Photometric Determination
of Traces of Metals

CHEMICAL ANALYSIS

A SERIES OF MONOGRAPHS ON
ANALYTICAL CHEMISTRY AND ITS APPLICATIONS

VOLUME 3, PART I
Fourth Edition

A WILEY-INTERSCIENCE PUBLICATION

JOHN WILEY & SONS

New York / Chichester / Brisbane / Toronto

Photometric Determination of Traces of Metals
General Aspects

Fourth Edition of Part I of
Colorimetric Determination of Traces of Metals

E. B. SANDELL
University of Minnesota (Retired)

HIROSHI ONISHI
University of Tsukuba

A WILEY-INTERSCIENCE PUBLICATION

JOHN WILEY & SONS
New York / Chichester / Brisbane / Toronto

Library of Congress Cataloging in Publication Data:

Sandell, Ernest Birger, 1906–
  Photometric determination of traces of metals.

  (Chemical analysis; v. 3)
  Includes bibliographical references and indexes.
  1. Trace elements—Analysis. 2. Metals—Analysis.
  3. Colorimetric analysis. I. Onishi, Hiroshi, joint
author. II. Title. III. Series.

QD139.T7S26 1977 546′.3 77-18937
ISBN 0-471-03094-5

Printed in the United States of America

10 9 8 7 6 5 4 3 2 1

FROM THE PREFACE TO THE FIRST EDITION

The colorimetric determination of traces of elements, especially of metals, has made great advances in recent years and it seemed to the writer that it would be useful to have available a collection of modern methods in this field of analysis. This book is the result of an attempt in this direction. It is not intended to be an encyclopaedia of methods for the colorimetric determination of small amounts of metals. The aim has been rather to present a limited number of methods which at the present time appear to be best suited for dealing with traces of metals. No one reagent is necessarily the best for the determination of an element in all kinds of samples or under all conditions, and consequently two or three methods are sometimes described in greater or less detail for a number of the metals. A few fluorimetric methods are included. The treatment is to a considerable extent based on the experience of the writer in testing or using various methods.

Anyone who surveys the methods of colorimetric trace analysis must experience a feeling of satisfaction arising from the many sensitive reactions available and, on the other hand, of something close to dismay at our imperfect knowledge of the application of these reactions. The effect of foreign elements on a particular color reaction is frequently poorly known and the prevention of the interference of foreign substances has, for the most part, been incompletely studied. Methods for the separation of traces are but poorly developed or even non-existent for many elements. The user of this book is likely to find many of his questions in this phase of trace analysis unanswered in the present treatment. It is to be hoped that the workers of the future will be willing to devote as much of their energies to this prosaic aspect of the subject as to the more inviting one of searching for new reagents.

v

PREFACE

As indicated on the title page, this book is a revision of Part I of *Colorimetric Determination of Traces of Metals* (1959), which deals with general aspects of inorganic photometric analysis (molecular absorptiometry and fluorimetry) and with metal separations. Since the late 1950s, thousands of papers germane to the subject matter of this volume have appeared. An exhaustive treatment of this mass of material might have led to the exhaustion of the reader—and certainly of the authors. An eclectic approach would seem to have advantages and that is what we have attempted in this revision. A sufficient number of references have been given to provide points of entry into the literature. One-third of the treatment is devoted to separations.

At this time we have not decided whether to revise Part II of *Colorimetric Determination of Traces of Metals*. We believe that the applied analyst will continue to find Part II of the third edition useful for information on the separation and determination of traces of individual metals.

E. B. SANDELL

HIROSHI ONISHI

Minneapolis, Minnesota
Ibaraki, Japan
January 1978

CONTENTS

Photometric Determination
of Traces of Metals

TRACE ANALYSIS:
ROLE OF PHOTOMETRIC METHODS

I. PRELIMINARY CONSIDERATIONS

Trace analysis (quantitative) may reasonably be defined as the determination of constituents making up less than 0.01% of a solid sample.[1] Obviously, there is no need for, nor can there be, a sharp boundary between trace and nontrace constituents. The lower limit of a trace constituent is zero, but practically the lower limit is set by the sensitivity of available analytical methods and, in general, is pushed downward with the progress of analysis. In this book, the trace constituents considered are inorganic, and quantities usually refer to the elements.

In Fig. 1-1, the relative constituent content of a sample is subdivided logarithmically along the Y-axis, with major (100–1%) and minor (1–0.01%) constituent classes included.[2] Along the X-axis in this figure, the sample weight classes are designated. Samples weighing more than 0.1 g may be called macro; those in the range 0.1 to 0.01 g, meso

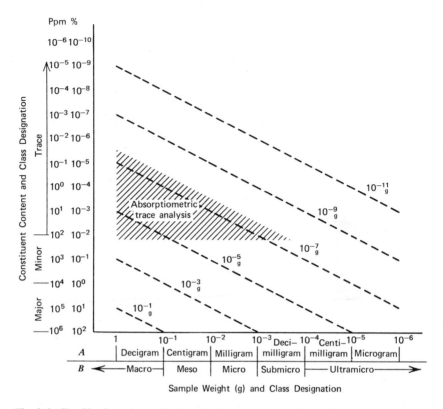

Fig. 1-1. Classification of analytical methods on the basis of sample size and constituent content. Major constituents make up 100–1% of a sample; minor constituents, 1–0.01%; and trace constituents, <0.01% (<100 ppm). The trace range is not subdivided into named classes, but these can be designated numerically, for example, 1–100 ppm, or as $-\log$ (concentration) $= p$ (concentration). Sample-size classes can be named according to A or B, or designated numerically, for example, 10^{-2}–10^{-3} g. *Macro* and *meso* will generally be understood, but there is no agreement as to the naming or the range of the classes here called micro, submicro, and ultramicro. The slanting dashed lines indicate constituent quantities in grams. The constituent content–sample weight field of molecular absorptiometric methods is indicated. The approximate lower absolute limit of solution absorptiometric methods (spectrophotometry) in terms of constituent weight is based on the use of the more sensitive color reactions (0.001–0.003 μg of element detectable instrumentally in a column of solution of 1 cm² cross section). This weight is taken to be ten times the limit of detection. The volume of solution is assumed to be 5 ml contained in a 5 cm cell (or 10 ml in a 10 cm cell). This diagram is based on Fig. 1 in Sandell, *Colorimetric Determination of Traces of Metals*, 3rd ed., 1959.

2

("semimicro"); and those below 0.01 g may be subdivided as follows:

10^{-2}–10^{-3} g   micro (or milligram samples)
10^{-3}–10^{-4} g   submicro (or decimilligram samples)
$<10^{-4}$ g   ultramicro

(The naming of the latter classes is rather arbitrary and there is no general agreement on these designations.)

The diagonal lines in Fig. 1-1 indicate absolute amounts of constituents (in grams) as a function of the relative content of constituent and the sample size (=weight). Depending on the sensitivity of the method and the constituent content, the sample size in a trace determination may vary from macro to ultramicro. Ordinarily in trace analysis, the sample size does not exceed 1–10 g (often not above 1 g), but occasionally much larger samples are taken. For easily handled materials such as water or ice, samples in the kilogram or ton range may be used.

The essential feature of a trace analysis is not so much the determination of a minute quantity of a constituent, as it is the determination of such a quantity in the presence of overwhelming quantities of other substances that may seriously affect the reaction of the trace constituent.

II. METHODS OF ELEMENTAL TRACE ANALYSIS

Our purpose in this section is not to treat the principles of the various types of methods applied in inorganic trace analysis. The reader is assumed to be familiar with them in a general way.[3] We consider instead the general sensitivity, precision, accuracy, and selectivity of common trace methods. The treatment is not intended to be detailed or comprehensive, but it may serve as a survey of possibilities, allow some comparisons to be made among the various methods, and permit the reader to draw some conclusions regarding the capabilities of solution absorptiometric and fluorimetric methods in trace analysis.

With a few exceptions, the classical stoichiometric (gravimetric and titrimetric) methods are not used in trace analysis, chiefly because they lack sensitivity. Titrimetric methods are applied occasionally in the upper trace range. The determination of gold and other precious metals by cupellation, in which the metal bead is finally weighed to 0.01 mg, is almost the only important application of a gravimetric method in trace analysis. The method used by Haber to determine gold in sea water involved microscopic measurement of the metal bead to obtain its mass. A similar method was applied by Stock to determine mercury in various materials. These micrometric methods are now rarely if ever used.

The required sensitivity for trace determinations can be obtained by using physical or physicochemical methods, in which manifestations of energy provide the basis of measurement. These methods are indirect in the sense that the emission or absorption of radiation or transformation of energy must be related in some way to the mass or concentration of the species that are being determined. The establishment of these relations almost invariably requires calibration, with the use of standards of known content of the constituent in question. For some types of methods, solutions of pure metals can be used. For others, solid standards of similar matrix composition are needed or are desirable. It is difficult to prepare such standards synthetically, and it may be necessary to use standards analyzed by another method; obviously, such methods have serious shortcomings as far as accuracy is concerned.

As a class, physical methods tend to be selective and may be rapid. When applied to trace constituents, they need not, as a rule, give results with errors less than a few percent—and such accuracy is not always readily attained in real analysis. Some physical methods—neutron activation, X-ray fluorescence, and flame or arc spectrography, for example—are more physical than others such as solution spectrophotometry and fluorimetry or electrochemical methods, which involve chemical reactions. The latter methods are best described as physicochemical.

We consider first the two types of methods that are the subject of this book, solution absorptiometry and solution fluorimetry, which for the want of a better term are here included under the term "photometric methods."

A. ABSORPTIOMETRY (SPECTROPHOTOMETRY) IN SOLUTION

Methods in this class are based on the absorption of radiation in the visible and UV (rarely the near infrared) portion of the electromagnetic spectrum by species in solution. The solutions are almost always aqueous or organic–solvent (rarely a melt), and the absorbing species are molecules or ions (hardly ever atoms).[4] The absorbing species are usually (and preferably) in true solution but may be solids in suspension or in a thin layer. The term "solution absorption spectrophotometry," while not entirely exact or all-inclusive, characterizes these methods fairly well, but is cumbrous. "Molecular (ionic) absorptiometry" conveys the meaning and distinguishes this type of method from atomic absorption. When, in this book, we refer to absorptiometry, we shall mean measurement of the absorption by species in solution, usually with a narrow band of wavelengths in the visible or the UV range. "Spectrophotometry" is another designation, which though not sufficiently restrictive, is usually understood to mean the methods considered here. The older term "colorimetry" does not accurately describe these methods because nowadays absorption is frequently measured in the UV as well as in the visible range, but analytical chemists will understand this term as referring to methods used to determine the concentrations of dissolved constituents by developing light-absorbing reaction products by chemical reaction followed by spectrophotometric (usually absorptiometric but also reflectance) measurement or comparison. Most absorptiometric reagents in use today form colored products with the metals to be determined. At times the use of "colorimetry" may be advantageous or necessary, as when attention is called to absorption in the visible range.

Color comparisons based on the use of such reagents as ammonia (for copper), thiocyanate (for ferric iron and cobalt), and stannous chloride (for gold) provided the earliest physicochemical trace determinations in the nineteenth century. The development of organic chemistry over the last hundred years or so has led to the discovery and synthesis of many reagents giving colored (and some fluorescent) chelates and ion-association complexes with metals, which have been put to use in analysis. This has been followed by a more or less systematic search (by synthesis) for new sensitive and selective organic reagents. The commercial availability of photoelectric spectrophotometers before the middle of the present century allowed accurate determinations of both metals and nonmetals to be made with the aid of such reactions.

In the last 25 yr or so, new physical methods (p. 10–21) have been developed (emission spectrography was in use earlier) that have revolutionized trace analysis. Absorptiometric analysis is not as important as formerly. However, it still has a useful, even valuable, role to play in trace analysis, especially in the upper range of trace analysis.

Lower Limits. The lowest solution concentration, c_{min}, at which an element is detectable absorptiometrically is given by the relation:

$$c_{min} = \frac{A_{min}}{ab}$$

in which

A_{min} = smallest detectable absorbance (log I_0/I).

a = absorptivity at some specified wavelength λ

= absorbance produced by a unit weight of element present as the absorbing species in a column of solution of unit cross sectional area (convenient units are $cm^2/\mu g$ or cm^2/ng).

b = light (radiation) path in centimeters.

(The assumption is made that A is rectilinearly proportional to b and c, so that a is constant; this is usually true in practice.)

a is typically 0.1–1 cm^2/μg for the color reactions used in inorganic trace analysis. A_{min} depends on a number of factors, particularly on the absorptiometer used, but 0.001 is a good average value. We then have

$$c_{min} = \frac{0.001}{ab}$$

It is convenient to replace a by $1/(1000\mathscr{S})$, where $\mathscr{S}$ is the sensitivity index of the reaction, defined as the number of micrograms of the element, converted to the absorbing species, in a column of solution of 1 cm^2 cross section giving an absorbance of 0.001:

$$(c_{min})_{\mu g/ml} = (c_{min})_{ppm} = \frac{\mathscr{S}}{b}$$

Values of $\mathscr{S}$ for many color reactions are available[5] or can be calculated from molar absorptivities ϵ if the reaction runs to completion, or from apparent molar absorptivities if it does not. For the most sensitive color reactions, $\mathscr{S} = 0.002$ μg/cm^2 on the average. Reactions having $\mathscr{S} = 0.002$–0.003 are available for at least 20 metals (Al, Sb, As, Be, Bi, Cd, Cr, Co, Cu, Ga, Ge, Au, Fe, Pb, Mg, Hg, Pd, Pt, Ru, Tl, and Zn). Very few reactions have $\mathscr{S} < 0.001$. It may be said then that with a 1 cm cell ($b = 1$) the lowest average c_{min} is ~ 0.002 ppm.

The lowest content of an element in a (solid) sample detectable absorptiometrically by a reaction of known $\mathscr{S}$ can be specified if the assumption is made that not more than 5 ml of solution is needed for the absorption measurement, the light path is 5 cm, and the sample size is 1 g. A final volume of 5 ml is quite realistic for most metal trace determina- -tions, because a liquid–liquid extraction is, or can be, made that brings the colored or absorbing species into this relatively small volume, or other concentration methods can be used, although less conveniently. If the final volume is 10 ml, the light path should be 10 cm. Although there is no maximum sample size[6] when an isolation (concentration) procedure is used, it is convenient to limit the (solid) sample to ~ 1 g. A sample of this size is handled without much difficulty, even if a fusion procedure, as of a silicate with sodium carbonate, is needed to decompose the material. On the same basis as before, the minimal detectable element content in the sample then is $\sim(1 \times 0.002)$ or ~ 0.002 ppm. Now suppose that the limit of detection represents approximately the standard deviation in the determination of a trace constituent in its lowest range and that a relative standard deviation of 10% is acceptable in this range. The content of constituent in the sample should then be $\sim 100/10 \times 0.002 \sim 0.02$ ppm. Of course, it is possible that the standard deviation of the method is not determined solely by the instrumental random deviation, here equated with the sensitivity. On the whole, in using colorimetric methods one would prefer to apply them to samples containing >0.1 ppm of constituent.

Generally, solutions giving an absorbance of ~0.4 provide the optimum precision if a transmittance scale of 100 divisions ($I_0 = 100$) can be read to 0.2 division ($\pm0.2\%$ I, absolute). (See Chapter 4.) Accordingly, under the conditions assumed (1 g sample, 5 ml final volume, 5 cm cell, $\mathscr{S} = 0.002$), no increase in photometric precision is to be expected above $0.4/0.001 \times 0.002 = 0.8$, or say 1 ppm of constituent. Therefore, if the constituent content is ~10 ppm, the sample size can be reduced to ~0.1 g, or if it is 100 ppm, it can be reduced to 0.01 g, and so on, for maximal photometric precision. With a 1 cm cell (5 ml volume), these weights would be multiplied by 5. Actually, in practical trace analysis, it is by no means necessary to strive for maximal photometric precision in all determinations, and, other factors remaining the same, smaller samples than indicated will often serve.

Precision and Accuracy. This topic is considered in Chapter 4, but some general statements are required in the present context. The random error in the measurement of absorbance, alluded to in the previous section, is only one of a number of errors afflicting absorptiometric determinations. There also may be random (indeterminate) errors in the color development step of the analysis. But determinate errors may be more important than indeterminate errors, and may be more difficult to evaluate. Absorptiometric analysis, like all other methods of analysis, is subject to positive and negative interferences. Hardly any metal color reaction is so selective that the possible reaction of other metals to give colored or absorbing species cannot occur. Negative interferences are of lesser importance. By adjustment of the hydrogen-ion concentration, use of differential complex formation or oxidation-reduction, it is often possible to eliminate or reduce the interference of foreign elements. But sooner or later such measures fail, especially in trace analysis, where the ratio of foreign elements to the element being determined is unfavorably large. Separations are then required to remove interfering elements and leave the desired element, preferably in a small volume of solution. Errors of two types can arise in separations: The recovery of the element in question may be incomplete and foreign elements may not be completely removed.

If blanks are large, precision may be impaired and determination limits may be raised.

Another source of error is in sampling. Because of heterogeneity the sample taken may not accurately represent the mass from which it is derived. This error is likely to be smaller in absorptiometric analysis than in some other methods of trace analysis in which a smaller sample is used.

In the absence of interfering elements, the precision and accuracy of a method can be determined without any particular difficulty. When interfering elements are present, this will not be so easy, because both precision and accuracy, especially the latter, are likely to depend on the number and the amounts of these elements. The accuracy of a method can be tested by using synthetic samples containing known amounts of the element in question. This is laborious, and it may be done only by the originator or the modifier of a method. A simpler way involves analyzing a standard sample by the method in question, and comparing the result with the results that have been obtained by other methods of presumed accuracy. Comparison of results by different methods is a powerful means of detecting determinate errors, if the methods are so different that the same determinate errors are not likely to be present in all. It may thus be possible to establish the accuracy of an absorptiometric method for the determination of a particular element in a class of materials whose composition is represented by the standard sample. See Section III for comparisons of trace metal results given by different analytical methods.

Selectivity. The great majority of determination forms in molecular absorptiometry have broad absorption bands, so that attainment of selectivity by optical means (as in atomic absorption) is of limited value. Instead, chemical properties of reagents and constituents must be exploited. It may perhaps be doubted that any color reaction for an element, more particularly for a metal, is strictly specific, meaning that all other metals do not yield reaction products absorbing at or near the wavelength of maximum absorption of the element in question. Specificity is especially hard to attain in practice when the ratio of foreign elements to the given element is very large, say 10^5 or 10^6. But many color reactions are selective—only a few elements give a positive reaction similar to that of the element in question. The use of masking agents is of great help in improving selectivity. In practice, limited specificity is, as a rule, acceptable. That is, most methods described for the determination of a given element are suitable for certain classes of material, in which the amounts of foreign elements will not exceed certain limits. Very few methods to be found in the literature can be applied to a sample of any composition.

Sooner or later—usually quite soon in trace analysis—the analyst using absorptiometric methods (and other methods as well) is driven to separations in order to attain requisite specificity. A combination of a selective separation with a selective color reaction may give virtual specificity. Separations, as already mentioned, may result in appreciable loss of the desired element. Moreover, they lengthen the analysis. But separations have one great advantage (if they are adequate): They make the determination independent of the composition of the sample. Matrix effects are virtually universal among physical trace methods for the elements, although the effects vary greatly from one method to another.

Some results are given later that allow judgments to be made concerning the accuracy and lower determination limits of absorptiometric methods in applied trace analysis.

B. FLUORIMETRY

Although the principles of fluorimetry and absorbance spectrophotometry are entirely different, it is convenient to consider fluorescence methods in conjunction with absorptiometric methods. Both are often included under "photometric methods." The apparative requirements are modest, and both methods are likely to require separations before they can be applied in trace analysis unless the samples are of the simplest type. Fluorescence methods are often more sensitive than the absorptiometric, sometimes by one or two orders of magnitude, but this is not always true. The sensitivity of a fluorescence method depends upon the intensity of the UV source, as well as other factors, and its objective expression is not as easy as for an absorptiometric method. Commonly, sensitivity is expressed in concentration terms, for example, the lowest concentration in ppm or $pp10^9$ (or even $pp10^{12}$) giving a detectable fluorescence under certain conditions, as with a specified instrument or solution volume. These concentrations are often in the $pp10^9$ range. For example, one commonly used fluorimeter in the USA allows a full-scale deflection (100 divisions) to be obtained with a 5×10^{-3}-ppm solution of quinine sulfate in water, and even a limit of a few $pp10^{12}$ has been claimed. A 10^{-4}-ppm solution of fluorescein gives a readily detectable fluorescence. With an esculin laser (to increase excitation intensity), the fluorescein detection limit is $\sim 2pp10^{12}$. With very few exceptions, organic reagents are used in inorganic fluorescence analysis, and the metal–organic compounds formed (often chelates) give sensitivities comparable to these. Bis-salicylalethylenediamine provides a sensitivity of 2×10^{-5} ppm for Mg, and 2-hydroxy-3-naphthoic acid a sensitivity of 2×10^{-3} ppm for Al and 2×10^{-4} ppm for Be in solution. Of course, not all fluorescence reactions are this sensitive.

Morin (Chapter 6B) is an example of a metal fluorescence reagent of demonstrated practical worth. It allows Be and Zr to be determined selectively in rather strongly basic and acid solutions, respectively. Instrumentally, 2×10^{-5} ppm of Be is detectable in solution, and visually, 4×10^{-4} ppm. Zirconium shows comparable sensitivities. With a 5-g biosample (leaves), 0.01–0.02 ppm Be can be determined; the detection limit is 0.002 ppm or less. With a 0.02 g sample of silicate rock, 1 ppm Zr is detectable, and 10 ppm Zr should be determinable to ~10%.

Some of the fluorescent metal complexes (Al and Ga 8-hydroxyquinolates, for example) can be extracted into immiscible organic solvents, thus increasing the concentration sensitivity.

In principle, fluorimetric methods are more selective than absorptiometric. Fluorescent complexes are generally given by diamagnetic ions, not the paramagnetic, and thus the number of positively interfering elements is reduced. As in absorptiometry, the effect of interfering ions can be eliminated or reduced by differential complexing and occasionally by a change in oxidation state.

Fluorescence determinations are widely subject to negative (fluorescence diminishing) effects. On the whole, fluorescence methods are not as precise as absorptiometric methods, and they are not likely to be chosen in preference to them unless a definite superiority in sensitivity or selectivity can be demonstrated.

A fluorescence trace determination of great value is that of uranium. In a matrix of sodium fluoride or sodium fluoride-carbonate, U(VI) fluoresces an intense yellow-green. As little as $0.001 \mu g$ U in ~1 g of this flux can be determined with reasonable precision. The fluorescence is specific or almost so under suitable excitation conditions, but is subject to quenching by foreign substances, so that separations must often be made.

The price range of fluorimeters suitable for inorganic trace analysis is roughly \$500–2000 (1970). Measurements can be made with 0.5 ml volumes of solution, but unless extractions can be made, such small volumes are not likely to be useful in trace analysis. See Chapter 4 for further discussion.

As far as measurement is concerned, nephelometry is very similar to fluorimetry, the intensity of scattered light being measured instead of emitted radiation. A suspension of very slightly soluble substance is the basis of nephelometry. It is used very little in trace metal determinations. Absorptiometric and fluorimetric analysis is almost always superior in sensitivity and reproducibility.

C. ATOMIC ABSORPTION, ATOMIC FLUORESCENCE, AND FLAME EMISSION SPECTROPHOTOMETRY

Today (1977) more trace determinations are possibly made by atomic absorption spectrophotometry than by any other method. This popularity is due to the generally good sensitivity for many elements, the frequent lack of serious interferences, and commonly no need for separations (except simple ones often made more for concentration than actual separation), rapidity, and satisfactory precision. For some elements, atomic fluorescence is more sensitive than atomic absorption. For still other elements, flame emission is more sensitive than either of these, though more susceptible to spectral interference ("... it is probably easier to obtain an erroneous result with a complex matrix in emission than with the other techniques."—Browner).

The detection limits of a number of metals by atomic absorption, fluorescence, and emission methods, all in flames, are given in Table 1-1. Customarily, the detèction limits are stated in terms of solution concentration. What counts in applied analysis is the detection limit (and thence the determination limit) of an element in (usually) a solid sample. Assume that the solution sprayed into the flame is a 2% solution of the sample. Taking Be as an example, the detection limit by atomic absorption or fluorescence in the solid sample then is $(100/2) \times 2 \times 10^{-9} \times 10^{6} = 0.1$ ppm. Multiplying 0.1 by 10, we obtain 1 ppm Be as the approximate limit of *determination.* This is very satisfactory sensitivity, but it is no better than that provided by solution fluorimetry with morin as reagent. Some metals are not easily determined sensitively by atomic absorption with conventional equipment. Included among these are Th, Ce, rare earth elements, Hf, Zr, Nb, Ta, U, W, and Re. The sensitivity may be inadequate (as in geochemical analyses) for Sb, Ge, Ti, and Mo. (Almost all of these elements can be determined sensitively by solution absorptiometric methods.) On the other hand, the sensitivity is very high for Zn, Cd, and Mg. Flame emission is generally less sensitive than atomic absorption and fluorescence in flames, but for some elements (Group 6 in Table 1-1), it is more sensitive and to some extent is complementary to absorption and fluorescence. Extraction of a metal as a chelate into a suitable immiscible organic solvent is commonly used to increase the sensitivity in flame atomic absorption methods; extraction also has the advantage of separating the metal from other substances in the aqueous solution that could affect its determination.

The sensitivity of atomic absorption and fluorescence methods for some elements can be improved by substituting a graphite furnace or other nonflame atomizer for the flame. The atomic absorption detection limits in solution for a few metals are[7]:

Ag   0.0025 ng/ml
Au   0.08 ng/ml
Cd   0.001 ng/ml
Mg   0.1 ng/ml
Sb   0.2 ng/ml

TABLE 1-1
Detection Limits in Solution by Atomic Flame Methods[a]

	Element	Detection Limit (ng/ml)		
		Atomic Absorption	Atomic Fluorescence	Flame Emission
Group 1: Atomic absorption	Be	2	10	1,000
(AA) more sensitive	Hf	2,000	100,000	20,000
than both atomic	Mg	0.1	1	5
fluorescence AF	Mo	20	500	100
and flame emis-	Pd	20	40	50
sion E	Sn	10	50	100
Group 2: (AA = E) > AF	Lu	700	3,000	1,000
	Rh	20	150	20
Group 3: (AA = AF) > E	As	100	100	10,000
	Fe	5	8	30
	Ni	2	3	20
	Pb	10	10	200
	Sb	40	30	600
Group 4: AA = AF = E	Cr	3	5	4
	Nb	1,000	1,500	1,000
Group 5: AF > (AA, E)	Ag	2	0.1	8
	Au	10	5	500
	Bi	25	5	2,000
	Cd	2	0.001	800
	Ce		500	10,000
	Co	10	5	30
	Cu	1	1	10
	Ge	200	100	400
	Hg	250	0.2	10,000
	Mn	2	1	5
	Tl	30	8	20
	Zn	1	0.02	10,000
Group 6: E > (AA, AF)	Al	20	100	5
	Ca	0.5	20	0.1
	Eu	20	20	0.5
	In	20	100	2
	Ru	70	500	20
	Sr	2	30	0.1
	V	4	70	10

[a] Selection of values from table given by R. F. Browner, *Analyst*, **99,** 617 (1974).

There is evidently a large decrease in the detection limit for Ag, Au, and Cd (compare with flame values in Table 1-1). However, precision is poorer than in flame methods, and interferences are reported to be greater. Determinations of elements in solid silicate samples by nonflame methods are especially subject to large matrix effects.

The precision of flame atomic absorption spectrophotometry is generally very satisfactory. Actual determinations may have standard deviations of 5% or less, if the amount of element lies sufficiently far above the detection limit. The precision parameters in Table 1-2 illustrate what may be expected in practice.

Spectral line interferences are uncommon in atomic absorption but can occur in flame emission spectrophotometry. More important are the negative errors that can arise from the presence in the sample of elements forming refractory compounds stable at high temperatures (phosphates, sulfates, aluminates, titanates, and others). Hotter flames minimize such

TABLE 1-2
Determination of Magnesium in Cast Iron by Atomic Absorption Spectrophotometry[a]

	Sample			
	A	B	C	D
Mg (%) (mean)	0.0022	0.0225	0.0522	0.0992
Repeatability (%)[b]	0.0003	0.0012	0.0019	0.0050
Coefficient of re-peatability (%)	13.6	5.3	3.6	5.0
Reproducibility (%)[c]	0.0006	0.0014	0.0029	0.0046

[a] From L. L. Lewis, "A Comparison of Atomic Absorption with Some Other Techniques of Chemical Analysis" in *Atomic Absorption Spectroscopy*, ASTM Special Technical Publication 443, p. 62, Am. Soc. Test. Mater., Philadelphia, 1969. Examples of results of determination of Mg in four samples of cast iron in ten laboratories.

[b] "... repeatability is a quantitative measure of the variability of the results by a single analyst in a given laboratory, using a given apparatus. It is defined as the greatest difference between two independent results that is to be expected due to random error on the basis of the 95 per cent confidence level."

[c] "Reproducibility is a quantitative measure of the variability associated with different analysts and equipment in two different laboratories. It is defined as the greatest difference between a result obtained in one laboratory and one obtained in another laboratory that is to be expected due to random errors on the basis of the 95 percent confidence level."

interference, as do "releasing agents" such as lanthanum that bind the interfering elements. Easily ionized elements (potassium, for example) can affect the proportion of atoms of the element being determined, and similar standards are needed to remedy this matrix effect. It has been reported that such measures are not always successful with complex matrices, and the element in question may need to be separated before determination. Thus barium cannot be accurately determined in silicate rocks unless first separated (by ion exchange). Physical differences (e.g., in viscosity) in sample and standard solutions can lead to differences in nebulization and cause errors. High salt content can cause scattering of radiation and produce absorption not due to the constituent determined.

Simultaneous multielement determinations by atomic absorption, fluorescence, and emission are difficult.

Atomic absorption by cold mercury vapor provides a deservedly popular method for the determination of this element. Mercury compounds in the sample solution are reduced to the metal with stannous chloride, and mercury is then volatilized into a stream of indifferent gas. The concentration of mercury in the gas phase is obtained by measurement of its absorption at 253.7 nm. The principle of this method was already applied in the 1930s to determine mercury in air. The molar absorptivity ϵ of atomic Hg is 4 or 5×10^6, which is much greater than ϵ of mercuric dithizonate ($\epsilon_{485} \sim 70,000$ in CCl_4), or any other absorbing mercury species in solution spectrophotometry.

D. EMISSION SPECTROGRAPHY (NONFLAME)

Solid samples are used and many elements (approximately 70) can be determined or estimated with greater or less reliability, a considerable number at one time. The detection limit varies greatly, but is quite low for many metals. The limit may be as small as 0.001 μg, but often is no better than 0.01–0.1 μg with d.c. carbon arc excitation, the technique commonly used in trace analysis. The sample size is usually small, say 10 mg, and consequently the concentration sensitivity is reduced, the limiting concentration frequently being greater than 1–5 ppm. Concentration methods are applied at times.

Emission spectrography is not noted for its precision. Accuracy also may be quite mediocre (see Fig. 1-6). But these shortcomings may sometimes be tolerated in favor of other advantages, particularly speed and the possibility of estimating many elements at one time. The best precision and accuracy are not always required in analysis, and emission spectrography may be an adequate method for some purposes. It has been, and still is to some extent, a favorite method for the simultaneous determination of traces of metals in geomaterials. However, a considerable number of metals occur in such low concentrations in rocks that they cannot be determined spectrographically without concentration.

E. X-RAY EMISSION (FLUORESCENCE) AND MICROPROBE ANALYSIS

Solid samples are ordinarily used. In X-ray emission, the absolute detection limit may be approximated as $0.1-1 \mu g$, sometimes worse, rarely better. The relative detection limit, in direct analysis, is roughly 1–10 ppm. It depends on the element, the matrix, and the thickness of the sample, as well as on the instrumentation. X-Ray fluorescence then is not one of the more sensitive techniques. The concentration sensitivity can be improved by applying separations and concentration in preliminary treatment of the sample. Ion exchange has found considerable use. In direct analysis, the matrix effect is likely to be large. The technique is rapid and multielemental.

Electron microprobe analysis is used for the nondestructive analysis of very small solid samples, for example, $1 \mu m^3$, and is of special value for heterogeneous solids. It is not a method for the determination of traces of elements, because the concentration sensitivity is rarely below 500 ppm.

F. MASS SPECTROMETRY

Ions for mass spectrometry can be produced in various ways. Commonly, in the analysis of a wide variety of solids, a spark source is used. Spark source mass spectrometry, like other types of mass spectrometry, is an exceedingly sensitive method of elemental trace analysis. The sensitivity does not vary much from one element to another, and for most elements, the sensitivity is comparable to the average or higher sensitivities of activation analysis. The sensitivity depends on the spectrometer and procedure, but the absolute detection limits lie in the nanogram–subnanogram range. The concentration detection limits lie in the range 0.001–0.1 ppm on the atomic basis.

In the present state of development (early 1970s), the accuracy and precision of spark source mass spectrometry tend to be mediocre, and the results are sometimes semiquantitative. With standards similar in matrix composition to the sample, results to within 10 to 20% have been reported. Sample preparation requires care. In spite of these deficiencies, and of the cost of the mass spectrometer and its maintenance, the method is important because of its sensitivity and the possibility of determining (or estimating) the concentration of almost all the elements simultaneously.

Isotope Dilution. This is the most sensitive and accurate mass spectrometric method for finding the concentration of an element in a sample. It is a kind of an internal standard method. The element in question must have at least two stable or long-lived isotopes. A known quantity of the element in which one of the isotopes has been artificially enriched (spike) is added to a known quantity of the sample. After equilibrium has been attained, the element is isolated from the sample (not necessarily quantitatively), and the new isotope ratio of the element is determined by mass spectrometry. Isotope ratios can be determined mass spectrographically with a reproducibility of 1–3 ppt, but the accuracy may not be this good. In geochemical applications (the chief inorganic use of the method), the error usually falls in the range 1–5% for routine determinations. Sensitivities of 10^{-6} to 1 μg are attainable. Usually the limiting factor in the accuracy is contamination. An example of the use of isotope dilution is the determination of uranium in stony meteorites (10^{-3} ppm range). Uranium in ammonium nitrate reagent ($\sim 10^{-4}$ ppm) can be determined to within $\sim 1\%$.

Another important type of isotope dilution makes use of radioactive isotopes (p. 19).

G. METHODS BASED ON RADIOACTIVITY

The most important of these is radioactivation. For a large number of elements, radioactivation (principally neutron activation, but also charged-particle and high-energy photon activation) is usually considered to provide the best means of determination in materials of complex composition when quantities in the lower trace ranges (say less than 0.1 ppm) are present. If separations are made after irradiation, interferences can virtually be eliminated. Even in nondestructive, wholly instrumental procedures (γ-ray spectrometry), results are usually reliable and of satisfactory precision. But a number of sources of error need to be guarded against.[8]

Although generally characterized by high sensitivity, neutron activation determinations do vary in sensitivity from element to element, as illustrated by the values in Table 1-3. A range of more than 10^6 in detection limits is shown.

Neutron activation, compared to other methods, has the advantage that contaminating foreign elements introduced after irradiation will not be counted with the elements being determined, but, of course, error can

TABLE 1-3
Examples of Sensitivities of Neutron Activation Determinations

		Limit of Detection
		Micrograms of Element Detectable with Thermal Neutron Flux of 10^{13} n/cm^2-sec and 1-hr Irradiation[a]
Low sensitivity	Bi	0.05
	Fe	5
	Pb	1
	Zr	0.1
Medium sensitivity	Cd	5×10^{-3}
	Cu	1×10^{-4}
	I	5×10^{-4}
	Sb	5×10^{-4}
High sensitivity	Au	5×10^{-5}
	Dy	1×10^{-7}
	In	5×10^{-6}
	Mn	5×10^{-6}

[a] From G. H. Morrison and R. K. Skogerboe, in Morrison, *Trace Analysis—Physical Methods*, Chapter 1.

arise from elements introduced before irradiation (crushing, destruction of organic matter).

For quantities of elements appreciably above the detection limits, neutron activation has been reported to have precisions corresponding to standard deviations of 1–3% and accuracies corresponding to errors of 2–4%; even exagerrated claims of ±1% for each have been made. In practice, results may not be this good, at least when samples of complex composition are analyzed. This conclusion is based on comparison of results obtained by different analysts in analyzing the same sample. In Table 1-4, we record analytical values obtained in as many as three different laboratories in the analysis of the standard granite G-2 and in as many as four laboratories in the analysis of the standard diabase W-1. For most elements the agreement is good or at least satisfactory. Only for Cr and Sb in G-2 is the agreement poor, and heterogeneity in the sample is perhaps not excluded. The difference in extreme values for most elements having contents above 1 ppm usually does not exceed 15–20%. In some neutron activation determinations of elements in plant material (standard sample of kale), different laboratories have occasionally obtained quite different results (p. 35), and a few results are less accurate than those by other methods.

TABLE 1-4
Comparison of Results (ppm) Obtained by Different Analysts in Neutron Activation Determination of Trace Elements in Silicate Rocks

Element	Sample[a] and Analysts[b]						
	G-2			W-1			
	A	*B*	*C*	*C*	*D*	*A*	*B*
Ba			1800	< 200			
			1627[c]	167[c]			
Ce		140	144	21	24		.21
Co	5.0	5.7	4.3	46	43	50	50
Cr	7.2	15	4.6	99	105	95	99
Cu						72.5	110
Eu	1.5	1.15	1.37	1.08	1.2	1.0	1.1
Ga		27			16	12.8	16
		22[d]					
Hf				3.0	3.4		
La		84	81	12	10.5	6.7	12
Lu		0.16	0.18	0.44	0.51		0.44
Sb	0.2	0.06	< 0.6	1	0.98	0.8	1
Sc	4.0	4.5	3.5	34.8	37	33	34
Sm				2.8	4.1		
Sr					186		
					193[c]		
Ta				0.67	0.4	0.5	
Th				2.6	1.9		
Zn					92	89	82
Zr				< 140	105		

[a] U.S. Geological Survey standard samples: G-2, granite; W-1, diabase.
[b] A: R. H. Filby and W. A. Haller, *Modern Trends in Activation Analysis*, Nat. Bur. Stand. (U.S.), Spec. Pub. 312, vol. I, p. 339, Washington, 1969. Co, Cr, Sb, Sc and Ta by γ-ray spectrometry and Cu, Eu, Ga, La and Zn by γ-ray spectrometry after removal of ^{24}Na.
B: S. F. Peterson, A. Travesi, and G. H. Morrison, *ibid.*, p. 624. γ-Ray spectrometry after group separations. C: G. E. Gordon et al., *Geochim. Cosmochim. Acta*, **32**, 369 (1968). γ-Ray spectrometry. D: O. Landström, K. Samsahl, and C. G. Wenner, *Modern Trends in Activation Analysis*, p. 353. Sequential separations, mainly by ion exchange.
[c] H. Higuchi, K. Tomura, H. Takahashi, N. Onuma, and H. Hamaguchi, *Modern Trends in Activation Analysis*, p. 334. γ-Ray spectrometry after precipitation of Sr and Ba carbonates.
[d] P. A. Baedecker and J. T. Wasson, *Science*, **167**, 503 (1970). Radiochemical separations.

18

Neutron activation results reported in the literature are not always free of gross errors. In Table 1-5, one set of neutron activation values for Mo in silicate rocks agrees well with solution spectrophotometric values, the other does not, and the probability is great that the second set of neutron activation values is in error.[9]

A similar example is provided by the determination of tin by spectrophotometric and neutron activation methods (Table 1-6). The spectrophotometric values agree well on the whole with the first set of neutron activation values but not with the second set.[10]

Activation analysis require facilities that put it out of reach of most analysts. It is hardly a routine method of analysis. Determinations are costly and may be time consuming. But sometimes neutron activation is the only method that will give the content of a low-trace element in a complex sample. Also, it is multielemental.

Substoichiometric Analysis. In a substoichiometric determination (a variety of isotope dilution analysis), a known amount of a radioisotope of the element being determined is added to the solution of the sample.

TABLE 1-5
Comparison of Determinations of Molybdenum in U.S.G.S. Standard Rocks by Neutron Activation and Spectrophotometry

| Rock | Mo, ppm | | |
	Spectrophotometry[a]	Neutron Activation[b]	Neutron Activation[c]
AGV-1	1.66	1.71	10.0
BCR-1	1.21	1.15	2.3
DTS-1	0.04	0.04	0.2
G-2	0.15	0.13	3.2
GSP-1	0.30	0.27	5.3
PCC-1	0.03	0.02	0.2
W-1	0.52	0.57	0.7

[a] K. Kawabuchi and R. Kuroda, *Talanta*, **17**, 67 (1970). Dithiol as reagent. Some of these rocks have also been analyzed for Mo by the spectrophotometric thiocyanate method by E. G. Lillie and L. P. Greenland, *Anal. Chim. Acta*, **69**, 313 (1974), who obtained results agreeing quite well with the dithiol values, namely AGV-1, 2.02 ppm; BCR-1, 1.42; G-2, <0.2; GSP-1, 0.37. The thiocyanate results are ~20% higher than the dithiol.
[b] E. Steinnes, *Anal. Chim. Acta*, **57**, 249 (1971).
[c] J. C. Laul, D. R. Case, M. Wechter, F. Schmidt-Bleck, and M. E. Lipschutz, *J. Radioanal. Chem.*, **4**, 241 (1970).

TABLE 1-6
Comparison of Determination of Tin in U.S.G.S. Standard
Rocks by Spectrophotometry and Neutron Activation

Rock	Sn (ppm)		
	Spectro-photometry[a]	Neutron Activation[b]	Neutron Activation[c]
G-2	1.32	1.2	2.0
GSP-1	4.35	4.5	8.6
AGV-1	3.03	3.0	5.65
PCC-1	0.88	0.5	1.68
DTS-1	0.74	0.5	0.68
BCR-1	1.81	1.5	2.71

[a] J. D. Smith, *Anal. Chim. Acta*, **57**, 371 (1971). Phenylfluorone as reagent.
[b] H. A. Das, J. G. van Raaphorst, and H. J. L. M. Umans, *J. Radioanal. Chem.*, **4**, 21 (1970).
[c] D. Schmidt, *Earth Planet. Sci. Lett.*, **10**, 441 (1971).

After isotope exchange equilibrium has been attained, a substoichiometric amount of reagent is added, that is, less than the equivalent amount required for precipitation, extraction, etc. Conditions must be such that the reagent reacts quantitatively. The radioactivity of the isolate (precipitate, extract, etc.) is compared against the radioactivity of the isolate from a standard solution of the radioactive isotope obtained in the same way. The same amount of the element in question is (ideally) isolated from the two solutions, so that if y = the amount of radioactive isotope added and A = radioactivity, the amount (x) of the element in the sample is

$$x = y\left(\frac{A_{standard}}{A_{sample}} - 1\right)$$

The sensitivity of this method is high (occasionally higher than that of neutron activation) if an isotope of high specific activity is used. As in all isotope dilution methods, the amount of isolated element is not required. Obviously the method is not free from pitfalls. The isolation reaction must be selective, indeed specific, for the sample in hand. Because of its selectivity, immiscible solvent extraction is frequently used for the isolation reaction. Masking agents can be used to increase selectivity. The substoichiometric method performs well in practice. The accuracy will of course vary with the element determined and the sample in which it is found. Results within ~5% are obtainable for some elements in a concentration range of 0.1–10 ppm.

The substoichiometric principle can also be applied in neutron activation determinations (to increase selectivity). Equal amounts of (inactive) carrier are added to sample and standard solutions after activation and before addition of reagent in substoichiometric amount.

H. ELECTROCHEMICAL METHODS[11]

More than 20 types of electroanalytical methods have been developed and some of these find application in trace analysis, for example, voltammetry, amperometry, coulometry, and anodic stripping. The sensitivity sometimes extends to 10^{-9} or 10^{-10} M. Electrochemical methods are not applied as widely in elemental trace analysis as some of the other methods mentioned in the preceding discussion, but they can be very useful at times. Anodic stripping analysis can be used for the simultaneous determination of a number of metals with good sensitivity. The range of elements amenable to determination by electrochemical methods is more limited than those that can be determined at low levels by methods based on absorption or emission of radiation.

Ion-selective electrodes have become important in trace determinations, especially of nonmetals.

I. REACTION RATE METHODS

Certain chemical reactions normally proceeding very slowly are markedly accelerated by low concentrations of catalysts. If the catalytic effect is sufficiently strong, it can be made the basis of an indirect determination of the catalyzing element by measurement of the reaction rate. If one of the reactants or one of the products is strongly absorbing, its concentration can conveniently and precisely be found as a function of the time by photometry. Rate methods not depending on catalysis (or induction) are also used in trace analysis. See Chapter 4, p. 204, for further discussion.[12]

J. MISCELLANEOUS METHODS

Reference was made earlier to trace determination of mercury and gold and other precious metals by isolating these elements as a microscopic globule or bead and obtaining the mass by measuring the diameter under the microscope. Although sensitive—as little as 0.01 ppm of the platinum metals can be determined in 2 g of silicate rock—such micrometric methods are now essentially obsolete. They have been replaced by radioactivation, which is more sensitive and has other advantages.

A nonchemical method for the trace determination of certain elements (Cu, Fe, Mn, Zn, Mo) that are micronutrients is based on the growth response of microorganisms toward them. Fungi, for example, *Aspergillus niger*, are the organisms commonly used. Determination is made by comparison or by weighing the dried organism. The sensitivity for some elements is high ($0.0001~\mu g$ Mo in 50 ml of solution is reported to be detectable), but the technique is infrequently used.

Determinations of traces of inorganic species can also be made by enzyme activation and inhibition.[13]

III. COMPARISON OF RESULTS BY DIFFERENT TYPES OF METHODS

At least in principle, the estimation of the precision of a photometric method is simple (Chapter 4). The assessment of the accuracy is a more difficult matter. The chief reason for this lies in the variable composition of the materials to which a given method is applied. No method is so specific that the determination of a given element is not affected in some degree by other elements in the material, but, of course, the effect varies greatly. In theory, the accuracy of a method can be studied by preparing synthetic samples containing known amounts of the element in question with varying amounts of foreign elements. A knowledge of the nature of the interference of foreign elements may make possible the prediction of error magnitudes by interpolation and extrapolation as the composition varies. Some foreign elements will have little if any effect and can, after preliminary investigation, be omitted. Nevertheless, this approach is laborious, and is likely to be limited to samples of certain compositional ranges. A difficulty in the preparation of synthetic standards for trace analysis is obtaining sufficiently pure matrix elements. Often synthetic standards are restricted to simple mixtures or even solutions. In the extreme case of a constituent in essentially pure solution, the standard is not useful for testing accuracy (except to the extent that this is determined by precision), but is used for testing the precision of a method in its various aspects (different analysts, different laboratories, different instruments) or for comparing the precision of different methods. We will call synthetic samples Type I standards.

Indirectly, the accuracy of a method can be judged by comparing it with other methods. For this purpose, what may be called Type II standards are used. These are simply suitably prepared and homogenized samples of important classes of materials, natural or man-made, to which the methods in question will be applied. Type II standards include rocks, minerals, ores, ferrous and nonferrous alloys, and some biomaterials

(Chapter 4). The elemental contents are not known and must be established analytically. When substantially the same numerical results are obtained by different analytical methods, which are unlikely to have the same systematic errors, we may reasonably conclude that the analytical values are accurate within certain limits and that the methods are correspondingly reliable. A method that does not give analytical values in substantial agreement with values by other methods that agree will be suspect. In such comparisons, it is desirable that methods be tested in various laboratories or by different analysts to exclude the possibility that lack of skill or sagacity may unduly affect a result. When the number of analysts reporting results by a specified method is small—especially if there is only one—it may be difficult to decide whether erroneous results are due to deficiencies in the method or in the analyst.

Some Comparisons of Results by Various Types of Methods Based on Type I Standards. The accuracy and precision attainable in the trace determination of four common metals, in the presence of each other, by emission spectrography, solution absorption spectrophotometry, polarography, and neutron activation have been studied in nine cooperating laboratories distributed internationally.[14] A solution containing approximately 5 ppm each of Cu and Mn and 15 ppm each of Cr and Hg (all concentrations accurately known) was analyzed in these laboratories, a total of 551 determinations being made. The accuracy and precision of the pooled results by each type of method[15] are shown in Figs. 1-2 and 1-3.

In Fig. 1-2 the accuracy is indicated by plotting the relative error in ppt in groups of 100 (0–100, 100–200, etc.) against the number of determinations in each group for each type of method. In Fig. 1-3, the precision is indicated similarly by plotting the deviations from the mean in groups against the number of determinations. The spectrophotometric determinations are seen to be the most accurate and precise; the activation and polarographic determinations follow. As expected, emission spectrography is the least accurate and the least precise method of the four. In practical trace analysis, in which samples are often complex, accuracy and precision (especially the former) might be expected to be worse than found here. Neutron activation might then be found to have relatively better accuracy. Because of the rather simple composition of the standard sample and the comparable amounts of the metals present, interferences are probably small, and what is being tested is chiefly precision or accuracy as determined by precision. Nevertheless, the results are instructive. Unless based on data such as these, analysts tend to overestimate the accuracy and precision of their results and their methods. For example, neutron activation results are sometimes claimed (as mentioned earlier) to

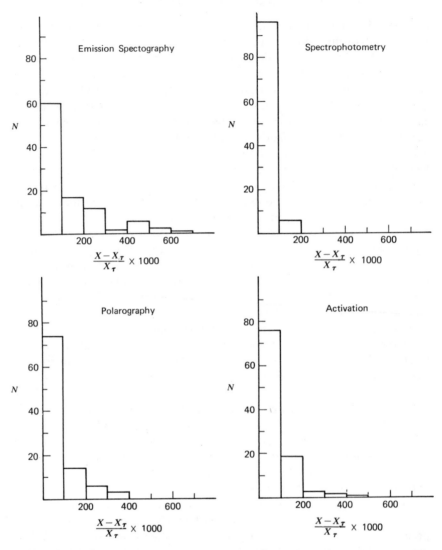

Fig. 1-2. Histograms showing accuracy of determinations of traces of Cu, Mn, Cr, and Hg by various methods. Mean error in parts per thousand, without regard to sign, plotted against number of results (N) in each group. From G. B. Cook, M. B. A. Crespi, and J. Minczewski, *Talanta*, **10,** 917 (1963).

be accurate and precise to 1–2%. The data given here do not support such statements.

Extensive data have been published on the results of the cooperative determination of trace elements in a "salted" uranyl nitrate solution by emission spectrography and absorption spectrophotometry.[16] Some results obtained by other methods (polarography, atomic absorption, flame photometry, and activation analysis) are included. Nineteen laboratories participated. The uranyl nitrate solution (50 mg U/ml) contained

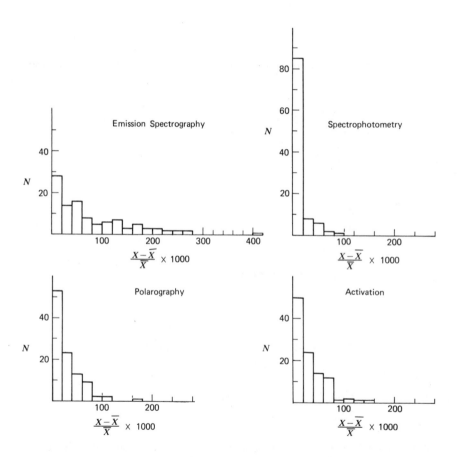

Fig. 1-3. Histograms showing precision of determinations of traces of Cu, Mn, Cr, and Hg by various methods. Mean deviation in parts per thousand plotted against number of results (N) in each group. From G. B. Cook, M. B. A. Crespi, and J. Minczewski, *Talanta,* **10,** 917 (1963).

0.50 ppm each of B and Cd; 5.0 of Co, Cu and Mo; ~10 of Cr and Mn, 20 of Ni and ~50 of Fe. It was concluded that the routine determination of these elements could be made more accurately by spectrophotometry than by other methods. By way of example, the results obtained for copper may be presented. The following (average) emission spectrographic values (ppm) were reported by individual laboratories, not all using the same technique:

5.3	5.4	1.7
5.8	4.0	4.8
4.9	2.5	4.8
5.0	4.9	5.6
		4.8

Avg. 4.58

The following spectrophotometric values (ppm) were obtained by various methods:

$\left.\begin{array}{c} 5.00 \\ 4.74 \end{array}\right\}$ 4.87   Oxalyldihydrazide-acetaldehyde

$\left.\begin{array}{c} 5.4 \\ 5.4 \\ 4.72 \\ 5.12 \\ 4.6 \\ 4.79 \end{array}\right\}$ 5.01   2,2'-Biquinoline

$\left.\begin{array}{c} 4.92 \\ 4.96 \\ 5.16 \\ 4.77 \end{array}\right\}$ 4.95   Neocuproine

$\left.\begin{array}{c} 4.95 \\ 4.99 \\ 4.82 \\ 5.00 \\ 4.96 \end{array}\right\}$ 4.94   Diethyldithiocarbamate ($CHCl_3$ or CCl_4 extraction)

5.09   Oxine ($CHCl_3$)

3.92   Dithizone

Similar comparisons could be made for the other elements. The good results obtained by the spectrophotometric procedures are to be expected. All these elements can be determined easily by spectrophotometric methods, and the concentrations are such that these methods are very suitable. Besides, uranium(VI) is a relatively harmless element in these determinations, so that separations other than those involved in the determination itself (extraction) can be avoided as a rule.

In the same publication,[16] some results of the determination of trace elements in nuclear-grade uranium are given; the uranium was that used to prepare the uranyl nitrate which was then "salted." The trace elements occur in such low concentrations in the pure uranium that they cannot be reliably determined by routine methods of spectrophotometric and other types of analysis, except activation analysis. Again we take copper to illustrate the variation of the analytical results (ppm):

Spectrography	Spectrophotometry	Polarography	Activation
<0.06	0.08		
<0.2	0.15		
<0.25	0.025		
<0.12	<1		
<0.025	Much below detection limit	0.13	
<0.012	0.1		0.008
0.024			
0.028	0.035	$\lesseqgtr 0.05$	
~0.5	0.15	~0.6	
<0.05			

(These determinations are actually on a Type II sample, but are treated here for convenience.)

Here the neutron activation value 0.008 ppm no doubt is close to the truth. All other values (not stated as upper limits) are badly in error. That so many results are reported above the detection limits of the methods is surprising and points to deficiencies in obtaining the blanks. The other possibility is the effect of uranium—in one way or another, it might give an absorbance that would be attributed to copper. This tendency to overestimate trace elements occurring in very low concentrations is not uncommon, and may be noted in the earlier determinations of trace elements in silicate rocks and meteorites carried out by various analytical methods. With samples of such complex composition, high results due to imperfect separations and final determination reactions of insufficient selectivity may be expected.

Some Comparisons of Results by Various Types of Methods Based on Type II Standards. A considerable number of such comparisons involving trace elements are available in the literature. The data are especially extensive for standard igneous rock samples. We will not attempt to review all this information, but some comparisons will be instructive, with special reference to solution spectrophotometric methods.

TABLE 1-7
Determination of Cu in Aluminum by Solution Spectrophotometry and Neutron Activation[a]

| Sample | Cu (ppm) | |
	Spectro-photometry	Neutron Activation
1	172	160
2	30	32
3	1.8	1.3
4	26	25
5	16	15
6	<0.5	0.3

[a] P. Urech, *Mitt. Lebensmitt. Hyg. Bern*, **49**, 443 (1958).

Table 1-7 compares determination of Cu in aluminum metal by activation and spectrophotometry. The agreement is good (hardly surprising). In four of five determinations, the spectrophotometric results are higher than the activation results, but no certain systematic difference is demonstrated.

In Table 1-8, results of the determination of tin in two U.S. Geological Survey standard rock samples by five or six different types of methods are

TABLE 1-8
Determination of Tin in Standard Granite G-1 and Standard Diabase W-1

| Method | Sn (ppm) | |
	G-1	W-1
Emission spectrographic[a]	5.2, 5.0, 9.1, 4.5, ~8	2.5, 2.3, 2.3
Neutron activation[a]	3.5, 3.1	3.4, 2.4
Mass spectrometric[a]	5, 2.3, 1.7	2, 1.0, 1.5
Spectrophotometric[b]	—	2.7
Colorimetric[c]	2.3	2.8
Fluorimetric[d]	2.7	

[a] Mostly from Fleischer (see Table 1-10). Each result from a different laboratory.
[b] A. J. MacDonald and R. E. Stanton, *Analyst*, **87**, 600 (1962). Gallein as reagent.
[c] H. Onishi and E. B. Sandell, *Geochim. Cosmochim. Acta*, **12**, 266 (1957). Dithiol as reagent.
[d] C. Huffman and A. J. Bartel, *U.S. Geol. Surv. Prof. Paper*, **501-D**, 131–133 (1964).

recorded. The spectrographic values for tin in the granite G-1 are so much higher than the values obtained by the other methods that a systematic error in the former, or sample inhomogeneity, is indicated. All methods agree much better on tin in the diabase W-1. One of the neutron activation results is appreciably higher than the other and higher than all the values by other methods, and is suspect.

Determinations of tin by various methods have been made in other standard rock samples and the values are shown in Fig. 1-4 (the spectrophotometric and neutron activation values are recorded in Table 1-8 and have already been discussed). Values by one set of emission spectrographic determinations agree well with those from spectrophotometry in four or five of six rocks and also with one set of neutron activation values, but the other set of spectrographic values do not agree (five out of six rocks). In the spectrophotometric method, tin was separated by anion-exchange chromatography and SnI_4 extraction into toluene and determined with phenylfluorone. The final solution had a volume of 14 ml, and absorbance was measured in a 4 cm cell. Not more than 1 g of rock sample was taken.

From the values tabulated in Table 1-9, it may be concluded that the Mo content of granite G-1 is close to 7 ppm and of the diabase W-1, to

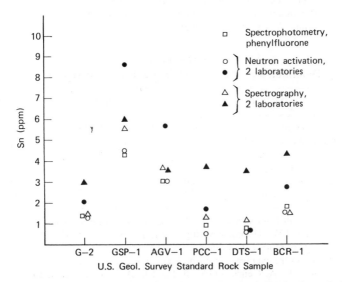

Fig. 1-4. Tin contents of U.S. Geological Survey Standard Rocks by solution spectrophotometry (Table 1-6), neutron activation (Table 1-6), and spectrography [△: Le Riche, quoted in F. J. Flanagan, *Geochim. Cosmochim. Acta*, **33**, 81 (1969); ▲: Blackburn, Griswold, and Dennen, *Chem. Geol.*, **7**, 143 (1971).].

TABLE 1-9
Determination of Molybdenum in Silicate Rocks by Various Methods

Method	Mo (ppm)	
	Granite, G-1	Diabase, W-1
Spectrophotometric, with thiocyanate[a]	6.6	0.5
Spectrophotometric, with dithiol[b]	6.3	0.48
Emission spectrographic[c]	7.3; 7; 8	—; 5; <1
Neutron activation[d]	7.0	1.3
Spark source, mass spectrometric[e]	<0.8; 9.5	<0.8; 0.7

[a] P. K. Kuroda and E. B. Sandell, *Geochim. Cosmochim. Acta*, **6**, 35 (1954). Lillie and Greenland, footrote a, Table 1-5, found 5.2 ppm Mo by the thiocyanate method.

[b] K. M. Chan and J. P. Riley, *Anal. Chim. Acta*, **36**, 227 (1966).

[c] N. Hertz and C. V. Dutra, *Geochim. Cosmochim. Acta*, **21**, 81 (1960); R. M. McKenzie, A. C. Oertel, and K. G. Tiller, *ibid.*, **14**, 68 (1958); M. C. Clark and D. J. Swaine, *ibid.*, **26**, 511 (1962).

[d] H. Hamaguchi, R. Kuroda, T. Shimizu, I. Tsukahara, and R. Yamamoto, *Geochim. Cosmochim. Acta*, **26**, 503 (1962).

[e] R. Brown and W. A. Wolstenholme, *Nature*, **201**, 598 (1964); S. R. Taylor, *Nature*, **205**, 34 (1965).

TABLE 1-10
Determination of Zirconium in Standard Granite G-1

Method	Zr (ppm)
Spectrographic[a]	210, 193, 165, 210, 155, 185; avg. 189
X-Ray fluorescence[a]	220, 214, 239; avg. 224
Mass spectrometric (spark source)[a]	185, 70, 206
Fluorimetric[b]	200
Neutron activation[c]	202

[a] Almost all from summary of M. Fleischer, *Geochim. Cosmochim. Acta*, **29**, 1277 (1965), covering results published in 1962–1965. Each value represents a different analyst.

[b] With morin; average of four determinations ranging from 192 to 207 ppm. R. A. Geiger and E. B. Sandell, *Anal. Chim. Acta*, **16**, 351 (1957). Includes the equivalent of Hf (5.6 ppm by neutron activation).

[c] O. Landström, K. Samsahl, and C. G. Wenner, *Modern Trends in Activation Analysis*, vol. I, p. 357.

30

0.5–1 ppm. Spectrography is erratic: good agreement for G-1, poor for W-1. Spectrophotometry by two different reagents gives closely agreeing values for both rocks and also agrees well with activation for the granite, but it does not agree as well for the diabase. The two different spectrophotometric methods give such closely agreeing values for W-1 that the higher value (by a factor of 2.5) by neutron activation seems to warrant confirmation. The mass spectrometric (spark source) values are in accord for W-1, but they are not at all in accord for G-1.

Determinations of Zr in granite G-1 by five different methods are listed in Table 1-10 and plotted in Fig. 1-5. Except for one erratic mass

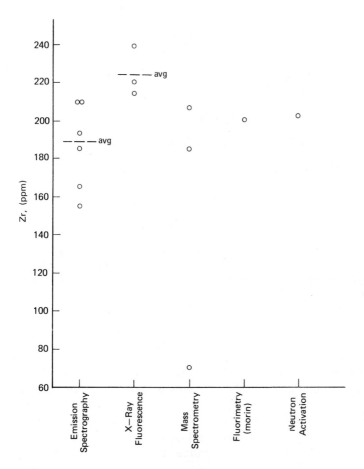

Fig. 1-5. Determinations of Zr in granite G-1 by various methods. Each point represents a different analyst. The mass spectrometric value of 70 ppm Zr is obviously grossly in error.

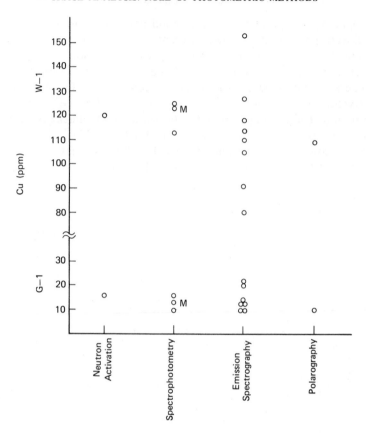

Fig. 1-6. Determinations of Cu in standard granite G-1 and standard diabase W-1 summarized in M. Fleischer and R. E. Stevens, *Geochim. Cosmochim. Acta,* **26,** 525 (1962). Each point represents the result of a different laboratory (some are averages). Spectrophotometric results obtained by Jean'ne Shreeve at the University of Minnesota (1956), not included in Fleischer and Stevens, are indicated by M. They agree closely with the neutron activation values.

spectrometric value, the agreement is fairly satisfactory. Neutron activation and fluorimetry (in solution) give results agreeing within 5% (the fluorimetric Zr value includes Hf, but the latter contributes only a few percent to the total). The emission spectrographic average is a little lower than the presumably correct or almost correct neutron activation value; the X-ray fluorescence average is about 10% higher.

Figure 1-6 brings out the pronounced scatter of the spectrographic determinations of copper in the two standard silicate rocks, the highest values being about twice the lowest in both.

Determinations of manganese in bone by spectrophotometry and activation are compared in Table 1-11.

A comparison of the determination of many elements in basalt (U.S. Geol. Survey Standard BCR-1) by mass spectrometry and neutron activation (both in the same laboratory) showed reasonably good agreement for most elements, better than might have been expected from earlier work.[17] A similar basalt (W-1) was used as standard.

Determination of mercury in a granite and a diabase (Table 1-12) gave good agreement between atomic absorption and solution spectrophotometry (dithizone) for both samples and in one with neutron activation also.

A number of metals can be determined with high sensitivity by solution fluorimetric methods. In Table 1-13 results obtained by the fluorimetric Rhodamine B method for gold are compared with results obtained by other trace methods. Heterogeneity in some of these samples can account for some differences. On the whole, agreement is good.

The results obtained in a considerable number of laboratories in the elemental analysis of a standard sample of plant material (kale, *Brassica oleracea acephala*) by various analytical methods are of much interest.[18] Some aspects of the findings are touched on here. Not unexpectedly, the agreement among results obtained by different methods depended on the particular element. Moreover, the precision of different methods varied from one element to another. Consistent results were obtained by more

TABLE 1-11

Determination of Manganese in Bone by Solution Spectrophotometry and Neutron Activation Analysis[a]

Method	Sample Weight (g)	Mn (ppm)
Spectrophotometric, dry ashing	1.0	0.95 ± 0.04
Neutron activation,		
Wet ashing $(HNO_3 + HClO_4)$	1.0	0.66 ± 0.04
Wet ashing $(HNO_3 + H_2SO_4)$	0.05	0.96 ± 0.06
Dry ashing	1.0	1.09 ± 0.05

[a] M. A. Rojas, I. A. Dyer, and W. A. Cassatt, *Anal. Chem.*, **38,** 788 (1966). Typical results. Spectrophotometric manganese by use of 4,4'-tetramethyldiaminotriphenylmethane.

TABLE 1-12
Results for Mercury in Silicate Rocks by Various Methods[a]

| | Hg found (ppm) | | |
Sample	Neutron Activation	Atomic Absorption	Spectrophotometry (dithizone)
U.S.G.S. G-1	0.34	0.11, 0.08; 0.13	0.090, 0.070, 0.095
U.S.G.S. W-1	0.17	0.18, 0.18; 0.34	0.17, 0.16

[a] W. R. Hatch and W. L. Ott, *Anal. Chem.*, **40**, 2085 (1968).

than one laboratory using the same type of method for the following elements (the method and element content in ppm are indicated).

Au (activation, 0.002)   I (catalysis, 0.08)
B (spectrophotometry, 51)   Rb (activation, 53)
Br (activation, 24)   Sc (activation, 0.008)
Ga (activation, 0.05)   W (activation, 0.06)

Consistent results were obtained by different laboratories, using various methods, for Ba, Ca, Cl, Co, Cr, Fe, Mn, Mo, N, P, and S. For example, the following means, with standard deviations, in ppm were obtained:

$$\text{Co} \begin{cases} 0.052 \pm 0.008 \text{ by activation (2 labs.)} \\ 0.0586 \pm 0.007 \text{ by spectrophotometry (5 labs.)} \end{cases}$$

$$\text{Fe} \begin{cases} 123 \pm 14 \text{ by activation (2 labs.)} \\ 118 \pm 12 \text{ by atomic absorption (4 labs.)} \\ 121 \pm 19 \text{ by spectrophotometry (10 labs.)} \end{cases}$$

$$\text{Mn} \begin{cases} 14.7 \pm 1.3 \text{ by activation (6 labs.)} \\ 15.5 \pm 1.8 \text{ by atomic absorption (5 labs.)} \\ 14.6 \pm 1.9 \text{ by spectrophotometry (9 labs.)} \end{cases}$$

$$\text{Mo} \begin{cases} 2.36 \pm 0.86 \text{ by activation (2 labs.)} \\ 2.33 \pm 0.32 \text{ by spectrophotometry (8 labs.)} \\ 1.00 \pm 0.5 \text{ by emission spectrometry (2 labs.)} \end{cases}$$

In this group of elements it may be noted that the precision of spectrophotometry was statistically the same as that of neutron activation for Co, and better for Mo. For Fe, atomic absorption was more precise than spectrophotometry and possibly more precise than activation. For Mn, activation was more precise (but not by much) than atomic absorption and spectrophotometry.

Small but significant differences among methods were found for Cu, K, Mg, Na, Pb, Se, Sr, and Zn. For example,

$$\text{Cu} \begin{cases} 4.12 \pm 0.73 \text{ by activation (3 labs.)} \\ 5.25 \pm 0.48 \text{ by atomic absorption (5 labs.)} \\ 4.65 \pm 0.52 \text{ by spectrophotometry (10 labs.)} \end{cases}$$

$$\text{Zn} \begin{cases} 32.3 \pm 1.9 \text{ by activation (3 labs.)} \\ 34.2 \pm 1.6 \text{ by atomic absorption (7 labs.)} \\ 30.7 \pm 5.5 \text{ by spectrophotometry (6 labs.)} \\ 24.6 \pm 8.0 \text{ by polarography (2 labs.)} \end{cases}$$

For copper, the precision of the three methods is not significantly different; for zinc, precision by atomic absorption and activation is better than by spectrophotometry and polarography. Polarography gave low results for zinc.

Gross differences among the methods were found for Al, As, Hg, Ni, and Ti. For example,

$$\text{Al} \begin{cases} 7.35 \text{ and } 78.5 \text{ by activation } (\gamma\text{-ray spectrometry; 2 labs.)} \\ 35.5 \text{ (4 labs.) and } 6.4 \text{ (1 lab.) by spectrophotometry} \\ 80.1 \text{ by emission spectrometry (2 labs.)} \end{cases}$$

$$\text{Hg} \begin{cases} 0.15 \text{ by activation (2 labs.)} \\ 0.01 \text{ by spectrophotometry (1 lab.)} \end{cases}$$

TABLE 1-13
Determination of Gold in Complex Materials by Various Trace Methods[a]

Sample	Rhodamine B (fluorescence) (ppm)	Atomic Absorption after Fire Assay (ppm)	Neutron Activation (ppm)	Neutron Activation After Fire Assay (ppm)
Quartz-pyrite (1 : 1)	3.6, 2.9	3.3, 4.0		
Quartz	2.2 ± 0.1	2.5		
Altered zone material	0.12			0.13
Mud, Red Sea	0.09, 0.085		0.13	
Rock (unspecified)	0.022		0.019	
Standard rock, W-1	0.0020, 0.0014, 0.0015		0.0048, 0.0049, 0.0034	

[a] J. Marinenko and I. May, *Anal. Chem.*, **40,** 1137 (1968). Sample weight, 10 g or greater. According to Fig. 1-1, the method can be described as $10 + g/10^{-3}$–1 ppm, meaning a sample weight of 10 g or greater, with the determination of Au in the range 0.001 to 1 ppm approximately.

Apparently loss of mercury by volatilization occurred in the decomposition of the sample for the spectrophotometric determination.

From these results, it appears that from the standpoint of precision and accuracy the colorimetric (solution spectrophotometric) determination of Co, Mo, Cu, Fe, Mn, and Zn in this sample is competitive with the neutron activation determination, or even better for Co and Mo.

In a study of the determination of lead (~ 15 ppm) in human rib bone by various methods, Hislop, Parker, Spicer, and Webb concluded that emission spectrography and absorptiometry with dithizone as reagent were the most suitable methods.[19] Atomic absorption (after dithizone or diethyl-dithiocarbamate extraction of Pb) was considered to be insufficiently sensitive. Absorptiometry based on measurement of the lead chloride complex in the UV range has the disadvantages of a somewhat high blank and a sensitivity lower than that of the dithizone method; moreover, it is not specific and may be subject to interference from traces of organic material. "... emission spectroscopy and absorptiometry using dithizone were selected since these were reliable, sensitive, comparatively simple to apply and were reasonably economic to operate. Gamma activation analysis, which is comparable in sensitivity with these methods and has the advantages of being independent of reagent and apparatus blanks, is a more costly method and was applied principally to the investigation of the loss of lead under various ashing conditions, for which purposes it is ideally suited."

IV. CHOICE OF METHOD

This topic is not considered in detail here—for this book deals with a restricted sector of trace analysis—but some general observations may be appropriate before conclusions are drawn concerning the part that solution photometry can play in trace analysis.

The prime requirement in any analytical determination is a certain degree of accuracy, which varies with the purpose of the determination. Accuracy obviously depends on the sensitivity of the method (especially in trace analysis), its reproducibility (precision), and its freedom from sizable determinate errors. The most serious determinate errors arise from the effect of other constituents of a sample: major, minor, and trace. Other important considerations in practice are the length of time required for an analysis and its cost. Other factors remaining the same, the use of expensive instrumentation obviously is not a disadvantage if the time of an analysis is reduced and many determinations are made.

One type of analytical method may be more selective and more sensitive than another in general, but all methods vary in these respects from one element to another. A particular type of method is not always the best choice for a given element in a given material. Moreover, what may be called local conditions—the availability of instruments and operator expertness—play an important role in the choice of an analytical method. Choice is often restricted in practice.

Some types of methods are clearly more suited for routine use than others. Atomic absorption, fluorescence, and emission are superior to neutron activation in this respect.

Turning now to the merits and demerits of solution absorptiometry and fluorimetry in trace analysis, we may first observe that the sensitivities of these methods are sufficiently good to allow the detection of 0.002–0.003 μg (less often ~0.001 μg) of many metals. As already discussed, such detection limits permit the determination of 0.05–0.1 μg of metal, or say down to 0.1 ppm with 0.5–1 g of sample. This sensitivity makes these methods competitive with neutron activation and flame atomic absorption for *some* metals. Absorptiometry and fluorimetry lack the selectivity of neutron activation and atomic absorption (and atomic fluorescence and emission). More often than not they require separations to be made before a metal can be determined in a complex matrix. Separations lengthen a procedure and tend to put it in an inferior position with respect to others not requiring separations or only simple ones. But methods requiring separations have an important advantage: They make the determination of a metal much less dependent on the matrix composition. Determinations then become absolute and are not dependent on the accuracy of standards used in some physical methods. For the most part, modern separation methods are very effective, with good recoveries, but the ease of separation varies from one element to another.

"By their fruits ye shall know them." The results recorded in the tables and figures of this chapter lead one to believe that Cu, Sn, Mo, Zr (+Hf), and Au can be determined by solution photometry in a complex material such as a silicate rock with accuracy (and precision) as good as, if not better than, by neutron activation, emission spectrography, and other physical methods as ordinarily practiced.[20] Other work indicates that as far as accuracy is concerned, determination of As, Be, Cd, Co, Cr, Ga, Ge, Ni, Pb, Sb, V, and Zn by solution photometry in silicate rocks is competitive with most of the physical methods. Metals such as these should also be determinable in biomaterial (if their content exceeds 0.1 ppm of the ashed sample), natural waters, and the like. On the other hand, metals such as the alkalies, alkaline earths, and the rare earths are not profitably determinable by molecular photometric methods. Atomic absorption, fluorescence, and emission are far superior.

The low cost and simplicity of absorptiometric and fluorimetric instruments are, at times, important factors in favor of their use. Portable spectrophotometers, weighing 0.5 kg or so, are convenient for use in the field. One such (grating) instrument, operating on Ni–Cd batteries (rechargeable), has a spectral range of 400–800 nm and a band width of 20 nm; it allows the use of 2.5 cm cells.

In fact, color and fluorescence methods may be applied in the field without any instruments. A series of tubes suffice for color and fluores-

TABLE 1-14
Trace Metals Advantageously Determined by Solution Absorptiometric or Fluorimetric Methods

Element	Reagent or Method
Ag	Dithizone or 5-(p-diethylaminobenzylidene)rhodanine and other reagents
As	As–Mo blue
Au	Rhodamine B (fluorescence)
Be	Morin (fluorescence)
Bi	Dithizone
Cd	Dithizone
Co	Nitroso-R salt and other nitroso reagents
Cr	Diphenylcarbazide
Cu	Dithiocarbamates, "cuproin" reagents, dithizone
Fe	1,10-Phenanthroline and analogs, thiocyanate
Ga	Rhodamine B (fluorescence) and other basic dyes, 8-hydroxyquinoline (fluorescence)
Ge	Phenylfluorone and other fluorones
Hg	Dithizone
Mo	Dithiol, thiocyanate
Ni	Oximes
Os	Catalysis (CeIV–AsIII)
Pb	Dithizone
Pd	p-Nitroso reagents, dioximes
Pt	p-Nitroso reagents
Re	Oximes, thiocyanate, basic dyes
Ru	Diphenylthiosemicarbazide, 1,10-phenathroline, catalysis
Sb	Rhodamine B and analogs
Sn	Phenylfluorone and others
Ta	Basic dyes ($+F^-$)
Th	Arsenazo III
Ti	Diantipyrinylmethane or thiocyanate-tri-n-octylphosphine oxide
Tl	Rhodamine B and analogs
U	Fluorescence of UO_2^{2+}, Arsenazo III
Zn	Dithizone
Zr	Morin (fluorescence)

cence comparisons. In this way, trace determinations can be made quickly with adequate accuracy at remote locations with minimal equipment. Such analyses are important in geochemical exploration.[21]

When the amount of sample available is limited, elements can be determined sequentially by photometric methods after suitable separations.[22]

Elements are not dissipated in the photometric methods considered here and can be recovered if necessary.

Finally, we list (Table 1-14) metals that can profitably be determined in trace concentrations in complex materials by color and fluorescence reactions. To some extent the listing is arbitrary. The metals included are those that can be determined sensitively and usually without undue difficulties in separation, because of the existence of selective separation reactions. For other metals, such as Al, numerous sensitive color reactions are available, but separations may leave something to be desired. Obviously, not all of these metals can be determined satisfactorily by these methods in every kind of material. On the other hand, metals not listed are also determinable, usually with good sensitivity (but not as sensitively as some of the others). Exceptions are the alkali and alkaline earth metals, the rare earths, Hf, and two or three others, as already mentioned.

What is contemporarily (early 1970s) the actual relative extent of use of various trace determination methods in the inorganic field? A good

TABLE 1-15
Types of Methods Applied in Determination of Inorganic Trace Constituents[a]

Type of Method	Number of Laboratories Using
Colorimetry and absorption spectrophotometry	78
Atomic absorption spectrophotometry	75
Emission spectrography (arc, spark)	43
Polarography	35
Radioactivity	26
X-Ray (emission and fluorescence)	25
Flame spectrography	17
Mass spectrometry (spark)	12
Electrical methods	10
Direct potentiometry (specific electrodes)	8
Chromatography (photometry)	6
Direct reading spectrometry	6

[a] M. Pinta, *Pure Appl. Chem.*, **37**, 483 (1974).

answer is provided by the results of a survey conducted by the Commission on Microchemical Techniques and Trace Analysis of the International Union of Pure and Applied Chemistry. One thousand questionnaires were sent worldwide to analysts in 600 laboratories. Replies were received from 208 laboratories. Of these, 20 laboratories indicated no interest in trace analysis. The remainder (188 laboratories) supplied information on the types of methods used by them in the determination of inorganic trace constituents (Table 1-15). Materials analyzed, in decreasing frequency order, were:

> Rocks, minerals, soils
> Water
> Metals, alloys
> Pure chemicals
> Plant materials
> Biological media
> Industrial products (e.g., glass)
> Air, atmosphere

NOTES

1. This definition has, in part, a historical basis. When, until the first part of this century, gravimetric and titrimetric methods were virtually the only analytical methods available, constituents below 0.01% were not easily determinable. There was only an occasional need to determine such constituents at that time. Hillebrand stated, "It may be said with regard to the use of the word 'trace' that the amount of a constituent thus indicated is supposed to be below the limit of quantitative determination in the amount of sample taken for analysis. It should, in general, for analysis laying claim to completeness and accuracy, be supposed to indicate less than 0.02 or even 0.01 percent." (W. F. Hillebrand, "The Analysis of Silicate and Carbonate Rocks", *U.S. Geol. Surv. Bull.*, **700**, 32 (1919); W. F. Hillebrand, G. E. F. Lundell, H. A. Bright, and J. I. Hoffman, *Applied Inorganic Analysis*, Wiley, New York, 1953, p. 807.) Trace does not refer to an absolute amount of a constituent.

2. A convenient unit for expressing the relative content (by weight) of a trace constituent is often parts per million (ppm, $\mu g/g$, mg/kg, or g/metric ton). The unit parts per billion (ppb) is ambiguous, being 1 part per 10^9 (ng/g) in the USA and 1 part per 10^{12} in Great Britain and on the Continent generally. A term long in existence for 10^9 is "milliard" (abbreviated M), but this seems not to be universally known in the USA. The abbreviation ppM may puzzle some persons and may be misinterpreted by others. This ambiguity is avoided easily enough by the use of $pp10^9$. If a solution is involved, and it is desired to base concentrations of elements on it, and not on the dissolved solid sample, this is made clear by using $\mu g/ml$ or ng/ml, instead of ppm or $pp10^9$. Occasionally, one sees in journals (usually considered reputable) ppt used for

parts per trillion. This usage is doubly ambiguous; ppt is well established as meaning parts per thousand.

3. For surveys of physical and physicochemical methods, see G. H. Morrison, Ed., *Trace Analysis: Physical Methods*, Interscience, New York, 1965, and J. D. Winefordner, Ed., *Trace Analysis: Spectroscopic Methods for Elements*, Wiley, New York, 1976.

4. Determination of Hg by the cold vapor method, in which Hg atoms in a gas phase are the absorbing species, comes close to being included here. Presumably, OsO_4 and a few other gaseous or volatile metal compounds could be determined in a gas phase by molecular absorption.

5. E. B. Sandell, *Colorimetric Determination of Traces of Metals*, 3rd ed.

6. Apparently the largest samples used so far in trace analysis are the 5000 and 10,000 metric ton samples of melted Antarctic ice taken for the determination of soluble and insoluble constituents [E. L. Fireman, *Antarc. J. U.S.*, 250 (1968)].

7. Calculated from figures tabulated by Browner for $100\,\mu l$ samples in a commercial graphite tube furnace. Even higher sensitivities are reported by others.

8. See, for example, V. P. Guinn and H. R. Lukens, Jr., in G. H. Morrison, Ed., *Trace Analysis. Physical Methods*, Chapter 9, Interscience, New York, 1965. (Because it is not our intention to go into the various physical methods in depth, literature references are kept to a minimum here and elsewhere.)

9. The average difference between the seven results by the spectrophotometric dithiol method and by the first activation set, without regard to sign, is only 0.03 ppm, and <0.01 ppm with sign taken into account. This is remarkably good agreement between methods of entirely different nature.

10. The average difference (spect. − neut. act.)/6 is $+0.15$ ppm for the first set of neutron activation results and -1.53 ppm for the second set. This difference is large enough to justify acceptance of the first set of neutron activation values in preference to the second. For the first set, the tin ratios spect./neut. act. are 1.1, 1.0, 1.0, 1.76, 1.5, 1.2. Four out of six of these values are satisfactorily close to unity.

11. For surveys of electrochemical methods with reference to trace analysis, see J. K. Taylor, E. J. Maienthal, and G. Marinenko, in G. H. Morrison, Ed., *Trace Analysis, Physical Methods*, p. 377.

12. Cf. H. B. Mark, Jr., and G. A. Rechnitz, *Kinetics in Analytical Chemistry*. Interscience, New York, 1968.

13. See, for example, A. Townshend, *Process Biochem.*, **8**, 22, 24 (1973).

14. G. B. Cook, M. B. A. Crespi, and J. Minczewski, *Talanta*, **10,** 917 (1963).

15. The accuracy and precision of the determinations of each element by each method in the nine laboratories individually are listed in Cook et al.[14]

16. *Analytical Chemistry of Nuclear Materials. Second Panel Report*, International Atomic Energy Agency, Vienna, 1966.

17. G. H. Morrison and A. T. Kashuba, *Anal. Chem.*, **41**, 1842 (1969).

18. H. J. M. Bowen, *Analyst*, **92**, 124 (1967). Results obtained by R. A. Greig, *Anal. Chem.*, **47**, 1682 (1975), are also of interest. He compared results of the determination of Ag, Cr, and Zn in a series of marine organisms by atomic absorption and neutron activation. According to our computations, the average differences between neutron activation and atomic absorption results were 29% for Ag (29 samples), 58% for Cr (19 samples), and 13% for Zn (13 samples).

19. J. S. Hislop, A. Parker, G. S. Spicer, and M. S. W. Webb, AERE-R7321 (1973).

20. Trace determinations of metals by physical methods in samples other than silicate rocks may also give results that do not agree well. See, for example, the comparison of results obtained in the analysis of coal by spark source mass spectrometry, DC arc spectrography, instrumental neutron activation, and the like in D. J. von Lehmden, R. H. Jungers, and R. E. Lee, Jr., *Anal. Chem.*, **46**, 239 (1974). Reported values, for the same coal sample, showed the following ranges (ppm): Hg, <0.02–<2; Be, <0.1–0.4; V, 5.5–10; Mn, 1.9–20; Sr, <30–160, Ba<2–500, and so on. Such large differences can hardly be accounted for by sample inhomogeneity.

21. A. A. Levinson, *Introduction to Exploration Geochemistry*, Applied Publishing, Calgary, 1974, p. 264. R. E. Stanton, *Analytical Methods for Use in Geochemical Exploration*, Halsted, New York, 1976.

22. B. Morsches and G. Tölg, *Z. Anal. Chem.*, **250**, 81 (1970), describe the determination of nanogram amounts of As, Be, Cd, Co, Cr, Cu, Fe, Hg, Mn, Mo, Ni, Pb, Sn, Tl, V, and Zn in 0.1 g samples of biomaterial after stepwise separations by liquid–liquid extraction. (Cd, Pb, Tl, and Zn were determined by inverse voltammetry but could also have been determined spectrophotometrically.) The volume of solution for absorption measurement was $\sim$1 ml and a 5 cm cell was used. Errors in analysis of synthetic mixtures of all these elements were usually less than 10%.

CONTAMINATION AND LOSSES
IN TRACE ANALYSIS

I. CONTAMINATION

The unintentional introduction of any substance during the preliminary treatment of a sample and in the course of its analysis constitutes contamination.[1] The most objectionable contaminant is, of course, the constituent being determined, but other contaminants may also be undesirable or objectionable. (Contamination in the preparation of samples for analysis is considered in Chapter 3.)

A. ATMOSPHERIC CONTAMINATION

Unfiltered laboratory air may contain enough suspended material to be a hazard in trace metal determinations. Airborne contaminants can be a serious source of high and irregular blanks in refined trace analysis. Even in what may be called unpolluted environments, the atmosphere near the earth's surface contains dust consisting of silicate particles (Na, K, Ca, Mg, Al, etc.), sodium chloride, and other salts, organic debris, and, in exceedingly low concentrations, cosmic material. To these are added particulate matter, widely dispersed, arising from the activities of man. Some elements may be present partly in gaseous form, for example, Hg, I_2, and various forms of C, N, S, Pb as $Pb(C_2H_5)_4$. In an urban environment, the concentration of dust and gaseous contaminants in the atmosphere normally increases. The values listed in Table 2-1 give an idea of the levels of certain elements (not all those present were determined) in the atmosphere of a large city. Table 2-2 supplies data for metals in urban air of the United States. The number and amounts of contaminants will vary greatly with local conditions (Table 2-3). Urban air is not likely to contain as much as 1 mg of solid material/m^3, but locally this value can be exceeded. Dust storms can increase normal values a hundredfold or more. Combustion of fuels and varied industrial operations pollute the atmosphere with a variety of elements. The use of leaded automotive fuels has led to a worldwide increase in atmospheric lead (and bromine) concentration. The air in the central sections of large cities may contain 1–10 μg Pb/m^3 (late 1960s), air over the mid-Pacific $\sim$0.001 μg Pb/m^3, and that in the Arctic and Antarctic regions <0.0005 μg Pb/m^3. Urban rain has been found to contain as much as 0.004 ppm lead alkyls. Clearly, the analyst in

TABLE 2-1

Trace Elements in Particulate Form in Surface Air of Chicago Metropolitan Area[a] (ng/m^3)

Al	480–3180	Cr	5.7–37	Na	80–670
Br	24–320	Fe	820–5130	Sb	1.4–55
Cl	610–5880	Hg	3.2–39	Sc	0.21–0.96
Co	0.32–4.84	Mn	100–900	V	2.2–121
				Zn	190–1700

[a] S. S. Brar, D. M. Nelson, E. L. Kanabrocki, C. E. Moore, C. D. Burnham, and D. M. Hattori, in J. R. DeVoe and P. D. LaFleur, Eds., *Modern Trends in Activation Analysis*, Nat. Bur. Stands. Spec. Pub., 312, Vol. 1, p. 43, Washington, 1969. Element range in samples collected at 22 localities is tabulated. The total particulate matter in these samples ranged from 27.8 to 144 μg/m^3. Sendai, Japan, atmospheric particulate material, analyzed by T. Kato, N. Sato, and N. Suzuki, *Talanta*, **23**, 517 (1976), has a similar composition.

TABLE 2-2
Heavy Metals in Urban Air, USA, 1964–1965[a]

Metal	Concentration (ng/m³)		Metal	Concentration (ng/m³)	
	Average	Maximum		Average	Maximum
As	20		Mo	< 0.5	780
Be	< 0.5	10	Ni	34	460
Bi	< 0.5	64	Pb	790	8,600
Cd	2	420	Sb	1	160
Cr	15	330	Sn	20	500
Co	< 0.5	60	Ti	40	1,100
Cu	90	10,000	V	50	2,200
Fe	1,580	22,000	Zn	670	58,000
Hg[b]					
Mn	100	9,980			

[a] G. B. Morgan, G. Ozolins, and E. C. Tabor, *Science*, **170**, 289 (1970). Average refers to contents found in biweekly sampling of particulate matter at 35 to 133 stations.
[b] R. S. Foote, *Science*, **177**, 513 (1972), found 3.25 ng Hg/m³ in Washington, D.C. air; 3.14, San Francisco; 3.38, Dallas.

an urban environment—and that is where he is likely to be found—needs to guard, for example, against lead contamination, to name but one metal.

In a general analytical laboratory, almost any element can become an airborne contaminant from the trace standpoint. Air quality may be much worse than in the external atmosphere. Common gaseous contaminants include hydrogen halides (especially HCl and HF), ammonia, hydrogen sulfide, organic solvents, and mercury.[2] A trace of hydrogen sulfide could ruin a spectrophotometric trace metal determination. Walls, ceilings, floors, the interior of hoods, and the like are all sources of solid contaminants: dust and other deposits, paint particles, and the like. These can carry calcium, and other light elements as well as heavy metals. Corrosion products may be contributed by ring stands, burners, hot plates, ovens, furnaces, and metal objects in general.

Frequent and thorough cleaning of the laboratory and its equipment will aid in reducing contaminants. Paints containing lead, zinc, and other heavy metals should be shunned. Metal apparatus should be avoided as far as possible when trace metals are being determined, or the metal should be covered with plastic, porcelain, or fused silica as the temperature requires. Metal Tirrill or Meker burners and water baths should not be used when trace metals such as copper, zinc, and nickel are to be determined. High temperature furnaces may be a source of volatilized

TABLE 2-3
Elements in Atmospheric Suspended Material from an Industrial $(A)^a$ and a Rural Locality $(B)^a$ and from the South Pole $(C)^b$

	Content (ng/m^3)				Content (ng/m^3)		
	A^c	B^d	C		A^c	B^d	C
Ag	2	< 1		La	6	1	5×10^{-4}
Al	2,200	1,200	6×10^{-1}	Mg	2,400	500	
As	10	5		Mn	260	60	1×10^{-2}
Br	80	40	6×10^{-1}	Na	500	170	7
Ca	7,000	1,000	5×10^{-1}	Ni	< 60		
Ce	10	1	2×10^{-3}	Pb			6×10^{-1}
Co	3	1	8×10^{-4}	S	13,000	11,000	
Cr	110	10	5×10^{-3}	Sb	30	6	2×10^{-3}
Cu	1,100	280	4×10^{-2}	Sc	3	1	1×10^{-4}
Eu	0.1	0.06·	2×10^{-5}	Se	4	3	6×10^{-3}
Fe	13,800	1,900	8×10^{-1}	Sm	0.4	0.2	6×10^{-5}
Ga	1	0.9		Th	1	0.3	6×10^{-5}
Hg	5	2		Ti	190	120	
In	0.1	0.04		V	20	5	2×10^{-3}
K	1,400	750	3×10^{-1}	W	6	0.4	
				Zn	1,500	160	3×10^{-2}

[a] R. Dams, J. A. Robbins, K. A. Rahn, and J. W. Winchester, *Anal. Chem.*, **42,** 861 (1970). Values rounded here.
[b] W. H. Zoller, E. S. Gladney, and R. A. Duce, *Science*, **183,** 198 (1974).
[c] East Chicago, Indiana (industrial).
[d] Niles, Michigan (rural location).

elements, as from the ceramic, the heating elements and occasionally from material that has been spilled in them.[3] Ovens also may be suspect. Electric motors (armatures) can contaminate the atmosphere with copper.

Blanks, properly designed and carried out, help to minimize errors due to aerial (and other) contamination, especially if the blank value is smaller than the content of trace constituent in the sample and is not erratic. Large blanks have an unfavorable effect on both accuracy and precision, so that contamination must be kept within bounds.

Reduction of aerial contamination requires exclusion of dust and gaseous contaminants. Depending on the content level of the trace constituents being determined, more or less elaborate (and correspondingly costly) measures are chosen. The requirements in photometric trace analysis are not as stringent as in more sensitive methods used for lower trace contents. Suppose we take 0.1 μg of an element as the lower limit of a determinable amount (representing, for example, 10^{-5}% or 0.1 ppm in a 1 g sample) and allow 1% contamination (0.001 μg). Assuming the air of the laboratory to hold the amounts of elements found for an industrial area as represented by column 1 in Table 2-3, we calculate the volumes of air (liters) containing 0.001 μg of some selected elements:

Al	~0.5	Fe	0.07
Co	~300	Mn	~4
Cu	0.9	Sb	~30

It may be surmised that if the usual precautions observed in ordinary analytical practice are taken,[4] the danger of contamination to the extent stipulated would be small for Co and Sb and probably so for Al and Mn, even if operations such as evaporations are rather prolonged. On the other hand, contamination by Fe is very likely, for example, dissolving a sample in boiling HCl in a 250 ml flask might be expected to bring into solution much of the iron in that volume of air (~ 0.0035 μg Fe). Copper might be a border-line element. An example such as this gives no particular assurances because local conditions conceivably could be quite different from those assumed (the assumption that the quality of the laboratory air is as good as the outside air is especially questionable), but it seems justifiable to conclude that when as little as 0.1 μg of an element is being determined, means to exclude dust and undesirable gaseous substances can often be simple ones.

A simple way of reducing atmospheric contamination is to cover the vessel, in which an evaporation or other operation is being carried out, with an inverted Pyrex crystallizing dish or beaker having a side arm through which a slow stream of carefully filtered air, nitrogen, or other gas is passed (see Fig. 3-3).[5] The protecting inverted beaker should rest on a base of Pyrex, porcelain, or the like so that it forms a chamber in which the inner vessel is contained. The whole is placed on a hot plate when an evaporation is made. Also, heating from above (infrared lamp) speeds up evaporation and prevents condensation.

A more elaborate arrangement may be needed for a wider range of operations in exacting work. Essentially what is required is a box of suitable material through which filtered air is passed. A description has been given of such a box, constructed of acrylic plastic and provided with polyethylene sleeves allowing hand manipulations within the box.[6] Only acrylic plastic and Teflon surfaces are exposed in the interior of the box; the heater has no exposed metal. Comparison of treatment of samples (evaporation of solutions) in the box and in the open air of the laboratory showed that as much as an order of magnitude more of some impurities was found in special purity materials when the preparation was carried out in the open air (Table 2-4).

TABLE 2-4
Impurities Introduced by Solution Evaporation in the Open Laboratory[a]

Sample	Treatment[b]	Impurities (ppm)									
		Al	Fe	Ca	Cu	Mg	Mn	Ni	Pb	Ti	Cr
H_2O	A	0.002	0.0003	0.002	6×10^{-5}	0.002	6×10^{-5}	$<2\times10^{-5}$	8×10^{-5}	4×10^{-4}	$<8\times10^{-5}$
	B	0.008	0.007	0.004	6×10^{-4}	0.01	4×10^{-4}	3×10^{-4}	8×10^{-4}	8×10^{-4}	3×10^{-4}
HCl	A	<0.004	0.003	0.005	2×10^{-4}	0.003	3×10^{-4}	$<1\times10^{-4}$	$<4\times10^{-4}$	$<1\times10^{-4}$	$<4\times10^{-4}$
	B	0.02	0.03	0.07	2×10^{-3}	0.04	4×10^{-4}	8×10^{-4}	1×10^{-3}	1×10^{-3}	7×10^{-4}

[a] Selection of values from Vasilevskaya et al.
[b] A, in box; B, in open laboratory. Determinations made spectrographically. Differences between B and A represent atmospheric contamination. Contents refer to original sample.

On a larger scale, clean-air, laminar-flow, hoods are commercially available. Special filters can supply air containing less than 100 particles (>0.5 μm in diameter)/ft^3 ($\sim$28 1).[7] The ultimate is a clean-air room supplied with such filtered air and kept under positive pressure, and having a laminar flow clean-air hood in which analytical operations are carried out.[8]

Gaseous contaminants may require removal from the air stream by adsorption or chemical reaction.[9]

Aside from airflow considerations, laboratory contamination can be diminished by proper laboratory design. This is a topic that will not be treated here, but a quotation will be instructive[10]:

The requirements for trace elements can best be illustrated by outlining some points in the design of the new spectrochemical laboratory of the Macaulay Institute now being built. The construction is conventional with plastered walls, linoleum-covered heated floor and teak benches with wooden cupboard and drawer units. The walls will be painted with an amide-cured epoxide resin paint, with the restriction that titanium will be the only metallic pigment. Lead, zinc and other biologically important metals are to be excluded. Metallic driers such as cobalt have also been eliminated. All exposed metal work has been avoided as far as practicable. Electric switches and sockets are plastic covered, as are the fluorescent electric light fixtures, under bench pyrotenax cables and copper hot water pipes. Other metal pipes, including those for gas and compressed air, are to be protected with a silicone-based aluminium paint, which has been found to be very useful for general laboratory use. Cold water supplies, sink fittings and waste pipes will be of polythene. Plastic-wheeled remote-control taps will control all water and gas services, so avoiding handling of metal as far as possible, the outlets being PVC covered. The fume cupboards are to be entirely lined with PVC, with hot plates and water baths mounted flush in a hard asbestos false floor. Corrosion-resistant water baths are desirable, and polypropylene would appear to be the most suitable material for the bath itself. For a number of years aluminium–5% magnesium alloy hot-plates have been employed in place of the usual readily-corroded iron hot-plates. They have given good service in an atmosphere of acetic acid, hydrochloric acid and nitric acid. The presence of Mg is undesirable, but more acceptable than Cu, a major constituent of dural-type alloys. Copper contamination also arises from the normal bunsen burner: it has been found impracticable to ash small amounts of organic materials in a crucible over a normal burner without increasing its Cu content appreciably. The remedy is to fit silica burner tubes or to use electric burners with silica sheaths over the heater elements. Silica-lined muffle furnaces and aluminium-lined ovens are the most suitable available at present. Equipment using nickel, zinc or cadmium plating should generally be avoided in trace element work.

Operator Contamination. The analyst himself may be a source of trace elements, conveyed to the sample or the utensils by way of the atmosphere or in other ways. The following metal contents (ppm) of human skin have been reported[11]:

Be	0.02	Pb	210
Co	0.7	Cd	1.1
Cu	0.75, 5.7	Zn	6, 19.5
Ag	0.035	As	0.07
Au	0.002		

Na and Cl, and other ions as well, occur in perspiration. Conceivably heavy metals could be introduced in the handling of samples, for example, mineral specimens. Because metals occur so frequently in the environment and metal objects are handled by the analyst, the metal content of his finger epidermis may be greater than indicated by these values. The silver content of some rocks is thought to have been overestimated by some spectrographers as a result of contamination from silver coins. Haber mentions introduction of gold, in a submicro determination of this element, from the gold-rimmed spectacles of the assayer.

More important than the skin per se as a source of contamination may be the varied powders, plasters, salves, ointments, lotions, antiperspirants, depilants, and antiseptics applied to the epidermis of *Homo sapiens.*

The byproducts of smoking—smoke and ash—are a fruitful source of some contaminating trace metals.

Laboratory contamination can also be introduced from the clothing and shoes of workers.

Particle masks, polyethylene sleeves, and PVC gloves have been worn to reduce contamination.[12]

B.   FROM APPARATUS

The contamination considered in this section arises mostly from the attack of containers by solutions.[13] Nowadays the analyst has a wide choice in the selection of the materials of his vessels. The sample level of the constituent being determined will be an important factor in his choice.

1.   Glass

Because of the generally favorable mechanical and thermal properties of glass and the availability of glass vessels in a great variety of forms and sizes, the trace analyst is likely to consider the use of such vessels at one stage or other of his work. To the analyst, "glass" means borosilicate glass, when not otherwise qualified, such as Pyrex and Jena (Table 2-5). These are quite resistant to the action of weakly acidic and weakly basic solutions, less so to strongly acid solutions, and still less to moderately alkaline solutions (Tables 2-6 and 2-7). Vycor, a high silica glass,[14] is

TABLE 2-5
Composition of Chemically Resistant Glasses
(n.d. = not determined)

	Pyrex[a] (%)	Jena[b] (%)	Vycor[c] (%)	Alkali-Resistant (%)
SiO_2	81.0	74.5	96.3	71
B_2O_3	13.0	4.6	2.9	<0.2
R_2O_3	2.2	8.5	0.4	16
(mostly Al_2O_3)		(Al_2O_3)		
ZnO	n.d.[d]		n.d.	
CaO	negligible	0.8	negligible	
	(0.1 Ca)			
BaO	n.d.	3.9	n.d.	
MgO	n.d.	0.1	n.d.	
	(0.06)			
Na_2O	3.6	7.7	<0.02	⎫
				⎬ 12
K_2O	0.2		<0.02	⎭
As_2O_3	0.002		0.005	
Sb_2O_3	n.d.		~0.1	
Undetermined			0.3	
Σ	100.0	99.9	100.0	

[a] Chemical Pyrex glass, Corning 7740; analysis by E. Wichers, A. N. Finn, and W. S. Clabaugh, *Ind. Eng. Chem. Anal. Ed.*, **13**, 419 (1941). D. E. Robertson, *Anal. Chem.*, **40**, 1067 (1968), reported the following metal contents (ppm) in Corning borosilicate glass: Fe, 280; Zn 0.7; Sb, 2.9; Co, 0.08; Sc, 0.1, Hf 0.6. According to W. Stross, *Talanta*, **9**, 740 (1962), British Pyrex contained as much as 0.3% As_2O_3 before 1950, <5 ppm in 1962. Duran 50 (Schott and Gen.) is sufficiently low in As and Sb to be satisfactory for use in determination of As by evolution as arsine.
[b] Geräteglass 20.
[c] Corning 7900. Robertson[a] reported 0.11% Sb.
[d] O. R. Alexander and L. V. Taylor, *J. Assoc. Offic. Agr. Chem.*, **27**, 325 (1944), found ~10 ppm Zn in Pyrex.

more resistant than the ordinary borosilicate glasses to the action of solutions and can be used at temperatures as high as 900°C for long periods of time.

When traces of such elements as Si, B, Al, Na, and K are to be determined, borosilicate glass of Pyrex composition should, in general, be avoided. Brief contact of solutions with such glass may be permissible, as in pipeting.[15] Heavy metals such as Fe, Sb, Zn, Pb, and Cu can be leached from Pyrex in amounts requiring attention in submicrogram, or even microgram, analysis (see Table 2-8).

Contamination of halide solutions by lead has been observed to occur in UV absorption cells constructed from fused silica or 96% silica glass

TABLE 2-6
Chemical Resistance of Glasses[a]

			Plate Weight Loss (mg/cm²)		
Liquid	Temp. (°C)	Time (hr)	Boro-silicate	Vycor	Alkali-Resistant
H_2O[b]	80	48	0.002	<0.001	0.002
5% HCl	100	24	0.005	0.0005	0.01
0.02 N Na_2CO_3	100	6	0.12	0.07	0.01
5% NaOH	100	6	1.4	0.9	0.04

[a] Mostly from J. P. Williams, *Microchem. J.*, **4,** 187 (1960). Other data may be found in E. Wichers, A. N. Finn, and W. S. Clabaugh, *Ind. Eng. Chem. Anal. Ed.*, **13,** 419 (1941). No simple relation is to be expected between the quantity of a particular trace element brought into solution and weight loss. For example, some elements may be brought into solution more easily in acid than in alkaline solution; nevertheless, the weight losses have some general interest.
[b] Other data indicate that very dilute acid solutions, for example, 0.001 M HCl, attack glass less than water.

TABLE 2-7
Leaching of Al and Fe from Jena Glass G20[a]
(30 ml of solution heated on boiling water bath and cooled 1 hr)

		Metal Leached (μg)	
Solution	Heating Time (hr)	Al	Fe
H_2O	1	0	0
HCl, 0.3 M	0.5	0.3	0
HCl, 0.3 M	1	0.6	0
HCl, 0.9 M	1	0.9	0
NaOH, 0.3 M	1	81	7
NH_4OH, 0.3 M	0.5	7	0.3
NH_4OH, 0.3 M	1	13	0.6
NH_4OH, 0.9 M	1	16	1.3

[a] Selection of values from W. Oelschläger, *Z. Anal. Chem.*, **154,** 329 (1957).

with the use of lead solder–glass. It has been proposed to store solutions without contamination from attack of the container walls by keeping them in the frozen state.

For drastic cleaning of glassware, a 1:1 mixture of concentrated sulfuric and nitric acid can be used. This mixture is better than the time-honored dichromate-sulfuric acid mixture (Cr is hard to remove in the subsequent rinsing with water). Warm hydrochloric or nitric acid often suffices for ridding glass surfaces of metals. Especially if fluorimetric determinations are to be made, detergents (likely to contain "optical brighteners") are best avoided. It is wise to keep glass vessels filled with dilute (say 0.1 M) hydrochloric or nitric acid when not in use.

2. Fused Silica (Quartz)

Vessels of this material have very low contents of most metals, and for this reason and also because silica undergoes only trifling attack by acidic or neutral solutions, they are much superior to borosilicate glass vessels from the contamination standpoint. Moreover, the possibility of using silica ware at high temperatures is sometimes an advantage.

It should be noted that, depending on the quality of the raw material, the trace metal content of fused silica varies (Tables 2-8 and 2-9). The occurrence of ~2 ppm Sb in some fused silica is unexpected. Cesium has been reported up to ~1 ppm, and zirconium, up to a few parts per million.

Silica is, of course, attacked by basic solutions. For example, a 75 ml fused silica flask filled with 10% ammonia solution at 18°C lost 0.8 mg in weight after 2 days. With stronger bases, especially hot, the attack is much more pronounced, and silica ware should not be used with such solutions; the introduction of silica may be objectionable, even if the foreign trace metals are insignificant. All forms of silica have a small intrinsic solubility in water, but equilibrium is reached extremely slowly. When powdered silica glass is shaken with water in polyethylene bottles, the concentration of SiO_2 attains a value of 1.5 mg/liter after 2 weeks and 129 mg/liter after 1 yr.[16]

A possible source of contaminating metals is adsorbed material from earlier use. Silica vessels—like all vessels used in the laboratory—should be heated with dilute HCl or HNO_3 before first use and now and then later. Such leaching can well be extended over 1 or 2 days.

Indirect evidence seems to indicate that HCl and HNO_3 can leach larger amounts of trace metals from fused silica dishes than from platinum and Teflon (see p. 59). This need not be generally true—the purity of the respective materials may vary.

TABLE 2-8
Content of Certain Trace Elements in Laboratory Ware[a]

	ppm							
	Ag	Co	Cr	Cu	Fe	Hf	Sb	Zn
Pyrex	$<10^{-6}$	0.08			280	0.6	2.9	0.73
Vycor[b]							1090	
Fused silica ("Spectrosil")	5×10^{-5}	4×10^{-4}	0.007	0.002	0.4	$<5\times10^{-6}$	5×10^{-5}	0.002
Fused silica ("Suprasil")	$<10^{-5}$	0.012	0.003	4×10^{-5}		$<5\times10^{-6}$	10^{-5}	0.001
Fused silica	$<10^{-4}$	0.001		9×10^{-5}		$<10^{-5}$	1.9	
Fused silica	$<10^{-4}$	0.002	0.2	2×10^{-4}		0.023	0.06	0.02
Polyethylene	$<10^{-4}$	7×10^{-5}	0.08	0.007	10.4	$<5\times10^{-4}$	2×10^{-4}	0.03
Polyethylene	$<10^{-4}$	3×10^{-4}	0.02	0.015	10.6	$<5\times10^{-4}$	8×10^{-4}	0.025
Teflon	$<3\times10^{-4}$	0.002	0.03	0.022	0.035		4×10^{-4}	0.009

[a] D. E. Robertson, footnote a, Table 2-5. See the original for contents of Cs and Sc, omitted here. Some of the values above have been rounded.

[b] Zr and Fe are leachable in small amounts by HCl from Vycor.

TABLE 2-9
Trace Elements in Fused Silica[a]

	ppm	
	"Vitreosil" (high purity quartz)	"Spectrosil" (synthetic quartz)
Al	50	<0.02–<0.25
As		<0.0002
B	0.1	<0.01–<0.5
Ca	0.4	<0.1
Cu	0.01	<0.0002
Fe	0.7	<~0.1
Ga		<0.004
K		<0.004
Mn	0.03	<0.001–0.04
Na	4	<0.04
P	0.01	<0.001
Sb	0.2	<0.0001

[a] R. Kleinteich, *Glas-Instrum.-Tech.*, **5**, 334 (1961); K. Jack and G. Hetherington, *ibid.*, **5**, 378 (1961). Silica glass of what might be called ordinary quality may contain 10 ppm or more of Fe and >100 ppm K and Na. Other elements may also be higher, but usually not exceeding 1 ppm.

3. Plastics

Polyethylene and Polypropylene. A great variety of laboratory ware, including separatory funnels and volumetric ware, is made from the thermoplastic polymers polyethylene and polypropylene, which are resistant to attack by most chemical reagents. Two general types of polyethylene are produced, differing in density and melting point. The lower density material is obtained by polymerization of ethylene at high pressures and temperatures of 150°–250°C without metal catalysts. The high-density (linear) polymers are produced at relatively low pressures and temperatures with the aid of metal catalysts. The lower-density (0.92) polymer melts at ~108°C and should not be heated above 70°; the higher density (~0.95) material, melting at ~135°, should not be heated above 100°. The lower-density polyethylene is more flexible and is used for squeeze-type wash bottles and the like, the higher-density type for bottles, flasks, and other containers requiring greater rigidity. Polypropylene is harder and more rigid than polyethylene and can be used at higher temperatures (melting point ~170°, maximum service temperature ~130°). Polyethylene and polypropylene are not appreciably attacked at room temperature by concentrated HCl, H_3PO_4, and HF, or strong hydroxide solutions. They are slowly attacked by strong H_2SO_4 ($>60\%$), concentrated HNO_3, Br_2, and other oxidizing agents, including $HClO_4$. Absorption of H_2S and NH_3 occurs, and their solutions should not be stored in plastic bottles.[17] In general, organic solvents should not be kept in them. Softening, swelling, and cracking are likely to occur, especially with hydrocarbons (including benzene), chlorinated hydrocarbons, ethyl ether, and others.

The lower density polyethylene, produced without catalysts, is generally lower in heavy metals than the higher density type.[18] Table 2-10 summarizes the content of a number of trace elements in polyethylene (the type is not always distinguished). Some common metals are present in concentrations of 0.1–~1 ppm. Considerable variation is shown by some elements, which is not surprising in view of variations in raw material, equipment, and manufacturing processes. More or less of the trace elements will be leached from polyethylene by neutral, acid, or alkaline aqueous solutions (see footnote ^{a}, Table 2-10). At times the quantities leached are not insignificant from the trace standpoint. For example, water (from polar ice) stored in a 19-liter polyethylene carboy for 2 yr, extracted 100 μg Cl.[19] The carboy had been carefully cleaned with nitric acid at the start. There seemed to be negligible leaching of sodium. Specks of partly exposed foreign material may sometimes be seen in polyethylene bottles, and these may contribute lead to a solution, in addition to that possibly contributed by more homogeneously distributed lead.[20] Lead derived from the walls of polyethylene bottles on long storage of water is believed to amount to ~0.06 μg Pb for 19–50 liters of water.[19] See Table 2-10A for elements extracted from polyethylene and other plastics by hot HCl.

Polypropylene vessels have been found to contribute essentially no heavy metals to acids evaporated or stored in them (Dabeka et al.[52]).

Polyethylene (and other plastics) usually contain additives (antioxidizers and UV absorbers, as well as other substances to aid mixing and fabrication) which can be very objectionable from the viewpoint of the trace analyst. Even the composition of the main components can vary, so that the material called polyethylene is not well defined and polyethylene vessels can vary a great deal in the inorganic and organic contaminants they yield and in their adsorptive properties. See especially Heiden and Aikens, Table 2-22, p. 86. Aqueous solutions extract small amounts of organic substances from polyethylene. Note has been made of significant extraction by acidic aqueous solutions; such organic matter resists oxidative degradation.[19] Polypropylene bottles have been observed to yield organic matter to acids stored in them for some time (Dabeka et al.[52]).

TABLE 2-10
Trace Elements in Polyethylene[a]

	A	B	C	D	E	F	G	H	I
Ag	0.02			<0.0001				<0.003–0.005	
Al	0.3	3.1	3				9	0.9–1.2	19
Au							0.002	0.002–0.007	
B	0.09								
Ba		0.05						<0.25	28
Br			0.1–0.2				0.2	0.8–1.3	0.1
Ca	0.2	20						<100	7.7
Cl			0.8–3			0.8	3.1	3.7–6.9	24
Co				~0.0001–0.0003			0.014	0.001–0.03	0.08
Cr	0.3	0.06		0.02–0.08			0.06	1.2–1.5	0.2
Cu	0.004			0.007–0.015	3–4		0.25		
Fe	0.6	2.1		~10				1–4.7	11
La							0.005	0.003–0.007	0.008
Mg	0.08	1.5						<60	12
Mn			0.1		0.07		0.5	0.14–0.3	0.1
Na		10	0.3–0.9			0.8	17	0.6–6.6	30
Pb	0.2								
Sb				0.0002–0.0008			0.0075	0.002–0.03	0.03
Sc							0.0024	0.0001–0.0007	0.007
Se								<0.003–0.006	0.04
Si	2								
Sr		0.8						<0.2	
V							0.84	0.2	0.15
Zn				0.03				0.15–0.8	1.6

[a] Sources of analyses: A and B: R. E. Thiers in D. Glick, Ed., *Methods of Biochemical Analysis*, Vol. 5, Interscience, New York, 1957, p. 296. A represents medians of analyses of six samples of high-pressure process (low-density) polyethylene bottles shipped in 1950–1953; B, medians of analyses of bottles manufactured in 1955. In A, K, Mn, Na, Ni, and Zn were not detected; in B, K, Mn, Ni, Pb, and Zn were not detected. C: V. P. Guinn et al., "Application of Neutron-Activation Analysis in Scientific Crime Detection," General Atomic GA-3491, 1962. D: Robertson, Table 2-5 (two containers). E: D. Brune, *Radiochim. Acta*, **5**, 14 (1966). F: Murozumi et al., reference 19. G: R. W. Karin, J. A. Buono, and J. L. Fasching, *Anal. Chem.*, **47**, 2296 (1975). These authors leached polyethylene used in the above determinations in cold 8 M HNO_3 for 3 days and found the following percentages of elements leached out: Al, 0; Sb, 34; Cl, 86; Cr, 45; Co, 19; Cu, 65; La, 100; Mn, 88; Na, 44; V, 90. Little additional leaching occurred after 3 days. Concentrated HNO_3 did not leach more than did 8 M acid. Other workers have found Al and Ti (Ziegler catalysts), as well as Zn, to be introduced when acids are evaporated in linear polyethylene vessels. H: Linear polyethylene film, 0.0025 cm, bag, soaked in 50% HNO_3 for 24 hr. S. H. Harrison, P. D. LaFleur, and W. H. Zoller, *Anal. Chem.*, **47**, 1685 (1975). Some values rounded. I: Conventional polyethylene film. Harrison et al.

TABLE 2-10A
Contamination of HCl Evaporated in Beakers of Various Materials (μg element/l)[a]

	LPE[b]	PP	TPX	Teflon	Vycor
Al	3	0.1	0.2	0.3	0.3
Co			0.002	0.02	
Cr	0.02	0.03	0.001	0.9	0.006
Fe	0.7	0.6	0.6	4	0.4
Mn	0.005	0.001	0.001	0.5	0.005
Ni	0.03	0.01	0.003	0.3	0.003
Si	0.8	0.4	0.8	1	6
Ti	0.2	0.07	0.03	0.04	0.04
Y					1
Zn	7	0.02	0.07	0.04	0.1
Zr					1

[a] A. Mykytiuk, D. S. Russell, and V. Boyko, *Anal. Chem.*, **48**, 1462 (1976). Beakers digested with 1:1 HNO_3 for 24 hr before use. 250 ml of high purity HCl evaporated to essential dryness at ~80°. Determinations by spark source mass spectrometry. According to these figures PP and TPX are superior to Teflon and Vycor.
[b] LPE = linear polyethylene, PP = polypropylene, TPX = poly(4-methyl-1-pentene).

Polypropylene bottles holding HCl have been reported to become brown after some time (HNO_3, yellow). So much organic matter is brought into solution that HCl and HF on evaporation leave a black residue. Whether the organic matter originates from the polypropylene itself or from unpolymerized material, releasing agents, antioxidants, or other substances is not clear. Presumably not all lots of polypropylene will show the same behavior. Organic matter derived from polyethylene is fluorescent (e.g., on excitation at 280 nm).[21] It is reducing in nature. Solutions containing Cr(VI), Ce(IV), OsO_4, and other strong oxidizing substances should not be brought into contact with polyethylene and other plastics.

Polyethylene apparently contains smaller quantities of trace elements than Tygon, methyl methacrylate, or polyvinyl chloride. Plexiglas seems to be equal or superior to polyethylene in purity. Polyethylene hose has been found to be higher in trace elements than polyethylene vials or bottles.

Polyethylene and polypropylene are chiefly of value to the trace analyst as materials for bottles for storing solutions and for separatory funnels. Although organic solvents should not be stored in polyethylene or polypropylene, extractions, as with $CHCl_3$ and CCl_4, can be carried out in them, but contact should not be prolonged.

Polypropylene separatory funnels soaked in 4 M HNO_3 have been observed to absorb the acid or nitrogen oxides that are not completely removed by thorough rinsing with water.[22] Similarly, funnels of this plastic soaked in 4 M HCl overnight, continue to release small amounts of chloride on washing for some time. These observations are in accord with the known permeability of polyethylene to HCl and other acids. It is also slowly permeable to NH_3, as already noted.

Several workers have noted that polyethylene and other plastic bags can give off volatile carbon compounds, permeating powdered material stored in them.

Teflon. This tetrafluoroethylene polymer (relatively high priced) is of special interest to the analyst because it can withstand temperatures approaching 300°C. Dishes, breakers, and other vessels of this plastic can, therefore, be used for evaporations of solutions above 100° and for decomposition with hydrofluoric acid. It is inert to essentially all analytical reagents and organic solvents. Above 250°, Teflon evolves small amounts of toxic tetrafluoroethylene, so that at higher temperatures Teflon vessels should be heated in a hood. The content of trace elements in Teflon seems to be comparable to that in polyethylene (Table 2-8). Comparative figures indicate that evaporation of acids in fused silica and platinum introduces larger quantities of metals than evaporation in Teflon. Thus 50 g of extra pure acids evaporated to dryness after mixing with 20 mg carbon powder (for spectrography) showed the following contents, in ppm of the original acids, of various metals[23]:

		Mg	Al	Mn	Fe	Ni	Cu	Ga	Pb
HCl	Teflon	0.003	<0.004	0.0001	0.003	<0.0004	0.0002	0.005	<0.0004
	Pt	0.006	0.002	0.0002	0.002	0.0006	0.001	0.01	<0.0004
	SiO$_2$("quartz")	0.01	0.01	0.0004	0.01	0.002	0.001	0.06	0.0005
HF	Teflon	<0.003	0.003	0.0001	0.003	<0.0004	<0.00004	0.001	<0.0001
	Pt	0.01	0.01	0.0002	0.01	0.0008	0.0004	0.01	0.0005

Dabeka and coworkers[52] report unfavorably on the use of Teflon–FEP for the evaporation of acids and their storage. The uptake of Fe, Ni, Cr, and Mn was especially significant, for example, 10 μg/liter of Fe and 0.4 of Ni after storage of concentrated HCl for 4 months. This contamination was found to result from the presence of minute particles of stainless steel (originating in the manufacturing process). Even after leaching with hot acids, these metals continue to be released. Whether such metallic contamination is common is apparently unknown.

Teflon is less prone than other plastics to yield fluorescent substances to liquids coming into contact with it.

Teflon stoppers and stopcocks are superior to glass[24] for separatory funnels and other glassware; cover glasses of Teflon are much to be preferred to glass, from which metals and SiO_2 can be leached. Teflon separatory funnels are now (1970) manufactured.

Teflon bottles have been preferred by some workers for the storage of acids and other solutions of the highest purity, but their use is suspect, as we have seen. The material is also used to line bombs for decomposition at higher temperatures and pressures.

Teflon soaked in 4 M HCl absorbs acid, which continues to be released for some time on washing with water (noted by T. Bidleman using Teflon stopcocks).

Other Plastics.   Polyvinyl chloride (PVC) is high in some trace metals, for example, 7 ppm Zn, 270 ppm Fe, 3 ppm Sb, and 0.6 ppm Cu (Robertson, *loc. cit.*); it has also been reported to contain Pb, Sn, Cd, and Ti.[25] On the other hand, Plexiglas seems to be low in metals, being as pure as, if not purer than, polyethylene and Teflon. Much Co (~0.15%) has been found in nylon.

All molded plastics can be surface contaminated with lubricants such as zinc stearate, and should be carefully cleaned before use.

Rubber.   It is advisable to exclude rubber tubing, stoppers, and other such articles from the analytical trace laboratory. Rubber contains additives such as zinc oxide, antimony compounds, and even lead compounds.

4. Platinum

Teflon can often be substituted for platinum in operations carried out below 200°–300°C. At higher temperatures, platinum is indispensable, or at least very useful, as in sodium carbonate fusions. Analysts are familiar with the substances that markedly attack platinum. But small amounts of platinum—enough to interfere in some photometric determinations[26]— may be brought into solution in such operations as evaporation of hydrochloric acid in platinum dishes or crucibles. In the presence of oxygen, halide ions form halo-platinum anions. For example, it has been found that evaporation to dryness on the water bath of 20 ml of concentrated HCl in a ~7 cm diameter platinum dish (50–60 g), with the residue being taken up in 10 ml 6 M HCl, brought from 10 to 85 μg Pt into solution, depending on the condition of the dish. Old dishes, used for many years in the ashing of biosamples at ~500°C, yielded larger amounts of Pt (28–85 μg) than new dishes (10–57 μg, average ~15 μg), the difference being explained by the larger available surface in old dishes.[27] Evaporation of 10 ml of concentrated HF in new Pt dishes, with the residue taken up in 6 M HCl, brought 8–11 μg Pt into solution, and evaporation of 2 ml of concentrated sulfuric acid over a burner introduced 7–10 μg Pt.

Sodium carbonate fusion, as of a silicate, always brings some platinum into solution, but the amount is variable, ranging from a few tenths of a milligram to milligram amounts, depending on the composition of the sample, the condition of the crucible, and other factors.[28] The attack of platinum seems to increase with the iron content of the sample and the introduction of as much as 10 mg of Pt has been claimed. Potassium pyrosulfate fusion also introduces platinum into the melt, usually not more than 1 mg. Fusion with KHF_2 attacks platinum little, figures of less than 0.2 mg Pt being reported.

Attack of platinum by molten sodium hydroxide is severe in the presence of air; for example, ~0.1 g Pt may be dissolved when 1 g NaOH is fused for a few minutes in a platinum crucible. In a nitrogen atmosphere, much less platinum is dissolved.[29] Sintering a sample with NaOH–Na_2O_2 at 440° for 30 to 60 min in air introduces little platinum, usually <1 mg.[28]

Fusion of iron-containing samples with sodium carbonate results in the uptake of iron by a platinum crucible, sometimes a milligram or more. The presence of iron in the ferrous state is said to increase the amount taken up. It appears that iron in the melt can be reduced to the metal, as by flame gases, which alloys with platinum. Later, treatment with acid (HCl) will dissolve some of the iron or the iron oxide formed on subsequent ignition. Likewise fusion of pyrosulfate in the crucible removes only part of the iron. Fusion of a sample with $K_2S_2O_7$ or KHF_2 does not lead to loss of iron from uptake by the crucible.[28] Other reducible metals can behave similarly to iron in sodium carbonate fusion.

Boron is reported to be a common contaminating element (~10 ppm) in new platinum crucibles. For contamination of biosamples by platinum in dry ashing, see Chapter 3.

5. Filter Paper

Small amounts of many elements are present in filter paper, even the acid-washed "ashless." The following elements have been found in quantitative filter paper[30]: Al, Ba, Ca, Cr, Cu, Fe, Ge, K, Mn, Na, Pb, Sb, Si, Ti, V, Zn, Zr, and the rare earths (especially Y). Other elements can be added to this list: B,[31] Cl, and Se. Some of these occur in minute amounts, others in appreciable amounts. For example, Rankama found approximately 1% PbO, 0.3% ZnO, and 0.3% SnO_2 in the ash of quantitative paper. Larger amounts of Ca, Mg, and others are always present (Table 2-11). As much as a few tenths of a percent TiO_2 may be

TABLE 2-11
Major Constituents of Ash of Schleicher and Schüll Blue Band Filter Paper No. 589[a]

	%		%
SiO_2	29.7	MgO	2.5
Al_2O_3	3.4	CuO	1.2
Fe_2O_3	12.1	SO_3	13.7
CaO	21.9	Σ	84.5

[a] A. Grüne, *Über Mineralstoffgehalte in Filtrierpapieren*, Schleicher and Schüll, 1954, as quoted by O. G. Koch and G. A. Koch-Dedic, *Handbuch der Spurenanalyse*, Springer, Berlin, 1964, p. 88. Values rounded to 0.1%. Composition of Black and White Band ash also given. Weight of ash in 7 cm paper ∼0.02 mg. H. Vogg and R. Härtel, *Z. Anal. Chem.*, **267**, 257 (1973), found the following amounts (ng) of elements in 20 cm² of S. and S. White Band filter paper 589/2: Fe 750, Na 1000, As 1, Br 20, Co 2, Cr 40, Cu 4, Mn 1, Zn 35.

present. An 11 cm paper has been found to contain ∼0.005 μg Se and ∼0.01 μg Sb. Lead contents of a few parts per million are not uncommon in filter paper.

Some of these elements are present in the original cellulose fibers, others may be introduced (as by adsorption) in the manufacturing process. The greater fraction of the trace elements cannot be washed out with acids. In general, it is best to use filter media other than paper, because of the possibility of adsorption of substances by paper and because contamination can occur, the latter being most serious when the filter paper is ignited and the ash becomes part of the sample. But there are times when the use of filter paper is convenient, if not indispensable. The smallest possible size should be used, and preliminary washing with solutions to be used in the determination may do some good. Filter paper should be carefully protected against laboratory dust and fumes.

Chromatographic paper can contain significant concentrations of trace metals as the following figures (parts per million of paper) show[32]:

Al	7	Co	1	Cu	7	Sb	3
Au	0.003	Cr	1	Zn	27		

The distribution of these elements was not homogeneous in the paper. This paper weighed 185–200 mg cm².

The practice of wiping pipet tips with absorbent paper can introduce Ca and other metals.

C. FROM WATER

Because water is used in large quantities compared to the reagents and to the sample analyzed—the ratio may exceed 100–particular attention must be paid to its purity. The ordinary distilled water of the laboratory is of variable quality; often it is not suitable for use in trace analysis.[33] It can be purified by redistillation or by ion exchange. As far as metals are concerned, there is little likelihood that they will be volatilized at 100°C (mercury is an exception), so that metals found in the distillate must have been carried over mechanically in droplets of spray in the current of vapor or by surface creepage, or have been leached from the condenser or container, if atmospheric contamination is excluded. Contamination by mechanical carryover can be reduced to negligible proportions by repeated distillation, but better by the use of some arrangement (e.g., superheater) that prevents droplets from being carried into the condenser. Distillation without boiling is the best method, but slow. The factor limiting the purity of distilled water is likely to be leaching from the condenser and container. Fused silica is usually considered to be the best condenser material,[34] but as far as metals are concerned, plastics, for example, polypropylene or polyethylene,[35] should be comparable, possibly better. Teflon is usually superior to polyethylene.

The amount of silica dissolved from a fused silica condenser is extremely small. A value of 1.6 μg SiO_2/liter has been reported for water triply distilled in silica apparatus.[36] Water tends to extract a minute, but not always negligible, amount of organic material from polyethylene.[37] Commercial fused silica stills are available.[38] In some models, an ion-exchange unit is incorporated, so that tap water can be used as the feed water.

Water distilled from Pyrex (the important feature is the Pyrex condenser) is sufficiently low in heavy metals to be satisfactory for some trace analyses. Two redistillations of ordinary distilled water in a simple still are usually sufficient, as may be seen from Table 2-12. No improvement is noted on distilling a third time, and the metals then present are presumably largely leached from the condenser or the storage container. A single distillation suffices if the still is well designed.[39]

In our experience, a single slow distillation (of ordinary distilled water) with a Pyrex flask and condenser can give water containing 1–1.5 μg/liter of dithizone-reacting metals expressed as Pb. No doubt the metal content can be higher if the glass has not been aged. Quite generally, borosilicate glass that has been exposed to hot water and acid solutions, subsequently furnishes lesser amounts of leachable substances than glass not so treated. Glass that has been well aged in this way should always be used for

TABLE 2-12
Metal Contents of Water Distilled in Various Ways[a]

Manner of Distillation	Micrograms of Metal per Liter				
	Cu	Zn	Mn	Fe	Mo
From tinned Cu still	10	2	1	2	2
Water from Cu still:					
Distilled once from					
Pyrex	1	0.12	0.2	0.1	0.002
Distilled twice from					
Pyrex	0.5	0.04	0.1	0.02	0.001
Distilled thrice from					
Pyrex	0.4	0.04	0.1	0.02	0.001

[a] D. J. D. Nicholas, *Analyst*, **77**, 632 (1952).
[b] Dabeka and coworkers[52] reported the following metal contents (μg/liter) of tin-distilled water: Cu, 12; Pb, 2.5; Zn, 33; Fe, 0.3; Al, 0.9; Mg, 0.8; Ni, 0.07; Cd, 0.02.

distillation and storage of water and acids. New glassware should at least be digested in hot HCl or other mineral acid before use.[40] Teflon or polyethylene bottles are better than borosilicate glass for storage of water (if a trace of organic matter is not objectionable).

In a few trace metal determinations, water free of organic reducing substances is needed. Such water is obtainable by distilling water treated with permanganate or by other oxidative pretreatment.[41]

TABLE 2-13
Removal of Heavy Metals from Distilled Water by Passage through a Column of Cation-Exchange Resin (Amberlite IR-100)[a]

	Metal Present (μg/liter)		
	Cu	Pb	Zn
Laboratory distilled water	200	55	20
Ion-exchange water,			
one passage	3.5	1.5	<10[b]
Ion-exchange water,			
five passages	0.0	1.0	<10[b]
Redistilled water (all-			
Pyrex still)	1.6	2.5	<10[b]

[a] G. F. Liebig, Jr., A. P. Vanselow, and H. D. Chapman, *Soil Sci.*, **55**, 371 (1943).
[b] Given as 0.00 ppm in the original.

TABLE 2-14
Trace Metals in Water Purified with Mixed Ion-Exchange Resins[a]

Element	Mg	Ca	Ba	Pb	Zn	Cr	Fe	Ni
μg/liter	2	0.2	0.006	0.02	0.06	0.02	0.02	0.002

[a] R. E. Thiers, in J. H. Yoe and H. J. Koch, *Trace Analysis*, Wiley, New York, 1957, p. 642. Boston tap water was passed through a 1.5 m column of mixed Amberlite IR 120 and IRA 410 resins. Other elements present: $Sr < 0.06$ μg/liter, $Mn < 0.02$ μg/liter, $Co < 0.002$ μg/liter, $Mo < 0.02$ μg/liter; B, Si, Sn, Cu, Ag, V (not determined). This water seems to be comparable to water doubly distilled in Pyrex (See Table 2-12). Water somewhat purer than Thiers was obtained by Dabeka et al. (reference 52) by passing tin-distilled water through two mixed-bed ion-exchange columns, for example, Pb, 0.01 μg/liter; Cu, 0.004 μg/liter; Cd, 0.0002 μg/liter; Ni, 0.001 μg/liter; Al, 0.05 μg/liter; Fe, 0.03 μg/liter. Thus this water is much lower in heavy metals than that obtained by subboiling distillation in silica or polypropylene apparatus. A commercially available (1977) water purification system embodying the use of semipermeable membranes (reverse osmosis), activated carbon, and ion-exchange resins, supplies water containing $< \sim 0.05$ μg/liter each of Al, Ca, Cu, Fe, Mg, Pb, and Zn, and < 0.003 μg/liter Cd. The lower limits here are set by the sensitivity of the flameless atomic absorption method used for determination.

The goal in purification is not necessarily the reduction of foreign elements to the lowest possible level, but rather to a level that is reasonably lower than that of the elements being determined. Blanks must always be run, and if water and reagents introduce, let us say, 0.1 as much of an element as being determined, accuracy is not impaired. Assuming that 100 ml of water is required for 1 g of sample, water twice-distilled from Pyrex, containing 0.5 μg Cu/liter, would be acceptable for use in the determination of $10 \times 0.5/10$ or 0.5 μg Cu or ≥ 0.5 ppm Cu in the sample. The blank/sample ratio of 0.1 used here for illustration is arbitrary, and a slightly larger ratio might be acceptable. Of course, the lowest possible impurity content is always desirable if it can be obtained without undue expense or expenditure of time.

The metal content of water can be reduced to very low levels by passage of distilled water, or even tap water, through a column of ion-exchange resin (Tables 2-13, 2-14, and 2-15). By use of a mixed bed of strong-acid cation- and strong-base anion-exchange resins, carbon dioxide and silica can be removed. Nonelectrolytes and colloidal substances present in the original water will not, in general, be removed. Resins always introduce small amounts of organic substances (dissolved, colloidal, or both), whose presence may be objectionable, as in radiochemical work or when reducing substances must be excluded. Microorganisms can grow in resins and contaminate the water. The

TABLE 2-15
Comparison of Trace Elements in Water Purified by Distillation and Ion Exchange[a]

	Element (μg/liter)			
	Ca	Al	Mg	Si
Doubly distilled in silica	0.07	0.5	0.05	5
Monobed deionization of distilled H_2O, polyethylene	0.03	0.1	0.01	1

[a] R. C. Hughes, P. C. Mürau, and G. Gundersen, *Anal. Chem.*, **43,** 691 (1971). A number of other metals, for example, Cu, Mn, Pb, Fe, and Zn, are reported as not detectable by emission spectrography in 1 or 2 liters of water. See this paper for references on the preparation of ultrapure water, especially on a large scale. F. Tera and G. J. Wasserburg, *Anal. Chem.*, **47,** 2214 (1975), obtained water containing 0.003 μg Pb/liter by distilling twice in silica, followed by passage through a cation-exchange resin column (AG-50W-X8, 100-200 mesh). Water distilled three times in silica contained 0.005 μg Pb/liter.

leaching of fluorescent substances from resins has been reported, as well as of nonpolymeric components in sufficient amounts to mask metals. No doubt resins will vary in the amounts of water-soluble substances they can contribute to solutions. Ion-exchange purified water is, therefore, not always the equal of suitably distilled water. It should not be used in preparing Hg, Ag, and Au solutions.

In distribution of distilled or deionized water from a central supply, possible contamination by the material of the tubing and valves is an important consideration. Polytetrafluoroethylene is recommended.[42]

D. FROM REAGENTS

1. Mineral Acids and Ammonia

Extra pure acids, very low in metals (Table 2-16), are commercially available, but are expensive.[43] These "ultrapure" or "electronic grade" chemicals are indeed lower in most metals, often by a wide margin, than the corresponding analytical reagent. However, the content of a particular trace element can vary considerably from one product to another, and the analysis supplied by the manufacturer does not always agree well with the actual content.[44] The discrepancy can arise in part from attack of the container by liquid reagents (acids).

Analytical reagent-grade acids vary considerably in their heavy metal contents, but are likely to be within the maximum limits specified on the label (*cf.* Table 2-16, column A). The following values in micrograms per liter, reported in years past by analysts (not the manufacturer), will give

TABLE 2-16
Illustrative Contents of Trace Elements in Reagent Grade and Extra Purity Acids[a]

	Content (μg/liter)									
	HCl		HClO$_4$		HF		HNO$_3$		H$_2$SO$_4$	
Element	A_{max}	B	A_{max}	B_{max}	A_{max}	B_{max}	A_{max}	B	A_{max}	B
Al		8	5		700		400	5	10	8
As	10	<1	50	5	50		10	1	500	<1
Ca		20		1000				7		10
Cd	10									
Co		<1		5				<1		<1
Cr		2								
Cu	500, 50	2	200	5	100	10	100	3, 10	500	3
Fe	100	7	500	100	1000	500	200	5	200	10, 3
Hg		<10						<10	<10	<10
Mg		4								5
Mn		1	500	5	1			1, 5	1, 5	1
Ni	500	<1		2	20		100	<1	500	1, 7
Pb	50	<1	100	5	20			2		5
Zn	100	<1		40				<1		<1

[a] A = analytical reagent, B = extra purity reagent ("electronic grade"). Maximum content indicated by *max*; otherwise actual analysis. Values are those of manufacturer (more than one company is represented). Selected elements. See Table 2-20 for metal contents of acids conforming to Am. Chem. Soc. specifications. For additional values of metals in reagent-grade and specially purified acids, see J. H. Oldfield and E. P. Bridge, *Analyst*, **85**, 97 (1960) and N. A. Kershner, E. F. Joy, and A. J. Barnard, *Appl. Spectrosc.* **25**, 542 (1971) (additional references in latter). The values above may be compared with those in still purer acids obtained by nonboiling distillation in silica apparatus (Table 2-20). C. D. Chriswell and A. A. Schilt, *Anal. Chem.*, **46**, 992 (1974), report the following Fe contents in ppm (w/w) in concentrated reagent grade acids: HCl, 0.05; HBr, 0.09; HClO$_4$, 0.4; H$_2$SO$_4$, 0.3; HNO$_3$, 0.1. C. Feldman, *ibid.*, **46**, 1608 (1974), found the following concentrations of Hg (ppm w/w) in reagent-grade acids: HClO$_4$, 0.00006–0.011; HNO$_3$, 0.0001–0.0002 (can be reduced to 0.00002 by distillation in quartz); H$_2$SO$_4$, 0.0002. Reagent grade H$_2$SO$_4$ has been reported to contain 5 ppm Al, HNO$_3 \sim$1 ppm, HF 0.2 ppm. Reagent grade H$_2$SO$_4$ may contain appreciable amounts of V, and H$_3$PO$_4$ may contain Sb (1–3 ppm). From 0.01 to 0.1 ppm Se has been reported in H$_2$SO$_4$.

some idea of the possible heavy metal contents of concentrated analytical reagent-grade acids:

HCl: 100–300 Cu,   10–30 (750) Pb,   500 Fe,   10–1000 Hg, 3 Cd
HNO_3:   100 Cu,   10–30 Pb,   300 Fe,   1 Cd,   1 Co,   50–350 Al
H_2SO_4:   100–600 Cu,   40–800 Pb,   100 Fe,   10–20 Cd,   200 Ca
HF: 50–800 Fe,   100–200 Pb,   1000 Ca,   700 Al

The zinc content of reagent-grade acids as found by M. Nishimura (University of Minnesota, 1960) is of interest:

	Zn (μg/liter)
HCl	120
HNO_3	14
H_2SO_4	24
$HClO_4$, 70%	36
HF	20
H_2O, deionized	1

The reagent acids now available may be purer than those marketed 10–25 years ago. Some reagent HCl and HNO_3 currently manufactured are much lower in heavy metals than these figures would indicate (Tables 2-17 and 2-20).

With the exception of H_3PO_4, the mineral acids can be purified by one or two distillations ($HClO_4$ under reduced pressure).[45] A fused silica condenser is preferable, in fact essential, for highest purity, but judging from the results in Table 2-17, a borosilicate condenser is often satisfactory for HCl and HNO_3. It would be expected that if atmospheric contamination is negligible, the metals in the distillate are derived mostly from the condenser or the receiver; exceptions would be elements forming volatile compounds under the conditions, for example, $GeCl_4$ and $AsCl_3$. Mechanical carryover should not be important in a double distillation, perhaps not in a single. Hydrofluoric acid requires platinum or Teflon distillation apparatus. In general, one or two distillations should give acids containing not more than a few parts per 10^9 of a heavy metal.

When cylinders of the gases are available, solutions of HCl and HF[46] extremely low in metals can be obtained by passing the filtered and washed gases into pure water (Tables 2-18 and especially 2-19). As may be seen from Table 2-19, HCl prepared in this way contains less Fe, Ca, Mg, and Cu than that obtained by distillation in fused silica. A slower and less efficient way of obtaining very pure solutions of HCl and other volatile substances is by the old procedure of isothermal distillation, in which vessels containing strong HCl and pure water are placed in a closed

TABLE 2-17
Purification of Acids by Distillation[a]

	Metal Content (μg/liter)						
Acid	Zn	Fe	Sb	Co	Cr	Ag	Cu
HCl, reagent (original)[b]	22	~1	0.2	~0.1	1	<0.1	82
HCl, doubly distilled	~1	~1	0.004	~0.1	6	<0.02	—[c]
HNO$_3$, reagent (original)	13	~2	~0.03	0.02	72	0.2	1
HNO$_3$, doubly distilled	~2	~1	~0.04	0.03	13	0.3	—

[a] Robertson, *loc. cit.* Borosilicate distillation apparatus.
[b] Thiers (Table 2-14) found (μg/liter) 0.6 Pb, 0.2 each Ni and Co, 0.02 Mo and 2 Cr in reagent-grade HCl.
[c] Other data indicate that 1:1 HCl (from reagent-grade acid) doubly distilled in borosilicate glass contains less than 5 μg/liter Cu, probably considerably less. HCl distilled in silica has been found to contain 0.05 μg/liter Pb (original acid 0.4 μg/liter Pb).

container such as a desiccator. After 3 or 4 days at room temperature equilibrium is reached.[47] HF can also be purified in this way if polyethylene ware is used.[48] If atmospheric contamination is negligible, the purity of the solution is limited only by the purity of the water and by the very small amounts of elements derived from the container (Teflon or polyethylene). Because the volatilization and absorption takes place at room temperature, the solution may be expected to be purer than that obtained by distillation at higher temperatures.

A simple distillation arrangement for purification of HF consists of two

TABLE 2-18
Purification of Acids by Distillation or Volatilization[a]

		Impurities (μg/liter)						
Acid	Purification	Al	Fe	Ca	Mg	Mn	Cu	Pb
HCl	Distillation in silica	2	6	30	20	0.1	1	0.7
	Saturation of H$_2$O in Teflon with HCl[b]	n.d.[c]	3	5	4	0.3	0.2	n.d.
HF	Distillation in Pt	50	30	n.d.	30	1	3	n.d.
	Distillation in Teflon	5	5	10	10	n.d.	0.05	n.d.

[a] Selection of data from Vasilevskaya et al., *Zh. Anal. Khim.*, **20,** 540 (1965).
[b] It may be noted that Thiers (Table 2-19) prepared HCl containing only 0.1 to 0.01 as much of most of these metals.
[c] Not determined.

TABLE 2-19
Trace Metals in Hydrochloric Acid Prepared from HCl Gas[a]

Contents (μg/liter)							
Mg	2	Pb	0.06	Fe	0.02
Ca	0.2	Zn	0.06	Co	<0.002
Sr	<0.06	Cr	<0.06	Ni	0.002
Ba	0.02	Mn	0.02	Mo	<0.02

[a] Thiers (footnote a, Table 2-14). HCl gas scrubbed with concentrated H_2SO_4 and absorbed in triply distilled water at 0°C. For most of the metals listed, the product is "10^{-10}–10^{-11} acid." J. H. Oldfield and D. L. Mack, *Analyst*, **87**, 778 (1962), found hydrochloric acid prepared by absorbing cylinder HCl gas in pure water, held in polyethylene, to contain (μg/liter) <2 each of Mn, Bi, Cr, Ni, Mo, In, Ti, Co, Ga, and V; <20 Fe and Pb; 5 Mg; 20 Ca; and 100 Si. T. E. Krogh, *Geochim. Cosmochim. Acta*, **37**, 485 (1973), prepared 6 M HCl containing only 0.002–0.005 μg/liter of Pb by using the Teflon bottle still of Mattinson (reference 49), but this method is very much slower than the gaseous HCl method. Tera and Wasserburg (*loc. cit.*, footnote a, Table 2-15) obtained 4M HCl containing 0.01 μg Pb/liter by absorbing HCl gas in water.

1000 ml Teflon bottles connected at right angles by a threaded Teflon block.[49] One bottle, containing the acid to be distilled, is heated from above with a 300 W heat lamp placed so that the liquid does not actually boil but condensation does not take place on the walls. The second bottle, the condenser, is cooled in a bath of water (preferably an ice-water bath). The collecting bottle is unscrewed a fraction of a turn to prevent pressure build-up. Distillation is slow. Three or four days are needed to distill 600–700 ml of HCl or HF and ~350 ml HNO_3 if ice cooling is used. The distillate is phenomenally pure as indicated by its lead content (other data for comparison):

	10^{-9} g Pb/g of acid		
	HF (48%)	HCl (6.2M)	HNO_3 (70%)
Two-bottle Teflon still	0.002, 0.005	0.0015	0.023, 0.049
Tatsumoto system	0.08		
Isothermal distillation, Kwestroo and Visser	0.2		
Distillation in Pt	0.2–1		
HF gas absorbed in H_2O	0.2–1		
Subboiling distillation in silica		0.12	0.18

The high purity is explained by avoidance of boiling in the distillation, use of Teflon, and thorough cleaning of the collecting bottle by a preliminary distillation procedure. Protection from the atmosphere may also be a factor.

A still similar to Mattinson's has been used for the purification of HCl as well as HF.[50] The purity of the product depends on the care taken in cleaning the bottles. They should be heated in aqua regia at 100°C for two 24 hr periods, followed by soaking in distilled water (100°) for four 24 hr periods (Cl is difficult to remove). Distillation is carried out slowly, so that collection of 600 ml of distillate requires 2 weeks. The effectiveness of the purification may be judged from the following figures (partial data[50]) for HCl:

	Feed (6 M reagent grade)	Distillate (6 M)
	(ng/g)	(ng/g)
Al	35	0.4
Ca	150	3
Cr	1	<0.05
Co	<0.5	<0.5
Cu	15	2
Fe	15	1
Pb	1	<0.5
Mg	4	0.2
Mn	0.4	0.05
Ni	<0.5	<0.5
Na	>500	5
Zn	1	<0.5

Hydrofluoric acid of similar purity is obtained.

The condenser bottle containing the distilled acid can be protected from external contamination by placing it in a polyethylene or polypropylene refrigerator pitcher.

A modification of the two-bottle distillation procedure for the preparation of high-purity concentrated HF has been described.[51] Two 1 liter polyethylene bottles connected with a 50 cm long Teflon tube, 1 cm in diameter, are used. One bottle, cooled in a dry-ice bath is three-quarters filled with 100% HF from a cylinder of HF gas. The dry-ice bath is switched to the collecting bottle and HF is allowed to distill into the latter. About 90% of the acid is distilled and the feed bottle is cleaned with high-purity water. The distilled acid is then allowed to distill into the cleaned bottle. The distillation is repeated in this way as many times as necessary. The following lead contents were found after repeated distillation:

No. of Distillations	ng Pb/g of HF
0	3.68
1	0.546
2	0.105
3	0.021
4	0.0048
5	0.0009

Five distillations require about 10 hr. The time can be shortened by using a heat lamp, but the acid is then less pure. The final distillate is diluted with high-purity water to give the desired concentration. (Rocks requiring up to 1 week for decomposition with 48% HF are reported to be decomposed in less than one day with 55–60% acid.)

Subboiling distillation has been applied on a larger scale for purification of mineral acids and water.[52] For HCl, HNO_3, H_2SO_4, and $HClO_4$, an all-silica ("quartz") still is used (commercially available), in which two radiators, enclosed in silica tubes, are used to heat the liquid. A tilted finger condenser (silica) allows condensed liquid to flow to the tip and be collected by a tube leading to an outside Teflon bottle. The capacity of the still is ~500 ml. For distillation of HF, a similar still constructed entirely of Teflon is used. Each still is housed in a clean-air chamber (Class 100 specifications). Production rates, liters/24 hr, are HCl, 10 M, 2.0; HNO_3 70% (~16 M), 0.5; $HClO_4$, 70% (11.5 M), 0.6; H_2SO_4, 96% (18 M), 0.3; HF 48% (~26 M), 0.3. The concentrations of the distilled acids are the same as the starting acids. Determinations of metals in the distilled acids and the starting acids were made by isotope dilution-spark source mass spectrometry (Table 2-20), which has detection limits of ~0.01 ng/g. The values in Table 2-20 refer to acids stored for at least 2 weeks in the Teflon collecting bottles. A few metals are volatile under the conditions of distillation (Sn in HCl and $HClO_4$, Cr in $HClO_4$ and HF). It appears that most of the metal found in the distilled acids must be leached from silica or Teflon surfaces. As far as HCl is concerned, a purer product can be obtained by absorbing HCl gas in pure water (Table 2-19). $HClO_4$ and H_2SO_4 distilled in this way are, in general, less pure than HCl and HNO_3, probably in large part because of the higher temperatures used and increased leaching of silica glass. In passing, it may be noted that the starting acids are surprisingly low in heavy metals except for iron. For many absorptiometric determinations, purification would not be necessary, though a blank correction would be required.

TABLE 2-20
Comparison of Heavy Metal Content of Acids Purified by Subboiling Distillation in Silica with That of Starting Acids (ACS Reagent Grade)[a]
(10^{-9} g metal/g acid)

	HCl		HNO$_3$		HClO$_4$		H$_2$SO$_4$		HF	
	Dis-tilled	ACS	Dis-tilled	ACS	Dis-tilled	ACS	Dis-tilled	ACS	Dis-tilled	ACS
Pb	0.07	0.5	0.02[b]	0.2	0.2	2	0.6	0.5	0.05	0.8
Te	0.01	0.1	0.01	0.1	0.05	0.05	0.1	0.1	0.05	0.1
Sn	0.05	0.07	0.01	0.1	0.3	0.3	0.2	0.6	0.05	11
Cd	0.02	0.03	0.01	0.1	0.05	0.1	0.3	0.2	0.03	2
Ag	0.03	0.05	0.1	0.03	0.1	0.1	0.3	0.6	0.05	0.1
Zn	0.2	2	0.04	4	0.1	7	0.5	2	0.2	4
Cu	0.1	4	0.04	20	0.1	11	0.2	6	0.2	3
Ni	0.2	6	0.05	20	0.5	8	0.2	0.5	0.3	12
Fe	3	20	0.3	24	2	330	7	6	0.6	110
Cr	0.3	2	0.05	6	9	10	0.2	0.2	5	20

[a] From Kuehner et al.,[52] where values for other metals may be found, and also metal contents of commercial high purity acids.
[b] Other workers have found the same Pb content after triple ordinary distillations in silica.

Distillation of water by subboiling in silica apparatus gives a product containing at most the following concentrations of (selected) metals (ng/g):

Pb	0.008	Zn	0.04
Tl	0.01	Cu	0.01
Sn	0.02	Ni	0.02
Te	0.004	Fe	0.05
Cd	0.005	Cr	0.02
Ag	0.002		

Four liters of water can be produced in 24 hr. The product is free from heavy metals at the 10^{-10} g/g level (though Fe and Zn come close to this) but not, generally, at the 10^{-11} level.

Ammonia. Aqueous solutions of ammonia in glass bottles will be contaminated with silica, aluminum, calcium, and other elements introduced by attack of glass by the alkaline solution. In part, these may be in suspended or colloidal state. Some analytical reagent ammonium hydroxide is surprisingly low in heavy metals, for example, <0.003 ppm Mn, <0.005 ppm Zn, 0.003 ppm Pb, and ~0.005 (~0.2) ppm Cu. Possibly adsorption on glass accounts for the low concentration of some trace metals. Ionic iron(III) could hardly be present in appreciable quantity; adsorption or occlusion on container walls or particles of insoluble silicates would be expected, or alternatively expressed, little leaching of the small amounts of iron in glass would be expected.[53] High-purity ammonium hydroxide can readily be obtained by absorbing suitably washed and filtered cylinder NH_3 in pure water in (ice-cooled) plastic containers. Isothermal distillation can be used when cylinder NH_3 is not available. Electronic grade ammonia has been reported to contain 0.03 ppm Na, 0.007 ppm Mn, and 0.004 ppm Cu; 0.03 ppm Ca.

2. Inorganic Solids

In some trace analyses, reagent quality solids may be sufficiently pure to be used without purification, but blanks must, of course, be run. Generally, blanks should be run first to learn whether the available reagents are of adequate purity. Because trace elements may not be homogeneously distributed in a solid reagent, sampling needs to be done with care. Since sodium hydroxide and especially sodium carbonate are often used as fluxes in decomposition of inorganic samples, and since alkaline substances are usually more difficult to purify than neutral salts or solid acids, we give examples of trace elements in these substances (Table 2-21) to convey some idea of what may be expected. Phosphate is not uncommon in sodium hydroxide. Considerable variation is to be expected from one manufacturer to another, or from one lot to another. Occasionally, an analytical reagent may have a lower content of a trace metal than a higher purity chemical.[54] For example, one sample of reagent quality Na_2CO_3 of Table 2-21 contains ~0.001 ppm Cr compared to 0.01 ppm in extra purity Na_2CO_3; "Spec-pure" Na_2CO_3 has been reported to contain 0.6 ppm Cr and 0.02 ppm Co. In our experience, some reagent quality Na_2CO_3 contained enough Mo to make it useless for determination of this element in silicate rocks, whereas other products were satisfactory.

Lead is an important trace metal. Its content in analytical reagent KCl, NaCl, NH_4NO_3, and KNO_3 is typically 0.1–0.3 ppm, but values greater than this may be expected. Bismuth is sometimes found to the extent of a few tenths of a ppm in KCl. As much as 4 ppm Cd been reported in NaCl (analytical reagent).

Purification. The time-honored method for the purification of soluble solids involves crystallization from a suitable solvent, usually water for inorganic solids, as by cooling a hot, filtered solution. When applicable, this method has the advantage of eliminating, or at least reducing, many impurities at one time. It often renders good service, but recrystallization does not always result in greater purity. When mixed crystal (solid solution) formation takes place between the macro and micro components, an increase in the micro component concentration in the crystals

TABLE 2-21
Trace Metals (ppm) in Hydroxides and Carbonates

Metal	NaOH Analytical Reagent (max)	NaOH Analytical Reagent[a]	KOH Analytical Reagent[a]	Na_2CO_3 Extra Pure[b]	Na_2CO_3 Analytical Reagent[a]
Ag		<0.0002	0.07	<0.001	<0.0001
Al	10				
Ca	5			17	
Cd	0.05				
Co		0.006		<0.01	0.002, 0.15
Cr		0.06	<0.01	0.01	~0.001, 0.02
Cu	0.5			0.1, 0.02	
Fe	10	<0.9	2.7	0.1	1.4, 0.05
Mn	0.1			0.003	0.1
Ni	1				
Pb	0.1			<0.02	
Sb		0.0003	0.002		0.005
V					0.1
Zn	0.1	<0.02	1.25	<0.2	0.07

[a] Robertson, *loc. cit.* (mostly).
[b] Manufacturer's analysis, "electronic grade" chemical.

occurs if its distribution coefficient is greater than unity (see Chapter 7). Crystallization of KCl or NaCl from an aqueous solution containing a trace of lead concentrates lead in the crystals.[55] Even if mixed crystal formation in the strict sense of the term does not occur, there is a possibility of incorporation of micro components in crystals by "coprecipitation"—a convenient term for the carrying down of foreign constituents by poorly understood or unknown processes. We largely lack definite information on the extent to which soluble solids can be freed from trace metals by crystallization.

Because of the formation of hydrolysis products or other slightly soluble compounds of trace metals (as with the anion of a salt or acid) in weakly acidic, neutral or basic solutions, soluble solids crystallized from such media may be contaminated with them. For example, oxalic acid crystallized from water may contain appreciable amounts of calcium oxalate, only slightly soluble under the conditions, or perhaps incorporated in the crystals of oxalic acid in some way. Crystallization from hydrochloric acid solution decreases the calcium content of the oxalic acid crystals.

Sometimes it is advantageous to crystallize out a solid by adding a suitable solvent or agent to the aqueous solution, instead of by lowering the temperature of the solution. For example, sodium chloride can be crystallized by adding strong hydrochloric acid (or HCl gas) to a solution. Apart from the fact that the temperature coefficient of solubility of NaCl is unfavorable for recrystallization, the high H^+ concentration should aid in keeping certain metals in solution. $AlCl_3$ is readily freed from $FeCl_3$ by HCl precipitation. Crystallization by addition of ethanol to an aqueous solution needs to be applied with caution. The solubility of the impurities may be reduced also, so that purification is ineffective. The concentration of the impurity is an important factor. Crystallization of beryllium sulfate from alcoholic solution is not a good way to eliminate aluminum sulfate, at least if the concentration of the latter is not small, and perhaps not then.

Any insoluble material suspended in a solution to be crystallized must be removed by filtration through a suitable medium; if not removed, it will be gathered in the crystals so that these will be more impure with respect to insoluble constituents than the original salt.[56]

The direct removal of a trace metal from a solution of a reagent may be the surest method. For example, many heavy metals can be extracted with a $CHCl_3$ or CCl_4 solution of dithizone from a weakly acidic, neutral or weakly basic aqueous solution.[57] Other general metal extraction reagents that can be used in this way are oxine and thiooxine.[58]

Extraction is usually to be preferred to precipitation for removal of trace metals, but the latter is sometimes applied in conjunction with a collector (e.g., H_2S precipitation with a suitable metal as a collector). After removal of the trace metals and the excess extractant, the solute could be crystallized out,[59] but usually the purified solution is used directly. It should be stored in a polyethylene or polypropylene bottle. Other separation methods discussed in Chapters 7–10 may be useful at times in purification of reagents.

It may be advantageous or necessary to prepare some pure reagents by synthesis, for example, $NaClO_4$ from Na_2CO_3 and $HClO_4$, $CaCO_3$ from Ca salt and $(NH_4)_2CO_3$, and so on. Dilute NaOH solution can be obtained by running NaCl solution through a column of pure anion-exchange resin in hydroxide form.

A few inorganic solid reagents can be purified by sublimation.

It may be noted in passing that so far as the requirements of trace analysis are concerned, reagent quality chemicals can be used to prepare standard solutions of the common metals if the compounds are anhydrous or of known water content. If the truth be told, no trace determination of lead will suffer in accuracy if anhydrous analytical reagent lead nitrate is used as standard in place of NBS Standard Reference Material 928 (100.01% lead nitrate). Suspicion is justified when standard solutions of less common metals such as Be, Ga, and Sc, for example, are to be prepared. It may be necessary to establish the purity of these metals or their salts by analysis or qualitative tests of known reliability.

3. Organic Reagents

Chelating reagents may contain a few ppm of metals, which are likely to be bound as complexes. The metals are introduced through the solvents and the synthesis reagents, as well as from the equipment. A few metal values (ppm) for common organic reagents may be quoted (Robertson, *loc. cit.*):

Dithizone: Zn 1.2, Fe <7, Cu 0.4
Cupferron: Zn 7, Fe <0.6, Cu 0.16
Sodium diethyldithiocarbamate: Zn 0.04, Fe <0.6
Thenoyltrifluoroacetone: Zn 0.3, Fe 11.3, Sb 0.006, Cr 0.2, Ag 0.2, Cu 0.02
8-Hydroxyquinoline: Zn <0.1, Fe 1; Zn <0.4, Fe 5.7; Zn <0.04, Fe <0.1.

Organic reagents are ordinarily used in small amounts so that they are unlikely to introduce much metal, but nevertheless such amounts may need to be considered, especially in separations.

Recrystallization of organic reagents from suitable solvents may be expected to reduce their heavy metal content as well as that of undesirable organic contaminants, but information on this is scanty. Fractional freezing is applicable to a few organic reagents. Purification of some organic reagents is touched upon in later chapters.

4. Organic Solvents

The metal content of such organic solvents as $CHCl_3$ and CCl_4 (reagent grade) is likely to be very low and distillation should reduce it further. Doubly distilled $CHCl_3$ has been reported to contain 0.002 ppm Zn, 0.002 ppm Fe, and 3×10^{-4} ppm Cu and doubly distilled CCl_4, 0.001 ppm Zn, 0.01 ppm Fe, and 1×10^{-4} ppm Cu.[60] Chloroform distilled once in Pyrex was found to have a Pb content of 5×10^{-5} ppm.

The trace analyst using photometric methods is often more concerned with the presence of undesirable organic substances in his organic solvents. Oxygen-containing organic solvents (ethers, esters, etc.) will contain peroxides that must be removed. CCl_4 and $CHCl_3$ are likely to contain, or to develop on standing, oxidizing substances such as $COCl_2$ that can be very harmful in dithizone procedures and in other procedures as well. Moreover, analytical reagent CCl_4 contains, to varying extents, substances inhibiting the extraction of some metals (Cu, Bi) from acidic solutions by dithizone. For purification of CCl_4 and $CHCl_3$, see Chapter 6G.[61] The presence of a small amount of a polar solvent in a nonpolar or weakly polar solvent can drastically alter the extraction properties of the latter. Ethanol is commonly added (~1%) to chloroform as a preservative (inhibition of phosgene formation). It can be removed by shaking chloroform with successive portions of water. Phosgene is removed by shaking with an alkaline aqueous solution.

II. LOSSES

In separations and determinations, trace elements can be lost by adsorption, volatilization, coprecipitation, and coextraction. Errors can also result from adsorption of the constituents of standard solutions on the walls of containers. Adsorption losses especially will be discussed in this section.

A. ADSORPTION

A considerable stock of information, largely empirical, has been accumulated on the adsorption of inorganic species on glass and plastics from aqueous solutions. The exact nature of this adsorption often is not clear. Some adsorption may be what is called molecular, but ionic adsorption of one kind or another might be expected to predominate. On (silicate) glass especially, one important type of ion adsorption is ion-exchange adsorption. Metal cations exchange with alkali or alkaline earth ions at the glass-solution interface. From the analytical standpoint, an important characteristic of this adsorption is the possibility of reducing it to negligible proportions for most metal cations by acidifying the solution. In acidic medium, often no more than $0.1 M$ in H^+, or even $0.01 M$, the cations are replaced by hydrogen ions. Anion exchange, as of $AuCl_4^-$, can take place on $\equiv$SiOH groups of glasses. An increase in OH^- concentration might then be expected to decrease exchange adsorption of the metal-containing anion, and this has indeed been observed to occur. Metal cations often are adsorbed maximally in a feebly acidic, approximately neutral, or feebly basic solution. Here, metal ion hydrolysis products are likely to be involved and these may be adsorbed molecularly. Unless the metal solution is extremely dilute the products may be insoluble, and the colloidal particles can attach themselves to the wall of the container.[62] Adsorption then passes into precipitation. In fact, vigorous shaking of the solution may bring some of the material on the walls back into (colloidal) solution.

Adsorption is by no means limited to glass surfaces. It can also take place on plastic surfaces (polyethylene, polypropylene, and Teflon). Such adsorption need not be wholly molecular, although this would be expected to be important. Ion-exchange adsorption on plastics is not excluded, because plastic surfaces can develop carboxyl and other groups by the action of oxygen, heat and light, which can function as ion adsorbers. Polyethylene and polypropylene adsorb or absorb molecular species such as hydrogen sulfide, ammonia, bromine, iodine, organic solvents, and organic molecules (for example, 8-hydroxyquinoline) in general. These substances penetrate into the plastic and are hard to remove.[63] Even HCl and HNO_3 penetrate into polyethylene to a slight extent. Mercury metal diffuses through polyethylene. Some plastics contain plasticizers and fillers that can exert an effect of their own.

Under some conditions—in approximately neutral solutions—polyethylene and polypropylene may adsorb some metals more strongly than does borosilicate glass (see Table 2-22). In the presence of thenoyltrifluoroacetone, Teflon adsorbs ^{234}Th strongly—about ten times more strongly than glass (Table 2-22). The similarity in the HTTA and the Teflon molecule may offer an explanation.

Dyes can be adsorbed appreciably on glass surfaces and this effect needs to be taken into account in some colorimetric determinations.[64] Contact of dyes with ground glass surfaces especially should be avoided. Metal chelates and ion associates can be adsorbed from organic solvent solutions, but little information is available on this. It is known that Be trifluoroacetylacetonate is adsorbed on glass from organic solutions but not on silica. Bidleman (Thesis, 1970) stored carbon tetrachloride solutions of bismuth dithizonate in stoppered Pyrex test tubes in the dark for 20 hours and found the following amounts of adsorption:

Original Concentration of Bi $(HDz)_3$, $\times 10^{-6}$ M	Decrease in Concentration at ~4°C due to Adsorption (%)	Decrease in Concentration at 20°C due to Adsorption (%)
2.1	6.4	
4.1	3.7	
6.3	3.7	
1.4		9.4
2.8		7.2
5.7		6.5

There was also some decrease in concentration due to decomposition of bismuth dithizonate, but adsorption accounted for 60–65% of the decrease in the dark. Adsorbed bismuth dithizonate can be removed from the glass with mineral acids (1 M HCl).

TABLE 2-22
Adsorption of Trace Elements on Glass, Plastic, and Other Surfaces[a]

Elements	Behavior	Reference
Cr(VI) (in H_2SO_4–$K_2Cr_2O_7$ cleaning mixture)	After thorough initial washing, Cr(VI) continues to be released from Pyrex and soda-lime glass surfaces for hours.	E. P. Laug, *Ind. Eng. Chem. Anal. Ed.*, **6**, 111 (1934).
Hg	Dilute solutions of Hg salts in glass or silica lose appreciable quantities of Hg on standing.	A. Stock, *Berichte*, **B72**, 1844 (1939).
Mo, V, Ni, Ti, Au, Pd, Pt, Ru (0.001% solutions in 6% mineral acid)	In Jena glass, Mo, V, Ni and Ti solutions decreased to 0.2–0.4 of original strength after 75 days; Au, Pd, Pt and Ru decreased to 0.1–0.3 of original after 230 days. In fused silica, little decrease in concentration, except Pd which decreased to 0.3 after 230 days.	F. Leutwein, *Zentr. Min. Geol.*, **A,** 129 (1940). E. S. Gladney and K. E. Apt, *Anal. Chim. Acta*, **85**, 393 (1976), contrarily found little change in 1 M HCl solutions of Ir, Os, and Ru stored for a month or more in glass and polyethylene bottles.
Au	~60% Au lost from 10 liters 3% NaCl solution containing 50 μg Au (as $AuCl_4^-$) when stored in glass for 8 weeks.	E. Bauer, *Helv. Chim. Acta*, **25**, 1202 (1942).
Na	Stronger adsorption on soft glass than on silica (0.0226 M Na, pH 8.6). Adsorption increases with pH in range 3–11.	J. W. Hensley, A. O. Long, and J. E. Willard, *Ind. Eng. Chem.*, **41**, 1415 (1949); A. O. Long and J. E. Willard, *ibid.*, **44**, 916 (1952).
Pb	See Figs. 2-1 and 2-2.	
Th (tracer concentrations)	Cation-exchange adsorption on glass, molecular adsorption (hydrolysis products) on polyethylene. Silicone coating on glass decreases adsorption.	J. Rydberg and B. Rydberg, *Svensk Kem. Tids.*, **64**, 200 (1952).
Th, U, Pb (0.1–1000 μg/ml)	Loss in strength (as much as 50–80%) of all solutions after 2.5 months in pH range in which hydrolysis occurs; usually greater loss in polyethylene than in borosilicate glass. No adsorption of Pb at pH < 5.6, of U < 4.2, of Th < 3. Decrease in concentration associated to large extent with formation of insoluble hydrolysis products.	R. G. Milkey, *Anal. Chem.*, **26**, 1800 (1954).
PO_4 in sea water	Almost all phosphate adsorbed on polyethylene in less than 6 days.	J. Murphy and J. P. Riley, *Anal. Chim. Acta*, **14**, 318 (1956).
Au in sea water	More than 75% of gold lost when sea water was stored in polyethylene 3 weeks.	R. W. Hummel, *Analyst*, **82**, 483 (1957).
Co, 0.01 ppm; Mn, 1 ppm; V, 7 ppm	Co at pH 3 in well-cleaned Pyrex showed no adsorption after 3 months. Mn and V in HCl (concentration not stated) did not decrease in concentration after 9 months in polyethylene.	R. E. Thiers, in J. H. Yoe and H. J. Koch, *Trace Analysis*, p. 646 (1957).

[a] This listing is not intended to be exhaustive. Order is approximately chronological.

TABLE 2-22 *(Continued)*

Elements	Behavior	Reference
Be in alkaline solution	Be appreciably adsorbed (up to 15%) on glass and (to a lesser extent) on plastic surfaces from quite strongly basic solutions. Presumably anion exchange on active surface sites. Adsorption decreases in presence of aluminate and stannite.	C. W. Sill, C. P. Willis, and J. K. Flygare, Jr., *Anal. Chem.*, **33**, 1671 (1961).
Ca	Ca adsorbed strongly on borosilicate glass previously soaked in EDTA solution, but not on glass previously soaked in $CaCl_2$ solution.	J. S. Brush, *Anal. Chem.*, **33**, 798 (1961).
As (V) in sea water	$\sim$16% As adsorbed on soda glass after 16 days, when equilibrium was attained. $\sim$6% As adsorbed on Pyrex and polyethylene within 10 days, when equilibrium was attained.	J. E. Portmann and J. P. Riley, *Anal. Chim. Acta*, **31**, 509 (1964).
Au (III)	No adsorption at pH 0 on borosilicate glass; cation-exchange adsorption occurs in less acidic solutions with maximum at pH 2–3. NaCl and $NaClO_4$ as well as HCl diminish adsorption.	P. Beneš, *Radiochim. Acta*, **3**, 159 (1964).
Ag, Cr, Co, Cs, Sb, Se, Zn in sea water	No significant adsorption in Pyrex in periods up to 6 months.	D. F. Schutz and K. K. Turekian, *Geochim. Cosmochim. Acta*, **29**, 259 (1965).
K, Th, U, Ag, Sr, Ba, Pb, Fe	Pretreatment of glass ("Normalglas") with acid and base diminishes adsorption. At $\sim 10^{-10}$ mole cation/cm^2, metal thought to be held by ion exchange, at higher concentrations by "adsorption" (then easier to remove). High adsorption values of Ag and Pb at low concentrations attributed to diffusion into glass. From 2×10^{-5} M Ag in 10^{-3} M HNO_3, 5×10^{-9} mole Ag adsorbed/cm^2 of glass treated with conc. HNO_3, 2×10^{-8} on untreated glass. From 1.3×10^{-4} M $Pb(NO_3)_2$ in 10^{-3} M HNO_3, 1×10^{-8} mole Pb/cm^2 adsorbed on glass treated with HNO_3, 5×10^{-8} on untreated.	K. H. Lieser, *Radiochim. Acta*, **4**, 225 (1965).
Cs, Sr, Y, Ce, Ba-La, Zr, I, Ru–Rh (as radioactive isotopes)	Less adsorption on borosilicate glass than on polypropylene, except for Cs, Ru, and Zr (pH mostly 4–8). ^{90}Y, for example, showed adsorption of 0.10%/cm^2 on polypropylene and 0.02%/cm^2 on Pyrex after 1 hr (10^{-12} N solution). Treatment of Pyrex with silicone coating agents of limited or moderate value in decreasing adsorption.	G. G. Eichholz, A. E. Nagel, and R. B. Hughes, *Anal. Chem.*, **37**, 863 (1965).
Mn	Adsorption on borosilicate glass (Sial). Percentage adsorption in pH range 6–13 decreases greatly as Mn concentration increases from 5×10^{-8} to 10^{-5} M.	P. Beneš and A. Garba, *Radiochim. Acta*, **5**, 99 (1966).

81

TABLE 2-22 (*Continued*)

Elements	Behavior	Reference
Ag	From 0.05 ppm Ag solution in distilled water, significant adsorption on Vycor, poly-propylene, polystyrene, and Teflon after 10 days, increasing on longer standing. No adsorption the first ~8–16 hr. When Cl⁻ is present induction period decreases and ad-sorption increases. Adsorption on walls very nonuniform.	F. K. West, P. W. West, and F. A. Iddings, *Anal. Chim. Acta*, **37**, 112 (1967); *Anal. Chem.*, **38**, 1566 (1966).
Ag	At pH 3.8, borosilicate glass adsorbs relatively little Ag, even less when treated with Desicote. Acid-washed borosilicate glass ad-sorbs less strongly than paraffin, Nalgene, and Teflon. Adsorption is less in more acidic solutions (pH 1.5). Na^+ suppresses adsorption of Ag. Very dilute (0.01 ppm Ag) solutions lost 5–10% Ag in 2–20 days at pH 1.5, or relatively more than 0.1 ppm solutions.	W. Dyck, *Anal. Chem.*, **40**, 454 (1968).
Ag, Co, Cs, Fe, In, Rb, Sb, Sc, Sr, U and Zn in sea water (pH 8)	Indium adsorbed >90% on polyethylene after 2 months, 20% on Pyrex. Fe and Sc similar to In on polyethylene, but adsorption on Pyrex stronger than of In. Ag hardly adsorbed at all on Pyrex, ~25% on polyethylene after 20–60 days. About 5% Co lost on Pyrex after 20–60 days, roughly twice as much on polyethylene. Little adsorption of Cs, Sb, Sr, and Zn on either material after as much as 75 days; ~10% Rb on Pyrex, none on polyethylene. Sea water acidified with HCl to pH 1.5 lost essentially no Co, Zn, In, Sr, Ag, Fe but about 50% Sc and 10% U after 2 months in polyethylene.	D. E. Robertson, *Anal. Chim. Acta*, **42**, 533 (1968).
Th $(10^{-10}\ M)$	On soda-lime glass, Th adsorbed maximally at pH 4–5, little at pH 7 and still less at pH 1. In presence of HTTA, maximal adsorption at pH ~3, with second maximum at pH 7. Adsorption of Th in presence of HTTA much stronger (~10 times) on Teflon than on glass at pH 5; also strong on polyethylene and polypropylene.	B. Edroth, *Acta Chem. Scand.*, **23**, 2636 (1969).
Sb(III)	0.01% Sb solution in 6.5 M HCl underwent no change in concentration after 1 month in polyethylene; in glass, effective concentration decreased ~10%/day. Solution of 2 ppm Sb is unstable in polyethylene also.	G. O. Kerr and G. R. E. C. Gregory, *Analyst*, **94**, 1037 (1969).
Ag	0.22 ppm Ag in neutral solution, various materials. For adsorption up to 10 days, see Fig. 2-3. At end of 46 days, adsorption (%): Pyrex 60, Desicoted Pyrex 51, Vycor 43, polyethylene 47, Teflon 50.	R. A. Durst and B. T. Duhart, *Anal. Chem.*, **42**, 1002 (1970).

TABLE 2-22 *(Continued)*

Elements	Behavior	Reference
Pb, Bi	Loss of ~15% Pb (6–9 μg) in dithizone extraction at pH 10.8 in Pyrex separatory funnels. Loss was ~5% in polypropylene funnels. Immediate extraction (<1 min) of Bi as dithizonate gave ~95% recovery, 85–90% after 0.5–1 hr, at pH 9.5 in Pyrex; with polypropylene funnels, recovery was ~99% (avg. of 40 determinations).	T. Bidleman, Ph.D. Thesis, University of Minnesota, 1970.
Y $(10^{-11}$– $10^{-12}\,M)$	Maximum adsorption on soft glass at pH 4–5 and on Teflon at pH 5–6. Amounts of Y adsorbed were less in the presence of NaCl (0.01 M).	P. Polansky and J. Bar, *Coll. Czech. Chem. Commun.*, **37**, 3024 (1972).
As in sea water	Up to 70% As lost from acidified sea water (9 ml HCl/liter) stored for 1 week in polyethylene.	B. J. Ray and D. L. Johnson, *Anal. Chim. Ácta*, **62**, 196 (1972).
Bi	Losses of 20–90% Bi found in decomposition of silicate rocks with HF–HClO$_4$ in platinum ware. Attributed to adsorption. Decomposition in Teflon vessels gave no losses; no adsorption on glass.	L. P. Greenland and E. Y. Campbell, *Anal. Chim. Acta*, **60**, 159 (1972).
Hg	Loss of Hg from distilled and natural waters stored in polyethylene containers is largely prevented by acidification to 0.1 M with HNO$_3$. Add HNO$_3$ first to bottle.	R. V. Coyne and J. A. Collins, *Anal. Chem.*, **44**, 1093 (1972).
P	Glass, polypropylene, and polycarbonate bottles adsorb phosphate. Treatment of polycarbonate vessels with KH$_2$PO$_4$ solution, followed by thorough washing with water, eliminated adsorption of inorganic phosphate from distilled and lake waters.	J. C. Ryden, J. K. Syers, and R. F. Harris, *Analyst*, **97**, 903 (1972).
Ag, Pb, Cd, Zn, Ni	~0.001 ppm Ag, 0.01 ppm Pb, 0.001 ppm Cd, 0.1 ppm Zn solutions. Ag: erratic, but little adsorption on polyethylene (PE) and borosilicate glass (BG) at pH 2 after weeks if protected from light. Pb: little adsorption on BG at pH 2 after 3 weeks; strong adsorption on PE (and polypropylene) at pH 2 in 5–10 days. Cd, Zn: no adsorption on BG and PE at pH 2 after weeks. Metals tend to be leached from polyproplene, Zn from BG but not PE, Ag (erratic) from BG. Adsorption losses greater above pH 2.	A. W. Struempler, *Anal. Chem.*, **45**, 2251 (1973).
Cr(III)	Sea water stored in Pyrex, polyethylene and polypropylene lost Cr(III) by adsorption at about same rate in all. Concentration of Cr reduced by ~50% after ~4 days in polypropylene. Equilibrium concentration not zero.	T. R. Gilbert and A. M. Clay, *Anal. Chim. Acta*, **67**, 289 (1973).

TABLE 2-22 (*Continued*)

Elements	Behavior	Reference
Hg(II)	At pH 7, loss of Hg by adsorption is greater on polyethylene and, especially on polyvinyl chloride, than on glass. Acidification of water with HNO_3 (not H_2SO_4) to pH 0.5 or less prevents appreciable decrease in Hg concentration for as long as 10–15 days storage in glass or plastics.	R. M. Rosain and C. M. Wai, *Anal. Chim. Acta*, **65**, 279 (1973).
Au, Ag, Pt	Au (10 ppm) in aqua regia stable for 6 months, Pt(IV) (1–5 ppm) in 0.5 M HCl for 2 months (glass container).	V. A. Pronin et al., *Zh. Anal. Khim.*, **28**, 2328 (1973).
Hg(II)	Distilled water solutions of Hg (0.1–10 μg/liter) 5% (v/v) in HNO_3 and 0.01% in $Cr_2O_7^{2-}$ showed no change after 5 months in borosilicate glass. Solutions 5% in HNO_3 and 0.05% in $Cr_2O_7^{2-}$ stored in polyethylene unchanged for at least 10 days. $KMnO_4$ should be avoided.	C. Feldman, *Anal. Chem.*, **46**, 99 (1974).
Pb	H_2O_2 prevents adsorption of Pb on borosilicate and polyethylene surfaces from neutral solution. Action not explained.	H. J. Issaq and W. L. Zielinski, Jr., *Anal. Chem.*, **46**, 1328 (1974).
Cd	When stored in bottles of polyethylene, polypropylene, and polyvinyl chloride, carrier-free ^{109}Cd showed less than 3% loss of activity from solution of pH 3–10 after 2 weeks. However, when the same solution at pH 9 was stored in a soft glass container, loss of Cd was 75% after 1 day. No loss observed at pH less than 7.	W. G. King, J. M. Rodriguez, and C. M. Wai, *Anal. Chem.*, **46**, 771 (1974).
Cr(III, VI)	Pyrex, flint glass, and polyethylene. Solution of 1 ppm Cr(III) at pH 6.95 showed the greatest instability, particularly in polyethylene, where losses were ~25% after 15 days. Cr(VI) showed losses less than 1%. In 0.5% HNO_3 and at pH 3.1, Cr(III) losses were less than 0.5% after 15 days.	A. D. Shendrikar and P. W. West, *Anal. Chim. Acta*, **72**, 91 (1974).
Hg	Hg (vapor) can diffuse through walls of polyethylene bottles and contaminate contained solution (sea water).	M. H. Bothner and D. E. Robertson, *Anal. Chem.*, **47**, 592 (1975).
Hg	Losses of Hg by adsorption on glass or PTFE greatly reduced by adding I^-, Br^-, or CN^-, which also reduce volatilization loss. Steaming of vessels with 65% HNO_3 superior to washing with 1 M HNO_3 in reducing losses.	G. Kaiser et al., *Talanta*, **22**, 889 (1975).

84

TABLE 2-22 (*Continued*)

Elements	Behavior	Reference
Hg	Demineralized water (5 μg Hg/liter) lost 77% of total Hg by adsorption on polyethylene walls and 18% by volatilization after 3 weeks. Adsorbed Hg could be removed with 16 M HNO_3. Water acidified to pH 0.5 lost 4% Hg by adsorption and 12% by volatilization. Solutions of pH 0.5 (HNO_3) containing 0.05% $K_2Cr_2O_7$ lost only 2% Hg after 3 weeks.	J. M. Lo and C. M. Wai, *Anal. Chem.*, **47**, 1869 (1975).
Hg	Order of decreasing adsorption of Hg: polyethylene > borosilicate glass > Teflon. 1 ppm Hg in 1 M HNO_3 showed no significant loss on Teflon after 1 month; 0.4 ppm Hg no loss on glass after 10 days, but a 0.015 ppm Hg solution lost ~10% in 24 hr. Washing glass with acid Cr(VI) solution reduces Hg loss. As much as 30% Hg lost in standard wet ashing (AOAC). Also serious losses of Hg in lyophilization of organic samples. No Hg loss in dry combustion of organic samples in O_2 stream with collection of products at liquid N_2 temperature.	R. Litman, H. L. Finston, and E. T. Williams, *Anal. Chem.*, **47**, 2364 (1975).
Hg	Bacterial action can result in volatilization of Hg as metal or organomercury compounds.	R. W. Baier, L. Wojnowich, and L. Petrie, *Anal. Chem.*, **47**, 2464 (1975).
Hg	HNO_3 and $K_2Cr_2O_7$ prevent loss of Hg from standard solutions and from waste water.	A. A. El-Awady, R. B. Miller, and M. J. Carter, *Anal. Chem.*, **48**, 110 (1976).
Hg	2,6-Di-*tert*-butylmethylphenol (antioxidant in polypropylene) can reduce Hg(II). Oxidizing agents and Cl^- reduce losses.	S. R. Koirtyohann and M. Khalil, *Anal. Chem.*, **48**, 136 (1976).
Hg as methylmercury	Loss processes obscure.	K. R. Olson, *Anal. Chem.*, **49**, 23 (1977).
U(VI)	U not adsorbed on polyethylene from neutral or acidic solutions, nor on glass from acidic solutions.	R. J. N. Brits and M. C. B. Smit, *Anal. Chem.*, **49**, 67 (1977).
Sn(IV)	Solutions (0.1 mg Sn/liter) in 0.5% HCl stable in glass or polyethylene for at least 80 days.	P. N. Vijan and C. Y. Chan, *Anal. Chem.*, **48**, 1788 (1976).
Hg	If polyethylene container is first soaked in (natural) water to be analyzed and the collected water is agitated to keep it saturated with air, Hg content (pp 10^9) does not decrease seriously in 10 hours, so that determination can be made within this period. (All natural waters may not show this behavior.)	K. I. Mahan and S. E. Mahan, *Anal. Chem.*, **49**, 662 (1977).

TABLE 2-22 (Continued)

Elements	Behavior	Reference
Hg	Some commercial linear polyethylene bottles were found to be a mixture of 45% polypropylene and 55% polyethylene. Hg solutions (1 pp 10^9) containing HNO_3 and $K_2Cr_2O_7$, stored for 10 days in such bottles, lost 43% Hg compared to a loss of 6% in 100% linear polyethylene bottles. The additives identified in the 100% polyethylene were a hindered alkylphenol (HAP) resembling 2,6-di-*tert*-butyl-4-methylphenol and an aliphatic ester which appeared to be a dialkylthio-propionate (obviously Hg-reactive). The polypropylene–polyethylene mixture contained an alkoxyhydroxybenzophenone, possibly a small amount of an HAP, an aliphatic ester, and a low molecular weight polyhydroxy compound. The presence of an arylphosphoric acid was indicated (can be formed by oxidation of an arylphosphorous acid, a protective agent commonly used with polypropylene). The polyhydroxy compound is believed to be responsible for the marked reduction of Cr(VI) by the mixed plastic. The relatively poor stability of Hg solutions in 100% polyethylene may be due to additives, surface oxidation products, and residual terminal unsaturation of polyethylene.	R. W. Heiden and D. A. Aikens, *Anal. Chem.*, **49,** 668 (1977).

Characteristically the adsorption of metals is slow, equilibrium or an approach to it being counted in days, even weeks, not minutes or hours (see Fig. 2-3). Another characteristic is the proportionately stronger adsorption from dilute than from less dilute solutions. In other words, adsorption follows the classical adsorption isotherm.

As long as the analyst can work in acid solutions he should not be much troubled as a rule by adsorption phenomena. Standard solutions of most metals should not undergo a significant decrease in strength in weeks or months if they are 0.1 M or stronger in mineral acid (solutions of easily hydrolyzable metals should be made 1 M or stronger) and are stored in polyethylene (polypropylene) or Pyrex (or similar borosilicate glass) bottles.[65] Plastic bottles may be preferred because of lesser leaching of trace elements. It is always wise to avoid storing very dilute standard solutions. Preferably, the stock solution is made 0.1% or thereabout in metal, and more dilute solutions, for example, 0.001%, are prepared from it by dilution as needed. Feebly acidic, neutral, or basic solutions of metals should not as a general rule be kept in contact with glass or plastic any longer than really necessary. Errors from adsorption occurring in such solutions during a determination are more or less cancelled if the standard photometric curve is established under the same conditions as exist in the determination itself. But errors can occur if adsorption takes place in dissolution or separation steps not duplicated in setting up the standard curve. It may be possible to remove adsorbed metal from the vessel walls (e.g., a separatory funnel) by acid treatment, but it is well not to rely on this.

Whenever permissible, sample solutions, natural waters for example, should be acidified to prevent or minimize adsorption of metals. Acidification is not always admissible. Thus acidification of a natural water may bring suspended matter into solution partially or totally and thus cause difficulties if the form of a trace element is of importance.

In connection with adsorption losses, attention should be called to the possibility of removal of metals from solutions by fungus and algae growth. Strontium, for example, is known to be ingested by fungi.[66] Bacteria may be implicated in the loss of mercury from solutions (volatilization).

The stability of dilute mercury solutions has received more attention than that of any other metal. The problem is complicated by possible

TABLE 2-23
Recovery of Elements after Solution Evaporation[a]

Element (μg in 10 ml of acid)	Residue Dissolved in 0.5 ml of Solvent	Recovery (%)
Co(II), 10^{-3}–1, 0.1 M HCl	H_2O	100
Co(II), 10^{-3}–1, 0.1 M HCl	0.1 M HCl	100
Zn(II), 10^{-2}–10, 0.1 M HCl	0.1 M HCl	100
Au(III), 3×10^{-2}–10, 0.9 M HCl −0.4 M HNO$_3$	0.1 M HCl	98
Cr(III), 2×10^{-2}–10, 0.1 M HCl	0.1 M HCl	93
Cr(III), 2×10^{-2}, 0.1 M HCl	3 M H$_2$SO$_4$	100
Cr(III), 2×10^{-2}, 0.1 M HCl −0.1 M H$_2$SO$_4$	0.1 M HCl	100
Ag(I), 1.5×10^{-2}–10^2, 1 M HNO$_3$	0.1 M HNO$_3$	30–100
Ag(I), 3×10^{-2}–5, 1 M HNO$_3$	0.1 M NH$_4$OH	86–89
Ag(I), 3×10^{-2}, 1 M HNO$_3$	0.2 M Na$_2$S$_2$O$_3$	96–100
Ag(I), 3×10^{-2}, 1 M HNO$_3$ −0.5 M NaNO$_3$	0.1 M HNO$_3$	98–100
Hg(II), 10^{-3}–10, 1 M HNO$_3$	0.1 M HNO$_3$	2
Hg(II), 10^3, 1 M HNO$_3$	0.1 M HNO$_3$	61

[a] A. Mizuike and Y. Takata, *Japan Analyst*, **12**, 1192 (1963). Evaporation in borosilicate glass dish on water bath. Radioactive nuclides for determination.

reduction of mercury to metal and its volatilization (see later). But adsorption can also take place. Loss of mercury increases markedly with a decrease in concentration, so that standard solutions should not be made unnecessarily dilute. Some authors state that a 0.001 ppm (1 μg Hg/liter) mercury solution in 0.2 M HNO$_3$ (containing a strong oxidizing agent to forestall reduction of Hg(II)) can decrease in strength by 5–10% in 24 hr when stored in borosilicate glass bottles. Others have found a solution of this mercury concentration in 1 M HNO$_3$ to show no loss of mercury in months when kept in borosilicate glass.

Solutions of photosensitive salts (Ag, Au, etc.) should be protected from light and should be prepared from water free from organic matter.

Finally, in connection with adsorption, the question may be asked, if a solution is evaporated to dryness, can the residue be brought quantitatively into solution in acid? Some metals, present in microgram or smaller amounts, are satisfactorily brought back into solution in 0.1 M acid, as may be seen from Table 2-23. Low recoveries of Hg(II) are explained by volatilization. In general, it is advisable when evaporating a very dilute

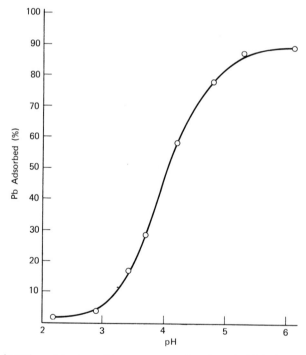

Fig. 2-1. Adsorption of lead on filter paper as a function of pH (0.06 g of filter paper, 30 ml of 10^{-5} M lead solution) according to E. Broda and T. Schönfeld, *Monatsh. Chem.*, **81**, 459 (1950).

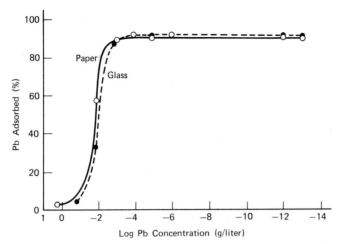

Fig. 2-2. Adsorption of lead by paper and glass at pH 6 as a function of the lead concentration according to T. Schönfeld and E. Broda, *Mikrochem. Mikrochim. Acta*, **36/37**, 537 (1951). Thirty milliliters of lead solution, 0.06 g of paper fibers (unsized, lignin-free) and 0.30 g of glass wool (250 cm^2 geometrical surface). The principal lead species in solution at pH 6 are Pb^{2+} and $PbOH^+$.

metal solution to add a soluble salt as a collector, for example NaCl to a heavy metal chloride solution. Subsequently, when the soluble salt is dissolved, the trace constituent is carried with it into solution.

Adsorption on Filter Media. Filtration of solutions through paper, sintered glass, glass wool, and other media presenting large surface areas can result in serious adsorption losses. Centrifugation is, in general, to be preferred to filtration, but the latter is often convenient and may be acceptable if the solution is sufficiently acidic.

The adsorptive properties of cellulose are partially due to the presence of $-COO^-$ groups, which allow cation exchange. The pH of the solution filtered is, therefore, an important variable. In feebly acidic, neutral, or feebly basic solutions, the adsorption of heavy metals may be so strong that most of the metal is removed from very dilute solution. Figure 2-1 illustrates the adsorptive behavior of lead on filter paper as a function of pH. Adsorption rises sharply above pH ~3; even at pH 2, slight adsorption of lead occurs. Other heavy metals show similar behavior, but adsorption is not limited to what are usually called heavy metals. Thus Ba is adsorbed by filter paper.

Adsorption of metals is proportionately stronger in more dilute solutions of the metal, as illustrated in Fig. 2-2, in which the adsorption of lead by paper fibers and glass fibers is shown as a function of its concentration in an aqueous solution of pH 6. It appears that the adsorption of a metal can often be represented by the Langmuir adsorption isotherm:

$$\frac{x}{m} = \frac{k_1 C}{k_2 C + 1}$$

where x = amount of adsorbed metal
  m = mass of adsorbent
  C = equilibrium concentration of metal in solution
  k_1 and k_2 = constants

  Adsorption equilibrium may be attained slowly (Fig. 2-3). The important general rule for the trace analyst is to filter only acidic solutions, and even with these caution is necessary. Weight for weight, sintered glass is to be preferred to filter paper or asbestos.[67] Adsorption of metals by finely divided platinum (Munroe crucible) seems to have received little attention, but it is known to occur.[68]

  Millipore filter discs and Whatman no. 4 filter paper have been reported to adsorb organic compounds (absorbance below 250 nm especially can be affected).

B. VOLATILIZATION

  Losses due to escape of metals as more or less volatile compounds are most likely to occur when the sample is decomposed (as in the destruction of organic matter) or brought into solution, especially at 100°C and higher. Metals are less likely to be lost from basic solutions or melts than from acid. Slight loss of Os is reported to occur from a Na_2O_2–NaOH melt. Mercury (boiling point 357°) and its compounds are easily lost and require special attention. Among the other metals, those of semimetallic character and those in their highest oxidation states tend to form volatile oxides and halides, for example, OsO_4, RuO_4, Re_2O_7, $GeCl_4$, $AsCl_3$, $SbCl_3$, $SbBr_3$, CrO_2Cl_2, $SnCl_4$ and TiF_4. In a strongly reducing environment, gaseous AsH_3 is formed, less easily SbH_3. Many metal chlorides and fluorides have rather low boiling points, but these refer to the anhydrous compounds that are unlikely to be present under analytical conditions. Besides, excess Cl^- or F^- can sometimes reduce volatility by complex formation. See Chapters 3 (destruction of organic matter) and 10 (separations by volatilization) for further information having a bearing on volatilization losses.

  Some loss of gold has been claimed to occur in the evaporation of chloride solutions, but this has been disputed and losses are attributed to creeping.[69] A few metals can be lost from solutions by volatilization at room temperature (OsO_4, RuO_4 and, from HCl, As(III) and Ge). It has been shown that the decrease in strength of very dilute Hg(II) solutions on standing is due to reduction of Hg(II) to Hg(I) by traces of reducing

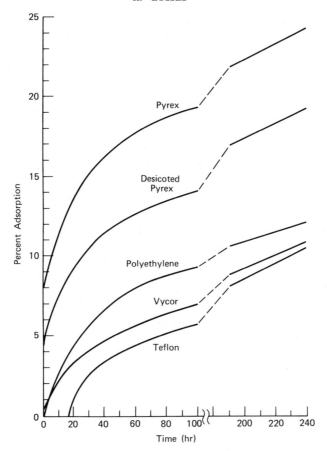

Fig. 2-3. Adsorption of silver on various materials from neutral solution (0.22 ppm Ag as AgNO₃) as a function of time according to Durst and Duhart, *loc. cit.* (Table 2-22). Reprinted with permission from R. A. Durst and B. T. Duhart, *Anal. Chem.*, **42**, 1003 (1970). Copyright by the American Chemical Society.

substances, followed by disproportionation of the latter to Hg(II) and Hg metal, which then slowly volatilizes.[70] The loss of Ca which has been reported to occur in fuming off ammonium salts must be due to mechanical loss. Ammonium salts can alternatively be removed by oxidation with $HNO_3 + HCl$.

C. OTHER LOSSES

Solid substances can react with silica glass (fused quartz), porcelain, and platinum at higher temperatures in ignitions (and sometimes in fusions) to give reaction products that may not be completely brought into solution on subsequent acid treatment. Thus SiO_2 can form silicates, especially with heavy metal oxides, on heating, and Pt can alloy with reduced heavy metals. This type of loss is most likely to occur in the destruction of organic matter and in the decomposition of refractory inorganic materials. See Chapter 3 for further discussion.

Insoluble material at any point in a procedure—whether a small amount of undissolved sample or a precipitate formed at a later stage—is a cause for concern, and it should be appropriately treated to recover any metal present unless experience has shown this to be unnecessary.

Losses can, of course, occur in separations, and these may outweigh those resulting from adsorption and volatilization. They fall in a different category and will not be discussed here.

Mention need hardly be made of the possibility of mechanical losses when solutions are boiled or gases are evolved, when extractions are made (leakage from separatory funnels, incomplete separation of phases), etc. Such errors can easily be avoided and will be of no importance except in the work of inexperienced or careless operators.

Apparent Losses. Even though the analytical concentration of a metal may not change on standing, changes in species may occur that must be taken into account in a colorimetric or fluorimetric procedure. Hydrolysis may occur, as with zirconium, leading to the formation of species not readily reacting with the reagent.[71] Oxidation [Fe(II) → Fe(III)] or reduction [Ce(IV) → Ce(III), Au(III) → Au, as by reducing impurities] may occur on standing. Usually procedures provide for such possible changes in oxidation state.

NOTES

1. Contamination problems are discussed in M. Zief and R. Speights, Eds., *Ultrapurity: Methods and Techniques*, Dekker, New York, 1972. T. J. Murphy in P. D. LaFleur, Ed., *Accuracy in Trace Analysis: Sampling, Sample Handling, Analysis*, Nat. Bur. Stands. Spec. Publ. 422, Washington, 1976, p. 509.

2. Mercury is adsorbed on filter paper and vessels, and solutions thus can become contaminated. Moreover, mercury can be present in laboratory dust. It has been reported that a vessel exposed uncovered in a chemical laboratory for 24 hr was contaminated to the extent of a few nanograms of Hg per $10\ cm^2$. Presumably dust contributed much of this mercury.

3. Contamination by silver in a furnace run at 900°C was traced to volatilization from a silver thermal fuse as the main source: P. J. Parle and G. A. Fleming, *Analyst*, **97**, 195 (1972). Contamination by copper (and some silver) was attributed to volatilization from the interior of the furnace, presumably contaminated at an earlier time.

4. Such as keeping vessels covered (cover glasses for beakers, inverted beakers for flasks). Covering vessels serves two purposes: preventing coarse foreign material from falling into solutions and hindering access of outside air to interiors of vessels.

5. Large evaporating chambers of Teflon or Pyrex (made from a 3 liter beaker) have been described by T. J. Chow and C. R. McKinney, *Anal. Chem.*, **30,** 1499 (1958); nitrogen flushing is used. According to results given by these workers, these chambers do not wholly prevent contamination by lead in ordinary laboratory air and they should be used in purified air (filtered and electrostatically precipitated).

6. L. S. Vasilevskaya, V. P. Muravenko, and A. I. Kondrashina, *Zh. Anal. Khim.*, **20,** 540 (1965). Similar glove boxes have been described by other workers.

7. U.S. Federal Standard No. 209a, *Clean Room and Work Station Requirements, Controlled Environment*, 1966, General Services Administration, defines three classes of air cleanliness in terms of particle density and other parameters. The class designation refers to the maximum number of particles > 0.5 μm in 1 ft^3 (English system) or in 1 liter (metric system):

English	Metric
100	3.5
10^4	350
10^5	3500

8. In a laminar airflow clean room the whole body of air moves through the room with uniform velocity, without turbulent flow characterized by eddies and pockets of more or less stationary air. To produce laminar flow, the inlet and exhaust areas must be equal to each other and equal to the cross section of the enclosed space. Thus filtered air could emerge from the entire ceiling of such a room and be withdrawn through a perforated floor. For an account of laminar airflow clean rooms, see D. Hughes, *Chem. Brit.*, **10,** 84 (1974). Also see J. A. Paulhamus on airborne contamination in Zief and Speights, *op. cit.*[1]

9. For example, lead determinations for geochronology demand very low blanks. Lead occurs in air (especially in urban areas) partly as lead alkyls. Filtration of air through activated carbon probably removes some of the gaseous lead compounds. The lead remaining can be removed by scrubbing a slow air stream with 16 M HNO$_3$ to 0.1 N KMnO$_4$. Air so purified can be led into vessels in which evaporations are being carried out. See J. W. Arden and N. H. Gale, *Anal. Chem.*, **46,** 2 (1974). Blanks as low as 0.002 μg Pb have been obtained in decomposition of, and (electrolytic) separation of lead from, 1–2 g of chondrites.

10. R. L. Mitchell, *J. Sci. Food Agr.*, **11,** 553 (1960).

11. H. Feuerstein, *Z. Anal. Chem.*, **232**, 196 (1967); B. Morsches and G. Tölg, *ibid.*, **250**, 81 (1970).

12. C. Boutron, *Anal. Chim. Acta*, **61**, 140 (1972).

13. Contamination can also result from the slow release from the container walls of substances previously stored in the container.

14. Not to be called silica glass, however, which refers to fused silica ("quartz").

15. F. B. Nesbett and A. Ames, *Anal. Biochem.*, **5**, 452 (1963), call attention to the release of sodium from acid-cleaned pipets (both soft-glass and Pyrex, but less from the latter) on dry storage. A carefully cleaned Pyrex pipet, showing no extractable sodium immediately after cleaning, yields easily determinable amounts of sodium to a KCl solution, less to water alone, after being kept dry for a day or more at room temperature. The amount of extractable sodium increases, but at a slowing rate, with time and with the temperature of dry storage. This behavior is attributed to diffusion of sodium ions in the glass. Therefore, pipets should be acid-cleaned immediately before use. (Plastic pipets are available.)

16. G. W. Morey, R. O. Fournier, and J. J. Rowe, *J. Geophys. Res.*, **69**, 1995 (1964).

17. L. S. Theobald, *Analyst*, **84**, 570 (1959).

18. High-density polyethylene may contain more than 1% Ti in its ash, as found by R. O. Scott and A. M. Ure, *Proc. Soc. Anal. Chem.*, **9**, 288 (1972).

19. M. Murozumi, T. J. Chow, and C. Patterson, *Geochim. Cosmochim. Acta*, **33**, 1247 (1969).

20. Dark particles embedded in polyethylene have been found to contain Al, Cu, Fe, Mg, Mn, Si, and Zn (D. H. Freeman, Ed., Nat. Bur. Stand. (U.S.) Tech. Note, **459**, 40 (1968); D. H. Freeman and W. L. Zielinski, Eds., *ibid.*, **509**, 56 (1970)). Similar particles have been found in Teflon, which also can be contaminated with steel particles.

21. R. F. Chen, *Anal. Lett.*, **5**, 663 (1972). Extraction of fluorescent substances (probably largely plasticizers and the like) is of common occurrence with plastics in general.

22. T. Bidleman, Thesis, University of Minnesota, 1970.

23. Vasilevskaya et al.[6] Cf. Table 2-10A. Boutron[12] mentions that more K, Ca, Mn, and Fe are taken up by evaporation in silica than in Teflon.

24. If glass stopcocks are used, they should not be lubricated with stopcock grease, which can dissolve in organic solvents and cause trouble, for example, give a fluorescence in fluorimetric determinations. Glass stopcocks can be lubricated well enough with a film of water.

25. Tributyl citrate acetate, a contaminant of many laboratory solvents, has been reported to originate as a plasticizer in polyvinyl chloride. Polyvinyl cap liners have been found to be a source of the plasticizer bis(2-ethylhexyl)phthalate contaminating dichloroethane and other solvents [R. A. de Zeeuw et al.,

Anal. Biochem., **67,** 339 (1975)]. The extracted substances can be fluorescent.

26. For example, determination of Mo and Re with $SnCl_2$–SCN^-. Decomposition of a silicate with sodium hydroxide in a gold crucible at 350° can be substituted for sodium carbonate decomposition in platinum. Fusions with pyrosulfate can be made in Vycor or silica.

27. W. Oelschläger, *Z. Anal. Chem.*, **246,** 376 (1969). The platinum dishes contained less than 1% Ir. As shown by H. von Wartenberg, *Z. Anorg. Allgem. Chem.*, **273,** 91 (1953), surprisingly large amounts of Pt are brought into solution when HCl solutions of KCl are evaporated. Thus evaporation to dryness (80°C) of 100 ml of 10% HCl containing 2 g of KCl in a platinum dish dissolved Pt equivalent to 2.5 and 3.6 mg $PtCl_6^{2-}$ in two trials; with 0.2 g KCl, 0.5 mg $PtCl_6^{2-}$ was dissolved. Apparently, precipitation of K_2PtCl_6 favors solution of Pt. Hydrochloric acid solutions of $FeCl_3$ attack Pt quite strongly and should not be heated in such dishes. Fuming $HClO_4$ in the presence of chloride also brings Pt into solution.

28. B. G. Russell, J. D. Spangenburg, and T. W. Steele, *Talanta*, **16,** 487 (1969), give extensive data.

29. R. Bock and A. Hermann, *Z. Anal. Chem.*, **248,** 180 (1969), fused 6 g of NaOH in a Pt crucible of 30 mm diameter in nitrogen and found 26 and 38 μg Pt dissolved after 10 min at 510°C, and 56 and 98 μg dissolved after 30 min at 690°. KOH and LiOH behave similarly. E. M. Dodson, *Anal. Chem.*, **34,** 966 (1962), investigated the use of crucibles of various materials for the decomposition of ruby and sapphire maser crystals by KOH fusion at low temperatures (400°–450°). Crucibles of carbon, commercial nickel, platinum, gold, and silver were all more or less unsatisfactory, either because of attack of the crucible or introduction or retention of trace elements. "Specpure" nickel was satisfactory but not readily available. Crucibles of pure zirconium used under controlled conditions were the most satisfactory. Incidentally, it may be noted that polyethylene is not recommended for holding the acidified solution of the melt because of possible adsorption of trace metals; glass is preferred. Because of insufficient sensitivity, a spectrophotometer is not used in the colorimetric determination of iron; visual comparison is carried out with standards on a tile or filter paper discs, with standards as low as 0.002 μg Fe.

30. K. Rankama, *Bull. Comm. Géol. Finlande*, No. 126, p. 14 (1939). For contents of elements in Millipore membrane filters, see T. Kato et al., *Talanta*, **23,** 517 (1976).

31. H. W. Windsor, *Anal. Chem.*, **20,** 176 (1948), found as much as 17 μg of easily extractable B in a 9 cm doubly acid-washed filter paper.

32. P. Patek and H. Sorantin, *Anal. Chem.*, **39,** 1458 (1967).

33. But some distilled water from a modern central distribution system is surprisingly low in heavy metals. The water in the new facilities of the National Bureau of Standards at Gaithersburg, Md., has been reported to

contain, 6 months after installation of the system, 0.5 μg Cu, 0.6 μg Pb, and <0.02 μg Cd/liter. Immediately after installation of the system the metal content was much higher.

34. Robertson, *loc. cit.* lists some trace element contents of "quartz" (fused silica)-distilled water (two samples) in μg/liter as: Zn~1, 9.5; Fe ~1, <0.2; Sb ~0.06, ~0.1; Co ~0.04, ~0.2; Cr ~2, ~10; and Ag <0.02, <0.02 (Boutron found Na<0.1, Mg<0.2, Ca$\leqq$0.5, Mn$\leqq$0.8, Fe$\leqq$1.) Particulars concerning the still and distillation procedure are not given. It seems difficult to believe that the values for Zn and Fe can be representative of water distilled in fused silica, because water distilled once in Pyrex may contain less (~0.1 μg/liter) of both elements, as may be seen from Table 2-12. Perhaps another factor should be taken into consideration: Is there adsorption of metals on borosilicate glass and not (or to a lesser extent) on silica? For heavy metal contents of water distilled in silica without boiling, see p. 73. This water, stored in Teflon bottles, seems to have as low heavy metal contents as any reported.

35. A laboratory still having a polyethylene condenser and receiver is described by J. V. Jordan and G. D. Wyer, *Chem.-Anal.*, **48,** 39 (1959).

36. W. Sonnenschein, *Z. Anal. Chem.*, **168,** 18 (1959). This value may be compared with 30 μg/liter SiO_2 for water distilled in Jena glass according to R. Wickbold, *ibid.*, **171,** 81 (1959/60). (Pyrex should be similar.)

37. R. Delhez, *Chem.-Anal.*, **49,** 20 (1960), observed that water stored in polyethylene for several days (even in bottles previously used to store water for several years) showed absorption at 270 nm. H_2O_2 destroys the absorbing substance. Whether all polyethylene behaves similarly does not seem to be known. Variations may be expected. Sodium hydroxide solutions kept in polyethylene bottles have been reported to reduce RuO_4 to $RuO_2 \cdot$ aq. Plasticizers are commonly used in the manufacture of plastic articles and may be the extractable substance. See polypropylene, p. 56.

38. J. S. McHargue and E. B. Offutt, *Ind. Eng. Chem. Anal. Ed.*, **12,** 157 (1940), describe a still having a silica condenser, which is capable of producing more than 10 liters of water/hr. Water collected in Pyrex contained dithizone-reacting heavy metals equivalent to 0.2 μg Zn/liter. This then is "10^{-11} water," that is, water containing a few parts of metal in 10^{11} parts of water. This water may be compared with water distilled from an all-silica still without boiling (p. 73), which contained ~0.1 μg/liter of metals reacting with dithizone (Pb, Tl, Cd, Ag, Zn, Cu, Ni). The latter water was stored in Teflon. Since nonboiling distillation yields water at a rate $\frac{1}{60}$ that of the McHargue–Offutt still and apparently not much purer, the use of nonboiling distillation for water is not always advantageous. Even if appreciable carry-over of nonvolatiles occurs in ordinary distillation, a second distillation will reduce this to negligible amounts. Suppose poor distilled water containing 100 μg heavy metals/liter is used as feed, and 1% carry-over occurs in a distillation. After two distillations ~0.01 μg heavy metals will be present per liter. A 1% carry-over seems excessive (see note 39).

39. Many laboratory stills have been described. R. Ballentine, *Anal. Chem.*, **26,** 549 (1954), constructed a still with a 5-ft Vigreux column to trap droplets and having a section of glass tubing preceding the condenser heated to 120°–130° to destroy any continuous film between the boiler and condenser. The distillate was found to give no dithizone test for metals, even when the water in the boiler was salted with 1000 ppm Cu. A test with radioactive Co showed that $\sim 10^{-7}$ had passed into the distillate when most of the charge had been distilled, that is, a carry-over of 0.0001 ml per 1000 ml water distilled.

40. D. C. Burrell and G. G. Wood, *Anal. Chim. Acta*, **48,** 47 (1969), found that leaching of zinc by water from coarse glass frit and other filter media was greatly reduced by soaking them overnight in 10% HCl, or better still in aqua regia.

41. Oxidation of organic matter resisting permanganate treatment by the use of a Pt–Rh catalyst at 800°C is described by B. E. Conway, H. Angerstein-Kozlowska, W. B. A. Sharp, and E. E. Criddle, *Anal. Chem.*, **45,** 1331 (1973).

42. N. A. Karamian, *Am. Lab.*, **5,** 11 (1973).

43. In 1970, 500 ml of "ultrapure" HCl was quoted at $40, or 20 to 100 times the price of analytical reagent, depending on the quantity.

44. J. W. Mitchell, C. L. Luke, and W. R. Northover, *Anal. Chem.*, **45,** 1503 (1973), who determined Mn, Cl, Na, and Cu in acids by neutron activation and these and other constituents in the insoluble matter of salts by X-ray fluorescence.

45. The azeotrope of HCl is 6.1 *M*, of HNO_3 15.5 *M* (the concentrated HNO_3 of the laboratory).

46. H. Stegemann, *Z. Anal. Chem.*, **154,** 267 (1957), has described a polyethylene purification and absorption train for HF. A more elaborate purification procedure is presented by M. Tatsumoto, *Anal. Chem.*, **41,** 2088 (1969), in which HF released from a cylinder is frozen with liquid nitrogen in a Kel-F tank, then distilled and frozen again, and finally absorbed in triply distilled water contained in Teflon. Hydrofluoric acid containing as little as 8×10^{-5} μg Pb/ml is obtained in this way. The acid obtained by absorbing HF gas, filtered through Teflon shavings, in water is reported to contain 2×10^{-4} to 1×10^{-3} μg Pb/ml.

47. H. Irving and J. J. Cox, *Analyst*, **83,** 526 (1958), place 500 ml of concentrated HCl in the bottom of a 4–6 liter desiccator and a polyethylene dish holding 250 ml of doubly distilled water on the desiccator plate. The desiccator cover is not greased. After 2 days at room temperature, 250 ml of 2 *M* HCl is obtained, after 4 days, 3 *M*. With only 50 ml of water, the concentration of HCl is about 10 *M* after 3 days. In the preparation of pure ammonia, starting with 500 ml concentrated NH_4OH, 250 ml of 8.7 *M* NH_4OH can be obtained after 4 days. A Teflon dish or beaker might be better than a polyethylene dish when highest purity is necessary.

48. W. Kwestroo and J. Visser, *Analyst*, **90,** 297 (1965), use two polyethylene bowls, one inverted over the other, with polyethylene beakers inside to hold

concentrated HF and pure water. The acid thus obtained contains only 2×10^{-4} μg Pb/ml. Earlier, Vasilevskaya et al., *Zav. Lab.*, **28,** 675 (1962), used a 1 liter Teflon vessel closed with a lid having an internal cone cooled with dry ice. Some 400 ml of HF was placed in the vessel, which was heated in a water bath or on a hot plate. The condensed acid drips from the tip of the cone into a dish within the vessel. The distillation rate is 15 ml/hr.

49. J. M. Mattinson, *Anal. Chem.*, **44,** 1715 (1972).

50. K. Little and J. D. Brooks, *Anal. Chem.*, **46,** 1343 (1974).

51. M. S. Lancet and J. M. Huey, *Anal. Chem.*, **46,** 1360 (1974).

52. E. C. Kuehner, R. Alvarez, P. J. Paulsen, and T. J. Murphy, *Anal. Chem.*, **44,** 2050 (1972). A subboiling still made of polypropylene has been used by R. W. Dabeka, A. Mykytiuk, S. S. Berman, and D. S. Russell, *Anal. Chem.*, **48,** 1203 (1976) for purification of water, HCl, and HF (but not HNO_3). The metal contents of the distillates were, in general, no greater than when a silica still was used.

53. But at 100°C, extraction of iron from Jena glass by dilute ammonia does occur (Table 2-7), presumably in suspended or colloidal form.

54. Greater contamination of supposedly purer salts may result, for example, from the use of stainless steel equipment not used for lower quality material. "V. M. Goldschmidt always insisted that the purest sodium carbonate from the trace-element aspect came from the stock of washing soda in the village shop!" (R. L. Mitchell[10]).

55. An interesting application of this fact was made by R. G. Clem, G. Litton, and L. D. Ornelas, *Anal. Chem.*, **45,** 1306 (1973), in reducing the lead content of KCl. Sufficient water is added to KCl crystals to dissolve about 95% at room temperature and the whole is stirred overnight in a Pyrex beaker. The supernatant liquid after 21 hr contains only a few parts Pb/billion (10^9) of solution, calculated on KCl basis, compared to 130 ppb Pb in the starting material. Presumably some recrystallization of KCl occurs, with incorporation of Pb in the crystals. Cu and Zn remain in the supernatant solution. When this is evaporated down, the separating KCl contains 5% or less of these metals present in the starting material. NaCl can be purified similarly.

56. Mitchell et al.[44] report contents of heavy metals in the insolubles of various salts of ultrapure grade.

57. R. L. Mitchell, *Spectrographic Analysis of Soils, Plants and Related Materials*, Harpenden, 1948, removes heavy metals from Na_2CO_3 by extracting a hot 20% solution with 0.01% H_2Dz in $CHCl_3$ (3–4 times with 50 ml for 1 liter Na_2CO_3). Excess H_2Dz is removed by repeated extraction with $CHCl_3$, and $Na_2CO_3 \cdot 10 H_2O$ is crystallized out by ice cooling, which is converted to $Na_2CO_3 \cdot H_2O$ at 25°C and to Na_2CO_3 at 300°.

58. Yu. A. Bankovskii, L. M. Chera, and A. F. Ievin'sh, *J. Anal. Chem. USSR*, **19,** 42 (1964).

59. A number of workers have noted that salts crystallized from aqueous solutions containing organic reagents (e.g., EDTA) can be significantly contaminated by these.

60. Robertson, *loc. cit.* R. Woodriff et al., *Anal. Chem.*, **45**, 230 (1973), found ~0.001 ppm (~1 μg/l) Ag in $CHCl_3$, and no way of purifying it (presumably including distillation) was successful. CCl_4, distilled once, was silver-free.

61. Percolation of acetone, ethers, chloroform, benzene, and the like, through a column of suitably activated alumina (commercially available) removes water and peroxides; ethyl alcohol is removed from $CHCl_3$ at the same time.

62. It is a common observation that finely divided suspended material in natural waters lodges on the walls of the container on storage.

63. This behavior can result in the contamination of another solution subsequently stored in a bottle that has been used to hold a more or less concentrated solution of a substance taken up by the material of the container. Even if carefully washed, such a bottle slowly releases the absorbed substance over periods ranging from minutes to days. Various workers have noticed the contamination of solutions in polyethylene by nitrate, fluoride, sulfide, and others derived from solutions previously kept in such bottles. This effect is not restricted to plastics. It can also be shown by glass (e.g., chromic acid cleaning solution), but to a smaller extent. In trace analysis, it is desirable to know the history of the vessels to be used.

64. The adsorption of the sulfonated azo dye Evans Blue on glass and fused silica has been studied by W. O. Caster, A. B. Simon, and W. D. Armstrong, *Anal. Chem.*, **26**, 713 (1954). As much as 0.06 μg of dye is adsorbed per square centimeter of glass after 1 min, 0.16 μg/cm^2 after 24 hr. Zephiran (an alkyldimethylbenzylammonium chloride) in sufficient concentration prevents the dye adsorption on glass and silica. Adsorption of Methylene Blue has been studied by D. T. Burns et al., *UV Spectrom. Group Bull.*, 23 (1974).

65. The large decreases in concentration observed by Leutwein (Table 2-22) do not seem to be characteristic of most metal solutions (>0.1 M in H^+) kept in Pyrex or plastic bottles.

66. S. Z. Lewin, P. J. Lucchesi, and J. E. Vance, *J. Am. Chem. Soc.*, **75**, 6058 (1953). A 0.06% solution of $SrCl_2$ underwent a 60% decrease in concentration in 3 months.

67. K. T. Marvin et al., *Limnol. Oceanogr.*, **15**, 320 (1970), found adsorption of Cu by glass filters in *neutral* solution.

68. T. Bidleman, University of Minnesota, 1968, observed that saturated solutions (15 ml) of bismuth in 0.5 M NaOH (~3 × 10^{-5} M or ~6 ppm Bi) decreased in concentration by as much as 75% on filtration through a Munroe crucible having a ~2.5 mm Pt mat. On the other hand, the same Bi solution showed no loss of Bi within the experimental error when filtered through a sintered Pyrex disk. These solutions were also 0.5 M in NaClO$_4$. Bismuth is present as Bi(OH)$_4^-$ and Bi(OH)$_3$ in 0.5 M NaOH.

69. D. Buksak and A. Chow, *Talanta*, **19,** 1483 (1972), who used [195]Au in their tests.

70. For example, T. Y. Toribara, C. P. Shields, and L. Koval, *Talanta*, **17,** 1025 (1970). Addition of $KMnO_4$ has been recommended to keep mercury in the II state but this is not advisable because any MnO_2 formed would adsorb Hg(II). $K_2Cr_2O_7$ is a better choice.

71. Antimony in 6 *M* HCl has been found to be converted on standing into forms not reacting completely with Brilliant Green. In the absence of chloride (in dilute sulfuric acid solution) this behavior is not observed (A. A. Al-Sibaai and A. G. Fogg, *Analyst*, **98,** 732 (1973)). The changes involved are not well understood. Hydrolysis and oxidation-reduction reactions seem to occur, and deterioration of the chloride-containing solutions is more rapid in light than in darkness.

SAMPLING AND SOLUTION OF SAMPLES

I. SAMPLING

Sampling is considered here from the standpoint of obtaining the average content of some constituent in a larger mass of material from which the sample is derived. Obviously analyses may be made for purposes other than obtaining average content, such as finding the distribution of a constituent in a larger mass (example: finding the content in the individual minerals of a rock, or even the distribution in the individual minerals). This is not the place to discuss the ramifications of sampling required for the latter purpose.

The aim in sampling is to obtain a portion of a large amount of material that will represent the composition of the latter within admissible limits of error. The problems arising in sampling result from the heterogeneous character of the material and the possible introduction of foreign substances in the sampling operation, and seldom the presence in the original material of extraneous substances that need to be excluded.

Both indeterminate (random) and determinate errors are involved in sampling, and it may not always be easy to distinguish between them.

Random errors can be reduced to acceptable values by taking sufficiently large samples of well-mixed material having sufficiently small grain size.

A. STATISTICAL INHOMOGENEITY

Even if sampling can be carried out without systematic errors—and that may not be easy—a limit to accuracy may be imposed by statistical fluctuations in the composition of a heterogeneous sample. Especially when a sample consists of more than one phase and the content of a particular constituent is not the same in all the phases, sizeable errors can occur if the material is not sufficiently finely divided or the sample weight is too small. The same is true if the sample consists of a single nonhomogeneous phase. The most unfavorable situation arises when the trace constituent is present in a small fraction of the particles of a sample. Thus gold metal may be scattered as fine particles in a matrix of quartz or other gangue, platinum metals may be concentrated in sparse grains of base-metal sulfide in an igneous rock, zirconium in a silicic rock may be present largely as grains of zircon occurring in limited numbers, and so on. The number of particles consisting of, or containing, a particular trace constituent in, say, a gram of sample, may be so small that statistical fluctuation can lead to a large percentage error. The approximate number of uniformly sized particles n having a constant content of trace constituent (other particles barren), that should be present in the sample taken (for a single analysis) to give a specified standard deviation is readily found (see p. 103). The relative standard deviation in percent is given by $100(n^{1/2}/n) = 100n^{-1/2}$. Thus to obtain a relative standard deviation not exceeding 1%, the sample weight must be such that at least 10^4 particles containing the trace constituent are present, if all the particles have the same size. If all the phases of a solid mixture contain the trace constituent in question, the deviation due to statistical inhomogeneity becomes smaller, other factors remaining the same (see below). Moreover, if the trace constituent occurs in a single phase making up a considerable fraction of the sample, sampling error becomes smaller. Trace constituents often replace a major element in a crystal lattice and the sampling problem then may be no greater than that for the major element. Thus little sampling error for gallium in most silicate rocks would be expected, since it largely replaces Al, which occurs in feldspars and other minerals making up most of the rock.

When very sensitive trace methods are used, the analyst may fall into the error of reducing the size of a sample that is a heterogeneous mixture below the weight that is needed to keep the error due to inhomogeneity below an acceptable value. Sometimes data may be available to the analyst that will enable him to estimate, even if only roughly, the deviation that can result from statistical fluctuations in the composition of a specified weight of sample. The discussion following is offered with this aim in mind.

CASE I. Trace Constituent is Present in One Phase of Mixture

Designate the phase containing the trace constituent as A and that containing none as B (B could also represent the sum of a number of barren phases). All particles are assumed to have uniform composition and the same size. Let

W = sample weight in grams
f_A = fraction by weight of (infinite) sample consisting of A
f_B = fraction by weight of (infinite) sample consisting of B
ρ_A = density (g/cm^3) of A
ρ_B = density of B
ρ = density of sample
v = volume (cm^3) of a particle
d = particle diameter (here taken as the edge of a cube, so that $d^3 = v$) in centimeters
N_A = number of particles of A in 1 g of sample
N_B = number of particles of B in 1 g of sample
$N = N_A + N_B = (f_A/\rho_A + f_B/\rho_B)v^{-1}$ = total number of particles in 1 g of sample

Then

$WN_A = f_A W/\rho_A v = f_A W/\rho_A d^3$ = number of particles of A in W g of sample
$WN_B = f_B W/\rho_B v = f_B W/\rho_B d^3$

According to the J. Bernoulli expression[1]:

s_A = particle *number* standard deviation of A for W grams of sample

$$= \pm\left\{W\left(N\times\frac{N_A}{N}\times\frac{N_B}{N}\right)\right\}^{1/2} = \pm\left\{\frac{WN_AN_B}{N}\right\}^{1/2} \tag{3-1}$$

The relative weight standard deviation of A in percent (= relative number standard deviation in percent) is given by

$$s_{A,\%} = \pm100\left\{\frac{WN_AN_B}{N}\right\}^{1/2}\times\frac{1}{WN_A} = \pm100\left\{\frac{N_B}{WNN_A}\right\}^{1/2} \tag{3-2}$$

The preceding expression can be applied when the sample size is stated in terms of particle number, $n = WN$, and N_A and N_B are known.

When N_B/N is near unity (e.g., >0.8), the following approximation is justified:

$$s_{A,\%} \sim \pm 100 \left(\frac{1}{WN_A} \right)^{1/2} \sim \pm 100 \left(\frac{\rho_A v}{Wf_A} \right)^{1/2} \qquad (3\text{-}3a)$$

$$\sim \pm 100 \left(\frac{\rho_A d^3}{Wf_A} \right)^{1/2} \qquad \text{(if particles are cubes)} \qquad (3\text{-}3b)$$

In practice, only rough values of $s_{A,\%}$ are to be expected. The particle size varies, the average size of A and B may not be the same, A and B may be intergrown, and it would be unusual for the particle size to be known with any accuracy.[2] But even rough values are helpful in assessing the precision of trace determinations, or finding the weight of sample to be taken for a stipulated precision.[3]

CASE II.   Trace Constituent Is Present in Both Phases of Binary Mixture.

C_A = content (e.g., in ppm by weight) of trace constituent in phase A
C_B = content of trace constituent in phase B
$C = f_A C_A + f_B C_B$ = content of trace constituent in sample

The relative standard deviation in percent by weight, $s_\%$, of the trace constituent in the sample is given by[4]:

$$s_\% = \pm 100 \frac{(C_A - C_B)}{C} \left\{ f_A f_B \frac{(\rho_A \rho_B)}{\rho} \frac{v}{W} \right\}^{1/2} \qquad (3\text{-}4)$$

Examples

1.   A sample consisting of pyrite (5% by weight) and quartz (95% by weight) contains a certain trace constituent in the pyrite only. If all the particles have a diameter (cube edge) of 0.015 cm and a 1 g sample is taken, what is the relative (%) standard deviation due to statistical heterogeneity? $\rho_{pyrite} = 5.0$, $\rho_{quartz} = 2.65$.

$$s_{A,\%} \sim \pm 100 \left(\frac{5 \times 0.015^3}{1 \times 0.05} \right)^{1/2} = \pm 1.8$$

This result is likely to represent the maximal[2] standard deviation to be expected from statistical heterogeneity. The probability of a standard deviation of $\pm 3.6\%$ ($= 2 \times 1.8$) is ~ 0.05. Evidently the sample size of this mixture should not be reduced much below 1 g.

It needs to be remembered that the weight of sample material from which the 100-mesh sample is prepared must be so large that it does not contribute significantly to the sampling error. Suppose that the original material (representing a larger body of material) has a particle size (cube edge) of 0.1 cm and its sampling error, s, is to be kept below 1%. The minimal weight of this coarser material to be taken for size reduction to 100-mesh is

$$W = 10^4 \left(\frac{5 \times 0.1^3}{1 \times 0.05} \right) = 1000 \text{ g}$$

If the original material consisted of 1 cm^3 lumps of pyrite and quartz, a 1000 kg sample would need to be taken! If sampling precision is relaxed so that a standard deviation of 1.8% in each step is acceptable (total sampling deviation of $1.8 \times 2^{1/2} = 2.5\%$), the required weight of 1 cm-lump material is reduced to ~ 300 kg. Evidently, the random sampling error can be large when the trace constituent is concentrated in a single phase making up a small fraction of a heterogeneous sample.

2. As before, the sample consists of 5% pyrite and 95% quartz by weight, but now the quartz holds 1.0 ppm of the trace element and the pyrite 10.0 ppm. The particle diameter is 0.015 cm as before. What is the relative (%) standard deviation due to statistical heterogeneity if a 1 g sample is taken?

From Eq. (3-4):

$$s_\% = \frac{100(10-1)}{(0.05 \times 10 + 0.95 \times 1)}$$

$$\times \left\{ 0.05 \times 0.95 \left[\frac{5.0 \times 2.65}{1/(0.05/5 + 0.95/2.65)} \right] \frac{0.015^3}{1} \right\}^{1/2}$$

$$= \pm 0.55$$

The quartz in the sample greatly exceeds the pyrite in weight and particle number and now contributes the larger fraction of the trace element to the sample. Quartz thus buffers the variation in the particle number of pyrite. A 0.2 g sample of this mixture would be large enough if a standard deviation of $\sim 1\%$ is acceptable (which usually it should be):

$$s_\% = 0.55 \times 0.2^{-1/2} = 1.2$$

B. SAMPLING IN TRACE ANALYSIS

Except for the greater precautions that must be taken against contamination, and the special attention that may need to be paid to sample and particle size, sampling for trace analysis is much the same as sampling for major constituents. The discussions of sampling to be found in standard references are also useful for the trace analyst.[5]

Occasionally, the analyst or his sampling surrogate needs to depart somewhat from the usual routine of starting with a mass of material and reducing its particle size, often in stages, until a relatively small test sample is secured.[6] When exceedingly pure materials (e.g., semiconductors) are dealt with, crushing may involve serious danger of introducing foreign elements, and fragments or lumps, as they are presented, may be dissolved as such after etching to remove surface impurities. Some degree of homogeneity is assumed. The sampling of plant material may require its selection and sometimes treatment. Adhering soil must be excluded. Brushing may be needed to remove soil and dust, not always satisfactorily. Resort may even be made to washing leaves with a detergent solution, although such treatment may affect the content of some elements.

The purpose of an analysis may, of course, influence the selection of a sample. Thus the natural content of lead in a biomaterial may be desired or the gross content (intrinsic + surface contaminant) may be required, and the sampling procedure will vary. Also, it may be noted that the distribution of trace elements in biomaterials may be far from uniform, and account often needs to be taken of this fact.[7]

Crushing and Grinding. Usually, in spectrophotometric analysis, reduction of inorganic sample material to particles passing through a 100-mesh (per inch) sieve will suffice. Occasionally, reduction to finer particle size is advisable, either for assuring a representative sample or for facilitating solution. More often than not, the size reduction will be carried out in stages, first crushing and then grinding, in agreement with the principle already stated that the number of separated particles in sample material at any stage should be comparable.

The initial stages in the sampling of ores and occasionally of rocks may require the taking of such large samples that suitable electrically-powered machinery (crushers, pulverizers, etc.) will be needed for crushing. Laboratory equipment of this type is commercially available but will not be described here.[8] Abrasion of the machinery will introduce fragments of steel, which may not do much harm unless alloying elements such as Cr and V are present, and they are to be determined in the sample. On a smaller scale, bypassing the use of machinery, pieces of compact material 1–2 cm in diameter may be obtained from a hand specimen, as of a rock, by striking with a geological hammer on an anvil or hardened steel plate. Or the specimen may be broken in the jaws of a manually operated rock splitting machine. The next stage in size reduction involves the use of a Plattner-type hardened (but not alloyed) steel mortar, in which a hardened steel pestle, struck with a hammer, crushes the material to ~1 mm size. A coarse sieve (nonmetallic) should be used to sift out the finer particles (< 1 mm), with return of the coarse material to the mortar for further crushing.[9] Finally, after all the material has been passed through the coarse sieve, it is carefully mixed[10] and a suitable fraction is taken (by successive quartering if necessary) for fine grinding. In all these operations care must be taken to avoid losses from flying fragments or dust, and no material may be rejected.

An agate mortar and pestle is commonly used in the final stage to grind the relatively coarse material to 100-mesh or finer.[11] Grinding vials also are often used, in which size-reduction is effected by the back and forth movement of a ball in an automatic shaker; these devices are made of tungsten carbide (WC), boron carbide, alumina, and so on. The ground powder is sifted at intervals through nylon or silk bolting cloth (100-mesh, or finer if required) held in a plastic or glass frame, and the fraction not passing through is returned to the mortar for further grinding. The sifted powder is collected on glazed paper, and when all the material has been ground and sifted, the powder is well mixed by successively raising the corners of the sheet to cause the powder to roll over itself. The mixed powder is transferred to a vial or bottle by means of a nonmagnetic spatula or small scoop (pouring may cause segregation). If the powder is transported, or stored for some time in a building subject to vibration, segregation due to particle density differences may occur, and the powder should then be remixed before a sample is taken.

The chief source of contaminating metals in the crushing–grinding procedure outlined in the preceding paragraphs is the steel mortar and pestle (or the steel plates of the hydraulic crusher or the steel rollers of the crusher–grinder if these are used). In trace analysis, it is the trace or minor elements in the steel that are undesirable. Introduction of iron is unavoidable and is chiefly objectionable because of the metals brought with it. The approximate extent of metal contamination can be judged from the values listed in Table 3-1 pertaining to the crushing of quartz crystals in a steel mortar. Contamination is lessened if the collar of the

TABLE 3-1
Amounts of Metals Introduced in Crushing Quartz in Hardened Steel (Plattner) Mortar[a]

Nine grams of quartz crystal (up to 1 cm in diameter) crushed with a total of 200 hammer blows; crushed material sifted through a screen with 0.7 mm holes after every 20 strokes, and coarse material returned to mortar.

Metal	Fe	Mn	Cr	V	Ni	Co	Cu
Contamination (ppm)	280	1.8	0.4	<0.1	0.25	<0.1	0.35

[a] E. B. Sandell, *Anal. Chem.*, **19**, 652 (1947). See this paper for further data. W. T. Schaller (private communication) found from 350 to 540 ppm of introduced iron in beryl crushed to 100-mesh in a Plattner mortar under conditions similar to those in the above experiments.

mortar is not used, and a cardboard screen is placed around the pestle to prevent loss of flying fragments. Occasionally, unexpected elements have been found in the steel of Plattner mortars, for example, Nb and Ta,[12] and Mo. Ordinarily, little metal will be introduced by grinding in agate. Traces of such metals as iron and magnesium occur in agate, but unless very hard material is ground, the abrasion of the agate is unlikely to introduce significant quantities of these metals.[13]

When crushing needs to be done without introducing iron, manganese, and possibly other metals found in steel, a buckboard and muller of high alumina ceramic may be used.[14] It consists of a heavy segment of a cylinder (muller) and a flat heavy base plate. When the muller is rocked, the fragments are crushed against the plate and can be reduced to 100-mesh if necessary. In crushing massive quartz to 100-mesh powder, the only elements found to be introduced were Ti (1 ppm) and Mg (30 ppm); some Al was doubtlessly introduced also, but its amount was not determined.

Various other pulverizers for inorganic materials have been described.[15]

Size reduction by crushing and grinding as described is only applicable to samples composed of brittle substances. Metals usually require drilling or milling, but ferro alloys (ferromanganese, ferrotungsten, etc.) can be crushed. Sampling of metals for chemical analysis may be difficult because of segregation and because small particle size is virtually impossible to obtain. Besides, contamination may be derived from the cutting tool. Without some knowledge of the distribution behavior of the trace elements in the matrix metal, intelligent sampling is hardly possible. In the sampling of iron meteorites, the metal phase must be free of silicate, sulfide, and phosphide inclusions (these may need to be analyzed separately if they can be isolated), and the drillings must be examined microscopically for them. Because of the coarse structure of many iron meteorites, a large sample will often be required if passably accurate trace element values are to be obtained. This is all the more true of silicate meteorites containing metal (e.g., chondrites). Crushing these in a steel mortar leaves the metal grains much the same as originally. The silicate and troilite fraction can be crushed or ground more easily if the metal fraction is (temporarily) separated with a weak magnet. The only effective method known at present for crushing the metal grains of chondrites involves chilling them and the mortar–pestle (sintered corundum) to liquid–air temperature. The metal ($<20\%$ Ni) then becomes so brittle that it can be crushed to 100-mesh.[16]

Metal contamination also occurs in the preparation of biomaterial for analysis, especially when grinding mills are used. Comparison of various methods of grinding plant material has demonstrated that little contamination results from hand grinding with a porcelain mortar and pestle, whereas ball-mill grinding introduces large amounts of Fe, Zn, Cu, and Co (Table 3-2).[17] Fe and Cu are introduced from Wiley mills and hammer mills (in part from screens).[18] Porcelain and mullite (3 $Al_2O_3 \cdot 2 SiO_2$) balls were found to be less satisfactory than flint balls. Abrasion in ball mills varies with the type of plant material, and contamination increases erratically with the time of grinding. Not surprisingly, contamination increases as the sample size is decreased. Flint balls contain appreciable amounts of trace metals, for example, 21 ppm Zn, 49 Cu, 99 Mn, 1.7 Co, and 0.4 Mo. Other work has shown an all-agate ball mill to be satisfactory for plant sample grinding.[19] Capacity is limited.

A pulverizer having knurled nylon slip-rollers, in which dried plant materials can be prepared in a few minutes, is reported not to contaminate samples.[20]

For preparation of samples of animal origin, chopping, grinding, and slicing are applied, and there is greater or less chance of metal contamination, depending on the equipment used in these operations and the degree of subdivision required. An unusual method, not often applied, is crushing the sample (animal tissue) with hammer blows after chilling in liquid nitrogen.[21]

TABLE 3-2
Contamination from Grinding Oat Grain Samples

Method	Contamination (%)[a]			
	Fe	Zn	Cu	Co
Porcelain mortar and pestle (23 cm)	0	0	5	0
Wiley mill, 10-mesh Cu alloy screens	35	0	58	0
Hammer mill, 10-mesh screen	43	0	14	0
Jar mill, 2-gal glazed porcelain, flint balls, ground 19 hr	29	87	31	67
Jar mill, 2-gal glazed porcelain, porcelain balls, ground 19 hr	84	89	59	99
Jar mill, 2-gal glazed porcelain, mullite balls, ground 19 hr	93	85	71	97

[a] Percentage refers to fraction of total metal in ground sample introduced by grinding. Original contents, unground, $\mu g/g =$ Fe, 24.9; Zn, 21.6; Cu, 3.8; Co, 0.01. Values have been rounded to nearest percent. See original paper for contamination by Na, Ca, S.

Lasers have been suggested for cutting soft and hard tissues, particularly bone samples.[22] As long as the section cut is not too narrow, possible volatilization of metals from the thin damaged section should not, it is believed, have much effect on the analytical values.

II. DECOMPOSITION OF SAMPLES

A. INORGANIC

In general, inorganic materials to be analyzed for trace metals are decomposed and brought into solution in the same way as samples to be analyzed for major constituents, by treatment with mineral acids or by fusion with fluxes. The acid or flux to be used depends principally on the composition of the sample,[23] but the possible effect of an element in the solvent on the subsequent determination or on volatilization needs to be taken into account.

Possible losses in decomposition of samples resulting from volatilization of metal compounds and their adsorption on, or reaction with, vessel walls were touched upon in Chapter 2. Volatilization losses are most likely to occur in hydrochloric (or hydrobromic) acid solutions (p. 90). Metals can also be lost in $HF-HClO_4$ or $HF-H_2SO_4$ decomposition. See Table 3-3. Losses can occur in fusions with acidic fluxes ($K_2S_2O_7$) in the presence of chloride and other anions forming volatile metal compounds (e.g., V is volatilized to a large extent in a pyrosulfate fusion in the presence of chloride). Alkaline fusions can result in losses of Hg, Tl, and As.[24]

Adsorption losses on walls are likely to be at a minimum in mineral acid solutions, including those resulting from fusions with acidic fluxes. From basic solutions adsorption may occur, and such losses need to be given consideration.

Any insoluble material remaining after acid decomposition[25] must be further treated, as by fusion, in order that all of the sample will be brought into solution, unless it is known from the nature of the sample that any small amount of such material may be safely discarded.

Fusions are likely to bring significant quantities of the crucible material into solution, and the effect of this (e.g., Pt) on the subsequent determination must be known. Traces of the metal being determined may also be introduced, as iron from a platinum crucible.[26] Traces of Pt are also likely to be introduced in hydrofluoric acid decompositions in Pt vessels. When the presence of Pt is objectionable, such decompositions can be carried out in Teflon vessels.[27] Fluxes, and acids for that matter, may have low contents of the element being determined and blanks must always be run. It may be worthwhile to consider alternative methods of decomposition if

TABLE 3-3
**Volatilization of Elements from Perchloric
and Hydrofluoric Acid Solutions[a]**

No Loss	Apparent Loss
Na, K	B, 100%
Cu, Ag, Au	Si, 100%
Be, Mg	Ge, up to 10%
Ca, Sr, Ba	As, 100%
Zn, Cd, Hg	Sb, up to 10%
La, Ce	Cr, varies greatly
Ti, Th	Se, varies greatly
Sn, Pb	Mn, up to 3%
V	Re, varies greatly
Bi	
Mo, W, U	
Fe, Co, Ni	

[a] Mixed $HClO_4$ and HF solutions of metals (un-specified amounts) evaporated in Pt to strong fumes of $HClO_4$ at 200°C. Reprinted with permission from F. W. Chapman, Jr., G. G. Marvin, and S. Y. Tyree, Jr., *Anal. Chem.*, **21,** 701 (1949). Copyright by the American Chemical Society.

sufficiently pure reagents are not available or if they cannot be purified readily.

Some inorganic samples may contain organic matter—their nature and origin may give a clue, for example, soils. Such organic matter, brought into solution or pseudosolution by acids, may be objectionable in the subsequent determination of a metal or it may contain metals that should be included in the analysis. Suitable oxidative treatment will then be needed to destroy the organic substances and render any combined metals ionic. In analysis of waters, when total metal contents are to be determined, care should be taken to see that metals are present in ionic form at the start.

Faster and more certain decomposition of refractory inorganic substances can be effected in acid decomposition by heating in a bomb that allows the use of high temperatures and pressures. In fact, some decompositions can only be carried out in this way. At the same time, possible volatilization of some elements is prevented. Several bombs or autoclaves have been described. They consist of a steel or other metal bomb lined with Teflon or having a Teflon crucible that can be suitably sealed. In one design, a thermostatted oven is used and the bomb can be rotated.[28]

Temperatures as high as 240° (all temperatures in °C) can be used. Decomposition can be carried out with H_2SO_4, H_3PO_4, HF, $HClO_4$, and KOH solutions. Because the sample need not be as finely ground as in other types of decomposition, contamination is lessened. Another bomb, commercially available, is designed for decomposition of silicates with HF at temperatures up to 150° and pressures to 1200 lb/in.^{2}.[29] In earlier years acid decompositions were carried out in sealed glass tubes.[30] Less refractory silicates can be decomposed with HF in a screwcapped polypropylene bottle at steam bath temperature.[31]

Bomb decomposition has also been applied to ores (as of U[32]) and steel.[33]

B. ORGANIC[34]

Organic materials include those of bio-origin and others, such as plastics and pharmaceuticals, that are man-made, as well as those having, it might be said, a double origin. The determination of trace metals is particularly important in biosamples. Trace elements occur in organic materials in varied forms, ranging from contaminants, perhaps in inorganic form, having no relation to the organic matrix, to those combined with the matrix material or closely associated with it. The mode of occurrence of a trace element may have a bearing on the method of sample decomposition. An element may occur in different forms in a given sample, whose relative amounts may be of interest and require determination, but here we are concerned only with total amounts of trace metals in organic samples. With some exceptions,[35] the determination of trace metals in organic materials by spectrophotometric methods requires the destruction of organic matter by oxidation, with the metals finally being brought into solution in ionic form. Even if the metals were not bound to the matrix so firmly that disruption of the organic molecules may be required for their release, the organic matter would likely interfere physically or chemically in the subsequent determination.

The requirements and chief desiderata for the destruction of organic matter in trace metal determinations are:

1. Oxidation of organic material in a reasonable time period without hazard and without need for undue attention or special apparatus.
2. Adequate recovery of metal.

   a.   No volatilization.
   b.   No retention by vessel.
   c.   No reactions leading to incomplete solution of the metal or to

formation of species interfering in separation and determination reactions.

3. No contamination by the material of the vessel.
4. Possibility of using large samples.
5. Applicability to a wide range of materials.

No method is wholly ideal, and in practice losses of metals and contamination will occur in greater or lesser degree. The task of the analyst is to choose a method or so modify it that these errors will be within permissible limits.

Some method can always be found for the complete destruction of almost any organic material, but the convenience and the time required vary. Some metals are subject to volatilization and retention losses and for them the conditions of decomposition may be fairly critical.

The commonly applied methods for destruction of organic matter may be classified as follows:

1. Wet oxidation (almost always in acid medium).
2. Dry oxidation (ashing).

a. Ordinary (open system).
b. Oxygen bomb or flask.
c. Excited oxygen.

3. Oxidative fusion.

Before considering the important destruction methods in greater detail, it may be of value to compare in a general way some features of the two main approaches, namely wet oxidation and dry oxidation by ordinary ashing in air:

Wet	*Dry*
Recovery of metals more certain because risk of volatilization is less at lower temperatures and because loss by retention on walls of vessel or by insoluble matter (silica) is generally less.	With care to regulate ashing temperature, most metals can be recovered quite satisfactorily in most types of organic material. Ashing aids may be helpful or necessary.
Use of acids and other reagents in relatively large quantities may result in a relatively large blank, especially when the metal content of the sample is very low (< 1 ppm). Large amount of acid at end of digestion often requires partial neutralization, possibly increasing blank.	Ash can usually be dissolved in small volume of acid.

Wet	Dry
Elapsed time may be shorter than in dry ashing, but more attention may be required.	Procedure often simple, but a certain amount of attention is required in initial stages; if much acid-insoluble material (silica) remains, treatment may be required, and procedure is then more troublesome than wet oxidation.
Ease of oxidation varies with sample and oxidizing mixture, but there are usually no problems in obtaining complete destruction of organic matter.	Some samples are ashed incompletely or with difficulty at safe ashing temperatures (e.g., some animal tissues, samples giving fusible ash coating carbonaceous particles).
Risk of atmospheric contamination less than in dry oxidation.	If amenable, large samples handled more conveniently than in wet oxidation.

Neither of these two general methods is so superior that it eliminates the other from consideration, and there is no procedure universal in the sense that it is always clearly superior to others. Depending on the type of sample and the metals in question, and sometimes on other factors, either wet or dry oxidation may be preferred. However, it is probably fair to say that wet oxidation, in one form or another, is more reliable than dry ashing (except at low temperature in excited O_2) for most metals and for a wider variety of organic materials, and it is likely that it would be

TABLE 3-4
Determination of Metals after Oxidation of the Alga *Scenedesmus acutus* by Different Methods (in % of dry sample)[a]

	Low-temperature Ashing (excited oxygen)	Muffle Furnace at 450°C		Wet Oxidation with HNO_3 and $HClO_4$
		In Pt Crucible	In Porcelain Crucible	
Na	0.029	0.031	0.030	0.031
K	1.43	1.47	1.41	1.41
Ca	0.364	0.354	0.354	0.351
Mg	0.32	0.36	0.34	0.33
Fe	0.26	0.24	0.23	0.24
Mn	0.014	0.015	0.015	0.015
Cu	0.0026	0.0025	0.0024	0.0027
Zn	0.024	0.025	0.025	0.026
Ni	0.0011	0.0010	0.0011	0.0012

[a] K. H. Runkel and I. Baak, *Z. Anal. Chem.*, **260**, 284 (1972). Determination by atomic absorption spectrophotometry. Because of ease and safety, simple dry ashing is preferred for routine determinations.

TABLE 3-5A
Recovery of Trace Elements after Wet Oxidation of Cocoa[a]

Element	Recovery (%)			
	Oxidation Mixture[b]			
	HNO_3-HClO_4	$HNO_3-HClO_4-H_2SO_4$	$HNO_3-H_2SO_4$	HNO_3
Ag	97	99	100	95
As	99	99	98	91, 98
Cd	101	102	102	99
Co	98	99	101	100
Cr	100	101	100	101
Cu	100	99	100	100
Fe	99	99	102	97
Hg[c]	79	89	93	
Mo	97	98	101	98
Pb	100	99, 93	90, 93	100
Sb	99, 94	100	100	97
Se	101	101	79	1
Sr	100	97	98	95
Zn	99	101, 94	100	99

[a] T. T. Gorsuch, *Analyst*, **84**, 135 (1959). Recoveries were determined by the use of radioactive nuclides. No retention of elements on glass was found. Trace element level 1–10 ppm.

[b] Oxidation mixtures respectively as follows: 15 ml concentrated HNO_3, 10 ml 60% $HClO_4$; 15 ml concentrated HNO_3, 5 ml concentrated H_2SO_4, 10 ml 60% $HClO_4$; 15 ml concentrated HNO_3, 10 ml concentrated H_2SO_4. Concentrated and fuming HNO_3, with a little H_2SO_4 according to Middleton and Stuckey (reference 43). A 500 ml flask was used; 2 g of cocoa.

[c] Hg in distillate of digest amounted to 11, 12, and 6%, respectively, to give total recoveries of $79 + 11 = 90$, $89 + 12 = 101$, and $93 + 6 = 99$%.

adopted for more or less untested samples. But once dry ashing has been shown to be satisfactory for a given metal in a given type of sample, it may be chosen in preference to wet oxidation by many analysts. For some biomaterials, ashing at low temperatures (excited oxygen), ashing at 450° and oxidation with HNO_3-HClO_4 all give essentially the same values for a number of common metals as demonstrated for a green alga (Table 3-4). But this generalization does not seem to apply to all types of biomaterial (see later). The level of the trace element in the sample may decide the method selected. When the level is very low, a dry oxidation method may be chosen in preference to wet oxidation, because of its generally lower blank.

Destruction of organic matter preparatory to the determination of trace

SAMPLING AND SOLUTION OF SAMPLES

TABLE 3-5B
Recovery of Trace Elements after Dry Oxidation of Cocoa[a]

	Recovery and Retention[b] (%)			
Element	Ashing Aid[c]			
	None	HNO$_3$	H$_2$SO$_4$	Mg(NO$_3$)$_2$
Ag	93, 99 (8, 1)	87 (7)	97 (1)	100 (0.5)
As	88	84	96	99
Cd	91 (6)	76 (6)	92 (2)	78 (2)
Co	99 (0.5)	96 (1.5)	98 (0)	99 (0)
Cr	98 (0.5)	99 (0)	99 (0)	92 (0)
Cu	86 (14)	94 (2)	96 (0.5)	98 (0.5)
Fe	99	101	101	100
Hg	0			
Mo	99	98	100	98
Pb, 450°	99 (1)	98 (0.8)	99 (0)	98 (0.3)
Pb, 550°	95 (3)	98 (1)	96 (0.3)	93 (2.5)
Pb, 650°	71, 83 (24, 15)	83, 69 (12, 19)	96, 90 (2, 2)	91, 96 (4, 2)
Sb	96 (1.5)	92 (1)	94 (0.5)	97 (0)
Sr	97	97	100	100
Zn	96 (2)	97 (0.5)	100 (0)	97 (0)

[a] T. T. Gorsuch, *Analyst*, **84**, 135 (1959). Ashing (2 g cocoa) carried out in 500 ml silica flask heated with burners at 550°C for 16 hr, except for Pb samples, which were ashed in silica crucibles in muffle furnace at various temperatures as indicated. Radioactive nuclides used.
[b] Retention, recorded in parentheses, is the percentage of element remaining on walls of vessel after solution of ash in acid.
[c] With HNO$_3$ as ashing aid, sample was charred and a few drops HNO$_3$ then added and ashing continued. Sulfuric acid (5 ml 5 N) or Mg(NO$_3$)$_2$·6 H$_2$O (10 ml 7% solution) as ashing aids were added before ashing.

elements has been extensively studied over the years, not always with the same results being obtained or the same conclusions being drawn. The availability of radioactive nuclides of most elements now makes it possible to determine accurately the recovery of trace metals in the destruction of organic matter and to find whether losses are due to volatilization or to retention on the walls of the vessel (chiefly a possibility in dry ashing) or both. See Tables 3-5A and 3-5B for recoveries of trace elements in the wet and dry oxidation of cocoa by various procedures. If we accept 95% recovery as being satisfactory in routine trace determinations, wet oxidation easily meets the requirements for 13 of the 14 elements in Table 3-5A (Hg requires special precautions). Dry oxidation without a fixer or ashing aid is unsatisfactory for three or four metals (omitting Hg), but with a suitable ashing aid all but two can be recovered satisfactorily. It will be realized that the same or similar recoveries need not necessarily be obtained with other organic samples, since the form in which an element is present and the composition of the sample may play an important role in element recovery. However, these results should serve as general indicators.

Values listed in Table 3-6 show some variation in the recovery of Au,[36] Fe, Hg, and Sb from different biosamples after oxidation with H_2SO_4–HNO_3–$HClO_4$. Because the use of a reflux condenser gave ~100% recovery for all these elements, volatilization loss (mostly of chlorides) is indicated. Additional data on the behavior of some elements in dry ashing with blood as the matrix are listed in Table 3-7. See p. 131 for further discussion.

Contrary to an earlier report (in which gas chromatography of Be trifluoroacetylacetonate was involved) that Be is lost in HNO_3–H_2SO_4 digestion of organic material (leaves) in an open beaker, no loss was found in acid digestion or in ashing at temperatures as high as 800°.[37]

In the following pages we touch on some more or less general aspects of various decomposition procedures.

TABLE 3-6
Recovery of Metals in Various Biomaterials after Wet Oxidation[a]

Metal	Recovery (%)				
	Blood	Urine	Leaves	Animal Tissue	Reflux Condenser
As	93	94	95	92	101
Au	77	100	77	65	100
Fe	98	92	95	85	100
Hg	24	87	45	30	100
Sb	99	95	94	94	101
Ag, Co, Cr Cu, Mn, Mo Pb, V, Zn	98–102% for all samples				

[a] J. Pijck, J. Hoste, and J. Gillis, *Proc. Int. Symp. on Microchem., Birmingham, 1958*, Pergamon, New York, 1960, p. 48. Sample (5 ml horse blood, 10 ml urine, 250 mg sugar beet leaves, or 1.5 g animal tissue) digested for not more than 1.5 hr with 5 ml of mixture of 98% H_2SO_4, 70% $HClO_4$, and 66% HNO_3 in the volume ratio 1:1:3; micro-Kjeldahl flask; radioactive nuclides. Amount of metal present: 0.05–0.3 mg. Digesting urine with concentrated nitric acid alone, A. J. Blotcky and coworkers, *Anal. Chem.*, **48,** 1084 (1976), found losses of Al above 80°C, attributed by them to volatilization of aluminum chloride.

TABLE 3-7
Recovery of Metals in Dry Ashing of Blood[a]

Metal	Recovery (%)		
	400°C	500°C	700°C
Ag	65	67	45
As	23	0	0
Au	19	0.5	0
Co	98	75	67
Cr	100	99	85
Cu	100	98	87
Fe	86	82	52
Hg	<1	0	0
Mn	99	96	85
Mo	100	97	85
Pb	103	69	32
Sb	67	82	35
V	101	99	70
Zn	100	98	69

[a] Pijck et al., *loc. cit.* 5 ml blood evaporated and ashed in porcelain crucible (24 hr, 400°C; 12 hr, 500°C); ash dissolved in 6 M HNO_3. Radioactive nuclides.

1. Wet Oxidation

Commonly this is carried out in a flask open to the atmosphere. Oxidation with fuming nitric acid in a sealed tube under pressure (Carius) is rarely applied in trace analysis, because samples larger than ~0.5 g are not easily handled and because the procedure is inconvenient. However, wet oxidation of small samples can be carried out in a Teflon-lined bomb.[38] Sample size should be limited with HNO_3 in a Teflon-lined bomb. A 0.1 g dry sample is specified for a 25 ml bomb. Another method calls for digestion of the sample with H_2SO_4, HNO_3, and $HClO_4$ under moderate pressure in a plastic bottle (linear polyethylene or polypropylene) closed with a screw cap.[39] After overnight predigestion, the bottle is placed in hot flowing water (60–70°) for 2–3 hr. The accuracy and precision obtainable appear comparable to those of standard digestion methods. Still another method involves high temperature and pressure destruction of organic matter with H_2O_2.[40] As much as 20 g of liquid or solid sample (1–2 g of fat or oil) can be completely digested in 10 min by 30% H_2O_2 at 370° in a high-pressure vessel (internal pressures exceed 100 atm). Losses of elements are small and due solely to adsorption by the vessel. Recoveries of Sr, Cs, Ru, and I were greater than 96%.

Wet oxidation is almost always effected in a strongly acid mixture. The strong oxidizing agent may be the acid itself (HNO_3 and $HClO_4$) or it may be a nonacid (H_2O_2) used in conjunction with a suitable acid. Because a high temperature greatly speeds up the oxidation of organic matter, the use of an acid of high boiling point ($H_2SO_4 \sim 330°$, $HClO_4 \sim 200°$) is advantageous. If the oxidizing agent is volatile at a lower temperature or is unstable, it must be replenished in the mixture as the digestion proceeds or retained by refluxing.

Both oxidation and hydrolysis are involved in the splitting of C–C bonds in wet oxidation.[41] A high hydrogen-ion concentration is clearly favorable. The dehydrating effect of sulfuric acid leading to double bond formation contributes to the hydrolysis of organic substances and their destruction.

a. Sulfuric and Nitric Acids

Sulfuric acid alone is not a sufficiently strong oxidizing agent for effective destruction of organic matter. An obvious choice of oxidant to be used with it is nitric acid, and this combination has been, and still is, much used. Hot nitric acid readily oxidizes many organic substances, especially those containing alcoholic –OH groups (often present in natural materials). At times nitric acid is a disadvantageous oxidant, namely when it forms nitro compounds, which are oxidized less easily than the parent compounds. Heterocyclic nitrogen compounds are not easily oxidized in general. When required, nitric acid can be removed by volatilization (b.p. $\sim 120°$) at the end of the digestion. (On evaporation to fumes of sulfuric acid, some nitrosyl sulfuric acid will remain, and it should be destroyed by adding a few milliliters of water and again evaporating to fumes; unless destroyed it can oxidize such reagents as dithizone in subsequent determination steps.)

This mixture oxidizes most organic samples, although some may require rather long digestion and a few cannot be successfully decomposed. Carbohydrates are more easily destroyed than proteins and fats. In general, plant materials are oxidized more easily than animal. Perchloric acid may be added in the final stages if the digest is hard to decolorize.

Destruction of organic matter with sulfuric–nitric acid at atmospheric pressure usually proceeds without dangerously violent reactions, but some exceptions have been observed. Explosions have occurred in heating hydroxy carboxylic acids with H_2SO_4–HNO_3 (R. W. Perlich, private communication); compare Note 84.

Some half-dozen elements are subject to volatilization loss during H_2SO_4–HNO_3 digestion, but for most of these, the losses are no worse than in other digestion mixtures. Special modifications of the procedure, such as the use of a condenser, are needed for Hg, Os, and Ru. The presence of Cl, either as chloride or bound covalently, can lead to volatilization of As(III) and Ge, and modified procedures are needed. Even in the absence of Cl, loss of As and Se can occur because of reducing conditions (charring) in the early stages of digestion.

Organomercuric compounds in fish are completely decomposed by the use of sulfuric acid-fuming nitric acid (methylmercuric chloride requires digestion for 1 hr).[42]

In general, the use of sulfuric acid in a digestion mixture is objectionable if slightly soluble sulfates are formed. The presence of these may be undesirable even if they do not contain the trace element in question. More serious is the possibility of loss of a few elements by coprecipitation. Thus the precipitation of sparingly soluble $CaSO_4$ will result in the loss of Pb because slightly soluble $PbSO_4$ is carried down by $CaSO_4$. Similarly, losses of Ba, Sr, and possibly Th are to be expected in the presence of much Ca. Much Ag is also lost.

With Little or No Sulfuric Acid. Oxidation by nitric or fuming nitric acid at a relatively high temperature (up to 300°–350°) has been recommended as a general method capable of dealing with a wide variety of biomaterials.[43] A small amount of sulfuric acid is used to prevent deflagration in the first evaporation.[44] Nitric acid is the only acid used in quantity, and it can be distilled if necessary to remove metal impurities. The method is considered to be particularly advantageous for nitrogenous animal materials. It is believed to be suitable for the determination of metals such as Cu, Fe, Zn, Mn, and Al; specifically, it has been demonstrated to be satisfactory for Pb and Cu.[45]

General Procedures

It is hardly possible to write a single procedure that can be followed for all materials, because the reactivity of the sample, the particular metals to be determined, and possibly other factors may require changes to be made. Nevertheless, some generalized procedures can be given that can be followed for most types of organic materials without major modifications. Directions for sulfuric–nitric acid decomposition have been subdivided here into Procedures A, B, C, and D, depending on sample reactivity in the initial stage, followed by a common procedure in the last oxidation stage. Directions are included for the use of $HClO_4$ or H_2O_2 in the latter, if organic matter proves hard to destroy.[46] For more reactive samples, Procedure A is better than B. Procedure C is suitable for samples that decompose quietly; this is a more rapid method than A or B. Despite the use of dilute acid, some samples (e.g., some basic dyes)

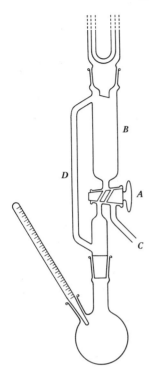

Fig. 3-1. Apparatus for wet oxidation of organic material according to T. T. Gorsuch, based on an earlier design by Bethge, *Anal. Chim. Acta*, **10**, 317 (1954). Stopcock *A*, having three positions, allows (1) refluxing, (2) distillation, (3) removal of distillate. Reproduced with permission from The Chemical Society, from T. T. Gorsuch, *Analyst*, **84**, 147 (1959).

subsequently cause trouble due to their liability to deflagrate violently during charring; appreciable losses of arsenic may be incurred in this way. In such instances, Procedure *D* must be used.

Digestions are usually carried out in 100–250 ml borosilicate Kjeldahl flasks (silica flasks are preferred for exacting work).[47] The special digestion flask illustrated in Fig. 3-1 can advantageously be used, especially if volatilized elements are to be recovered.[48] It is also possible to use conical (Erlenmeyer) flasks for digestions satisfactorily.[49]

Note is to be made of the volumes of all reagents added so that the same volumes can be used in the blank run.

Some workers use a microwave oven for heating to speed up the digestion.

Procedure A. Weigh 5 g (Note 1) of sample into a 100 ml or larger Kjeldahl flask, and add 10 ml of 1:2 nitric acid. As soon as any initial reaction subsides (Notes 2 and 3), heat gently until further vigorous reaction ceases, and then cool the mixture. Add, gradually, up to 10 ml of concentrated sulfuric acid (sp. gr. 1.84) at such a rate as not to cause excessive frothing or heating (5–10 min are usually required), and then heat until the liquid darkens appreciably.

Continue as in "Continuation for Procedures A to D."

Procedure B. Weigh 5 g (Note 1) of sample into a 100 ml Kjeldahl flask, and add 5 ml of concentrated nitric acid (sp. gr. 1.42). As soon as any initial reaction subsides, heat gently until further vigorous reaction ceases, and then cool the mixture. Add, gradually, 8 ml of concentrated sulfuric acid at such a rate as not to cause excessive frothing or heating, and then heat until the liquid darkens appreciably.

Continue as in "Continuation for Procedures A to D."

Procedure C. Weigh 5 g (Note 1) of sample into a 100 ml Kjeldahl flask, and add a mixture of 8 ml of concentrated sulfuric acid and 10 ml of concentrated nitric acid (Note 4). Warm cautiously until the reaction subsides, and then boil rapidly until the solution begins to darken because of incipient charring.

Continue as in "Continuation for Procedures A to D."

Procedure D. Treat 5 g (Note 1) of sample in a 100-ml Kjeldahl flask with 20 ml of 1:2 nitric acid, and warm until the initial vigorous reaction is over. At this point, a spongy, tarry cake is formed. Cool the mixture, pour off the acid into a beaker, and wash the residue with four 1 ml portions of water, adding the washings to the beaker. Add 8 ml of concentrated sulfuric acid to the tarry residue, agitate to disperse the cake, and add concentrated nitric acid dropwise, with warming if necessary, until vigorous reaction ceases. Return the original acid solution to the flask, and boil until the solution just begins to darken.

Continue as in "Continuation for Procedures A to D."

Continuation for Procedures A to D. Add concentrated nitric acid slowly in small portions (1–2 ml), heating after each addition, until darkening takes place. Do not heat so strongly that charring is excessive, or loss of arsenic may occur; a small, but not excessive, amount of free nitric acid must be present throughout. Continue this treatment until the solution fails to darken on heating to fuming (5–10 min) and is, at most, only pale yellow in color. Cool somewhat, dilute the solution with 10 ml of water, and boil gently to fuming. Cool again, add 5 ml of water, and boil to fuming. Cool and dilute with 5 ml of water.

With Final Addition of $HClO_4$ *or* H_2O_2. If color tends to persist in this procedure, it may be destroyed more easily with these reagents than by continued addition of nitric acid. With perchloric acid as the auxiliary oxidant, run 0.5 ml of the 60% acid (Note 5) and a little more nitric acid into the flask, and heat for about 15 min; then add a further 0.5 ml of perchloric acid, and heat for a few minutes longer. Allow to cool somewhat, and dilute the mixture with 10 ml of water. The solution should be quite colorless, except when much iron is present, when it may be faintly yellow. Boil gently until white fumes appear. Allow the solution to cool, add a further 5 ml of water, and again boil gently to fuming. Finally, cool and dilute the solution with 5 ml of water.

Hydrogen peroxide may be used in place of perchloric acid. Add 100 volume (30%) reagent-grade hydrogen peroxide in small quantities (1–2 ml) with a few drops of concentrated nitric acid. Heat to fuming after each addition until the solution is colorless or no further reduction in color can be obtained. Cool and dilute the solution with 10 ml of water, evaporate to fuming, add 5 ml of water, and again evaporate to fuming. Finally dilute with 5 ml of water.

Notes. **1.** The size of the sample depends on the level of the metal content and on the methods to be used for the determination of the metal. A 5 g sample is suitable for 10–100 ppm of metal. If a very sensitive method is available, 2 g is often enough.

For biomaterials, 10–15 g of blood or tissue or 50 ml of urine or plasma are generally sufficient for all determinations; they should be boiled down to small bulk with nitric acid before sulfuric acid is added.

For other liquids, for example, fruit juices or beverages, take 20–50 g of the sample, containing less than 5 g of solids, and boil down to small bulk with nitric acid before adding sulfuric acid.

2. With certain materials such as tea, or when much carbohydrate is present, the initial reaction may be violent and heating should be delayed, if necessary even overnight. More nitric acid may then be added as necessary.

With some extremely reactive organic compounds, it is necessary to carry out the preliminary treatment with dilute nitric acid in a 500 ml beaker, heating on a water bath until the initial reaction ceases.

3. If excessive frothing occurs in the earlier stages, a drop or two of 2-octanol may be added or the preliminary treatment may be carried out in a 500 ml beaker with the addition of glass beads to prevent bumping.

4. With some organic materials, for example, rubbers, coated fabrics, and polymers, wet decomposition sometimes takes place more readily if the material is first charred by warming gently with concentrated sulfuric acid, followed by nitric acid; but this procedure must not be used if arsenic is to be determined. Wet oxidation of petroleum is carried out by charring with sulfuric acid, then adding nitric acid (or H_2O_2) and, if color remains, a little $HClO_4$.[50]

5. Many oxidations have been carried out safely by the use of perchloric acid, but some explosions have been reported with this reagent (see Section B.1.c.). Full precautions should be taken against possible injury in the event of such an explosion, even in well-tried procedures.

b. *Sulfuric Acid and Hydrogen Peroxide*

The availability in recent years of strong H_2O_2 solutions (~50% w/w) of high purity has led to increased use of this oxidizing agent for the destruction of organic matter.[51] Most often it is used in conjunction with sulfuric acid (probably the active agent is permonosulfuric acid). It is a more vigorous oxidant than nitric acid, and has been found to be of special value in the decomposition of resistant plastics. The only reduction product in oxidations with hydrogen peroxide is water, but unless unreacted reagent is destroyed at the end, there can be interference in subsequent steps of the procedure (separations, photometric determination). Hydrogen peroxide is also useful in the final stages of H_2SO_4–HNO_3 digestion to destroy any remaining small amount of difficultly oxidizable substances.[52]

There is little explosive hazard in the use of H_2O_2 if H_2SO_4 is present in excess. "A Sub-Committee member produced explosions deliberately by evaporating large volumes (about 50 ml) of peroxide with small volumes (less than 5 ml) of sulphuric acid. Though noisy, the explosions did not produce mechanical fracture of the glassware involved."[51] Still it is best to use a safety shield.

Most organic materials, even plastics, can be destroyed in 0.5 hr digestion with H_2SO_4–H_2O_2.[53] Copper phthalocyanine can be decomposed with H_2SO_4–H_2O_2–HNO_3. Most elements are quantitatively recovered in H_2O_2–H_2SO_4 digestion of organic material. Exceptions are Ru, Os, Ge, As, and Se.[54] The loss of Ge and As is attributable to the volatilization of the chlorides. Mercury requires special measures. Precipitated $CaSO_4$, if not brought into final solution may retain Pb and Ag. Elements showing good recoveries include Cd, Bi, V, Cr, Sn, Zn, Sb, Zr, Mn, In, and Te, even in the presence of chloride.[55]

After decomposition by the following procedures, the sulfuric acid solution, diluted with water, should be boiled gently for 10 min to decompose any remaining H_2O_2, which might interfere in subsequent determination steps.

Procedures[51]

Oxidations are conveniently carried out in Kjeldahl flasks (100 ml for 2 g sample), with hydrogen peroxide added from a small separatory funnel having a Teflon plug.[56]

Procedure A (for plastics and other materials).[57]   Transfer 2 g of sample to the flask, and add 5–25 ml of concentrated sulfuric acid. Heat until charring begins and then add 50% H_2O_2 dropwise until the solution becomes colorless. Some samples require alternate charring and addition of H_2O_2, but care should be taken not to reduce greatly the volume of

H_2SO_4. Avoid excessive charring, lest the carbon be difficult to oxidize. Finally, heat to fuming to destroy H_2O_2.

Procedure B (for readily oxidizable materials). Add 20 ml of concentrated sulfuric acid to 2 g of sample and then, to the cold mixture, 50% H_2O_2 dropwise until the reaction slows down or until the solution becomes colorless. Heat until fumes of sulfuric acid are evolved, adding more peroxide as necessary until a colorless solution is obtained. Heat at fuming to destroy H_2O_2.

This procedure can be used for solid fruits, vegetables, meat, and milk products. If reaction does not begin within 2 min after addition of H_2O_2, the mixture may be heated gently to start it. A few glass beads may be added to lessen the danger of bumping.

Procedure C (for soft drinks, beer, etc.). To a 250 ml flask add 5 ml of concentrated sulfuric acid and a few glass beads, followed by not more than 50 ml of sample. Add 20 ml of 50% H_2O_2, and heat gently until the initial reaction subsides. Then heat until fumes of sulfuric acid appear. If charring occurs at any point, add further 1 ml portions (not larger) of H_2O_2. Digestion is complete when the fuming sulfuric acid remains colorless. If at any stage sulfuric acid tends to approach dryness, add 2–3 ml more of this acid before continuing.

Procedure D (for syrups). Transfer not more than 50 ml of sample to a 500 ml flask and add a few glass beads, 50 ml of water, 5 ml of concentrated sulfuric acid and 20 ml of 50% H_2O_2. Heat gently until the initial reaction has ceased. If the solution is dark, repeat the addition of peroxide, several times if needed, and when reduced to a small bulk, heat the solution strongly to fumes of sulfuric acid. If charring occurs, add further 1 ml portions (not larger) of H_2O_2 until the solution remains colorless when fumes are evolved. If it seems that the sulfuric acid may approach dryness at any stage, cool, add 2–3 ml of sulfuric acid, and continue. Excess H_2O_2 is destroyed by fuming the mixture.

Procedure E (for herbs, spices, gums, etc.).[58] Mix 2–5 g of sample with the minimum amount of water to form a slurry and add 3–5 ml of concentrated sulfuric acid. Heat the mixture gently and add 50% H_2O_2 dropwise, ensuring that all peroxide has decomposed before making further additions. When most of the sample has been decomposed, add more peroxide with continued heating until a clear liquid is obtained on fuming.

c. Perchloric Acid Mixtures

Perchloric acid is never used alone in oxidations of organic matter, for this could result in violent explosions. Compounds containing hydroxy groups react with perchloric acid to give perchlorate esters that decompose explosively. If perchloric acid is mixed with nitric acid and gradually heated with hydroxy-containing compounds, the latter are likely to be destroyed by the nitric acid before reacting with perchloric acid.[59] In cold dilute solution, perchloric acid has little oxidizing power, but in hot, strongly acid solutions it is a vigorous oxidizing agent, destroying organic matter more rapidly than nitric acid. On dilution with water its oxidizing properties are lost and the digest is not oxidizing except for the effect of free chlorine or chlorine oxides that may be present in small amount. Perchloric acid is stable indefinitely as the 60 or 72% aqueous solution in which the reagent-grade acid is usually supplied. The azeotropic mixture contains 72% $HClO_4$ and boils at 203°C. By addition of sulfuric acid, the temperature and oxidation potential of the digest can be increased and a very potent oxidizing mixture is obtained.[60] Sulfuric acid should be omitted when the presence of sulfate is objectionable, as in determination of lead.

Because few samples resist oxidation with perchloric acid mixtures and because digestions take less time than with most other oxidants, perchloric acid is extensively used to destroy organic matter. But it must always be borne in mind that perchlocic acid is a potentially dangerous reagent that may destroy more than the sample.[61] Procedures involving perchloric acid must be carefully followed and precautions taken to lessen the effects of explosions should they occur, small though the number may be. Some analysts look upon the use of perchloric acid with suspicion and avoid it.

There is no doubt that, handled with knowledge and care, this reagent is extremely efficient in the destruction of organic material, and from a technical rather than a safety standpoint, a mixture of nitric and perchloric acids probably has less disadvantages than any other oxidation mixture. However, its degree of tolerance toward mishandling is probably less than is the case with the other common wet oxidation reagents, and for use in unskilled or inexperienced hands avoidance is perhaps better than regret. · · · for the newcomer in the field, most of the other oxidation techniques will allow him to make more mistakes in safety.[62]

The initial nitric acid oxidation mentioned above must be carried out in such a way that the more easily oxidized constituents are actually destroyed before perchloric acid oxidation begins: Sufficient nitric acid must be present and it must be allowed to react for an adequate length of time. Most of the oxidation in the mixture should be done by the nitric acid. Samples remaining immiscible with the digestion mixture—fats and waxes particularly—are a source of hazard. Reaction takes place only at the interface, and the nitric acid oxidation may not be adequate. Milk fat seems to be the most dangerous (Gorsuch). Possibly high temperature

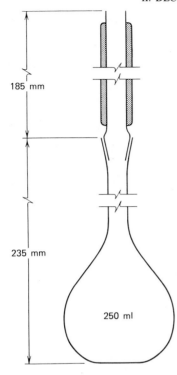

185 mm

235 mm

250 ml

Fig. 3-2. Reaction flask with asbestos-jacketed air condenser according to C. Feldman. The flask is a 250 ml borosilicate glass volumetric flask. A condenser is made from ST 19/38 ground glass joint, with jacket of woven asbestos tubing held in place by wrapping with Teflon sealing tape. Reprinted with permission from C. Feldman, *Anal. Chem.*, **46,** 1607 (1974). Copyright by the American Chemical Society.

hydrolysis of fats can lead to formation of alcohols at temperatures where reaction and decomposition are almost simultaneous. It is always wise practice to test a small quantity of a sample before proceeding with a full-size sample, but hazard may increase with sample size.

The only elements likely to be recovered incompletely in perchloric acid digestions are those whose chlorides are volatile at relatively low temperatures (chloride results from reduction of perchlorate). The recovery of Hg is low unless provision is made for collecting the distillate from the digest (Table 3-5A) or other measures are taken (p. 128). Germanium tetrachloride would be expected to be lost unless the distillate is caught. Tin and Sb(?) may be lost from a digest heated to fuming with sulfuric acid. Arsenic is not volatilized appreciably if charring does not occur. Osmium and ruthenium will be volatilized as the tetroxides, rhenium as Re_2O_7.

A recent (1973) method for the destruction of plant material uses concentrated nitric acid vapor to oxidize about 90% of the organic matter in 5–6 min, followed by destruction of the remainder with 70% perchloric acid.[63] The advantages of this method are the shorter decomposition time (a total of 45 min) and very low blanks. Metal recoveries are comparable to those obtained in other methods.

Perchloric acid digestion with special reference to the determination of Hg has been studied in some detail.[64] The essential feature of the procedure worked out is the use of a 250-ml borosilicate volumetric flask with an asbestos-insulated air condenser (Fig. 3-2). Depending on the nature of the sample, digestion is carried out with HNO_3–$HClO_4$ or HNO_3–H_2SO_4–$HClO_4$ (+ small amount of $K_2Cr_2O_7$). The insulated condenser, in conjunction with hot plates maintained at certain temperatures, allows H_2O, HNO_3, some $HClO_4$, and others, to be distilled off without loss of Hg, so that a sufficiently high temperature can be reached to assure complete destruction of organic material. The rate of boiling—and frothing—must be controlled to prevent volatilization of Hg compounds. The time required for decomposition ranges from 1.5 to 2.5 hr, the latter when H_2SO_4 must be used. Attention is called to hazards. For samples low in lipids, explosion danger is avoided if the sample is first boiled with HNO_3 or a mixture of HNO_3 and $HClO_4$, provided that sufficient HNO_3 (~15 ml for up to 5 g of wet tissue) is used and that, by refluxing, HNO_3 remains in the system until easily oxidized material has been destroyed. (A beaker covered with a watch glass might allow much of the HNO_3 to escape before this oxidation has been completed, and an explosion might follow when the temperature is raised in the presence of $HClO_4$.) Preliminary boiling with HNO_3 under total reflux does not guarantee complete oxidation of gram amounts of lipids and such samples should first be heated with concentrated H_2SO_4 alone. Final oxidation with HNO_3 and $HClO_4$ can then be made smoothly. For details see the original paper.

Recovery of Hg in this decomposition procedure is within a few percent of 100%. Diphenyl mercury, phenyl mercury acetate, and ethyl mercury chloride are recovered within 2% of 100.

Procedure[65]

It is advisable to use a safety shield in all oxidations with perchloric acid. Transfer 2 g or more of finely divided sample to a 300 ml Kjeldahl flask. Add 15 ml of concentrated nitric acid (7 ml more for each gram of sample above 2 g) and 4 ml of 60% perchloric acid.[66] After mixing, add 2–5 ml of concentrated sulfuric acid (the larger volume is preferable if it will not interfere in the subsequent metal determination). At least 2 ml of sulfuric acid should always be present to ensure a sufficiency of an acid of high boiling point to prevent over-heating of the digest in the later stages after the nitric acid has been expelled; in the absence of sulfuric acid, local overheating of the digest may lead to decomposition of ammonium perchlorate with explosive violence.

Heat gently the mixed contents of the flask for 3–5 min or to the first appearance of dense brown fumes. Remove the flask from the heat for ~5 min to allow the initial vigorous reaction to subside. Then continue the digestion slowly at low heat until thick white fumes of sulfuric acid appear. If bumping tends to occur, add two small glass beads.

The rate of digestion is important. If the temperature is too high, loss of nitric acid occurs before oxidation of organic matter is completed, whereas if it is too low the time required for digestion is unduly prolonged. When 25–30 ml of nitric acid is used, ~90 min should be required to reach the fuming stage. If the nitric acid is boiled off too rapidly, the liquid in the flask becomes black with charred organic matter. In that event, add 1–2 ml of nitric acid and continue the digestion at a slower rate.

When dense fumes of sulfuric acid appear, continue the digestion for 5–10 min at low heat and finally for a couple of minutes at higher heat. If digestion is complete, the liquid will be colorless at this stage. Any sign of carbonization indicates incomplete digestion, and 1–2 ml of nitric acid should then be added, followed by digestion to fuming. When the digest is cold, dilute with 50 ml of water.

Notes. **1.** If perchloric acid of adequate purity is not available, sodium perchlorate plus an equivalent amount of sulfuric acid can be substituted. Sodium perchlorate can be purified by shaking its slightly basic solution (pH ~9 or 10) with dithizone in CCl_4 to extract most heavy metals. Of course, nothing is gained by this purification unless the sulfuric acid is free from heavy metals.

2. Silica separating in the digestion contains inappreciable amounts of Fe, Co, Mn, and the like and need not be hydrofluorized except in the most refined work.

Procedures for Decomposition of Organic Material for Mercury Determination[64]

Preliminary Test. Test small (0.1 g) samples to determine which of the procedures described below is most suitable.

Treat the test sample first according to the initial steps of Procedure A; use at least the quantity of each acid specified. Because the reaction of some substances with concentrated nitric acid has a fairly long induction period, wait 5 min before heating the sample. If the reaction of unheated mixture is mild, heat the flask (Fig. 3-2) at 230°C. If the reaction appears to be too vigorous to permit scaling up to a larger sample, use Procedure B for this material instead. If the reaction is smooth, heat the flask at 265°. If the foam produced by the test sample indicates that more than 1 in. of foam would be produced by 5 g of sample, use Procedure C for this material instead. If the amount of foam produced is moderate, carry Procedure A to completion. In doubtful cases, test 0.5 g of sample as before.

It is convenient to maintain three separate hot plates at the surface temperatures indicated. Perform all addition and heating of acids in a hood which is equipped for handling perchloric and nitric acid fumes.

Procedure A (*for muscle and other low-lipid tissue, vegetation, and soil*). Transfer $\sim$20 mg granular $K_2Cr_2O_7$ and 5 g of sample to the reaction flask. Add 10–15 ml of concentrated nitric acid and 15 ml of concentrated (70%) perchloric acid. Swirl the flask gently and connect the condenser. Place the assembly on the first hot plate (230 ± 5°), and allow it to remain there until evolution of NO_2 fumes has essentially ceased (20–30 min). Transfer the assembly to the second hot plate (265 ± 5°) and allow it to remain for approximately 30 min. Next transfer the assembly to the third hot plate (340 ± 5°), and allow it to remain there until the color of the solution changes to bright orange ($\sim$45 min). Remove the assembly from the hot plate, and allow it to cool to room temperature. Add water through the condenser to rinse any condensate into the flask. Remove the condenser, and dilute the solution to volume.

Procedure B (*for light tars, wheat flour, gasoline, and ethanol*). Transfer 5 g of sample to the reaction flask, and add 15 ml of water and 15 ml of concentrated nitric acid. Connect the condenser, swirl the assembly, and allow it to stand until the initial reaction has subsided. Place the assembly on the 230° hot plate, and allow it to remain there until evolution of NO_2 fumes has essentially ceased. Add 15 ml of concentrated perchloric acid, heat the mixture at 230°, and continue as in Procedure A.

Procedure C (*for high-lipid materials such as fatty tissues, brain, and liver*). Transfer 5 g of sample to the reaction flask, and add 15 ml of concentrated sulfuric acid. Connect the condenser, swirl the assembly, and place it on the 230° hot plate for 15 min. Remove the assembly from the hot plate, and allow it to cool to room temperature. Cool the lower part under cold water, and slowly, with swirling, add 25 ml of concentrated nitric acid through the condenser. Replace the assembly on the 230° hot plate for 60 min. Add 40 ml of concentrated perchloric acid, and heat the mixture at 265° for 30 min. Transfer the assembly to the 340° hot plate, and allow it to remain there for approximately 45 min. Allow the assembly to cool, rinse the condenser walls into the flask, and dilute to volume. Use caution in the rinsing and diluting, because the heat of mixing is high.

2. Dry Oxidation

a. In Air

In the common method of dry oxidation (ashing), the sample is gradually heated in air to a temperature usually not exceeding ~500°C. If the sample is an aqueous solution, or if it is moist, water is first driven off slightly below the boiling point, and the temperature is then slowly raised to char the material. In this stage, carbon-containing products of decomposition escape, which must not be allowed to catch fire for fear of mechanical losses or because the temperature may increase unduly.[67] The carbonaceous residue must then be destroyed by continued heating at as low a temperature as feasible. This is a slow process, and it is best carried out in a muffle furnace overnight. If the residue is not free from particles of carbonaceous material, it can be moistened with nitric acid and reheated. A few workers have used oxygen, passed into the furnace, to hasten oxidation.

As indicated earlier, the recovery of metals is not as certain in dry oxidation as in wet oxidation, and although satisfactory recoveries can be obtained for most metals in most organic samples, care is required. The two chief causes of losses are volatilization and retention by the container in which the ashing is done. Metals can be volatilized as such after reduction by organic matter, as inorganic compounds (chlorides) and as metal–organic complexes. Mercury, as metal or any compound, will be volatilized completely at the usual ashing temperature in the vicinity of 500°C. Very few other metals will be volatilized as the elements; cadmium probably can be lost in this way. Arsenic is likely to be volatilized in combined forms. A larger number of metals can be volatilized as inorganic compounds, mostly as the chlorides.[68] Even if a metal is not present originally as a chloride, this may be formed from covalent chlorine in the sample on decomposition (e.g., polyvinyl chloride) or from inorganic chloride. For example, NH_4Cl, which gives HCl on heating, can react with metal oxides, carbonates, and the like; NaCl and other alkali metal chlorides are unlikely to react. Volatilization of Fe, As, Sb, Pb, and Zn (and possibly some other metals) as chlorides is a possibility in dry ashing.[69] Even alkali metal chlorides can be lost above 500°.[70] Dry ashing of liver and muscle at 550° results in the loss of 8–15% Cs, primarily by volatilization of the chloride; in ashing of bone there is no loss.[71] When metals subject to volatilization are present, closer temperature control and lower ashing temperatures than otherwise may be needed, or an ashing aid may be desirable (see later). Sulfuric acid is sometimes added to convert inorganic chloride to HCl.

Occasionally, samples may contain metal–organic complexes that are volatile at comparatively low temperatures and can be lost in the early stages of dry oxidation, for example, nickel and vanadium porphyrin complexes in petroleum[72] and arsenic and other elements in some biomaterials.[73] See the individual elements for particulars concerning volatilization and other losses of trace metals in the destruction of organic matter.

Loss of elements from retention by the vessel in which dry ashing is carried out is probably as important a source of error as volatilization loss in general. Ashing vessels are composed of fused silica, porcelain, or platinum. With the first two, reaction of metal oxides, carbonates, and other salts to form silicates at the surface of the dish must be considered a possibility and does in fact occur. The extent of such reactions depends on a number of factors, among them the nature of the metal, the composition of the ash and its quantity, and the temperature of ashing. In general, retention increases with temperature. The amount of ash is very important. If the amount is small so that an appreciable fraction can come into contact with the walls of the vessel and react with it, losses are greater than when the ash is bulky. Such loss can be greatly reduced by adding an ashing aid to increase the amount of ash. For example, magnesium nitrate is often a good ashing aid. The magnesium oxide formed is a good physical buffer, dissolves readily in acids, and the nitrate aids the oxidation of organic matter. The presence of much magnesium is sometimes objectionable in the subsequent separations and determinations, and another ashing aid may then be substituted. Calcium carbonate has been recommended.

If the silicate formed at the surface of a silica or porcelain dish were soluble completely in acid no harm would be done, but as a rule metal is retained. Perhaps the metal diffuses some distance into the dish, so that it cannot be reached by acid, or if exposed to acid, it may not be completely released from the silicate framework. Adsorption of metal on the surface of the dish from a rather strongly acid solution will, in general, tend to be small. Metal oxides of weakly ionic character may be expected to combine more readily with SiO_2 than those which are strongly ionic. Lead oxide combines readily with SiO_2. Lead is one element that may give trouble in dry ashing. If the temperature of ashing is above 450–500°, some retention of lead by a silica (or porcelain) dish is likely to occur (Table 3-5B and 3-7)[74]; volatilization of the chloride may also contribute to loss of lead. It has been found that the retention of lead on a silica dish is greatly increased if much sodium chloride is present.[75] The silicate structure is thought to be weakened by sodium chloride attack, leading to increased reaction of lead and greater difficulty in leaching it from the dish.

According to Gorsuch, losses of Au, Ag, and Cu in silica vessels can result from reduction to metal, followed by diffusion into the silica. This type of loss will be greater, in general, in platinum dishes when easily reducible metals are involved and alloy formation takes place. For example, evaporation of an organic solvent solution of Cu(II) dithizonate in a platinum dish followed by gentle ignition to destroy organic matter and treatment with acid brings very little copper into solution. Other elements that can be reduced by organic matter or flame gases and then alloy with platinum include the noble metals, Pb, Bi, As, Ni, Fe, Mn, and P. Ashing aids may be desirable with such elements and temperatures should be kept as low as feasible.

Another factor calling for consideration in ashing is attack of the vessel, including that by the acid when the ash is dissolved. When silica vessels are used, metal contamination is minimal and probably need rarely be feared—only losses are important—but with porcelain dishes, there is some danger of introduction of trace metals from attack of the glaze. For example, traces of copper and zinc have been reported to be derived from porcelain. Platinum gives the greatest danger. Introduced platinum can interfere in the spectrophotometric determination of Mo with SCN^-— reducing agent, of Co with nitrosonaphthol and in some other determinations. The following figures illustrate the amounts of platinum that can be introduced in ashing 10-g samples of biomaterial for 15 hr[76]:

	Pt (μg)	
	450°C	600°C
Grass	145	103
Clover	5	10
Blood	130	305
Liver	68	555

The high phosphorus content of liver is believed to be the cause of much Pt introduced at 600°.

Finally, loss of metals can occur by retention in or on insoluble material remaining when the ashing residue is treated with acid. The insoluble material is most likely to be silica. Iron and aluminum are held tenaciously by silica (Table 3-8). Such retention is not as serious as that by the vessel, since the silica can be filtered off, hydrofluorized, and the metals brought into solution, but it is, nevertheless, inconvenient. In wet oxidation, metal retention by silica is much less. It seems that material

TABLE 3-8
Retention of Iron, Aluminum and Manganese by Silica in Ashing[a]

In each case 5 g of sample were ashed as indicated; the ash was treated with dilute hydrochloric acid, the solution evaporated to dryness, and the residue extracted with dilute hydrochloric acid; iron, aluminum, and manganese were then determined in the washed residue after hydrofluorization. Samples as follows: No. 1, hay low in silica (0.4% crude SiO_2), No. 2, hay with normal silica (1.5%); No. 3, straw (2.4% SiO_2), No. 4, hay high in silica (~4%).

		Amount of Metal in Silica					
		Fe_2O_3		Al_2O_3		MnO	
Method of Ashing	Sample No.	mg	Percent of Total	mg	Percent of Total	mg	Percent of Total
Over Teclu burner	1	0.61	59	0.42	48	<0.005	<3
	2	0.30	72	0.54	57	0.03	3.5
	3	0.95	73	1.10	56	0.04	6
	4	1.35	78	2.13	85	0.51	17
In electric muffle furnace	1	0.39	38	0.20	23	<0.005	<3
	2	0.17	40	0.29	30	0.01	1
	3	0.51	39	0.90	45	<0.005	<1
	4	0.98	56	1.59	63	0.09	3

[a] Selection of data from T. Wijkstrom, *Metoder för Undersökning av Fodermedels Mineralbeståndsdelar.* Svanbäck, Stockholm, 1935.

containing much silica is best wet-oxidized when metals such as iron are to be determined.

General Procedures

It is impossible to give a set of directions that can be followed in detail for the ashing of all types of organic material. The exact procedure (actually there are alternatives) will depend on the composition of the sample—whether much or little ash is present, the content of chlorine and its state of combination, and so on. Moreover, the metal that is being determined may have a bearing on how a procedure is carried out. Nevertheless, an attempt is made here to give some general directions. These may be of some aid to those who are newcomers in the field.

Silica (fused quartz) dishes are perhaps the most suitable for ashing. Platinum dishes are also very good (and preferred by some workers), except when the introduction of a small amount of platinum may cause trouble in the subsequent determination. Of course, platinum should not be used when easily reducible elements (Au, Ag, etc.) are present. In less exacting work, porcelain dishes may serve. All dishes should have a smooth, unetched interior,[77] and should be acid-cleaned (hot 1:1 HCl) before use.

From what has been said in the preceding discussion, it is evident that the ashing temperature should be no higher than really needed. If it is too low, oxidation of organic matter is unduly slow. A good compromise is 450°C, and at this temperature, heating overnight (say 15 hr) is often required. If an ashing aid is used, a temperature of 500° may be permissible, but an increase above this value is risky.

Dry ashing is contraindicated for Se and of course Hg. It may be used for As with a suitable ashing aid.[78]

Procedure A (*without ashing aid*). Transfer 5 g (or more or less) of sample that will yield relatively much ash (e.g., many solid foodstuffs) to a silica dish of suitable size. Dry if necessary and then char the material by heating with an infrared lamp or on a high temperature hot plate or slowly by means of a charring burner. The arrangement shown in Fig. 3-3 is convenient for preliminary charring. Do not allow the sample to ignite. When it is well charred and volatile matter has been largely expelled, place the dish in a muffle furnace at about 300°. It is advisable to place the dish on a triangle to avoid local overheating. Gradually increase the temperature to about 450°, as ashing proceeds. When the door of the muffle is partly opened, the charred material should glow only faintly. If the temperature is raised too rapidly or too much air is allowed to enter, oxidation may be so rapid that the temperature of the sample will exceed that of the furnace, and volatilization of metals and other undesirable effects will result. The temperature of the furnace should be monitored during the ashing or a temperature calibration should be carried out beforehand in terms of furnace settings and time of heating. Care should be taken to obtain the temperature at the place where the dish is located—temperatures inside a furnace can vary significantly. It is often convenient to allow ashing to continue overnight at 450° (or at 500° if this is known to be permissible). A grayish ash should usually be obtained when ashing is complete. If more than a few specks of carbon are present, moisten the cold residue with a few drops of 6 M HNO$_3$, evaporate, and again heat in the furnace until all carbonaceous matter has been destroyed.

Add, from beneath a cover glass, 5 or 10 ml of 1:1 HCl[79] to the cold dish[80] and heat on the steam bath for 15 min or longer to bring all soluble material into solution. Dilute with 10 ml of water. If more than a minute amount of insoluble material remains (most likely silica), filter it off on a very small filter paper previously washed with HCl, and wash with a little

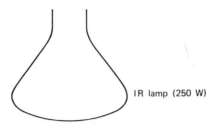

IR lamp (250 W)

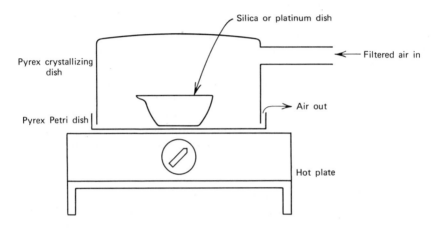

Silica or platinum dish

Filtered air in

Pyrex crystallizing dish

Air out

Pyrex Petri dish

Hot plate

Fig. 3-3. Arrangement for preliminary charring of organic materials. (According to R. E. Thiers in J. H. Yoe and H. J. Koch, Jr., Eds., *Trace Analysis*, Wiley, New York, 1957, p. 640; R. E. Thiers in D. Glick, Ed., *Methods of Biochemical Analysis*, Vol. 5, Interscience, New York, 1957, p. 279.)

dilute HCl. Combine filtrate and washings, and dilute with water to a convenient volume in a volumetric flask.

If the amount of silica is appreciable, and especially if there is reason to believe that it may contain Fe, Al, Cu, or other trace elements, ignite the paper in a platinum crucible and hydrofluorize the residue.

Procedure B (*with ashing aid*). If the sample is low in ash, addition of an ashing aid is desirable for reasons already mentioned. Magnesium nitrate often serves well as a general ashing aid, but other substances also have their uses (see Note).

Wet the sample in a silica dish with magnesium nitrate solution (5 to 10 ml of 7% $Mg(NO_3)_2 \cdot 6\,H_2O$ solution for 5 g of sample), evaporate to dryness, and continue with the ashing according to Procedure A.

Note. The use of magnesium oxide (for raw rubber and rubber latex), sodium carbonate (for dyestuffs, intermediates, medicinals, silk, wool, etc.) and sulfuric acid (foodstuffs of low ash content and compounded rubbers) has been described.[81] Sulfuric acid is sometimes used to volatilize ionic chloride before oxidation of organic matter, with the intent of forestalling metal chloride volatilization. Magnesium acetate has been recommended in place of magnesium nitrate to avoid deflagration.[82]

Magnesium salts may be objectionable as ashing aids if at a later stage insoluble magnesium salts can be formed, for example, $MgNH_4PO_4 \cdot 6 H_2O$ in the presence of phosphate in ammoniacal solution.

b. With Excited Oxygen

A comparatively new method of ashing is based on the use of a stream of excited oxygen (plasma) obtained by subjecting oxygen at low pressure (2–5 mm Hg) to a high-frequency electrodeless discharge.[83] The oxidation of organic matter by atomic and ionic species of oxygen takes place at 100°–200°C, with obvious advantages to be gained from the low temperatures.[84] At low power input levels, good recovery of arsenic and selenium has been reported, but the composition of the sample plays an important role. Mercury is not recovered satisfactorily. The equipment is commercially available.

The apparatus has been modified to allow recovery of volatile elements.[85] By incorporation of a cold trap in the train, recoveries of volatile elements of 98% or better can be obtained (nanogram amounts of elements up to 1 g of sample). Further modification allows >91% recovery of mercury.

c. In Closed Oxygen Systems

The use of an oxygen bomb for destroying organic materials in trace analysis has been examined.[86] The bomb (capacity 300 ml) is made of 18-8 Cr–Ni stainless steel, and the interior of the bomb is preferably platinum plated so that it can be rinsed with acid. A platinum sample cup and firing wire are placed inside the bomb. Recoveries of various elements were determined photometrically after ignition of filter papers that had been loaded with known amounts of these elements (Table 3-9). The bomb is electrically fired at an oxygen pressure of 25 kg/cm^2. On ignition of the sample, the elements enter both the liquid and gas phases inside the bomb, although they tend to concentrate in the liquid. The liquid phase contains determinable concentrations of nitric and nitrous acids. Samples as large as 1 g can be burned. Polyethylene can be ashed in this way. The use of this oxygen bomb for the determination of mercury in

TABLE 3-9
Recovery of Metals After Sample Combustion in Oxygen Bomb

Element	Treatment of Ignition Product	Recovery (%)
Al	Fused with $KHSO_4$	80
Cu	Digested with dil. aqua regia	95–97
Fe	Digested with H_2SO_4 and HNO_3	95–97
Hg	None	96–100
Pb	Digested with 1:1 HNO_3	93–95
Ti	Fused with $KHSO_4$	80
V	Digested with $HClO_4$	92–96
Zn	Digested with 1:1 HNO_3	90

various samples of organic origin has been described.[87] The combustion is carried out in the presence of 1 M HNO_3; in its absence, reproducibility is unsatisfactory.[88]

The oxygen combustion flask provides a simpler closed oxygen system. Here the sample, wrapped in filter paper and placed in a platinum gauze holder, is burned in a closed flask filled with oxygen at atmospheric pressure, together with an absorbing solution. But this system finds little use in the determination of trace constituents, because a large sample, not well suited for this approach, is required. Nevertheless, attempts have been made to use it for mercury, with samples as large as ~1 g. An oxygen flask of 500 ml capacity is needed for a 0.1 g sample.

What may be called a semiclosed oxygen system is a combustion tube fitted with traps to catch volatilized elements when the sample is burned in a stream of oxygen. Nonvolatilized element in the residue can be combined with the volatilized fraction to give the total. Such a system might have value for metals (mercury) volatilized in whole or in part in other methods. Limitation on the size of sample is a drawback. A recent combustion tube system is fitted with a liquid–air-cooled trap and reflux condenser; the combustion products are frozen and then dissolved in mineral acid.[89] Good recoveries can be obtained for As, Ba, Cd, Co, Cu, Hg, Mo, Sb, and Zn but not for Cr and Sn. The amount of sample is limited to 0.3 g.

3. Oxidative Fusion

A well-known method of this type is fusion with sodium peroxide or other powerful oxidizing agent in a Parr bomb. However, it is not well suited for trace elements, which usually demand the use of fairly large samples.

Fusion with an equimolar mixture of $NaNO_3$ and KNO_3 in an open vessel has been proposed for the destruction of organic samples such as polyethylene, flour, blood, and kale.[90] A final temperature of $390 \pm 10°$ is used and oxidation of organic matter is rapid. A drawback of the method is the necessity for using the flux in a ratio of 20:1 to the sample (a small amount of nitrate would give what might be called low-grade gunpowder). Thus the use of a sample weighing more than a decigram or two would be unattractive and the method is not likely to find much use in ordinary trace analysis. The presence of much nitrite in the fusion mixture at the end could also be objectionable. On the other hand, the oxidation requires only a few minutes and can be carried out in a borosilicate beaker or flask. Radioactive tracers show that few elements are lost in the fusion. Mercury is essentially completely volatilized, As and Se (and S and halogens) appreciably, other elements not at all or not more than 1–2%.

A combination of burning and fusion with a 3:1 mixture of K_2SO_4 and KNO_3 has been used to destroy polyethylene.[91]

Fusion of vanadium 8-hydroxyquinolate with Na_2CO_3 in air (platinum crucible) provides an easy way of destroying small amounts of this complex, obtained by chloroform extraction, and bringing vanadium into solution as vanadate.

NOTES

1. If p and q represent the fractions by number of two kinds of units in an infinite population, the standard deviation in drawing n units randomly is $s = (npq)^{1/2} = \{np(1-p)\}^{1/2}$. That is, the number of expected units is, respectively, np and $n(1-p)$ (the most probable values), and s is the standard deviation for each unit.

2. The maximum particle size of approximately equidimensional material can be estimated roughly from the mesh size of the screen or bolting cloth through which the crushed sample has been sifted. The openings in U.S. standard sieves are respectively 0.03, 0.015, and 0.0075 cm for 50-, 100-, and 200-mesh. Assuming that particles are either spherical or cubic with a diameter (sphere) or edge (cube) equal to the sieve opening, we find the following number of particles of that size in 1 g of sample of density = 1:

	Diameter (cm)	Number (rounded)	
		Spheres	Cubes
50-mesh	0.03	7×10^4	4×10^4
100-mesh	0.015	6×10^5	3×10^5
200-mesh	0.0075	5×10^6	2.5×10^6

Of course, many particles will be smaller than the sieve opening but most of the mass will be in the larger particles. In general, the sieve opening gives the maximum particle size. Elongated particles could have an average diameter which is larger than this.

3. F. E. Wickman, *Ark. Min. Geol.*, **3**, 131 (1963).

4. The derivation is given by A. D. Wilson, *Analyst*, **89**, 18 (1964). A similar equation was first obtained by B. Baule and A. A. Benedetti-Pichler, *Z. Anal. Chem.*, **74**, 442 (1928). For mixtures consisting of more than two phases, each holding the constituent in question, the equations become quite complex. See Wilson, who also considers nonuniform grain size. C. O. Ingamells and P. Switzer, *Talanta*, **20**, 547 (1973), and Ingamells, *ibid.*, **21**, 141 (1974) and **23**, 263 (1976), have proposed the use of a sampling constant (K_s) in geochemical analysis. It can be defined by

$$s_\% = \left(\frac{K_s}{W}\right)^{1/2} \qquad (W = \text{sample weight in grams})$$

It is the weight in grams needed to ensure (68% confidence) a result which differs from the mean by no more than 1%. The square root of the sampling constant is numerically equal to the relative standard deviation of a set of results obtained from 1-g samples by a precise analytical method. The sampling constant is specific for each constituent of a sample *in a given state of subdivision*. See further Ingamells, *Geochim. Cosmochim. Acta*, **38**, 1225 (1974); *cf.* H. Jaffrezic, *Talanta*, **23**, 497 (1976). For other types of sampling constant, see Ingamells. W. E. Harris, and B. Kratochvil, *Anal. Chem.*, **46**, 313 (1974), give plots of the minimum number of particles required for $s \gtrless 0.1$ and 1% as a function of the 'difference in content of a given constituent in a solid mixture of two components having the same density and particle size.

5. See, for example, W. W. Walton and J. I. Hoffman, Chapter 4, Volume 1 of Part I of I. M. Kolthoff, P. J. Elving, and E. B. Sandell, *Treatise on Analytical Chemistry*, Interscience, New York, 1959. In addition to a general treatment of sampling, these authors supply sources of specific sampling information. A similar treatment is given by R. C. Tomlinson in Chapter II.3, Volume IA, of C. L. Wilson and D. W. Wilson, *Comprehensive Analytical Chemistry*, Elsevier, Amsterdam, 1959. Also see Chapter 2, Volume IIA, in F. J. Welcher, *Standard Methods of Chemical Analysis*, Van Nostrand, Princeton, N.J., 1963. Sampling of various commercial materials is described in A.S.T.M., *Annual Book of ASTM Standards;* of metals in G. E. F. Lundell, J. I. Hoffman, and H. A. Bright, *Chemical Analysis of Iron and Steel*, Wiley, New York, 1931, and in A.S.T.M., *Annual Book of ASTM Standards, Chemical Analysis of Metals;* of foods and fertilizers in Assoc. Offic. Agr. Chem., *Official Methods of Analysis*. Books on the analysis of particular materials will usually give information on their sampling.

6. If a material is a mixture, the composition of one or more of its component phases may be desired, and this will require the separation of the phases if

chemical methods—or many physical methods—are to be applied. This topic, as already mentioned, lies outside the scope of this volume.

7. See B. G. Davey and R. L. Mitchell, *J. Sci. Food Agr.*, **19,** 425 (1968), for the distribution of 24 trace, minor and major elements in cocksfoot at flowering.

8. See, for example, F. E. Beamish, *The Analytical Chemistry of the Noble Metals*, Pergamon, Oxford, 1966, pp. 558, 559.

9. Some workers prefer to use a hydraulic rock crusher with steel plates, between which the large pieces of material are reduced to ~1 mm size, instead of a percussion mortar. This coarse powder is then pulverized to 100- or even 200-mesh by running it through a mechanically driven roller–crusher in which the steel rollers are adjustable. See L. R. Wager and G. M. Brown, in A. A. Smales and L. R. Wager, Eds., *Methods in Geochemistry*, Interscience, New York, 1960, p. 13. Miniature jaw crushers have become available that can handle 0.5 kg of material in pieces of 2–3 cm diameter, reducing it to ~4 mm size. A miniature pulverizer can then be used to give 200-mesh powder. A special hood in which this equipment can be used, so that the laboratory remains dust free, is described by P. L. Beaulieu, *Anal. Chem.*, **43,** 798 (1971). Ceramic and iron–tungsten carbide crushing and grinding equipment for use in the laboratory are commercially available (1970).

10. See, for example, Maxwell (reference 23) for mixing, coning, and quartering.

11. Mortars made of sintered or vitreous alumina, boron carbide, and titanium carbide are available, all of which are harder than agate and are particularly valuable when substances harder than quartz are to be ground. As shown by J. F. Boulton and R. P. Eardley, *Analyst,* **92,** 271 (1967), boron carbide mortars are more resistant than alumina mortars to such hard substances as alumina and sillimanite. Thus $\frac{1}{16}$ in. grains of alumina ground for 15 min in boron carbide contained 0.18% boron carbide, representing about one-third of the abrasion produced by similar grinding in an alumina mortar. A boron *nitride* mortar and pestle has been shown to contaminate a ground sample of silicon carbide with ~120 ppm Fe, 780 ppm Co, 7 ppm Mn, as well as other elements (L. F. Lowe, H. D. Thompson, and J. P. Cali, *Anal. Chem.*, **31,** 1951 (1959)).

12. K. Rankama, *Bull. Comm. Géol. Finlande*, No. 133, p. 11 (1944).

13. G. Thompson and D. C. Bankston, *Appl. Spectrosc.*, **24,** 210 (1970), have determined trace element contamination produced by grinding samples of "Specpure" silica and calcium carbonate in alumina, boron carbide, and agate mortars and in WC, alumina, and lucite vial-and-ball. A WC grinding vial introduced much Co and Ti. An alumina mortar introduced Al, Cr, Fe, Ga, and Zr. An alumina–ceramic vial and ball resulted in contamination by Al, Fe, Cu, Ga, Li, Ti, B, Ba, Co, Mn, Zn, and Zr. There was little or no contamination from a boron carbide mortar (except from B) and an agate mortar or from nylon sieves. Metal (brass and stainless steel) sieves led to

serious contamination as could have been predicted. Cross contamination from previously ground material can be significant.

14. H. Bloom and P. R. Barnett, *Anal. Chem.*, **27**, 1037 (1955). Contamination from crushing with a steel buckboard and muller was also studied by these workers; much iron and its accompanying elements were found to be introduced. Also see P. R. Barnett, W. P. Huleatt, L. F. Rader, and A. T. Myers, *Am. J. Sci.*, **253**, 121 (1955). In this connection, it may be mentioned that surface contamination can result from material previously crushed in a steel mortar. Soft substances can leave smears difficult to remove. If the history of a mortar is unknown, it is advisable to crush quartz crystal in it and run a blank for the element being determined. It is well to reserve a mortar for a given type of sample; for example, a mortar used to crush stony meteorites should not be used for silicate rocks.

15. L. Lykken, F. A. Rogers, and W. L. Everson, *Ind. Eng. Chem. Anal. Ed.*, **17**, 116 (1945), constructed a small high-speed hammer mill having internal surfaces of nitrided steel. Contamination by iron is reported to be less than 0.01% for materials having a hardness less than 6.

16. H. Berry and R. Rudowski, *Geochim. Cosmochim. Acta*, **29**, 1367 (1965).

17. S. L. Hood, R. Q. Parks, and C. Hurwitz, *Ind. Eng. Chem. Anal. Ed.*, **16**, 202 (1944).

18. R. L. Mitchell, *J. Sci. Food Agr.*, **11**, 553 (1960), states that with a good mild-steel hammer-mill, the amount of iron introduced with material of the texture of pasture herbage is no more than 1–2 ppm in a total of 50–100 ppm. In general trace work, scoops, trays, and screws of brass or copper should be replaced by Al, steel, or plastic, and the material used for bearings should be checked. Molybdenum sulfide lubricants should be guarded against. A. W. Palm and R. S. Beckwith, *Anal. Chem.*, **28**, 1637 (1956), have modified a Christy–Norris mill by chromium-plating the beater and plastic-coating the interior to eliminate introduction of iron. Attention is drawn to the possibility of contamination by dusty air drawn through the mill at high speed.

19. W. J. A. Steyn, *Agr. Food Chem.*, **7**, 344 (1959), using a ball mill of this type, found that dried citrus leaves ground therein 2 hr showed the same Fe, Mn, Cu, and Zn contents as a sample that was hand ground with an agate mortar and pestle.

20. A. E. Kretschmer and J. W. Randolph, *Anal. Chem.*, **26**, 1862 (1954).

21. A. Purr, *Biochem. J.*, **28**, 1907 (1934).

22. J. S. Hislop and A. Parker, *Analyst*, **98**, 694 (1973).

23. A useful tabulation of solvents and fluxes for ores, refractories, minerals, and metals may be found in G. E. F. Lundell and J. I. Hoffman, *Outlines of Methods of Chemical Analysis*, Wiley, New York, 1938. Also see J. Dolezal, P. Povondra, and Z. Sulcek (transl. by D. O. Hughes, P. A. Floyd, and M. S. Barrett), *Decomposition Techniques in Inorganic Analysis*, Iliffe, London, 1968; R. Bock, *Aufschlussmethoden der anorganischen und organischen*

Chemie, Verlag Chemie, Weinheim/Bergstr., 1972. Recent review: H. Flaschka and G. Meyers, *Z. Anal. Chem.*, **274**, 279 (1975). The rationale of the decomposition of inorganic materials is discussed by H. H. Willard and C. L. Rulfs in *Treatise on Analytical Chemistry*, Vol. 2 of Part I, Interscience, New York, 1961, pp. 1027–1050. See J. A. Maxwell, *Rock and Mineral Analysis*, Interscience, New York, 1968, p. 88, for the decomposition of silicates. G. Tölg, *Talanta*, **19**, 1489 (1972), tabulates acid decomposition methods for various high-purity inorganic materials. (Contamination problems are also considered.) Occasionally, decomposition in a vapor phase is of interest, as of silicates in HF, for which a Teflon apparatus is described by J. W. Mitchell and D. L. Nash, *Anal. Chem.*, **46**, 327 (1974), and J. F. Woolley, *Analyst*, **100**, 896 (1975). C. Feldman, *Anal. Chem.*, **49**, 825 (1977) describes the decomposition of siliceous materials by $HF-HNO_3$ in the vapor phase at 70°C for the determination of As. The blank is reduced by this decomposition method.

24. R. Bock and Syed Laik Ali, *Z. Anal. Chem.*, **258**, 12 (1972), found that fusion of the following compounds (0.1 g) with Na_2CO_3 (1 g) resulted in losses (%) as follows: HgO and Tl_2O, 100; $As_2O_3 > 20$; $As_2O_5 \sim 0.2$. In_2O_3, Sb_2O_3, and Sb_2O_5 showed very small losses.

25. For example, zircon, tourmaline, garnet, pyrite, and other minerals less commonly encountered may be left after $HF-HClO_4$ treatment.

26. Another possible source of contamination is the muffle furnace when used for fusions or destruction of organic matter. Substances can be volatilized from the ceramic, the heating element or from material that has been spilled in the furnace and can then be taken up by the charge in a crucible (Chapter 2, p. 45). R. J. Hall, *Analyst*, **88**, 76 (1963), mentions that the inside of a covered Pt crucible became contaminated with several micrograms offluorine after heating in a muffle furnace at 400°C for a few hours.

27. Because of the poor heat conductivity of Teflon, decomposition in a Teflon beaker placed on the top of a steam bath is not feasible. The beaker or other vessel needs to be placed in the steam. L. Shapiro, *Chem.-Anal.*, **48**, 46 (1959), has described the design of a Teflon vessel, having a tightly fitting cover, for $HF-H_2SO_4$ decompositions. The removal of the last amounts of fluoride by fuming with sulfuric acid after most of the hydrofluoric acid has been evaporated is best done by transferring the decomposed sample with a minimum of water to a Vycor or silica beaker and heating on a hot plate. Any small amount of organic matter present can be destroyed by evaporation with nitric acid.

28. J. Dolezal, J. Lenz, and Z. Sulcek, *Anal. Chim. Acta*, **47**, 517 (1969). Even α-Al_2O_3, corundum, can be brought into solution (H_2SO_4, $2 + 1$, 3 hr, 240°). Bombs for HF decomposition: F. J. Langmyrh and P. E. Paus, *ibid.*, **49**, 358 (1970); P. Knoop, *Anal. Chem.*, **46**, 965 (1974).

29. B. Bernas, *Anal. Chem.*, **40**, 1682 (1968); Y. Hendel, *Analyst*, **98**, 450 (1973).

30. E. Wichers, G. W. Schlecht, and C. L. Gordon, *J. Res. Nat. Bur. Stand.*, **33**, 363 (1944).

31. W. J. French and S. J. Adams, *Anal. Chim. Acta*, **62**, 324 (1973).

32. P. Pakalns, *Anal. Chim. Acta*, **69**, 211 (1974).

33. J. B. Headridge and A. Sowerbutts, *Analyst*, **98**, 57 (1973) (determination of Al); A. Ashton, A. G. Fogg, and D. T. Burns, *ibid.*, **99**, 108 (1974) (Zr).

34. See the monograph by T. T. Gorsuch, *The Destruction of Organic Matter*, Pergamon, Oxford, 1970 (151 pages). A literature survey may be found in O. G. Koch and G. A. Koch-Dedic, *Handbuch der Spurenanalyse*, Vol. 1, Springer, Berlin, 1974, p. 183 et seq. G. Middleton and R. E. Stuckey, *Analyst*, **78**, 532 (1953), critically reviewed earlier methods.

35. For example, organic samples that are liquids immiscible with water can at times be shaken with a more or less strongly acid solution to transfer a trace metal to the aqueous phase in ionic form (extraction of lead from naphtha with hot hydrochloric acid). Zinc in urine can be determined by direct extraction with dithizone in CCl_4, as shown by J. H. R. Kägi and B. L. Vallee, *Anal. Chem.*, **30**, 1951 (1958). Iron and copper in serum can be determined after precipitation of protein with trichloroacetic acid. Methods of this kind require careful testing for reliability.

36. Much gold is volatilized in the decomposition of organic matter (filter paper) with HNO_3–$HClO_4$, as shown by H. A. Moore and R. A. Wessman, *Anal. Chem.*, **44**, 2398 (1972). No tin is lost in H_2SO_4–HNO_3 decomposition.

37. T. M. Florence, Y. J. Farrar, L. S. Dale, and G. E. Batley, *Anal. Chem.*, **46**, 1874 (1974).

38. A. M. Hartstein, R. W. Freedman, and D. W. Platter, *Anal. Chem.*, **45**, 611 (1973). Coal (50 mg) is decomposed with fuming HNO_3–HF at 150°. The Bernas bomb (p. 112) is used. A similar procedure for biomaterial in general, with HNO_3 or HNO_3–HF, was described by L. Kotz, G. Kaiser, P. Tschöpel, and G. Tölg, *Z. Anal. Chem.*, **260**, 207 (1972). Hg, Se, I, and Be recoveries were >98%. Bombs should not be used for H_2SO_4–HNO_3 decomposition of organic materials, especially not fatty substances. Explosions can result (formation of nitroglycerin and other nitrated products). See *Chem. Eng. News*, July 30, 1973, p. 32.

39. W. J. Adrian, *Analyst*, **98**, 213 (1973). In general, sample size is limited to 1 g in a 4 oz bottle. The test material was a sage, *Artemisia tridentata*.

40. G. Denbsky, *Z. Anal. Chem.*, **267**, 350 (1973).

41. M. M. Shemyakin and L. A. Shchukina, *Quart. Rev.*, **10**, 261 (1956).

42. P. Kivalo, A. Visapää, and R. Bäckman, *Anal. Chem.*, **46**, 1814 (1974). For extraction of Hg from sediments and soils containing HgS, see H. Agemian and A. S. Y. Chau, *Analyst*, **101**, 91 (1976). They digest with HNO_3–H_2SO_4 and a small amount of HCl, followed by treatment with permanganate and persulfate. M. R. Hendzel and D. M. Jamieson, *Anal. Chem.*, **48**, 926 (1976), find that fish samples can be satisfactorily decomposed for Hg determination by HNO_3–H_2SO_4 digestion at 180°C.

43. G. Middleton and R. E. Stuckey, *Analyst*, **79**, 138 (1954). Gorsuch assesses

this method as follows "[It] gave reasonable results for most of the trace elements, although the recoveries obtained were often a little lower than those by wet-oxidation methods. Selenium and mercury were lost completely, and in use it proved somewhat tedious and time consuming."

44. Explosions have occurred in the decomposition of organic matter with HNO_3 + metal nitrate. Their likelihood increases in the order Mg, Ca, and Na nitrate, as shown by T. Greweling, *Anal. Chem.*, **41,** 540 (1969). These explosions occurred in the range 300°–470°C.

45. Compare the procedure for decomposing organic matter on p. 435 of Sandell, *Colorimetric Determination of Traces of Metals*, 3rd ed.

46. These procedures are taken from a report by the Analytical Methods Committee, *Analyst*, **85,** 643 (1960). A later report, *Analyst*, **100,** 54 (1975), deals with determination of As in organic matter after sulfuric–nitric acid digestion, followed by distillation of As as the bromide.

47. A modified Kjeldahl flask having an extension to the neck by means of a standard ground joint is illustrated in the reference 46. This modification increases refluxing of the acids. Also see p. 128.

48. Directions are given by Gorsuch, *The Destruction of Organic Matter.*

49. H. Ssekaalo, *Lab. Prac.*, **19,** 720 (1970), describes and illustrates a unit in which 12 250 ml conical flasks provided with Quick-fit reflux condensers are used. This simple arrangement condenses and returns volatile acids to the digest and reduces the oxidation time by 50%. Digestion with H_2SO_4–HNO_3–$HClO_4$ is described.

50. O. I. Milner, *Analysis of Petroleum for Trace Elements*, Pergamon, Oxford, 1963, p. 21. Decomposition for determination of Hg: H. E. Knauer and G. E. Milliman, *Anal. Chem.*, **47,** 1263 (1975).

51. Analytical Methods Committee, *Analyst*, **92,** 403 (1967). A second report of this committee [*Analyst*, **101,** 62 (1976)], appearing too late for detailed consideration here, deals with destruction of fats, oils, foods with high-fat contents, syrups, and fruit concentrates. Potential hazards of peroxide decomposition are pointed out.

52. Hydrogen peroxide in combination with ferrous iron has been recommended for the destruction of certain types of organic material by B. Sansoni and W. Kracke, *Z. Anal. Chem.*, **243,** 209 (1968). The active oxidizing agent is the hydroxyl free radical:

$$Fe^{2+} + H_2O_2 \rightarrow Fe^{3+} + OH^- + \cdot OH$$

The sample is heated with 20–30% H_2O_2 and enough $FeSO_4$ to give an iron concentration of 0.001–0.01 M. The mixture is evaporated to a small volume at 100°C, and the residue is further treated with 30% H_2O_2. The oxidation is effected under mild conditions (relatively low temperatures), and the danger of loss of metals is minimized. No undesirable reduction products remain in the solution, and the oxidation is relatively rapid. However, the presence of relatively much iron in the final solution may be objectionable and traces of heavy metals may be introduced with the ferrous sulfate. Oils and fats cannot be destroyed in this way and must be removed by extraction with trichloroethylene or in some other way. Samples that can be wet oxidized by this procedure include meat, farinaceous materials and ion-exchange resins. See, further, R. Berner and H. R. Baechli, *Mitt. Geb. Lebensmitt. Hyg.*, **65**, 229 (1974).

53. C. Whalley in P. W. West, A. M. G. Macdonald, and T. S. West, Eds., *Analytical Chemistry 1962*, Elsevier, Amsterdam, 1963, p. 397. I. L. Marr, *Talanta*, **22**, 387 (1975), successfully decomposed organotin compounds and PVC plastics with $H_2SO_4 + H_2O_2$.

54. J. L. Down and T. T. Gorsuch, *Analyst*, **92**, 398 (1967).

55. Adrian[39] reports low results for Cu, Fe, and Zn by H_2SO_4–H_2O_2 digestion in his pressure method (atomic absorption determination).

56. The special digestion flask illustrated on p. 121 can be used to advantage. Directions are given in reference 54.

57. Polypropylene and Perspex (granules) are hard to decompose, and butyl rubber is not completely decomposed after 60 min. Carbon black and Teflon are not attacked. Liquid paraffin reacts so vigorously that its oxidation is considered dangerous. See, further, R. P. Taubinger and J. R. Wilson, *Analyst*, **90**, 429 (1965).

58. With certain oils and balsams, e.g., oil of copaiba balsam and Peru balsam, 50% H_2O_2 is reported to react too vigorously for safe and accurate work.

59. G. F. Smith, *Anal. Chim. Acta*, **8**, 397 (1953); **17**, 175 (1957). Perchloric–nitric acid decomposition has been further examined by G. D. Martinie and A. A. Schilt, *Anal. Chem.*, **48**, 70 (1976). They found that "the following reacted violently with explosive force or with sparks and flame production" (mainly in the vapor phase): cottonseed oil, lanolin, lecithin, cholesterol, squalene, Tygon, latex rubber, Amberlite XAD-2, and others. The following substances, among others, undergo vigorous but not pyrophoric oxidation: Na tetraphenylborate, tetraphenylarsonium chloride, 2-hydroxyquinoline, Amberlite CG-120, quinoxaline, and anthranilic acid. These are mostly fatty materials or those not miscible with the acid mixture. About half of the 85 substances tested left some organic matter after digestion. Compounds having *N*-methyl, *S*-methyl, *C*-methyl, and pyridyl moieties were resistant to perchloric–nitric acid oxidation. Such substances can be completely destroyed with nitric–sulfuric–perchloric acid, with catalysts if need be. For destruction of animal tissue with HNO_3–$HClO_4$ preliminary to determination of Ag by atomic absorption, see R. C. Rooney, *Analyst*, **100**, 471 (1975).

60. H. Diehl and G. F. Smith, *Talanta*, **2**, 209 (1959).

61. For precautions in the use of perchloric acid, see Analytical Methods Committee, *Analyst*, **84**, 214 (1959).

62. Gorsuch,[34] p. 22.

63. A. D. Thomas and L. E. Smythe, *Talanta*, **20**, 469 (1973).

64. C. Feldman, *Anal. Chem.*, **46,** 1606 (1974).

65. C. S. Piper, *Soil and Plant Analysis,* Interscience, New York, 1947, p. 272. The directions given are for plant material but can be extended without much change to most other biomaterials. Destruction of polyethylene is described by W. T. Bolleter, *Anal. Chem.*, **31,** 201 (1959).

66. If the sample consists of seeds rich in fats and oils, a preliminary oxidation with sulfuric–nitric acid alone is advisable.

67. An exception occurs in some methods for decomposing petroleum, in which the sample is burned.

68. T. T. Gorsuch, *Analyst,* **87,** 112 (1962).

69. According to P. F. Reay, *Anal. Chim. Acta,* **72,** 145 (1974), dry ashing of plant material with $NaHCO_3$ plus Ag_2O at 450°C in Pyrex retains As without loss.

70. According to T. Y. Kometani, *Anal. Chem.*, **38,** 1596 (1966), 20 μg quantities of NaCl, $NaNO_3$, and NaF showed volatilization losses when heated in platinum above 500°. Chlorides and nitrates of Rb, Cs, and Li may be expected to behave similarly. Apparently, $CaCl_2$ in small amounts can volatilize significantly above 700°C, $MgCl_2$ above 600°. Sodium sulfate does not volatilize below 700°. Volatilization losses may be expected to be relatively greater for small amounts of substances than for larger, other factors remaining the same, because of the greater surface/weight ratio for the former.

71. C. Blincoe, *Anal. Chem.*, **34,** 715 (1962). Likewise, R. M. Green and R. J. Finn, *Anal. Chem.*, **36,** 692 (1964), found loss of Cs-137 at temperatures above 550°, but none after 20 hr at 450°. But G. C. Goode, C. M. Howard, A. R. Wilson, and V. Parsons, *Anal. Chim. Acta,* **58,** 367 (1972), found Na to begin to volatilize at 500° and to be 50% lost after 2 hr at 600°. Dry ashing of animal tissues at 550° was found to give no loss of Na or K by E. L. Grove, R. A. Jones, and W. Matthews, *Anal. Biochem.,* **2,** 221 (1961). Sr was found to volatilize slightly in ashing of blood at 450° by E. I. Hamilton, M. J. Minski, and J. J. Cleary, *Analyst,* **92,** 257 (1967). J. J. Cleary and E. I. Hamilton, *ibid.,* **93,** 235 (1968), recovered Po-210 incompletely from rat tissue on ashing at temperatures as low as 200–300°.

72. Losses of Ni, V, Cu, and Fe can be prevented by adding concentrated sulfuric acid to the petroleum sample before carbonizing, as by heating with infrared lamp and hot plate: O. I. Milner, J. R. Glass, J. P. Kirchner, and A. N. Yurick, *Anal. Chem.*, **24,** 1728 (1952); J. T. Horeczy, B. N. Hill, A. E. Walters, H. G. Schutze, and W. H. Bonner, *ibid.,* **27,** 1899 (1955). Ashing is completed in a muffle at ~500°. Metal losses can also be prevented by adding benzene sulfonic acid or sulfur before burning and ashing: J. E. Shott, Jr., T. J. Garland, and R. O. Clark, *ibid.,* **33,** 506 (1961); E. J. Agazzi, D. C. Burtner, D. J. Crittenden, and D. R. Patterson, *ibid.,* **35,** 332 (1963).

73. P. Strohal, S. Lulic, and O. Jelisavcic, *Analyst,* **94,** 678 (1969), allowed a mollusc, *Mytillus galloprovincialis Lam.*, to take up radioisotopes of Ce, Co, Mn, Pa, Ru, and Zn from sea water. The soft tissues were then heated at

various temperatures and the loss in radioactivity was determined. Even after heating at 110°, losses of 9–14% were found, except for Ce. At 250°, losses ranged from 10 to 21% and at 450°, from 15 to 33%. Presumably, the losses arise from the volatilization of chelate compounds. Wet digestion with H_2O_2 as oxidizing agent according to Down and Gorsuch gave recoveries of 97–101%. Contrarily, J. G. van Raaphorst, A. W. van Weers, and H. M. Haremaker, *Analyst*, **99,** 523 (1974), found no significant loss of Co or Zn by volatilization in the dry ashing of marine mussels and brown seaweed in porcelain up to 1000°. Leaching with dilute HCl quantitatively removed both metals from the crucibles after ashing at 450° and 550°. Cobalt could not be completely recovered by HCl treatment of residue from dry ashing of mussels at 1000°, but was recovered within ~2% from seaweed residue left at 1000°. Results obtained by G. A. Knauer, *Analyst*, **95,** 476 (1970), in analyzing shrimp for Mg, Mn, Fe, Cu, and Zn indicate that ashing in a muffle furnace at 550° for 16 hr in porcelain crucibles is satisfactory. Hydrochloric acid (1:1), nitric acid (1:1), and aqua regia were equally effective in dissolving the metals from the residue (atomic absorption determination). H. O. Fourie and M. Peisach, *Analyst*, **102,** 193 (1977), found serious losses of Se, Cd, and Pb on freeze drying and oven drying of oysters at low temperatures, whereas Cr, Fe, Mn, Co, and Zn probably are not lost.

74. M. D. Webber, *Can. J. Soil Sci.*, **52,** 282 (1972), found that in ashing of plant tissues, Pb was lost above 430°. Ashing at 300–430° gave complete recovery of Pb, but much carbon remained. Ashing at 430° (with Na_2CO_3 fusion) gave results agreeing closely with those obtained by wet oxidation.

75. T. T. Gorsuch, *Analyst*, **84,** 135 (1959).

76. W. Oelschläger, *Z. Anal. Chem.*, **246,** 376 (1969).

77. E. F. Dalton and A. J. Malanoski, *J. Assoc. Off. Anal. Chem.*, **52,** 1035 (1969), obtained seriously low results for Cu and Pb when old Vycor crucibles were used.

78. G. M. George et al., *J. Assoc. Off. Anal. Chem.*, **56,** 793 (1973), obtained good recoveries of As from animal tissues by heating with MgO and cellulose powder and then with $Mg(NO_3)_2$.

79. Occasionally another acid such as HNO_3 may be needed, as when a trace element may be present in metallic form (copper). Certain elements, Ti for example, may require $K_2S_2O_7$ fusion of the ash.

80. If a platinum dish is used and nitric acid was previously added, transfer the ash to a beaker before HCl treatment to avoid possible liberation of chlorine and attack of the platinum.

81. Reference 46, p. 654.

82. Piper, *Soil and Plant Analysis*, p. 268.

83. C. E. Gleit and W. D. Holland, *Anal. Chem.*, **34,** 1454 (1962); C. E. Gleit, *ibid.*, **37,** 314 (1965) and *Microchem. J.*, **10,** 7 (1966); J. R. Hollahan, *J. Chem. Educ.*, **43,** A401 (1966); J. Fabry and P. Nangniot, *Analusis*, **1,** 117 (1972).

84. R. Coomber and J. R. Webb, *Analyst*, **95,** 668 (1970), determining sodium in acrylic fiber, found better reproducibility by this method than by dry ashing or wet oxidation (H_2SO_4–HNO_3 and H_2SO_4–H_2O_2). Fabry and Nangniot, *Analusis*, **1,** 117 (1972), obtained good recoveries of Cu and Co in grass. Unsatisfactory recoveries of As in atmospheric particulate samples were reported by P. R. Walsh and coworkers, *Anal. Chem.*, **48,** 1012 (1976).

85. G. Kaiser, P. Tschöpel and G. Tölg, *Z. Anal. Chem.*, **253,** 177 (1971).

86. S. Fujiwara and H. Narasaki, *Anal. Chem.*, **40,** 2031 (1968). In an earlier application of this method, R. Anduze, *ibid.*, **29,** 90 (1957), decomposed polyethylene for the determination of Ti.

87. M. Fujita, Y. Takeda, T. Terao, O. Hoshino, and T. Ukita, *Anal. Chem.*, **40,** 2042 (1968).

88. In another bomb method, a 1.8 liter bomb is used and Hg-203 is added to correct the recovery. A 10 g sample (alfalfa) is burned at an oxygen pressure of 17 atm. Recoveries of added inorganic and organic Hg ranged from 78 to 91%. See E. W. Bretthauer, A. A. Moghissi, S. S. Snyder, and N. W. Mathews, *Anal. Chem.*, **46,** 445 (1974).

89. B. Morsches and G. Tölg, *Z. Anal. Chem.*, **219,** 61 (1966).

90. H. J. M. Bowen, *Anal. Chem.*, **40,** 969 (1968). Teflon can be decomposed: S. Sherken and E. J. Friedman, *ibid.*, **45,** 2399 (1973).

91. W. T. Bolleter, *Anal. Chem.*, **31,** 201 (1959).

ABSORPTIOMETRY AND FLUORIMETRY IN INORGANIC TRACE ANALYSIS

Absorptiometry in the title above refers to spectrophotometry and colorimetry in solution, the absorbing species being molecules and ions.[1] The spectral range for inorganic trace analysis is predominantly the visible and near ultraviolet. The reader is assumed to have a general knowledge of spectrophotometers (and filter photometers)[2] and of absorption spectrophotometry.[3]

I. SPECTROPHOTOMETRY

A. GENERAL CONSIDERATIONS

The trace analyst should have available a photoelectric spectrophotometer of good precision that can be used in the range 200–900 nm. The advantage of a spectrophotometer over a filter photometer lies in the choice of wavelengths and the narrow band of wavelengths possible with the former. The precision of a spectrophotometer may be no better—it may be considerably worse—than that of a filter photometer. Other factors remaining the same, a spectrophotometer furnishes better sensitivity than a filter photometer, but the difference may be small. Most chromogenic species that are made the basis of absorptiometric methods have broad absorption bands, the half-band width being 50–100 nm or more. The absorbance curves of permanganate ion (Fig. 4-1 *a*) and of metal dithizonates in organic solvents (Figs. 6G-8 to 6G-12) illustrate the curves of such substances. Filters having half-band widths of 50–100 nm give sensitivities only slightly inferior to those of a spectrophotometer of 1-nm band width, if centered close to the absorbance peak of these substances. On the other hand, species having narrow absorption bands, for example rare earth salts (Fig. 4-1*b*), give poor sensitivity with filters.[4]

Filter photometers, in contrast to spectrophotometers, may show apparent deviations from Beer's law because of the heterochromatic radiation, especially when the transmission peak of a filter lies at the side of the absorption band of the substance whose absorbance is being measured. The absorbance–concentration curve then deviates from linearity, tending to become horizontal with the concentration axis at higher concentrations. This effect is more likely to be observed at the UV edge of the visible spectrum than inside the visible. At low concentrations of colored species, such as often worked with in trace analysis, the effect may not be perceptible, and even at higher concentrations it may be insignificant.

When two absorbing species are to be determined simultaneously in a solution by measurement of absorbance at two suitable wavelengths, a

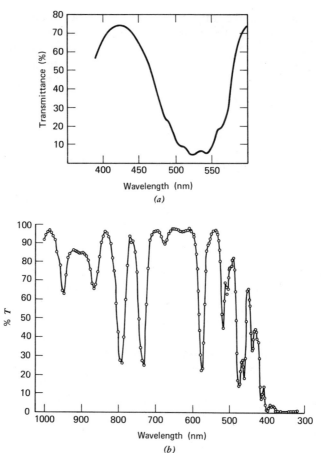

Fig. 4-1. *a.* A portion of the transmission curve ($100I/I_0$ versus λ) of potassium permanganate in water ($\sim 0.002\%$ Mn, 1.3 cm cell, band width of light $= 5$ nm). Broad absorption bands such as this are characteristic of most absorptiometric determination forms of metals.

b. Transmission curve of samarium chloride solution. (According to W. C. Miller, G. Hare, D. C. Strain, K. P. George, M. E. Stickney, and A. O. Beckman, *J. Opt. Soc. Am.*, **39,** 377 (1949).) Measurement made with Fery prism spectrophotometer. Few absorptiometric determination forms have such narrow absorption bands.

spectrophotometer is practically a necessity. Likewise, a spectrophotometer may be superior to a filter photometer when the reagent is colored or foreign absorbing substances are present.

In many trace determinations, a filter photometer having a transmission scale graduated to 0.1% would be chosen in preference to a spectrophotometer graduated to 1%.

In most inorganic trace determinations, final volumes are relatively large, say from 5 to 25 ml. For the smaller volumes, absorption cells of 1 or 2 cm light path are usually satisfactory, but longer paths may be needed for larger volumes to provide adequate sensitivity. The 1 or 2 cm cells should not require more than 5 ml of solution for measurement. For larger volumes of solution, cells of greater stratum thickness (and a photometer to accommodate them) are desirable, such that 1 (or 2) cm corresponds to 5 or 10 ml of solution. A great many cells of microsize have been described, and some are commercially available, but they find more use in microanalysis than in trace analysis.[5] In most trace analyses, the final solution volume is not so small that microcells are required. However, the absorbance of ~1 ml of solution may need to be measured, as in an immiscible solvent extraction involving a relatively large volume of aqueous phase and a small volume of the extractant.

B. PRECISION AND ACCURACY

1. Precision in Transmission and Absorbance Measurements

The analyst applying absorption methods for the determination of constituents in solution is ultimately concerned with the precision and accuracy of a *concentration* determination, and these depend in part on the precision and accuracy of transmission (or absorbance) measurements. For a single absorbing species, the Lambert–Beer law holds for a narrow band of wavelengths ("monochromatic" light):

$$A = \log \frac{I_0}{I} = abc \qquad (4\text{-}1)$$

where A = absorbance.

I_0 = intensity (radiant power) of radiation incident perpendicularly on solution of absorbing species in cell having plane parallel walls.

I = intensity of radiation leaving solution (absorption and reflection of radiation by glass is essentially cancelled out in the usual way of making the measurements so that A is due only to absorbing species).

a = absorptivity (depends on wavelength λ, temperature, and solvent) of a given species.

b = light path in centimeters.

c = concentration of (single) absorbing species; when c is expressed in moles per liter, a becomes ϵ, the molar absorptivity (or molar extinction coefficient), whose units are liters mole^{-1} cm^{-1} or 1000 cm^2 mole^{-1}.

(Beer's law refers to the relation between A and c at constant b.)

In the following discussion, it is assumed that deviations (indeterminate errors) in I_0 and I, and therewith in A, are due to the photometer and possibly to the observer (in reading the scale, e.g., in estimating tenths of a division).

A meter spectrophotometer or a filter photometer may have two scales, one graduated in absorbance, the other in percentage transmission. Because the latter scale is divided into equidistant divisions (usually at 1% or, less often, at 0.1% intervals), it can be read more precisely and is preferably used in the most exact work. First of all, in a series of measurements the photometer is adjusted so that when no light strikes the photoelectric cell, the pointer reads exactly 0% (dark current). Then, with a single beam photometer, the instrument is adjusted so that the light beam (of specified λ) passing through the working cell containing water or a blank solution gives a reading of 100.0 on the transmission scale ($I_0 = 100.0\%$, $A_0 = 0.000$). Finally the solution whose transmittance, I/I_0 or $100I/I_0$ in percent, is desired is transferred to the cell, and the I reading is obtained. Because of possible drift in readings, these are usually obtained in pairs, I_0 and I (corresponding to A_0 and A). From a series of N pairs of readings, estimates of the absolute standard deviation of the (percentage) transmittance and absorbance are found:

$$s_{I/I_0} = \left\{ \frac{\sum (I/I_0 - \overline{I/I_0})^2}{N-1} \right\}^{1/2} \times 100 \tag{4-2}$$

$$s_A = \left\{ \frac{\sum (A - \bar{A})^2}{N-1} \right\}^{1/2} = \left\{ \frac{\sum (\log I_0/I - \overline{\log I_0/I})^2}{N-1} \right\}^{1/2} \tag{4-3}$$

When estimates can be obtained of the standard deviation of the readings at the ends of the I interval ($I_0 = 100$ and I):

$$s_{I/I_0} = (s_I^2 + s_{I,100}^2)^{1/2}$$

and

$$s_A = \left\{ \left(\frac{s_I \times 0.434}{I} \right)^2 + \left(\frac{s_{I,100} \times 0.434}{100} \right)^2 \right\}^{1/2} \tag{4-4}$$

s_{I/I_0}, and therewith s_A, can depend on many factors, one or two of which may be limiting over a certain range of conditions: various types of noise in the photoelectric system[6]; scale reading error (estimation of tenth divisions, parallax, resolvability); and possibly irreproducibility of placement of absorption cell. s_{I/I_0} can vary with I but sometimes, in some instruments, is determined by the scale reading error (including resolvability of the readout device) and then may be constant, or approximately so, over the whole I scale. When s_{I/I_0} is constant, s_A necessarily varies with I.

The relative standard deviation of an absorptiometric concentration determination of an element is given by the relative standard deviation of an absorbance measurement,[7] namely

$$s_{A,\%} = \frac{100 s_A}{A}$$

where s_A is the absolute standard deviation of a (single) A measurement. It may be surmised that $s_{A,\%}$ will be larger at low A values than at higher values. In fact, it is found by error analysis[6] as well as experimentally that usually there is an optimal value or range of values of A at which $s_{A,\%}$ has a minimal value (see Fig. 4-2). The minima depend on instrumental

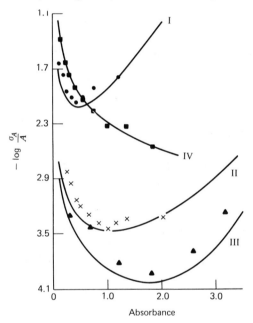

Fig. 4-2. Theoretical and experimental relative standard deviations in spectrophotometry as determined by limiting instrumental parameters [L. D. Rothman, S. R. Crouch, and J. D. Ingle, *Anal. Chem.*, **47**, 1226 (1975)]. Curves give theoretical calculated values. $K_2Cr_2O_7$ solutions. I, precision limited by dark current noise. Represents low light levels and narrow slit widths. Typically observed with limited readout resolution (most common cause), with excessive amplifier or dark current noise or very low photocurrent. II, precision limited by photocurrent shot noise. Photocurrent $\gg$ dark current. Source flicker noise contributes. III, precision limited by source flicker noise. At low transmittances, dark current noise and photocurrent shot noise become increasingly significant and curve passes through minimum. IV, imprecision due to cell positioning (for particular spectrophotometer used). Reprinted with permission from L. D. Rothman, S. R. Crouch, and J. D. Ingle, Jr., *Anal. Chem.*, **47**, 1231 (1975). Copyright by the American Chemical Society.

TABLE 4–1
Precision of a Filter Photometer Having a Transmission Scale Graduated in Tenths of a Percent^a

Solution	Filter	I (%) ($I_0 = 100$)	A (rounded)	s_{I/I_0} (absol. %)	s_A (absol.)	$s_{A,\%}$
H_2O	S53	100	0	0.01	<0.0001	
K_2CrO_4	S38E	97.4	0.01	0.02	0.0001	1
$CuSO_4$	S72E	93.1	0.03	0.02	0.0001	0.3
		71.3	0.15	0.02	0.0001	0.07
		50.0	0.30	0.01	<0.0001	<0.03
		4.0	1.40	0.02	0.0021	0.15
$KMnO_4$	S53E	97.1	0.01	0.02	0.0001	1
		57.8	0.24	0.01	<0.0001	<0.04
		5.0	1.30	0.02	0.0017	0.13

^{a} Zeiss Elko II filter photometer; measurements by R. S. Alberg, University of Minnesota. The half-band width of the filters is ~20 nm.

variables (mentioned in the legend of this figure) and occur at relatively high A. These minima are of secondary interest to the trace analyst because he generally works with solutions too dilute to come near them. Besides, it may not be necessary for him to obtain the highest photometric precision. At trace concentrations, the permissible relative deviation is greater than when higher concentrations are determined.

An important case in practice is that in which $s_{A,\%}$ is largely determined by scale-reading error. Some filter photometers having transmission scales divided at 1% intervals are so reproducible that the error in scale reading by estimating tenths of a division is the predominant source of irreproducibility. Even with the less often encountered filter photometers graduated to 0.1% transmission (allowing estimation to 0.01%), noise in the photoelectric system may contribute less than does the reading error to $s_{A,\%}$. The precision of one such photometer was tested with three filters for I values ranging from 97 to 4% (A ranging from 0.01 to 1.4) with the results shown in Table 4-1. The estimated standard deviation in transmittance expressed in percent ($I/I_0 \times 100$, $I_0 = 100$) was ~0.02 and did not change significantly with I or with the filters used (Table 4-1). Because an I measurement requires two readings, the estimated standard deviation of a single I reading in this series is ~0.02/1.4~0.014% or a little more than 0.1 division (1 division = 0.1%). This is hardly more than the scale-reading error. Our experience with several filter photometers graduated to 1% transmission (instead of 0.1%) indicates that with them a standard deviation not exceeding 0.2% I absolute is obtainable. Of course, this is not a generalization applying to all filter photometers or spectrophotometers for that matter.

Suppose that the (estimated) absolute standard deviation of an I scale *reading* or pointer setting[8] is ± 0.1 division or $\pm 0.1\%$ (scale graduated at 1% intervals from 0 to 100%). (It is likely to be a little less than this for an experienced observer with most photometer scales.) The absolute standard deviation of the absorbance as a function of I then is

$$s_A = \left\{ \left(\frac{0.1 \times 0.434}{I}\right)^2 + \left(\frac{0.1 \times 0.434}{100}\right)^2 \right\}^{1/2} \qquad \text{[from Eq. (4-4)]} \quad (4\text{-}5)$$

Values of s_A are plotted in Fig. 4-3, which also shows the relative standard deviation of A in percent, $s_{A,\%}$. The latter values are of particular interest to the analyst, because they are also the relative standard deviations of the concentration determination if the photometric deviations are the significant deviations. Table 4-2 lists values of s_A and

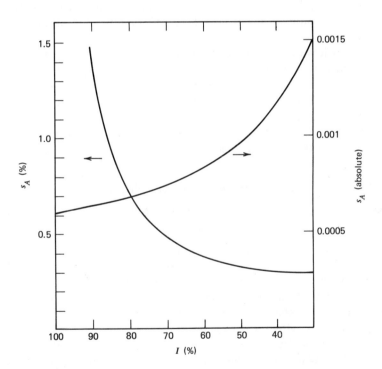

Fig. 4-3. Absolute standard deviation s_A and relative standard deviation in percent $s_{A,\%}$ of absorbance corresponding to a standard deviation of 0.1% absolute in a single transmission reading or 0.14% $(= 0.1 \times 2^{1/2})$ in transmittance.

TABLE 4-2

Absolute and Relative Standard Deviations of Absorbance Corresponding to Standard Deviation of 0.1% Absolute in $I_0 (= 100)$ and I

I (%)	A	s_A	$s_{A,\%}$
100	0	0.00061	
90	0.046	0.00065	1.4
80	0.097	0.00069	0.71
70	0.155	0.00076	0.49
60	0.222	0.00084	0.38
50	0.301	0.00097	0.32
40	0.398	0.00117	0.29
30	0.523	0.00150	0.287
20	0.699	0.00221	0.316

$s_{A,\%}$ plotted in Fig. 4-3. Even if the absolute standard deviation in a transmission reading is not 0.1%, the graph or table is useful for finding s_A or $s_{A,\%}$.

When I is less than 60 or so, the setting or reading error at $I_0 = 100$ can be disregarded in approximate calculations. For all practical purposes, down to $I \sim 50\%$, the approximation $100 s_{I/I_0}/(100 - I)$ can be used to obtain $s_{A,\%}$. Thus at $I = 50\%$, when $s_{I/I_0} = 0.14\%$, the approximation gives 0.28 instead of the accurate value 0.32%.

At times it may be convenient to use water in a separate cell as reference, and the following sequence of transmission readings might be made:

$$H_2O, \ I_0 = 100 \ (\text{reference cell}); \ I_{blank};$$

$$H_2O, \ I_0 = 100 \ (\text{reference cell}); \ I_{sample}$$

A considerable period of time could elapse between the first and second pair of readings and the water reference would be reset to 100 before obtaining I_{sample} (i.e., I of sample + reagent). The difference $A_{sample} - A_{blank}$, giving A due to the constituent converted to the absorbing species, requires four independent I readings. For relatively low absorbances, where the standard deviation of a single absorbance reading does not change much with I, s_I, or s_A, the standard deviation for four readings is $\sim 2/2^{1/2}$ or 1.4 times s_I or s_A for two readings. The loss in precision is about 30%. Water is frequently used in the reference cell when a standard curve is established (see p. 160 for a discussion of associated random errors).

Irreproducibility in placement of absorption cells in a photometer can result in appreciable indeterminate absorbance errors.[9] If the light beam traverses slightly different sections of the cell, variations in light intensity can occur because of variation in glass absorption and reflection. This may arise from heterogeneity in the glass, contamination of cell surfaces, and slight changes in the angle between cell faces and the light beam. The stratum thickness may vary somewhat from one point to another. Such errors can be eliminated by carefully emptying and filling a cell without removing it from its holder, but such a procedure may be troublesome. If the cell fits snugly in its holder and the cell faces traversed by the beam are not wetted so that wiping is not necessary, the placement error should be negligible.

Scale expansion or digital readout devices in modern spectrophotometers increase resolvability and lessen or eliminate scale-reading error. Noise in the photoelectric system then becomes the limiting instrumental factor and the relative precision is determined by the ratio noise/signal (or signal/noise in the usual phraseology).[6] Imprecision due to noise can be kept well below $0.001A$ or even below $0.0001A$ (see Table 4-1 and Figs. 4-2 and 4-6). Instrumental imprecision is unlikely to set a limit to precision obtainable in absorptiometric trace analysis unless available absorptiometers are of low quality.

The effect of noise on photometric precision depends on the type of detector used in a filter photometer or spectrophotometer. Evidently, the relative effect of noise increases as I decreases. For a photoconductive type of detector, the noise/signal ratio in terms of $\Delta C/C$ is given by

$$\frac{\Delta C}{C} = \frac{const.}{AI}$$

and for a photomultiplier detector (used in most spectrophotometers) by

$$\frac{\Delta C}{C} = \frac{const.'}{AI^{1/2}} \qquad \text{(photomultiplier shot noise)}$$

For the photoconductive detector, the minimal value of $\Delta C/C$ is obtained at $I = 37\%$, $A = 0.43$, but for the photomultiplier, the minimum falls at $I = 14\%$, $A = 0.85$. (Again, see Fig. 4-2.) But as already pointed out, such high A values are of no interest in trace analysis.

2. In Concentration Determinations

a. Precision

Conversion of an absorbance value to the concentration of the absorbing substance requires, at a minimum, the value of the absorptivity a, or its equivalent. Almost always the analyst establishes a standard curve for a particular constituent by taking known amounts of the constituent and converting it, under prescribed conditions, into the absorbing species contained in a known volume of solution. The absorbances of the series of solutions, determined at a fixed wavelength or with a suitable filter, are plotted versus the concentration (total or analytical concentration) to give the standard (calibration, analytical, or reference) curve. Ideally, and often in practice, the standard curve is a straight line.

When the absorbance of the standards is measured against water in the reference cell, the intercept β (see following discussion) will have an appreciable value if the reagent absorbs more or less at the wavelength used for the absorbing species of the constituent. If the measurements are made against a blank solution (containing reagent) in the reference cell, β will be small and could result from a slight difference in stratum thickness between reference and working cells, as well as from experimental error. Measuring the absorbance of a standard against water saves time. Otherwise, a blank solution should be prepared for each standard measured, unless it is known that the absorbance of the blank does not change with time.

The standard curve gives information other than the effective value of a:

i.   Whether a rectilinear relation exists between the analytical concentration C of the constituent and the absorbance, that is, whether the relation

$$A = \alpha C + \beta \qquad (\lambda,\ b,\ \text{solvent, and temperature constant}) \qquad (4\text{-}6)$$

is followed.[10] Adherence of A to this relation is frequently described as "Beer's law is followed."[11]

ii.   The precision of the measurements.

As soon as the constituent being determined photometrically must be converted to an absorbing species (or possibly a mixture of several species) a new set of determinate and indeterminate errors are added to the determination. The extent of formation of the colored species, or the relative amounts of more than one, may depend upon the concentration of the chromogenic reagent, the pH of the solution, and the time of color development. Moreover, fading may occur as a result of decomposition or by the action of light. Factors such as these, not always as easy to control as might be desired, rather than the precision of photometric measurements, may determine the precision and accuracy of the determination.

As already mentioned, an important reason for not obtaining a rectilinear relation between concentration and absorbance is the use of too broad a band of wavelengths, so that departure from rectilinearity is most likely to be encountered when filter photometers are used. However, departure from rectilinearity may also be the result of changes in equilibrium concentrations of absorbing species as the total concentration of the constituent changes. (See Section I.D.4.)

Standard curves that are straight lines are, of course, desirable, and most photometric methods, even when a filter photometer is used, give them. Usually a straight line can be drawn by eye through the points with sufficient accuracy for the purpose. In the most exacting work, the best straight line should be drawn by the use of least squares. We shall show how this is done, and how the precision is evaluated, by using a specific example.

The Rectilinear Standard Curve as Illustrated by the Determination of Iron with Thioglycolic Acid. Various amounts of Fe(II) were measured

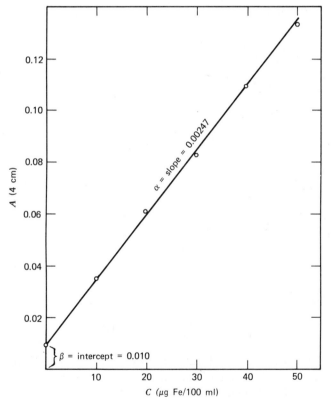

Fig. 4-4. Standard curve for the absorptiometric determination of iron with thioglycolic acid. Data of Higgins, *loc. cit.*

out accurately into 100 ml flasks and treated with excess reagent and ammonia, the solutions were diluted to volume and the absorbance of each in a 4 cm cell was determined by measuring against water (605 nm filter)[12]:

C (μg Fe in 100 ml)	0	10	20	30	40	50
A ($\log I_0/I$)	0.009	0.035	0.061	0.083	0.109	0.133

A plot of these points appears to give a straight line (Fig. 4-4). We wish to test these points for rectilinearity and to obtain the best straight line passing through them, namely the line for which the sum of the squares of the deviations in the A direction is a minimum (C is assumed errorless). The values of the constants in the equation for this line,

$$A_f = \alpha C + \beta \quad (A_f \text{ are the fitted values, those on the line})\qquad(4\text{-}7)$$

are found from the relations

$$\alpha = \frac{\sum\limits_{i=1}^{n}(A_i - \bar{A})(C_i - \bar{C})}{\sum\limits_{i=1}^{n}(C_i - \bar{C})^2}\qquad (i = 1, 2, \ldots, n)\qquad(4\text{-}8)$$

$$\beta = \bar{A} - \alpha\bar{C}\qquad(4\text{-}9)$$

The following values are obtained:

$\bar{C} = 25(\mu$g Fe/100 ml)
$\bar{A} = 0.072$
$\sum(C - \bar{C})(A - \bar{A}) = 4.32$
$\sum(C - \bar{C})^2 = 1750$
$\sum(A - \bar{A})^2 = 0.01067$
$\alpha = 4.32/1750 = 0.002469$ or 0.00247
$\beta = 0.072 - 0.00247 \times 25 = 0.010$

Accordingly,

$$A_f = 0.00247C + 0.010\qquad(4\text{-}10)$$

This is the line drawn in the figure. (It may be noted that subtracting the blank absorbance from each of the others before finding the best straight line is not admissible because the corrected values then are no longer statistically independent.)

We find the absolute deviations (and their squares) of the measured A

values from the best straight line:

C	A	A_f	$A - A_f$	$(A - A_f)^2$
0	0.009	0.0100	−0.0010	100×10^{-8}
10	0.035	0.0347	0.0003	9
20	0.061	0.0594	0.0016	256
30	0.083	0.0841	−0.0011	121
40	0.109	0.1088	0.0002	4
50	0.133	0.1335	−0.0005	25
				$\Sigma 515 \times 10^{-8}$

There is no trend in the magnitudes or signs of $A - A_f$, verifying the visual impression that the points lie along a straight line.[13] The absolute standard deviation of A is

$$s_A = \left\{ \frac{\Sigma (A - A_f)^2}{N - 2} \right\}^{1/2} = \left\{ \frac{5.15 \times 10^{-6}}{4} \right\}^{1/2} = 1.14 \times 10^{-3} \qquad (4\text{-}11)$$

The number of degrees of freedom, $N - 2$, is $6 - 2 = 4$, because two parameters have been calculated, that is, α and β.

It is interesting to compare the observed s_A with s_A calculated on the assumption that the deviations are due entirely to a standard deviation of 0.1% in transmission reading. In the range $A = 0.00$–0.13, $s_I = 0.14$ (i.e., $2^{1/2} \times 0.1$) corresponds to an s_A of ~0.0007 (varies from 0.0006 to 0.0008), leaving the remainder of the total s_A, approximately the same, to be explained by random deviations in the color system or elsewhere (small errors in measuring out the iron solution are possible). Therefore, the color system has good reproducibility over a concentration range from 0 to 50 μg Fe in 100 ml. It is, of course, possible that s_I is a little greater than 0.14% absolute, so that the reproducibility of the color system is even better than that obtained on this assumption.

We next obtain an expression for the standard deviation of C determined by the use of the standard curve in Fig. 4-4. From a measured value A, the iron concentration of the sample solution (μg Fe in 100 ml) is given by

$$C = \frac{A - \beta}{\alpha} = \frac{A - 0.010}{0.00247} \qquad (4\text{-}12)$$

A, α, and β all have indeterminate errors that combine to give the indeterminate error in C. The standard deviation of C expressed in percentage is

$$\frac{100 s_C}{C} = 100 \left\{ \left(\frac{s_{A-\beta}}{A - \beta} \right)^2 + \frac{s_\alpha^2}{\alpha^2} \right\}^{1/2} \qquad (4\text{-}13)$$

As an approximation assume[14] that $s_\beta = s_A$. Then

$$\frac{100 s_C}{C} = 100 \left\{ \frac{2 s_A^2}{(A - \beta)^2} + \frac{s_\alpha^2}{\alpha^2} \right\}^{1/2}$$

$$= 100 \left\{ \frac{2 s_A^2}{\alpha^2 C^2} + \frac{s_\alpha^2}{\alpha^2} \right\}^{1/2}$$

$$= \frac{100}{\alpha} \left\{ \frac{2 s_A^2}{C^2} + s_\alpha^2 \right\}^{1/2}$$

$$s_\alpha = \left\{ \frac{s_A^2}{\sum (C_i - \bar{C})^2} \right\}^{1/2} = \left\{ \frac{N s_A^2}{N \sum C_i^2 - (\sum C_i)^2} \right\}^{1/2} = \frac{6^{1/2} \times 1.14 \times 10^{-3}}{(6 \times 5500 - 22500)^{1/2}}$$

$$= 2.7 \times 10^{-5} \tag{4-14}$$

Therefore, the estimated standard deviation of C in percent is

$$\frac{100 s_C}{C} = \frac{100}{2.47 \times 10^{-3}} \left\{ \frac{2 \times 1.29 \times 10^{-6}}{C^2} + 7.4 \times 10^{-10} \right\}^{1/2}$$

$$= 4.05 \times 10^4 \left\{ \frac{2.58 \times 10^{-6}}{C^2} + 7.4 \times 10^{-10} \right\}^{1/2} \tag{4-15}$$

The first term within the braces combines, as we have seen, the effect of indeterminate errors in absorbance measurement of the sample solution and of the intercept of the standard curve; the second term involves only the error in the slope of the line. When C is small, the term 7.4×10^{-10} can be neglected in comparison with $2.58 \times 10^{-6}/C^2$, that is, the precision (relative) of the determination depends largely on the precision of the intercept β (as well as on A), very little on the slope of the line. The estimated percentage standard deviations in the determination of 1, 10, and 50 μg of Fe, together with the error contributions of the two terms inside the braces, are the following:

C (μg Fe/100 ml)	$2 s_A^2/C^2$	s_α^2	$100 s_C/C$
1	2.6×10^{-6}	7.4×10^{-10}	65
10	2.6×10^{-8}	7.4×10^{-10}	6.6
50	1.03×10^{-9}	7.4×10^{-10}	1.7

If there were no indeterminate error in the standard curve—a state of affairs that could be approached by basing it on a large number of points[15]—the relative standard deviations of C for 1, 10, and 50 μg Fe would be 46, 4.6, and 0.9%.

The range in which the "true" value of iron will lie at a specified probability level is found with the aid of t values. Suppose that a single absorbance measurement gives $C = 10.0 \ \mu$g Fe in 100 ml. The percentage standard deviation for this amount of iron is 6.6, as already calculated. We wish to find the range of iron concentrations in which the (statistically) true value lies with a probability of 95%. The range is given by

$$C_\mu \pm \bar{C} = \pm \frac{ts}{N^{1/2}} \qquad (4\text{-}16)$$

where C_μ = true value, $\bar{C}$ is average of N observed values, and $s = s_C$. Here $N = 1$ and $\bar{C} = C$, and t for a confidence level of 95% and 4 degrees of freedom $(6-2)$ is found to have the value 2.78 from a t table (or Fig. 4-5). Therefore,

$$C_\mu \pm C = \pm 2.78 \times 6.6 \times \tfrac{10}{100} = \pm 1.8 \ \mu\text{g Fe} \qquad (4\text{-}17)$$

The probability is 0.95 or 95% that the statistically true value of iron lies in the range 8.2 to 11.8 μg Fe, or within $\pm 18\%$ of 10.0 μg. At the 50 μg Fe level, 0.95 probability corresponds to ± 2.36 μg Fe or $\pm 4.7\%$. If there were no error in the standard curve, the range of iron values for 95% probability at the 50 μg Fe level would be $50 \pm 1.96 \times 0.85 = 48.3 - 51.7$ μg or $\pm 3.4\%$.

Probably most trace analysts are willing to accept a confidence level of 95% for the majority of their results. Figure 4-5 gives t values as a function of the number of degrees of freedom for a confidence level of 95%. It is seen that t decreases only slowly above $n = 4$ or 5 at this probability level. In other words, increasing the number of points (or their replication) above 6 or 7 $(= n+2)$ in a rectilinear calibration graph brings only a slow increase in its precision.

For values of t at other confidence levels, see Fig. 4-10.

Ordinarily in absorptiometric determinations, absorbances are not measured closer than 0.001. In large part, this limit is set by the use of

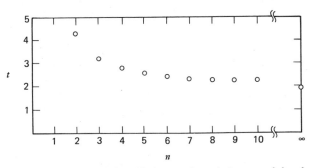

Fig. 4-5. The value of t as a function of the number of degrees of freedom n for a confidence level of 95%.

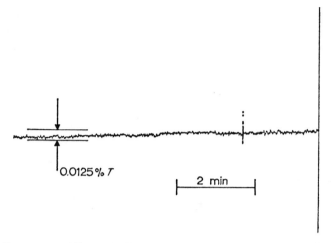

Fig. 4-6. Short-term stability and noise of spectrophotometer with optical feedback stabilization (Pardue and Deming, *loc. cit.*). Reprinted with permission from H. L. Pardue and S. N. Deming, *Anal. Chem.*, **41,** 987 (1969). Copyright by the American Chemical Society.

spectrophotometers that cannot be read closer than 0.1% in transmittance, and, for the rest, because of convention. Today, however, commercial instruments are available that by scale expansion in the range 0–0.1 A allow readings to be made with a reproducibility of $\sim0.0002\ A$ (standard deviation of a single reading). This is by no means the present day limit. Specially designed instruments can give reproducibilities of $5\times10^{-5}\ A$ with average drifts of 0.0001 A/hr (cf. Fig. 4-6).[16]

In trace analysis, a fairly large relative error is often allowable, so that the usual reactions with good photometric reproducibility should also be suitable at low absorbances. Reagents that are themselves absorbing at the working wavelength or which decompose into absorbing species are less desirable than colorless reagents. As a rule, temperature control would not need to be more rigorous than ordinarily. Photometric errors caused by placement of absorption cells and cleaning of cell faces could be eliminated by filling and emptying cells without removing them from the photometer. The most serious source of error will arise from foreign substances giving absorbing species with the reagent, so that reagents of high selctivity coupled with good separation methods are prime requisites for determinations based on measurements of low absorbance values.

b. Accuracy

The chief sources of error in photometric analysis arise in the chemistry of the systems and especially from the presence of foreign substances.

These are discussed later (Section I.B.4). Here we touch on possible (determinate) errors that are more related to the photometric measurement itself.[17]

In present day photometers it is unusual to find that transmittance scale readings are not closely proportional to true transmittance, or that A scale readings are not proportional to $\log I_0/I$.[18] Photometric linearity could be tested for, with monochromatic light, by measuring I/I_0 or A as a function of the concentration of a substance known to "obey Beer's law." As a matter of fact, this test is in a sense made by the analyst when he obtains a straight line for his standard curve.[19] Failure to see that the photometer reads zero at the beginning ($=$ dark current) results in a determinate error. For example, if the dark current reading is 1.0% transmittance instead of 0 as assumed, $\log I_0/I = (100-1)/(90-1)$ instead of $\log 100/90$, or 0.0462 instead of 0.0458. But when a standard curve is used to obtain the concentration, there is no error (if the dark current remains constant). A standard curve also cancels other determinate errors. Thus, if the same absorption cell is used for both standards and sample, the exact stratum thickness b is immaterial. The cell should always be placed in the instrument in the same orientation (same side toward the light source), because the cell walls may not be strictly parallel, or possibly not of the same reflectivity, so that the light transmitted to the photocell may not have the same intensity from both sides. It is important that the cell fit snugly in its holder, else random fluctuations in transmittance may result. In the most accurate work, cells should not be wiped between I_0 and I readings, which means that solutions or water may not come in contact with the outside faces of the walls through which light passes.

A small error in the wavelength scale is usually of small importance analytically. But the wavelength must be the same when standards and unknown are measured, especially if the absorption maximum is sharp. Likewise the slit width must remain constant. Lack of precision in wavelength setting can cause irreproducibility in transmittance readings; temperature changes can conceivably lead to small wavelength changes. The wavelength scale can be checked for accuracy when required by use of a mercury arc or rare earth (e.g., Ho and Sm) solutions or more conveniently of a rare earth glass. When a filter photometer is used, as it usually is, with a tungsten lamp as a light source, the standard curve may show a change with time, because deposition of tungsten on the glass will change the spectral characteristics of the light, especially in the blue and violet range.

Systematic errors in absorbance measurements can result from multiple reflections and stray light in spectrophotometers,[20] but they are not of great importance in trace analysis. When a standard curve is used—as it always should be—errors in absorbance measurement of the sample solution will be cancelled out by errors in measurement of the standard solutions, if conditions remain the same and the composition of the solutions are the same. A small error in concentration determination can result if the standard and sample solutions do not have the same refractive index. The greater the difference in the refractive indices of glass and solution, the larger the fraction of reflected light at the interface. At normal incidence, the reflected light fraction f at a glass–solution interface is

$$f = \left\{ \frac{(n_g - n_s)}{(n_g + n_s)} \right\}^2$$

n_g being the refractive index of glass and n_s that of solution. The index of dilute solutions is so close to that of water that the fraction of light reflected at a glass–water interface is usually almost the same as at a glass–solution interface, and compensation is achieved. For example, the transmission of a glass cell containing $1\,M$ HCl ($n = 1.342$) is 0.1% greater than the transmission of a cell holding water ($n = 1.333$). When necessary, the electrolyte content of the standard solutions can be made approximately the same as that of the sample solution.

It may be noted in passing that the accurate determination of absorptivities a or ϵ at particular wavelengths is not always easy because of the presence of determinate errors in the absorbance measurements. The analyst is spared these difficulties. He does not require absolute values. But he must see that conditions are the same in the absorbance measurement of the sample solution as in the measurement of the solutions for the standard curve. This requirement should not, in general, be difficult to fulfill—except for the "foreign" constituents of the sample.

The standard curve requires checking at reasonable intervals, preferably daily but possibly more often if reagents are unstable or if danger of contamination of solutions is great. Both the intercept and the slope of the line are subject to change. Deterioration of the photometric reagent may cause a change in slope, as may the aging of the tungsten lamp in a filter photometer, as already noted. A significant systematic error in wavelength setting when a spectrophotometer is used is unlikely but not excluded. The intercept may show a change because of a change in absorbance of the reagent (slow fading of a colored reagent, development of a color in a colorless reagent). Moreover, contamination of the reagent by traces of metals from the container or the atmosphere may bring about small changes in absorbance that can be of importance in trace analysis.

The absorbance of a cell may change because of deposition of material on the walls. When the standard curve is a straight line, only two points at the ends of the line need be verified. Often the reference cell ($I_0 = 100\%$) is filled with a reagent blank solution, instead of water, and absorbance measurements are made versus this solution; the absorbance of the working cell filled with reagent blank then gives the intercept. However, if the standard curve is not a straight line, absorbances should be measured against water in the reference cell.[21]

Factors mentioned earlier as contributing to imprecision can, of course, also be the source of determinate errors. The pH at which a color reaction is carried out is often important and may need to be controlled quite closely. Bringing a sample into solution may require the use of strong acids or alkaline fluxes, the excess of which often needs to be largely neutralized. Variables such as reagent concentration and time and temperature of color development may need to be regulated, and to be kept the same in the determination as in the establishment of the standard curve. Attention also needs to be paid to the effect of temperature on the absorbance of colored products, some of which have appreciable temperature coefficients.[22] In some photometers, the cell compartment becomes quite warm (35°C) during operation. A thermostated compartment is probably not necessary for most trace analyses, but cells should not be kept in the instrument longer than really necessary and the absorbance of sample and standard solutions (calibration curve) should be obtained at very nearly the same temperature.

The stability of the products of color reactions varies over a wide range. Inorganic species such as permanganate and chromate may show no change in solution for months or years in a suitable environment (not in plastic bottles), whereas certain organic products, illustrated by the evanescent substance formed when benzidine is oxidized by permanganate and other strong oxidizing agents, decompose in minutes or seconds. As a rule, organic reagents give less stable products than do the inorganic, but often the organic product is sufficiently stable to furnish results of high accuracy. For example, Fe(II)–1,10-phenanthroline ion is a stable species. Metal complexes of dithizone, thiocyanate, and dithiocarbamates are less stable.

Some metal chelates decompose more or less rapidly on exposure to light, especially UV radiation or the shorter wavelengths of the visible range (dithizonates, for example). Sometimes the decomposition results from the photodecomposition of the solvent, for example, chloroform. A number of metal–organic complexes, as well as the reagents themselves, are more stable in the absence of oxygen (or oxidizing agents in general). Lower temperatures may also aid stability.

3. Precision in Analysis

The heading is meant to indicate that there is a difference between the determination of a constituent in a simple solution and its determination in a solid sample that may have a complex composition and that may contain interfering substances. The precision and accuracy of a simple determination may depend largely or entirely on the precision of the photometer and the reproducibility of the color system, whereas these may not be the limiting factors in a real analysis. In applied analysis, determinate and indeterminate errors may result from heterogeneity of the sample, losses of constituent in separations, introduction of constituent from reagents, vessels or atmosphere, and other sources.

Especially in trace analysis, the blank is important. It is obtained by carrying the reagents, in the same amounts as used for the sample, through all the steps of the procedure and finding the amount of the constituent in question that is present in the final solution. The average of a number of blank values (bl) is subtracted from the *analytical* value (an) to obtain the *sample* value, that is, if $\mathbf{C}$ represents the content (say in ppm) of the constituent:

$$\bar{\mathbf{C}}_{samp} = \mathbf{C}_{an} - \bar{\mathbf{C}}_{bl}$$

($\bar{\mathbf{C}}_{bl}$ is reduced to sample basis.) $\mathbf{C}$ is found from A and the standard curve.

If s is the standard deviation of $\mathbf{C}$:

$$s_{samp} = (s_{an}^2 + s_{bl}^2)^{1/2} \tag{4-18}$$

The values of s_{an} and s_{bl} are not necessarily the same (taking differences in I into account) but are usually assumed so. In practical work, it is unusual for the analyst to run more than one, or at most two, determinations of a constituent in a particular sample, so that s_{samp} may not be known except roughly, perhaps from statements of a worker who has studied a method for a certain class of materials, in which the sample in question may be included. The composition of the sample may, of course, affect s_{samp} (as well as the accuracy of $\mathbf{C}_{samp}$), but in a good analytical method the dependency should be small. If indeterminate errors in an analysis were due largely to the photometer and the color system, their combined value could be found easily enough by replicating one or more points on the standard curve, but such an estimate is more likely than not to be too favorable in a real analysis.

Although reproducibility may have a significant effect on accuracy, especially if the number of replicates are small, errors are more likely to arise from other constituents of a sample. But again, much depends on the particular method and, of course, on the composition of the sample. See further, p. 171.

4. Accuracy in Analysis

One scheme for the classification of determinate errors in chemical analysis divides them into (1) additive, and (2) proportional (or multiplicative). Additive errors exert the same absolute effect, independently of the constituent content of a sample, other factors remaining the same. The effect of proportional errors is relatively the same (ideally) for varying amounts of constituent (absolute error is proportional to the amount of constituent at constant concentration of interfering substance). Some additive errors in photometric analysis are easily corrected or allowed for, others not at all. Proportional errors in the determination of a constituent can usually be corrected for, if they are actually proportional over the range in which the constituent can vary. This condition is not always fulfilled.

The limiting factor in the accuracy of real analyses is likely to be the effect of what is sometimes called foreign substances—though they are no more foreign to the sample than the constituent being determined. The interference of other constituents of a sample can be of two types: positive and negative. Negative interference can result from the reaction of the interfering constituent with the constituent being determined, so that the latter is not completely converted into the absorbing (or fluorescing) determination form. The common type of such interference is complexing or other masking of the constituent by the interfering substance, for example, Fe(III) by F^-. Another mode of interference is by combination of the interfering constituent with the color reagent to form a (colorless) complex, thus preventing or hindering the reaction between the desired constituent and the reagent. For example, much Hg(II) prevents the reaction between Cu(II) and diethyldithiocarbamate by forming the more stable, essentially colorless, Hg(II) diethyldithiocarbamate. Errors of this kind are likely to be of the proportional type.

In positive interference, constituents other than the desired one react with the reagent to give absorbing (or fluorescing) species, so that high results are obtained. This is the more common type of interference in the photometric determination of metals. These errors are likely to be of the additive type. Even if the reagent does not react with the other constituents to yield absorbing products, these constituents, may, in the form in which they are present, absorb to a greater or less extent and lead to positive errors unless their effect is nullified or corrected for.

In Table 4-3, general methods for preventing or minimizing both positive and negative interferences are listed.

TABLE 4-3

General Methods for Preventing or Minimizing Errors Due to Foreign Substances in Absorptiometric Analysis

Type of Interference	Method	Remarks
I. Positive		
A. The foreign substance (species) is colored (or absorbs in the UV) but undergoes no change on addition of reagent (simple additive error).	1. Measure absorbance of solution at a wavelength at which foreign substance shows practically no absorption, but at which reaction product of constituent absorbs strongly, or use method in Section I.B.1.	Occasionally applicable. Also used when reagent is colored (or, more generally, absorbing).
	2a. Measure absorbance of solution before addition of reagent and deduct (proportionately) from final absorbance.	The most generally applicable method.
	b. Measure absorbance of solution after reagent addition against original solution (suitably diluted) in reference cell.	Monochromatic light should be used. Apply cell correction if necessary.
	3. Add a reagent (usually complexing or reducing) to destroy color of foreign substance at beginning, e.g., H_3PO_4 to decolorize Fe(III).	
	4. Develop color, measure absorbance, discharge color with a reagent not affecting color of foreign substances and again measure absorbance.	Example: In determination of Mn as MnO_4^- in presence of Fe(III), H_2O_2 may be used to destroy MnO_4^-.
	5. Add a solution of foreign substance to standards to duplicate color of sample before adding reagent.	For colorimetry by the series and colorimetric titration methods. Approximate amount of absorbing foreign substance may be known (Fe in steel). Infrequently used.

172

6. Extract colored reaction product into an immiscible solvent and determine its absorbance. (Actually a separation.)

Example: Extraction of Co(II)–thiocyanate complex with amyl alcohol, Ni being left in aqueous solution.

B. The foreign substance may or may not be colored, but it gives a colored product with the reagent. Error usually additive.

1a. If overlapping of the two absorption curves is not too great, measure absorbance at two suitable wavelengths and calculate concentration from two simultaneous equations (simultaneous determination). Let

λ_1 = a wavelength at which constituent M, as colored product, absorbs strongly, but at which N absorbs only slightly

λ_2 = a wavelength at which N absorbs strongly but M weakly

a_1^M, a_2^M = absorptivities of colored product of M at wavelengths λ_1 and λ_2

a_1^N, a_2^N = absorptivities of colored product of N at wavelengths λ_1 and λ_2. (Units of a can be made whatever desired; conveniently let a represent the absorbance of a solution containing 1 μg of M or N in 1 ml solution, i.e., 1 ppm, and having a depth of 1 cm.)

C_M, C_N = concentrations of M and N in the solution

At wavelength λ_1, absorbance of the solution is sum of absorbances due to M and N:

$$A_1 = a_1^M C_M + a_1^N C_N$$

At wavelength λ_2:

$$A_2 = a_2^M C_M + a_2^N C_N$$

From these two equations:

$$C_M = \frac{a_1^N A_2 - a_2^N A_1}{a_2^M a_1^N - a_1^M a_2^N}$$

(and an analogous expression for C_N)

Values of a are found from standard curves (A versus C) at λ_1 and λ_2; a is the slope, A/C. Absorbances must be proportional to concentrations for this method to be applied. A spectrophotometer, not filter photometer, should be used, although sometimes a filter photometer may serve. The accuracy may be unsatisfactory if the foreign substance preponderates or gives a strongly colored product.

173

TABLE 4-3 (*Continued*)

Type of Interference	Method	Remarks
	1b. A component M absorbing strongly at wavelength λ_1 can be determined by measuring its absorbance at this wavelength versus its absorbance at a wavelength λ_2, the latter being chosen so that the absorbance of N is the same at λ_1 and λ_2. A variation in the concentration of N does not affect the difference. This is the principle of the automated dual-wavelength method of S. Shibata, M. Furukawa, and K. Goto, *Anal. Chim. Acta*, **46**, 271 (1969); **53**, 369 (1971); also see T. J. Porro, *Anal. Chem.*, **44** (4), 93A (1972).	
	2. Convert interfering substance into a nonreactant by (a) complex formation or (b) oxidation or reduction.	A powerful method, convenient to use, frequently applicable.
	3. Determine absorbance of sample solution, in which color has been developed before and after addition of a reagent that destroys color of constituent without affecting that of foreign substance; difference is absorbance due to constituent. A variant consists in treating two aliquots of sample solution with the color-producing reagent, adding to one aliquot a reagent destroying the color of the constituent and measuring absorbance of one against the other.	Suitable only when color (or fluorescence) of the foreign substance is weak. Example: Morin develops strong fluorescence with Zr, weak fluorescence intensity of Zr+Al, add EDTA to destroy fluorescence of Zr, and find residual fluorescence due to Al. See Geiger and Sandell, *Anal. Chim. Acta*, **16**, 346 (1957).
	4. Develop color under two conditions: first, such that only one constituent M reacts; second, such that both M and N react. Determine N from difference in absorbances (consecutive determination).	Example: determination of Th and rare earths with Arsenazo I (H. Onishi, H. Nagai, and Y. Toita, *Anal. Chim. Acta*, **26**, 528 (1962)).

5. Make separations, preferably applying a method that effects a direct isolation of constituent to be determined.

The method that is applied when others fail; applicable to most constituents; likely needed when minute traces are to be determined. See Chapters 7–10.

II. Negative

Foreign substance prevents or hinders reaction between constituent and reagent, thus reducing color intensity. Interference may be due to formation of a complex between substance and constituent or reagent, destruction of reagent, etc. Error often of the proportional type. Foreign "indifferent" substances rarely increase color intensity.

1. Increase excess of reagent.

Of limited value. Helpful when interfering substance forms a relatively weak complex with constituent. Thus, a large excess of thiocyanate reduces error due to phosphate in determination of Fe(III).

2. Find absorbance A_x, given by x g of constituent in V ml of solution. Add a g of constituent in a volume of v ml and obtain absorbance A_{x+a} of the mixture (a should be two or three times larger than x, and v should be very small). If there are no complications:

$$\frac{A_x}{A_{x+a}} = \frac{x/V}{(x+a)/(V+v)}$$

(in which A is corrected for blank).

$$x = aV \frac{A_x}{A_{x+a}} \Big/ \left(V+v - V\frac{A_x}{A_{x+a}}\right)$$

Assumption of constant relative error can be tested to some extent by varying a.

Not often applied in spectrophotometry, but it can be used profitably in checking results obtained with solutions of unknown composition to avoid gross errors. A linear relation between A and constituent concentration is assumed for the range covered. Statistical aspects of the (general) standard addition method are discussed by G. Ehrlich and R. Gerbatsch, *Z. Anal. Chem.*, **209**, 35 (1965). The addition method has been applied to a system containing a (known) foreign constituent giving an absorbing product with the color reagent, by absorbance measurement at two wavelengths; see L. P. Adamovich, *J. Anal. Chem. USSR*, **17**, 893 (1962).

TABLE 4-3 (*Continued*)

Type of Interference	Method	Remarks
	3. Prepare standards having same composition as sample solution.	Occasionally applicable, as in determination of trace elements in sea water, in chemical reagents or other samples of known composition, especially of major constituents. It may be possible to remove completely the constituent in question from a portion of the sample and then prepare standards by adding known amounts of the constituent (example: sea water).
	4. As in Section I.B.2.	
	5. Apply separations.	

Expression of Photometric Interference. On the basis of certain assumptions, to be mentioned shortly, the relative photometric effect of a constituent B in the amount Q_B on the determination of a constituent A in the amount Q_A is given by

$$T_B \frac{Q_B}{Q_A}$$

in which T_B represents the ratio of the absorbance given by B to the absorbance given by an equal weight of A reacting with a certain reagent under specified conditions.[23] The absorbance produced by B depends on the fraction of B in the form of the absorbing species and its absorptivity at a specified wavelength. Conditions in analysis are, of course, so arranged, if possible, that as little of B, C, D, . . . , and as much as possible of A are present in the form of absorbing species. Selection of favorable acidities and wavelength and the use of suitable differential masking agents are obvious ways of reducing interferences.

As a first approximation, T_B may be assumed to be independent of absolute concentrations, that is, that the absorbances produced by B and A are proportional to their analytical concentrations under specified conditions. This is not always true. There may be interaction of A and B. For example, B and A could react with the photometric reagent to yield a mixed complex. Thus alizarinsulfonic acid gives a stronger color with aluminum in the presence of calcium than in its absence. If such interaction does not occur, the absorbance produced by B at a known total concentration could be calculated if ϵ or a of its colored product and all the equilibrium constants (especially the formation constant of the colored complex) were known. Occasionally, especially when the hydrogen-ion concentration is the chief determinant in the formation of the absorbing species, the required absorbance can be calculated. For example, such calculations can be made for a considerable number of metals with dithizone as reagent, the formation constants and absorptivities (from an absorbance curve) being known. The absorbance of constituent B under prescribed conditions can be found experimentally and therefrom the value of T_B, the absorbance of A always being known from the standard curve.

If Q_B/Q_A in the sample is approximately known, the calculated ratio of absorbances produced by B and A is $T_B(Q_B/Q_A)$ and the positive error in percent is given by $100 T_B(Q_B/Q_A)$. Thus if $T_B = 0.002$ and $Q_B/Q_A = 100$, the error is $100 \times 0.002 \times 100 = 20\%$.

Testing the Accuracy of Methods. Analytical results vary a great deal in accuracy, in large part because samples vary greatly in composition Separations may be necessary and these then are an essential part of the method. The final photometric step of the analysis may be the quickest and easiest one of all.

Suppose it is desired to test a new reagent for the determination of a metal A. The color system A + reagent would first be studied: the spectral characteristics (reagent and colored species), the effect of pH and reagent excess, rate of color formation, stability of the colored product, effect of temperature, and so on.[24] Such investigation can establish the nature of the colored complex(es) and the value of the formation constant. Information of this kind is fundamental, and is needed for rationally evaluating the reaction, for predicting the effect of varying conditions, and so on.[25] The standard curve for the determination of A would then be established, and the reproducibility would be found in doing this. Next the behavior of other metals under the conditions of the determination would be examined, and if colored species are formed, their absorbances would be determined. The predicted interference might then be checked by determining the absorbances of known mixtures of A and other metals. Also the effect of anions would be studied. It would be very laborious, indeed impossible, to test all possible interferences at different levels, so that a selected list of possibly interfering substances would generally be studied.[26] The selection would be made on the basis of the reacting group in the reagent (similarity to other reagents of like type) and the composition of samples that might profitably be analyzed by use of the reagent. If positive interferences are found, the use of masking agents might be tried, with and without A being present.[27] Although complexing or other types of masking (e.g., oxidation or reduction) is very useful in preventing or minimizing interference it is unlikely by itself to solve all problems of interference. Sooner or later, separations will be required, especially when the ratio of the interfering element to A is large.

In working out or testing the suitability of an analytical method, the analyst may modify an existing method. The modification may be slight or it may be radical. The spectrophotometric stage of the method may remain largely unchanged, but new methods of separation might be introduced. Often it is a matter of modifying or adapting an existing method to the determination of a particular metal in a certain type of sample. Rarely has a more or less general method been studied in sufficient detail to allow it to be applied in practice to a given sample or class of samples, for example, silicate rocks, meteorites (iron and stone), ferrous alloys, nonferrous alloys (many types), biomaterials. In the first stage of testing, a solution containing the elements in the average amounts present in the material in question can render good service. It would be put through all the steps of the method, and the amounts of element A found would be compared with the amounts added, the latter being corrected for the small amounts of A introduced as an impurity with the other components added as determined from blank runs.[28] Values for the recovery (p. 699) of A would thus be obtained from which conclusions concerning the accuracy would be drawn.[29] It would be desirable, if at all possible, to test a method on a solid sample, much like, if not identical with, the material for which the method is intended. After all, bringing the sample into solution is a part of the method, and may be a step that entails errors not detected by the use of a synthetic solution. Finding a suitable solid sample is not always easy. It may be possible to obtain a sample of the class of materials in question that has a low content of A. Additional amounts of A could then be added, and their recovery determined. Synthetic solid samples are occasionally prepared.

For some materials *standard samples* are available, and these are valuable. Standard samples have been carefully analyzed by a number of analysts, often using different methods. If essentially the same values are obtained by radically different methods, which can hardly have the same determinate errors, they are likely to be accurate within certain limits. The U.S. National Bureau of Standards prepares, analyzes (cooperatively), and sells standard reference materials, some of which are of interest to the trace analyst.[30] A considerable number of trace elements have been determined in a wide variety of metals: steels, ingot iron, steelmaking alloys, and nonferrous alloys (Al-, Co-, Cu-, Pb-, Mg-, Ni-, Sn-, Zn-, and Zr-base). Recently, the Bureau has prepared a series of glasses containing trace elements at four levels, nominally 0.02, 1, 50, and 500 ppm.[31] From other sources, samples of geomaterials (especially silicate rocks) are available (Table 4-4). Some of these have been extensively analyzed for trace elements, in various laboratories and by a variety of methods, including neutron activation, which supposedly should furnish the most reliable values for the lowest trace contents. Agreement is not always as good as desirable, but some values seem to be well established.[32]

A few samples of biomaterial have been prepared and analyzed cooperatively (see p. 33). A reference material compounded from gelatin as a matrix and containing roughly 50 ppm of trace elements has been analyzed in a number of laboratories. [See D. H. Anderson, J. J. Murphy, and W. W. White, *Anal. Chem.*, **48,** 116 (1976).] A botanical material (SRM 1571), orchard leaves, is available from the U.S. National Bureau of Standards. This standard sample is certified (1970) for Fe, Ni, Cu, Pb and other metals. More recently (1976) other botanical samples have become available from the Bureau of Standards. Tomato Leaves (SRM 1573) and Pine Needles (SRM 1575) are certified for the trace elements

TABLE 4-4
Geochemical Standards[a]

Sample Name	Sample Number	Sample Name	Sample Number
Adularia	PSU-Or-1	Gold-quartz	USGS-GQS-1
Albite	PSU-Ab-1	Granite	GIB-G-B
Andesite	USGS-AGV-1		QMC-I-1
Anhydrite	ZGI-AN		ANRT-GS-N[b]
Basalt	CRPG-BR		CRPG-GR, GA, GH
	ZGI-BM		ZGI-GM
	USGS-BCR-1		USGS-G-1[c]
	USGS-BHVO-1		USGS-G-2
	GSJ-JB-1		NIM-G
Bauxite	ANRT-BX-N	Granodiorite	USGS-GSP-1
	NBS-69a		GSJ-JG-1
Biotite	CRPG-Mica-Fe	Hematite	GFS-453
	USGS-Btt-1[b]	Hornblende	Basel-1-H
Brick	Basel-1-B		USGS-Hbl-1[b]
Fire	BCS-269,315	Jasperoid	USGS-GX-1[b]
Silica	BCS-267,314	Lepidolite	NBS-183
Calcsilicate	QMC-M-3	Limestone	ZGI-KH
Cement, Portland	BCS-372		GFS-401
	NBS-633 to 639	Argillaceous	NBS-1b
Clay		Dolomitic	NBS-88a
Flint	NBS-97a	Lujavrite	NIM-L
Plastic	NBS-98a	Magnesite	UNS-MK
Coal	NBS-1630,1631	Meteorite, Allende	USNM-3529
Deposit, hot spring	USGS-GX-3[b]	Monazite sand	NBL-7A
Diabase (dolerite)	QMC-I-3	Mud, marine	USGS-MAG-1
	USGS-W-1[c]	Nepheline syenite	Len X
Diorite	ANRT-DR-N		USGS-STM-1
Disthene (kyanite)	ANRT-DT-N	Norite	NIM-N
Dolomite	BCS-368	Ores	
	GFS-400	Cr	BCS-308
Dunite	USGS-DTS-1[c]	Cu	NBS-330,331
Dunite, chrysolite	NIM-D		USGS-GX-4[b]
Feldspar, potash	BCS-376	Cu–Mo	CSRM-HV-1[b]
	ANRT-FK-N[b]	Fe	BCS-175/2 and
	NBS-70a		others
Feldspar, soda	BCS-375		NBS-27e
	NBS-99a	Mn	BCS-176/1
Fluorspar	NBS-79a, 180		NBS-25c
Gabbro	CSRM-MRG-1[b]	Mo	CSRM-PR-1
Galena	GF-1		NBS-333
Glasses	NBS-89,91,92,93a	Pt	CSRM-PTA-1,
	NBS-620,621		CSRM-PTC-1
Trace elements	ANRT-VS-N	Ni–Cu matte	CSRM-PTM-1
	NBS-608 to 619	Sulfide	CSRM-MP-1
	USGS-GSA to GSE		ASK-3

180

TABLE 4-4 (*Continued*)

Sample Name	Sample Number	Sample Name	Sample Number
Zn	NBS-113a	Mica	USGS-SDC-1
Peridotite	USGS-PCC-1	Sediment, lake	IAEA-SL-1
Petalite	NBS-182	Serpentine	ANRT-UB-N
Phlogopite	CRPG-Mica-Mg	Shale	KnC-Shp-1
Phosphate	NBS-120a		JOS-1
	NBL-1		ZGI-TS
Pitchblende ore	IAEA-S-12,S-13		USGS-SCo-1
Pyrite	PS-1		USGS-SGR-1
Pyroxene	PSU-Px-1	Slate	ZGI-TB
Pyroxenite	NIM-P	Soils	CSRM-^{b}
Quartz, Brazilian	GEM-430		IAEA-Soil-5
Quartz latite	USGS-QLO-1	Sphalerite	SF-1
Refractories (and	NBS-103a and	Spodumene	NBS-181
ceramics)	others	Sulfide	CSRM-SU-1
	BCS-313 and	Syenite	NIM-S
	others		CSRM-SY2,3
Rhyolite	USGS-RMG-1	Tonalite	MRT-T-1
Sand, glass	NBS-81a,165a	Ultramafic rocks	CSRM-UM-1,2,4
	UNS-SpS	Zircon	BCS-388
Schist	QMC-M-2		
	ASK-2		

^{a} From F. J. Flanagan, *Geochim. Cosmochim. Acta*, **38,** 1731 (1974). For earlier lists, see F. J. Flanagan and M. E. Gwyn, *ibid.*, **31,** 1211 (1967); F. J. Flanagan, *ibid.*, **34,** 121 (1970).
^{b} In preparation.
^{c} Exhausted.

Issuing organizations:

ANRT: same as CRPG
ASK: University of Oslo
Basel: Universität Bernoullianum, Basel, Switzerland
BCS: Bureau of Analysed Samples, England
CRPG: Centre de Recherches Petrographiques et Geochimiques, France
CSRM: Canadian Standard Reference Materials Project
GEM: Engineered Materials, New York
GF: see SF
GFS: G. Frederick Smith Chemical Company, Ohio
GIB: Geological Institute, Bulgarian Academy of Sciences
GSJ: Geological Survey of Japan
IAEA: International Atomic Energy Agency
JOS: Natural Resources Authority, Amman, Jordan
KnC: Knox College, Illinois
Len: Leningrad State University, USSR
MRT: Mineral Resources Division, Dodoma, Tanzania
NBL: New Brunswick Laboratory, U.S. Atomic Energy Commission
NBS: National Bureau of Standards, Washington, D.C.
NIM: South African Bureau of Standards
PS: see SF

PSU: Pennsylvania State University
QMC: Queen Mary College, England
SF: Bergakademie Freiberg, Deutsche Demokratische Republik
UNS: Institute of Mineral Raw Materials, Czechoslovakia
USGS: U.S. Geological Survey
USNM: Smithsonian Institution, Washington, D.C.
ZGI: Zentrales Geologisches Institut, Deutsche Demokratische Republik

As, Cr, Cu, Fe, Mn, Pb, Rb, Sr, Th, U, and Zn (the latter in 1573) and Al and Hg (in 1575). Values for many other trace elements are given for information. Spinach (SRM 1570) is certified for K, Ca, P, Al, Fe, Mn, Sr, Zn, Rb, Cu, Cr, Pb, As, Th, U, and Hg, with values for other trace elements. Brewer's Yeast (SRM 1569) is certified for Cr.

C. ABSORPTIOMETRIC SENSITIVITY

The term sensitivity is used in several different but related senses by analytical chemists. In a general sense, it refers to that characteristic of analytical instruments that enable them to detect weak signals and to that characteristic of analytical reactions and methods permitting the detection and hence the determination of very small amounts or very low concentrations of a constituent. In a narrower or more technical sense, it is often used to refer to the slope of the line relating signal strength Y to the amount or concentration X of a constituent:

$$Y = f(X)$$

Frequently in practice, a rectilinear relation exists between Y and X:

$$Y = const. \ X + const'. \qquad (const. \text{ and } const'. = \text{constants})$$

and the first constant, the coefficient of X, is the sensitivity (or a measure of the sensitivity) provided by the system. The larger the slope, the greater is the sensitivity and the lower the limit of detection. The detection limit increases with insensitivity, but the latter term is rarely used.

The sensitivities of absorptiometric reactions can be compared on the basis of a or ϵ, but these values in themselves do not provide analytical sensitivity. To the analyst, sensitivities are meaningful only in terms of detection limits. These are conventionally expressed as concentrations (mass/volume), but they can also be expressed in terms of mass in a column of solution of unit cross-sectional area. In the first part of this discussion, we use the conventional mode of expression.

Designate the minimum concentration of substance that can be detected (which itself requires specification) as C_d. Assume a rectilinear relation between C and the absorbance A. Then the lowest concentration of C detectable is determined by the smallest difference in A that is detectable at low A values:

$$C_d = \frac{\Delta A}{\alpha b} \qquad (b = \text{stratum thickness}) \qquad (4\text{-}19)$$

(If $C = c$ and light is monochromatic, α becomes a; see p. 232.) ΔA is likely to be determined by the sensitivity and reproducibility of the spectrophotometer. Conceivably, ΔA might be determined by the reproducibility of the absorptiometric reaction, but when only the reproducibility of the instrument is to be tested, a preformed stable solution of the absorbing species can be used. Evidently, ΔA and C_d are not fixed quantities for a given instrument. They depend on probability levels and are a function of the number of replicate absorbance measurements.

Consider Fig. 4-7, representing the standard curve (usually rectilinear) for the determination of a constituent by some reagent under given conditions. The uncertainty in the slope of the curve can be made so small that it is of no importance for the present considerations. $\bar{A}_{bl}$ is the mean absorbance of a blank solution, measured either against water or an identical blank solution in the reference cell. Let A_d (or the mean $\bar{A}_d$) be the minimum absorbance $(A_d > \bar{A}_{bl})$ that can be distinguished from $\bar{A}_{bl}$ with a specified probability. Since $A_d > \bar{A}_{bl}$, the stipulated probability

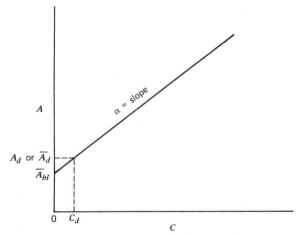

Fig. 4-7. Illustration of limit of detection in absorptiometry (see text). Solution stratum thickness (path length) $b = 1$ cm (unless otherwise specified). C is the analytical concentration of the element in question. Wavelength is specified. α becomes absorptivity a when $C = c$ (see p. 232) and reagent does not absorb. If the standard curve is not rectilinear, α is the initial slope. (If the reagent absorbs at the wavelength used for the absorbing species of the constituent, α and a will not be the same.)

refers to the positive value $A_d - \bar{A}_{bl} > 0$. Corresponding to $\Delta A = A_d - \bar{A}_{bl}$, we have $\Delta C = C_d - 0 = C_d$, which is the concentration detection limit under the conditions. For a given reaction, C_d will depend on the photometer and possibly to some extent on the analyst in reading the scale.

CASE 1. σ_A is known. σ_A is the (true) absolute standard deviation of A_{bl} or A_{samp}. A_{samp} is a measured absorbance of a sample solution (constituent + reagent) in which the presence or absence of the constituent at a specified concentration is to be demonstrated at stipulated probability levels. $\bar{A}_{samp}$ will therefore have a small value, in an absorbance range over which σ_A will be constant or essentially so. If the reproducibility of A_{samp} depends on the reproducibility of the absorptiometer and not on the reproducibility of the absorbing solution, σ_A can be found from repeated measurements of either A_{bl} or A_{samp}. Often this assumption is valid. Suppose then that σ_A is found from a large number of measurements of A_{bl}, which at the same time furnish an average close to the true blank value, $A_{bl,\mu}$ (measured against water). The Gaussian distribution of blank absorbance deviations is plotted as curve 1 in Fig. 4-8, with $\bar{A}_{bl,\mu}$ and σ_A indicated. The probability of occurrence of one-sided deviations, in terms of σ_A, is given by:

$+(A_{bl} - \bar{A}_{bl,\mu})/\sigma_A = +x$	Probability that $+x$ will be Exceeded
1.65	0.05
2.0	0.023
2.3	0.011
2.5	0.006
3.0	0.0013
4.0	0.00003

The probability that a single deviation equal to, or larger than, $+2\sigma_A$ will be observed is ~ 0.02 (more exactly 0.023); it is represented by the cross-hatched area under curve 1 on the right. There is thus a reasonably good chance that a deviation larger than $+2\sigma_A$ will not be obtained, and we adopt this value as a criterion in the following discussion.

Now suppose we determine the absorbance, by a single measurement, of a sample solution (treated with reagent) that may or may not contain the constituent in question. If we find that $A_{samp} - \bar{A}_{bl,\mu} \geqq 2\sigma_A$, there is only a probability of ~ 0.02 that a deviation of this size is due to chance alone if no constituent is present, and we conclude that the sample solution does contain the constituent with a probability of $1.00 - 0.02 \sim 0.98$. This is the detection limit, A_d, based on a single absorbance

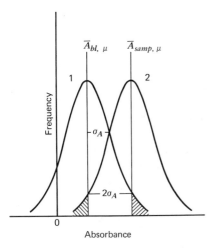

$\overline{A}_{bl,\,\mu}$ $\overline{A}_{samp,\,\mu}$

Frequency

σ_A

$2\sigma_A$

0

Absorbance

Fig. 4-8. Gaussian distribution of deviations $A_{bl} - \overline{A}_{bl,\mu}$ of blank solution (curve 1) and $A_{samp} - \overline{A}_{samp,\mu}$ of a sample solution having a concentration of constituent such that $\overline{A}_{samp,\mu} - \overline{A}_{bl,\mu} = +2\sigma_A$ (curve 2). A sample solution of constituent having this concentration will give an absorbance reading equal to, or greater than, $\overline{A}_{bl,\mu}$ 98% of the time. This is a reasonable criterion for detection, but an arbitrary one. Only ~2% of the time will an absorbance reading on such a solution give a value $< \overline{A}_{bl,\mu}$ (cross-hatched area under curve 2 on the left), if Gaussian distribution holds. And only ~2% of the time will a solution not containing the constituent (blank solution) give a reading $> \overline{A}_{bl,\mu} + 2\sigma_A$ (cross-hatched area under curve 1 on the right).

measurement of the sample solution and the $2\sigma_A$ criterion. The constituent concentration corresponding to A_d is $C_d = 2\sigma_A/\alpha$. A solution containing the constituent at this concentration will give an absorbance $\geq \overline{A}_{bl,\mu}$ 98% of the time if Gaussian distribution of deviations holds. If the criterion of detectability is taken as $A_{samp} - \overline{A}_{bl,\mu} \geq +3\sigma_A$, a solution having a concentration of $3\sigma_A/\alpha$ will give an absorbance greater than $\overline{A}_{bl,\mu}$ approximately 999 times in 1000. If error distribution is not Gaussian, estimates of detectability become less favorable.

If more than one absorbance measurement of the sample solution is made and the readings are averaged, A_d (and C_d) at the 0.98 probability level becomes smaller as shown in the following tabulation, which also shows the effect of basing $\overline{A}_{bl}$ not on a large number of readings ($\rightarrow \overline{A}_{bl,\mu}$) but on a small number:

No. of A Measurements		$\geq +(\overline{A}_{samp} - \overline{A}_{bl})/\sigma_A$
Blank	Sample	for Probability ≥ 0.98
∞	1	2.0
∞	2	$2/2^{1/2} = 1.41$
∞	3	$2/3^{1/2} = 1.16$
∞	4	1.0
1	1	2.9
2	2	2.05
3	3	1.7

Thus if triple absorbance readings are made of both blank and sample solutions, the presence of the constituent in the sample solution is indicated at the 0.98 probability level if

$$(\overline{A}_{samp} - \overline{A}_{bl}) \geq 2.05(\tfrac{1}{3} + \tfrac{1}{3})^{1/2}\sigma \qquad \text{or} \qquad \geq 1.7\sigma.$$

Case 1 is not often encountered in practice because the number of replicate absorbance measurements is too small to establish $\bar{A}_{bl}$ and σ_A with sufficient precision to enable probabilities to be obtained in the manner described. (σ_A might be known quite precisely from sets of previous absorbance measurements, but not $\bar{A}_{bl}$.) Therefore, a more realistic state of affairs must be confronted, and that is Case 2.

CASE 2. $\bar{A}_{bl}$, $\bar{A}_d$, and s_A are obtained from a limited number of replications.

This situation is represented in Fig. 4-9. Because of the small number of absorbance measurements, the observed mean of the blank, $\bar{A}_{bl}$ will not coincide with the true blank mean, $\bar{A}_{bl,\mu}$, nor will $\bar{A}_d$ coincide with $\bar{A}_{d,\mu}$. Moreover, the observed standard deviation s_A will be smaller than the true standard deviation σ_A. Account is taken of these differences in calculating probabilities by using t statistics (see any book on applied statistics or a comprehensive textbook on quantitative analysis). t is the analog of x (p. 184) used for a large number of measurements.

We wish to determine whether $+(\bar{A}_{samp} - \bar{A}_{bl})$ is sufficiently greater than s_A to indicate the presence of the constituent in the sample solution. The t of the quotient $(\bar{A}_{samp} - \bar{A}_{bl})/s_A$ is obtained from the expression

$$+t = \frac{\bar{A}_{samp} - \bar{A}_{bl}}{s_A} \left(\frac{N_{samp}N_{bl}}{N_{samp} + N_{bl}} \right)^{1/2} \tag{4-20}$$

in which

s_A = estimated standard deviation of individual absorbance measurements.

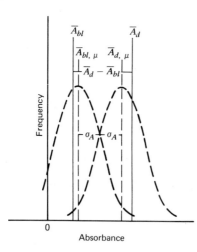

Fig. 4-9. Diagram illustrating means derived from a limited number of absorbance measurements. The hypothetical Gaussian distribution curves of deviations from the theoretical blank mean $\bar{A}_{bl,\mu}$ and from some specified theoretical mean $\bar{A}_{d,\mu}$ taken to represent the difference $\bar{A}_{d,\mu} - \bar{A}_{bl,\mu}$ on which probabilities of detection are to be based. $\bar{A}_{bl}$ represents the mean derived from a limited number of absorbance measurements of a blank, and $\bar{A}_d$, the mean of a limited number of absorbance measurements of A_d. σ_A is the standard deviation (based on a large number of measurements) of both sets. $\bar{A}_{d,\mu}$ is here taken as $\bar{A}_{bl,\mu} + 2\sigma_A$ (as in Fig. 4-8).

N_{samp} = number of absorbance measurements of sample solution made to obtain the average $\bar{A}_{samp}$.

N_{bl} = number of absorbance measurements made to obtain the average $\bar{A}_{bl}$.

In the usual repartition of absorbance measurements, $N_{bl} = N_{samp} = N$ and Eq. (4-20) becomes

$$+t = \frac{\bar{A}_{samp} - \bar{A}_{bl}}{s_A} \left(\frac{N}{2}\right)^{1/2}$$

s_A may usually be taken as the same in the two sets of absorbance measurements, so that it is calculated from the pooled deviations:

$$s_A = \left\{ \frac{\sum (A_{bl} - \bar{A}_{bl})^2 + \sum (A_{samp} - \bar{A}_{samp})^2}{N_{bl} + N_{samp} - 2} \right\}^{1/2} \qquad (4\text{-}21)$$

Example

Suppose that $\bar{A}_{bl}$ and $\bar{A}_{samp}$ are obtained from 3 absorbance measurements each, and s_A is estimated from the pooled observations. What is the minimum value of $\bar{A}_{samp} - \bar{A}_{bl}$ at which the presence of the constituent is indicated with a probability of 0.98?

$$\frac{(\bar{A}_{samp} - \bar{A}_{bl})}{s_A} \left(\frac{3}{2}\right)^{1/2} = \frac{1.22(\bar{A}_{samp} - \bar{A}_{bl})}{s_A} = +t = 3.0$$

($n = 6 - 2 = 4$ and for the probability $1 - 0.98 = 0.02$, $+t \sim 3.0$ from Fig. 4-10.)

$$\therefore \quad (\bar{A}_{samp} - \bar{A}_{bl})/s_A \geqq 3.0/1.22 \geqq 2.5$$

If, for example, $\alpha = 0.5$ ml/μg (1 cm light path) and $s_A = 0.0010$, the limit of detection in concentration terms, C_d, is 0.005 μg/ml at the 0.98 probability level. α and C_d are usually expressed in terms of the *element* of interest. For *determination* of the element, $\sim 10 \times C_d$ would usually be considered the minimal desirable concentration. This matter has already been considered more exactly in the discussion of the standard curve (section I.B.2 of this chapter).

The choice of detection limit must, of course, be left to the analyst. He decides what the consequences will be if he chooses a criterion that leads him to a wrong conclusion. Is the increased probability of not exceeding a specified deviation worth the extra expenditure of labor and time? The criterion can vary from one analysis to another. The detection limit is chiefly of importance to the analyst because it enables him to assess the precision of results as affected by absorptiometer reproducibility and

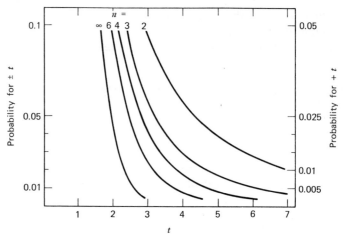

Fig. 4-10. Probabilities corresponding to $\pm t$ and $+t$. n is the number of degrees of freedom. (The curve for $n = 100$ is barely distinguishable from that for $n = \infty$ on this scale.)

possibly other factors. When the reagent absorbs appreciably at the wavelength used for the absorbing species of the constituent, random errors in measuring out the reagent can conceivably lead to larger deviations than result from absorptiometer irreproducibility.

The detection limit is not necessarily the same with real samples as with simple solutions considered above. Other constituents of a sample, even if not leading to a determinate error, could affect reproducibility unfavorably. Even so-called indifferent electrolytes might exert an effect. If separations must be made, reproducibility can hardly be expected to equal that when none are required.

Probability considerations are usually based on Gaussian distribution of deviations. The assumption that Gaussian distribution holds for a particular set of measurements is not necessarily valid. Evidence has been presented that Gaussian distribution is more or less followed in some absorptiometric measurements.[33] If a final result is subject to a combination of relatively many errors of not greatly different magnitude, a better approach to Gaussian distribution may be expected than otherwise. If the photometer and the absorptiometric reaction are highly reproducible, the precision of the absorbance measurements might be determined by the operator in estimating tenths of a division of a meter instrument. Such deviations may be expected to crudely approximate "normal" distribution.

Another factor involved here is the quantization, so to speak, of scale readings. Ordinarily these are made to the nearest 0.1% and smaller variations would not be detected. When the scale is sufficiently expanded, noise becomes the determining factor in precision. If noise could be eliminated, limited response of the photometer could become a deciding factor, that is, a small increase in signal might not produce a proportional change in instrument reading.

Some papers dealing with, or bearing on, absorptiometric sensitivity are mentioned here.[34]

Finally, attention should be called to a more or less special case in formulating sensitivity of photometric reactions. In the previous discussion it was assumed that the standard curve (blank readings subtracted) passes through the origin. This is not necessarily true. Occasionally it has been found that the curve (usually a straight line) does not pass through the origin but cuts the concentration axis at a small positive value, as shown in the adjoining diagram. Such a curve would be obtained if an impurity in the reagents masked, or otherwise inactivated, a small amount of the constituent being determined, so that it gives no absorbing (or fluorescing) species. This behavior was observed by T. Bidleman (Thesis, 1970) in determining bismuth with dithizone in basic medium in the presence of cyanide to mask most other dithizone-reacting metals. Blank absorbance readings of excess dithizone in CCl_4 at 590 nm were subtracted. A rectilinear curve was obtained, but it intersected the Bi concentration axis at a small positive value, the intercept corresponding to ~0.05 μg Bi in 5 ml of CCl_4. The failure of a small fraction of Bi to react with dithizone was attributed to formation of Bi sulfide from a trace of sulfide (or possibly some other impurity) in the KCN used. Recrystallization of KCN reduced the value of the intercept.

If, as before, ΔA is the smallest detectable absorbance difference for a specified confidence level, C_d in the diagram represents the minimal

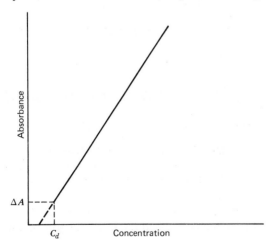

detectable concentration of the constituent in question, which of course is larger than the corresponding value if the standard curve passed through the origin. Also, see p. 229 of this chapter.

Absorptiometric Sensitivity Index. We have found it convenient and advantageous to express the sensitivity of absorptiometric reactions by the use of a sensitivity index $\mathscr{S}$, defined as the number of micrograms of an element, present as the absorbing species, in a column of solution of 1 cm^2 cross section giving an absorbance of 0.001 at a specified wavelength or a defined band of wavelengths (filter).[35] The absorbance is referred to a blank solution containing the same amount of reagent as the test solution. All factors affecting $\mathscr{S}$ (pH, reagent concentration, etc.) should be specified. $\mathscr{S}$ is best found from a standard (calibration) curve but can be derived from a transmission curve if direct proportionality between analytical concentration and absorbance exists, as it usually does. The units of $\mathscr{S}$ are mass/area since

$$bC = \frac{A}{\alpha} \quad \text{or} \quad \frac{\text{cm } \mu\text{g}}{\text{cm}^3} = \frac{\mu\text{g}}{\text{cm}^2}$$

($\mathscr{S}$ could also be expressed as ng/cm^2.)

$\mathscr{S}$ is an experimental quantity and has advantages over ϵ or a for expressing the sensitivity of reactions. It does not require a knowledge of the composition of the absorbing species, whether more than one species is formed, nor the extent to which reaction has occurred, all of which need to be known to express ϵ and a. However, if only one absorbing species is formed, if the constituent is completely converted into this species, and if its true ϵ or a is known, $\mathscr{S}$ can be calculated:

$$\mathscr{S} = \frac{1}{1000a} \tag{4-22}$$

when a is expressed as square centimeters per microgram. In terms of ϵ,

$$\mathscr{S} = \frac{n(\text{at. wt.})}{\epsilon} \tag{4-23}$$

where $n =$ number of atoms of element in formula of absorbing molecule representing a mole.

at. wt. = atomic weight of element in which $\mathscr{S}$ is expressed.

It needs to be noted that if the reagent absorbs at the wavelength used to measure the absorbance of the species formed, account must be taken of the amount of reagent consumed in calculating $\mathscr{S}$. This is done automatically when $\mathscr{S}$ is found from a standard curve.

Results of trace analyses are usually reported on a simple weight basis, not on a mole basis, and sensitivity indexes based on ϵ are not convenient in practice. The range in atomic weight of elements leads to greatly different $\mathscr{S}$ values for the same ϵ value. Thus, if a complex of Be and of U each have an ϵ value of 10,000, $\mathscr{S}$ for Be is 0.0009 and for U approximately 0.024 $\mu g/cm^2$.

The principal reason for selecting an absorbance of 0.001 as the basis for the sensitivity index is a practical one. An absorbance of 0.001 is *roughly* the minimum detectable absorbance with a spectrophotometer or filter photometer having a transmission scale graduated to 1%, and therewith $\mathscr{S}$ is *roughly* the minimum weight of an element detectable in a column of solution of 1 cm^2 cross-sectional area. From tabulated values of $\mathscr{S}$, the analyst can, by simple calculation, find the weight of sample to be taken for stipulated photometric sensitivity or precision, or the relative photometric error at low absorbances. As a ratio, $\mathscr{S}$ is, of course, as exact as a or ϵ when calculated from them, and analytically more realistic when determined experimentally. In practice, it is more useful than these to the analyst because of the way in which it is expressed and because it is an effective value. Values of ϵ and a (if they can be found) may not agree closely with values holding in a particular determination. $\mathscr{S}$ can be used to find photometric detectability more exactly defined on probability considerations (p. 187).[36]

For the more sensitive metal reactions, $\mathscr{S}$ usually ranges from 0.01 to 0.001 $\mu g/cm^2$. Values much less than 0.001 are uncommon, and such reactions are especially prized by the trace analyst, but he will be content to have available reactions with $\mathscr{S}$ approaching 0.001. Even reactions having $\mathscr{S} \sim 0.01$ are not to be disdained. For the determination of manganese as MnO_4^-, $\mathscr{S} \sim 0.03$ at the optimal wavelength, but this is, nevertheless, a useful reaction.

It may be noted that the upper limit of ϵ lies near 2×10^5, more or less in accord with theoretical predictions. The maximum value of ϵ is reported to be given approximately by

$$\epsilon_{max} = 9 \times 10^{19} P\mathbf{a},$$

where P is the probability of electronic (or other) transition and $\mathbf{a}$ is the chromophore area.[37] Values of $\mathbf{a}$ are of the order 10^{-15}–10^{-16} cm^2, so that with $P = 1$, $\epsilon_{max} = 10^4 - 10^5$. For zinc di-2-naphthylthiocarbazonate in chloroform, ϵ at 560 nm is slightly greater than 1×10^5 (ϵ for the reagent itself has been reported to be 4.6×10^4 at 645 nm), corresponding to $\mathscr{S} \sim 7 \times 10^{-4}$ for Zn. Dithizone is almost as sensitive for zinc. The rosocyanin complex of boron[38] has a maximum ϵ of $\sim 180,000$, corresponding to $\mathscr{S} = 6 \times 10^{-5}$. The unusually small value of $\mathscr{S}$ results from the combination of high ϵ and low atomic weight. "The sensitivity is so great that it can prove an embarrassment unless elaborate precautions are used to reduce contamination." For Re as Methyl Violet perrhenate in toluene, $\mathscr{S} = 8 \times 10^{-4}$ (corresponding to $\epsilon = 230{,}000$).

Indirect photometric methods sometimes provide sensitivity indexes smaller than 1×10^{-3}. For example, Mn(VII) oxidizes 4,4'-tetramethyldiaminotriphenylmethane to a yellow product, $\mathscr{S}$ for the reaction being $1 \times 10^{-4}\ \mu\text{g Mn/cm}^2$ at 475 nm. The exceptional sensitivity is largely explained by the five oxidizing equivalents of Mn(VII), and such reactions are not typical. Phosphate can be determined extremely sensitively by an amplification reaction in which phosphomolybdic acid, $H_3PO_4 \cdot 12\ MoO_3$, is extracted into butanol–chloroform, and molybdenum in the extract (washed free of excess molybdic acid) is determined with 2-amino-4-chlorobenzenethiol.[39] The effective ϵ (at 710 nm) for P is 359,000 or $\mathscr{S} = 9 \times 10^{-5}\ \mu\text{g/cm}^2$.

Another type of indirect method, the catalytic (p. 201), can often provide unusual sensitivity.

D. GENERAL DESIDERATA FOR COLOR OR ABSORPTIOMETRIC REACTIONS FOR TRACE ANALYSIS

These may be summarized as follows.

1. High Sensitivity

This is clearly an important desideratum when minute amounts of elements are to be determined, and other desiderata may be relaxed in its favor in choosing a reagent or method. Within limits, reproducibility and stability of the colored product may be sacrificed to obtain it. Thus, the stability of some dithizonates in solution is not as good as might be desired, but the great sensitivity of dithizone as a heavy metal reagent more than makes up for this deficiency.

With a fixed minute quantity of an element, the precision may be determined by the sensitivity, other factors being equal. As pointed out earlier, the sensitivity index $\mathscr{S}$ roughly represents the minimal detectable amount of an element under certain conditions, and it may therefore determine the relative precision for small amounts of an element.

It is desirable that the reaction product absorb strongly in the visible portion of the spectrum, rather than in the ultraviolet. In general, a greater variety of substances are likely to interfere by absorption in the UV than in the visible spectrum. Nitrate, iodide, perchlorate, and the like absorb in the UV, as do traces of organic substances derived from filter paper, ion-exchange resins, stopcock grease, and plastic vessels. To be sure, when their amounts are not too large and do not change in the subsequent procedure, they can be corrected for by finding the absorbance of the sample solution before addition of the reagent forming the absorbing species. An off-hue in a *colored* solution may give warning of the presence of foreign reacting substances. In general, greater care is needed when using UV absorption methods than when using visible absorption methods. But measurements in the UV increase the number of available methods for metals (and nonmetals) and some UV methods are very sensitive, more so than visible spectrum methods.

As a class, organic reagents are more sensitive than the inorganic. Often, organic complexes of metals, whether chelates or ion-association complexes, can be extracted into organic solvents immiscible with water. In this way, the sensitivity of the determination can be increased if the aqueous sample volume is relatively large, as it often is when ~1 g samples are taken. Sometimes an advantage adduced for a method is the convenience of nonextraction, the absorbing species being water soluble. That may be, but this advantage is greatly outweighed by the increase in sensitivity gained by extraction. Almost all the most sensitive trace metal methods entail extraction.

2. Specificity or Selectivity

There are three levels of specificity: of reagent, of reaction, and of determination (or method). A spectrophotometric (absorptiometric) reagent may be called specific for a particular element if it forms a reaction product with that element absorbing at some wavelength in the visible or UV spectral range, at which the reaction products (if any) of all other elements do not absorb.[40] Such reagents are very few, if indeed they exist at all. Biquinoline and related compounds are described as specific for Cu(I), because no other element has apparently been found to yield a colored complex extractable by suitable immiscible solvents, although a few elements give colored nonextractable complexes. A reagent reacting with relatively few elements is called selective.

A reagent that reacts with a considerable number of elements can often be made to give more selective reactions, possibly even a specific reaction, for an element by suitably complexing other elements or altering their oxidation states (these measures are the common ways of *masking*) or even by adjusting the pH. In practice, it is more useful to speak of selectivity of reactions than of reagents, and the former is usually understood when no qualification is made.[41] Dithizone is an example of an unspecific reagent (it reacts with more than a score of metals) that can be made to give selective reactions with a number of metals by masking and adjustment of the hydrogen-ion concentration.

We are still so far from the ideal of specific reactions—to say nothing of specific reagents—that to obtain a specific or even selective determination it is often necessary to get rid of interfering constituents in a sample. This is especially true in trace analysis, where the ratio of interfering constituents to the trace constituent is often very unfavorable. Sooner or later, separations become necessary. The choice of a separation method may or may not depend on the color reagent to be used later—or conversely. Masking may, of course, be combined with separations. In combination, they may provide a method of determination that is specific. But that term is subject to reservations. The specificity is often conditional. A method may be specific for a constituent in a certain class of materials, with a composition falling in a certain range. The presence of an unexpected constituent may cause trouble and require procedure changes that are not always minor.

If good, easy methods of separation are available, the selectivity of the color reaction becomes less important, but the greater the selectivity the greater the assurance.

It is advantageous to have available several sensitive color reagents for a given element, if these differ in their reactivity with other elements. It may be found that one reagent may be much more suitable for a certain type of sample than another reagent, which may react with one or more of the other elements present in the sample.

3. Good Reproducibility and Stability of the Color

A certain latitude is always desirable in the variables (pH, reagent concentration, possibly others) determining the color intensity. Methods requiring careful adjustment of pH are undesirable but may need to be used; adequate buffering is then essential. If there is a choice, methods that can be carried out at fairly high acidities are preferred; difficulties due to hydrolysis of metals are then minimized, and the selectivity may be greater. Most of the chelating reagents are weak acids, and the stability of their metal chelates is such that weakly acidic or even weakly basic reaction media are often necessary. Complexing agents such as citrate may then be needed to prevent precipitation of hydrolyzable metals. On the other hand, ion-association complex formation, for example, Rhodaminium B chloroantimonate, may require mineral acid media.

Reagents that are colorless are preferred to those that are colored. But many colored reagents are used perforce. Sometimes a choice can be made from a group of similar colored reagents. It is obviously desirable that the wavelengths of maximum absorption by the complex and by the reagent not lie too close together. Other factors remaining the same, the reagent chosen would be the one for which $\Delta\lambda_{max}(=\lambda_{max, complex} - \lambda_{max, reagent})$ is the largest. Dithizone is an example of a reagent showing favorable $\Delta\lambda_{max}$ for many metals.

The reaction product should be soluble either in aqueous solution or in an organic solvent. Occasionally, suspensions or colloidal solutions of strongly absorbing reaction products are used in absorptiometric analysis. The reproducibility of such methods varies. On the whole, they are best avoided, if other methods are available. The formation of insoluble products may sometimes be prevented by adding miscible organic solvents in sufficient amount, but this is not always desirable when other (inorganic) constituents are present, because precipitation of these may occur. Extraction of the colored product into an immiscible organic solvent is often to be preferred, but occasionally chelates not easily extractable are encountered.

Because of its otherwise favorable characteristics, the analyst may choose a reagent whose stability or that of its reaction product is not as good as desirable. Greater care is then obviously needed to obtain reproducible results. Special attention must be paid to the time of standing and exposure to light. Refrigeration of reagent solutions and the use of solvents free from oxidizing substances may be helpful.

4. Rectilinear Relation Between Analytical Concentration of Constituent and Absorbance

The term "adherence to Beer's law" and equivalent expressions are often used loosely to indicate a rectilinear relation between the analytical concentration of an element in a solution and its absorbance. As we have seen, failure of this relation is most unlikely to be due to real failure of Beer's law, which merely specifies a rectilinear relationship between absorbance and the concentration of a single, definite dissolved species at a given wavelength.[42]

With a sufficient excess of reagent and the formation of a soluble reaction product, a rectilinear relation is almost always obtained in trace metal determinations when the wavelength band is sufficiently narrow. A filter photometer occasionally gives an initially straight line for absorbance versus concentration, curving at higher concentrations so that it tends to run parallel to the concentration axis. Better precision is, of course, attainable when the standard curve is a straight line, but slight curvature is acceptable, although more time is then required to establish the curve.

It may be of interest to examine reasons for nonlinearity between absorbance and analytical concentration of an element arising from equilibria in the color system. Suppose a colorless metal ion M (charges omitted) reacts with a colorless reagent R to give a soluble, colored complex according to the reaction

$$m\text{M} + n\text{R} \rightleftharpoons \text{M}_m\text{R}_n$$

The formation constant of the colored complex is

$$\frac{[M_m R_n]}{[M]^m [R]^n} = K$$

The solution absorbance is proportional to $[M_m R_n]$. If virtually all M is converted to $M_m R_n$, absorbance is proportional to the original (analytical) concentration of M. If M is incompletely converted to $M_m R_n$, A will not be proportional to the analytical concentration of M except when $m = 1$, and then only if $[R]$ is practically constant, which means that the original concentration of R must be so large that its concentration does not change appreciably on reaction. In the general case, intermediate species of M and R will be formed, and these would not be expected to have the same ϵ, although they might be similar. If such species are present, the concentration of species having $m > 1$, would depend on the analytical concentration of M and A might not vary rectilinearly with the analytical concentration. Most chromogenic chelating agents furnish singly charged anions, so that the complexes $MR, MR_2 \cdots$ can be formed; A should be proportional to the analytical concentration of M if $[R]$ is practically constant, unless association occurs:

$$p(MR_n) \rightleftharpoons (MR_n)_p$$

and the absorptivities of the two species are different. An inorganic example of association is Cr(VI) in acid solution:

$$2\,HCrO_4^- \rightleftharpoons Cr_2O_7^{2-} + H_2O \qquad (K = 43)$$

At a fixed pH, the relative amounts of the two Cr species will depend on the total Cr concentration, and A will not exactly vary rectilinearly with that concentration.

When the reagent is colorless, there is usually no reason for not adding a reasonably liberal excess, so that its concentration in the solution remains essentially unchanged, and the same fraction of M is converted to a complex such as MR_n, no matter what its concentration. When the reagent is colored and absorbs more or less at the wavelength used for measuring the absorbance of the complex, it may be undesirable to add a large excess of reagent and then conversion of M to the colored complex may be incomplete when the analytical concentration of M is relatively high. This effect can be observed with strongly colored extraction reagents, for example, dithizone. When the concentration of dithizone in the organic phase is small at equilibrium, the reacting metal may not be extracted completely and the standard curve departs from linearity.

Although, as said, rectilinear relations between A and analytical concentration are the rule in absorptiometric trace analysis, exceptions are encountered, as shown in Fig. 4-11.

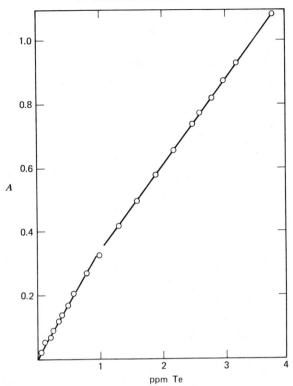

ppm Te

Fig. 4-11. Unusual standard curve (two straight line segments) for the determination of Te(IV) with iodide. The iodo species present have not been determined. From K. Marhenke and E. B. Sandell, *Anal. Chim. Acta*, **38**, 421 (1967).

5. Simplicity

Finally, if the choice is offered, methods that are simple and easy to carry out—and therefore less time consuming—will obviously be preferred to those that do not have these advantages, provided that more important requirements are not sacrificed.

E. INDIRECT ABSORPTIOMETRIC METHODS

If these are defined as including all methods in which the colored (or absorbing) determination form does not contain the element being determined, the following classification can be made:

1. The constituent transforms the reagent into a colored product. The latter is almost always an oxidation or reduction product. *Example:* determination of thallium(III) by liberation of iodine on treatment with iodide.

2. The constituent reacts with a colored reagent to form a colorless product, and the extent of bleaching is a measure of the amount of constituent. *Example:* determination of fluoride by bleaching of red zirconium alizarinate. The colorless substance is usually a complex of the constituent and a component of the reagent. However, it can be an oxidation or reduction product. Thus cerium(IV) can be determined by its oxidation of red ferrous phenanthroline to almost colorless (pale blue) ferric phenanthroline.

3*a*. The constituent is precipitated as a slightly soluble compound, which, after being washed, is dissolved and

(*1*) The color intensity of the solution (if the precipitate contains a colored component) is measured.[43] *Example:* Magnesium 8-hydroxyquinolate is dissolved in mineral acid and absorption due to $HOx \cdot H^+$ is measured.

(*2*) A reagent is added to develop a color by reaction with a component of the precipitate other than the constituent. *Example:* Sodium uranyl magnesium acetate is treated with thiocyanate to give yellow uranyl thiocyanate. Or the precipitate can be made to bleach a colored substance. *Example:* calcium oxalate is oxidized with excess ceric sulfate.

3*b*. The constituent is precipitated as a slightly soluble compound by a reagent, and the excess of the latter is found by measurement of its own color or of the color formed by treatment with a suitable reagent. *Examples:* calcium by chloranilic acid; potassium by treating the excess of chloroplatinic acid with stannous chloride to develop a yellow or orange coloration.

4. The constituent catalyzes a reaction (or otherwise alters the reaction rate) between two substances, one of which is colored or can be converted into a colored species, and its concentration is found from the reaction rate, i.e., by measuring the color intensity as a function of time.

Disadvantages and Advantages. On the whole, indirect methods are inferior to the direct, and they are used more by necessity than by choice. Principally, except for catalytic methods, they fill a certain need in making possible a colorimetric determination of the alkali metals and some of the alkaline earths, as well as a number of nonmetals.

The main objection that can be raised against the oxidation-reduction methods of class (1) is their limited selectivity. Difficulties can also arise from nonstoichiometric reactions and in the removal of excess oxidizing or reducing agent used to put the constituent in reactive form. It is true that similar difficulties can be encountered in analogous direct methods.

Reactions of class (2) are also likely to be lacking in selectivity, and there is no way of telling from the absorption curve whether substances other than the constituent in question are reacting. Attention must be paid to the proper photometric technique in this type of determination. Since the reagent solution is likely to have a fairly high absorbance and the change produced by a trace of constituent may be small, it is best to measure against unbleached reagent solution or standards in the reference cell.[44]

In indirect methods for metals, reactions of class (3) play the major role. The solubility of the precipitate is of course the most important factor in determining the accuracy and sensitivity. The range is very great. At one extreme are precipitates such as sodium uranyl zinc (or magnesium) acetate, which in spite of its low sodium content can be made to serve for small amounts (not traces) only by special measures, and at the other, precipitates such as silver p-diethylaminobenzylidenerhodanine, whose solubility can be reduced to less than $10^{-7} M$ (<0.01 ppm Ag). Especially when organic precipitants are used, account must be taken of the molecular as well as the ionic solubility of the precipitate. Often it is possible to reduce the ionic solubility to negligible values by adjustment of pH, but the molecular solubility remains unchanged and may have an appreciable value. For example, nickel(II) dimethylglyoximate has a molecular solubility of $9.7 \times 10^{-7} M$, corresponding to 0.057 ppm Ni in water at 0.05 ionic strength. In alcoholic solution, such as would result from using an alcoholic solution of precipitant, the molecular solubility will be greater.

Even if the total solubility of the precipitate is satisfactorily low, difficulties may arise from formation of supersaturated solutions. Occasionally, the reagent itself, if slightly soluble, can be made to function as collector. This is illustrated by the rhodanine reagent in the indirect method for silver. When the precipitate has been collected, the reagent can be washed out with alcohol, in which the silver rhodanate is virtually insoluble. This favorable combination of properties will not often be encountered.

Another factor limiting the accuracy of precipitation methods is the possible nonstoichoimetric composition of the precipitate. For example, reagent or other elements may be coprecipitated, even if not precipitated. Some methods of this type are very indirect, with possibilities of error in each step. Thus one method for calcium consists in its precipitation as oxalate, reaction of this with ceric sulfate, and determination of the excess cerium by reaction with iodide to liberate iodine which is determined finally as I_3^-.

Sometimes a precipitation method will provide a separation from other constituents in the first step, and may thus be preferred to a direct method which will require an extra separation step.

The main reason for the use of a catalytic method is the high sensitivity it may afford. Such a method is not likely to be applicable without a preliminary separation of the constituent, because neutral salts and the like will affect the rate, even if the reaction is selective. At the present time, three elements at least, osmium, ruthenium, and iodine, can advantageously be determined by catalytic methods in consequence of great sensitivity and the availability of effective methods of isolation.

We consider catalytic—or more generally, reaction rate—methods in the following section.

Reaction Rate Methods.[45]   In nonrate methods, the mainstay of analytical photometry, the reaction between a constituent and a photometric reagent is allowed to proceed to equilibrium (or to a point close to it if the reaction is slow), when a large fraction, if not essentially all, of the constituent will have been converted to the absorbing or fluorescing product. Another approach is sometimes profitable. The rate of formation (or disappearance) of some absorbing species is related to the concentration of the constituent being determined, so that measurement of the rate gives the concentration. In a sense, all rate methods are indirect, but some are more indirect than others. An analytically useful distinction in considering rate methods is whether a stoichiometric relation exists between the amount of constituent that has reacted and the amount of absorbing species formed or disappearing. In a catalyzed reaction, the concentration of the constituent in its active forms remains ideally constant with time and the ratio (amount of absorbing species formed/amount of constituent) can attain very large values. Accordingly, such reactions can be very sensitive and are of special value in trace analysis. On the other hand, if there is a stoichiometric relation between the two quantities, that is, if there is no catalysis, the sensitivity may be no better than in a corresponding nonrate method. Since such stoichiometric rate methods are not characterized by exceptional sensitivity and since their precision can hardly be better than the corresponding equilibrium method, their use must be attended by some other gain. The advantage lies mostly in the greater selectivity that they may be able to provide. Two metals M and M' may react with a reagent to give similarly absorbing products, but it may be possible to determine one in the presence of the other if they react at different rates with a suitable reagent. Greater speed may also be an advantage of such methods. More or less elaborate and sophisticated instrumentation is usually required in the application of these methods.

The following discussion touches on the two types of rate methods. It is intended to be illustrative, not exhaustive.

1. Catalytic Methods

a. Oxidation–Reduction Reactions

Probably the greater number of proposed determinations based on catalysis involve oxidation–reduction systems, and most of those likely to be useful in practice seem to fall in this class.

An oxidation–reduction reaction

$$Ox + Red \rightarrow Ox' + Red'$$

is likely to be slow, even though the oxidation potentials are favorable, if an unequal number of electrons are involved in the half reactions. When this is so, it has been observed that some element can often be found that will effectively catalyze the reaction. A reduced ionic form (or forms) of the catalyst must be capable of rapid oxidation by Ox, and this oxidized form must then be rapidly reduced by Red to its reduced form. The repetition of this cycle many times sooner or later brings the system to equilibrium. Evidently, the reaction rate will be some function of the catalyst concentration. Often a simple proportionality exists (Fig. 4-12). For this system to be useful in trace analysis, Ox and Red when alone must show very little interaction, and the catalyzed reaction must be rapid at very low concentrations of catalyst.

An exemplar of oxidation–reduction systems useful in trace determinations by catalysis is the As(III)–Ce(IV) system

$$As(III) + 2\ Ce(IV) \rightarrow As(V) + 2\ Ce(III)$$

  colorless   yellow   colorless   colorless

For a long time, this reaction has been known to be very slow.[46] We have observed that a mixture of the two reactants in equivalent amounts [or with an excess of As(III)] in sulfuric acid medium remains yellow for days if sufficiently pure (not all reagent products are). Powerful catalysts for reaction are $I(I_2-I^-)$, Os, Ru, and Ir. The determination of iodine by this catalysis (described in 1934) is probably the only catalytic method that has found extensive use. Osmium, Ru, and Ir are not elements whose trace determination is often required, but when so, they can be determined even more sensitively than I_2 by this reaction.[47] As little as 0.01 μg Os can be determined.

The sensitivity and the precision of the Os and Ru determinations may be judged from Fig. 4-12.[48] The equation for the Ru line in this figure is

$$[Ru] = 0.00574 \left(\frac{1}{T} - 0.0015 \right) \qquad ([Ru] \text{ in ppm})$$

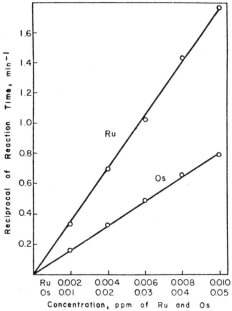

Fig. 4-12. Reciprocal of reaction time as a function of Ru and Os concentration in catalysis of ceric–arsenite reaction. Original concentrations of Ce(IV) and As(III) each 0.040 N; 2 M sulfuric acid; 25°C. Reaction time is defined as time required for percentage transmission to change from 22 to 60 at 488 nm. From E. B. Sandell, *Colorimetric Determination of Traces of Metals*, 3rd ed., p. 787.

T is the reaction time (arbitrarily defined) in minutes. The term 0.0015 (corresponding to ~700 min) represents the rate of the Ce(IV)–As(III) reaction in the absence of (added) catalyst. Since the blank is very small, $0.00574 \times 0.0015 = \sim 10^{-5}$ ppm Ru, minute amounts of Ru can be determined. A reaction time of 10 min corresponds to 6×10^{-4} ppm Ru.

Few catalytic methods are likely to be of much value in applied trace analysis unless selective separation methods are available for the elements being determined. In this respect, Os and Ru (as well as I) are exceptional. As the tetroxides, Os and Ru are easily isolated and separated from essentially all other elements by volatilization or CCl_4 extraction. Also, Os and Ru can be separated from each other by differential oxidation: Os(IV) is more easily oxidized to Os(VIII) than Ru(IV) to Ru(VIII). Besides, Os and Ru can be determined in the presence of each other by altering the conditions in the reaction solution so that the two metals exert different catalytic effects.[49] Silver or Hg(II) inactivates I^-.

An example of a combination of *induced* and *catalytic* reactions in trace analysis is provided by the system Cr(VI)-Te(IV)-1,10-phenanthroline. Chromium(VI) is reduced by Te(IV) to a lower oxidation state thought to be Cr(IV):

$$Cr(VI) + Te(IV) \rightarrow Cr(IV) + Te(VI) \tag{1}$$

When Cr(VI) is present in excess, this reaction can be used to determine Te by the induced reaction

$$\underset{\text{pale blue}}{Cr(IV)} + 2\,FePhen_3^{3+} \rightarrow Cr(VI) + \underset{\text{orange-red}}{2\,FePhen_3^{2+}} \tag{2}$$

The amount of $FePhen_3^{2+}$ formed, found spectrophotometrically, corresponds to an equivalent amount of Te(IV), and no unusual sensitivity is to be expected. On the other hand, if Te(IV) is present in excess, Cr(VI) formed in reaction (2) is again reduced by Te(IV) to Cr(IV), which reduces $FePhen_3^{3+}$ and the cycle continues. Chromium (VI) is detectable in this way at a concentration of 5×10^{-6} ppm, an unusual sensitivity.[50] Presumably this induction–catalysis system could be used for the determination of Cr. The cyclical reactions eventually come to an end because of side reactions, such as that between Te(IV) and Cr(IV), which yields Cr(III).

b. Exchange Reactions

Relatively recently, catalytic reactions not involving oxidation–reduction have been made the basis of rate determination of metals. An important type of reaction in this category is that in which one ligand exchanges with another in a metal complex according to the general scheme

$$ML + L' \rightleftharpoons ML' + L \tag{1}$$

or a double exchange takes place:

$$ML + M'L' \rightleftharpoons ML' + M'L \tag{2}$$

An example of (1) is the catalysis by Hg^{2+} of the slow reaction

$$Fe(CN)_6^{4-} + \underset{\text{nitrosobenzene}}{C_6H_5NO} \rightleftharpoons \underset{\text{violet}}{Fe(CN)_5(C_6H_5NO)^{3-}} + CN^-$$

The detection limit is $1 \times 10^{-7}\,M$ Hg.[51] Silver, Au, and Pt also catalyze.

An example of double exchange (2) is provided by the slow reaction between triethylenetetramine-Ni and ethylenediaminetetraacetato-cuprate(II), which is catalyzed by free EDTA:

$$NiT^{2+} + CuY^{2-} \rightleftharpoons CuT^{2+} + NiY^{2-}$$

Such exchange reactions are often slow because a series of coordinate bonds must be broken in succession (six with EDTA). Any substance stabilizing the intermediate species (thus blocking reformation of the chelate ring) will increase the rate of dissociation of CuY^{2-} and formation of CuT^{2+}. Addition of a small amount of free EDTA catalyzes the above reaction by a chain propagating mechanism:

$$Y_{tot} + NiT^{2+} \rightarrow NiY^{2-} + T_{tot} \qquad (tot = \text{total})$$
$$T_{tot} + CuY^{2-} \rightarrow CuT^{2+} + Y_{tot}$$

Metals that are complexed by EDTA can be determined indirectly by their inhibition of the catalysis resulting from removal of free EDTA.[52] The rate measurement gives the concentration of free EDTA, and since the total amount added is known, the initial amount of metal can be found. The exchange rate is found from the absorbance of CuT^{2+} (bright blue) as a function of time. Metal concentrations as low as 10^{-8} M can be determined. Evidently one of the variables that must be controlled is the pH.

2.   Uncatalyzed Rate Methods

The success of some photometric equilibrium methods rests on the substantially complete formation of the absorbing species of the constituent before undesired constituents have reacted appreciably, but otherwise little use is made of a difference in reaction rates in such methods. With modern instrumentation, differences in reaction rates of rather rapid reactions can be exploited and made the basis of useful determination methods.

For example, an automated fast reaction-rate system allows determination of phosphate by stopped-flow in the millisecond range as 12-molybdophosphoric acid.[53] At HNO_3 concentrations below 0.5 M, the rate law for formation of yellow 12-MPA is

$$\frac{d[12\text{-MPA}]}{dt} = k[H_3PO_4][Mo(VI)]$$

If Mo(VI) is kept in large excess, the initial reaction rate is proportional to the phosphate concentration. Digital rate data proportional to phosphate concentration are obtained 0.1–0.2 second after mixing. Concentrations are found from phosphate standards.[54] Concentrations as low as 0.1 ppm P can be determined. In the range 0.2–0.75 ppm, averages of 10 results show a standard deviation of less than 1%. Arsenic in the ratio 1000:1 P produces an error of $\sim +1\%$. The interference of Si is greater, but 50–100 times as much Si as P can be tolerated.

Uncatalyzed ligand exchange between metals by the gross reaction

$$ML + M' \rightleftharpoons M'L + M$$

may take place with greatly different rates depending on the metals (and ligand) involved, and then determination of one or more metals in a metal mixture becomes possible. This reaction system is illustrated by the use of

CyDTA (*trans*-1,2-diaminocyclohexane-*N,N,N',N'*-tetraacetate) chelates.[55] The exchange of M' for M in M–CyDTA does not take place
directly, because the rigid cage structure of CyDTA cannot accommodate
more than one metal ion at a time.

Hydrogen ion reacts however, displacing M, which can be replaced with a
suitable metal such as Pb or Cu(II):

$$MCyDTA^{2-} + H^+ \xrightleftharpoons{\text{rate-detmng.}} HCyDTA^{3-} + M^{2+} \qquad (1)$$

$$HCyDTA^{3-} + Pb^{2+} \xrightarrow{\text{fast}} PbCyDTA^{2-} + H^+ \qquad (2)$$

The rate-determining step is the first reaction, with a characteristic rate
for each metal. The second reaction is rapid and the ultraviolet absorbance of $PbCyDTA^{2-}$ (or $CuCyDTA^{2-}$) is used to monitor the exchange
reactions, the rate of which is accordingly given by $d[PbCyDTA^{2-}]/dt$.
When an excess of Pb^{2+} (scavenger) is present, each metal CyDTA
complex reacts independently of others and of metal ions in the solution.
Stopped-flow mixing is usually required.

Reaction rate constants for Mg:Ca:Sr:Ba are in the ratio
1:6.5:96:1660. Binary and ternary mixtures in which the metal ratios
are not too unfavorable can be analyzed with errors usually not greater
than 10–15% down to 10^{-5} M. The absorbance (or transmittance)–time
data are analyzed graphically or by computer[56] to obtain metal concentrations. The metals being determined are run as pure components before
and after a series of measurements on the sample. Only ~0.2 ml of
solution is required for a measurement. The transition metals hardly
interfere with the determination of the alkaline earths because their
CyDTA complexes react much more slowly with H^+. Other metals react
even more slowly than the transition metals.

Reaction (1) (p. 205) is first-order with respect to $[H^+]$ and to [M–
CyDTA] and the rate constant $k_{H^{MCy}}$ or $k_{H^{MHCy}}$ varies over an enormous
range for different metals, for example, 1×10^8 M^{-1} sec^{-1} for Ba (pH 7–7.6)
to 3.4×10^{-4} for Ni (pH 4–5.2) (if the half life is 0.01 sec for the Ba
complex, that of the Ni complex would be a century). The rate constants
for the transition metals do not differ so much, however; some are close
together. Reaction rates can be varied by changing the pH so as to bring
the rates into a favorable observational range. Some binary mixtures

(comparatively unfavorable cases), as Zn–Mn, Zn–Cd, and Cd–Pb at the 10^{-4}–10^{-5} M level can be analyzed with errors usually not exceeding 10% when the metal ratios are favorable ($\sim 1:1$).

3. Enzyme Methods

The rate of an enzyme-catalyzed reaction may increase (activation) or decrease (inhibition) in the presence of ppm- or ppb-concentrations of a metal ion.[57] The metal can be determined by monitoring the reaction spectrophotometrically or fluorimetrically. Zinc and calcium have been determined by their activation[58] of the apoenzyme of calf-intestinal alkaline phosphatase and beryllium and zinc by their inhibition[59] of calf-intestinal alkaline phosphatase. Beryllium and bismuth have been determined by their inhibition of alkaline phosphatase.[60] Enzyme methods are not very selective. Preparation and purification of enzyme and the pH and temperature of the reaction must be carefully controlled.

II. COLORIMETRY

The essence of colorimetry—this term being used here in its narrow or specialized sense—is comparison of the color[61] of the sample solution with the color of a standard solution containing the same colored substance. The comparison is effected by arranging matters so that the color intensities (or hues) of the two solutions appear identical to the observer. Ordinarily, white light is used for illumination. Methods of arriving at the balance point are mentioned below.

The precision of colorimetric determinations is inferior to that of absorptiometric determinations, and the sensitivity is usually less. But these disadvantages are outweighed by certain advantages under some circumstances. Colorimetric methods provide quick, inexpensive determinations of trace metals with a minimum of apparatus. They are of special value in the field, perhaps hundreds of miles from a laboratory with sophisticated instrumentation. Analyses can be made on the spot with easily transported apparatus. Colorimetric analysis finds extensive use in geochemical exploration (prospecting),[62] away from inhabited localities. Frequently, in such applications high precision is not necessary—it may be a waste of time and money—and it is more important to be able to obtain approximate results without delay. The standard series method is usually applied, and for rough work color comparisons can be made in test tubes.

A. SENSITIVITY

Visual sensitivity is usually determined by using identical flat-bottomed tubes, one containing the blank solution, and the other variable amounts of the substance yielding the colored product together with the reagent. White light (daylight reflected from a white surface) is used for illumination. Such factors as the acidity, the excess of reagent, and the time of standing must be specified. The quantity of substance ($\mu g/cm^2$) required to give a perceptible difference in appearance from the blank can be defined exactly only on the basis of probability. Thus the minimum amount might be defined as that which allows a detectable difference in appearance from the blank in 95 out of 100 comparisons. The determination of the limit thus defined would be laborious, and a less exactly defined and somewhat larger limit may be taken for all practical purposes, namely that which in practically every comparison will show a difference from the blank. A subjective factor is involved to a greater or less extent, and different observers (especially the inexperienced or those suffering from some visual impairment) may obtain quite different results. See Table 4-5 for some values obtained by a supposedly normal observer.

TABLE 4-5
Some Visual Sensitivity Values

Element	Reagent	Sensitivity (minimum detectable amount, $\mu g/cm^2$)
Al	Aurintricarboxylate (pH 5.5)	0.01
Au	p-Diethylaminobenzylidenerhodanine	0.05–0.1
Be	Morin (fluorescence)	0.002
Bi	Iodide	0.5
Cd	Dithizone	0.03
Co	Nitroso-R salt	0.01
Co	Thiocyanate-acetone	1
Cr	As CrO_4^{2-}	1.0
Cr	Diphenylcarbazide	0.03
Cu	Ammonia	10
Cu	Hydrochloric acid (9 M)	0.5
Cu	Diethyldithiocarbamate (H_2O)	0.05
Fe	o-Phenanthroline	0.05
Fe	Thiocyanate	0.1
Ga	8-Hydroxyquinoline ($CHCl_3$, fluorescence)	0.1
Mg	Titan Yellow	0.05
Mn	As MnO_4^-	0.1
Ni	Bromine-dimethylglyoxime	0.05–0.1
Pt	Iodide	0.5
Sb	Iodide	1
V	Hydrogen peroxide	2
V	Phosphotungstic acid	1
W	Thiocyanate–stannous chloride	0.3

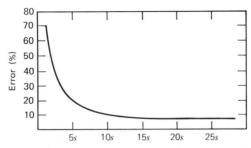

Fig. 4-13. Maximum error in color comparison by standard series method as a function of the amount of colored substance in a column of unit cross section (1 cm^2) in terms of the visual sensitivity s of the reaction. The error becomes essentially constant at $15s$.

For low values of mass per square centimeter, the sensitivity value thus defined determines the precision of the visual color duplication by the standard series method. The probability is great that the error of a single color duplication will not exceed $(s/q) \times 100\%$ (where s is the sensitivity value and q is the weight of substance in micrograms per square centimeter cross section of solution) as long as q is less than 15 or $20s$. When q/s is greater than 15 or 20, the precision remains essentially constant over a wide range of q values.[63] (See Fig. 4-13.) In other words, no useful purpose is served in working under conditions such that q/s is larger than about 20.

For substances having broad absorption bands in the part of the visible spectrum in which the human eye is not too insensitive, the minimum amount of substance visually detectable is 5 or 10 times the spectrophotometric sensitivity index $\mathscr{S}$ defined on p. 190. Accordingly, at a stratum thickness of, say, 25 cm (Nessler tube), the minimum *concentration* of a substance visually detectable might be less than that detectable spectrophotometrically in a 1 cm cell.

B. METHODS

As already mentioned, the essential feature of colorimetric methods is the matching of the color of the sample solution with that of a comparison solution, so that when duplication is attained both solutions contain the same amount of the colored substance in columns of equal cross section:

$$c_x l_x = c_s l_s,$$

where c_x is the concentration of sample solution, l_x is the depth of sample solution (in line of sight), c_s is the concentration of standard solution, and l_s is the depth of standard solution. Since $c = \text{mass}/l^3$, the dimensions of cl are mass/l^2 or mass per unit cross section.

If $c_x l_x$ is fixed, there are evidently three ways in which this product can be duplicated if one starts with a standard solution of concentration c_s (Fig. 4-14): (1) standard series and colorimetric titration methods, (2) balancing method (Duboscq colorimeter), (3) dilution method.

Standard Series. A number of known solutions is prepared, each having the same volume as the sample solution and contained in identical tubes with flat bottoms. The sample and comparison solutions are treated with reagent at as nearly the same time as possible. This method of color comparison is the most sound in principle, because color intensity need not be directly proportional to constituent concentration, error from fading or slow development of color is reduced to a minimum, and any effect of variable concentration of reagent is eliminated. The method is not as precise as some nor as convenient to carry out, for a comparatively large number of standards may have to be prepared. Sometimes the amount of constituent in the sample solution is roughly determined with a preliminary series (or by a colorimetric titration) and then more exactly by the use of a second series in which adjacent members differ by smaller increments.

Nessler tubes are generally less convenient to use in color comparisons in trace analysis than shorter and somewhat narrower tubes such as the one illustrated in Fig. 4-15. These tubes are provided with ground-glass

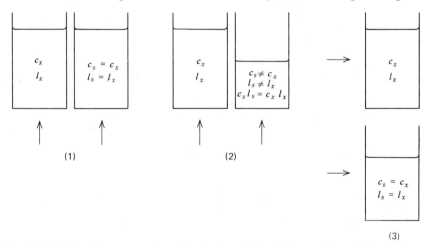

Fig. 4-14. Schematic representation of methods used in arriving at the state of equicoloration in colorimetry.

	Vary	Keep Constant
Method (1)	c_s, l_s	—
Method (2)	l_s	c_s
Method (3)	c_s	l_s

Fig. 4-15. Tube for color comparisons.

stoppers to permit their use in extraction procedures with immiscible solvents, and to facilitate mixing by inversion. It is convenient to have available two sizes of these tubes, namely 1.8×15 cm (25 ml) and 1.2×8 cm (5 ml).

Comparison of the solutions should be made in daylight by examining the tubes axially against a white background such as obtained by inclining a white card at an angle of about 45° in front of a window receiving light from the sky, but not directly from the sun. When layers of immiscible solvents are to be compared, the tubes are viewed at right angles to their axes against a white vertical background. The tubes should be interchanged when the colors appear to match, because to some observers the tube on one particular side will always appear a little more strongly colored than that on the other when the color intensities are actually the same.

An observer can usually differentiate without hesitation between two solutions when the amounts of colored substance in columns of equal cross-section differ by about 1/15 (or 7%) in the optimum range. As a rule no useful purpose is served in having adjacent members of the standard series differ by less than 10 or 15%. As a matter of fact, standards differing by 20% will serve satisfactorily. With such a series, an experienced observer can usually find the content of the sample within

5%. Suppose X (the sample) contains 11.2 μg of constituent, and the two adjacent standards contain 10 and 12 μg. The observer will probably have difficulty in deciding whether X is closer to one or the other of the standards, or at least will see that it is very close to being midway between the two, and he will call the content 11 μg. If X contains 11.6 μg he probably will be unable to distinguish X from the 12 μg standard and will call the amount of constituent present 12 μg, and the error will still be less than 5%. If the amount of constituent is very small, then the standard solution may contain respectively 0, 2s, 4s, 6s, 8s, and 10s μg of the constituent per square centimeter of cross-section (s is the sensitivity as defined on p. 207).

Sometimes a series of artificial standards, prepared from a substance, or more generally a mixture of substances, as similar as possible in transmission characteristics to the constituent being determined, is used for comparison in place of standards prepared from the constituent itself.[64] Such artificial standards are convenient when natural standards are difficult to prepare, as for instance in the determination of free chlorine in water and in certain routine work.

In the colorimetric titration method, a standard solution of the constituent being determined is added to a tube or other suitable vessel containing water and reagent until the color is the same as that of a second tube containing the sample solution which has been treated with reagent. The tubes are examined axially. The standard solution of suitable strength is conveniently added from a microburet subdivided to 0.01 ml. The final volumes of the two solutions should be approximately the same. A difference of 10% or so is usually of no importance. The tubes should be checked for equal diameter. The tubes previously mentioned may be used to advantage in colorimetric titrations if the volume of solution is not too great. Mixing may be accomplished simply by shaking the tubes or, if necessary, by the use of a glass rod. It is hardly necessary to slip a cylinder of black paper around the tubes to exclude extraneous light as sometimes recommended.

In order that this method may yield accurate results, it is necessary that the manner of mixing the reagent and the solution of the reacting constituent have no effect upon the color. Moreover the color development must be very rapid,[65] and the colored product must be stable. Thus the colorimetric titration method is not to be recommended for the determination of iron by the thiocyanate method, because the color of the thiocyanate complex fades fairly rapidly on standing. Even if the conditions mentioned are not entirely fulfilled, colorimetric titration may at times be used to advantage in connection with the standard series method. The approximate amount of the constituent in the sample solution can first be found by colorimetric titration, and then a series of standards covering a narrow range can be used to fix the amount more precisely, a new aliquot of the sample solution being used if necessary. The titration technique is generally to be preferred to the standard series method whenever it is applicable and only one or two determinations are to be made.

The following procedure is advantageous in finding the amount of constituent required for equicoloration in colorimetric titration. The standard solution is added to tube S (standard) until it is seen that the color intensities are nearly the same. The standard is now added in small increments with comparison after each addition, and the amount noted which corresponds to the last definite difference between X and S. The addition of standard is continued until S is definitely stronger in color than X. If a_1 is the amount of standard for which S is just weaker than X, and a_2 is the amount for which S is just stronger than X, the amount of constituent in the sample solution is taken to be $(a_1 + a_2)/2$. The error should not exceed 5%.

Balancing Method. A Duboscq colorimeter is ordinarily used. This instrument consists of two cups holding solution which can be moved up and down so that the depth of the solutions can be varied to obtain a balance. The observer sees the solutions as two halves of a circular field. This juxtaposition enables the color intensities to be compared more exactly than in the standard series method.

A Duboscq colorimeter is not suitable for very faintly colored solutions because the maximum depth of solution that can be examined is about 5 cm. It cannot be used if the reagent or sample solution is colored. Color intensity is assumed to be directly proportional to the analytical concentration of the constituent being determined.

In using a colorimeter, the following precautions should be taken. It should be ascertained whether each scale reads zero when the plunger touches the bottom of the cup. Cups must not be changed from one side to the other. Cups and plungers must be scrupulously clean. The optical perfection of the instrument should be tested by filling both cups with a standard solution of the colored substance being determined, setting one cup at a height which gives a suitable color intensity and permits the depth to be read with small percentage error, then moving the other cup until the two halves of the field appear the same, and repeating the balancing to obtain a total of about 10 readings. The average reading should agree with the setting of the stationary cup within 1 or 2%. If it does not, the reading of the unknown solution can be corrected with the aid of the ratio that has been determined. Alternatively, if the optical perfection of the instrument is in doubt, one can proceed in the following manner. One of the cups, say L (the left), is filled with standard solution and set at a convenient height. Cup R is filled with the same standard solution and the average of 10 settings of R found; cup R is emptied, rinsed with the unknown (sample) solution, and then filled with the latter. The point of balance is found by moving R and averaging a sufficient number of readings. The concentration of the unknown solution then is $c_x = R_s/R_x \times c_s$, R_s and R_x being the average readings given by the standard and unknown solutions in the right-hand cup, the other cup being kept stationary.

Either daylight or artificial light of proper color quality may be used for illumination. The white glass plate beneath the cups should be set at the angle giving the optimum illumination when daylight is used; this plate must be kept clean, else the two halves of the field may appear of different color intensities when a balance should exist, or it may be impossible to match two solutions because of a difference in hue. The same difficulties may be encountered if any other surfaces in the optical paths become dirty or dusty.

The precision of color comparison with a Duboscq colorimeter may be specified by giving the standard deviation of a single setting of a cup. Table 4-6 contains some data on the precision of comparison for solutions of different colors and, in the case of potassium chromate and the bismuth

TABLE 4-6

Precision of Color Intensity Duplication with Duboscq Colorimeter[a]

Solution	Q (μg/cm^2)	Standard Deviation of a Single Setting (%)
K_2CrO_4	10 Cr	7.5
	20	3
	60	1.3
	100	1.0
	300	2.2
$Bi^{3+} + I^-$	2 Bi	3.2
	4	1.8
	6	1.1
	8	1.4
	12	1.2
$Fe^{3+} + CNS^-$	10 Fe	1.0
$NiCl_2$	35,000 Ni	0.8
$Cu^{2+} + NH_4OH$	100 Cu	1.2

[a] Data of a supposedly normal observer (one of the authors).

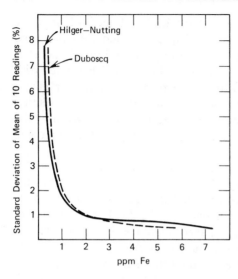

Fig. 4-16. Illustration of precision of readings with a Duboscq colorimeter and a visual (Hilger–Nutting) spectrophotometer on a green solution of the ferric complex of 8-hydroxyquinoline-5-sulfonic acid. [From data of J. Molland, *Archiv. Math. Naturvid.,* **B43**, 67 (1940).] Stratum thickness not reported.

iodide complex, for different concentrations. The second column of the table shows the quantity of the element in a column of solution having a cross-section of 1 cm². The standard deviations given in the third column of the table were obtained in each instance from a series of 10 settings. Daylight (cloudy western sky in the afternoon) was used for illumination. So far as these results permit generalization, it may be concluded that the hue of the solution plays a minor role in the precision of matching. This is in accord with observations of others. At suitable color intensities the standard deviation is substantially the same for red, yellow, green, and blue colors, and is ~1% on the average (Table 4-6) or less (Fig. 4-16). The brightness of the field is an important factor in the precision of matching. At low light intensities the standard deviation becomes greater. For this reason the use of filters with a Duboscq colorimeter in the comparison of faintly colored solutions is of dubious value.

Over a three- or fourfold range of concentrations where the color intensity is neither too weak nor too strong, the standard deviation of Duboscq settings remains roughly constant as we have seen.[66] Accepting 1% as the relative standard deviation of a setting in the favorable concentration range, we will find the best precision that can be expected in the use of a Duboscq colorimeter. Usually two sets of values are obtained as noted earlier: standard versus standard, sample versus standard. Suppose 10 readings ($=N$) are made in each set, that the scale reading error is not a factor in the precision, and that standard and sample have about the same concentration of the colored substance. The relative standard deviation of the ratio of the set averages, and therewith the relative standard deviation in the concentration of the sample is

$$1 \times \left(\frac{1}{N_{stand}} + \frac{1}{N_{samp}} \right)^{1/2} = (\tfrac{1}{10} + \tfrac{1}{10})^{1/2} = \pm 0.45\%$$

The probability that the random error (deviation) will not exceed 1% (for example) is obtained from the t value $1/0.45 = \pm 2.22$, which at 18 degrees of freedom $(10 + 10 - 2)$ corresponds to a probability of ~95% that this error will not be exceeded.

If a solution containing 10 μg Cr/cm^2 (representing ~50 μg Cr in ~10 ml of solution required to fill the cup) is compared against a similar solution in the same way as above, the standard deviation of a single setting is $\pm 7.5\%$ (Table 4-6) and the standard deviation of the ratio is

$$7.5(\tfrac{1}{10} + \tfrac{1}{10})^{1/2} = \pm 3.35\%$$

The probability that the ratio is in error by more than 10% ($t = \pm 10/3.35 = \pm 3.0$) is less than 0.5%.

Nowadays the Duboscq colorimeter is not likely to be used except in the most primitive of laboratories—that is to say in the field.

Dilution Method. This comparison method is not often used, but it can be useful at times. Application may be made of it in extraction methods if the reagent is colorless (or is not extractable if colored). For example, the aqueous solution can be shaken with a small volume of the immiscible solvent in a glass-stoppered tube, and more of the solvent then added from a buret until the layer of the organic solvent shows the same color intensity as that in the unknown tube when the tubes are examined at right angles to their axes.

Since the concentration of the reagent is not the same in the sample and standard solutions at the point of balance, the concentration of the constituent may not be the same in the two solutions in spite of the equal color intensities if the excess of reagent affects the color intensity. This must, therefore, be used with caution.

Ring Oven Colorimetry. The colorimetric methods discussed in the preceding pages are carried out wholly in solution. Another technique developed by Weisz in 1954 employs filter paper for separating constituents of a minute sample and concurrently determining one or more of them.[67] A quantitative filter paper is placed on an electrically heated flat aluminum ring (usually 22 mm inside diameter) and one or more drops of sample are transferred to the center of the paper. The addition of an appropriate reagent precipitates one or more of the constituents. Non-precipitated constituents are then washed out of the precipitate area by slow addition of a suitable solvent and are carried to the inner periphery of the ring by capillary action of the paper. Because of the rapid evaporation of the solvent as the solution approaches the heated surface of the ring, the solutes are deposited, forming a ring of narrow width (0.1–0.3 mm). A colored ring is obtained by applying a suitable color reagent. The actual area of the ring, which is as thin as a pencil line, is smaller than that of the original spot. In other words, the concentrations of the solutes in the ring are much greater than in the original spot. Determination of an element can be made by comparing the intensity of the colored ring with that of standard rings produced from known amounts of the element (ring colorimetry). The color intensity or size of the ring is independent of the number of drops used to form the ring. The rings should be observed in both transmitted and direct light. Photometry can also be applied.

In the ring oven technique, uniform diffusion of solutes to the ring area is necessary. The ring formed should be regular and the deposited solutes should be in the form of a sharply defined ring. Under favorable conditions, an element in the microgram to nanogram range can be determined with an error of less than 10%.

Because the sample used in the ring oven method is very small, the method is not suitable for trace analysis of a solid sample. However, it finds applications in air pollution studies where it is used for the analysis of airborne particulates. Ring oven procedures have been described for the determination of Al, Fe, Be, Cu, Pb, Sb, V, Zn, and other elements.

III. FLUORIMETRY

We assume that the reader is familiar in a general way with luminescence processes. Excluding atomic fluorescence (as in flames), the type of luminescence of greatest importance to the inorganic analyst is molecular or ionic fluorescence in solution at room temperature.[68] Phosphorescence has some applications in inorganic analysis but it is of slight importance. Chemiluminescence finds occasional use in applied inorganic trace analysis.[69]

The trace analyst is primarily interested in fluorimetric methods because of their great sensitivity. Almost all useful fluorimetric metal determinations (uranium is a striking exception) are based on the formation of fluorescent chelate or organic ion-association complexes. Many of these can be extracted into immiscible organic solvents with further lowering in the detection or determination limit. Fluorescence methods vary in sensitivity, but the more sensitive ones can provide a detection limit 0.01 or less that of absorption photometric methods for metals in solution. For reasons to be touched on later, fluorescence reagents for some metals are more selective than absorption reagents. However, fluorimetric determinations are subject to a variety of interferences, including those operative in absorption photometry, and others in addition, that can lead to low results. On the whole, fluorescence methods are not as precise or accurate as absorption methods in the range where they can be compared, but the sensitivity often more than makes up for this. For some trace metals, fluorimetric methods are clearly superior to the colorimetric (spectrophotometric).

The inorganic trace analyst understandably prefers to apply fluorimetry to aqueous or organic solvent solutions at room temperature—as he does when using absorptiometry. But there are other possibilities. In the last 10–12 y, the luminescence (fluorescence and phosphorescence) of frozen solutions at low temperatures (as that of liquid nitrogen, −196°C) has been investigated from the analytical viewpoint.[70] Phosphorimetry becomes possible in viscous solutions, glasses and solids. Fluorescence intensity may increase at low temperatures; structureless luminescence spectra may be converted to fine-structure spectra, which is analytically advantageous. At −196°C, in frozen HCl or HBr solutions, Tl, Pb, Bi, Sb, and Te are detectable at concentrations of 0.001–0.01 μg/ml by their luminescence, other metals less sensitively.

Traces of certain elements (Cu, Ag, Mn, Sb, Pb, and others) produce luminescence in solid polycrystalline phosphors (Ca tungstate, ZnS, KCl, many others) at room temperature and are occasionally determined in this way with greater or less success, as are the rare earth elements. The fluorimetric determination of U(VI) in a (cold) NaF melt is of great practical importance.

A. PRINCIPLES OF ANALYTICAL FLUORIMETRY

1. Absorption, Excitation, and Emission Spectra

Consider a spectrofluorimeter having a monochromator A for providing a narrow band of wavelengths for excitation over the spectral range (e.g., 200–700 nm) and another monochromator B for selecting a narrow band of wavelengths from the emitted (fluorescent) light of longer wavelength. Let the solution of the fluorescent species be irradiated at a wavelength giving strong fluorescence, and find the wavelength at which the fluorescence is most intense (the wavelength of maximum emission intensity is independent of excitation wavelength). Now scan the spectral range, using monochromator A, to obtain the excitation spectrum, that is, a plot of fluorescence intensity versus the exciting wavelength. Then, with monochromator A set at the wavelength of maximum fluorescence excitation, with B scan the spectral range to record the emission (fluorescence) spectrum (fluorescence intensity versus wavelength).

If the intensity of the source were the same for all irradiating wavelengths and other possible complications did not arise, the excitation curve should correspond to the absorbance curve, and the wavelength of maximum emission should be the wavelength of maximum absorbance. If the detector (phototube or electron multiplier) were equally sensitive to all wavelengths and showed a linear response, the emission spectrum should be similar to the absorbance curve,[71] but would be shifted to longer wavelengths (Stokes' law). (The amount of shifting between corresponding absorption and fluorescence peaks depends primarily on solvent composition and molecular structure; if solvation or molecular geometry is different in the excited state relative to the ground state, no direct correlation between the two spectra will be apparent. Moreover, the correlation holds only if the vibrational levels of the ground and excited states are identical.) See Fig. 4-17 for the absorbance, excitation and emission curves of the beryllium–morin chelate as an example.

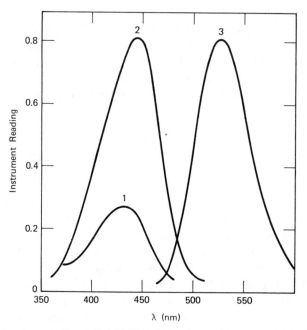

Fig. 4-17. Absorbance (1), excitation (2), and emission (3) curves of beryllium–morin chelate in aqueous solution ($1 \times 10^{-5}\,M$ morin, $45 \times 10^{-5}\,M$ Be). The excitation curve is uncorrected for source (tungsten lamp) intensity at various wavelengths; when corrected, the peak of curve 2 falls at 432 nm (instead of at 446 nm as in the figure), thus coinciding with the peak of the absorbance curve (432 nm). The heights of curves 2 and 3 are arbitrary, and not comparable. Secondary filter Corning 3384 for curve 2; primary filter 5850 for curve 3. From C. W. Sill, *Anal. Chem.*, **33**, 1579 (1961).

If the same spectrofluorimeter is to be used in the determination of the fluorescing species in a sample, the uncorrected excitation and emission curves provide the wavelengths for maximum sensitivity. If another fluorimeter—a filter instrument, for example—is to be used, the corrected curves are needed, together with the characteristics of the second instrument and its radiation source, for the choice of wavelengths or filters. To be sure, the latter can be selected empirically. If known, the absorption (absorbance) curve of the fluorescing species (not including the excess reagent) should be an acceptable substitute for the excitation curve. The possibility of error from fluorescence in obtaining the absorbance curve should be remembered.

The wavelengths for maximum sensitivity are not always the best for a particular determination, important as sensitivity is in trace analysis. Fluorescence of the reagent or of compounds of the reagent with other elements may make other wavelengths preferable. Excitation and emission spectra of other species will aid in the proper selection.

2. Instruments

Details of fluorescence instrumentation will not be treated here,[72] but some observations of a general nature may be made. The double monochromator spectrofluorimeter mentioned in the last section is an essential instrument for obtaining the characteristics of a fluorescence system. Usually a high-pressure xenon arc lamp is used to supply a continuum of radiation in the UV and visible range. However, a simpler instrument may often be used in applied analysis, and may in fact have advantages at times. It may be possible to increase the sensitivity by dispensing with the monochromators and using filters.[73] The use of a mercury arc with a filter to isolate required wavelengths for excitation can provide a more intense source than obtainable from a monochromated continuous source. Even if the wavelength thus obtained is not the optimum one, the greater intensity of the radiation may more than make up for this. Filters may also be used to isolate a suitable band of the fluorescent light. See Fig. 4-18 for a schematic representation of a filter fluorimeter. Of course, monochromators (usually gratings) are superior to filters when mixtures of fluorescent species are present, which it may be possible to resolve by critical selection of wavelengths for excitation and for impingement on the detector.

Glass cells for holding solutions are suitable for use with a filter fluorimeter having a high-pressure mercury vapor lamp (emission at 365, 398, 436, 546, etc., nm). When excitation by UV radiation is suitable, visible radiation from the Hg lamp is readily absorbed with a Ni–Co oxide glass filter. Then, with fluorescence in the visible spectrum, error due to scattered primary radiation is much decreased.

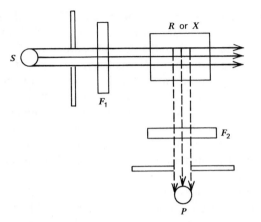

Fig. 4-18. Schematic representation of the elements of a simple filter fluorimeter.
  S, radiation source (usually a low-pressure Hg-vapor lamp).
  P, phototube (a balanced circuit may be used to compensate fluctuations in source).
  F_1, filter to absorb undesired wavelengths from S (often a filter to absorb visible radiation and transmit UV).
  F_2, filter to transmit desired wave band of fluorescent light and absorb stray UV.
  X, cell for sample solution.
  R, cell for reference solution (e.g., quinine bisulfate).
(In a double phototube fluorimeter, the fluorescence of X is compared directly against the fluorescence of R.)

Occasionally, fluorescence determinations by visual comparison are worth considering (see p. 207). The standard series or titration procedure (see discussion of colorimetric titration, Section II.B) can be applied. In a favorable concentration range, in which the fluorescence is not too strong, a difference of ~7% (1:15) in fluorescence intensities can be detected. The beginning of the optimum concentration range may be taken to lie at 10–15 times the concentration giving minimal visible fluorescence.

3. Fluorescence Intensity as a Function of Concentration

Consider a rectangular cell of length b (cm) holding a solution of a fluorescent substance of concentration C, through which a narrow beam of exciting radiation is passed (Fig. 4-19). The intensity of the beam decreases as it traverses the solution because some of its energy is transformed into that of the fluorescent light. Assume that at any point in the path the fluorescence intensity is proportional to the intensity of the exciting radiation, that the latter decreases in accordance with the

Lambert–Beer law, and that the fluorescent light is not absorbed by the solution. Then the fluorescent light energy dF radiated from the element dx to the detector (photoelectric cell, photomultiplier) is given by

$$dF = KI_x \, dx = KI_0 \times 10^{-aCx} \, dx \qquad (4\text{-}24)$$

where

$K =$ a constant depending on the fluorescence yield (quantum efficiency) of the substance and on instrumental factors (cross-sectional area of beam, size of detector face intercepting light, distance between beam and detector, etc.).

$a =$ absorptivity of the fluorescing substance (depends on solvent, temperature, and wavelength of excitation for specified units of C).

The total energy, from the entire length of the beam, reaching the detector is given by

$$F = KI_0 \int_b^0 10^{-aCx} \, dx = KI_0(1 - 10^{-aCb}) \qquad (4\text{-}25)$$

If aCb is small ($< \sim 0.01$), signifying that the intensity of the exciting radiation has not been much diminished in traversing the solution, the preceding expression reduces to

$$F \sim 2.30 KI_0 aCb \qquad (4\text{-}26)$$

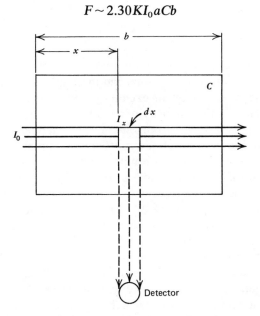

Fig. 4-19. Principle of fluorimetry of solutions.

and F (and usually the meter deflection of the instrument) is directly proportional to the concentration of the fluorescent substance.[74] As the value of C increases at constant b, F does not increase linearly, and a plot of F against C eventually gives a curve tending to run parallel to the C-axis.

Another reason for failure of the linear relation may be self-quenching. As the concentration of a fluorescent substance increases, the efficiency of conversion of absorbed radiation to fluorescent radiation may fall, probably largely because the excited molecules collide with themselves, and energy is dissipated in forms other than fluorescence. The absorption of exciting radiation is not affected. Self-quenching may become so strong at relatively high concentrations of the fluorescing substance that the fluorescence decreases with concentration after having reached a flat maximum. This behavior is unlikely to be encountered in trace analysis.

For simplicity, we considered the exciting beam to be narrow in the preceding treatment. In practice it is likely to be quite broad, but this does not alter the relationships found. Other factors remaining the same, F increases as the cross-sectional area of the beam is increased, but not necessarily in direct proportion.

A more detailed examination of the relation between the fluorescence signal and the concentration of the fluorescing substance may be found in the literature.[75]

B. SENSITIVITY

Expression of the sensitivity of fluorescence methods (or reactions) is not as simple as that of absorptiometric methods. For a given reaction, with a particular wavelength for excitation and of emission, the detector response depends on the intensity of radiation and the amplifier gain, as well as on the apparative arrangement, for example, the volume of fluorescent solution from which the emitted light impinges on the detector (see Fig. 4-19). For a given fluorimeter, cell, etc., the sensitivity can be expressed arbitrarily but reasonably as the minimum concentration of an element, present as the fluorescent species, producing a reading perceptibly greater than that of the blank at a specified probability, for example, a concentration corresponding to twice the standard deviation ($\sim$95% confidence level) of the measurements.[76] It is evident that fluorimetric detection limits reported in the literature are likely to be highly approximate, and will depend on the instrumentation.[77] A high blank (fluorescence of reagent itself) can, of course, increase the detection limit. Moreover, if the reagent absorbs the exciting or the emitted radiation, the sensitivity will depend on the excess of reagent in the solution. The blank fluorescence can often be reduced appreciably by careful choice of wavelengths (or filters), the use of optimal concentrations of fluorogenic reagent, and purification of auxiliary reagents and solvents. These measures are obviously especially important when determinations of metals are being made near the detection limit.

Equations have been derived that relate the minimum detectable concentration of a luminescing substance in a determination to apparative and spectral parameters and noise factors.[78]

The sensitivities of fluorescence reactions can be compared on the molar basis by using the sensitivity factor

$$S = \frac{\epsilon\phi}{H} \tag{4-27}$$

where ϵ is the molar absorptivity of a solution of the fluorescing substance at a specified wavelength of excitation, ϕ is the quantum efficiency of the fluorescence, and H is the half-band width (nm) of the emission spectrum.[79] (Comparison on the weight basis, which may be more useful analytically, can be made by replacing ϵ with a.) This expression is chiefly of theoretical interest. Values of S (ranging from 71 to 130) have been obtained for seven reagents for Mg and Ca (Dagnall et al.[79]). For these reagents, ϕ varied from ~0.2 to 1.0. Values of $\epsilon\phi$ have been reported for Al reagents.[80] S is the intrinsic sensitivity of a reaction (assuming that the constituent in question is converted completely into the fluorescing compound) and is, of course, independent of the instrument used. It is only one of the factors involved in obtaining the practical sensitivity of a reaction and therewith of a determination, which is ordinarily stated in terms of the minimum amount or concentration of constituent that can be detected at a specified probability level.

If the fluorescent substance can be extracted into an immiscible organic solvent from an aqueous solution a great increase in sensitivity can often be obtained. The fluorescence of the substance may be more intense in an organic solvent than in water, and the extraction may favor the formation of the fluorescing species. Also, the concentration of the element into a small volume of solvent is important. If the extraction coefficient is favorable, the sensitivity can become essentially independent of the volume of the aqueous phase. Sometimes the blank can be reduced by extraction, if the weakly fluorescing reagent largely remains in the aqueous phase.

An undue increase in the reagent concentration can result in an appreciable sensitivity decrease. This is likely to be due to absorption of exciting radiation or the emitted light by the excess reagent.

It will be evident that fluorescence reactions are inherently more sensitive than absorptiometric reactions, at least if ϕ is not much less than 1, which is often true. The absorbance of a very dilute solution must be found as the difference between two transmission values that are almost equal. On the other hand, very weak fluorescences can be detected, if the blank fluorescence is even weaker, by suitable amplification. Moreover, increasing the intensity of the exciting radiation, can, within limits, increase the fluorescence intensity. The fluorescence resulting from absorption of radiation corresponding to an absorbance of 10^{-6} is detectable, if ϕ is not much less than 1, whereas the detection of an absorbance of 10^{-6} ($I/I_0 = 99.9998\%$) spectrophotometrically would be impossible.

C. SELECTIVITY

Most of the useful fluorimetric methods for the determination of traces of metals are based on the formation of metal–organic reagent complexes (chelate or ion-association complexes) that fluoresce in aqueous or organic solvent solution. Normally, paramagnetic ions (usually colored) do not form fluorescent chelate complexes. For example, in $1:1$ CHCl$_3$–C$_2$H$_5$OH solution the 8-hydroxyquinoline chelates of Al, Y, Sc, Zr, Mg, Sn(II), Be, Zn, and Ag fluoresce, but those of the colored ions Fe, Ni, Co, Cr, V, and Cu (among others) do not.[81] There is accordingly some gain in selectivity by using a fluorescence method, since the colored ions all give absorbing complexes. Of course, the colored hydroxyquinolates interfere with the fluorimetric determination of, say, aluminum by absorbing the exciting and emitted radiation. Fe(III) hydroxyquinolate is black and interferes seriously, but Fe(III) can be reduced to Fe(II), which does not interfere in reasonable amounts.

It may be possible to increase selectivity among fluorescing metals by selection of suitable exciting and emitted wavelengths if the spectra of the complexes are sufficiently different. Even when the spectra are similar, some decrease in interference may be obtained by selection of wavelengths.[82] But usually the specta overlap to such an extent that this approach is only partly successful.

Of course, fluorescence reactions, like colorimetric reactions, can often be made more selective by adjustment of pH and by masking (including a change in oxidation state). Extraction of the fluorescent compound into an immiscible solvent can often be combined with these measures to increase selectivity. In fact, the extraction is often a type of painless separation, a by-product of the concentration of the fluorescent compound into a small volume of organic solvent. For example, by extraction of gallium 8-hydroxyquinolate into chloroform at pH ~3, ~0.1 μg Ga can be detected in the presence of 30 mg Al (1:300,000).[83] This approach to increasing selectivity, when applicable, is much superior to the selection of wavelengths mentioned in the preceding paragraph.

Occasionally, highly selective fluorescence methods for a metal can be developed by making use of what might be called special properties. For example copper(II), which forms no fluorescent compounds, can be determined fluorimetrically by an indirect reaction: The ion-association complex of Cu(II), 1,10-phenanthroline and Rose Bengal is extracted into chloroform, and the extract is treated with an ammoniacal acetone solution to dissociate the complex and bring about fluorescence of the Rose Bengal.[84] No interference from 56 cations was reported; the sensitivity is 0.001 ppm (570 nm).

Uranium(VI) fluoresces strongly in sodium fluoride solid solution (obtained by fusion). As little as 0.001 μg U is determinable in a gram of flux (excitation at 365 nm). More conveniently, but less sensitively, U can be determined fluorimetrically in a phosphoric acid solution (excitation at 280 nm, emission at 493 nm); the lowest detectable U concentration is 0.0003 ppm. Both procedures are essentially specific for U, but many substances quench the U fluorescence.

Some generalizations concerning organic fluorimetric reagents for metals may be offered. Ideally, the reagent itself should be nonfluorescent. As a rule, organic molecules having a nonrigid, nonplanar structure are not fluorescent. If a rigid planar structure results on chelation with a metal, the chelate is likely to be fluorescent. 8-Hydroxyquinoline in chloroform solution is extremely weakly fluorescent, but its chelates with Al, Ga, Zr, and some other colorless cations fluoresce strongly in the same solvent. If an organic compound is itself fluorescent, it is not necessarily disqualified as a fluorimetric reagent for metals. The respective fluorescence intensities and fluorescence emission curves of reagent and chelate may be sufficiently different for a distinction to be made. Even if the fluorescence properties are similar, it may be possible to extract the metal complex into an immiscible organic solvent which does not extract the reagent (Rhodamine B and other basic dyes are examples).

Evidently, fluorescence reagents are to be looked for among aromatic and heterocyclic compounds. In later chapters, a number of organic compounds of definite value as fluorescence reagents are mentioned. These include, in addition to 8-hydroxyquinoline, certain azo dyes, flavonols (morin is one), Schiff bases, basic dyes (such as the rhodamines) forming ion associates with complex metal anions, as well as others. Some β-diketones are reagents for rare earth metals. They show the unusual behavior of transferring absorbed energy to the cation of the chelate, from which it is emitted with the characteristic spectrum of the cation.

D. ERRORS

The important errors in fluorimetry are those arising from the presence of other constituents of the sample, considered below. A standard curve is always established over the range in which the determination is to be made, and such variables as pH, reagent concentration, organic solvent concentration (if any), time of reading after addition of reagent, and temperature are controlled as closely as needed for good reproducibility. Temperature differences of 1° or 2°C are usually not important (fluorescence intensity normally decreases with temperature: Fig. 4-20). However, in any determination the standard curve should be verified by fluorescence readings on one or two standards and the reagent blank at the time the fluorescence of the sample is read. Unnecessary exposure of solutions to UV radiation should be avoided because photochemical decomposition can occur. A solution of quinine in sulfuric acid serves well as a fluorescence reference. Solutions should be entirely free from suspended or colloidal material that can scatter the exciting radiation. With a properly chosen secondary filter, much of the scattered radiation will be

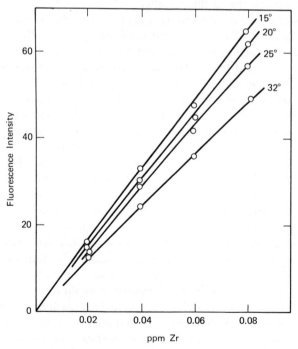

Fig. 4-20. Fluorescence intensity of Zr–morin system as function of temperature (°C). (Data of Richard Geiger.)

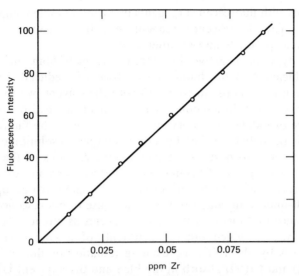

Fig. 4-21. Fluorimetric determination of Zr ($+$ Hf) with morin. Aqueous solution 2.0 M in HCl; 0.20 mg morin in 25 ml; Coleman 12B fluorimeter. A blank reading of 3 has been subtracted. (Data of Richard Geiger, University of Minnesota, 1955.)

absorbed, but it should be kept at a minimum from the start. Care is needed to exclude foreign fluorescent substances of organic origin (from filter paper, stopcock grease, dust, lint, detergents, stoppers, plastic ware, the skin, etc.); organic solvents will often require special purification, such as percolation through a suitable adsorbent, followed by distillation.

The precision attainable in fluorimetric determinations obviously can vary a great deal, depending on the analytical fluorescence system and the fluorimeter. It is not uncommon to find that fluorimetric precision approaches the absorptiometric as far as trace analysis is concerned. The precision that can often be expected is represented by the standard curve for the zirconium–morin system in Fig. 4-21, which was obtained with an inexpensive filter fluorimeter. A standard deviation of less than 2–3% is indicated in the range 0.05–0.1 ppm Zr. In passing, it may be noted that the sensitivity expressed as the limiting concentration is ~0.002 ppm Zr.

Most fluorimetric reactions are far from specific, so that there is risk of positive errors when they are applied to real samples. Familiarity with the reaction—it may have been studied fairly thoroughly—and with the likely composition of the sample material is needed to avoid possible gross errors. Negative errors can also be serious, but often they can be detected, and corrections can be made. Interference can arise from chemical reactions among the constituents of a sample or with the reagent. Fluoride, for example, can prevent or impede the reaction of heavy metals with fluorescence reagents by forming stable complexes with the metals. The fluorescence reagent may be consumed in forming nonfluorescent products and in other ways.

Another type of negative interference is *quenching*, in which the quantum efficiency of the fluorescence process is reduced. The excited molecules can lose energy in collisions with the solvent, with solute ions, and with each other. In aqueous solution, quenching can be observed in the presence of iodide and thiocyanate. For example, the fluorescence of uranyl ion is strongly quenched by iodide. Oxygen in solution can quench fluorescence and also destroy the metal complex. Ions of the transition metals can also quench. Chelates may fluoresce less in strongly polar solvents. Organic solvents having bromo, iodo, and nitro groups tend to quench. Self-quenching may result in a decrease of fluorescence intensity at higher concentrations of a fluorescing species as already mentioned.

Another way in which other constituents of a sample can weaken fluorescence is by absorbing the exciting radiation or the emitted light. Thus Fe(III) and Cr(VI) absorb in the blue and the adjacent UV and can lead to serious negative errors, even if they are otherwise innocuous. Colored solutions are to be looked upon with suspicion; they may act as internal filters for the exciting or the emitted wavelengths.

Negative interferences of the types mentioned can be detected and often corrected by adding a known amount of the constituent in question to the sample and comparing the increase in fluorescence obtained with that which should be obtained. This procedure requires that the negative effect (quenching or other) of the foreign constituents be independent of the absolute amount of the fluorescing constituent, that is, that it be proportionately the same for different amounts of the fluorescing constituent.[85] Moreover, the fluorescence intensity must be proportional to the concentration of the fluorescing substance.

Let

x = amount of constituent in sample.

r = amount of constituent introduced from reagents and utensils.

a = amount of constituent added to sample.

F_{x+r} = fluorescence of sample, including contribution from reagents.

F_{x+r+a} = fluorescence of sample with added constituent, including reagents (corrected for any volume change).

F_r = fluorescence of reagent blank.

k = fluorescence of unit weight of constituent (unquenched).

Then, if the preceding assumptions hold,

$$\frac{F_{x+r}}{F_{x+r+a}} = \frac{x+r}{x+r+a}$$

and

$$x = \frac{aF_{x+r}}{F_{x+r+a} - F_{x+r}} - r$$

Since $r = F_r/k$ (no quencher present), the final expression is

$$x = \frac{aF_{x+r}}{F_{x+r+a} - F_{x+r}} - \frac{F_r}{k}$$

Obviously the amount a added should be large compared to x, so that the experimental error in the difference $F_{x+r+a} - F_{x+r}$ will not be too great. However, it should not be so large that the range of constant proportional quenching or the linear fluorescence range is exceeded.

This "spiking" approach to correcting for negative interferences has been applied extensively in the fluorimetric determination of uranium (sodium fluoride melt and H_3PO_4 solution). The spiked and unspiked volumes should not be greatly different.

An increase in ionic strength of a solution may have a slight effect on the fluorescence intensity, sometimes increasing it. The addition method should correct this error.

IV. TURBIDIMETRY AND NEPHELOMETRY

In turbidimetry, the light transmitted by a suspension is measured, and the light scattered or absorbed is thus found by difference. In nephelometry, the light scattered is measured directly. A spectrophotometer can be used for turbidimetry. On the other hand, a nephelometer does not differ essentially from a fluorimeter except in the character of the radiation.

Turbidimetry and nephelometry are seldom used in the trace determination of metals and an examination of these methods is not worthwhile here. It is clear that turbidimetry of colorless suspensions would not be profitable—the sensitivity is poor. Colored suspensions or colloidal solutions are made the basis of absorptiometric methods, but only because of the absorption of light, and such systems are avoided if possible. Nephelometry can be made very sensitive but problems in reproducibility of suspensions impose limitations on the attainable precision. Also, all substances have an intrinsic solubility, and accordingly there is a lower limit to nephelometric methods.

ADDENDUM

Extremely dilute solutions of quinine sulfate have been observed to show anomalous fluorescence behavior (unexplained), in that the standard

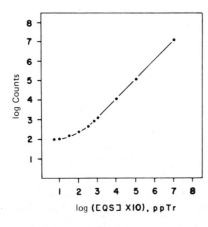

Fig. 4-22. Molecular fluorescence standard curve for quinine sulfate in $0.1\,N\ H_2SO_4$. Analytical signal corrected for blank fluorescence. Reprinted with permission from J. D. Ingle, Jr. and R. L. Wilson, *Anal. Chem.*, **48**, 1642 (1976). Copyright by the American Chemical Society.

(calibration) curve levels off at these concentrations (Fig. 4-22).[86] To find the detection limit from such a curve, it has been suggested that the blank signal be taken as the extrapolated value of the curve and not as the measured value of the blank solution.[86]

"The reasons for this behavior are not clear but it may be due to equilibria involving the analyte at low concentrations and sorption and desorption of the analyte and contaminants from the sample containers and the sample cell. If the blank solution is first transferred to a scrupulously cleaned beaker and then to the sample cell, the blank signal increases."[86]

NOTES

1. Spectrophotometry and colorimetry by *reflection* are sometimes applied in analysis. A solid phase is usually involved. In one general method, a colored precipitate is formed or collected on a disk of filter paper, or other material, having a constant diameter. See, for example, K. W. Franke, R. Burris, and R. S. Hutton, *Ind. Eng. Chem. Anal. Ed.*, **8**, 435 (1936) (selenium collected with barium sulfate on a small disk of paper); H. Yagoda, *ibid.*, **9**, 79 (1937), and *Mikrochemie*, **24**, 117 (1938) (confined spots); Gutzeit test for arsenic. A special apparatus, called the chromograph, has been devised for forming confined spots of colored precipitates by filtration on paper for visual comparison: R. E. Stevens and H. W. Lakin, *U.S. Geol. Survey Circ.*, No. 63 (1949); *U.S. Geol. Survey Bull.*, **992**, 71 (1953). Colored spots of constant area can also be obtained by applying a small measured volume of solution to absorbent papers and adding reagent. If the color of the reaction product is sufficiently intense and of suitable hue, spectrophotometric measurement may be made; see, for example, C. W. Ayres, *Mikrochim. Acta*, **1956**, 1333; and E. H. Winslow and H. A. Liebhafsky, *Anal. Chem.*, **21**, 1338 (1949). Visual (or sometimes photometric) comparison of stains is applied in paper strip

chromatography. E. C. Hunt, A. A. North, and R. A. Wells, *Analyst*, **80,** 172 (1955), use this technique for approximate determination of Cu, Co, Ni, Nb, Ta, Pb, and U in geochemical prospecting. Reflectance measurements or comparisons are also applied in thin layer chromatography and in the use of the Weisz ring oven (Section II.B). For reflectance spectroscopy of Cu, Ni, and Zn on cellulose chromatographic plates, see M. M. Frodyma, D. F. Zaye, and V. T. Lieu, *Anal. Chim. Acta*, **40,** 451 (1968). D. Kealey, *Talanta*, **19.** 1563 (1972), has studied the technique of quantitative reflectometry, using the determination of nickel with dimethylglyoxime as the model. A recording densitometer was used, and standard deviations as low as 2–4% could be obtained. For general treatments of reflectance measurements, see W. W. Wendlandt and H. G. Hecht, *Reflectance Spectroscopy*, Wiley-Interscience, New York, 1966; G. Kortüm (translated by J. E. Lohr), *Reflectance Spectroscopy. Principles, Methods, Applications*, Springer, Berlin, 1969.

2. See, for example, S. Z. Lewin, *J. Chem. Educ.*, **37,** A197, A271, A341, A401, A455, A507, A797 (1960). P. F. Lott, *ibid.*, **45,** A89, A169, A182, A273 (1968). Useful information will be found in American Society for Testing and Materials, *Manual on Recommended Practices in Spectrophotometry*, ASTM, Philadelphia, 1966, especially p. 9 et seq., dealing with description and measurement of the performance of spectrophotometers, and p. 82 et seq., dealing with photometric methods for chemical analysis of metals.

3. See the more comprehensive works on analytical chemistry for the rudiments. On a more advanced level: M. G. Mellon, Ed., *Analytical Absorption Spectroscopy*, Wiley, New York, 1950; J. R. Edisbury, *Practical Hints on Absorption Spectrophotometry*, Hilger and Watts, London, 1966. G. Kortüm, *Kolorimetrie, Photometrie und Spektrometrie*, Springer, Berlin, 1962. E. J. Meehan, *Optical Methods of Analysis*, Interscience, New York, 1964.

4. Interference filters having half-band widths of 10–15 nm or less, much narrower than that of ordinary absorption filters, may be suitable for some substances whose solutions show sharp bands. A limited number of very narrow wave bands can be obtained by using suitable filters in conjunction with a mercury arc to absorb the unwanted wavelengths.

5. D. S. Howell, J. C. Pita, and J. F. Marquez, *Anal. Chem.*, **38,** 434 (1966), have described an ultramicro cell that allows the spectrophotometric determination of $0.01–0.0001\ \mu$g of substance. This cell is made from Pyrex or polyethylene capillary tubing of ~0.6 mm diameter (volume about 2.8 μl/cm) and may have a length of 3 cm. Such a cell, used with a suitable spectrophotometer and holder permits measurement of absorbances of ~0.1 with a standard deviation of ~0.4%. Calcium has been determined in 5×10^{-4} ml of biological fluid (30–120 ppm Ca) by precipitation as oxalate, with final photometric determination with Chlorophosphonazo III in a volume of 0.015 ml. The standard deviation of the recovery of calcium was 1–2%.

6. A detailed analysis of the effect of noise on the precision of absorbance measurements has been presented by J. D. Ingle, Jr., and S. R. Crouch, *Anal.*

Chem., **44**, 1375 (1972); L. D. Rothman, S. R. Crouch, and J. D. Ingle, Jr., *ibid.*, **47**, 1226 (1975). Their main conclusions are summarized in Fig. 4-2. For photometric precision and accuracy of the "GeMSAEC fast analyzer," see G. Goldstein, W. L. Maddox and M. T. Kelley *ibid.*, **46**, 485 (1974).

7. Absorbance is assumed to be directly proportional to the analytical (total) concentration of the element.

8. The assumption that the error in setting to a line is the same as the error in estimating tenths of a division is an approximation. If the distance between graduations is relatively large, the setting can be made more precisely than the estimation, but usually the distance between marks is small enough to make the respective errors comparable.

9. A surprisingly large standard deviation of ~ 0.005 in A (equivalent to $\sim \pm 1\%$ absolute in transmittance) from this source was reported by Rothman, Crouch, and Ingle, *loc. cit.*; *cf.* Fig. 4-2. We have not observed such large deviations in the use of filter photometers. In one such instrument, error from this source could not definitely be demonstrated. Also pertinent are the results of A. C. Docherty, S. G. Farrow, and J. M. Skinner, *Analyst*, **97**, 36 (1972), using a digital single-beam spectrophotometer (Unicam SP3000 system), who determined the reproducibility of absorbance measurements of potassium dichromate solutions for which A at 235 and 420 nm varied from approximately 0.2 to 1.1. The sample cell (1 cm) was filled for each measurement. Standard deviations ranging from ~ 0.0005 to $0.0007\ A$ were found. Relative standard deviations varied from 0.2–0.3% at $A \sim 0.2$ to 0.05–0.07% at $A \sim 1$.

10. We use capital C to indicate analytical concentration, as of various species of metal in solution, and lower case c to represent the concentration of a particular absorbing species. When C is c, α becomes a (at 1 cm stratum thickness if the reagent does not absorb at the wavelength λ). Also, when a filter is used, α should be used in place of a, since the latter refers to "monochromatic" radiation. Note that when the reagent converting the constituent into absorbing species also absorbs at λ, α becomes smaller. The limiting case $\alpha = 0$ is reached when the absorbance of the reagent consumed is equal to the absorbance of the product formed (i.e., if ϵ of reagent and product are the same and the constituent is itself nonabsorbing and reacts in a $1:1$ ratio with the reagent). When the product is nonabsorbing, or absorbs less strongly than an equivalent amount of reagent, α becomes negative.

11. Curiously, in both papers and books in the analytical field, failure to obtain a rectilinear relation between absorbance and analytical concentration is more often than not ascribed to failure of Beer's law. This is taking the name of Beer in vain. Beer's law has nothing to do with *analytical* concentrations. It applies only to the absorbance of a single species at a "single wavelength." Failure of the rectilinear relationship can result from the presence of a number of absorbing species of the constituent being determined (p. 195) and the use of insufficiently monochromatic light (less often from scattered light or fluorescence). There is little reason to believe that Beer's law ever fails by

a detectable amount at concentrations used in analytical absorptiometry. The essence of Beer's law is absorption of radiation by an ion or a molecule independent of other ions and molecules. Only at very high concentrations is there a possibility of such interference (shielding of a molecule by other molecules in the light path). Failure of Beer's law in aqueous solution would be difficult to prove in any event. Even in a solution supposedly having a single absorbing ionic species, a change in the hydration of the ions with a change in their concentration and therewith a small change in absorptivity is a possibility. In practice, the great majority of standard curves are closely rectilinear if sufficiently monochromatic light is used.

12. J. Higgins, *Analyst*, **89**, 211 (1964).

13. See, further, on testing for deviation from a straight line, M. J. Maurice, *Z. Anal. Chem.*, **204**, 401 (1964). It is shown how k_1 and k_2 can be calculated when A is not proportional to C, but the relation between the two can be satisfactorily approximated by $A = k_1 C + k_2 C^2$. For error analysis when standard curves are fitted quadratic or cubic curves (automated analysis), see D. Whitehead, *Talanta*, **20**, 193 (1973). For a criticism of straightforward least squares prodedure and use instead of weighted regression, see C. Marmasse, *Appl. Opt.*, **4**, 163 (1965). Nonlinear calibration curves (general): L. M. Schwartz, *Anal. Chem.*, **48**, 2287 (1976).

14. This assumption makes s_β a little larger than it really is, because β is not based on one calibration point at $C = 0$, but on other points as well, more or less near the blank point ($C = 0$), and on the slope of the curve. We are making the situation slightly more unfavorable than it actually is, but perhaps it is better to err on the pessimistic side. The assumption would be correct if the standard curve (line) were based on two points, one at $C = 0$, the other at the end of the line. It may be noted that α and β are negatively correlated: an increase in α decreases β. This correlation is ignored here. For an equation (quadratic) taking this correlation into account in calculation of the confidence interval, see J. Mandel, *The Statistical Analysis of Experimental Data*, Wiley, New York, 1964. Also it should be noted that s_A is not strictly constant. As pointed out earlier, if s_A depends mostly on transmission reading deviation, it becomes slightly larger at higher absorbances, but the difference is not large for the range in Fig. 4-4.

15. The precision of the standard curve in Fig. 4-4 could be improved most efficiently by replication of points at $C = 0$ and 50 μg Fe. The six points already plotted give sufficient assurance that the plot is a straight line. The slope would be found most precisely from points at 50 μg, the intercept from the blank absorbances. In this connection—for the effect of repartition of standards on precision—see A. Hubaux and G. Vos, *Anal. Chem.*, **42**, 849 (1970).

16. For example, see H. L. Pardue and S. N. Deming, *Anal. Chem.*, **41**, 986 (1969); H. V. Malmstadt, R. M. Barnes, and P. A. Rodriguez, *J. Chem. Educ.*, **41**, 263 (1964).

17. For example, A. Reule, *Appl. Opt.*, **7,** 1023 (1968), tested the photometric scale of a recording spectrophotometer by a light addition method and found the errors (deviation from the straight line 0–100% T) not to exceed 0.1% T (transmittance) for wavelengths in the range 350–665 nm. The Elko II filter photometer (p. 156) was found to show a maximal error of 0.07% T (filter not specified).

18. However, even if a straight line is obtained in a plot of concentration of absorbing substance versus A, the absorbance readings are not necessarily the true values. For example, if for any reason, $T_{observed} = T^a$, $A_{observed} = aA$. See A. Reule, *loc. cit.* But, in general, the analyst is interested only in finding the relation between concentration and observed A. When ϵ is to be reported, assurance is required that observed A is an accurate value. This also requires that the light (radiation) be sufficiently monochromatic and that stray light be absent. The wavelength scale must be checked.

19. R. Mavrodineanu, J. I. Shultz, and O. Menis, Eds., *Accuracy in Spectrophotometry and Luminescence Measurements*, Nat. Bur. Stand. Spec. Pub. 378, Washington, D.C., 1973. Some of the chapters contain information of interest to the trace analyst.

20. For an analysis of the effect of stray light, see M. N. States and J. C. Anderson, *J. Opt. Soc. Am.*, **32,** 659 (1942); W. Slavin, *Anal. Chem.*, **36,** 561 (1963); R. E. Poulson, *Appl. Opt.*, **3,** 99 (1964). A related error is the fluorescence error, to which attention has been drawn by K. S. Gibson and H. J. Keegan, *J. Opt. Soc. Am.*, **28,** 180 (1938). In recent spectrophotometers, the arrangement is such that the radiant energy is dispersed before it reaches the sample, and only a narrow band of wavelengths is incident on the latter. In this way, undue heating of the sample and possible resultant change in transmittance are avoided. However, this arrangement may introduce a significant error if the sample is activated to fluorescence by this narrow band, since the fluorescent light is added to that normally transmitted. Thus, a sample illuminated with green light may show a red fluorescence and the transmittance in the green will then be erroneously high, especially if the photocell is highly sensitive to red light, as is often the case. T. C. J. Ovenston, *Photoelectric Spectrometry Group Bulletin*, **6,** 133 (1953), investigated experimentally the fluorescence error in Unicam SP 500, Uvispek, and Beckman model DU instruments and concluded that in the range log I_0/I 0.2 to 0.8, accurate absorption measurements could be made on even strongly fluorescent solutions.

21. The blank correction in the case of nonrectilinear standard curves has been discussed by W. E. van der Linden, *Z. Anal. Chem.*, **269,** 26 (1974). It is recommended that the correction be omitted and that the standard curve be constructed by carrying a blank and known amounts of standard solution through the entire procedure and plotting the amounts taken against the absorbances measured against water.

22. For example, the transmittance of the red Os-thiourea system increases by about 0.1% per °C in the range 25–45°C according to G. H. Ayres and W. N.

Wells, *Anal. Chem.*, **22,** 317 (1950). The absorbance of the germanium-*p*-dimethylaminophenylfluorone system in ~90% ethanol varies with temperature (°C) according to the relation

$$A_{t°} = A_{20°} + 0.0317(20 - t°)$$

as reported by A. Campe and J. Hoste, *Talanta*, **8,** 453 (1961). Errors can also arise if the reagent is colored and the absorbance of the excess varies with temperature.

23. V. A. Malevannyi and Y. L. Lel'chuk, *J. Anal. Chem. USSR*, **23,** 1372 (1968), have proposed an index for interference from foreign cations in spectrophotometry, which is defined as the ratio of the molar absorptivity of the complex of the test ion to the molar absorptivity of the complex of the interfering ion with a particular reagent. This index then is the inverse of T above if the molar basis is changed to the straight weight basis. The molar absorptivity of these authors is actually the apparent (or we could say the *effective*) molar absorptivity. T, or its reciprocal, is a ratio of sensitivities of two elements with a given reagent. Using the sensitivity index $\mathscr{S}$ (p. 190), we have $T_B = \mathscr{S}_A / \mathscr{S}_B$ for a particular wavelength.

24. If the dissociation constant of the reagent HR, the formation constant of the complex of A with HR, and the hydrolysis constants of A cation are known beforehand, the optimum pH for the reaction, the required concentration of reagent, and so on, can be calculated. See L. P. Adamovich, *J. Anal. Chem. USSR*, **20,** 1317 (1965).

25. G. F. Kirkbright, *Talanta*, **13,** 1 (1966), has outlined the steps in the study of spectrophotometric methods and the presentation of the results.

26. If the method is quite selective, the effect of many elements can be tested at one time by adding them all in small amounts. For example, J. O. Hibbits and S. Kallmann, *Talanta*, **11,** 1443 (1964), tested the effect of ~70 elements on the determination of a few micrograms of a number of metals, adding 10 mg each of these elements at one time. Prior separations were made in some of these methods.

27. Much information is now available on masking (mostly by complex formation) in analysis. See the review of the field by D. D. Perrin in *Masking and Demasking of Chemical Reactions*, Wiley-Interscience, New York, 1970; also K. L. Cheng, *Anal. Chem.*, **33,** 783 (1961). For quantitative treatment of complexing, see especially A. Ringbom, *Complexation in Analytical Chemistry*, Interscience, New York 1963; the tables of constants given are very useful for calculating the extent of metal complexing with common complexing reagents, as far as known.

28. Not foolproof. A blank of zero, or almost so, might be obtained when A is not zero. For example, this could occur if the solution contained a trace of a substance (sulfide in reagents) that would inactivate A originally present as a trace but not much more. Poor recoveries of small amounts of A would be obtained, but recoveries would improve with an increase in A added.

29. For statistical tests (based on F and t) to detect an effect of foreign substances on the precision and the accuracy of determination of an element, see M. J. Maurice and K. Buijs, *Z. Anal. Chem.*, **244**, 18 (1969). An example is given involving the effect of phosphate on the spectrophotometric determination of plutonium(VI).

30. *Catalog of NBS Standard Reference Materials*, Nat. Bur. Stand. Spec. Pub. 260, Washington, D.C., 1975–1976.

31. "This series ... represents one of the most complex and comprehensive trace standards ever produced and issued. ... Well over 11,000 man hours of analytical time has been expended in the work done thus far. An equal or even greater expenditure of time will be required over the next few years before most of the elements are measured and certified" (NBS leaflet, July 24, 1970). Altogether 61 elements were added to the standard. At the present time (1970), the following elements have been certified at one or more levels: Cu, Fe, Pb, Mn, Rb, Ag, Th, Tl, Ti, U. "The target precision for homogeneity testing, and accuracies for assay were set at: 2% or better for the 500 ppm materials, 5% or better for the 50 ppm materials, and the state of the art for the 1 ppm and 0.02 ppm materials." The standards are made available as wafers 1 or 3 mm in thickness cut from ~1.2 cm rods.

32. Data on G-1 (granite; exhausted) and W-1 (diabase) have been summarized by H. W. Fairbairn et al., *U.S. Geol. Surv. Bull.*, **980** (1951); R. E. Stevens et al., *ibid.*, **1113** (1960); M. Fleischer and Stevens, *Geochim. Cosmochim. Acta*, **26**, 525 (1962); Fleischer, *ibid.*, **29**, 1263 (1965); Fleischer, *ibid.*, **33**, 65 (1969). Data on new U.S. Geological Survey rocks, G-2 (granite), GSP-1 (granodiorite), AGV-1 (andesite), PCC-1 (peridotite), DTS-1 (dunite), and BCR-1 (basalt) have been compiled by F. J. Flanagan, *Geochim. Cosmochim. Acta*, **33**, 81 (1969). Data on CAAS (Canadian Association for Applied Spectroscopy; now Spectroscopy Society of Canada) syenite rock-1 (exhausted) are given by N. M. Sine, W. O. Taylor, G. R. Webber, and C. L. Lewis, *Geochim. Cosmochim. Acta*, **33**, 12 (1969). More recent data on various geochemical reference samples are evaluated by F. J. Flanagan, *Geochim. Cosmochim. Acta*, **37**, 1189 (1973) and **38**, 1731 (1974); S. Abbey, *Geol. Surv. Canada Paper* **73–36** (1973); and K. Randle, *Chem. Geol.*, **13**, 237 (1974). O. H. J. Christie, *Talanta*, **22**, 1048 (1975), lists values for trace elements obtained in the cooperative analysis of samples of larvikite, schist, and a sulfide ore.

33. K. Behrend, *Z. Anal. Chem.*, **235**, 391 (1968), made 400 absorbance measurements of a copper sulfate solution, using a filter photometer and filling the absorption cell each time. A Gaussian distribution of deviations seemed to be obtained, with a standard deviation of 0.00117 in absorbance. Curiously, a plot of the averages of four successive points gave an irregular curve broader than the theoretical, that is, a standard deviation of 0.93×10^{-3} instead of $(1.17 \times 10^{-3})(1/4)^{1/2} = 0.59 \times 10^{-3}$.

34. J. D. Ingle, Jr., *J. Chem. Educ.*, **51**, 100 (1974); R. Püschel, *Mikrochim. Acta*, **1968**, 82; H. Kaiser, *Z. Anal. Chem.*, **209**, 1 (1965) and *Anal. Chem.*, **42** (4),

56A (1970); A. L. Wilson, *Analyst*, **86**, 72 (1961); J. B. Roos, *ibid.*, **87**, 832 (1962); L. A. Currie, *Anal. Chem.*, **40**, 586 (1968).

35. E. B. Sandell, *Colorimetric Determination of Traces of Metals*, Interscience, New York, 1st ed., 1944, p. 40 and subsequent editions.

36. For example, the filter photometer of Table 4-1 gives $s_A \sim 0.0001$ at low absorbance values. If $3s_A$ is accepted as representing the absorptiometric detection limit (p. 187), the detection limit of Sb as Rhodamine B chloronatimonate in benzene ($\mathscr{S} = 0.0012~\mu g~Sb/cm^2$) with this instrument is $3 \times 0.0001/0.001 \times 0.0012 \sim 0.0004~\mu g~Sb/cm^2$. Incidentally this value illustrates the low detection limits attainable by using an absorptiometer of good reproducibility with a sensitive reaction. Since 2–3 ml of benzene suffices for extraction of Rhodamine B chloroantimonate, the sample detectability in a 1 cm cell would be $\sim 0.001~\mu g$ Sb as far as instrumental reproducibility is concerned. In a determination, this probably will not be the limiting factor, which may be the reproducibility of the reaction or of the method (involving separations).

37. E. A. Braude, *J. Chem. Soc.*, 379 (1950).

38. G. S. Spicer and J. D. H. Strickland, *Anal. Chim. Acta*, **18**, 231 (1958). In this connection, it may be mentioned that F. Umland et al., *Z. Anal. Chem.*, **215**, 400 (1966), have obtained a detection limit of $2 \times 10^{-4}~\mu g$ B with curcumin as reagent by photometry on chromatographic paper. For a list of $\mathscr{S}$ values, see P. A. St. John, Chapter 7A, J. D. Winefordner, Ed., *Trace Analysis: Spectroscopic Methods for Elements*, Wiley-Interscience, New York, 1976.

39. V. Djurkin, G. F. Kirkbright, and T. S. West, *Analyst*, **91**, 89 (1966). A similar method, in which thiocyanate is used to determine Mo is described by F. Umland and G. Wünsch, *Z. Anal. Chem.*, **213**, 186 (1965).

40. It may be noted that a specific reagent, as defined above, may not allow a particular element to be determined directly in the presence of all others. Other elements might give a precipitate under the required conditions. They might consume reagent by forming colorless reaction products, they might complex the desired element, and so on. Besides, the specific reagent would need requisite sensitivity to be of practical value.

41. For proposals concerning the expression of selectivity, as by the use of a selectivity index, see R. Belcher, *Talanta*, **12**, 129 (1965); R. Belcher and D. Betteridge, *ibid.*, **13**, 535 (1966); D. Betteridge, *ibid.*, **12**, 129 (1965).

42. A true deviation from Beer's law would occur, if a solution were şo densely populated that absorbing ions or molecules would "hide behind one another in the path of the beam." As already noted, apparent failure of Beer's law can result from photometer deficiencies (particularly insufficiently monochromatic light and, less likely in trace analysis, stray radiation and nonparallel beams), in addition to chemical interactions touched on above. For a review of Beer's law and its use in analysis, see G. F. Lothian, *Analyst*, **88**, 678 (1963). Apparent deviations are considered by K. Buijs and M. J. Maurice, *Anal. Chim. Acta*, **47**, 469 (1969), who show how the different effects (instrumental and chemical) can be distinguished.

43. The distinction between a direct and an indirect method can become tenuous here. If the coloration is due to the undissociated compound in the solvent, the method must be called a direct one. Thus cobalt has been determined by precipitation as 2-nitroso-1-naphtholate followed by solution of the precipitate in chloroform. This method has been improved by extracting the cobalt complex and excess reagent with toluene or chloroform, removing the reagent from the organic phase by shaking with sodium hydroxide, and obtaining the absorbance. This is certainly a direct method for cobalt. The precipitation method does not differ from the extraction method in principle.

44. J. J. Lothe, *Anal. Chem.*, **27**, 1546 (1955), has examined the precision of indirect spectrophotometry by the differential technique. Also see C. N. Reilley and G. P. Hildebrand, *Anal. Chem.*, **31**, 1763 (1959).

45. K. B. Yatsimirskii, *Kinetic Methods of Analysis*, Pergamon, Oxford, 1966 (2nd Russian edition, 1967); H. B. Mark, Jr., and G. A. Rechnitz, *Kinetics in Analytical Chemistry*. Interscience, New York, 1968. Reviews: H. B. Mark, *Talanta*, **19**, 717 (1972); Z. Gregorowicz and T. Suwinska, *Chem. Anal.*, **11**, 3 (1966); G. A. Rechnitz, *Anal. Chem.*, **40**, 455R (1968); G. G. Guilbault, *ibid.*, **42**, 334R (1970); R. A. Greinke and Mark, *ibid.*, **44**, 295R (1972); H. V. Malmstadt, E. A. Cordos, and C. J. Delaney, *ibid.*, **44** (12), 26A, 79A (1972); G. Svehla, *Analyst*, **94**, 513 (1969) (Landolt reactions); R. P. Bontchev, *Talanta*, **17**, 499 (1970); **19**, 675 (1972); I. P. Alimarin, *Pure Appl. Chem.*, **34**, 1 (1973).

46. P. B. Browning and W. O. Cutler, *Z. Anorg. Chem.*, **22**, 303 (1900).

47. Os: R. D. Sauerbrunn and E. B. Sandell, *Mikrochim. Acta*, **1953**, 22. The kinetics and mechanism of the Os-catalyzed reaction have been studied by R. L. Habig, H. L. Pardue, and J. B. Worthington, *Anal. Chem.*, **39**, 600 (1967). They find that As(III) reduces Os(VIII) to Os(VI) and Os(IV) in two-electron steps, and reduced forms of Os are oxidized to Os(VIII) by Ce(IV) in one-electron steps. Rate expressions holding over a wide range of conditions are given. Ru: C. Surasiti and E. B. Sandell, *Anal. Chim. Acta*, **22**, 261 (1960). J. B. Worthington and H. L. Pardue, *Anal. Chem.*, **42**, 1157 (1970), investigated the kinetics of the Ru-catalyzed reaction and developed a method for the simultaneous determination of Ru and Os based on their different catalytic activities at different Ce(IV) and As(III) concentrations (see note 49).

48. H. L. Pardue and R. L. Habig, *Anal. Chim. Acta*, **35**, 383 (1966), have described an instrument which measures the rate of the catalyzed Ce(IV)–As(III) reaction photometrically and automatically prints the Os concentration on tape. Concentrations of 0.001–0.002 ppm Os can be determined with an average error of ~3%.

49. Worthington and Pardue[47] found the initial rates of the catalyzed reactions to depend on Ce(IV) and As(III) concentrations as follows in $2\,M\,H_2SO_4$:

$$V_1 = 240\,[RuO_4] + 90.4\,[OsO_4] \quad (0.005\,M\,\text{CeIV and }0.001\,M\,\text{AsIII})$$

$V_2 = 2.69\,[\text{RuO}_4] + 78.0\,[\text{OsO}_4]$

(0.0005 *M* CeIV and 0.001 *M* AsIII, latter added first)

By measuring the rates under these two conditions, they were able to analyze Os + Ru mixtures (10^{-10}–10^{-7} M) with generally good results, even at unfavorable ratios.

50. R. Lang, *Mikrochim. Acta*, **3**, 116 (1938).

51. S. Asperger and I. Murati, *Anal. Chem.*, **26**, 543 (1954). The method has been used for determining Hg in air. An automatic method is described by T. P. Hadjiioannou, *Anal. Chim. Acta*, **35**, 351 (1966).

52. D. C. Olson and D. W. Margerum, *J. Am. Chem. Soc.*, **85**, 297 (1963); Margerum and R. K. Steinhaus, *Anal. Chem.*, **37**, 222 (1965).

53. A. C. Javier, S. R. Crouch, and H. V. Malmstadt, *Anal. Chem.*, **41**, 239 (1969). J. D. Ingle, Jr., and S. R. Crouch, *Anal. Chem.*, **43**, 7 (1971), determine phosphate in a mixture with silicate by forming molybdophosphoric acid (rapid reaction), reducing this with ascorbic acid and measuring the rate of formation of the P-heteropoly blue (Si reacts very slowly). β-Molybdosilicic acid is then formed in another aliquot (slow reaction) and the rate measured after the rapid formation of molybdophosphoric acid. Not more than 5 min is required for these determinations.

54. Is such a photometric method indirect? In one sense of the word perhaps not. The rate measurement is equivalent to finding the concentration of reacted P at various times after time 0 instead of at a time equivalent to ~∞ in an equilibrium method by comparing against a standard. But if reaction rates are not the same in sample and standards—always a possibility when foreign substances are present—an error may result that would not be present in an equilibrium method.

55. D. W. Margerum, J. B. Pausch, G. A. Nyssen, and G. F. Smith, *Anal. Chem.*, **41**, 233 (1969) (determination of transition metals, lanthanides, etc.); J. B. Pausch and D. W. Margerum, *ibid.*, **41**, 226 (1969) (Mg, Ca, Sr, Ba).

56. For computer-assisted rate analyses using this type of reaction, see, for example, B. G. Willis, J. A. Bittikofer, H. L. Pardue, and D. W. Margerum, *Anal. Chem.*, **42**, 1340 (1970); B. G. Willis, W. H. Woodruff, J. R. Frysinger, D. W. Margerum, and H. L. Pardue, *ibid.*, **42**, 1350 (1970); J. B. Worthington and H. L. Pardue, *ibid.*, **44**, 767 (1972). For kinetic determination of iron with an automatic stopped-flow spectrophotometer, see P. M. Beckwith and S. R. Crouch, *Anal. Chem.*, **44**, 221 (1972).

57. G. G. Guilbault, *Enzymatic Methods of Analysis*, Pergamon, Oxford, 1970.

58. A. Townshend and A. Vaughan, *Talanta*, **17**, 289 (1970).

59. A. Townshend and A. Vaughan, *Talanta*, **16**, 929 (1969).

60. G. G. Guilbault, M. H. Sadar, and M. Zimmer, *Anal. Chim. Acta*, **44**, 361 (1969).

61. Usually it is the intensity of the color that is duplicated, but sometimes it is the *hue* (e.g., in mixed color dithizone methods). As an item of interest, it

may be mentioned that comparison of hues in reflected light formed the basis of one of the earliest (if not the earliest) methods of chemical analysis, namely, the use of the touchstone in obtaining the proportion of gold in its alloys. The color of a streak made by the sample on a black surface of stone was compared with the color of streaks made by gold alloys of known composition. Theophrastus (ca. 371–288 B.C.), writing on the touchstone, says, "... it now serves not only for the trial of refined gold, but also of copper or silver coloured with gold; and shows how much of the adulterating matter by weight is mixed with gold ..." The sample in this earliest of physicochemical methods of analysis was of micro size, or possibly even of submicro size according to the terminology we have adopted (Fig. 1-1, p. 2). More than two millenia later, this colorimetric method was converted to a photometric method by measuring the reflectivity of an Au–Ag alloy at 550 nm (the reflectivities of Au and Ag are very dissimilar in the 450–600 nm range). Silver contents determined in this way compare favorably with contents found by electron-probe microanalysis. See H. Squair, *Trans. Instn. Min. Met.*, **74,** 917 (1965).

62. See A. A. Levinson, *Introduction to Exploration Geochemistry*, Applied Publishing, Calgary, 1974. Chapter 6 is devoted to analytical methods used in exploration geochemistry. Procedures are given in the appendix, pp. 549–559, for determination of Zn, Cu, total heavy metals, Mo, and W by the standard series method. The chromogenic reagents used are dithizone, biquinoline, and dithiol. Literature references to methods can be found in this work.

63. Symbolically expressed, $\Delta q/q$ = constant, where Δq is the increase or decrease in the amount of q required to give a just perceptible difference in color intensity in the optimum concentration range. In strongly colored solutions, the value of $\Delta q/q$ again increases. In comparing color intensities in tubes according to the procedure described, the value of $\Delta q/q$ is usually found to be 0.05–0.07. See, for example, J. H. Yoe and F. H. Wirsing, *J. Am. Chem. Soc.*, **54,** 1866 (1932). R. C. Wells, *J. Am. Chem. Soc.*, **33,** 504 (1911), found $\Delta q/q$ equal to 0.065 in the determination of titanium with hydrogen peroxide. D. W. Horn and S. A. Blake, *Am. Chem. J.*, **36,** 202 (1906), obtained the respective values 0.07 and 0.08 for potassium chromate and copper sulfate in the favorable concentration range.

64. The popular Lovibond system of colorimetric analysis uses colored discs for comparison. It finds use in the analysis of water, foodstuffs, metals, plastics, and others. Many tests are described in L. C. Thomas and G. J. Chamberlin, *Colorimetric Chemical Analytical Methods*, 8th ed., Tintometer, Salisbury, England, 1974.

65. This requirement can often be circumvented by using a standard solution containing the constituent in the form of the colored product. Thus in determining manganese as permanganate, a solution of potassium permanganate can be used as the standard.

66. Over a somewhat wider range the minimum detectable amount of the colored

substance does not vary much. This Duboscq minimum is smaller than the sensitivity value s (p. 207) but apparently not by much. As Q decreases below the optimal range, the standard deviation increases, and the product remains roughly constant. Thus for CrO_4^{2-} (Table 4-6) at $Q = 10, 20, 60$, and $100 \ \mu g \ Cr/cm^2$, $Q \times$ standard deviation $= 75, 60, 78$, and $100 \ \mu g \ Cr/cm^2$. This product is only slightly smaller than the value $100 \ \mu g \ Cr$ given for s in Table 4-5, which admittedly is only approximate. If a more exact way of specifying visual sensitivity were needed, it could be based on the minimum standard deviation.

67. H. Weisz, *Microanalysis by the Ring Oven Technique*, 2nd ed., Pergamon, Oxford, 1970. Before the advent of the ring oven, colorimetry based on comparison of colors of spot reaction products was (and still is) applied (see note 1).

68. Some post-1960 books dealing with the underlying principles or with analytical applications may be noted: C. A. Parker, *Photoluminescence of Solutions with Applications to Photochemistry and Analytical Chemistry*, American Elsevier, New York, 1968; D. M. Hercules, Ed., *Fluorescence and Phosphorescence Analysis*, Wiley, New York, 1966; J. D. Winefordner, S. G. Schulman, and T. C. O'Haver, *Luminescence Spectrometry in Analytical Chemistry*, Wiley-Interscience, New York, 1972; C. E. White and R. J. Argauer, *Fluorescence Analysis: A Practical Approach*, Dekker, New York, 1970; G. G. Guilbault, Ed., *Fluorescence*, Dekker, New York, 1967; G. G. Guilbault, *Practical Fluorescence: Theory, Methods and Techniques*, Dekker, New York, 1973; E. L. Wehry, Ed., *Modern Fluorescence Spectroscopy*, 2 vols., Plenum, New York, 1976; M. A. Konstantinova–Shlezinger, Ed., *Fluorimetric Analysis*, Moscow, 1961, transl. by Israel Program for Scientific Translations, Jerusalem, 1968. Accounts of fluorimetry are to be found in the more comprehensive reference works on analytical chemistry. A useful discussion, with particular reference to trace analysis is presented by P. A. St. John in Winefordner, Ed., *Trace Analysis: Spectroscopic Methods for Elements*, Wiley, New York, 1976, Chapter 7B. Reviews are numerous , for example, C. A. Parker and W. T. Rees, *Analyst*, **87**, 83 (1962) (a useful general survey). C. E. White, *Ind. Eng. Chem. Anal. Ed.*, **11**, 63 (1939); *Anal. Chem.*, **21**, 104 (1949); **22**, 69 (1950); **24**, 85 (1952); **26**, 129 (1954); **28**, 621 (1956); **30**, 729 (1958); **32**, 47R (1960). C. E. White and A. Weissler, *Anal. Chem.*, **34**, 81R (1962); **36**, 116R (1964); **38**, 155R (1966); **40**, 116R (1968); **42**, 57R (1970) (biennial literature reviews). General accounts of fluorescence and phosphorescence analysis are presented by D. M. Hercules, *Anal. Chem.*, **38** (12), 29A (1966) and T. S. West in *Trace Characterization, Chemical and Physical*, Nat. Bur. Stand. Monograph 100, U.S. Government Printing Office, Washington, D.C., 1967. Review of luminescence analysis, 1966–1973 (1300 papers): D. P. Shcherbov and R. N. Plotnikova, *Zav. Lab.*, **41**, 129 (1975).

69. Chemiluminescence refers to the emission of light in a chemical reaction. Certain chemiluminescence reactions are catalyzed by metals and traces of

these can then be determined in this way, that is, the intensity of the emitted light is proportional to the concentration of the catalyzing metal. Such methods can be highly sensitive but are likely to be unselective. Occasionally they can be made sufficiently selective by masking to be of practical use. For example, Cr(III) can be determined down to concentrations of ~ 1 in 10^{10} by its catalysis of the light emitting reaction between Luminol (5-amino-2,3-dihydrophthalazine-1,4-dione) and H_2O_2 in the presence of EDTA, as shown by W. R. Seitz, W. W. Suydam, and D. M. Hercules, *Anal. Chem.*, **44**, 957 (1972); application to biosamples: R. T. Li and D. M. Hercules, *ibid.*, **46**, 916 (1974). Other heavy metals such as Cu also catalyze the reaction, but they are complexed in basic solution with EDTA, which reacts extremely slowly with Cr(III) at room temperature. However, Fe(II, III) and Co(II) still catalyze and require application of a different method; Sn(IV) leads to low results. Some metals can be determined by their inhibiting effect on chemiluminescence; see a review of metal determination by catalysis and inhibition: U. Isacsson and G. Wettermark, *Anal. Chim. Acta*, **68**, 350 (1974).

70. E. A. Solov'ev and E. A. Bozhevol'nov, *J. Anal. Chem. USSR*, **27**, 1648 (1972).

71. Commercial spectrofluorimeters are available that record corrected fluorescence.

72. See some of the reviews mentioned in note 68 for continuing developments in instrumentation and P. F. Lott and R. J. Hurtubise, *J. Chem. Educ.*, **51**, A 358 (1974). Parker and Rees[68] touch on some important features of spectrofluorimeters.

73. Transmittance curves of many filters have been given by C. W. Sill, *Anal. Chem.*, **33**, 1584 (1961).

74. If $aCb = 0.010$, $F = KI_0 \times 0.02303$ according to Eq. (4-26), compared to the true value $F = KI_0 \times 0.02277$ according to Eq. (4-25). In other words, the linear relation fails by about 1% $[(0.02303 - 0.02277)/0.023 \times 100 = 1.1\%]$. If $aCb = 0.05$, the difference is 5.5%.

75. R. van Slageren, G. den Boef, and W. E. van der Linden, *Talanta*, **20**, 501 (1973).

76. Not surprisingly, visual fluorescence sensitivity can be expressed more simply, provided the excitation source is specified. Consider a sample solution contained in a tube exposed to the exciting radiation along its whole length to be compared against a blank solution in an identical tube. The sensitivity can then be expressed as the minimum amount of the element in a column of solution of 1 cm^2 cross section showing a fluorescence perceptibly stronger than the blank fluorescence (if any) when the tubes are examined axially. For example, the visual sensitivity of the morin reaction for beryllium in full daylight is $\sim 0.02 \ \mu\text{g Be/cm}^2$. With long-wave ultraviolet, the limit is $\sim 0.002 \ \mu\text{g Be/cm}^2$. With 10 ml of solution, a volume easily accommodated in 25 ml color comparison tubes (Fig. 4-15), the concentration sensitivity is $(2 \times 0.002)/10 = 4 \times 10^{-4}$ ppm of Be (2 is the cross-sectional area of the tube in square centimeters). The instrumental sensitivity has been reported as

2×10^{-5} ppm Be. This detection limit is roughly 0.01 that of sensitive absorptiometric methods for Be.

77. For a compilation of fluorimetric methods for the determination of inorganic substances with sensitivities, see A. Weissler and C. E. White in L. Meites, Ed., *Handbook of Analytical Chemistry*, p. 6–176, McGraw-Hill, New York, 1963, and St. John.[68]

78. P. A. St. John, W. J. McCarthy, and J. D. Winefordner, *Anal. Chem.*, **38**, 1828 (1966).

79. C. A. Parker and W. T. Rees, *Analyst*, **85**, 587 (1960); R. M. Dagnall, S. J. Pratt, R. Smith, and T. S. West, *ibid.*, **93**, 638 (1968).

80. A. K. Babko, A. I. Volkova, and T. E. Get'man, *J. Anal. Chem. USSR*, **22**, 842 (1967).

81. H. M. Stevens, *Anal. Chim. Acta*, **20**, 389 (1959). The generalization stated has exceptions. For instance, the Ru(II) chelate of 1,10-phenanthroline is fluorescent, as are Cu(I) pyridine complexes and the Co(III)–PAN chelate. The fluorescence of the last is explained by Co(III) becoming diamagnetic when chelated with PAN, a strong field ligand. Zinc fluoresces with morin but not with PAN. For the relevant underlying principles see some of the references cited in note 68, for example, Winefordner et al. and Hercules.

82. J. W. Collat and L. B. Rogers, *Anal. Chem.*, **27**, 961 (1955), were able to determine Al and Ga in the presence of each other, when present in not greatly different amounts, by using excitation of the hydroxyquinolates in $CHCl_3$ at different wavelengths.

83. E. B. Sandell, *Anal. Chem.*, **19**, 63 (1947).

84. B. W. Bailey, R. M. Dagnall, and T. S. West, *Talanta*, **13**, 1661 (1966).

85. The addition method can be used for up to 1 mg Fe in 1 ml in the fluoride procedure for U. See B. S. Osipov, *Zh. Anal. Khim.*, **21**, 70 (1966).

86. J. D. Ingle, Jr., and R. L. Wilson, *Anal. Chem.*, **48**, 1642 (1976).

5

INORGANIC PHOTOMETRIC REAGENTS
AND DETERMINATION FORMS

This chapter provides some information of a general nature on inorganic photometric determination forms for the metallic elements. The inorganic reagents converting metals into these forms may supply the ions for coordination complex formation or they may act as oxidizing or reducing agents to put the metals into the required oxidation state. They may also function as indirect reagents. In the present context, *photometric* includes *absorptiometric* (loosely *colorimetric*) and *fluorimetric*, as well as *turbidimetric* and *nephelometric*, although the latter two methods are of minor importance.

A number of elements can be determined sufficiently sensitively in inorganic chromogenic and fluorigenic forms to make these useful in trace analysis. Some of these compete with, and a few are even superior to, organic photometric determination forms because they provide better selectivity, greater convenience, and comparable sensitivity. As a class, inorganic photometric reagents are much less sensitive than the organic, perhaps by almost an order of magnitude on the average. On the other hand, the stability (as a function of time) of the inorganic determination forms is usually greater.

I. ELEMENTS

With the exception of mercury, and possibly iodine, photometric determination of elements as such is not of great importance (atomic absorption at high temperatures is excluded here). Mercury is unique in being easily reducible and volatile at room temperature, with a sufficiently high vapor pressure to enable small amounts to be determined in a gas phase (air, nitrogen, etc.) by strong absorption at $\sim$254 nm. Iodide is easily oxidized to I_2 (I_3^- in aqueous iodide solution). Occasionally, indirect determination of metals [e.g., Tl(III)] as oxidizing species is based on the formation of an equivalent amount of I_3^-, or of I_2 extracted into an organic solvent. For I_3^- in 0.049 M I^-, $\epsilon = 4 \times 10^4$ at 288 nm and 2.64×10^4 at 353 nm. I_2 in CCl_4 has $\epsilon_{517} = 930$, in benzene $\epsilon_{500} = 1,030$. Iodine can also be determined as the blue reaction product with starch, $\epsilon \sim 1 \times 10^5$ at 575 nm. Conversely, an obsolete method for Mg was based on the adsorption of I_3^- by $Mg(OH)_2$ (Schlagdenhauffen reaction, 1878, brown hydrosol).

Colloidal suspensions of easily reducible elements such as Au, Te, Se, and As can serve for their turbidimetric or nephelometric determination, but the sensitivity is mediocre and the usual problems in reproducibility of suspensions are encountered.[1]

II. SIMPLE CATIONS

With one or two exceptions, simple (hydrated) cations are not sufficiently strongly colored (or absorbing in the UV range) to be useful determination forms in trace analysis. The prominent exception is Ce(IV), which actually is probably largely complexed in perchloric acid, and certainly in sulfuric acid, solution. Its absorbance peak in 1 N sulfuric acid lies at 320 nm, and the sensitivity index $\mathscr{S} = 0.025$ (= the number of micrograms of Ce in a solution of 1 cm^2 cross section giving an absorbance of 0.001). This sensitivity allows the determination of a few micrograms of cerium in a volume of 10 ml when a 1 cm absorption cell is used in a conventional spectrophotometer. At 390 nm, $\mathscr{S} \sim 0.13$. Because the determination can be made moderately selective, it has some value in trace analysis. Protactinium(V) in sulfate solution has $\epsilon_{220} = 7700$; Pa(IV), $\epsilon_{290} = 1500$ (whence $\mathscr{S}_{290} = 0.15$ μg Pa/cm^2).

Some of the trivalent lanthanides have sharp absorption bands in the visible range (see Fig. 4-1b), but the sensitivity is so poor (e.g., for Pr at 444 nm in perchloric acid solution $\mathscr{S} \sim 13$) that their determination as M^{3+} is of little, if any, value in trace analysis. For Am^{3+}, $\mathscr{S}_{504} = 0.6$. On the other hand, some of the rare earth elements fluoresce sufficiently strongly in aqueous solution when excited by UV radiation to make their trace determination of interest. The fluorescence of Ce(III) is detectable in 0.1 ppm solution (excitation at 254 nm, emission at 350 nm); Tb can also be determined in this way. See further the chapter on rare earth elements in *Colorimetric Determination of Traces of Metals*, 3rd ed. Fluorescence in a solid matrix will not be considered here.

III.   OXO ANIONS AND OXIDES—OXO CATIONS

CrO_4^{2-} and MnO_4^- are old determination forms that are still useful in colorimetry when the highest sensitivity for the determination of chromium and manganese is not required ($\mathscr{S}_{Cr} = 0.012$ at 366 nm, $\mathscr{S}_{Mn} = 0.027$ at 522 nm). In acid solutions, Cr(VI) is present as $Cr_2O_7^{2-}$ and $HCrO_4^-$ (mostly as the latter in the dilute solutions encountered in trace analysis), so that oxidation in basic medium to give only CrO_4^{2-} is preferable when not objectionable on other grounds. Oxidizing agents are discussed in the chapters on these elements (*Colorimetric Determination of Traces of Metals*).

In basic solution, MoO_4^{2-} absorbs quite strongly at 230 nm. This property has been made the basis of an indirect method for phosphorus.[2] Phosphomolybdic acid is extracted from an acidic aqueous solution into a mixture of ethyl ether and isobutyl acetate (excess molybdic acid is not extracted or is washed out), and the separated extract, after washing, is shaken with an ammonia buffer to decompose the heteropoly acid and bring MoO_4^{2-} into the aqueous phase. Presumably, other elements forming heteropoly molybdic acids could be determined in the same way.

Perrhenate ion absorbs maximally at 226 nm (sensitivity $\sim$1 ppm). V(V) in $1\ M\ H^+$ absorbs maximally at 222 nm.

Absorption by UO_2^{2+} and PuO_2^{2+} is too weak to be of use in trace determinations. But the fluorescence of U(VI), as UO_2^{2+} or its complex species, especially in fluoride melts, provides the most sensitive photometric method for uranium. In aqueous solutions (as in H_3PO_4–H_2O), the fluorescence is much weaker, but still of analytical value.

Osmium and ruthenium in higher oxidation states give species absorbing strongly enough to be of marginal interest to the trace analyst: OsO_4 in $CHCl_3$, $\epsilon_{304} = 1400$, $\epsilon_{289} = 1760$; RuO_4^{2-}, $\epsilon_{465} = 1700$; RuO_4^-, $\epsilon_{385} = 2200$; RuO_4 in H_2O, $\epsilon_{310} = 3000$, $\epsilon_{385} \sim 1000$.

IV.   PEROXO COMPLEXES

A number of congeneric elements react with H_2O_2 in acid solution to form peroxo ions or acids of moderate or mediocre absorptivity (Table

TABLE 5-1
A Portion of the Periodic Table Showing Elements Known to Form Peroxo Complexes in Mineral Acid Solution Absorbing in the Visible and Ultraviolet[a]

Sc	Ti	V	Cr	Mn
	400 nm	450 nm		
	0.07	0.17		
Y	Zr	Nb	Mo	Tc
		365 nm	330 nm	
		0.11	0.1	
La	Hf	Ta	W	Re
		285 nm	295 nm	
		0.19	0.33	
Ac	Th	Pa	U	

[a] The wavelengths of maximum absorption and sensitivity indexes (μg of element/cm$^2 \equiv \log I_0/I = 0.001$) are shown. Values for Nb, Ta, and W are for 96% H_2SO_4. Trans-U elements are not included.

5-1).[3] Because of their predominant absorption in the visible spectrum, the peroxo complexes of Ti(IV) (yellow) and V(V) (reddish brown) have long been used as determination forms for these elements (Fig. 5-1). From the standpoint of applied analysis, the possibility of forming these complexes in mineral acid solution is of great importance (H_2O_2 is a very weak base) and provides an advantage over the more sensitive organic reagents that often require solutions of low acidity. H_2O_2 reacts mole for mole with titanium over a range of acidities.[4] Apparently the equimolar molybdenum species is formed only at high acidities and low metal concentrations. TBP extracts the peroxo Mo complex, $HMoO_6^-$, the

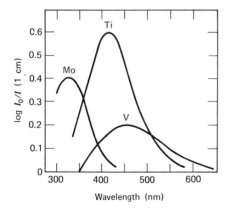

Fig. 5-1. Absorbance curves of the peroxo complexes of Mo(VI), Ti(IV), and V(V). The W-peroxo complex curve (not shown) lies in the ultraviolet, with a peak at 262 nm in 20% H_2SO_4. Reprinted with permission from A. Weissler, *Ind. Eng. Chem. Anal. Ed.*, **17,** 698 (1945). Copyright by the American Chemical Society.

extract showing weak absorption at 340 nm ($\epsilon \sim 1000$); Fe, W, Ti, V, and Nb interfere. The quantitative formation of the peroxo complexes requires an appreciable excess of hydrogen peroxide. Vanadium(V) forms two peroxo complexes[5]:

$$VO_2^+ + H_2O_2 \rightleftharpoons VO(O_2)^+ + H_2O \qquad K_1 = 3.5 \times 10^4 \ (\mu = 1, 25°)$$

reddish, $\epsilon_{455} = 280$

$$VO(O_2)^+ + H_2O_2 \rightleftharpoons VO(O_2)_2^- + 2 H^+ \qquad K_2 = 1.3 \ (1 \ M \ HClO_4)$$

yellow, $\epsilon_{350} = 610$, $\epsilon_{455} = 25$

(Isosbestic point of the two complexes at 404 nm).

In sulfuric acid solution, hydrogen peroxide forms three niobium(V) peroxo complexes, which absorb in the ultraviolet and the blue region of the spectrum[6] (Fig. 5-2). Their relative amounts depend on the concentration of sulfuric acid:

$$H_2O_2/Nb \quad 4:1 \ \rightleftharpoons \ 1:1 \ \rightleftharpoons \ 2:1 \ \rightleftharpoons \ 3:2 \ \rightleftharpoons \ none$$

Basic soln.	Neutral and dilute sulfuric acid	Inter- mediate sulfuric acid	Concen- trated sulfuric acid	Free SO_3

Tantalum forms peroxo complexes absorbing in the ultraviolet. Both niobium and tantalum can be determined absorptiometrically, but the sensitivities are mediocre. Titanium is a troublesome interfering element in spite of the fact that it gives less color in strong sulfuric acid than in dilute. Phosphoric acid bleaches the titanium color.[7]

The peroxo complexes of molybdenum, tungsten, and rhenium are not of importance in the determination of these elements because of the availability of better photometric determination forms. The unstable,

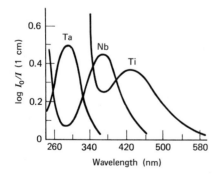

Fig. 5-2. Absorbance curves of Nb, Ta, and Ti peroxo acids in 96% H_2SO_4. Molar concentrations: Nb, 0.00054; Ta, 0.00055; Ti, 0.0021. (After Palilla, Adler, and Hiskey, *loc. cit.*)

blue, peroxo chromium(VI) compound (said to be $CrO_5 \cdot H_2O$) finds occasional use in separation of chromium.

In basic medium, H_2O_2 yields a reddish brown color with Mo(VI), a yellow with Ce(III or IV) and U(VI), and a pale yellow with W(VI). In the presence of excess bicarbonate, cobalt is oxidized to Co(III) with the formation of a green color. Np(V) forms a yellow $1:1$ peroxo complex in $1\ M$ NaOH; $\epsilon_{430} = 5000$.

In mineral acid solution, H_2O_2 is present as such since

$$\frac{[H^+][HO_2^-]}{[H_2O_2]} = 10^{-11.6} \qquad (\mu = 0.1)$$

In a $0.1\ M$ OH^- solution, most of the peroxide is present as HO_2^-, that is,

$$\frac{[HO_2^-]}{[H_2O_2]} = 10^{-11.6+13} = 10^{1.4}$$

and the peroxo complexes might be expected to be different from those formed in mineral acid medium. Uranyl ion does not react with H_2O_2 in strongly acid solution but does react in sodium hydroxide or carbonate solution to give yellow species; it also reacts in weakly acidic solution. Various species are formed.

V. HETEROPOLY ACIDS AND HETEROPOLY BLUES

The heteropoly molybdic acids formed when excess molybdate reacts in acidic solution with phosphate, silicate, arsenate, germanate, and other oxo anions are made the basis of sensitive methods for the determination of the corresponding elements. The heteropoly anion formed in the common series of the 12-molybdoheteropoly acids is of the type

$$\{PMo_{12}O_{40}\}^{3-} \qquad (\text{or } \{P(Mo_3O_{10})_4\}^{3-})$$

The central atom in the anion is P, to which the Mo atoms are coordinated through O atoms. In the 12-heteropolymolybdic acids there are 12 Mo atoms arranged around the P, or other central, atom.

These anions (or acids) have an absorption band in the near ultraviolet extending into the visible, so that their solutions are yellow. These complexes are chiefly of importance in the determination of P and Si, although those with As or Ge as the central atom can be used for their determination. Tungsten forms heteropoly acids analogous to the heteropoly molybdic acids, but less frequently used analytically, tungstic acid being only slightly soluble. Vanadium(V) can also sometimes function as the coordinating atom.[8] More than 30 elements can serve as the

TABLE 5-2
Extraction of Heteropoly Acids into 1-Butanol[a]
(Optimum conditions and wavelength maximum)

Central Atom	Na_2MoO_4 (M)	HCl (M)	λ_{max} (nm)
P	0.025	0.96	315
As	0.013	0.67	310
Si	0.010	0.15	300
Ge	0.010	0.15	290

[a] C. Wadelin and M. G. Mellon, *Anal. Chem.*, **25**, 1668 (1953). $\mathscr{S}_{As}$
~0.003 at 310 nm, $\mathscr{S}_{Ge}$ ~0.005 at 290.

hetero (central atom) element, and some can be determined in this way. The heteropoly acids are extractable into oxygenated solvents, such as esters and higher alcohols, immiscible with water.[9] Extraction is a consequence of bonding of H of the heteropoly acid to O of the organic solvent. Most or much of the excess molybdic acid remains in the aqueous phase, and the remainder can be washed out of the organic phase. Because molybdic acid also absorbs in the UV range in which the heteropoly acids absorb, it is advantageous to extract the latter and measure the absorbance of the organic phase to keep the blank low.

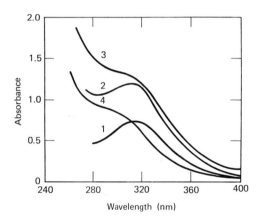

Fig. 5-3. Absorbance curves of heteropoly molybdic acids in 1-butanol (25 μg P, 100 μg each As, Si, and Ge in 25 ml butanol). (1) Molybdophosphoric acid, (2) Molybdoarsenic acid, (3) Molybdosilicic acid, (4) Molybdogermanic acid. Extraction conditions as in Table 5-2. Reprinted with permission from C. Wadelin and M. G. Mellon, *Anal. Chem.*, **25**, 1670 (1953). Copyright by the American Chemical Society.

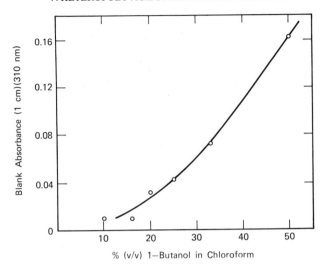

Fig. 5-4. Extraction of molybdic acid by 1-butanol–chloroform mixtures as indicated by reagent blank. 25 ml aqueous solution (0.96 M in HCl and 0.6% in $Na_2MoO_4 \cdot 2\,H_2O$) extracted with 10 + 10 ml organic solvent mixture and this diluted to 25 ml. From data of Wadelin and Mellon, *loc. cit.*

Molybdic acid does not absorb in the visible, but there the sensitivity becomes poor for the hetero elements. See Table 5-2 for optimal conditions of extraction of common heteropoly acids into 1-butanol and Fig. 5-3 for their absorption spectra. In general, phosphomolybdic acid (more accurately called molybdophosphoric acid) is extracted more easily than the other heteropoly acids. Differential extractions have been worked out for the determination of P, Si, As, and Ge in the presence of each other by choice of extractant and acidity.[10] Mixed organic solvents find use and may be preferable to a pure extractant. Thus a mixture of 1-butanol and chloroform[11] allows a more selective extraction of phosphorus than does butanol alone, and does not extract much of the reagent if the proper ratio is used (Fig. 5-4).

Some use has been made of masking agents for molybdenum to increase selectivity of heteropoly molybdate reactions. Thus oxalic acid and other organic acids destroy phosphomolybdic acid but not silicomolybdic acid (stability of hetero acids is a function of free molybdate concentration).

The heteropoly acids are not particularly useful for the determination of molybdenum and tungsten. Some mixed heteropoly acids have analytical value. Thus V(V) can be determined in mineral acid solution by adding (colorless) phosphotungstic acid to form yellow phosphotungstov-anadic acid, as already noted.[12] A mixed Bi–P dimeric heteropoly-molybdate has been shown to exist, in which the molar ratio Mo:P:Bi is 18:1:1.[13] It is more stable than 12-molybdophosphoric acid in perchloric acid solutions but is reduced more easily by ascorbic acid to heteropoly blue.

Titanium, Zr, and Th form ternary heteropoly acids in which $M:P:Mo = 1:1:12$.[14] The absorption spectra of the ternaries are very similar to the spectrum of phosphomolybdic acid, with $\lambda_{max} \sim 315$ nm, but the former show higher peaks (roughly double absorbances). These three elements can be determined as the ternary complexes,[15] with phos-phomolybdic acid as reagent, if the excess of the latter is extracted with n-butyl acetate, leaving Ti, Zr, or Th in the aqueous phase ($\mathscr{S}_{\text{Ti,330}} \sim 0.01$). Other ternary complexes are involved in heteroblue methods (p. 255).

The course of the reaction between phosphate and molybdate in acidic solution has not yet been completely elucidated. Stoichiometric studies[16] indicate that in nitric acid solution the reaction can be represented by

$$H_3PO_4 + 6\{Mo(VI)\}_{tot} \rightleftharpoons 12\text{-molybdophosphate} + 9\ H^+$$

Apparently, molybdenum is largely dimerized[17] in HNO_3 solutions (whence the coefficient 6 in the above equation). An increase in the hydrogen-ion concentration above $\sim 0.1\ M$ markedly decreases the equilibrium concentration of phosphomolybdic acid (with a restricted excess of molybdate). In sulfuric acid solution, a 5:1 combining ratio of $Mo(VI)_{tot}$ to P is reported, with 7 moles of H^+ reacting with 1 mole of phosphomolybdate in the reverse reaction. The heteropoly acid is formed to a smaller extent in sulfuric than in nitric acid medium and complexing by sulfate is indicated.[18]

The formation of phosphomolybdic acid is rapid under analytical conditions, only a few seconds (or less) being required. The kinetics has been investigated.[19] Phosphate and a Mo(VI) dimer react initially, and polymerization to 12-Mo–P heteropoly acid follows. The rate law for the formation of 12-molybdophosphoric acid is

$$\frac{d[12\text{-MPA}]}{dt} = \frac{k_1 k_2 [H_3PO_4][Mo(VI)]^2}{k_{-1}[H^+]^4 + k_2[Mo(VI)]}$$

which at acid concentrations $< 0.5\ M$ becomes

$$\frac{d[12\text{-MPA}]}{dt} = k_1[H_3PO_4][Mo(VI)]$$

The rate of formation of As heteropoly acid is much slower than that of P heteropoly acid, so that there is little interference from As in the determination of P by a fast reaction procedure. Silicate interferes in such a method, but much less than in conventional colorimetric methods (Chapter 4, Section I.E).

Strickland[20] has demonstrated that two forms of molybdosilicic acid exist in solution, both having the composition $H_4SiMo_{12}O_{40}$. The acid designated α is formed at low acidities and the β-acid at high acidities. The α-form is the stable form and the β-acid changes more or less rapidly into the α-form. Major factors controlling the molybdosilicic acid reaction are the state of molybdate ions, the concentration of molybdate, the ratio of acid to molybdate, and the type of acid used in the reaction.[21] Absorption spectra of unreduced and reduced (with tin(II) chloride) α- and β-molybdosilicic acids are given by Strickland. The reduced α-molybdosilicic acid is greenish blue and the reduced β-acid is royal blue.

The existence of the α- and β-forms of heteropolymolybdates of P, As, and Ge has been reported, and a structure for the β-form has been proposed.[22] The unstable β-forms can be stabilized by addition of comparatively large amounts of polar organic solvents, especially acetone, presumably because the solvent displaces water from the surface of the heteropoly anion and so inhibits any reorganization of the structure of the anion. The stable α-form of molybdogermanic acid is obtained by adding Ge solution to an excess of molybdate at pH 1.5[23] The pH is critical.

Procedures have been described for the simultaneous determination of two or three of the elements in a mixture of P, As, Si, and Ge by varying acid and Mo concentration, and wavelength.[24]

Heteropoly Blues. The heteropoly molybdic and tungstic acids have an important analytical property: They are reduced more easily than the parent molybdic and tungstic acids. With a suitable reducing agent used under proper conditions, as of acidity, the heteropoly molybdic (and tungstic) acid can be reduced to a soluble blue compound, known as heteropoly blue, without reduction of the excess molybdic acid. If the reducing agent is too strong or the acidity too low, the uncombined molybdic acid will also be reduced to a blue compound (in colloidal solution), called simple molybdenum blue. The heteropoly blues contain Mo(V) as well as Mo(VI) and are mixed-valence[25] complexes of indefinite composition, which can be described as containing electrons trapped on certain Mo atoms. The heteropoly blues and molybdenum blue do not have the same absorption curves, and the heteropoly blue curves, though similar, differ among themselves, even those of the same central element, depending on the conditions of reduction. Absorbance peaks usually range from ~700 to ~840 nm for the various elements. Molybdenum blue does not show this band.

The heteropoly blues provide determination forms of about the same sensitivity as the heteropoly acids from which they are derived if the absorbance of both is measured at λ_{max}. They are extractable into oxygenated organic solvents, butanol for example,[26] but usually extraction procedures are not applied (whereas they are needed for the heteropoly acids if maximal sensitivity is required). The heteropoly blues serve for the determination of As and Ge, as well as (and more importantly) for P and Si. A standard method for arsenic involves its isolation as $AsCl_3$ or AsH_3, formation of arsenimolybdic acid and reduction of the latter to heteropoly blue. The determination of germanium as heteropoly blue is no longer as important as formerly, now that sensitive organic chelating reagents are available for this element.

Various reducing agents find use in heteropoly blue methods. Stannous chloride and certain organic reducing agents such as 1-amino-2-naphthol-4-sulfonic acid + sulfite and hydroquinone have been, or are, frequently used in the determination of phosphorus. Hydrazine sulfate is very suitable for the reduction of arsenimolybdate in boiling solution and can also be used for phosphomolybdate. More recently ascorbic acid has been used and is very good. Ferrous sulfate has been used for the reduction of germanomolybdate.

General studies of heteropoly blue methods have been made, largely empirical in nature, and these papers may be consulted for the effect of a number of variables and for transmission curves.[27]

The kinetics of the formation of phosphomolybdenum blue has been investigated in nitric and sulfuric acid solutions (necessarily in narrow concentration ranges) with ascorbic acid and 1-amino-2-naphthol-4-sulfonic acid as reducing agents.[16] The following rate law is suggested for nitric acid solutions with ascorbic acid as reductant:

$$\frac{d[\text{P--Mo blue}]}{dt} = K[H_3PO_4][\text{Mo(VI)}_{tot}]^6 \left\{ \frac{K'[H^+]^9}{[\text{ascorb. acid}]} + 1 \right\}^{-1}$$

For sulfuric acid solutions the expression is the same, except $[H^+]^{10}$ replaces $[H^+]^9$. The rate of P--Mo blue formation is thus strongly decreased by an increase in acidity. At room temperature, with $[\text{Mo(VI)}] = 0.0085$ and $[\text{aminonaphtholsulfonic acid}] = 0.00017$, the reaction is completed in ~2 minutes when $[HNO_3] = 0.5$ and in ~10 minutes when $[HNO_3] = 0.9$. The amount of P--Mo blue formed is independent of acidity in the range $0.3–1\ M\ HNO_3$. It has been noted that freshly prepared sodium or ammonium molybdate solutions give different kinetic (and equilibrium) results than solutions allowed to stand for a few hours.

The heteropoly blue methods require fairly close control of conditions (as do the heteropoly acid methods). At a given molybdic acid concentration, reduction of the excess reagent to molybdenum blue will occur if the

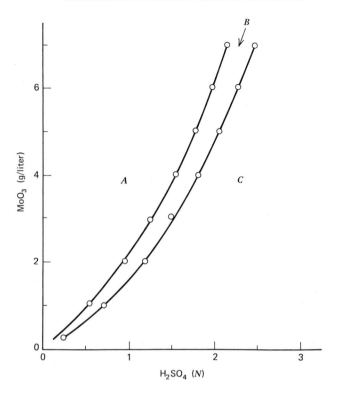

Fig. 5-5. Analytical fields in the reduction of arsenimolybdic acid with ascorbic acid, hydrazine sulfate, or stannous chloride in sulfuric acid solution. *A* is the zone of reduction of excess molybdic acid, *B* is the zone of formation of stable heteropoly As blue, *C* is the zone in which fading of heteropoly blue occurs. After Duval, *loc. cit.*

acidity is too low. On the other hand, if the acidity is too high, the heteropoly blue becomes unstable (color fades). The zone in which a satisfactory reaction is obtained is relatively narrow, as illustrated for arsenic in Fig. 5-5.[28]

Heteropoly blue methods based on the reduction of heteropoly (molybdate) or mixed heteropoly (ternary molybdate) complexes have been proposed for a number of metals (reagents in parentheses, $SnCl_2$ or ascorbic acid the usual reductant):

Bi	$(Mo, P)^{29}$	Th	$(Mo, P)^{35}$
Bi	$(Mo, sulfate)^{30}$	Ti	$(Mo)^{36}$
Ce	$(Mo)^{31}$	W	$(Mo, P)^{37}$
Nb	$(Mo, P)^{32}$	Y	$(Mo, P)^{38}$
Nb	$(Mo)^{33}$	Zr	$(Mo, sulfate)^{39}$
Ni	$(Mo, P)^{34}$	Hf	$(Mo, sulfate)^{40}$

Because of inadequate selectivity, the careful control of conditions required, or for other reasons, most of these methods have limited utility.

A study of the reduction of 12-tungstophosphate and 12-tungstosilicate anions has shown that each anion can accept two electrons without decomposition. Further electrons can be added in acidic solutions where protonation can accompany reduction and keep the overall ionic charge low.[41]

$$\{PW_{12}O_{40}\}^{3-} \rightarrow \{PW_{12}O_{40}\}^{4-} \rightarrow \{PW_{12}O_{40}\}^{5-} \rightarrow \{HPW_{12}O_{40}\}^{6-}$$

$$\{SiW_{12}O_{40}\}^{4-} \rightarrow \{SiW_{12}O_{40}\}^{5-} \rightarrow \{SiW_{12}O_{40}\}^{6-} \rightarrow \{H_2SiW_{12}O_{40}\}^{6-}$$

The central atoms (P and Si) play essentially no part in the reduction of these heteropoly anions. The blue species formed by the addition of one and two electrons to 1:12- and 2:18-heteropolytungstates are best described as class II mixed-valence complexes.[42] This description implies the trapping of the added electrons on tungsten atoms.[43]

VI. HALIDE COMPLEXES

A considerable number of heavy metals form anionic halide (Cl^-, Br^-, I^-) complexes absorbing moderately strongly in the ultraviolet and visible regions of the spectrum. Except for Fe(III), Co, Cu(II), Au, and the platinum metals, the absorption bands of the chloro and bromo complexes lie in the ultraviolet, whereas those of the iodo complexes lie partly in the visible (Figs. 5-6a, b, c). In general, the iodo complexes provide a little greater sensitivity. With any one metal, a series of halide complex species is formed and even at high halide concentration, for example, 6 M HCl, complete conversion to the highest species may not occur. The sensitivity varies with the halide (and hydrogen-ion) concentration. The sensitivity indexes $\mathscr{S}$ are of the order of 0.01 μg of metal/cm^2 for an absorbance of 0.001 (Figs. 5-6a, b, c), with a range from 0.002 to 0.04 or greater.

A. CHLORIDES

The chloro anions of iron(III), copper(II), cobalt, and the platinum metals absorb sufficiently strongly in the visible spectrum to allow moderately small amounts of these metals to be determined colorimetrically, as they were before the introduction of photoelectric spectrophotometers and filter photometers. The determination of iron[44] and copper with hydrochloric acid finds occasional use when the highest sensitivity is not required. The chloro complexes of these metals are yellow. Copper(II) gives the maximum color intensity with hydrochloric acid when the acid concentration is 28% or higher.[45] Iron interferes seriously since, weight for weight, it produces twice as strong a color as copper. The coloration given by cobalt is much less intense than that of copper. Copper and cobalt in strong hydrochloric acid can be determined more selectively by absorption measurements at 970 and 625 nm, where chloroferric acid absorbs but little. In 6 M HCl, $\mathscr{S}_{940}$ for Cu(II) = 0.7, too high for trace determinations. Addition of acetone (to decrease dissociation of chloro complexes) permits the use of lower HCl concentrations in the determination of Cu and Co. Determination of Ir(IV) as IrCl$_6^{2-}$ in HCl medium in the visible has low sensitivity ($\mathscr{S}_{487} \sim 0.05$) but conceivably might be of use at times. Care needs to be taken that the absorbing species is actually

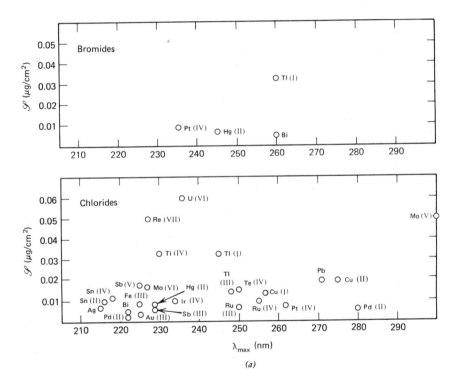

Fig. 5-6. $\mathscr{S}$ (μg of element/cm^2 of solution corresponding to log $I_0/I = 0.001$) at λ_{max} of complex metal chlorides, bromides, and iodides. Values, from various sources, are to be looked upon as approximate, because of variable Cl$^-$, Br$^-$, and I$^-$ concentrations (Cl$^-$ usually = 6 M, Br$^-$ = 4 M) and acidity. Sn(II)–halide complexes of platinum metals not included here.

    a. Spectral range 210–300 nm
    b. Spectral range 300–400 nm
    c. Spectral range >400 nm

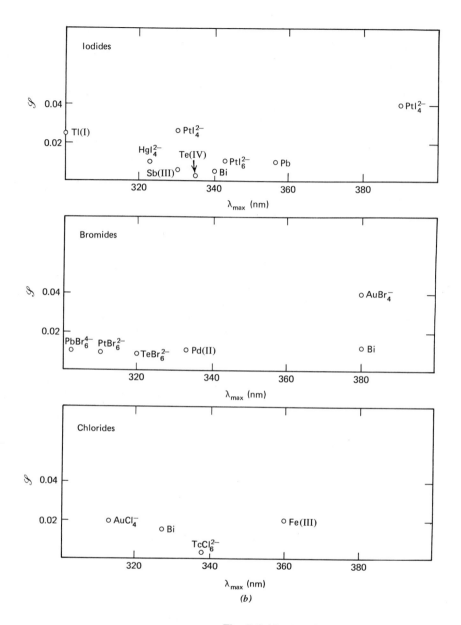

Fig. 5-6 (*Continued*)

258

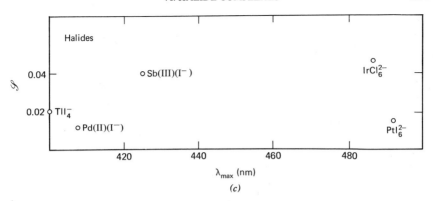

Fig. 5-6. (*Continued*)

$IrCl_6^{2-}$; equilibria in chloride solutions are complex.[46] Rhenium has been determined by reducing ReO_4^- to $ReCl_6^{2-}$ in 8 M HCl and measuring the absorbance at 370 or 282 nm.

Ultraviolet absorption measurement in 6 M hydrochloric acid has been proposed for the determination of Tl(I) (245 nm), Pb (271 nm), and Bi (222 and 327 nm).[47] No absorption is reported for As(III), Ga, In, Fe(II), Se(IV), and W(VI) in the UV range at 10 ppm concentration. Absorption is shown by Cu(I), Mo(VI), Sn(II), and Ti(III, IV) and other metals as well. Re(IV) in strong HCl absorbs markedly at 282 nm. Nb [possibly as $Nb(OH)_2Cl_4^-$] in 12 M HCl absorbs maximally at 281 nm, and its determination in this way has been proposed, but numerous metals, particularly V, Cr, Pb, Fe(III), Cu(II), Mo(VI), and Ti also absorb. $TcCl_6^{2-}$ (obtained by reduction of Tc(VII) with hot concentrated HCl) absorbs maximally at 240 and 338 nm; at the latter wavelength $\epsilon = 32,000$ ($\mathscr{S} = 0.003$), and Re does not absorb.[48]

Determination of heavy metals as chloro complexes is attractive because of its simplicity and, sometimes, adequate sensitivity, and no doubt is useful for some samples,[49] but interferences are numerous, especially in the UV range.

With half-band widths of 20 nm or more in the UV, there is bound to be overlapping of absorption bands. Also it should be remembered that traces of organic substances can cause trouble by absorbing in the UV. Water stored in polyethylene bottles has been reported to show absorption at 270 nm. Solutions run through ion-exchange columns may also absorb in the UV.

Thallium(I) in solutions of high chloride concentration gives a blue fluorescence when excited by UV radiation, which can be made the basis of a sensitive fluorimetric method. A number of metals—Sb(III), Bi,

Ce(III), Pb, Te(IV), Tl(I)—fluoresce strongly in HCl glass at liquid nitrogen temperature.[50] The limiting concentrations are 10^{-6}–10^{-8} M.

Some of the chloro (and other halo) acids are extractable into solvents of the oxygen type. As anionic partners of bulky organic cations, such as tetraphenylarsonium, some are extractable into inert organic solvents— but then we are using an organic reagent.

Stannous Chloride Complexes of Platinum Metals. The reactions of stannous chloride with Pt(IV or II) in hydrochloric acid solution to give yellow to orange soluble species, and with Rh(III) to give a reddish one, have long been used for the sensitive colorimetric determination of these metals. Some light has been thrown on these reactions in recent years but obscurities still remain. Evidence has been presented[51] to show that these colored species (and analogous species formed by Ru and Ir) are complex anions in which $SnCl_3^-$ groups are coordinated to Pt metal chlorides, the platinum metal being in a lower oxidation state:

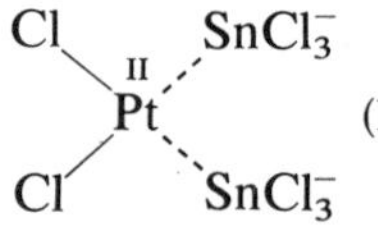

(Ru(II) forms a similar anion)

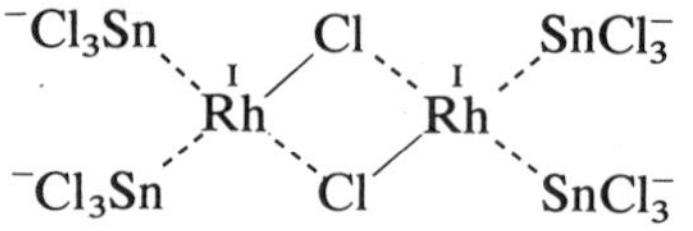

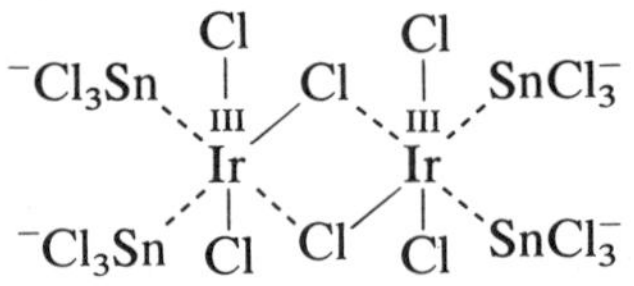

These formulas were established chiefly by adding tetramethylammonium chloride, or other salt having a large cation, to the reaction mixture to precipitate the salt of the complex anion, which was analyzed.

The complex platinum anion, $PtCl_2(SnCl_3)_2^{2-}$, forms two isomers, one red, the other yellow. The red form is obtained at high Sn/Pt ratios, so that when this ratio is >8, it alone exists in solution, and this then should be the isomer existing under analytical conditions. The spectra of the yellow and red forms are greatly different, λ_{max} for the former falling at 254 and 278 nm and for the latter at 310 and 401 nm as determined by the use of NMe_4^+ salts. The latter maxima agree with those observed in analytical solutions, namely at 310 and 403 nm. The red isomer is believed to be *trans*, the yellow *cis*.

Investigation of the Pt–SnCl$_2$ reaction by Elizarova and Matvienko[52] have led to some conclusions at variance with those of Young et al. The former authors find platinum to be present as Pt(0) in the SnCl$_2$ compounds.[53] They report the formation of two platinum complexes, having Pt:Sn = 1 and 2, in 1–2 M HCl (with HClO$_4$ to give $\mu = 2$). With a slight excess of SnCl$_2$, a yellow 1:1 species is formed ($\lambda_{max} = 275$ nm), with a large excess the 1:2 species is formed (absorption maxima at 255, 310, and 400 nm). When the chloride concentration of the solution is 0.25–0.5 M, an orange complex (Pt:Sn = 2) is obtained, whose absorption spectrum remains unchanged until [Sn]:[Pt] ~10. If the latter ratio is made greater than 20, the orange color changes to yellow, and the absorption spectrum becomes identical with that of the yellow complex in 2 M HCl with a large excess of Sn(II). The color change is interpreted as being due to *cis–trans* isomerism. The orange complex is considered to be the *trans* form ($\lambda_{max} = 275$ nm), the yellow complex the *cis* ($\lambda_{max} = 255$, 310, 400 nm). The *cis* form is obtained in procedures specifying HCl concentrations ~2 M ($\mathscr{S}_{400} = 0.025$, $\mathscr{S}_{310} = 0.0054$).

Young et al. do not mention different forms of anionic rhodium–stannous chloride complexes, but it is known that red and yellow solutions of these complexes can be obtained, depending on the hydrochloric acid concentrations. The 1:1 HCl solution of the NMe$_4^+$ salt of the rhodium anion represented above shows λ_{max} at 303 and 425 nm, whereas analytical solutions (~2 M HCl, color development by heating) have λ_{max} at 330–340 and 470–475 nm, so that the species are not identical. In fact, the existence of four or five Rh–SnCl$_2$ species has been claimed.[54] One of these, supposedly isolated in pure form as a yellow compound ($\lambda_{max} = 425$ nm), has a molar ratio Rh:Sn = of either 1:4 or 1:2 (according to the reference). Another, formed in 2 M HCl with 0.06–0.25 M SnCl$_2$ ($\lambda_{max} = 475$ nm) has Rh:Sn = 1:2.[55] In 0.05 M HCl– 6 M HClO$_4$, $\mathscr{S}_{330} = 0.002$. The red species formed in 0.5–6 M HCl with 0.06–0.25 M SnCl$_2$ is recommended for the determination of Rh. In 2 M HCl, the color develops fully in 1 hr at room temperature or in 10–20 minutes on heating at 100°C. $\lambda_{max} = 475$ nm. The color is stable for 24 hr and is said to be relatively reproducible.

The reactions occurring in the Rh–SnCl$_2$ system are evidently complex and still far from being well understood.[56]

The platinum metal–stannous chloride complexes can be extracted into oxygenated immiscible organic solvents (esters and alcohols). In the organic solvents, the SnCl$_3^-$ groups in these species presumably are replaced by SnCl$_2$ or SnCl$_2$·HCl.

Palladium(II) also reacts with $SnCl_2$ in HCl–$HClO_4$ solution.[57] Soluble orange to green products are formed. Chloride is not necessary for a color reaction between Pd(II) and Sn(II). Tin(II) phosphate yields a soluble red product with palladium in phosphoric–perchloric acid, which serves for the sensitive determination of palladium (see Pd chapter, *Colorimetric Determination of Traces of Metals*). The sensitivity of the Sn(II)–H_3PO_4 reaction is more than a magnitude greater than that of the Sn(II)–HCl reaction (for the former $\mathscr{S}_{487} = 0.003$, for the latter $\mathscr{S}_{635} = 0.04$). The species formed in these reactions have not been identified.[58]

A separation of Rh from Ir is based on extraction of the yellow Rh–$SnCl_2$ complex into isopentyl alcohol. Ir remains in the aqueous phase.[59]

B. BROMIDES

Substitution of HBr for HCl brings about a slight analytical gain in sensitivity and a shift in the maximum absorbance to longer wavelengths for most reacting metals.[60] There is a definite advantage in determining gold as $AuBr_4^-$ ($\lambda_{max} = 380$ or 400 nm in butyl acetate) instead of as $AuCl_4^-$ ($\lambda_{max} = 312$). Pd(II) absorbs quite strongly in bromide solution at 333 nm.

Stannous Bromide Complexes of Platinum Metals.   Substituting $SnBr_2$ for $SnCl_2$ gives greater sensitivity in determination of the platinum metals. Presumably the reaction products are analogous to those formed in the $SnCl_2$ reactions (p. 260). The determination of Pt, Rh, Pd, and Ir in this way has been investigated.[61] The determination of Rh and Ir is of special interest because their reactions are sensitive, and these metals do not have many good color reactions. (Gold is reduced to the metal by stannous bromide, and a violet sol can be obtained with gelatin as protective colloid, but this reaction is of no great interest.) The reactions of the four platinum metals can be summarized as follows:

Pt   Red color develops immediately at room temperature and is stable for several days. Acidity should be 1 M or greater. $\lambda_{max} = 463$ nm.

Rh   Red-orange color develops at room temperature ($\lambda_{max} = 475$ nm), but this changes to yellow ($\lambda_{max} = 427$ nm) after several hours. Perchloric acid stabilizes the yellow color. Heating gives less stable yellow color. Concentrations of $SnBr_2$ and acids require control; optimal HBr concentration $\geqq 0.1$ M, ~1.5 M each HBr and $HClO_4$.

Pd   Yellow-brown color, the intensity of which is dependent on Sn(II) and HBr concentrations; color stable for an hour (HBr~3 M, $SnBr_2$ >0.1 M).

Ir   No color at room temperature after several days. Yellow color produced on heating ($\lambda_{max} = 403$ nm), which is stable for several hours in 0.1–0.2 M $SnBr_2$ and 2–3 M HBr.

All these colored products can be extracted into isoamyl alcohol (Fig. 5-7). The Pt and Rh complexes are reasonably stable in this solvent, but

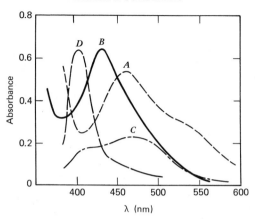

Fig. 5-7. Absorbance curves of $SnBr_2$–Pt–metal complexes in isoamyl alcohol. *A*, 10.6 ppm Pt; *B*, 2.3 ppm Rh; *C*, 10.1 ppm Pd; *D*, 6 ppm Ir. F. Pantani and G. Piccardi, *Anal. Chim. Acta*, **22**, 232 (1960).

the Ir complex is not. Pt, Rh, and Pd can be determined in the presence of Ir (which does not react at room temperature), with or without extraction. Pt and Rh can be determined in the presence of each other, if the ratios are not too unfavorable, by absorbance measurements at 463 and 427 nm. Sensitivity indexes in aqueous solution are Pt, 0.021 (463 nm); Rh, 0.0034 (427 nm); Pd, 0.016 (385 nm); Ir, 0.008 $\mu g/cm^2$ (403 nm). The values are much the same for Pt and Rh in isoamyl alcohol. Stannous bromide itself absorbs below 400 nm.

If an organic base such as diantipyrinylmethane, diphenylguanidine or tribenzylamine is added, chloroform, which is a more convenient solvent than isoamyl alcohol, can be used to extract the stannous bromide complexes of Rh and Ir.[62] Sensitivities are good, for example, with diphenylguanidine, $\mathscr{S}_{Ir,410nm} = 0.0032$ $\mu g/cm^2$ and $\mathscr{S}_{Rh,430nm} = 0.0027$ $\mu g/cm^2$.

C. IODIDES

Because of their strong absorption in the visible range, the anionic iodo complexes of Bi (yellow), Pt (PtI_6^{2-}, brownish red), Pd (PdI_4^{2-}, brown), and Te (reddish yellow) have been used as photometric determination forms for a long time. They are still of value when the highest sensitivity is not required. Sb(III) gives a sensitive reaction at high iodide concentrations, comparable to the bismuth reaction (at low iodide concentrations).[63] Some of the iodo complexes have absorption bands in the UV as well as in the visible range. Thus Bi shows a higher absorption peak at ~340 nm than at 465 nm. HgI_4^{2-} absorbs strongly at 323 nm, but many metals interfere. Because of the liberation of free iodine from iodide solutions by air oxidation, it is desirable or necessary to have a suitable reducing agent present, especially when the iodide concentration is high. Ascorbic acid is a good choice. It can be used down to about 300 nm, below which its absorbance increases rapidly. At 300 nm, a 0.005 M solution of ascorbic acid gives an absorbance of $\sim$0.005 in a 1 cm cell, changing little in 1 or 2 hr.

Extraction of iodides or iodo acids into organic solvents is advantageous at times for the colorimetric determination of a few metals. Antimony(III) is extractable as SbI_3 from aqueous solutions of low iodide concentration, for example, 0.01 M, into nonoxygen solvents such as CCl_4, $CHCl_3$, or benzene. The absorbance of the benzene solution of SbI_3 can be measured at 300 nm ($\epsilon \sim 9400$). Bismuth can be extracted as iodobismuthous acid into oxygenated solvents such as esters and higher alcohols. Determination of Sn(IV), As(III), and Ge(IV) by extraction of their iodides into cyclohexane has been proposed.[64] At the absorption maxima, $\epsilon_{Sn,364} = 8700$, $\epsilon_{As,282} = 9700$, $\epsilon_{Ge,360} = 6600$. In benzene, $\epsilon_{Sn,365} = 9120$, corresponding to $\mathscr{S} = 0.013\ \mu g\ Sn/cm^2$. Though the sensitivity is fair, a high iodide concentration and moderately high acidity are needed for the successful extraction of SnI_4.[65]

Palladium(II) forms a triphenylphosphine (or –arsine, or –stibine) adduct $PdI_2 \cdot 2(C_6H_5)_3P$ extractable into cyclohexane ($\mathscr{S}_{325} = 0.005$).[66]

Certain metals [Cu(II), Au(III), Ir(IV)] are reduced to lower oxidation states by iodide, so that they cannot be determined as iodides, and increasing the sensitivity of their determination must stop at the bromo complexes.

Determination of Tl as TlI_4^- in alcoholic solution has been proposed ($\mathscr{S}_{400} = 0.02$).[67]

Iodide in Stannous Halide Methods for Platinum Metals.   Rhodium can be determined more sensitively with stannous chloride in the presence of iodide than with stannous chloride alone (by about a factor of three over the "red" rhodium–$SnCl_2$ method). The formation of a mixed Sn(II) chloride–iodide–Rh(I) complex anion may be suspected, but the composition has not been determined.[68]

In ethanol solution, $SnBr_2$ with I^- reacts with Pd(II) to give a purple species which is anionic and supposedly has the structure in which the bridging halogen is thought to be I.[69] $\mathscr{S}_{555} = 0.005$. Platinum and rhodium give red and yellow complexes respectively.

D. FLUORIDES

As mentioned earlier, a solid solution of U(VI) (as uranyl) in sodium fluoride fluoresces strongly (yellow) in UV radiation and provides the most sensitive means of determining uranium. (For thiocyanates, see Chapter 6F, p. 536).

VII. MISCELLANEOUS: SULFIDES, AZIDES, CYANIDES, AND AMMINES

Under suitable conditions, H_2S or a soluble sulfide can be used as reagents to form colloidal suspensions of various heavy metal sulfides of limited usefulness for photometric determination (strong light absorption), and this was once a general method for traces of Cu, Pb, and other metals when more sensitive and selective reagents were not available. Now, dithizone and other organic reagents provide far superior methods for the heavy metals. Occasionally, sulfide is used as a semiquantitative reagent for heavy metals as a group in materials such as certain chemical reagents. Selectivity can be increased by adding cyanide to mask metals such as Cu, Hg, and Ag. Sulfides of Pb, Tl, and Bi still form in basic cyanide medium.

The brown to black stain formed by reaction of AsH_3 with Hg(II) salt in the Gutzeit method for arsenic presumably is an arsenide of mercury or some mixed species (if not arsenide, then metallic mercury).

Azide ion, N_3^-, shows similarities to thiocyanate in giving colored complexes with Fe^{3+} (red), UO_2^{2+} (yellow), Cu^{2+} (yellow-green to brown),[70] and other cations. Azide is not often used as a colorimetric reagent, though it has been recommended for the determination of U,[71] and also Pd, Au,[72] and Rh.[73] Whereas HSCN is a strong acid, HN_3 is a weak acid with a pK_a of 4.6 ($\mu = 0.1$). The hydrogen-ion concentration will thus have a greater effect on azide reactions than on thiocyanate reactions.

Some cyanide complexes absorb quite strongly in the UV, for example, that of Cu at 238 nm ($\epsilon = 12,800$) and of Ni at 267 nm ($\epsilon = 11,700$).[74] Cyanide ion itself does not absorb above 240 nm.

The ammine complexes formed by excess ammonia with Cu(II), Ni, and Co(III) have been used for their photometric determination, but the sensitivity is so inferior to that provided by the many excellent organic reagents for these metals that ammonia is hardly a trace reagent. Various amines give stronger colors than ammonia with copper.

VIII. INDIRECT METHODS—CATALYSIS

Some observations on indirect photometric methods are made in Chapter 4. A number of such methods can be carried out with inorganic reagents alone. This is true of some important catalytic methods, as those based on the reaction $2\,Ce(IV) + As(III) \rightarrow 2\,Ce(III) + As(V)$, which can be followed by measuring the absorbance of Ce(IV) as a function of the time. I$^-$, Os and Ru especially catalyze this reaction.

A Backward Look at Inorganic Photometric Determination Forms. Table 5-3 is an appraisal of the status of inorganic determination forms for metals in traces at the present time. There is no need for confrontation of inorganic and organic reagents as competitive—we choose whatever reagent best meets a particular need. For the metals

TABLE 5-3
Metals (enclosed) for Which Photometric Inorganic Determination Forms Often Are of Value

Li	Be														
Na	Mg											Al			
K	Ca	Sc	Ti	V	Cr	Mn	Fe	Co	Ni	Cu	Zn	Ga	Ge	As	
Rb	Sr	Y	Zr	Nb	Mo	Tc	Ru	Rh	Pd	Ag	Cd	In	Sn	Sb	Te
Cs	Ba	RE*	Hf	Ta	W	Re	Os	Ir	Pt	Au	Hg	Tl	Pb	Bi	Po
	Ra	Ac	Th	Pa	U										

*Ce

indicated, determination in inorganic forms will often—hardly ever always—be advantageous. A table such as this will necessarily be arbitrary to a certain extent and probably subject to the prejudices of the one who draws it up.[75] It will be seen that few reagents are represented: H_2O_2 for Ti, heteropoly acids for V and As, iodide for Sb, Bi, and Te, oxidizing agents for Cr, Mn and Ce, stannous halides for Pt, Rh, and Ir, and one or two others. U is included because of its strong fluorescence in fluoride melts; Ru and Os, because they can be determined by established catalytic methods; and Hg, because it can be determined in atomic form at room temperature by absorption at 253.7 nm.

Some metals are on the border line for inclusion: Nb with H_2O_2, Tl by fluorescence in chloride solution, and the like, but most of the time organic reagents will be chosen instead. Reactions involving chloride and bromide, except for some of the platinum metals, are not sufficiently distinctive to have wide use, although, as already mentioned, they are of value at times.

NOTES

1. The red-brown color produced when AsH_3 is passed into a solution of silver diethydithiocarbamate in pyridine is believed to be due to colloidal silver: H. Bode and K. Hachmann, *Z. Anal. Chem.*, **241**, 18 (1968). This is the basis of a popular method for the colorimetric determination of arsenic. The production of 6 Ag from 1 AsH_3 explains the fairly high sensitivity of the method.

2. D. F. Boltz, *Colorimetric Determination of Nonmetals*, Interscience, New York, 1958, p. 40; C. H. Lueck and D. F. Boltz, *Anal. Chem.*, **30**, 183 (1958).

3. It appears that the species formed can contain H_2O_2, HO_2^-, and O_2^{2-}. The reactions are not simple, and the products are often not certainly established. See L. G. Sillén, *Stability Constants of Metal–Ion Complexes, Section I: Inorganic Ligands*, The Chemical Society, London, 1964, for a survey of probable species and their formation constants (La, Ti, Zr, V, Nb, Ta, Cr, Mo, W, U, Pu, etc.), and also J. A. Connor and E. A. V. Ebsworth, *Peroxy Compounds of Transition Metals* in *Advances in Inorganic Chemistry and Radiochemistry*, vol. 6, Academic, New York, 1964. Mixed ions, as with SO_4^{2-} and EDTA, can be formed.

4. Y. Schaeppi and W. D. Treadwell, *Helv. Chim. Acta*, **31**, 577 (1948); see also *Anal. Chem.* **25**, 928 (1953). In sulfuric acid solutions, $Ti\begin{smallmatrix}O\\ |\\ O\end{smallmatrix}(SO_4)_2^{2-}$ may be formed.

5. M. Orhanović and R. G. Wilkins, *J. Am. Chem. Soc.*, **89**, 278 (1967).

6. N. Adler and C. F. Hiskey, *J. Am. Chem. Soc.*, **79**, 1827, 1831, 1834(1951). For absorption curves of the peroxo complexes of Nb, Ta, Ti, W, Mo, and Re in 96% sulfuric acid, see F. C. Palilla, N. Adler, and C. F. Hiskey, *Anal. Chem.*, **25**, 927 (1953). V. F. Gorlach, *Zh. Anal. Khim.*, **26**, 2372 (1971), found peroxidized Nb and Ta to be extracted from H_2SO_4 solution into a benzene solution of dibutyl phosphate (L) as $NbO(OH)_2(H_2O_2)L^-$ and $Ta(OH)_4(H_2O_2)L^-$ (negative charge neutralized by H^+).

7. G. Thanheiser, *Mitt. Kaiser-Wilhelm-Inst. Eisenforsch. Düsseldorf*, **22**, 255 (1940).

8. Vanadium forms a yellow complex ion with tungstate in which the molar ratio W:V is 3, serving for the determination of V, but not as satisfactory as phosphotungstovanadic acid. The latter determination form was proposed by A. P. Vinogradov, *C. R. Acad. Sci. URSS*, **1931A**, 249, resuscitated by E. B. Sandell, *Ind. Eng. Chem. Anal. Ed.*, **8**, 336 (1936), and further studied by E. R. Wright and M. G. Mellon, *ibid.*, **9**, 251 (1937). The sensitivity is not great, but a mineral acid solution can be used and the method has been found useful in applied analysis. Phosphotungstovanadic acid (yellow) can be extracted satisfactorily into hexyl alcohol, and the effective sensitivity thus increased.

9. F. Umland and G. Wünsch, *Z. Anal. Chem.*, **225**, 362 (1967), studied the extraction of molybdic and phosphomolybdic acid from HCl and $HClO_4$

solutions into $CHCl_3$–butanol, isobutyl alcohol–ethyl ether, ether, and butyl acetate as a function of acidity and time of extraction. F. Wünsch and G. Umland, *ibid.*, **247**, 287 (1969), found arsenomolybdic acid to be extracted well by isobutyl alcohol, MIBK, and ethyl ether–butanol. H_2SO_4 medium minimizes extraction of molybdic acid. For extraction of heteropoly tungstates of P, Si, Ge, and B, see F. Wünsch and G. Umland, *Z. Anal. Chem.*, **250**, 248 (1970).

10. A discussion of these extractions lies beyond the field of this book, but reference may be made to a procedure of J. Paul, *Anal. Chim. Acta*, **35**, 200 (1966), for determining As, Ge, and P in the presence of each other. Phosphomolybdic acid is first extracted with isobutyl acetate at pH 1–0.8 and then germanomolybdic acid is extracted from the remaining aqueous phase with 2-ethylhexanol at pH 0.4, leaving As in that aqueous phase. All three elements are actually finally determined as the heteropoly blues, not the parent complex acids. Also see Wünsch and Umland, reference in note 9, and Chapter 9A.

11. A 25% v/v solution of 1-butanol ($+CHCl_3$) was used by R. I. Alekseev, *Zav. Lab.*, **11**, 122 (1945), and N. A. Filippova and L. I. Kuznetsova, *ibid.*, **16**, 536 (1950). Wadelin and Mellon, *loc. cit.*, recommend a 20% solution as providing the best compromise between selectivity and sensitivity for P. For various other extractants, optimum pH and *E* values, see S. J. Simon and D. F. Boltz, *Anal. Chem.*, **47**, 1758 (1975).

12. An insensitive method for W utilizes phosphovanadic acid as reagent: G. J. Lennard, *Analyst*, **74**, 253 (1949).

13. H. D. Goldman and L. G. Hargis, *Anal. Chem.*, **41**, 490 (1969). The stoichiometry of complex formation and the kinetics of reduction to heteropoly blue are considered.

14. K. Murata, Y. Yokoyama, and S. Ikeda, *Anal. Chim. Acta*, **48**, 349 (1969). Also see A. K. Babko and Y. F. Shkaravskii, *Zh. Neorg. Khim.*, **6**, 2091 (1961); Y. F. Shkaravskii, *ibid.*, **11**, 120 (1966).

15. Methods for Ti and Zr were described earlier by R. M. Veitsman, *Zav. Lab.*, **25**, 408 (1959); **26**, 927 (1960).

16. S. R. Crouch and H. V. Malmstadt, *Anal. Chem.*, **39**, 1084 (1967).

17. Dimerization has been known for some time. P. Souchay, *Pure Appl. Chem.*, **6**, 61 (1963), has proposed an equilibrium of the type $PO_4(MoO_3)_{12}^{3-} + 9\ H^+ \rightleftharpoons H_3PO_4 + 6\ HMo_2O_6^+$.

18. It may be noted that Goldman and Hargis, reference 13, found the reaction in $HClO_4$ or HNO_3 solution to be represented by $H_3PO_4 + 6\{Mo(VI)\}_{tot} \rightleftharpoons$ 12-molybdophosphate $+ 6\ H^+$. The difference in H^+-dependency has not been explained.

19. A. C. Javier, S. R. Crouch, and H. V. Malmstadt, *Anal. Chem.*, **40**, 1922 (1968); *ibid.*, **41**, 239 (1969) (the last reference dealing with a fast-reaction method for P). Further work on the kinetics of molybdophosphate formation in perchloric, sulfuric, and nitric acid solutions has been done by P. M.

Beckwith, A. Scheeline, and S. R. Crouch, *ibid.*, **47**, 1930 (1975). The initial reaction between Mo(VI) and phosphate is followed by several polymerization steps.

20. J. D. H. Strickland, *J. Am. Chem. Soc.*, **74**, 862, 868, 872 (1952). The nature of the α- and β-molybdosilicic acids is still obscure. See further, V. W. Truesdale and C. J. Smith, *Analyst*, **100**, 203, 797 (1975); V. W. Truesdale, C. J. Smith, and P. J. Smith, *ibid.*, **102**, 73 (1977). Kinetics of formation of α- and β-molybdosilicic acids: L. G. Hargis, *Anal. Chem.*, **42** 1494, 1497 (1970).

21. A. Ringbom, P. E. Ahlers, and S. Siitonen, *Anal. Chim. Acta*, **20**, 78 (1959) (determination of silicon as α-molybdosilicic acid); G. J. S. Govett, *ibid.*, **25**, 69 (1961) (as β-molybdosilicic acid).

22. R. A. Chalmers and A. G. Sinclair, *Anal. Chim. Acta*, **33**, 384 (1965).

23. R. Jakubiec and D. F. Boltz, *Anal. Chem.*, **41**, 78 (1969). Determination of Ge without extraction (measurement at 315 nm), with extraction (1-pentanol + ethyl ether, 295 nm) and indirectly by transferring Mo in the washed extract to ammoniacal solution as molybdate ($\sim$225 nm). The last method is the most sensitive.

24. A. Halász, E. Pungor, and K. Polyak, *Talanta*, **18**, 577 (1971).

25. P. Souchay and S. Ostrowetsky, *Compt. Rend.*, **250**, 4168 (1960), obtained a molybdenum blue, formulated as $H_2Mo_2^VMo_4^{VI}O_{18}$, by mixing Na_2MoO_4 and Mo(V) in 0.06 M HCl.

26. V. I. Klitina, F. P. Sudakov, and I. P. Alimarin, *J. Anal. Chem. USSR*, **20**, 1197 (1965). Extraction of ascorbic acid-reduced molybdophosphoric and molybdoantimonylphosphoric acids with oxygenated solvents (alcohols, esters, and carbonyl-containing solvents) has been studied by S. J. Eisenreich and J. E. Going, *Anal. Chim. Acta*, **71**, 393 (1974). Acetophenone appears to be a potentially useful solvent for either heteropoly acid, based on minimal mutual solubility, wide acidity range, and extractability. Butyl acetate also appears to be suitable for extraction of reduced molybdoantimonylphosphoric acid. R. A. Karanov and A. N. Karolev, *J. Anal. Chem. USSR*, **20**, 594 (1965), extract molybdoarsenic acid into butanol + ether and form Mo blue in the latter by shaking with aqueous $SnCl_2$.

27. J. T. Woods and M. G. Mellon, *Ind. Eng. Chem. Anal. Ed.*, **13**, 760 (1941): P, Si, As; effect of other elements on P determination; study of various reducing agents. R. E. Kitson and M. G. Mellon, *ibid.*, **16**, 466 (1944): AOAC method for P. D. F. Boltz and M. G. Mellon, *Anal. Chem.*, **19**, 873 (1947): P, Si, Ge, As. Although not directly pertinent to metals, the following papers are of interest: J. D. H. Strickland, in note 20. I. R. Morrison and A. L. Wilson, *Analyst*, **88**, 88, 100 (1963), dealing with determination of silica in water. Reviews: M. Jean, *Chim. Anal.*, **37**, 125, 163 (1955); **44**, 195, 243 (1962).

28. L. Duval, *Chim. Anal.*, **51**, 415 (1969). Earlier, similar diagrams were

presented for P, As, Si and Ge in HCl solution, with 1-amino-2-naphthol-4-sulfonic acid and SnCl₂ as reductants, by H. Levine, J. J. Rowe, and F. S. Grimaldi, *Anal. Chem.*, **27**, 258 (1955).

29. R. H. Campbell and M. G. Mellon, *Anal. Chem.*, **32**, 54 (1960). A spectrophotometric rate method is described by L. G. Hargis, *Anal. Chem.*, **41**, 597 (1969).

30. J. C. Guyon and L. J. Cline, *Anal. Chem.*, **37**, 1778 (1965).

31. Z. F. Shakhova and S. A. Gavrilova, *Zh. Anal. Khim.*, **13**, 211 (1958).

32. G. Norwitz and M. Codell, *Anal. Chem.*, **26**, 1230 (1954).

33. J. C. Guyon, G. W. Wallace, Jr., and M. G. Mellon, *Anal. Chem.*, **34**, 640 (1962).

34. R. L. Heller and J. C. Guyon, *Talanta*, **17**, 865 (1970).

35. B. L. Madison and J. C. Guyon, *Anal. Chem.*, **39**, 1706 (1967).

36. J. C. Guyon and M. G. Mellon, *Anal. Chem.*, **34**, 856 (1962).

37. J. C. Guyon and J. Y. Marks, *Anal. Chem.*, **40**, 837 (1968).

38. B. L. Madison and J. C. Guyon, *Anal. Chim. Acta*, **42**, 415 (1968).

39. G. C. Dehne and M. G. Mellon, *Anal. Chem.*, **35**, 1382 (1963).

40. C. C. Clowers, Jr., and J. C. Guyon, *Anal. Chem.*, **41**, 1140 (1969).

41. M. T. Pope and G. M. Varga, Jr., *Inorg. Chem.*, **5**, 1249 (1966).

42. G. M. Varga, Jr., E. Papaconstantinou, and M. T. Pope, *Inorg. Chem.*, **9**, 662 (1970). For the classification, see M. B. Robin and P. Day, *Advan. Inorg. Chem. Radiochem.*, **10**, 248 (1967).

43. However, P. Stonehart, J. G. Koren, and J. S. Brien, *Anal. Chim. Acta*, **40**, 65 (1968), state that the reduction electron may be shared with 12 tungsten atoms.

44. In 6 *M* hydrochloric acid, ferric iron shows maximum absorption at 343 nm according to M. A. Desesa and L. B. Rogers, *Anal. Chim. Acta*, **6**, 534 (1952). Absorption curves for various metals in hydrochloric acid are given in this paper. In hydrochloric acid solutions stronger than 1 *M*, polonium absorbs strongly at 418 nm (D. J. Hunt, *Chem. Abst.*, **50**, 4697 (1956)). Tellurium(IV) in 10–12 *M* hydrochloric acid gives a yellow solution with an absorption maximum at 375 nm.

45. C. Hüttner, *Z. Anorg. Chem.*, **86**, 351 (1914).

46. Review: E. A. Bus'ko, K. A. Burkov, and S. K. Kalinin, *J. Anal. Chem. USSR*, **29**, 286 (1974).

47. C. Merritt, Jr., H. M. Hershenson, and L. B. Rogers, *Anal. Chem.*, **25**, 572 (1953). Absorption curves of chloro, bromo, and iodo complexes of thallium, lead, bismuth, mercury, and other metals are shown. For determination of lead as chloro complex, see K. E. Kress, *Anal. Chem.*, **29**, 803 (1957).

48. A. A. Pozdnyakov, *J. Anal. Chem. USSR*, **20**, 439 (1965).

49. For example, M. J. Maurice and S. M. Ploeger, *Z. Anal. Chem.*, **179**, 246 (1961), determine Pb, Fe, and Cu in ZnSO₄ as the chloro complexes (Pb at

270 nm after precipitation as the sulfate, Fe at 278 nm, and Cu at 360 nm; a correction is needed for absorption by Zn). M. Ishibashi, Y. Yamamoto, and Y. Inoue, *Bull. Inst. Chem. Res. Kyoto Univ.*, **37**, 38 (1959), determine Sn(IV) in $MgCl_2$ (2.3 M, 1.5 N in HCl) at 215 nm; As, Sb, Fe, V and Ti interfere. R. M. Sherwood and F. W. Chapman, *Anal. Chem.*, **27**, 88 (1955), found HCl satisfactory for determining Pb in petroleum catalysts after dithizone separations. A. Glasner and P. Avinur, *Talanta*, **11**, 761 (1964), give ϵ for Bi, Cu(I), Cu(II), Fe(III), and Pb in KCl (4 M) and other chloride solutions.

50. G. F. Kirkbright, C. G. Saw, and T. S. West, *Talanta*, **16**, 65 (1969).

51. J. F. Young, R. D. Gillard, and G. Wilkinson, *J. Chem. Soc.*, 5176 (1964).

52. G. L. Elizarova and L. G. Matvienko, *J. Anal. Chem. USSR*, **25**, 254 (1970).

53. G. L. Elizarova and L. G. Matvienko, *Zh. Neorg. Khim.*, **15**, 1606 (1970).

54. S. K. Kalinin, K. P. Stolyarov, and G. A. Yakovleva, *Zh. Anal. Khim.*, **25**, 132 (1970).

55. S. K. Kalinin and G. A. Yakovleva, *Zh. Anal. Khim.*, **25**, 312 (1970).

56. Review: E. A. Bus'ko, K. A. Burkov, and S. K. Kalinin, *J. Anal. Chem. USSR*, **25**, 839 (1970).

57. In a procedure proposed by A. T. Pilipenko and I. P. Sereda, *Ukr. Khim. Zh.*, **27**, 524 (1961), Pd and Pt are determined by developing an orange color in H_2SO_4–$HClO_4$–$SnCl_2$ (+HCl), then diluting with water to give a blue-green color and measuring the absorbance at 630 nm (Pt does not absorb) and 310 nm (both Pd and Pt absorb). For Pd, $\epsilon_{630} = 4700$, $\epsilon_{300} = 31{,}000$; for Pt, $\epsilon_{310} = 36{,}000$. Ir, Rh, and Ru interfere, but Os does not.

58. In this connection, mention may be made of a qualitative test for Pd reported by W. B. Pollard, *Analyst*, **67**, 184 (1942), in which Sn(II) chloride and Hg(II) chloride give a strong orange-red color with Pd in HCl solution. Evidently the Pd(II)–Sn(II) system requires further study.

59. G. G. Tertipis and F. E. Beamish, *Anal. Chem.*, **34**, 623 (1962).

60. Determination of Cu: W. Nielsch and G. Böltz, *Z. Anal. Chem.*, **142**, 94 (1954). The red-violet color given by strong HBr with Cu(II) was already utilized for its determination by G. Denigès and E. Simonot, *J. pharm. chim.*, **11**, 186 (1915).

61. F. Pantani and G. Piccardi, *Anal. Chim. Acta*, **22**, 231 (1960). Rh: S. S. Berman and R. Ironside, *Can. J. Chem.*, **36**, 1151 (1958); Ir: S. S. Berman and W. A. E. McBryde, *Analyst*, **81**, 566 (1956).

62. A. T. Pilipenko, V. N. Danilova, and S. L. Lisichenok, *Zh. Anal. Khim.*, **25**, 1154 (1970).

63. As shown by B. Julin, Ph.D. Thesis, University of Minnesota, 1965, a series of complexes from SbI^{2+} to SbI_6^{3-}, if not higher, is formed, the relative amounts depending on the iodide concentration.

64. K. Tanaka and N. Takagi, *Anal. Chim. Acta*, **48**, 357 (1969).

65. D. D. Gilbert and E. B. Sandell, *J. Inorg. Nucl. Chem.*, **24,** 989 (1962). The extraction is of greater interest for the separation of tin.

66. P. Senise and F. Levi, *Anal. Chim. Acta*, **30,** 509 (1964).

67. D. Betteridge and J. H. Yoe, *Anal. Chim. Acta*, **27,** 1 (1962).

68. J. J. Markham, Ph.D. Thesis, University of Minnesota, 1957, first gave a tin(II)–iodide procedure for Rh. E. W. Berg and H. L. Youmans, *Anal. Chim. Acta*, **25,** 366, 470 (1961), studied the determination of Rh and Ir in this way.

69. J. A. W. Dalziel, J. D. Donaldson, and B. W. Woodget, *Talanta*, **16,** 1477 (1969).

70. F. Maggio and F. P. Cavasino, *Ann. Chim. Roma*, **51,** 1392 (1961). With excess azide, $Cu(N_3)_6^{4-}$ forms.

71. F. G. Sherif and A. M. Awad, *Anal. Chim. Acta*, **26,** 235 (1962).

72. Tetraazidogold(III) ion is formed ($\lambda_{max} = 325$ nm), extractable into *n*-butyl alcohol ($\lambda_{max} = 330$ nm). See R. G. Clem and E. H. Huffman, *Anal. Chem.*, **37,** 1155 (1965).

73. A. K. Majumdar and B. K. Mitra, *Anal. Chim. Acta*, **33,** 670 (1965). Rose red color; insensitive.

74. R. P. Buck, S. Singhadeja, and L. B. Rogers, *Anal. Chem.*, **26,** 1240 (1954). See this paper for UV absorption curves of nitrate, nitrite, thiocyanate, acetate, oxalate, citrate, tartrate, bromate, iodate, periodate, persulfate, and other ions that might be present in a solution in which a constituent is determined by UV absorption. H_3PO_4, SO_4^{2-}, and ClO_4^- absorb very little above 220 nm. In 2 M H_3PO_4 solution, Hg(II) absorbs strongly ($\epsilon = 1.4 \times 10^4$) at 235 nm. A. Glasner, S. Sarig, D. Weiss, and M. Zidon, *Talanta*, **19,** 51 (1972), find the isosbestic point for $Cu(CN)_3^{2-} + Cu(CN)_4^{3-}$ at 235 nm with $\epsilon = 11,300$. Gold cyanide also absorbs at this wavelength. M. W. Scoggins, *Anal. Chem.*, **42,** 301 (1970), determined Ni as $Ni(CN)_4^{2-}$. Co, Cu, Fe, Pt, Rh, and Mn interfere. He finds $\epsilon = 12,100$ or $\mathscr{S} = 0.005$.

75. An earlier table, in the third edition of *Colorimetric Determination of Traces of Metals*, included indirect methods (excluded here except for two metals advantageously determined catalytically) and was somewhat more permissive than the present one.

6A

ORGANIC PHOTOMETRIC REAGENTS
GENERAL ASPECTS—
CHELATING REAGENTS

Almost all metals can be determined photometrically by the use of organic reagents,[1] the alkali metals forming the chief exception (and this only partly). For most metals, organic photometric reagents are superior to the inorganic because of better sensitivity or selectivity or both. Chelate formation especially makes organic compounds effective metal reagents. The formation of ion-association complexes is also important, though for a smaller number of metals. The wide choice of chromogenic and chelating groups and the possibility of suitably modifying organic molecules account for the versatility of organic reagents. The almost universal extractability of metal–organic complexes into organic solvents contributes much to the sensitivity and selectivity of organic reagents. Some metal–organic complexes are fluorescent and allow metals to be determined very sensitively by fluorimetric methods.

Many of the organic reagents giving color reactions with metal ions were first used qualitatively; their application to photometric determinations was an easy transition. A considerable number of early organic reagents were introduced as precipitants and became valuable trace reagents only when it was realized that their metal complexes could be extracted into immiscible organic solvents. 8-Hydroxyquinoline is an example. Cupferron (useful as a separatory reagent) was found to be an effective precipitant for Fe(III) and other metals before 1910, but was not used extractively until the 1920s and 1930s and then only sparingly.

273

Dimethylglyoxime was first used for the determination of nickel by precipitation in 1907, but the extractability of nickel dimethylglyoximate by chloroform was not demonstrated until 1939 and the isolation of trace quantities of nickel then became a simple matter. The slow development of extraction methods, whether for separation or photometric determination, was largely a consequence of little interest in the determination of traces of heavy metals before ca. 1930. The use of dithizone as an extractive reagent for heavy metals (introduced in the mid-1920s) received impetus in the 1930s from determination of lead in foods and in geo samples (Cu, Zn, and Pb in silicate rocks). Shortly thereafter, the discovery of nuclear fission produced a tidal wave of extractive separatory methods in which organic reagents played a predominant role.

Surprisingly, the simple relation between the extractability of a metal and the concentration of the complexing (chelating) reagent in the immiscible organic solvent and the hydrogen-ion concentration of the aqueous phase was not formulated until 1939 (p. 964), as far as we know.

I. GENERAL ASPECTS

A. TYPES OF ORGANIC PHOTOMETRIC REAGENTS

Most organic reagents reacting with metals to give colored, UV-absorbing, or fluorescent products fall into one of the following classes:

1. Those forming chelate complexes.
2. Those forming ion-association complexes.
3. Those oxidizing or reducing a metal species and being converted into an absorbing or fluorescent product not containing the metal in the process, or reacting in some other way not included in (1) or (2).

Reactions of the first two classes are the most important. Chelate formation is usually thought of as providing the majority of useful photometric determination forms for metals. This is no doubt true, but it should not be forgotten that ion-association complexes are also important in trace metal determinations (Chapter 6H). Strictly, reagents in (1) should be classified as coordination-complexing reagents, not chelating reagents, because ring formation is not involved in all of these complexing reactions, though in most of them. The colored products obtained by reaction of metals with organic reagents unable or unlikely to form chelates are in part π-complexes, as for example the yellow complex formed when salicylic acid reacts with TiO^{2+} in concentrated sulfuric acid (Chapter 6B, Section II.D).

If it is to be useful in photometric analysis, the reaction product in all of these reactions must either (a) differ markedly in some photometric property (absorption spectrum, fluorescence) from that of the excess reagent, or (b) if the optical properties of the two are much the same, reagent and product must be easily separable. In most ion-association reactions, the photometric properties of the two are very similar, but determination of the metal complex is possible because it can be extracted into some organic solvent that does not extract the excess of reagent.

In most chelation reactions, an excess of the reagent reacting as L^- with M^{n+} forms the neutral chelate ML_n. More likely than not, ML_n will be quite insoluble in aqueous solution unless the reagent contains a solubilizing group such as $-SO_3H$. At low metal concentrations, insoluble metal chelates form dispersions, which, if kept from flocculation by a protective colloid, can be used for absorptiometric determination. However, a true solution of the metal complex is much to be preferred. A miscible organic solvent, such as ethanol, is occasionally used to give an aqueous mixture that keeps both reagent and chelate in solution. An undesirably large volume of miscible organic solvent may be needed, however, to increase the solubility sufficiently and, besides, inorganic salts may be precipitated. The alternative is to extract the chelate into an immiscible organic solvent.[2] This is almost always the preferred procedure in trace analysis. An important advantage thus gained is the concentration of the chelate into a small volume of organic solvent and therewith an increase in effective sensitivity. At the same time, other light-absorbing species, as of other metals, and sometimes the reagent itself, may be left in the aqueous phase. Also when foreign metals are masked with complexing agents, unextractable colored species may be formed that will not interfere in extraction methods.

Reactions of class (2) above involve cations and anions of large size. Likely candidates for such reactions are a bulky organic cation (colored) and a complex anion of the metal. Thus the cation might be a bulky dye cation, R^+ or RH^+ and the anion a metal chloro complex, preferably both of single charge. Rhodaminium B (RH^+) chloroantimonate, (RH^+)-($SbCl_6^-$), is a typical example. The cation could be a chelated metal species having a positive charge and the anion could be a chelated metal having a negative charge. In most ion-association type determinations, extraction is essential. Thus in the determination of Sb(V) as (RH)($SbCl_6$), an immiscible solvent extracting the latter but not the excess of Rhodamine B is required. See further discussion in Chapter 6H dealing with ion-association reagents.

Extraction of metal–organic complexes in the wider context of separations is discussed in Chapters 9A–9D, to which the reader is referred for principles that also hold in extraction of absorbing or fluorescing determination forms.

Reactions based on color formation by oxidation (less often reduction) of an organic compound by a metal species (class 3) are of occasional value. For example, V(V), Ir(IV), and Mn(VII), as well as other metals in higher oxidation states, oxidize suitable organic compounds to strongly colored products. Such methods are indirect and are generally less desirable than direct methods. The colored reaction products may not be stable. Another type of indirect method involves conversion of the anionic organic component of a metal compound (usually as a separated precipitate) to a colored or more strongly colored product. For example, the oxine in a washed precipitate of an oxinate can be coupled with a diazo compound to give a strongly colored dye. Such methods find infrequent use.

B.   SPECIFICITY AND SELECTIVITY

Every determination laying claim to accuracy must be specific in a qualified sense at least, namely that the other elements in a particular sample do not react similarly to the element in question in the final (determination) step.[3] In theory, and to some extent in practice, an absorptiometric reagent could react with a number of metals and still give a specific reaction with one or more of them if the respective reaction products had sharp, well separated absorption peaks. But in practice, absorption peaks are likely to be broad and to lie so close together that overlapping occurs. Specificity—and selectivity—must realistically be based on the nonreaction of most or all of the other constituents of a sample. As noted in an earlier chapter, it is convenient to look upon the specificity (or selectivity) of a determination as lying at three levels:

1.   Reagent.
2.   Reaction.
3.   Procedure or method (separations).

To speak of the specificity of a reagent is dubious in an exact sense because its reactions are necessarily carried out under certain conditions: in a particular solvent or solvent system and at some pH. Nevertheless, it is convenient to speak loosely of the specificity or selectivity of a reagent. Certainly, if a reagent reacts characteristically with only one element, no matter what the conditions, it is specific for that element. This state of affairs is rarely encountered or even approached in practice. Almost always the best that can be expected is a *selective* reagent—one that reacts with a small number of elements. Reagents of poor selectivity can often be made to give selective reactions by using differential complexing agents to mask other metals,[4] as well as by adjusting the pH, the latter measure often affecting the concentration of the reactive form of the reagent and sometimes that of the metal. Also, a change in the oxidation state of the metal to be determined or of one (or more) of the accompanying metals may increase selectivity. See discussion of dithizone (Chapter 6G) for a good example of a reagent reacting with numerous metals that can be made selective for one (lead) by adjustment of pH, complexing with cyanide, and changes in oxidation state of some metals.

These measures have their limits, and when the limits are reached, we are left with separations as the sole remaining way of achieving the required specificity in a determination. Separation methods are rarely truly specific, and again high selectivity is usually all that can be expected. A combination of a selective photometric reaction and a selective separation may result in a specific determination—within limits. It is evident that a knowledge of the approximate composition of a sample may be needed before it can be analyzed successfully, especially if trace constituents are to be determined.

The existence of a specific reaction for a metal does not necessarily mean that the metal can be determined forthwith in any sample. The required reaction conditions may bring about interference from other elements, not connected with reaction with the reagent. Such situations can occur in practice. 2,9-Dimethyl-1,10-phenanthroline appears to be specific for Cu(I) in the sense that no other element is known to give a color reaction under the specified conditions. Copper(II) must be reduced to Cu(I) with a reducing agent such as hydroxylammonium chloride or ascorbic acid. In the presence of easily reducible metals such as gold, a precipitate would be obtained that would disturb the determination.

Factors contributing to the selectivity of photometric chelating reagents are examined in Section II.B of this chapter. For ion-association reagents see Chapter 6H.

C. SENSITIVITY

Briefly, the presence of highly resonating structures in a molecule gives rise to light absorption. At least some change in an absorption spectrum occurs when a proton in an organic reagent is replaced by a metal ion. Chelate formation usually results in a marked change in the visible or ultraviolet absorption spectrum of the reagent.[5] An originally colorless reagent may give a colored chelate (or one more strongly colored than the metal ion if it is colored) or a chelate of different hue, or one that is colorless, but showing a change in the ultraviolet spectrum. Changes in the infrared spectrum are of less importance in inorganic trace analysis.

It was mentioned in Chapter 4 that the maximum attainable molar absorptivity is given approximately by

$$\epsilon_{max} = 9 \times 10^{19} P\mathbf{a}$$

where P is the transition probability and $\mathbf{a}$ is the chromophore area. The upper value of $\mathbf{a}$ is usually $\sim 10^{-15}\ cm^2$, so that with $P = 1$, $\epsilon_{max} \sim 10^5$, which agrees quite well with observed values. As already noted, this approximate value sets a limit to the intrinsic sensitivity of absorptiometric methods. In order that this limit may be approached in practice, there must not be too much overlapping of the absorption curves of the reagent and its chelate, that is, there must be a considerable wavelength spread in λ_{max} of the two components. Dithizone fulfills this requirement. Not all colored chelating reagents are satisfactory in this respect. As already mentioned, ion-association reagents may show much the same absorption characteristics as the metal complex, but the two can be separated by extracting the complex. The sensitivity of ion-association reactions can be fully as good as those of chelating reagents. Molar absorptivities of 50,000–100,000 are not uncommon in both classes, but 200,000 is rarely exceeded.

Ligands of satisfactory chelating power, but unsatisfactory spectral characteristics can often be modified by introducing chromophoric groups so that absorption takes place at longer wavelengths (i.e., in the visible range), or greater contrast in the colors of reagent and chelate is provided or the color intensity of the chelate is increased. Groups such as

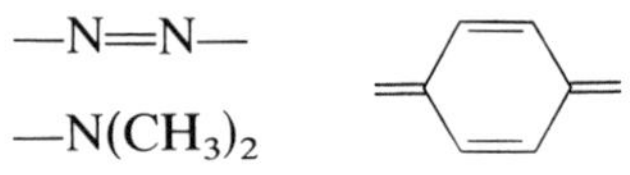

$$-N{=}N-$$
$$-N(CH_3)_2$$

and many others confer such effects. EDTA is colorless, but the introduction of various chromophoric groups converts it into colorimetric or absorptiometric reagents. Rhodanine forms yellowish white precipitates

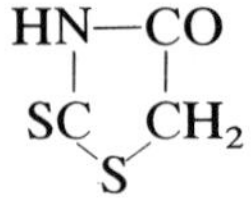

$$
\begin{array}{c}
HN\!-\!CO \\
|\quad\ |\ \\
SC\quad CH_2 \\
\diagdown\ S\ \diagup
\end{array}
$$

with Ag and Hg by N and S bonding, of no value as photometric determination forms. Feigl (1928) transformed rhodanine into a useful colorimetric reagent by introducing the chromophoric group

$$=CH\langle\bigcirc\rangle N(CH_3)_2$$

into the molecule to form dimethylaminobenzylidenerhodanine

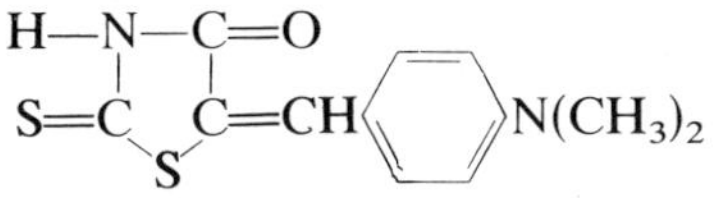

$$\begin{array}{c} H-N-C=O \\ | \quad\quad | \\ S=C \quad C=CH\langle\bigcirc\rangle N(CH_3)_2 \\ \diagdown S \diagup \end{array}$$

The complexes of Ag, Au, Hg(I), Cu(I), and Pd(II) formed by this reagent have intense red-violet colors (reagent yellow).

Less drastic modifications of a molecule, such as replacing H by CH_3- or C_6H_5- can produce favorable changes in absorption spectra, including increase in sensitivity. At the same time, changes (not always analytically favorable) are likely to be produced in the solubility of a reagent, its pK_a, the formation constants of its chelates and their solubility and extractability. The effect of substitutions has been studied especially with 1,10-phenanthroline and related compounds.[6] A simple substitution of CH_3- for H– can have a profound effect on the reactivity of a reagent if made at a point in the molecule where the coordination of a metal atom can be blocked (steric effect, p. 282).

II. CHELATING REAGENTS

A. GENERAL

For a chelating reagent to be of photometric value it must contain chromogenic or fluorigenic as well as chelating groups (they may be the same).

Most chelating reagents of interest to the analyst are of the type HL, in which H can be replaced by an equivalent of metal. The metal atom is coordinatively bonded to a functional group of basic character to form a ring structure which characterizes a metal chelate. The acidic hydrogen may be present in such groups as

—OH	—NH$_2$	—SH
—COOH	=NOH	—PO$_3$H$_2$

Typical coordinating groups are

=O	—NH—	=S
—NH$_2$	=N—	

The chelate ring is five- or six-membered; less often it is four-membered, as when both donors are S atoms. Tautomeric rearrangements in the reagent molecule frequently occur in chelate formation (see discussion of acetylacetone, p. 926). Most chelating reagents used in photometry contain one replaceable hydrogen (they are monobasic) and one coordinating group. Such a reagent is bidentate and occupies $2n$ coordination positions of the metal (n is the charge on metal ion) when the chelate is uncharged. Some dibasic (even tribasic) and polydentate chelate reagents find use in photometric analysis.

Generally a series of reaction products ML^{n-1}, $ML_2^{n-2} \ldots ML_n$ can be formed when L^- reacts with M^{n+}. Under analytical conditions—an excess of reagent and a sufficiently high pH—ML_n is usually the predominant species, but some reagents can give anionic species such as ML_{n+1}^-.

The stability of metal chelates varies over a very wide range. It depends on many factors: the nature of the metal (its electronic structure) and its oxidation state, the nature of the coordinating and acidic groups in the reagent, the stereochemistry of the chelate (number of atoms in the ring, the number of rings of which the metal is a member). The effect of these factors cannot adequately be examined in a brief treatment here and the reader is referred elsewhere for an introduction.[1]

The stability of a chelate is related to the acidity of the reagent. In general, other factors remaining the same, a chelating reagent HL with a small value of K_a will complex metals more strongly than others with larger acid dissociation constants. This is understandable from the similarity in the strength of binding of H and a metal atom. In a series of reagents of similar structure, the dissociability of the metal complex generally decreases as K_a decreases, that is, the weaker the acidity of the reagent, the stronger the complexing. It should be remembered, however, that in acidic solution, at a fixed pH, the less acidic reagent will give a lower concentration of L^-, and the greater complexing ability is counteracted.

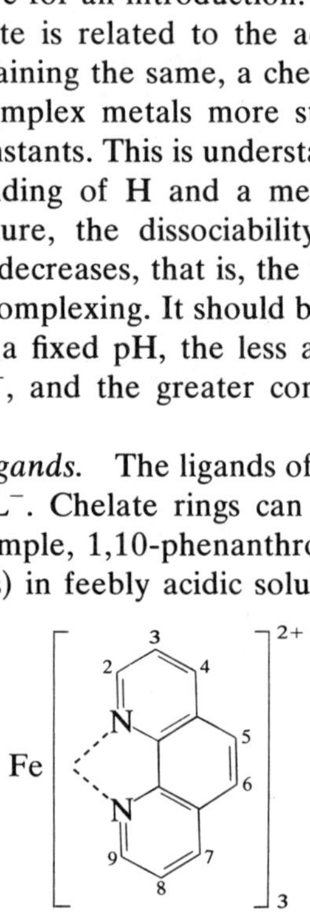

Uncharged Chelating Ligands. The ligands of the preceding discussion are anionic, of the type L^-. Chelate rings can also be formed with an uncharged ligand. For example, 1,10-phenanthroline combines with Fe^{2+} (and other divalent metals) in feebly acidic solution to give the chelated ion in which Fe^{2+} is bonded to N. The architecture of the 1,10-phenanthroline molecule, with the N atoms suitably positioned to form a stable five-membered ring with the coordinated metal atom, is favorable for chelation. 1,10-Phenanthroline is colorless in solution, the complex ferrous cation is red; the color is ascribed to a charge-transfer band. Hydrogen ions are not directly involved in the formation of this chelate, yet it is dissociated in acid solution. The reason for this lies in the basic properties of 1,10-phenanthroline. Hydrogen ions add to the N atoms in the reagent and compete with Fe^{2+}. The extraction of the ferrous complex requires neutralization of the charge with a bulky anion such as ClO_4^-, or better, for easy extraction, with an organophilic anion.

Another example of a chelating agent of this type is 2,2'-biquinoline and analogous reagents (see Chapter 6C), which are virtually specific for Cu(I). The violet ion $CuBq_2^+$ is extractable into amyl alcohol in the presence of chloride and other ions forming ion-pair compounds. Still another example is thiourea, which gives a red color with osmium (IV or VIII), due to $Os(Thi)_6^{3+}$ (thiourea also acts as a reducing agent). This complex is formed in mineral acid solution, because the basic properties of thiourea are extremely weak. Actually, $Os(Thi)_6^{3+}$ is not a chelate but a coordination complex of the ammine type.

Exceptionally, a protonated base, RH^+, chelates a metal cation.

B. FACTORS IN SELECTIVITY

The old hope of analysts was to find reactive groupings in reagents that would be specific for each element. The fulfillment of this hope may have a better chance than a successful search for the philosopher's stone or the fountain of youth, but they all fall in much the same category. Realistically, the objective should be more limited: specificity or a close approach to it for a few elements (metals) and good selectivity for many. Considerable progress has been made toward this goal.

In general, chelating (more accurately, coordinating) reagents react not with one or two metals but with groups of metals.[7] The number of reactive metals varies from one reagent to another—the selectivity varies—and also depends on the reaction conditions. An important factor in reactivity is obviously the strength of the bond between the metal atom and the coordinating atoms in the reagent. Binding of a metal in a ring structure with bonding to several atoms increases the stability of a complex. Stereochemical factors are also of importance in reactivity and stability of reaction products. Ionic sizes and symmetry relations come into play here. Detailed treatment of such factors must be sought elsewhere, but some examples are given subsequently.

The important bonding elements in organic reagents are O, N, and S. Reagents in which O is the sole bonding element can react with many metals, but are likely to give distinctive reactions with oxyphilic elements, namely those elements tending to bond more strongly to O than to N or S. Thus alizarin and morin, which form chelates through O, furnish characteristic reactions with such metals as Al, Ga, Be, and Zr. The higher valent oxyphilic elements tend to form the more stable complexes and greater or less selectivity can be obtained by regulation of pH. The reactivity of the O atom depends on the functional group of which it is a part, and indeed on the whole reagent molecule. There are many possibilities and experiment is required to find favorable reagents.

Reagents in which bonding to both O and N occurs show varying selectivity. Some such reagents—8-hydroxyquinoline, for example—react with most metals, even the alkaline earths, but others—the nitrosonaphthols are examples—give characteristic color reactions with a limited number of metals, namely those showing an affinity for N [Pd, Co, Ni, Fe(II)]. To be sure, such reagents may form complexes with metals other than those showing a color change. Reagents having only N as the bonding element are usually more selective than reagents having only O or O and N. Certain elements, including Cu(I), Ag, Cu(II), Cd, Ni, Co, and Hg, bond more strongly to N than to O.

An excellent example of a reagent bonding only through N is 1,10-phenanthroline, already mentioned. It chelates many metals [among them, Cd, Co, Cu(I,II), Fe(II,III), Ni, Zn, Hg(II), Tl(III)], yet it is a selective photometric reagent for ferrous iron. Only the Fe(II) and Cu(I) chelates absorb strongly in the visible range, giving orange-red colors. Zinc and Cd do not absorb in the visible, Co and Ni rather weakly. Copper(II), as $CuPhen_2^{2+}$, also absorbs weakly ($\epsilon_{700} \sim 60$, still less at 510 nm, where the ferrous complex has $\epsilon = 11,100$). Reduction of Fe(III) to Fe(II) without reduction of Cu(II) to Cu(I) is possible so that iron can be determined in the presence of considerable amounts of copper. Other reagents containing the "ferroin" grouping

$$-N{=}C{-}C{=}N-$$

in a suitable framework react similarly. 1,10-Phenanthroline is of no interest for the determination of Cu(I), because of the reaction of Fe(II). However, 1,10-phenanthroline can be converted into a specific reagent for Cu(I) by introducing methyl groups in the 2- and 9-positions adjacent to the reactive grouping. Because of the blocking by the two methyl groups, 2,9-dimethyl-1,10-phenanthroline cannot form 3 : 1 chelates or

2:1 *planar* chelates with any cation. However, the methyl groups do not interfere with the coordination of Cu(I), which forms a 2:1 *tetrahedral* chelate, $(CuL_2)^+$, in which the two ligand molecules lie in planes at right angles.[8] The chelated copper cation (orange) can be extracted into amyl alcohol and other solvents as $(CuL_2)X$, where X^- is Cl^- and others. Apparently, no other element gives a color under these conditions, and if so, this reaction is that of great rarity, a specific reaction. It appears that the most promising way of obtaining a highly selective reagent is by combining selective coordinating groups with groups imposing steric hindrance. Such combination may lead to an approach to, sometimes possibly attainment of, the old ideal of specific atomic groupings. Few such reagents are known at the present time.

Another example of steric blocking is provided by 8-hydroxyquinaldine (2-methyl-8-hydroxyquinoline). The parent reagent 8-hydroxyquinoline (Chapter 6D) forms chelates with most metals by bonding through O and N. Replacement of H next to N with CH_3 makes difficult the arrangement of three 8-hydroxyquinaldinate radicals around the small Al^{3+} to form the uncharged trischelate. Substituting H in the 2 position by C_6H_5 instead of CH_3 is even more effective in preventing the formation of the trischelate.

A structural factor is also involved in the extractability (and therewith the determination) of nickel as the dimethylgloximate in the presence of cobalt. Dimethylglyoxime (H_2Dx) forms $Ni(HDx)_2$ having planar configuration with the groups –OH and –O– strongly hydrogen bonded internally and, therefore, not easily hydrated, leading to chloroform extractability. Cobalt(II) [and also Cu(II)] dimethylglyoximate does not have a planar structure, the chelate rings do not lie in the same plane, the OH groups bond to water molecules, and as a result the complex is water soluble and is not extracted by chloroform. A sharp separation of Ni and Co thus becomes possible.

Turning now to reagents bonding through S or S and N (S and O bonding is less important), we note that they tend to be more selective than O- and O,N-bonding reagents. Generally, they do not react readily, if at all, with oxyphilic metals: the alkaline earths, the rare earths, Al, Ti, and some others. Not surprisingly, there is close similarity between reactivity of S-chelating organic reagents and H_2S in aqueous solution. Roughly, metals forming sulfides insoluble in acidic solutions tend to form S-chelates not easily dissociated in acidic solutions. The range in dissociation constants (or inversely, of formation constants) of S-chelates can be greater than with other chelates. Dithizone is the outstanding example of an S,N-bonding reagent (Chapter 6G). It reacts with well over a score of metals, but in mineral acid essentially only with Hg, Ag, Au, Pd(II), Pt(II), and to some extent Cu. The stability of the Hg(II) complex is indicated by the hydrogen-ion concentration which would allow the extraction of 99% of $Hg(HDz)_2$ when a 0.01% solution of dithizone in CCl_4 is shaken with an equal aqueous volume of Hg^{2+}. This H^+ concentration is found by calculation to lie between 10^8 and 10^9 M. For the extraction of 99% of Ni as $Ni(HDz)_2$, a pH of 6 would be required, for extraction of 99% Pb, a pH of 4. Evidently, pH adjustment will allow selective extraction (determination) of some metals.

S-Chelators such as dithizone are especially valuable for the extractive separation of sulfophilic trace metals from geo samples (silicate and carbonate rocks) and biomaterials, which are composed largely of non-reactive elements, $Z \leq 26$.

It may be observed that for some metals better selectivity is secured by not using chelating reagents at all but ion-association reagents instead. For example, a trace of Ga in the presence of much Al is more easily determined by the use of a basic dye to form $RGaCl_4$ than by any chelating reagent. Aluminum shows little tendency to form a chloro complex and does not react. Other metals usually better determined by ion-association reactions than by chelation include Sb, Tl, Ta, Re, and possibly Au.

Reaction Rate. It may be noted that occasionally selectivity can be obtained for some metals by making use of differences in reaction rate. The best example is furnished by various inert complexes of Co(III). Once formed, in approximately neutral solution, the isonitrosonaphtholates of Co(III) are not (or essentially not) decomposed in mineral acid solution, whereas the complexes of associated metals are decomposed rapidly and a selective determination of Co is secured. Chromium(III) often reacts sluggishly with chelating reagents at room temperature, so that it does not interfere in the reaction of other metals.

C. MIXED LIGAND AND MIXED METAL CHELATES[9]

These are respectively represented by the formulas MLL' and MM'L, omitting subscripts indicating the number of atoms and charges if any. In other words, a mixed ligand chelate, charged or uncharged, consists of a single metal combined stoichiometrically with two or more different ligands, anionic or neutral; the mixed metal chelate consists of two or more different metals combined stoichiometrically with the same ligand. The term ternary complex is often used for a mixed complex having a single metal and two different anions or vice versa.[10]

In mixed ligand chelates, L and L' can both be anionic or one can be uncharged (e.g., H_2O, H_2O_2, NH_3, and various other neutral inorganic or organic ligands coordinated to M). Such mixed chelates are not uncommon. Some simple chromogenic examples are:

$FeOHY^{2-}$, $Fe(OH)_2Y^{3-}$ ($H_4Y = EDTA$)

TiO_2Y^{2-} ($O_2^{2-} = $ peroxo anion)

$Pd(C_6H_5NHC_6H_5NO)_2Cl_2$ ($C_6H_5NHC_6H_5NO = p$-nitroso-diphenylamine)

$Co(HDx)_2I_2^{2-}$ ($H_2Dx = $ dimethylglyoxime)

$Hg(HDz)Cl$ ($H_2Dz = $ dithizone)

Many other mixed ligand chelates are mentioned elsewhere in this book (especially in the chapters on chelating reagents). Some mixed ligand chelates having an inorganic component such as Cl^- may be formed only with an excess of metal in solution, so that the concentration of organic L^- is low. Chelates (uncharged) having neutral organic molecules in their structure are important in extraction separations (Chapter 9B). The neutral molecule can be the acid of the anionic component of the chelate as in

$$SrOx_2 \cdot 2\,HOx \qquad (HOx = 8\text{-hydroxyquinoline})$$

(which is often not considered a mixed-ligand chelate), or it can be different from the anionic ligand as in

$$Th(TTA)_4 \cdot TBP \qquad (TBP = \text{tributyl phosphate})$$

Such chelates are commonly known as adducts. Other examples of this type of mixed chelate are mentioned elsewhere in this chapter.

An example of mixed chelate in which both organic ligands are anionic is provided by the compound formed by tantalum with oxalic acid and o-dihydroxybenzene. The chelates formed separately by tantalum with

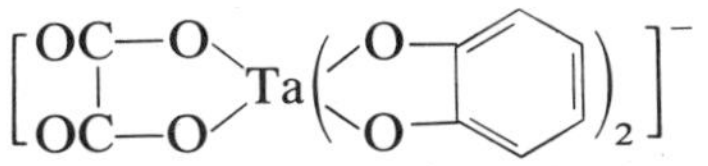

o-dihydroxybenzene and oxalic acid absorb in the ultraviolet, whereas the mixed chelate absorbs in the visible range and is yellow. Mixed ligand chelates of this kind are important in the photometric determination of higher-valent metals.

In addition to favorable photometric and extractive properties, mixed ligand chelates may have the advantage of being formed more rapidly than the corresponding simple chelate. Thus nickel reacts comparatively slowly with dithizone to form $Ni(HDz)_2$. In the presence of pyridine or other N-base, the reaction is much more rapid and a mixed nickel pyridine dithizonate is formed. It appears that dithizone replaces pyridine coordinated to Ni^{2+} more rapidly than it does coordinated H_2O molecules.

The formation of mixed ligand chelates results from the possibility of fitting different ligands simultaneously into the coordination sphere of a metal cation. The strength of the bonding of the respective ligands, their concentrations in the solution, steric factors, and sometimes kinetic factors determine whether mixed ligand chelates can be formed, and if so, their composition and stability.

Mixed metal chelates are much less common than mixed ligand chelates. An example is the secondary dithizonate $Hg(AgDz)_2$ (see discussion of dithizone for suggested structure). The strongly colored complexes produced by interaction of stannous chloride and Pt(IV), Rh(III), and other platinum metals contain both Sn and platinum metal. Cobalt and iron(III) alone give no precipitate with dimethylglyoxime in basic solution, but when both are present a red-brown precipitate is formed having $Co:Fe:H_2Dx = 1:1:3$.

Lumogallion (p. 451) forms a colored $1:1$ complex with scandium that cannot be extracted into ethyl ether. However, when Y, a rare earth element, or Fe(III) is present, a binuclear extractable complex, $Sc:M':L = 1:1:2$, is formed.[11] The apparent molar absorptivity of the Sc–Y–Lumogallion complex at 480 nm is 3.41×10^4, whereas that of the Sc–Lumogallion complex at 530 nm is 1.0×10^4.

The sensitivity of the Alizarin Red S reaction for Al is doubled in the presence of Ca and a mixed Al–Ca chelate is believed to be formed, for which the following structure has been proposed:

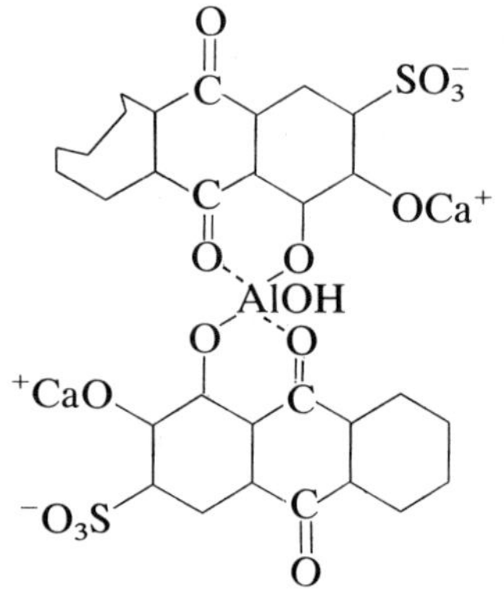

The intensification of the color of Mo(V) thiocyanate by Fe(II), or Cu(I), has been explained by the formation of a mixed Mo–Fe thiocyanate complex.

In general, formation of mixed metal chelates is to be looked for among multidentate chelating agents. Their formation may have an unfavorable effect on photometric determinations. For example, low results for copper with 2,2'-biquinoline are obtained in citrate solution when Cr(III) is present. The explanation is the formation of a kinetically inert mixed Cu(II)–Cr(III) citrate complex in the procedure. Copper can be displaced from this complex by Fe(II), Zn, and other divalent transition metals. Aluminum and U(VI) also form a mixed complex in citrate medium, as do Al(Zr, Bi) and Co in tartrate medium. Also see Chapter 9B,II.A.4.

Inorganic complexes containing a metal in two different oxidation states usually have a color stronger than that of a complex containing either alone (examples, molybdenum blue, Prussian blue). There seem to be no good examples among metal organic chelates.

As already indicated, some seemingly mixed cation chelates are actually ion-association complexes.

NOTES

1. No single book gives a comprehensive account of organic photometric reagents. D. D. Perrin, *Organic Complexing Reagents*, Interscience, New York, 1964, though not devoted exclusively to photometric reagents, is very useful for its treatment of the theoretical basis of the chemical reactions involving organic reagents as indicated by its subtitle "Structure, Behavior, and Application to Inorganic Analysis." Also see A. Holzbecher, L. Divis, M. Kral, L. Sucha, and F. Vlacil, *Handbook of Organic Reagents in Inorganic Analysis*, Halsted, New York, 1976. K. Burger, *Organic Reagents in Metal Analysis*, Pergamon, Oxford, 1973, includes spectrophotometric applications. A number of monographs are devoted to certain classes of organic reagents useful in photometry; these are noticed in the following chapters. Several books deal with immiscible-solvent extraction of metal–organic complexes, especially chelates (See Chapter 9A). F. Feigl, *Chemistry of Specific, Selective and Sensitive Reactions*, Academic, New York, 1949, written from the standpoint of qualitative analysis, is also of interest to the quantitative analyst. Information useful to the photometric analyst can of course also be found in books and papers not primarily concerned with analytical aspects, such as those on coordination and chelate chemistry, bonding, and the like. Books devoted primarily to applied photometric analysis are not mentioned here. In passing, it may be noted that some organic compounds used as photometric reagents are health hazards for one reason or another. A few are demonstrably carcinogenic (notably benzidine, also tolidine); some organic solvents can also be dangerous (Chapter 9A).

2. There is still another possibility, sometimes applicable: The slightly soluble

organic substance is solubilized by micelles of a surfactant. A number of examples are given elsewhere in this book (cf. p. 530).

3. It is difficult to give a foolproof definition of specificity. In practice, the term is used in a relative rather than an absolute sense. In nontrace analysis, we might be willing to concede that a photometric reagent giving an absorbance with any foreign element of, say, 10^{-5} that obtained with an equal weight of the desired element is to all intents specific, yet the foreign element would interfere if trace amounts of the element in question were being determined in a sample composed $\sim 100\%$ of the interfering element.

4. Both inorganic and organic auxiliary complexing agents find extensive use. Included among the former are halides (Cl^-, Br^-, I^-, F^-), OH^-, CN^-, S^{2-}, $S_2O_3^{2-}$, phosphate (PO_4^{3-}, $P_2O_7^{4-}$), H_2O_2. Organic complexing agents include EDTA and other aminopolycarboxylic acids, citrate, tartrate, oxalate, thioglycolic acid, and others. See Đ. D. Perrin, *Masking and Demasking of Chemical Reactions*, Wiley-Interscience, New York, 1970. From the effective constants for complexing reactions given in that book or more comprehensively in A. Ringbom, *Complexation in Analytical Chemistry*, Interscience, New York, 1963, at least rough values for the extent of complexing of common metals by some of the masking agents mentioned can be calculated and conclusions drawn concerning the likelihood of success of various complexing possibilities.

5. D. D. Perrin, *Organic Complexing Reagents*, Interscience, New York, 1964, Chapter 8, deals with the ultraviolet and visible spectra of metal complexes. Also see Burger.[1]

6. Early work: W. H. McCurdy and G. F. Smith, *Analyst*, **77**, 846 (1952); G. F. Smith, *Anal. Chem.*, **26**, 1534 (1954). Subsequent work is touched upon in Chapter 6C.

7. Metals within a group (such as certain horizontal or vertical sequences in the periodic table) may form complexes whose stabilities vary more or less in the same order irrespective of the particular ligand. Thus in the periodic series of elements Mn $\cdots$ Zn (all in the 2+ state) the stability sequence of high-spin complexes with a given ligand is Mn $<$ Fe $<$ Co $<$ Ni $<$ Cu $>$ Zn (Irving-Williams series, *Nature*, **162**, 746 (1948); *Analyst*, **77**, 813 (1952)). This behavior evidently bears on attainable selectivity. If the stability constants lie sufficiently far apart (the range can vary from one reagent to another), Cu, and sometimes Zn, can be determined in the presence of the other metals by adjusting the pH, that is, by changing the concentration of L^-. The stability order can be broken by changing the oxidation state of one or more metals (Mn, Fe, Co, possibly Ni, Cu) or by the use of a reagent effecting a change in the stereochemistry of one of the metal complexes. Even if all the metals in a group react, there may be differences in absorption or fluorescence that can contribute to selectivity. Auxiliary complexing can change the order.

8. H. M. Irving and R. J. P. Williams, *Analyst*, **77**, 813 (1952).

9. A. K. Babko, *Talanta*, **15**, 721 (1968); A. T. Pilipenko and M. M.

Tananaiko, *ibid.*, **21,** 501 (1974) (review covering 1966–1972); I. P. Alimarin and V. I. Shlenskaya, *Pure Appl. Chem.*, **21,** 461 (1970).

10. Certain ion-association compounds are often called ternary complexes. For example Rhodaminium B (RH^+) chloroantimonate, $(RH)SbCl_6$, is sometimes described as a ternary complex of R, Sb, and Cl. Such practice is dubious. $(RH)SbCl_6$ is fundamentally a binary complex, $(RH^+)(SbCl_6^-)$, even as $(RH^+)(ClO_4^-)$ is a binary complex (ion-pair complex). Referring to such ion-association complexes as ternary complexes conceals their true nature. Admittedly it is not always known whether an apparent mixed anion or cation chelate is an ion associate. $Ca(ScOx_4)_2$ is not a mixed cation complex but an ion-pair compound with Sc in the anion, as shown by electrophoresis.

11. I. P. Alimarin, I. M. Gibalo, and A. K. Pigaga, *Zh. Anal. Khim.*, **25,** 2336 (1970).

6B

ORGANIC PHOTOMETRIC REAGENTS
OXYGEN-BONDING CHELATION

Photometric reagents bonding to metals only through oxygen are discussed in this chapter. The products are mostly chelates. One or more OH groups are involved, an H of which is replaced by an equivalent of metal with bonding to =O, >O, OH, or COOH. Five- or six-membered rings are formed. The reactive grouping can occur in a variety of frameworks (mostly aromatic) that have an important effect on the stability, solubility, extractability, and color (sometimes fluorescence) of the chelate. These reagents are chiefly of interest for their reaction with the higher-valent oxyphilic metals in periodic groups IIIB, IVA, IVB, VA, and VIA. Other metals also react, but with a few exceptions, are not advantageously determined with these reagents. Some of these reagents react with such metals as Zr(Hf), Sn(IV), and Ge in mineral acid medium and, therefore, are in some measure selective. Moreover, not all reagents of this type show uniform reactivity with oxyphilic metals. There is a rough correspondence between the strength of the metal–oxygen bond and the tendency of the metal cation to hydrolyze and give hydroxides of low solubility in water. Cations of small size and high charge tend to form firm complexes with the reagents considered here.

The reactive oxygen atoms need not be attached to carbon. In cupferron (p. 942) the reactive grouping is

$$-N\begin{array}{c}{}^{\nearrow\text{OH}}\\{}_{\searrow\text{NO}}\end{array}$$

Replacement of H in OH by metal with bonding to NO through O gives a five-membered ring. The resulting cupferrate is usually not colored (unless the metal is) and is not used photometrically, but is important in separations by precipitation and extraction.

Hydroxamic acids have the reacting grouping

$$\begin{array}{c}-\text{C}=\text{O}\\ |\\ -\text{N}-\text{OH}\end{array}$$

and metals are bonded to O. Reagents with this grouping are of some interest in absorptiometry (determination of vanadium).

The reagents in the following discussion are mostly those having O bonded to C. Reagents in which O is attached to N and other elements are considered in Section IX.

Reagents in which a metal replaces H in an OH group but is also bonded to N or other nonoxygen element are not included in this chapter.

I. NONAROMATIC REAGENTS

Examples of metal-reactive groups are

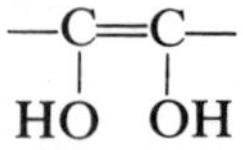

$$\begin{array}{cc} -C{=}C- \\ \ |\ \ \ | \\ HO\ \ OH \end{array}$$

and

$$\begin{array}{ccc} & & H \\ & & | \\ -C-CH_2-C- & \rightleftharpoons & -C{=}C-C- \\ \ \|\ \ \ \ \ \ \ \ \| & & \ \ |\ \ \ \ \ \| \\ O\ \ \ \ \ \ \ \ \ O & & HO\ \ \ \ O \end{array}$$

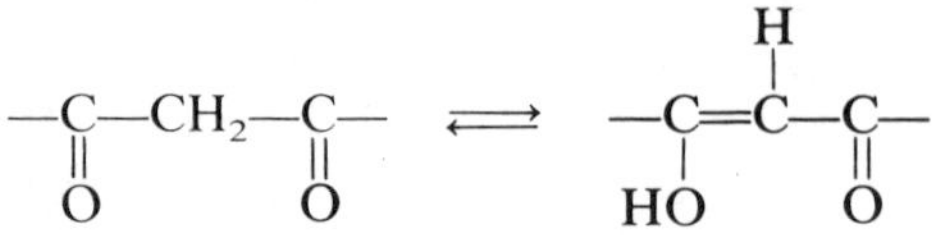

The first occurs in dihydroxymaleic acid and ascorbic acid (A below), the second in β-diketones (B below).

A. DIHYDROXYMALEIC ACID AND ASCORBIC ACID

The first named compound

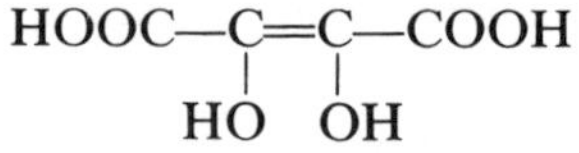

$$\begin{array}{c} HOOC-C{=}C-COOH \\ \ \ \ \ \ \ \ |\ \ \ | \\ \ \ \ \ \ \ HO\ \ OH \end{array}$$

is an old (1908) qualitative reagent for Ti, with which it gives yellow to orange chelate species. It is interesting to note that Th, Zr, and Ce give no color; V(V) and W are reduced to blue products; Mo(VI) and U(VI) give red or brown colors that are said to be destroyed by acids. The dihydroxylamine salt of dihydroxymaleic acid has also been used as reagent for Ti at pH 2.5–3.

Ascorbic acid

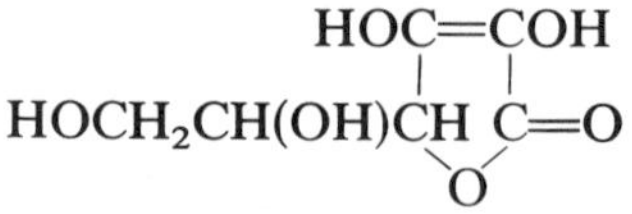

$$\begin{array}{cc} HOC{=}COH & \text{L-Ascorbic acid} \\ \ \ \ \ |\ \ \ \ \ | & \\ HOCH_2CH(OH)CH\ \ C{=}O & \text{Mol. wt. 176.1} \\ \ \ \ \ \ \ \ \ \ \ \backslash O \diagup & \text{M. p. 192°.} \end{array}$$

likewise contains adjacent hydroxy groups and gives similar color reactions, yellow or orange with Ti, orange or brown with U(VI, IV). Other metals giving colors at pH 4–5 include Cu, Te, Bi, Nb, W, Mo, V(IV, V).[1] The titanium complex has been formulated as

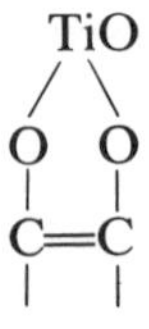

$$\begin{array}{c} TiO \\ \diagup\ \diagdown \\ O\ \ \ O \\ |\ \ \ \ \ | \\ C{=}C \\ |\ \ \ \ | \end{array}$$

but a 3 ligand: 1 Ti complex (TiL_3^{2-}) formed at pH 4 is used analytically.[2] Probably various species can be formed depending on pH. Bismuth forms the species $Bi(H_2A)^{3+}$, $Bi(OH)A$, and BiA^+ (ascorbic acid $= H_2A$).[3]

The yellow species formed with Ti (λ_{max} reported as 360 nm) absorbs less strongly than the Ti–H_2O_2 complex. No advantage in the determination of Ti with ascorbic acid is evident, but separation applications have been made (Chapter 9C).

Ascorbic acid has been suggested as a spectrophotometric reagent for Nb.[4] Ta gives a weaker color. Ascorbic acid also finds use as a reducing agent for Fe(III) and other metals in photometric procedures.

B. β-DIKETONES

The extractive properties of these reagents are discussed in Chapter 9B. The extracts, as in chloroform, of the metal diketonates may absorb in the UV or visible range of the spectrum. Usually the colors are not strong and the sensitivity is mediocre or poor. The selectivity is low unless masking agents are used. Occasionally the reaction is carried out in a mixture of miscible organic solvent and water. The complexes formed have the general structure

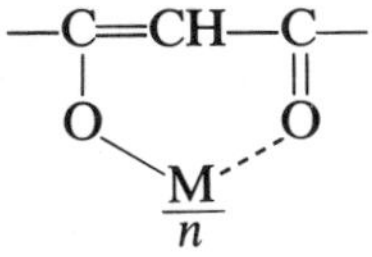

$$-C=CH-C-$$

Acetylacetone (2,4-pentanedione)

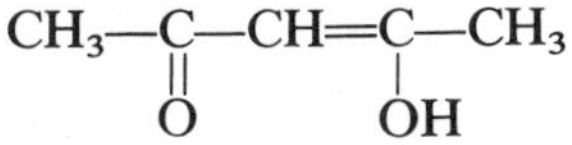

$$CH_3-\underset{\underset{O}{\|}}{C}-CH=\underset{\underset{OH}{|}}{C}-CH_3$$

as a photometric reagent is chiefly of interest because its Be chelate absorbs strongly in the UV ($\epsilon_{295} = 31{,}600$ in $CHCl_3$; $\mathscr{S} = 0.0003\ \mu g/cm^2$).[5] Many other metals can be masked more or less successfully with EDTA. The excess reagent, which also absorbs, is removed by shaking the $CHCl_3$ extract with NaOH, but it seems difficult to reduce the blank to a low value. Ti(III) gives a red color, not very sensitive.[6] Fe(III) can also be determined, but the reaction is of negligible interest.

Dibenzoylmethane, $C_6H_5COCH_2COC_6H_5$, finds use in the determination of uranium ($\epsilon_{400} = 20{,}000$ in butyl acetate)[7] and iron(III).

2-Thenoyltrifluoroacetone (HTTA, Chapter 9B) has been used as reagent for Cu(II), Fe(III), Ce(IV), and other colored cations, but it is chiefly of value in extractive separations. In the presence of pyridine, Fe(II) is extracted synergically into benzene ($\mathscr{S} \sim 0.01\ \mu g\ Fe/cm^2$).[8]

β-Diketones form fluorescent chelates with some of the rare earth elements and are used for their determination, for example, 2-thenoyltrifluoroacetone for Eu.[9]

II. HYDROXYBENZENES AND RELATED REAGENTS

The reaction of Fe(III) with phenol and substituted phenols to give colored products was known at least as early as 1834 (Runge). Later it was noted that other oxyphilic metals (Ti, V, Mo, etc.) also reacted with phenolic reagents, giving less lively colors (usually yellow) than ferric iron (often red or violet). The preferred reagents for iron today are not found in the phenolic group, but some phenolic reagents are nonetheless respectable reagents for Fe(III). Before considering some of the individual reagents of this class, it may be of interest to glance at the reaction of Fe(III) with phenolic compounds.[10] The complexes formed are not all necessarily chelates. Thus Fe(III) gives colors with phenol, C_6H_5OH, chlorophenols, bromophenols, and nitrophenols, which do not contain coordinating groups. The colors are weak (apparent $\epsilon_{max} < 50$) and often unstable. Iron(III) has been found to give a 1:1 species with phenol in acid solution. It has a violet color and is presumably a charge-transfer complex. The formation constant is approximately[11]

$$K = \frac{[C_6H_5OFe^{2+}][H^+]}{[Fe^{3+}][C_6H_5OH]} = 1 \times 10^{-2}$$

This reaction has no photometric importance. The color is very unstable; iron(III) is partially reduced by phenol. Other workers[10] report a reacting ratio of 3 C_6H_5OH:1 Fe(III), so that $Fe(OC_6H_5)_3$ (blue-violet, apparent $\epsilon_{555} = 19$) is also formed. 2,4,6-Tribromophenol gives no color with Fe(III). Steric hindrance must be the reason for this. On the other hand, phenolic compounds having a coordinating group *ortho* to the OH group give Fe chelates, and these have large or moderately large values of apparent ϵ_{max}.[12] Examples are

Salicylic acid (*o*-hydroxybenzoic acid), $\epsilon_{max} = 3780$, compared to *m*-hydroxybenzoic acid, 9.5, and *p*-hydroxybenzoic acid, 13.0.

o-Aminophenol, 8400, compared to *m*-aminophenol, 2.8, and *p*-aminophenol 253.

Catechol, 435, compared to resorcinol, $C_6H_4(OH)_2$ (1,3), 32.

o-Methoxyphenol (guaiacol) gives an iron chelate having apparent $\epsilon_{max} = 2160$ and 2-hydroxy-3-naphthoic acid, 2160.

A. CATECHOL (PYROCATECHOL)

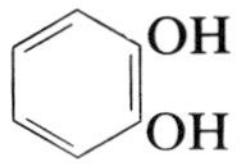

Mol. wt. 110.1   M. p. 105°.

$K_1 = 3.5 \times 10^{-10}$ $K_2 = 2 \times 10^{-13}$

Itself colorless when pure, catechol forms colored chelates with a considerable number of metals, of which those with Fe(III) (violet with excess reagent, insensitive), Mo(VI) (orange-red), Ti (orange), V (blue), Nb, and Ta (yellow) have been used as photometric determination forms. Most of these are of little practical importance, but the reactions are of interest.

The formation of variously colored products in the reaction of catechol with Fe(III), depending on reaction conditions, shows that a number of complexes must be formed. In fact, some have been isolated in the solid state. A red complex is

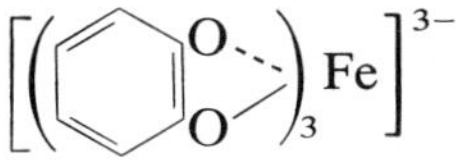

$$\left[\left(\begin{array}{c}\text{O}\\\text{O}\end{array}\right)_3 Fe\right]^{3-}$$

as shown by Weinland[13] and others. A blue complex has the composition[14]

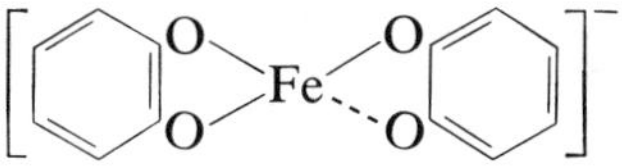

$$\left[\begin{array}{c}\text{O}\quad\text{O}\\Fe\\\text{O}\quad\text{O}\end{array}\right]^{-}$$

A complex containing chloride has also been isolated,[15] which can be formulated

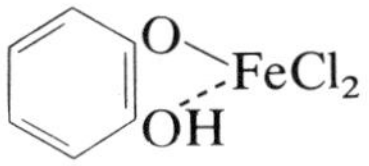

(catechol with O / OH)–FeCl$_2$

Molybdenum(VI)[16] reacts in a 1:1 ratio in acid medium and in a 1:2 ratio in neutral. The 1:2 chelate is considered[17] to have the structure

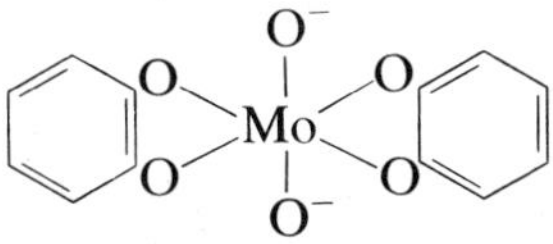

$$\begin{array}{c}\text{O}^-\\ \text{O}\quad|\quad\text{O}\\ \quad\text{Mo}\\ \text{O}\quad|\quad\text{O}\\ \text{O}^-\end{array}$$

with molybdenum in octahedral coordination.

The reaction of catechol or its derivatives with W has been studied, not always from the analytical viewpoint.[18]

Niobium forms a 1:1 complex with catechol at pH 2.[19] Germanium also reacts.[20] It is known from early work (Causse, 1892) that Sb(III) forms a colorless complex SbOHL (H_2L = catechol); this can be looked upon as an ester of antimonous acid and catechol. The reaction of V with catechol has been extensively studied.[21]

The reaction of titanium with catechol has been studied in some detail.[22] Depending on the pH and the concentration of reagent, sundry species are formed, for example, $TiO(HL)^+$, $TiO(HL)_2$, $Ti(HL)^{3+}$, TiL_3^{2-}. The analytical conditions should be such that the species TiL_3^{2-} is predominantly formed, namely pH ~ 3.8 (formate buffer) and $\Sigma[L]/\Sigma[Ti] > 400$. Ferric iron can be reduced to Fe(II), which gives no color, with ascorbic acid. EDTA is reported to prevent the interference of a number of other color-producing elements. Titanium–oxalate–catechol mixed complexes have been studied.[23]

Tin(IV) is known to form a 1(Sn):3 complex with catechol (λ_{max} ~ 285 nm) at pH 6–7.

At least some of the anionic catechol chelates can be extracted into nonoxygen solvents such as chloroform, if an organic base is added that furnishes a cation under the extraction conditions. The structure of the extracted products does not seem to have been established. Apparently the OH group in catechol whose H is not replaced by metal becomes sufficiently acidic on chelation to provide anionic species. *n*-Butyltriphenylphosphonium bromide and *n*-propyltriphenylphosphonium bromide have been found to be good extracting reagents for the catechol complexes of Fe(III), Mo(VI), Ti, Nb, V(IV), U, and W.[24] Extraction of Fe, V (reduction to IV), Ti, and U (into chloroform) requires a weakly acidic solution, but Nb, Mo (partial reduction to V), and W are extractable from rather strongly acid solutions (H_2SO_4); the latter extractions are of analytical interest.[25] Extraction is slow (10–15 min). The Mo complex of catechol and diantipyrinylmethane (1:2:1) is extractable from 0.6–4 *M* HCl into $CHCl_3$ ($\epsilon_{440} = 5000$).[26]

3,4-Dinitrocatechol forms an anionic complex with V(IV or V) at pH ~2, which is extractable into $CHCl_3$ if diphenylguanidinium chloride is added.[27] The extracted complex has the composition 1 V:2 dinitrocatechol:2 diphenylguanidinium. The nitro groups increase color intensity and enable the reaction to be carried out in more acidic solutions. (These products are often called ternary complexes, a term which obscures their nature. They are ion-association compounds.)

The anionic complexes formed by reaction of Ge and Sn(IV) with nitro derivatives of catechol can be extracted as partners with basic dyes or ferrous-2,2'-bipyridyl cation into nonoxygen solvents.

Catechol (and pyrogallol) are subject to air oxidation, which is prevented by sulfite and other reducing agents. Oxidizing ions such as Fe^{3+} and VO_4^{3-} can oxidize these and other hydroxybenzenes and give colored products (quinones and other compounds). Catechol solutions in water can be kept for a day or two if stored in the cold and dark; they are more stable at low pH than high.

Adrenaline (*Epinephrine*)

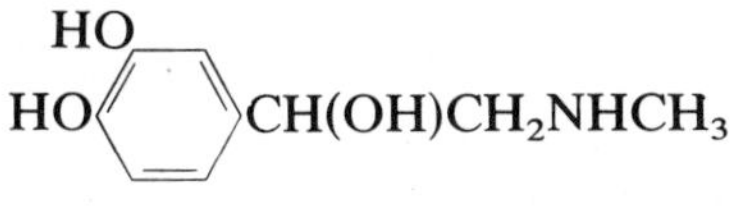

$$\text{HO}-\underset{\text{HO}}{\bigcirc}-\text{CH(OH)CH}_2\text{NHCH}_3$$

forms an orange or red chelate with Ti in strong sulfuric acid.[28] Nb and Ta react less sensitively. No doubt the vicinal OH groups are involved in chelate formation.

B. 1,2-DIHYDROXYBENZENE-3,5-DISULFONIC ACID

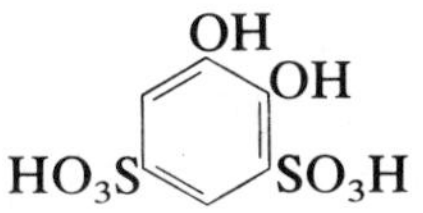

$$\underset{\text{HO}_3\text{S}}{\bigcirc}\overset{\text{OH}}{\underset{\text{SO}_3\text{H}}{\text{OH}}}$$

K_a values of OH groups $= 2.2 \times 10^{-8}$, 2.5×10^{-13}.

(the disodium salt is known as Tiron) forms colored soluble chelates with Ti (orange), Mo (yellow), Nb, V, Fe(III), Ce(III), Cu, as well as with others.[29] Tungsten gives a complex absorbing mainly in the ultraviolet, as does Sc. Tiron has had some vogue in the determination of titanium[30]; better reagents are now available. It reacts with Nb in hydrochloric acid solution to form a yellow species; Ta gives no color (in 6 M HCl, both Nb and Ta are reported to react).

Tiron is a sensitive reagent for iron(III), but the selectivity leaves something to be desired (interference from Ti, Mo, Cu), so that it is not a recommended reagent for this metal. Its reactions with iron are of interest, however, from the viewpoint of chelate composition. Depending on the acidity of the solution, three complex species at least can be formed: red in basic (pH > 7), violet in feebly acidic, and blue in more acidic solutions (apparently mixed with the violet form, pH 4 and thereabouts).[31] The red species has a reagent/Fe ratio of 3 and may be formulated

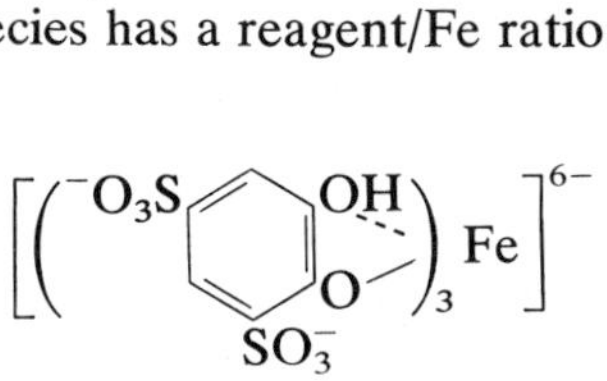

$$\left[\left(\,{}^{-}\text{O}_3\text{S}-\underset{\text{SO}_3^-}{\bigcirc}\overset{\text{OH}}{\underset{\text{O}^-}{}}\right)_3 \text{Fe}\right]^{6-}$$

[Copper(II) forms an analogous species in neutral solution.] Conceivably the –OH in the formula might become –O⁻ in more strongly basic solutions. The blue species is believed (by Yoe and Jones) to have a reagent/Fe ratio of 2 from analogy with a blue complex formed by unsulfonated catechol, which, in solid form, has the composition

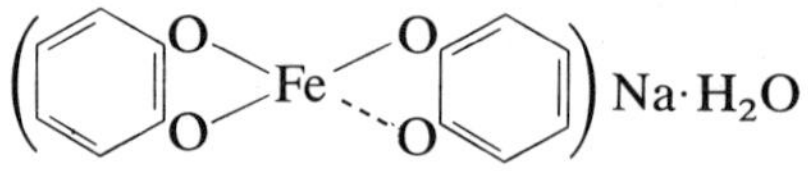

The composition of the violet form (supposing it is not a mixture of the blue and red) has not been established.

The Ti chelates are analogous to those formed by catechol.[22] An ion-associate with diphenylguanidine (1 Ti:2 Tiron:4 diphenylguanidine) is extractable into butanol. Chloroform extracts the associate with tributylamine ($\epsilon_{380} = 1.6 \times 10^4$).

Terbium gives a 2:3 complex with Tiron at pH ~ 9, which is fluorescent.[32] Apparently both OH groups in the reagent react to form this chelate. At pH 2.5–3, Tiron forms a 1:1 complex with Ga which absorbs at 310 nm,[33] and at pH ~ 5 a 1:3 complex with Sn(IV) (~ 305 nm). Tiron, EDTA, and the rare earths form mixed complexes.[34] At pH ~ 4, Tiron forms a green complex with Cu and Ca whose stoichiometry is 1 Cu:1 Ca:2 Tiron.[35]

Tiron increases the heights of the absorbance peaks of Nd, Ho, and Er by a factor of 4 to 9 over the heights shown by the chlorides in aqueous solution and this effect has been put to analytical use.[36] The effect is even stronger with Eu (1:1 complex).[37]

Osmium(IV) forms a 1:1 red-violet complex at pH 5 that may be[38]

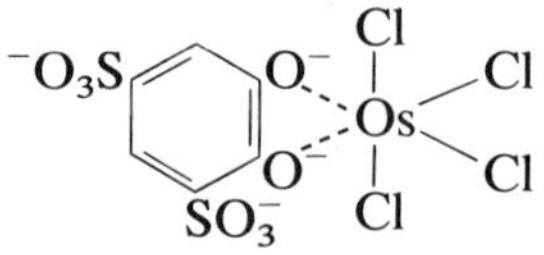

Monosulfonated catechol (3, 4-dihydroxybenzene-1-sulfonate) can be used for the determination of Al in the UV range.[39] It forms 1:1, 2:1, and 3:1 Al chelates with increasing pH.

C. PYROGALLOL

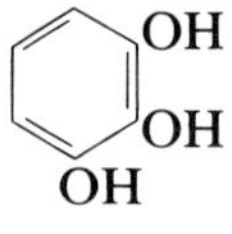

Mol. wt. 126.1.   M. p. 133°–134°.
$K_1 = 1 \times 10^{-9}$.

Pyrogallol reacts with Ti, V(V), Mo(VI), Ta, Nb, and other metals, in much the same way as catechol, to give colored complex species. It has real value for the determination of tantalum in mineral acid solution, though of limited selectivity (Ti interferes, Nb less so); a mixed complex may be formed in the oxalate or oxalate–tartrate solution in which the reaction is carried out.

The reaction product with Mo(VI) is reported to be $MoO_2(HL)_2^{2-}$, pyrogallol being H_3L.[40] One hydrogen of each H_3L is replaced by metal. The complex of niobium and pyrogallol formed in neutral solution must be anionic, since it is extracted into ethyl acetate in the presence of tetrabutylammonium cation.

Osmium(VIII) gives a blue color with pyrogallol; the colored species are anionic and can be extracted into amyl alcohol.

It is known from early work that bismuth and antimony are precipitated from mineral acid solutions by pyrogallol as 1:1 compounds. The precipitates are said to be polynuclear. The absence of adjacent OH groups, as in 1,3,5-trihydroxybenzene or in resorcinol (1,3-dihydroxybenzene), results in no reaction (at least no precipitation) with bismuth and antimony.[41]

Tribromopyrogallol is an extractive photometric reagent for V(IV) and V(V); Fe and Mo are masked with mercaptoacetic acid, Zr with F^-.[42] Nb reacts in neutral tartrate medium.[43]

The orange 1:3:2 complex of Ti(IV) with tribromopyrogallol and antipyrine is extractable into isoamyl alcohol–$CHCl_3$ at pH 1.[44]

D. HYDROXYBENZENE CARBOXYLIC ACIDS

Chelates can be formed by OH in a benzene ring if this group is adjacent to –COOH or other coordinating group containing O. Reactions can occur without chelation. Thus p-hydroxybenzoic acid, which according to current concepts should not chelate metals, reacts with them, nevertheless, through the –COOH group, but such reactions are insensitive and have no analytical value.

Salicylic Acid (o-Hydroxybenzoic Acid)

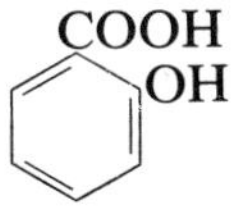

reacts with Fe(III), Ti, U(VI), and other metals to give colored products, but is used only occasionally as a photometric reagent. Like many phenolic reagents it reacts with Ti in concentrated sulfuric acid to form a 1:1 complex,[45] which is suitable for the determination of Ti in the presence of Zr, Mo, and Nb; V(IV) gives a strong blue color. Titanium also reacts in weakly acidic solution.[46]

The yellow product obtained when salicylic acid reacts with Ti(IV) in strong sulfuric acid is considered to be a π-complex, formed by reaction of protonated salicylic acid with TiO^{2+} and formulated as[47]

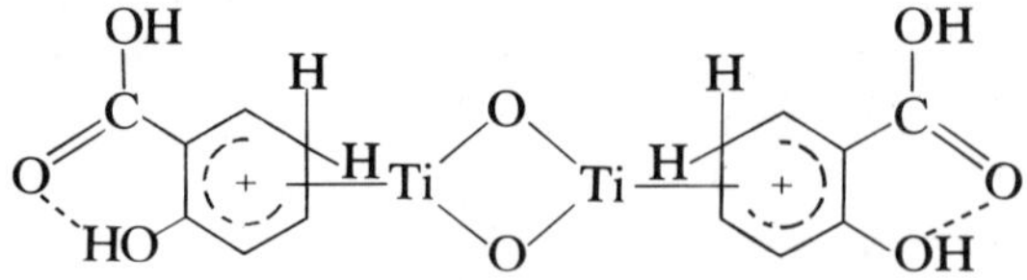

5-Sulfosalicylic acid is perhaps best known as a masking agent, but it is sometimes used as a color reagent, as for Ti and U. Its complex with titanium[48] can be extracted as the tributylammonium salt into organic solvents. As expected, its color reactions are much the same as those of salicylic acid. It forms a 2:1 complex with beryllium in basic solution, absorbing maximally at 317 nm. Iron(III)[49] and Al react with sulfosalicylic acid in weakly acidic medium to form 1:1, 1:2, and 1:3 complexes with increasing pH and reagent concentration. The Al complexes absorb in the ultraviolet, but the sensitivity is not exceptional ($\epsilon = 4000–7000$).

Terbium can be determined selectively in basic solution by the fluorescence of its sulfosalicylic acid–EDTA complex (1:1:1).[50] A wavelength of ~320 nm (absorbance maximum of sulfosalicylic acid) is used for excitation, and the sharp emission line of Tb lies at 545 nm. Energy is transferred intramolecularly from coordinated sulfosalicylic acid to terbium ion.[51] EDTA strengthens the fluorescence a thousandfold, presumably by protecting Tb^{3+} and sulfosalicylate ion from collisional interference by solvent and solute molecules. The other rare earth elements do not give a fluorescence (probably because of mismatching of their energy levels with the triplet level of sulfosalicylic acid) and do not interfere. Since an excess of EDTA does not affect the fluorescence intensity, it can be used to mask other metals. Most metals do not interfere at a ratio 50: 1 Tb. Less than 0.1 μg Tb is determinable.

2,4- and 3,4-Dihydroxybenzoic acids give color reactions with various metals. The latter (protocatechuic acid) forms a 1:1 chelate with V(V) in neutral solution, absorbing in the UV; not selective.

Pyrogallol Carboxylic Acid (2,3,4-Trihydroxybenzene-1-carboxylic Acid)

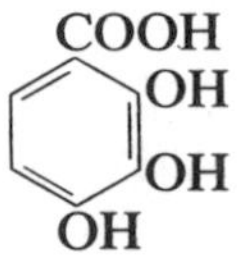

reacts with alkaline earth metals in basic solution to form blue products. Oxidation (air) of the reagent is thought to be necessary for this reaction.

Gallic Acid (3,4,5-*Trihydroxybenzoic Acid*)

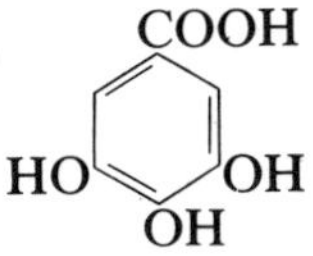

Mol. wt. 170.1   M. p. 220° (decomp.).

gives much the same color reactions as pyrogallol. It has been recommended for the determination of Mo.[52] The chelate formed with Mo(VI) is analogous to that which pyrogallol forms. Gallic acid can be used in place of pyrogallol for the determination of Ta. Niobium can be determined by extracting the three-component complex, Nb(V):gallic acid:aniline = 1:1:1, with a mixture of amyl alcohol and dichloroethane at pH ~ 4.5.[53] Interference from Ta is slight.

Dibromogallic acid has been recommended as better than catechol for the determination of niobium and tantalum.[54] Species such as NbO_2H_2L, NbO_2HL^-, and $TaO_2L_2^{5-}$ seem to be formed (H_3L = dibromogallic acid). Mixed ligand complexes are formed in the presence of oxalate and EDTA.

5,5'-Thiodisalicylic Acid

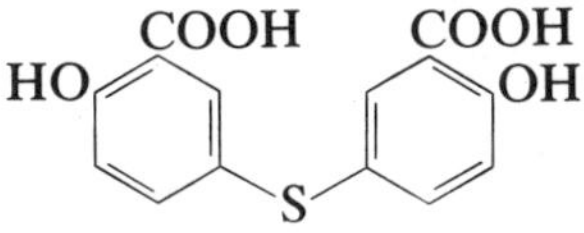

forms colored chelates with Fe, U, Pd, Rh, and other metals.[55] The chelating center is the phenolic OH group. The S atom is not directly involved in the complexing.

E.   MISCELLANEOUS REAGENTS

Hydroquinone (1,4-dihydroxybenzene) and thymol (3-hydroxy-4-isopropyltoluene) react with Ti and W (and some other metals) to form red or orange products in strong sulfuric acid. Niobium and Ta give yellow colors with hydroquinone in H_2SO_4. The nature of these products is not definitely known. The reagents would not be expected to form chelates. (See p. 294.)

Many other aromatic compounds having –OH groups give color reactions with metals, for example kojic acid (actually a heterocyclic compound), chloranilic acid,[56] and other acids. Their reactions show little of general interest. Chloranilic acid is used for indirect determination of metals, for example, calcium, by precipitating the metal with a known amount of reagent and obtaining, by absorbance measurements, the amount remaining in the centrifugate. This is hardly an ideal trace method.

It may be noted that simple o-hydroxyquinones do not react with Sn(IV) in acidic solution.

2,3-Dihydroxypyridine forms a purple-red 2:1 complex with Fe(III) above pH 5,[57] to which the formula

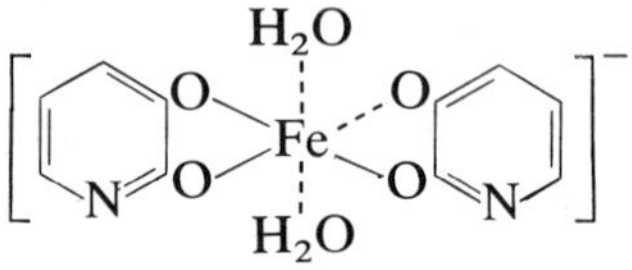

is assigned. In acid solutions a blue 1:1 complex forms, which is stable in 1 M mineral acid.[58]

III. CHROMOTROPIC ACID AND DERIVATIVES

Hydroxynaphthalenes, especially sulfonated hydroxynaphthalenes, have been extensively studied as reagents for metals reacting with phenolic OH. We discuss here the use of chromotropic acid and its derivatives, the most popular of the hydroxynaphthalene reagents. Incidentally, hydroxynaphthoic acids are fluorimetric reagents for Al and Be.

Chromotropic acid, 1,8-dihydroxynaphthalene-3,6-disulfonic acid,

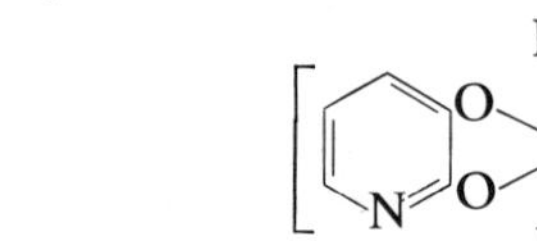

Mol. wt. (+2 H_2O) 356.3; di-Na salt, 346.3.
pK_1 0.6, pK_2 0.7, pK_3 5.45, pK_4 15.5 (the latter two for ionization of OH groups). Solutions not very stable.

is best known as a reagent for titanium, and particulars are given in the chapter on that element (in *Colorimetric Determination of Traces of Metals*). From the general standpoint, the composition of its chelates is of interest. As exemplified by titanium, various complex species can be formed, depending on the acidity. Titanium forms a yellow 1:3 complex above pH 5, to which the following structure is assigned:

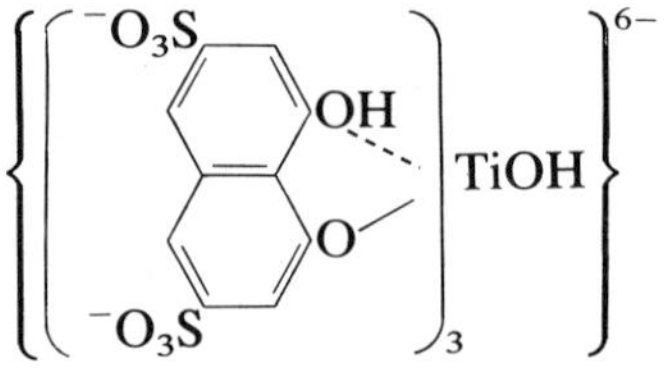

and a red 1:2 complex below pH 3[59]:

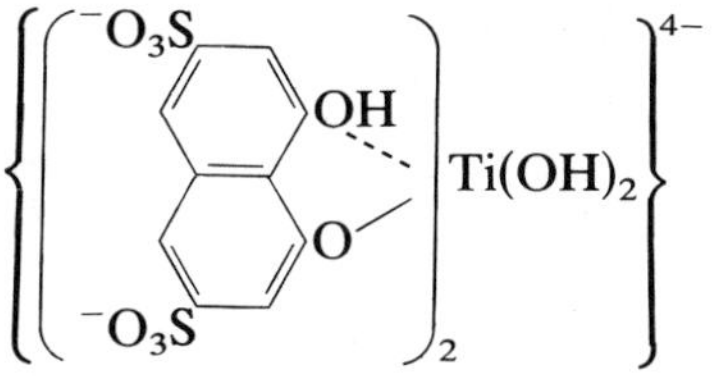

$$\left\{ \left[\begin{array}{c} ^-O_3S \\ OH \\ \\ O \\ ^-O_3S \end{array} \right]_2 Ti(OH)_2 \right\}^{4-}$$

In strong sulfuric acid (>75%) solution, a purple 1:1 complex is formed. Its structure is uncertain. Titanium-hydroxyl ions would not be expected in such solutions. Many metals give colors in weakly acidic solution.[60] The rare earth metals form 1:1 complexes at pH ~ 5. In strong sulfuric acid, only a few metals react: Cr(VI), Ta, Nb, and V(V), although the titanium reaction is less sensitive. The color formation by Cr(VI) and V(V) involves oxidation of the reagent, and does not take place in the presence of reducing agents (Fe^{2+}).

A ternary (presumably ion-association) complex is formed by titanium, chromotropic acid and diantipyrinylmethane.[61] The composition is 1:2:2.

1,8-Dihydroxynaphthalene reacts much the same as the sulfonic acid (1:3 complex with Ti above pH 6, 1:2 complex below pH 4). 2,3-Dihydroxynaphthalene with EDTA gives a fluorescent ternary complex with terbium.

2,7-Dichlorochromotropic acid (ionization constants of $OH = 7.5 \times 10^{-4}, > 10^{-12}$) is a slightly better titanium reagent than chromotropic acid. The composition of its iron(III) complexes has been established[62]:

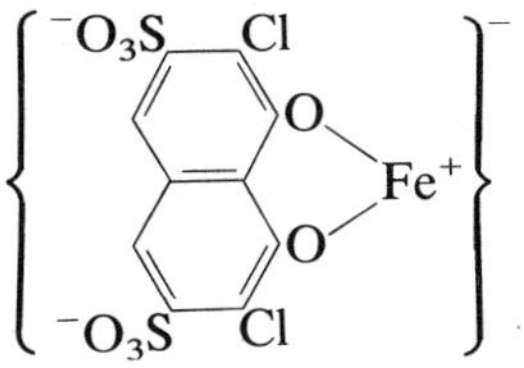

$$\left\{ \begin{array}{c} ^-O_3S \quad Cl \\ O \\ Fe^+ \\ O \\ ^-O_3S \quad Cl \end{array} \right\}^{-} \quad \text{Yellow-green } (\lambda_{max} = 760 \text{ nm}). \\ \text{pH 2.}$$

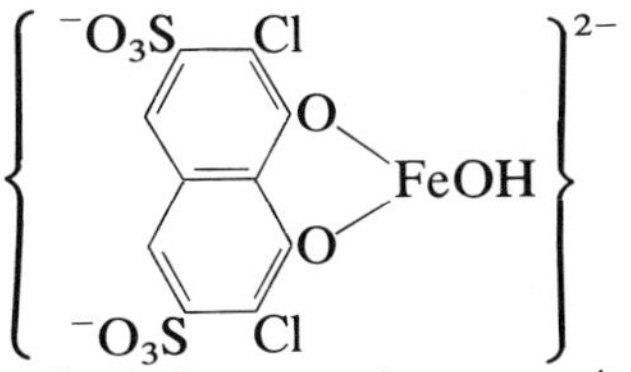

$$\left\{ \begin{array}{c} ^-O_3S \quad Cl \\ O \\ FeOH \\ O \\ ^-O_3S \quad Cl \end{array} \right\}^{2-} \quad \text{Blue-green } (\lambda_{max} = 660 \text{ nm}). \\ \text{pH 5.}$$

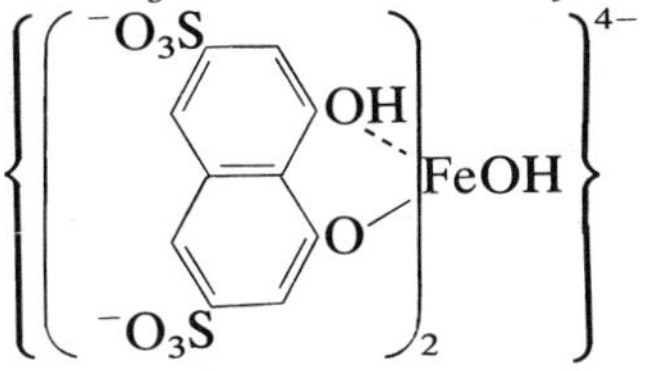

$$\left\{ \left[\begin{array}{c} ^-O_3S \\ OH \\ \\ O \\ ^-O_3S \end{array} \right]_2 FeOH \right\}^{4-} \quad \begin{array}{l} \text{(Chromotropic acid complex)} \\ \lambda_{max} = 660 \text{ nm}. \\ \text{pH} > 7. \end{array}$$

The reagent Khimdu[63] is the disodium salt of 1,8-dihydroxy-2-[N,N-bis(carboxymethyl)-aminomethyl]naphthalene-3,6-disulfonic acid. Its dissociation scheme is[64]

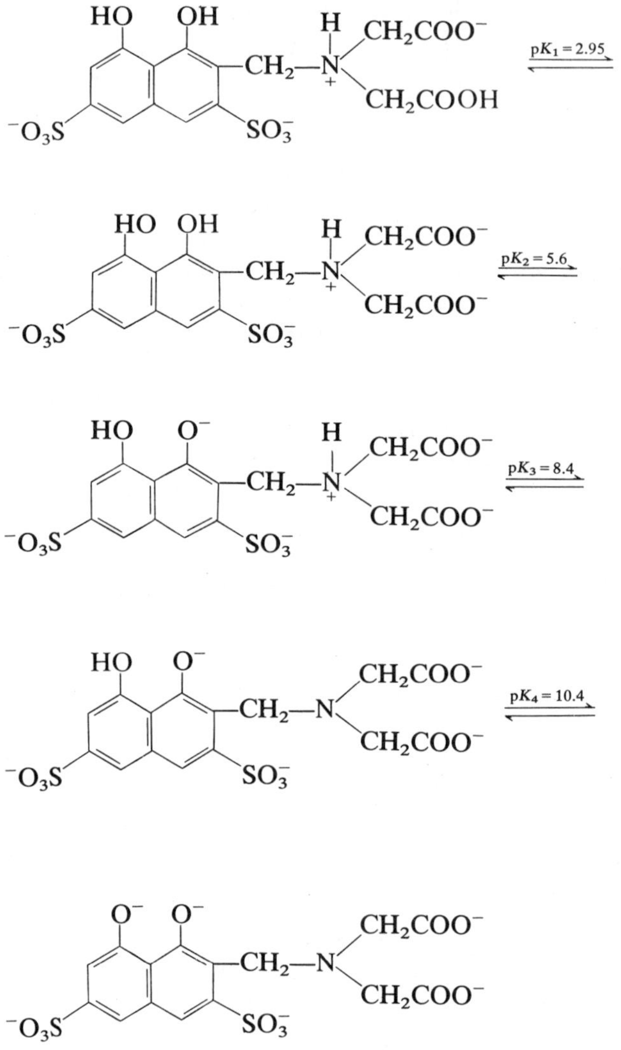

HO   OH   H   CH$_2$COO$^-$
                 N
$^-$O$_3$S   —CH$_2$—N$^+$
         SO$_3^-$ CH$_2$COOH $pK_1 = 2.95$

HO OH   H   CH$_2$COO$^-$
                 N
$^-$O$_3$S   —CH$_2$—N$^+$
         SO$_3^-$ CH$_2$COO$^-$ $pK_2 = 5.6$

HO   O$^-$ H   CH$_2$COO$^-$
                 N
$^-$O$_3$S   —CH$_2$—N$^+$
         SO$_3^-$ CH$_2$COO$^-$ $pK_3 = 8.4$

HO   O$^-$ CH$_2$COO$^-$
$^-$O$_3$S   —CH$_2$—N
         SO$_3^-$ CH$_2$COO$^-$ $pK_4 = 10.4$

O$^-$ O$^-$ CH$_2$COO$^-$
$^-$O$_3$S   —CH$_2$—N
         SO$_3^-$ CH$_2$COO$^-$

Khimdu is reported to give a selective reaction with titanium (2:1 Ti complex, brownish red, pH ~2). Iron(III), zirconium, and thorium are stated to react with the iminodiacetate group in preference to –OH. Niobium gives a yellow color.

Another chromotropic acid derivative (colorless) is Tichromin[65]:

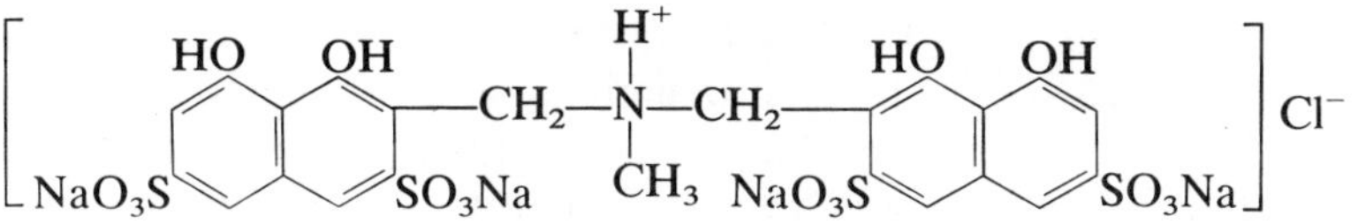

N-methyl-N,N-bis(methylene chromotropic acid)amine (Na salt)

It reacts sensitively with titanium in mineral acid solution ($\sim 1\ M$ HCl), giving a red-brown chelate. Iron(III) can be reduced with ascorbic acid. Molybdenum reacts less sensitively. Niobium also reacts and can be determined in $2–4\ M$ HCl or $2–3\ N\ H_2SO_4$; Ta, Ti and Mo interfere (Mo can be masked with tartaric acid).[66]

Dibromotichromin is a more stable reagent than Tichromin.[67] In conjunction with diphenyl- or triphenylguanidine, it gives an orange-red Ti complex extractable into butanol.

IV. HYDROXYANTHRAQUINONES

A. ALIZARIN

Alizarin (1,2-dihydroxyanthraquinone)

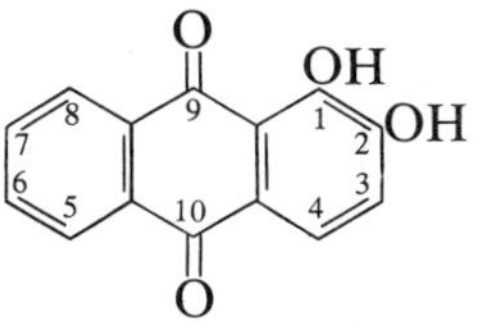

Mol. wt. 240.2. M. p. $287°–289°$. K_1 (ionization of OH in 2-position) reported as 6.6×10^{-9} and 2×10^{-7}; the latter value is in better accord with pH color interval $5\cdot5–6.8$. $K_2 = 1.1 \times 10^{-11}$. H_2Az is yellow, HAz is pink, Az^{2-} is violet.

and Alizarin Red S (sodium alizarin-3-sulfonate)[68]

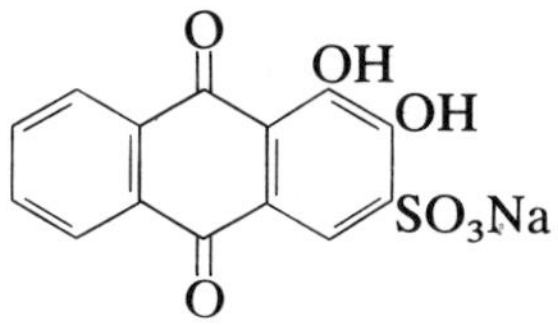

Mol. wt. ($+H_2O$) 360.3. M. p. $> 300°$. K_2 (corresponding to K_1 of alizarin) $= 3.2 \times 10^{-6}$, $K_3 = 1.4 \times 10^{-11}$ ($\mu = 0.5$, 25°).

have been in use a long time for the photometric determination of tri- and quadrivalent metals, particularly aluminum (Atack, 1915), zirconium, and the rare earths. Earlier, the colored reaction products were referred to as lakes, which were considered to be metal hydroxides containing adsorbed dye. However, it has often been shown that it is possible to obtain reaction products in which metal (aluminum and zirconium) and alizarin are combined in integral, or almost integral, ratios, from which it follows that chelates of rather definite composition are formed, at least under certain conditions.

These chelates, usually red in color,[69] may be charged, and the composition varies with the acidity of the solution, which determines the alizarinate ion concentration and the form of the metal (hydroxo species). In dilute mineral acid solution, zirconium (presumably as ZrO^{2+} or an equivalent ion) reacts with alizarin (or sulfonated alizarin) very nearly in a 1:1 ratio. Aluminum reacts in the ratios 1:1 alizarin, 1:2, and 1:3, depending on the pH, the higher alizarin/Al ratios being obtained at higher pH. The colored solutions, as of aluminum, are not particularly stable (even with sulfonated alizarin) and tend to form precipitates, especially in the pH range in which aluminum hydroxide is but little soluble. Such precipitates may be considered to be aluminum hydroxide with adsorbed alizarin. Given the complexity of solutions of metal–hydroxide species and the slow attainment of equilibrium in such solutions, it is not surprising to find that alizarin methods are not ideal in every particular.

Although the reacting ratios of metal and alizarin may be known, the structures of the complex species formed are less certain. Probably the structure of 1:1 chelates of alizarin or sulfonated alizarin is represented by

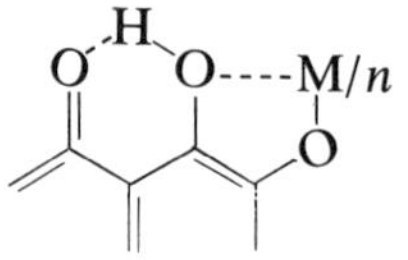

(M/n may be M-hydroxo species)

though replacement of H of OH in the 1-position with bonding of the metal to the quinoid O is not positively excluded.[70] Evidence has been adduced for the following structure for the 1:1 Zr–Alizarin Red complex[71]:

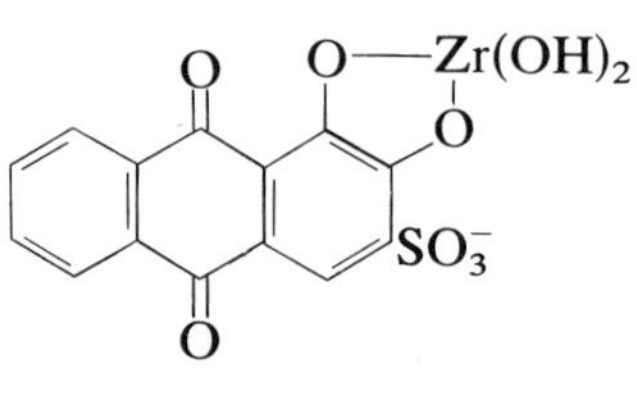

Vanadium(IV) is believed to form a chelate of analogous structure.[72] The H in OH at position 2 has been postulated to be replaced by Ca/2 in the formation of a mixed Al–Ca alizarinsulfonate chelate. (See the structure on p. 286.)

The effective formation constants of the 1:1 alizarinsulfonic acid complexes of quadrivalent metals have been determined.[73] The values of K in

$$K = \frac{[M(IV)Az]}{[M(IV)][H_2Az]} \qquad (H_2Az = \text{alizarinsulfonic acid})$$

are: Ti, 3.7×10^4; Zr, 3.6×10^5; Hf, 5×10^5; Th 3.6×10^6, based on spectrophotometric measurements in 1.6×10^{-4}, 0.1, 0.05, and 10^{-4} N H_2SO_4, respectively. Tin(IV) also forms a 1:1 complex. The constant

$$K = \frac{[Zr(OH)_2Az][H^+]^4}{[Zr^{4+}][H_2Az]}$$

has been determined[71] and found to have the value 4.2×10^4.

Alizarin reacts with many cations and is not a selective reagent. However, at relatively high acidity the number of reacting metals decreases, so that the determination of zirconium (and hafnium) can be carried out in the presence of most divalent metals and reasonable concentrations of trivalent metals. Iron(III) can be reduced to Fe(II). Niobium reacts slowly at pH 3–5 in the presence of EDTA, oxalate, and the like.

Scandium can be determined at pH 3.5 with sulfonated alizarin.[74] Other metals reacting at this pH include Fe(III), V(V), Th, Al, Ga, Cu, Zr, Ti(III and IV), Mo, Sn(II), and U(VI). Metals not reacting include Pb, Bi, Zn, Co, Ni, As, Sb(V), Mg, and Ca. Some of the complexes, as of Al and Ga, are fluorescent.

Lanthanum forms two complexes with Alizarin Red S, in which La:reagent ratio is 1:1 (acid or neutral solution) and 1:2 (alkaline medium).[75] The red color of the reagent can be removed with H_3BO_3, which forms a yellow compound with Alizarin Red S. In neutral medium, Ca participates in the reaction, possibly giving a complex in which La:reagent is 1:2; it serves for the determination of La.

In the presence of $Na_2S_2O_4$, Alizarin Red S gives a green product with Ti.[76] Both reagent and Ti(IV) are partly reduced.

The alizarin or alizarinsulfonic acid complexes are usually not extracted for determination, but the negatively charged complexes of the latter are extractable with high molecular weight cationic partners.[77] At least some alizarin chelates (as of Th) can be extracted into butanol.

The following hydroxyanthraquinones, in addition to alizarin, give colored precipitates with zirconium in dilute mineral acid solution[78]:

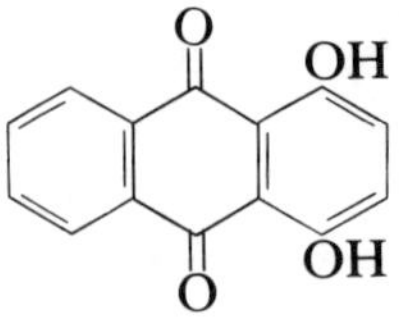

Quinizarin 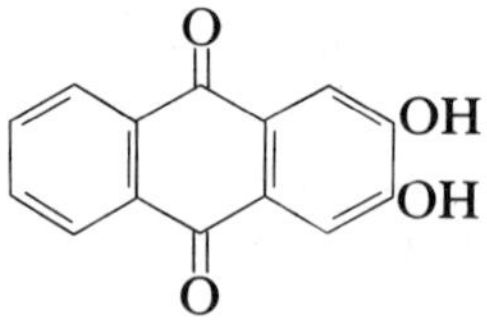   Hystazarin 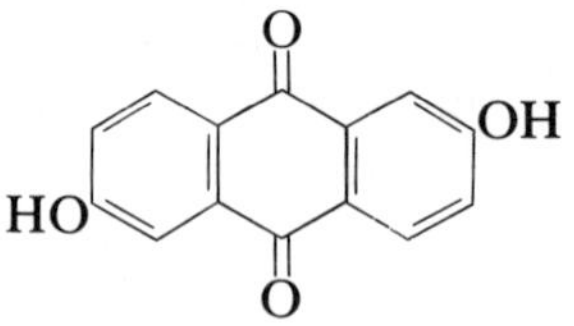   Anthraflavic acid

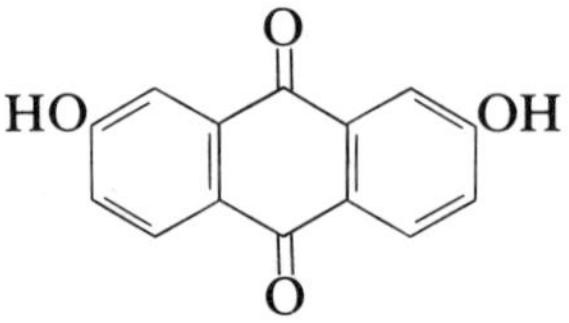

Isoanthraflavic acid 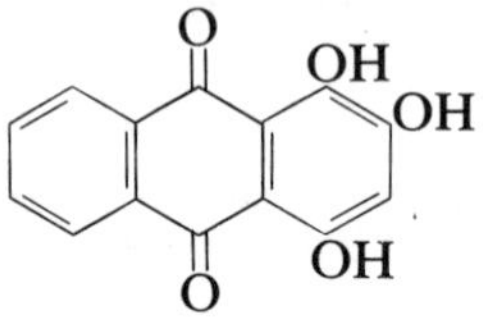   Purpurin

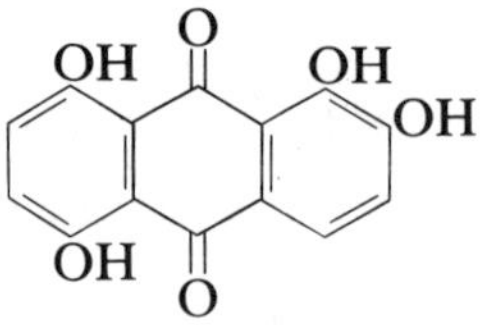

Quinalizarin

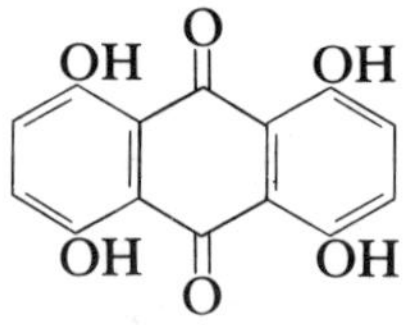

The following hydroxyanthraquinones do not give precipitates:

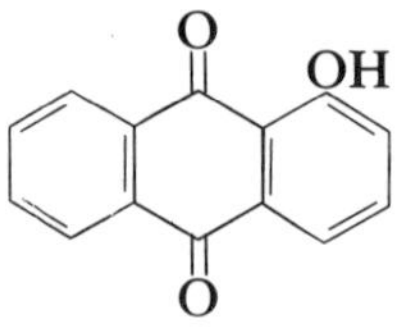

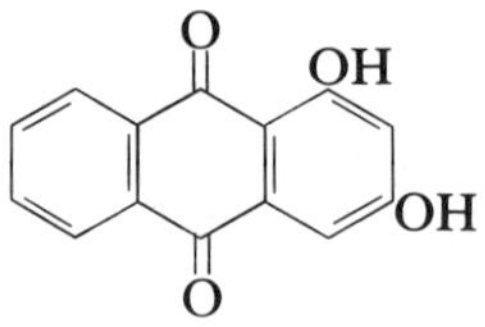

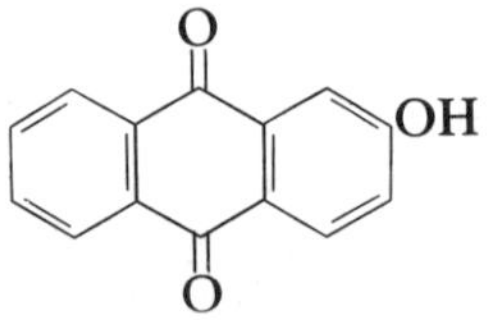

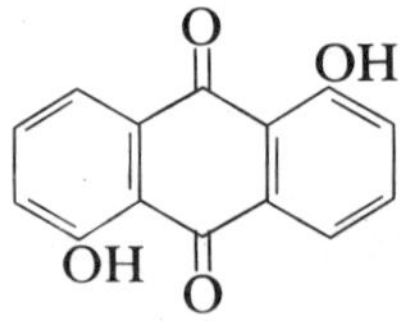

The Zr chelate of quinizarinsulfonic acid is relatively easily dissociated.

The nonreaction of monohydroxyanthraquinones with zirconium is noteworthy. Moreover, according to Feigl, aluminum does not react with 1-hydroxyanthraquinone, but beryllium does react, though not sensitively. It is not clear why 2,8- or 2,7-dihydroxyanthraquinone should react with zirconium (or other cation). Evidently, the chelating properties of the hydroxyanthraquinones require further study.[79]

Alizarin Complexan (*Alizarin Fluorine Blue*). This compound, 3-dicarboxymethylaminomethyl-1,2-dihydroxyanthraquinone (or 3-amino-methylalizarin-N,N-diacetic acid), is more important for the determination of fluoride than of metals[80]:

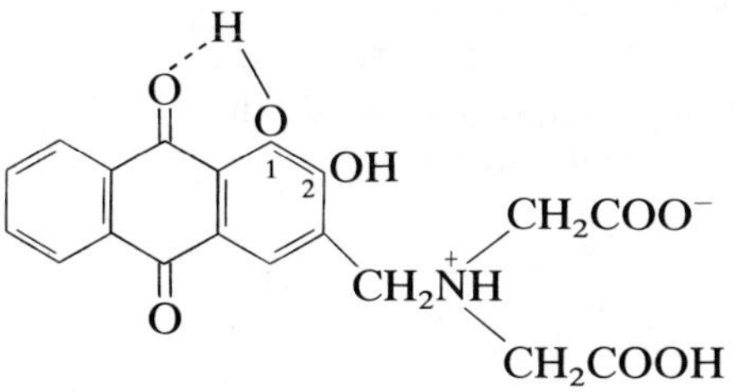

H_4L

With increasing pH, four protons are successively lost from H_4L and the following ions are formed:

H_3L^-(carboxyl group H removed), yellow, pH ~ 4
H_2L^{2-}(phenolic proton in position 2 removed), red, pH ~ 8
HL^{3-}(imino proton removed), red, pH ~ 11
L^{4-}(phenolic proton in position 1 removed), blue, pH 13–14

In addition, the species $H_5L_2^{3-}$ has been postulated to be formed at pH 5–6 (Ingman). The following values have been found for log K (stability constant) of the species at $\mu = 0.1$[81]: HL^{3-}, 11.98; H_2L^{2-}, 10.07; H_3L^-, 5.54; H_4L, 2.40.

Alizarin Complexan reacts with Al, the lanthanides, and other hexacoordinating metals to give red 1:1 complexes at pH ~4.[82] These complexes have the composition $M^{III}HL$ when the metal cation is triply charged, and their structure is analogous to that of H_2L^{2-}; the absorption spectra are similar. In the presence of substoichiometric amounts of fluoride, the La or Ce(III) chelate gives a blue ternary 1:1:1 complex whose structure is analogous to that of L^{4-}. The binary complex of Ce(III) or La is represented as

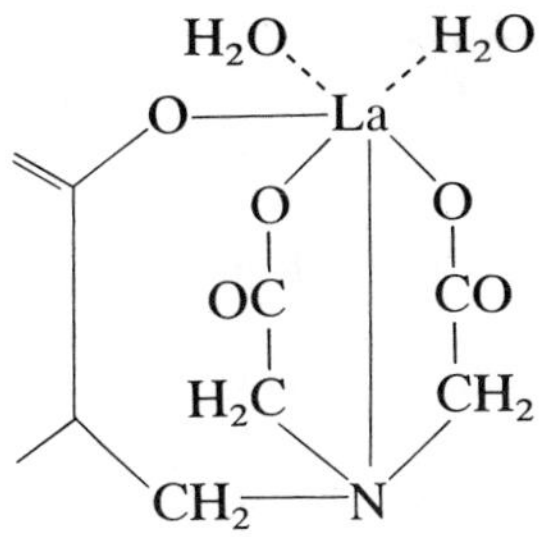

Introduction of F^- (replacing H_2O) weakens the proton–oxygen bond so that the OH group in position 1 dissociates at pH ~ 4, whereas in the absence of F^- it does so only at pH > 12. Accordingly, the fluoride complex at pH ~ 4 shows a spectrum similar to that of the reagent alone in strongly basic solution, and hence the color change from red to blue produced by fluoride. Later work[83] suggests that the ternary complex containing F consists of a ring-formed dimer $(LaHL)_2$ with one F attached to each of the two metal atoms:

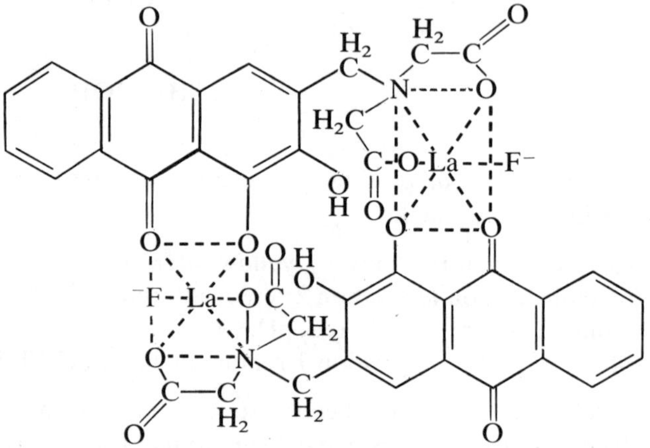

Still later work suggests the composition $La(LaL)_4F_2$ for the fluoride complex.[84]

Aluminum can be determined reproducibly and fairly sensitively ($\mathscr{S}_{455} = 0.01\ \mu g$ Al/cm^2) with Alizarin Complexan.[85] A solution containing 20% dioxane is used to keep Al(HL) dissolved. A pH of 4.1–4.3 is suitable (below pH 4 the reagent precipitates, above pH 5 Al(OH)$_3$ may precipitate or the reaction becomes very slow). Aluminum can be determined in the presence of Fe(III) by masking Al (but not Fe) with F^- and finding the difference in absorbance, $A_{Al} = (A_{Al+Fe} - A_{Fe})$. Titanium does not interfere in this procedure. Heating for an hour at 70° is required to form the Al chelate.

Alizarin Complexan gives a ternary complex with La and Ni having the composition 2 reagent:1 La:1 Ni, suggested for the determination of nickel.[86]

B. OTHER HYDROXYANTHRAQUINONES

Quinizarin (*1,4-Dihydroxyanthraquinone*). As might be expected, quinizarin gives color reactions with many metals. These have not received much attention. Analysis of the solid chelates of Ni and Co indicates a composition 1 mole quinizarin, 2 moles Ni or Co, and 4 moles H_2O.[87] The Cu and Mn chelates, prepared in the presence of acetate, contained quinizarin, Cu(or Mn), and acetate in the molar ratios 1:2:4. 1,5-Dihydroxyanthraquinone is reported to yield chelates of analogous composition.

Quinizarin forms a red fluorescent chelate with Be in basic medium. The 2-sulfonic acid[88] has been used as a colorimetric reagent for the same element. Aluminum forms a 1:1 fluorescent complex at $pH \sim 5$.

1,4-Dihydroxy-2-(2-pyridylmethyl)anthraquinone (yellow alcoholic solution) forms a blue, fluorescent, 2:1 reagent-UO_2^{2+} chelate in neutral solution.[89]

1-*Amino*-4-*hydroxyanthraquinone.* An alcoholic solution of this compound produces a purple colloidal precipitate with thorium in weakly acidic solution (pH 2), which shows a reddish fluorescence in ultraviolet light.[90] The sensitivity is mediocre, approximately 1:125,000.

Gallium and praseodymium[91] also fluoresce under these conditions but not as strongly as thorium (0.1 mg Th $\equiv$ 1.5 mg Ga $\equiv$ 10 mg Pr). Zirconium and iron(III) decrease the thorium fluorescence, and cerium(IV), silver, gold, and mercury(I), as well as metals of the platinum group, are stated to destroy the reagent. Phosphate, fluoride, and sulfate in small amounts destroy the thorium fluorescence. In basic medium, the reagent produces a fluorescence with beryllium.

Purpurin (1,2,4-*trihydroxyanthraquinone*) is a reagent for zirconium. It gives a sensitive fluorescence reaction with stannic tin at pH 4.5; gallium also reacts sensitively at this pH. Also a reagent for Ge[92].

Germanium has been determined by extracting the Ge–purpurin complex as an ionic associate with diphenylguanidinium into chloroform at pH 3. The associate can be represented by the following formula[93]:

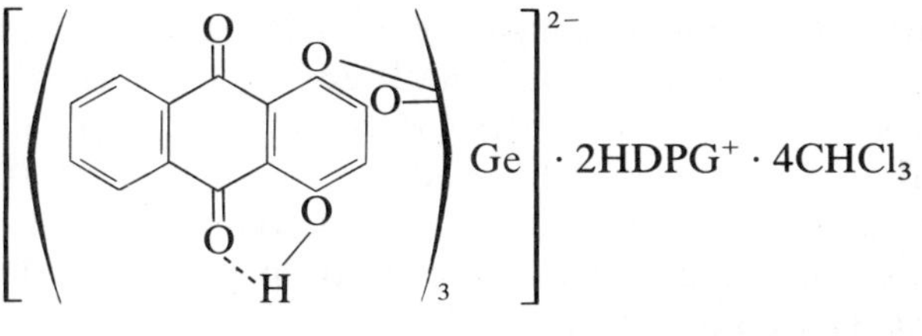

$$\left[\left(\text{structure} \right)_3 Ge \right]^{2-} \cdot 2HDPG^+ \cdot 4CHCl_3$$

1,2,7-*Trihydroxyanthraquinone.* Hydroxyanthraquinones with a hydroxy group in the ortho position are stated to give sensitive color reactions with stannic tin.[94] One of the most suitable of these is the 1,2,7-compound.

Quinalizarin (1,2,5,8-*tetrahydroxyanthraquinone*). This dye dissolves in alkalies to give a red-violet solution that is turned yellow by acids. It is also soluble in alcohol and acetone. Quinalizarin has been used for the colorimetric determination of beryllium, gallium, magnesium, aluminum, germanium, molybdenum, and uranium. No doubt it could be used for other metals.

The addition of quinalizarin to a sodium hydroxide solution containing beryllium gives a blue color which is quite different from the violet color obtained in the absence of beryllium.[95] Aluminum does not react in strongly basic solution. A method for the determination of beryllium has been based on this behavior. Other metals giving a sensitive reaction (blue color) in sodium hydroxide solution are magnesium, scandium, rare earths, nickel, and cobalt.[96] Various other metals react less sensitively. Attempts to determine scandium in this way did not lead to reproducible results.

Gallium at pH 4.5–6 yields a pink to amethyst color with quinalizarin, and has been determined in this way,[97] but the method is no longer used.

Magnesium hydroxide adsorbs quinalizarin and gives a blue lake.

In faintly acidic solution, aluminum produces a violet color. The reaction has found little quantitative application.

In strong sulfuric acid, boric acid gives a blue color with quinalizarin, which is due to the formation of a complex boronhydroxyanthraquinone compound. Niobium and tantalum also react in sulfuric acid medium.

Many metals will react with quinalizarin in slightly acidic medium, giving color changes. Thus at pH 5, iron(III) and lead produce a blue color; tin(II), antimony(III), copper(II), indium(III), germanium, vanadium(IV and V), and molybdenum(VI), a pink color. These reactions are not prevented by the addition of fluoride. The following metals also react at pH 5 but fluoride prevents the reaction: zirconium, thorium, rare earths (all blue colors); tin(IV), beryllium, aluminum, thallium(III), titanium(IV), arsenic(III), antimony(V) (all pink colors). The alkalies, alkaline earths, magnesium, manganese, iron(II), mercury(II), thallium(I), cadmium, uranium(VI), and tungsten(VI) give no color at pH 5, whereas other metals, such as silver, mercury(II), bismuth, and tantalum, are precipitated as chlorides or hydrolyzed out.

Quinalizarin and benzoic acid form 1:2:1 complexes with the yttrium subgroup elements, extractable into butanol at pH ~ 10.[98] The Ce subgroup elements are said not to interfere.

Carminic acid

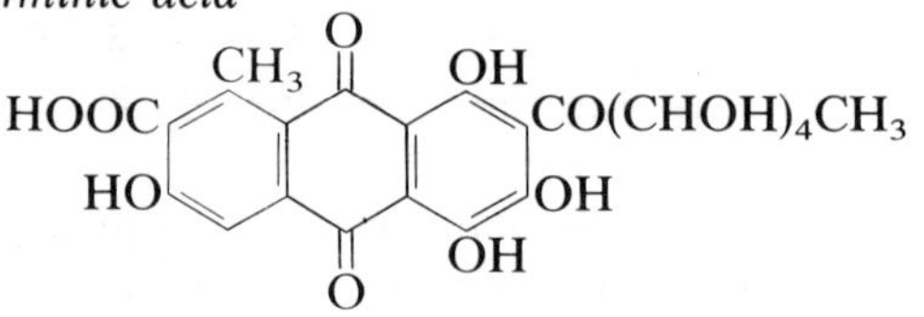

Mol. wt. 492.4
M. p. 136° (decomp.).

is evidently a reagent capable of giving color reactions with a multitude of metals. It provides a fluorescence reaction with Sn(IV) and has been used in the colorimetric determination of scandium, but it is best known as a boron reagent. The fluorimetric determination of Mo and W (1:1 chelates) has been described.[99] The rare earth metals can be determined (2:1 metal complexes).[100]

V. PHENYLFLUORONE AND OTHER TRIHYDROXYFLUORONES[101]

The compound 2,3,7-trihydroxy-9-phenyl-6-fluorone (or 2,6,7-trihydroxy-9-phenylisoxanthene-3-one) is

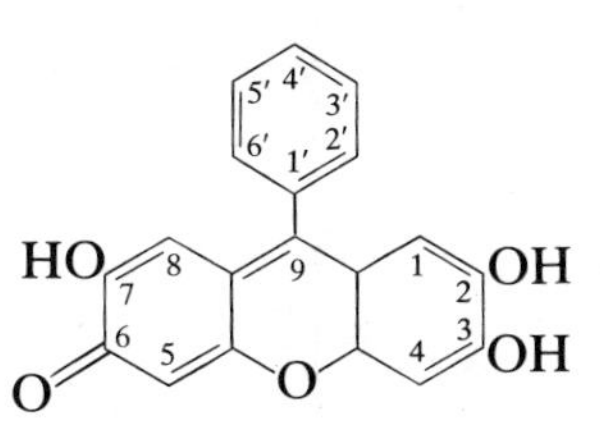

Mol. wt. 320.3. Orange powder. Some commercial products have unsatisfactory purity. Can be recrystallized from warm acidified alcohol by addition of ammonia. An impurity interfering in the determination of Ge has been identified as 4-benzylidene-2, 5-dihydroxycyclohexa-2, 5-dien-1-one.[102] It can be removed by extracting commercial phenylfluorone with ethanol in a Soxhlet apparatus for 10 hr.

usually called phenylfluorone by analytical chemists. It is best known as a reagent for germanium, with which it gives a sensitive reaction (red color) in quite strongly acid solutions. Like other reagents with oxygen coordinating groups, it also reacts sensitively in acid solution with oxyphilic elements of higher valence such as molybdenum(VI), titanium(IV), tin(IV), zirconium, niobium, and tantalum.[103] Vanadium(IV) reacts at pH ~5, but analytically this reaction is unimportant. Aluminum reacts at pH ~6, Bi in acidic solution (pH 1.5).

Phenylfluorone is an ampholyte. If the neutral substance is represented as H_3Pf, the ionization constants for the reactions

$$H_4Pf^+ \rightleftharpoons H^+ + H_3Pf$$
(xanthylium ion)

yellow   yellow-orange

$$H_3Pf \rightleftharpoons H^+ + H_2Pf^-$$
orange-pink

in 20% ethanol are $K_1 = 5 \times 10^{-3}$ and $K_2 = 1 \times 10^{-7}$ as determined by Schneider.[104] The color changes result from a change in the structure of the compound with pH. In strongly acid solution, phenylfluorone exists as the xanthylium ion with relatively simple triphenylmethane-like structure. As the pH is increased, the neutral molecule is formed in increasing quantities; it has a quinoid conjugated system. In alkaline solution, the anion is formed; the ionic structure facilitates resonance, and the absorptivity increases markedly. Above pH 8, in 20% alcoholic solution, a greenish fluorescence appears. The equilibria may be represented as follows with increasing pH (0–~9):

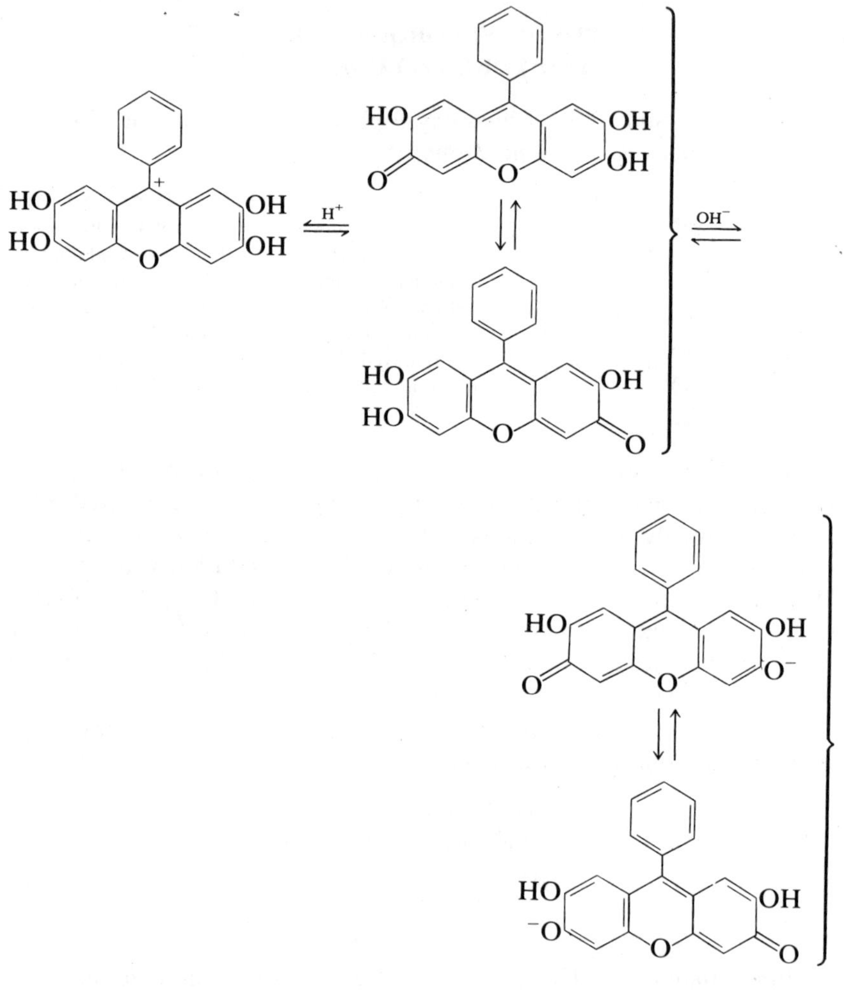

In more strongly basic solutions, doubly charged anions are formed. The reagent decomposes slowly in strongly alkaline solutions.

The reagent is very slightly soluble in water ($3 \times 10^{-7} M$ in 20% ethanolic solution, 25°), as are its metal chelates. The reacting ratios of a number of metals with phenylfluorone have been determined. Germanium especially has been studied. As shown by Schneider and others, the combining ratio in ~0.1 M H^+ solution is 1 Ge:2 H_3Pf, and it seems reasonable to believe that the chelate

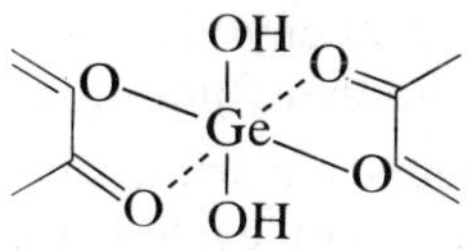

is formed. This is red. In spite of difficulties arising from the insolubility of the reaction product and slow attainment of equilibrium, Schneider was able to show that an approximate equilibrium constant could be determined for the reaction

$$2\,H_3Pf + Ge(OH)_4 \rightleftharpoons Ge(OH)_2(H_2Pf)_2 + 2\,H_2O$$

namely

$$K = \frac{[Ge(OH)_2(H_2Pf)_2]}{[H_3Pf]^2[Ge(OH)_4]} = 8(\pm4) \times 10^{12} \qquad (25°, 20\%\ ethanol,\ \mu = 5)$$

([H^+] varied from 0.06 to 0.34 M.)

The phenylfluorone complexes of Sn(IV), Ge, Sb(III), Ta, Nb, Zr, Hf, Ti, Mo(VI), W, V(V), and Fe(III) (all in 0.1 mg amounts) can be extracted with strong color into methyl isobutyl ketone from 1:100 HCl solution.[105] In the presence of H_2O_2, Sn(IV), Ge, Sb(III), Hf, and Fe(III) are extracted with strong colors at the same acidity; Zr gives a light color, and the other elements, little or no color. In the presence of oxalic acid, strongly colored extracts are given by Ge and Sb(III) and lightly colored by Sn(IV), Ta, Ti, and Mo(VI). Only Sn(IV), Ge, and Sb(III) give a strong color in the presence of both H_2O_2 and oxalic acid. H_2O_2 increases the color due to Sn(IV). Zr and Hf diminish the color due to Sn, but not that due to Ge. The Sn–phenylfluorone complex can be extracted quantitatively into MIBK; the reagent divides between the two phases, and, as expected, the fraction remaining in the aqueous phase increases with the acidity. The color intensity of the MIBK solution of Sn decreases slowly on standing, the more so the higher the acidity of the aqueous phase.

It has been shown that Ti forms two types of complexes with trihydroxyfluorones, in each of which Ti:reagent is 1:2. Below pH 6 the complex-forming ion is $Ti(OH)^{3+}$, above pH 6 it is $Ti(OH)_2^{2+}$; the reagent reacts as an o-hydroxyquinone.[106] In pyridine buffer media, mixed complexes are formed (Py:Ti:reagent = 1:1:2), which are more strongly colored than the binary complexes.[107]

Various derivatives of phenylfluorone have been prepared and tested as metal reagents. Among these are

2,3,7-trihydroxy-9-(4'-dimethylaminophenyl)fluorone[108]
2,3,7-trihydroxy-9-(3'-nitrophenyl)fluorone[109]
2,3,7-trihydroxy-9-(4'-nitrophenyl)fluorone[109]
2,3,7-trihydroxy-9-(2'-sulfophenyl)fluorone[109]

The spectra of these compounds in acidic solution differ only slightly from that of phenylfluorone. It is not evident that these reagents have any advantage over the parent compound in the determination of germanium. For zirconium, 3'-nitrophenylfluorone has an advantage in greater stability of the chelate and lower blank readings. The sulfophenylfluorone gives chelates of lower stability than the nonsulfonated reagents and so, in spite of the greater solubility of its chelates, is apparently not a useful reagent. All these reagents require the use of a protective colloid to give a stable suspension of the chelates.

Still other derivatives include 9-(3-pyridyl)fluorone[110] and o-carboxy- and p-(methyloxycarbonyl)phenylfluorone.[111] The latter two reagents give sensitive reactions in 0.5 M HCl with W(VI), Mo(VI), Fe(II and III), Nb and some other elements, but not with Sn(IV), Zr, or Ge. 2-Quinolylfluorone forms a 4:1 complex with Zr, useful for the determination of this metal.[112] 9-(2'-Quinolyl)-, 9-(8'-hydroxy-2'-quinolyl)-, 9-(2'-quinoxalyl)-, and 9-(2'-carboxy-3'-pyridyl)- derivatives of trihydroxyfluorone have been prepared.[113]

Bis-fluorones such as p-phenylene-bis-fluorone

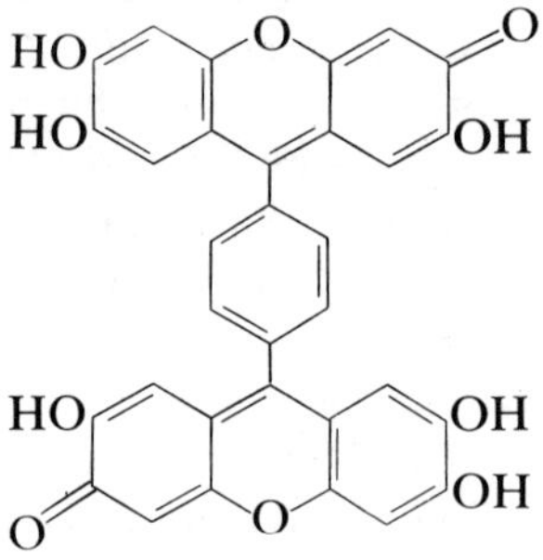

and the ester of ethylene glycol and dihydroxyfluorescein

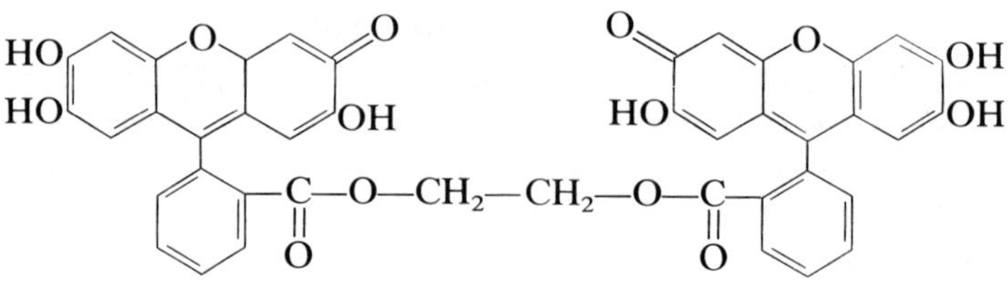

have been prepared[114] and their reactions with the metals of groups III–VI examined.[115] p-Phenylene-bis-fluorone has been proposed as a reagent for gallium.[116]

Phenylfluorone and derivatives (nitro-, bromo-, etc.) form 1:1 complexes with Nb in ~1 M HCl that are extractable into $CHCl_3$. The extracts have remarkably high molar absorptivities, for example, phenylfluorone $\epsilon_{515} = 131,000$, 3-nitrophenyl derivative $\epsilon_{527} = 169,000$.[117] Sn, Sb(III), Ti, Ta, W, and Mo, but not Zr and Fe, are also extracted.

Propylfluorone (2,3,7-trihydroxy-9-propyl-6-fluorone) is very similar to phenylfluorone in its reactions, as would be expected. It combines as a singly charged anion with $Ge(OH)_2^{2+}$ to give a 2:1 (Ge) chelate, which is very slightly soluble ($K_{sp} \sim 10^{-23}$).[118] Propylfluorone has also found use in the determination of Mo.[119] Either of two chelates can be formed, depending on the pH:

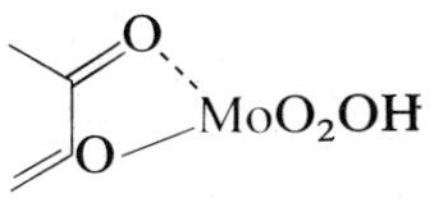

pH 1.8–3.0.

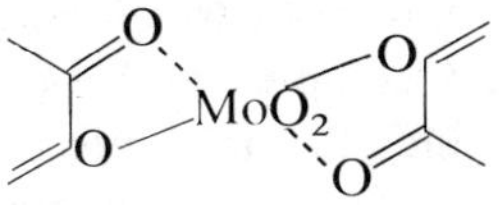

pH 0–1.5 ($\lambda_{max} = 565$ nm).

(2,3,7-Trihydroxy-9-trichloromethyl-6-fluorone forms a 1:1 complex in the pH range 0–3.5.)

Particulars concerning the analytical properties of salicylfluorone complexes are available (Tataev et al.[119]), and are partially tabulated here:

	Optimal pH	λ_{max} (nm)	$\epsilon \times 10^{-3}$	Composition
Zr	1	490	23	1:2
Hf	1	490	25	1:2
Sn(IV)	2	490	18	1:2
Bi	2.5	490	25	1:2
W	2.5	530	25	1:2
Mo	3	530	28	1:2
Ti	3	530	92	1:2
Th	3.7	530	43.5	1:2
Ga	4	530	47	1:1
Al	6	530	50	1:1
Sc	6	530	36	1:1
U(VI)	6	530	12	1:2
La	6.5	530	18	1:1

Trihydroxyfluorones, antipyrine, and the anion of a strong acid (A^-) form an ion-pair complex with Ge^{120}:

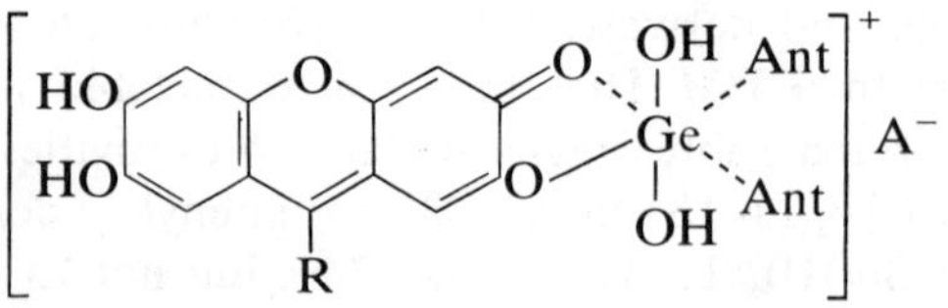

This complex is extractable into chloroform (6 $CHCl_3$ are stated to be bound by a molecule of the ion-pair) at pH 0.8–2. (Phenylfluorone and salicylfluorone have been mainly studied as the trihydroxyfluorones; $\epsilon_{515} = 1.1 \times 10^5$.) Molybdenum(VI) gives an analogous complex (with propylfluorone)

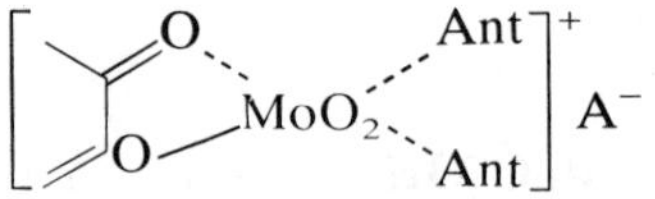

that is extractable into $CHCl_3$ at pH 2.0–2.6 $(\epsilon_{515} = 1.46 \times 10^5)$.[121] Tin(IV) can be determined similarly with trihydroxyfluorones.[122]

Titanium(IV) forms a mixed-ligand complex with benzoylphenylhydroxylamine (BPHA) and phenylfluorone (PF) in dilute hydrochloric acid solution.[123] The complex can be extracted into chloroform in the presence of methanol $(\lambda_{max} = 550$ nm, $\epsilon = 7.5 \times 10^4)$. The complex is reported to have the composition: $Ti(IV)(BPHA)_2 \cdot PF \cdot Cl \cdot 3\ CH_3OH \cdot 4\ CHCl_3$.

Gallein (4,5-dihydroxyfluorescein)

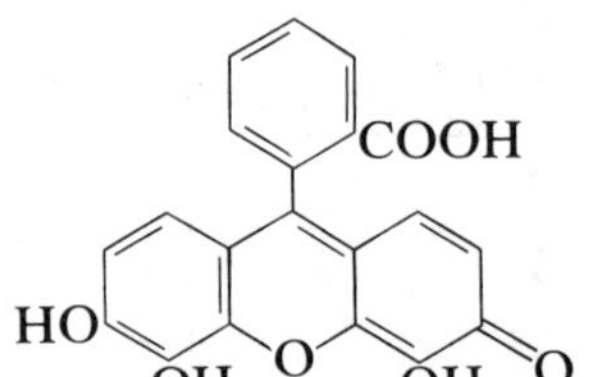

Brownish red powder (1.5 H_2O). Almost insoluble in H_2O, $CHCl_3$; soluble in ethanol and alkaline aqueous solutions. $pK_1(H_2L^+) = 1.8$; $pK_2 = 6.3$ [see further, *Chemia Anal.*, **20**, 339 (1975)].

has vicinal OH groups in its structure and may be expected to chelate the metals that are complexed by phenylfluorone in acidic solution. Zirconium has been reported to form a violet 2:3 complex at pH 2–3. This reaction is of little analytical value. But gallein finds practical use in the photometric determination of tin. It has also been suggested as a Ge and W reagent. Oxidizing substances destroy the reagent.

The isomer, 2,7-dihydroxyfluorescein,

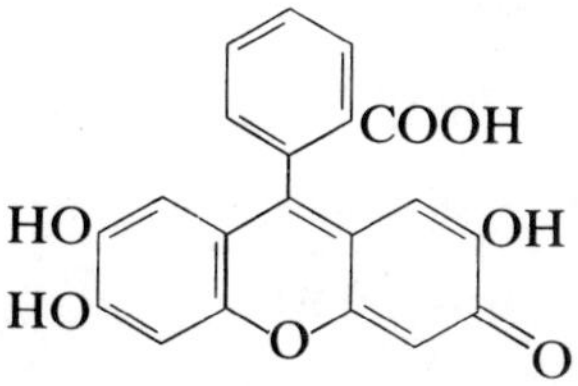

$$pK_1(H_2L^+) = 2.5; \ pK_2 = 6.0.$$

reacts with Sn(IV) at an optimal pH of 2.4, giving a complex 1(Sn):2, extractable into cyclohexanone ($\epsilon_{510} = 108,000$).[124] Zr, Ge, and a number of other metals interfere.

VI. SULFONEPHTHALEINS

Numerous dyes having vicinal OH groups (or OH adjacent to —COOH or C=O) in a chromophoric framework form chelates whose colors are usually different from the dyes themselves and which can serve as photometric reagents. Dyes of the triarylmethane and phthalein series provide many reagents of this type. Aluminon, Catechol Blue, and Gallein are examples of such dyes

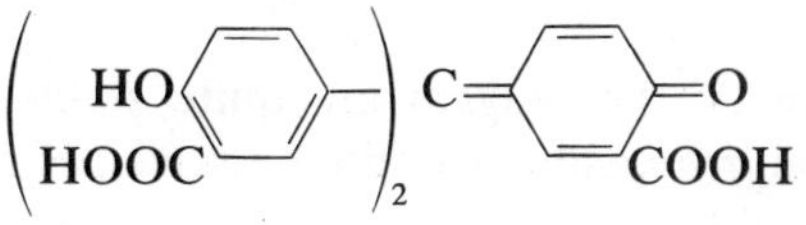

Aurintricarboxylic acid
(Ammonium salt is Aluminon)

that are unsulfonated. Sulfonated dyes are more frequently used, however, principally because of their greater solubility in water. The sulfonephthaleins furnish many useful chelating agents. They are best known as metal indicators in compleximetric (EDTA) titrations, but some have been adapted to photometry. They react with most metals in weakly acidic or weakly basic solution. Their unselective character allows them to be used as reagents for many metals but is a trial for the analyst, and all the resources of differential masking—and finally separations—must be called upon to make them useful in practice.

Xylenol Orange and Methylthymol Blue are examples of sulfonephthaleins containing the group –CH$_2$N(CH$_2$COOH)$_2$, and the metal in their chelates is bonded to N as well as O. (See p. 398.) The presence of N does little to alter selectivity.

A. CATECHOL (OR PYROCATECHOL) VIOLET

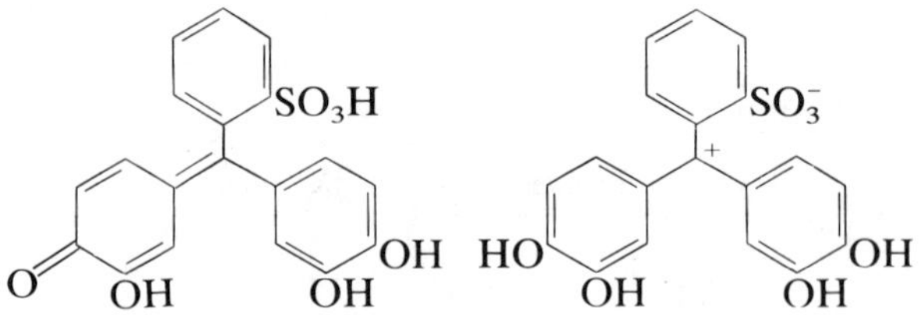

Zwitter-ionic form
Mol. wt. 386.4. Dark red solid.

This reagent, 3,3′,4′-trihydroxyfuchsone-2″-sulfonic acid (catecholsulfonephthalein), forms colored chelates (predominantly blue) with more than 30 metals, mostly in weakly acidic and basic solutions.[125] The adjacent hydroxy groups explain the strong chelating properties of the reagent.

In an aqueous solution of reagent alone, the following color changes occur as the pH is increased (approximate λ_{max} indicated)[126]:

$$H_4L \xrightarrow{\text{pH}\sim1} H_3L^- \xrightarrow{\text{pH}\sim7} H_2L^{2-} \xrightarrow{\text{pH}>10} HL^{3-}, L^{4-}$$

Violet (560 nm) or red (550 nm)	Yellow (460 nm) (loss of proton from one of OH groups *para* to central C)	Violet (620 nm) (loss of proton from second *p* OH)	Blue

Weakly acidic or neutral solutions of the reagent are quite stable, but alkaline solutions are unstable. The irreversible oxidation above pH 5 can be prevented by adding ascorbic acid.

As usual with reagents of this type, with bonding to O, cations of higher charge or metals forming oxocations react in acidic solution. Thus Nb(V), ZrO^{2+}, TiO^{2+}, Sn(IV), Th^{4+}, Bi^{3+}, Sb^{3+}, Fe^{3+}, Ga^{3+}, and MoO_2^{2+} form chelates below pH 5. Divalent cations such as Cu, Zn, Ni, and Co react above pH 7, the rare earth metals maximally above pH 8. The alkaline earths react only slightly at, or below, pH 8. The chelates can have various ratios of reagent and metal. For example, Sn(IV) at pH 2.5 is reported to form at least two species, in which the mole ratios L:Sn are 2 and 0.5. Later work[127] has shown the formation of three tin species: SnH_2L^{2+}, $Sn(H_2L)_2$, and $Sn_2H_2L^{6+}$. H_3L^- is the reacting reagent species but H_2L^{2-} forms the complexes, possibly because its stabilization is induced by highly charged Sn(IV); resonance stabilization may explain the high molar absorptivity of $Sn(H_2L)_2$ ($\epsilon_{555} = 7.96 \times 10^4$).

Bismuth is said to form a doubly charged cation having the composition

2 Bi:1 L. Some of the complexes are insoluble, and a protective colloid is needed to disperse them (e.g., Y). Ion-association complexes can be formed between some Catechol Violet chelates and high molecular-weight organic cations, especially surface-active compounds that increase the sensitivity (see Chapter 6H).

Catechol Violet seems to have utility in the determination of tin after separations have been made. Zirconium has also been determined. Yttrium can be determined sensitively but unselectively at pH 8.5. Germanium reacts even in 1 M HCl. Aluminum is determinable at pH 6.[128] The Al–Catechol Violet–diphenylguanidinium complex, with a ratio of 1:3:3, can be extracted at pH 6.3–7.9 by isobutyl alcohol, giving a blue-green color ($\epsilon_{590} = 5.6 \times 10^4$).[129]

The Catechol Violet on the market varies in quality and some products are grossly impure. One impurity is reported to be Catechol Green. A solution of the reagent in acetate buffer (pH 5.2–5.4) should be lemon yellow in color with no green tinge.[130]

B. BROMPYROGALLOL RED

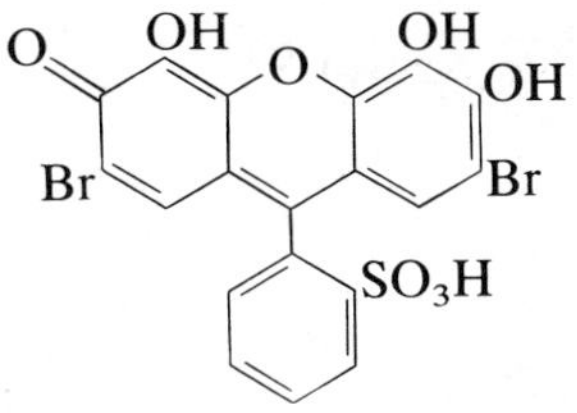

$pK_1 \cdots pK_4 = 0.16$, 4.4 (3.9), 9.1, 11.3. Orange-yellow in acid solutions, red in neutral, and violet to blue in basic. Resists oxidation better than Catechol Violet.

This reagent is seen to resemble phenylfluorone and to have the reactive grouping we have been discussing. It forms a 3:1 chelate with germanium ($\epsilon_{540} = 38,000$) at pH 0.5–4.5 (preferably 2).[131] An ion associate (through the sulfonic acid group) with diphenylguanidinium cation can be extracted into higher alcohols. Many heavy metals react with Brompyrogallol Red in weakly acidic solution. Niobium forms a blue complex; EDTA and tartrate are of help in reducing interferences.[132] Other metals determinable with Brompyrogallol Red include Sb(III), Sc, Ti, Ag, Th, Sn(IV), and the rare earths. For the determination of Sn, this reagent and Pyrogallol Red are less sensitive than phenylfluorone.

For the use of Brompyrogallol Red in ion-association reactions, see Chapter 6H, Part I (determination of Ag).

C. ERIOCHROME CYANINE R AND CHROME AZUROL S

Eriochrome Cyanine R (Mordant Blue 3) is a well-known reagent for determining Al, and has also been used for Be, Ga, R.E., and Zr. In acid

solution it probably has the structure

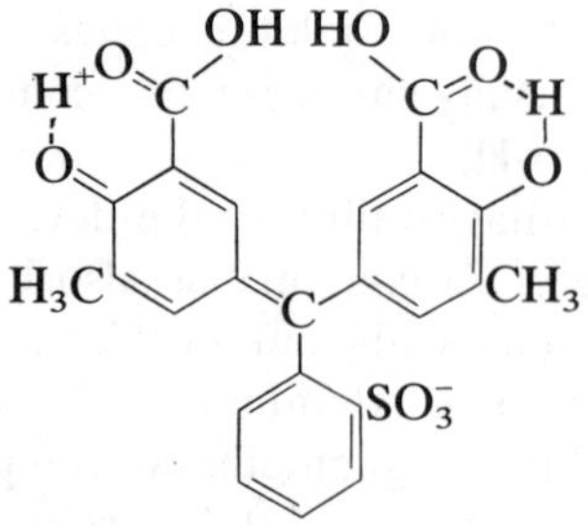

Iron(III) forms a dimeric cation $Fe_2L_2^{2-}$ as well as two monomers in which Fe:L ratios are 2:1 and 1:1.[133] The reagent is of no interest for the determination of iron.

In media of pH 2–5 above 70°C, Cr(III) forms a 1:1 complex, $\lambda_{max} = 540$ nm (at pH 4), $\epsilon = 26,000$.[134] No reaction occurs at room temperature. If, after formation of the complex, the solution is made 1 N in HNO_3, complexes of all metals other than Cr are destroyed at once or (for Al and Zr) within 10 min. $\lambda_{max} \sim 500$ nm.

A reagent similar to the preceding is Chrome Azurol S

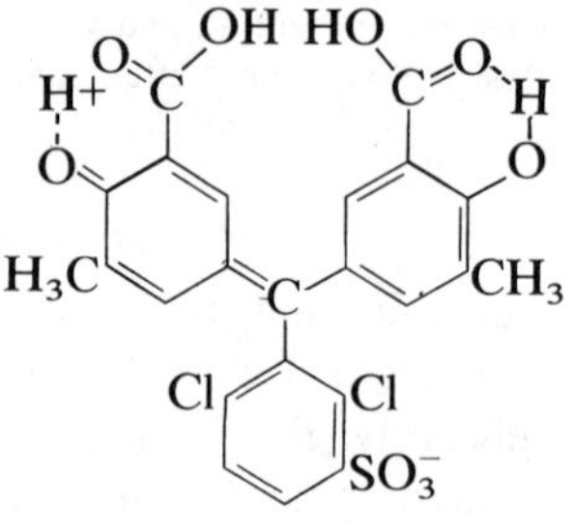

$pK_1 < 0.0, \quad pK_2 = 2.25, \quad pK_3 = 4.71, \quad pK_4 = 11.81.$
Colors $(H_3L^- \cdots L^{4-})$ = orange, red, yellow, violet.

Zwitterion (acid soln.)

It has found use in the determination of Al, Be, Th, U, and other metals. The reagent can be purified by adding HCl to a solution of the impure commercial trisodium salt to precipitate H_4L. The dissociation constants $pK_2 \cdots pK_4$ have been determined.[135] The reaction of Chrome Azurol S with Fe(III) in the pH range 2–4 has been studied by Langmyhr and Klausen,[135] who found four complexes:

$[Fe(H_2O)_2]_2L_2^{2-}$ (ring structure)
$Fe(H_2O)_4HL$ or $Fe(H_2O)_4L^-$
$[Fe(H_2O)_4]_2L^{2+}$
$Fe(H_2O)_2H_xL_2^{x-5}$

The formation constants of these species have been determined. Hydroxo complexes of the dimer are formed at higher pH values.

Beryllium forms a 1:1 complex.[136]

The joint addition of Chrome Azurol S and cetyltrimethylammonium bromide to a Th solution at pH ~4.5 gives a blue color that is stronger than that given by Chrome Azurol alone.[137] A bathochromic shift in absorption maximum occurs (554 → 635 nm) in this ion-association type of reaction. The reaction is sensitive ($\mathscr{S}_{635} = 0.002\ \mu g$ Th/cm^2, corresponding to $\epsilon = 1.18 \times 10^5$), but many metals interfere.

VII. HYDROXYFLAVONES

The parent compound, flavone, has the structure

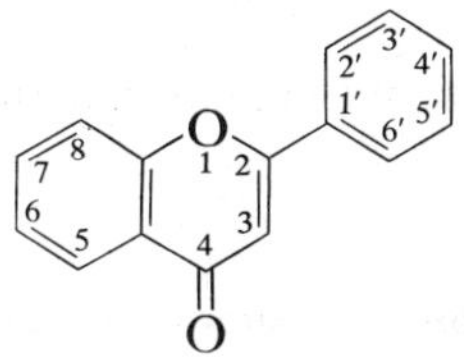

Flavonols contain an OH group at position 3. Most of the analytically useful hydroxyflavones are flavonols, and these usually also have an −OH in the 5-position. The following hydroxyflavones, among others, have been used analytically[138]:

Flavonol	3-hydroxyflavone
	3,7,4′-trihydroxyflavone
Kaempferol	3,5,7,4′-tetrahydroxyflavone
Morin	3,5,7,2′,4′-pentahydroxyflavone
Quercetin	3,5,7,3′,4′-pentahydroxyflavone
Rutin	Quercetin-3-rutinoside
Galangin	3,5,7-trihydroxyflavone
Robinetin	3,7,3′,4′,5′-pentahydroxyflavone
Myricetin	3,5,7,3′,4′,5′-hexahydroxyflavone
Quercetagetin	3,5,6,7,3′,4′-hexahydroxyflavone

Hydroxyflavones having −OH in the 3- or 5-position or two *ortho* OH groups form chelates with what may be called oxyphilic elements such as Al, Be, Ga, Hf, In, Fe, Mo, Nb, rare earths, Sc, Ta, Th, Sn, Ti, U, V, W, and Zr. Univalent and divalent cations [alkaline earths, Hg, Tl(I), Ag, Pb, Co, Ni, Cu, Cd, Zn] usually form less stable chelates, and small amounts of these may be tolerated in weakly acidic solutions when trivalent or quadrivalent elements are determined fluorimetrically or spec-trophotometrically. The chelates of colorless cations are usually yellow (alcoholic solutions of the reagents are pale yellow, with absorption bands in the 320–380 and 240–270 nm regions). Most of these metals give fluorescent chelates, which provide very sensitive determination forms. Selectivity can be improved by pH adjustment and, of course, by differential complexing. Also, the number and arrangement of the OH groups in the reagents can affect reactivity.

5-Hydroxyflavones form six-membered rings

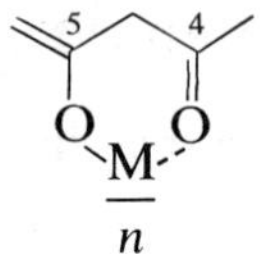

$$ \underset{n}{\overset{\displaystyle 5 \qquad 4}{\underset{\displaystyle O \diagdown M \diagup O}{}}} $$

and 3-hydroxyflavones form five-membered rings[139]:

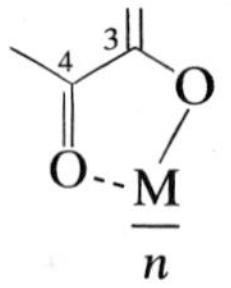

(M can be an oxo- or hydroxo-metal cation; ZrO^{2+} can form a 1:2 chelate

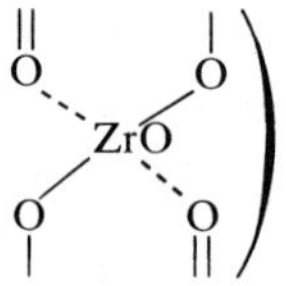

By the use of various flavones in metal reactions, the centers of chelation and the binding strengths can be found. Thus zirconium forms a 1:1 or 1:2 (Zr/reagent) chelate with 3-hydroxyflavone that is ~100 times more stable than the chelate formed with 5-hydroxyflavone.[140] 3',4'-Dihydroxyflavone does not react with zirconium in acidic solution. This order of decreasing chelating power seems to hold for most metals, two *ortho* OH groups having the lowest chelating ability. However, niobium does not chelate with 5-hydroxyflavones; it reacts in acid solution with an OH group and the carbonyl group *ortho* to each other, and in neutral solution it can react with *ortho* OH groups, for example, 4' OH and 5' OH.[141]

Dimethylkaempferol, in which the 3- (and 4'-)positions are blocked, does not react with Ga and In, whereas kaempferol does. Hence the conclusion is drawn that OH in the 3-position participates in the chelation of these (and many other) metals.[142]

The relations between structures of hydroxyflavones, their metal reactivity, and the fluorescence of the products require further study. Present knowledge indicates that metals (including Zr, Th, Ga, Al, Sc, Ge, Nb, and Ta) are chelated by the *o*-hydroxy-carbonyl group, and hydroxyflavones not having this group are unreactive. In neutral and alkaline solution, the reactive group can be perihydroxy-carbonyl or vicinal OH. The *o*-hydroxyquinone group is believed to be responsible for fluorescence, that is, bonding of metal to oxygen attached to C in the 3- and 4-positions gives rise to fluorescence (but note the intense fluorescence produced by reaction of Be with morin in sodium hydroxide solution).

The ratio of metal to hydroxyflavone in the chelate species is often 1:1 or 1:2, but can be higher, and, for a particular metal, can vary with the acidity. Some of these species must be charged. A number of them are extractable into oxygenated solvents. Even if uncharged, chelates formed by morin, quercetin, and other polyhydroxyflavones would not be extractable into inert solvents such as $CHCl_3$.[143] Since some polyhydroxyflavones have two chelating centers, it is possible for 2:1 metal/reagent chelates to be formed. Thus robinetin

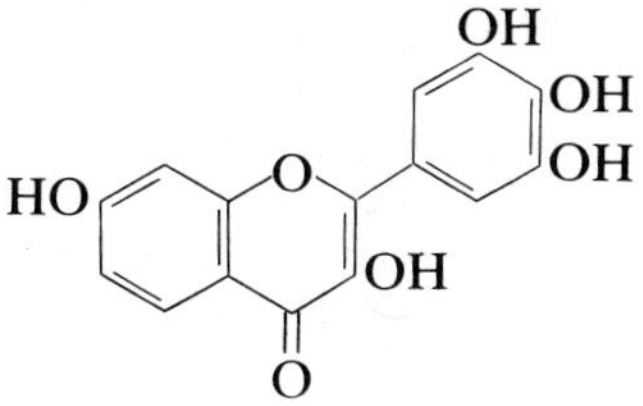

forms a 2 U:1 Rob (as well as a 1:1) complex with U(VI).[138] One U atom is chelated by $-\overset{\|}{\underset{O}{C}}-\overset{|}{\underset{OH}{C}}=$, the second by $-\overset{|}{\underset{OH}{C}}=\overset{|}{\underset{OH}{C}}-$. Polynuclear species are also a possibility and could account for 2:1 metal/reagent ratios.[144]

Investigation of the morin and quercetin chelates of scandium has revealed that several 1:1 species can be formed, some fluorescent, some not[145]:

	Coordinating Ion	pH Range	Charge	Fluorescence
Morin	$ScOH^{2+}$	1.2–2.8	+	Yes
	$Sc(OH)_2^+$	6.0–7.6	0	No
Quercetin	$ScOH^{2+}$	5.5–6.0	+	Yes
	$Sc(OH)_2^+$	6.0–6.5	0	Yes
	$Sc(OH)_2^+$	7.6–8.8	0	No
Rutin	$Sc(OH)_2^+$	6.5–7.1	0	No

The following structures have been assigned to these chelates:

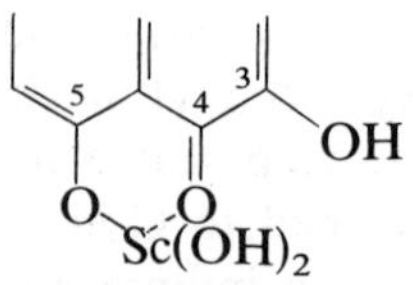

Uncharged, nonfluorescent complexes of Sc with morin, quercetin, and rutin (latter has —OR in 3-position).

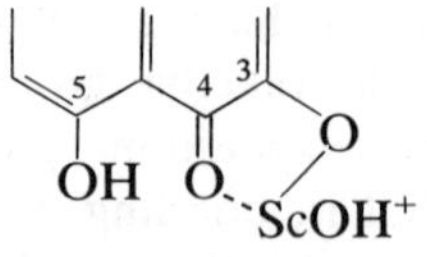

Positively charged, fluorescent complexes of Sc with morin and quercetin. Uncharged, fluorescent complex with quercetin probably has analogous structure.

Hydroxyflavone reagent solutions are usually prepared in ethanol. Occasionally, sulfonated hydroxyflavones are used as reagents.[146]

A.   MORIN

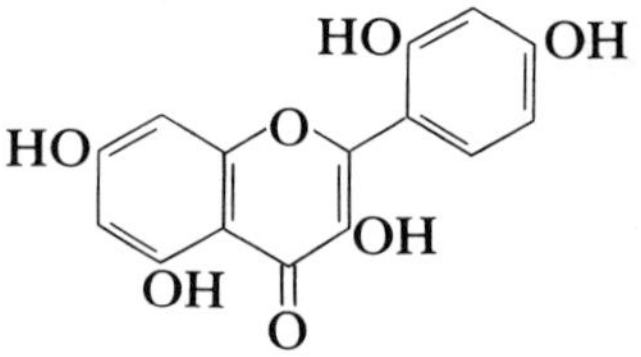

Mol. wt. ($+2H_2O$) 338.3. M.p. 285°. Can be recrystallized by forming HBr derivative in acetic acid and decomposing this with boiling water. Products >99% pure are commercially available. Morin solutions in alcohol are stable for months. Approximate pK values ($pK_1 \cdots pK_5$) are -1, 4.8, 7, 9, 12 (M. H. Fletcher).

This is probably the polyhydroxyflavone finding the most frequent use. It is of particular interest for the fluorimetric determination of Be, Zr, and Th[147]. Various other metals also give fluorescent morinates (Table 6B-1). Most metals react most sensitively in weakly acidic solution. Zirconium is distinguished by its sensitive reaction in mineral acid solution (2 M HCl), and traces can be determined very satisfactorily. A 1:2 morin complex is predominantly formed.[148] Less acidic solutions are required for the determination of thorium and scandium, and they are determined less satisfactorily in acidic solution. Aluminum requires a weakly acidic medium, and doubtless other metals could be determined in such medium. Magnesium reacts at pH ~10. W(VI) forms a 1:1 complex at pH 5.

The determination of Be rests on its ability to react with morin in alkaline solution. In rather strongly basic solution (~0.02–0.3 M NaOH), morin, with the aid of masking agents, becomes an almost specific reagent for Be. Amphoteric metals such as Al, Ga, and In, and some others, do not fluoresce if the hydroxyl ion concentration is sufficiently high. Zinc fluoresces weakly under these conditions, but cyanide prevents all reaction. Lithium fluoresces very weakly. The weak fluorescence of Ca is destroyed by EDTA, which also prevents the feeble fluorescence of other elements (cf. footnote *a*, Table 6B-1). Diethylenetriaminepentaacetic acid (DTPA) is also used as a masking agent and is more effective than EDTA for some metals. Separation of Be by acetylacetone-chloroform extraction is a useful preliminary step in the analysis of many materials for Be.

Three principal species are reported to be formed in the Be–morin reaction in basic solution: BeL, Be_2L_2, and BeL_2 (charges omitted).[149]

Thorium can be determined sensitively by fluorescence in basic medium if DTPA, triethanolamine, and citrate are added to prevent or reduce the interference of most other elements. Beryllium interferes, Zr(Hf), Ti, U, and some other elements less so, but separations can be made. The determination of Th is less satisfactory than that of Be, but nevertheless has value in practice.

Morin has also found use for the *absorptiometric* determination of Th (yellow color), Ga, In, Zr, U (reddish brown in ammonial solution), and other elements, but fluorimetric determinations are more sensitive when applicable. Titanium forms a yellow-to-orange product in acid solution, which is not fluorescent. Tin can be determined absorptiometrically at pH 2. In 25% by volume of sulfuric acid, Ta gives an orange precipitate and Nb a red precipitate with morin.

Ternary complexes of morin and metals have been investigated. Morin in methanolic solution reacts with V(IV) and antipyrine to give a complex having the composition 1V(IV): 2 Mor:1 Ap, extractable into di-chloroethane at pH 4.5.[150] In the presence of anions of strong acids (preferably ClO_4^-), Sc at pH 3.3 forms a chloroform-extractable complex with morin and phenazone and can be determined absorptiometrically or fluorimetrically in this way, though the method lacks selectivity.[151] Lutetium can be determined fluorimetrically in the presence of Yb, Tm, Er, or Ho by extraction into chloroform from a perchlorate solution containing morin and diantipyrinylmethane.[152] In the presence of an-tipyrine and ClO_4^-, indium reacts with morin to form a mixed-ligand complex that is extracted by chloroform at pH 4.3–4.7 and can be determined fluorimetrically or absorptiometrically.[153] The following formula is suggested for the complex:

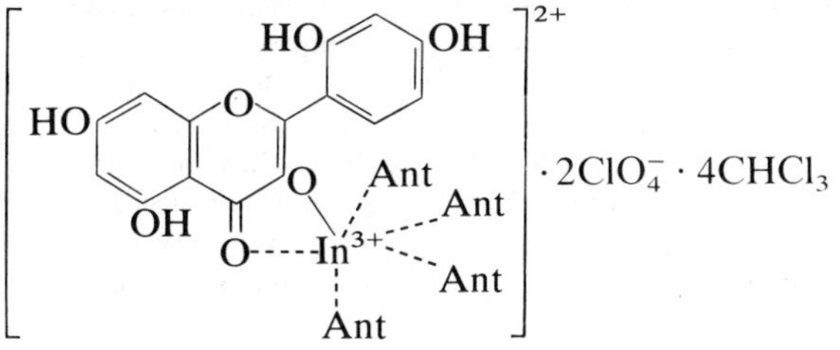

TABLE 6B-1
Visual Sensitivity (mole/liter) of Fluorescence Reactions of Morin with Various Metals[a]

	1 N HCl	pH 3	pH 4.5
Zr	10^{-4}	10^{-4}	10^{-4}
Ti		No fluorescence	
Th	10^{-3}	10^{-3}	10^{-4}
Ga	10^{-2}	10^{-3}	10^{-5}
Al	10^{-1}	10^{-3}	10^{-4}
In	—	10^{-4}	10^{-4}
Sc[b]	10^{-3}	$10^{-4.3}$	$10^{-4.3}$
Y	—	10^{-2}	10^{-3}
Ce(III)	—	10^{-2}	10^{-4}
Be	—	1	10^{-1}
Sb(III)	10^{-3}	—	—
Sn(IV)	—	10^{-4}	10^{-5}

[a] G. Charlot, *Anal. Chim. Acta*, **1**, 233 (1947). Values pertain to qualitative conditions: about 10 drops of solution. See Zirconium (*Colorimetric Determination of Traces of Metals*) for other comparative fluorescence intensities. Relative fluorescence intensities in basic solution containing EDTA may be judged from some figures tabulated by I. May and F. S. Grimaldi, *Anal. Chem.*, **33**, 1251 (1961). In ~0.035 M NaOH containing 0.5% EDTA, the following weights of oxides (μg) give the same fluorescence intensity as 1 μg Be with the filter combination used: Th, 600; Lu, 700; Y, 700; Zr, 5000; La, 5000; Sc, 7000. It may be noted in passing that Be fluorescence is weakened in the presence of Ce(III), Cr(III, VI), Er, Tb, Sm, Pt(IV), and U(VI) (weak effect). Sill, Willis, and Flygare[147] compared the reactivity of metals in alkaline solution in the presence of EDTA and DTPA, respectively, and with DTPA as masking agent obtained the following relative fluorescence intensities based on Be = 1: Th, 0.02; Zr, 0.009; Sc, 2×10^{-5}; La, 8×10^{-6}; Y, 6×10^{-7}; Li, 4×10^{-7}. DTPA is the more effective masking agent (in determination of Be) for Sc, La, and Y. The rare earth elements decrease the fluorescence of Be when EDTA is used as masking agent, presumably by consuming morin to give complexes fluorescing slightly or not at all. DTPA allows Be to be determined in the presence of the rare earth metals.

[b] Sensitivity in 2 N NaOH is 10^{-4}. This and other sensitivities in the table are of interest more from the relative than the absolute standpoint. With UV radiation (instead of daylight as here), limits are much lower.

There is good reason to believe that the 1 Th:1 morin fluorescent complex formed in the presence of DTPA (to mask other metals as mentioned in the foregoing pages) is a mixed chelate in which DTPA is also bound to Th. DTPA is known to bind Th firmly, and it seems unlikely that morin could displace all the DTPA ligands.

Most (simple) metal–morin chelates can be extracted into isoamyl alcohol. Some of the extracts are fluorescent. Differences in the pH of extraction are of aid in increasing selectivity. Metals that can be extracted from acid solution ($\sim 1\,M$ HCl) include Sn (II, IV), Zr, Hf, and Mo. See Table 6B-2.

Morin alone shows practically no fluorescence below pH ~ 3 in 50% alcohol[154]; at pH ~ 4 it shows an emerald green fluorescence, which remains unchanged with increasing pH until a value of 8–9 is reached, when the fluorescence becomes yellow-green.[155] These fluorescence intensities are much weaker than those of the metal complexes.

The slow decrease in fluorescence intensity of the Be–morin complex on standing has been attributed to air-oxidation of morin catalyzed by Cu(II) and other metals as well. Copper can be inactivated by EDTA (or DTPA) or stannite. Morin is oxidatively decomposed more rapidly in basic than in acid or neutral solutions.

B. QUERCETIN

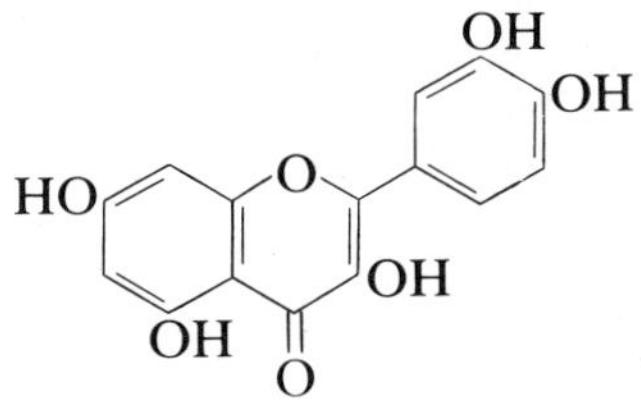

Mol. wt. ($+\ 2H_2O$) 338.3.
M. p. $> 300°$ (316–317°).

Quercetin differs from morin only in having an OH group in the 3′ position instead of in the 2′. It would be expected that the two reagents would be quite similar in their metal reactions, which for the most part involve 3- and 5-OH. This is true to a large extent, and it seems that in many reactions one could be used in place of the other without much difference. However, there appear to be differences in some reactions.

Quercetin, like other hydroxyflavones, is a feebly acidic substance. With strong acids, it can form orange salts, but this equilibrium is of no importance analytically except perhaps in reactions with zirconium, carried out at low pH, in which it may need to be taken into account in formulation of the reaction system.

Quercetin has been used for the spectrophotometric or fluorimetric

Extraction of Morin Chelates with Isoamyl Alcohol[a]

	pH for 80–100% extn.	Organic Phase[b]	
		Color	Fluorescence
Be	6.3–7.6	Yel	Gr (int)
Al	5.1–6.2	Yel (int)	Gr-yel (int)
Sc	4.0–9.0	Yel (int)	Yel (pale)
Y	5.3–7.1	Yel (int)	Gr-yel (int)
Eu	7.5	Yel (int)	
Ge		Yel (pale)	
Sn(II)	<0(4 M HCl)–6.5	Yel	
Sn(IV)	~0(1 M HCl)–2.0	Yel (int)	Gr-yel (pale)
Pb	4.4–7.1	Yel	Gr-yel
Sb(III)		Yel	
Sb(V)		Yel	
Bi	5.2–7.2	Yel (int)	
Cu(II)	5.5–7.5	Yel	
Ag	7.0–10.3	Yel (pale)	Yel (pale)
Au(III)		Yel–gr	
Cd	5.6–8.5		
Hg(I)	5.4–6.5		
Hg(II)	4.1–5.8		
Ga	2.7–6.4	Yel	Gr
In	4.9–7.5	Yel	Gr
Ti	2.6–7.0	Yel (int)	Yel (pale)
Zr	<0(3.5 M HCl)–9.2	Yel (int)	Gr-yel (int)
Hf	~0–9.0	Yel (int)	Gr-yel (int)
Th	1.7–9.1	Yel (int)	Gr (int)
V	3.0–3.8	Yel-brown	Yel (pale)
Nb	6.7	Yel (int)	Gr-yel
Ta	3.0–6.7	Yel (int)	Gr (pale)
Cr(III)		Yel	Yel (pale)
Cr(VI)	3.4–4.4	Yel (int)	Yel (pale)
Mo(VI)	<0(6 M HCl)–3.4	Yel	
W(VI)		Yel-gr	Gr
U(VI)	7.5	Yel	
Mn(II)		Yel (pale)	
Fe(III)	1.5–8.4	Brown (int)	
Co	7.5	Yel (pale)	
Ni		Yel (pale)	Yel (pale)

[a] A. B. Blank, I. I. Mirenskaya, E. S. Zolotovitskaya, and E. I. Yakovenko, Zh. Anal. Khim., **26**, 656 (1971). Blank and coworkers, Zh. Anal. Khim., **28**, 1331 (1973) and **31**, 703 (1976), use isopentyl alcohol to extract the 1:2 Zr–morin complex from 0.1 M HCl ($\epsilon = 70{,}600$) and TBP with isopentyl alcohol to extract the 1:2 complex of Eu at pH 3–5.5 ($\epsilon \sim 4700$).

[b] $V_{org}:V_{aq} = 0.5$. Yel = yellow, Gr = green, int = intense. Alkaline earth metals, Cu(I), Fe(II), and Tl(I) essentially not extracted.

determination of Zr, Nb, Sn, Th, V, Mo, and U. A method for Hf in the presence of Zr is based on the differential complexing of the metals with hydrogen peroxide.[156]

It is interesting that quercetin reacts with Mo α-benzoinoximate in $CHCl_3$ if alcohol is added.[157]

Thorium forms a 1:1 complex in weakly acidic solution and can be determined spectrophotometrically (425 nm) and fluorimetrically (541 nm).[158]

Scandium gives two fluorescent species at pH 5.5–6.5.[159] This element can also be determined colorimetrically (pH 4.4, $\epsilon_{435} = 12,800$).[160] Borate fluoresces at pH 7–11.

Titanium can be determined spectrophotometrically in weakly acidic solution ($\epsilon_{420} \sim 16,000$) with quercetin sulfonic acid. Tl(III) gives a violet complex in faintly acidic solution.

Pure quercetin can be prepared from dihydroquercetin, which is commercially available.[161]

C. OTHER HYDROXYFLAVONES

3,7,4'-Trihydroxyflavone is a better fluorimetric reagent than morin for thorium, because it gives about 1/400 as strong a fluorescence with beryllium as does morin, while having much the same sensitivity for thorium. Evidently, the OH in the 5-position in morin contributes, directly or indriectly, to the beryllium reaction. Trihydroxyflavone has also been used for the fluorimetric determination of Sb(III), Sn(IV)[162], and Zr,[163] all of which give sensitive fluorescence reactions in dilute H_2SO_4 solutions. Other metals fluorescing quite sensitively include W, Ga, and Al. Sn(IV) is reported to fluoresce in $\sim 0.1\ M$ NaOH (difference from morin).

Kaempferol has been suggested as a reagent for antimony(III).[164] The sensitivity is good ($\epsilon_{420} = 10,900$) but the selectivity is poor, even in 0.1 M HCl. Interferences arise especially from the presence of Zr, Ti, W, Mo, and Fe(III) in very small quantities. Kaempferol has also been used as a reagent for scandium and for Sn(IV), the latter in 0.1 M HCl.[165]

Flavonol has been used for the spectrophotometric determination of tin(IV)[166] and fluorimetrically for W.[167]

VIII. OTHER REAGENTS CHELATING METALS THROUGH OXYGEN BOUND TO CARBON

A number of compounds not readily falling into the classes of the preceding discussion are collected here. With some exceptions, such as hematoxylin, they have not seen much use in photometric metal determinations. There may be meritorious reagents among them, and some may be worth further study.

A. HEMATOXYLIN

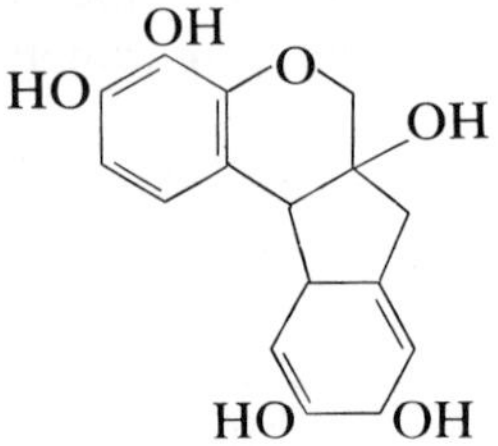

Mol. wt. 302.3. M. p. 148°–150°C (monohydrate).

Hematoxylin has been used for the colorimetric determination of Al, Hf, Sn, Nb, V(V), and some other metals.[168] The reacting substance is actually hematein (hematin). Hematoxylin (colorless) is easily oxidized in air to reddish brown hematein, which forms blue, violet, or red insoluble

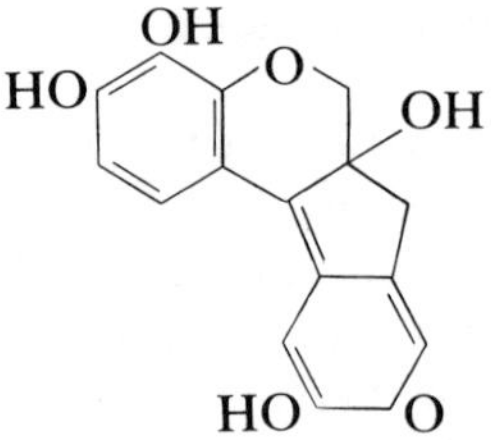

chelates with many metals. As usual with phenolic reagents, chelates of hematein with elements in groups III, IV, and V of the periodic table are more stable than others and can be formed in weakly acid solution.

Hafnium can be determined sensitively ($\mathscr{S}_{520} = 0.003$ μg Hf/cm^2) at pH $\sim$2 as red-brown HfO(HL)$_2$ (H$_2$L = hematein).[169] Zirconium absorbs much less strongly so that Hf can be determined in its presence if Hf > Zr (a situation not often encountered in practice). Niobium forms a 1:1 violet complex (pH 1).

Hematoxylin forms 1:1 and 2:1 complexes with Sn(IV) in methanol–0.6 M HCl.[170] V(V) in 2 N H$_3$PO$_4$ gives a blue color.

Hematein is readily prepared by oxidizing hematoxylin with hydrogen peroxide in alkaline solution.[171] It is a dibasic acid (pK_1 = 6.58, pK_2 = 9.67), with a color change from yellow to red-purple. Its aqueous solutions are unstable. In basic solution, it undergoes further air-oxidation. In acid solution, an H$^+$-catalyzed reaction produces dihydroxyhematein. Organic solvent solutions (alcohol) are more stable. Molar absorptivities of hematein chelate suspensions have high values (e.g., ϵ_{max} of tin is

76,000) but except for sensitivity, hematein has drawbacks as a colorimetric reagent.

B.   HYDROXYCHROMONES

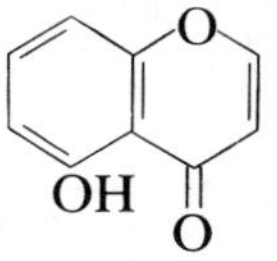

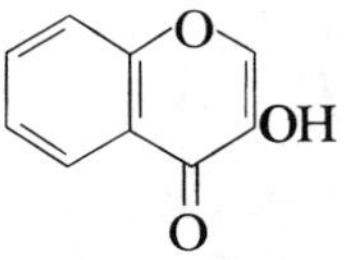

5-Hydroxychromone   3-Hydroxychromone

5-Hydroxychromone (pale-yellow crystals, m. p. $125.5° - 127.5°$, $pK_a =$ 10.75)[172] and 3-hydroxychromone (pale-yellow crystals, m. p. 178.5°, $pK_a = 8.37$)[173] react with many metal ions at various pH values. The reactions with iron(III) have been studied in some detail. 5-Hydroxychromone (HL) forms three complexes, FeL^{2+}, FeL_2^+, and FeL_3·HL, depending on pH[174]; with 3-hydroxychromone 1:1, 1:2, and 1:3 metal/reagent complexes are formed.[175] 5-Hydroxychromone has been used for the extraction-spectrophotometric determination of iron, beryllium,[176] and palladium.[177] The palladium(II) complex has a metal/reagent ratio of 1:2. Iron(III) and titanium interfere with the determination of both beryllium and palladium. 2-Ethyl-3-methyl-5-hydroxychromone can be used for the extraction-fluorimetric determination of beryllium; the method is sensitive ($0.01-0.25$ μg Be in 10 ml of carbon tetrachloride) but unselective.[178]

C. o-DIHYDROXYCHROMENOLS

These compounds form chelates with Ge, W, and other metals in which the latter (often as a hydroxo cation) replaces an H of the adjacent hydroxy groups and forms a five-membered ring.

6,7-Dihydroxychromenol (6,7-dihydroxy-2,4-diphenylbenzopyranol or, in salt form, 6,7-dihydroxy-2,4-diphenylbenzopyrilium chloride) is a reagent of this class. The following equilibria exist in aqueous solution ($R = C_6H_5$):

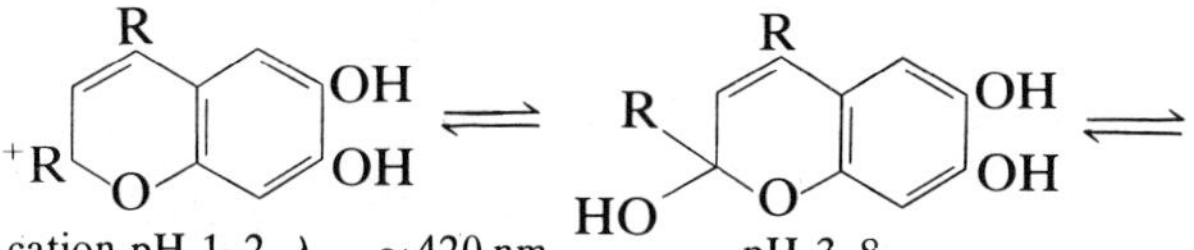

Benzopyrilium cation pH 1–2. $\lambda_{max} \sim 420$ nm.   pH 3–8.

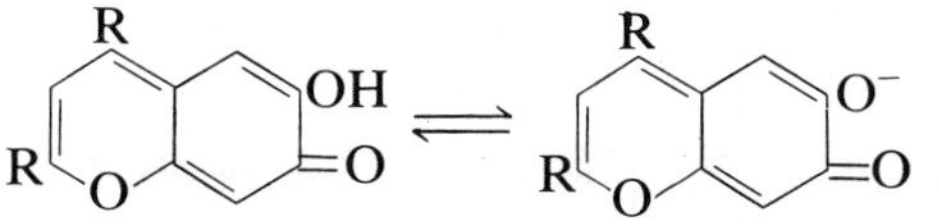

Anhydro base   Anion pH 10–13.

Germanium forms two chelates,[179] having Ge:reagent = 1:3 at pH <3 and 1:2 at pH >5:

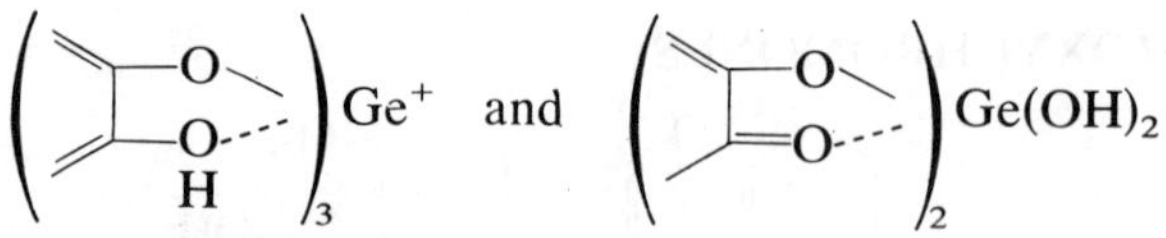

The first chelate can be extracted into $CHCl_3$ (λ_{max} = 535 nm, ϵ = 12,000) as an ion-pair compound with ClO_4^- as anionic partner of the cation. Excess reagent in $CHCl_3$ shows low absorbance. 6,7-Dihydroxy-2,4-dimethylbenzopyrilium reacts similarly, as does 7,8-dihydroxy-2,4-diphenylbenzopyrilium chloride.

Tungsten(VI), as WO_2OH^+, reacts with 6,7-dihydroxychromenol to form a 1:1 complex.[180]

The o-dihydroxychromenols and 2-thenoyltrifluoroacetone form ternary complexes with the rare earth metals.[181]

6,7-Dihydroxy-2,4-diphenylbenzopyrilium chloride

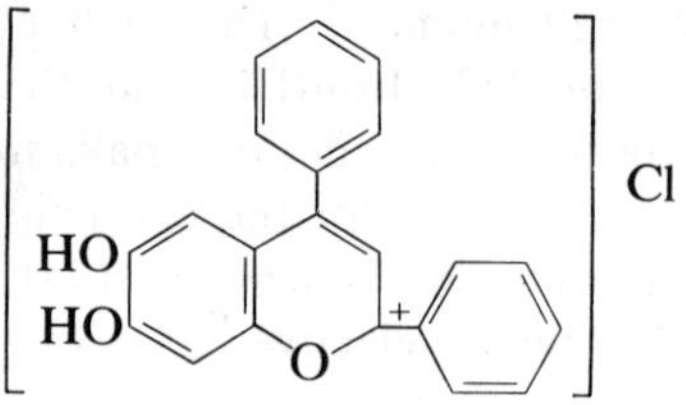

gives a sensitive reaction with germanium.[182] A 0.1 M HCl solution is used, with gelatin as protective colloid. $\mathscr{S}_{500}$ = 0.0013 μg Ge/cm^2.

D. HYDROXYCOUMARINS[183]

Esculetin, 6,7-dihydroxycoumarin,

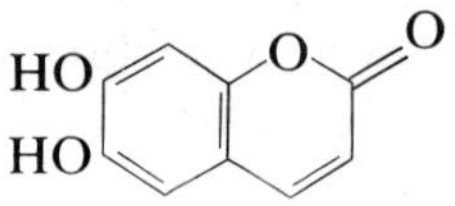

and daphnetin, 7,8-dihydroxycoumarin,

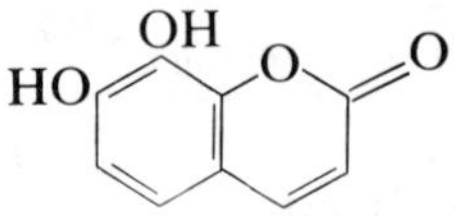

and their derivatives have received some attention as absorptiometric reagents. The vicinal –OH groups lead to chelate formation with such metals as Ge, Fe, Mo, Nb, Ti, and U. A 1:4 Ge complex is formed at pH 7; a 1:3 complex, at pH 10. The chelates are often quite strongly colored, but the reactions are not selective (nor would they be expected to be). It does not seem that these reactions have much analytical value.

A related compound is 6,7-dihydroxycoumarone-3, which chelates

$>$Ge(OH)$_2$ to form

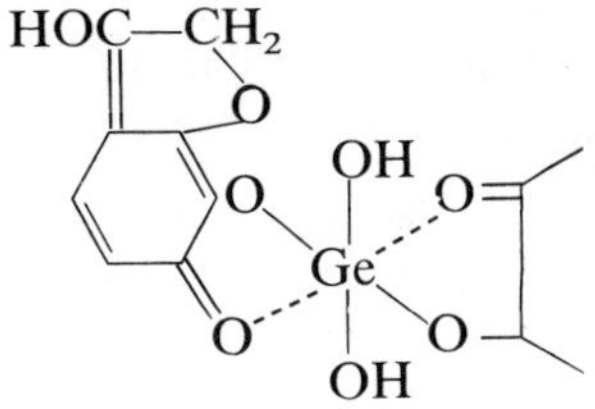

at pH 3.5–6.[184] The complex is colorless, having maximum absorbance at 320 nm ($\epsilon_{330} = 16{,}700$).

E.   TROPOLONE AND DERIVATIVES

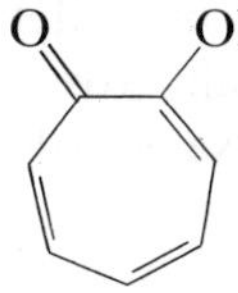

The parent compound has been suggested for the colorimetric determination of a number of metals,[185] but does not seem to be of great practical interest or potential except for vanadium (blue complex, extractable into CHCl$_3$ from 6 M HCl, $\mathscr{S}_{590} = 0.011 \ \mu g/cm^2$).[186]

β-Isopropyltropolone (separatory extraction reagent, Chapter 9B) is more stable than tropolone. It has been used for the extraction determination of V in 1 M HCl, but Fe in any amount must be separated.[187]

Purpurogallin (2,3,4,6-tetrahydroxy-5H-benzocyclohepten-5-one)

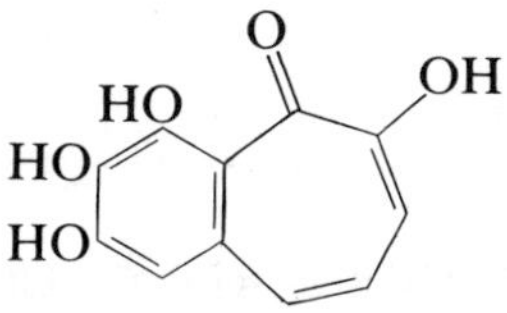

$pK_a = 6.35.$

forms a rose-colored 2:1 (Ge) chelate (insoluble) with germanium in mineral acid solution.[188] Germanium, as $>$Ge(OH)$_2$, is bonded to hydroxy O and carbonyl O of the tropolone ring. This reagent is less sensitive than trihydroxyfluorones for germanium, but more selective, because the acidity can be higher. In 3 M HCl, Nb, Ta, Zr, much Ti, and much Fe(III) react, but Sn and Sb in amounts below 4 μg/ml do not interfere (ferric iron can be reduced with ascorbic acid). $\mathscr{S}_{340} = 0.002$. Purpurogallin is also a reagent for Zr.

F. MISCELLANEOUS

3,4-Dihydroxynaphthyldiphenylmethanol[189]

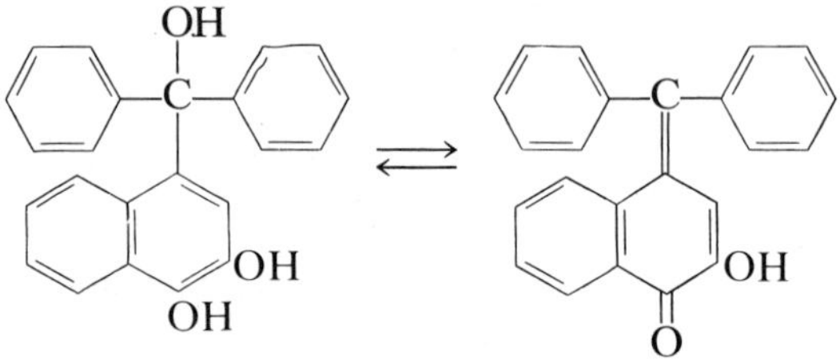

forms a 1:1 complex with germanium in strongly acid solutions by reacting as the orthohydroxyquinone. This reagent does not seem to have been studied from the viewpoint of analytical application.

Gossypol, 1,1',6,6',7,7'-hexahydroxy-5,5'-diisopropyl-3,3'-dimethyl-[2,2'-binaphthalene]-8,8'-dicarboxaldehyde,

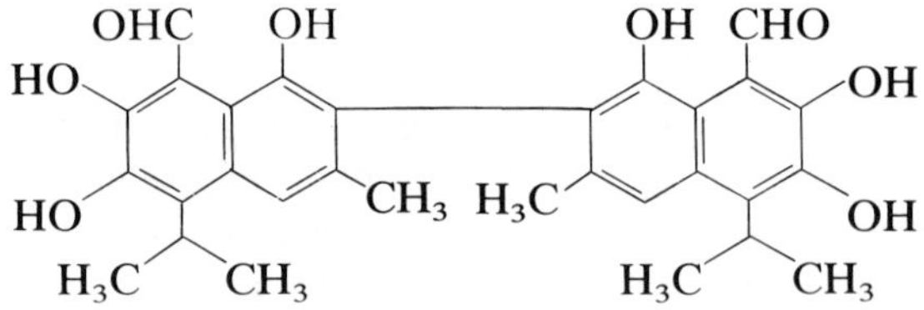

is of interest as being a reagent that forms a germanium complex extractable into chloroform from 2.8–6.4 M sulfuric acid solutions.[190] It is also a tin(IV) reagent; Sb(III) and Mo give colors.[191]

1,2-Dihydroxycyclobutenedione (or 3,4-dihydroxy-3-cyclobutene-1,2-dione), known as squaric acid,

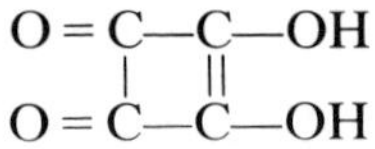

$K_1 = 0.5, \ K_2 = 1.3 \times 10^{-3}$.

forms colored chelates with Fe(III)(red-violet), U(VI) (yellow-green), Cu(II) (yellow-green), and other metals.[192] It has not received much study as a colorimetric reagent, but it would seem to be suitable for determination of iron. But for this metal there is already a surfeit of reagents. It has been suggested for the determination (insensitive) or U(VI) ($\epsilon_{400} = 1450$),[193] which forms a 1:1 chelate at pH 2.5.

IX. REAGENTS CHELATING METALS THROUGH OXYGEN BOUND TO ELEMENTS OTHER THAN CARBON

Metals can be bonded to oxygen bound to P (Chapter 9C), As (Chapter 6E), or N. It this section, some photometric reagents are considered in which one of the chelating oxygens is bound to nitrogen. Chelation to O–N groups usually takes place through N, not O, as in –NO groups. Hydroxamic acids are an exception. At least the oxyphilic metals bond to them through oxygen.

A. HYDROXAMIC ACIDS

Compounds containing the grouping

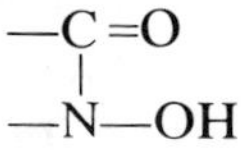

$$
\begin{array}{c}
-C=O \\
| \\
-N-OH
\end{array}
$$

react with heavy metals (especially the oxyphilic of higher valence) to form chelates of the type

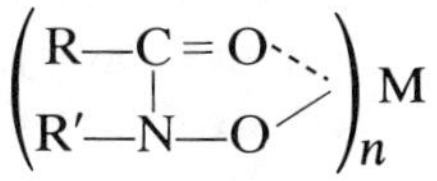

$$
\left(
\begin{array}{c}
R-C=O \\
| \\
R'-N-O
\end{array}
\right)_n M
$$

The most useful hydroxamic acids are those in which both R and R' of

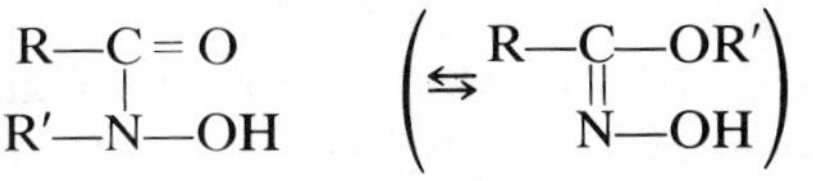

$$
\begin{array}{cc}
R-C=O & \left(\leftrightarrows \; R-C-OR' \right) \\
| & \\
R'-N-OH & \left(\quad\quad\; N-OH \right)
\end{array}
$$

are organic groups (alkyl, aryl, thienyl, etc.), instead of R' being H (as it is in benzohydroxamic acid and salicylhydroxamic acid). The former give chelates generally extractable into chloroform, whereas the latter give chelates usually requiring the less desirable oxygen-type solvents for extraction.

Benzohydroxamic acid

$$
\begin{array}{c}
C=O \\
| \\
NHOH
\end{array}
$$

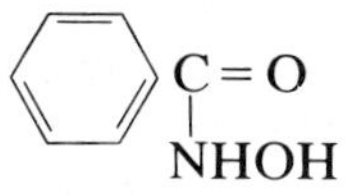

has been used for the determination of V, Mn, Nb[194] and U,[195] the latter giving an orange complex at pH ~6, extractable into 1-hexanol. Salicylhydroxamic acid is a similar reagent. Both benzohydroxamic and salicylhydroxamic acid have been studied as reagents for Mo(VI).[196]

Nicotinhydroxamic has been suggested for the determination of Mo, Ti, and U.

N-Benzoyl-N-phenylhydroxylamine   (or   N-Phenylbenzohydroxamic Acid)

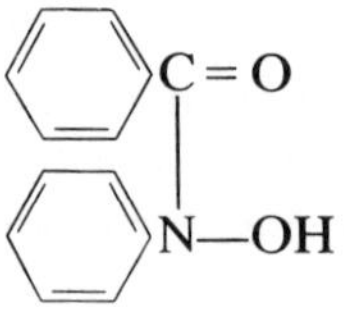

Mol. wt. 213.2.
M. p. 121°–122°.

This is a more important reagent.[197] It may be regarded as an analog of cupferron, to which it is superior in a number of respects. It is a valuable reagent for the extraction separation of certain metals (see Chapter 9B) and for the photometric determination of V (violet 2:1 V chelate in $CHCl_3$). In mineral acid medium (~6 M HCl), it becomes a selective reagent for vanadium; few other metals give colored chloroform extracts at high acidity. Titanium(IV) is also extracted from strongly acid solutions (yellow $CHCl_3$ extract). In 10 M HCl in the presence of $SnCl_2$, iron and vanadium do not react, and niobium is the chief interfering element. The absorbance of the titanium extract depends on the acid and its concentration, and apparently mixed complexes are formed.[198] The chloroform solution of the reagent alone does not absorb appreciably above 375 nm.

Titanium, Nb, Ta, Mo(VI), and W form mixed-ligand complexes with *N*-phenylbenzohydroxamic acid and phenylfluorone, extractable into $CHCl_3$ from ~0.1 M HCl.[199]

Other Hydroxamic Acids. More than 150 N-arylhydroxamic acids have been synthesized by Indian workers,[200] with particular reference to the determination of vanadium. Those derived from N-arylhydroxylamines are preferred because the V complexes are easily extracted into inert solvents. Among these, those derived from aliphatic and substituted aliphatic acids are relatively less sensitive. Length of conjugation increases sensitivity, as does methoxy or methyl substitution. Electronegative substituents (Cl, Br, I, NO_2) offer no special advantage. *N-m*-Tolyl-*p*-methoxybenzohydroxamic acid is one of the new reagents. In ~4 M HCl, with $CHCl_3$ as extractant, $\mathscr{S}_{530} = 0.009$ μg V/cm^2. The selectivity is much the same as with N-phenylbenzohydroxamic acid. Interference is caused by Mo, Ti, and Zr. Iron(III), Cu, Sn(IV), Th(IV), and U(VI) can be tolerated at a ratio of 200 or greater.

N-Cinnamoyl-*N*-phenylhydroxylamine[201] provides $\mathscr{S}_{540} = 0.008$. *N*-3-Styrylacryloylphenylhydroxylamine is equally sensitive.[202] *N*-Furoylphenylhydroxylamine yields a $CHCl_3$-extractable complex (4 M HCl) believed to have the composition VOL_2Cl.[203]

N-Phenylacetylhydroxamic acid in benzene extracts Nb from 10 M HCl; $\epsilon_{370} = 9800$.[204] PAHA:Nb = 2. N-Substituted acetyl-salicylohydroxamic acids with SCN^- form mixed complexes with Nb in which Nb:reagent:$SCN^- = 1:2:1$.[205] They are extractable from 4–8 M HCl into CCl_4.

B. MISCELLANEOUS

It has been demonstrated that syn-α-oximinophenylacetamide chelates Cu(II) through two oxygens, one of which is bonded to C (CuL_2 formed above pH ~8.5)[206]:

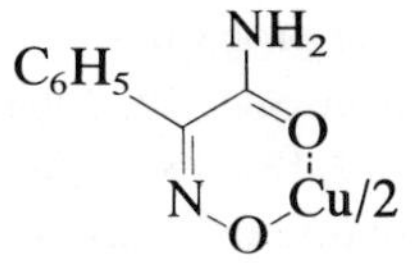

$$C_6H_5 \quad \overset{NH_2}{\underset{\underset{O}{N}}{\overset{\|}{\underset{\|}{C}}}} \quad \overset{O}{\underset{Cu/2}{}}$$

The reaction is of no analytical value, but is mentioned here because of the structure of the chelate.

An indirect method for zirconium (of little importance) is based on precipitation with p-dimethylaminoazophenylarsonic acid to form

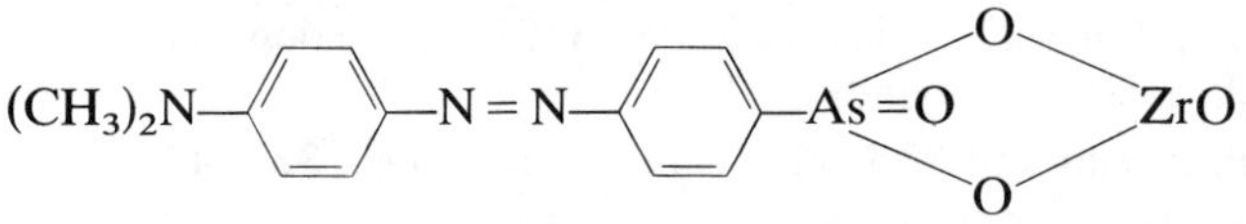

$$(CH_3)_2N-\langle \rangle-N=N-\langle \rangle-As=O \quad \overset{O}{\underset{O}{\diagdown \diagup}} ZrO$$

or an analogous compound with Zr^{4+} (salt formation). No organic chelation occurs, and the color of the precipitate is much the same as that of the reagent. But if the $-AsO_3H_2$ group were adjacent to $-N{=}N-$ with an $ortho$-OH on the other benzene ring (Arsenazo-type reagent, Chapter 6E), chelate formation would occur with a striking color change.

NOTES

1. For absorptivities of ascorbinate complexes of Ti, U, Fe, Ni, Eu, and others at pH 9, see K. P. Stolyarov and I. A. Amantova, *Talanta*, **14**, 1237 (1967). Review of analytical applications of ascorbic acid: A. T. Pilipenko and M. B. Kladnitskaya, *Zav. Lab.*, **32**, 3 (1966).

2. L. Sommer, *Coll. Czech. Chem. Commun.*, **28**, 449 (1963).

3. N. Elenkova, C. Palasev, and L. Ilceva, *Talanta*, **18**, 355 (1971).

4. G. E. Janauer and J. Korkisch, *Anal. Chim. Acta*, **24**, 270 (1961).

5. J. A. Adam, E. Booth, and J. D. A. Strickland, *Anal. Chim. Acta*, **6**, 462 (1952). Determination of Be in sediments and natural waters: J. R. Merrill, M. Honda, and J. R. Arnold, *Anal. Chem.*, **32**, 1420 (1960).

6. W. E. van der Linden and G. den Boef, *Anal. Chim. Acta*, **37**, 179 (1967).

7. T. Shigematsu and M. Tabushi, *J. Chem. Soc. Japan Pure Chem. Sec.*, **81**, 265 (1960). Benzoylacetone is a reagent for Nb in 7 M HCl, $\epsilon_{377} = 12,360$ (benzene): G. P. Ozerova, N. V. Mel'chakova, L. I. Nosova, and V. M. Peshkova, *Anal. Abst.*, **17**, 2074 (1969).

8. H. Akaiwa, H. Kawamoto, and M. Hara, *Anal. Chim. Acta*, **43**, 297 (1968). Determination of Cu: H. Akaiwa, H. Kawamoto, and M. Abe, *Bull. Chem. Soc. Japan*, **44**, 117 (1971); without pyridine: *Talanta*, **23**, 403 (1976).

9. R. Belcher, R. Perry, and W. I. Stephen, *Analyst*, **94**, 26 (1969). Compare K. De Armond and L. S. Forster, *Spectrochim. Acta*, **19**, 1393, 1403, 1687 (1963) (fluorescence process).

10. P. H. Gore and P. J. Newman, *Anal. Chim. Acta*, **31**, 111 (1964), have surveyed the light-absorbing properties of ferric complexes of 38 phenols.

11. R. M. Milburn, *J. Am. Chem. Soc.*, **77**, 2064 (1955). A more accurate value is 7×10^{-3} [G. Limb and R. J. Robinson, *Anal. Chim. Acta*, **47**, 451 (1969)].

12. It is unclear whether the reason for weak coloration is incomplete formation of the complex or a low true ϵ_{max}. Probably both factors are involved.

13. R. Weinland and E. Walter, *Z. Anorg. Chem.*, **126**, 148 (1923).

14. H. Reihlen, *Z. Anorg. Chem.*, **123**, 173 (1922).

15. R. F. Weinland and K. Binder, *Berichte*, **45**, 2498 (1912).

16. L. Havelkova and M. Bartusek, *Coll. Czech. Chem. Commun.*, **34**, 2919 (1969); Bartusek, *ibid.*, **38**, 2255 (1973) (reactions of Mo and W with o-dihydroxy compounds).

17. G. P. Haight and V. Paragamian, *Anal. Chem.*, **32**, 642 (1960).

18. D. N. Brown, *J. Inorg. Nucl. Chem.*, **17**, 146 (1961). N. V. Chernaya et al., *Chem. Abst.*, **75**, 155,439x (1971). J. Halmekoski, *Suom. Kemistil.*, **B36**, 29, 46 (1963).

19. V. Patrovsky, *Coll. Czech. Chem. Commun.*, **23**, 1774 (1958), studied the Nb reaction.

20. Chelates with o-dihydric phenols: V. Stejskal and M. Bartusek, *Coll. Czech. Chem. Commun.*, **38**, 3103 (1973). Catechol: M. Nadasy and K. Jonas, *Acta Chim. Acad. Sci. Hung.*, **34**, 339 (1962); 1:3 complex.

21. For references see the paper by Nardillo and Catoggio.[25] Spectrophotometric determination: V. Patrovsky, *Coll. Czech. Chem. Commun.*, **20**, 1328 (1955); **31**, 3392 (1966).

22. L. Sommer, *Coll. Czech. Chem. Commun.*, **28**, 2102 (1963). This paper also treats the reaction of titanium with Tiron and pyrocatechuic acid.

23. A. T. Pilipenko and V. V. Lukachina, *Zh. Anal. Khim.*, **25**, 2125 (1970).

24. M. Vrchlabsky and L. Sommer, *Talanta*, **15**, 887 (1968); B. Grebenova and Vrchlabsky, *Coll. Czech. Chem. Commun.*, **38**, 379 (1973).

25. The extraction of Nb complexes of catechol and substituted derivatives from essentially neutral solution was studied by A. M. Nardillo and J. A.

Catoggio, *Anal. Chim. Acta*, **66**, 359 (1973), who also describe the extraction of the ion associates formed with quaternary ammonium ions. A. G. Ward and O. Borgen, *Talanta*, **24**, 65 (1977), found that sparteine gives an ion-association complex, thought to be [NbO(Cat)$_2$]Sp, extractable into chloroform. It serves as a fairly selective determination form for Nb ($\mathscr{S}_{342} =$ 0.008 μg Nb/cm^2). λ_{max} for the ion associate is different from λ_{max} of the Nb-catechol complex (300 nm) and ϵ_{max} is larger.

26. M. M. Tananaiko, L. I. Gorenshtein, and L. G. Zhidik, *Ukr. Khim. Zh.*, **36**, 703 (1970). W reacts similarly: *Anal. Abst.*, **19**, 3051. Extraction of Mo complex with tetraphenylarsonium cation: R. Bock and S. Strecker, *Z. Anal. Chem.*, **248**, 157 (1969).

27. A. I. Busev and Z. P. Karyakina, *J. Anal. Chem. USSR*, **22**, 1266 (1967). Rare earth metals of the Y subgroup react with 3,5-dinitrocatechol and Rhodamine B to form an ion associate (1:4:1) extractable into benzene-isobutyl alcohol, as shown by N. S. Poluektov and coworkers, *Zh. Anal. Khim.*, **30**, 1513 (1975).

28. L. Jerman and F. Polacek, *Anal. Chim. Acta*, **36**, 240 (1966).

29. For complex formation of Tiron (and chromotropic acid and sulfosalicyclic acid) with some trivalent metals, see V. T. Athavale et al., *J. Inorg. Nucl. Chem.*, **30**, 3107 (1968). The Al chelates (2:1 and 3:1, reagent/Al) absorb in the UV, the former maximally at 258 and 310 nm (T. Yotsuyanagi et al., *Anal. Abst.*, **16**, 1794).

30. L. J. Clark, *Anal. Chem.*, **42**, 694 (1970). Fe(III) is reduced with dithionite, Cu is masked with thiourea, and the Ti reaction is carried out in oxalic-citric acid solution to prevent hydrolytic precipitation of Ti and to mask other metals.

31. J. H. Yoe and A. L. Jones, *Ind. Eng. Chem. Anal. Ed.*, **16**, 111 (1944). Later work by M. Morin and J. P. Scharff, *Anal. Chim. Acta*, **60**, 101 (1972), indicates that in addition to the three simple species FeL$_n$ ($n = 1$, 2, 3), a protonated complex FeHL is formed.

32. N. S. Poluektov, L. A. Alakaeva, and M. A. Tishchenko, *Zh. Anal. Khim.*, **25**, 2351 (1970). Later, M. A. Tishchenko, L. A. Alakaeva, and N. S. Poluektov, *Ukr. Khim. Zh.*, **39**, 475 (1973), determined Tb and Dy fluorimetrically in basic solution by forming a 1:1:1 complex of these metals with Tiron and EDTA. Also see *Anal. Abst.*, **29**, 4B93 (1975); **28**, 4B61 (1975).

33. W. Reksc, *Chemia Anal.*, **14**, 795 (1969).

34. B. K. Afghan and J. Israeli, *Talanta*, **16**, 1601 (1969).

35. S. Arribas, M. L. Alvarez Bartolome, and M. Gonzalez Alvarez, *An. Quim.*, **69**, 633 (1973).

36. T. Taketatsu and N. Toriumi, *Talanta*, **17**, 465 (1970). The combining ratios Tiron/R.E. were found to be 3:2 and 2:1.

37. T. Taketatsu and T. Yamauchi, *Talanta*, **18**, 647 (1971).

38. A. K. Majumdar and C. P. Savariar, *Anal. Chim. Acta*, **21,** 146 (1959).

39. T. Yotsuyanagi, Y. Kudo, and K. Aomura, *Japan Analyst*, **18,** 619 (1969).

40. J. Halmekoski, *Ann. Acad. Sci, Fenn.*, **1959,** A, II, 96.

41. In ammoniacal solution, resorcinol gives blue or violet colors with Fe(III), Cu, Pb, and other metals. The composition of the products is apparently not known, but it may be suspected that an oxidation product of resorcinol is the reactant.

42. A. I. Busev and Z. P. Karyakina, *Zh. Anal. Khim.*, **22,** 1350 (1967).

43. G. Ackermann and S. Koch, *Talanta*, **9,** 1015 (1962).

44. A. I. Busev and N. G. Solov'eva, *Zh. Anal. Khim.*, **27,** 1529 (1972).

45. A. E. Hultquist, *Anal. Chem.*, **36,** 149 (1964).

46. A yellow pyridine adduct (1:1:*n* pyridine), formed at pH ~ 4 and extracted into $CHCl_3$, serves for the determination of Ti ($\epsilon_{366} = 7300$): R. S. Ramakrishna and H. D. Gunawardena, *Talanta*, **20,** 21 (1973).

47. R. S. Ramakrishna, V. Paramasigamani, and M. Mahendran, *Talanta*, **22,** 523 (1975).

48. The system Ti–sulfosalicylic acid was investigated by L. Sommer, *Coll. Czech. Chem. Commun.*, **28,** 2716 (1963). With a large excess of reagent in the pH range 2–5, the dominant chelate species is TiL_3^{5-}. U(VI) forms 1:1 and 1:2 chelates with the reagent in excess (J. Havel and L. Sommer, *ibid.*, **33,** 529).

49. Formation constants of complexes are given by M. Morin, M. R. Paris, and J. P. Scharff, *Anal. Chim. Acta*, **57,** 123 (1971). Mixed chelates of Fe(III) with nitrilotriacetic acid and sulfosalicylic acid or catechol-3,5-disulfonic acid have been investigated by M. Morin and J. P. Scharff, *ibid.*, **66,** 113 (1973).

50. R. M. Dagnall, R. Smith, and T. S. West, *Analyst*, **92,** 358 (1967); R. G. Charles and E. P. Riedel, *J. Inorg. Nucl. Chem.*, **28,** 527 (1966).

51. Review of transfer process: G. A. Crosby et al., *J. Chem. Phys.*, **34,** 743 (1961).

52. H. Buchwald and E. Richardson, *Talanta*, **9,** 631 (1962).

53. T. D. Ali-Zade, G. A. Gamid-Zade, and G. A. Akhmedova, *Zh. Anal. Khim.*, **29,** 735 (1974).

54. G. Ackermann and S. Koch, *Talanta*, **16,** 95, 284, 288 (1969). Spectra are illustrated. Determination in presence of masking agents: *ibid.*, **17,** 757 (1970).

55. M. L. Good and S. C. Srivastava, *Talanta*, **12,** 181 (1965).

56. Chloranilic acid is 2,5-dichloro-3,6-dihydroxy-*p*-benzoquinone. It has found use for the direct determination of Mo and Zr. Review: W. C. Broad, K. Ueno, and A. J. Barnard, Jr., *Japan Analyst*, **9,** 257 (1960). It forms a 2:1 complex with Sn(IV) at pH 1–1.5, absorbing maximally at 540 nm.

57. M. Katyal, D. P. Goel, and R. P. Singh, *Talanta*, **15,** 711 (1968).

58. K. E. Curtis, J. A. Thomson, and G. F. Atkinson, *Anal. Chim. Acta*, **49**, 351 (1970).

59. E. A. Biryuk and V. A. Nazarenko, *Zh. Anal. Khim.*, **23**, 1018 (1968), A. Okac and L. Sommer, IUPAC XV Congress, Analytical Chemistry, Vol. II, p. 925 (1958), write the composition of the 1:2 and 1:3 complexes as TiL_2 and $TiOH(HL)_3$, if H_2L = chromotropic acid. L. Sommer, *Coll. Czech. Chem. Commun.*, **28**, 3057 (1963), investigated 2,3-dihydroxynaphthalene-6-sulfonic acid as a titanium reagent. At pH 3 with a large excess of reagent, the composition of the orange Ti chelate is reported as TiL_3^{5-}, if the reagent is written as H_3L.

60. Mo(V) forms a yellow 1:2 complex ($\lambda_{max} = 370$ nm) at pH 2–8, as shown by I. A. Tserkovnitskaya and N. A. Kustova, *Zh. Anal. Khim.*, **23**, 72 (1968).

61. L. I. Ganago and L. V. Kovaleva, *Zh. Anal. Khim.*, **25**, 1517 (1970). See V. M. Peshkova et al., *ibid.*, **29**, 592 (1974).

62. V. A. Nazarenko and E. A. Biryuk, *J. Anal. Chem. USSR*, **24**, 35 (1969); see V. I. Kuznetsov and N. N. Basargin, *Zh. Anal. Khim.*, **16**, 573 (1961).

63. N. N. Basargin, M. K. Akhmedli, and M. M. Shirinov, *J. Anal. Chem. USSR*, **23**, 1590 (1968). Y. K. Lee, K. J. Whang, and K. Ueno, *Talanta*, **19**, 1665 (1972), have raised doubts about the nature of Khimdu. The product synthesized by Basargin and coworkers seems to be a mixture of compounds.

64. According to N. N. Basargin, N. A. Panarina, and P. Ya. Yakovlev, *Zh. Anal. Khim.*, **29**, 979 (1974).

65. N. N. Basargin, M. K. Akhmedli, and M. M. Shirinov, *J. Anal. Chem. USSR*, **24**, 276 (1969); N. N. Basargin et al., *Zh. Anal. Khim.*, **30**, 177 (1975). Tipyrogin is a similar reagent. It is N-methyl-N,N-bis(methylenepyrogallol)amine. Niobium gives a yellow color in 0.5 M HCl. See N. N. Basargin, P. Ya. Yakovlev, and N. A. Panarina, *Zh. Anal. Khim.*, **25**, 746 (1970).

66. P. Ya. Yakovlev, N. N. Basargin, and N. A. Panarina, *Zh. Anal. Khim.*, **25**, 505 (1970).

67. N. N. Basargin, P. Ya. Yakovlev, and R. S. Deinekina, *Zav. Lab.*, **39**, 1043, 1305 (1973).

68. Partial purification of commercial products has been effected by extracting with methyl alcohol in a Soxhlet extractor. Better purification is obtained by applying a chromatographic procedure with a column of Sephadex G-10 as demonstrated by H. G. C. King and G. Pruden, *Analyst*, **93**, 601 (1968). The commercial products examined by them contained large amounts of Na or K as sulfates or chlorides. Free alizarinsulfonic acid, in addition to Na or K alizarin sulfonate, was present. Organic impurities were very small in amount. The effect of the inorganic salts was to shorten the linear portion of the Al calibration curve, but not its slope.

69. There is, as might be expected, a general similarity between the absorption spectra of the metal chelates and of HAz^-.

70. M. Otomo and K. Tonosaki, *Talanta*, **18**, 438 (1971), concluded from the infrared spectrum of the 1:1 complex of In and Alizarin Red S that the metal is probably bonded to the phenolic oxygen atoms as shown above.

71. H. E. Zittel and T. M. Florence, *Anal. Chem.*, **39**, 320 (1967).

72. P. Sanyal and S. P. Mushran, *Anal. Chim. Acta*, **35**, 400 (1966).

73. Y. Dorta-Schaeppi, H. Hürzeler, and W. D. Treadwell, *Helv. Chim. Acta*, **34**, 797 (1951).

74. A. W. Ashbrook, *Analyst*, **88**, 113 (1963).

75. L. S. Serdyuk and U. F. Silich, *Anal. Abst.*, **10**, 1372 (1963).

76. V. I. Kuznetsov and A. I. Zabelin, *Zh. Anal. Khim.*, **17**, 318 (1962).

77. For example, N. Ishibashi and H. Kohara, *Japan Analyst*, **15**, 1137 (1966), extract the reaction product of Mo(VI) and Alizarin Red S into dichloroethane in the presence of tetradecyldimethylbenzylammonium chloride. Similarly, M. Otomo, S. Masuda, and K. Tonosaki, *J. Chem. Soc. Japan Pure Chem. Sec.*, **92**, 739 (1971), extract the Ga chelate of Alizarin Red S into butanol–benzene after adding 1,3-diphenylguanidinium salt. The 1:1 In chelate is extractable into *n*-butyl acetate.[70] Germanium can also be determined: V. A. Nazarenko, G. V. Flyantikova, and T. N. Selyutina, *Zh. Anal. Khim.*, **27**, 2369 (1972). Both the Alizarin and Alizarin Red S complexes of Ge are extracted into chloroform at pH 6–7 in the presence of diphenylguanidine. With the former reagent the extracted compound is reported to be $(GeAz_3)^{2-}(2DPG)^{2+} \cdot 4\ CHCl_3$.

78. Mostly according to F. Feigl, *Chemistry of Specific, Selective and Sensitive Reactions*, Academic, New York, 1949.

79. Formation constants of the chelates of 1-hydroxyanthraquinone and 1,8-dihydroxyanthraquinone with U(VI), Cu, Be, Ni, Co, Zn, Cd, Mn in dioxane (75%)–H_2O are given by H. Kido, W. C. Fernelius, and C. G. Haas, *Anal. Chim. Acta*, **23**, 116 (1960). (Also in this paper, formation constants of chelates of these metals with hydroxynaphthoquinones and hydroxy-γ-pyrones.) Jackson and Leonard[87] found that the product formed by reaction of Ni (or Co) with 1-hydroxyanthraquinone (HL) in methyl alcohol in the presence of acetate had the molar ratio 1 Ni:1 HL:2 acetate. For Cu and Mn, metal–HL ratios are 1:2 with no acetate in the chelate. Dimethylformamide can also coordinate in the Ni (Co) chelate.

80. R. Belcher, M. A. Leonard, and T. S. West, *J. Chem. Soc.*, 3577 (1959); *Talanta*, **2**, 92 (1959). Alizarin Fluorine Blue sulfonated in the 5-position has been prepared: M. A. Leonard and G. T. Murray, *Analyst*, **99**, 645 (1974).

81. F. Ingman, *Talanta*, **20**, 135 (1973). Values for $\mu = 0.5$ also are reported. Later, M. A. Leonard, *Analyst*, **100**, 275 (1975), working with sulfonated Alizarin Fluorine Blue, obtained similar values for these constants and determined two more (for H_6L^+ and H_5L). He also supplies solubilities for K salts and purification procedures.

82. M. A. Leonard and T. S. West, *J. Chem. Soc.*, 4477 (1960).

83. F. J. Langmyhr, K. S. Klausen, and M. H. Nouri-Nekoui, *Anal. Chim. Acta*, **57**, 341 (1971). They found La or Ce(III) to form the 1:1 chelate, $M^{III}(HL)$, at pH 4.5; at pH 5.5 and 6.5 mixtures of the 1:1 complex and complexes containing two or more metal ions per ligand molecule are formed.

84. T. Anfält and D. Jagner, *Anal. Chim. Acta*, **70**, 365 (1974).

85. F. Ingman, *Talanta*, **20**, 999 (1973).

86. M. A. Leonard and F. I. Nagi, *Talanta*, **16**, 1104 (1969).

87. M. Jackson and M. A. Leonard, *Proc. Soc. Anal. Chem.*, **9**, 192 (1972).

88. The purification of the acid has been studied by J. A. Thomson and G. F. Atkinson, *Talanta*, **18**, 817 (1971). The commercial sodium salt can be purified through precipitation by adding NaCl to the aqueous solution. The free acid can be obtained by running a solution of the commercial Na salt through a column of cation exchanger in the H-form, concentrating the eluted solution and crystallizing out the acid by addition of HCl. The pK values of the OH groups are 12.14 and 8.71. Alkaline solutions of the reagent are unstable (color fades).

89. K. Al-Ani and M. A. Leonard, *Proc. Soc. Anal. Chem.*, **8**, 190 (1971).

90. C. E. White and C. S. Lowe, *Ind. Eng. Chem. Anal. Ed.*, **13**, 809 (1941).

91. A. K. Jain, V. P. Aggarwala, P. Chand, and S. P. Garg, *Talanta*, **19**, 1481 (1972), studied the reaction with the rare earth metals.

92. *Anal. Abst.*, **21**, 1010 (1971).

93. V. A. Nazarenko, G. V. Flyantikova, and A. M. Tetereva, *Zh. Anal. Khim.*, **29**, 284 (1974).

94. E. Eegriwe, *Z. Anal. Chem.*, **120**, 81 (1940).

95. H. Fischer, *Wiss. Veröffent. Siemens-Konzern*, **5**, 99 (1926); *Z. Anal. Chem.*, **73**, 54 (1928).

96. W. Fischer and J. Wernet, *Angew. Chem.*, **A60**, 729 (1948).

97. H. H. Willard and H. C. Fogg, *J. Am. Chem. Soc.*, **59**, 40 (1937).

98. N. S. Poluektov, O. P. Makarenko, A. I. Kirilov, and R. S. Lauer, *Zh. Anal. Khim.*, **28**, 285 (1973).

99. G. F. Kirkbright, T. S. West, and C. Woodward, *Talanta*, **13**, 1637 (1966). Many interferences.

100. N. S. Poluektov, R. S. Lauer, and M. A. Sandu, *Zh. Anal. Khim.*, **25**, 2118 (1970).

101. V. A. Nazarenko and V. P. Antonovich, *Trihydroxyfluorones*, Nauka, Moscow, 1973, 180 pp. (in Russian). Chapter III, the largest chapter, is on the use of trihydroxyfluorones for determining the chemical elements, mainly in photometric analysis.

102. G. S. Petrova, Yu. S. Ryabokobilko, A. M. Lukin, N. N. Vysokova, and R. V. Poponova, *Anal. Lett.*, **5**, 695 (1972).

103. Phenylfluorone was preceded by methylfluorone as a qualitative reagent. P.

Wenger, R. Duckert, and C. R. Blancpain, *Helv. Chim. Acta*, **20**, 1427 (1937), were led to investigate the latter compound as a metal reactant (especially for antimony) by the similarity of its reacting group with that of catechol and pyrogallol. J. Gillis, J. Hoste, and A. Claeys, *Anal. Chim. Acta*, **1**, 291 (1947), found methylfluorone to be a selective color reagent for Sb, Sn, Ge, Mo, V, Fe(III), Ti, and Zr. The same workers, *Med. Vlaamsche Acad.*, **9**, 1 (1947), tested fluorones having C_6H_5, C_6H_4OH (*o*, *m*, *p*) and $C_6H_3(OH)_2$ (*m*, *p*) in place of the CH_3 group, and found them to react with the preceding metals in 1 *M* HCl solutions. For trihydroxyfluorones as reagents for Nb, see V. A. Nazarenko and G. Ya. Yagnyatinska, *Anal. Abst.*, **12**, 1144 (1963).

104. W. A. Schneider, M. S. Thesis, University of Minnesota, 1953. O. V. Sivanova, G. S. Ivankovich, and G. S. Lagutina, *J. Anal. Chem. USSR*, **26**, 1675 (1971), have determined all four ionization constants: pK_1 2.3, pK_2 5.8, pK_3 11.3, pK_4 12.3 (25% ethanol). These authors also report the ionization constants of o^-, m^-, and p^- nitro derivatives of phenylfluorone.

105. C. L. Luke, *Anal. Chim. Acta*, **37**, 97 (1967).

106. V. A. Nazarenko and E. A. Biryuk, *Zh. Anal. Khim.*, **22**, 57 (1967).

107. Ternary and quaternary complexes containing phenazone: M. B. Shustova and V. A. Nazarenko, *Ukr. Khim. Zh.*, **33**, 623 (1967); *Anal. Abst.*, **15**, 5294 (1968).

108. K. Kimura, H. Sano, and M. Asada, *Bull. Chem. Soc. Japan*, **29**, 640 (1956); K. Kimura and M. Asada, *ibid.*, **29**, 812 (1956). The germanium chelate is soluble in ethanolic–HCl solution. N. F. Kazarinova and N. L. Vasilieva, *Zh. Anal. Khim.*, **13**, 677 (1958), prefer this reagent to the parent compound for the greater solubility of its Ge chelate (stable solutions up to 1.2 ppm Ge). L. S. Serdyuk, A. A. Novikova, and A. V. Fedin, *J. Anal. Chem. USSR*, **28**, 1138 (1973), have determined the dissociation constants of the reagent (H_5L^{2+}, H_4L^+, H_3L, H_2L^-, HL^{2-}).

109. H. Sano, *Bull. Chem. Soc. Japan*, **31**, 974 (1958).

110. E. Asmus and U. Kossmann, *Z. Anal. Chem.*, **245**, 137 (1969). Sn in steel.

111. J. Vrbsky and J. Fogl. *Coll. Czech. Chem. Commun.*, **35**, 2497 (1970).

112. E. Asmus and W. Klank, *Z. Anal. Chem.*, **265**, 260 (1973). This reagent is inferior to methylfluorone for Mo.

113. V. A. Nazarenko, T. P. Yakovleva, E. M. Nevskaya, and V. P. Antonovich, *Zh. Anal. Khim.*, **29**, 1478 (1974). The spectral characteristics and ionization constants of these compounds are similar to those of the trihydroxyfluorones containing aliphatic and aromatic radicals in position 9. V. P. Antonovich, T. P. Yakovleva, E. I. Shelikhina, and E. M. Nevskaya, *Zh. Anal. Khim.*, **29**, 2348 (1974), have examined reactions between quinolylfluorone or quinoxalyfluorone and metals of periodic groups III, IV, and VI.

114. V. P. Antonovich, T. P. Yakovleva, E. M. Nevskaya, and V. A. Nazarenko, *Zh. Anal. Khim.*, **29**, 2341 (1974).

115. V. P. Antonovich, T. P. Yakovleva, E. I. Shelikhina, and E. M. Nevskaya, *Zh. Anal. Khim.*, **29**, 2348 (1974).

116. A. V. Fedin and S. I. Kravchuk, *Zh. Anal. Khim.*, **29**, 1734 (1974).

117. V. A. Nazarenko and G. Ya. Yagnyatinskaya, *Zav. Lab.*, **38**, 1427 (1972). In *Zh. Anal. Khim.*, **29**, 1977 (1974), these values are reduced to 130,000 and 140,000 for ethanolic $CHCl_3$, in which the chelate molecule is said to be solvated with 2 C_2H_5OH and 4 $CHCl_3$.

118. V. A. Nazarenko, G. V. Flyantikova, and A. M. Andrianov, *Zh. Neorg. Khim.*, **12**, 3072 (1967).

119. V. P. Antonovich, E. I. Shelikhina, B. V. Zhadanov, and V. A. Nazarenko, *Zh. Anal. Khim.*, **27**, 100 (1972). Reactions of salicyl- and other fluorones with Mo and other metals: O. A. Tataev et al., *Anal. Abst.*, **28**, 5A9 (1975).

120. V. A. Nazarenko, N. I. Makrinich, and M. B. Shustova, *Zh. Anal. Khim.*, **25**, 1595 (1970). For reaction of Sb(III), see *ibid.*, **28**, 1104 (1973).

121. V. A. Nazarenko, E. I. Shelikhina, and V. P. Antonovich, *Zh. Anal. Khim.*, **27**, 307 (1972).

122. A. Sh. Shakhabudinov and O. A. Tataev, *Zh. Anal. Khim.*, **27**, 2382 (1972).

123. A. T. Pilipenko, E. A. Shpak, and O. S. Zul'figarov, *Zh. Anal. Khim.*, **29**, 1074 (1974). Titanium also forms a mixed complex with thiocyanate and salicylfluorone (and other fluorones) having the composition 1:1:2, extractable by $CHCl_3$–isobutyl alcohol ($\mathscr{S} = 0.0003 \, \mu$g Ti/cm^2). See V. P. Antonovich, E. I. Shelikhina, and S. A. Lozovaya, *ibid.*, **28**, 1506 (1973).

124. G. Ackermann and H. Heegn, *Talanta*, **21**, 431 (1974). Of a number of hydroxyxanthene dyes tested as Sn(IV) reagents, this compound and gallein are recommended.

125. V. Suk and M. Malat, *Chem.-Anal.*, **45**, 30 (1956); O. Ryba, J. Cifka, M. Malat, and V. Suk, *Coll. Czech. Chem. Commun.*, **21**, 349 (1956).

126. A closer examination of Catechol Violet equilibria by I. Mori, M. Shinogi, E. Falk, and Y. Kiso, *Talanta*, **19**, 299 (1972), indicates that this scheme may be over-simplified and the system may require further study. They find the reagent solution to be red up to pH $\sim$1.5 ($\lambda_{max} = 550$ nm), yellow to pH $\sim$7 (443 nm), greenish yellow to pH 9 (500 nm), blue to pH $\sim$10.5 (590 nm) and brown-violet at pH $>$10.5. Absorbance curves are given. $pK_1 = 0.1$, $pK_3 = 9.75$, $pK_4 = 11.7$ (ionic strength not specified, but low). E. A. Biryuk and R. V. Ravitskaya, *J. Anal. Chem. USSR*, **25**, 494 (1970), determined K_2, K_3, and K_4 at ionic strengths from 0.1 to 1 M. At $\mu = 0.1$, $pK_2 = 7.9$, $pK_3 = 9.94$, $pK_4 = 11.82$. They show that pK varies linearly with μ. Thus $pK_2 = 8.01 - 1.10\mu$.

127. W. D. Wakley and L. P. Varga, *Anal. Chem.*, **44**, 169 (1972). The formation constants are given. For a critique of this work, see B. W. Budesinsky, *Analyst*, **97**, 909 (1972).

128. A. D. Wilson and G. A. Sergeant, *Analyst*, **88**, 109 (1963). The extent of interference by common metals is listed. On the basis of experimental

comparison, W. K. Dougan and A. D. Wilson, *ibid.*, **99**, 413 (1974), concluded that Catechol Violet was superior to Eriochrome Cyanine R, Stilbazo and other photometric reagents for determination of Al in water. $\mathscr{S}_{580} = 4 \times 10^{-4} \mu g$ Al/cm^2.

129. M. M. Tananaiko, O. P. Vdovenko, and V. M. Zatsarevnyi, *Zh. Anal. Khim.*, **29**, 1724 (1974).

130. V. F. Luk'yanov and E. M. Knyazeva, *J. Anal. Chem. USSR*, **23**, 455 (1968).

131. V. A. Nazarenko and N. I. Makrinich, *Zh. Anal. Khim.*, **24**, 1694 (1969).

132. R. Belcher, T. V. Ramakrishna, and T. S. West, *Talanta*, **12**, 681 (1965); T. V. Ramakrishna, S. A. Rahim, and T. S. West, *ibid.*, **16**, 847 (1969). For determination of W with Pyrogallol Red, see Y. Shijo and T. Takeuchi, *Japan Analyst*, **22**, 1341 (1973).

133. F. J. Langmyhr and T. Stumpe, *Anal. Chim. Acta*, **32**, 535 (1965). The preparation of pure reagent in the form of the tetrabasic acid by precipitation with HCl from impure trisodium salt is described. Purification of the commercial dye was further investigated by E. J. Dixon, L. M. Grisley, and R. Sawyer, *Analyst*, **95**, 945 (1970), who preferred a chemical method to paper, thin-layer, and column chromatography. In this method, the free acid, obtained by adding concentrated HCl to an aqueous solution of the sodium salt, was dissolved in chloroform, then back-extracted into water, and crystallized therefrom. N. G. Elenkova and Ek. Popova, *Talanta*, **22**, 925 (1975), found 1:1 complexes formed by the reagent with Mg and Al in basic solution, and report the formation constants for MgOHL, AlH$_2$L, AlHL, and AlL.

134. R. R. Abdulaev, O. A. Tataev, and A. I. Busev, *Zav. Lab.*, **37**, 389 (1971).

135. E. J. Langmyhr and K. S. Klausen, *Anal. Chim. Acta*, **29**, 149 (1963); M. Malát, *ibid.*, **25**, 289 (1961). Chromatographic purification of reagent: V. Malanik and M. Malát, *ibid.*, **76**, 464 (1975).

136. S. C. Srivastava and A. K. Dey, *J. Inorg. Nucl. Chem.*, **25**, 217 (1963).

137. B. Evtimova, *Anal. Chim. Acta*, **68**, 222 (1974). Y. Nakamura, H. Nagai, D. Kubota, and S. Himeno, *Japan Analyst*, **22**, 1156 (1973), found the complex formed by Fe(III) with Chrome Azurol S and hexadecyltrimethylammonium chloride at pH 3.5 to have $\epsilon = 147,000$ or $\mathscr{S} = 0.0004 \mu g$ Fe/cm^2. Beryllium can be determined sensitively if a micelle-forming cationic reagent, such as a quaternary amine having a long carbon chain (Zephiramine, for example), is added with Chrome Azurol S: H. Nishida, *Japan Analyst*, **20**, 1080 (1971); R. Ishida and K. Tonosaki, *J. Chem. Soc. Japan*, 1077 (1974); L. Sommer and V. Kuban, *Anal. Chim. Acta*, **44**, 333 (1969). F. W. E. Strelow, R. G. Böhmer, and C. H. S. W. Weinert, *Anal. Chem.*, **48**, 1550 (1976), were able to increase the sensitivity of the Be reaction four- to fivefold by using benzyldimethyltetradecylammonium chloride as the micelling reagent ($\mathscr{S}_{620} = 0.0001 \mu g$ Be/cm^2). The determination is preceded by separatory cation-exchange in water-organic solvent medium; Ca-EDTA is used as

masking agent. The method is applied to silicate rocks, for which the lower limit is 0.02 ppm with a 1 g sample.

138. Reviews of hydroxyflavones as analytical reagents: M. Katyal, *Talanta*, **15,** 95 (1968); E. M. Nevskaya and V. A. Nazarenko, *Zh. Anal. Khim.*, **27,** 1699 (1972), 150 references.

139. For some formation constants of 3-hydroxyflavone chelates see reference in note 79.

140. L. Horhammer, R. Hänsel, and W. Hieber, *Z. Anal. Chem.*, **148,** 251 (1955).

141. T. Kanno, *Japan Analyst*, **10,** 8 (1961).

142. B. S. Garg and R. P. Singh, *Talanta*, **18,** 761 (1971).

143. However, the gallium and indium complexes of kaempferol are reported[142] to be extractable into chloroform from perchlorate medium; possibly perchlorate ion-association complexes are formed.

144. F. L. Urbach and A. Timnick, *Anal. Chem.*, **40,** 1269 (1968), found that flavonol forms a series of polynuclear chelates with aluminum in absolute ethanol. Fluorimetric observations indicate the formation of a 6 Al:1 flavonol species as well as others.

145. V. A. Nazarenko and V. P. Antonovich, *Zh. Anal. Khim.*, **22,** 1812 (1967).

146. Flavonol-2'-sulfonic acid has been used to determine Bi absorptiometrically by K. Yamamoto and T. Nishio, *J. Chem. Soc. Japan Pure Chem. Sec.*, **89,** 1214 (1968). A 1 Bi:2 reagent chelate is formed by excess reagent in the acid range pH 5 to ~0.2 M HClO$_4$.

147. Beryllium: E. B. Sandell, *Ind. Eng. Chem. Anal. Ed.*, **12,** 674, 762 (1940); C. W. Sill and C. P. Willis, *Anal. Chem.*, **31,** 598 (1959); Sill, Willis, and J. K. Flygare, Jr., *ibid.*, **33,** 1671 (1961). The determination of Be in a variety of sample materials has been thoroughly studied by Sill and coworkers. Zirconium: R. A. Geiger and Sandell, *Anal. Chim. Acta*, **16,** 346 (1957). Thorium: Sill and Willis, *Anal. Chem.*, **34,** 954 (1962).

148. A. T. Pilipenko, T. U. Kukibaev, A. I. Volkova, and T. E. Get'man, *Ukr. Khim. Zh.*, **39,** 813 (1973), report that Hf forms a 1:2 complex with maximal absorbance in 0.1 N H$_2$SO$_4$ and a 1:1 complex with maximal fluorescence in 1 N H$_2$SO$_4$. The same authors, *Zh. Anal. Khim.*, **27,** 1787 (1972), determine Hf in the presence of an equal amount of Zr by fluorimetry in 7 M HCl in the presence of acetone.

149. M. H. Fletcher, *Anal. Chem.*, **37,** 550 (1965). Approximate formation constants are given. Effect of time is reported.

150. S. Ya. Shnaiderman and G. N. Prokof'eva, *Zh. Anal. Khim.*, **25,** 2368 (1970).

151. V. A. Nazarenko and V. P. Antonovich, *Zh. Anal. Khim.*, **24,** 358 (1969).

152. N. S. Poluektov, R. S. Lauer, and O. F. Gaidarzhi, *Zh. Anal. Khim.*, **26,** 898 (1971).

153. N. L. Olenovich, L. I. Koval'chuk, and E. P. Lozitskaya, *Zh. Anal. Khim.*, **29,** 47 (1974).

154. But morin fluoresces in mineral acid solution, the intensity increasing with the hydrogen-ion concentration. R. A. Geiger, M.S. Thesis, University of Minnesota, 1955, found the fluorescence intensity of morin to increase as follows with the HCl concentration: 0.5 M, 0; 1.4 M, 0.5; 2.2 M, 1; 4 M, 4; 6 M, 18; 8 M 55; 10 M, 105. The fluorescence intensities are in arbitrary units; an intensity of 30 corresponds to that given by an 0.08 ppm Zr solution + morin in 2 M HCl (approximately optimal acidity). The fluorescence of morin in stronger acid solutions is presumably due to the formation of a morinium ion or salt (addition of H^+ to $-O-$).

155. E. A. Kocsis and G. Zádor, *Z. Anal. Chem.*, **124,** 42 (1942). Fletcher[149] gives absorption and fluorescence spectra of morin species $H_5L \cdots L^{5-}$.

156. E. Cerrai and C. Testa, *Energia Nucleare (Milan)*, **7,** 477 (1960). More recently it has been shown that in 7 M HCl (38% CH_3OH) Hf fluoresces strongly, Zr rather less so, so that twice as much Zr as Hf does not interfere in the determination of the latter: A. T. Pilipenko, T. U. Kukibaev, and A. I. Volkova, *Zh. Anal. Khim.*, **28,** 510 (1973). A better method, which can be used to determine >0.2% Hf in Zr, is based on the fluorescence of Hf (498 nm, excitation at 396 nm) in 55% $HClO_4$ that is 0.15 mM in quercetin: T. Kouimtzis and A. Townshend, *Analyst*, **98,** 40 (1973). Morin gives a more sensitive reaction with Hf, but Zr then fluoresces weakly under the conditions used. Quercetin-6'-sulfonic acid can be used to determine Zr and Hf fluorimetrically in 9–10 M HCl; formation of 1:1 complexes (termed ion associates) is indicated [A. T. Pilipenko, T. U. Kukibaev, and A. I. Volkova, *Zh. Anal. Khim.*, **29,** 710 (1974)]. Also see Y. Hoshino, *J. Chem. Soc. Japan*, **81,** 1273 (1960).

157. Kh. Kuus and A. Lust, *Anal. Abst.*, **22,** 2271 (1972).

158. A. K. Babko, T. H. Chan, A. I. Volkova, and T. E. Get'man, *Ukr. Khim. Zh.*, **35,** 292 (1969).

159. V. A. Nazarenko and V. P. Antonovich, *Zh. Anal. Khim.*, **22,** 1812 (1967).

160. H. Hamaguchi, R. Kuroda, R. Sugusita, N. Onuma, and T. Shimizu, *Anal. Chim. Acta*, **28,** 61 (1963). 1:1 complex.

161. L. E. Dowd, *Anal. Chem.*, **31,** 1184 (1959).

162. T. D. Filer, *Anal. Chem.*, **43,** 1753 (1971).

163. T. D. Filer, *Anal. Chem.*, **43,** 469 (1971). It is implied that this reagent is superior to morin in giving less fluorescence with Al, Ga, Sb, and other metals. Actually the reverse is true, though the difference may be due largely to the difference in acidities that are specified in procedures with the two reagents.

164. B. S. Garg, K. C. Trikha, and R. P. Singh, *Talanta*, **16,** 462 (1969).

165. B. S. Garg and R. P. Singh, *Microchem. J.*, **18,** 509 (1973). Also a fluorimetric reagent for Ga (1:1 complex at pH 3–4) and In (pH 6).

166. Y. Oka and R. Tanaka, *J. Chem. Soc. Japan Pure Chem. Sec.*, **81**, 1846 (1960).

167. R. S. Bottei and B. A. Trusk, *Anal. Chem.*, **35**, 1910 (1963).

168. References may be found in K. Kodama, *Methods of Quantitative Inorganic Analysis*, Interscience, New York, 1963.

169. Y. P. Dick and N. Vesely, *Anal. Chim. Acta*, **73**, 377 (1973).

170. C. Yoshimura and H. Noguchi, *J. Chem. Soc. Japan Pure Chem. Sec.*, **86**, 399 (1965).

171. E. Asmus, H. J. Altmann, and E. Thomasz, *Z. Anal. Chem.*, **216**, 3 (1966).

172. A. Murata, T. Suzuki, and T. Ito, *Japan Analyst*, **14**, 630 (1965).

173. A. Murata, T. Ito, K. Fujiyasu, and T. Suzuki, *Japan Analyst*, **15**, 143 (1966).

174. A. Murata, T. Ito, and T. Suzuki, *J. Chem. Soc. Japan Pure Chem. Sec.*, **89**, 54 (1968). A. Murata, T. Ito, and T. Suzuki, *Japan Analyst*, **17**, 1284 (1968), have obtained 1:1 and 1:3 metal/reagent complexes of alkyl-substituted 5-hydroxychromones with iron. For the reactions of other 5-hydroxychromone derivatives with iron, see A. Murata, F. Hirano, and T. Suzuki, *Japan Analyst*, **19**, 1346 (1970).

175. A. Murata and T. Ito, *Japan Analyst*, **18**, 1131 (1969). Also a fluorimetric reagent for Zr: T. Ito and A. Murata, *ibid.*, **23**, 274 (1974).

176. A. Murata and T. Suzuki, *Japan Analyst*, **16**, 248 (1967).

177. A. Murata and K. Mizoguchi, *Japan Analyst*, **18**, 1471 (1969). The molar absorptivity in the benzene extract is 7.7×10^3 at 430 nm.

178. T. Ito and A. Murata, *Japan Analyst*, **20**, 335 (1971). 2-Ethyl-5-hydroxy-3-methylchromone is a Sc reagent: M. Nakamura and Murata, *ibid.*, **22**, 1474 (1973).

179. V. A. Nazarenko and N. I. Makrinich, *Zh. Anal. Khim.*, **25**, 719 (1970). Indium forms insoluble red $In(HL)_3$, which as a suspension in aqueous ethanol has $\epsilon_{550} = 45,700$, as shown by N. L. Olenovich, A. A. Bazilevich, V. A. Nazarenko, and O. D. Dira, *ibid.*, **29**, 2287 (1974).

180. V. A. Nazarenko and E. N. Poluektova, *J. Anal. Chem. USSR*, **23**, 1461 (1968).

181. N. S. Poluektov, M. A. Sandu, and R. S. Lauer, *Zh. Anal. Khim.*, **25**, 899 (1970).

182. L. I. Kononenko and N. S. Poluektov, *J. Anal. Chem. USSR*, **15**, 63 (1960).

183. Review: M. Katyal and H. B. Singh, *Talanta*, **15**, 1043 (1968).

184. V. A. Nazarenko and E. N. Poluektova, *J. Anal. Chem. USSR*, **22**, 755 (1967).

185. Review of metal complex formation with tropolone and derivatives: Y. Dutt, R. P. Singh, and M. Katyal, *Talanta*, **16**, 1369 (1969). A table of formation constants is included.

186. G. H. Rizvi and R. P. Singh, *Talanta*, **19**, 1198 (1972). Ru(III) forms a yellow complex in feebly acidic solution ($CHCl_3$– extractable): G. H. Rizvi,

B. P. Gupta, and R. P. Singh, *Anal. Chim. Acta*, **54,** 295 (1971). Rh(III) can also be determined: *Mikrochim. Acta*, 459 (1972).

187. O. Menis and C. S. P. Iyer, *Anal. Chim. Acta*, **55,** 89 (1971).

188. V. A. Nazarenko and E. N. Poluektova, *J. Anal. Chem. USSR*, **19,** 1359 (1964).

189. V. A. Nazarenko and G. V. Flyantikova, *J. Anal. Chem. USSR*, **18,** 156 (1963). 3,4-Dihydroxyazobenzene-4'-sulfonic acid is a similar reagent.

190. S. T. Talipov, R. K. Dzhiyanbaeva, A. Inoyatov, L. V. Chaprasova, and M. Ziyaeva, *Zh. Anal. Khim.*, **25,** 1420 (1970).

191. A. Vioque-Pizarro, *Anal. Chim. Acta*, **5,** 529 (1951).

192. See R. E. Stevens, *Am. Lab.*, **7,** 57 (1975); S. Cohen et al., *J. Am. Chem. Soc.*, **81,** 3480 (1959).

193. P. H. Tedesco and H. F. Walton, *Inorg. Chem.*, **8,** 932 (1969).

194. I. M. Gibalo, V. S. Voskov, and F. I. Lobanov, *Zh. Anal. Khim.*, **25,** 1918 (1970). Nb is extracted into $CHCl_3$ from 10 M HCl. Mo, Zr, Ti, W, Fe, and Ta are said not to interfere. As shown earlier by E. N. Poluektova and V. A. Nazarenko, *ibid.*, **22,** 746 (1967), W is extracted from 1–5 M HCl ($\epsilon_{290} = 6500$). Benzohydroxamic acid has also been used as a separatory reagent for Be preparatory to determination with Beryllon II. Beryllium is extracted from basic solution into cyclohexanone. EDTA prevents extraction of Ca, Mg, and Al, but Ti, Mo, Nb, and V interfere. See N. P. Borzenkova and L. A. Burmistrova, *ibid.*, **27,** 682 (1972).

195. C. E. Meloan et al., *Anal. Chem.*, **32,** 791 (1960).

196. I. P. Alimarin and N. P. Borzenkova, *Anal. Abst.*, **19,** 3872 (1970); Nb and Ta, *ibid.*, **12,** 630 (1965).

197. Introduced by S. C. Shome, *Analyst*, **75,** 27 (1950). The homolog N-phenyl-2-naphthohydroxamic acid is very similar (Y. K. Agrawal, *Anal. Chem.*, **47,** 940). The most vanadium-sensitive derivative of phenylnaphthohydroxamic acid is N-[(p-N,N-dimethylanilino)-3-methoxy-2-naphtho] hydroxamic acid, synthesized and studied by S. A. Abbasi, *Anal. Chem.*, **48,** 714 (1976). In chloroform, $\mathscr{S}_{570} = 0.004$ μg V cm^{-2} (2–6 M HCl). As much as ~50 mg Ti, Zr, Mo, W, Fe, or Pd in 25 ml of sample solution do not interfere appreciably. "... the presence of strong electron-donating substituents in the para position of the N-phenyl fragment of PBHA, an increase in conjugation at the benzo site and introduction of electron-donating groups ortho to the functional $-C{=}O$ group help in enhancing the sensitivity and selectivity of a hydroxamic acid towards vanadium(V) relative to PBHA."

198. B. K. Afghan, R. G. Marryatt, and D. E. Ryan, *Anal. Chim. Acta*, **41,** 131 (1968); M. Ishii and H. Einaga, *Japan Analyst*, **17,** 976 (1968).

199. A. T. Pilipenko, E. A. Shpak, and O. S. Zul'figarov, *Zav. Lab.*, **40,** 241 (1974). Determination of Ti is described ($\epsilon_{550} = 75,000$).

200. U. Tandon and S. G. Tandon, *J. Indian Chem. Soc.*, **46,** 983 (1969); V. K.

Gupta and S. G. Tandon, *Anal. Chim. Acta*, **66**, 39 (1973), where references to other work may be found.

201. U. Priyadarshini and S. G. Tandon, *Analyst*, **86**, 544 (1961).

202. D. C. Bhura and S. G. Tandon, *Anal. Chim. Acta*, **53**, 379 (1971).

203. A. T. Pilipenko, E. A. Shpak, and G. T. Kurbatova, *Zh. Anal. Khim.*, **22**, 1014 (1967).

204. I. M. Gibalo, V. S. Voskov, V. F. Kamenev, and F. I. Lobanov, *Zh. Anal. Khim.*, **27**, 2405 (1972). Mixed complexes are formed with pyrocatechol and thiocyanate.

205. C. P. Savariar and J. Joseph, *J. Indian Chem. Soc.*, **50**, 528 (1973).

206. H. M. El Fatatry, A. W. von Smolinski, C. E. Gracias, and D. A. Coviello, *Talanta*, **20**, 923 (1973).

6C

ORGANIC PHOTOMETRIC REAGENTS
NITROGEN AND
NITROGEN–OXYGEN CHELATION

The reagents discussed in this chapter are those that bond metals through N alone or jointly through N and O. The order is roughly: nonheterocyclic (or partly heterocyclic) N, heterocyclic N, and N, O. Precise classification is impossible because not all metals are necessarily coordinated the same way nor is the type of bonding always known.

Departure from a strictly logical order is sometimes expedient from the analytical viewpoint. Separate chapters are devoted to 8-hydroxyquinoline (Chapter 6D) and azo reagents (Chapter 6E).

In general, reagents binding metals through N only are more selective than those binding through N and O. Metals such as Fe(II), Ni, Co, Cu, Zn, Cd, and those in the Pt group form more or less strong bonds with N. On the other hand, 8-hydroxyquinoline coordinates metals through both N and OH, the latter group in large measure accounting for the complexing of oxyphilic metals.

I. AROMATIC AMINES AND RELATED COMPOUNDS

A. o-PHENYLENEDIAMINE AND OTHER COMPOUNDS CONTAINING NH$_2$ GROUPS

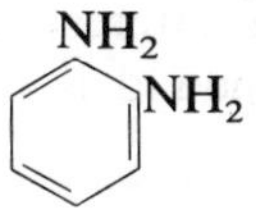

The platinum metals especially bond to the vicinal N atoms in o-phenylenediamine and form colored complexes. Platinum in particular reacts sensitively.[1] A blue chelate of Pt(II) is formed in hot neutral solution, soluble in dimethylformamide. EDTA is of some use in masking other Pt and non-Pt metals. $\epsilon_{703} = 98,000$; $\mathscr{S}_{703} = 0.002 \, \mu g$ Pt. Platinum(IV) is reduced to Pt(II) by the reagent, which in the form $C_6H_4N_2H_2^{2-}$ reacts in the ratio 2:1(Pt) according to Golla and Ayres. It may be noted in passing that various heavy metals can coordinate to N in o-phenylenediamine without replacing H. Zinc, Ni, and Co form complexes in which 2, 4, or 6 molecules of o-phenylenediamine are coordinated to an atom of metal. Phenylenediamines are easily oxidized by metals in higher oxidation states (e.g., Fe(III) and Cu(II)) and by O$_2$. o-Phenylenediamine thus gives yellow solutions absorbing maximally at 430 nm.

3,4-Diaminobenzoic acid is a reagent for Pt(IV)[2] and Ru(III)[3]. Platinum(IV) is reduced by the reagent and Pt(II) reacts with the reagent in the ratio 1(Pt):2. No doubt Pt is coordinated to vicinal N atoms, but the composition of the (blue) complex remains undetermined. Ruthenium(III) reacts in 1:3 and 1:2 ratios; the composition of the colored species (brown to purple-red) has not been definitely established. Many other aromatic compounds with vicinal NH$_2$ groups reacts similarly.[4]

It may be noted that many compounds containing –NH$_2$ can coordinate with heavy metal salts. Thus benzidine (carcinogenic) and thiocyanate form a blue precipitate with Cu(II), which may be

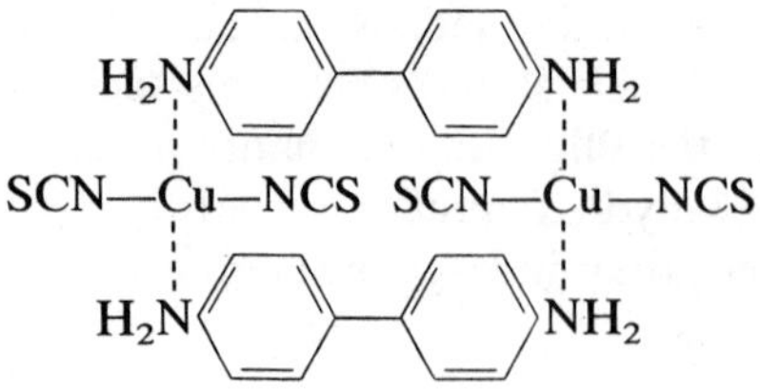

$$
H_2N\!-\!\langle\rangle\!-\!\langle\rangle\!-\!NH_2
$$
$$
SCN\!-\!Cu\!-\!NCS \quad SCN\!-\!Cu\!-\!NCS
$$
$$
H_2N\!-\!\langle\rangle\!-\!\langle\rangle\!-\!NH_2
$$

Tolidine reacts similarly.

B.   8-AMINOQUINOLINE

A sensitive reagent for Pd ($\mathscr{S}_{590} = 0.0029\ \mu g/cm^2$ in $CHCl_3$). The manner of reaction with Pd is interesting. In dilute HCl (pH $<\sim 2$) a yellow insoluble addition compound is formed, which is not soluble in chloroform:

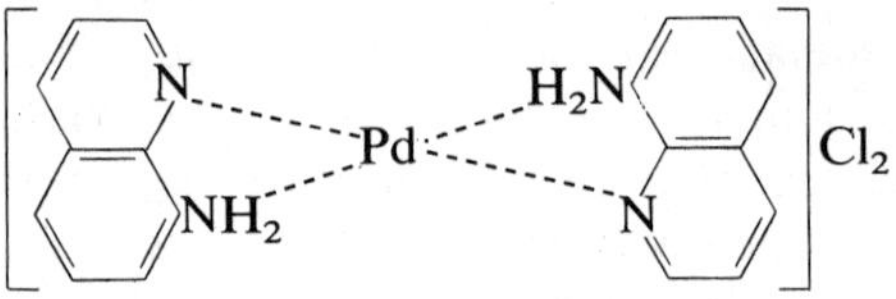

$$
\left[\ \text{(quinoline)}\,N\cdots Pd\cdots H_2N\,\text{(quinoline)}\ \right]Cl_2
$$

On addition of base to make the pH >10, the insoluble violet chelate

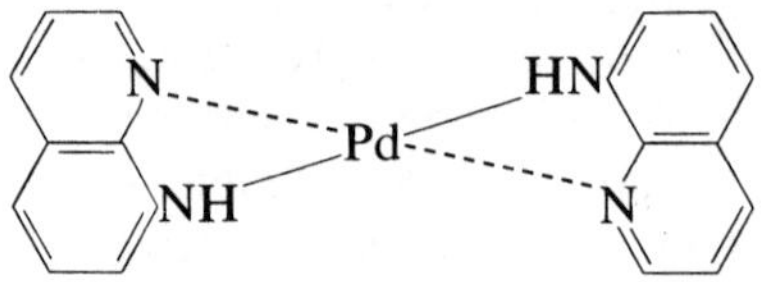

$$
\text{(quinoline)}\,N\cdots Pd\cdots HN\,\text{(quinoline)}
$$

is formed, which is soluble in chloroform.[5] Ir(IV) reacts less sensitively than Pd. Cu(II) and Co(II?) also give extractable complexes, but EDTA prevents their reaction.

An indirect method for Fe(III) has been suggested in which 8-aminoquinoline is oxidized to a red-violet product. There are too many good reactions for iron to make this reaction of any interest.

A class of reagents closely related to 8-aminoquinoline is the 8-sulfonamidoquinolines, which are said to form chelates only with Ag, Hg(II), Cu(II), Pb, Co, and Zn.[6] The chelates are colored and insoluble in water and in common polar and nonpolar solvents. Their composition is

given by

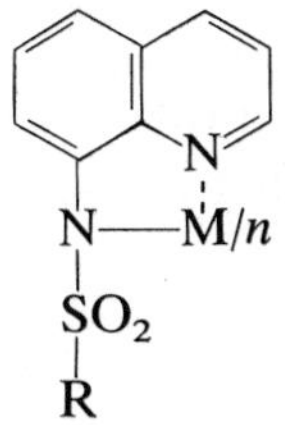

$$N \text{---} M/n$$
$$\underset{R}{\overset{SO_2}{|}}$$

Silver forms an adduct AgL·HL.

The much reduced reactivity of H bound to N compared to H attached to O or S (8-hydroxyquinoline and 8-mercaptoquinoline) is evident.

1-Naphthylamine-4,6,8-trisulfonic acid reacts at pH ~2 with Os(VI) (violet) and Ru(VI) (red), forming 2:1 complexes of undetermined structure.[7]

C. *p*-NITROSOPHENYLAMINO COMPOUNDS

These reagents are placed here somewhat arbitrarily. Compounds containing the group p-NOC$_6$H$_4$N< form strongly colored, slightly soluble complexes with Pd(II), Pt(II), Rh(III), Ru, and some other metals in weakly acidic (chloride) solution. *p*-Nitrosodimethylaniline

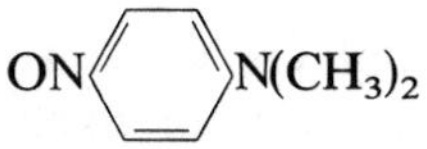

ON⟨ ⟩N(CH$_3$)$_2$ Dark green solid, m. p. 85°–87°. Mol. wt 150.2. Soluble in alcohol.

and *p*-nitrosodiphenylamine (and analogous compounds)

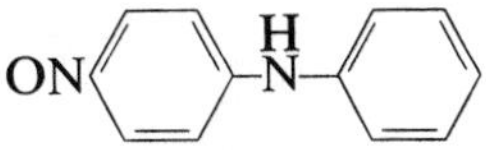

ON⟨ ⟩$\overset{H}{N}$⟨ ⟩   Dark green or blue solid, m. p. 144°–145°. Mol. wt. 198.2. Soluble in alcohol, CHCl$_3$.

are sensitive Pd and Pt(II) reagents. The Pd complex formed by nitrosodimethylaniline has the composition Pd[NOC$_6$H$_4$N(CH$_3$)$_2$]$_2$Cl$_2$, with uncharged reagent molecules and Cl filling four coordination positions around Pd. Whether Pd is bound to N of the nitroso group or to N of the amino group is not known (but presumably to the former). The NO group is at least necessary for a chromophoric effect. Dimethylaniline gives no color with Pd. The reagent itself is colored, but there is a difference of about 50 nm between λ_{max} of reagent and chelate.

At least the Pd complex of nitrosodiphenylamine is extractable into chloroform and other solvents.

The color given by *p*-nitrosodiethylaniline with Ru is weaker than that given by p-nitrosodimethylaniline, and p-nitrosodiphenylamine gives little if any color. The difference may be due to steric effects. p-Nitrosodimethylaniline is a sensitive reagent for Rh ($\mathscr{S}_{510} = 0.0015$), but not selective; heating is required (pH ~4.5).

In passing it may be noted that the group

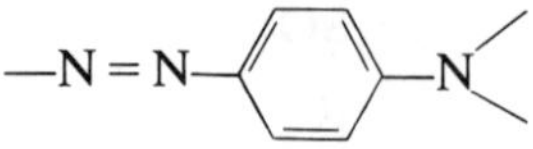

$$-N = N-\!\!\!\bigcirc\!\!\!-N\big\langle$$

is reactive with Pd.[8] Thus Methyl Orange reacts with Pd.

II. HYDRAZONES AND RELATED COMPOUNDS

Hydrazones are characterized by the grouping $\rangle C{=}N{-}N\langle$. Various substituted hydrazones have been examined as photometric reagents for metals, especially the divalent transition metals.[9]

Compounds of the type

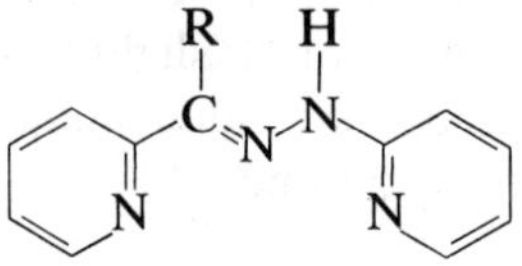

(—N= is protonated in acid solution. For R = H, $pK_1 = 2.9$, $pK_2 = 5.7$.)

functioning as terdentate ligands form cationic complexes with many metals, which can lose the imino H in each ligand in less acid or alkaline solution to give uncharged complexes extractable into organic solvents.[10]

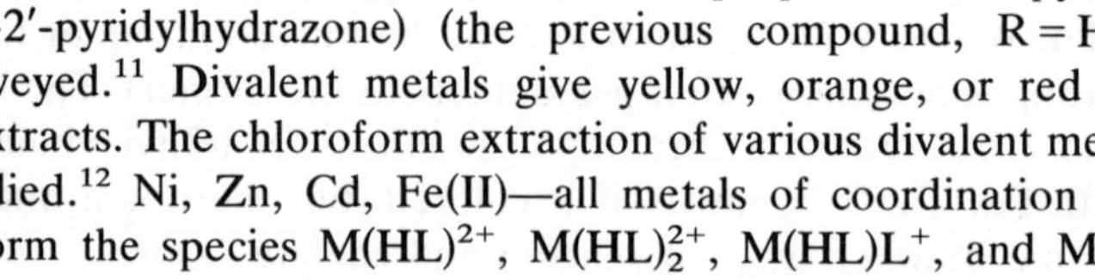

The reactions of 1,3-bis(2'-pyridyl)-1,2-diazaprop-2-ene (or pyridine-2-aldehyde-2'-pyridylhydrazone) (the previous compound, R = H) have been surveyed.[11] Divalent metals give yellow, orange, or red organic solvent extracts. The chloroform extraction of various divalent metals has been studied.[12] Ni, Zn, Cd, Fe(II)—all metals of coordination number 6—can form the species $M(HL)^{2+}$, $M(HL)_2^{2+}$, $M(HL)L^+$, and ML_2, and the last species extracts into chloroform. Palladium and Cu(II), having a coordination number of 4, form PdL^+ and CuL^+ and these can be extracted as ternary complexes such as PdLCl and CuLCl. Copper can also be extracted as CuL_2. The Pd reaction has analytical value.[13] PdLCl is extracted into o-dichlorobenzene from aqueous solution of pH ~3 (red extract). Though sensitive ($\mathscr{S} = 0.0065$ μg Pd/cm^2), the method is subject to interference from many metals. Other divalent metals can be determined even more sensitively.(as ML_2), for example, $\mathscr{S}_{470}$ for Zn (CHCl$_3$) = 0.001 μg/cm^2, but selectivity is poor.

   2-(3'-Sulfobenzoyl)pyridine-2-pyridylhydrazone

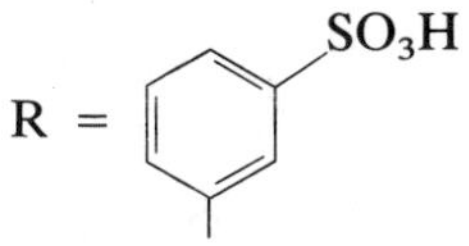

$$R = $$ SO$_3$H

in formula on p. 358) forms soluble, intensely colored (yellow) chelates with Zn, Cd, and Hg in basic solution and an orange chelate with Co(II → III).[14] These complexes are deprotonated above pH 11, and their charge is 2− (from the sulfonic acid groups) for a divalent metal. They are retained on anion-exchange resins and may be useful for isolation of traces of the complexed metals. The protonated complexes (uncharged) are formed in neutral or slightly acidic solution and are essentially colorless. The deprotonated chelates of Co, Zn, Cd, and Hg have ϵ_{max} ranging from ~3×10^4 to 5×10^4.

Pyridine-2-aldehyde-2-quinolylhydrazone

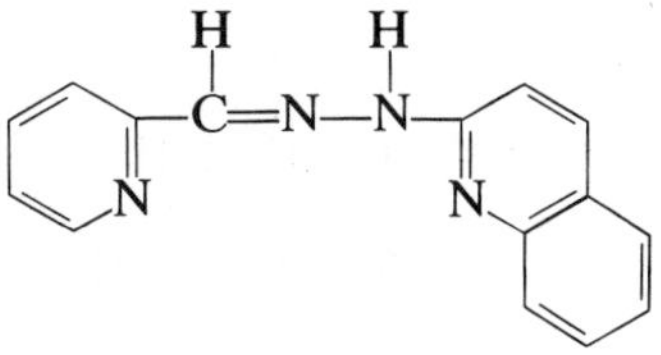

forms colored chelates with Pd, Cu, Ni, Co(III), Fe(II), Zn, and Cd.[15] The H attached to −N− is metal-replaceable and many divalent metals are extractable as ML_2 into inert and active solvents from weakly acidic solutions; Co is extracted as CoL_2X (X = ClO$_4^-$, etc.). The reagent itself forms the species H_2L^+, HL, and L^- in aqueous solution. Isoamyl alcohol extracts some $H_2L \cdot ClO_4$ as well as HL. Benzene is not as good an extractant as isoamyl alcohol and MIBK. The reagent is of practical value in the determination of nickel; with masking of iron and other metals, nickel can, for example, be determined sensitively in sea water. The chloroform extract of the zinc complex is fluorescent.[16]

Similar reagents include quinoline-2-aldehyde-2-quinolylhydrazone, phenanthridine-6-aldehyde-2-pyridylhydrazone and phenanthridine-6-aldehyde-2-quinolylhydrazone, which form red chelates with divalent transition metals, mostly in weakly basic solution, extractable into sundry organic solvents.[17] The first named reagent (H_2L) gives (CuOH)HL extractable into nitrobenzene.[18]

2,2'-Dipyridyl-2-pyridylhydrazone

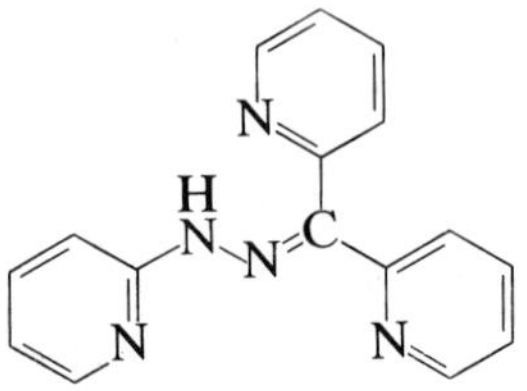

forms colored complexes with many metal ions including cobalt. After complexation, if a strong acid (perchloric acid) is added, the reaction with cobalt(III) becomes selective.[19]

N-Heterocyclic hydrazones are fluorescence reagents for zinc. Benzimidazole-2-aldehyde-2-quinolylhydrazone is the most sensitive of a number tested.[20] At pH 8 in ethanol solution, it allows the determination of as little as 0.001 ppm Zn ($\lambda_{excit} = 470$ nm, $\lambda_{emis} = 520$ nm).

The *anti* isomer of 2-benzoylpyridine-2-pyridylhydrazone chelates Cu, Zn, Ni, and other metals, forming strongly colored complexes.[21]

Biscyclohexanoneoxalyldihydrazone

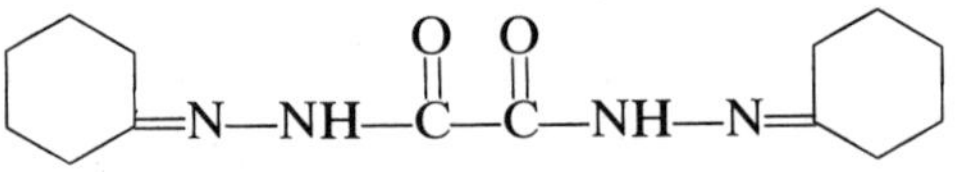

is a sensitive Cu(II) reagent (blue color, pH ~8),[22] $\mathscr{S}_{595} \sim 0.004 \ \mu g$ Cu/cm^2.

2-Hydroxy-1-naphthaldehyde benzoic acid hydrazone is a fluorescence reagent for Al.[23] It is not especially sensitive. 8-Hydroxyquinaldehyde-8-quinolylhydrazone forms a 1:1 fluorescent complex with Ca in 0.1 M KOH.[24]

Bis(6-methyl-2-pyridyl)glyoxal dihydrazone

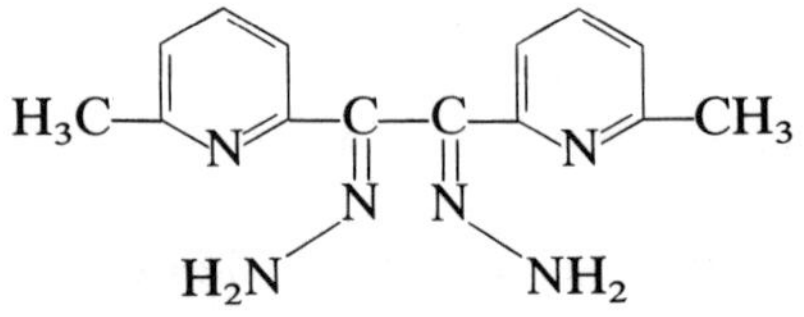

is a cuproine-type reagent for Cu(I); Pd reacts less sensitively.[25]

Bis(4-hydroxybenzoylhydrazones) of glyoxal, methylglyoxal, and di-methylglyoxal have been proposed for the determination of Ca, Cd, and other metals (color and fluorescence):[26]

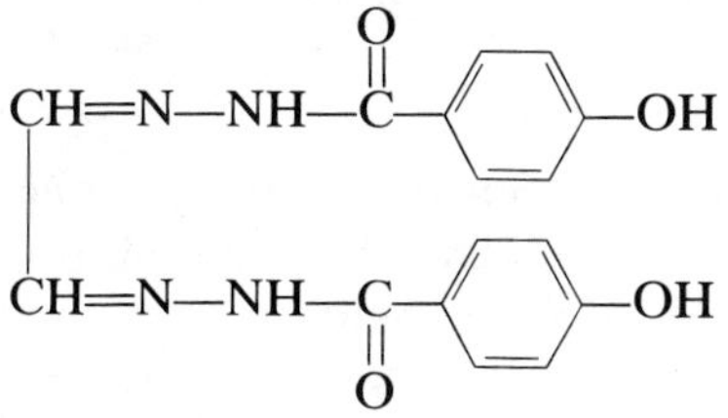

Glyoxal bis(4-hydroxybenzoylhydrazone)

Formazans

These contain the grouping

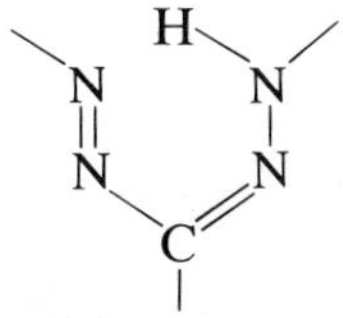

Those having hydroxy, carboxy, arsono, and heterocyclic radicals have been synthesized, and their color reactions with metal ions (divalent and Ag⁺) and the extractive properties of the chelates have been examined.[27] Formazans having heterocyclic radicals are potential extraction-photometric reagents. Reactions are usually sensitive, for example, Cd with 1,3-diphenyl-5-(2-thiazolyl)formazan gives $\epsilon_{625} = 57,500$, but selectivity is poor.

3-Acetyl-1,5-bis(3,5,6-trichloro-2-hydroxyphenyl)formazan forms a green 2:1 (Sc) complex with Sc, used in an extraction photometric method ($\epsilon_{675} = 27,000$).[28]

Acyclic aromatic azines have been proposed for the determination of Co.[29] 3-Hydroxypicolinealdehyde azine

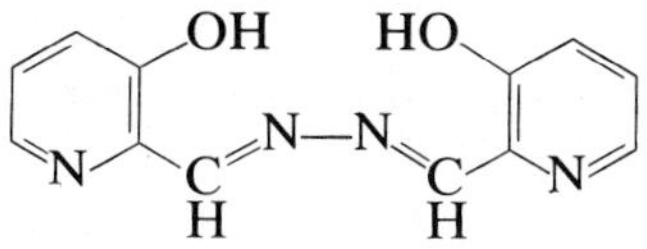

gives a sensitive reaction ($\mathscr{S}_{545} = 0.002 \ \mu g \ Co/cm^2$), but the selectivity is poor.

Hydroxytriazenes[30]

3-Hydroxy-1-*p*-sulfonatophenyl-3-phenyltriazene   reacts   sensitively with Pd(II) at pH 2–4 ($\mathscr{G}$ = 0.0036 μg/cm^2).[31] The following structure is suggested for the chelate

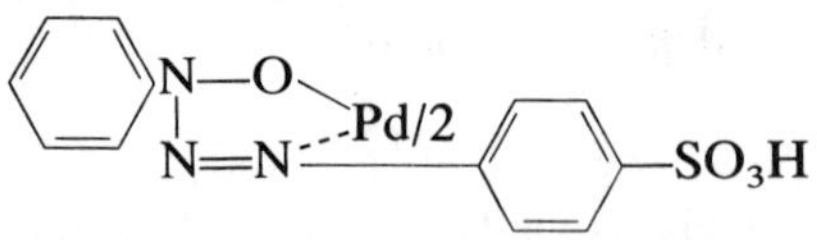

1-(4-Sulfonato-5-methylphenyl)-3-hydroxy-3-phenyltriazene is used for determination of vanadium; the method is not selective.[32] 1-(2-Carboxyphenyl)-3-hydroxy-3-methyltriazene is an extractive photometric reagent for Fe and Ti,[33] and 1-(2-carboxyphenyl)-3-hydroxy-3-phenyltriazene for Fe, Ti, and V.[34] 1-(2-Carboxy-4-sulfonatophenyl)-3-hydroxy-3-phenyltriazene is used for determination of Fe, Mo, and V in aqueous solution.[35] Titanium does not give a color. The 5-sulfonato derivative is used for determination of Fe and V.[36] Cu, Pd, Ti, and other metals interfere.

III.   REAGENTS HAVING $-N{=}\overset{|}{C}{-}\overset{|}{C}{=}N-$ AS REACTIVE GROUP

A.   GENERAL—1,10-PHENANTHROLINE   AND   COGNATE   COMPOUNDS

F. Blau reported the preparation of 2,2-bipyridine in 1888 and 1,10-phenanthroline in 1898, and discovered their color reactions with Fe(II). Analytical chemists were slow to make use of these compounds, the first applications as chromogenic reagents for iron and as indicators occurring around 1930. G. F. Smith and coworkers prepared and studied many derivatives and compounds related to 1,10-phenanthroline. A monograph covers the field comprehensively up to 1967.[37]

The compounds of this class are characterized by the presence of the α,α'-diimine grouping (ferroin chain)

$$-N{=}\overset{|}{C}{-}\overset{|}{C}{=}N-$$

in an aromatic system. They form five-membered chelate rings with metals. The chelates of Fe(II) and Cu(I) are especially noteworthy for serving as the basis of selective (essentially specific for Cu) chromogenic methods for the determination of these metals. Cobalt(II) also reacts sensitively with some of these reagents. The complexes formed are of the ammine type, the metal being bound to nitrogen. Unlike most useful organic reagents, hydrogen atoms are not replaced and extraction into an immiscible solvent requires the participation of an anion to form an ion associate. The color (red, violet, orange) of the Fe(II), Cu(I), and other metal complexes results from charge-transfer.

1,10-*Phenanthroline*

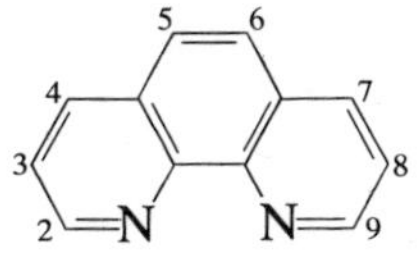

Mol. wt. (monohydrate) 198.2. M. p. 98° (anhydrous 117°). Sparingly soluble in water (~ 0.3 g/100 ml, room temperature); more soluble in alcohol, ether, benzene, etc. Can be purified by recrystallization from $C_2H_5OH{-}H_2O$.

forms phenanthrolinium ion by protonation of one of the N atoms

$$Ph + H^+ \rightleftharpoons PhH^+ \qquad \frac{[Ph][H^+]}{[PhH^+]} = 1.4 \times 10^{-5} \ (25°, \ \mu = 0)$$

so that the extent of the formation of its metal complexes, for example, MPh_p^{n+}, is a function of the hydrogen-ion concentration. Above $1 \, M \, H^+$, PhH_2^{2+} begins to be formed.

Phenanthroline forms cationic complexes with a great many metals, some of which are quite stable (Table 6C-1). This means that when Fe(II) is determined in the presence of such metals a sufficient excess of reagent must be added to chelate these metals as well as iron. The rather slow reaction of some metals [e.g., nickel and vanadium (IV)] with the reagent allows larger amounts to be present than would be permissible under equilibrium conditions. In the determination of iron, the reduction of Fe(III) must be carried out under conditions such that Cu(II) is not reduced also. Once formed, the phenanthroline (and bipyridine) complexes of Fe(II), Ni, the Pt metals, and V(IV) tend to dissociate slowly if the acidity is increased; the Zn, Cd, and Cu chelates dissociate more rapidly.

With an excess of 1,10-phenanthroline, Fe(II) forms the *tris* chelate $FePh_3^{2+}$ (orange-red). The lower chelates are only weakly colored.

Ruthenium(II), obtained by reducing Ru(III or IV) with hydroxylammonium chloride, reacts slowly with 1,10-phenanthroline in hot neutral solution to give a yellow tris chelate. The chelates of Ru(II) and Ir(III) are luminescent in aqueous solution. Vanadium has been determined in ammoniacal solution, after reduction to V(II) with dithionite, as VPh_3^{2+} ($\epsilon_{645} = 8000$).[38]

2,2'-*Bipyridine* (*or* 2,2'-*dipyridyl*)

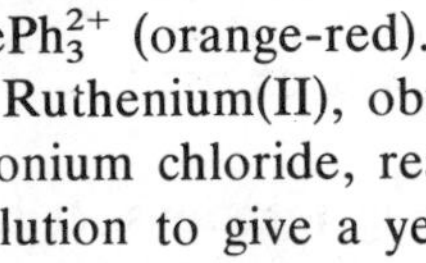

Mol. wt. 156.2. M. p. 70°. Sublimes (a method of purification). Sparingly soluble in water ($\sim 0.5\%$, room temperature); soluble in alcohol and other organic solvents. K_a of Bipy·$H^+ = 4.5 \times 10^{-5}$ ($25°$, $\mu = 0$).

TABLE 6C-1
Overall-Stability Constants of Some Phenanthroline Chelates and Related Chelates[a] ($25°$, $\mu = 0.1$ usually)

Metal	Chelating Agent	$\log \beta_3$
Cd	Ph	14.3
	2,9-Dimeth-Ph	~10.4
Co	Ph	19.8
	Bipy	15–16
Cu(I)	Bipy	14.2 ($\log \beta_2$)
Cu(II)	Ph	21
	2-Cl-Ph	13.9
	Bipy	17
Fe(II)	Ph	21.2
	5-Meth-Ph	21.9
	2-Meth-Ph	10.8
	2-Cl-Ph	11.6
	5-Cl-Ph	19.7
	Bipy	17.5
	Ferrozine	15.5
Fe(III)	Ph	14.1
	Bipy	12.0
Hg(II)	Ph	23.3
Mn(II)	Ph	10.3
Ni(II)	Ph	24
	2-Cl-Ph	13.6
Zn	Ph	17.0
	2-Meth-Ph	12.7
	Bipy	13

[a] From various sources. For a more extensive list see Schilt, *op. cit.* Ph = 1,10-phenanthroline, Bipy = 2,2′-bipyridine, Meth = methyl. Stepwise stability constants ($k_1 \cdots$) are also tabulated in Schilt's monograph. For stabilities of Cl-, OH-, CN-, and carboxy-substituted 1,10-phenanthroline chelates of Fe(II), see C. J. Hawkins, H. Duewell, and W. F. Pickering, *Anal. Chim. Acta*, **25**, 257 (1961).

is very similar to 1,10-phenanthroline in its chelating reactions. It is a little less sensitive as an iron(II) reagent, and the formation constant of its iron complex is a little smaller than the constant of the iron complex of 1,10-phenanthroline. It can usually be substituted for 1,10-phenanthroline, but has no advantages save one—it is considerably lower in price.

1,10-Phenanthroline and bipyridine form mixed chelates, which are mentioned elsewhere under other chelating agents. Also it may be noted in passing that the protonated bases LH^+ can form slightly soluble ion-pair compounds with ClO_4^-, $TeCl_6^{2-}$, and similar ions.

$FePh_3^{2+}$ forms ion pairs, especially with anions of large size. The ion associate with ClO_4^- is extractable into nitrobenzene. With more organophilic anions, for example, SCN^-, organic solvents of lower dielectric constant can be used as well. Methyl Orange anion (MO^-) forms $FePh_3(MO)_2$, extractable into $CHCl_3$. The sensitivity of the iron determination is thus increased ($\epsilon_{420} \sim 48{,}000$) but the selectivity is much decreased (Cu, Zn, etc., give extractable ion associates),[39] so that the reaction is of little practical value.

1,10-Phenanthroline (and bipyridine) reacts with Mo(VI) in $HCl–SnCl_2$ solution to form a red-violet complex of uncertain composition. It has been reported to contain Mo(VI) and Mo(V), but Mo(III) is thought more likely. A mixed chelate forms in the presence of thiocyanate.[40]

Effect of Substituents. Replacement of H in 1,10-phenanthroline (and related compounds) by alkyl and aryl groups, Cl, Br, OH, NO_2, etc. leads to a change in the electron distribution in the molecule and therewith in the basicity of N, its chelating ability and in the spectral characteristics of the chelates (λ_{max}, ϵ). Substitution of phenyl groups for H in the 4- and 7-positions (or in the 4- and 4'-positions in 2,2'-bipyridine) increases ϵ_{max} of the Fe(II) and Cu(I) chelates roughly twofold. In general, $\log \beta$ of the chelates increases with pK_a. The introduction of alkyl and aryl groups decreases the aqueous solubility and increases the extractability of phenanthroline and its complexes. Polar groups such as OH, NH_2, and especially SO_3H[41] increase the solubility of the reagent and its chelates.

Replacement of H by OH in positions 4 and 7

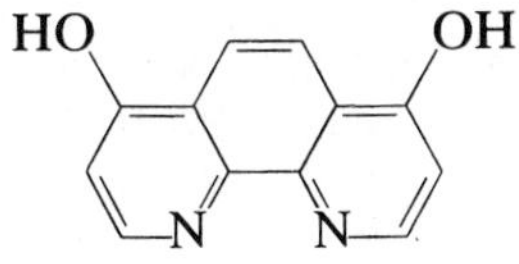

4,7-Dihydroxy-1,10-phenanthroline

gives a reagent with weakened complexing power for Fe(II) in neutral or feebly acidic solution. Very likely a tautomeric equilibrium exists, in which the amide form

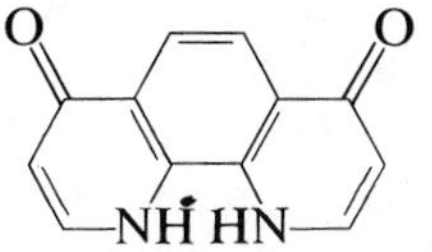

is favored and this has only weak complexing power. However, in strongly alkaline solution the OH groups of 4,7-dihydroxy-1,10-phenanthroline are ionized, and the dianion reacts sensitively with

Fe(II).[42] Incidentally, 6-hydroxy-1,7-phenanthroline forms colored complexes with Fe(III) and V(V), but these reactions are not of the type considered in this section. The chelates formed are analogous to those given by 8-hydroxyquinoline (five-membered ring involving N, O, and Fe or OH–V$=$O, H of OH being replaced).

The steric effect of substituents is of the greatest analytical interest. The introduction of a bulky group next to N in 1,10-phenanthroline makes difficult[43] or impossible the coordination of three reagent molecules to a cation such as Fe^{2+} requiring this number to fill all its coordination positions (weakly colored mono or bis complexes can be formed, however). Thus 2,9-dimethyl-1,10-phenanthroline gives no significant color with Fe(II). Three molecules of this reagent (Neocuproine) cannot be arranged around Fe^{2+} to fill the six coordination positions required for octahedral coordination of this ion. On the other hand, 2,9-dimethyl-1,10-phenanthroline reacts with Cu(I) to give orange CuL_2^+ (or an equivalent ion pair). Only two molecules of the reagent are required to fill the four coordination positions of Cu(I) (tetrahedral configuration). Similarly, 2,9-dimethyl-4,7-diphenyl-1,10-phenanthroline gives a colored chelate with Cu(I) but not with Fe(II).

The same steric effect is shown by 2,2'-biquinoline

Mol. wt. 256.3. M. p. 193°–196°. Solubility in water $2.5 \times 10^{-7} M(\mu = 1)$, in isoamyl alcohol 0.0078 M. Can be recrystalized from ethanol. pK_a of Bq·H$^+ = 1.0 \times 10^{-4}(26°)$.

(sometimes called Cuproine[44]), which is a good absorptiometric reagent for Cu(I), but which does not react with Fe(II).

The grouping

$$R—\overset{\|}{C}—N=\overset{|}{C}—\overset{|}{C}=N—\overset{\|}{C}—R' \qquad (R \text{ and } R' \neq H)$$

(cuproine chromophore) has been called specific for Cu(I). This statement requires qualification. Apparently no metal other than Cu(I) has been found to give strongly colored chelates extractable into immiscible organic solvents (such as amyl alcohol). However, some metals can form mono or bis complexes extractable into organic solvents, though the colors are not strong. Thus in $CHCl_3$, $CoCl_2$·(DMP) has $\epsilon_{660} = 495(max)$, and little contribution would be made to the absorbance of Cu(I) at 455 nm (DMP is 2,9-dimethyl-1,10-phenanthroline).

2,2'-Biquinoline reacts with Cu(I) in very weakly acidic solution, whereas 4,4'-dihydroxy-2,2'-biquinoline reacts in strongly basic solution (even 12 M NaOH).[45] The colored species CuL_2^{3-} is extracted into isoamyl alcohol from hydroxide solution, presumably as M_3CuL_2 (M = Na or K). The reaction is termed specific for Cu(I). The apparent pK_a for $HL^- \rightleftharpoons L^{2-} + H^+$ is ~12.

The spectrophotometric properties of the cuprous and ferrous chelates of 1,10-phenanthroline, some substituted 1,10-phenanthrolines and some related reagents (mostly pre-1967) are listed in Table 6C-2. In the following paragraphs, a number of other reagents (mostly post-1967) of this class are noted.

TABLE 6C-2
Spectral Characteristics of Cu(I) and Fe(II) Chelates of 1,10-Phenanthroline and Some Related Reagents[a]

Reagent	Cu(I)		Fe(II)	
	λ_{max}(nm)	$\epsilon \times 10^{-3}$	λ_{max}(nm)	$\epsilon \times 10^{-3}$
1,10-Phenanthroline	435	7.0(7.25)	508	11.1
2,9-Dimethyl- (Neocuproine)	455	7.95	—	—
		(isoamyl alcohol)		
4,7-Diphenyl- (Bathophenanthroline)	457	12.1	533	22.4
2,9-Dimethyl-4,7-diphenyl-(Bathocuproine)	479	14.2	—	—
2,2'-Bipyridine	435	4.5	522	8.7
4,4'-Diphenyl-	463	9.6	552	21.1
2,2',2''-Terpyridine[b]	—	—	552	12.5
4,4',4''-Triphenyl-	—	—	583	30.2
2,2'-Biquinoline	545	6.5	—	—
4,4'-Diphenyl-	559	9.0	—	—
4,4'-Dihydroxy-	525	6.9	—	—
		(isoamyl alcohol)		
4,4'-Bis-(4-ethoxycarbonylanilino)-	556	17.4		
		(butanol)		
2,4,6-Tripyridyl-1,3,5-triazine (TPTZ)			594	22.6
3-(2-Pyridyl)-5,6-diphenyl-1,2,4-triazine (PDT)	488	8.0	555	23.5
3-(2-Pyridyl)-5,6-bis(4-phenylsulfonic acid)-1,2,4-triazine (Ferrozine)	470	4.3	562	27.9

[a] From various authors. A more extensive list is given by Schilt, *op. cit.* Spectral data for $Cu(TMB)_2X_2$ in chloroform and isoamyl alcohol and for $FeCl_2(TMB)$, $CoCl_2(TMB)$, and $NiCl_2(TMB)$ in chloroform are reported by J. R. Hall, M. R. Litzow, and R. A. Plowman, *Anal. Chem.*, **35**, 2124 (1963). (TMB is 4,4',6,6'-tetramethyl-2,2'-bipyridine.) These authors point out that 2,9-dimethyl-1,10-phenanthroline (DMP) forms mono and bis complexes with Fe^{2+}, Ni^{2+}, and Co^{2+}, and that $Cu(DMP)_2Cl$ (or bromide or iodide) is dissociated to CuCl(DMP) in chloroform, unless excess DMP is present; isoamyl alcohol is a better extractant.

[b] Protonated 2,2',2''-terpyridine, RH^+, also combines with Fe^{2+}, forming $R_2H_2Fe^{4+}$.

B. SUNDRY $-N{=}\overset{|}{C}{-}\overset{|}{C}{=}N-$ REAGENTS

A swarm of other compounds having this reactive group have been synthesized and investigated as reagents, chiefly for iron and copper. No attempt is made here to give a complete listing.

A series of hydroxy-substituted 1,10-phenanthrolines having CH_3- in the 2-position and CH_3- or C_6H_5- in the 9-position have been evaluated as Cu(I) reagents in strongly alkaline solution.[46] The most sensitive of these is 2,9-dimethyl-4,7-dihydroxy-1,10-phenanthroline, which gives a yellow chelate, probably $(CuL_2)^{3-}$, above pH 11 ($\epsilon_{400} = 11{,}500$). The chelate is stable in concentrated NaOH solutions. It is not extractable by organic solvents (unless bulky organic cations are added). Co, Zn, and Fe interfere above 10 ppm. The alkaline reaction medium is of advantage when Cu is to be determined in NaOH, Na_3PO_4, NH_4OH. Cu(II) gives a much less sensitive reaction (blue chelate, presumably CuL_2^{2-}).

3,3'-Bipyridazyl

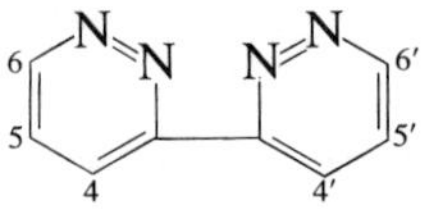

and its derivatives are reported to react more sensitively (about twofold) than 2,2'-bipyridyl with Fe(II).[47]

Bis-3,3'-(5,6-dimethyl-1,2,4-triazine)[48]

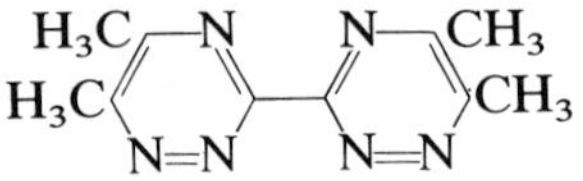

forms orange FeL_3^{2+} ($\log \beta_3 = 9.6$) as well as colored chelates with Cu(I), Co, and Ru(II). The metal is bonded to the 2,2'-nitrogen atoms. The Fe reaction is sensitive ($\epsilon_{493} = 15{,}000$). Ni, Cu, and Cr(III) interfere. Prolonged heating is required to form the Ru complex, and it is sensitive to pH.

2,4,6-Tris(2'-pyridyl)-s-triazine (TPTZ)

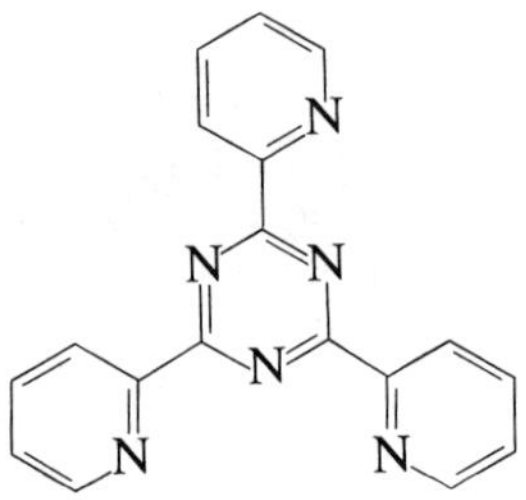

is a reagent for Fe(II)[49] and Ru(III).[50] It has also been suggested for Co,[51] which forms CoL_2^{2+}.

3-(2-Pyridyl)-5,6-diphenyl-1,2,4-triazine

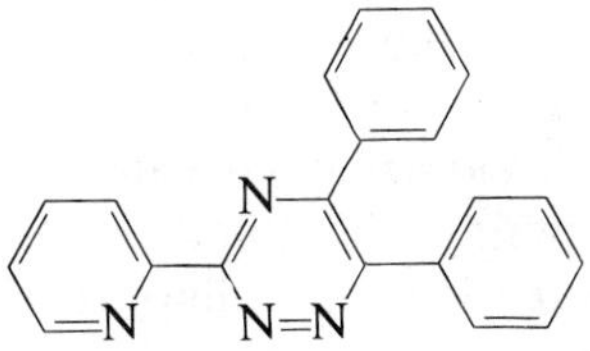

Mol. wt. 310.3. M. p. 189°–190°. $[H^+][L]/[HL^+] = 1.1 \times 10^{-3}$ (25°)

is of interest as a reagent that allows determination of iron in mineral acid solution by an extractive procedure.[52] In the presence of thiocyanate as anionic partner, the ferrous chelate is extracted into $CHCl_3$ as the ion associate $(FeL_3)(SCN)_2$. The success of this extraction rests on the very large partition constant of this complex, $P_{CHCl_3/H_2O} > 10^5$. At equilibrium the concentration of L in the aqueous phase is very low (LH^+ is formed in the acid solution; moreover, L partitions into $CHCl_3$ as the uncharged molecule, and probably partly as (LH)(SCN)). In spite of this, enough $(FeL_3)(SCN)_2$ is formed in the aqueous phase to lead (multiplication by 10^5) to the partitioning of most of the iron into chloroform from 1 to 3 M HCl. Perchloric acid and H_2SO_4 solutions, up to 4 M, can also be used. The concentration of SCN^- (as NaSCN) in the aqueous phase should be ~0.5 M. Chloroform was found to be the best extractant of a number of solvents tested. $\mathscr{S}_{555} = {\sim}0.002\ \mu g\ Fe/cm^2$ ($\epsilon = 24{,}500$). Thiocyanate reduces Fe(III), but it is best to add ascorbic acid. The highest tolerable concentrations of some interfering ions are (ppm): Cr 10, Co 2, Cu 50, Mn 50, Ni 2, V 2.

2,4-Bis(5,6-diphenyl-1,2,4-triazino-3-yl)pyridine

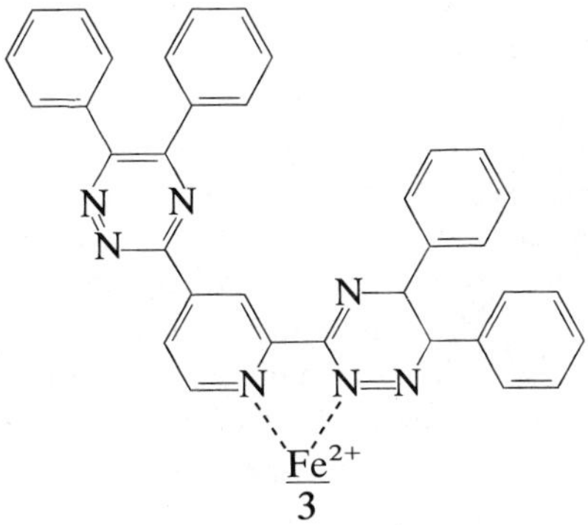

forms a violet Fe(II) chelate having $\epsilon_{563} = 3.2 \times 10^4$ ($\mathscr{S} \sim 0.002$) in $CHCl_3$, the highest ϵ value found for Fe(II)–ferroin chelates[53].

Investigation of a number of substituted triazines related to 6-cyano-2,2'-bipyridine did not reveal any compound superior to the iron and copper reagents then known.[54] The results lead to the conclusion that no benefit accrues from extending the ferroin grouping beyond that of tridentate functionality. Later work showed that pyridyl and isoquinolyl derivatives of phenyl- or pyridyl-substituted 1,2,4-triazine have analytical utility (p. 369).[55] Synthesis of 3-(4-phenyl-2-pyridyl)-5,6-(diphenyl-4,4'-disulfonic acid)-1,2,4-triazine diammonium salt for determining iron(II) has been described.[56] Cu(I) and Co(II) form colored chelates but with significantly lower intensities: $\epsilon_{570}^{Fe} = 29,300$. $\epsilon_{470}^{Cu} = 5200$, $\epsilon_{530}^{Co} = 3500$.

Examination of pyrido and pyridyl derivatives of quinoxaline and phenazine showed some of the quinoxaline derivatives to be promising reagents for Fe(II), but they are difficult to synthesize.[57] They have no advantages over earlier reagents for the determination of Cu(I). Similar conclusions were reached concerning some 26 derivatives of 2,3-bis(2-pyridyl)quinoxaline as copper reagents.[58]

6,7-Dimethyl-2,3-di(2-pyridyl)quinoxaline

has been shown to be a good reagent for the determination of traces of Cu in the presence of much Fe, Ni, Co, Zn, and other metals.[59] CuL_2X is extracted into nitrobenzene ($\lambda_{max} = 514$ nm). In the presence of Methyl Orange, a yellow-green ion-association complex is formed, which can be extracted into the more pleasant solvent ethylene dichloride at pH 5. No doubt the extracted compound is $(CuL_2^+)(MO^-)$. $\epsilon_{418} = 29,300$. The absorption is now due mostly to MO and the sensitivity is increased.

A study of pyridyl-substituted pyrazines demonstrated that 2,3,5,6-tetrakis(2'-pyridyl)pyrazine is a sensitive reagent for Fe(II) ($\epsilon_{570} = 2 \times 10^4$), and 2,3-bis(2'-pyridyl)-5,6-dihydropyrazine and 2,3-bis[2'-(6'-methyl pyridyl)]-5,6-dihydropyrazine are sensitive reagents for Fe(II) and Cu(I), respectively.[60] These reagents are not superior to some of those in use earlier, but the ease of preparation is a point in their favor.

A study of derivatives of benzimidazole or imidazole in which pyridyl, quinolyl, 2,2'-bipyridyl or 1,10-phenanthrolyl groups are substituted for

H at the 2-position of the imidazole ring showed that they are inferior to earlier reagents for Cu(I) and Fe(II).[61]

Finally, some compounds may be mentioned in which the $-\overset{|}{N}=\overset{|}{C}-\overset{|}{C}=N-$ chain is present, but in which one or both of the nitrogens forms another functional group as well in combination with other groups (e.g., =NOH). The sensitivity for Fe(II) (and sometimes for Cu) may be retained, but selectivity may be impaired. Thus the oximes of substituted methyl and phenyl 2-pyridyl ketones are of considerable interest as analytical reagents, especially for iron.[62] It appears that only the *syn*-isomer of ferroin-type oximes is capable of metal chelation.

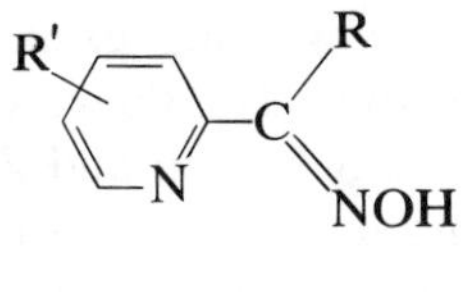

syn 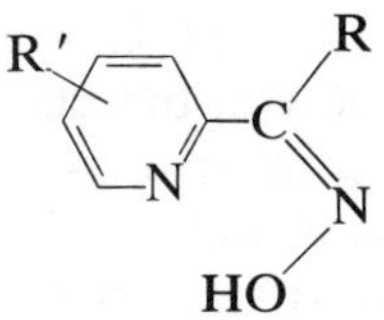   *anti*

The *syn*-methyl isomer of the oxime of 2-acetyl-4-phenylpyridine and the *syn*-phenyl isomer of 2-benzoyl-4-phenylpyridine oxime are sensitive Fe(II) reagents ($\epsilon_{568} = 23,400$ and $\epsilon_{566} = 25,700$, respectively, in isoamyl alcohol extract from NaOH solution). The chelation of iron in strongly basic solution is noteworthy. These two reagents have the drawback of being expensive and difficult to synthesize. As Cu reagents they (and others studied) are not superior to bathophenanthroline.

2-(2-Pyridyl)-benzimidazole (PB) and 2-(2-pyridyl)-imidazoline (PI)

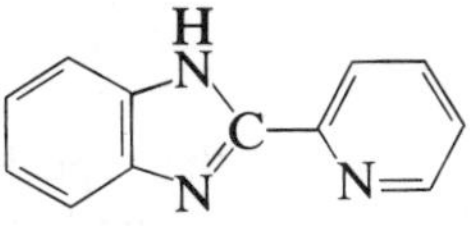

PB 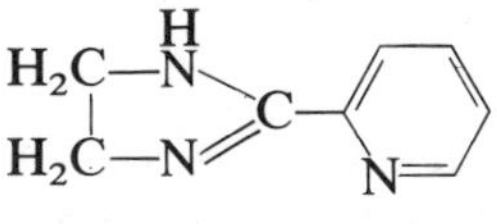   PI

have the reactive grouping discussed here and both give red-purple colors with Fe(II).[63] 2-(2-Pyridyl)-benzoxazole gives no color, however, presum-

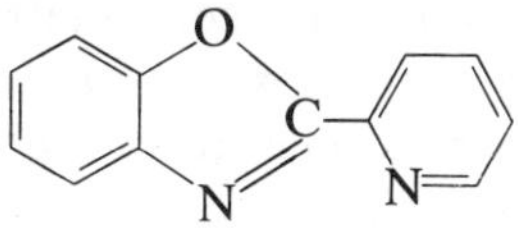

ably because of the low basicity of oxazole nitrogen. ϵ_{max} of the Fe(II) complexes of PB and PI are, respectively, 3800 (2600 at pH 5.5) and

7800. PB reacts with Cu(II) and Co to give brown to orange colors [white precipitate with Hg(II)]. A pH of 5–6 is suitable for the determination of Fe(II) with PB. The iron complex is extractable into isoamyl alcohol. The colored product is FeL_3^{2+}. Apparently, more than one colored species can be obtained because λ_{max} is not the same at pH 5.5 ($\sim$500 nm) as at pH 11.5 (550 nm). The reagent itself is colorless. PI gives a yellow color with Fe(III) and a blue with Cu(II). Its Fe(II) complex is not extracted into isoamyl alcohol unless perchlorate is present.

Pyridylbenzimidazole has also found use in the fluorimetric determination of Ga, In, and Zn.[64] Under the conditions (faintly acidic solution), Cd also fluoresces, but apparently not Al, Be, alkaline earths, rare earths, Th, and Zr. Colored ions interfere by absorbing exciting or emitted radiation (roughly in the range 300–400 nm). The fluorescent complexes have the composition reagent: metal of 1:1 for Ga and Zn, 2:1 for In. The metals can be determined in concentrations as low as 0.1–1 ppm. The reaction of Ga with this reagent shows that it can bond to N as well as to O, and that it is less oxyphilic than Al.

IV.   α-DIOXIMES

Dimethylglyoxime and other 1,2-dioximes having the oxime groups in *cis* configuration, with the OH groups in *anti* orientation (α),

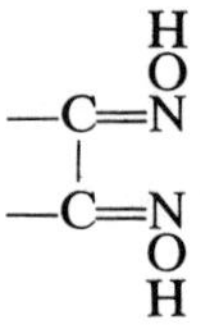

$$
\begin{array}{c}
\text{H} \\
\text{O} \\
-\text{C}=\text{N} \\
| \\
-\text{C}=\text{N} \\
\text{O} \\
\text{H}
\end{array}
$$

form chelates with many of the transition metals of the type $M^{II}(HDx)_2$ (H_2Dx = dimethylglyoxime or other α-dioxime). M^{2+} is bonded to N with simultaneous loss of equivalent H^+ from −OH. See the structure depicted on p. 374. The bright red precipitate produced by reaction of dimethylglyoxime with nickel (Tschugaeff, 1905) remains a striking example of a slightly soluble metal chelate. The α-dioximes, with dimethylglyoxime as their best known representative, are colorimetric reagents principally for Ni, Pd, and Re.[65] *Amphi* (γ) *cis* dioximes can form 1:1 metal complexes, whereas the *syn* (β) isomers usually form[66] no complexes (steric effect; intramolecular hydrogen bonding).

Although the dioximes are generally thought of as being selective for Ni, Pd, and Pt(II) in the sense of forming colored slightly soluble, and hence extractable, chelates with these metals, this is too sweeping a generalization. For example, the following dioximes

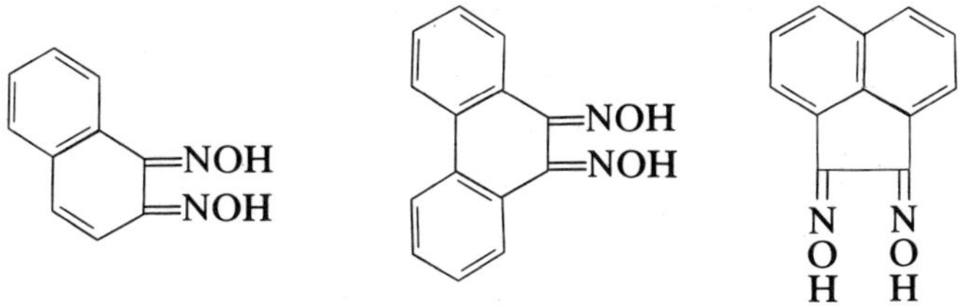

give slightly soluble chelates also with Co, Fe(II and III), and Cu(II), which are extractable into $CHCl_3$, CCl_4, or nitrobenzene.[67] These extracts absorb strongly in the UV or blue region of the spectrum. Clearly, such reagents are of no interest for the determination of Ni. Generally, an increase in the molecular weight of the dioxime reduces selectivity for Ni.

Even dimethylglyoxime, which is more selective than most other dioximes for Ni, extracts some Cu into $CHCl_3$. It does not extract Fe(III) and only a trace of Co(II).

A.   DIMETHYLGLYOXIME

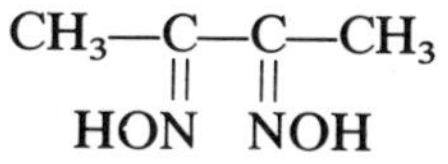

$$CH_3\!-\!\underset{\underset{\displaystyle HON}{\|}}{C}\!-\!\underset{\underset{\displaystyle NOH}{\|}}{C}\!-\!CH_3$$

Mol. wt. 116.1. M. p. 235°–237° (or 238°–240°) (decomposition). The commercial material is usually quite pure, and purification by recrystallization will rarely be required. Solubility in water 0.632 g/liter (5.4×10^{-3} M, 25°). More soluble in ethanol. Chloroform solubility is 4.5×10^{-4} M (25°). $P_{CHCl_3/H_2O} = 0.083$ (25°). Slowly decomposed (hydrolyzed) in 0.1 M H^+ at room temperature (various products formed).

Dimethylglyoxime (H_2Dx) is a weak acid:

$$\frac{[H^+][HDx^-]}{[H_2Dx]} = 2.6 \times 10^{-11}$$

($\mu = 0.05$, $[H^+]$ expressed as hydrogen-ion activity, 25°)

$$\frac{[H^+][Dx^{2-}]}{[HDx^-]} = 9 \times 10^{-13}$$

It also shows weakly basic properties (protonation of N in NOH)

$$\frac{[H^+][H_2Dx]}{[H_3Dx^+]} = 8.7 \qquad (25°)$$

(In a 1.0 M H^+ solution, ~90% of the reagent is present as H_2Dx, ~10% as H_3Dx^+.)

Dimethylglyoxime forms chelates especially with the transition metals. The nickel(II) complex may be represented

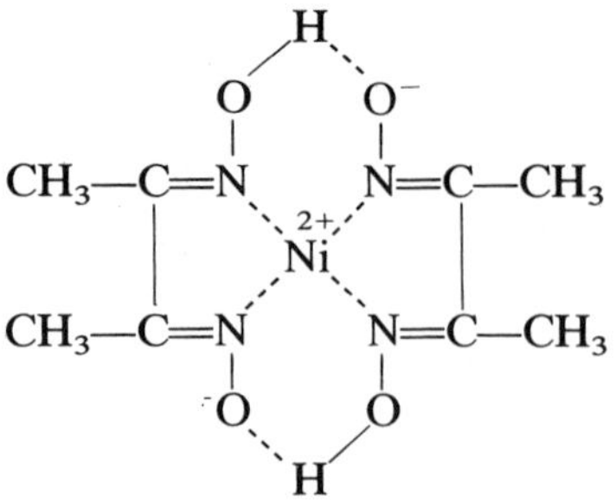

The Ni or other metal atom is bonded to 4 N atoms. Formation constants have the following values (rounded) in water solution (20 or 25°, $\mu = 0.1-0.3$):

	$\log K_1$	$\log K_2$	$\log K_3$	$\log \beta_2 (= \log K_1 K_2)$
Co(II)	8.7	9.0		17.7
Co(III)	15.0	10.0	7.2	25
Cu(II)				19.2
Fe(II)				7.25
Ni	8.9	8.6		17.5
Pd(II)				34.3

Pb, Zn, and Cd form less stable complexes than do Ni or Cu(II).

From the analytical standpoint, the most important property of dimethylglyoxime (and other α-dioximes) is its ability to form water-insoluble, colored chelates with Ni, Pd(II), and Pt(II). Since the dimethylglyoximates of these metals are extractable into immiscible organic solvents, selective methods for the separation and determination of Ni, Pd, and Pt become available. The reason(s) for the water-insolubility of Ni (Pd, Pt) dimethylglyoximate has been much discussed. A structural factor seems to be of major importance. The molecule of $Ni(HDx)_2$ has a planar configuration and the groups $-OH$ and $-O^-$ are strongly hydrogen-bonded internally and are not readily solvated. Copper(II) and cobalt(II) dimethylglyoximates on the other hand do not have a planar structure, and since the chelate rings do not lie in the same plane, the $-OH$ groups can be bonded to water molecules, therewith conferring water solubility. Also, nickel tends to form weak metal to metal bonds in the solid chelate, and this is believed by some to contribute to the water-insolubility of $Ni(HDx)_2$; metal to metal bonding does not take place in copper and cobalt chelates.

Chloroform is generally the best solvent for the extraction of the insoluble dimethylglyoximates.[68] From the solubilities of the dimethylglyoximates in chloroform and in water saturated with chloroform (M), the partition constants are calculated to be

$$P_{\text{CHCl}_3/\text{H}_2\text{O}} \text{ of Ni(HDx)}_2 = 4.8 \times 10^{-4}/1.2 \times 10^{-6} = 4.0 \times 10^2$$

$$P_{\text{CHCl}_3/\text{H}_2\text{O}} \text{ of Pd(HDx)}_2 = \frac{2.0 \times 10^{-3}}{6.4 \times 10^{-6}} = 3.1 \times 10^2$$

Similarly, from the chloroform and water solubilities of Cu(HDx)$_2$, its $P \cong 10^{-3}/10^{-2.2} \cong 0.14$. Co(II) dimethylglyoximate is not appreciably extracted by chloroform. Copper, as well as any trace of Co, can be washed out of the chloroform phase with dilute ammonia solution. See discussion of nickel in *Colorimetric Determination of Traces of Metals* for further information (E as a function of pH, separation of Ni from Cu and Pd).[69] The Ni and Pd chelates have a yellow color in chloroform and can be determined in this way, though not with the highest sensitivity in the visible range. For Ni, $\mathscr{S} = 0.0024\ \mu\text{g Ni/cm}^2$ at 260 nm, 0.013 at 330 nm, and 0.016 at 374 nm.

A more sensitive method for Ni is based on the formation of what appears to be a dimethylglyoxime complex of Ni(IV) in basic aqueous solution in the presence of a strong oxidizing agent such as bromine or persulfate. The course of this reaction is still obscure, and a number of products can apparently be formed. For earlier work on this method see discussion of nickel in *Colorimetric Determination of Traces of Metals*, 3rd ed. A separation of nickel from most other metals, conveniently by chloroform extraction of Ni(II) dimethylglyoximate, must generally precede the application of this method. The wine red or brownish oxidized Ni-dimethylglyoxime species can be extracted as an ion-association complex.[70]

Dimethylglyoxime and other α-dioximes can form a variety of mixed complexes with a number of metals, especially with Co(II and III), Cu(II), and Fe(II). The addends can be ions (OH$^-$, Cl$^-$, Br$^-$, I$^-$, CN$^-$, etc.) or uncharged molecules (NH$_3$, pyridine, and other nitrogen bases). The resulting species can be charged or uncharged. A few of these are of some analytical interest. Cobalt(III) is especially prone to form mixed (composite) species. The arrangement of three dimethylglyoxime ligands around Co(III) is sterically impossible, but two can be bound to form the species Co(HDx)$_2$(H$_2$O)$_2^+$. By introduction of an anion X$^-$ (such as Cl$^-$) to replace a water molecule, the $+$ charge is neutralized, and a mixed chelate is formed.

Copper(II), Co(II and III), and Fe(II) dimethylglyoximates form various adducts, as with ammonia and pyridine (Ni, Pd, and Pt dimethylglyoximates do not form adducts or do so only weakly). $Fe(HDx)_2$-(pyridine)$_2$ is extractable into $CHCl_3$. Pyridine can replace the water molecules in $Co^{III}(HDx)_2(H_2O)_2^+$, and the resulting cation is extractable as a partner of Cl^- and other anions into chloroform–isoamyl alcohol. Basic molecules other than pyridine form adducts of this type.

Cobalt(II), unlike Ni(II), can coordinate axially two monofunctional ligands, X^-, in addition to the two HDx^- ligands occupying four coordination positions. The red-brown ion $Co(HDx)_2I_2^{2-}$ formed in this way has been proposed for the photometric determination of cobalt.[71] Copper(II) and Co(II) form $M(HDx)_2(OH)_2^{2-}$ in basic solution. Palladium(II) can form $Pd(HDx)_2OH^-$, unlike $Ni(HDx)_2$, which does not add OH^-. Iron(III) is reported to combine very slowly with dimethylglyoxime in acidic solution, equilibrium being attained in ~ 24 hr at pH 2.5–3.[72] The cationic species $Fe(OH)(HDx)^+$ is formed (instability constant 3×10^{-12}). In basic solution, $Fe(OH)Dx$ is probably formed.

Addition of dimethylglyoxime to an ammoniacal tartrate solution of Fe(III) and Co(II) results in the formation of a red-brown precipitate thought to be a polynuclear complex of Fe and Co ($Fe:Co:H_2Dx = 1:1:3$). Either metal alone gives no precipitate. The polynuclear complex is partially extractable into chloroform.

Dimethylglyoxime and other α-dioximes give a red color with rhenium in $SnCl_2$–HCl solution (Tougarinoff reaction). The identity of the complex formed is not clear. It apparently contains Re(IV) (see discussion of furildioxime). Mo(VI) or Mo(V) gives a red-violet complex extractable into alcohols.[73] (Nioxime and Heptoxime are reported to form $MoO(HL)_2^+$.)

B. 2-FURILDIOXIME

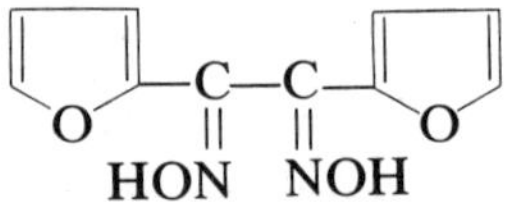

Mol. wt. = 220.2.

There are three isomers, α, β, and γ, of 2-furildioxime. Pure α-furildioxime melts at 192°–193°, whereas the β and γ compounds melt at 150°–152° and 182°–183°, respectively. The α, β, and γ isomers probably correspond to the *anti-*, *syn-*, and *amphi-*stereoisomers of 2-furildioxime, respectively. The so-called α-furildioxime or 2-furildioxime (colorless, m.p. 166°–169°) prepared by the usual procedure is a mixture of α and γ isomers.[74] The α isomer, for which the structural formula is shown above, is analytically important.

2-Furildioxime is slightly soluble in water (more so than dimethylglyoxime), but easily soluble in alcohol and ether. The chloroform solubility is 0.048 M; $P_{CHCl_3/H_2O} = 0.35$. The first acid dissociation constant of 2-furildioxime is $(5.5 \pm 0.2) \times 10^{-11}$ at 20°.

2-Furildioxime reacts with nickel(II) to form a 2:1 (ligand:metal) chelate extractable into organic solvents (yellow color; in $CHCl_3$, $\mathscr{S}_{436} = 0.003$ $\mu g/cm^2$). The reagent also has been used for the determination of cobalt,[75] palladium, rhenium,[76] and technetium. It has some advantages over dimethylglyoxime (such as greater solubility of its chelates in organic solvents and higher sensitivity) in photometric determinations, but Co and Cu are also extracted into $CHCl_3$ (the Cu chelate absorbs at 480 nm).

From ammoniacal solution, a hexyl alcohol solution of α-furildioxime (H_2L) extracts Fe(II) as $Fe(HL)_2(NH_3)_2$.[77] (The α-benzildioxime chelate can be extracted with $CHCl_3$.) From HCl solution, Fe(II)–Sn(II) complexes of furildioxime and other dioximes are said to be extracted into various solvents.

C. OTHER DIOXIMES AND TRIOXIMES

Nickel methylethylglyoximate has two crystal modifications: α-Ni(MEG)$_2$, monoclinic, red, and β-Ni(MEG)$_2$, orthorhombic, yellow. Their properties, including absorption spectra, stability constants, and partition constants (between chloroform and water), have been measured and compared with those of nickel dimethylglyoximate.[78]

1,2-*Cyclohexanedionedioxime* (or cyclohexane-1,2-dioxime),

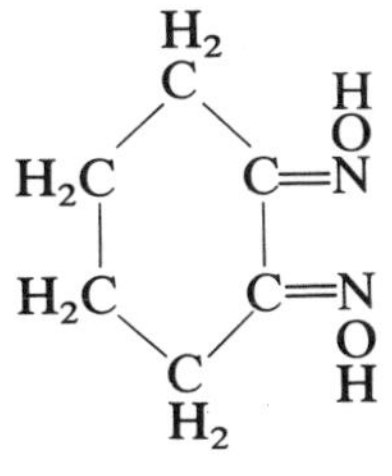

called Nioxime, gives much the same photometric reactions as dimethylglyoxime. The reagent is more soluble in water (8 g/liter, room temperature).

4-Methyl-1,2-cyclohexanedionedioxime has been used for extractive photometric determination of Ni, Pd, and Re[79], and 4-isopropyl-1,2-cyclohexanedionedioxime for determination of Ni and Pd. Determination of nickel with 4-*tert*-butyl-1,2-cyclohexanedionedioxime has been described.[80] 1,2-Cycloheptanedionedioxime (Heptoxime) has been used for extractive photometric determination of nickel (chloroform, 377 nm). Cu, Co, and Fe(II) give colored chloroform extracts also. It is of interest that Heptoxime and Nioxime form the species $Mo^VO(HL)_2^+$ with Mo in the presence of $SnCl_2$.[81] 1,2-Cyclodecanedioxime behaves much like dimethylglyoxime and other 1,2-dioximes in its reactions with Ni, Pd, and other metal ions.[82]

Benzildioxime

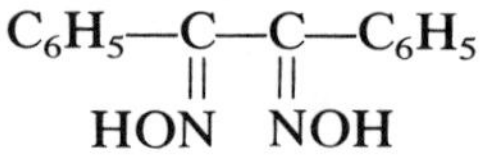

$$C_6H_5-\underset{\underset{HON}{\|}}{C}-\underset{\underset{NOH}{\|}}{C}-C_6H_5$$

has been used for extraction photometric determination of Ni, Pd, and Re. Chloroform extraction of the Ni chelate provides markedly greater sensitivity than does that of the dimethylglyoximate; Co and Cu(II) are also extracted but their chelates do not absorb in the visible range. A number of Cl-, Br-, and NO_2-substituted benzildioximes have been studied as extraction reagents.[67] It is not evident that they have any great advantage over the parent reagent (sensitivity is a little better).

1,2,3-Cyclohexanetrionetrioxime has been proposed as reagent for Co and Ni. 5-Methyl-1,2,3-cyclohexanetrionetrioxime is inferior to furildioxime as a reagent for rhenium.

Alicyclic *vic*-dioximes containing from 5 to 12 C atoms give yellow precipitates with Bi in basic solutions, which are insoluble in organic solvents.[83] Their stoichiometric composition and organic solvent insolubility suggest a polymeric structure (**I** or **II**). Infrared spectra are in accord with structure **II**.

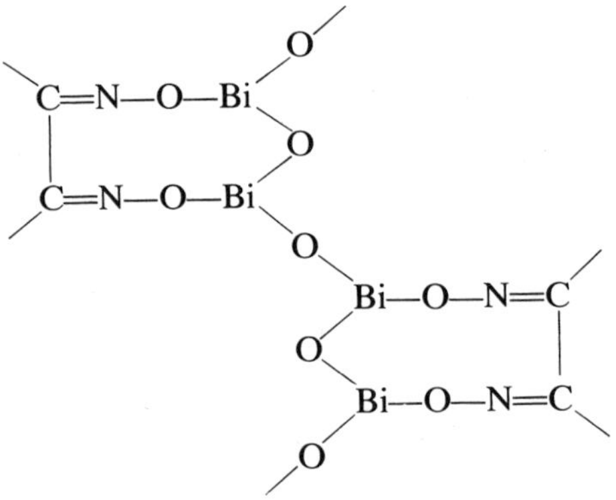

I 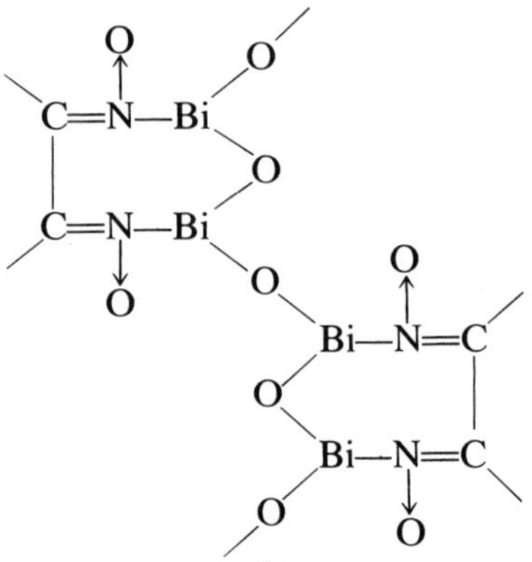   II

Dimethylglyoxime forms a similar precipitate, to which the following structure was assigned[84]:

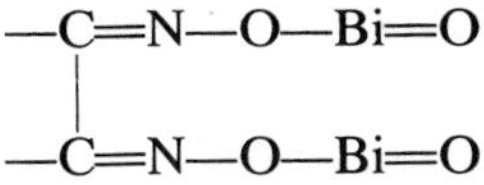

$$-C=N-O-Bi=O$$
$$-C=N-O-Bi=O$$

2,2'-Dipyridyl-α-glyoxime

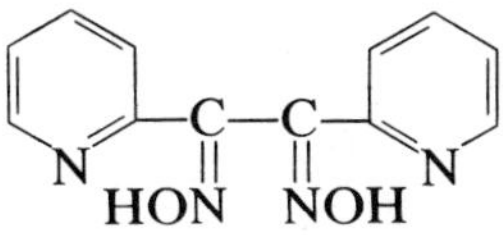

has two functional groups: $=NOH$ and the Fe(II) selective chain $-N=C-C=N-$. It provides a sensitive determination of Fe(II) at pH 10.5 ($\mathscr{S}_{534} = 0.004$ μg Fe).[85] Large amounts of Cu and Ni may be present. 2,2'-Dipyridyl-β-glyoxime reacts similarly. In CHCl$_3$, $\mathscr{S}_{558} = 0.003$ μg Fe.[86] Gold can be determined with the α isomer as a yellow complex extracted into dichloromethane-*n*-amyl alcohol.[87]

Dimedone Dioxime

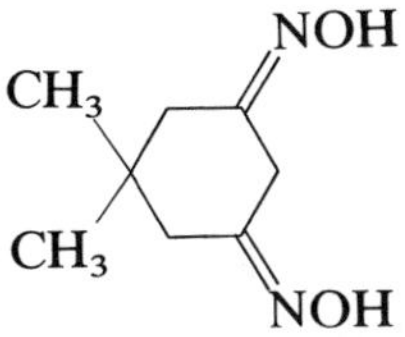

is a 1,3 dioxime.[88] It forms colored complexes with Co, Ni, and Cu.[89] An extractive absorptiometric method ($\mathscr{S} \sim 0.006$ μg/cm^2) for Co has been described.[89] The colored Co species, formed in basic solution, is believed to be CoL_6^{4-}. Acidification seems to give CoL_2.

V. MONOXIMES

Reagents having a single $=NOH$ group form metal complexes of varied nature depending on the second reactive group. Compounds in Section V.A chelate metals through two nitrogens (usually); those in Section V.B, through nitrogen and oxygen.

A. PYRIDINE KETOXIMES

The pyridine nitrogen in conjunction with the oxime group forms the reactive grouping

$$-N\!\!=\!\!\overset{|}{C}\!\!-\!\!\overset{|}{C}\!\!=\!\!N-$$

which is selective for Fe(II), even though not part of a wholly aromatic system here. These compounds are principally Fe(II) reagents, but they also react with metals binding strongly to ligands with N donor atoms and have been suggested for the determination of some of these.

Pyridine-2-aldoxime

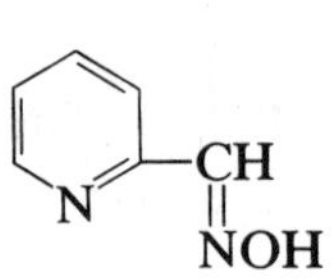

$pK_{a1} = 3.55; pK_{a2} = 10.17.$

forms strongly colored chelates with many transition metals.[90]

Methyl-2-pyridyl ketoxime
Iron(II) reacts with this reagent in basic solution, giving a red-violet color ($\mathscr{S}_{525} = 0.004 \ \mu g$ Fe) which is stable even in 6 M KOH.[91] Very likely the colored species is

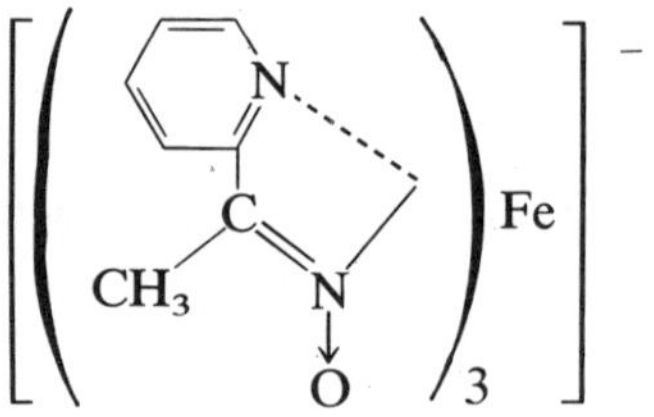

In acidic solution, an orange-red species is obtained

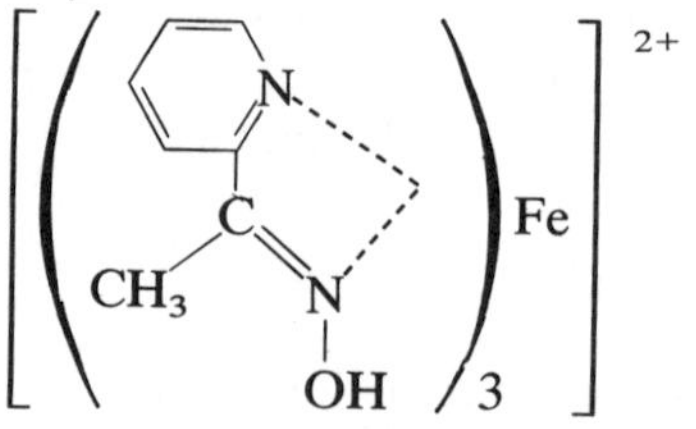

This species slowly changes to the anionic species represented above when the solution is made alkaline. The slow transformation is believed to be explained by a slow conversion of the unstable cis-Fe(HL)$_3^{2+}$ to stable $trans$-Fe(HL)$_3^{2+}$, which then gives $trans$-FeL$_3^-$ in basic solution.

Copper, Ni, Co, and Mn interfere seriously in the iron determination. Copper(I) can be determined in basic solution as a yellowish green species, probably CuL$_2^-$.

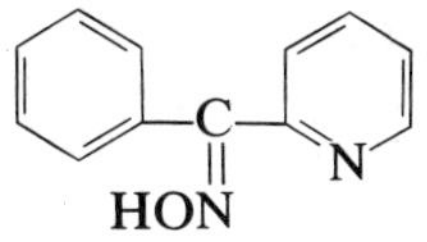

Mol. wt. = 198.2.

Phenyl-2-pyridyl ketoxime has two stereoisomers, *syn-* (above) and *anti*-forms. The *syn*-form (m.p. 151°–153°), a colorless crystalline substance, is analytically important. It is insoluble in water, slightly soluble in cold ethanol and soluble in benzene, chloroform, dioxane, and isoamyl alcohol. Its solutions are stable when protected from direct sunlight.[92]

Phenyl-2-pyridyl ketoxime behaves as a weak acid in basic solutions

$$>C{=}NOH \rightleftharpoons >C{=}NO^- + H^+$$

The ionization constant K_a at 20° and ionic strength 0.1 in 40% acetone[93] is 6.5×10^{-13}. The ionization constant of the protonated form HLH$^+$ is 14.5 (20°, $\mu = 0.1$).

The reagent forms soluble colored chelates with Ni, Co, Fe(II, III), Cu(I, II), and others, which are extractable into CHCl$_3$. It has the interesting property of reacting with iron(II) in strongly alkaline solutions to form a red 1:3 (metal:reagent) complex that can be extracted by isoamyl alcohol.[92] It also has been used for the determination of cobalt, gold, palladium, and rhenium.[94] The formation constants of phenyl-2-pyridyl ketoximates of Ni(II), Zn(II), Cd(II), Hg(II), and Pb(II) have been determined. The bonding of metal to N, rather than to O, of =NOH has been regarded as not conclusively proved.

Phenyl-2-(6-methylpyridyl) ketoxime

This reagent can be used for the determination of copper in alkalies by extraction of an orange 1:2 (Cu(I)-reagent) chelate into isoamyl alcohol.[95] The following structure is assigned to the chelate formed in alkaline solution:

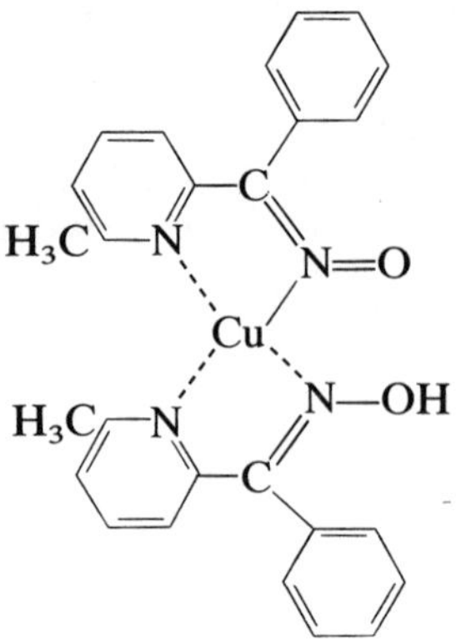

In weakly acidic solution, a $1:3$ chelate (orange-yellow) forms:

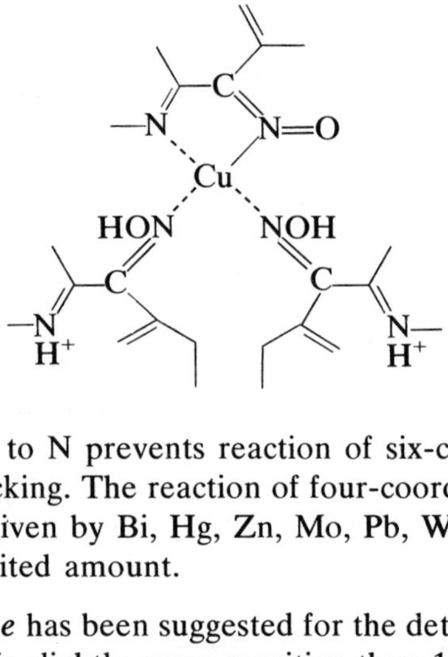

The CH_3 group *ortho* to N prevents reaction of six-coordinated iron(II) by stereochemical blocking. The reaction of four-coordinated Cu(I) is not affected. No color is given by Bi, Hg, Zn, Mo, Pb, W, Pd, or Pt. Cobalt may be present in limited amount.

Di-2-pyridylketoxime has been suggested for the determination of Fe[96], Au, Pd, and Co.[97] It is slightly more sensitive than 1,10-phenanthroline for Fe(II), but requires a basic solution (pH 10.5–13.5) for color development. Ni and Cu can be masked with cyanide after the iron color has been developed. The Pd complex is PdL_2 (reagent = HL), the H of each =NOH being replaced, with Pd bonded to 4 N and two 5-membered rings being formed.

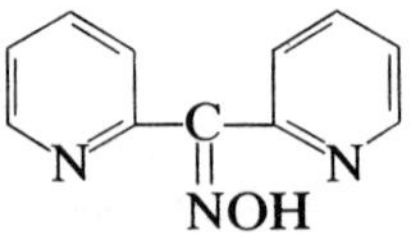

Miscellaneous Monoximes

The following monoximes are not pyridine ketoximes but are mentioned here as being reagents chelating metals through two nitrogens.

Diacetyl monoxime benzothiazolehydrazone forms a red-violet chelate with Pd(II) in acid solution, extractable into $CHCl_3$.[98] The complex probably is

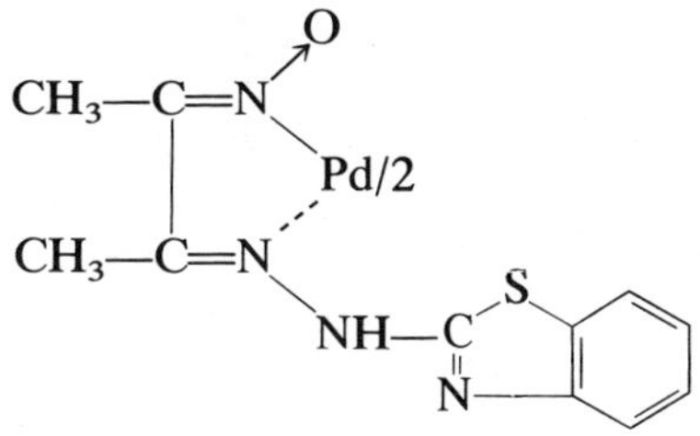

Acenaphthenequinone monoxime has been proposed for the determination of Pt(IV) in the presence of Pd(II),[99] and of Ru[100] and Rh,[101] as well as other metals.

B. MONOXIMES CHELATING THROUGH NITROGEN AND OXYGEN

Salicylaldoxime

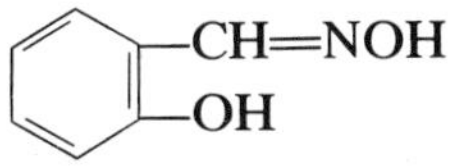

forms a slightly soluble 2:1 chelate with Cu(II) at pH ~3 and with Pd(II) in more acidic solution. The hydrogen of the —OH group is replaced by Cu, and bonding to N with formation of a six-membered ring occurs:

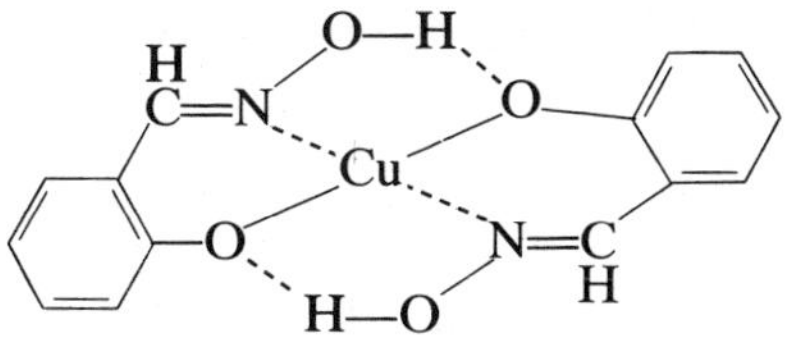

At higher pH values, 1:1 complexes can be formed by Cu, Pb, and other metals, H of =NOH as well as of —OH being replaced by metal. These chelates are extractable into $CHCl_3$ and other solvents.[102] It is interesting that Mg forms a chelate and is extracted incompletely at pH>11. Salicylaldoxime is not often used as a photometric reagent.

α-Benzoinoxime, an α-acyloinoxime, gives a chloroform-extractable green Cu(II) chelate but its chief analytical use is for the extractive separation of Mo, W and Nb from mineral acid solutions.

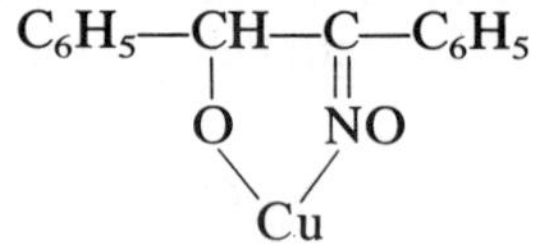

$$C_6H_5-CH-C-C_6H_5$$

1,3-*Dimethyl-4-imino-5-oximino-alloxan*

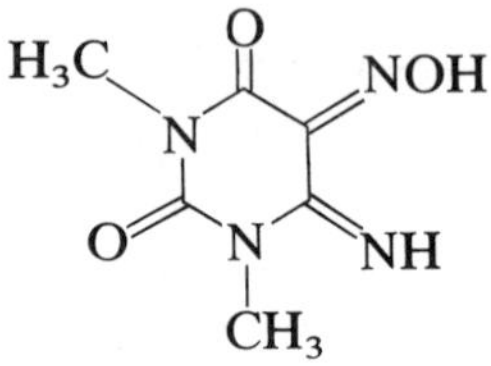

(violet solid, red-brown in alkaline solution) has two metal-reactive groups,

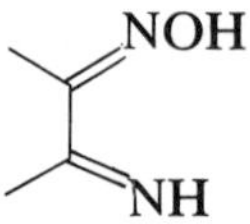

and

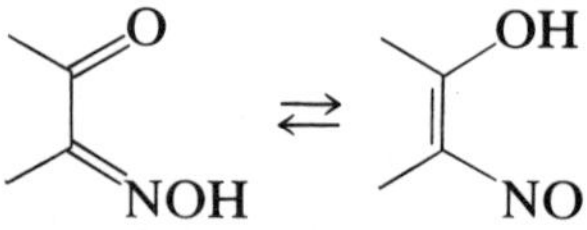

The first group is responsible for chelate formation with Ni and Pd (colored, insoluble), the second for reaction with Cu(II), Fe(II), and Co(II).[103] In ammoniacal solution the green complex is formed, but the reaction is of little practical importance for determination of copper.

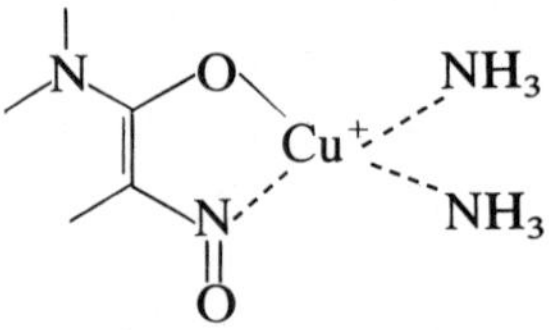

Formaldoxime

$$CH_2{=}NOH$$

is included here, though the constitution of its complexes with Mn, Ni, Fe(III), V(V), and Ce is not known. Since these metals, in the oxidation states that they are present in the complexes, are predominantly oxyphilic, bonding to oxygen seems likely. The metals mentioned form strongly colored formaldoxime complexes in basic solution (mostly brownish), and can be determined in this way.[104] All polyvalent metals are reported to be complexed in alkaline solution, and the products are colored if the metals are chromophoric. The chelate complexes are anionic. The composition of the Mn and Ni complexes is given as $Ni^{IV}(L_3)_2^{2-}$ and $Mn^{IV}(L_3)_2^{2-}$ (oxidation of metals by O_2); the Ce, V, and Fe complexes have an analogous composition, with 6 CH_2NO- for each atom of metal. The sensitivities are quite high: Mn, $\epsilon_{455} = 11,200$; Ni, $\epsilon_{473} = 18,400$; V, $\epsilon_{403} = 6,600$; Ce, $\epsilon_{340} = 4,700$.

The reagent is customarily used in the form of its trimer, $(CH_2NOH)_3 \cdot HCl$. An aqueous solution rapidly loses its activity, apparently by formation of solvated molecules, an equilibrium that is established in a few minutes. A 1 M solution is reported to contain 55% of the active reagent, a 0.1 M solution 5%. The reagent solution should, therefore, be quite concentrated, and the sample solution should be made alkaline immediately after addition of reagent. A solution of the trimer in $NaHCO_3$ can be kept for 24 hr. Oxidizing agents convert formaldoxime to colored products.

Reagents with the Group —CO—C(NOH)—
Compounds having the isonitrosoketone grouping in an aliphatic chain

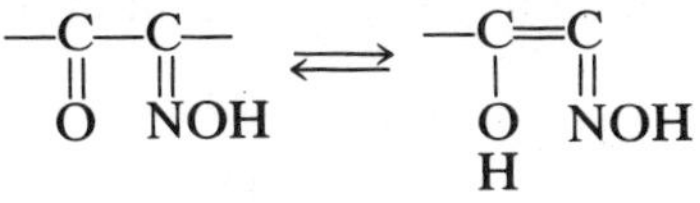

react with Cu, Hg(II), Cd, Zn, Ni, Co, Fe(II), and other divalent metals to give colored chelates extractable into organic solvents[105]:

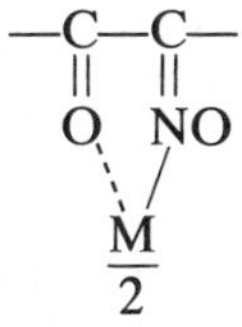

It was observed at the beginning of this century (Whitely, 1903) that isonitrosoacetone and analogous compounds reacted with Fe(II) to form blue-colored products. The ferrous complex of isonitrosoacetophenone gives a blue chloroform extract (Kröhnke, 1927). The Pd(II) complex is extractable into benzene (yellow).[106] Isonitrosoacetylacetone reacts similarly. Isonitrosoacetophenone has also been used for the determination of nickel by extraction of the chelate (yellow) with $CHCl_3$.[107]

VI. NITROSOPHENOLS AND NITROSONAPHTHOLS

The reacting group is $=C(NO)-C(OH)=$, which forms chelates of the type

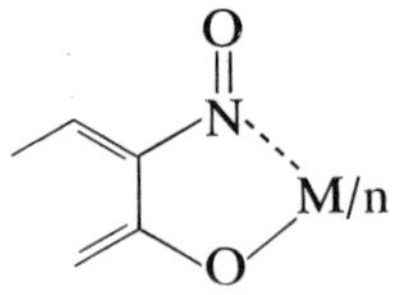

Metals such as Co(II, III), Fe(II), Pd, and Ni coordinate to N and O, with replacement of H of OH by an equivalent of metal. Other metals, not coordinating readily to N, can also react with this group, possibly bonding to O of NO and forming six-membered rings. Thus zirconium gives slightly soluble compounds of the type $ZrOL_2$. The reactions of Co, Pd, and to some extent Fe(II) with these reagents are of chief interest. The reagents themselves are colored (yellow in solution), but the chelates are more intensely colored, with absorption peaks shifted into the blue region.

The premier analytical reagents in this group are the nitrosonaphthols.

A. NITROSOPHENOLS

o-Nitrosophenol is not an important reagent in practice, partly because

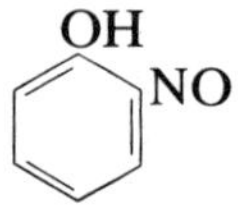

of its poor stability and that of its chelates, but it is of some general interest. In weakly acidic solution, it gives chelates soluble in water and some polar solvents (ethyl ether) with Cu(II) (red-violet),, Hg(II) (red-violet), Ni (red), and Fe(II) (green).[108] The chelates of Pd (green), Co(III) (gray-brown), and Fe(III) (brown) are extractable into petroleum ether. Apparently chelates of the type $M^{II}L^+$ can be formed together with $M^{II}L_2$. Probably coordination of water molecules is the reason for the nonextraction of the uncharged chelates of Cu(II), Fe(II) and Ni into nonpolar solvents.

o-Nitrosocresol (1-methyl-3-hydroxy-4-nitrosobenzene) reacts with metals in much the same way as a nitrosophenol. It has been used as an extraction reagent for cobalt.[109] Although a sensitive reagent, it has some drawbacks: The reagent is not obtainable in solid form, the solution must be prepared by the analyst and is not stable, and special precautions must be taken to prevent the interference of traces of iron.

3-Nitrososalicylic acid is of some interest as a reagent that can be used for the simultaneous determination of Co and Ni.

o-Nitrosoresorcinol monomethyl ether (3-methoxy-6-nitrosophenol) has been proposed as a colorimetric reagent for Fe(II) and Co.[110] The former metal gives a water-soluble green complex (reaction said to be more sensitive than with 1,10-phenanthroline) at pH 1.5–2.5, the latter a benzene-extractable, red-brown complex at pH 6–8.

2-Nitroso-5-dimethylaminophenol[111] forms a 3:1 chelate with Co(III), extractable into 1,2-dichloroethane. This chelate shows the same characteristics as the Co(III) chelates of the nitrosonaphthols. It is stable to the action of strong acids, so that 4 M HCl (or NaOH) can be used to wash out the excess reagent from the extract and decompose most other metal complexes. This reagent is of interest because it is a little more sensitive ($\mathscr{S}_{456} = 0.001$ μg Co) than other nitroso reagents. It is quite stable in acid aqueous solution. The high sensitivity is presumably attributable to the strongly electron-donating dimethylamino group in $para$-position. 2-Nitroso-5-diethylaminophenol provides a trifle higher sensitivity than the methyl analog.[112]

2-Nitroso-4-chlorophenol forms the complex species $FeL_2 \cdot HL$ with Fe(II), that is, a reagent adduct. This ionizes to give a bulky anion that forms an ion pair, $(RH^+)(FeL_2 \cdot L^-)$, with Rhodamine B at suitable pH ($\sim$4.8), extractable into benzene or toluene. It has been proposed to determine iron in natural waters in this way (ferric iron can be reduced to ferrous with hydroxylammonium chloride).[113] The absorptivity of the ion pair ($\epsilon_{558} = 9 \times 10^4$) is greater than that of the adduct ($\epsilon \sim 3 \times 10^4$). Interfering metals include Co, Ni, Cu, and Sn(II).

In general, o-nitrosophenols and o-nitrosonaphthols with electron-attracting substituents (Cl, Br, COOH, for example) give extractable ion pairs similar to the above. The presence of electron-donating substituents (dialkylamino, OH, methoxyl, methyl) in the nitroso reagents results in decreased or no extraction of iron.

3-Nitroso-2,6-pyridinediol (included at this point for convenience)

reacts with Ru(III) (purple chelate), Os(VIII) (purple), Rh(III) (yellow), Pd (yellow), and Au(III) (blue). It has been used to determine Os in weakly acidic solution.[114] An anionic chelate is formed ($\epsilon = 24{,}000$).

Many metals interfere, including Cu and Co. It is not clear what the reacting group is.

B. NITROSONAPHTHOLS

1-Nitroso-2-naphthol was one of the earliest organic precipitants for

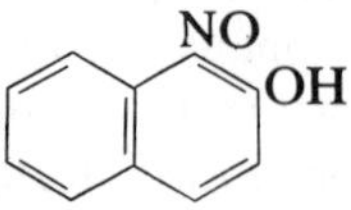

Mol. wt. 173.2. M. p. 109°–110°. Yellowish brown solid. Solubility in water 1.06×10^{-3} M, in $CHCl_3$ 1.35 M. $K_a = 2 \times 10^{-8}$. Commercial material often of low purity. A "certified" reagent has been found to be grossly impure. Can be purified by recrystallization, as from petroleum ether. Chromatography should also be an effective means of purification.

metals to be discovered (Ilinski and v. Knorre, 1885), and its cobalt precipitating power was soon noted. It can be used for a highly selective absorptiometric extraction determination of this element. Once formed, the reddish Co(III) chelate is exceptionally resistant to dissociation by acids (inert stability), whereas most other metal chelates[115] are decomposed. The formation of the Co(III) complex requires a weakly acidic (or basic citrate) solution and the reaction is slow. The aqueous reagent is added to the cobalt solution and 0.5 hr is allowed for reaction at room temperature. The reaction is more rapid in hot solution. Cobalt(II) is oxidized to Co(III) by the reagent.[116] Probably atmospheric oxygen would also oxidize cobalt. Chloroform is a good extractant for the cobalt complex. When the chloroform extract is shaken with mineral acid (e.g., 2 M HCl), the nitrosonaphtholates of such metals as Cu and Ni are decomposed, but the cobalt chelate is not affected. The Fe(II) and Pd(II) chelates are also resistant to mineral acids.[117] The extracted excess reagent can be removed from the chloroform with alkaline aqueous solution.

1-Nitroso-2-naphthol can form adducts. Thus $Co(C_{10}H_6NO \cdot O)_2 \cdot C_{10}H_6NO \cdot OH$ and its Na and K salts are known.[118] In basic solution, Ni gives $Ni(C_{10}H_6NO \cdot O)_2 \cdot (C_{10}H_6NO \cdot O)^-$.

2-Nitroso-1-naphthol

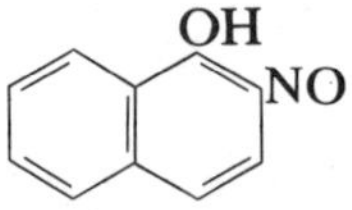

Mol. wt. 173.2. M. p. 165° (decomposition). Yellow solid. Solubility in water, 8.4×10^{-4} M; in $CHCl_3$, 0.096 M. $K_a = 6 \times 10^{-8}$.

The reactions of this compound are much the same as those of its isomer. Apparently, it oxidizes Co(II) more easily in weakly acidic solution than does 1-nitroso-2-naphthol.

Ruthenium reacts in hot mineral acid (3 M HCl) solution containing SO_2, giving a blue complex. Palladium(II) and Rh react in less acidic solutions.

Nitroso-R salt, 1-nitroso-2-hydroxy-3,6-naphthalene disodium sulfonate,

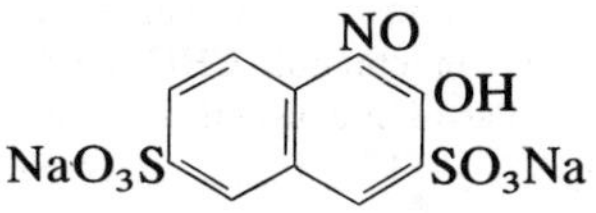

Mol. wt. 377.3. M.p. > 320°. Can be recrystallized from water. Reagent solution (aqueous) not particularly stable.

is an important reagent for cobalt and it has also been used for Fe(II) (green color in basic medium), V(IV) (red-brown in acidic solution), Ru(III) (blue-violet), Rh(III) (red), and Pd (red).[119] Its reactions with metals are much the same as those of the unsulfonated reagent, the main difference being the solubility of the chelates. They are not extractable into solvents of the inert type.

An analogous reagent is sodium 2-nitroso-1-naphthol-4-sulfonate. The Co(III) and Pd chelates are extractable into isoamyl alcohol.

3-Nitroso-4-hydroxy-5,6-benzocoumarin reacts slowly with Ru(III) at pH 5.5–8 to give a pink-violet 1 (metal):2 complex species extractable into *n*-butanol.[120] Other Pt metals are reported to interfere relatively little.

VII. DIPHENYLCARBAZIDE AND DIPHENYLCARBAZONE

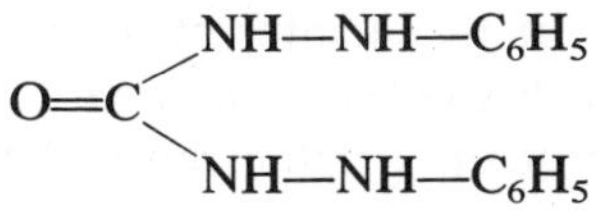

sym-Diphenylcarbazide (1,5-Diphenylcarbohydrazide)

Mol. wt. 242.3. M.p. 172°. White solid, gradually becoming pinkish in air. Only slightly soluble in water (~0.001 M), but soluble in alcohol, acetone and other organic solvents. Solutions gradually become yellow and brown by oxidation. Protect from light. Many products sold as diphenylcarbazide are (or have been) an approximately equimolar mixture of diphenylcarbazide and phenylsemicarbazide.[121]

The acidic properties of diphenylcarbazide are very weak and acid dissociation constants do not seem to have been reported.

$$O=C \overset{N=N—C_6H_5}{\underset{NH—NH—C_6H_5}{<}}$$

Diphenylcarbazone (Phenylazoformic acid 2-phenylhydrazide)

Mol. wt. 240.3. M.p. 126°–127° (decomposition). Orange-red solid. Very slightly soluble in water ($3.8 \times 10^{-4} M$); more soluble in ethanol, $CHCl_3$, CCl_4 ($2.8 \times 10^{-3} M$), and other organic solvents. $K_{a1} = 2.9 \times 10^{-9}$. Partition constant of uncharged substance is $P_{CCl_4/H_2O} = 7.5$ ($\mu = 0.1$, perchlorate); $P_{benzene/H_2O} = 126$; $P_{toluene/H_2O} = 39$.

The product sold as diphenylcarbazone is grossly impure. The products of nine manufacturers (sold in the early 1970s) are predominantly mixtures of diphenylcarbazone and diphenylcarbazide with smaller amounts (usually) of phenylsemicarbazide and diphenylcarbadiazone.[122] Pure diphenylcarbazone can be obtained from such mixtures by chromatographic[122] or extraction[123] procedures.

Diphenylcarbazone is formed by oxidation of diphenylcarbazide, even by air. Further oxidation yields diphenylcarbadiazone, $OC(N=N—C_6H_5)_2$, by removal of H attached to N.

Absorption spectra of diphenylcarbazone species in water, carbon tetrachloride, and toluene are available.[124] The spectra of diphenylcarbazone in methyl isobutyl ketone in equilibrium with aqueous solutions of various pH values indicate that the reagent exists in keto and enol forms in the MIBK phase.[125] An increase in pH shifts the equilibrium to the enol form:

$$\text{keto} \underset{}{\overset{OH^-}{\rightleftharpoons}} \text{enol}$$
$$(\lambda_{max} = 460 \text{ nm}) \qquad (\lambda_{max} = 505 \text{ nm})$$

In the aqueous phase, λ_{max} of HL^- (H_2L = diphenylcarbazone) is 495 nm.

Diphenylcarbazide is chiefly of interest because of the very sensitive color reaction it gives with Cr(VI) in mineral acid solution (Cazeneuve, 1900). The composition of the soluble red-violet product remained unknown for more than 50 yr. Relatively recent work has shown that the colored species is a chelate of Cr(III) (formed by reduction of Cr(VI) by diphenylcarbazide) and diphenylcarbazone, namely $Cr(HL)_2^+$, diphenylcarbazone being H_2L (and diphenylcarbazide H_4L).[126] The reaction occurring can be written: $2 CrO_4^{2-} + 3 H_4L + 8 H^+ \rightarrow Cr^{III}(HL)_2^+ + Cr^{3+} + H_2L + 8 H_2O$. The intensely colored cation can be extracted as chloride (from a chloride solution) into isoamyl alcohol. That Cr(III) should form such a strongly colored complex (true $\epsilon \sim 84,000$ or higher) seems odd. The direct reaction of Cr(III) with H_2L in aqueous solution is difficult to effect. The typical inertness of Cr(III) comes into play in these reactions. Diphenylcarbazide is also a reagent for Re, Tc(IV) (CCl_4 extraction, $\epsilon_{520} = 48,600$), Au(III), and Se(IV) [but not Te(IV)].[127]

Diphenylcarbazone reacts with many heavy metals (Hg, Cu, Cd, Rh, etc.) to form blue, violet, or red chelates in weakly acidic solution.[128] It may be assumed that metal is bonded to N and O and that the structure of the chelates is analogous to that of the dithizonates (Chapter 6G). From the practical viewpoint, these reactions are of little importance, but they are of general interest. It appears that diphenylcarbazide does not react with these metals unless it is first oxidized to the carbazone [as by Fe(III), Cu(II), or air]. Diphenylcarbazone complexes of Hg(I, II) and Cu(II) are unstable in toluene solution; oxidation to diphenylcarbadiazone occurs and UV and visible radiation accelerates the decomposition.[129]

Hg(I, II) and Ni are reported to form both 1:1 and 1:2 chelates with diphenylcarbazone.[129] Pb and Zn give only the 1:2 complexes. The reaction with mercury has been examined further, and two chelates each of Hg(I) and Hg(II) are recorded[130]: Hg_2L and $Hg_2(HL)_2$, and HgL and $Hg(HL)_2$. These are counterparts of the primary and secondary dithizonates. The diphenylcarbazonates are very much less stable than the dithizonates, as illustrated by the extraction constant for mercuric carbazonate,[129] $Hg(HL)_2$:

$$\frac{[Hg(HL)_2]_{toluene}[H^+]^2}{[Hg^{2+}][H_2L]^2_{toluene}} \sim 2 \times 10^5$$

This is to be compared with the value 10^{27} for the extraction constant of mercuric dithizonate, $Hg(HDz)_2$, in CCl_4.

Extraction constants for other metals in water–toluene have also been obtained.[131] Only Sn(II) diphenylcarbazonate has a larger extraction constant than the corresponding dithizonate. The extraction constants of $Cu(HL)_2$ and $Zn(HL)_2$ in the system methyl isobutyl ketone–water have been determined (logs are -1.1 and -7.1, respectively).[125]

Diphenylcarbazone has been used as an extractive photometric reagent for copper (extraction with benzene solution, pH 4–5).[132] The sensitivity is high ($\lambda_{550} = 75,000$). Copper forms not only the extractable complex $Cu(HL)_2$ but also the species $CuHL^+$ and $Cu(HL)_2 \cdot HL^-$, so that the extractability decreases with increasing reagent concentration and pH after attaining an optimum value.[133] See Fig. 6C-1. Because the partition constant of the uncharged reagent is not very large (reported as 12.6 and

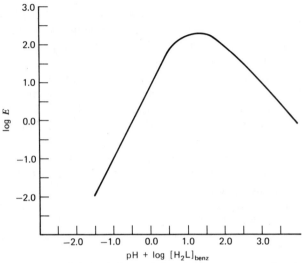

Fig. 6C-1. Extraction coefficient of Cu as diphenylcarbazonate as a function of pH and diphenylcarbazone concentration in benzene. After Geering and Hodgson, *loc. cit.* Plotting $\log E$ versus $(pH + \log [H_2L]_{benz})$ is equivalent to plotting versus $(\log [HL^-] - \log K_a + \log P_{H_2L}) = \log [HL^-] + 10.6$.

126 for benzene–water), sufficiently high reagent concentrations can be obtained in the aqueous phase to form appreciable amounts of the anionic chelate. Maximal extraction occurs when $[H_2L]_{benzene}/[H^+] = 20$ (or practically, between 10 and 30). The value of K_{ext} is roughly 0.5, compared to $\sim 10^{10}$ for K_{ext} of Cu(II) dithizonate. The great difference is due mostly to the much stronger bonding of Cu to S than to O.

Diphenylcarbazone is also a reagent for Ru(III), Rh, and Ir(III, IV), as well as Re.[134]

Tributyl phosphate and diphenylcarbazone form ternary benzene-extractable complexes with Mn(II), Cd, and Ni.[135]

A reagent showing structural similarity to diphenylcarbazone is 2-carboxy-2′-hydroxy-5′-sulfoformazylbenzene (Zincon):

Available as Na salt. Moderately stable in alkaline solution, but decomposes rapidly in acidic solutions (pH <4).

Equilibria in water:

$$H_4L \rightleftharpoons H_3L^- \rightleftharpoons H_2L^{2-} \rightleftharpoons HL^{3-} \rightleftharpoons L^{4-}$$

red	pink	pK	yellow	pK	red-orange	violet
	(red-violet)	4.0 or 4.5		7.9 or 8.3	(orange-yellow)	

(Data are incomplete and not in good agreement. Lack of purity may be suspected.)

which reacts with numerous metals, including Cu, Hg, Zn, Ni and Co. It forms blue 1:1 chelates (slightly soluble) with Cu(II) and Zn and has been used for the colorimetric determination of these elements.[136] The determinations are sensitive ($\mathscr{S}_{600} = 0.003 \ \mu g/cm^2$ for both Cu and Zn). The structure of the chelates formed is not stated. Probably the metals are bound by two anionic oxygens and two nitrogens. Zinc reacts optimally at pH 9, Cu in the pH range 5–9.5. Since zinc is relatively easily separable from many metals by ion-exchange, this reagent is of interest for the determination of this element in spite of poor selectivity. The metal complexes of Zincon are sufficiently soluble in water (sulfonic acid group) not to require extraction.

VIII.   GLYOXAL BIS(2-HYDROXYANIL) AND LIKE REAGENTS

Glyoxal bis(2-hydroxyanil) or 2,2′-(ethanediylidenedinitrilo)diphenol is not the correct name for the colorless or pale yellow crystalline substance used for the determination of calcium. The solid reagent, m. p. 204°, is 2,2′-bibenzoxazoline(I), which rearranges to glyoxal bis(2-hydroxyanil) (II) and then reacts with calcium in an alkaline medium.[137]

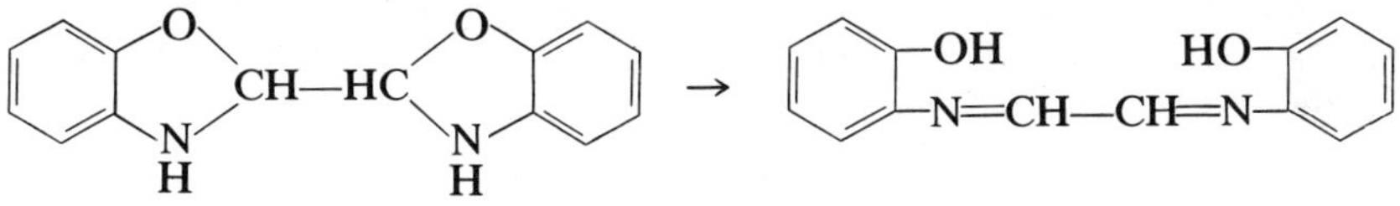

I   II

(Mol. wt. = 240.3.)

The solid reagent is nearly insoluble in water but soluble in aqueous alkaline solution. It is not very soluble in alcohol. It can be recrystallized from methanol. The so-called glyoxal bis(2-hydroxyanil) ($1 \times 10^{-3} \ M$) in 0.04 M sodium hydroxide in 50% methanol shows an absorption maximum at approximately 450 nm (absorbance in a 1 cm cell = 1.0).[138]

Glyoxal bis(2-hydroxyanil) is a useful reagent for the determination of calcium. A red 1:1 chelate is formed in an alkaline tartrate medium.

Presumably the chelate has the structure

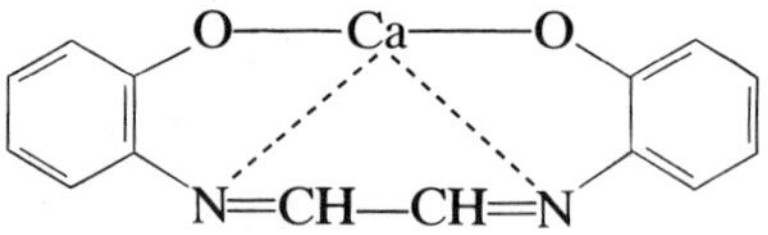

A mixture of chloroform with a higher alcohol (hexanol or isopentanol[139]) is a suitable extractant for the chelate. Chloroform alone and other inactive solvents are not satisfactory. In the presence of a sufficient amount of tetradecyldimethylbenzylammonium chloride (Zephiramine), a complex having Ca:ligand = 1:2, probably $CaL(H_2L)$, can be extracted into the inactive solvent 1,2-dichloroethane.[140] (A small amount of ethanol is also needed for good extraction.) Apparently, the extracted complex does not contain Zephiramine. The action mechanism of Zephiramine is not clear, though it seems to cause replacement of water molecules in the 1:1 complex by H_2L molecules and thus leads to extraction into inactive solvents. Nishimura and coauthors regard the phenomenon as a kind of synergism. Cadmium behaves the same as Ca.

Glyoxal bis(2-hydroxyanil) also has been proposed as a reagent for the determination of cadmium and uranium.[141] In 2 M hydrochloric acid, tin(II) gives a violet color, and the colored product is extractable into chloroform.[142] Scandium forms a 1:2 complex at pH ~5.[143]

Derivatives of glyoxal bis(2-hydroxyanil) have been studied as calcium reagents.[144] The reaction with glyoxal bis(2-hydroxy-4-methylanil) is more sensitive than that with the parent compound. Acenaphthene-quinone-(2-hydroxyanil) is another possible reagent for calcium.[145]

The sulfur analog of glyoxal bis(2-hydroxyanil), glyoxal bis(2-mercaptoanil), has been synthesized, and its reactions have been studied.[146] It is more selective.

Schiff Bases
These are best known as fluorimetric reagents. The fluorescence reactions of many Schiff bases (derived mostly from salicylidene-o-aminophenol) with Al in 95% alcohol[147] and with Al and Ga in 8% dimethylformamide[148] have been studied.

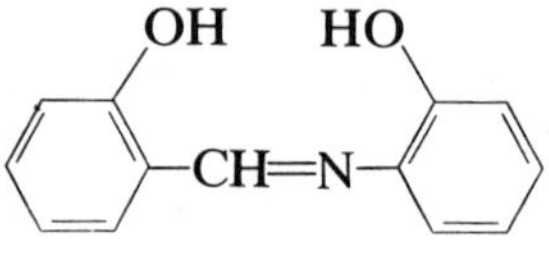

Salicylidene-o-aminophenol (2-hydroxyaniline-N-salicylidene) is a fluorescence reagent for Al, Ga, and In.[149] In weakly acidic solution, the chelates of these metals have the composition 2 reagent:1 metal.[149] At pH 5.6 a 1:1 complex is formed.[150]

This reagent has also found use for the extractive absorptiometric determination of some metals, as of Cu.[151] It forms insoluble chelates with divalent cations (Co, Ni, Fe, etc.) in alkaline solution.[152] Apparently, H of both OH groups is replaced by $M^{2+}(= Zn^{2+}, Ni^{2+})$, but these uncharged chelates are not extractable into inactive solvents. If Zephiramine is added, the complex M:2 reagent:2 Zeph is extracted into dichloroethane.[140] Evidently, the —OH of each ligand is now ionized and the anion of charge 2—forms an extractable ion associate with two $Zeph^+$. Copper forms a 1:1 chelate extractable into inactive solvents and the addition of Zephiramine has no effect.

2-Hydroxy-5-sulfoaniline-N-salicylidene and 2-hydroxy-5-carboxy-aniline-N-salicylidene are more sensitive reagents than salicylidene-o-aminophenol for Al and Ga.[148] The former reagent gives a 1:1 complex with Be at pH 10.[153]

N,N'-Bis-salicylidene-2,3-diaminobenzofuran reacts with Mg at pH

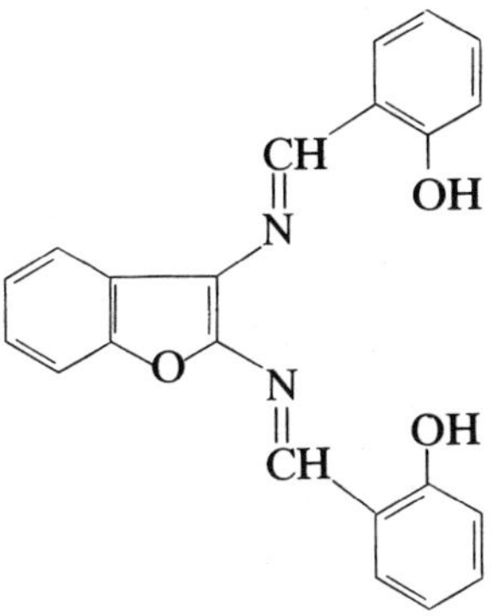

10.5 in methanolic solution to form a fluorescent 1:1 complex.[154] It can be extracted into MIBK. Most other metals can largely be made harmless by masking.

Salicylaldehyde semicarbazone forms a blue fluorescent 1:1 complex with Sc at pH 6.0.[155] 2,4-Dihydroxybenzaldehyde semicarbazone also reacts with Sc at pH 6.[153] 2-Hydroxy-5-methylbenzaldehyde thiosemi-carbazone and 2-hydroxy-5-chlorobenzaldehyde thiosemicarbazone form fluorescent 1:1 complexes with Ga but not Al.[148]

IX. REAGENTS WITH IMINODIACETIC
ACID GROUPS

Ethylenediaminetetraacetic acid (*EDTA*) forms colored chelates with

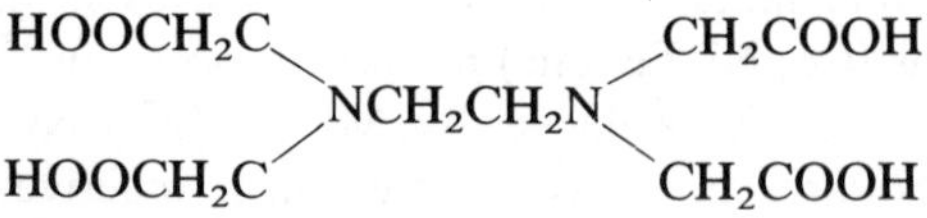

$$\begin{array}{ccc} HOOCH_2C & & CH_2COOH \\ & \diagdown NCH_2CH_2N \diagup & \\ HOOCH_2C & & CH_2COOH \end{array}$$

metal cations that are themselves colored. It has been used for the colorimetric determination of Bi, Cr, Co, Cu, Fe, Ir, Mn, Ni, and some other metals.[156] These methods are not likely to be of much value to the trace analyst because of their general lack of sensitivity and selectivity.

Useful colorimetric trace reagents are obtained by introducing the chelating group $—CH_2N(CH_2COOH)_2$ into a chromophoric framework. Such reagents react with numerous metals, but some selectivity can be realized by adjusting the acidity and using masking reagents. Sensitivity is likely to be good. A number of these reagents are discussed in the following pages.

Xylenol Orange (XO),

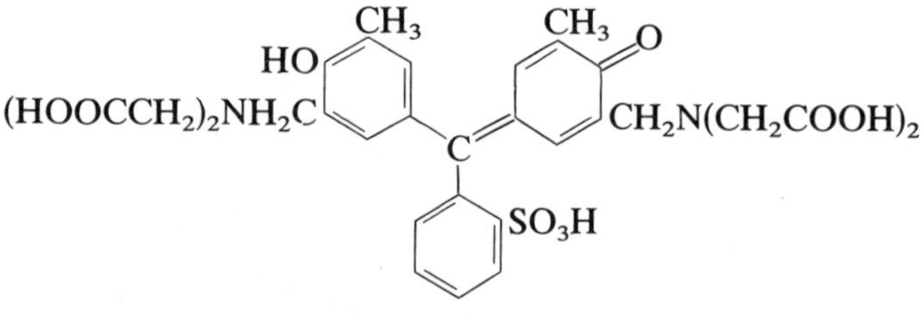

Mol. wt. (free acid) 672.7. 4 Na and 2 Na salts commercially available.

3',3''-Bis{[bis(carboxymethyl)amino]methyl}-5',5''-dimethylphenolsulfon(e)-phthalein, or its sodium salt, was initially proposed as a metal indicator in EDTA titrations. It is soluble in water, and its color is yellow below pH 6 and red-violet above pH 7. XO in aqueous solutions decomposes slowly.[157] The compound with one $—CH_2N(CH_2COOH)_2$ group is known as Semixylenol Orange.

Values of the acid dissociation constants of XO are not in good agreement.[158] Presumably this is mostly due to lack of purity of the XO reagents used. However, the values obtained by Rehak and Körbl[159] have been confirmed by Otomo[160]: $pK_1 = -1.74$, $pK_2 = -1.09$, $pK_3 = 0.76$, $pK_4 = 1.15$, $pK_5 = 2.58$, $pK_6 = 3.23$, $pK_7 = 6.40$, $pK_8 = 10.46$, and $pK_9 =$

12.28. These ionization constants refer to the following groups:

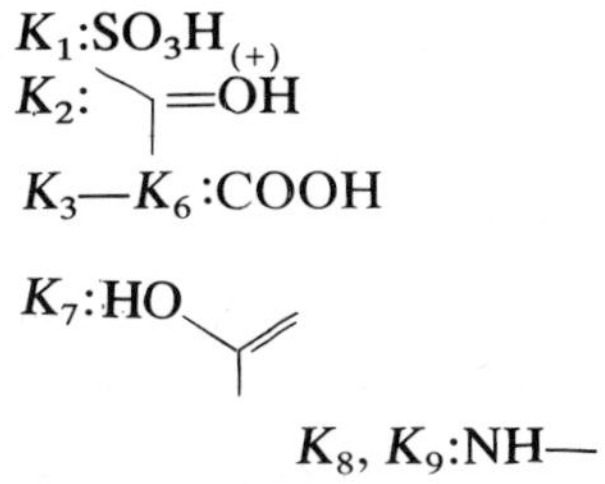

$$K_1\text{:}SO_3H_{(+)}$$
$$K_2\text{: }\diagdown\!\!=\!\overset{..}{O}H$$

$$K_3\text{—}K_6\text{:}COOH$$

$$K_7\text{:}HO$$

$$K_8, K_9\text{:}NH\text{—}$$

Commercial products of XO are usually impure. A major impurity is Semixylenol Orange, 3'-[bis(carboxymethyl)amino]methyl-5',5″-dimethyl-phenolsulfon(e)phthalein. XO and Semixylenol Orange can be separated by chromatography on a cellulose column.[161] Cresol Red may also be an impurity.

XO forms strongly colored red complexes with many metal ions in the pH range 0–6.0. Palladium(II) reacts in 1.5 M perchloric acid solution. Zirconium, hafnium, bismuth, niobium, and iron(III) react in solutions of fairly high acidity ($\sim$0.1 N). These ions can be determined fairly selectively with the help of masking agents.[162] XO is also used for the determination of the rare earths,[163] scandium, and aluminum.

The metal complexes often have 1:1 composition, but other ratios are not uncommon. Nickel forms a 1:1 complex at pH 3 and two complexes at pH 6.4, one having a Ni/reagent ratio of 1:2, the other the ratio 2:2 (Langmyhr and Paus[161]). Bismuth reacts in 1:1 ratio at pH 1 but in a ratio 1 (Bi):2 above pH 1.[164] Scandium forms a violet 1:1 complex at pH 1.2–5.5 and a red 1:2 complex at pH 3.5–6.0, the former being used for determination of Sc. Although earlier work indicated that XO forms a 1:1 complex with zinc at pH 6, later work[165] with purified XO has shown that it forms a 1:2 (XO:Zn) complex at the same pH. Thorium forms a ternary complex with XO and EDTA. Stability constants of some XO complexes are available. The alkaline earth metals form three 1:1 and two 2 (metal):1 complex species with Xylenol Orange. The structures of three Th–XO complexes have been represented as shown at the top of page 398.[166]

Methylthymol Blue (MTB)

MTB (formula on following page) and SMTB (Semimethylthymol Blue, which has one —$CH_2N(CH_2COOH)_2$ group) are difficult to obtain pure. Even after chromatography on cellulose and ion-exchange columns, they may be only 97% (MTB) and 90% (SMTB) pure.[167]

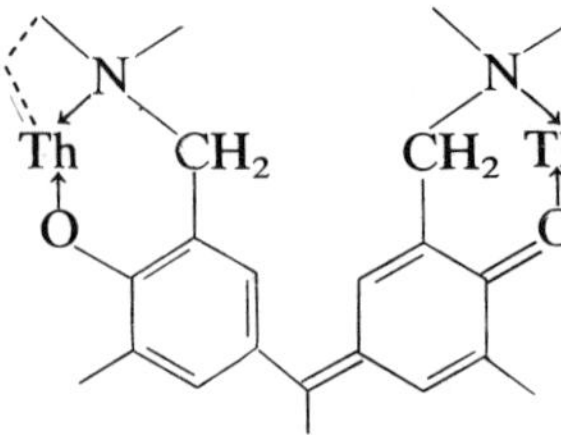

Th_2H_2L

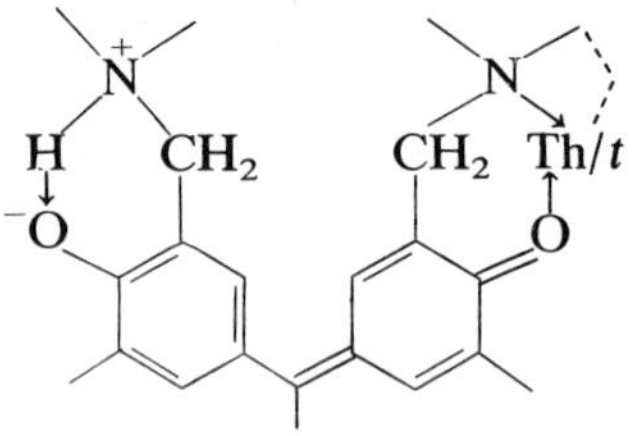

ThH_2L $(t = 1)$, ThH_4L_2 $(t = 2)$

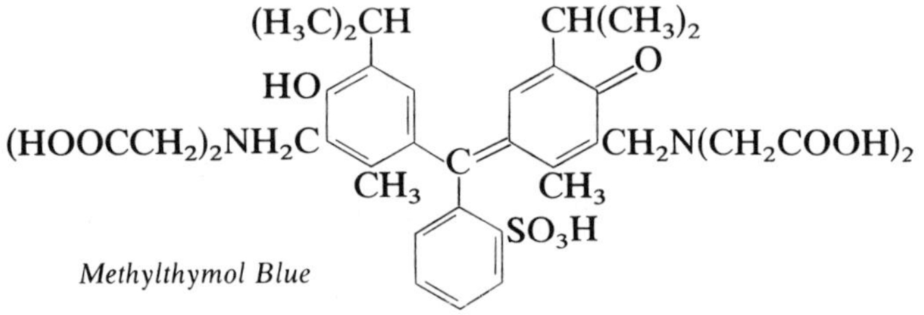

Methylthymol Blue

Mol. wt. 866.7 for 5 Na salt, 756.8 for free acid. M. p. 252°–256° for free acid (decomp.) (99.7% pure product). Na salts less stable than acid.

Methylthymol Blue is yellow in acidic solution, yellow to purple-blue in the pH range 5.4–~11, the color becoming gray around pH ~12 and changing to deep blue above pH 12.5. pK values for $H_6L \cdots L^{6-}$ are 1.8, 2.0, 3.04, 6.9, 11.14, and 12.94.[168] Cobalt, Ni, Cu, and Zn form various charged species with the reagent[168]: ML^{4-}, MHL^{3-}, MH_2L^{2-}, MH_3L^-, M_2L^{2-}, M_2HL^-, $M_2(OH)_2L^{4-}$, the formation constants of which have been determined.[169] The structure of alkaline earth metal complexes is represented in Fig. 6C-2. Because of the presence of two identical chelating groups in the reagent molecule, 1:1 or 2(metal):1 chelates can be formed. Thus Ba in alkaline solution forms both complexes, depending on the ratio of Ba to reagent. The two complexes show maximal absorption at much the same wavelength, but ϵ is larger for the 2:1 complex. The reactions of MTB with metals are similar to those of Xylenol Orange. (It may be noted that aqueous solutions of MTB are less stable than those of Xylenol Orange.)

MTB has been used as colorimetric reagent for Zr, Hf, Th, Nb, Ti, Ga, Al, Fe, Hg, rare earths, and Mg,[170] but these methods are not of great interest except for Zr(Hf), which reacts in mineral acid solution (0.1–1 M HCl or $HClO_4$). At pH 9–10, Th forms a violet 1:2 complex, extractable as a red-orange ion-association complex with trioctylmethylammonium chloride.[171] EDTA masks other metals, except Be, Ti, U, Ag, and Sb.

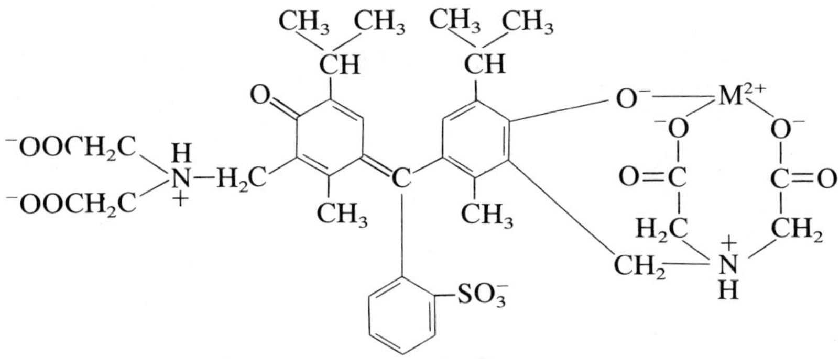

MH_2L^{2-}

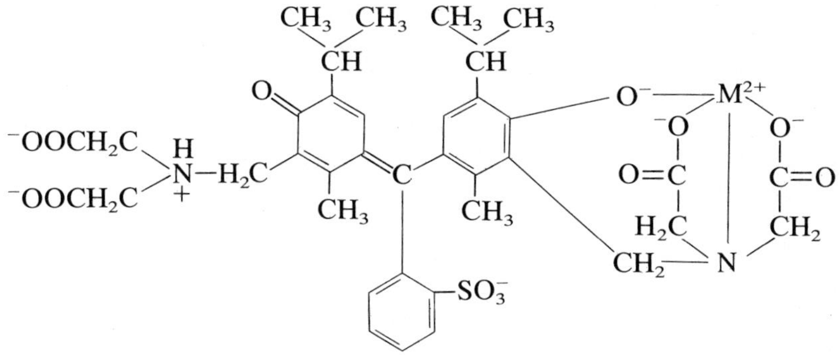

MHL^{3-}

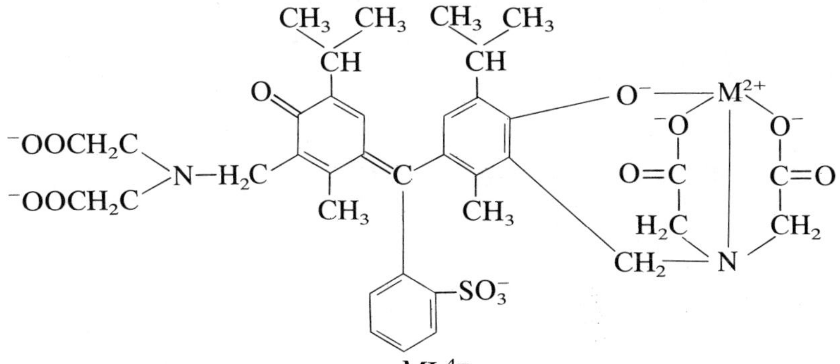

ML^{4-}

Fig. 6C-2. Structures of alkaline earth metal complexes of Methylthymol Blue according to T. Yoshino, H. Okazaki, S. Murakami, and M. Kagawa, *Talanta*, **21,** 678 (1974).

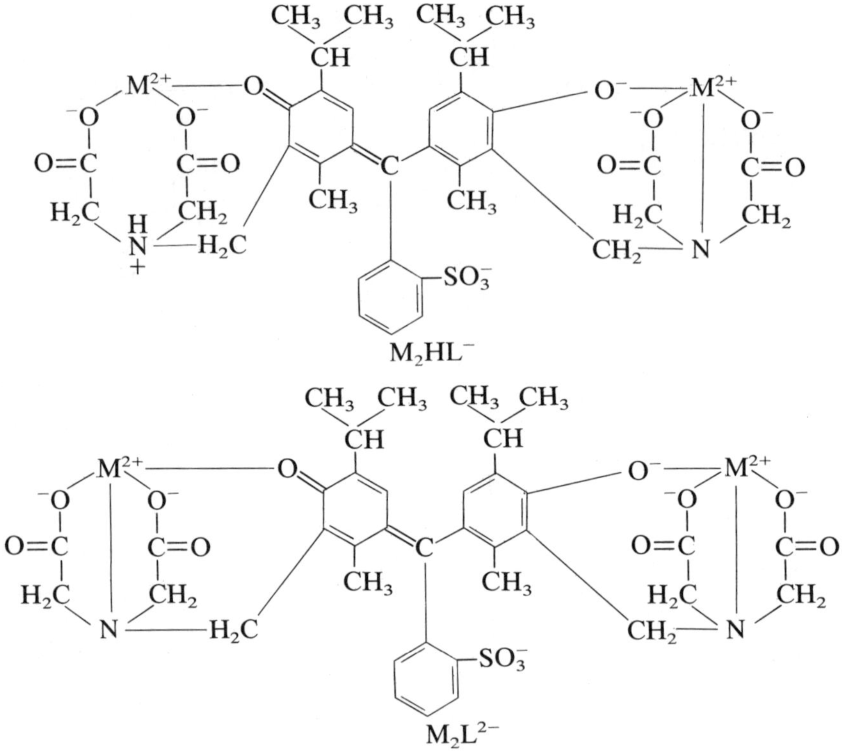

M_2HL^-

M_2L^{2-}

Fig. 6C-2. (*Continued*)

Formation constants of the species of **SMTB** are available, as are those of its Co, Ni, Cu, and Zn complexes.[172]

Methylxylenol Blue. This substance is the same as Methylthymol Blue except that CH_3 replaces $(CH_3)_2CH$. It has found use in the determination of Sc at pH 2.3–2.7 and of Y at pH 6.6–9.3 in the presence of hexadecyltrimethylammonium chloride.[173]

Other Iminodiacetate Reagents. Many of these are fluorescence reagents for metals, obtained by introducing the iminodiacetate group into a fluorophoric framework.

1,5-Bis(dicarboxymethylaminomethyl)-2,6-dihydroxynaphthalene

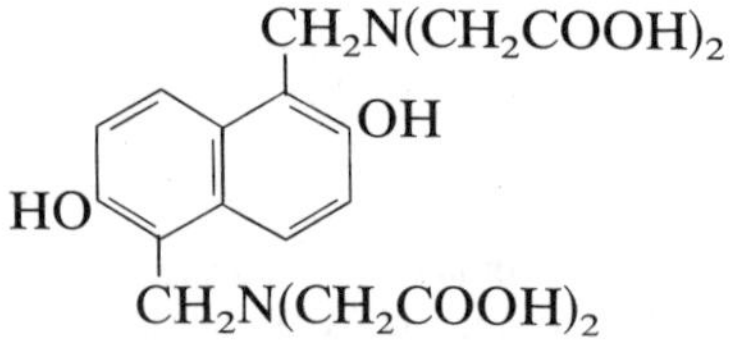

forms a strongly fluorescent 2:1 (Ca/reagent) chelate at pH 11.7 with calcium.[174] Mg, Ba, and Sr and other metals likewise give fluorescence. 1-Dicarboxymethylaminomethyl-2-hydroxy-3-naphthoic acid has been suggested as a fluorimetric reagent for Be, La, and Lu.[175] (2-Hydroxy-3-naphthoic acid, the parent compound, is itself a fluorimetric reagent for various metals.)

3,5′-Bis(dicarboxymethylaminomethyl)-4,4′-dihydroxy-*trans*-stilbene forms fluorescent complexes (2 metal:1 reagent) with many metals and has been proposed for the determination of Cd in the presence of Zn and for La, Gd, and Lu.[176]

2,6-Dibromoindo-3′-methyl-5′-N,N-dicarboxymethylaminomethylphenol, called Indoferron, has been used as a nonselective reagent for Sc[177] and Zr and Fe(III).[178]

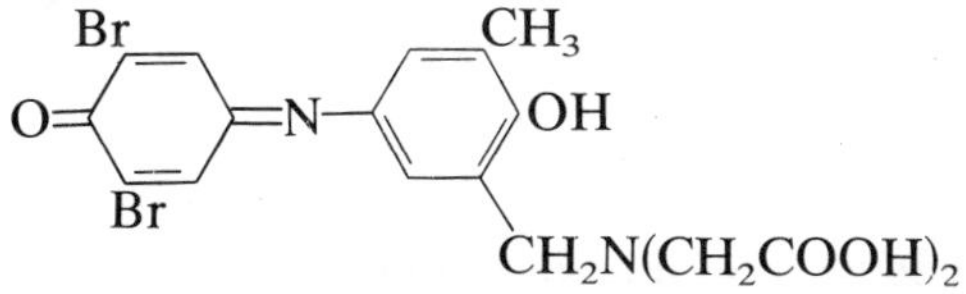

Isocein

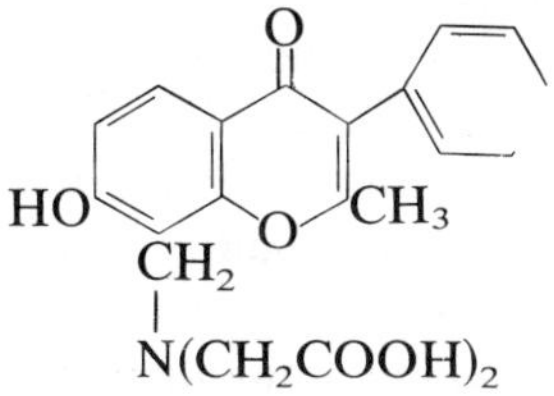

a methyleneiminodiacetic acid derivative of 2-methyl-7-hydroxyisoflavone, has been proposed as a spectrofluorimetric reagent for calcium.[179] It forms a 1:1 complex in 0.8 M KOH. Fluorescence intensities are in the order: Ca > Sr > Ba > Mg.

Bis{[bis(carboxymethyl)amino]methyl}fluorescein

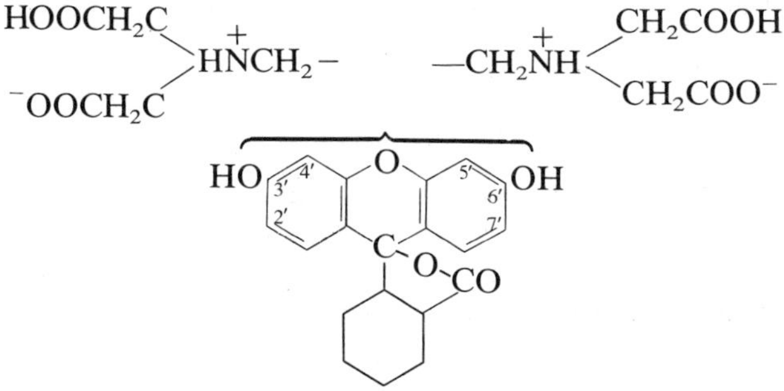

can be prepared by condensation of fluorescein, formaldehyde, and iminodiacetic acid. The positions of the two

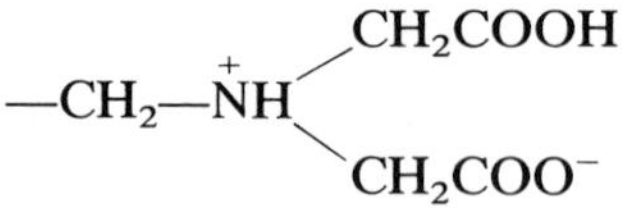

$$-CH_2-\overset{+}{N}H\underset{CH_2COO^-}{\overset{CH_2COOH}{<}}$$

groups are not certain: $(2', 7')$, $(2', 4')$, $(4', 5')$, and $(2', 5')$ are possible.[180] Trivial names such as Calcein[181] and Fluorescein Complexon(e)[182] have been proposed.

Dissociation constants of purified Calcein and fluorescein have been reported as follows:

	Carboxyl groups		Phenol groups		Imino groups	
	pK_{a1}	pK_{a2}	pK_{a3}	pK_{a4}	pK_{a5}	pK_{a6}
Calcein (Wallach)	<4		5.4	9.0	10.5	>12
Calcein (Iritani)	2.1	2.9	4.2	5.5	10.8	11.7
Fluorescein (Iritani)	—	—	5.1	6.8	—	—

Calcein is difficultly soluble in water but readily soluble in alkaline solutions. It fluoresces strongly in the pH range 6–10. At pH >12 it shows only weak residual fluorescence, but the fluorescence intensity increases in the presence of Ca, Sr, Ba, Cd, and other metals. Methods have been described for the fluorimetric determination of Ca[183] and Cd[184] in KOH solutions. These methods suffer from the instability of the reagent solution and the complexes.

For still other reagents containing iminodiacetic acid groups, see discussions of Khimdu, Tichromin, and Alizarin Complexan (Chapter 6B).

X. MISCELLANEOUS COMPOUNDS WITH N- AND O-BONDING GROUPS

A few compounds not readily fitting into the previous classes are mentioned here.

1,2,3-Phenyloxyamidine

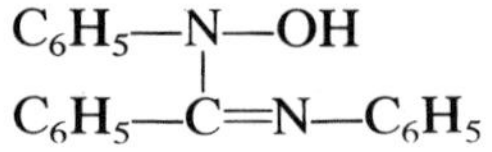

$$C_6H_5-\underset{|}{N}-OH$$
$$C_6H_5-C=N-C_6H_5$$

has been proposed as a colorimetric reagent for V(V), which gives a blue-violet precipitate in acetic acid solution, extractable into $CHCl_3$.[185] $\mathscr{S}_{560} = 0.011\ \mu g\ V/cm^2$. Iron(III), Cu(II), Ni, and Mo(VI) are listed among noninterfering metals.

2-Pyridyl-2'-hydroxymethane sulfonic acid

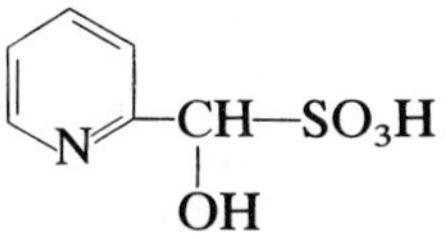

chelates Cu, Ni, Co, Zn, and Fe(II). It is a reagent for Fe (sensitivity comparable to that with 1,10-phenanthroline), but an ammoniacal medium (pH ~10) is required, which is a disadvantage. Reaction is slow. Hydroxylammonium chloride is used to reduce Fe(III), but in addition to its reducing function it is an essential component (as NH_2OH) of the wine red chelate adduct.[186] Apparently, both $FeL_2(NH_2OH)_2$ and $Fe(HL)_4(NH_2OH)_2$ can be formed. NH_2OH probably replaces water molecules in $Fe(H_2O)_6^{2+}$. Hydrazine sulfate or hydroquinone in place of NH_2OH give no color.

ADDENDUM

Dibenzothiazolylmethane[187]

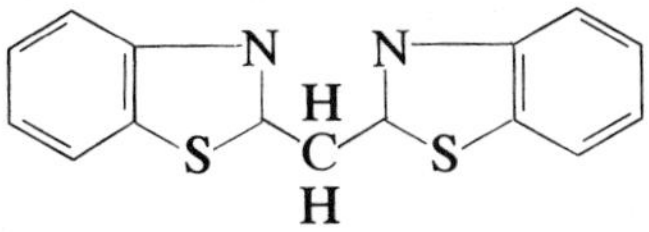

is a fluorimetric reagent for Zn(and Li). Approximately 0.001 ppm Zn in solution is detectable, and Cd does not interfere. Zinc is bonded to the two N atoms; S is not involved in chelate formation.

NOTES

1. E. D. Golla and G. H. Ayres, *Talanta*, **20**, 199 (1973); J. G. Sen Gupta, *Anal. Chim. Acta*, **23**, 462 (1960).

2. L. D. Johnson and G. H. Ayres, *Anal. Chem.*, **38**, 1218 (1966); R. Keil, *Z. Anal. Chem.*, **254**, 191 (1971).

3. G. H. Ayres and J. A. Arno, *Talanta*, **18**, 411 (1971). The reaction is sensitive: $\mathcal{S}_{550} = 0.0024$ μg Ru.

4. For example, 2,3-diaminopyridine reacts with Ru in III, IV, VI, VII, and VIII oxidation states to give variously colored products of unknown structure: G. H. Ayres and D. T. Eastes, *Anal. Chim. Acta*, **44**, 67 (1969).

5. V. K. Gustin and T. R. Sweet, *Anal. Chem.*, **35**, 44 (1963).

6. J. H. Billman and R. Chernin, *Anal. Chem.*, **34**, 408 (1962); see *ibid.*, **36**, 552 (1964).

7. E. L. Steele and J. H. Yoe, *Anal. Chim. Acta*, **20**, 205, 211 (1959).

8. Gh. Baiulescu, C. Greff, and F. Danet, *Analyst*, **94**, 354 (1969). See azo reagents, Chapter 6E.

9. Review: M. Katyal and Y. Dutt, *Talanta*, **22**, 151 (1975).

10. F. Lions and K. V. Martin, *J. Am. Chem. Soc.*, **80**, 3858 (1958). J. F. Geldard and F. Lions, *ibid.*, **84**, 2262 (1962); *Inorg. Chem.* **2**, 270 (1963) and **4**, 414 (1965).

11. A. J. Cameron, N. A. Gibson, and R. Roper, *Anal. Chim. Acta*, **29**, 73 (1963).

12. M. A. Quddus and C. F. Bell, *Anal. Chim. Acta*, **42**, 503 (1968).

13. A. J. Cameron and N. A. Gibson, *Anal. Chim. Acta*, **40**, 413 (1968).

14. J. E. Going and C. Sykora, *Anal. Chim. Acta*, **70**, 127 (1974); Going, G. Wesenberg, and G. Andrejat, *ibid.*, **81**, 349 (1976).

15. M. L. Heit and D. E. Ryan, *Anal. Chim. Acta*, **32**, 448 and **34**, 407 (1966); S. P. Singhal and D. E. Ryan, *ibid.*, **37**, 91 (1967); R. W. Frei, G. H. Jamro, and O. Navratil, *ibid.*, **55**, 125 (1971).

16. R. E. Jensen and R. T. Pflaum, *Anal. Chem.*, **38**, 1268 (1966).

17. Among other reports: G. G. Sims and D. E. Ryan, *Anal. Chim. Acta*, **44**, 139 (1969); V. Zatka, J. Abraham, J. Holzbecher, and D. E. Ryan, *ibid.*, **54**, 65 (1971).

18. R. E. Jensen, N. C. Bergman, and R. J. Helvig, *Anal. Chem.*, **40**, 624 (1968).

19. G. S. Vasilikiotis, Th. Kouimtzis, C. Apostolopoulou, and A. Voulgaropoulos, *Anal. Chim. Acta*, **70**, 319 (1974).

20. D. E. Ryan, F. Snape, and M. Winpe, *Anal. Chim. Acta*, **58**, 101 (1972).

21. J. E. Going and R. T. Pflaum, *Anal. Chem.*, **42**, 1098 (1970).

22. G. Nilsson, *Acta Chem. Scand.*, **4**, 205 (1950).

23. T. Uno and H. Taniguchi, *Japan Analyst*, **20**, 1123 (1971).

24. E. A. Bozhevolnov et al., *Zh. Anal. Khim.*, **24**, 531 (1969).

25. M. Valcarcel and F. Pino, *Analyst*, **98**, 246 (1973).

26. M. Lever, *Anal. Chim. Acta*, **65**, 311 (1973).

27. A. Kawase, *Japan Analyst*, **16**, 1364 (1967); also **20**, 156. 1,3-Diphenyl-5-(2-thiazolyl)-formazan: A. Uchiumi, *J. Chem. Soc. Japan Pure Chem. Sec.*, **90**, 1133 (1969).

28. V. M. Dziomko, V. M. Ostrovskaya, and O. V. Kon'kova, *Zh. Anal. Khim.*, **25**, 267 (1970). 1,5-Di(2-hydroxy-5-sulfophenyl)-3-cyanoformazan has been used to determine U in the presence of Th by V. V. Sergovskaya, M. I. Ermakova, and S. P. Onosova, *Zh. Anal. Khim.*, **29**, 2216 (1974).

29. A. Garcia de Torres, M. Valcarcel, and F. Pino-Perez, *Anal. Chim. Acta*, **68**, 466 (1974); *Talanta*, **20**, 919 (1973). The related compound di-2-pyridyl ketone azine, $(C_5H_4N)_2C{=}N{-}N{=}C(C_5H_4N)_2$, has been proposed as

a photometric reagent for Fe(II) by M. Valcarcel, M. P. Martinez, and F. Pino, *Analyst*, **100,** 33 (1975). Many heavy metals react with this reagent.

30. Review: D. N. Purohit, *Talanta*, **14,** 353 (1967).

31. N. C. Sogani and S. C. Bhattacharyya, *Anal. Chem.*, **29,** 397 (1957). Also a reagent for Mo: *ibid.*, **33,** 1273 (1961).

32. D. Chakraborti, *Anal. Chim. Acta*, **71,** 196 (1974).

33. A. K. Majumdar, B. C. Bhattacharyya, and B. C. Roy, *Anal. Chim. Acta*, **67,** 307 (1973).

34. A. K. Majumdar and S. C. Saha, *Anal. Chim. Acta*, **44,** 85 (1969).

35. A. K. Majumdar and D. Chakraborti, *Anal. Chim. Acta*, **53,** 127 (1971).

36. D. Chakraborti, *Anal. Chim. Acta*, **70,** 207 (1974).

37. A. A. Schilt, *Analytical Applications of 1,10-Phenanthroline and Related Compounds*, Pergamon, Oxford, 1969. Earlier review: F. Vydra and M. Kopanica, *Chem.-Anal.*, **52,** 88 (1963).

38. A. K. Bhadra, *Talanta*, **20,** 13 (1973).

39. A. Hulanicki and J. Nieniewska, *Talanta*, **21,** 896 (1974).

40. A. K. Bhadra and S. Banerjee, *Talanta*, **20,** 342 (1973).

41. 1,10-Phenanthroline cannot be used in acidic perchlorate solutions because of the low solubility of phenanthrolinium perchlorate. Sulfonation of phenanthroline remedies this defect: D. E. Blair and H. Diehl, *Anal. Chem.*, **33,** 867 (1961).

42. A. A. Schilt, G. F. Smith, and A. Heimbuch, *Anal. Chem.*, **28,** 809 (1956).

43. $\log \beta_3$ of the Fe(II) chelate of 2-methyl-1,10-phenanthroline is 10.8 compared to 21.2 for 1,10-phenanthroline [H. Irving and D. H. Mellor, *J. Chem. Soc.*, 5237 (1962)]. The steric effect of Cl in the 2-position is much the same as of the CH_3 group, if the difference in ligand basicities is taken into account ($\log \beta_3 = 11.6$) as shown by H. Irving and P. J. Gee, *Anal. Chim. Acta*, **55,** 315 (1971).

44. *The Copper Reagents: Cuproine, Neocuproine, Bathocuproine*, G. Frederick Smith Chemical Co., Columbus, Ohio, 1972.

45. A. A. Schilt and W. C. Hoyle, *Anal. Chem.*, **41,** 344 (1969).

46. W. E. Dunbar and A. A. Schilt, *Talanta*, **19,** 1025 (1972).

47. M. Maeda, A. Tsuji, H. Igeta, and T. Tsuchiya, *Chem. Pharm. Bull. Tokyo*, **18,** 1548 (1970).

48. R. E. Jensen and R. T. Pflaum, *Anal. Chim. Acta*, **32,** 235 (1965).

49. P. F. Collins, H. Diehl, and G. F. Smith, *Anal. Chem.*, **31,** 1862 (1959). See Table 6C-2.

50. W. A. Embry and G. H. Ayres, *Anal. Chem.*, **40,** 1499 (1968).

51. M. J. Janmohamed and G. H. Ayres, *Anal. Chem.*, **44,** 2263 (1972).

52. C. D. Chriswell and A. A. Schilt, *Anal. Chem.*, **42,** 992 (1974). Earlier used for simultaneous determination of Fe and Cu: A. A. Schilt and P. J. Taylor,

ibid., **42**, 220 (1970). As an Fe reagent, it is as sensitive as bathophenanthroline and easier to prepare (therefore cheaper). The sulfonated derivative (Ferrozine) is available and used for determination of Fe ($\mathscr{S} = 0.002$ μg Fe). See L. L Stookey, *Anal. Chem.*, **42**, 779 (1970). S. K. Kundra, M. Katyal, and R. P. Singh, *ibid.*, **46**, 1605 (1974), used Ferrozine to determine Co ($\epsilon_{500} = 4600$, $\mathscr{S} = 0.013$ μg/cm^2). They found commercial Ferrozine (disodium salt) to be grossly impure (three recrystallizations from water were required before use). The preparation of purer reagent. has been described by C. R. Gibbs, *Anal. Chem.*, **48**, 1197 (1976).

53. A. A. Schilt, C. D. Chriswell, and T. A. Fang, *Talanta*, **21**, 831 (1974). This paper treats various ferroin-type reagents based on triazine, triazoline, and triazole.

54. A. A. Schilt and K. R. Kluge, *Talanta*, **15**, 475 (1968). A. Bergh, P. O. Offenhartz, P. George, and G. P. Haight, *J. Chem. Soc.*, 1533 (1964), studied Fe(II) chelates of 2,2′, 2″, 2‴-quaterpyridine without finding better reagents.

55. A. A. Schilt, W. E. Dunbar, B. W. Gandrud, and S. E. Warren, *Talanta*, **17**, 649 (1970); see L. Stookey, *ibid.*, **17**, 644.

56. E. Kiss, *Anal. Chim. Acta*, **72**, 127 (1974).

57. A. A. Schilt and W. C. Hoyle, *Talanta*, **15**, 852 (1968).

58. W. I. Stephen and P. C. Uden, *Anal. Chim. Acta*, **39**, 357 (1967).

59. F. C. Trusell and W. F. McKenzie, *Anal. Chim. Acta*, **40**, 350 (1968); **47**, 154 (1969) (in presence of Methyl Orange).

60. W. I. Stephen, *Talanta*, **16**, 939 (1969). For the first-named reagent, see R. T. Pflaum, C. J. Smith, Jr., E. B. Buchanan, Jr., and R. E. Jensen, *Anal. Chim. Acta*, **31**, 341 (1964); Jensen, J. A. Carlson, and M. L. Grant, *ibid.*, **44**, 123 (1969) (protonation of FeL$_2^{2+}$).

61. A. A. Schilt and K. R. Kluge, *Talanta*, **15**, 1055 (1968).

62. A. A. Schilt and P. J. Taylor, *Talanta*, **16**, 448 (1969). Earlier reagents of the oxime type include phenyl 2-pyridyl ketone oxime and methyl 2-pyridyl ketone oxime (references in this paper). (See p. 380.)

63. J. L. Walter and H. Freiser, *Anal. Chem.*, **26**, 217 (1954).

64. L. S. Bark and A. Rixon, *Anal. Chim. Acta*, **45**, 425 (1969).

65. The following general reviews are available: C. V. Banks, *Analytical Chemistry* 1962, Elsevier, Amsterdam, 1963, p. 131; Banks, *Rec. Chem. Progr.*, **25**, no. 2, 85 (1964); K. Burger in H. A. Flaschka and A. J. Barnard, Jr., Eds., *Chelates in Analytical Chemistry*, vol. 2, Dekker, New York, 1969, pp. 179–212; D. Dyrssen, *Svensk Kem. Tids.*, **75**, 618 (1963); B. Egneus, *Talanta*, **19**, 1387 (1972). The last reference surveys the literature comprehensively since 1963.

66. Exceptions are known. β-Diphenylglyoxime reacts with Pd(II); see A. Warshawsky, *Talanta*, **21**, 624 (1974).

67. S. Kuse, S. Motomizu, and K. Toei, *Anal. Chim. Acta*, **70**, 65 (1974).

68. Chloroform extraction of nickel dimethylglyoximate: E. B. Sandell and R. W. Perlich, *Ind. Eng. Chem. Anal. Ed.*, **11**, 309 (1939); H. Christopherson and Sandell, *Anal. Chim. Acta*, **10**, 1 (1954).

69. The effect of tartrate and citrate on the extraction of Ni with dimethylglyoxime and other α-dioximes has been investigated by V. M. Peshkova, V. M. Savostina, E. K. Astakhova, and N. A. Minaeva, *Anal. Abst.*, **13**, 2970 (1966). All dioximes are suitable for tartrate solution, but benzil α-dioxime should be used for citrate solutions.

70. Yu. A. Zolotov and G. E. Vlasova, *Zh. Anal. Khim.*, **28**, 1540 (1973), use iodine as oxidizing agent at pH ~ 12.5 and extract what is thought to be $NiDx_3^{2-}$ with diphenylguanidinium as cationic partner into chloroform-isopentyl alcohol. $\mathscr{S}_{490}$ calculated from their data is $1.6 \times 10^{-4} \mu g\,Ni/cm^2$, a value so small that confirmation is desirable.

71. K. Burger and I. Ruff, *Acta Chim. Acad. Sci. Hung.*, **45**, 77 (1965).

72. Yu. V. Morachevskii, L. I. Lebedeva, and Z. G. Golubtsova, *J. Anal. Chem. USSR*, **15**, 541 (1960).

73. A. K. Babko, P. B. Mikhel'son, and V. F. Mikitchenko, *Zh. Neorg. Khim.*, **11**, 817 (1966).

74. F. A. Fryer, D. J. B. Galliford, and J. T. Yardley, *Analyst*, **88**, 188 (1963).

75. J. L. Jones and J. Gastfield, *Anal. Chim. Acta*, **51**, 130 (1970).

76. D. Thierig and F. Umland, *Z. Anal. Chem.*, **240**, 19 (1968); V. W. Meloche, R. L. Martin, and W. H. Webb, *Anal. Chem.*, **29**, 527 (1957); $\mathscr{S}_{530} = 0.005 \mu g\,Re/cm^2$. According to V. M. Peshkova and N. G. Ignat'eva, *J. Anal. Chem. USSR*, **22**, 642 (1967), the red complex, formed in the presence of HCl and $SnCl_2$, has furildioxime : Re(IV) = 4. It is $CHCl_3$ extractable.

77. P. B. Mikhel'son, A. K. Boryak, and L. T. Moshkovskaya, *Zh. Anal. Khim.*, **26**, 787 (1971).

78. B. Egneus, *Anal. Chim. Acta*, **48**, 291 (1969). Properties of the reagent: *ibid.*, **43**, 53 (1968).

79. J. L. Kassner, S.-F. Ting, and E. L. Grove, *Talanta*, **7**, 269 (1961).

80. M. M. Barling and C. V. Banks, *Anal. Chem.*, **36**, 2359 (1964). Absorbance of the nickel complex in xylene is measured at 386 nm. Co, Cu, and Fe(III) interfere if not masked.

81. I. N. Marov, E. S. Gur'eva, and V. M. Peshkova, *Zh. Neorg. Khim.*, **15**, 3039 (1970).

82. P. Collins and H. Diehl, *Anal. Chim. Acta*, **18**, 384 (1958).

83. J. Bassett, G. B. Leton, and A. I. Vogel, *Analyst*, **92**, 279 (1967).

84. P. F. Lott and R. K. Vitek, *Anal. Chem.*, **32**, 391 (1960).

85. D. Soules and W. J. Holland, *Mikrochim. Acta*, **1972**, 247. Preparation of α and β isomers of reagent: D. Soules, W. J. Holland, and S. Stupavsky, *ibid.*, **1970**, 787.

ORGANIC PHOTOMETRIC REAGENTS

86. H. R. Notenboom, W. J. Holland, and D. Soules, *ibid.*, **1973**, 187.

87. D. Soules and W. J. Holland, *Mikrochim. Acta*, 565 (1971).

88. "Compounds with the 1,3-dioxime grouping · · · have received very little attention, probably because the dioxime groups are too far apart to form chelates with metal ions. It should be appreciated, however, that although the formation of a chelate ring confers added stability on a complex, complexes that do not contain chelate rings are not necessarily unstable, and may also possess many other desirable analytical qualities such as high molar absorptivity, rapid and selective formation, and solubility in water or organic solvents."[89]

89. R. Belcher, S. A. Ghonaim, and A. Townshend, *Talanta*, **21**, 191 (1974).

90. B. Kirson, *Bull. Soc. Chim.*, 1032 (1962). For structure of complexes, see C. F. Liu and C. H. Liu, *Inorg. Chem.*, **2**, 706 (1963). In connection with pyridine-2-aldoxime, pyridine-2,6-dialdoxime may be mentioned. S. P. Bag, Q. Fernando, and H. Freiser, *Anal. Chem.*, **35**, 719 (1963), determined the formation constants of the Cu(II), Ni, Zn, and Mn chelates of this reagent. The colors of the Cu, Co, Mn, Ni, and Zn chelates in water are, respectively, green, yellow, blue-violet, pale yellow and colorless; ϵ_{max} lies in the range 1000–10,000. Cu(II) forms a 1:1 chelate; the other metals, 1:2 reagent. The second oxime group is not involved in chelation.

91. D. K. Banerjea and K. K. Tripathi, *Anal. Chem.*, **32**, 1196 (1960). The formation constant of the complex is reported and observations of color reactions of other metals (notably V, Mo, and Re) are recorded.

92. H. Diehl and G. F. Smith, *The Iron Reagents: Bathophenanthroline*, 2,4,6-*Tripyridyl-s-triazene, Phenyl-2-pyridyl Ketoxime*, G. F. Smith Chemical Co., Columbus, Ohio, 1965. F. Trusell and H. Diehl, *Anal. Chem.*, **31**, 1978 (1959).

93. D. C. Shuman and B. Sen, *Anal. Chim. Acta*, **33**, 487 (1965).

94. Pd: B. Sen, *Anal. Chem.*, **31**, 881 (1959). The reaction is more sensitive ($\epsilon_{410} = 30,000$ in $CHCl_3$) than that with dimethylglyoxime but apparently a higher pH is required for extraction. Re: J. Guyon and R. K. Murmann, *Anal. Chem.*, **36**, 1058 (1964), carried out the reaction at room temperature in a solution containing HCl, $SnCl_2$, and methanol, and isolated a purple anionic complex to which the composition $[Re(V)(HL)O_2(OH)Cl]^-$ was assigned.

95. J. R. Pemberton and H. Diehl, *Talanta*, **16**, 393 (1969).

96. W. J. Holland, J. Bozic, and J. T. Gerard, *Anal. Chim. Acta*, **43**, 417 (1968).

97. W. J. Holland and J. Bozic, *Anal. Chem.*, **39**, 109 (1967); **40**, 433 (1968); *Talanta*, **15**, 843 (1968).

98. D. Goldstein and E. K. Libergott, *Anal. Chim. Acta*, **51**, 126 (1970). Qualitative only.

99. S. K. Sindhwani and R. P. Singh, *Talanta*, **20**, 248 (1973).

100. *Anal. Chim. Acta*, **55**, 409 (1971).

101. *Microchem. J.*, **18**, 686 (1973). The Rh complex (1:3) is formed on heating, and Co, Ni, and Pd can first be extracted at room temperature into $CHCl_3$.

102. G. Gorbach and F. Pohl, *Mikrochemie*, **38**, 258 (1951). The extraction of metals with a benzene reagent solution has been studied by I. Dahl, *Anal. Chim. Acta*, **41**, 9 (1968). Cu Ni, Co can be determined.

103. K. Burger, *Talanta*, **8**, 77 (1961).

104. Z. Marczenko and J. Minczewski, *J. Anal. Chem. USSR*, **17**, 19 (1962); Z. Marczenko, *Anal. Chim. Acta*, **31**, 224 (1964); A. Okac and M. Bartusek, *Z. Anal. Chem.*, **178**, 198 (1960).

105. F. Feigl, *Chemistry of Specific, Selective and Sensitive Reactions*, Academic, New York, 1949, p. 213.

106. U. B. Talwar and B. C. Haldar, *Anal. Chim. Acta*, **39**, 264 (1967). $\mathscr{S} = 0.01 \ \mu g$ Pd.

107. Talwar and Haldar, *Anal. Chim. Acta*, **51**, 53 (1970).

108. G. Cronheim, *Ind. Eng. Chem. Anal. Ed.*, **14**, 445 (1942).

109. G. H. Ellis and J. F. Thompson, *Ind. Eng. Chem. Anal. Ed.*, **17**, 254 (1945).

110. S. M. Peach, *Analyst*, **81**, 371 (1956); T. Torii, *J. Chem. Soc. Japan Pure Chem. Sec.*, **76**, 328, 333, 675, 680, 707, 825 (1955); *Japan Analyst*, **4**, 177 (1955).

111. S. Motomizu, *Anal. Chim. Acta*, **56**, 415 (1971); *Analyst*, **97**, 986 (1972). Also an Fe(II) reagent ($FeL_3 \cdot HL^-$ or FeL_4^{2-} formed): K. Toei, S. Motomizu, and T. Korenaga, *Analyst*, **100**, 629 (1975).

112. S. Motomizu, *Talanta*, **21**, 654 (1974); Co in sea water: *Anal. Chim. Acta*, **64**, 217 (1973).

113. T. Korenaga, S. Motomizu, and K. Toei, *Anal. Chim. Acta*, **65**, 335 (1973); *Talanta*, **21**, 645 (1974). Also see *Analyst*, **101**, 974 (1976).

114. C. W. McDonald and R. Carter, *Anal. Chem.*, **41**, 1478 (1969).

115. 1-Nitroso-2-naphthol (and its isomer) forms 1:1 and 1:2 complexes with numerous divalent metals. Even Mg reacts in dioxane solution. Most of these complexes have small formation constants and are easily decomposed by acids. Thorium, Zr, and other higher valent metals also react, but the chelates are not important photometrically. The presence of the –OH group with replaceable H explains the reactivity of the reagent with many metals.

116. C. M. Callahan, W. C. Fernelius, and B. P. Block, *Anal. Chim. Acta*, **16**, 101 (1957), have shown that in basic medium, both 1-nitroso-2-naphthol and 2-nitroso-1-napthol become good oxidizing agents and oxidize the first-formed $Co(ORNO)_2$ to $Co(ORNO)_3$. One mole of 1-amino-2-naphthol is formed for 4 moles of the cobaltic complex. Formation constants of a number of nitrosonaphtholates are given in this paper. For a survey of reactions of nitrosonaphthols with metals, see H. Fischer, *Mikrochem. ver. Mikrochim. Acta*, **30**, 47 (1942).

117. Probably also Ru(III) chelates. RuL$_3$ and, in citrate solution, RuLCit⁻ have been shown to be formed by C. Konecny, *Anal. Chim. Acta*, **29**, 423 (1963). 2-Nitroso-1-naphthol forms analogous complexes. Nitrosyl-ruthenium(III) forms 1:1 and 1:2 chelates (*ibid.*, **31**, 352). G. Kesser, R. J. Meyer, and R. P. Larsen, *Anal. Chem.*, **38**, 221 (1966), form the Ru complex of 1-nitroso-2-naphthol in CCl$_4$ solution (after isolation of RuO$_4$ by distillation) allowed to stand overnight. $\mathcal{S} = 0.006 \ \mu\mathrm{g \ Ru/cm}^2$.

118. H. Iinuma, *Res. Rept. Fac. Engin. Gifu Univ.*, **1**, 35 (1951).

119. Ru(III): D. J. Miller, S. C. Srivastava, and M. L. Good, *Anal. Chem.*, **37**, 739 (1965). The complex (1 Ru:2 reagent) is formed at pH 3–7 (acetate buffer containing chloride and iodide) on heating. Rh(II): O. W. Rollins and M. M. Oldham, *Anal. Chem.*, **43**, 146 (1971). The chelate, formed slowly at boiling temperature, has the composition 1 Rh:3 reagent (H of –OH replaced). The solid Na$_6${Rh[C$_{10}$H$_4$NO$_2$(SO$_3$)$_2$]$_3$} has been isolated. The Pd chelate [*ibid.*, **43**, 262 (1971)] has the expected 1:2 (reagent) composition. It forms rapidly at boiling temperature. For later work on the determination of Rh see L. S. Markova, V. M. Savostina, and V. M. Peshkova, *Zh. Anal. Khim.*, **29**, 1378 (1974).

120. N. Kohli and R. P. Singh, *Talanta*, **21**, 638 (1974).

121. G. J. Willems, R. A. Lontie, and W. A. Seth-Paul, *Anal. Chim. Acta*, **51**, 544 (1970), examined ten commercial products sold as diphenylcarbazide and found only two (BDH Analar; Matheson, Coleman, and Bell) to be pure (m.p. 171.2°–172.0°). The other eight, from European and U.S. suppliers (some purportedly "reagent grade" and "analyzed reagent"), contained from 31 to 40% phenylsemicarbazide (m.p. 162.1°–164.9°). The pure substance can be obtained by reaction of diphenyl carbonate (not urea) with phenylhydrazine according to the original method of Cazeneuve and Moreau, *Bull. Soc. Chim. Fr.*, 51 (1900). The two components can be separated chromatographically with polyamide as adsorbent.

122. G. J. Willems and C. J. De Ranter, *Anal. Chim. Acta*, **68**, 111 (1974). The content of diphenylcarbazone ranged from 16 to 58%, of diphenylcarbazide from 37 to 70%, of phenylsemicarbazide from 6 to 34%, of diphenylcar-badiazone from ~0.5 to 2.5%, as determined by elution chromatography. Other compounds are present in traces. An Eastman product contained 58% diphenylcarbazone, an Aldrich (Milwaukee) product 55%, and one from Merck (Darmstadt) 24%. Diphenylcarbazone and diphenylcarbazide are not present in equimolar proportions as claimed in earlier work: P. Krumholz and E. Krumholz, *Monatsh.*, **70**, 431 (1937); J. J. R. F. da Silva, J. C. G. Calado, and M. L. de Moura, *Talanta*, **11**, 983 (1964).

123. J. J. R. F. da Silva, J. C. G. Calado, and M. L. de Moura, *Talanta*, **11**, 983 (1964). One gram of commercial diphenylcarbazone is dissolved in 100 ml five percent NaOH containing a few milligrams of KCN and the solution is extracted continuously with 100 ml of ethyl ether in a suitable extractor (e.g., that of Ashley and Murray) for 36 h. Because of its acidic properties,

diphenylcarbazone remains in the basic aqueous phase, whereas diphenyl-carbazide is extracted into the ether, from which it crystallizes (m.p. 172°–173°). Diphenylcarbazone is recovered from the aqueous phase by adding sulfuric acid, and recrystallizing the precipitate by dissolving in hot alcohol and adding water (orange crystals, m.p. 126°–127°, rapid heating). The solubility of diphenylcarbazone is roughly 10^{-3} M in ethanol and chloroform. In carbon tetrachloride, $\epsilon_{290} = 12{,}750$; $\epsilon_{466} = 3300$; $\epsilon_{563} = 800$.

124. S. Balt and E. van Dalen, *Anal. Chim. Acta*, **27**, 188 (1962).

125. H. Einaga and H. Ishii, *Analyst*, **98**, 802 (1973).

126. H. Sano, *Anal. Chim. Acta*, **27**, 398 (1962); E. V. Kovalenko and V. I. Petrashen', *J. Anal. Chem. USSR*, **18**, 645 (1963); E. Najdeker, *Proc. Soc. Anal. Chem.*, **8**, 194 (1971); H. Marchart, *Anal. Chim. Acta*, **30**, 11 (1964).

127. Se: S. G. Sushkova and V. I. Murashova, *Anal. Abst.*, **16**, 1233. Diphenyl-thiocarbazide is a more sensitive Se reagent. Tc: F. J. Miller and H. E. Zittel, *Anal. Chem.*, **35**, 299 (1963). Tc(VII) is reduced to Tc(IV) by the reagent, and this oxidation state forms the red complex.

128. F. Feigl and A. Lederer, *Monatsh.*, **45**, 63, 115 (1924); H. Fischer, *Mikrochem. ver. Mikrochim. Acta*, **30**, 44 (1942); S. Balt and E. van Dalen, *Anal. Chim. Acta*, **29**, 466 (1963). For qualitative use of derivatives of diphenylcarbazone and of the dinaphthyl compounds, see P. Krumholz and F. Honel, *Mikrochim. Acta*, **2**, 177 (1937).

129. E. van Dalen and S. Balt, *Anal. Chim. Acta*, **25**, 507 (1961).

130. S. Balt and E. van Dalen, *Anal. Chim. Acta*, **27**, 416 (1962).

131. S. Balt and E. van Dalen, *Anal. Chim. Acta*, **30**, 434 (1964). Values of P (partition constant) are also given. They range from 0.5–0.8 for $Mn(HL)_2$ and $Ni(HL)_2$ to $\sim 10^3$ for $Hg(HL)_2$ and $Fe(HL)_2$.

132. L. N. Lapin and N. V. Reis, *Zh. Anal. Khim.*, **13**, 426 (1958). Diphenylcarbazide in benzene (pH ~ 10) has also been used: R. E. Stoner and W. Dasler, *Anal. Chem.*, **32**, 1207 (1960).

133. H. R. Geering and J. F. Hodgson, *Anal. Chim. Acta*, **36**, 537 (1966). The stability constants of the copper complexes are $\log \beta_1 = 9.8$, $\log \beta_2 = 19.5$, $\log \beta_3 = 29$; $\log P_{Cu(HL)_2} = 2.82$ (benzene, 28°, $\mu < 0.01$).

134. V. K. Akimov, A. I. Busev, and L. Ya. Kliot, *Zh. Anal. Khim.*, **28**, 1014 (1973), extract perrhenate with 1,1-diantipyrinylbutane in $CHCl_3$, treat the extract with diphenylcarbazide in acetone $+ 6$ M HCl and measure the absorbance of the violet complex of Re(V) and diphenylcarbazone (ϵ_{540} $\sim 20{,}000$).

135. L. G. Gavrilova and Yu. A. Zolotov, *Zh. Anal. Khim.*, **25**, 1054 (1970).

136. R. M. Rush and J. H. Yoe, *Anal. Chem.*, **26**, 1345 (1954). Review: M. B. Johnston, A. J. Barnard, Jr., and W. C. Broad, *Rev. Univ. Indust. Santander*, **4**, 43 (1962).

137. I. Murase, *Bull. Chem. Soc. Japan*, **32**, 827 (1959); E. Bayer and G. Schenck, *Berichte*, **93**, 1184 (1960); Bayer, *Berichte*, **90**, 2325 (1957).

ORGANIC PHOTOMETRIC REAGENTS

Preparation of reagent: J. Drewry, *Analyst*, **87**, 827 (1962). The slow fading of the Ca color at high pH has been attributed by F. Lindstrom and C. W. Milligan, *Anal. Chem.*, **39**, 132 (1967), to conversion of the reagent to glycolic acid, which precipitates or complexes Ca. A. V. Kuczerpa, *ibid.*, **40**, 581 (1968), has shown that the stability of the colored product is increased by reducing the temperature, excluding CO_2, and making other modifications in the procedure.

138. F. Umland and K.-U. Meckenstock, *Z. Anal. Chem.*, **176**, 96 (1960).

139. H. G. C. King and G. Pruden, *Analyst*, **94**, 39 (1969). The interference of small amounts of Mg and Fe(III) is prevented by addition of mannitol. J. R. W. Kerr, *ibid.*, **85**, 867 (1960), develops the color in ethanol–butanol–water mixture. K. T. Williams and J. R. Wilson, *Anal. Chem.*, **33**, 244 (1961), state that Mg does not interfere (10 Mg : 1 Ca); $CHCl_3$–C_2H_5OH was used for extraction.

140. M. Nishimura and S. Noriki, *Japan Analyst*, **21**, 640 (1972); M. Nishimura, S. Noriki, and S. Muramoto, *Anal. Chim. Acta*, **70**, 121 (1974).

141. A. D. Wilson, *Analyst*, **87**, 703 (1962).

142. A. Okac and M. Vrchlabsky, *Z. Anal. Chem.*, **182**, 425 (1961).

143. A. Okac and M. Vrchlabsky, *Z. Anal. Chem.*, **195**, 338 (1963).

144. F. Lindstrom and C. W. Milligan, *Anal. Chem.*, **36**, 1334 (1964); **44**, 1822 (1972).

145. T. Pelczar, *Mikrochim. Acta*, **1968**, 1295.

146. W. Ch'i, S. Chang, and M. Hsu, *Acta Chim. Sin.*, **31**, 179 (1965); *Anal. Abst.*, **13**, 4007 (1966).

147. R. J. Argauer and C. E. White, *Anal. Chem.*, **36**, 2141 (1964). The effect of substituent groups in some 30 2,2'-dihydroxyazomethines was investigated. Review of Schiff bases: K. Dey, *J. Sci. Ind. Res.*, **33**, 76 (1974).

148. K. Morisige, *Anal. Chim. Acta*, **72**, 295 (1974).

149. J. H. Saylor and J. W. Ledbetter, *Anal. Chim. Acta*, **30**, 427 (1964), where earlier work is cited. This reagent was already used for Al by Z. Holzbecher, *Coll. Czech. Chem. Commun.*, **19**, 241 (1954); **24**, 3915 (1959). *N*-Salicylidene-2-amino-3-hydroxyfluorene is also an Al fluorescence reagent [*Anal. Chem.*, **39**, 367 (1967)].

150. R. M. Dagnall, R. Smith, and T. S. West, *Talanta*, **13**, 609 (1966).

151. H. Ishii and H. Einaga, *Japan Analyst*, **18**, 230 (1969).

152. K. Isagai, *J. Chem. Soc. Japan Pure Chem. Sec.*, **82**, 1172 (1961).

153. K. Morisige, *Anal. Chim. Acta*, **73**, 245 (1974).

154. R. M. Dagnall, R. Smith, and T. S. West, *Analyst*, **92**, 20 (1967). This reagent has been used for extraction-photometric determination of Cu (H. Ishii and H. Einaga, *Analyst*, **94**, 1038) and Ni (Ishii and Einaga, *Japan Analyst*, **19**, 1351). Both Cu and Ni form 1 : 1 complexes.

155. G. F. Kirkbright, T. S. West, and C. Woodward, *Analyst*, **91**, 23 (1966). Both Al and Y, among others, also react.

156. Details may be found in F. J. Welcher, *The Analytical Uses of Ethylenediaminetetraacetic Acid*, Van Nostrand, New York, 1957, pp. 276–285.

157. R. Pribil, *Talanta*, **3**, 200 (1959).

158. V. P. Antonovich and V. A. Nazarenko, *Zh. Anal. Khim.*, **23**, 1143 (1968); M. Murakami, T. Yoshino, and S. Harasawa, *Talanta*, **14**, 1293 (1967); I. Mori et al., *ibid.*, **19**, 299 (1972).

159. B. Rehak and J. Körbl, *Coll. Czech. Chem. Commun.*, **25**, 797 (1960).

160. M. Otomo, *Japan Analyst*, **14**, 45 (1965); **21**, 436 (1972).

161. D. C. Olson and D. W. Margerum, *Anal. Chem.*, **34**, 1299 (1962); Murakami et al.[158] According to Olson and Margerum the first fraction (of three) is Semixylenol Orange, but F. J. Langmyhr and P. E. Paus, *Acta Chem. Scand.*, **20**, 2456 (1966), on chromatographing the first fraction again obtained three bands and they conclude that the three components are only three protonated forms of the reagent, probably $H_4(XO)^{2-}$, $H_3(XO)^{3-}$, and $H_2(XO)^{4-}$, which predominate at pH 5. Semixylenol Orange has been proposed as a sensitive reagent for zirconium ($\epsilon = 59,000$ at 535 nm).

162. R. Pribil and V. Vesely, *Talanta*, **17**, 801 (1970), list methods up to the late 1960s.

163. V. Svoboda and V. Chromy, *Talanta*, **13**, 237 (1966); *ibid.*, **12**, 431 (1965). W. J. de Wet and G. B. Behrens, *Anal. Chem.*, **40**, 200 (1968). Addition of cetylpyridinium bromide to an alkaline XO solution results in a color change from red-violet to grayish pink. If a rare earth ion solution is now added, a blue complex ($\lambda_{max} = 605-615$ nm) forms (Ni also gives a blue color). The composition of the complexes is 1 M : 2 XO : 4 CPB. Determination of Sc: V. P. Antonovich and V. A. Nazarenko, *Zh. Anal. Khim.*, **23**, 1143 (1968); Y. Shijo, *J. Chem. Soc. Japan*, 889 (1974).

164. D. Kantcheva, P. Nenova, and B. Karadakov, *Talanta*, **19**, 1450 (1972).

165. Murakami et al.[158]

166. B. W. Budesinsky and J. Svec, *Anal. Chim. Acta*, **61**, 465 (1972).

167. T. Yoshino, H. Imada, T. Kuwano, and K. Iwasa, *Talanta*, **16**, 151 (1969). Later (*ibid.*, **21**, 211 and 199), purer products were obtained. Commercial products may contain as little as 50% MTB. Inorganic salts are often present.

168. T. Yoshino, H. Imada, S. Murakami, and M. Kagawa, *Talanta*, **21**, 211 (1974). $\mu = 0.1$, 25°.

169. Reference in note 168. For formation constants of alkaline earth complexes see T. Yoshino, H. Okazaki, S. Murakami, and M. Kagawa, *Talanta*, **21**, 676 (1974). For complexes with Fe, V, Ti, Zr, Th, In, Ga, see V. Tikhonov, *J. Anal. Chem. USSR*, **21**, 1043 (1966).

170. Literature references in Adam and Pribil.[171] Be forms a 1 : 1 complex at pH 5: K. L. Srivastava and S. K. Banerji, *Analusis*, **1**, 132 (1972). Thymolphthalexon S for alkaline earths: *Anal. Abst.*, **29**, 4B50 (1975).

171. J. Adam and R. Pribil, *Talanta*, **16**, 1596 (1969). Solvent not mentioned. V. N. Tolmachev et al., *J. Anal. Chem. USSR*, **22**, 797 (1967), extract MTB and Xylenol Orange metal complexes into butanol in the presence of diphenylguanidine.

172. T. Yoshino, S. Murakami and M. Kagawa, *Talanta*, **21**, 199 (1974). Alkaline earth metal complexes: *ibid.*, **21**, 673.

173. J. Ueda, *J. Chem. Soc. Japan*, 1467 (1973). Lanthanoids: *ibid.*, 724 (1973).

174. B. Budesinsky and T. S. West, *Talanta*, **16**, 399 (1969).

175. B. Budesinsky and T. S. West, *Anal. Chim. Acta*, **42**, 455 (1968).

176. B. Budesinsky and T. S. West, *Analyst*, **94**, 182 (1969).

177. T. Shimizu and K. Ogami, *Talanta*, **16**, 1527 (1969).

178. T. Sakai, *Bull. Chem. Soc. Japan*, **43**, 3171 (1970). The Fe(III)–Indoferron complex can be extracted into $CHCl_3$ in the presence of diphenylguanidine, as shown by Y. Wakamatsu and M. Otomo, *Japan Analyst*, **19**, 537 (1970).

179. G. M. Huitnik, *Anal. Chim. Acta*, **70**, 311 (1974).

180. For example, see D. F. H. Wallach, D. M. Surgenor, J. Soderberg, and E. Delano, *Anal. Chem.*, **31**, 456 (1959); N. Iritani and T. Miyahara, *Japan Analyst*, **22**, 174 (1973).

181. H. Diehl and J. L. Ellingboe, *Anal. Chem.*, **28**, 882 (1956); Diehl, *Calcein, Calmagite, and o,o-Dihydroxyazobenzene*, G. F. Smith Chemical Co., Columbus, Ohio, 1964.

182 J. Körbl and F. Vydra, *Coll. Czech. Chem. Commun.*, **23**, 622 (1958); J. Körbl, F. Vydra, and R. Pribil, *Talanta*, **1**, 138 (1958).

183. Wallach et al., *loc. cit.*; F. H. Wallach and T. L. Steck, *Anal. Chem.*, **35**, 1035 (1963); B. L. Kepner and D. M. Hercules, *ibid.*, **35**, 1238 (1963). For determination of Ca from a decrease in absorbance see C. Robinson and J. A. Weatherell, *Analyst*, **93**, 722 (1968).

184. A. J. Hefley and B. Jaselskis, *Anal. Chem.*, **46**, 2036 (1974).

185. K. Satyanarayana and R. K. Mishra, *Anal. Chem.*, **46**, 1609 (1974).

186. E. E. H. Pitt and V. M. Stanway, *Anal. Chem.*, **41**, 981 (1969).

187. R. R. Trenholm and D. E. Ryan, *Anal. Chim. Acta*, **32**, 317 (1965); A. E. Pitts and Ryan, *ibid.*, **37**, 460 (1967); Ryan and B. K. Afghan, *ibid.*, **44**, 115 (1969).

ORGANIC PHOTOMETRIC REAGENTS
8-HYDROXYQUINOLINE

This compound, also known as 8-quinolinol, is commonly called oxine (HOx) by analysts.

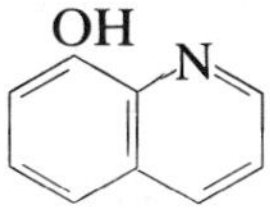

Mol. wt. 145.15. M. p. 73°–74°. Can be recrystallized from water–alcohol or ligroin. Quite stable in solution (especially acidic), but protect from light.

Oxine forms chelates with most metals. The uncharged chelates MOx_n are very slightly soluble in water but dissolve in organic solvents. If unhydrated, they can be extracted by chloroform and other "inert" solvents. If the coordination number of the metal is greater than $2n$ (n is the charge on the cation), the otherwise empty coordination sites are filled with H_2O, HOx, or some neutral ligand. Hydrated oxinates require an oxygenated organic solvent for extraction; if the water is replaced by HOx or an organophilic ligand, they can be extracted into chloroform and similar solvents. Primary and secondary alkylamines and alcohols form chloroform-extractable adducts that are analytically important; the amines are especially useful. Mixed chelates, such as $AuCl_2Ox$, are sometimes formed. More complicated species can be formed under certain conditions. Thus, Ni and Zn at relatively low pH can form ion-association complexes $[M_2Ox_3(HOx)_3]X$, where X^- is ClO_4^-, NO_3^-, Cl^-, among others.

The extracted oxinates or their adducts can be determined absorptiometrically or fluorimetrically. Chloroform is the favored extracting solvent. Oxine is also of value as a separatory reagent, with the final determination being made by another chromogenic reagent.

I. PROPERTIES OF REAGENT

Oxine is an ampholyte, forming oxinium (8-hydroxyquinolinium) ion by protonation of N in acid solutions

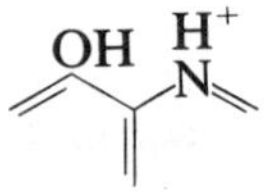

and oxinate ion in basic solutions

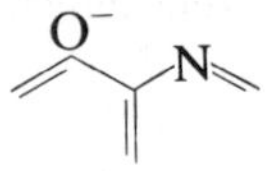

Less than 4% of the neutral species exists as the zwitterion

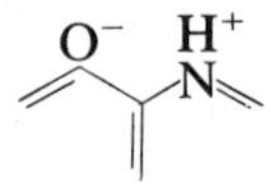

Absorbance curves of Ox^- and $HOx \cdot H^+$ ions are shown in Fig. 6D-1. Table 6D-1 gives $\bar{\nu}_{max}$, λ_{max}, and ϵ_{max} for the different oxine species. The molar absorptivity curve of HOx in $CHCl_3$ is represented in Fig. 6D-2.

Oxine fluoresces very weakly or not at all in aqueous solutions and organic solvents at room temperature, but does fluoresce in strong perchloric and sulfuric acid.[1] The ionization constants of oxine have the values

$$K_1 = \frac{aH^+[HOx]}{[HOx \cdot H^+]} = 9 \times 10^{-6} \quad \text{at} \quad \mu = 0.1(25°)^2 \quad (6D\text{-}1)$$

$$K_2 = \frac{aH^+[Ox^-]}{[HOx]} = 2 \times 10^{-10} \quad \text{at} \quad \mu = 0.1(25°)^3 \quad (6D\text{-}2)$$

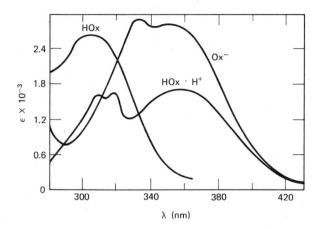

Fig. 6D-1. Molar absorptivity curves of HOx, Ox⁻, and HOx · H⁺ in water (pH, respectively, 7.1, 12.3, and 1.7). T. D. Turnquist, Ph.D. Thesis, 1965.

TABLE 6D-1
Absorption Maxima and ϵ_{max} of Oxine in Aqueous Solution[a]

Form	$\bar{\nu}_{max}(\mu m^{-1})$	λ_{max}(nm)	ϵ_{max}
Neutral (pH 7.6)			
HOx	4.18	239	3.24×10^4
	3.28	305	2.63×10^3
HOx± (zwitterionic form)	3.71	270	2.84×10^3
	2.32	431	64
Cation (pH 1)			
HOx · H⁺	3.98	251	3.16×10^{4} [b]
	3.25, 3.13	308, 319	1.48×10^3
			1.55×10^3
Anion (pH 12)			
Ox⁻	3.97	252	3.02×10^4
	2.99, 2.84	334, 352	2.88×10^3,
			2.82×10^3

[a] As listed by P. D. Anderson and D. M. Hercules, *Anal. Chem.*, **38,** 1702 (1966).
[b] 4.39×10^4 according to J. Fresco and H. Freiser, *Anal. Chem.*, **36,** 631 (1964).

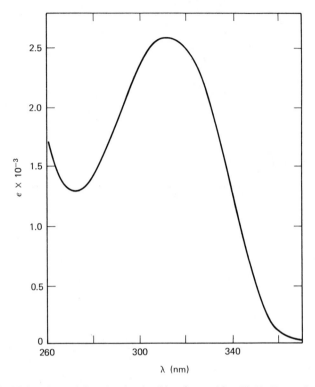

Fig. 6D-2. Molar absorptivity of oxine in chloroform. After T. D. Turnquist, Thesis.

The partition constant (of molecular oxine) for $CHCl_3/H_2O$ is

$$P = \frac{[HOx]_{CHCl_3}}{[HOx]} = 3.4 \times 10^2 \qquad at \qquad 25^{o4} \qquad (6D-3)$$

(Concentrations without subscripts refer to aqueous phase.) If the mixed ionization constants given in the foregoing are to be used, aH^+ is substituted for $[H^+]$. Between pH 4 and 11, $[HOx]_{CHCl_3}$ remains almost constant and most ($>90\%$) of the oxine is found in the chloroform phase (if the volumes of the two phases are equal). Turnquist (Thesis) found equilibrium to be attained rapidly; values of P after various times of shaking were: 2 min, 334; 4 min, 337, 335; 10 min, 329, 341; 20 min, 339.

The chloroform solubility of oxine is 2.63 M (25°)[5] and the water solubility is 4–5×10^{-3} M (20°).[6]

With the aid of these constants, the extraction coefficient of oxine for

$CHCl_3/H_2O$ can be obtained at any pH. The total concentration (i.e., the sum of all forms) of 8-hydroxyquinoline in the aqueous phase is given by

$$\Sigma[HOx] = [HOx] + [HOx \cdot H^+] + [Ox^-]$$

$$= [HOx] + \frac{[H^+][HOx]}{K_1} + \frac{K_2[HOx]}{[H^+]} \qquad (6D\text{-}4)$$

Since $[HOx]_{CHCl_3} = 3.4 \times 10^2[HOx]$,

$$E_{\Sigma HOx} = \frac{3.4 \times 10^2}{1 + [H^+]/K_1 + K_2/[H^+]} \qquad (6D\text{-}5)$$

where $E_{\Sigma HOx}$ is the extraction coefficient (the distribution coefficient of oxine in all its forms) between chloroform and water (Fig. 6D-3).[7]

The preceding expression enables us to calculate the total concentration of hydroxyquinoline in the aqueous phase (if the original amount of reagent is known), from which the concentration of Ox^- is found at any pH:

$$[Ox^-] = \frac{\Sigma[HOx]}{1 + [H^+]^2/K_1K_2 + [H^+]/K_2} \qquad (6D\text{-}6)$$

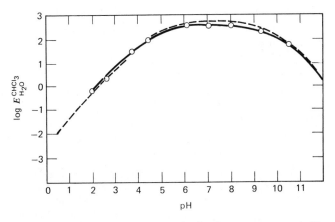

Fig. 6D-3. Extraction coefficient of 8-hydroxyquinoline between water and chloroform as a function of pH. --- Calculated, with $P = 5 \times 10^2$, $K_1 = 8 \times 10^{-6}$, $K_2 = 2 \times 10^{-10}$. —— Experimental. [According to T. Moeller and F. L. Pundsack, *J. Am. Chem. Soc.*, **75**, 2258 (1953). Ionic strength not specified.] The use of the newer value of P (3.4×10^2) would bring the calculated and experimental curves closer together.

From perchloric acid solution, MIBK extracts some oxinium perchlorate, $(HOx \cdot H^+)ClO_4^-$, and for this solvent and acid

$$\frac{[HOx]_{org}}{[HOx]} = 140$$

$$\frac{[H_2OxClO_4]_{org}}{[H_2OxClO_4]} = 0.59 \ (0.5 \ M \ NaClO_4) \ \text{(Zolotov and Kuz'min)}$$

$$= 0.11 \ (0.1 \ M \ ClO_4^-) \ \text{(Dyrssen}[8])$$

II. PROPERTIES OF OXINATES

The chloroform solubilities of some oxinates are listed in Table 6D-2. Table 6D-3 summarizes the chloroform extraction of oxinates with special reference to pH (some pH precipitation data are included).[9] Chloroform is the preferred extracting solvent for unhydrated oxinates and those forming HOx adducts. Hydrated oxinates are very slightly soluble in chloroform (the solubility of $ZnOx_2 \cdot 2 H_2O$ has been reported as $2 \times 10^{-6} M$). Nevertheless, when a solution of Cd, Zn, Ni, or Mg is shaken with a chloroform solution of oxine at the proper pH, these metals are extracted as the anhydrous oxinates, but in a minute or two they become hydrated and precipitate out of the chloroform phase.[10] Some oxinates forming hydrates at low oxine concentrations will form HOx adducts at higher oxine concentrations by replacement of H_2O. Hydrated oxinates can be extracted into oxygenated solvents. Methyl isobutyl ketone is used for this purpose.[11] Hydrate water can sometimes be replaced by adducts other than HOx. Thus primary (and secondary) amines replace a molecule of water in $ZnOx_2 \cdot 2H_2O$ and $CdOx_2 \cdot 2H_2O$ to form $MOx_2 \cdot RNH_2 \cdot H_2O$, and the metals can then be extracted into chloroform at low ($< 0.001 \ M$) oxine concentrations.[12] The chloroform extraction of magnesium from basic solution becomes analytically feasible in the presence of butylamine.[13] The extracted substance is $MgOx_2 \cdot C_4H_9NH_3Ox$ (more properly written as $C_4H_9NH_3[MgOx]_3$). Calcium[14] forms an analogous amine adduct. Adducts with pyridine[15] and other compounds containing basic N are known, for example, $ZnOx_2 \cdot Py$. In the presence of 1,10-phenanthroline and ClO_4^-, zinc forms the chloroform-extractable mixed-ligand complexes $ZnOx_2 \cdot Phen$ and $ZnOxPhen_2^+ \cdot ClO_4^-$ as well as $ZnOx_2 \cdot HOx$ and $ZnPhen_3^{2+} \cdot 2 ClO_4^-$; in the presence of acetate (Ac^-), $ZnPhenAc_2$ can also be extracted.[16]

It has been found that zinc can be extracted into a chloroform solution of oxine when a relatively high concentration of perchlorate ion is present.[17] An ion-pair compound, to which the formula $Zn_2Ox_3(HOx)_3ClO_4$ is assigned, is the extracted species. In the absence of

TABLE 6D-2
Solubilities of Oxinates in Chloroform

Oxinate	Solubility ($\times 10^4$ M)
AlOx$_3$	435 (19°)[a]
	450 (18°)[b]
	380[d]
CuOx$_2$	27 (19°)[a]
	34[d]
FeOx$_3$	90.5(19°)[a]
	190 (25°)[c]
	150[d]
GaOx$_3$	920 (18°)[b]
InOx$_3$	2400 (18°)[b]
NbOOx$_3$	207[e]
ThOx$_4$	8.8(25°)[f]; 34[d]
UO$_2$Ox$_2 \cdot$HOx	28[d]
ZrOOx$_2$	495 (19°)[a]
Rare earths[g]	

[a] F. G. Zharovskii and V. L. Ryzhenko, *J. Anal. Chem. USSR*, **22**, 963 (1967). Al, Fe, Cu, and Zr oxinates are less soluble in oxygenated solvents than in inert solvents. Solubility increases in the order $CCl_4 < C_6H_5CH_3 < C_6H_6 < C_2H_4Cl_2 < CHCl_3$. FeOx$_3$ is almost insoluble in butyl and amyl acetates.

[b] S. Lacroix, *Anal. Chim. Acta*, **1**, 260 (1947).

[c] Turnquist and Sandell, *loc. cit.*

[d] M. Oosting, *Anal. Chim. Acta*, **21**, 505 (1959).

[e] L. Kosta and M. Dular, *Talanta*, **8**, 265 (1961).

[f] D. Dyrssen, *Svensk Kem. Tids.*, **68**, 212 (1956).

[g] I. V. Pyatnitskii, E. F. Gavrilova, and L. I. Gorenshtein, *Zh. Anal. Khim.*, **28**, 2241 (1973), determined the solubilities of the oxinates of these elements at 18° (10^{-4} M): La, 2.7; Nd, 25; Sm, 100; Eu, 33; Gd, 270; Dy, 770; Ho, 370; Er, 540; Tm, 380; Yb, 830.

perchlorate (chloride or nitrate solution), extraction of zinc is poor, and a precipitate appears, which is ZnOx$_2 \cdot 2$ H$_2$O. The constant for the extraction of the binuclear zinc-oxine-perchlorate is

$$K_{ext} = \frac{[Zn_2Ox_3(HOx)_3ClO_4]_{org}[H^+]^3}{[Zn^{2+}]^2[HOx]^6_{org}[ClO_4^-]} = 10^{1.1}$$

The extraction coefficient increases with the zinc concentration. At pH 4,

Metal	Composition[a]	pH Range for Complete Precipitation	pH for 90% Extraction (CHCl₃)[b]	Remarks
Ac		>7.5		
Ag	AgOx·HOx	6.1–11.6		~90% max. extn. at pH 8–9 (0.1 M HOx). Amines do not help extn.
Al	AlOx₃	4.2–9.8	4.0(0.01 M HOx)	Complete extn. pH 4.5–10 (0.01 M HOx); 4.5–11 (Kambara and Hashitani).
As(III), As(V)				Not extd.
Au(III)[c]	AuCl₂Ox			Greenish CHCl₃ soln., $\lambda_{max} = 400$ nm
Ba	$\begin{cases} \text{BaOx}_2 \cdot 2\,\text{H}_2\text{O} \\ \text{BaOx}_2 \cdot 2\,\text{HOx} \end{cases}$	$>8?$		Incomplete extn. at pH ~11 (1 M HOx).
Be	BeOx₂			Incompletely extd. (max. 87%) at pH 6–10 (0.5 M HOx).
Bi	BiOx₃·H₂O (ppt)	5.0–8.3	2.5 (0.1 M)	Quant. extn. pH >2.5–11 (0.1 M HOx).
Ca	$\begin{cases} \text{CaOx}_2 \cdot 2\,\text{H}_2\text{O} \\ \text{CaOx}_2 \cdot \text{HOx} \end{cases}$	9–13	~10 (0.5 M)	Quant. extn. pH ~11 (0.5 M HOx).
Cd[d]	$\begin{cases} \text{CdOx}_2 \cdot 2\,\text{H}_2\text{O} \\ \text{CdOx}_2 \cdot 2\,\text{HOx} \end{cases}$	5.7–14.6	5.1 (0.1 M)	Quant. extn. pH 5–9 (0.1 M HOx). Amines aid extn. log $P_{\text{CdOx}_2 \cdot 2\,\text{HOx}} = 4.1$.
Ce(III)	CeOx₃	9–10		Extd. at pH ~10, citrate soln.
Co(II)[e]	$\begin{cases} \text{CoOx}_2 \cdot 2\,\text{H}_2\text{O} \\ \text{CoOx}_2 \cdot \text{HOx} \end{cases}$	4.3–14.5	5.7 (0.01 M)	Quant. extn. pH 4.5–10.5 (0.1 M HOx).
Co(III)	CoOx₃			By air oxidation of CoOx₂·HOx. Extracts into CHCl₃.
Cr(III)	CrOx₃			Boiling required to form chelate; then extd. in faintly acidic soln. Little extn. at room temp.[f]
Cu(II)	CuOx₂ CuOx₂·2 H₂O	5.3–14.6	1.6 (0.1 M)	Quant. extd. pH 2–12 (0.1 M HOx). $P_{\text{CHCl}_3/\text{H}_2\text{O}}$ of anhydrous CuOx₂ = 2.4×10^3.
Fe(II)	$\begin{cases} \text{FeOx}_2 \cdot 2\,\text{H}_2\text{O} \\ \text{FeOx}_2 \cdot \text{HOx} \end{cases}$			Not extd. below pH 4.
Fe(III)	FeOx₃	2.8–11.2	1.7 (0.01) M	Quant. extn. pH 2–~10 (0.01 M HOx). P_{FeOx_3}(CHCl₃/H₂O) = 1×10^5.
Ga	GaOx₃	3.6–11	1.7 (0.01 M)	Quant. extn. pH 2–12 (0.01 M HOx).

Metal	Composition[a]	pH Range for Complete Precipitation	pH for 90% Extraction (CHCl$_3$)[b]	Remarks
Ge[g]	GeOOx$_2$			
Hf	HfOx$_4$(CHCl$_3$)	4.5–11.3		Quant. extn. pH ~2 (0.1 M HOx).
Hg(I)	Hg$_2$Ox$_2$	5.2–>8.2		Incompletely extd. in basic soln. Some disproportionation.
Hg(II)	HgOx$_2$·2 H$_2$O	4.8–7.4		Extd. incompletely. pH>3 (0.1 M). Cl$^-$ interferes.
Ho	HoOx$_3$			Extd. pH 8–10, 0.1 M HOx.
In	InOx$_3$	2.5–3.0(?)	2.4 (0.01 M)	Quant. extn. pH 3–11 (0.01 M HOx); log P ~3.2.
La	LaOx$_3$	6.5–>10.3	6.7 (0.1 M)	Quant. extn. pH 7–10 (0.1 M HOx); log P = 2.6. LaOx$_3$ slightly sol. CHCl$_3$.
Mg	{ MgOx$_2$·2 H$_2$O MgOx$_2$·4 H$_2$O	9.4–12.7	8.5 (0.5 M)	Hydrated MgOx$_2$ insol. in CHCl$_3$ but sol. in MIBK. Butylamine forms CHCl$_3$-extractable adduct (λ_{max} = 380 nm). In presence of quaternary ammonium salt, Zephiramine, MgOx$_3$(Zeph) is extracted into nitrobenzene, 1,2-dichloroethane, CHCl$_3$, and CCl$_4$ at pH ~9.5.[h]
Mn(II)	{ MnOx$_2$·2 H$_2$O MnOx$_2$·HOx	5.9–10	~7 (0.01 M)	Apparently metastable anhydrous MnOx$_2$ extd. into CHCl$_3$. Oxidation to Mn(III) in basic soln.
Mn(III)	MnOx$_3$			Selective extn. at pH 9 in presence of NH$_4$F.
Mo(V)[i]	Mo$_2$O$_3$Ox$_4$·H$_2$O (solid)	~4		Purple or red color in CHCl$_3$.
Mo(VI)	MoO$_2$Ox$_2$	3.6–7.3	0.7 (0.01 M)	Quant. extn. pH 1–5.5 (0.01 M HOx). Slow.
Nb[j]	NbOOx$_3$			When HOx is added to tartrate (0.3%) soln. and extn. made with CHCl$_3$, recovery almost complete in pH range 4–9 (separation from Ta).
Ni[k]	{ NiOx$_2$·2 H$_2$O NiOx$_2$·HOx	4.3–14.6	3.5 (0.01 M) or lower	Very slow equilibrium. Faster extn. with 0.1 M soln. (quant. at pH ~ 5). Max. extn. at pH 8.5.
Np(V)				Hardly extd. by 0.1 M CHCl$_3$ soln., but partly extd. by butyl alcohol soln. at pH 10.

423

Metal	Composition[a]	pH Range for Complete Precipitation	pH for 90% Extraction $(CHCl_3)$[b]	Remarks
Pa		>6		Incompletely extd. by amyl alcohol soln. from basic medium.
Pb	$PbOx_2$	8.4–12.3	5.5 (0.1 M)	Quant. extn. pH 6–10 (0.1 M HOx).
Pd	$PdOx_2$	3–>11.6		Quant. extn. at pH 0 (0.01 M HOx). Very slow equilibrium at low pH.
Pm(III)				Partly extd. from basic solns.
Po				Partly extd. at pH 3–4.
Pu(IV)				Extd. into amyl acetate, pH < 8 (purplish brown).
Pu(VI)				Extd. into amyl acetate at pH 4–8 (orange-brown).
Rh(III)				Chelate formed on boiling. Quant. $CHCl_3$ extn. pH 6–9.
Ru[l]				Green chelate forms in hot soln. Ru(III–IV?)
Sb(III)	$SbOOx \cdot 2\,HOx$ (pH 9)			
Sc[m]	$ScOx_3$	6.5–8.5	4.0 (0.1 M)	Quant. extn. at pH 4.5–10 (0.1 M HOx).
Sm[n]	$SmOx_3$			Quantitative extn. at pH 6–8.5 (0.5 M HOx).
Sn(II)	$SnOx_2$			Quant. extn. weakly basic soln.
Sn(IV)[o]	$SnOx_2Cl_2$			Can be extracted from H_2SO_4 soln. (pH 0.85) containing halide; no extn. in absence of halide (sepn. from Mo).
Sr	$\begin{cases} SrOx_2 \cdot 2\,H_2O \\ SrOx_2 \cdot 2\,HOx \end{cases}$		11.3 (0.5 M)	Quant. extn. at pH 11.3 with 5% HOx in presence of 4% $BuNH_2$.
Ta[p]				$CHCl_3$ does not extract well. Oxygenated solvents better.
Tb	$TbOx_3$			Almost complete extn. pH > 7 (0.1 M HOx).
Th[q]	$ThOx_4 \cdot HOx$	4.4–8.8	3.2 (0.1 M)	Quant. extn. pH 4–10 (0.1 M HOx). $\log P = 2.4$.
Ti(IV)	$TiOOx_2$	<3.7–8.7	1.8 (0.1 M)	Quant. extn. pH 2.5–9 (0.1 M HOx).
Ti(III)	(Green ppt)			$\lambda_{max} = 660$ nm, C_2H_5OH–H_2O.
Tl(I)	$TlOx \cdot HOx$	> ~4		Incompletely extd. by $CHCl_3$, isobutyl alcohol, etc., from basic soln.

TABLE 6D-3 (*Continued*)

Metal	Composition[a]	pH Range for Complete Precipitation	pH for 90% Extraction (CHCl$_3$)[b]	Remarks
Tl(III)	TlOx$_3$	6.5–7.0	3.0 (0.1 M)	Quant. extn. pH 3.5–>9 (0.01 M HOx).
U(VI)	UO$_2$Ox$_2\cdot$HOx[r] (UO$_2$Ox$_2\cdot$2 HOx)	5.7–9.8	4.2 (0.01 M)	Quant. extn. pH 5–9 (0.1 M HOx).
V(V)[s]	VO(OH)Ox$_2$, VO(OH)Ox$_2\cdot$HOx	2–5	1.7 (0.01 M)	Quant. extn. pH 2–4. No extn. above pH 9. VOOx$_2$ extd. incompletely.
W	WO$_2$Ox$_2$(?)	5–6		Quantitative extn. (slow) at pH 2.5–3.5 (0.01–0.14 M HOx) in presence of 0.01 M EDTA. Hardly extd. pH>7.
Y[t]	YOx$_3$ YOx$_3\cdot$HOx	5.9–>9.3		Almost completely extd. pH 7–10 (0.2 M HOx).
Zn[u]	$\begin{cases} ZnOx_2\cdot2\,H_2O \\ ZnOx_2\cdot HOx \end{cases}$	4.6–13.4	3.6 (0.1 M)	Hydrate precipitates out of CHCl$_3$ on continued shaking at higher pH. Amines aid extraction.
Zr	ZrOOx$_2$	4.7–12.5	1.3 (0.1 M)	Quant. extn. pH 1.5–4 (0.1 M HOx) or at pH 9. ZrOx$_4$ has also been reported.

[a] The composition of stable solid phases and of extracted species, as far as known, are indicated. Some oxinates, normally hydrated, may be extracted into chloroform and similar solvents as metastable anhydrous oxinates. Water of hydration may be replaceable by HOx, and the resulting adduct is extractable into chloroform. Adducts other than HOx may be useful in chloroform extractions, but the composition of these is not indicated in this table.

[b] Figures in parentheses refer to original concentration of HOx in CHCl$_3$. Equal volumes of CHCl$_3$ and aqueous phases. In Stary's work, on which the values in this column are based, the initial metal concentration in the aqueous phase was 1 or 2×10^{-4} M. Apparently pH of aqueous solution was adjusted before shaking with CHCl$_3$–oxine.

[c] H. M. Stevens, *Anal. Chim. Acta*, **18**, 359 (1958).

[d] Extraction of CdOx$_2\cdot$2 HOx: H. Hellwege and G. K. Schweitzer, *Anal. Chim. Acta*, **28**, 236 (1963). Effect of temperature on equilibrium constants. B. Magyar and P. Wechsler, *Talanta*, **23**, 95 (1976), found log solubility of CdOx$_2\cdot$2 H$_2$O in CHCl$_3$ = −6.88. CdOx$_3^-$ is formed (log $\beta=17.96$). β of CdOx$_2\cdot$2 HOx in CHCl$_3$ = 2000. P (CHCl$_3$/H$_2$O) of dihydrate ~10. $K_{sp}=4\times10^{-22}$.

[e] S. Oki, *Anal. Chim. Acta*, **50**, 465 (1970), found Co(II) oxinate to be extracted into CHCl$_3$ in two forms, predominantly as CoOx$_2\cdot$HOx. Oxidation of Co(II) to Co(III) can occur in basic solution, with extraction of Co(III) oxinate. Large excess of HOx prevents air-oxidation of Co(II) oxinate. Co(III) oxinate not back-extracted with EDTA solution. Maximal extraction of Co(II) oxinate at pH ~5.5. CoOx$_3^-$ can be formed in basic soln. Apparently, polymerized Co(II) species extracted in basic solutions. Formation and extraction constants given. Also see N. M. Kuz'min and Yu. A. Zolotov, *Zh. Neorg. Khim.*, **11**, 2316 (1966).

[f] Once formed and extracted into CHCl$_3$, CrOx$_3$ can hardly be stripped from the extract. K. Beyermann, H. J. Rose, and R. P. Christian, *Anal. Chim. Acta*, **45**, 51 (1969), found less than 1% Cr to be removed from the extract by aqueous solutions varying in acidity from pH 11 to 3 M HCl. This behavior provides a means

(Table footnotes continued)

of separating Cr from many metals. Co(III) oxinate shows similar inert behavior. These authors found the optimal pH to be 4 for the formation and extraction of CrOx₃. Heating for 5 min at 100° is required.

^{g} N. P. Rudenko and L. V. Kovtun, *Anal. Abst.*, **11**, 5413 (1964).

^{h} S. Noriki and M. Nishimura, *Anal. Chim. Acta*, **72**, 339 (1974).

^{i} A. I. Busev and C. Fan, *Zh. Anal. Khim.*, **15**, 455 (1960); H. Berge and H.-L. Kreutzmann, *Z. Anal. Chem.*, **210**, 81 (1965).

^{j} K. Motojima and H. Hashitani, *Anal. Chem.*, **33**, 48 (1961), found extraction of Nb almost complete in pH range 3–10.5 when HOx was added to acid solution, pH adjusted, and extraction made with CHCl₃. Also see R. Keil, *Z. Anal. Chem.*, **229**, 267 (1967), who extracts at pH ~ 10 after other metals have been precipitated in presence of H₂O₂ at pH ~8.

^{k} S. Oki, *Anal. Chim. Acta*, **49**, 455 (1970), found NiOx₂·HOx extracted into CHCl₃. NiOx₃⁻ formed in basic solution. Equilibrium constants reported. S. Oki and I. Terada, *ibid.*, **66**, 201 (1973), found that from perchlorate solution, CHCl₃ extracts Ni₂Ox₃(HOx)₃ClO₄ at low pH and Ni₂Ox₄(HOx)₂ at high pH. The latter species is also extracted from sulfate solution. Constants given. Other salt solutions: *ibid.*, **69**, 220 (1974).

^{l} H. Hashitani, K. Katsuyama, and K. Motojima, *Talanta*, **16**, 1553 (1969), oxidize Ru to ruthenate, digest with HOx at pH 4–6.5 and 85° for 30 min, and then extract (quantitative, pH 4–7).

^{m} J. N. Petronio and W. E. Ohnesorge, *Anal. Chem.*, **39**, 460 (1967). T. J. Cardwell and R. J. Magee, *Anal. Chim. Acta*, **36**, 180 (1966), report (ScOx₃)₂·HOx by homogeneous precipitation at pH 8.8 and ScOx₃·HOx at pH 6.5.

^{n} D. Dyrsssen, *Svensk Kem. Tids.*, **68**, 212 (1956), found quantitative extraction above pH 7 with 0.05 *M* HOx in CHCl₃.

^{o} A. R. Eberle and M. W. Lerner, *Anal. Chem.*, **34**, 627 (1962).

^{p} I. P. Alimarin et al., *Trudy Komis. Anal. Khim.*, **8**, 152 (1958); *Zh. Neorg. Khim.*, **7**, 2725 (1962). L. Kosta and M. Dular, *Talanta*, **8**, 265 (1961).

^{q} Orange adduct, approximately ThOx₄·HOx]: A. Corsini and J. Abraham, *Talanta*, **17**, 439 (1970).

^{r} More properly written H[UO₂(C₉H₆NO)₃]; the anion forms salts in basic solutions. See E. P. Bullwinkel et al., *J. Am. Chem. Soc.*, **80**, 2955, 2959 (1958). S. Oki, *Anal. Chim. Acta*, **44**, 315 (1969), believed UO₂Ox₂·2 HOx to form in CHCl₃ soln. in presence of HOx. UO₂Ox₂·HOx is formed when precipitation is made in aqueous solution, from which UO₂Ox₂ is obtained by heating. Apparently, (UO₂Ox₂)₂·HOx also exists, as well as UO₂OxOH and UO₂Ox·OC₂H₅, the latter two formed by reaction with water and ethanol.

^{s} M. Tanaka and I. Kojima, *Anal. Chim. Acta*, **36**, 522 (1966). [VO(OH)Ox₂·HOx]/[VO(OH)Ox₂] = 1.8[HOx] in CHCl₃. Also see A. J. Blair, D. A. Pantony, and G. J. Minkoff, *J. Inorg. Nucl. Chem.*, **5**, 316 (1958); H.-J. Bielig and E. Bayer, *Liebigs Ann.*, **584**, 96 (1953). V. Yatirajam and S. P. Arya, *Talanta*, **23**, 596 (1976), reduce V(V) to V(III) by adding Na dithionite to an aqueous solution of pH 3–10 and extract V(III) oxinate with carbon tetrachloride (yellow solution, $\lambda_{max} = 420$ nm) for determination.

^{t} T. J. Cardwell and R. J. Magee, *Anal. Chim. Acta*, **43**, 321 (1968).

^{u} Stary gives composition of extracted species as ZnOx₂·2 HOx, but F. Chou, Q. Fernando, and H. Freiser, *Anal. Chem.*, **37**, 361 (1965), found Zn to be extracted as ZnOx₂·HOx when working with tracer concentrations of Zn. With 0.1 *M* HOx in CHCl₃, pH 4.56, log E_{Zn} ~ 1.5. (It should be noted that at higher concentrations of Zn and low HOx concentrations, ZnOx₂ is unstable in CHCl₃ solution and precipitates out as ZnOx₂·2 H₂O.) G. K. Schweitzer and W. van Willis, *Anal. Chim. Acta*, **30**, 114 (1964), also assign the composition ZnOx₂·HOx to the extracted species. In the presence of perchlorate, an ion-association oxine complex containing that ion is obtained (see p. 420). The special effect of perchlorate was not known in earlier work, and some erroneous conclusions were drawn.

with 0.1 *M* oxine (chloroform), 0.1 *M* ClO₄⁻, and $10^{-4.4}$ *M* Zn, *E* is approximately 100. At low zinc concentrations, mononuclear zinc oxinate is also present. The difference in behavior of a perchlorate and chloride solution of zinc is explained by the large ionic size of ClO₄⁻. Findings differing in some respects from the preceding have been reported.[18] In the presence of excess oxine and a hydrophobic anion (X⁻ = ClO₄⁻ or tetraphenylborate), Zn is readily extracted into dichloroethane as

$(Zn \cdot 3HOx)^{2+} \cdot Ox^- \cdot X^- (\lambda_{max} = 360$ nm, log $\epsilon = 3.88)$. (Analogous extractable complexes are formed by Ca and Ba in basic solution.) In the presence of an excess of Zn, $(Zn \cdot 3HOx)^{2+} \cdot 2X^-$ is extracted.

The ion associate of Sn(IV) hydroxyquinolate and the Methyl Orange anion (X^-), $(SnOx_3)^+ X^-$, is extracted into 1, 2-dichloroethane at pH 3.7 (critical).[19]

The pH for 90% extraction (at the specified HOx concentration) is listed in Table 6D-3, instead of pH for 50% extraction, as being somewhat more convenient for use (two extractions at $pH_{90\%}$ with equal volumes of phases correspond to 99% extraction of metal). To be sure, $pH_{90\%}$ values can hardly be relied upon closer than a few tenths of a unit, but they are of use in estimating possibilities of separations by pH adjustment. The $pH_{90\%}$ values in the table have been read off pH–R (recovery) curves which tend toward $R = 100\%$ at higher pH values, and have the normal S-shaped form. A few metals, including Ag and Be, never attain $\sim 100\%$ extractability at any pH.

The chloroform extraction of $AlOx_3$ has been investigated in detail.[20]

Oxine shows some tendency to form double or complex oxinates; for example, the rare earth metals M with Ca form $CaOx_2 \cdot 2MOx_3$ or $Ca(MOx_4)_2$, extractable as such into chloroform.[21] The formation of such complexes invalidates separations devised on the behavior of metals when present alone.

Oxine with H_2O_2 at pH 3 forms $TiOOx_2 \cdot H_2O_2$, extractable into $CHCl_3$ ($\lambda_{max} \sim 265$ nm).

Values of what may be called effective extraction equilibrium constants $(CHCl_3/H_2O)$

$$K_{ext} = \frac{[MOx_n \cdot aHOx]_{org}[H^+]^n}{\Sigma[M][HOx]_{org}^{n+a}} \qquad (org = CHCl_3) \qquad (6D-7)$$

are listed in Table 6D-4. Most of these constants are effective constants in the sense that they are based on the analytical concentration, $\Sigma[M]$, of the metal, not on $[M^{n+}]$. Hydrolyzed metal species and metal present as charged (intermediate) oxine complex ions contribute to $\Sigma[M]$.[22] In general, the various components of $\Sigma[M]$ will depend on pH and $[Ox^-]$, so that only rough constancy of the effective K_{ext} is to be expected. Nevertheless, the effective constants have analytical value, if they are used in the pH region (say $pH_{10\%}$–$pH_{90\%}$) in which they have been determined.

The values of a (Eq. 6D-7) on which the calculation of K_{ext} has been based are indicated in Table 6D-4. As noted, other workers have found different values of a for three or four metals, but differences in extraction conditions may account for the discrepancy.

TABLE 6D-4
Effective Extraction Constants of Oxinates $(CHCl_3/H_2O)^a$

	a	$\log K_{ext}$		a	$\log K_{ext}$
Al	0	-5.2	Ti	$-^g$	0.9
Ba	2	-21	Tl(I)		(-8.8)
Bi	0	-1.2	Tl(III)	0	$\sim 5^h$
Ca	1	-17.9	U(VI)	1	-1.6
Cd	2	$-5.3; -6.1$	V(V)	$-^i$	1.65
Co	2^b	-2.15	Zn	$2^b(?)$	-2.4
Cu(II)	0	1.75	Zr	$-^j$	2.7
Fe(III)	0	$4.1; 3.9^c$			
Ga	0	3.7			
Hg(II)	(?)	~ -3			
In	0	$0.9; 1.35^d$			
La	0	-16.35			
Mg	0	-15.15			
Mn(II)	0	-9.3			
Mo(VI)	$-^e$	9.9			
Ni	2^b	-2.2			
Pb	0	-8.05			
Pd	0	~ 15			
Sc	1^f	-6.65			
Sr	2	-19.7			
Th	0	-7.2			

^{a} J. Stary, *Anal. Chim. Acta,* **28,** 132 (1963). Values rounded to nearest ± 0.05.

^{b} Does not agree with results of other workers (see Table 6D-3).

^{c} Zolotov and Kuz'min, *loc. cit.*

^{d} Yu. A. Zolotov and V. G. Lambrev, *J. Anal. Chem. USSR,* **20,** 1204 (1965). This is the constant based on In^{3+}. $\log \beta_1 = 12.0$, $\log \beta_2 = 24.0$, $\log \beta_3 = 35.5$, $\log P_{InOx_3} = 3.09$. G. K. Schweitzer and G. R. Coe, *Anal. Chim. Acta,* **24,** 311 (1961), report effective $\log K_{ext} = 0.6$.

^{e} MoO_2Ox_2.

^{f} $a = 0$ seems more likely.

^{g} $TiOOx_2$.

^{h} Hydrolysis of Tl^{3+} taken into account.

^{i} VO_2Ox (doubtful).

^{j} $ZrOOx_2$.

428

The formation constants of all M^{n+}–HOx complex species have been determined for a number of metals in *aqueous* solution. For Fe^{3+}–HOx, the following constants have been obtained by two different methods:

	By Extraction[23]	By Spectrophotometry[24]
$\log \beta_1$	12.85	13.69
$\log \beta_2$	25.45	26.3
$\log \beta_3$	36.95	36.9
pK_{sp}		43.5

The distribution diagram of Fe(III)–oxinate is represented in Fig. 6D-4. If the hydrolysis of a metal ion and the formation of hydroxo-oxinate metal species can be neglected and polynuclear species are not formed, the ratios of the species are a function only of Ox^- concentration. Figure 6D-4 holds for a pH of 1.52, at which $\sim 10\%$ of iron not complexed by oxine is present as $FeOH^{2+}$ and $\sim 90\%$ as Fe^{3+}.

Suppose a $0.010\,M$ solution of oxine in chloroform is shaken with an equal volume of Fe(III) solution buffered at pH 1.5 ($[H^+] = 0.03$). The iron solution is taken to be so dilute that the concentration of HOx is unchanged. From Eq. (6D-5), $E_{\Sigma HOx} = 0.10$. Therefore, in the aqueous phase, $\Sigma[HOx] = 0.009$ (approximately); it is present almost entirely as $HOx \cdot H^+$. The concentration of Ox^- in the aqueous phase at pH $1.5 = 1.8 \times 10^{-13}$ (from Eq. 6D-6), and $pOx^- = 12.75$. From Fig. 6D-4, at this pOx^- we read the approximate percentages of iron present in the various

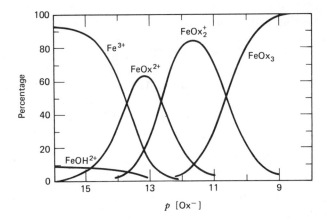

Fig. 6D-4. Distribution diagram, Fe(III)–oxine system, pH 1.52. From T. D. Turnquist, Thesis.

species as: $Fe^{3+}6$, $FeOx^{2+}53$, $FeOx_2^+40$, $FeOx_3 < 1$. There is thus a large difference between K_{ext} based on Fe^{3+} and on $\Sigma[Fe]$ in the aqueous phase.[25] In practice, the effective extraction constant is more useful, and is easily obtained from extraction data.

Some metals (Cd is one) can form anionic oxinate species at a relatively high Ox^- concentration.

The maximal attainable value of E for an oxinate is $P = [MOx_n]_{org}/[MOx_n]$. For $FeOx_3$, the solubility in $CHCl_3$ is 0.019 M, and the intrinsic solubility in water is 1.6×10^{-7} M (Turnquist) so that $P_{CHCl_3/H_2O} = 1 \times 10^5 (25°)$.

Experimental pH ranges for quantitative extraction of various metal oxinates have been reported by many workers.[26] The manner in which the extraction is made can have a great effect on the pH range, particularly on the upper limit. The soundest way is to add oxine solution (e.g., in very dilute HCl) to the metal solution, then to adjust the pH and finally to extract with chloroform or other solvent. This procedure is also desirable when the metal (e.g., V) is somewhat slowly extracted by a $CHCl_3$–oxine solution. Extraction of a metal solution with a chloroform solution of oxine may be entirely satisfactory in the acidic range. But if, for example, an aluminum solution is adjusted to pH 7 before extracting with oxine in chloroform, extraction of aluminum will be incomplete because of precipitation of aluminum hydroxide, which, once formed, is not quickly transformed to $AlOx_3$, the stable phase under the conditions.

The pH range for complete precipitation (Table 6D-3) and chloroform extraction of oxinates is quite wide for most metals. Eventually, at high pH values, formation of hydroxo complexes may gain the upper hand, and precipitation or extraction decreases. Formation of anionic oxinate species, occurring with some metals, will also decrease E. In slightly basic solutions, mixed oxinate–hydroxide complexes may be formed. The formation of $FeOx_2(OH)_2^-$ and $FeOx(OH)_4^{2-}$ (probable composition) has been demonstrated[27] and solid $FeOx_2OH$ and $FeOx(OH)_2$ have been prepared (brown, soluble in $CHCl_3$). The equilibrium constant for the reaction

$$FeOx_3 + 2\,OH^- \rightleftharpoons FeOx_2(OH)_2^- + Ox^-$$

has the value

$$\frac{[FeOx_2(OH)_2^-][Ox^-]}{[OH^-]^2} = 7.7 \times 10^{-5}$$

The choroform extraction of the mixed oxinate-hydroxo complexes has not been studied.

Thallium (III) forms the hydroxo complex $TlOx(OH)_2$.[28] Zinc shows a slight tendency to form $ZnOx_3^-$ at high $[Ox^-]$.

The effect of high electrolyte concentrations (KCl and NaI) on the extraction of various oxinates into chloroform and isopentyl alcohol has been studied.[29]

III. DETERMINATION OF METALS WITH OXINE

A. DIRECT ABSORPTIOMETRIC METHOD

If only a single extractable metal is present in the sample solution, its determination is simple enough, at least if a high concentration of oxine is not needed for quantitative extraction and if equilibrium is rapidly attained. A chloroform solution of oxine absorbs very little above 370 nm (see Fig. 6D-2), in which region metal oxinates have an absorption band. With the exception of Fe(III) (green-black), V(V) (magenta-black), U(VI), Ce(III), Ru(III?), and possibly some of the other platinum metals, the oxinates give yellow chloroform solutions (λ_{max} and ϵ_{max} for selected metals in Table 6D-5). In general, if the metal is extractable from acidic solution, the sample solution is buffered to the required pH and is shaken with a fixed volume and concentration of oxine in chloroform.[30] Adducting agents are added when necessary. If the metal is hydrolyzed to insoluble or polymerized products at the pH of extraction, the acidic solution is first treated with oxine[31], the pH is adjusted to the required value, and the extraction is then made with pure chloroform. Tartrate is added to prevent precipitation of hydrolyzed zirconium.

Equilibrium is usually attained within a few minutes of shaking. Metals requiring long shaking, especially at lower pH values and in the presence of complexing agents, include Ni, Pd, Mo, and W, and hours may be required for equilibrium according to Stary. Extraction is generally more rapid at higher oxine concentrations and higher pH values. Complexing agents such as tartrate, oxalate, and cyanide usually have little effect on extraction rate, but aminocarboxylic acids can reduce it seriously, depending on the strength of the complex formed with the metal. When chloroform solutions of oxine extract slowly, better results may be obtainable by adding oxine to the aqueous solution, adjusting the pH, allowing it to stand to form the oxinate, and then extracting with chloroform.

Because iron(III)[32] and vanadium(V) oxinates in chloroform absorb at longer wavelengths than most other oxinates, these metals can be determined in the presence of reasonable amounts of the others. Vanadium(V) is reported not to be extracted above pH 9, whereas iron(III) is still extracted at pH 10, so that it should be possible to determine iron in the presence of vanadium in this way.

Other possibilities in preventing interference of other metals in the determination of a particular metal by pH adjustment will be evident

TABLE 6D-5
Photometric Determination of Some Metals with Oxine after Chloroform Extraction[a]

Metal	Determination Form	Method[b]	λ_{max}(nm)	$\epsilon_{max} \times 10^{-3}$
Al	$AlOx_3$	A, F	386	6.64
Ca	$CaOx_2 \cdot BuNH_3Ox$	A	375	5.61
Cr(III)	$CrOx_3$	A	420	7.28
Cu(II)	$CuOx_2$	A	410	5.7–6.4
Fe(III)	$FeOx_3$	A	580	3.75
			470	5.5
			375	4.73
Ga	$GaOx_3$	A, F	390	6.97
In	$InOx_3$	A, F	400	6.9
Mg	$MgOx_2 \cdot BuNH_3Ox$	A, F	380	5.6
Rare earths[c]		A	~375	
Sc	$ScOx_3$	A, F	378	6.4
U(VI)	$UO_2Ox_2 \cdot HOx$	A	380	7.1
V(V)	$VO(OH)Ox_2 \cdot HOx$	A	370, 550	~3.5
Y	YOx_3	F(A)	380	5.7
Zr	$ZrOOx_2$	F	398(excit.)	
		A	386	14.7

[a] Mostly from F. Umland, *Z. Anal. Chem.*, **190**, 186 (1962). λ_{max} values for some other oxinates are: Be (MIBK), 400; Bi 395; Ce(III), 395, ~500; Co(II), 420; Hf, 385; La, 380; Mo(VI), 375; Nb, 385; Ni, 380; Pb, 400; Pd, 425; Rh(III), 425; Sn(IV), 385; Ta, 388; Th, 378; Tl(III), 400; Ti, 385; W, 360, 380; Zn (MIBK) 395; Zr, 393 nm. These are from various sources.
[b] A = absorptiometric, F = fluorimetric.
[c] I. V. Pyatnitskii and E. F. Gavrilova, *Zh. Anal. Khim.*, **25**, 445 (1970). Extraction from citrate or tartrate solution at pH 8.5–9.5.

from the data in Tables 6D-3 and 6D-4. The trio Al, Ga, In provides an illustration, for which the log K_{ext} values are, respectively, -5.2, 3.7, and 0.9. The Ga constant is much greater than that of Al, and considerably greater than that of In, so that Ga can be selectively extracted in the presence of these two elements. The separation from aluminum is of special importance because of an Al/Ga ratio of 10^4 or greater in natural materials, such as silicate rocks. If 2 ml of a 0.0008 M CHCl$_3$ solution of HOx is shaken with 12 ml of an aqueous solution of pH 3 (these conditions correspond approximately to those existing in a fluorimetric method for gallium),[33] we calculate that $E_{Ga} = 320$ and $E_{Al} = 4 \times 10^{-7}$. A few micrograms of Ga can be determined in the presence of 50 mg of Al.

Differential complexing is of aid in increasing the selectivity of oxine extractions.[34] The extraction of iron(III) can be prevented in acidic solution by reduction with hydroxylammonium chloride or ascorbic acid.

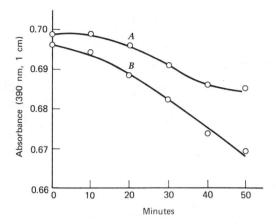

Fig. 6D-5. Stability of $AlOx_3$ solutions in chloroform (5 mg/100 ml). A, analytical-reagent chloroform (0.6% ethanol), kept in dark; B, analytical-reagent chloroform (0.6% ethanol), exposed to 60-W incandescent bulb at distance of 18 in. After R. H. Linnell and F. H. Raab, *Anal. Chem.*, **33**, 154 (1961).

Chloroform solutions of oxinates show some photodecomposition (Fig. 6D-5). Slight decrease in absorbance (390 nm) also occurs in the dark. The greatest effect is shown by exposure to light in the presence of oxygen. Probably oxygenated decomposition products (phosgene) of chloroform are responsible for the decomposition. The rate of photochemical decomposition of oxinates is reported to increase with the size and the equivalent weight of the central metal atom.[35] Thallium(I) oxinate is unstable in chloroform solution, even in the dark. Thallium(III) oxinate is markedly light sensitive. See further discussion under fluorescence determination (p. 434).

B. INDIRECT ABSORPTIOMETRIC METHODS

These are inferior to the direct method, and are not often used. All require precipitation of the metal as oxinate, which then must be washed free of excess oxine.

1. The oxine in the precipitate can be coupled with a diazo compound to yield a dye of strong color. Thus Alten, Weiland, and Loofmann,[36] in the colorimetric determination of aluminum, dissolve the oxinate in a mixture of equal volumes of dilute hydrochloric acid and alcohol, add a mixture of sulfanilic acid and sodium nitrite in acetic acid, allow to stand for 10 minutes, and then make alkaline with sodium hydroxide.

2. Phosphotungstic, molybdotungstic, and phosphotungstomolybdic acid are reduced by oxine in basic medium and the resulting blue solution containing tungsten or molybdenum in a lower valence can be compared with a standard obtained in the same way.[37]

3. The precipitate may be dissolved in dilute hydrochloric acid and the absorption of the solution measured at 358 nm (more sensitively at 252 nm), where 8-hydroxyquinolinium ion absorbs strongly (see Fig. 6D-1). At 358 nm $\epsilon = 1660$; at 250 nm, 43,000. This is the best indirect method. Relatively small amounts of magnesium have been determined in this way.[38]

C. FLUORIMETRY

In chloroform (and other immiscible solvents), oxinates of diamagnetic metals fluoresce strongly.[39] Oxine itself shows only an extremely weak fluorescence. Compared to absorptiometric determination, the fluorimetric has the advantages of greater selectivity and sensitivity and the disadvantage of lower precision and accuracy, arising in part from instability of oxinates exposed to UV radiation. The fluorescence properties of the oxinates of the second, third, and fourth group elements have been studied.[40] Approximately 20 metal oxinates fluoresce, half of these (Mg, Zn, Cd, Al, Ga, In, Sc, Y, Zr, Hf, and Sn) strongly enough to be of analytical interest. Yttrium (as little as $10^{-5}\%$) can be determined in a mixture of the rare earths if such elements as Al, Zr, Sc, and Ga are first separated. In $CHCl_3$ solution, emission maxima (nm) are Al, 510; Ga, 526; In, 528.

Extraction of Zr oxinate with $CHCl_3$ from tartrate-fluoride solution at pH 9 allows fluorimetric determination of Zr in the presence of limited amounts of Ti, Mo, and W (emission at 520 nm).[41] Molybdenum and Ti (but not W) are partly extracted and interfere by absorbing UV radiation.

If the fluorescence intensity of $ZnOx_2 \cdot BuNH_2 \cdot H_2O$ in chloroform is taken as 100 (emission at 599 nm), that of oxinates or oxinate adducts under 365 nm excitation is

$ZnOx_2 \cdot 2ROH$ (alcohol adduct) 87.2 (593 nm)
$AlOx_3(+HOx^{42})63.2(533$ nm)
$YOx_3(+10\%$ $CH_3OH)$ 34.6
$ScOx_3(+HOx)$ 21.7
$MgOx_2 \cdot 2ROH$ 21.5
$ZrOx_4(+HOx)$ 11.3

Yttrium oxinate in chloroform is very sensitive to UV radiation; the fluorescence intensity decreases appreciably in <1 min. On the other hand, $AlOx_3$ is quite stable, showing only a slight intensity decrease in 5 min. The fluorescence intensity of $CdOx_2 \cdot 2\,ROH$ decreases on irradiation, whereas that of $CdOx_2 \cdot 2\,BuNH_2$ increases, the effect being attributable to the fluorescence of decomposition products.

Substances absorbing UV radiation or the emitted light will, of course, produce error. $FeOx_3$ is one of these. The effect of Fe(III) is largely eliminated by reduction to Fe(II).

Niobium oxinate (extracted from citrate solution at pH 9.4) phosphoresces at liquid nitrogen temperature and can be determined in this way.

IV. DERIVATIVES AND ANALOGS OF OXINE

The more important derivatives analytically are those in which hydrogen in the 2- or 7-position is replaced by a bulky group and selectivity is increased.

A. 8-HYDROXYQUINOLINE-5-SULFONIC ACID

Usually in trace analysis, reagents yielding metal complexes extractable into organic solvents are preferred, but occasionally those forming soluble complexes find use, as when the highest sensitivity is not required. Sulfonation of oxine provides a reagent forming soluble, colored chelates with Fe(III) and V(V).[43] The stability constants of its chelates do not differ much from those of the oxine chelates. The fluorescence of the Mg, Zn, and Cd chelates has been studied.[44] Zinc shows maximal fluorescence at pH > 5.5, Cd at pH > 6, and Mg at pH > 6.5 (approximate values). Magnesium can be determined in the presence of Ca in a solution buffered at pH 7, with activation at 380 nm and measurement of fluorescence at 510 nm.[45] Tin (II or IV) fluoresces strongly in aqueous solution at pH 4–5. The reagent itself does not fluoresce appreciably in the pH range 1–12.

Depending on the pH, Al, Ga, and In form predominantly 1:1, 1:2, or 1:3 complex species, all fluorescent. The formation constants K_1 and K_2 have been determined roughly[46]:

	$\log K_1$	$\log K_2$
Al	9.7	9.0
Ga	13.3	11.2
In	~11	8

These values indicate that Ga can be determined in acid solution in the presence of limited amounts of Al. The maximum fluorescence of Al is reported to be shown at pH 4, that of Ga at pH 2.5.

Chloroform extracts ion-association complexes of metal-8-oxine-5-sulfonates and Zephiramine (tetradecyldimethylbenzylammonium chloride, ZCl). The extracted complexes have the composition[47] $Co^{2+}(L^{2-})_2(L^-)(Z^+)_3$ (analogous composition for Ni), $Cu^{2+}(L^{2-})_2(Z^+)_2$ and $Fe^{3+}(L^{2-})_3(Z^+)_3$.

8-Hydroxyquinoline-7-sulfonic acid and its 5-halo derivatives form green 1:1 complexes with Fe(III) at pH 2–4.[48]

B. HALO-OXINES

Introduction of Cl, Br, or I into the oxine molecule increases its acid strength (Table 6D-6), which is analytically advantageous, in that, other factors remaining the same, the extraction of a metal can be made in more acidic solutions. However, this effect is generally counteracted to a greater or less extent by increased dissociation of the metal chelate. The partition constants (P) of the metal chelates of the di-substituted oxines seem to be greater than those of oxine for chloroform and other common organic solvents; the same is true for the reagents themselves. In general, the chelates of the dihalo oxinates are extractable at somewhat lower pH values than the corresponding chelates of oxine, as far as the available data allow generalization to be made.

5,7-Dichloro-8-Hydroxyquinoline

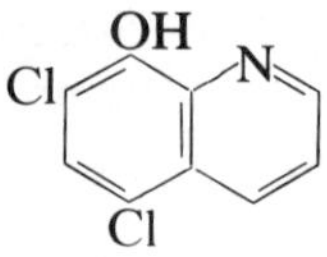

Mol. wt. 214.1. M. p. 179°–180°.

has some advantages over oxine in the extraction of Th, U, and the rare earths. Quantitative extraction of Th is obtained at pH ~2.5 (0.05 M reagent solution in $CHCl_3$), compared to pH ~4.5 for oxine. For U and La the pH difference is ~1.[49] The log K_{ext} values for Th, U, and La are -0.9, -1.25, and -13.45 (rounded to nearest 0.05).

Gallium and indium can be determined by extraction from slightly acidic solution (pH 3–4) by a chloroform solution of the reagent, followed by absorbance measurement at 409 and 412 nm ($\epsilon = 2540$ and 2380, respectively).[50] Fluorescence measurement is also possible. Yttrium is determinable fluorimetrically by extraction of its chelate from weakly basic solution into $CHCl_3$ (excitation at 365 nm, maximum emission at 530 nm).[51] Among other metals determined by this reagent, scandium may be mentioned.[52] Tin(IV) forms $SnCl_2L_2$, extractable into CCl_4 (405 nm).[53]

TABLE 6D-6
Constants of Dihalo-Oxines[a]

	pK_1	pK_2	Soly. in $CHCl_3(M)$	log P $(CHCl_3/H_2O)$
5,7-Dichloro-8-hydroxyquinoline	2.9	7.4	0.0587	3.86
5,7-Dibromo-8-hydroxyquinoline,	2.6	7.3	0.0302	4.15
$\mu = 3.0$[b]	3.3	7.2		4.35
5,7-Diiodo-8-hydroxyquinoline	2.7	8.0	0.0109	4.15
5-Chloro-7-iodo-8-hydroxyquinoline	2.7	7.9	0.0585	3.88

[a] D. Dyrssen, M. Dyrssen, and E. Johansson, *Acta Chem. Scand.*, **10**, 341 (1956). Temp. 25°, $\mu = 0.1$. Dissociation constants are also available for 5-halo-oxines (Sillén and Martell, *op. cit.*; M. R. Chakrabarty et al., *Anal. Chem.*, **39**, 238 (1967); R. G. Beimer and Q. Fernando, *ibid.*, **41**, 1003 (1969)). The last named authors also give formation constants of transition metal chelates of 5-halo-8-hydroxyquinolines and 5-halo-2-methyl-8-hydroxy-quinolines in 75% dioxane, together with proton magnetic resonance spectra of the reagents.
[b] D. A. Knyazev, *Zh. Anal. Khim.*, **19**, 273 (1964).

5,7-Dibromo-8-Hydroxyquinoline. The determination of tin(IV) by extraction with this reagent (pH ~ 1, isobutanol, 410 nm) is of some interest.[54] The extraction and determination of various other metals have been described.

Tungsten(VI) at low concentrations (10^{-5}–10^{-7} M) is extracted by a $CHCl_3$ solution of the reagent from 1–4 M HCl as well as at lower acidities.[55] The extracted species has not been identified. Molybdenum is also extracted from acid solution (up to 2 M HCl).

Lithium forms a fluorescent complex in basic alcoholic solution (pH ~ 9).

Yttrium is extractable at a lower pH (6.5, benzene) than La, and can thus be determined fluorimetrically in the presence of the latter.[56] The limit of determination is ~ 20 ppm Y_2O_3 in La_2O_3. Al also gives a fluorescence, but its effect can be corrected for by measuring the fluorescence before and after adding EDTA to destroy the Y fluorescence.

Thorium forms a complex with 5,7-dibromo-oxine and Rhodamine B, extractable into benzene at pH 4.5–7.5 ($\epsilon_{552} = 88,000$); moderate amounts of the rare earths may be present.[57]

This reagent has also found use in the isolation of trace metals from sea water by cocrystallization.

5,7-Diiodo-8-Hydroxyquinoline. This reagent has been used for the extraction of a few metals but does not seem to have any particular value.[58]

7-Iodo-8-Hydroxyquinoline-5-Sulfonic Acid

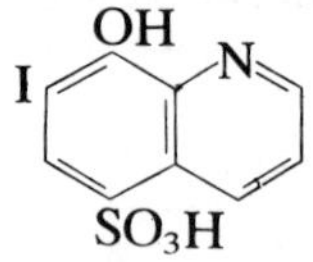

$pK_1(H_2L^+) = 2.58$
$pK_2(HL) = 7.11 \ (\mu = 0.1)$

The presence of the sulfonic acid group in the molecule results in the formation of soluble chelate species, as with Fe(III) and V(V), which metals are determinable by this reagent.[59] No great differences in the behavior of 7-iodo-8-oxine-5-sulfonic acid and 8-oxine-5-sulfonic acid are to be expected. The sulfonated metal chelate anions form ion-pair compounds with tributylammonium cations (and other bulky organic cations), extractable into isobutyl alcohol.[60] Aluminum is determinable in the presence of cetyltrimethylammonium chloride ($\lambda = 385$ nm).[61]

The iodine in the reagent is of little significance, and the analytical use of this compound seems to be a result of its availability as a medicinal agent.

In spite of the presence of the group –SO$_3$H, the reagent is only sparingly soluble in water (~0.2%, room temperature).

The formation constants of Ga and Ga–OH complexes with the reagent have been determined.[62]

Fluoro-oxines

5-Fluoro-oxine has been tested qualitatively as a precipitant for heavy metals.[63] It does not appear to show any significant differences in behavior from the parent reagent.

6-Trifluoromethyl-8-hydroxyquinoline shows much the same precipitation behavior as oxine.[64] 5-Trifluoromethyloxine gives rather less sensitive reactions than either. 7-Trifluoromethyloxine gives no precipitates with metals (Al, Fe, Cu, Zn, and many others), probably because of *ortho*-effects (steric factors, intramolecular hydrogen bonding between hydroxy and trifluoromethyl groups).

C. 2-METHYL-8-HYDROXYQUINOLINE (8-HYDROXYQUINALDINE) AND COGNATES

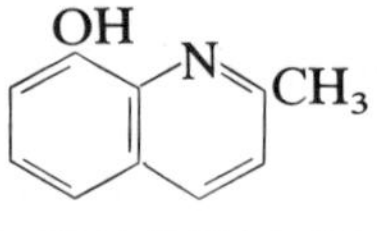

Mol. wt. 159.2. M. p. 74°. Soly. in H$_2$O (25°), 0.0025 M; in CHCl$_3$ (25°), 0.51 M. $K_1 = 1.7 \times 10^{-6}$, $K_2 = 9.1 \times 10^{-11}$. $P_{\text{CHCl}_3/\text{H}_2\text{O}} = 1.7 \times 10^3$ (25°). Can be recrystallized from alcohol.

The interest in this[65] methyl-substituted oxine arises from the effect of the CH_3- group adjacent to $-N=$ on the formation of chelates of metals having a relatively small ionic radius and high charge. Aluminum is such a metal. It seems reasonable to expect that the presence of the methyl group in the 2-position will, at least, make the formation of the aluminum *tris* chelate more difficult because of the steric effect of the methyl group. Early observations seemed to indicate that 2-methyl-oxine does not precipitate aluminum or give a chloroform-extractable chelate. Indeed, it was even asserted that soluble complex aluminum species were not formed either. The early image of 2-methyl-oxine as a classical example of a reagent displaying the effect of steric hindrance vis-à-vis aluminum has become tarnished. More recent work has shown that 2-methyl-oxine forms a *tris* chelate with aluminum, and this chelate can be precipitated from aqueous solution by careful pH adjustment.[66] Further work[67] has shown that the bi-substituted oxines 2,5-dimethyl- and 2-methyl-5-chloro- form *tris* chelates (as precipitates) with aluminum in aqueous solution; 2,3-dimethyl- and 2,3,4-trimethyl-oxine form *bis* chelates containing the linkage Al–O–Al; 2-phenyloxine does not react. Apparently then, CH_3- in the 2-position makes *tris* chelate formation with aluminum difficult, C_6H_5- in the same position makes it impossible.

Contrary to earlier reports, 2-methyl-oxine in chloroform extracts aluminum partially.[68] According to Dyrssen,[69] thorium is not extracted by 2-methyl-oxine in chloroform. Other metals are extracted, the minimum pH for quantitative extraction usually being 1 or 2 pH units higher than for oxinates in chloroform.[70] The absorption spectra of the metal 2-methyl-oxinates are much the same as for the oxinates. There is usually no advantage in the determination of metals as the 2-methyl-oxinates except when aluminum and thorium are present. Beryllium can be determined spectrophotometrically[71] or fluorimetrically with 2-methyl-oxine in the presence of *small* amounts of aluminum, but usually a separation must first be made. Chloroform solutions of the reagent hardly absorb above 370 nm.[72]

Unlike oxine, 2-methyl-oxine does not form an adduct with Zn, presumably for steric reasons. Because of the absence of water in its Zn chelate, methyl-oxine allows chloroform extraction of Zn at low reagent concentrations.

The structures of the 2-methyl-oxinates, VOL_2, and $GaClL_2$ have been elucidated by X-ray analysis.[73] The metals are five-coordinated.

Compounds in which the 2-methyl group of 8-hydroxyquinaldine has been modified to contain a donor atom have been investigated as chelating agents for Cu, Ni, Co, Zn, Mn, and Mg.[74] 2-Aminomethyl-8-hydroxyquinoline behaves as a tridentate ligand.

2-(1-Ethylpropyl)-8-hydroxyquinoline does not precipitate aluminum, but does give a precipitate with indium.[75]

2-Phenyl-8-Hydroxyquinoline

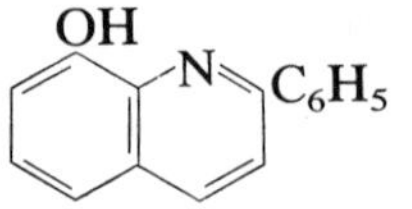

Mol. wt. 221.3. M. p. 59°, 72° (polymorphic). Soly. in H_2O (22°), 1.7×10^{-5} M. $pK_1 = 3.7$, $pK_2 = 10.4$ (20% ethanol).

Little is known about the extraction of metals with organic solvent solutions of this reagent. It does not precipitate Al, Cr(III), or Mg.[76] The nonprecipitation of chromium is probably explained by its inertness; the nonprecipitation of magnesium seems to show a greater steric effect of C_6H_5- than of CH_3- in the 2-position.

2-(2'-Hydroxyphenyl)-8-hydroxyquinoline forms 1:1 complexes with Cu, Zn, Cd, Co, and Ni, which are more stable than the corresponding complexes of 8-hydroxyquinoline.[77] With Al, 1:1 and 1:2(reagent) complexes are formed.

D. 7-(α-[o-CARBOMETHOXYANILINO]-BENZYL)-8-HYDROXYQUINOLINE

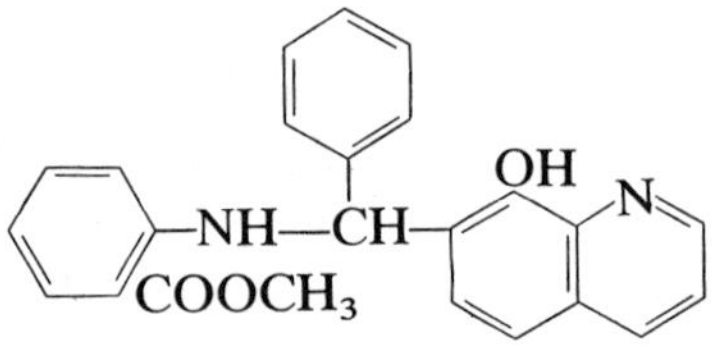

M. p. 133°–134°

This reagent and other oxine derivatives with substituents, containing one or more nitrogen atoms, in the 7-position show greatly diminished reactivity with trivalent metals in consequence of steric hindrance. It neither precipitates nor extracts (chloroform solution) Al, Fe, Cr(III), Ga, and La.[78] However, Bi, which has a larger ionic radius than these cations, is chloroform-extractable (quantitatively at pH 12.2 ± 0.4).[79] Indium is also extracted, but apparently only partially. Quadrivalent metals are not extracted. The divalent metals, such as Cu, Pb, Zn, Ni, and Mg, are extracted, as is Tl(I). A selective method for the photometric determination of magnesium has been developed with the aid of differential complexers.[80] This reagent and other 7-substituted oxines have been suggested for the determination of V, Cu, Pb, Zn (in Cd), and Ni (in Co).

This reagent (and other oxines substituted in the 7-position) is very slightly soluble in water. As a consequence, extractions (as with chloroform solutions) are much slower than with oxine. It may also be supposed that the blocking effect of the bulky substituent leads to decreased reaction rates with metal ions in aqueous solution. Extraction of magnesium requires 15–20 minutes of machine shaking for attainment of equilibrium, and that of Be and U(VI) is so slow that practically they are not extractable. The presence of tartrate, citrate and phosphate, but not ammonia, further decreases the extraction rate.

E. COMPOUNDS WITH TWO OXINE FUNCTIONS

Examples of such reagents[81] are the following:

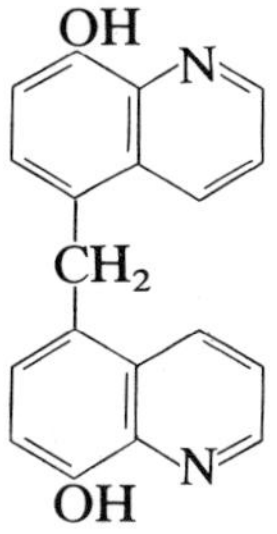

8,8'-Dihydroxy-5, 5'-biquinolylmethane

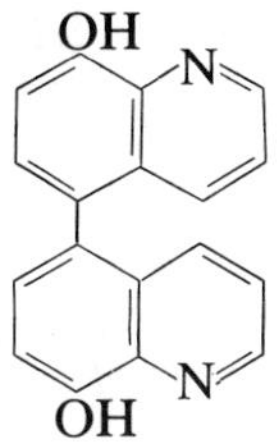

8,8'-Dihydroxy-5, 5'-biquinolyl

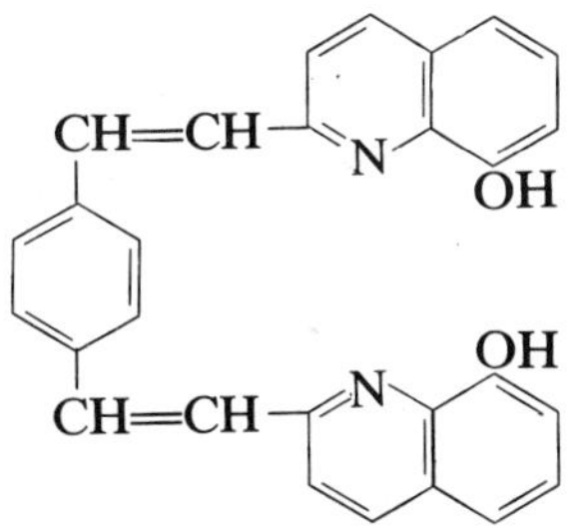

2,2'-(p-Phenylenedivinylene)di-8-quinolinol

Only one of the oxine functions forms chelates, with the possible exception of 8,8'-dihydroxy-5,5'-biquinolyl.

F. MISCELLANEOUS OXINE DERIVATIVES

A number of other derivatives or analogs of oxine have been examined for the extraction or determination of metals without much of analytical interest being found. Among these are 5,7-dinitro-, 5- and 7-nitroso-, and 7-(1-piperidylmethyl) oxine; 8-hydroxyquinazoline; 8-hydroxy-2,4-dimethylquinazoline; 8-hydroxycinnoline; various azo derivatives (Chapter 6E); oxine N-oxide.

In connection with 7-nitroso-oxine, a point of theoretical interest is the suggestion[82] that its Hg(II) chelate has two coordination isomers as indicated by the equilibrium

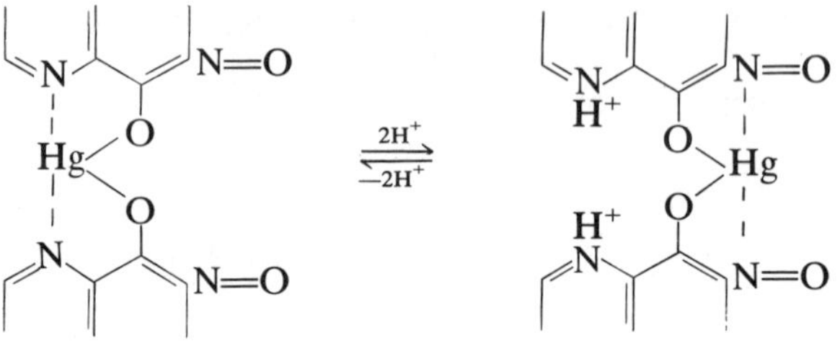

Isobutyl alcohol

Aqueous phase
(pH 5)

Azomethine and aminomethyl derivatives of oxine have recently been synthesized and their metal reactions studied.[83]

Indo-oxine,

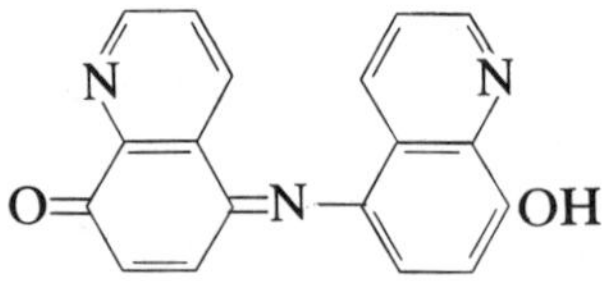

which is easily prepared,[84] gives red acidic solutions (pH <6) and green basic solutions (pH >8). It forms slightly soluble blue or violet chelates with most metals, analogous to the oxinates.[85] Zinc has been determined colorimetrically with this reagent, and other metals probably can be determined, but little attention has been paid to it.

G. 1-HYDROXYACRIDINE ("NEOOXINE")

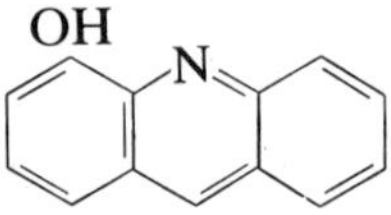

Brownish yellow solid, slightly soluble in water, easily soluble in ethanol or $CHCl_3$. M. p. 115°–116°. $K_1 = 4.9 \times 10^{-6}$, $K_2 = 1.4 \times 10^{-10}$. Can be purified by recrystallization from 75% ethanol.

This compound is similar to oxine in its properties and reactions.[86] A noteworthy difference is the nonprecipitation of Al. Cu(II) is essentially completely extractable by $CHCl_3$ in the pH range 7–>13, Co 10–12, Ni ~9, Zn 10–12.

NOTES

1. M. Goldman and E. L. Wehry, *Anal. Chem.*, **42,** 1178 (1970); M. P. Bratzel, J. J. Aaron, J. D. Winefordner, S. G. Schulman, and H. Gershon, *Anal. Chem.*, **44,** 1240 (1972).

2. J. G. Mason and I. Lipschitz, *Talanta*, **18,** 1111 (1971). R. Näsänen, P. Lumme, and A. Mukula, *Acta Chem. Scand.*, **5,** 1199 (1951), reported 9.61×10^{-6} for the thermodynamic constant at 20°, from which the mixed constant 8×10^{-6} is calculated. Slightly higher values have also been reported: 1.2×10^{-5} by B. Magyar and P. Wechsler, *Talanta*, **21,** 539 (1974), and D. Dyrssen, *Svensk Kem. Tids.*, **64,** 213 (1952): 1.0×10^{-5} by S. Paljk, C. Klofatar, F. Krasovec, and M. Suhac, *Mikrochim. Acta*, **II,** 485 (1975). K_1 is not greatly affected by a change in ionic strength.

3. From thermodynamic value of 1.5×10^{-10} of Näsänen et al. Other recent workers have reported values from 1.5×10^{-10} to 2.6×10^{-10}.

4. T. D. Turnquist and E. B. Sandell, *Anal. Chim. Acta*, **42,** 239 (1968). The P value is for $\mu = 0.1$ (phosphate buffer), 0.01–0.02 M HOx in $CHCl_3$ (25°). Y. A. Zolotov and N. M. Kuz'min, *J. Anal. Chem. USSR*, **20,** 442 (1965), report the value 3.53×10^2. Mason and Lipschitz, *loc. cit.*, found log $P = 2.63$ for $CHCl_3$, 2.25 for toluene, 2.36 for benzene, and 2.59 for nitrobenzene, all at 25°. For effect of ionic strength on P (benzene–water), see the same authors, *Talanta*, **13,** 1462 (1966). For benzene, log P varies from 2.42 ($\mu = 0.1$, NaCl) to 2.58 ($\mu = 1.5$).

5. D. Dyrssen, M. Dyrssen, and E. Johansson, *Acta Chem. Scand.*, **10,** 341 (1956). L. S. Berges, Thesis, Zaragoza, Spain (1952), found 2.55 M at 20°, and D. C. Spindler, Thesis, University of Minnesota, 1948, a higher value, 2.97 M at 25°.

6. Spindler found 4.75×10^{-3} M (25°) in water saturated with $CHCl_3$; Berges, 3.6×10^{-3} M; Paljk et al., 4.9×10^{-3} M.

7. If the aqueous phase contains a relatively high concentration of an acid, such as acetic, which dissolves in the chloroform phase, some 8-hydroxyquinolinium ion will be formed in the latter, and its presence must be taken into account in a more accurate expression. From the practical standpoint, the presence of this species in the chloroform is objectionable because it increases the absorption in the vicinity of 390 nm (peak absorption at 368 nm), where the absorption of aluminum and other hydroxyquinolates is measured. Attention has been called to this by D. W. Margerum, W. Sprain, and C. V. Banks, *Anal. Chem.*, **25,** 249 (1953), who noted the effect when working with 1 M acetic acid–acetate buffer of pH 5. At higher pH values and lower acetate buffer concentrations, the absorbance due to $HOx \cdot H^+$ will, of course, decrease. Washing the separated chloroform phase with a little water removes the acetic acid.

8. D. Dyrssen, *Svensk Kem. Tids.*, **64,** 213 (1952).

9. The data in this table are drawn largely from J. Stary, *Anal. Chim. Acta*, **28,** 132 (1963), and F. Umland, *Z. Anal. Chem.*, **190,** 186 (1962), the latter a

review of oxinate extractions. Other sources of information on the analytical use of oxine include K. Motojima and H. Hashitani, *Japan Analyst,* **9,** 151 (1960);A. K. De, S. M. Khopkar, and R. A. Chalmers, *Solvent Extraction of Metals,* Van Nostrand Reinhold, London, 1970 (bibliography of more than 200 references on oxine and derivatives); R. G. W. Hollingshead, *Oxine and Its Derivatives,* Butterworths, London, 1954–1956 (4 vols.). Precipitation ranges are mostly from H. R. Fleck and A. M. Ward, *Analyst,* **58,** 388 (1933) and R. Bock and F. Umland, *Angew, Chem.,* **67,** 420 (1955). The extremes are not always accurately located.

10. An illustration of the Oswald phase rule: The metastable phase forms first. The chloroform solution can be stabilized by addition of polar organic solvents (alcohols, glycols, amines) forming adducts with the oxinates. E. E. Rakovskii, B. L. Serebryanyi, and N. D. Klyueva, *Zh. Anal. Khim.,* **29,** 1710 (1974), explain the precipitation of $ZnOx_2 \cdot 2\,H_2O$ from solvents of low coordination capability by polymerization of extracted monomeric $ZnOx_2 \cdot 2H_2O$.

11. See J. W. Edwards, G. D. Lominac, and R. P. Buck, *Anal. Chim. Acta,* **57,** 262 (1971), who extract Mg, Mn, Fe, Zn, and Cu from ammonia solution.

12. F. Umland and W. Hoffmann, *Z. Anal. Chem.,* **168,** 268 (1959). Alcohols and glycols are less effective, but they function similarly, forming $MOx_2 \cdot 2\,ROH$ and, with amines, $MOx_2 \cdot ROH \cdot RNH_2$.

13. F. Umland and W. Hoffmann, *Anal. Chim. Acta,* **17,** 234 (1957); F. Umland, W. Hoffmann, and K.-U. Meckenstock, *Z. Anal. Chem.,* **173,** 211 (1960). 2-Butoxyethanol was used earlier for the same purpose by C. L. Luke and M. E. Campbell, *Anal. Chem.,* **26,** 1778 (1954). S. J. Jankowski and H. Freiser, *ibid.,* **33,** 776 (1961), studied the use of 2-butoxyethanol, isopentyl alcohol, and ethanolamine as solvating addends enabling magnesium oxinate to be extracted by $CHCl_3$. Dioxane, ethylene glycol, methyl amine, and pyridine are ineffective. Tetrabutylammonium ion forms the undoubted ion-pair compound $(C_4H_9)_4N^+MgOx_3^-$, quantitatively extracted into $CHCl_3$ at pH 11.4–11.6. The solution shows an absorption peak at 388 nm.

14. F. Umland and K.-U. Meckenstock, *Z. Anal. Chem.,* **165,** 161 (1959); F. Umland, W. Hoffmann, K.-U. Meckenstock,[13] p. 221 and p. 222 (Sr).

15. F. Chou and H. Freiser, *Anal. Chem.,* **40,** 34 (1968). Magnesium oxinate forms a 1:1 adduct with pyridine, extractable into 1, 2-dichloroethane; see S. Nakaya and M. Nishimura, *Japan Analyst,* **22,** 733 (1973).

16. C. Woodward and H. Freiser, *Anal. Chem.,* **40,** 345 (1968).

17. S. Oki and I. Terada, *Anal. Chim. Acta,* **61,** 49 (1972).

18. E. E. Rakovskii, B. L. Serebryanyi, and N. D. Klyueva, *Zh. Anal. Khim.,* **29,** 1086 (1974).

19. E. E. Rakovskii and T. D. Krylova, *Zh. Anal. Khim.,* **29,** 910 (1974).

20. T. Kambara and H. Hashitani, *Anal. Chem.,* **31,** 567 (1959).

21. R. Keil, *Z. Anal. Chem.,* **245,** 362 (1969). Sc forms $Ca(ScOx_4)_2$.

22. Stary, *loc. cit.*, with some exceptions [Tl(III)], appears to consider the unextracted metal to be present as M^{n+} and disregards hydrolyzed species and charged oxine complexes; K_{ext} is calculated on the assumption that $\sum[M] \sim [M^{n+}]$.

23. Zolotov and Kuz'min, *loc. cit.* $\mu = 0.50$.

24. Turnquist and Sandell, *loc. cit.* $\mu = 0.10, 25°$.

25. Using Stary's value for log K_{ext} (4.1) based on total iron, we calculate log K_{ext} based on Fe^{3+} as 5.3.

26. Among others, C. H. R. Gentry and L. G. Sherrington, *Analyst*, **75,** 17 (1950); T. Moeller, *Ind. Eng. Chem. Anal. Ed.*, **15,** 346 (1943); D. Dyrssen and V. Dahlberg, *Acta Chem. Scand.*, **7,** 1186 (1953).

27. Turnquist and Sandell, *loc. cit.*

28. E. A. Biryuk, V. A. Nazarenko, and N. I. Zabolotnaya, *Zh. Anal. Khim.*, **23,** 853 (1968).

29. Yu. A. Zolotov, N. M. Kuz'min and V. G. Lambrev, *Anal. Abst.*, **13,** 2849 (1966).

30. MIBK can be used to extract hydrated oxinates, which cannot be extracted satisfactorily with chloroform unless an adducting agent such as butylamine is added. Y. Kakita and H. Goto, *Sci. Rept. Res. Inst. Tohoku Univ.*, **12,** 334 (1960), treat the metal solution with oxine, adjust the pH (Zn > 9.4; Mg > 10.2; Be ammoniacal tartrate) and extract with MIBK.

31. A very dilute hydrochloric acid solution of oxine is often suitable. The acidity need be no higher than required to convert HOx to HOx·H^+. Solutions in acetic acid and alcohol may be objectionable.

32. Iron has also been determined as the green intermediate complexes $FeOx^{2+}$ and $FeOx_2^+$, or more accurately expressed as a mixture of these (Fig. 6D-4), without extraction. This method has little to recommend it. The sensitivity is poor, and the pH must be precisely fixed to maintain the ratio of the species constant. Turnquist (*loc. cit.*) found $\epsilon_{625} \cong 750$ and $\epsilon_{620} = 2190$ for $FeOx^{2+}$ and $FeOx_2^+$, respectively (these are λ_{max}).

33. E. B. Sandell, *Anal. Chem.*, **19,** 63 (1947).

34. Stary, *loc. cit.*, lists stability constants of metal complexes with oxalate, tartrate, EDTA, and other masking agents, by the aid of which separation possibilities can be deduced. The tables in Ringbom, *Complexation in Analytical Chemistry*, are also useful for this purpose.

35. T. Moeller and A. J. Cohen, *J. Am. Chem. Soc.*, **72,** 3546 (1950).

36. F. Alten, H. Weiland, and H. Loofmann, *Angew. Chem.*, **46,** 668 (1933).

37. M. Teitelbaum, *Z. Anal. Chem.*, **82,** 366 (1930). R. Berg, W. Wölker, and E. Skopp, *Mikrochemie (Emich Festschrift)*, 18 (1930).

38. H. C. Deterding and R. G. Taylor, *Ind. Eng. Chem. Anal. Ed.*, **18,** 127 (1946).

39. H. M. Stevens, *Anal. Chim. Acta*, **20,** 389 (1959). The fluorescence is yellowish green.

40. R. Haar and F. Umland, *Z. Anal. Chem.*, **191,** 81 (1962). Also see Y. Nishikawa et al., *Japan Analyst*, **25,** 459 (1976).

41. H. O. Schneider and M. E. Roselli, *Analyst*, **96,** 330 (1971); R. T. Van Santen, J. H. Schlewitz, and C. H. Toy, *Anal. Chim. Acta*, **33,** 593 (1965).

42. Oxine in $CHCl_3$ has been reported to affect the fluorescence properties of $AlOx_3$, possibly pointing to a complexing interaction between the two. Or is the effect due to absorption of radiation?

43. J. Molland, *Tids. Kjemi Bergvesen*, **19,** 119 (1939); *Arch. Math. Naturvidens-kab.*, **43,** 67 (1940). For stability constants of a number of chelates of oxine-5-sulfonic acid, see Sillén and Martell, *Stability Constants*.

44. J. A. Bishop, *Anal. Chim. Acta*, **29,** 172 (1963); (Cd) D. E. Ryan, A. E. Pitts, and R. M. Cassidy, *ibid.*, **34,** 491 (1966).

45. D. Schachter, *J. Lab. Clin. Med.*, **58,** 495 (1961).

46. J. A. Bishop, *Anal. Chim. Acta*, **63,** 305 (1973).

47. T. Kambara and M. Sugawara, *Bull. Chem. Soc. Japan*, **45,** 1430 (1972) and earlier papers. Effect of Zephiramine on fluorescence of Zn: K. Kina et al., *Japan Analyst*, **23,** 1404 (1974).

48. K. Balachandran and S. K. Banerji, *J. Prakt. Chem.*, **312,** 266 (1970).

49. D. Dyrssen, footnote *f*, Table 6D-2.

50. A. P. Golovina and I. P. Alimarin, *Vestn. Moskov. Univ. Ser. Mat. Mekh. Astron. Fiz. Khim.*, **12,** 211 (1957); T. Moeller, F. L. Pundsack, and A. J. Cohen, *J. Am. Chem. Soc.*, **76,** 2615 (1954); Y. Nishikawa, *J. Chem. Soc. Japan Pure Chem. Sec.*, **79,** 631 (1958).

51. T. Shigematsu, Y. Nishikawa, and K. Hiraki, *Japan Analyst*, **15,** 493 (1966).

52. Y. Nishikawa, K. Hiraki, S. Goda, and T. Shigematsu, *J. Chem. Soc. Japan Pure Chem. Sec.*, **83,** 1264 (1962); T. Shigematsu et al., *ibid.*, **84,** 336 (1963); Y. Nishikawa et al., *ibid.*, **90,** 483 (1969).

53. T. Matsuo and K. Funayama, *J. Chem. Soc. Japan Pure Chem. Sec.*, **87,** 433 (1966).

54. E. Ruf, *Z. Anal. Chem.*, **162,** 9 (1958).

55. K. Awad, N. P. Rudenko, V. I. Kuznetsov, and L. S. Gudyn, *Talanta*, **18,** 279 (1971). (Thiooxine behaves similarly.)

56. A. I. Kirillov, R. S. Lauer, and N. S. Poluektov, *Zh. Anal. Khim.*, **22,** 1333 (1967).

57. V. T. Mishchenko and T. V. Zavarina, *Zh. Anal. Khim.*, **25,** 1533 (1970). Analogous complexes of rare earths and Y: N. S. Poluektov and Mishchenko, *ibid.*, **24,** 1434 (1969).

58. A. K. Mukherjee and B. Banerjee, *Naturwissenschaften*, **42,** 416 (1955).

59. For complex formation of the reagent with Ga, see A. Massoumi, P. Over-voll, and F. J. Langmyhr, *Anal. Chim. Acta*, **68,** 103 (1974) and with In and Tl(III), *ibid.*, **71,** 205 (1974).

60. M. Ziegler, O. Glemser, and N. Petri, *Z. Anal. Chem.*, **153,** 415 (1956); **154,** 170 (1957); *Angew. Chem.*, **69,** 174 (1957); *Mikrochim. Acta*, 215 (1957). N. Kurmaiah, D. Satyanarayana, and V. P. R. Rao, *Anal. Chim. Acta*, **35,** 484 (1966). T. Singh and A. K. Dey, *Talanta*, **18,** 225 (1971), found that the Pd chelate can be extracted into *n*-butanol as the sulfonic acid, best at pH ~ 1. The extracted acid contains six molecules of solvating *n*-butanol per molecule of the chelate. Fe(III), V(V), and some other metals are also extracted.

61. K. Goto, H. Tamura, M. Onodera, and M. Nagayama, *Talanta*, **21,** 183 (1974). The surfactant leads to the formation of a 1 (Al):3 complex at low reagent concentrations (in its absence, mixtures of complexes are obtained). Stability constants of Al complexes: F. J. Langmyhr and A. R. Storm, *Acta Chem. Scand.*, **15,** 1461 (1961). The interference of iron is circumvented by absorbance measurements at two wavelengths.

62. A. Massoumi, P. Overoll, and F. J. Langmyhr, *Anal. Chim. Acta*, **68,** 103 (1974).

63. R. G. W. Hollingshead, *Chem. Indust.*, 344 (1954). J. C. Tomkinson and R. J. P. Williams, *J. Chem. Soc.*, 2010 (1958), found log β_3 of the Fe(III) 5-fluoro-oxinate to have the value 35.6 (50% dioxane, 0.3 M NaCl), which is similar to the value 38 for Fe(III) oxinate in the same medium.

64. R. Belcher, A. Sykes, and J. C. Tatlow, *J. Chem. Soc.*, 376 (1955).

65. Methyl oxines with CH_3- in the 4-, 5-, 6-, and 7-positions do not seem to have any properties of analytical interest compared to oxine itself. Values of formation constants of a number of metals in dioxane solutions may be found in compilations. Sundry alkyl oxines: T. Rudolph, J. P. Phillips, and H. Puckett, *Anal. Chim. Acta*, **37,** 414 (1967).

66. T. J. Cardwell, *Inorg. Nucl. Chem. Lett.*, **5,** 409 (1969). Earlier, P. R. Scherer and Q. Fernando, *Anal. Chem.*, **40,** 1938 (1968), prepared a solid adduct of Al methyl-oxinate and dimethylsulfoxide, $Al(Ox \cdot CH_3)_3 \cdot DMSO$, by reaction of $AlCl_3$ with 2-methyl-oxine in DMSO containing diethylamine. Reaction in $CHCl_3$ solution gave $Al(Ox \cdot CH_3)_2OH$ and a $CHCl_3$ adduct of the latter. These authors attribute the apparent nonreaction of 2-methyl-oxine with Al in aqueous solution to the inability (actually the difficulty) of methyl-oxine to replace water molecules of hydrated Al^{3+}. W. E. Ohnesorge and A. L. Burlingame, *Anal. Chem.*, **34,** 1086 (1962), have shown the formation of a 1:1 fluorescent complex between Al an 2-methyl-oxine in absolute ethanol. (Oxine forms an analogous monoligand complex in alcohol, also fluorescent.)

67. M. A. Khwaja, T. J. Cardwell, and R. J. Magee, *Anal. Chim. Acta*, **53,** 317 (1971). Earlier work: L. L. Merritt and J. K. Walker, *Ind. Eng. Chem. Anal. Ed.*, **16,** 387 (1944); J. P. Phillips and H. P. Price, *J. Am. Chem. Soc.*, **73,** 4414 (1951). Indium forms mono and triligand complex species with 2-methyl-oxine (Ohnesorge, *Anal. Chem.*, **35,** 1137); the former is more strongly fluorescent than the latter. A diligand species is also believed to be formed. Zinc fluorescence: *Anal. Chem.*, **36,** 327 (1964).

68. R. M. Dagnall, T. S. West, and P. Young, *Analyst*, **90,** 13 (1965). According to Yu. A. Zolotov, L. A. Demina, and O. M. Petrukhin, *J. Anal. Chem.*

USSR, **25,** 1283 (1970), the presence of acetate in solution should be avoided when the aim is to leave aluminum in solution, because acetate forms an ion-association complex with Al-2-methyl-oxine. These authors extract aluminum as a mixed complex of 2-methyl-oxine and caproic acid into TBP–CHCl$_3$. N. V. Shakhova, O. A. Kiseleva, and Yu. A. Zolotov, *Zh. Anal. Khim.*, **29,** 1229 (1974), extract Nd and Eu, as ion-association complexes of 2-methyl-oxine and trichloroacetate or ClO$_4^-$, into TBP–benzene at pH ~ 4.

69. D. Dyrssen, *Svensk Kem. Tidskr.*, **68,** 231 (1956).

70. pH ranges and other particulars are given by K. Motojima and H. Hashitani, *Japan Analyst,* **9,** 151 (1960); K. Motojima, *Progress in Nuclear Energy,* Series IX, Vol. 8, Part I, Anal. Chem., Pergamon, New York, 1967. Determination of Cr in U; *Anal. Chem.,* **33,** 239 (1961). Effect of masking agents (CN$^-$, H$_2$O$_2$) on extraction: R. J. Hynek, *ASTM Spec. Tech. Publ.* No. 238, 36 (1958).

71. R. Keil, *Mikrochim. Acta,* 919 (1973).

72. M. Borel, Thesis, Lyon, 1952, gives the wavelengths of maximum absorption of 2-methyl-oxine in CHCl$_3$ as 251 nm ($\epsilon = 5.3 \times 10^4$) and 307 nm ($\epsilon = 2.95 \times 10^3$).

73. M. Shiro and Q. Fernando, *Anal. Chem.,* **43,** 1222 (1971).

74. R. L. Stevenson and H. Freiser, *Anal. Chem.,* **39,** 1354 (1967).

75. H. Irving and D. J. Clifton, *J. Chem. Soc.,* 288 (1959).

76. G. Bocquet and R. A. Pâris, *Anal. Chim. Acta,* **13,** 508 (1955); **14,** 1, 201 (1956).

77. A. Corsini and R. M. Cassidy, *Talanta,* **21,** 273 (1974).

78. F. Umland and K.-U. Meckenstock, *Z. Anal. Chem.,* **177,** 244 (1960). This and analogous reagents appear to react as monobasic tridentate complexers with most divalent metals.

79. G. Röbisch, *Anal. Chim. Acta,* **48,** 161 (1969).

80. F. Umland, B. K. Poddar, and K.-U. Meckenstock, *Z. Anal. Chem.,* **185,** 362 (1962).

81. J. P. Phillips, J. F. Deye, and T. Leach, *Anal. Chim. Acta,* **23,** 131 (1960). Ionization constants and λ_{max} given.

82. A. I. Cherkesov and V. S. Tonkoshkurov, *Zh. Anal. Khim.,* **26,** 1301 (1971).

83. T. Hata and T. Uno, *Bull. Chem. Soc. Japan,* **45,** 477, 2497 (1972).

84. R. Berg and E. Becker, *Berichte,* **73,** 172 (1940).

85. R. Berg and E. Becker, *Z. Anal. Chem.,* **119,** 81 (1940).

86. M. Ishibashi, Y. Yamamoto, and H. Yamada, *Bull. Chem. Soc. Japan,* **32,** 1064 (1959). Absorption spectra are given.

6E

ORGANIC PHOTOMETRIC REAGENTS
AZO REAGENTS

The group $-N{=}N-$ is strongly chromophoric and in combination with other groups, such as aryl $-OH$ and $-AsO_3H_2$, furnishing replaceable H, is likely to give chelating agents of value in spectrophotometric analysis. Azo dyes having metal chelating properties are numerous and varied, and for purposes of discussion we classify them broadly as follows:

1. Those having two aryl groups linked by $-N{=}N-$, that is, monoazo compounds.

2. Those having three aryl groups linked by $-N{=}N-$, that is, bisazo compounds.

3. Those having a heterocyclic system as one of the linked components.

I. ARYL MONOAZO REAGENTS

Compounds of this class are very numerous, and they include reagents that have been used for a long time, and continue to be used, in photometric determinations. They are likely to show pronounced color changes with metals and to give sensitive reactions. In general, their selectivity is poor or mediocre. Most of them have one or, more usually, two OH groups. No useful purpose would be served by attempting to review all such reagents here. Some general characteristics of this group of reagents are touched on, as illustrated by a number of typical members.

A. SOME REPRESENTATIVE MONOAZO REAGENTS

The chelating grouping is

$$HO—C{=}C—N{=}N— \qquad or \qquad HO—C{=}C—N{=}N—C{=}C—OH$$

For a chelate to be formed, the OH group needs to be in the *ortho* position or to yield this arrangement by tautomerism. (There seem to be a few exceptions to this rule.) Most metals can be chelated by reagents having this grouping. The higher valent (III, IV, V) metals generally form more stable complexes than the divalent metals, and so react in more acidic solutions than the latter and can be determined in the presence of certain amounts of these. Some divalent metals (Mg, for example) are more reactive than others (Ca). Some of the chelates are fluorescent. Broadly, in these respects, the azo reagents are very similar to others bonding metals through N and O, or even O alone. But the azo reagents are likely to be more sensitive.

Depending on the metal and its ionic charge, the pH, the excess of reagent, and other factors, chelate species having various ratios of ligand/metal or ligand/metal–hydroxo can be formed, for example,

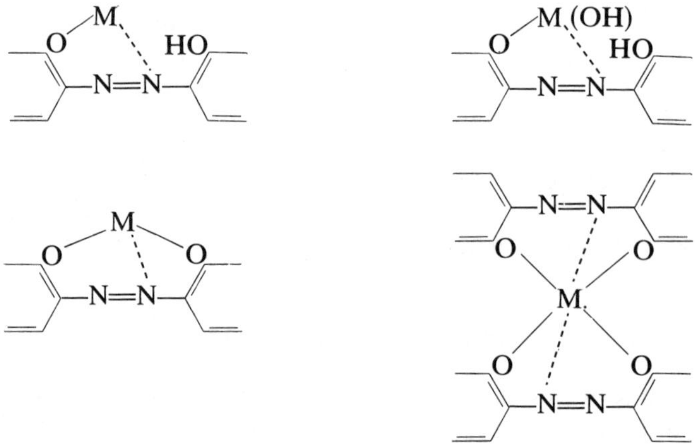

(Charges not indicated above, nor coordinated H_2O.) Ca and Mg often form 1:1 complexes in aqueous solution. The reagent has analytical utility only when λ_{max} of the chelate is considerably different from λ_{max} of the reagent (similarly for fluorescence). Groups such as Cl and NO_2 affect λ_{max} of reagent and chelate and can serve as auxochromes. In addition, such groups affect the chelating power of the functional group (see the discussion on p. 473). Sulfonic groups promote solubility of both the dye itself and the chelate complexes. Although the second OH group need not be directly involved in chelation, its presence is favorable to chelate formation. (Chelate formation with Mg and Ca requires two o-OH groups, with a few exceptions.) In general, the group

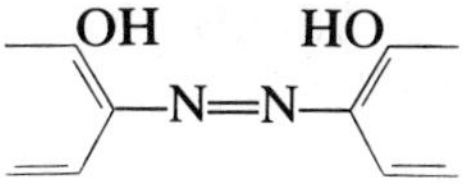

chelates better than

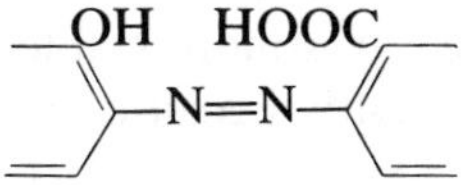

or

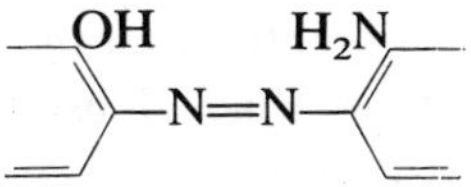

o,o'-Dihydroxyazobenzene

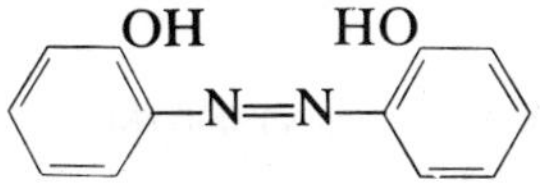

   $pK_1 = 7.8$, (also reported as 9.3); $pK_2 = 11.5$ (?).

is the simplest of the useful monoazo reagents. It forms 1:1 and 1:2 M/reagent complexes with Al, Ga, and In.[1] The 1:1 complex is formed by replacement of both phenolic hydrogens:

$$M^{3+} + H_2L \rightleftharpoons ML^+ + 2H^+$$

The 1:2 chelates, formed at higher pH values, apparently have the composition ML_2^-. Zirconium forms ZrOL, copper $CuLOH^-$.

Lumogallion (C. I. Mordant Red 72)

This is the sodium salt of 4-chloro-6-(2,4-dihydroxyphenylazo)-1-hydroxybenzene-2-sulfonic acid

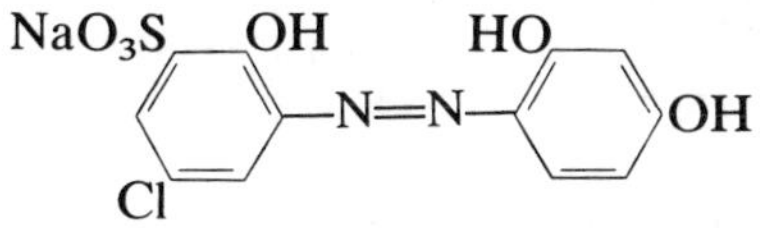

The $-SO_3H$ group confers water solubility on the reagent and its chelates, yet allows extraction of the chelates if need be with a suitable organic

solvent such as a higher alcohol. If a bulky organic cation (quaternary ammonium salt, diphenylguanidinium salt, etc.) is added, the ion associate can be extracted into $CHCl_3$ and similar solvents. (The presence of two $-SO_3H$ groups in the reagent is less favorable for extraction.)

Gallium reacts at pH ~3,[2] In at pH 4–5,[3] Al at pH ~5 and Sn(IV) at pH 1 (butyl alcohol extraction of latter).[4] Indium, for example, can be determined absorptiometrically at pH 4–5 in the presence of limited amounts of Ca, Mg, Zn, and Fe(II). Zirconium reacts in dilute HCl.

Calmagite

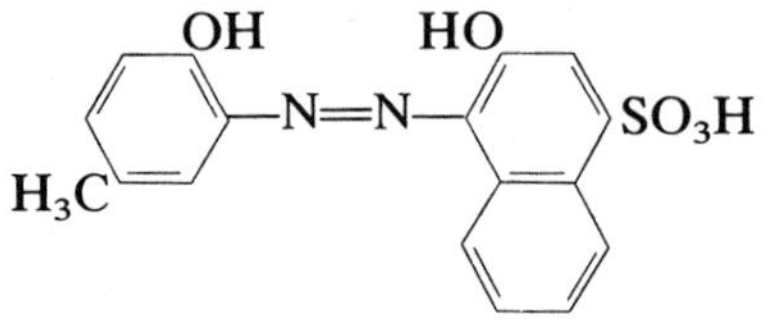

Mol. wt. 358.4. M.p. > 330°

This indicator for the titration of Ca and Mg with EDTA is also of value in the absorptiometric determination of Mg.[5] It is more stable in solution than Eriochrome Black T used earlier (apparently because of the absence of the $-NO_2$ group).

SPADNS

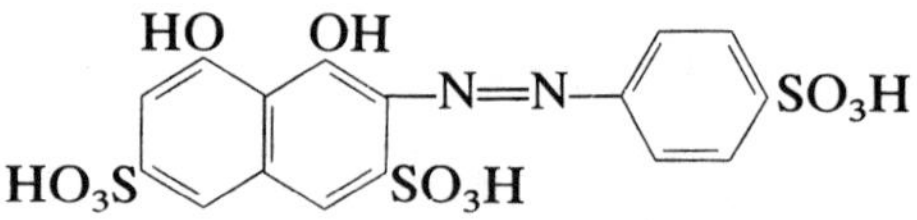

Mol. wt. 570.4 (tri-Na salt). M. p. > 300°.

This compound, 2-(4-sulfophenylazo)-1,8-dihydroxy-3,6-naphthalene-disulfonic acid, is a reagent for Th,[6] Zr,[7] and the rare earth elements,[8] which form 1:1 complexes. The similar dye Eriochrome Blue SE can be used to determine Ca in 1 M NaOH solution. At this high alkalinity Mg does not react because it is precipitated as the hydroxide, or at very low concentrations forms undissociated $Mg(OH)_2$.

Gallion

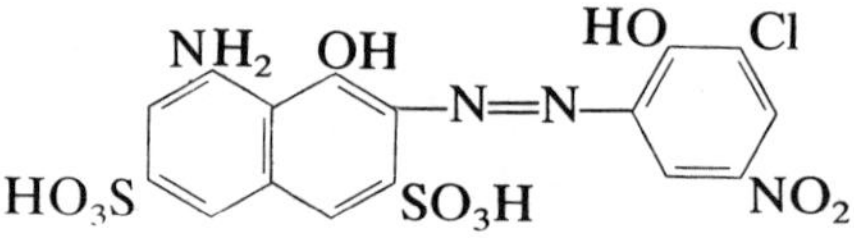

Mol. wt. 518.9. M. p. ~268° (decomp.).

is a reagent similar to Lumogallion. Gallium reacts at pH 2.4–3.2.

Another *o,o'*-dihydroxyazo compound, 4-hydroxy-3-(2-hydroxy-5-nitro-3-sulfophenylazo)naphthalene-1-sulfonic acid, reacts with Ga at pH 1.6, forming a 1:1 complex ($\epsilon_{540} = 15,400$); it allows determination of Ga in the presence of Zn, Pb, In, and Mn.[9]

Picramines. Numerous azo dyes derived from picramic acid have been prepared and tested as heavy metal reagents. The nitro groups increase the acidity of the OH group and the sensitivity of metal reactions.

Picramine R[10]

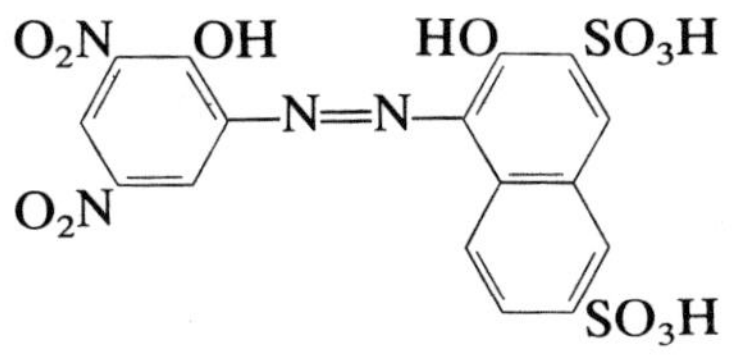

3-Hydroxy-4-(2-hydroxy-3, 5-dinitrophenylazo)naphthalene-2, 7-disulfonic acid

finds use in the determination of Zr, and is reported to be superior to Arsenazo III, since it can be used in the presence of Th, Ti, rare earths, and U(VI).

Picramine CA

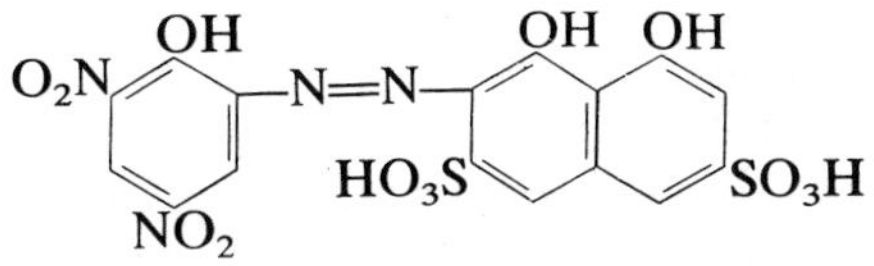

Disodium salt is stable, violet powder. Acidic solution (pH <1.5) red, neutral and alkaline, violet.

is a slightly more sensitive Zr reagent than Picramine R and can be used in the presence of limited amounts of the elements mentioned.[11] Zr, Sc, Nb, and Cu(II) react in 0.5 M HCl (color change red to violet). The Zr chelate has the composition 1 Zr:2 reagent.

Picramine-epsilon

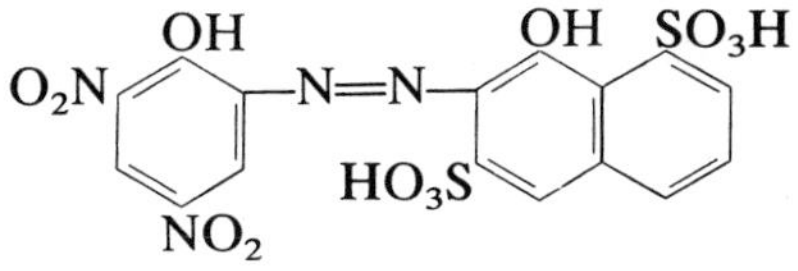

reacts with Cu(II) in 0.1–0.2 M HCl and with Zr in 1 M HCl.[12] Though more sensitive than Picramine R for Zr, Nb, and Cu, it is hardly more selective.

Sulfonitrazo,

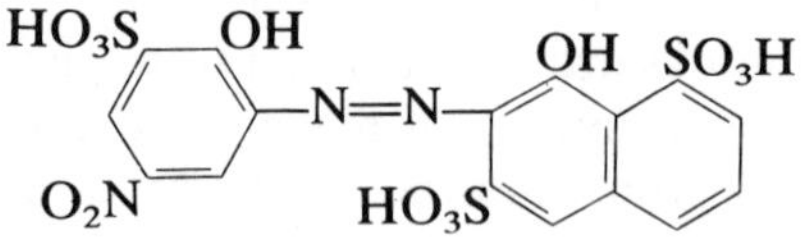

a similar reagent with a single $-NO_2$ group has been used to determine V(IV) at pH ~2.

Beryllon II

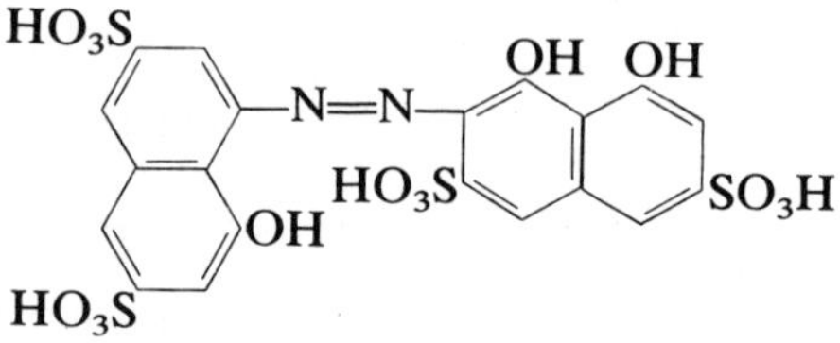

Used as di-Na salt. Mol. wt. 694.6.

which is 4,5-dihydroxy-3-(8-hydroxy-3,6-disulfo-1-naphthylazo)-2,7-naphthalenedisulfonic acid, has enjoyed some popularity for the determination of Be. Separations and masking are essential. It is also a reagent for the rare earth metals (1:1 chelates, pH 2).

Beryllon III, which is 5-(4-diethylamino-2-hydroxyphenylazo)-4-hydroxynaphthalene-2,7-disulfonic acid, is superior to Beryllon II as a beryllium reagent.[13] In ~0.1 M NaOH solution in the presence of triethanolamine and EDTA, it allows a fairly selective determination of Be. Some 500–1000 times as much Al, Fe(III), and Mg may be present. $\mathscr{S}_{526} = 0.0005 \ \mu g$ Be.

Eriochrome Black T (Solochrome Black T)

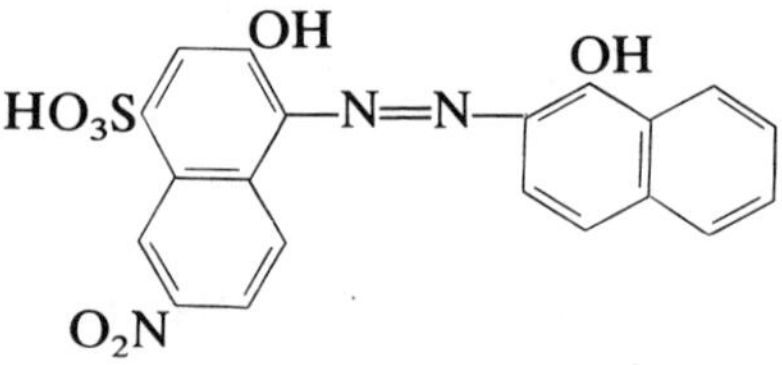

Used as Na salt. Mol. wt. 461.4.

is a reagent for Mg and Ca. It is not stable in solution. Eriochrome Blue–Black B lacks the $-NO_2$ group; it reacts similarly, as does its isomer Eriochrome Blue-Black R (Calcon).

Calcichrome. This compound, 1-(3,6-disulfo-8-hydroxy-1-naphthylazo)-2,8-dihydroxynaphthalene-3,6-disulfonic acid, was formerly thought to be a *bis-* or *tris*-azo compound having a cyclic structure, but is now known[14] to be represented by the formula

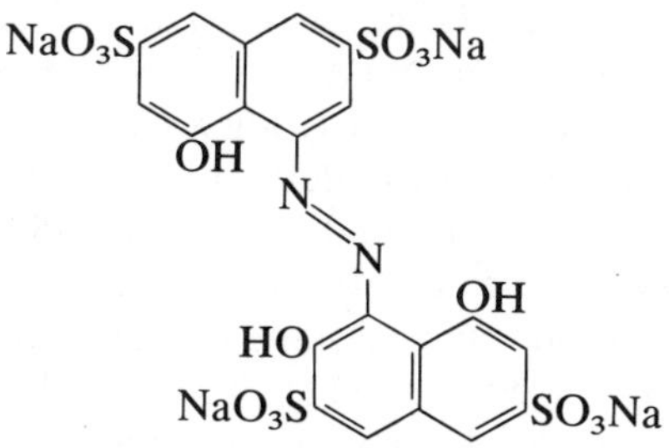

Calcichrome and Calcion are the same substance.[15]

At pH 12–13, Ca forms a red chelate; Sr and Ba do not react. Calcium ($\epsilon_{615} = 7600$) can be determined in the presence of several thousandfold amounts of Sr and Ba, if account is taken of their electrolyte effect.[16]

B. MONOAZO REAGENTS WITH –AsO(OH)$_2$ OR –PO(OH)$_2$ GROUPS

More selective azo reagents are obtained by introducing the above groups into the molecule, so that the reacting metal can be bound to the arsono (or phosphono) group, azo N and O of an OH group (replacement of H). Oxyphilic metals of high valence tend to react in quite strongly acid solutions with such reagents.

Thorin, Thoron, APANS. These trivial names[17] are used to designate *o*-(2-hydroxy-3,6-disulfo-1-naphthylazo)benzenearsonic acid, or its sodium salt,

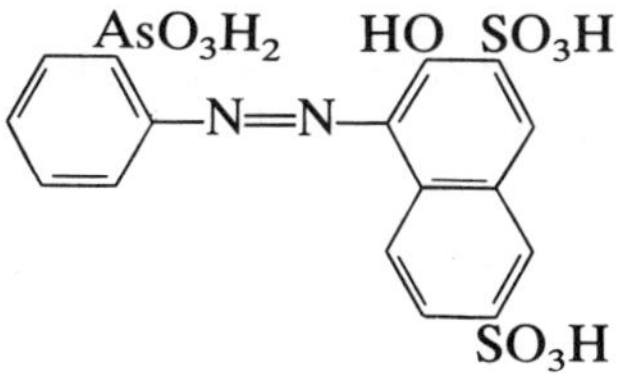

Disodium salt, $(NaO_3S)_2^-$, commercially available (impure). Mol. wt. 576.3. M.p. $> 300°$.

which was synthesized and its reactions with uranium, thorium, and other elements studied as early as 1941.[18]

The acid dissociation constants are $pK_3 = 3.7$ and $pK_4 = 8.3$ for the arsono hydrogens and $pK_5 = 11.8$ for the naphtholic OH.[19] The noteworthy characteristic of this reagent is its reaction with such metals as Zr, U(IV), Th, Np(IV), and Pu(IV) in fairly acid solution ($\sim 0.1\ N$). That the reagent should show selectivity for these metals is not unexpected, considering the low solubility of the arsenates (and phosphates) as indicating strong bonding between the metals and As(V). The chelates are orange or red (acid reagent solution is yellow, $\lambda_{max} = 445$ nm). Thorium and U(IV) form 1:2 chelates. Probably tetradentate complexes are formed by coordination of Th or U through the oxygen of the arsono group, phenolic O and an azo N (see p. 463). The effective value of the formation constant of ThL_2 at pH 1, $\log K_1 K_2$, is 10.15.[20] The reagent has found considerable use in the determination of thorium. Interfering elements include Zr, Ti, Nb, Ta, and Sn [Fe(III) can be reduced to Fe(II), and U(IV) can be oxidized to U(VI)], so that the selectivity for thorium is only moderate. Separation and masking of these elements are required. Bismuth can be determined in 0.02–0.04 M $HClO_4$.

As the acidity is decreased, many other metals begin to react, but such determinations are of no great interest. Uranium (VI) forms a blue 1:1 complex (pH 8 optimally). The reagent has even been suggested for the determination of lithium in a strongly alkaline, acetone-containing solution.

The reagent can be purified by recrystallization (addition of HCl to sodium salt) from water. No doubt chromatographic purification is possible.

Arsenazo I,

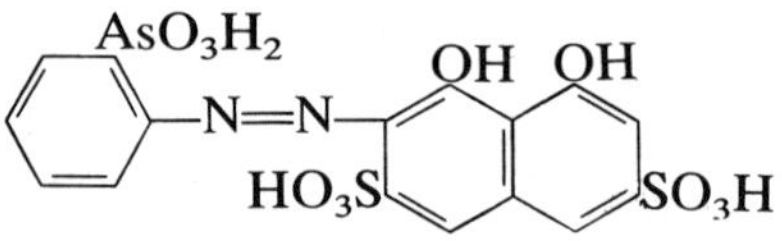

Mol. wt. 548.3. Commercial products (available as di- or tri-Na salt), often of low purity, can be purified by slowly dropping a saturated aqueous solution of the disodium salt into an equal volume of concentrated HCl. The precipitate is filtered off, washed with acetonitrile, and dried at 100° for 1–2 hr.

o-(1,8-Dihydroxy-3,6-disulfo-2-naphthylazo)benzenearsonic   acid,   is based on chromotropic acid and is the simplest member of the series known as Arsenazos.

The ionization constants have been reported as[21]:

$$pK_1 = 0.6 \qquad pK_3 = 3.5 \qquad pK_5 = 11.6$$
$$pK_2 = 0.8 \qquad pK_4 = 8.2 \qquad pK_6 = 15$$

The species of Arsenazo I present in strong sulfuric acid, 0.1 N sulfuric acid and weakly basic solutions are stated to be[22]:

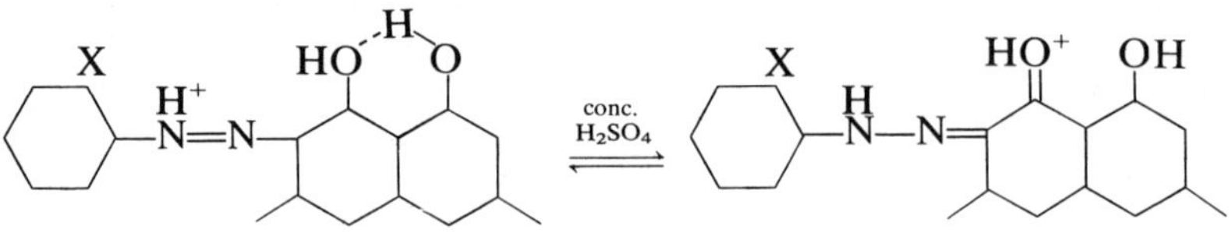

Quinone–hydrazone form

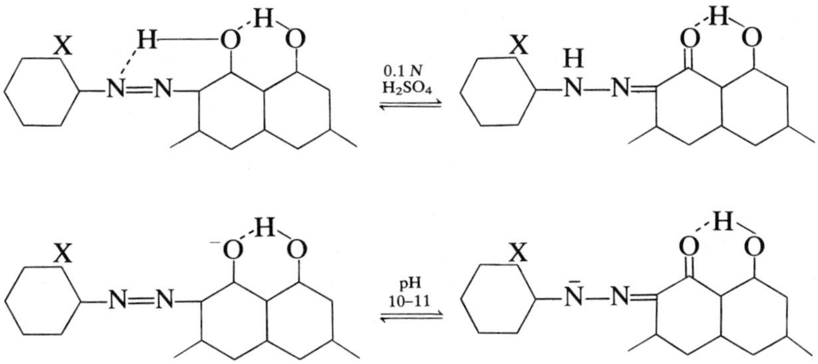

(X = arsono group)

Aqueous solutions of the reagent are red; λ_{max} (595 nm) changes little with pH in the range 4–8.

Arsenazo I forms violet chelates [U(VI), blue] with many metals in the pH range 0–8, but, in general, only those reactions occurring in quite acidic solutions are of much interest to the analyst.[23] Thorium, zirconium, plutonium(IV), niobium, and tantalum can be determined at pH 1–2. Scandium, the rare earths, uranium(VI), and other elements can be determined in approximately neutral solutions. Calcium and strontium react in basic solution.

The complexes formed with most metals have a 1:1 ratio of reagent to metal. In acid solution, with a metal such as zirconium, we would expect replacement of one H in $-AsO_3H_2$ with bonding to an azo N and to O of OH (see Arsenazo III). According to Kuzin et al.,[24] the structure of the Arsenazo I–Sc complex is

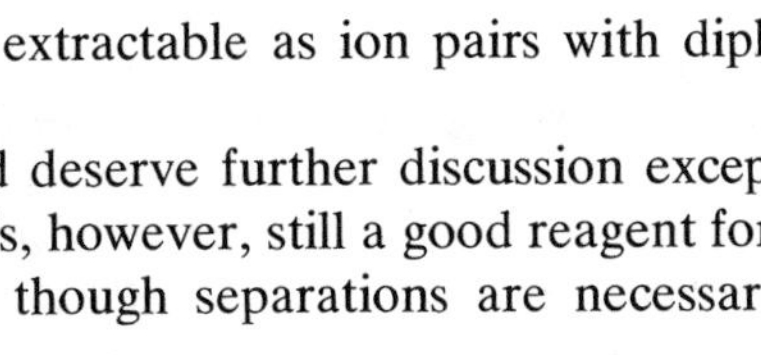

The chelates are extractable as ion pairs with diphenylguanidine into amyl alcohol.

Arsenazo I would deserve further discussion except that it is eclipsed by Arsenazo III. It is, however, still a good reagent for Th in the presence of the rare earths, though separations are necessary when R.E./Th is unfavorable.[25]

A reagent that has been given the name Arsenazokhimdu[26]

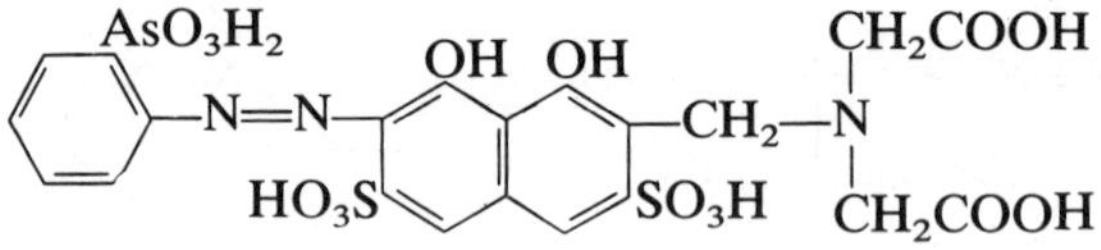

is Arsenazo I with the group $-CH_2-N(CH_2COOH)_2$ introduced. It is reported to be more selective than the parent compound and is recommended for the rare earth metals, which form 1:1 chelates. Cu, Al, Zr, Pb, Ti, Sc, and In react less sensitively with Arsenazokhimdu than with Arsenazo I. No color reactions are observed with Li, Zn, Bi, Cr(III), W, Mo, Cd, or Hg(II). The reactions with Th (pH 2) and with Be, La, and V (pH 4.5–5) are pronounced.

Another analog of Arsenazo I is Beryllon IV

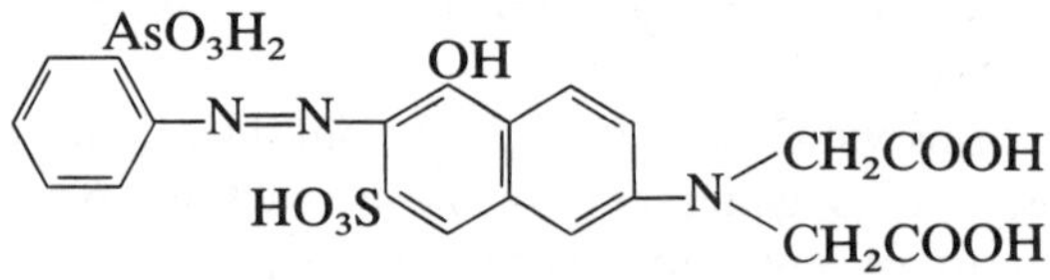

which as its name suggests is a photometric reagent for beryllium.[27] It allows a few micrograms of Be to be determined in basic solution in the presence of 1 mg of Al and other metals if EDTA is added. The $-N(CH_2COOH)_2$ group does not participate in chelate formation with beryllium.

Chlorophosphonazo R

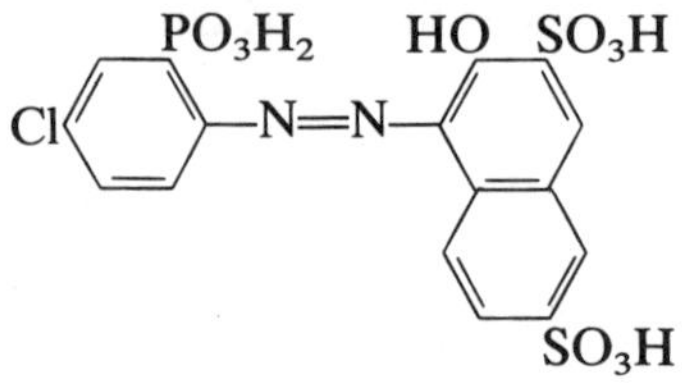

has found use in the determination of Zr, Nb, Sc, and other metals. The closely similar reagent Phosphonazo R serves for the determination of Li (see note 45).

II. ARYL BISAZO REAGENTS

In recent years, bisazo compounds based on chromotropic acid have been extensively synthesized and studied,[28] especially for metals whose cations are strongly hydrolyzed in dilute (some even in quite strong) mineral acid medium. Bisazo reagents containing arsonic acid groups, $-AsO_3H_2$, are the most valuable in general. Other bisazo (less often trisazo) compounds—not based on chromotropic acid—find photometric application, for example, Diamond Black F (C.I. 26695) and Congo Corinth (C.I. 22145), to name two at random, but they are of lesser interest and value than the reagents discussed in the following pages.

In this section we consider first some important reagents of this class and then survey some work dealing in a general way with bisazo reagents for niobium and for the alkaline earth metals. Bisazo chromotropic acid derivatives also find use in the determination of the rare earth elements.

A. ARSENAZO III

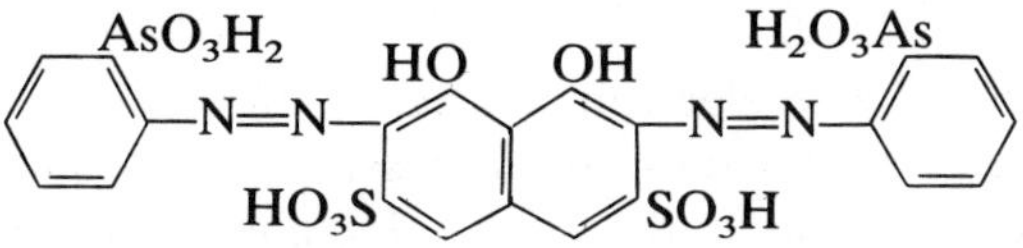

2,2'-[1,8-Dihydroxy-3,6-disulfo-2,7-naphthylenebis(azo)]dibenzenearsonic Acid

Dark red powder. Mol. wt. 776.4. M. p. >320° (decomposition). The commercial product is usually impure, containing Arsenazo I and other substances. Precipitation of Arsenazo III by addition of HCl to an alkaline solution (NaOH or NH_4OH) has been used for its purification, but this method is not very effective.[29] A better method[30] consists of a double HCl precipitation of the dye from ammoniacal solution, its solution in 1:1:1 n-propanol–concentrated $NH_4OH–H_2O$ mixture (0.5 g in 7 ml, ~50°) and chromatography on a 25× 120 mm column of microcrystalline cellulose (elution with a 3:1:1 volume ratio of the solvents named). After concentration to 10–15 ml by evaporation <80°, the eluate is treated with HCl to precipitate Arsenazo III, which is washed with alcohol and air-dried. λ_{max} in water = 540 nm, $\epsilon = 3.6 \times 10^4$. Arsenazo III solutions are stable (though deterioration after 2 weeks has been reported). Oxidizing agents in the sample solution should be avoided. In concentrated H_2SO_4, ϵ_{675} of Arsenazo III is 52,800; Arsenazo I does not absorb at this wavelength.[31]

This reagent[32] and its structural analogs[33] are of special value because of selectivity for the acid-hydrolyzing metals combined with high sensitivity.

Arsenazo III is moderately soluble in water. In the acid range pH 3– 10 M HCl its solutions are pink or red-crimson in color, depending on the concentration (Table 6E-1). See Fig. 6E-1 for $A - \lambda$ curves, pH 2.4–12.7. At a pH above 4–5, a violet or blue color is observed, and in concentrated sulfuric acid (Fig. 6E-2) a green color (protonation of azo group). The reported values of the acid dissociation constants are not in good agreement (Table 6E-2), probably largely because of impure reagent. pK_1 and pK_2 correspond to the dissociation of sulfonate groups, $pK_3 \cdots pK_6$ to the dissociation of arsono groups, and pK_7 and pK_8 to the dissociation

TABLE 6E-1
Absorption Maxima of Arsenazo III as a
Function of Acidity^a

pH Range	λ_{max} (nm)
Concd. H_2SO_4	675
0.5–2.8	540
2.8–6.3	545
6.3–8.0	560
8.0–10.7	$\lambda_1 = 595, \lambda_2 = 640$
10.7–11.7	585
11.7–13.0	580

^{a} P. K. Spitsyn and V. S. Shvarev, *Zh. Anal. Khim.*, **25,** 1503 (1970). Also see Fig. 6E-1.

of hydroxy groups. The electronic structure of Arsenazo III at various acidities has been elucidated.[34] In contrast to the asymmetric monoazo compounds, which exist in the azo and quinone–hydrazone forms (p. 456), bisazo compounds in moderately acid solutions (pH 1, up to pH 8) exist only in the azo form. In alkaline solution (pH >9) or in very strong acid (85% sulfuric acid), ionization destroys the symmetry of the Arsenazo III molecule and azo and quinone–hydrazone forms coexist. In general, bisazo derivatives of chromotropic acid have violet or blue colors as a result of the conjugation of both parts of the molecule and a planar

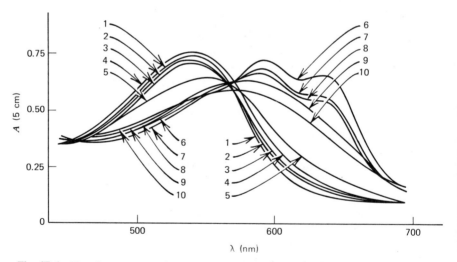

Fig. 6E-1. Absorbance curves of Arsenazo III solutions (3.4×10^{-6} *M*) as a function of pH: *1,* 2.40; *2,* 2.6; *3,* 3.45; *4,* 5.52; *5,* 7.30; *6,* 9.68; *7,* 10.2; *8,* 10.7; *9,* 11.3; *10,* 12.7. P. K. Spitsyn and V. S. Shvarev, *Zh. Anal. Khim.*, **26,** 1314 (1971).

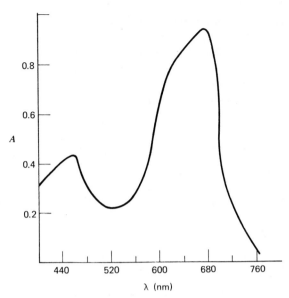

Fig. 6E-2. Absorption curve of Arsenazo III in concentrated sulfuric acid. P. K. Spitsyn and V. S. Shvarev, *Zh. Anal. Khim.*, **25**, 1503 (1970).

TABLE 6E-2
Acid Dissociation Constants of Arsenazo III

pK_n	Budesinsky (1963)[a]	Budesinsky (1965)[b]	Budesinsky (1969)[c]	Palei et al. (1967)[d]	Spitsyn and Shvarev (1970)[e]
1	−2.6	−1.3	0.6	1.9	0.8
2	0	−0.6	0.8	3.0	2.9
3	2.4	3.8	1.6	3.6	3.5
4	2.4	5.2	3.4	5.1	5.0
5	5.3	7.1	6.3	6.8	7.0
6	5.3	9.0	9.1	7.6	8.4
7	7.5	11.7	12.0	9.3	10.2
8	12.3	14.6	15.1	11.9	12.4

[a] B. Budesinsky, *Coll. Czech. Chem. Commun.*, **28**, 2902 (1963).
[b] B. Budesinsky, *Z. Anal. Chem.*, **207**, 247 (1965).
[c] B. Budesinsky, *Talanta*, **16**, 1277 (1969). pK values for the dissociation of H_9L^+ and $H_{10}L^{2+}$ are −2.7 and −2.7, respectively.
[d] P. N. Palei, N. I. Udal'tsova, and A. A. Nemodruk, *Zh. Anal. Khim.*, **22**, 1797 (1967).
[e] P. K. Spitsyn and V. S. Shvarev, *Zh. Anal. Khim.*, **25**, 1503 (1970). S. B. Savvin and R. F. Propistsova, *Zh. Anal. Khim.*, **23**, 653 (1968), give pK 7.6 for the dissociation of AsO_3H^-.

461

structure. The pink color of Arsenazo III in acidic solution is attributed to a breakdown in the planar structure and conjugation brought about by the presence of the bulky $-AsO_3H_2$ in the *ortho* position with respect to the azo group.

According to Budesinsky, the species of Arsenazo III and similar compounds in very strongly acid solutions are

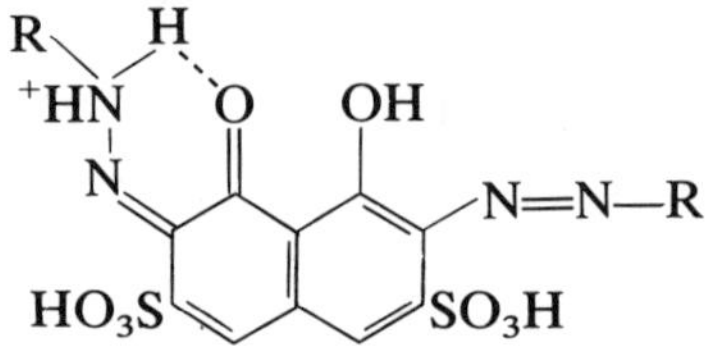

$$H_9L^+ \qquad\qquad\qquad H_{10}L^{2+}$$

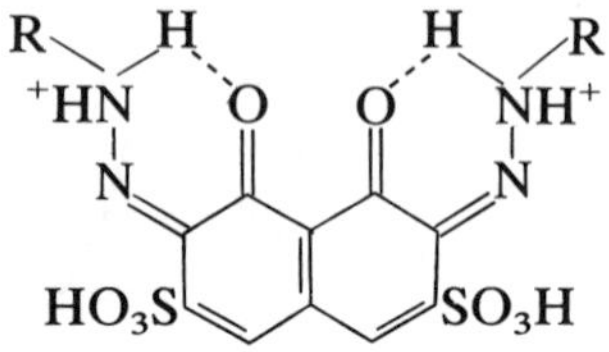

The following maximal molar absorptivities have been reported for various forms of Arsenazo III (Budesinsky, 1969):

Form	$\epsilon_{max} \times 10^{-4}$	λ_{max} (nm)	Color
L^{8-}	4.52	600	Blue
HL^{7-}	3.18	585	Blue
H_2L^{6-}	3.98	600	Violet
H_3L^{5-}	3.78	560	Wine red
H_4L^{4-}	4.05	535	Red
$H_{10}L^{2+}$	5.10	670	

Arsenazo III is noteworthy for reacting with Zr, Hf, Th, Pa, Np, Pu, and U(IV and VI)[35] in strongly acid solutions, for example, in 3 M or stronger HCl. These reactions are highly sensitive as may be judged from the large ϵ values[36] of the chelate species formed. Thus for the 1:1 complex of U(VI) in 5.6 M $HClO_4$, $\epsilon_{655} \sim 7.5 \times 10^4$ ($\mathscr{S}_{655} = 0.003$ μg U/cm^2) and for the 1(U):2 complex of U(IV) in 6 M HCl, $\epsilon_{655} = 1.2 \times 10^5$ ($\mathscr{S}_{665} = 0.002$). Protactinium, Np(IV), and Pu(IV) can also be determined sensitively in strongly acid medium. Determination of Sc, Y, the rare earth elements, Pd(II), and some other elements requires lower acidity, namely a pH of 2–3 or a little higher. Such metals as Cu, Cd, Pb, and Ca react in still less acidic solution, but these reactions are comparatively unimportant analytically. Cations having a radius less than 0.7–0.8 Å are said not to give a color reaction with Arsenazo III.[37] Included among these are Be, Zn, Al, Ga, In, Ge, Ti, and Sn. Actually, Zn, Al, and Be[38] react. Anyway, Arsenazo III is a more selective reagent than Arsenazo I (or II). The optimal acidity for the determination of U (VI, 1:1 complex) in perchloric acid medium has been reported as

5.6 M.[39] Interference from Al, Fe, Sc, rare earths, and Ca is a little less in $HClO_4$ than in HNO_3 solution, but that from Th, Zr, U(IV), and Pu(IV) remains unchanged. The U(VI) complex can be extracted into benzyl alcohol.[40] The reaction of Zr (but not of Th) is masked by oxalic acid.

Most of the Arsenazo III chelates have a green color (Pd purple; Sc violet; Bi violet-blue; Pb, Zn, Cd, and Ca blue). See Fig. 6E-3 for absorption curves of the reagent and the Zr and La chelates. Pd(II) absorbs maximally at 630 nm.

Arsenazo III forms 1:1 complexes with doubly and triply charged cations, for example, UO_2^{2+} and La^{3+}. With quadruply charged cations [Th, Zr, Hf, and U(IV)], it forms both 1:1 and 1:2 complexes, the latter when it is present in large excess.[41] Higher ratios are hardly to be expected (steric hindrance), though some have been reported. A closer study of the U(VI) reaction in acid solution has revealed the existence of two complexes of the type ML,[42] the ratio of these depending on the hydrogen-ion concentration. Thorium forms complexes of the type M_2L, ML (two), and ML_2. The last complex is formed in strong nitric acid solution and has a high molar absorptivity, $\epsilon_{658} = 1.2 \times 10^5$.

Metals reacting in acid solutions no doubt replace a hydrogen in the AsO_3H_2 group. Only one of the functional groups reacts. The following structures have been suggested for 1:1 complexes[43]:

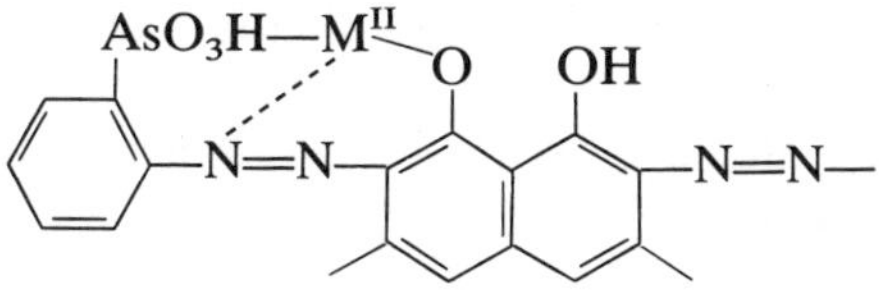

1

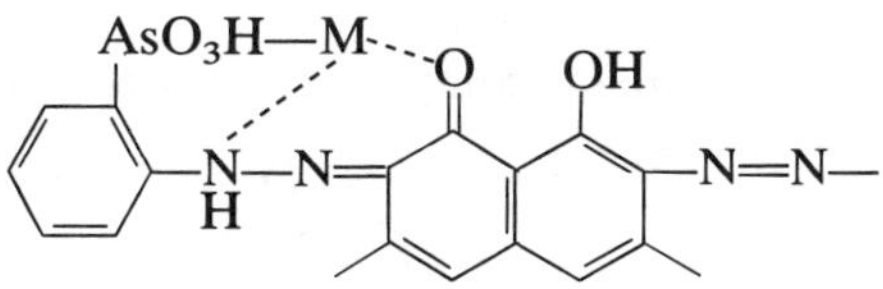

2

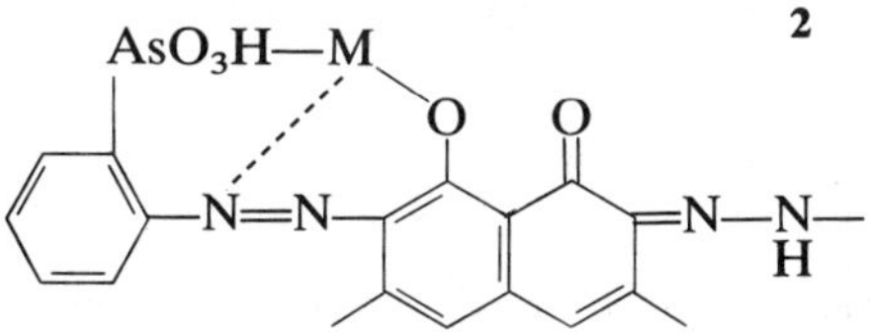

3

For 1:2 (metal:ligand) complexes another molecule of Arsenazo III should be added. Structures **4** and **5** are proposed by Budesinsky[44] for metal complexes (chelates) of Arsenazo III, Phosphonazo III, and

Sulfonazo III:

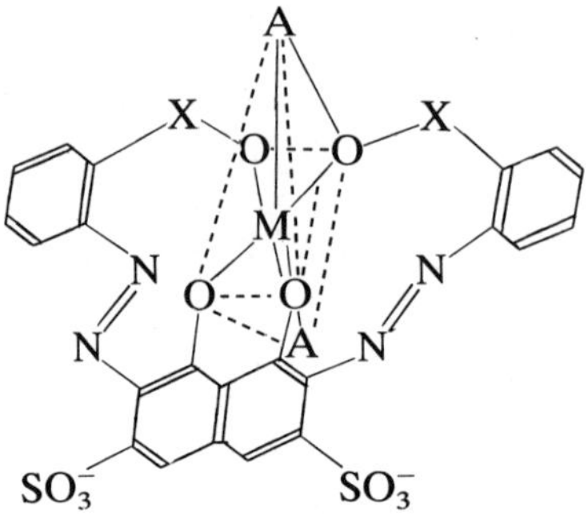

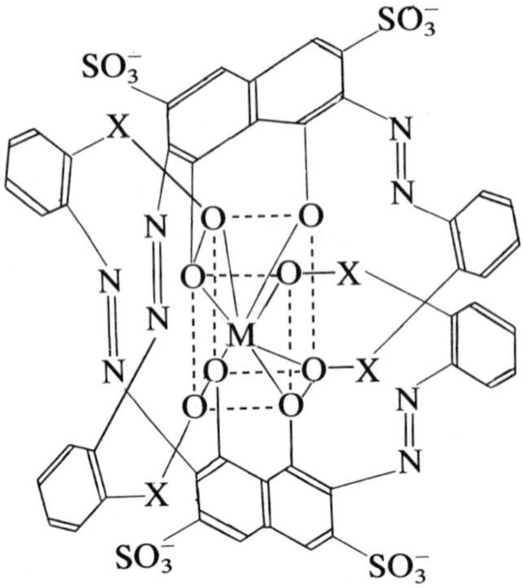

4   X = AsO·OH, PO·OH, SO₂   **5**

Octahedral structure **4** corresponds to the composition metal:ligand 1:1 and coordination number 6 (or 4-planar tetragonal complexes). Cubic structure **5** corresponds to the molar ratio 1:2 and coordination number 8.

In weakly basic solutions, the alkaline earth metals react with Arsenazo III according to the equation[45]

$$M^{2+} + H_3L^{5-} \rightleftharpoons MH_2L^{4-} + H^+$$

and the metal complex species is considered to have the structure

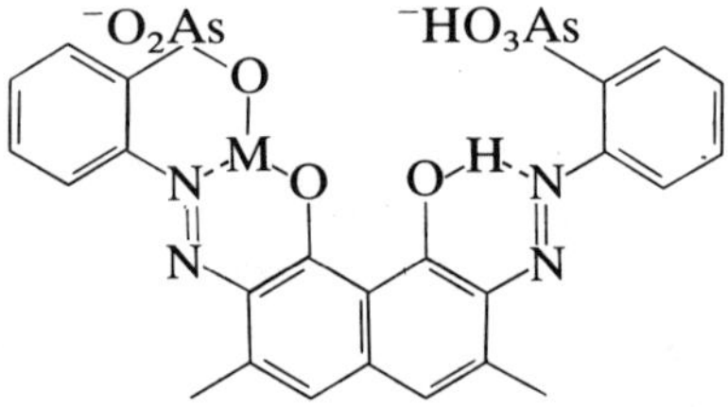

Zinc and Cd also react in basic solution, maximally at pH 10–11 ($\mathscr{S}_{590} = 0.0023$ μg Zn/cm², $\mathscr{S}_{600} = 0.0043$ μg Cd/cm²).[46] Triethanolamine decreases the interference of Al, Fe(III), and Mn(II), but hardly that of Ti. Fluoride is reported to mask Cd so that it does not interfere in the determination of Zn. Iodide forms a mixed chelate with Arsenazo III and Zn and increases the sensitivity. It is evident that the determination of Zn and Cd is better carried out with dithizone, which forms no chelates with Al, Fe, Ti, Zr, U, and other metals, than with Arsenazo III.

It is possible that the arsono group may not be involved in the chelation of some metals reacting at low acidities. The *peri* hydroxy groups of chromotropic acid in combination with an azo group would then be chelate forming.

Arsenazo III complexes of metals, such as of U(VI), can be extracted into butyl alcohol and like solvents as the anionic partner of diphenyl-guanidinium cation.[47]

It is interesting that the absorption spectra of Arsenazo III chelates (Zr, Th, U, La, etc.) show two maxima, located at ~665 and ~610 nm (see Fig. 6E-3). These maxima result from the presence of two weakly interacting chromophoric systems in the chelate molecule. Chelate formation (through one of the complexing groups) destroys the symmetry of the reagent molecule. The longer wavelength peak is ascribed to the functional group directly connected with the metal, the other peak to the

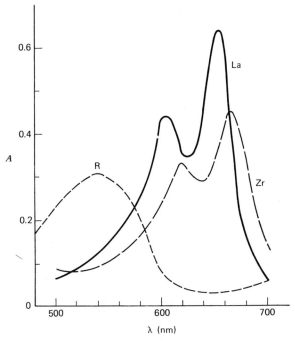

Fig. 6E-3. Absorption spectra of Zr (9 *M* HCl) and La (pH 3) Arsenazo III chelates and reagent R (9×10^{-6} *M* in 9*M* HCl). Concentration of Zr and La not specified. After S. B. Savvin, *Zh. Anal. Khim.*, **17**, 785 (1962); *Talanta*, **11**, 10 (1964).

chromophoric center (conjugated system) not directly connected to the metal, which, under the influence of the functional group changes to another ionic state.[37] It is thought that chelation may partly restore the planarity of the Arsenazo III molecule and contribute to an increase in absorbance.

Particulars concerning the extent of interferences in Arsenazo III determinations will not be entered into here.[48]

Palladiazo[49]

This is the name given to 2,7-bis(4-azophenylarsono)-1,8-dihydroxy-naphthalene-3,6-disulfonic acid. It differs from Arsenazo III only in having the arsono groups in the p,p'-positions. It is much less reactive with metals, since chelation through the arsono group is no longer possible. Apparently some reactions take place in HCl solution with Fe(III), Th, and a few other metals. Palladium slowly gives a bluish reaction product at pH 2–4. The reaction is selective. The product has been assigned the formula Pd_2L_3.[50]

B. ANALOGS OF ARSENAZO III

All the following reagents resemble Arsenazo III in having the same chromotropic acid–bisazo kernel, but the substituents on the flanking benzene rings differ. One or both $-AsO_3H_2$ groups is replaced by H, OH, PO_3H_2, or SO_3H, and additional substituents may appear in these rings.

Arsenazo M

This compound has been recommended as a reagent for the rare earth metals.[51] It lacks one of the $-AsO_3H_2$ groups of Arsenazo III and has an $-SO_3H$ group m to $-N{=}N-$ in that benzene ring. It was found to be the most sensitive of a series of 2,7-bisazo derivatives of chromotropic acid having a single arsono group. Lanthanum gives a blue-green 1:1 complex at pH 3.5 having $\epsilon_{640} = 8.6 \times 10^4$ (other rare earths similar). The sensitivity with Arsenazo III at the same pH is not much less (about 25% smaller). But Arsenazo M is the better reagent when the rare earths are to be determined in the presence of U(VI), because the optimal pH for masking the latter with H_2O_2 is 4–5, at which pH the sensitivity drops sharply with Arsenazo III, but not with Arsenazo M, as reagent.

Arsenazo M can be used for the determination of Np(IV) in 1.5–3 M HCl; a 1(Np):2 complex is formed having $\epsilon_{664} = 1.1 \times 10^5$.[52]

pK_a values (pK_1 to pK_7) of the reagent are: 0.7, 3.5, 6.5, 8.0, 9.5, 10.6, and 13.0.[53]

Arsenazo-p-NO₂

A sensitive reagent for La and other rare earths.[54]

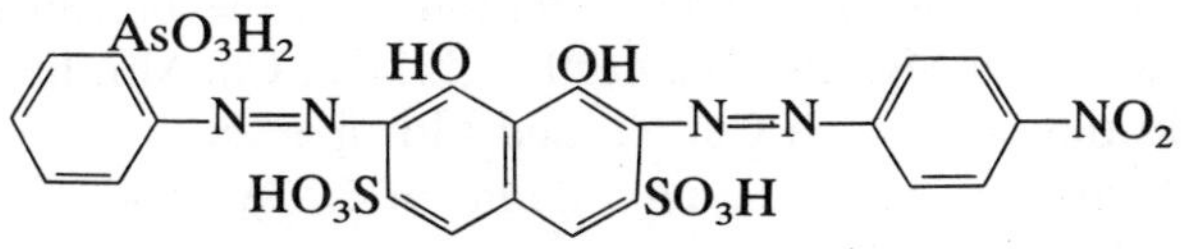

The molar absorptivities of the rare earth complexes formed at pH 3.6–3.8 decrease from 58,000 for La to 21,000 for Lu (at 660 nm). The lighter rare earth elements are believed to form more stable complexes than the heavier ones and to react at lower pH; more than one complex is formed. (Arsenazo III and other reagents of this type form rare earth metal complexes whose ϵ values do not vary much from one rare earth to another.)

Chlorophosphonazo III[55]

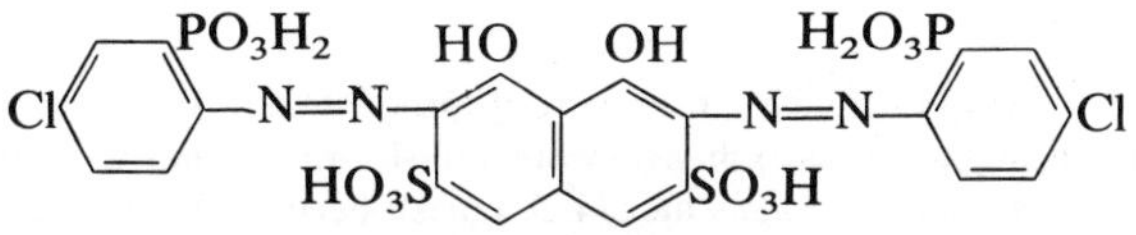

2,7-Bis-(4-chloro-2-phosphonophenylazo)-1,8-dihydroxynaphthalene-3,6-disulfonic acid

This reagent has $-PO_3H_2$ in place of the $-AsO_3H_2$ groups of Arsenazo III. pK_a values $(pK_1 \cdots pK_{10})$ are: $-1.1, -1.1, 0.6, 0.8, 1.5, 2.5, 5.5, 7.2, 12.15,$ and 15.3 at room temperature (Budesinsky, 1969).

Its solution is raspberry red in color and it forms green chelates with Zr and Th and blue with Ti and Sc:

	Acidity for Reaction	M:L in Chelate	$\epsilon_{max} \times 10^{-4}$
Th	1–2 M H⁺	1:2	4.3
Zr	2 M H⁺	1:2	3.3
Pu(IV)	1 M H⁺		3.7
Ti	pH 1–2	1:1	1.1
Sc	pH 2–3	1:1	1.25

U(VI), Y, and the rare earths react at pH ~2 (usually 1:2 species). The rare earth elements have been determined by extracting their chelates into *n*-butanol.[56] Later work[57] has shown that U(VI) can be determined in more strongly acid solutions (2.5–3.5 M HCl), in which other elements give less interference; the solution is made ~30% in isopropyl alcohol to prevent precipitate formation. Still greater sensitivity can be obtained by reducing uranium to U(IV) and extracting its complex with the reagent from ~2 M HCl into 3-methyl-1-butanol.[58]

Thorium and zirconium can be determined in the presence of 1000–5000 times as much Mg, alkaline earths, Zn, Cd, Co, Mn, Fe(II), Pb, Al, and Be; 100 times as much Cu, Ni, and Mo; five times as much Fe(III), Sc, Ta, and rare earths; an equal amount of Ce(IV) and Ti; and 0.1 as much Nb and U(VI).

Scandium can be determined in the presence of small amounts of Zr, Ti, Fe, and Al when these are masked with tartaric acid. The rare earth metals have also been determined,[59] as have Ca and Mg.[60,61]

Sulfonazo III

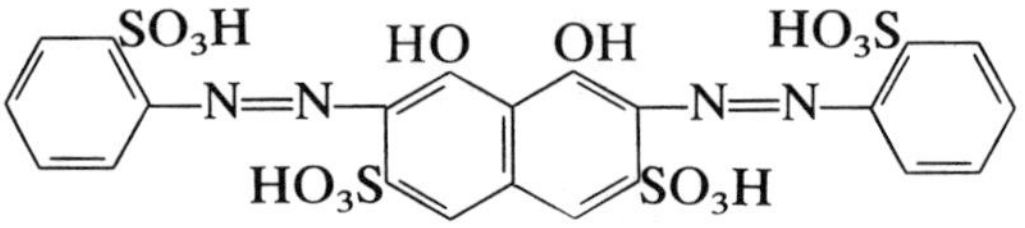

Mol. wt. 688.6; decomposes 222°. Commercial product can be purified by running aqueous solution through cation-exchange resin (H⁺-form) column a number of times. Product dried at 50° contains hydrate water (4–5 moles); becomes anhydrous when dried in vacuum over NaOH at 120° (99.7% purity by titration). $\epsilon_{570} = 37,500$. Moderately stable in solution (~1% decomposition in a week). Protect from light.

This azo compound, 2,7-bis(2'-sulfophenylazo)-1,8-dihydroxy-3,6-naphthalenedisulfonic acid, has received special attention as a reagent for Ba and Sr.[62] These metals are determined at pH ~6. By the use of EGTA (ethyleneglycol bis(β-aminoethyl ether)-N,N'-tetraacetic acid) and CyDTA (*trans*-1,2-diaminocyclohexane-N,N'-tetraacetic acid) as complexing agents, the two metals can be determined simultaneously in the presence of moderate amounts of calcium and most heavy metals. Alkali metal salts decrease the color intensity given by Ba and Sr.

Sulfochlorophenol S[63] (*Chlorosulfophenol S*) (S = symmetrical)

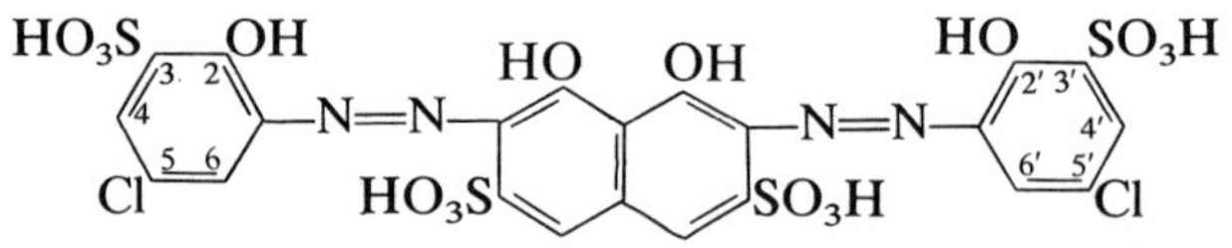

This reagent has been used to determine the elements listed in Table 6E-3. The reactions occurring in weakly acidic medium are of little interest, but those with Nb, Zr, Hf, and possibly Sc are analytically useful.

The reagent (dark red powder) is sparingly soluble in water (0.25%), easily soluble in alkaline solutions, and sparingly soluble in strongly acid

TABLE 6E-3
Reactions of Sulfochlorophenol S

Metal	λ_{max} (nm)	Optimal Acidity for Determination	$\epsilon \times 10^{-3}$
Nb	610, 645	1 M HCl	33
Zr	640	0.5 M HCl	50
Hf	640	0.5 M HCl	46
Sc	630	pH 2.2	12
Mo	600	0.25 N H$_2$SO$_4$	13
Cu	670, 630	pH 3.5	49
V(IV)	620	pH 3.5	30
V(V)	620	pH 3.5	40
Al	600 (660)	pH 3.5	34
Ga	620 (660)	pH 3.0	41
In	640	pH 3.5	40

solutions and most organic solvents. Aqueous solutions are stable for a month or more. The solid can be purified by repeated recrystallization; it should be dried at $<30°$. The tetrasodium salt contains 4 H$_2$O. pK_5 (OH of first benzene ring) = 6.6, pK_6 (OH of second benzene ring) = 8.6; the first OH of the naphthalene ring ionizes at pH >10, the second in strongly basic solution. The color of the reagent deepens when the OH of the benzene rings ionize, likewise when the naphthalene ring OH groups ionize (hypsochromic effect). The red-violet color of the reagent changes to blue when metal chelates are formed.

The alkali and alkaline earth elements and cationic organic compounds (diphenylguanidine) in relatively high concentration (0.5 M) shift the absorption curve to longer wavelengths in weakly alkaline medium. Presumably ion associates are formed.

The acidity range for the determination of niobium is 1–3 M HCl or HClO$_4$. Sulfuric acid tends to have a bleaching effect. The niobium reaction is slow. Maximum absorbance is observed after 1 hr at room temperature and after 10 min at 40°.

Sulfonitrophenol M (Nitrosulfophenol M)

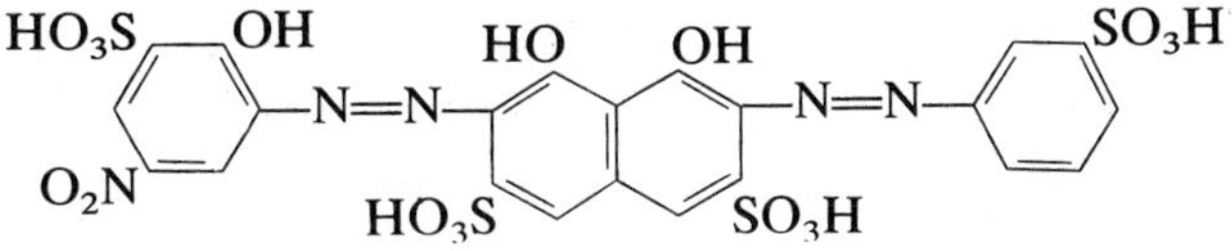

This is one of the best of a series of 2,7-bisazo derivatives of chromotropic acid tested for the determination of palladium.[64] It is sensitive

($\epsilon_{620} = 82,000$) and can be used in 3–5 N H_2SO_4 (HCl reduces color intensity). Heating (20 min at 80°) is required for full development of color. Ten milligrams of Fe, Ni, Co, and Al may be present with 20 μg of Pd without much error, but Cu and especially Zr and Mo interfere in milligram quantities. The selectivity is thus only moderate, even though the solution is highly acidic. The absorption curves of palladium complex and reagent overlap considerably, though $\Delta\lambda_{max} = 80$ nm.

This reagent is also used for the determination of Nb.[65]

C. BISAZO DERIVATIVES OF CHROMOTROPIC ACID AS REAGENTS FOR NIOBIUM AND FOR THE ALKALINE EARTH METALS

At this point it will be instructive to consider the effect of structure and the nature of substituents on the reactivity of these reagents with Nb and with the alkaline earth metals, as interpreted by Russian investigators.

Niobium.[66] Monoazo derivatives of chromotropic acid such as

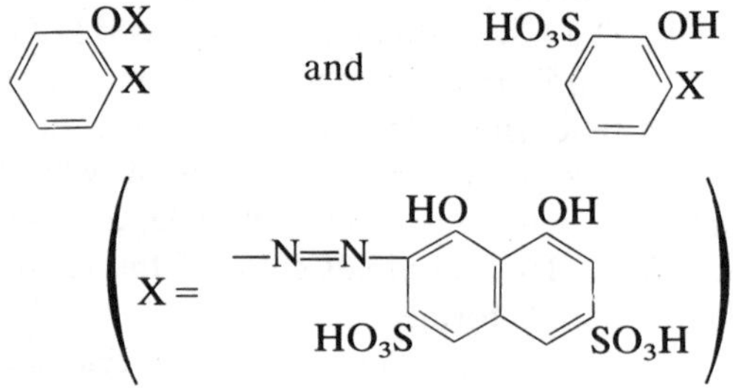

give no color with niobium, whereas the derivatives

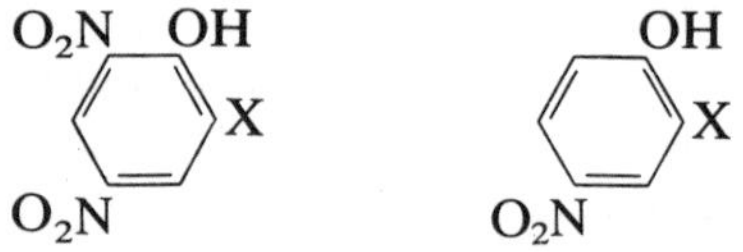

give chelates with so little color contrast that they are not useful analytically. Bisazo compounds without hydroxy groups in the benzene ring do not give noteworthy color reactions with niobium, for example,

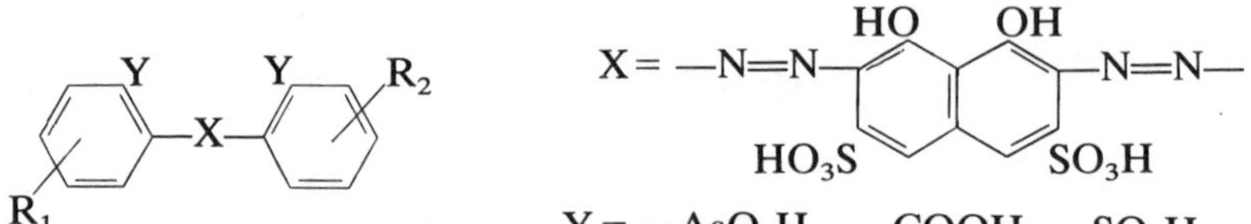

The functional group for niobium (and other metals reacting similarly) is the *o,o'*-dihydroxy grouping in 2,7-bisazo derivatives of chromotropic acid. Only one of these groups is chelatingly active. A typical useful reagent for niobium is Sulfochlorophenol S. It has been found that almost all the best reagents for niobium[67] have two electron-withdrawing substituents *ortho* and *para* to the reacting OH group; introduction of donor groups into the reacting benzene ring impairs the chelating properties of the reagent. Although the second diazo component is not directly involved in metal chelation, it affects the reactivity of the functional group and the absorption spectrum of the metal chelate. The reagent includes two partially conjugated systems, each of which includes the naphthalene nucleus. The effect of the second component is determined by its spatial location, electron-withdrawing properties, and the possibility of internal hydrogen-bonding and of tautomerism. The best reagents usually have a symmetrical structure or have an anthranilic, metanilic, or ε-amino acid as the second coupling component, which do not have an OH group in the *o*-position. When the second component contains an *o*-hydroxy group, the maximum in the absorption curve of the reagent is shifted to longer wavelengths and there is less contrast in color with that of the niobium chelate. The unfavorable effect of the second OH group is attributed to hydrogen bonding:

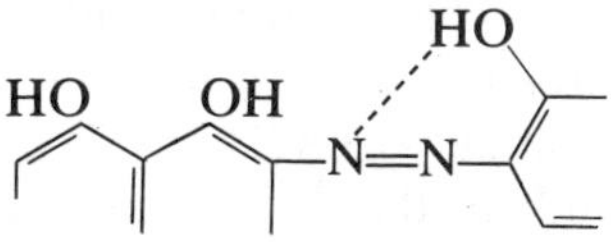

Symmetrical reagents having an OH group in the *o*-position in the second coupled component with an SO₃H– or NO₂– group ortho to OH have intermediate properties. It is postulated that hydrogen bonds may then be formed between the hydroxy group and the O atom in these substituents

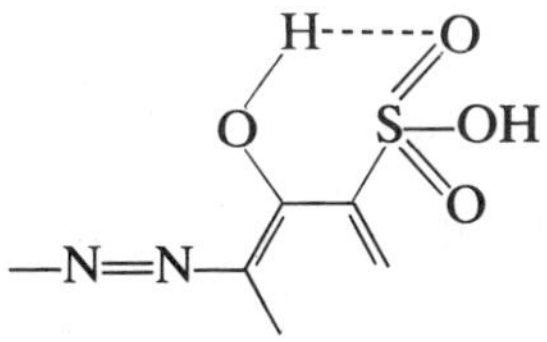

and this structure is not as unfavorable to contrast as hydrogen bonding to N. Reagents of this type have reasonably good contrast ($\Delta\lambda = 50$–80 nm, compared with 20–50 nm for reagents with hydrogen bonding to N and as high as 90–95 nm for reagents without a second OH).

Most of the 2,7-bisazo derivatives of chromotropic acid form monomeric 1:1 chelates with niobium and other metals reacting in mineral acid solution. Picramine S, which is the 3,3',5,5'-tetranitro reagent, forms 1:2 (metal:L) complexes.[68] In the formation of a complex with one of the functional groups, a positive charge is induced on the azo group of the second functional group, and this hinders formation of a second nitrogen–metal bond. Some transition metals (Cu, V, Mo), reacting in weakly acidic solution, can form 2:1 (metal:L) chelates. The presence of a labile H in the OH group on the benzene ring seems to be necessary for the reaction of the o,o'-dihydroxyazo grouping with niobium. It is believed that the chelate formed is of the type

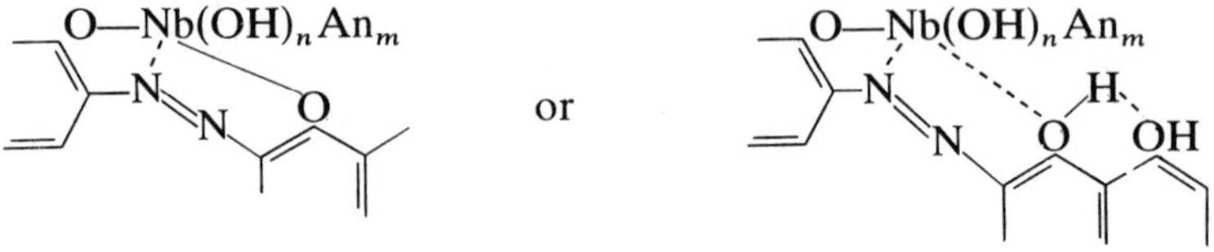

                         or

where An is the anion of a mineral acid or an organic acid (tartrate, citrate). Niobium exists as hydroxo species in 1–3 M, and even stronger, hydrochloric acid.

Tantalum reacts less sensitively than Nb with these reagents so that Nb can usually be determined in the presence of ~5 times as much Ta.

Alkaline Earth Metals. A study has been made of the reactions of the alkaline earths with bisazo chromotropic acid derivatives which do not have salt-forming substituents in the o-position to the azo group in the benzene ring, or which contain –COOH in this position.[69] Three types of color reaction were found. In the first, complexes of the alkaline earths formed by reaction with these reagents show two absorption maxima, the second at 640–650 nm; $\epsilon_{max} = 4$–7×10^3. Maximal color intensity is usually obtained at pH 6–7, and is greater in aqueous–organic solutions than in water. It is believed that the chelate formed is analogous to that obtained when a salt-forming group such as $-AsO_3H_2$ and $-PO_3H_2$ occupies the o-position, namely

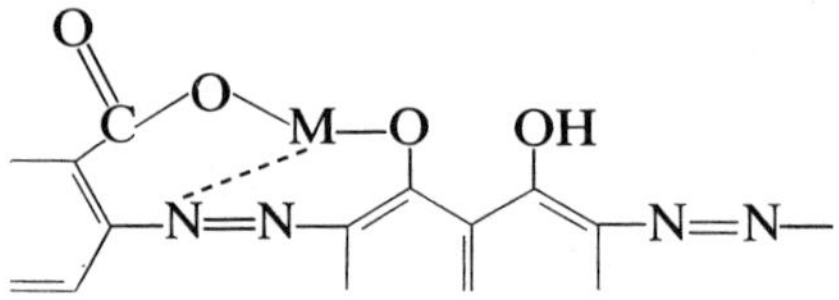

or, with no substituent in this position

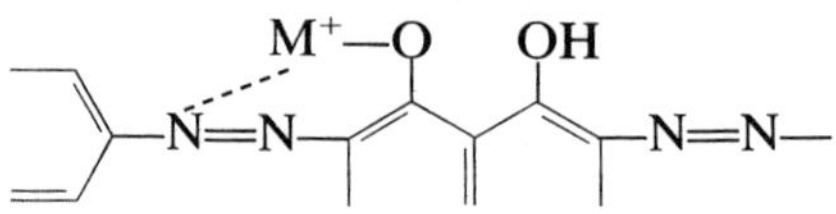

In the second type of reaction, occurring in alkaline solutions (pH 10–12, ionization of OH groups in the naphthalene ring), the product is

probably

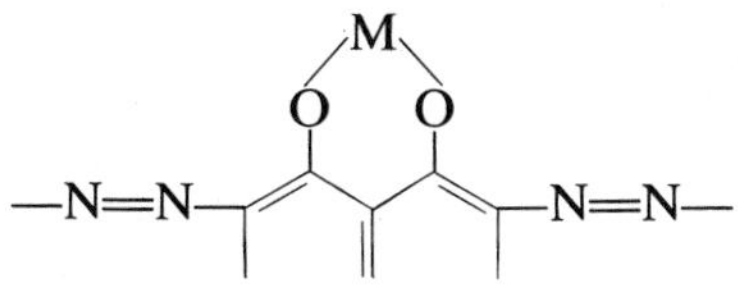

There is only one absorption maximum, separated by 20–40 nm from the maximum of the reagent, and ϵ_{max} is not greater than 4×10^4.

The third type of reaction occurs in solutions of low pH (2.5–3 and a little higher); the chelate formed has exceptionally high ϵ values, up to 1.8×10^5. Two maxima are observed in the absorption curves (the longer wavelength maximum lies at 690–710 nm). λ_{max} in aqueous solutions is greater than in water–organic solutions. The following structure is suggested for these chelates:

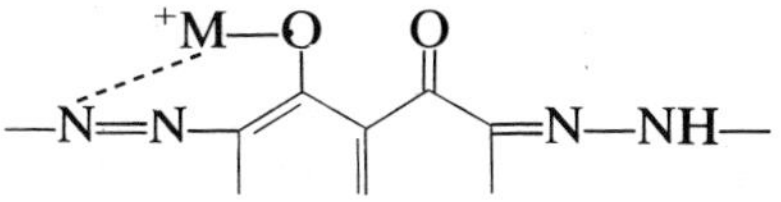

The reagent reacts in the quinonehydrazone tautomeric form. The presence of NO_2 groups in the p-position to the azo group facilitates this rearrangement. The OH group ionizes more readily, and reaction with the alkaline earths can occur in weakly acidic solutions.

Carboxynitrazo, incorporating these desiderata,

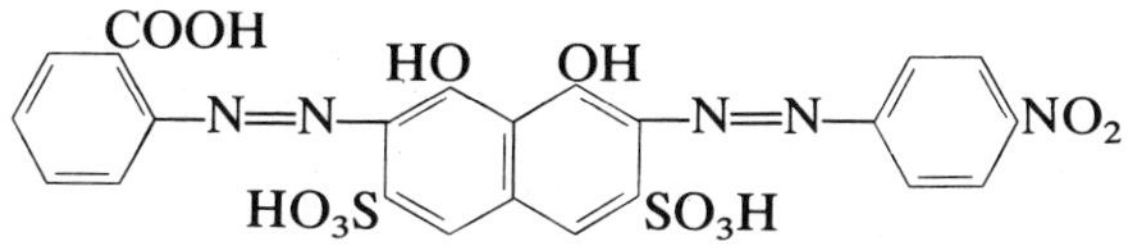

has been used for the determination of Ba and Sr in ethanolic solution at pH 4–4.5.[70] The molar absorptivity is high, $\epsilon_{701} = 142,000$ for Ba and 140,000 for Sr. Similar sensitivity is obtained in determination of the rare earth metals in weakly acidic solution.[71] La, Ce, Pr, Nd, Sm, Eu, and Gd are reported to react (pH 1.8–4.8), but not Tb, Dy, Ho, Er, Yb, or Y. Lanthanum can be determined in the presence of 40 times as much Yb.

D. SOME OTHER ARYL BISAZO REAGENTS

Arsenazo II

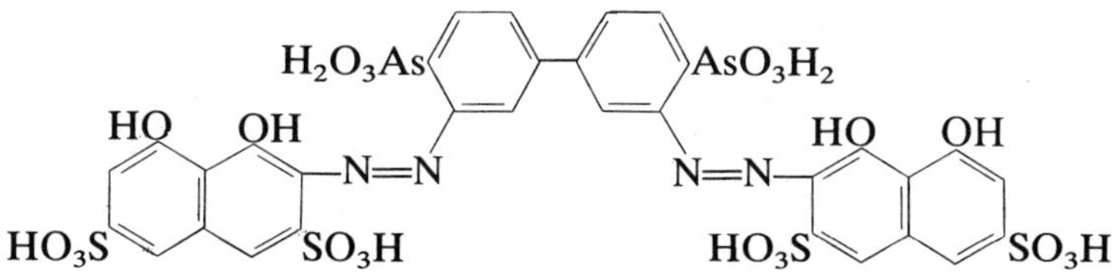

is less useful than Arsenazo III. It has been used to determine Th, U, and Sc. The selectivities and sensitivities of these determinations are much the same as with Arsenazo I.

Stilbazokhimdu

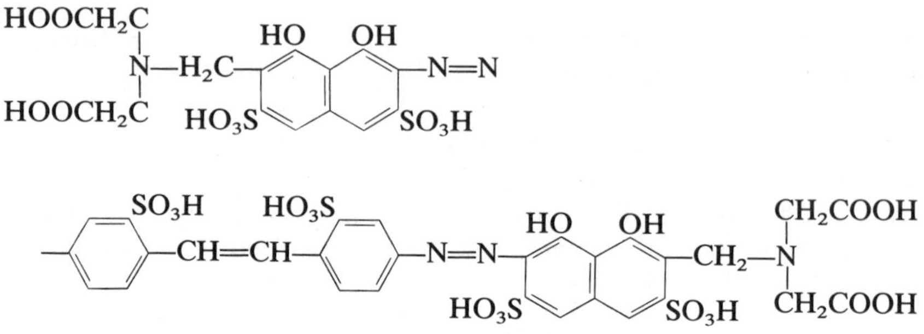

is a bisazo compound based on chromotropic acid that contains iminodi-acetate groups.[72] Compared to its analog Stilbazochrome, which does not have the iminodiacetate groups, it reacts with the yttrium subgroup of the rare earth elements to form 1:1 complexes in more acidic solutions (pH 3 versus pH 6). The analytical functional group consists of the peri OH groups in the naphthalene ring, not the iminodiacetate group. The latter group affects the analytical properties of the reagent in much the same way as do negative substituents such as $-Br$ and $-NO_2$. It may be supposed that small amounts of metals that might otherwise react with chelating $-OH$ and $-N{=}N-$ may be bound instead by the iminodiacetate group.

Fast Sulfon Black F

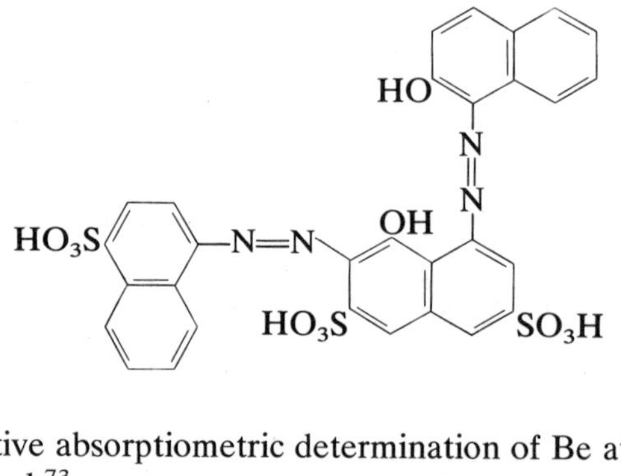

provides a sensitive absorptiometric determination of Be at pH 11. A 1:1 complex is formed.[73]

For other bisazo reagents (containing heterocyclic components) see Section III.C of this chapter.

III. HETEROCYCLIC AZO REAGENTS

A. PYRIDYLAZO

The most widely used reagents of this class are 1-(2-pyridylazo)-2-naphthol (commonly designated PAN or β-PAN) and 4-(2-pyridylazo)-resorcinol (PAR).[74] The former was first used by Cheng and Bray in 1955 as a reagent for many metals and as an indicator in EDTA titrations. The latter reagent was used as indicator in EDTA titrations by Wehber in 1957 and as photometric reagent for Co, Pb, and U(VI) by Pollard, Hanson, and Geary in 1959.

Both reagents are unselective and each reacts with some 40 or 50 metals. However, the color reactions are very sensitive. PAR has the possible advantage over PAN that it is readily soluble in water as the sodium salt and forms more strongly colored chelates, at least with some metals.

1. 1-(2-Pyridylazo)-2-naphthol (PAN)

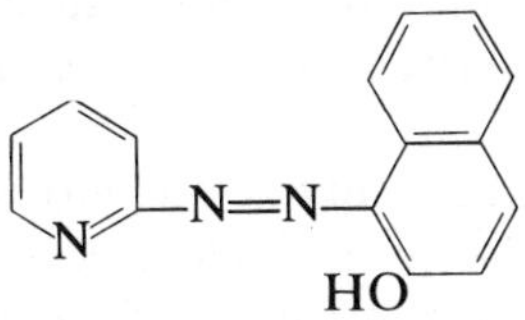

Orange-red solid, m. p. 137°. Mol. wt. 249.3. Essentially insoluble in water and in alkaline solutions in which $[OH^-] < 0.01$. Soluble in alcohols, $CHCl_3$, CCl_4 ($\sim 0.01\ M$), etc. (yellow color). Can be purified by recrystallization from alcohol. Commercial products often impure. Solutions are stable.

In acid solutions the pyridine nitrogen is protonated and in basic solutions the H of the OH group is ionized:

$$H_2L^+ \; \rightleftharpoons \; HL \; \rightleftharpoons \; L^-$$

yellow-green	yellow	red
pH < 2.5	pH > 2.5	pH > 12
	−12	

The absorption peaks of the three forms lie at 425, 470, and 495 nm. pK_1 ($= pK_{NH}$) $= 2.9$, $pK_2 = 11.5$ (1.9 and 12.2 in 20% dioxane, $\mu = 0.08$, 25°).[75] PAN dissolves in concentrated sulfuric acid to give a red-violet solution.

The metal chelates of PAN have the metal atom (or oxo- or hydroxo-metal cation) bonded to O of the OH group (H replaced), to pyridine N and to azo N as illustrated by the following structures:

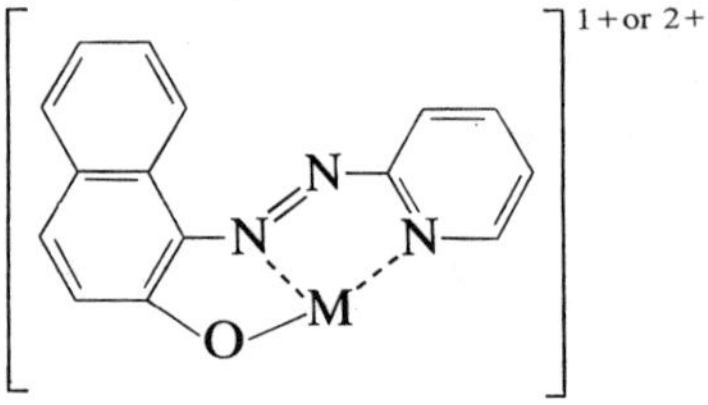

M(II or III) 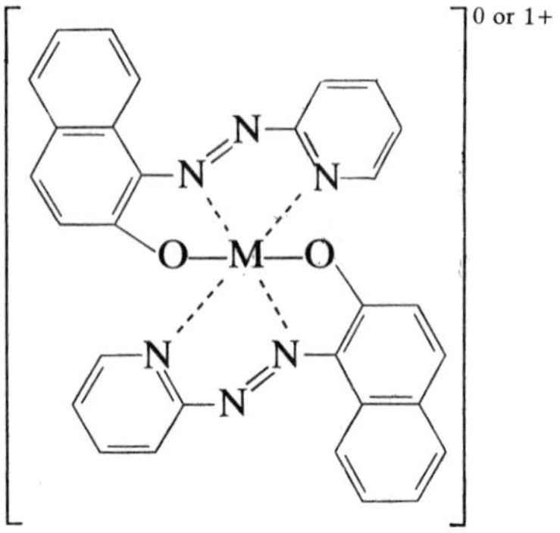   M(II or III) (M may be $M(OH)_2^+$, etc.)

(Other coordinating groups that may be required to fill the coordination sphere are not shown.) Structures other than the above are possible. Conceivably, PAN (and similar reagents) could react as a hydrazone tautomer.

The composition of species extracted into immiscible solvents is not known for all metals. The normal chelate is not always the extracted species. For example, a mixed PAN–acetate complex of indium is extracted into chloroform.[76] The extraction of indium is poor from a biphthalate or citrate buffer, but satisfactory from an acetate buffer. Acetate can be replaced with chloride or perchlorate in high concentration. At pH 10, Cu(II) is extracted as CuL_2. At pH 8, violet CuL^+ is formed which is not extracted unless Cl^-, Br^-, SCN^-, or acetate is present.[77]

Table 6E-4 lists the properties of some metal–PAN chelate species.

The uncharged chelates are extremely slightly soluble in water,[78] but can be extracted into various immiscible organic solvents; chloroform is usually preferred. PAN (as HL) has a yellow color in chloroform ($\lambda_{max} \sim$ 470 nm, little absorption >560 nm); $P_{CHCl_3/H_2O} = 10^{5.2}$. Chelates of the following metals have a red color in chloroform (λ_{max} mostly 540–570 nm): Cd, Zn, Hg(II), Bi, Cu, La, In, Ga, Fe(III, brown-red), Sn(II), Pb, U(VI), Ni, Th, Ir; those of Pd(II), Rh(III), and Co(III) are green; those of V(V) and Tl(III) are violet. Immiscible solvent extraction is usually preferred for the photometric determination of metals.[79] Occasionally, water-miscible organic solvent solutions are used. The alkali metals, Be, As, Ge, Se, and Te do not give color reactions with PAN.

TABLE 6E-4
Properties of Some Metal–PAN Complexes[a]

	M:PAN	λ_{max} (nm)	$\epsilon \times 10^{-3}$
Bi	1:1	580	21
Cd	1:2	~550	50
Co(II)	1:2	525	~30
Co(III)	1:2	580	23.1
		630	19.6
Cu(II)	1:2 (CHCl$_3$)	~550	45
	1:1 (H$_2$O)	~550	~22
Fe(III)		775	16
Ga	1:1, 1:2	550	21.7
Hg(II)		560	
In	1:2	560	19.6
Ir	1:2	550	10.3
Mn(II)	1:2	550	40
Ni	1:2	575	51
Os		~560	11
Pd	1:1	620 (675)	16
		626, 678	14.1 (678 nm)
Pt(II)	1:1	460, 690	4.9
Rare earths	1:2	530, 560	57–80
Rh	1:2	598	19.5
Tl(III)	1:1 (H$_2$O)	560	21.7
U(VI)		570	23 (dichlorobenzene)
V(V)	1:1	615	16.9
Y		560	
Zn	1:1	515, 514	22.6, 20.8
		550	
Zr		555	32

[a] From Busev and Ivanov, *loc. cit.*; Anderson and Nickless, *loc. cit.*; W. Berger and H. Elvers, *Z. Anal. Chem.*, **171**, 185 (1959), and other authors. λ_{max} and ϵ usually for CHCl$_3$ except ethyl ether for rare earths. Results of different workers often not in good agreement (differences in conditions can result in extraction of different species). Formation of Rh and Ir complexes requires heating at the boiling point for an hour or more at pH 4–5; J. R. Stokely and W. D. Jacobs, *Anal. Chem.*, **35**, 149 (1963).

The extraction constants

$$K_{ext} = \frac{[ML_n]_{org}[H^+]^n}{[M^{n+}][HL]^n_{org}}$$

of PAN chelates have comparatively small values as illustrated by K_{ext} for MnL$_2$ and ZnL$_2$ with CCl$_4$ as the extractant, namely 3×10^{-11} and 2×10^{-5} (calculated from extraction data of Betteridge, Fernando, and

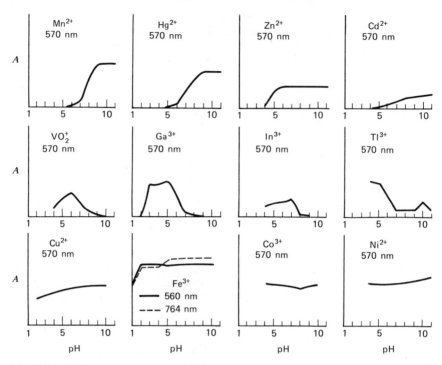

Fig. 6E-4. Effect of pH on extraction of PAN chelates into chloroform according to R. Püschel. A in arbitrary units. R. Püschel, *Z. Anal. Chem.*, **221**, 138 (1966).

Freiser, reference in Table 6E-5). With $CHCl_3$ as extractant, K_{ext} for Cd is $\sim 1 \times 10^{-9}$ (Galik). The partition constant

$$[ZnL_2]_{CCl_4}/[ZnL_2]_{H_2O}$$

has a value of $\sim 1 \times 10^4$.

Most metals require a feebly acidic or basic solution for extraction and careful adjustment of pH is needed (Fig. 6E-4). Variation of pH is of limited value in increasing the selectivity of PAN reactions. Masking, as with EDTA, is sometimes useful. For example, if a chloroform solution of metal–PAN chelates is shaken with an EDTA solution at pH 5, all are decomposed except those of Fe, Co, Ni, V (and Pd); V can be complexed with H_2O_2, leaving Fe, Co, and Ni to be determined simultaneously by absorbance measurements at 565, 628, and 764 nm.[80] Zinc can be determined in the presence of Cd, Cu, Hg, Pb, and Ag by masking these metals with I^-; Ni and Co can be masked with cyanide.[81]

The extraction of metals with a solution of PAN in chloroform and similar solvents may be slow. Nickel has been reported to require 40-hr shaking. It may be preferable to add the reagent in a small volume of alcohol to the aqueous solution and allow the chelate to form over a period of some minutes and then extract with chloroform.[82] Püschel and coworkers heat the solution to 80° after adding PAN and buffer (pH 5), cool, and then extract Fe, Co, and Ni.

Aluminum has been determined by the fluorescence of its PAN complex in ethanolic solution.[83]

Determination of Co in aqueous solution by solubilization of the Co(III)–PAN complex with a neutral surfactant (Triton X-100) in combination with Na dodecylbenzene sulfonate has been proposed.[84]

The formation constants of some metal–PAN species are listed in Table 6E-5.

TABLE 6E-5
Stability Constants of Some PAN Complexes[a]

Ion	$\log K_1$	$\log K_2$	$\log K_1 K_2$	Reference
Co(III)	>12			Corsini, Yih, Fernando, and Freiser, *Anal. Chem.*, **34**, 1090 (1962).
Cu(II)	12.6			Betteridge, Fernando, and Freiser, *Anal. Chem.*, **35**, 294 (1963).
	15.5	8.4	23.9	Wada and Nakagawa, *J. Chem. Soc. Japan Pure Chem. Sec.*, **85**, 549 (1964).
Mn(II)	8.5	7.9	16.4	Corsini et al. (1962).
Ni(II)	12.7	12.6	25.3	Corsini et al. (1962).
	14.0	13.5	27.5	Nakagawa and Wada, *J. Chem. Soc. Japan Pure Chem. Sec.*, **84**, 639 (1963).
Zn(II)	11.2	10.5	21.7	Corsini et al. (1962).
			21.8 (H_2O)	Betteridge et al. (1963).
	10.9	10.0	20.9	Nakagawa and Wada (1963).

[a] In 50% dioxane for the most part; $\log K_1$ of Cu(II) by Wada and Nakagawa in 5% dioxane.

4-(2-Pyridylazo)-1-naphthol (p-PAN)

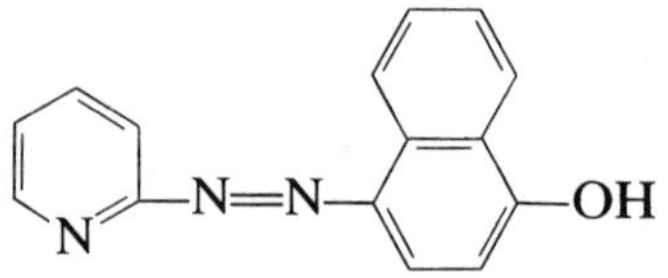

$pK_{NH} = 3.0$, $pK_{OH} = 9.1–9.5$ (H_2O)

has been synthesized and its metal complexing properties examined.[85] Its reactions with most heavy metals (colored precipitates) are apparently more sensitive than the reactions of the o-isomer. In faintly acidic solution, the composition of the chelates of Cu(II), Zn, and Ni is given by ratio 1 M : 2 reagent. The reagent is believed to react in the tautomeric form c:

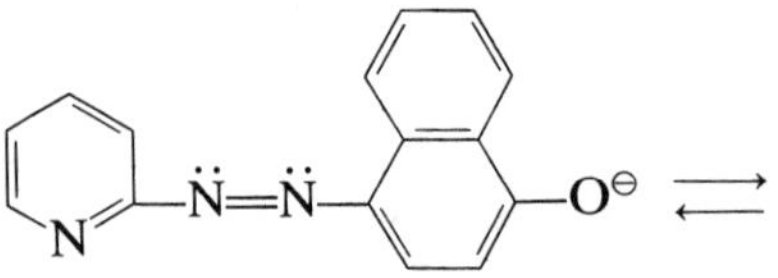

a

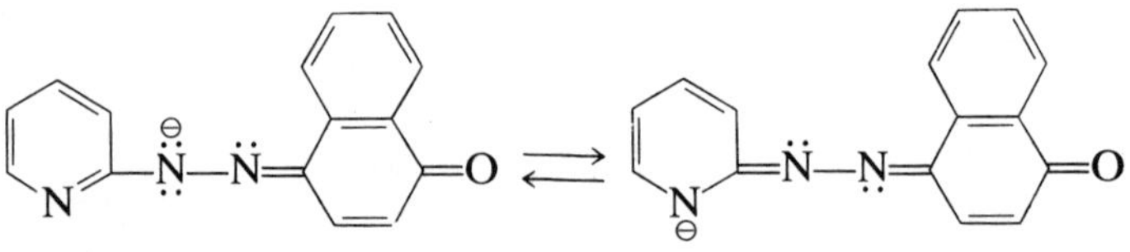

b   c

Substituted pyridylazonaphthols. The compound 1-[(5-chloro-2-pyridyl)azo]-2-naphthol has been synthesized and its photometric metal reactions studied.[86] It gives more sensitive reactions than PAN; for example, ϵ_{569} for the Zn chelate of the chloro-substituted reagent is 71,000, ϵ_{550} of the Zn–PAN chelate is 56,000. Chloroform, benzene, and other solvents are suitable for the extraction of the chelates. The reactivity of tri- and quadrivalent metals is reported to be decreased by the introduction of Cl. As with the parent compound, some increase in selectivity is obtained by the use of such complexing agents as EDTA, cyanide and citrate. For the most part, the chelates are of the type ML and ML_2. In acidic solution, Cu(II) forms CuLX (X = Cl), extractable into $CHCl_3$. In basic solution CuL_2 is formed.

1-[(5-Bromo-2-pyridyl)azo]-2-naphthol has also been prepared.[87] It gives more sensitive reactions than PAN, but otherwise seems to have no noteworthy analytical properties. pK_a for H_2L^+, OH, and H_3L^{2+} in 1:1 dioxane–H_2O are ~1, 11.8, and ~4. ϵ of chelates does not reach 1×10^5.

2-[(5-Nitro-2-pyridylazo)]-1-naphthol gives more sensitive reactions than α-PAN,[88] thus $\mathscr{S}_{598} = 0.00074\ \mu g/cm^2$ for Cu $(CHCl_3)$.[89]

2. 4-(2-Pyridylazo)-resorcinol (PAR)

The reagent on the market is usually in the form of its sodium salt (+1 or 2 H_2O), an orange-red solid. Mol. wt. (+2 H_2O). 295.2. Can be purified by recrystallization from 50% ethanol.[90] Aqueous solutions are said to be stable for a year or more (but see note 96).

Depending on the acidity, PAR can exist as six species in solution[91]:

$$H_5L^{3+} \rightleftharpoons H_4L^{2+} \rightleftharpoons H_3L^+ \rightleftharpoons H_2L \rightleftharpoons HL^- \rightleftharpoons L^{2-}$$

90% H_2SO_4 50% H_2SO_4 pH <2   pH 2.1–4.2   pH 4.2–7   pH >10.5

The ionization constants for the last three stages are[92]:

$$H_3L^+ \rightleftharpoons H_2L + H^+ \qquad K_1 = 10^{-3.1}$$
$$H_2L \rightleftharpoons HL^- + H^+ \qquad K_2 = 10^{-5.6}\ (\text{ionization of } p\ OH)$$
$$HL^- \rightleftharpoons L^{2-} + H^+ \qquad K_3 = 10^{-11.9}\ (\text{ionization of } o\ OH)$$

The absorption maxima of the six forms, H_5L^{3+} to L^{2-}, lie at 433, 390, 395, 385, 413, and 490 nm. In weakly acidic or basic solutions, the reagent has an orange color.

The chelates formed by PAR are similar in structure to PAN chelates, the H of the *ortho* OH group being replaced by an equivalent of metal, with bonding of the metal to pyridine N and azo N (two five-membered rings). Charged (+ or −) as well as uncharged complex species can be formed, and the composition can vary with the concentrations of HL^- and L^{2-}, that is, with the pH. The reacting ratios of metal and PAR may be known, but the composition of the product is not always clear. Evidence has been presented to show that Ni combines with PAR in the ratio 1:2 in both acidic and basic solutions, but different species are present.[93] In slightly acidic solutions (pH 3.3), $Ni(HL)_2$ is present, having a red color ($\epsilon_{520} = 37,200$). When the solution is made basic, the color changes to orange ($\epsilon_{496} = 79,400$, pH 8). The color change is attributed to the formation of NiL_2^{2-}. Cobalt forms $Co(HL)_2$ in acidic solution, and the monoprotonated species $Co(HL)L^-$ in basic solution (both forms red).[93] Manganese(II) is said to form MnL_2^{2-} in slightly acidic as well as basic solutions.[94] Zinc forms ZnL_2^{2-} in alkaline solution. Some metals form complexes whose M:PAR ratio depends on the pH (see Table 6E-6). Thallium(III) is reported to form a 1:1 chelate at pH 1–2 and a mixture of 1:1 and 1:2 chelates at higher pH values.[95] One of the chelates formed by Pd(II) does not involve replacement of a hydrogen in PAR (note 99).

Absorptiometric determinations of most metals are carried out in weakly acidic or weakly basic solutions[96]:

	Approx. pH
Bi	1
Cd	10
Co	7–8
Cu(II)	2; 7; 10
Ga	4; 7
In	7
Fe(III)	9
Mn	10
Ni	9
Nb	5–6
Pb	10
Pd	6; H_2SO_4 soln.
Pt(II)	3
Rare earths	6–7
Rh	5
Sc	3–5
Ta	5
Th	7
U(VI)	7–8
V(V)	6.5
Zn	8

Zr, Ti, and Tl(III) (as well as Bi and Pd) react in quite strongly acid solution (pH ~1). PAR gives no color reactions with As, Sb, Mo,[97] W, and Be. It is oxidized by Fe(III) (brown-colored solution or precipitate). The 1 metal:2 ligand chelates of many divalent metals have been obtained by precipitation from aqueous solution.

Most PAR chelates are red or red-violet in color.[98] Palladium chelates are green in acid and red in neutral and basic solutions.[99] Colorimetric determinations can be carried out in aqueous solution.[100] Extraction is rarely applied (ethyl acetate for the palladium chelate, isoamyl alcohol for the Hf complex), except when a cationic reagent such as a quaternary ammonium salt is used to form an ion associate with an anionic metal–PAR complex, which can be extracted into $CHCl_3$. Apparently the p-OH group accounts for the slight solubility of uncharged PAR chelates (and reagent) in aqueous solution, which is greater than that of the chelates of

TABLE 6E-6
Properties of Some Metal–PAR Species[a]

	M:PAR	λ_{max} (nm)	$\epsilon \times 10^{-3}$
Au(III)	1:1	540	8.3 (*tert*-butanol + xylene)
Bi	1:1	515	10.7
Cd	1:2	495	57.8
Co(III?)	1:2	510	5.5
Cu(II)[b]	1:1 (pH 2.3–5)	522	12.1
	1:2 (pH > 5)	505–510	58.9
Ga	1:1 (pH 1.5–3)	490–95	21.2
	1:2 (pH 3–5)	500–505	9.9
Hf	1:4 (pH 2.5)	510	37.5 (isoamyl alcohol)
In	1:1	500–510	32.8
Mn	1:2	496	86.5
Nb	1:1 (0.1–0.2 N H$_2$SO$_4$)	530	18
	1:1 (pH ~6)	555	38.7, 31.2
Pb	1:1	512	10.8
	1:2	522	50.2
Pd	1:1 (H$_2$SO$_4$ solution)	440	18.4
Pt(II)	1:1	450, 660	22.9
Rare earths	1:2	515	~16 to 50
Sc	1:1 (pH 2)	505 (515)	14.7 (22.1)
Ta	1:1	515	20.4
Th	1:4	500	38.9
Tl(III)[c]	1:1	520	18–19.4
U(VI)	1:1	510 (530)	38.5
V(V)	1:1	~550	
Zn		495	81
Zr	1:1 (pH 4)	535	21

[a] Condensed from Busev and Ivanov, *loc. cit.*, with later data from other authors. Solvent: H$_2$O.
[b] O. A. Tataev, S. A. Akhmedov, and Kh. A. Akhmedova, *Zh. Anal. Khim.*, **24**, 834 (1969), report formation of a 1:1 complex at pH 1.5–2.5, $\epsilon_{540} = 27,000$.
[c] As shown by E. A. Biryuk and R. V. Ravitskaya, *Zh. Anal. Khim.*, **26**, 1767 (1971), Tl(III) forms Tl(OH)$_2$L (HL = PAR and PAN).

PAN. Although very sensitive, PAR reactions in weakly acidic or weakly basic solutions have the drawback of poor selectivity. Heavy reliance must be placed on masking and separations.[101] The reactions taking place in dilute mineral acid solution are really of greater interest, but determinations in such medium have not received much attention.

Vanadium(V) forms the anionic species VO$_2$L$^-$ with PAR at pH 4–7, which is extractable into CHCl$_3$ with (C$_6$H$_5$)$_4$As$^+$ as cationic partner.[102] In the presence of antipyrine and the anion of a suitable monobasic acid, the In–PAR complex is extracted into $CHCl_3$[103]:

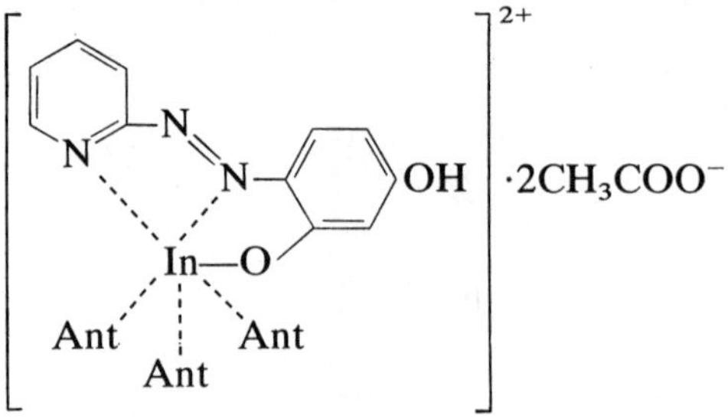

An analogous reagent, 5-(2-pyridylazo)-2-monoethylamino-p-cresol, gives the following complex with Sb(III) in acid iodide solution (extractable into benzene)[104]:

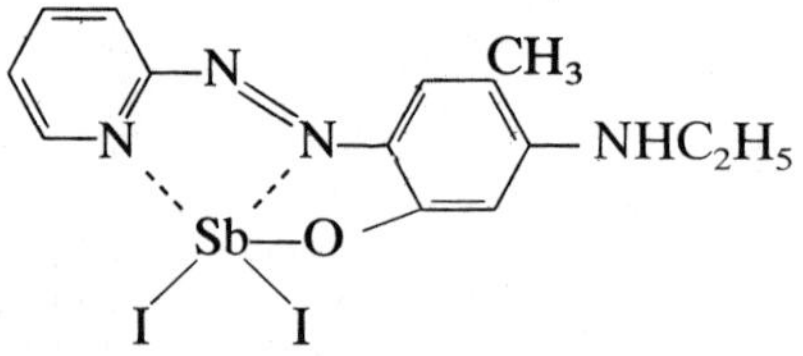

3. Other Pyridylazo Reagents

Many other reagents of the type

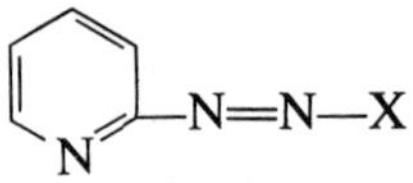

have been prepared, where X is phenol,[105] orcinol, cresol, dimethylaniline, 2,3-dihydroxynaphthalene-6-sulfonic acid, 8-hydroxyquinoline, and the like.[106] In general, they react with metals similarly to PAN and PAR. The sensitivities of their metal reactions may be higher, but otherwise they do not seem to have any noteworthy features compared to PAN and PAR. In the following paragraphs an account is given of some recent studies of these (and similar) reagents that have general interest.

A series of azo compounds containing the m-dimethylaminophenol group have been synthesized and examined as photometric reagents[107]:

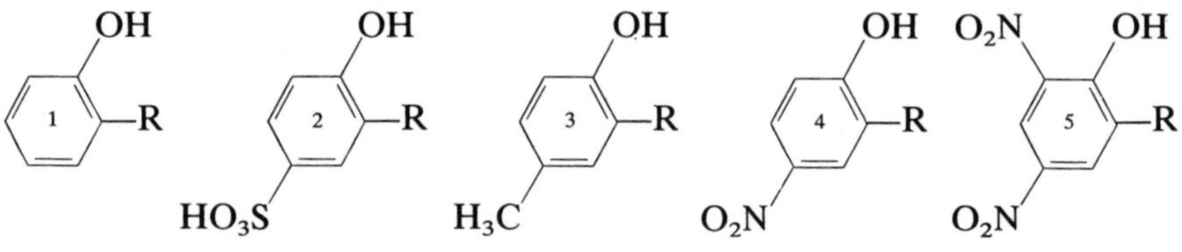

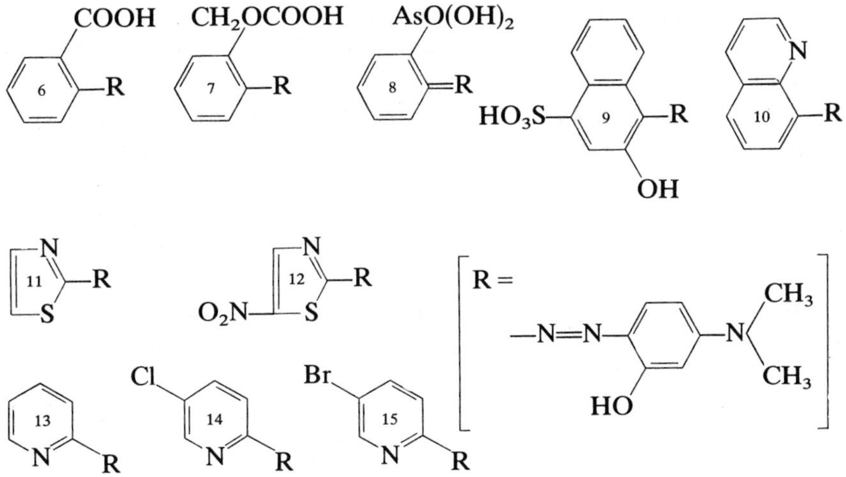

COOH   CH$_2$OCOOH   AsO(OH)$_2$

6 —R   7 —R   8 =R   HO$_3$S— 9 —R   10 —R

OH

11 —R   12 —R   O$_2$N—S

$$R = \quad -N=N-\!\!\!\!\!\!<\!\!\!\!\!\!>\!\!\!-N\!\!\begin{array}{c}CH_3\\CH_3\end{array}$$

HO

Cl   13   14   Br   15

N R

Numbers 6, 7, and 8 give no visible reaction with metals. The others react with Cd, Co, Cu, Fe, Mn, Ni, Pb (except numbers 9, 10, 11), and Zn; some react with Al, La, and Mg. The heterocyclic azo derivatives are the most suitable chromogenic reagents, the 2-pyridylazo compounds being the best. The latter give $\mathscr{S}$ values in the range 0.0005–0.001 μg/cm^2 for Co, Ni, Cu, and Zn. 2-(2-Pyridylazo)-5-diethylaminophenol is about 70% more sensitive than PAR for the determination of U(VI).[108]

2-[2-(5-Bromopyridyl)azo]-5-dimethylaminophenol (no. 15) is a good example of these pyridylazo reagents. Depending on the acidity, it exists in various forms:

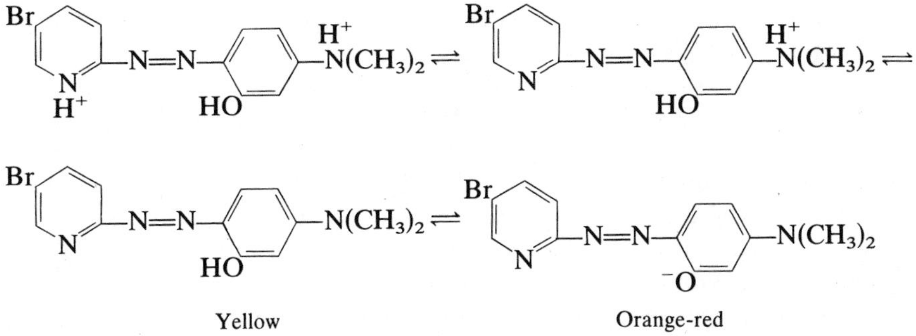

Yellow   Orange-red

The uncharged form is sparingly soluble in water but dissolves in most organic solvents. In 40% (v/v) methanol, $pK_{a1} < 1$, $pK_{a2} = 3.4$, $pK_{a3} = 11.6$ (25°, $\mu = 0.1$). Zinc, Ni, and Co form 1:2 (reagent) chelates, Cu both 1:1 and 1:2 chelates; they are soluble in 50% ethanol. Chelate formation is essentially complete above the following pH values: Co, 3.0; Ni, 4.5; Zn, 7.5; Cu, 2.0 (1:1), 9.0 (1:2). Optimal λ varies, with the metal, from ~550 to ~590 nm, in which range there is little absorption by the reagent. The chelates are reddish purple. The reagent is unselective but the high sensitivity is noteworthy ($\mathscr{S}$ varies from 0.00046 μg/cm^2 for Ni to 0.00067 for Co).

The structure of these chelates must be analogous to those formed by PAR. The greater sensitivity is attributable to the presence of the $-N(CH_3)_2$ group.

Substituted pyridylazodiaminobenzenes have been investigated as reagents for Co.[109] The sensitivities given by the parent compound 4-(2-pyridylazo)-1,3-diaminobenzene (PADAB)

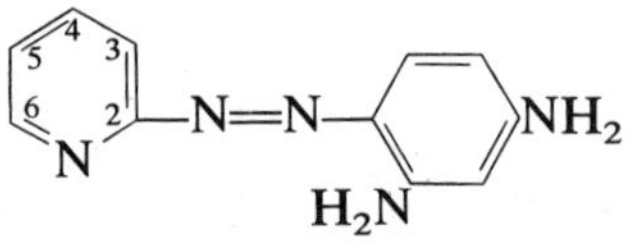

and the halogen-substituted (pyridyl ring) derivatives are

Reagent	$\mathscr{S}$ (μg Co/cm^2)	λ_{max} (nm)
PADAB	0.00059	565
5-Cl-PADAB	0.00052	570
5-Br-PADAB	0.00051	575
5-I-PADAB	0.00050	580
3,5-Cl-PADAB	0.00050	585
3,5-Br-PADAB	0.00048	590

It is seen that introduction of halogen increases the sensitivity slightly, but there is little difference between Cl, Br, and I. Nor does substitution by two atoms of Cl or Br appreciably increase the effect of one. All these compounds react with Fe(II and III), Cu(I and II), Hg(II), Ni, and Zn, as well as Co, to give colored chelates at pH 4–7. Most of the other common metals (including Be, Al, Pb, Sn, Cr, Mn, Zr, Ti, V, Th, La, Mo, and U) give no color reaction at pH 4–7 or 8–10. The Co species formed in weakly acidic solution changes into another (identity not stated) on acidification with HCl to 2–5 M. In a later paper,[110] acidification is shown to protonate the $-NH_2$ groups. It appears that these reagents function as tridentate ligands, forming two five-membered rings through coordination of Co to the o-amino group, heterocyclic N and azo group. The N of the pyridine ring is essential for the formation of Co chelates. (Benzeneazo-m-phenylenediamine does not react with Co, but pyridine-2-azo-p-dimethylaniline does.) The Co chelates of PADAB and its halogen derivatives are extractable as $CoL_2 \cdot X_2$ (X = univalent anion) into 3-methyl-1-butanol and TBP above pH 5. Cobalt is back-extracted into an acid aqueous solution as $CoL_2H_2^{2+}$.

4-[(5-Chloro-2-pyridyl)azo]-1,3-diaminobenzene reacts sensitively with Pd(II) in HCl solution ($\mathscr{S}_{572} = 0.002$ μg/cm^2).[111]

B. THIAZOLYLAZO REAGENTS[112]

These are represented by the general formulas

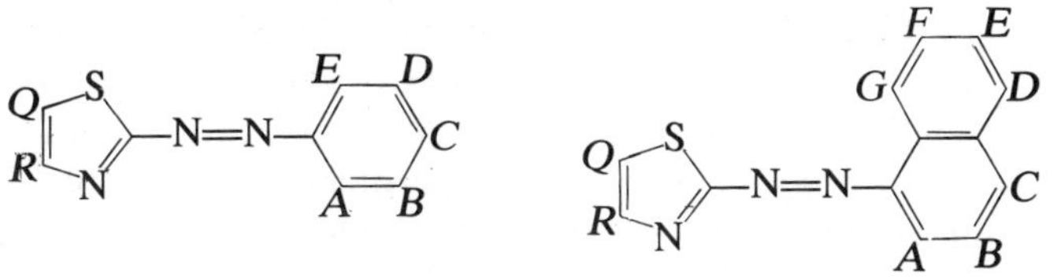

An OH group is usually present in position A.

Metals replace H of ortho OH groups in thiazolylphenols and thiazolylnaphthols and bond to N of the azo group to form chelates. Presumably two five-membered rings are formed, the second involving metal, azo N, and thiazole N (analogous to PAN complexes). The commonest reagent of the thiazolylphenols is TAR, 4-(2'-thiazolylazo)-resorcinol; of the thiazolylazonaphthols, TAN, 1-(2'-thiazolylazo)-2-naphthol. A swarm of derivatives have been prepared, some with substitutions in the thiazole ring. Both TAR (mol. wt. 221.2, m.p. ~210°, decomp.) and TAN (mol. wt. 255.3, m.p. ~137°) are available commercially. For purification see Hovind.[112]

TAR is only slightly soluble in water, but it dissolves in acidic and basic solutions:

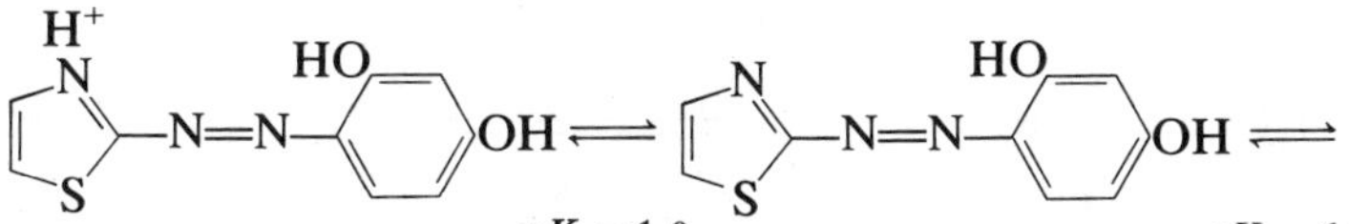

Red, $\lambda_{max} = 488$ nm   Yellow, $\lambda_{max} = 410-440$ nm

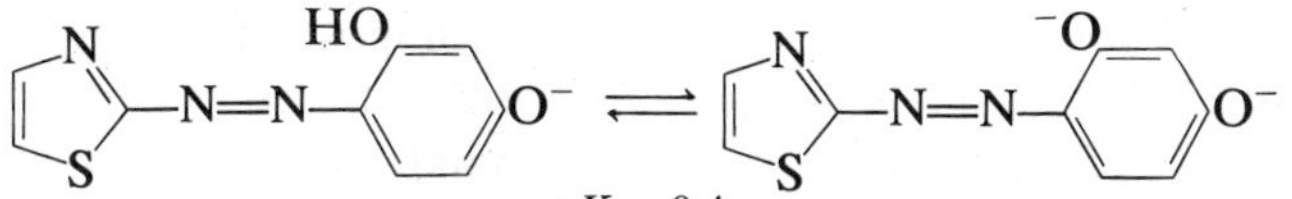

Orange-red, $\lambda_{max} = 470$ nm   Red-violet, $\lambda_{max} = 510$ nm; $\epsilon = 34,500$

The presence of two OH groups in TAR makes it more soluble in water than it would otherwise be, soluble enough to be used as reagent for some metals in water solution without extraction. It may be noted that a single OH in p position allows complex formation with some metals, but such complexes have low stability.

In weakly acidic medium, most reacting metals form 1:1 or mixtures of 1:1 and 1 M:2 complexes with TAR and its derivatives, with increasing tendency to form 1:2 complexes in basic solution. In organic solvents, complexes of the type ML_2 (reagent = HL) are usually present. Protonated complexes are formed by a few metals. Mixed anion complexes, as with OH^- and Ac^- as well as with others are quite common. Some metals can give negatively charged species with excess TAR. Ion-association complexes, as of SbI_4^-, or with cationic partners can be formed.

As photometric reagent, TAR is very similar to PAR. Owing to the lower basicity of the thiazole nucleus, TAR complexes are formed in slightly more acidic solutions, but only a few metals react in more than weakly acidic medium, for example, Tl(III) at pH 1–2. The complexes are extractable into organic solvents (often chloroform is suitable). Numerous procedures for the photometric determination of metals, with or without extraction, have been described. ϵ values usually range from 20,000 to 50,000, occasionally higher with some derivatives. Sensitivity is, therefore, good, but selectivity is poor. TAR and its derivatives may have advantages over PAR for some metals. The TAR dyes are more strongly colored than the pyridylazo dyes, but often there seems to be less contrast between reagent and chelate. However, the bathochromic shift on complexation with some metals is satisfactory, as with Th. TAR is of interest for the determination of Sc, Th, U, Zn, and a few other metals. 2-(2-Thiazolylazo)-5-dimethylaminophenol (TAM) is a sensitive U and Nb reagent.

Dissociation constants of TAN and colors of the species are:

$$H_2L^+ \rightleftharpoons HL \rightleftharpoons L^-$$

yellow	$pK_1 =$	red	$pK_2 =$	violet
$\lambda_{max} = 440$ nm	2.4	(also in org. soln.)	8.7	$\lambda_{max} = 530$ nm
		$\lambda_{max} = 490$ nm		

Like TAR, TAN forms complexes with most transition metals. In acidic solution 1:1 chelates are predominantly formed, whereas in basic solutions 1:2 chelates are formed. The complexes are essentially insoluble in water, and a medium of organic solvent miscible with water or extraction into chloroform is required for determination. On the whole, there seems to be nothing particularly noteworthy about these determinations. Copper, Zn, Ni, Co, and other metals have been determined.

Numerous reagents have been synthesized in which benzothiazolyl replaces thiazolyl in both the phenol and naphthol azo series.[113]

Chromotropic acid derivatives of thiazolylazo and benzothiazolylazo dyes have also been examined as metal reagents (water soluble), for example[114]:

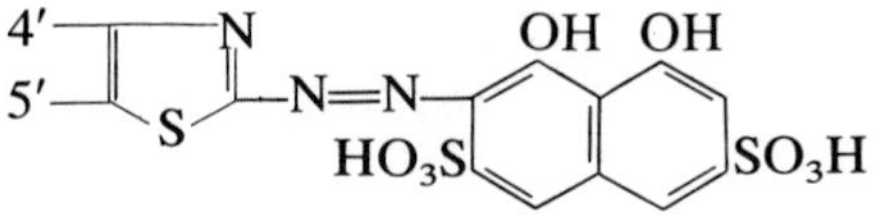

In H_2SO_4 solution, palladium(II) reacts selectively with[115]

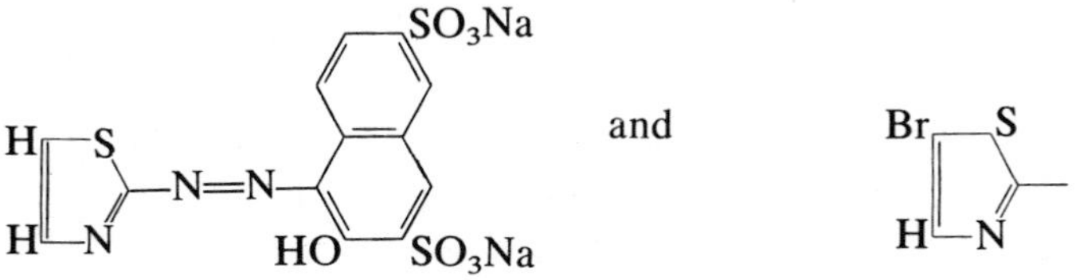

and

Titan Yellow

This dye, C.I. 19540, containing the thiazole grouping, may be included here. The commercial material (also known as Clayton Yellow, Thiazole Yellow, and by other names) is a mixture, whose magnesium-active component is[116]

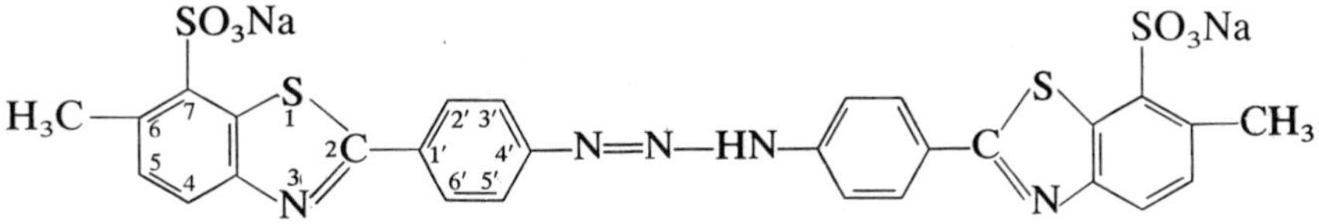

The impure product also contains inorganic salts (NaCl) and gives reactions of varying sensitivity with magnesium, for which it is chiefly used. The purification has been studied.[117] By use of the dextran gel Sephadex 10, the commercial product can be split into several fractions and separated from NaCl. Two yellow bands of Mg-reactive dye are obtained, one of which (with NaCl) is washed out of the column with water. The more active component is washed out of the column with water–acetone. Unfractionated Titan Yellow contains a colorless fluorescent substance, believed to be sulfonated dehydro-p-toluidine. From one commercial product about 30% of the most active fraction could be isolated.

In Mg solutions sufficiently basic to precipitate $Mg(OH)_2$, a color change from yellow to red occurs. The dye is adsorbed on the surface of the precipitate, but the mechanism of the color change is not well understood.

C. OTHER HETEROCYCLIC AZO REAGENTS

A limited number of these are mentioned here.

2-(2-Pyrimidylazo)-1-naphthol (α-MAN)

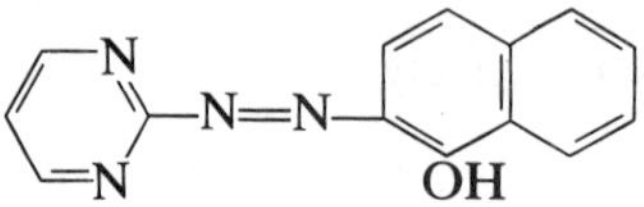

$pK_{NH} = 0.92$
$pK_{OH} = 9.66$ } 50% CH_3OH, $\mu = 0.1$

and its sulfonated derivatives form chelates of lower stability than its pyridyl analog α-PAN, but the pH range for the formation of its chelates is larger, which should make it a slightly more selective reagent.[118] The bathochromic shifts on chelation are greater with α-MAN than with α-PAN, which is an advantage in spectrophotometric determinations. These reagents have found little use so far. The high cost of preparation is a drawback.

Quinolinazo R

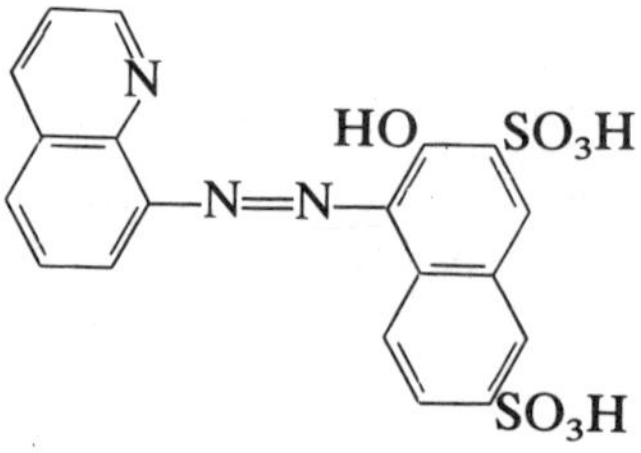

has $pK_{NH^+} = 1.5$ and $pK_{OH} = 11.5$ ($\mu = 0.1$, 20°).[119] It reacts with a multitude of metals, but the violet 1:3 chelate with Co (presumably III), once formed at pH 1.5, is stable to HCl and H_2SO_4 as strong as 6 N.[120]

5- and, more particularly, 7-arylazo-8-hydroxyquinolines form colored chelates with many metals but find little use in photometric determinations. From the standpoint of the structure of the chelates formed, it is interesting to note that 7-arylazo-8-hydroxyquinolines can chelate metals through the hydroxy group and heterocyclic N in neutral solution (yellow chelates) and through the hydroxy group and azo N in acidic solution (red chelates)[121]:

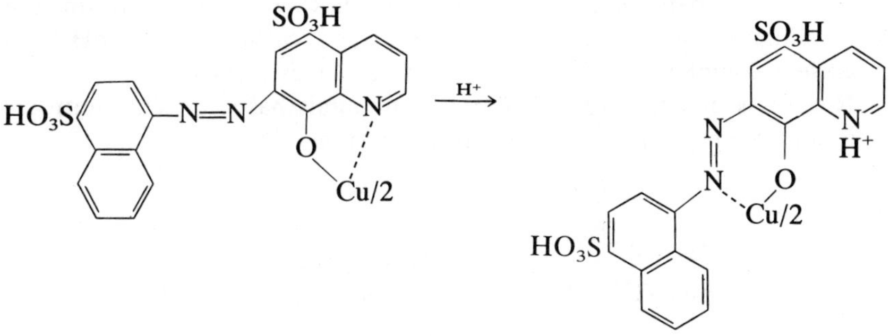

Yellow   Red

The 7-(4,8-disulfo-2-naphthylazo)-derivative of 8-hydroxyquinoline forms a yellow 1:1 complex with Sc, useful for its determination.[122]

5-Arylazo-8-hydroxyquinolines give only yellow chelates because the metal cannot coordinate with the azo group.

The reagents

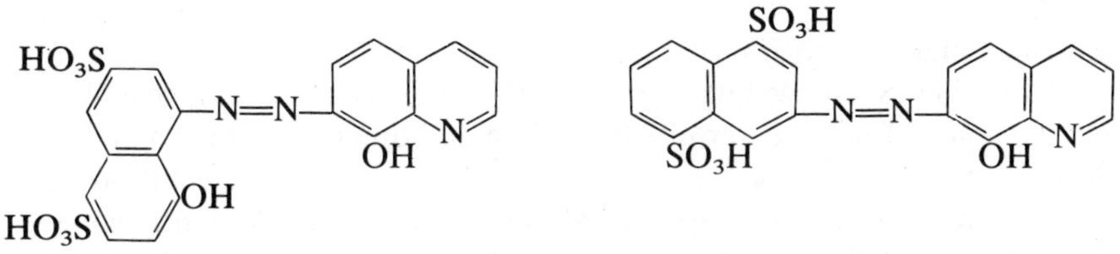

Azoxin-H   Azoxin-C

$K_1 = 2 \times 10^{-9}$ $K_1 = 3.85 \times 10^{-9}$
$K_2 = 1.65 \times 10^{-13}$

have been proposed for the determination of Hg.[123]

Quinazolinazo

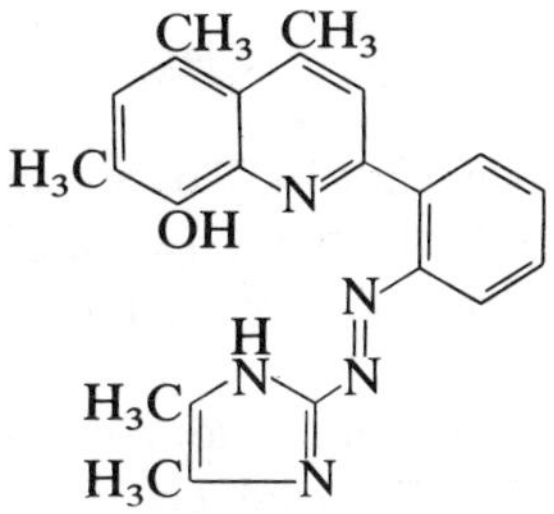

is possibly the best photometric reagent for lithium but is difficult to synthesize.[124]

Antipyrylazo III or Diantipyrylazo (3,6-bis-(4-antipyrinylazo)-4,5-dihydroxy-2,7-naphthalenedisulfonic acid) has been proposed for the determination of rare earth elements[125] and calcium.[126] At pH 13 a binuclear complex, in which Ca:reagent is 2:1, is formed.

Reactions of zinc with 11 bisazo compounds based on chromotropic acid and 4-aminoantipyrine have been examined.[127]

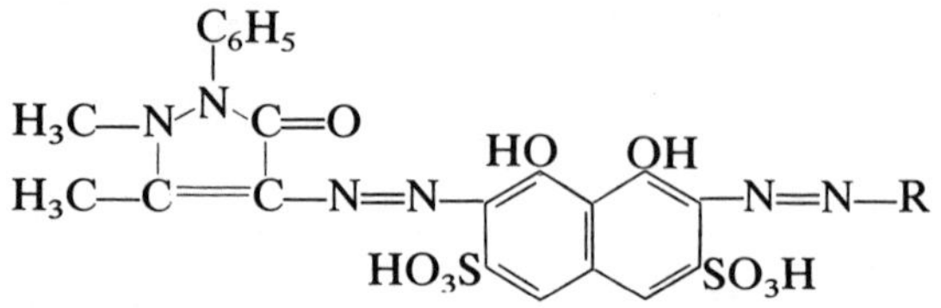

Some of the compounds give sensitive reactions ($\epsilon \sim 4 \times 10^4$ at $\lambda = 600$–625 nm). It is stated that

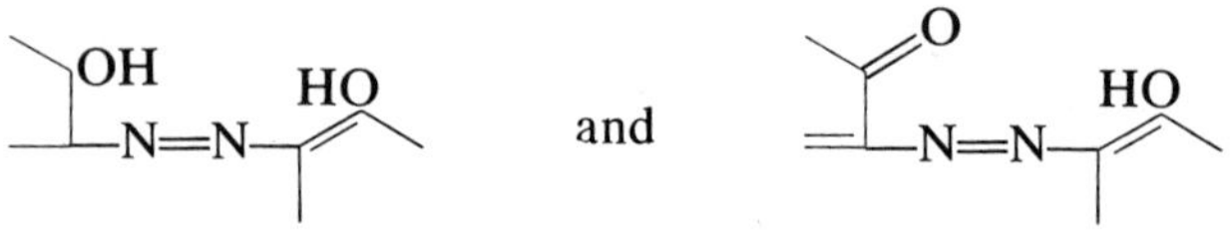

and

are functional groups for zinc ions.

Antipyrine S (preceding formula, R = antipyrine radical) has been recommended for the determination of La and other rare earth metals.[128] At pH 2.5, LaL_4 is formed, having $\epsilon_{max} = 1 \times 10^5$. Many metals interfere.

3-(4-Antipyrinylazo)-2,4-pentanedione (Azonol A) contains the β-diketone structure in combination with the chromogenic $-N{=}N-$ group. It gives red complexes with many heavy metals.[129] The following structure has been suggested for the complexes:

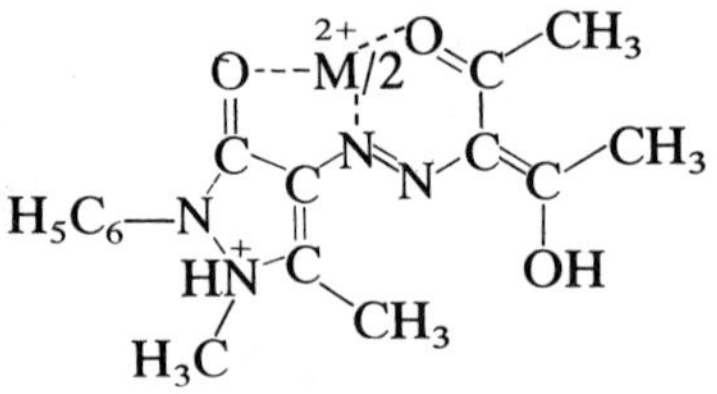

In particular, the reactions with Bi, Th, Sc, and Co are sensitive (for Sc, $\mathscr{S}_{610} = 0.002 \ \mu g/cm^2$), and the latter two metals have been determined with this reagent. The selectivity is mediocre, and the reagent is easily oxidized and reduced.

1-(2-Quinolylazo)-2-acenaphthylenol has been suggested for the determination of Zn, Cd, and Hg (red chelates in weakly basic solution, extractable into carbon tetrachloride).[130] For Zn the sensitivity index is $0.0009 \ \mu g/cm^2$ at 540 nm.

NOTES

1. J. R. Kirby, R. M. Milburn, and J. H. Saylor, *Anal. Chim. Acta*, **26**, 458 (1962); J. H. Saylor and J. W. Ledbetter, *ibid.*, **32**, 398 (1965). Apparent formation constants are reported, namely $pK_1 = 3.1, 0.6, 5.2$; $pK_2 = 7.4, 5.2,$ 8.0 (in the order Al, Ga, In). In basic solution, the reagent can be used to determine Mg fluorimetrically in the presence of much Ca: H. Diehl, R. Olsen, G. I. Spielholz, and R. Jensen, *Anal. Chem.*, **35**, 1144 (1963); also see *ibid.* **35**, 1142, in which various *o,o'*-dihydroxyazo compounds were tested as fluorimetric reagents for Mg and Ca (generally Mg fluoresces more strongly). Diehl and Ellingboe, *ibid.*, **32**, 1120 (1960), showed that two OH groups in *ortho* position (or *o*-OH and *o*-COOH) are needed for reaction with Ca and Mg. Introduction of a sulfonic group and another OH group (in the *p* position) gives the similar reagent 2,2',4'-trihydroxyazobenzene-5-sulfonic acid, which has been carefully studied as a reagent for Zr by M. H. Fletcher, *Anal. Chem.*, **32**, 1822, 1827 (1960). Zr forms two chelate complexes, in which dye:reagent is 1:1 and 1:2, the former having a net charge of 1+, the latter of 2−. Zr is bonded to O and N. The two *ortho* OH groups are involved in the bonding.

2. Y. Nishikawa et al., *Japan Analyst*, **16**, 692 (1967). Fluorimetric (Al similarly at pH 5; see especially D. J. Hydes and P. S. Liss, *Analyst*, **101**, 922 (1976)).

3. V. D. Salikhov and M. Z. Yampolskii, *Zh. Anal. Khim.*, **22**, 998 (1967).

4. P. V. Marchenko and N. V. Obolonchik, *Anal. Abst.*, **16**, 2388 (1969).

5. F. Holz, *Landw. Forsch.*, **26**, 56 (1973). Auto-Analyzer procedure for Mg in plant materials. Ca can be masked. Results agree well with those by atomic-absorption flame photometry.

6. G. Banerjee, *Anal. Chim. Acta*, **16**, 56 (1957); J. A. Cooper and M. J. Vernon, *ibid.*, **23**, 351 (1960).

7. G. Banerjee, *Anal. Chim. Acta*, **16**, 62 (1957).

8. S. Zelinski and V. Radecka, *Zh. Anal. Khim.*, **28**, 67 (1973); **29**, 1150, 1224 (1974).

9. Yu. M. Dedkov, L. M. Khokhlov, and V. D. Salikhov, *Zh. Anal. Khim.*, **26**, 2350 (1971).

10. Yu. M. Dedkov, D. I. Ryabchikov, and S. B. Savvin, *Zh. Anal. Khim.*, **20**, 574 (1965). Cu(II) forms a 1:1 chelate in solutions as acid as 0.7 M: V. I. Bogdanova and Yu. M. Dedkov, *Zav. Lab.*, **34**, 688 (1968).

11. S. S. Goyal and J. P. Tandon, *Talanta*, **15**, 895 (1968).

12. Yu. M. Dedkov and A. V. Kotov, *Zh. Anal. Khim.*, **25**, 650 (1970) and later papers.

13. V. I. Kuznetsov, L. I. Bol'shakova, and M. Fang, *Zh. Anal. Khim.*, **18**, 160 (1963); P. Pakalns and W. W. Flynn, *Analyst*, **90**, 300 (1965).

14. C. V. Stead, *J. Chem. Soc.* (*C*), 693 (1970). The structure is very similar to that of Beryllon II.

15. A. E. Mendes-Bezerra and W. I. Stephen, *Analyst*, **94**, 1117 (1969).

16. M. Herrero-Lancina and T. S. West, *Anal. Chem.*, **35**, 2131 (1963); P. Pakalns and T. M. Florence, *Anal. Chim. Acta*, **30**, 353 (1964). H. Ishii and H. Einaga have published a series of papers on the use of this reagent for the determination of Cu, Fe, Al, Ti, Mn, Mg, Ni, and Co (*Japan Analyst* and *Bull. Chem. Soc. Japan*, 1966–1967).

17. In passing, it may be noted that objections have been raised by S. B. Savvin and others to the practice of naming reagents for the element for which they are supposed to provide a selective determination. We agree. Hardly any reagent is specific for the element for which it is named. Such names as Thorin, Aluminon, Ferron, Beryllon, Cupferron, Bismuthiol, Cadion, Calcein, Calcion, Calcon, Calcichrome, Cupron, and Zincon are quite misleading. Moreover, this system does not satisfactorily provide for naming reagents that may later be discovered or developed, which are found to be more selective for the element than the earlier reagent. On the other hand, names such as Arsenazo, dithizone, PAN, and dithiol are much more satisfactory: they do not arouse false hopes, and they give some indication of the composition or nature of the reagent.

18. V. I. Kuznetsov, *Dokl. Akad. Nauk SSSR*, **31**, 895 (1941).

19. D. W. Margerum, C. H. Byrd, S. A. Reed, and C. V. Banks, *Anal. Chem.*, **25**, 1219 (1953). Mori et al. (1972) found the values 3.6, 7.4, 10.2. Such discrepancies, which are common, are most likely due to unsatisfactory purification of reagents.

20. A. R. Palmer, *Anal. Chim. Acta*, **19**, 458 (1958).

21. B. Budesinsky, *Talanta*, **16**, 1277 (1969). A. E. Klygin, V. A. Lekae, and N. A. Nikolskaya, *Zh. Anal. Khim.*, **25**, 76 (1970), found $pK_3 = 3.5$, $pK_4 = 6.6$; S. Nakashima, H. Miyata, and K. Toei, *Bull. Chem. Soc. Japan*, **41**, 2632 (1968), found $pK_3 = 3.0$, $pK_4 = 7.6$, $pK_5 = 10.0$. Such differences are not unusual for reagents of this kind, and arise in part, if not largely, from lack of purity.

22. S. B. Savvin, E. L. Kuzin, and L. A. Gribov, *J. Anal. Chem. USSR*, **23**, 1 (1968).

23. Dissociation constants of the chelates have been reported by A. F. Kuteinikov, *Zav. Lab.*, **28**, 1179 (1962).

24. E. L. Kuzin, S. B. Savvin, and L. A. Gribov, *Zh. Anal. Khim.*, **23**, 490 (1968). A similar structure is proposed for the Be complex.

25. D. Rajkovic, *Analyst*, **97**, 114 (1972). For U: H. P. Holcomb and J. H. Yoe, *Anal. Chem.*, **32**, 612 (1960).

26. N. N. Basargin, M. K. Akhmedli, and Sh. U. Islamov, *Zav. Lab.*, **37**, 269 (1971); *Anal. Abst.*, **24**, 2736 (1973).

27. V. I. Kuznetsov, L. I. Bol'shakova, and M. Fang, *Zh. Anal. Khim.*, **18**, 160 (1963).

28. Some papers dealing in a general way with these reagents and their

interrelationships may be mentioned here. T. V. Petrova, S. B. Savvin, and N. Khakimkhodzhaev, *J. Anal. Chem. USSR*, **25**, 1098 (1970), consider the protonation of chromotropic acid derivatives as affected by seven substituents (SO_3H, AsO_3H_2, COOH, etc.); substituents in the *m*- and *p*-positions to the azo groups exert inductive and conjugation effects, whereas steric effects are more pronounced with substituents in *o*-positions. Also see A. Muk and S. B. Savvin, *Anal. Chim. Acta*, **44**, 59 (1969). The effect of substituents on the dissociation of hydroxy groups in the naphthalene ring is reported in *J. Anal. Chem. USSR*, **25**, 188 (1970). The effect of sulfonate groups in the benzene ring on protonation and dissociation constants is considered by V. N. Nikolich, A. A. Muk, and T. V. Petrova, *ibid.*, **26**, 1497 (1971). *Ortho*-effects of carboxyl, sulfo, and arsono groups on the protonation and dissociation of monoazo and bisazo chromotropic acid derivatives are discussed by A. A. Muk, V. N. Nikolich, and T. V. Petrova, *Talanta*, **21**, 1296 (1974). Quantum mechanical interpretation of the structures of bisazo and monoazo reagents and their chelates is discussed by S. B. Savvin and E. L. Kuzin, *J. Anal. Chem. USSR*, **26**, 367 (1971); E. L. Kuzin, E. A. Likhonina, and S. B. Savvin, *ibid.*, **27**, 301 (1972). This listing is far from complete, but the papers mentioned and the references contained in them will serve as a point of entrance for those interested in the theoretical aspects.

29. H. Onishi and K. Sekine, *Japan Analyst*, **18**, 524 (1969).

30. J. Borák, Z. Slovák, and J. Fischer, *Talanta*, **17**, 215 (1970). For the use of paper chromatography in testing the purity of mono- and bisazo derivatives of chromotropic acid, see J. Siemroth and I. Hennig, *Talanta*, **15**, 765 (1968). Paper electrophoresis is superior to paper chromatography for separation of the components of impure azo reagents for their estimation as demonstrated by S. B. Savvin, T. G. Akimova, E. P. Krysin, and M. M. Davydova, *J. Anal. Chem. USSR*, **25**, 369 (1970). Arsenazo III reagents manufactured in England, Czechoslovakia, Switzerland, and Russia were found to be grossly impure, with as many as six to eight components. The purest Arsenazo III reagents are obtained by carrying out the synthesis in the presence of Li or Ba salts. The electrophoretic method reveals the presence of isomers of mono- and bisazo derivatives of chromotropic acid (T. G. Akimova, S. B. Savvin, and T. V. Petrova, *ibid.*, **25**, 680), and is also effective for testing the purity of heterocyclic azo compounds (S. B. Savvin et al., *ibid.*, **26**, 145). Purification of Sulfonazo III by treatment of the solution with active carbon and passage through a Dowex 50W (H^+-form) column has been reported (note 62).

31. A. A. Nemodruk, *Zh. Anal. Khim.*, **22**, 629 (1967).

32. S. B. Savvin, *Talanta*, **8**, 673 (1961); **11**, 7 (1964). Savvin, *Arsenazo III. Methods for Determining Rare Elements and the Actinides* [in Russian], Atomizdat, Moscow, 1966. Arsenazo III was first prepared in 1959.

33. More than a hundred compounds based on 2,7-bisazo chromotropic acid are

known. S. B. Savvin, *Talanta*, **15**, 1494 (1968), has suggested a nomenclature of trivial names for these reagents, as has B. Budesinsky, *Talanta*, **17**, 361 (1970). A few of these, or the series to which they belong, are mentioned in this section. See, among others, S. B. Savvin and Yu. M. Dedkov, *Zh. Anal. Khim.*, **19**, 21 (1964); Savvin's book; B. Budesinsky, K. Haas, and D. Vrzalova, *Coll. Czech. Chem. Commun.*, **30**, 2375 (1965); Budesinsky in H. A. Flaschka and A. J. Barnard, Jr., Eds., *Chelates in Analytical Chemistry*, Vol. 2, Dekker, New York, 1969, p. 1.

34. S. B. Savvin and E. L. Kuzin, *Talanta*, **15**, 913 (1968).

35. J. A. Pérez-Bustamante and F. Palomares Delgado, *Analyst*, **96**, 407 (1971), cite numerous papers relating to reaction of U with Arsenazo III. Zr can be determined in the presence of much Fe in 6 M HClO$_4$ if hydroxylammonium chloride is added.

36. However, since the reagent can absorb at the optimal wavelength, the sensitivity is better expressed in terms of the sensitivity index $\mathscr{S}$ (p. 190).

37. S. B. Savvin, *J. Anal. Chem. USSR*, **17**, 776 (1962).

38. A procedure for Be has been published: Sh. T. Talipov et al., *Anal. Abst.*, **23**, 89. Moreover, Al can be determined at pH 3.5, as shown by V. Mikhailova, *Acta Chim. Hung.*, **76**, 221 (1973). Zinc: see note 46.

39. A. A. Nemodruk, P. N. Palei, and L. P. Glukhova, *Zh. Anal. Khim.*, **23**, 214 (1968).

40. A. A. Nemodruk and L. P. Glukhova, *Zh. Anal. Khim.*, **23**, 552 (1968).

41. S. B. Savvin[32]; A. Muk and S. B. Savvin, *Zh. Anal. Khim.*, **26**, 98 (1971). Onishi and Sekine[29] found a 1:2 chelate of U(IV) in 6 M HCl. Am(V) forms a 1:1 complex (pH 5.2).

42. Borak et al.[30]

43. S. B. Savvin and E. L. Kuzin, *Zh. Anal. Khim.*, **22**, 1058 (1967).

44. B. Budesinsky, *Talanta*, **15**, 1063 (1968).

45. V. Michaylova and N. Kouleva, *Talanta*, **21**, 523 (1974). Formation constants are reported. It may be noted that in more strongly basic solution (~0.2 M KOH), Arsenazo III reacts with Li but not sensitively (miscible organic solvents increase the sensitivity). Phosphonazo R is a little more sensitive as shown by A. I. Lazarev and V. I. Lazareva, *Zh. Anal. Khim.*, **23**, 36 (1968). Still more sensitive is 2-(3-methyl-5-oxo-1-phenylpyrazolin-4-ylazo)-5-nitrobenzoic acid, called Nitroanthranilazo. It gives a 1:1 complex in aqueous–acetone solution ($\epsilon_{530} = 16,500$). See V. M. Dziomko, S. L. Zelichenok, and I. S. Markovich, *ibid.*, **23**, 170 (1968).

46. V. Michaylova and L. Yuroukova, *Anal. Chim. Acta*, **68**, 73 (1974).

47. V. I. Kuznetsov and S. B. Savvin, *Radiokhim.*, **2**, 682 (1960). Mixed-ligand complexes are reported to be formed by Y-subgroup rare earths, Arsenazo III, and diphenylguanidine (1:2:8). See M. K. Akhmedli et al., *Zh. Anal. Khim.*, **28**, 1304 (1973).

48. For determination of microgram quantities of Zr, U, Th, and the rare earth

metals after systematic separations, see H. Onishi and K. Sekine, *Talanta*, **19**, 473 (1972).

49. J. A. Pérez-Bustamante and F. Burriel-Marti, *Anal. Chim. Acta*, **37**, 49 (1967); Arsenazo III gives the complexes Pd_2L and PdL at pH 2–4: *ibid.*, **37**, 62 (1967). Dissociation constants of Palladiazo: *Talanta*, **18**, 183 (1971); **18**, 717 (1971).

50. Later work by L. Bocanegra Sierra, J. A. Pérez-Bustamante, and F. Burriel-Marti, *Anal. Chim. Acta*, **59**, 231 (1972), points to formation of 1:2 and 1:3 (Pd:L) complexes. See also, J. A. Pérez-Bustamante, C. Morell Garcia, and F. Burriel-Marti, *ibid.*, **44**, 95 (1969).

51. S. B. Savvin, R. F. Propistsova, and R. V. Strel'nikova, *J. Anal. Chem. USSR*, **24**, 24 (1969). Dissociation constants of the rare earth complexes have been reported by P. K. Spitsyn, V. S. Shvarev, and M. E. Korepina, *ibid.*, **26**, 1894 (1971), who also investigated differential masking by EDTA.

52. Yu. P. Novikov, M. N. Margorina, B. F. Myasoedov, and V. A. Mikhailov, *Zh. Anal. Khim.*, **29**, 698 (1974); Yu. P. Novikov, M. N. Margorina, B. F. Myasoedov, A. N. Usolkin, and V. A. Mikhailov, *ibid.*, **29**, 705 (1974).

53. P. K. Spitsyn, V. S. Shvarev, and M. E. Korepina, *Zh. Anal. Khim.*, **26**, 2121 (1971).

54. N. U. Perisic-Janjic, A. A. Muk, and V. D. Canic, *Anal. Chem.*, **45**, 798 (1973).

55. V. I. Fadeeva and I. P. Alimarin, *J. Anal. Chem. USSR*, **17**, 987 (1962); A. A. Nemodruk, Yu. P. Novikov, A. M. Lukin, and I. D. Kalinina, *ibid.*, **16**, 187 (1961).

56. T. Taketatsu, M. Kaneko, and N. Kono, *Talanta*, **21**, 87 (1974). $\epsilon_{668} \sim 2 \times 10^5$.

57. V. F. Luk'yanov, A. M. Lukin, E. P. Duderova, and T. E. Barabanova, *Zh. Anal. Khim.*, **26**, 772 (1971).

58. T. Yamamoto, *Anal. Chim. Acta*, **65**, 329 (1973). Similar method for Th: *ibid.*, **63**, 65 (1973).

59. J. W. O'Laughlin and D. F. Jensen, *Talanta*, **17**, 329 (1970). A. M. Lukin et al., *Anal. Abst.*, **23**, 146 (1972): 1:1 complexes in 0.2 M HCl; oxalic acid increases selectivity.

60. J. W. Ferguson, J. J. Richard, J. W. O'Laughlin, and C. V. Banks, *Anal. Chem.*, **36**, 796 (1964). Also see *Anal. Abst.*, **18**, 1515.

61. P. Ya. Yakovlev and M. P. Zhukova, *Zav. Lab.*, **36**, 1169 (1970), prefer this reagent to glyoxal bis-(2-hydroxyanil) for Ca. The Ca complex (660 nm) is formed at pH 2.5.

62. P. J. Kemp and M. B. Williams, *Anal. Chem.*, **45**, 124 (1973). Also earlier papers by B. Budesinsky and others mentioned in this reference.

63. S. B. Savvin, I. P. Alimarin, T. Ya. Belova, and L. A. Okhanova, *J. Anal. Chem. USSR*, **23**, 983 (1968).

64. S. B. Savvin, R. F. Propistsova, and L. A. Okhanova, *Talanta*, **16,** 423 (1969).

65. L. I. Gerkhardt, L. A. Okhanova, S. B. Savvin, and V. F. Vagan, *Zav. Lab.*, **39,** 769 (1973).

66. I. P. Alimarin, S. B. Savvin, and L. A. Okhanova, *Talanta*, **15,** 601 (1968); S. B. Savvin, I. P. Alimarin, T. Ya. Belova, and L. A. Okhanova, *J. Anal. Chem. USSR*, **23,** 983 (1968); S. B. Savvin, I. P. Alimarin, L. A. Okhanova, and T. Ya. Belova, *Trudy Kom. analit. Khim.*, **17,** 163 (1969).

67. See a listing of 17 reagents in the first-named reference in note 66. Most of these react with niobium in mineral acid solution (1 to 3–6 M HCl); $\epsilon_{max} = 3$–5.3×10^4; $\Delta\lambda$ (difference in absorption maxima of Nb chelate and reagent) = 50–90 nm.

68. The authors (Alimarin, Savvin, and Okhanova) cite Pfitzner's principle in explanation. This principle (*Angew. Chem.*, **62,** 242) states that reagents containing smaller numbers of salt-forming groups, whether participating in complex formation or not, form complexes having a higher proportion of reagent.

69. S. B. Savvin and T. V. Petrova, *J. Anal. Chem. USSR*, **24,** 80 (1969).

70. T. V. Petrova, N. Khakimkhodzhaev, and S. B. Savvin, *Izv. Akad. Nauk SSR Ser. Khim.*, No. 2, 259 (1970).

71. S. B. Savvin, T. V. Petrova, and P. N. Romanov, *ibid.*, No. 6, 1257 (1972); *Talanta*, **19,** 1437 (1972).

72. M. K. Akhmedli, N. N. Basargin, and S. U. Islamov, *J. Anal. Chem. USSR*, **24,** 94 (1969).

73. A. M. Cabrera and T. S. West, *Anal. Chem.*, **35,** 311 (1963).

74. Reviews: A: I. Busev and V. M. Ivanov, *Zh. Anal. Khim.*, **19,** 1238 (1964) [*J. Anal. Chem. USSR*, **19,** 1150 (1964)]; R. Püschel, *Z. Anal. Chem.*, **221,** 132 (1966); R. G. Anderson and G. Nickless, *Analyst*, **92,** 207 (1967); S. Shibata in H. A. Flaschka and A. J. Barnard, Jr., Eds., *Chelates in Analytical Chemistry*, vol. 4, Dekker, New York, 1972, pp. 1–232; M. N. Desai and M. H. Gandhi, *Rec. Chem. Progr.*, **30,** 223 (1969) (for PAR). General review of heterocyclic azo compounds: V. M. Ivanov, *Zh. Anal. Khim.*, **31,** 993 (1976). Purity (chromatographic analysis): *Zh. Anal. Khim.*, **29,** 2447 (1974).

75. B. F. Pease and M. B. Williams, *Anal. Chem.*, **31,** 1044 (1959); G. Nakagawa and H. Wada, *J. Chem. Soc. Japan Pure Chem. Sec.*, **84,** 639 (1963). More recently, R. G. Anderson and G. Nickless, *Analyst*, **93,** 13 (1968), have reported the respective values 2.32 and 12.00 in 50% methanol at $\mu = 0.1$. Data on isomers: S. I. Gusev, I. N. Glushkova, L. A. Ketova, and A. S. Pesis, *Zh. Anal. Khim.*, **25,** 260 (1970).

76. Yu. A. Zolotov, I. V. Seryakova, and G. A. Vorobyeva, *Talanta*, **14,** 737 (1967). In this connection it may be noted that V. A. Nazarenko, E. A. Biryuk, and R. V. Ravitskaya, *Zh. Anal. Khim.*, **30,** 74 (1975), concluded

that Ga, In, and Tl complexes of hydroxypyridylazo reagents are extracted into chloroform as $1 M:2 L:Cl$ (or other inorganic anion) compounds.

77. A. Galik, *Talanta*, **16**, 201 (1969).

78. R. Püschel, E. Lassner, W. Martin, and D. Martin, *Mikrochim. Acta*, 145 (1969), have suggested the isolation of microgram quantities of Ti, V, Mn, Fe, Co, Ni, Cu, Zn, and Pb from 100 ml of aqueous solution by precipitation at the proper pH. The excess reagent, being insoluble, acts as a collector. This separation method is of interest when the metal determination is to be made by emission spectrography or X-ray fluorescence. Also see *Talanta*, **16**, 351 (1969).

79. Among others, the extraction has been studied by S. Shibata, *Anal. Chim. Acta*, **22**, 479 (1960); **23**, 367, 434 (1960); **25**, 348 (1961); W. Berger and H. Elvers, *Z. Anal. Chem.*, **171**, 185 (1959). See Anderson and Nickless,[74] for an extensive tabulation of PAN spectrophotometric methods. Extraction for separation: I. H. Qureshi and M. N. Cheema, *J. Radioanal. Chem.*, **22**, 75 (1974).

80. R. Püschel, E. Lassner, and K. Katzengruber, *Z. Anal. Chem.*, **223**, 414 (1966).

81. H. Flaschka and R. Weiss, *Microchem. J.*, **14**, 318 (1969).

82. See I. V. Seryakova, Yu. A. Zolotov, and G. A. Vorobeva, *Zh. Anal. Khim.*, **24**, 1613 (1969).

83. P. R. Haddad, P. W. Alexander, and L. E. Smythe, *Talanta*, **21**, 123 (1974). Likewise Ga: L. Sommer et al., *Anal. Abst.*, **29**, 3B 60 (1975). Co(III) is also determinable fluorimetrically: G. H. Schenk, K. P. Dilloway, and J. S. Coulter, *Anal. Chem.*, **41**, 510 (1969).

84. H. Watanabe, *Talanta*, **21**, 295 (1974).

85. D. Betteridge, P. K. Todd, Q. Fernando, and H. Freiser, *Anal. Chem.*, **35**, 729 (1963). According to Betteridge and D. John, *Talanta*, **15**, 1238 (1968) and *Analyst*, **98**, 390 (1973), 2-(2-pyridylazo)-1-naphthol (α-PAN) is a better metal reagent than β-PAN; it reacts more rapidly with metals and at lower pH. An extraction (CCl_4)–photometric determination of Cu has been described by Betteridge, John, and F. Snape, *Analyst*, **98**, 512 (1973). It is sensitive, but the limiting concentrations of other metals are not sufficiently favorable to make the reagent competitive with older Cu reagents. The same authors, *ibid.*, **98**, 520, have investigated the reaction of Ti. Sulfonated derivatives of α-PAN were prepared by Anderson and Nickless, *loc. cit.*

86. S. Shibata, M. Furukawa, E. Kamata, and K. Goto, *Anal. Chim. Acta*, **50**, 439 (1970). Determination of Zn: *ibid.*, **51**, 271 (1970) (cyanide complexing).

87. Shibata, Goto, and Kamata, *Anal. Chim. Acta*, **45**, 279 (1969).

88. A. Kawase, *Japan Analyst*, **16**, 569 (1967).

89. I. Dahl, *Anal. Chim. Acta*, **62**, 145 (1972).

90. The purity of PAR (and PAN) can be tested by thin-layer chromatography: F. H. Pollard, G. Nickless, T. J. Samuelson, and R. G. Anderson, *J. Chromatogr.*, **16**, 231 (1964).

91. M. Hnilickova and L. Sommer, *Coll. Czech. Chem. Commun.*, **26**, 2189 (1961).

92. Somewhat different values are reported by Corsini et al., *Anal. Chem.*, **34**, 1090 (1962), for 50% dioxane. Also see R. W. Stanley and G. E. Cheney, *Talanta*, **13**, 1619 (1966); R. A. Chalmers, *Talanta*, **14**, 527 (1967); F. Lindstrom and A. E. Womble, *Talanta*, **20**, 594 (1973).

93. D. Nonova and B. Evtimova, *Anal. Chim. Acta*, **62**, 456 (1972).

94. D. Nonova and B. Evtimova, *Talanta*, **20**, 1347 (1973). Mn can be determined very sensitively ($\mathscr{S}_{496} = 0.0006 \ \mu g \ Mn/cm^2$). Cyanide is used to mask Ni, Cu, and others. Fe interferes.

95. M. Hnilickova and L. Sommer, *Talanta*, **16**, 83 (1969).

96. Mainly from Anderson and Nickless[74] and Shibata.[74] A careful study of the determination of Zn and Mn has been made by S. Ahrland and R. G. Herman, *Anal. Chem.*, **47**, 2422 (1975), who found pH 8 to be suitable for the determination of Zn. The determination of Zn is straightforward ($\mathscr{S}_{495} = 0.00081 \ \mu g/cm^2$), that of Mn less so. The absorbance of Mn–PAR solutions varies markedly with pH (and kind of buffer), excess of reagent, time of standing and sometimes on the absence or presence of reducing agent (ascorbic acid). Good results can be obtained at pH 10 (borate buffer) with a sufficient excess of PAR, without reducing agent; $\mathscr{S}_{500} = 0.0007 \ \mu g/cm^2$. It was observed that an old PAR solution (2 months) gave lower absorbance readings with Mn than one freshly prepared.

97. But PAR and hydroxylamine form a ternary complex with Mo(VI) having the composition 1:1:1, which has been proposed as a determination form for the element by E. Lassner, R. Püschel, K. Katzengruber, and H. Schedle, *Mikrochim. Acta*, 134 (1969). Vanadium forms analogous complexes: V. V. Lukachina et al., *Zh. Anal. Khim.*, **28**, 86 (1973).

98. It may be noted that the presence of pyridine N is crucial for obtaining chelates whose λ_{max} differ by 100 nm or so from λ_{max} of H_2L and HL^-. As shown by W. J. Geary, G. Nickless, and F. H. Pollard, *Anal. Chim. Acta*, **26**, 575 (1962), λ_{max} of the copper chelate of benzeneazoresorcinol is not significantly different from λ_{max} of that reagent.

99. Pd(II) forms four kinds of stable 1:1 chelates with PAR: below pH 4, a green chelate $[Pd(H_2L)H_2O]^{2+}$; at pH 4–7 two species of red chelate $[Pd(HL)H_2O]^+$ and $[Pd(L)H_2O]^0$; and at pH 8–11 a red chelate $[Pd(L)OH]^-$. At pH 6–11, the red Pd(II)–PAR chelates react with tetradecyldimethylbenzylammonium chloride to form an ion-association complex. $[Pd(L)(OH)]^- [TDBA]^+$, which can be extracted into chloroform. These facts have been elucidated by T. Yotsuyanagi, H. Hoshino, and K. Aomura, *Anal. Chim. Acta*, **71**, 349 (1974).

100. Co, U, Pb: F. H. Pollard, P. Hanson, and W. J. Geary, *Anal. Chim. Acta,* **20,** 26 (1959). Anderson and Nickless, *loc. cit.,* tabulate many methods.

101. Masking can be quite successful for some elements. For example, T. M. Florence and Y. J. Farrar, *Anal. Chem.,* **35,** 1613 (1963), were able to determine U(VI) selectively by using a mixture of complexing agents, namely CyDTA, sulfosalicylate, and fluoride, at pH 8. (Cyanide should be an effective complexer for many heavy metals in basic solution.) Once formed, the PAR complexes of Fe(II), Co(III), and Pd(II) are stable against EDTA on long boiling, as shown by T. Yotsuyanagi, R. Yamashita, and K. Aomura, *Anal. Chem.,* **44,** 1091 (1972). The stability of the Co(III) complex is no doubt due to kinetic inertness, but that of Fe(II) and Pd complexes seems to be thermodynamic [PAR is believed to stabilize Fe(II)]. Cr(III) requires boiling with PAR at pH 5 for about 2 hr for quantitative reaction. The reaction product is $Cr(HL)_3$, which ionizes sufficiently to $Cr(HL)_2L^-$, to enable its extraction into $CHCl_3$ as an ion pair with tetra-decyldimethylbenzylammonium cation ($\mathcal{S}_{540} = 0.0011$ μg Cr/cm^2). See T. Yotsuyanagi, Y. Takeda, R. Yamashita, and K. Aomura, *Anal. Chim. Acta,* **67,** 297 (1973). Determination of Nb becomes quite selective in the presence of EDTA and tartrate (rock analysis): L. P. Greenland and E. Y. Campbell, *J. Res. U.S. Geol. Surv.,* **2,** 353 (1974). Also (for Nb): J. B. Bodkin and D. A. Rogowski, *Analyst,* **102,** 110 (1977); R. Belcher, T. V. Ramakrishna, and T. S. West, *Talanta,* **10,** 1013 (1963), who first described the procedure.

102. M. Siroki and C. Djordjevic, *Anal. Chim. Acta,* **57,** 301 (1971). V(III) and V(IV) also react with PAR. Zephiramine can also be used as cationic reagent.

103. E. A. Biryuk and R. V. Ravitskaya, *Zh. Anal. Khim.,* **26,** 735 (1971).

104. A. I. Fomina, N. A. Agrinskaya, Yu. A. Zolotov, and I. V. Seryakova, *Zh. Anal. Khim.,* **26,** 2376 (1971).

105. D. Betteridge and D. John, *Analyst,* **98,** 377, 390 (1973).

106. See Anderson and Nickless, *loc. cit.,* and S. Shibata, *Japan Analyst,* **21,** 551 (1972) (tables in English) for reviews. Also see S. Shibata in *Chelates in Analytical Chemistry.*[74] Numerous papers by S. I. Gusev, Sh. T. Talipov, and their coworkers are noted in *Anal. Abst.,* **16–27** (1969–1974); they are surveyed by Shibata.

107. S. Shibata, M. Furukawa, and K. Toei, *Anal. Chim. Acta,* **66,** 397 (1973). E. Kiss, *ibid.,* **77,** 205 (1975), has shown that 2-(3,5-dibromo-4-methyl-2-pyridylazo)-5-diethylaminophenol forms a blue 1:1:1 complex with V(V) and H_2O_2 extractable into $CHCl_3$ ($\mathcal{S} = 0.001$ μg V/cm^2).

108. T. M. Florence, D. A. Johnson, and Y. J. Farrar, *Anal. Chem.,* **41,** 1652 (1969). A little later it was found that 2-(5-bromo-2-pyridylazo)-5-diethyl-aminophenol (Br–PDAP) is even more sensitive for U(VI): D. A. Johnson and T. M. Florence, *Anal. Chim. Acta,* **53,** 73 (1971). Continuing investigations by the same authors (*Talanta,* **22,** 253) demonstrated the formation of a 1:1 U complex at pH 2–5 (possibly $ML(OH)_2ML$), an unstable 1:1:1 OH

complex at pH 5.5–8 and, in the presence of fluoride, a 1 : 1 : 1 F complex at pH 6–8. All metal chelates of Br–PDAP show a double maximum in their absorption spectra. See the reference cited for a theoretical discussion of the spectra. Br–PDAP can also be used for the determination of Cu, Bi, Ni, Co, etc. The sensitivity is usually about twice that given by PAR.

109. E. Kiss, *Anal. Chim. Acta*, **66**, 385 (1973). A procedure for determining Co in silicate rocks and meteorites with 4-[(5-bromo-2-pyridyl)azo]-1,3-diaminobenzene is described. This compound, and similar ones, were earlier synthesized and studied as heavy metal reagents by S. Shibata, M. Furukawa, Y. Ishiguro, and S. Sasaki, *ibid.*, **55**, 231 (1971). 3-[(5-Chloro-2-pyridyl)azo]-2,6-diaminopyridine has been used as a Co reagent by S. Shibata, M. Furukawa, and K. Goto, *Talanta*, **20**, 426 (1973).

110. S. Shibata, M. Furukawa, and K. Goto, *Anal. Chim. Acta*, **71**, 85 (1974). S. Shibata and M. Furukawa, *Japan Analyst*, **22**, 1077 (1973) and S. Shibata, M. Furukawa, and E. Kamata, *Anal. Chim. Acta*, **73**, 107 (1974), have studied the reactions of Co with 5-(2-pyridylazo)-2,4-diaminotoluene and its derivatives. 5-[(3,5-Dichloro-2-pyridyl)azo]-2,4-diaminotoluene gives $\mathscr{S}_{590} = 0.00042 \ \mu g$ Co/cm^2 in aqueous solution.

111. S. Shibata, Y. Ishiguro, and R. Nakashima, *Anal. Chim. Acta*, **64**, 305 (1973).

112. A comprehensive review covering the literature in the period 1966–1974 has been presented by H. R. Hovind, *Analyst*, **100**, 769 (1975). Earlier reviews: O. Navratil, *Chem. Listy*, **60**, 451 (1966); Anderson and Nickless.[74] Because of the easy availability of the review by Hovind, a minimal number of references are given in the following discussion, which should be looked upon as a brief overview.

113. For example, S. I. Gusev, M. V. Zhvakina, and I. A. Kozhevnikova, *Zh. Anal. Khim.*, **26**, 859 (1971). A. Kawase, *Japan Analyst*, **17**, 56 (1968), found 2-(2'-benzothiazolylazo)-1-naphthol to be a sensitive zinc reagent ($\epsilon_{621} = 70,000$ in CHCl$_3$).

114. S. B. Savvin, Yu. G. Rozovskii, R. F. Propistsova, and E. A. Likhonina, *Zh. Anal. Khim.*, **25**, 423 (1970). Complex formation with Al, Ga, Th, Zr investigated. Al reacts sensitively with such reagents in rather acidic media (pH ~2, $\mathscr{S}_{630} = 0.5 - 1 \times 10^{-3} \ \mu g/cm^2$).

115. A. I. Busev, L. S. Krysina, T. N. Zholondkovskaya, G. A. Pribylova, and E. P. Krysin, *Zh. Anal. Khim.*, **25**, 1575 (1970). A. I. Busev, T. N. Zholondkovskaya, L. S. Krysina, and O. A. Barinova, *ibid.*, **29**, 1758 (1974), have determined nickel with 1-(2-thiazolylazo)-2-naphthol-3,6-disulfonic acid at pH 9.

116. H. G. C. King, G. Pruden, and N. F. Janes, *Analyst*, **92**, 695 (1967). Directions for synthesizing the pure compound are given.

117. H. G. C. King and G. Pruden, *Analyst*, **92**, 83 (1967). Earlier work: E. G. Bradfield, *ibid.*, **85**, 666 (1960); R. J. Hall, G. A. Gray, and L. R. Flynn, *ibid.*, **91**, 102 (1966). Also see W. Werner, *Anal. Abst.*, **28**, 3A 9 (1975).

118. R. G. Anderson and G. Nickless, *Analyst*, **93,** 20 (1968); F. H. Pollard, G. Nickless, and R. G. Anderson, *Talanta*, **13,** 725 (1966).

119. N. N. Basargin, A. V. Kadomtseva, and V. I. Petrashen, *Zh. Anal. Khim.*, **25,** 34 (1970).

120. *Ibid.*, **25,** 285 (1970).

121. A. I. Cherkesov, *Dokl. Akad. Nauk SSR*, **142,** 1098 (1962); *Zh. Anal. Khim.*, **17,** 16 (1962).

122. A. I. Busev and G. E. Lunina, *Zh. Anal. Khim.*, **20,** 1069 (1965).

123. A. I. Cherkesov, V. S. Tonkoshkurov, A. I. Postoronko, and V. N. Ryzhov, *Zh. Anal. Khim.*, **25,** 466 (1970). The second reagent forms HgL_2 at pH 1.6, $\epsilon_{540} = 4.4 \times 10^4$.

124. V. M. Dziomko, S. L. Zelichenok, and I. S. Markovich, *Zh. Anal. Khim.*, **18,** 937 (1963).

125. B. Budesinsky and D. Vrzalova, *Anal. Chim. Acta*, **36,** 246 (1966). 5-(4-Antipyrinylazo)-2-monoethylamino-*p*-cresol and 6-(4-antipyrinylazo)-3-diethylaminophenol form $CHCl_3$-extractable 1:1 complexes with U(VI) according to S. I. Gusev, V. A. Agilov, and L. M. Shchurova, *Zh. Anal. Khim.*, **29,** 2427 (1974).

126. B. W. Budesinsky, *Anal. Chim. Acta*, **71,** 343 (1974).

127. S. B. Savvin, L. A. Sokolovskaya, L. M. Trutneva, and Yu. G. Rozovskii, *Zh. Anal. Khim.*, **27,** 2354 (1972).

128. S. B. Savvin, L. A. Sokolovskaya, and Yu. G. Rozovskii, *Zh. Anal. Khim.*, **27,** 1263 (1972).

129. B. Budesinsky and J. Svecova, *Anal. Chim. Acta*, **49,** 231 (1970).

130. I. Singh, Y. L. Mehta, B. S. Garg, and R. P. Singh, *Talanta*, **23,** 617 (1976).

ORGANIC PHOTOMETRIC REAGENTS SULFUR, SULFUR–OXYGEN, AND SULFUR–NITROGEN CHELATION

Reagents having chelating sulfur atoms may bond metals through S alone, through S and N or, less often, through S and O, or S, N, and O. Definite assignment of a reagent to one subclass or another is not always possible, because not all metals react in the same way with a given reagent, and tautomeric rearrangements may occur, so that a particular reagent may react differently depending on the existing conditions. In the following discussion, reagents are roughly classified in the order S, S–O, and S–N bonding. Most of the reagents bonding through S have an –SH group in which H can be replaced by an equivalent of metal, but metal can also be bonded to S in some reagents without replacement of H. Less frequently, H attached to N is replaced, with bonding to S. An –SH group may be attached to atoms other than C, as in thiophosphoric acids, but in most reagents S is bound to C. Not all of the metal complexes formed are chelates. Some can be simple coordination complexes.

Sulfur-containing reagents vary a great deal in their analytical value, but some of the most important photometric reagents for heavy metals are found in this class. They allow easy determination of sulfophilic elements in the presence of the nonsulfophilic, that is, alkaline earths, Sc, Y, rare earth elements, Ti, Zr, V, Nb, Ta, Th, U, and Al. Some are sufficiently selective to enable certain sulfophilic elements to be determined in the presence of each other.

Some S-reagents are unstable. Those that have –SH groups tend to be unstable in air (oxidation to disulfide), but the stability varies. Some of these reagents may be hydrolyzed in aqueous solution, forming H_2S or other products.

Note on Sulfophilic Elements. We use the term sulfophilic to denote elements that have considerable affinity for S(II) in aqueous solution.[1] This affinity or tendency to bond with S(II) is made evident inorganically by the formation of metal sulfides, most of which are only slightly soluble. Elements forming sulfides stable in the presence of water and slightly soluble in aqueous solution are shown in Table 6F-1. The elements V, W, and U do not readily form sulfides under these conditions and are not

TABLE 6F-1
Sulfophilic Elements Defined as Those Forming Slightly Soluble Sulfides

(Elements within solid lines can be quantitatively precipitated, those within broken lines incompletely.)

V		(Mn)	Fe	Co	Ni	Cu	Zn	(Ga)	Ge	As	Se
	Mo	Tc	Ru	Rh	Pd	Ag	Cd	(In)	Sn	Sb	Te
	W	Re	Os	Ir	Pt	Au	Hg	Tl	Pb	Bi	Po
	U										

quantitatively precipitable. They may be called semisulfophilic. The elements in the main block of the table vary in their sulfophilicity. In general, those elements that can be precipitated as sulfides in more acid solutions are the more sulfophilic.

We are primarily concerned here with the reactivity of metals with organic reagents having chelating or coordinating S atoms. To what extent is that reactivity correlative with the formation of insoluble sulfides shown in Table 6F-1? The correlation is close. This is hardly unexpected with reagents containing –SH groups. However, exceptions or apparent exceptions occur. Dithizone (having an –SH group) reacts with most of the elements in the sulfophilic block, but not, as far as now known, with Mo and Ge (nor with V, W, and U), and the extent of its reactions with As and Sb is unclear. Dithiol (two –SH groups) reacts with most elements in the block, including Mo, As, and Sb, as well as W. The presence of bonding N or O as well as bonding S in the reagent molecule can confuse the issue. But usually the sulfophilic character predominates. Some S-containing reagents unequivocally react with metals lying well outside the sulfophilic block. For example, SCN$^-$ reacts with Ti and Nb as well as with the border elements U and V, in addition to Fe(III), Co, Mo, W, Re, and some of the platinum metals, to give colored coordination complexes.

Weakly sulfophilic elements such as Mn and Ga may not give notable reactions with these reagents. The range in the values of the formation constants of the metal complexes of S-bonding reagents can be very great. For some, the span is known to be as large as 10^{30} among the reacting metals in the sulfophilic block. Evidently, such reagents can be quite selective for metals forming the more stable complexes. Usually the most stable complexes are formed by Cu, Ag, Au, Hg, and the platinum metals. The latter may not react rapidly [except for Pd(II)] so that heating for some time may be required, but once formed their complexes can be very stable. There may be marked differences in the reactivity of different

S-reagents with the metals within the sulfophilic block, especially when bonding N is present in the molecule. For example, in a weakly acidic solution, diethylaminobenzylidenerhodanine reacts with Pd(II), Pt(II), Ag, Hg, and Cu(I), but not with Cu(II), whereas dithizone reacts with these metals as well as with Cu(II) and Bi and to some extent with Zn and other metals. Thiocyanate is useful for determination of metals not giving characteristic reactions with dithizone. Few of the sulfophilic elements will bond to a lone S (not –SH) in a molecule not also containing bonding N (or O).

I. REAGENTS BONDING THROUGH S ALONE

A. PHENYL-SUBSTITUTED 1,2-DITHIOLE-3-THIONES

PdCl$_2$ and PtCl$_2$ can bond to suitable S atoms, in the manner they bond to N (p. 356), without replacing H. The above compounds have been examined as reagents for the determination of palladium[2] and platinum.[3] PdCl$_2$ forms a 1:2 complex with the 5- or 4-phenyl (or the 4-methyl-5-phenyl) derivatives of 1,2-dithiole-3-thione in acid medium, and the complexes can be extracted into chloroform. The absorption maximum of each reagent falls at 420 nm and that of the three complexes at 440 nm. The molar absorptivities for the complexes range from ~10,000 to ~15,000. The proposed formula of the complex of PdCl$_2$ with 4-methyl-5-phenyl-1,2-dithiole-3-thione is

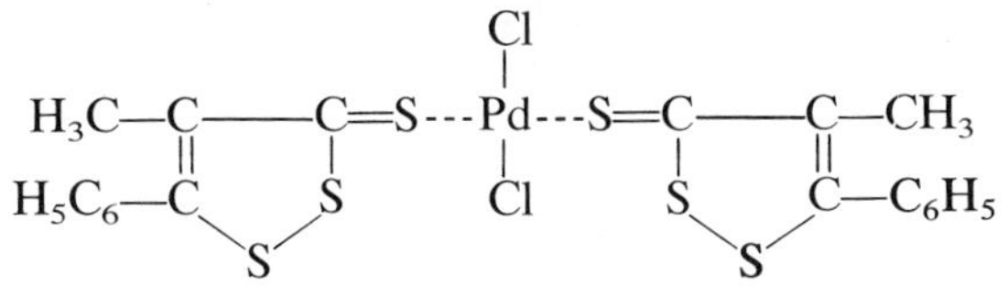

(HgCl$_2$ forms an analogous complex.) Platinum(II) and gold(III) interfere with the determination of palladium. PtCl$_2$ forms yellow-orange water-insoluble complexes with 5-phenyl-1,2-dithiole-3-thione and 4-methyl-5-phenyl-1,2-dithiole-3-thione in neutral and acid media, but does not react with 4-phenyl-1,2-dithiole-3-thione. The complexes are extracted by chloroform, and platinum can be determined by measuring the absorbance of the extracts at 470 nm. The composition of the complexes is PtR$_2$Cl$_2$. Palladium and gold interfere with the determination of platinum.

A complex analogous to the preceding is formed by PdCl$_2$ with

phenoxthine

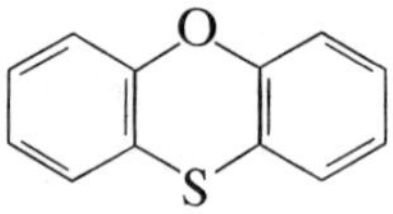

Au also reacts.

It is of interest that $HgCl_2$ forms $HgCl_2 \cdot C_4H_8S_2$ with dithiane

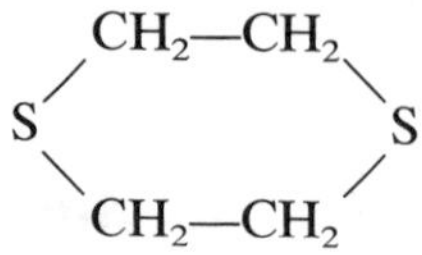

B. COMPOUNDS WITH VICINAL –SH GROUPS OR –SH AND =S GROUPS

Vicinal –SH groups in an aromatic framework provide useful photometric reagents. Such groups in an aliphatic chain are of lesser value. 2,3-Dimercaptopropanol (BAL: British Anti-Lewisite, Dimercaprol)

$$CH_2SHCHSHCH_2OH$$

is a well-known masking agent for the sulfide-forming metals. It forms colored complexes with Co, Cu, Fe(III), and other metals but is not used as a colorimetric reagent. 2,3-Dimercaptopropane sulfonic acid (Na salt) (Unithiol) gives a red iron complex in basic solution.[4]

Compounds having –SH and –S– suitably spaced in an aliphatic chain react with sulfophilic metals. Thus 2,2'-dimercaptodiethylsulfide (2,2'-thiodiethanethiol)

$$HSCH_2CH_2SCH_2CH_2SH$$

forms a red 1:1 chelate with Ni in faintly acidic or weakly basic solutions.[5] The Ni chelate is dimeric in dichloroethane solution. In acid solutions palladium forms a yellow-green 1:1 complex (probably trimeric in solution) and other species. This reagent appears promising for the determination of Pd. The chelates formed with Cu(I), Ag, Cd, Hg, and Zn are colorless.

1. Toluene-3,4-dithiol

Aromatic compounds containing the grouping

form slightly soluble, mostly colored, chelates with heavy metals, namely those that give slightly soluble sulfides with H_2S. 1,2-Dimercaptobenzene is one such reagent. The toluene analog (dithiol) is more often used as reagent:

CH₃ (structure)   Toluene-3,4-dithiol or 4-methyl-1,2-dimercaptobenzene

Colorless liquid or solid (when pure). Mol. wt. 156.3. M. p. 31°. Supplied in glass ampoules. Very easily air-oxidized to the disulfide. Soluble in the usual organic solvents and in sodium hydroxide solution. The latter solution is often stabilized with mercaptoacetic acid. $K_{a1} = 4 \times 10^{-6}$, $K_{a2} = 1 \times 10^{-11}$. $P_{CHCl_3/H_2O} (= [H_2L]_{CHCl_3}/[H_2L]_{H_2O}) = 1.3 \times 10^4$.

The reactions of dithiol with tin, molybdenum, tungsten, rhenium, and technetium form the basis of colorimetric methods for these metals. Tin(II) gives a red dithiolate in acid medium (yellow in organic solvent solutions). Molybdenum(VI) and tungsten(VI) form slightly soluble yellow-green and blue-green chelates, respectively, in strongly acid solution, which are extractable into benzene and various other organic solvents. The Mo, W, and Re chelates have the composition[6]

$$M\left(\begin{array}{c} S \\ S \end{array} \text{---} \bigcirc \text{---} CH_3 \right)_3$$

Zinc and apparently other metals reacting above pH ∼ 4 form uncharged chelates in which the metal: reagent ratio is 1:1 so that the composition must be $M^{II}L$, both H atoms of H_2L being replaced.

The colors of some other dithiolates are: Bi, red; Sb(III) and Pb, yellow; Cu, Ni, Co, Fe(II), black; Pd(II), red; Pt(II), violet. The solubilities of the dithiolates roughly parallel those of the sulfides. Pd, Pt, and other platinum metals, Au, Te, Se, As, Ge, Cu, Hg, Ag, W, and Re are precipitable (and extractable in general into nonpolar solvents) from 4 M and stronger acid solutions. In alkaline aqueous pyridine solution, dithiol gives color reactions with Cu, Co, Fe, Tl, Ni, Mn, V, Sb, and Ru; the reaction with Co (blue color) is sensitive, but none of these reactions is of practical importance quantitatively.[7] Silver dithiolate (yellow) can be dispersed with dodecyl sodium sulfate for determination.

Solid zinc dithiolate is much more stable than dithiol and can be used as reagent for dithiol reactions occurring in acid solutions.[8] The derivatives diacetyldithiol and dibenzoyldithiol likewise are more stable than dithiol and can be used as dithiol reagents after hydrolysis to dithiol in basic solution.[9]

Dithiol in excess has a strong tendency to form adducts or charged anionic complexes with divalent and trivalent metals, for example, $ZnL \cdot 2 H_2L$ or ZnL_2^{2-}. Early workers (Hamence and others) noted that metal dithiolates dissolved in ammonia in the presence of excess reagent. Uncharged zinc dithiolate is not extractable into "inert" solvents such as $CHCl_3$, presumably because of the presence of coordinated water molecules ($ZnL \cdot 2 H_2O$). With excess dithiol in weakly acidic solution, zinc and other metals forming anionic chelates can be extracted into nonoxygenated solvents ($CHCl_3$) as a partner of a bulky cation such as tetra-*n*-hexylammonium cation (R^+), for example, as the ion associate R_2ZnL_2.[10]

Hydrated dithiolates can also be made extractable into $CHCl_3$ by replacing water with a bidentate ligand such as 1,10-phenanthroline.[10] Zinc can thus be extracted in the approximate pH range 4–8. Many other metals can be extracted at pH 4.5 and higher.

It may be noted in passing that thiophenol, C_6H_5SH, does not give pronounced color reactions with heavy metals. Thus Hg mercaptide is colorless; Ag and Pb mercaptides are light yellow. A second –SH group, *ortho* to the first, is needed for the formation of strongly colored products, though not all reacting metals give strong colors.

o-Xylene-4,5-dithiol is a reagent much like toluene-3,4-dithiol and has been used for the determination of Mo and W.[11]

It may be mentioned at this point that disodium *cis*-1,2-dicyanoethylene-1,2-dithiolate

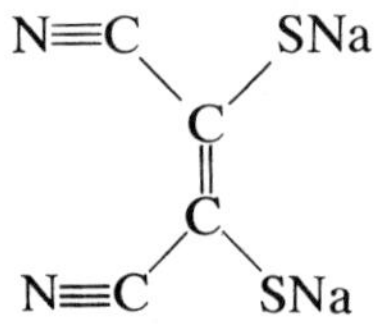

has been proposed for the determination of Mo.[12] An insoluble green complex is formed in dilute HCl (1 Mo:4 reagent). W(VI) and V(V) also form colored complexes.

2. Quinoxaline-2,3-dithiol

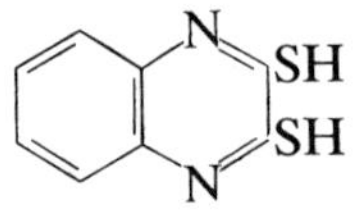

Mol. wt. 194.3. M. p. 345° (decomposition). Reddish brown solid, essentially insoluble in water, sparingly soluble in common organic solvents and in ammonia. Reagent solutions can be prepared in ethanol, 1:1 ethanol–dimethylformamide or dimethylformamide alone. Solutions unstable (presumably oxidation to disulfide occurs) and should be prepared daily.[13] Protect against light. $pK_1 = 6.84$, $pK_2 = 9.95$ (both at $20°$, $\mu = 0.01$). $K_{sp} = [H^+][HL^-] = 1.2 \times 10^{-11}$ ($20°$). $[H_2L]_{satd.}$ = intrinsic solubility = 7.2×10^{-5} ($20°$). Sodium salt, $Na_2L \cdot 2H_2O$, is readily soluble in water, and solution is relatively stable; can be purified by precipitation (addition of ether to aqueous ethanol solution).

Various tautomeric forms of the reagent can be written. The structure

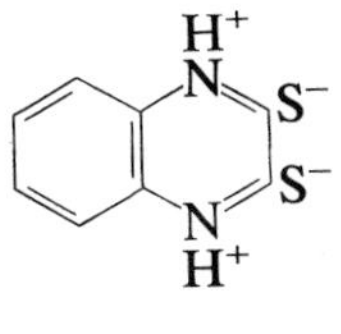

is considered the most probable.[14] It reacts with numerous sulfur-bonding metals to give insoluble colored chelates. These reactions have not been studied methodically, but it is known that Ni and Pd(II) react with excess reagent in weakly acidic solution to form 1:2 chelates, probably of the type

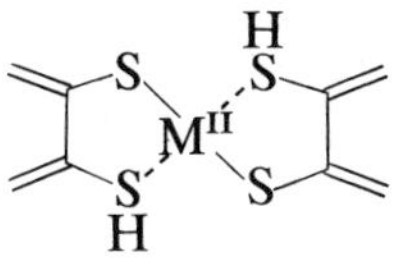

(H of —SH is ionizable)

In basic solutions, anionic complexes can be formed. Thus at pH 10 Pd(II) gives bright yellow PdL_2^{2-}. The green nickel chelate is extractable from acidic solutions into organic solvents. It can also be extracted as a quaternary onium ion associate. Nickel gives a pink color in ammoniacal solution (evidently an anionic chelate) and can be determined in this way.[15] In alcoholic solution at pH 3, Ni produces a blue color; Co (air-oxidized to III), a pink color; and simultaneous photometric determination has been suggested.[16] The reagent has also been suggested for the determination of Pd (orange-red color in a dimethylformamide medium),[17] Os (green),[18] and Pt(II) (blue color).[19] A superior method for Pd is carried out at pH 10 with S-2-(3-mercaptoquinoxalinyl)thiuronium chloride.[20] Bismuth has been reported to form a 1:5 complex in H_2SO_4 solution, and a determination method has been described.

Cobalt reacts in solutions of high acidity (10 N H_2SO_4), presumably as Co(III).[21] $\mathscr{S}_{540} = 0.0011$ μg Co/cm^2. The determination is possible in the presence of Ni ($150\times$) and Cu ($15\times$). Bismuth also reacts in acid solution (4 N H_2SO_4), forming a 1:5 complex, but many metals interfere.[22]

C. COMPOUNDS WITH –CS(SH) GROUP

1. Xanthates

In reaction with xanthates

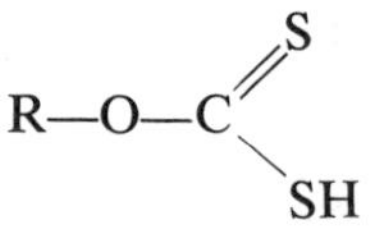

$$R-O-C \overset{\displaystyle S}{\underset{\displaystyle SH}{\big<}} \quad \text{(as alkali metal salt)}$$

a heavy metal replaces H of the –SH group and is also bonded to =S. Potassium ethyl xanthate and potassium benzyl xanthate are commonly used xanthate reagents. A number of metals, including Bi, Co, Cu, Mo, Ni, Pd, and Zn(UV), have been determined colorimetrically as xanthates by extraction procedures.[23]

Xanthates are probably of greater interest as separatory extraction reagents.[24] Dithiocarbamates, having the group $>$NC(S)SH, are more important reagents than xanthates.

2. Dithiocarbamates

Dithiocarbamates are valuable reagents for the separation and photo-metric determination of heavy metals, and the literature on analytical applications is extensive.[25] These compounds

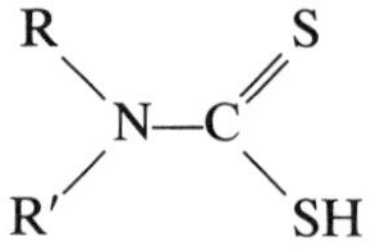

$$\overset{\displaystyle R}{\underset{\displaystyle R'}{>}}N-C \overset{\displaystyle S}{\underset{\displaystyle SH}{\big<}} \quad \text{(R and R}' = \text{H, Alk, Ar)}$$

(used in the form of their salts) form four-membered chelate rings, predominantly with sulfophilic elements, the H of –SH being replaced by an equivalent of metal with bonding to =S. Most of the analytically useful dithiocarbamates are disubstituted. The monosubstituted com-pounds (RNHCSSH) and their chelates tend to be unstable, decomposing to H_2S and other products, and are little used, though precipitation and extraction reactions of several have been studied.[26]

The dithiocarbamates of the heavy metals are only slightly soluble in water when uncharged[27] and hydrophilic groups are not present in the molecule. They are soluble in the usual organic solvents, often to a large extent in chloroform. The extracts absorb at the blue end of the spectrum or in the ultraviolet. The molar absorptivities are often large. Dithiocar-bamates are not selective *reagents*, but adjustment of the hydrogen-ion concentration and reagent concentration (by metal-exchange extraction) and the use of masking agents allow a number of metals to be determined selectively and likewise to be separated extractively.

Di-substituted dithiocarbamates tend to be unstable in acid solutions, some markedly so. The dithiocarbamate ion is protonated in acidic solution and decomposes into carbon disulfide and protonated amine. The decomposition rate can vary by a factor as great as 10^4 from one dithiocarbamate to another, and the choice of dithiocarbamate reagent is of importance when extractions are carried out in acid solutions. (See p. 514 for further discussion of the decomposition.)

It will be convenient to discuss the dithiocarbamates further by examining the reactions of a number of these reagents that are commonly used.

a. Sodium Diethyldithiocarbamate

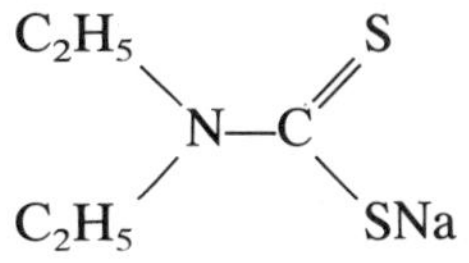

Mol. wt. (+3H$_2$O) 225.3. M. p., anhydrous salt, 94°–96°. Freely soluble in water (35 g in 100 ml). Solid is stable. Acidic aqueous solutions decompose rapidly, but slightly basic solutions (from hydrolysis) are fairly stable. A 1% stock solution in water shows little change in concentration in a few weeks. A 0.1% solution can be kept for a week (use a dark-glass bottle). Slow oxidation to diethyldithiouram

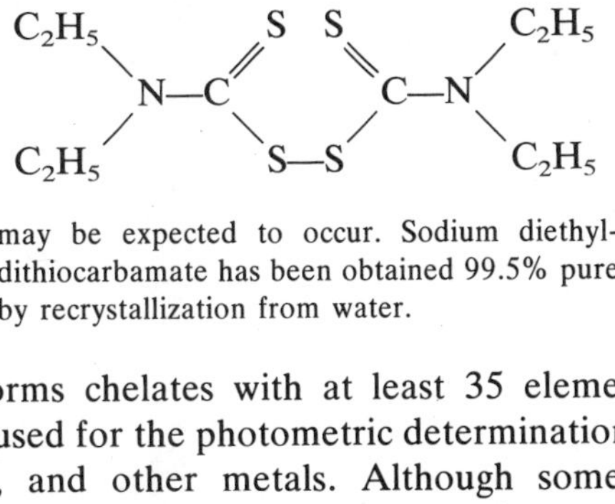

may be expected to occur. Sodium diethyldithiocarbamate has been obtained 99.5% pure by recrystallization from water.

Diethyldithiocarbamate forms chelates with at least 35 elements. In addition to separations, it is used for the photometric determination of Cu and, less often, of Bi, Ni, and other metals. Although some other dithiocarbamates are superior copper reagents because they are more selective in consequence of greater stability in acid solution, diethyldithiocarbamate still serves. Besides, much work has been done on diethyldithiocarbamate as analytical reagent, and it exemplifies the reactions of the dithiocarbamates in general. For these reasons, a fairly detailed treatment of this reagent seems called for.

Diethyldithiocarbamic acid (HDt) is a moderately weak acid, having $K_a = 4.5 \times 10^{-4}$ in H$_2$O at 0°, 4.1×10^{-4} ($\mu = 0.01$) in H$_2$O at 25° and

$\sim 3 \times 10^{-6}$ in 60% v/v ethanol at 25°. The following partition constants have been reported:

$$P_{CCl_4/H_2O} = \frac{[HDt]_{CCl_4}}{[HDt]_{H_2O}} = 3.4 \times 10^2 \text{ (Bode)}, 2.4 \times 10^2 \text{ (Stary)}$$

$$P_{CHCl_3/H_2O} = 2.30 \times 10^3 \text{ (Bode)}, 2.35 \times 10^3 \text{ (Stary)}$$

The extraction coefficient is given by

$$E_{HDt} = \frac{[HDt]_{org}}{[HDt]+[Dt^-]} \sim \frac{[HDt]_{org}}{[Dt^-]} = \frac{P}{K_a}[H^+] = K_{ext}[H^+]$$

(Concentrations without subscripts refer to aqueous phase.) K_{ext} has been determined for H_2O–CCl_4:

$$\left.\begin{array}{l} 1.6 \times 10^6 \text{ (Bode)} \\ 1.6 \times 10^6 \text{ (Stary)}, 20°, \mu = 0.1 \end{array}\right\} \log K_{ext} = 6.2$$

The extraction of the reagent into CCl_4 has been shown to follow the expected relation in the tested pH range 5.2–7.[28] For the $CHCl_3$–H_2O system (0.1 M phosphate), $K_{ext} = 1.0 \times 10^7$ (25°).[29]

Acid Decomposition. Dithiocarbamic acids are more or less unstable in acidic solutions:

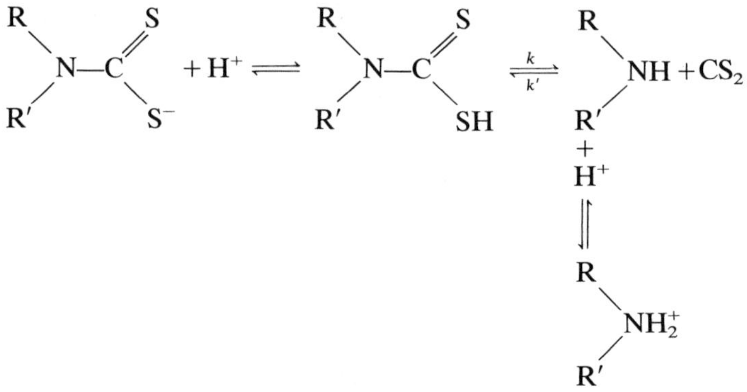

The rate of decomposition depends on the substituents R and R' and, for a given dithiocarbamate, on the pH.[30] The rate equation is

$$V = \frac{k[H^+]}{K_a + [H^+]} C$$

where V is the decomposition rate, k is the limiting rate constant, K_a is the acid dissociation constant of the dithiocarbamic acid, and C is the analytical concentration of dithiocarbamate. When $pH \gg pK_a$,

$$V = \frac{k[H^+]}{K_a} C$$

and when $pH \ll pK_a$,

$$V = kC, \text{ independent of pH.}$$

(Most dithiocarbamic acids, at least those in common analytical use, have pK_a values in the range 3–4.) When the first relation is applicable, the decomposition rate is proportional to $[H^+]$.

Bode[28] found the following half lives of diethyldithiocarbamate ion as a function of pH:

pH	
8.0	3.5 days (extrapolated)
7.0	8.3 hr = 498 min
6.0	51 min
5.0	4.9 min
4.0	30 sec
3.0	3 sec
2.0	0.3 sec (extrapolated)

(The decomposition rate follows closely that given by the first rate equation above.) Therefore, diethyldithiocarbamic acid is unstable even in weakly acidic solutions. Other dithiocarbamic acids are more stable in acidic solution and are to be preferred when metals are to be extracted from acid medium. Thus tetramethylenedithiocarbamate (pyrrolidine-dithiocarbamate) has a half life of 270 min at pH 3.0. Dibenzyldithiocarbamic acid is also much more stable than diethyldithiocarbamic acid and can be used, for example, for the extraction of Cu(II) from 1 M mineral acid.

Dithiocarbamic acids are more stable in organic solvents than in acidic aqueous solution. Diethylammonium diethyldithiocarbamate is soluble in organic solvents (in contradistinction to the sodium salt) and can be used for the extraction of metals from acid solutions.

Composition of Diethyldithiocarbamates; Intrinsic Solubilities.   Most of the solid dithiocarbamates have the composition MDt_n (n is the cation charge), and this usually represents the species extracted into organic solvents. However, mixed complexes containing OH and other anions can be formed. For example, bismuth gives $BiOHDt_2$ as well as $BiDt_3$.

Gold(III) forms mixed chloro chelates

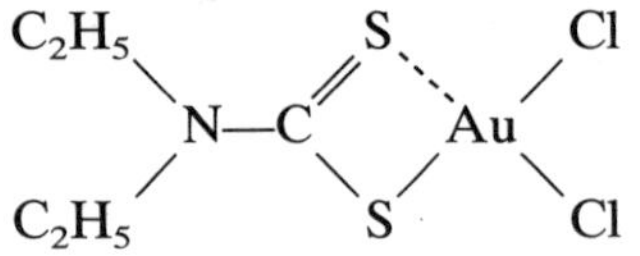

(with a deficiency of dithiocarbamate)

and

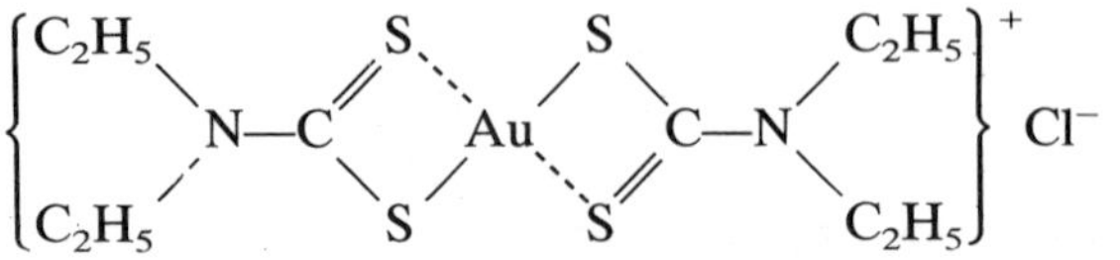

(with an excess of dithiocarbamate)[31]

Because of its polar character, the latter chelate does not extract well into nonpolar solvents. Extraction can be aided by increasing the HCl concentration and adding ClO_4^-. Palladium(II) forms PdClDt as well as $PdDt_2$.

Niobium forms $NbODt_3$; molybdenum apparently, MoO_2Dt_2. Since diethyldithiocarbamate is a reducing agent, higher oxidation states of elements can be reduced. Vanadium is known to form the chelate $VODt_2$. Iron(III) is reduced and forms $FeDt_2$. Copper(II) normally forms $CuDt_2$, but the Cu(I) complex is also known. Under some conditions, cobalt is extracted as $CoDt_3$ or is transformed into this chelate; Stary maintains that $CoDt_2$ is extracted.[32] There is also dispute concerning the oxidation state of extracted manganese. It appears that both $MnDt_2$ and $MnDt_3$ or $MnDt_4$ (oxidation by air) can be formed and extracted into organic solvents.

Tellurium reacts both in the IV and II oxidation states. Not all the ligands are bound in the same manner in $TeDt_4$; two are bidentate, and two are unidentate.[33]

The intrinsic (molecular) solubilities of most diethyldithiocarbamates in water are very low, whereas the solubilities in organic solvents, especially in chloroform, are large (Table 6F-2). It is evident that P_{CHCl_3/H_2O} of MDt_2 will have large values, even if the solubility of MDt_2 in water saturated with $CHCl_3$ is considerably greater than in water alone.

Some diethyldithiocarbamates are associated in organic solvents (C_6H_6 or $CHCl_3$). Åkerström found m in $\{(C_2H_5)_2NCS_2M\}_m$ to be close to 4 for Cu(I), 6 for Ag, and 2 for Au(I) (apparently some dissociation occurs).[34] Tl(I) dialkyldithiocarbamates dimerize in benzene.[35]

Solubility Products, Formation Constants, and Extraction Constants of Diethyldithiocarbamates. Besides being of interest in themselves, these

TABLE 6F-2
Solubilities of Metal Diethyldithiocarbamates in Water, Chloroform, and Carbon Tetrachloride[a]

	H_2O (M)	$CHCl_3$ (M)	CCl_4 (M)
Bi	2.1×10^{-10}	0.21	0.01
Cd	1.8×10^{-7}	0.14	0.12
Co	1.3×10^{-7}	0.51	0.03
Cu(II)	3.2×10^{-10}	0.70	0.07
Mn	1.5×10^{-4}	0.36	0.20
Ni	2.3×10^{-7}	0.54	0.11
Pb	3.4×10^{-8}	0.89	0.10
Zn	2.5×10^{-6}	0.48	0.16

[a] Y. I. Usatenko, V. S. Barkalov, and F. M. Tulyupa, *J. Anal. Chem. USSR*, **25**, 1257 (1970), who also give solubilities of dibutyldithiocarbamates in H_2O, $CHCl_3$, and CCl_4. (Earlier literature data included.) Solubilities of MDt_n in various organic solvents are also given by V. Sedivec and J. Flek, *Chem. Listy*, **52**, 545 (1958). Chloroform solubilities from these two sources do not agree well. Y. I. Usatenko and N. P. Fedash, *Anal. Abst.*, **11**, 5481 (1964), give water solubility of $MnDt_3$ as 3.3×10^{-4}, of $MnDt_4$ as 8.5×10^{-5} M.

constants, together with others, determine the extraction coefficient of a metal (*E*, Chapter 9B, p. 890). A considerable number of solubility products have been determined (Table 6F-3). A number of formation constants are known for aqueous solution and for alcoholic solutions. For Hg^{2+}–Dt^-, $\log \beta_1 = 22.3$, $\log \beta_2 = 38.1$, $\log \beta_3 = 39.1$, all for aqueous solution.[36] For Tl^+–Dt^-, $\log \beta_1 = 4.3$ and $\log \beta_2 = 5.3$ (cited by Hulanicki). $\log \beta_2$ for Zn = 11.4–11.6, for Ni = 12.9–12.1 and for Pb 18.3–17.7 ($\mu = 0.01$–1).[37] The tendency for Dt^- to form anionic metal complexes is not strong. Uranium(VI) can form $UO_2Dt_4^{2-}$.

If the extraction reaction of a metal is written

$$M^{n+} + nDt^- \rightleftharpoons MDt_n$$

$$\text{H}_2\text{O} \quad\quad \text{H}_2\text{O} \quad\quad\quad org$$

the extraction equilibrium constant in terms of the diethyldithiocarbamate ion concentration in the aqueous phase is

$$K_{ext}^* = \frac{[MDt_n]_{org}}{[M^{n+}][Dt^-]^n} = \frac{[MDt_n]_{org}}{[MDt_n]} \times \frac{[MDt_n]}{[M^{n+}][Dt^-]^n}$$

$$= \frac{P_m}{K_m} = \beta_n P_m$$

TABLE 6F-3
Solubility Product Constants (K_{sp}) of Some Diethyldithiocarbamates in Aqueous Solution

	Conditions	pK_{sp}
AgDt	$\mu = 0.3$	20.66^a
		20.47^b
	0.1 M KNO$_3$, 25°	19.6^c
CdDt$_2$	$\mu = 0.3$	19.2^a
		20.0^b
	0.1 M KNO$_3$, 25°	22.0^c
CoDt$_2$	$\mu = 0.3$	20.06^a
CuDt$_2$	0.1 M KNO$_3$, 25°	29.6^c
	0.2 M NaCl, 25°	30.2^d
GaDt$_3$	$\mu = 0.3$	24.5^a
HgDt$_2$	0.1 M KNO$_3$, 20°	43.5^d
InDt$_3$	$\mu = 0.3$	27.06^a
	0.1 M KNO$_3$, 20°	25.0
NiDt$_2$	$\mu = 0.3$	19.26^a
	0.1 M KNO$_3$, 25°	23.1^c
PbDt$_2$	0.1 M KNO$_3$, 25°	21.7^c
	0.1 M KNO$_3$, 25°	21.9^d
TlDt	0.1 M KNO$_3$, 25°	10.1
ZnDt$_2$	$\mu = 0.3$	16.8^a
		17.1^b
	0.1 M KNO$_3$, 25°	16.9^c
	0.1 M KNO$_3$, 25°	17.3^d

[a] G. A. Pevtsov, T. G. Manova, and L. K. Raginskaya, *J. Anal. Chem. USSR*, **22**, 134 (1967).
[b] E. Still, *Finska Kemistsamfundets Medd.*, **73**, 90 (1964).
[c] A. Hulanicki, *Acta Chim. Acad. Sci. Hung.*, **27**, 41 (1961).
[d] W. Kemula and Hulanicki, *Bull. Acad. Polon. Sci. Ser. Sci. Kem.*, **9**, 477 (1961).

in which P_m is the partition constant of metal diethyldithiocarbamate, K_m is the ionization constant of metal diethyldithiocarbamate, β_n is the overall formation constant of MDt$_n$ ($\beta_n = 1/K_m$); K_{ext}^* varies with μ. (The asterisk indicates that K_{ext} is based on [Dt$^-$], not [HDt]$_{org}$.) Also,

$$K_{ext}^* \sim \frac{S_{org}}{K_{sp}}$$

(S_{org} molar solubility of MDt$_n$ in organic extracting solvent.)

The more conventional extraction constant K_{ext} is based on the concentration of HDt in the organic phase at equilibrium:

$$K_{ext} = \frac{[MDt_n]_{org}[H^+]^n}{[M^{n+}][HDt]^n_{org}} = K^*_{ext} \times \left(\frac{K_{HDt}}{P_{HDt}}\right)^n$$

(K_{HDt} is the ionization constant of HDt, P_{HDt} is the partition constant of HDt). For H_2O–CCl_4,

$$K_{ext} = K^*_{ext}\left(\frac{1}{1.6 \times 10^6}\right)^n = (6.25 \times 10^{-7})^n K^*_{ext}$$

and for H_2O–$CHCl_3$

$$K_{ext} = (1.0 \times 10^{-7})^n K^*_{ext}$$

TABLE 6F-4
Extraction Constants of Metal Diethyldithio-carbamates with CCl_4 as Extractant[a]

	$\log K_{ext}$	$\log K^*_{ext}$
Ag	11.9	18.1
Bi	16.8	35.4
Cd	5.4	17.8
Co(II)	2.3	14.8
Cu(II)	13.7	26.1
Fe(II)	1.2	13.4
Hg(II)	32	44.5
In	10.35	29.0
Mn(II)	−4.4	8.0
Ni[b]	11.6	24.0
Pb	7.8	20.2
Pd[c]	>32	
Tl(I)	−0.5	5.7
Zn	3.0	15.4

[a] Values rounded to nearest 0.1. The ionic strength is 0.1.
[b] These values do not agree with the observed incomplete extraction of nickel (diethylammonium diethyldithiocarbamate as reagent) from 0.3 M H^+ solution.
[c] $\log K^*_{ext}$ for Pd(II) in H_2O–$CHCl_3$ is given as 69.8 by G. B. Briscoe and S. Humphries, *Talanta*, **16**, 403 (1969). The constant $[PdDtCl]_{CHCl_3}/[Pd^{2+}][Dt^-][Cl^-]$ is $10^{47.2}$.

K_{ext} or K_{ext}^* can, of course, be determined directly by suitable experimental arrangement. Values of K_{ext} and K_{ext}^* for various metals with CCl_4 as the extracting solvent, determined experimentally, are listed in Table 6F-4.[38]

K_{ext}^* is a convenient constant to use when extractions are made from basic solution (pH > 8 for CCl_4), in which most of the diethyldithiocarbamate is in the aqueous phase. When a metal is extracted from approximately neutral or basic solution—and this is usually the condition existing in extractions with NaDt because of its decomposition in acid solutions—metal species other than M^{n+} and MDt_n must be taken into account. These other species include hydrolyzed metal species, charged metal–diethyldithiocarbamate species and metal complexes formed by foreign anions present in the sample solution or intentionally added. In general, citrate or tartrate will be added to prevent precipitation of metals in basic medium. EDTA and CN^- may be added as differential masking agents. Formation constants of the species mentioned will be needed for a calculation of the extraction coefficient of a metal. More convenient is the use of α-coefficients, derived from formation constants, for this purpose. The α-coefficient of a metal for a particular complexing agent is the ratio of the concentration of metal in its complexes (more accurately its total concentration) to its concentration as the simple metal ion in the aqueous solution. Values of α for many metals and ligands (OH^-, NH_3, citrate, tartrate, CN^-, EDTA, etc.) are tabulated by Ringbom.[39] An example follows.

We wish to find the extraction coefficient of Cu(II) as diethyldithiocarbamate from an aqueous solution, which is 0.01 M in NH_3, 0.1 M in tartrate, 0.1 M in EDTA, and 0.001 M in Dt^- and which has a pH of 9. CCl_4 is the extracting solvent. First, we find $[CuDt_2]_{CCl_4}/[Cu^{2+}]$:

$$\frac{[CuDt_2]_{CCl_4}}{[Cu^{2+}]} = K_{ext}^* \times [Dt^-]^2 = 10^{26.1} \times 10^{-6} = 10^{20.1}$$

(at pH 9, very little HDt is transferred to the CCl_4 phase). The α values are the following:

NH_3	$10^{3.3}$	($\mu = 0.1$)
Tartrate	$10^{3.2}$	($\mu = 1$)
EDTA	10^{16}	($\mu = 0.1$)

It is evident that Cu complexes of NH_3 and tartrate can be neglected, so that

$$E_{Cu} = \frac{[CuDt_2]_{CCl_4}}{\sum[Cu]} = 10^{20.1-16} \sim 10^4$$

The magnitude of E is such that quantitative extraction of Cu is indicated. Formation of intermediate Cu^{2+}–Dt^- complexes has been neglected, which probably is justifiable.

Certain metal diethyldithiocarbamates can be dissociated and removed from an organic solvent phase by shaking with dilute acid, whereas others are not dissociated. For example, when a chloroform solution of the dithiocarbamates of Zn, Cd, and Pb is shaken with 0.16 M hydrochloric acid, these metals are transferred to the aqueous phase, leaving other metal carbamates (e.g., Cu, Ni, Co, and Mn) in the chloroform.[40] Because of rapid decomposition of diethyldithiocarbamic acid in mineral acid solution, the time of shaking may be expected to affect the separations and separations in this way may not be as satisfactory as in the dithizone system.

Extractability of Metals. The metals known to be extractable into chloroform or other organic solvents (not necessarily quantitatively) by diethyldithiocarbamate are

Ti	V	Cr	Mn		Fe	Co	Ni	Cu	Zn	Ga			As	Se
	Nb	Mo			Ru	Rh	Pd	Ag	Cd	In	Sn		Sb	Te
	(Ta)	W	(Re)		Os	Ir	Pt	Au	Hg	Tl	Pb		Bi	Po
		U, Pu												

With some exceptions (Ti, U, Pu, Cr, Nb), these elements form sulfides in aqueous solution. The extraction of Ru, Rh, Os, Ir, and Pt(IV) is slow and incomplete. Manganese and cobalt tend to be oxidized (air) to Mn(III) and Co(III). The existence of Ti^{41} and Ta^{42} chelates is asserted by some authors. It has been claimed that Mg is extracted as diethyldithiocarbamate to the extent of ~10% at pH 5.2–5.4 by $CHCl_3$ or CCl_4.[43] This observation requires confirmation. It seems clear enough that Ti can be extracted quantitatively at pH 2; little is extracted above pH 7.

Because of the acid instability of diethyldithiocarbamic acid, increase in selectivity by extraction from acid solutions is not feasible, and extraction is made from weakly acidic or basic solutions in the presence of masking agents. Bode investigated in detail the extractability of diethyldithiocarbamates by CCl_4.[44] His findings are summarized in Table 6F-5. Al, Be, Ca, Mg, as well as other elements not appearing in the foregoing listing are not extracted in the pH range 4–11. Below pH 6 the reagent decomposes rapidly, but it is quite stable in alkaline solutions as already noted. By the use of cyanide and ethylenediaminetetraacetate, various separations become possible, as can be seen from the table. It appears that it should be possible to determine bismuth in the presence of other elements at pH 11–12 in a solution containing tartrate, cyanide, and sodium ethylenediaminetetraacetate. Thallium must be in the univalent

TABLE 6F-5
Extractability of Diethyldithiocarbamates with Carbon Tetrachloride

Elements extracted at pH > 11 as well as at lower pH (>8):
Ag,[b] Bi, Cd,[a] Co(III),[a,b] Cu(II),[b] Hg(II),[b] Ni,[a] Pb,[a]
Pd,[b] Tl(I),[a] Tl(III), Zn[a,b] (tartrate present).

Elements extracted at pH 9 but incompletely at higher pH:
Fe(III),[a,b] In,[a] Mn(III),[a] Sb(III), Te(IV),[c] (tartrate present).

Elements extracted at pH 6, very little above pH 7:
As(III), Se(IV), Sn(IV), V(V).

Elements incompletely extracted:
Au, Ga, Ir, Nb, Os(IV), Rh, Ru, Pt, U, Te(VI).[d]

---

[a] Elements complexed by ethylenediaminetetraacetate and not extracted at pH 9 or 11.

[b] Elements complexed by cyanide and not extracted at pH 9 or 11. (According to D. Weiss, *Z. Anal. Chem.*, **204,** 102 (1964), Tiron complexes Fe, Mn, and Cu at pH 10, without hindering the extraction of Pb and Bi.)

[c] In presence of Tl(I), Te(IV) gives Te(II) diethyldithiocarbamate on extraction with CCl_4 at pH 8–8.5 (H. Bode, *Z. anorg. u. allgem. Chem.*, **289,** 207 (1957)).

[d] Te(VI) does not react.

state. By first extracting bismuth and other elements reacting at pH 11–12, antimony can be determined in the presence of all elements except tellurium.

Chromium(III) is reported to be extracted into $CHCl_3$ at pH > 5 after the aqueous solution containing diethyldithiocarbamate is heated to 80–90°.[45] Cr(VI) is reduced to Cr(III) and extracted.

Iridium is extracted into CCl_4 at pH 5–7 if first heated for 0.5 hr.

Most diethyldithiocarbamate extractions with $CHCl_3$ or CCl_4 as extracting solvents proceed rather rapidly and usually extraction equilibrium is reached after a few minutes of shaking. But there are exceptions. Complexing agents such as EDTA tend to decrease the extraction rate. Copper in microgram amounts can be extracted as diethyldithiocarbamate from a 2% EDTA solution at pH 8.5 in 1–2 min with $CHCl_3$, but the extraction becomes slow above pH 9 and with higher concentrations of EDTA.

When sodium diethyldithiocarbamate is used as reagent, it is necessarily added to the aqueous phase and heavy metal diethyldithiocarbamates may be precipitated. As a rule, metal chelates dissolve quickly in organic extractants ($CHCl_3$, CCl_4, etc.), and precipitate formation increases the rate of metal extraction. The opposite effect has been found with heavy metal diethyldithiocarbamates by a number of workers. For example, 110 ml of aqueous phase of pH 6, 0.01 M in NaDt, containing 50 μg of

Cu(II), when shaken vigorously with 5 ml of CCl_4 gave only 75% extraction of the copper after 2 hr.[46] Extraction was about 90% complete after 10 min shaking with a 1:1 mixture of chloroform and ethyl acetate. Other diethyldithiocarbamates (except silver) show similar behavior. The slow extraction into nonpolar and weakly polar solvents has been explained by their failure to wet the surface of the insoluble dithiocarbamate particles. Oxygenated solvents, such as esters, wet the particles and increase the solution rate. It should be noted that in many procedures this slow extraction into $CHCl_3$ or CCl_4 is not observed. Also, the rate of extraction is apparently greater in the presence of (suitable) electrolytes.[47] If necessary, precipitate formation can be avoided by using a $CHCl_3$ solution of diethylammonium diethyldithiocarbamate in place of aqueous NaDt.

Absorptiometric Metal Determination.   Chloroform and carbon tetrachloride are the usual solvents for the extraction of metal diethyldithiocarbamates. Unless the pH is high, some HDt will also be extracted. Thus at pH 9, $[HDt]_{CCl_4}/[Dt^-] \sim 0.002$. Diethyldithiocarbamic acid in organic solvents does not absorb in the visible range, so that colored diethyldithiocarbamates are easily determined photometrically. If the absorption peak of the metal chelate lies in the ultraviolet, there may be interference from the excess reagent, which begins to absorb appreciably below ~390 nm. Values of ϵ_{max} in CCl_4 are listed in Table 6F-6. Wavelengths of maximal absorption in $CHCl_3$ are almost the same as in CCl_4[48]: Bi, 370 nm; Co, 650; Cu(II), 440; Fe(II and III), 515; Ni, 395; U(VI), 390 (plateau). Mo(VI), Cr(VI), and Sn(II, IV) give red, green, and orange extracts, respectively.

Simultaneous determination of Cu, Ni, and Co with diethyldithiocarbamate has been proposed.[49]

Some diethyldithiocarbamates (e.g., Cu) tend to be photochemically unstable in CCl_4, usually less so in $CHCl_3$. Probably decomposition products of the solvent act as reducing agents. Mo(VI) diethyldithiocarbamate in $CHCl_3$ gives Mo(V) on standing.

When required, diethyldithiocarbamates can be decomposed with nitric acid or aqua regia and the metal thus obtained in ionic form.

Exchange (Replacement) Extractions.   The differences in the values of K_{ext} for various metals can be put to use in obtaining more selective separations and determinations of metals. The use of $PbDt_2$ as an extraction reagent for Cu(II) furnishes an illustration. When an organic solvent solution of $PbDt_2$ is shaken with an aqueous solution of Cu^{2+} the following equilibrium is established:

$$Cu^{2+} + \underset{org}{PbDt_2} \rightleftharpoons \underset{org}{CuDt_2} + Pb^{2+}$$

TABLE 6F-6
Wavelengths of Maximal Absorption and Molar Absorptivities of Metal Diethyldithiocarbamates in Carbon Tetrachloride[a]

Metal[b]	Color[c]	λ_{max} (nm)	ϵ
Bi*	yellow	366	8620
Co(II)*	green	323	23,300
		367	15,700
		650	549
Cr(VI)	green		
Cu(II)*	brown	365	1180
		436	13,000
Fe(III)*	brown	340	12,700
		515	2490
		600	2050
Mo(V)	red-violet	~450[d]	
Mn(II)*	red (brown-violet)	355	9520
		505	3710
Ni	yellow-green	326	34,200
		393	6110
Pd(II)*	yellow	305	54,800
Sb(III)*		350(shoulder)	3370
Sn(IV)*	orange		
Te(IV)*		428	3160
Tl(III)*		426	1330
U(VI)	red-brown[d]	400–420[d]	
V(V)*		400(plateau)	3790

[a] Mostly from H. Bode, Z. Anal. Chem., **143**, 182 (1954); **144**, 165 (1955). Metals with asterisks can be extracted satisfactorily into CCl_4.

[b] Original oxidation states indicated; Co(II) and Mn(II) form Co(III) and (sometimes) Mn(III) diethyldithiocarbamates by air oxidation. The Mn(II) chelate is colorless. Cr(VI) presumably is reduced to Cr(III). Fe is not extracted if reduced to Fe(II), with thioglycolic acid, in acid solution.

[c] Diethyldithiocarbamates of Pb, Zn, Cd, Ga, In, Tl(I), Se(IV) are colorless, those of Ag and Hg(II) almost so. The last absorbs maximally at 278 nm, with $\epsilon = 33,000$. In $CHCl_3$, ϵ_{max} of $Cd = 3.3 \times 10^4$ (at 262 nm).

[d] $CHCl_3$ solution.

the constant for which is the exchange constant. With CCl_4 as extractant:

$$\frac{[CuDt_2]_{CCl_4}[Pb^{2+}]}{[Cu^{2+}][PbDt_2]_{CCl_4}} = \frac{(K_{ext})_{Cu}}{(K_{ext})_{Pb}} = 10^{13.7-7.8} = 10^{5.9}$$

Evidently there is a strong tendency for Cu^{2+} to be converted to $CuDt_2$, replacing Pb in $PbDt_2$:

$$\frac{[CuDt_2]_{CCl_4}}{[PbDt_2]_{CCl_4}} = 10^{5.9}\frac{[Cu^{2+}]}{[Pb^{2+}]}$$

At equilibrium,

$$\frac{[CuDt_2]_{CCl_4}}{[Cu^{2+}]} = 10^{5.9} \frac{[PbDt_2]_{CCl_4}}{[Pb^{2+}]}$$

In an analytical extraction of Cu, $PbDt_2$ will be in excess of the amount required for reaction with Cu. With a trace amount of Cu, the equilibrium concentration of $PbDt_2$ in CCl_4 will be much the same as the original concentration. $[Pb^{2+}]$ includes Pb^{2+} originally present as well as Pb^{2+} formed by replacement. If much Pb^{2+} is originally present with Cu^{2+}, the extraction of Cu may be incomplete. Suppose we require $[CuDt_2]_{CCl_4}/[Cu^{2+}]$ to be 1000 (99.9% extraction of Cu with equal phase volumes). Then

$$\frac{[PbDt_2]_{CCl_4}}{[Pb^{2+}]} = \frac{10^3}{10^{5.9}} \sim 10^{-3}$$

If this ratio is less than 10^{-3}, $[CuDt_2]_{CCl_4}/[Cu^{2+}]$ drops below 1000. This extraction is suitable for the photometric determination of Cu(II), because remaining $PbDt_2$ in the organic phase does not absorb in the visible range, whereas $CuDt_2$ absorbs strongly at 436 nm. The use of $PbDt_2$ instead of HDt (NaDt) for determining Cu has the advantage that metals having definitely lower K_{ext} values than that of Pb (Fe, Mn, Zn), as well as Pb itself, do not interfere, at least not in reasonable ratios.

If metals are present in forms other than the simple ions, these must be taken into account, that is, $[MDt_n]_{org}/\sum[M]$ must replace $[MDt_n]_{org}/[M^{n+}]$, and effective K_{ext} (or K_{ext}^*) must replace K_{ext}. This requires constants for hydrolysis of metals and complex formation or the equivalent α-values (p. 520). Exchange constants for specified conditions can, of course, also be obtained experimentally. The order of extractability of metals with an organic solvent solution of HDt has often been determined.[50] For example, at pH 6 with $CHCl_3$ as extractant, the order of extraction is (Eckert):

Hg(II) > Ag > Cu(II) > Bi > Pb > Cd > Sn(II), Fe(III) >
Zn, As(III) > Mn(II)

Bode and Tusche, using CCl_4 at pH 8.5 or 11, found the order to be

Hg(II) > Pd(II) > Ag > Cu(II) > Tl(III) > Ni > Bi > Pb >
Co(III) > Cd > Tl(I) > Zn > In > Sb(III) > Fe(III) >
Te(IV) > Mn

(This order is much the same as that given by log K_{ext}/n, Table 6F−4.) The metals In $\cdots$ Mn are extracted only at pH ~8.5, the others also at pH 11.

A considerable number of dithiocarbamate procedures involving exchange extraction are to be found in the literature.[51] Metals such as lead, zinc, and cadmium, whose dithiocarbamates are colorless can be determined indirectly as copper dithiocarbamate. The organic solvent solution of zinc dithiocarbamate, for example, is washed free of excess dithiocarbamate with a weakly alkaline aqueous solution, and is then shaken with an aqueous solution of cupric salt in excess to transform zinc dithiocarbamate into cupric dithiocarbamate.

In practice, deviations from the expected replacement order may occur because of rate factors. For example, Ni diethyldithiocarbamate (and some other carbamates) once formed is more stable than the extractability of Ni from aqueous solution would suggest. Cobalt(II) dithiocarbamate is transformed by air oxidation to Co(III) dithiocarbamate, which is very stable, comparable to Hg(II) dithiocarbamate.

b. Diethylammonium Diethyldithiocarbamate

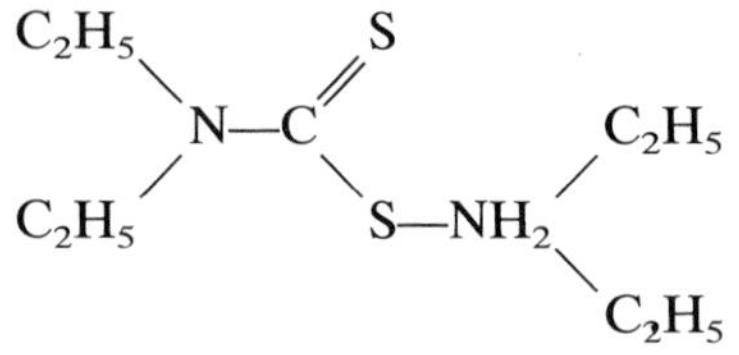

Mol. wt. 222.4. M. p. 82°–83°. Soluble in water, chloroform, carbon tetrachloride and other organic solvents. The solution in chloroform is quite stable if protected from light, but it is preferable to prepare fresh solutions every few days. The commercial reagent sometimes contains yellow-colored impurities that interfere in photometric determinations.

The anion of this reagent is the same as that of sodium diethyldithiocarbamate, but there is a marked difference in the extraction behavior of these two reagents in acid solution. As already pointed out, sodium diethyldithiocarbamate cannot, in general, be used for extraction of metals from mineral acid solutions because of the rapid decomposition of diethyldithiocarbamic acid in aqueous solution. When extractions are made with the diethylammonium salt, the reagent is present in the immiscible organic solvent at the beginning. On shaking with an acidic aqueous solution, more or less of the salt is converted to the dithiocarbamic acid, but this remains in the organic solvent for the most part because the partition coefficient favors distribution into the organic solvent, in which it is quite stable. After a CCl_4 solution of diethylammonium diethyldithiocarbamate has been shaken with an equal volume of 0.1–4 M HCl, 0.1–1.5 M HNO$_3$ or 0.1–4 M H$_2$SO$_4$ for 10 min at 20°, approximately 75% of the reagent still remains undecomposed in the

CCl_4.[52] Accordingly, the decomposition is so slow that metal extractions can be made from quite strong mineral acid solutions. (Above 4 M HCl, decomposition of the reagent in CCl_4 increases rapidly, and almost complete decomposition takes place in 1 min in 7 M HCl; above 5 M H_2SO_4, decomposition rises rapidly.)

The extractability of metals from neutral and basic aqueous solutions would be expected to be much the same for sodium and for diethylammonium diethyldithiocarbamates, and this seems to be true as far as available data go.[53] The extractability from acid solutions, with a carbon tetrachloride solution of diethylammonium diethyldithiocarbamate, is

TABLE 6F-7
Extractability of Metals from Strongly Acid Solutions with a Carbon Tetrachloride Solution of Diethylammonium Diethyldithiocarbamate (0.04%)[a]

Solution	Metals Quantitatively Extracted[b]
6 M HCl	Pd(II), Pt(II), Cu, Hg(II), Tl(III), As(III), Sb(III), Se
10 M HCl	Pd(II), Pt(II)
0.5 M H_2SO_4	Pd(II), Pt(II), Cu(II), Ag, Hg, In, Tl(III), Sn(II), As(III), Sb(III), Bi, Se(IV), Mo(VI),[c] Cd, Pb, Te(IV)
5 M H_2SO_4	As from 0.5 M H_2SO_4, except Cd, Pb, and Te are not extracted quantitatively

[a] H. Bode and F. Neumann, *Z. Anal. Chem.*, **172**, 1 (1960). For extraction of metals with a $CHCl_3$ solution of reagent, see C. L. Luke, *Anal. Chem.*, **28**, 1276 (1956). From 1:9 H_2SO_4, extractable elements are As(III), Sb(III), Sn(II), Hg, Ag, Cd, In, Cu, Bi, Se, Te, Pd, Mo, Cr(VI), Fe(III), and Pb ($HClO_4$ solution), not all quantitatively. As(V), Sb(V), and Sn(IV) are not extractable.

[b] Most of these metals will also be extracted quantitatively from weakly acidic and weakly basic (pH 10) solutions. In addition, such metals as Co(II), Fe(II and III), Mn(II), Ni, and Zn are extracted from weakly acidic and basic solutions; partial extraction of Co, Ni, and Zn takes place in solutions having $[H^+]$ ~0.3 M; Fe(III) is extracted incompletely from 0.1–7 M HCl solutions. Au(III), Os(IV), Pt(IV), Nb(V), and Sn(IV) do not seem to be quantitatively extracted at any pH. Metals not extracted at all include Al, As(V), Cr(III), (but Cr(VI) is partly extracted in acid medium), Sb(V), Ta, Th. U(VI) and W(VI) are hardly extracted by a 0.04% solution of reagent in CCl_4.

[c] 4:1 volume ratio of CCl_4–amyl alcohol for extraction. With this mixed extractant, Mo is completely extracted in the range 1.8 N–pH 4.7 in HCl solution, and in the range 10 N–pH 4.7 in sulfuric acid solution. The extractability of molybdenum into amyl alcohol may indicate the formation of "yl" dithiocarbamate, which would be expected to be very slightly soluble in CCl_4 alone. Mo(VI) is reported to be extractable into $CHCl_3$ from 1.5–4 N HCl.

summarized in Table 6F-7. With a chloroform solution, the extractabilities are much the same, except that with some metals slightly more acid solutions apparently can be used. One difference is the quantitative extraction of U(VI) from a chloride or sulfate solution of pH 6.2–8 and of W(VI) (may not be quantitative) from 5 M HCl. Germanium is partially extractable from ~6 M HCl.

A separation frequently required in trace analysis is that of small amounts of sulfophilic elements from samples composed largely of elements having atomic numbers less than 26 (= Fe). Diethylammonium diethyldithiocarbamate and dithizone (Chapter 6G) are useful extractive reagents for this purpose. There are differences in behavior between the two reagents. In general, for the same metal, K_{ext} is larger with diethyldithiocarbamate than with dithizone. This means that certain metals essentially not extracted by dithizone from dilute mineral acid medium can be extracted with diethyldithiocarbamate. The data already given indicate that In, Sn(II), Cd, and Pb, for example, can be extracted by a CCl_4 solution of diethylammonium dithiocarbamate from ~0.1 M H$^+$, whereas dithizone in the same solvent will not extract these metals from such an acidic solution. Sb(III) and As(III) do not form dithizonates readily if at all, but are easily extracted by diethyldithiocarbamate, even from strongly acid solutions. Mo(VI) diethyldithiocarbamate is extracted by CCl_4 + amyl alcohol; dithizone does not react with Mo. Ag, Cu, Pd, Pt(II), and Hg are extracted by both reagents from mineral acid medium. The extractability of metals from more acid solutions by diethyldithiocarbamate may favor its use for certain purposes. For example, the extraction of lead by dithizone requires an alkaline (citrate or tartrate) solution. This requirement may cause difficulties when a sample contains calcium and phosphorus, which may not be kept in solution by citrate (precipitation of calcium phosphate). On the other hand, lead can be extracted as the dithiocarbamate from a solution of pH ~1, in which calcium phosphate remains in solution. The extraction of iron by dithiocarbamate is somewhat a disadvantage of that reagent. [dithizone does not extract Fe(III)]. A fairly strong reducing agent will apparently keep iron in the Fe(II) state during a dithiocarbamate extraction, so that it will not be extracted from an acid solution. Moderate amounts of Fe(III) can be complexed with citrate so that it is not extracted from a basic solution (pH >9). The extraction of Ti by diethyldithiocarbamate at pH ~2 needs to be borne in mind.

Some metal diethyldithiocarbamates, in contrast to dithizonates, do not absorb appreciably in the visible range. This makes the determination of the absorbing metals more selective but that of the other metals less simple.

Metal dithiocarbamates extracted into an organic solvent can be converted into ionic form by evaporating the organic solvent in a silica dish, igniting the residue for a few minutes below 500° in the presence of a small amount of KCl, and dissolving in 6 M HCl.[54] There is no loss of such elements as Cu, Pb, Cd, Zn, Ni, Co, Mn, and Fe. It would be expected also that metals in the evaporation residue could be recovered after destruction of the complexes by oxidizing acid attack according to standard procedures. Ammoniacal H_2O_2 oxidation, followed by aqua regia treatment has been used.[55]

c. Ammonium Tetramethylenedithiocarbamate (Ammonium Pyrrolidine Dithiocarbamate or More Correctly Ammonium 1-Pyrrolidine-carbodithioate)

$$
\begin{array}{c}
CH_2-CH_2 \\
| \quad\quad\quad\quad\quad N-C \\
CH_2-CH_2 \quad\quad\quad\quad S-NH_4
\end{array}
$$

Mol. wt. 164.3. M. p. 142°–144°. Can be purified by recrystallization from ethanol. The acid has $K_a = 5.1 \times 10^{-4}$ (H_2O, $\mu = 0.01$, 25°); $P_{CHCl_3/H_2O} = 1.1 \times 10^3$.

As noted earlier, tetramethylenedithiocarbamic acid is more stable than diethyldithiocarbamic acid in acidic solutions. Half lives in aqueous citrate solution at pH 2 and 3 are 26 and 40 min.[56] At pH ~1, Bi, Cu, As, Sb, Sn, Pb, Ni, Co, and V are virtually completely extracted into $CHCl_3$ from a solution that is ~0.01 M (~0.2%) in the reagent.[57] (See Table 6F-8.) Quantitative extraction of Cu, Sb and Sn is said to be possible from 6 M or stronger hydrochloric acid. Niobium can be extracted from 8–10 M HCl.[58] It may be expected that, in general, metal extractabilities with tetramethylenedithiocarbamic acid will parallel extractabilities with diethylammonium diethyldithiocarbamate (or with sodium diethyl-dithiocarbamate in neutral or basic solutions). Photometric determinations should also be very similar.[59] Extractions are rapid from weakly acidic solutions. Solubilities of metal tetramethylenedithiocarbamates in $CHCl_3$ and CCl_4 are listed in Table 6F-9. In general, they are less soluble than the corresponding diethyldithiocarbamates.

Determinations of Cd, Co, Bi, and Mo by absorbance measurements in the ultraviolet range have been described.[60] Many metals interfere.

Values of log K_{ext} for a number of metals in the system $CHCl_3$–H_2O have been reported[61]:

Cu(II)	11.4	Zn	0.4
Co(II)	−0.8	Bi	15.5
Cd	1.0	Ga	4.5

These values hold for an ionic strength of 0.1 (24°).

TABLE 6F-8

Chloroform Extractabilities of Some Metals with Ammonium Tetramethylenedithiocarbamate from HCl Solution[a]

| | Metal Extracted (%) | | | |
	6 M HCl	1 M HCl	0.1 M HCl	pH 2
As(III)	93	100	100	99
Bi	90	100	100	98
Cd		20	36	20
Co(II)	16	41	100	100
Cr	0	0	0	0
Cu(II)	99	100	100	100
Fe	45	66	73	100
Mn	trace	0	0	0
Ni	90	100	100	100
Pb		96	100	100
Sb(III)	99	99	100	100
Sn(II?)	96	96	97	96
V(V)	22	70	97	100
Zn	22		80	95

[a] Solution (26 ml) containing ~1 mg metal treated with 5 ml 1% NH_4 tetramethylenedithiocarbamate and extraction made with 20 ml $CHCl_3$ after 2 min. Other data, obtained by S. Gomiscek and H. Malissa, *Anal. Chim. Acta*, **58**, 484 (1972), are relevant. After shaking a $CHCl_3$ solution of metal tetramethylenedithiocarbamates with an equal volume of 1 M HCl for 5 min, the following percentages of metal were found in the aqueous phase: Bi, Cu, Ni, all 0; Fe, 5; Co, 6; Sn, 3; Pb, 26; V, 12; Cd, 96; Zn, 100. Nitric acid, 1 M, decomposes the complexes.

Te(IV) and Te(II) react to form water-soluble complexes in the presence of Triton X-100 (polyethyleneglycol alkyl phenyl ether, a nonionic surfactant). It is considered that the complexes dissolve in micelles of Triton X-100.[62]

Tetramethylenedithiocarbamic acid and catechol form a ternary complex with Nb, which is extractable from 8 M HCl into $CHCl_3$ and serves for the determination of the element ($\lambda = 440$ nm).[63] Titanium and Ta can be masked with F^-, and Fe(III) can be reduced with $SnCl_2$.

Hexyl- and cyclopentyldithiocarbamic acids are reported to be even more stable than tetramethylenedithiocarbamic acid in mineral acid solution.[64]

d. Other Dithiocarbamates

Dibenzyldithiocarbamic acid is relatively stable in acid solution. In the

form of its zinc complex

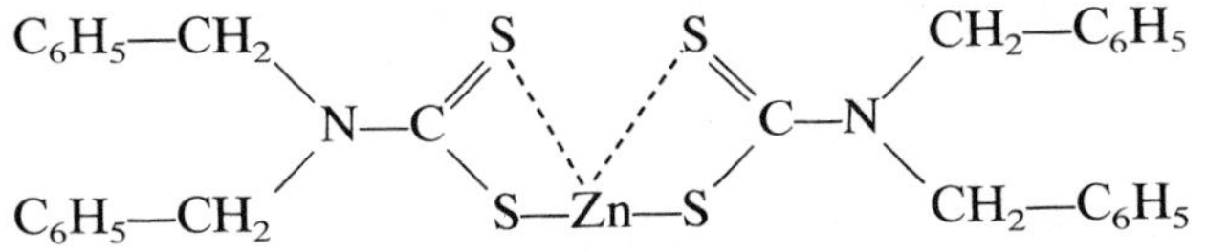

$$C_6H_5-CH_2 \diagdown \qquad \diagup CH_2-C_6H_5$$

it is a well-known reagent for copper. It is also used for determination and separation of Pd and Pt in mineral acid solution.[65] Pb, Ag, Tl(III), Sb(III), and Bi can be extracted from mineral acid solution into CCl_4.[66] It is interesting to compare this reagent with dithizone. The latter does not react with Sb(III) and requires a less acid solution than does dibenzyldithiocarbamic acid for extraction of Bi, nor does it extract Pb and Tl from mineral acid solution.

Other dithiocarbamic acids, having hydrophobic substituents other than those mentioned in the preceding discussion, have been investigated, but their chelating properties and the photometric properties of their chelates do not show anything of great analytical interest. If substituents of hydrophilic character, containing −OH and −COOH groups, are introduced into the dithiocarbamic acid molecule, the reagent and its chelates

TABLE 6F-9

Solubilities of Tetramethylene- dithiocarbamates in $CHCl_3$ and CCl_4[a]

	Solubility (g in 100 ml (20°))	
	$CHCl_3$	CCl_4
NH_4	0.38	0.12
Na	trace	0.003
Bi	26.37	0.06
Cd	0.02	trace
Co	0.39	3.64
Cu(II)	0.32	0.01
Fe(III)	14.49	0.23
Ni	0.09	0.02
Pb	0.03	0.02
Sb(III)	5.28	0.06
Sn	1.77	0.19
V	5.55	0.15
Zn	0.70	0.04

[a] H. Malissa and S. Gomiscek, *Anal. Chim. Acta,* **27,** 402 (1962).

may become soluble in water and little soluble in organic solvents such as carbon tetrachloride. Diethanoldithiocarbamic acid[67] is an example of such a reagent. It forms a Cu(II) chelate soluble in water. Occasionally, the use of a reagent of this type is of value. Various other dithiocarbamates giving soluble metal complexes have been prepared.[30]

Dimethyl- and diethyldiselenocarbamates have been studied as metal reagents, but no advantages over the corresponding sulfur compounds are evident.[68]

Tetraethylthiuram Disulfide

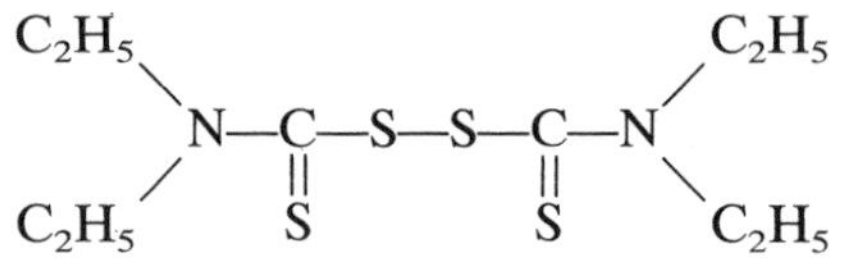

Mol. wt. 296.5. M. p. 71°–72°.

is formed on oxidation of diethyldithiocarbamate. It is reported to form yellow or brown complexes with Pd(II), Cu(II), Te(IV), and Bi and a colorless complex with Hg(II), all extractable into $CHCl_3$ from mineral acid solution.[69] Actually, the complexes formed are the respective diethyldithiocarbamates.[70]

3. Miscellaneous

Dimercaptothiopyran-4-one (Dimercaptothiopyrone)

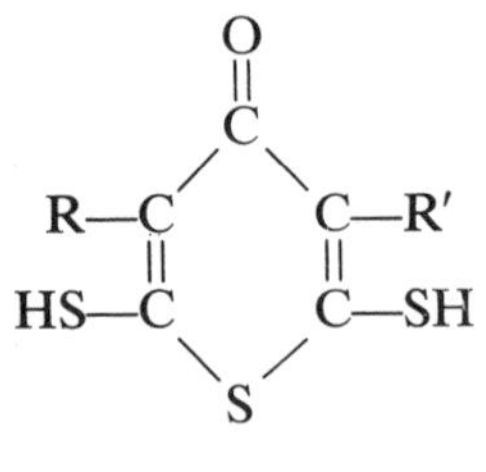

Derivatives of this compound have been prepared and tested as reagents for Bi, which gives colored precipitates in dilute HCl.[71] A surface-active agent is used to keep the precipitates in colloidal suspension. The best reagents have R and R′ = Et or Ph. With these, $\epsilon_{Bi} - \epsilon_{H_2L} = 25{,}400$ and 30,000 (360 nm, 0.5 M HCl). Bi(HL)$_3$ is formed with excess reagent. They have also been suggested for the determination of Sn(IV).[72] Many other metals react.

II. REAGENTS BONDING THROUGH S AND O

Compounds in this division usually have a single S as –SH or in an arrangement tautomerizing to –SH. Oxygen is (usually) the second bonding element. Few of these reagents are analytically meritorious.

A. MERCAPTO ACIDS AND RELATED COMPOUNDS

Mercaptoacetic (Thioglycolic) Acid

$$HSCH_2COOH$$

($pK_{COOH} = 3.68$, $pK_{SH} = 10.55$, 25°) is a respectable reagent for iron. In ammoniacal medium it reacts with Fe(III) to form the red-purple product

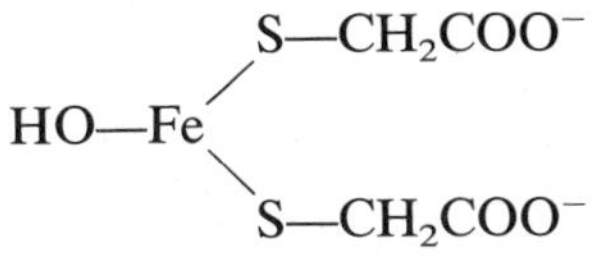

$$HO—Fe \Big\langle \begin{array}{l} S—CH_2COO^- \\ S—CH_2COO^- \end{array}$$

(Iron is probably bonded to oxygen also.) It gives a yellow color with Mo(VI) in acid solution and finds occasional use in its determination. The product is $H_4[Mo_2O_3(SCH_2COO)_4]$.[73] It has also been proposed as a reagent for Ni, Bi, U, Re,[74] Tc, Cr,[75] and other metals, but many elements interfere. Palladium reacts at room temperature, Rh only on heating, so that the two elements can be determined in the presence of each other; other platinum metals are said not to interfere.[76] Colorless complexes are formed by Ag, Bi, Pb, Cd, Hg, Tl, Zn, In, and Sn.

Mercaptoacetic acid reduces V(V) to V(IV), and the dithioglycolic acid (formed mostly by air oxidation) reacts with the latter to give a blue-colored anion

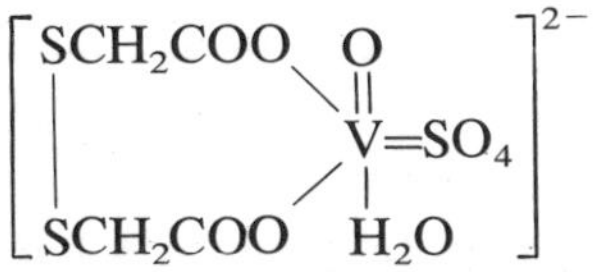

$$\left[\begin{array}{l} SCH_2COO \\ | \qquad\qquad\quad \overset{O}{\underset{|}{\overset{\|}{V}}}{=}SO_4 \\ SCH_2COO \quad H_2O \end{array} \right]^{2-}$$

which can be extracted by the primary amine Primene 81-R ($C_{13}H_{27}NH_2$) into a 1:1 mixture of n-octanol and benzene. However, the sensitivity is so poor ($\epsilon_{630} = 57.5$) that the method is of little interest.[77] Perhaps vanadium is not S-bound, but the reaction is mentioned here as an item of interest.

β-Mercaptopropionic acid has been suggested as a reagent for Ni and Co; mercaptosuccinic acid for Mo and Co; and mercaptobenzoic acid (thiosalicylic acid) for Ni, Bi, Fe(III) (red color in acid, green color in ammoniacal solution), and Re in the presence of Sn(II). Since better reagents are available for these metals, the mercapto acids are of small importance for them.

Thioacetic acid, $CH_3CO(SH)$, forms a red charge-transfer complex with Fe(III), but is of little interest as a colorimetric reagent. Thioacetic acid and analogous thio acids are easily hydrolyzed to hydrogen sulfide and for this reason, if none other, are not suitable as color reagents.

Dithiooxalic acid (as the potassium salt)

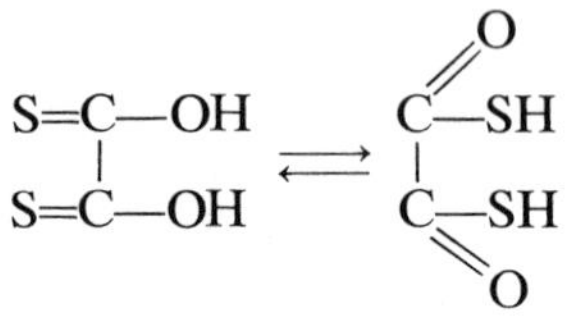

$$
\begin{array}{ccc}
\text{S=C—OH} & & \text{C—SH} \\
| & \rightleftharpoons & | \\
\text{S=C—OH} & & \text{C—SH}
\end{array}
$$

gives a fine red color with Ni, but many other metals (Fe, Cu, Co, etc.) form colored complexes also, so that it is of little importance as a nickel reagent. The red species is anionic, and the H of only one –SH group is replaced by Ni as shown by the extraction into chloroform-acetophenone of $(Ph_3MeAs)_2[Ni(C_2S_2O_2)_2]$.[78] Platinum and Pd have also been determined with dithiooxalate. Dithiosalicyclic acid has also been suggested as an extractive reagent for Ni.[79]

B. β-THIOXOKETONES

Compounds of this class

$$R_1\text{—C(OH)}=\text{CH—C(S)—}R_2$$

form strongly absorbing chelates with sulfophilic elements. The metal complexes are quite stable. Monothiodibenzoylmethane has been considerably studied as a photometric reagent.

Monothiodibenzoylmethane (m.p. 83°–84°) is essentially insoluble in water (7.5×10^{-6} M) but easily soluble in organic solvents[80]:

	Soly. (25°, M)	log P
n-Heptane	0.01	5.6
CCl_4	0.84	5.8
C_6H_6	1.3	6.1
$CHCl_3$	2.0	6.3

The reagent is present in tautomeric forms:

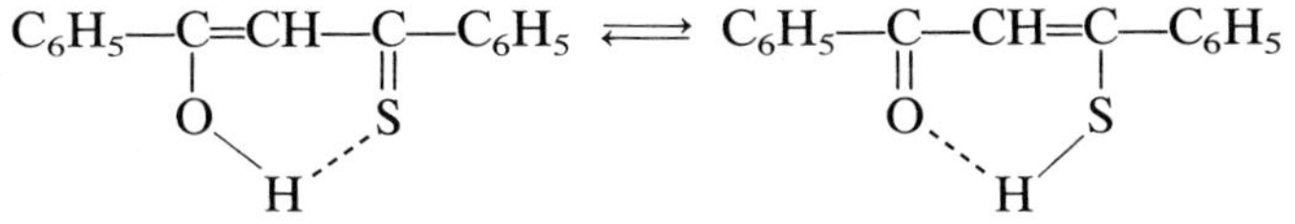

$$C_6H_5-\underset{\underset{H}{\overset{|}{O}}}{C}=CH-\underset{\overset{\|}{S}}{C}-C_6H_5 \;\rightleftharpoons\; C_6H_5-\underset{\overset{\|}{O}}{C}-CH=\underset{\underset{H}{\overset{|}{S}}}{C}-C_6H_5$$

A cyclohexane, *n*-heptane, or benzene solution is often used analytically (prepare fresh daily and store in the dark). The disulfide is easily formed in the light.

The pK_a value is 6.3. When $[H^+]$ is small compared to K_a, the extraction coefficient of the reagent is given by

$$E = [H^+] \times \frac{P}{K_a}$$

(Experimental value of $P/K_a = 10^{12.1}$ for *n*-heptane–H$_2$O; for CHCl$_3$, $P/K_a = 10^{13.0}$.)

Monothiodibenzoylmethane in organic solvents absorbs maximally in the 400–450 nm region, where its chelates also absorb (yellow or brownish colors). The excess reagent must, in general, be removed from the organic solvent before the absorbance of the metal complex can be determined. This can be done by washing out the reagent with dilute sodium hydroxide solution. The extraction of metals is more rapid from basic than from acidic solutions. Only those metals whose chelates resist this treatment and remain in the organic solvent can be determined, namely, principally Cu,[81] Ni,[82] Co,[83] Hg,[84] Cd,[85] and Pd. With 0.001 *M* reagent in *n*-heptane, quantitative extraction of Cu is obtained at pH >2, of Ni and Co at pH >7. Extraction is very slow. Masking agents (e.g., EDTA in the determination of Hg) find use.

Thiothenoyltrifluoroacetone

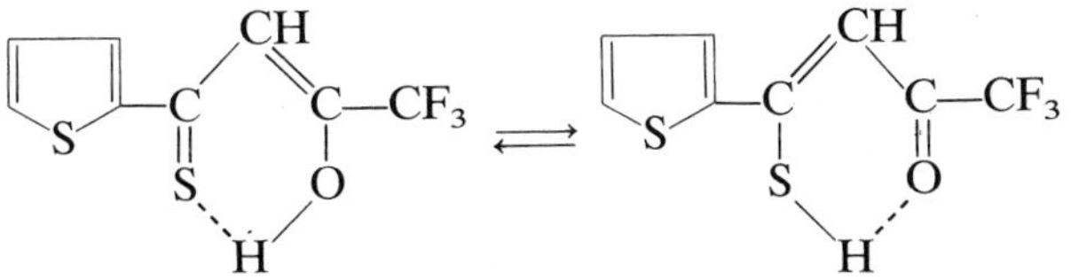

$$\text{S}\underset{\underset{H}{\overset{\cdot}{\underset{\overset{|}{S}}{}}}}{}\overset{\overset{\text{CH}}{\|}}{C}\underset{\overset{|}{O}}{C}-CF_3 \;\rightleftharpoons\; \text{S}\underset{\underset{H}{\overset{\cdot}{\underset{\overset{|}{S}}{}}}}{}\overset{\overset{\text{CH}}{\|}}{C}\underset{\overset{\|}{O}}{C}-CF_3$$

has been used as an extractive photometric reagent for Cu (olive-brown complex in CCl$_4$, $\epsilon_{490} \sim 6000$, pH 4).[86] Many metals interfere. With so many good copper reagents available, it is difficult to see that this reagent can have any practical value. Mercury can be determined by extraction from acid solution.[87] Excess reagent is removed from the extract with basic solution (pH 11). With CHCl$_3$ as the extractant, $K_{ext} = 4 \times 10^{19}$

(Yokoyama et al.[84]). Cobalt can be determined by extraction at pH ~6 into cyclohexane, followed by washing the extract with 1–12 M HCl to remove various reacting metals.[88]

Monothiotrifluoroacetylacetone has been suggested as a reagent for nickel.[89]

C. MISCELLANEOUS

3-Hydroxypyridine-2-thiol

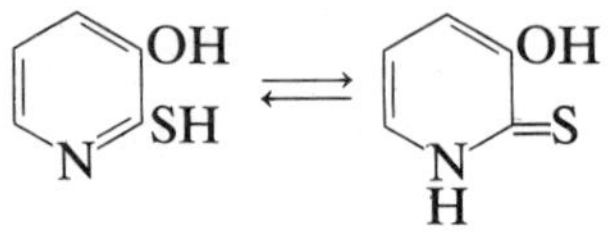

is of slight practical importance, but the complexes it forms with iron are of interest from the general viewpoint.[90] It is an example of a reagent bonding to metal through O and S. It forms three soluble complexes with Fe(III) [Fe(II) apparently reacts similarly], all having a reagent/iron ratio of 2. A green chelate is formed at pH 2, a blue at pH 3.5–11 and a red above pH 11. The complexes are thought to have the compositions $Fe(HL)_2X_2^-$ (H_2L = reagent, X = Cl^-, ClO_4^-, etc.), $Fe(HL)_2^+$ and $Fe(OH)_2L_2^{3-}$.

Thiotropolone has been used as an extractive ($CHCl_3$) photometric reagent for Ni and Co at pH 7.[91]

III. REAGENTS BONDING THROUGH S AND N

This division has numerous members. Some of the most useful reagents of the inorganic trace analyst are counted among them.

Diphenylthiocarbazone (dithizone) is treated in Chapter 6G.

A. THIOCYANATE

Thiocyanic acid is reported to exist as

$$H—S—C \equiv N$$

in water, as

$$S=C=NH \quad \text{(isothiocyanic acid)}$$

in CCl_4, and in both forms in ethyl ether. We write thiocyanic acid as HSCN with the understanding that this can represent either tautomer when not specified. Replacement of H by an equivalent of metal gives thiocyanates, many of which are useful as absorptiometric determination forms or in extractive separations. The metal is bonded to S or N or both. For example, silver thiocyanate forms a polymeric structure $(-Ag-S-C \equiv N-)_n$. In Co thiocyanate, Co is bonded to N.

At least 11 metals—Fe(III), Co, Ti(IV), Nb, Mo(V), W(V), Re(IV,V), U(VI), Tc(V), Ru(III), and Os—form thiocyanate complexes sufficiently strongly colored and stable to make them useful in colorimetric trace analysis (Table 6F-10).[92] Thiocyanate was the earliest colorimetric reagent for iron (detection, 1790; determination, 1852) and it still finds use for this element. Thiocyanate is a standard reagent for Mo, W, and Nb. These elements are readily determined geochemically (as in silicate rocks) in amounts as low as ~1 ppm by use of extraction methods. For Mo, $\mathscr{S}_{480} = 0.0075 \ \mu g/cm^2$ (butyl acetate); for Nb, $\mathscr{S}_{385} = 0.004 \ \mu g/cm^2$ (ethyl acetate).

In the presence of pyridine, chloroform extracts a blue V(IV) thiocyanate [V(V) is reduced by thiocyanate] from HCl solution, whose composition is reported as $VO(C_5H_5N)_2 \ (SCN)_2 \cdot C_5H_5N$; not sensitive. Vanadium-(II) forms a yellow to brown pyridine–thiocyanate complex, chloroform extractable.

In general, thiocyanate forms a series of consecutive complexes with a given metal. This has been demonstrated especially for iron,[93] bismuth,[94] molybdenum, and tungsten.[95] Thiocyanate in the Mo and W complexes is reported to be bonded to metal through N. At low thiocyanate concentrations, Fe(III) forms $FeSCN^{2+}$ predominantly, then $Fe(SCN)_2^+$, $Fe(SCN)_3$ and anionic species as the thiocyanate concentration is increased. These species are all red and have similar spectral curves; the absorption peaks are shifted to longer wavelengths with increasing thiocyanate concentration.

TABLE 6F-10
Elements Giving Colored Thiocyanates

Ti	V	Cr	Mn	Fe	Co	Ni	Cu	Zn	Ga	Ge	As
Zr	Nb	Mo	Tc	Ru	Rh	Pd	Ag	Cd	In	Sn	Sb
Hf	Ta	W	Re	Os	Ir	Pt	Au	Hg	Tl	Pb	Bi
	U										

Solid lines enclose elements forming strongly colored species suitable as determination forms.

Broken lines enclose elements giving more or less strong colors, generally not well suited for determination and infrequently used in trace analysis.

Other thiocyanato complexes can absorb strongly in the UV. For example, $Hg(SCN)_4^{2-}$ has a sensitivity index of $\mathscr{S} = 0.012 \ \mu g \ Hg/cm^2$ for $A = 0.001$. $Pd(SCN)_4^{2-}$ absorbs maximally at 310 nm.

The reaction between thiocyanate and Mo(VI) in the presence of a reducing agent, the basis of an important trace determination of molybdenum, has often been investigated, but with conflicting conclusions and interpretations. Until recently it was generally agreed that the amber to orange-red color produced when Mo(VI) reacts with thiocyanate in the presence of stannous chloride or a weaker reducing agent was due to a species of Mo(V). This belief has now been questioned.[96] It is suggested that the colored complex produced by Cu(I) and ascorbic acid in reaction with Mo(VI) is formed by an isomerization of $MoO_2(SCN)_2$ and hence that it may be an Mo(VI), not an Mo(V), species. Stannous chloride is considered to reduce Mo(VI) to a polynuclear Mo(V)·Mo(VI) complex, then to Mo(V).

The complex formed by W in its determination with thiocyanate and reducing agent is considered[97] to be $W(OH)_2(SCN)_4^-$, and this seems to be in accord with the ion-association complex, $(\phi_4As)[W(OH)_2(SCN)_4]$, formed with tetraphenylarsonium cation and extracted into $CHCl_3$.[98]

Blue $Ru(SCN)^{2+}$ forms on reaction of Ru(III) and Ru(IV), as well as RuO_4, with thiocyanate in $HClO_4$ (and HNO_3) solution.[99] The reaction is sensitive, ϵ_{590} reported as 40,000. The composition of the complexes was later shown to be represented by $RuO(SCN)_n^{1-n}$, n being 1, 2, or 3.[100]

When Re(VII) is reduced with $SnCl_2$ in the presence of SCN^-, a yellow-green thiocyanate complex of Re(V) or an orange-red complex of Re(IV) can be obtained.[101] The latter is reported to be ReO_2SCN^-. The composition $ReO(SCN)_4^-$ is also claimed.

From the practical standpoint, it is of significance that thiocyanic acid is highly ionized. This means that thiocyanate methods can be applied in mineral acid medium, which is always a great advantage. Thiocyanic acid is one of few organic reagents of the type HL forming complexes at high acidities.

All of the metals mentioned give thiocyanate complexes that can be extracted by oxygen-containing organic solvents such as ethers, ketones, and esters. Probably the species extracted are predominantly the thiocyanato acids, for example, $HFe(SCN)_4$. With strongly bonding solvents definite solvates are formed. Tributyl phosphate extracts iron as $Fe(SCN)_3 \cdot 3$ TBP. The yellow Ti(IV) thiocyanate complex is extractable into TOPO–cyclohexane. A 15% solution of TBP in $CHCl_3$ extracts Nb thiocyanate but not Fe(III) and Ti thiocyanates.

A factor requiring attention in most thiocyanate methods is the instability of the colored solutions. A change in color intensity may result from the reducing properties of thiocyanate or its decomposition products (causing fading of ferric thiocyanate color) or, more often, from the slow decomposition of thiocyanate in acidic medium. This decomposition, usually described as a polymerization of thiocyanic acid, results in the formation of yellow products, especially in fairly strongly acidic solutions.[102] Usually better color stability is obtained by using organic solvent media. This is one reason for extracting metal thiocyanate complexes with immiscible organic solvents. Acetone is frequently added to aqueous solutions to improve color constancy. The stability of ethyl ether solutions of niobium thiocyanate is increased by addition of an equal volume of acetone, the chief function of which appears to be the inhibition of the decomposition of extracted thiocyanic acid. Strong light accelerates decomposition in thiocyanate systems.

The use of an acetone–water medium in the determination of metals as thiocyanates has been extensively investigated.[103] In addition to stabilizing the color, acetone increases the color intensity by repressing the dissociation of the thiocyanato complex and aiding the replacement of water molecules around the metal ion by thiocyanate ions. At low thiocyanate concentrations, cobalt does not give a blue color (cobaltothiocyanate species) unless an adequate amount of acetone, ethyl alcohol, or similar solvent[104] is present. It appears that in some metal systems, for example, niobium, acetone exerts an additional favorable effect by minimizing hydrolysis of the metal.

The use of acetone is not without some drawbacks. Metals that do not interfere in aqueous medium may, by giving colors, do so in acetone medium (compare discussions of molybdenum and tungsten). Also, to obtain adequate sensitivity, the sample solution may have to be reduced to an inconveniently small volume. Frequently, immiscible solvent extraction will be superior to the use of water-miscible organic solvent.

Care should be taken that all organic solvents are peroxide-free.

Thiocyanate and Organic Bases. Certain organic bases, usually of high molecular weight, which in acidic medium form the cation RH^+, will give slightly soluble salts with thiocyanato anions such as of zinc and copper. The zinc thiocyanato salts of suitable organic cations are but slightly soluble and are at times utilized in colorimetric, nephelometric, and turbidimetric determinations. Methyl Violet and Rhodamine B are examples of such bases. Rhodamine B has been reported to form $(RH)_2MoO(SCN)_5$ with Mo(V). Tetraphenylarsonium cation allows extraction of Mo(V), W(V), and Nb(V) thiocyanato anions into chloroform. It appears that Nb is extracted as the mixed species $[(C_6H_5)_4As]^+[Nb(SCN)_2X_4]^-$, where X may be Cl^-, OH^-, among others. Extraction into an inert solvent may have advantages over extraction into an O-type solvent. Triisooctylamine in CCl_4 extracts blue $(R_3NH)_2[Co(SCN)_4]$. Tricaprylmethylammonium can be used as well. Niobium has also been determined by extraction of the ion associate

Nb(V)–SCN$^-$–1,10-phenanthroline with a 2% solution of TBP in benzene from 2 M H_2SO_4.[105] The composition of the extracted complex is probably $(HPhen^+)$ $[NbO(OH)(SCN)_3^- \cdot H_2O \cdot TBP]$. $\epsilon_{400\,nm} = 4 \times 10^4$.

In another type of slightly soluble complex thiocyanate compound, the organic base functions as a neutral component coordinated to the metal. For example, pyridine forms slightly soluble compounds of the composition $[Me^{II} \cdot Py_n](CNS)_2$ with copper(II), cadmium, zinc, nickel, cobalt, and other metals. These precipitates are soluble in chloroform; the copper compound $(CuPy_2)(CNS)_2$ gives a green solution, the nickel compound $(NiPy_4)(CNS)_2$, a blue solution, and the cobalt compound $(CoPy_4)(CNS)_2$, a pink solution. The old Biazzo method for copper was based on the extraction of the copper compound with chloroform and comparison or measurement of the color intensity of the latter. Occasionally, traces of the metals mentioned are isolated by extraction of their pyridine–thiocyanate complexes preparatory to determination by other methods. It will be evident that the formation of compounds of this type will, in general, require the use of weakly acidic solutions.

B. THIOUREA, DERIVATIVES, AND RELATED COMPOUNDS

1. Thiourea

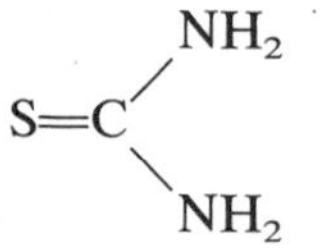

$$S{=}C\Big\langle\begin{array}{l}NH_2\\[4pt]NH_2\end{array}$$ Mol. wt. 76.1. M. p. 180°.

Thiourea forms complexes of the simple coordination type with numerous heavy metals,[106] most of which are colorless and some only slightly soluble in the presence of Cl$^-$ (e.g., Ag, Hg, Tl). Metal is bonded to S, sometimes N, or possibly both jointly. Elements giving colored cationic complexes in mineral acid solution include Bi (yellow), Te (yellow),[107] Os (red), Ru (blue). Procedures have been developed for all of these elements. Selenium(IV) is reduced to the element (red). Antimony gives a weak yellow color; Pd(II), a stronger one. Rhenium develops a yellow coloration in the presence of stannous chloride and can be determined by this reaction.

The species formed with Os(VIII or IV)[108] is Os^{III} $(NH_2CSNH_2)_6^{3+}$. On the other hand, Ru(III,IV) is reported to form complexes in which an H of an NH_2 group is replaced by an atom of metal: Ru(III) $(NHCSNH_2)^{2+}$ and Ru(III) $(NHCSNH_2)_3$.[109] Under analytical conditions, Bi reacts in the ratio 1 Bi : 3 thiourea.[110] The yellow Te complex extracted from a mineral acid solution containing ClO_4^- into MIBK is believed to have the composition Te(II) $Thi_4(ClO_4)_2$. Tellurium(IV) is reduced to II by thiourea.[111]

Thiourea is a very weak base, as shown by an inappreciable change in pH of a dilute strong acid solution to which thiourea has been added. From this observation, K_a of (Thi)H^+ is likely not smaller than 1,[112] and unprotonated Thi will be present to a large extent in moderately acid solutions. This conclusion is in accord with the reaction of metals with the species Thi in mineral acid medium and the relatively small effect of a change in acid concentration.

Thiourea is stable in solution at room temperature, but decomposes slowly at higher temperatures. One of the hydrolysis products is hydrogen sulfide.

Thiourea is a fairly strong reducing agent. In acidic medium, it is oxidized to formamidine disulfide which is protonated.

$$H_2N\diagdown \qquad NH_2\diagup$$
$$C\text{--}S\text{--}S\text{--}C$$
$$HN\diagup\diagup \qquad \diagdown\diagdown NH$$

Substituted Thioureas. The qualitative metal reactions of a large number of these have been studied; some give more sensitive reactions for bismuth than does the parent compound.[113] *s*-Diphenylthiourea, *s*-di-*o*-tolylthiourea and *s*-di-*p*-tolylthiourea have been used as reagents for ruthenium[114]; the sensitivity is much the same as with thiourea (all give blue colors). In general, the substituted thioureas give less sensitive reactions with osmium than does thiourea itself; the osmium complexes absorb only weakly at ~650 nm, where ruthenium absorbs maximally (this is especially true for *s*-diphenylthiourea). The other platinum metals and rhenium hardly absorb at ~650 nm. The substituted thioureas may be said to be selective reagents for ruthenium, at least within the platinum group.

s-Diphenylthiourea reacts more sensitively than thiourea with Te(IV) and its tellurium complex, unlike the thiourea complex, is extractable into chloroform.[115] The reaction proceeds at high acidities (up to 7 *M* HCl, 8 *M* or stronger $HClO_4$). The Re·complex is also chloroform-extractable. 1-Phenyl-2-thiourea finds use in the determination of Re.[116] Diphenyl- or ditolylthiourea reacts with Bi in $HClO_4$ solution to give $BiR_4(ClO_4)_3$, extractable into dichloroethane.[117]

N-Phenylthioureas and thiocyanate form ion-association complexes with Bi that are extractable into $CHCl_3$.[118] ϵ(440–465 nm) ~8400 (in dichloroethane).

2. Thiohydroxamic Acids

A few derivatives have been studied as photometric reagents. N,N'-Diphenylthiocarbamoylhydroxamic acid (N-hydroxythiocarbanilide)

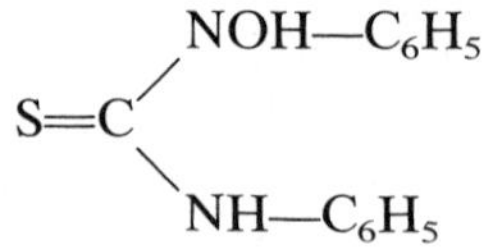

$$S=C\begin{cases} NOH-C_6H_5 \\ NH-C_6H_5 \end{cases}$$

can be used to determine Cu(II) by extraction of its complex from 0.1–0.2 M H^+ by $CHCl_3$.[119] Apparently, H in the group NOH is replaced by metal. Zn, Co, Ni, Fe(II), V(IV), and Mn do not react at this acidity.

Molybdenum can be determined with the same reagent.[120] Mo(VI) is reported to be reduced to Mo(V) by the reagent, which then reacts as follows (0.5 M HCl):

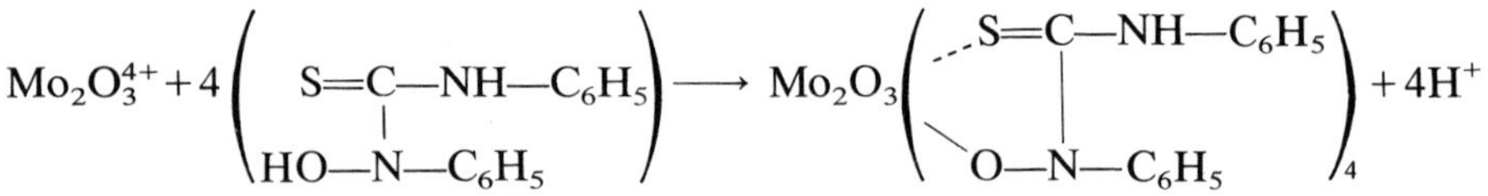

$$Mo_2O_3^{4+} + 4 \left(\begin{array}{c} S=C-NH-C_6H_5 \\ | \\ HO-N-C_6H_5 \end{array} \right) \longrightarrow Mo_2O_3 \left(\begin{array}{c} ..S=C-NH-C_6H_5 \\ | \\ O-N-C_6H_5 \end{array} \right)_4 + 4H^+$$

Fe(III) and V(V) are reduced with ascorbic acid; Cu(II) is masked with thiourea.

3. Thiosemicarbazide

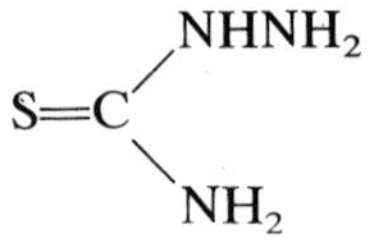

$$S=C\begin{cases} NHNH_2 \\ NH_2 \end{cases}$$

This reagent (m.p. 181°–183°) and its derivatives are chiefly of interest for their chelating reactions with the platinum metals; ruthenium is of special interest. From the standpoint of sensitivity for ruthenium, thiosemicarbazide (green-violet Ru complex)[114] is inferior to 4-phenylthiosemicarbazide[114] (rose violet), 2,4-diphenylthiosemicarbazide (red-violet),[121] and 1,4-diphenylcarbazide (red-violet).[122] The last three give about the same sensitivity ($\mathscr{S} \sim 0.01$ $\mu g/cm^2$ Ru). 1,4- and 2,4-Diphenylthiosemicarbazide both give $CHCl_3$-extractable Ru(III) chelates and very little color with osmium. In our experience, the 1,4-isomer (m.p. 176°) is the preferred reagent. Pd, Pt,[123] and Rh give colored chelates, extractable into chloroform; Ir does not react.

Thiosemicarbazide is also a Cu reagent (pH 3), but it can hardly compete with the better-known reagents for this metal.

Thiosemicarbazide decomposes to S^{2-} in alkaline solution.

1-Phenylthiosemicarbazide and 1,4-diphenylthiosemicarbazide have been proposed as photometric reagents for rhenium in HCl solution.[124] The reaction between the former and Re(VII) in 8 M HCl proceeds in two stages; first the reagent reduces Re(VII) to Re(V), and then it forms a complex with Re(V) ($\epsilon_{365} = 1.59 \times 10^4$). The structure of the complex can be represented by the formula

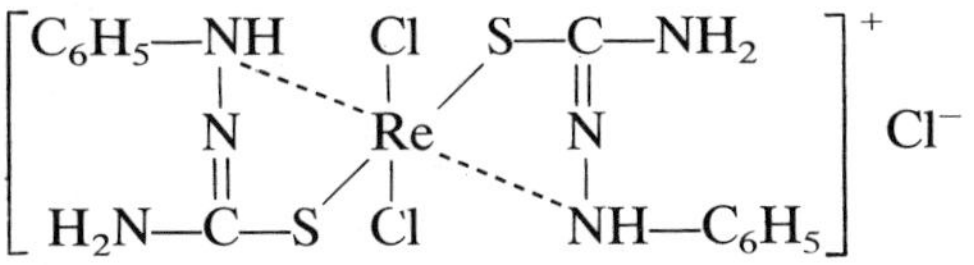

4. Thiosemicarbazones

Biacetyl bis(4-phenyl)thiosemicarbazone forms strongly colored chelates with many heavy metals.[125] These are extractable into CHCl₃ and other solvents.

Observations have been made on the metal reactions of picolinaldehyde thiosemicarbazone (and its Se analog), nicotinaldehyde thiosemicarbazone, and benzaldehyde thiosemicarbazone.[126]

Glyoxal dithiosemicarbazone

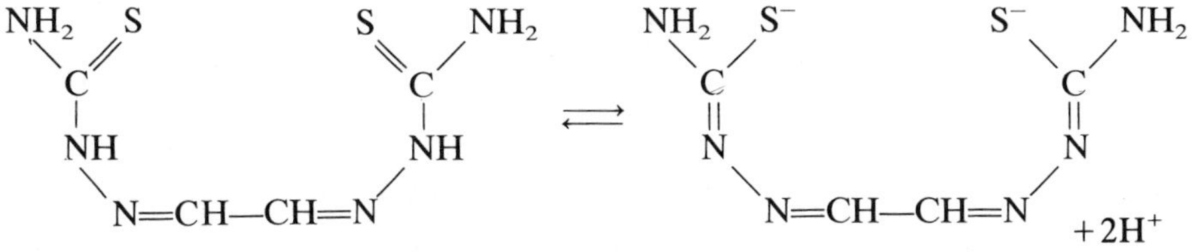

forms 1:1 chelates with Ag, Hg(II), Cu(II), and Pd and has been suggested for the determination of Ag and Hg ($\epsilon_{335} \sim 43{,}000$) at pH $\sim$ 1.[127] EDTA is required for selectivity. Presumably divalent metals give the chelate structure

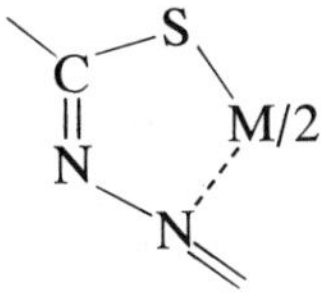

5. o-Hydroxythiobenzhydrazide[128]

$$HO \cdot C_6H_4CS \cdot NHNH_2$$

This compound forms a purple chelate with Ru(III) in 1–8 M HCl on heating. It is soluble in 20% ethanol; $\mathscr{S}_{540} = 0.007$ μg Ru/cm². Ru(III) is reported to be reduced to Ru(II) in the reaction. The reagent reacts with numerous other metals, but small amounts of these are said to be tolerated [Fe(III) and Cu(II) reduced with SO_2].

Platinum forms two complexes, one colorless (in 1–6 M HCl), the other blue (pH 5.5–6.5, hot solution, extractable into MIBK).[129] The blue complex has the composition Pt:reagent $= 1:2$ (apparently two H in the reagent are replaced by one Pt). The colorless complex is believed to be $(C_7H_7N_2OS)_2PtCl_2$.

6. Dithiooxamide

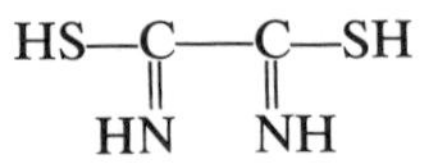

$$HS-\underset{\underset{HN}{\|}}{C}-\underset{\underset{NH}{\|}}{C}-SH$$

Orange-red solid. Mol. wt. 120.2. M. p. 142° (decomposition). Soluble in ethanol.

This compound, also called rubeanic acid, is weakly acidic (p$K = 10.4$, $\mu = 0.1$, 25°) and is assumed to have various tautomeric forms:

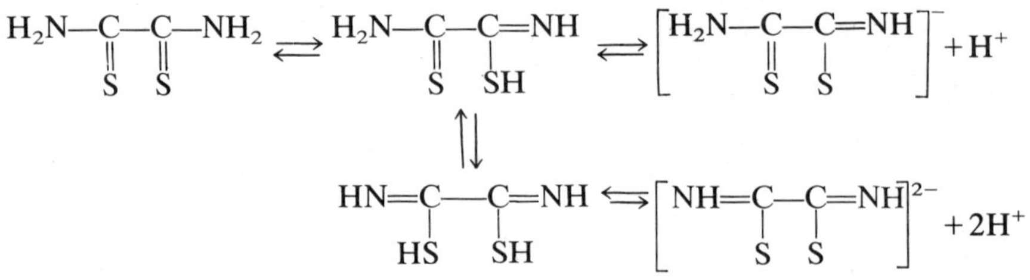

$$H_2N-\underset{\overset{\|}{S}}{C}-\underset{\overset{\|}{S}}{C}-NH_2 \rightleftharpoons H_2N-\underset{\overset{\|}{S}}{C}-\underset{\overset{|}{SH}}{C}=NH \rightleftharpoons \left[H_2N-\underset{\overset{\|}{S}}{C}-\underset{\overset{|}{S}}{C}=NH\right]+H^+$$

$$HN=\underset{\overset{|}{HS}}{C}-\underset{\overset{|}{SH}}{C}=NH \rightleftharpoons \left[NH=\underset{\overset{|}{S}}{C}-\underset{\overset{|}{S}}{C}=NH\right]^{2-}+2H^+$$

It forms slightly soluble colored chelates with numerous sulfophilic metals, including Cu(II), Pd, Pt, Os, Au, Mo, Hg, Pb, Ni, and Co, and has been suggested for the simultaneous photometric determination of Cu (olive green color), Ni (blue), and Co (yellow).[130] These metals give the uncharged complexes CuL, NiL, and $Co_2(III)L_3$, which must be kept in suspension with a protective colloid. Probably polymeric chelates are formed by bonding of metal to S at each end of the ligand. For the Cu(I) complex (which is soluble, unlike the divalent Cu complex), the following structure has been suggested (Paul[130]):

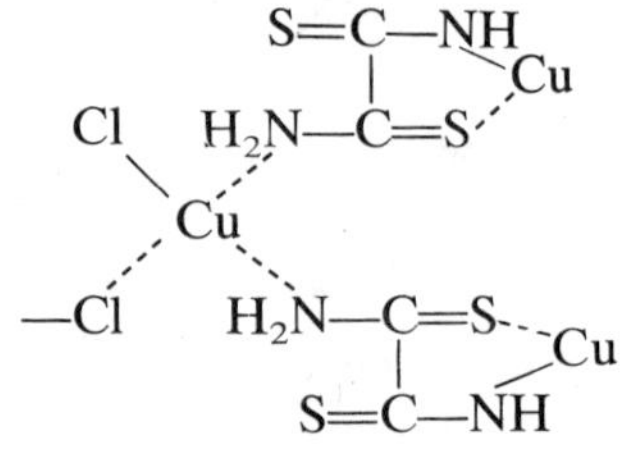

$$
\begin{array}{c}
S{=}C{-}NH \\
| \qquad\;\; {}_\diagdown Cu \\
Cl \quad H_2N{-}C{=}S\diagup \\
\diagdown \qquad\qquad\qquad \\
\;\,Cu \\
{-}Cl \quad H_2N{-}C{=}S\diagdown \\
\qquad\qquad\qquad |\qquad {}_\diagup Cu \\
\qquad\qquad S{=}C{-}NH
\end{array}
$$

The polynuclear nature of the chelates accounts for their insolubility in organic solvents (they do not seem to be easily extractable into the usual immiscible organic liquids).[131]

Of greater interest than the Cu, Ni, Co reaction is the reaction of dithiooxamide with the platinum metals, particularly ruthenium. In acid solution, Ru(III or IV) forms RuL^{2+}, RuL_2^+, and RuL_3, the last having a blue color.[132] Osmium (VIII and VI) reacts in $6\,M$ HCl containing pyridine to give a brownish violet complex having the composition 1:2 reagent.[133] The Pd, and Pt(II), chelate may have the structure

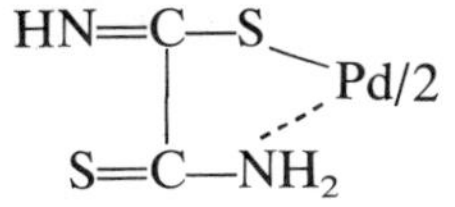

$$
\begin{array}{c}
HN{=}C{-}S \\
| \qquad\qquad Pd/2 \\
S{=}C{-}NH_2
\end{array}
$$

Substituted Dithiooxamides. N,N'-Bis(3-dimethylaminopropyl)dithiooxamide is a reagent for Co, Ni, and Cu[134] and for Pd,[135] Ru,[136] and Rh.[137] N,N-Bis(2-sulfoethyl)dithiooxamide has been used to determine Pd.[138] Strong HCl solutions are used for Pt metal reactions. N,N'-Bis-(m-sulfobenzyl)dithiooxamide forms a red 1:1 complex with Ni in basic solution.[139]

Dimethyl-, diethyl-, and diphenyldithiooxamide have found use in the determination of Cu, Co, Ni, and Pd.

It may be noted that N,N'-bis(2-hydroxyethyl)dithiooxamide forms a series of polynuclear complexes with Ag in acid solution, the composition of which can be represented by two general formulas, $H_{2q}Ag_pX_q$ and $H_{2q-1}Ag_pX_q$, with $q=1, 2, 3$, and $p=q$ or $q-1$.[140]

C. 5-(p-DIMETHYLAMINOBENZYLIDENE)RHODANINE

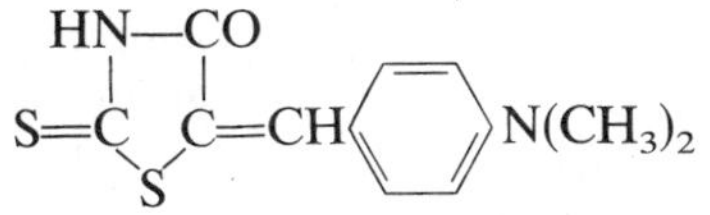

$$
\begin{array}{c}
HN{-}CO \\
| \quad\; | \\
S{=}C \quad C{=}CH{-}\!\!\!\bigcirc\!\!\!{-}N(CH_3)_2 \\
S
\end{array}
$$

Mol. wt. 264.4. M. p. 285°–288°. Red-violet solid, soluble in alcohol and other organic solvents.

This compound was introduced as a qualitative reagent for silver by Feigl in 1928. It, or the ethyl homolog, forms slightly soluble red or violet precipitates with Ag, Hg(I and II), Cu(I), Au(III → I), and Pd(II) in dilute mineral acid solution ($\sim0.05\,M\,H^+$). Platinum(IV) also reacts, but not sensitively, probably by reduction to Pt(II).

These reagents are ampholytic:

$$H_2L^+ \rightleftharpoons HL + H^+ \qquad HL \rightleftharpoons H^+ + L^-$$

very pale   pink   yellow
yellow

(H^+ adds to $-N(CH_3)_2$ or $-N(C_2H_5)_2$.) The dissociation constants

$$K_{H_2L^+} = \frac{[HL][H^+]}{[H_2L^+]} \qquad K_{HL} = \frac{[L^-][H^+]}{[HL]}$$

have been determined for both the methyl (DMABR) and the ethyl (DEABR) reagents[141]:

	K_{H_2L}	K_{HL}	$K_{H_2L}K_{HL}$
DMABR (20% ethanol, $\mu = 0.1, [H^+] = aH^+$)	4×10^{-2}	2×10^{-7}	8×10^{-9}
DEABR (20% ethanol, $\mu = 0.05, [H^+] = aH^+, 20°$)	2×10^{-4}	7×10^{-7}	1.4×10^{-10}

The concentration of undissociated DEABR in a saturated 20% ethanol solution at 20° is 2×10^{-7} M. Because of the low solubility of DEABR and DMABR, they may be present in supersaturated solution in some procedures for the determination of metals, but this does not lead to difficulties as long as the supersaturation is not too great. Thus the calculated solubility of DEABR in 0.05 M HNO_3 (20% ethanol) is about 15 mg/liter. Nevertheless, in the determination of silver, concentrations of 20 and even 40 mg of reagent/liter may exist in the final solution without any abnormalities being evident an hour after mixing.

The insoluble chelates formed by reaction of DMABR and DEABR with Ag, Cu(I), Hg(I), and Au(III→I) have the composition ML. An atom of metal replaces H of the imino group and is bonded to S of the thioketo group (cf. below the functional group of rhodanine and related compounds); Pd forms insoluble PdL_2. It is unlikely that four-membered rings are formed. Probably the structure is polymeric[142]:

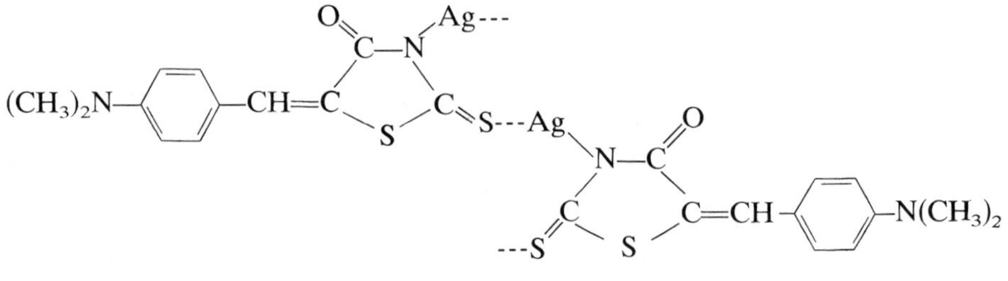

The essential insolubility of the silver complex in the common organic solvents is in accord with its polymeric structure.

Species other than the 1:1 chelates (or 1:2 for Pd) can be formed.[143] Silver also forms AgL_2^-. According to Borissova et al. (*loc. cit.*) the formation constants of the silver complex species are:

$$\frac{[AgL][H^+]}{[Ag^+][HL]} = 9 \times 10^9 \qquad (HL = DMABR)$$

$$\frac{[AgL_2^-][H^+]}{[AgL][HL]} = 2 \times 10^8$$

Pd(II) forms three mononuclear species: PdL^+, PdL_2, and an uncharged species having Pd:DMABR = 1:4, in which two DMABR dimers are supposedly bound to one Pd ion (Borissova et al.[141]). PdL_2 is the predominant species when pH > 1, Pd < 1.8×10^{-6} M, and DMABR < 10^{-5} M; the 1:4 complex is formed at higher Pd and DMABR concentrations. In 0.25 M Cl$^-$ (20% ethanol), the apparent formation constants (β_1, β_2) of PdL^+ and PdL_2 are, respectively, $10^{3.2}$ and $10^{7.8}$; these are the important constants under the conditions of photometric determination of Pd. Au(III) is reduced to Au(I) by DMABR (and DEABR), which reacts to give slightly soluble AuL. The oxidation product of DMABR also absorbs to an extent depending on the pH (see further, Borissova et al.[141]).

These reagents provide sensitive determinations of Ag, Au, and Pd.[144] The methods are not ideal because of the insolubility of the products AgL, AuL, and PdL_2. The Ag chelate of DEABR has $K_{sp} = 8 \times 10^{-19}$ in 20% ethanol, and the molecular solubility is extremely low (not determined but < 10^{-7} M). Possibly supersaturated solutions of the chelates of these metals are obtained at low concentrations, but usually colloidal suspensions are formed. These are fairly stable in very dilute solution, and at higher concentrations can be made so by addition of a protective colloid. The acidity (0.05–0.15 M H$^+$ depending on the metal) must be closely controlled, not only to transform the metals reproducibly into the absorbing species, but also to maintain the excess of reagent at constant absorbance. Differently absorbing species are formed by the metals and the reagent, the relative concentrations of which are a function of the acidity as we have already seen. These methods are valuable in applied analysis because of their sensitivity and selectivity. Silver does not react in acidic chloride solution so that gold can be determined in its presence;

dimethylglyoxime prevents the reaction of Pd. Copper(II) and Pb do not react.

Probably because of their polymeric structure, the insoluble Ag, Au, and Pd chelates of DMABR and DEABR generally do not dissolve as readily in the usual organic extractants as do other chelates. Still, extraction procedures of determination have been described.[145]

Strong oxidizing agents such as chlorine (sometimes found in distilled water) convert DMABR and DEABR into a reddish oxidation product.

Compared to other N–S bonding reagents (many of which form five-membered rings with heavy metals), dimethylaminobenzylidenerhodanine is more selective. Thus, as mentioned, Cu(II) and Pb(II) in relatively high concentrations do not react. A steric effect may be involved in the greater selectivity of the rhodanine reagent.

Numerous derivatives of rhodanine have been prepared and tested as spectrophotometric reagents.[146] These included various 5-substituted rhodanines, 3-substituted 5-p-dimethylaminobenzylidenerhodanine and others. Although many of these give sensitive reactions with Ag, Au, Pd, etc., the metal complexes are insoluble in water and, with one exception, cannot be extracted into organic solvents such as $CHCl_3$, MIBK, isoamyl acetate, and nitrobenzene. A derivative containing a sulfo group gave soluble, colored complexes with Ag and other reacting metals, but the sensitivity was poor.

Various other compounds containing the –HN–CS– group react similarly to dimethylaminobenzylidenerhodanine. Azo compounds based on rhodanine and thiorhodanine have been prepared and tested as reagents for Au, Ag, Pt, and other platinum metals.[147] One of these reagents is Sulfochlorophenolazorhodanine

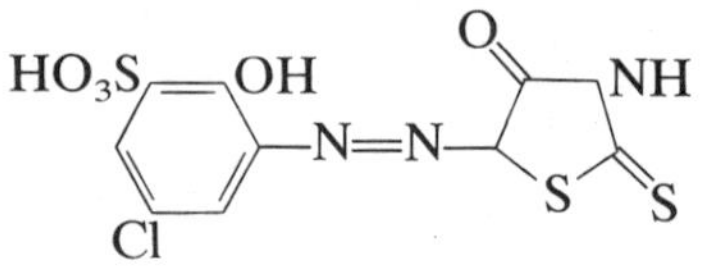

The reactions with Pt (+reducing agent) and Au take place in quite strong mineral acid solutions and are sensitive ($\mathscr{S}_{520} = 0.002\ \mu g$ Pt) but slow. The composition of the products formed is not stated. All the reagents prepared and tested contain –OH *ortho* to –N=N–. The composition of the complexes and the function of the rhodanine grouping are obscure.

The Silver-reacting Group in Rhodanine and Related Compounds. Stephen and Townshend[142] have tested the reactivity of

Ag^+ with compounds differing in part of the molecule from rhodanine to locate the group complexing silver. Their findings may be summarized as follows:

Replacement of S in position 1 of the rhodanine

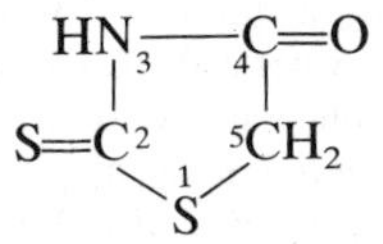

$$\underset{\underset{\displaystyle S}{\diagdown\;\diagup}}{\overset{\displaystyle HN\underset{3}{\rule{0.8cm}{0.4pt}}\overset{4}{C}{=}O}{S{=}\overset{2}{C}\qquad\overset{5}{C}H_2}}$$

molecule by $=NH$ or O (Feigl) does not prevent the formation of an insoluble silver compound.

Replacement of S in position 2 by O destroys silver reactivity, whereas $C{=}NOH$ allows some reactivity and $C{=}NH$ gives as good reactivity as $C{=}S$.

If H in position 3 is missing (replacement by C_6H_5 for example), the rhodanine type of reaction does not occur. A slow reaction does occur, but it is believed to result from decomposition products.

Replacement of CO by CH_2 in position 4 does not prevent reaction, though the product is more dissociated (an excess of reagent then favors solution of the silver precipitate, forming $Ag(LH)_2^+$).

Substitutions in position 5 have no effect on the rhodanine type of reaction.

It follows from these observations that the reactive grouping in rhodanine and related compounds is

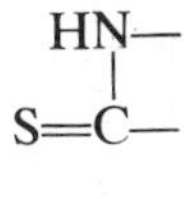

$$\begin{array}{c} HN{-}\\ |\\ S{=}C{-} \end{array}$$

Formation of the silver precipitate by simple replacement of acidic imino H by Ag cannot be the explanation, because then $O{=}C$, in place of $S{=}C$, should promote reaction because of its greater electronegativity. Bonding of Ag to S in position 2 must take place. Bonding of Ag by N in this position also occurs; bonding by $=NOH$ is less effective because O draws electrons away from N and makes them less available for coordination. Substitutions in other parts of the molecule can affect the acidity of the imino group (CH_2 in positions 1 and 4 reduces the acidity and therewith the extent of the precipitation reaction). When NH becomes sufficiently weakly acidic, the H may not be replaced, but Ag^+ can still bond to this group, forming a charged complex. [The stable Ag–thiourea complexes are of this type, e.g., $Ag(thiourea)_3^+$.]

D. 8-MERCAPTOQUINOLINE (THIOOXINE)

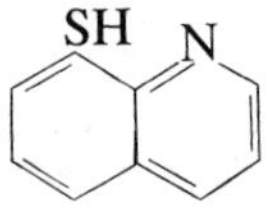

Mol. wt. $(+2H_2O) \doteq 197.2$. The dihydrate (dark red) melts at $58°$–$59°$ with loss of water to a blue liquid. Solubility of dihydrate in water is $0.0034\ M$ (0.67 g in 1 liter, $20°$). More soluble in alcohol (125 g/liter). Both the solid and its solution are easily air-oxidized to $8,8'$-diquinolyl disulfide (major product), more rapidly in alkaline than acidic solution. Mercapto-quinolinium chloride (yellow crystals) is more stable than mercaptoquinoline in air. Hydrated sodium salt of mercaptoquinoline is stable. Reagent can be purified by recrystallization from dilute HCl. pK_{a1} (pK_{NH}) $= 2.16$; pK_{a2} (pK_{SH}) $= 8.38$ (both at $25°$, $\mu = 0.1$).

Properties of Reagent in Solution. 8-Mercaptoquinoline (HTx) can exist in aqueous solution in the forms HTx, HTx$^{\pm}$ (zwitterion), H_2Tx^+, and Tx$^-$. The isoelectric point lies at about pH 5.2. Below this pH, HTx is converted to H_2Tx^+, and above to Tx$^-$:

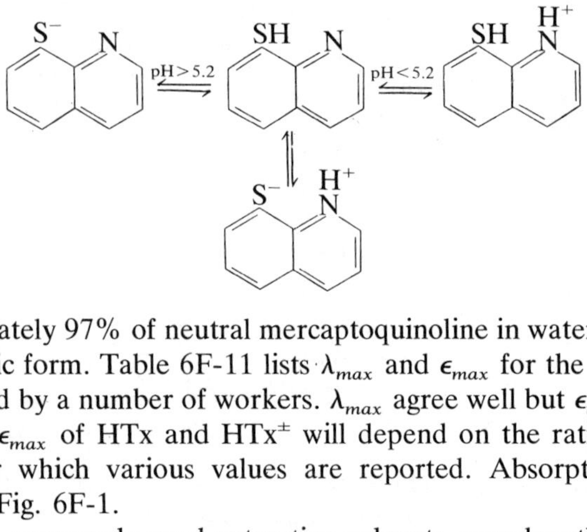

Approximately 97% of neutral mercaptoquinoline in water consists of the zwitterionic form. Table 6F-11 lists λ_{max} and ϵ_{max} for the various species as reported by a number of workers. λ_{max} agree well but ϵ_{max} do not. The values of ϵ_{max} of HTx and HTx$^{\pm}$ will depend on the ratio of these two forms, for which various values are reported. Absorption curves are shown in Fig. 6F-1.

In some commonly used extracting solvents, wavelengths of maximum absorption and molar absorptivities of 8-mercaptoquinoline are[148]:

	λ_{max} (nm)	ϵ
CHCl$_3$	250	22,500
	325	4400
CCl$_4$	333	4540
C$_6$H$_6$	328	4460

TABLE 6F-11

Molar Absorptivities of 8-Mercaptoquinoline at Absorption Maxima (Water Solution)

	Anderson and Hercules[a]		Albert and Barlin[b]		Lee and Freiser[c]		Bankovskii et al.[d]	
	λ(nm)	ϵ	λ(nm)	ϵ	λ(nm)	ϵ	λ(nm)	ϵ
H_2Tx^+ (pH -1)	243	16,570	242	20,000			240	20,870
	316	4280	321	5260			315	5250
HTx (pH 5.2)	252	14,400	254	14,800				
	316	820	323	980				
$HTx^{\pm}$ (pH 5.2)	278	17,970	280	19,100			277	26,400
	446	1600	461	1740	448	2032	466	2460
Tx^- (pH 13)	260	19,290	263	19,100			260	22,300
	367	3250	375	3550			367	4320

[a] P. D. Anderson and D. M. Hercules, *Anal. Chem.*, **38**, 1702 (1966).
[b] A. Albert and G. B. Barlin, *J. Chem. Soc.*, 2385 (1959).
[c] H.-S. Lee and H. Freiser, *J. Org. Chem.*, **25**, 1277 (1960). For λ_{max} and ϵ_{max} of methyl- and phenyl-substituted 8-mercaptoquinolines, see A. Kawase and Freiser, *Anal. Chem.*, **39**, 22 (1967), who also give acid dissociation and zwitterion–thiol tautomer equilibrium constants.
[d] Yu. A. Bankovskii, L. M. Chera, and A. F. Ievin'sh, *J. Anal. Chem. USSR*, **18**, 577 (1963). They report that 79% of the mercaptoquinoline in aqueous solution at the isoelectric point is in the zwitter-ionic form. Kawase and Freiser (*loc. cit.*) find a higher value.

There is hardly any absorption in the visible range. In nonpolar organic solvents, the molecular form HTx predominates over $HTx^{\pm}$. As little as ~0.1% of the latter has been reported in chloroform solutions, accounting for a very weak absorption band at ~560 nm ($\epsilon \sim 1.5$). In ethanol, this band is shifted to ~500 nm and strengthened ($\epsilon \sim 20$), in accordance with increasing solvent polarity and hydrogen bonding.

Chloroform solutions of mercaptoquinoline fluoresce (emission maximum 395 nm), carbon tetrachloride solutions less strongly ($\lambda_{max} = 390$ nm). Because of hydrogen bonding, approximately neutral aqueous solutions do not fluoresce, alcoholic solution extremely weakly. Moderately concentrated perchloric acid solutions of the reagent show blue fluorescence, which is quenched on dilution.[149]

The disulfide, 8,8'-diquinoline disulfide, formed on oxidation of the reagent, absorbs in the ultraviolet but not in the visible. It does not fluoresce in aqueous solution, but other oxidation products formed in small amounts are fluorescent.

The uncharged form of mercaptoquinoline is hydrated in water, but is

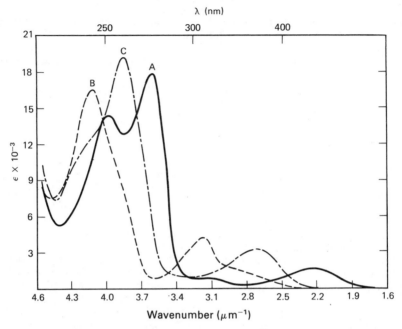

Fig. 6F-1. Absorption curves of 8-mercaptoquinoline (HTx) species: A, $HTx^{\pm}$ (mostly), pH 5.2; B, H_2Tx^+, pH -1; C, Tx^-, pH 13. Reprinted with permission from P. D. Anderson and D. M. Hercules, *Anal. Chem.*, **38**, 1703 (1966). Copyright by the American Chemical Society.

extracted as the anhydrous form into chloroform and similar solvents. Partition constants are as follows at 20°:

	P
$CHCl_3$	320
CCl_4	81
C_6H_6	160

P_{CHCl_3/H_2O} increases by approximately 50% on going from an ionic strength of 0 to 2 M. Since H_2Tx^+ and Tx^- are not extracted into solvents such as $CHCl_3$, the extraction coefficient E decreases from P on either side of the isoelectric point at pH ~ 5.2.

Chelates. 8-Mercaptoquinoline forms slightly soluble colored chelates with the metals of the hydrogen sulfide and ammonium sulfide analytical groups, that is, with the sulfophilic elements.[150] Most of these are extractable into the common organic solvents, chloroform being the favored extractant. A considerable number of metals can be extracted from quite strongly acidic solutions, whereas others require weakly acidic or approximately neutral solutions, so that various metal separations are possible. For example, Cu(II) can be completely extracted from 1 M HCl, whereas Zn hardly begins to be extracted until pH 1 (or higher) is reached. The extractability as a function of acidity is determined largely by the magnitude of the formation constant $[MTx_n]/([M^{n+}][Tx^-]^n)$. The mercaptoquinolates have larger formation constants than do the 8-hydroxyquinolates, which is explainable by the donor π-bonds in the former. It appears that the stability of some mercaptoquinolates depends mainly on the strength of the $M \rightarrow S$ bond, whereas for others the $M \leftarrow N$ bond is more important. The order of decreasing stability of the mercaptoquinolates is stated to be[151]: Re, Au, Ag, Hg, Pd, Pt, Ru, Os, Mo, Cu, W, Cd, In, Zn, Fe, Ir, V, Co, Ni, As, Sb, Sn, Bi, Pb, Mn, and Tl. The decreasing order of extractability of metals into a chloroform solution has been given as: Hg(II), Au(III), Ag, Pd(II), Re(VII), Cu(II), Ni, Co(III), Mo(VI), Bi, Fe(III), (In, Zn, Pb), Mn, Ga, Sb(III), V(IV), Tl(I), Sn(II).[152] It is interesting that W, an element with weak sulfophilic properties, forms an uncharged chelate with mercaptoquinoline having the composition WO_2OHTx ($+2H_2O$ in the solid state).[153] This complex is formed at pH 0.5–3 and can be quantitatively extracted by a 1 : 1 mixture of $CHCl_3$ and isobutyl alcohol. Tellurium(IV) and Se(IV) also react with mercaptoquinoline. Chromium(VI) is reduced in acid solution; the Cr(III) formed reacts at pH 4–7 to form a 1 : 3 complex insoluble in water.[154]

Partition constants of mercaptoquinolates have been reported as follows for $CHCl_3-H_2O$[155]: $HgTx_2$, 1.4×10^6; $ZnTx_2$, 8.4×10^5; $FeTx_3$, 1.3×10^4.

Values of the extraction equilibrium constant based on $[Tx^-]$ in the aqueous phase

$$\frac{[M^{n+}][Tx^-]^n}{[MTx_n]_{CHCl_3}}$$

(this constant is the reciprocal of the extraction constant K_{ext}^*, p. 517) are listed in Table 6F-12.

Most mercaptoquinolates are of the normal type MTx_n but chelates having HTx as addend are also formed. Gallium forms $GaTx_2^+$ (extractable as $GaTx_2Cl$ or $GaTx_2OH$). Since the reagent is a strong reducing agent, metals in higher oxidation state may be reduced to lower. Thus Cu(II) is reduced to Cu(I) and gives the chelate $CuTx \cdot HTx$. Thallium(III), Pt(IV), and V(V) are reduced, but Fe(III) is apparently extracted as such. Ion-pair compounds can also be formed and extracted. For example, Sb(V) is reported to form $H_2TxSbCl_6$ in 6–12 M HCl, $TxSbCl_4$ in acetic acid solution; Sb(III) forms $H_2TxSbCl_4$, $HTxSbCl_3$, and $TxSbCl_2$ under various conditions.[156] Heterocyclic nitrogen bases (e.g., pyridine) are reported to form adducts with $NiTx_2$.

TABLE 6F-12
Values of the Constant $\{[M^{n+}][Tx^-]^n\}/[MTx_n]_{CHCl_3}$[a]
($\mu = 0.1$ in aqueous phase except as noted)

	Constant		Constant
$BiTx_3(2\ M\ HClO_4)$	2×10^{-46}	$MoO_2Tx_2(1.5\ M\ HCl)$	6×10^{-31}
$CoTx_2 \cdot 2\ HTx$	5×10^{-28}	$NiTx_2$	6×10^{-26}
$FeTx_3$	2×10^{-40}	$PbTx_2$	5×10^{-22}
$GaTx_2^+$	5×10^{-27}	$SbTx_2^+(5\ M\ HClO_4)$	2×10^{-32}
$HgTx_2(3\ M\ HCl)^b$	3×10^{-48}	$TlTx$	3.5×10^{-7}
$InTx_3$	5×10^{-42}	$VOTx_2$	2×10^{-24}
$MnTx_2 \cdot 2\ HTx$	8×10^{-29}	$ZnTx_2$	2×10^{-26}

[a] Yu. A. Bankovskii, L. M. Chera, and A. F. Ievin'sh, *J. Anal. Chem. USSR*, **23**, 1134 (1968).
[b] The Hg(I) chelate is reported to be insoluble in $CHCl_3$, but (like AgTx) soluble in pyridine and quinoline.

TABLE 6F-13
Wavelengths of Maximum Absorption, Molar Absorptivities, and Minimum Extraction pH of Mercaptoquinolates[a]
(Chloroform solutions except as indicated)

	λ_{max} (nm)	ϵ	Minimum pH[b]
Cr(VI → III)	450	12,000	3.5
Cu(II → I)	431	7530	0
Fe(III)	444	7000	3
Ga(toluene)	397	8400	6
In(toluene)	407	11,000	4
Ir(IV → III)	485	9950	7
Mn(II)	413	~7000	7
Mo(VI)[c]	420	8600	2
Ni	332, 390, 538	6400 (390 nm)	
Os(III)	558	11,200	4
Pd(II)	485	7750	6 M HCl
Pt(II)	567	7600	3
Re	438	8470	6 M HCl
Rh	465	11,600	5
Ru	555	7300	5
V(V → IV)	412	7400	4

[a] Mostly culled from the papers of Bankovskii and coworkers.
[b] Only approximate. Will vary with concentration of reagent and possibly other factors.
[c] Said to be extractable from ~2 M HCl with a toluene solution of reagent. R. J. Magee and A. S. Witwit, *Anal. Chim. Acta*, **29**, 27 (1963), state that Mo(VI) is reduced and extracted as $MoTx_4$ into $CHCl_3$ at pH 1.3; $\lambda_{max} = 425$ nm, $\epsilon_{max} = 8000$.

Because the mercaptoquinolates are quite strongly colored in organic solvent solution (most are brown or red, Mo is green), a considerable number can be determined photometrically after extraction (Table 6F-13). It does not seem that mercaptoquinoline has any advantages over dithizone—or can even equal it—for the determination of most of the sulfophilic metals. It is a less stable reagent and ϵ_{max} of its chelates is 1/5–1/10 that of the dithizonates. That it is a colorless reagent is a point in its favor, though not a weighty one. Also, it reacts with Re, some platinum metals (Os, Ru, Ir, Rh), Mo, and V, for which no dithizone procedures have been developed, nor are likely to be, since these metals (or most of them) do not react with dithizone as far as now known.

Certain mercaptoquinolates (cf. discussion of 8-hydroxyquinoline), as of Zn and Cd, fluoresce strongly in chloroform solution, but as far as we know mercaptoquinoline has no advantage over 8-hydroxyquinoline for fluorimetric determinations.

Derivatives and Analogs. 2-Methyl-8-mercaptoquinoline, various other methyl- and phenyl-substituted 8-mercaptoquinolines, and various halogen derivatives of 8-mercaptoquinoline have been examined as metal extracting reagents, but it is not apparent that they have any special value in practice. The same may be said of 8-quinolineselenol (8-selenoquinoline).[157]

The allyl ether of 5-sulfo-8-mercaptoquinoline has been used to determine Rh.[158] Allthiox (allyl quinolin-8-yl solfide)

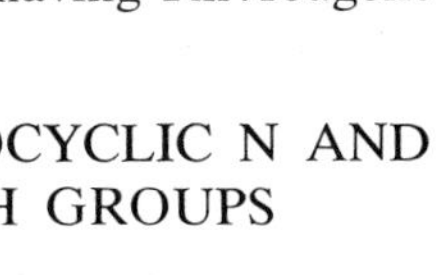

reacts with Rh in hot dilute H_2SO_4 in the presence of ascorbic acid to give a compound ($\epsilon_{400} = 7000$) having Rh:reagent = 1:2.[159]

E. VARIOUS HETEROCYCLIC N AND N–S COMPOUNDS, MOST HAVING –SH GROUPS

2-Mercaptobenzoxazole

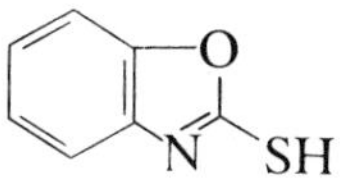

has been tried as a reagent for the platinum metals, but its worth has not been demonstrated.

2-Mercaptobenzimidazole

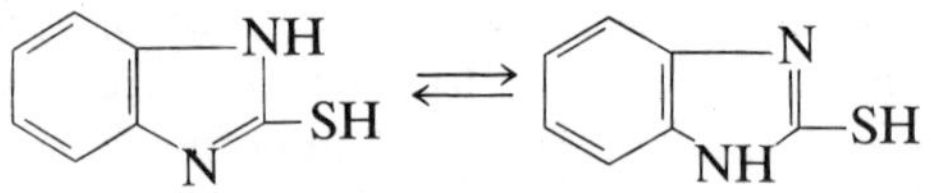

forms $1:1$, $1:2$, and $1:3$ (Ru:reagent) complexes with ruthenium(III).[160] The metal can be determined in aqueous HCl–ethanol medium ($\lambda_{max} = 680$ nm, $\epsilon = 3900$) or by chloroform extraction ($\lambda_{max} = 660$ nm, $\epsilon = 5500$). Osmium has been determined by extracting the $1:2$ complex from dilute HCl solution into a $3:1$ mixture of butanol and benzene ($\lambda_{max} = 550$ nm, $\epsilon = 10,500$).[161] 2-Mercaptobenzimidazole and 2-mercaptobenzothiazole have been used for the spectrophotometric determination of palladium at pH 3.0–6.5 (10% v/v ethanol). Both show similar analytical characteristics.[162] In a medium 0.5 to $1\,M$ in HCl, palladium forms with 2-mercaptobenzimidazole a bright-red compound extractable into butanol.[163]

A "ternary" complex of bismuth with 2-mercaptobenzimidazole and perchlorate can be extracted into chloroform from $2\,M$ HClO$_4$–ethanol solution. The complex has the composition $[\text{Bi}(\text{C}_7\text{H}_5\text{N}_2\text{S})_2 \cdot 2\,\text{C}_7\text{H}_6\text{N}_2\text{S}]$-ClO$_4$. The reaction is sensitive ($\epsilon_{350} = 3.4 \times 10^4$ and $\epsilon_{490} = 1.0 \times 10^4$) and selective.[164]

5-Amino-2-mercaptobenzimidazole forms an orange-yellow $1:3$ (M:reagent) complex with rhodium and an orange $1:2$ complex with palladium. The two metals can be determined by developing the Pd color in the cold and the Rh color by heating.[165]

2-(Mercaptomethyl)benzimidazole (isoamyl alcohol extraction) is claimed to be a more selective reagent for palladium than 2-mercapto-benzimidazole, -benzoxazole, or -benzothiazole.[166]

2-Mercaptobenzothiazole

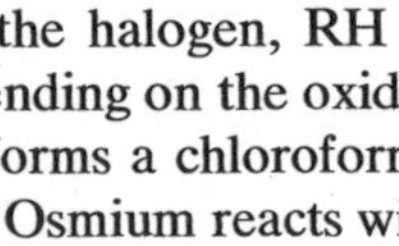

reacts with Ru, Rh, Ir, and Pd in acid halide solution as follows:

$$\text{MX}_j + 2\,\text{RH} \rightarrow \text{M(RH)}_2\text{X}_j$$

where M is the metal, X is the halogen, RH is the reagent, and j may have the value 2, 3, or 4 depending on the oxidation state of the metal.[167]

2-Mercaptobenzothiazole forms a chloroform-extractable blue precipitate with osmium at pH 4.[168] Osmium reacts with the reagent in a ratio of $1:2$. Nickel has been determined by extraction of its yellowish brown complex into chloroform at pH 8.[169] Many metals interfere.

2-Mercapto-4,5-dimethylthiazole gives colored complexes with Rh (yellow) and other platinum metals and has been used for the determination of the former.

Bismuthiols

Bismuthiols are reagents having the group

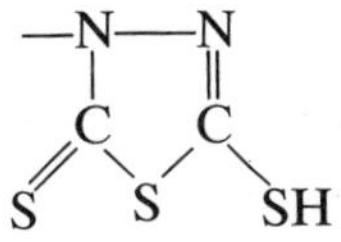

Bismuthiol I (H is attached to N) has been used for the determination of palladium. Bismuthiol II is the phenyl compound; the potassium salt is commercially available.[170] It reacts with sulfide-forming ions to form colored, water-insoluble complexes. The formula of the complexes is probably

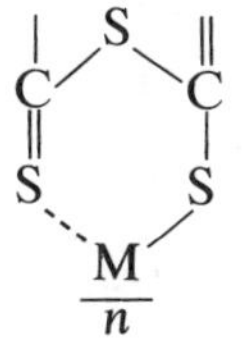

Bismuth and palladium have been determined in the presence of a dispersing agent.[171] Osmium forms a complex with Bismuthiol II that can be extracted by ethyl acetate; the reaction is not selective.[172] Bismuthiol II is a useful reagent for extraction-photometric determination of tellurium.

Several Bismuthiols have been synthesized, and their analytical utility has been examined. For example, 5-mercapto-3-(2-naphthyl)-1,3,4-thiadiazole-2-thione

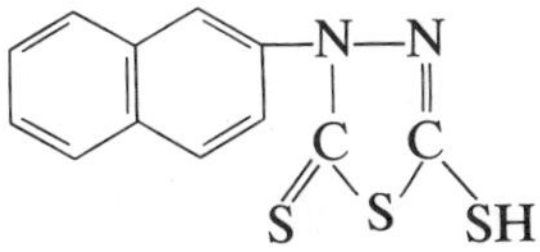

has been used for the extractive determination of Au[173] and Bi.[174]

1,2,4-Triazoline-3-thione

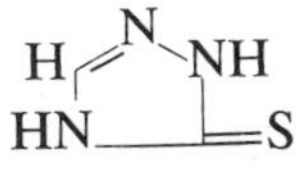

presumably can yield –SH tautomerically, but the reported complexes of Rh and Pd are cationic with the uncharged molecule in the structure, that is, coordination complexes, similar to some thiourea complexes. Rhodium(III) is reported[175] to form RhL_6^{3+}, Pd(II) to form PdL_4^{2+}.

F. MISCELLANEOUS REAGENTS BONDING THROUGH N AND S

1. 2-Diethylaminoethanethiol

$$(C_2H_5)_2N—CH_2—CH_2—SH \text{ (as HCl salt)}$$

A reagent for Rh and Pd (yellow colors).[176] Given the propensities of these metals to bond to S and N, it may be supposed that five-membered ring complexes are formed.

2. Thioglycolic-β-aminonaphthalide (Thionalide)

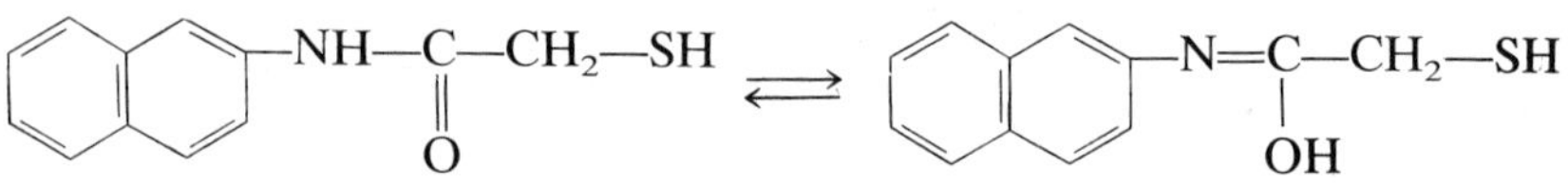

Mol. wt. 217.1. M. p. 111°–112°.

Thionalide yields very slightly soluble white or pale-colored chelates of the type ML_n with most of the metals of the hydrogen sulfide group.[177] It can be used for indirect determination of these metals (reduction of phosphotungstomolybdic acid by the thionalide in the metal complex).

Thionalide, like hydrogen sulfide, precipitates copper, silver, gold, mercury, tin, arsenic, antimony, bismuth, platinum, palladium,[178] rhodium, and ruthenium from dilute mineral acid solutions. The precipitates are extremely slightly soluble (Table 6F-14). Unlike hydrogen sulfide, thionalide does not precipitate lead and cadium in dilute mineral acid medium. In a tartrate solution made basic with sodium carbonate, copper, gold, mercury, cadmium, and thallium are precipitated; in an alkaline cyanide–tartrate solution, gold, thallium, tin, lead, antimony, and bismuth are precipitated; in a sodium hydroxide–cyanide–tartrate solution only thallium is precipitated. Thionalide is superior to hydrogen sulfide as a precipitant because coprecipitation is less and the precipitates have the theoretical composition.

Thionalide is readily soluble in the common organic solvents, but only slightly soluble in water or dilute mineral acid medium. In water, the solubility in 100 ml is ~0.01 g at 20°, and 0.08 g at 95°. Acetic acid increases the solubility. Usually a 1% solution of the reagent in ethyl

TABLE 6F-14
**Sensitivity of Precipitation of Metals by
Thionalide**[a]
(in 0.2 N mineral acid solution)

Metal	Limiting Concentration
Cu	$1:10^7$
Ag	$1:5\times10^6$
Au	$1:2.5\times10^6$
Hg(II)	$1:1.5\times10^7$
Sn(II)	$1:1.25\times10^7$
As	$1:10^8$
Sb(III)	$1:4\times10^7$
Bi	$1:10^7$
Pt	$1:10^7$
Pd	$1:10^7$

[a] According to R. Berg. Some thionalates are chloroform extractable (As, Sb, Bi, Pd, and others).

alcohol or glacial acetic acid is used. The solution decomposes in a few hours and should be prepared fresh before use.

The solvent effect of alcohol or acetic acid on the metal–thionalide complexes is not of importance as long as the volume of the organic solvent is not greater than 10–15%. Oxidizing agents, including ferric iron, destroy the reagent, and hydroxylamine sulfate should be added before precipitation when these are present.

3.   Sodium-p-(mercaptoacetamido)benzene Sulfonate

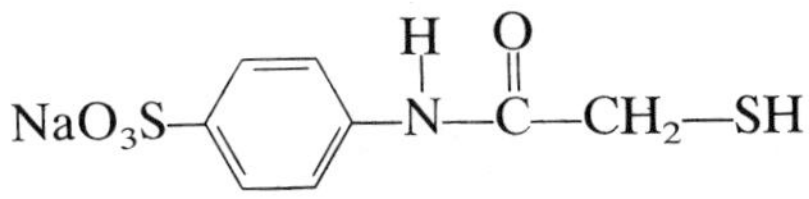

This compound, whose functional group is the same as that of thionalide, is known as a Co reagent.[179] In the pH range 6.5–7.5, it gives a soluble brown-red complex with Co (III, by air oxidation)

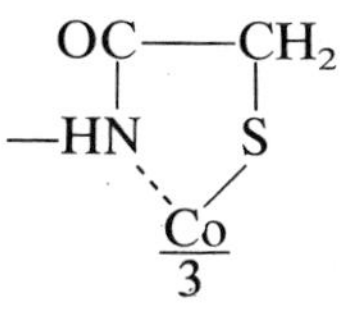

and a brown complex at pH 8.5–11, apparently

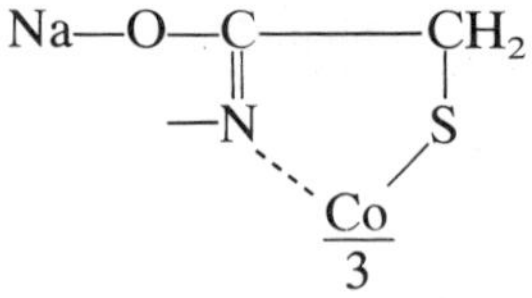

$$\underset{\substack{\displaystyle | \\ \displaystyle \underset{\displaystyle Co}{\underset{\displaystyle 3}{N\cdots}}}}{Na-O-\underset{\|}{C}\text{------}\underset{\displaystyle S}{CH_2}}$$

The reagent is mentioned here because most other metals give no color or pale colors (thionalide also gives pale-colored chelates). Pd(II) produces a yellow color, Ni and Fe(II) light brown colors. The bluish black color given by Cu disappears on addition of excess reagent.

4. 4,4'-Bis(dimethylamino)thiobenzophenone (Thio-Michler's Ketone)

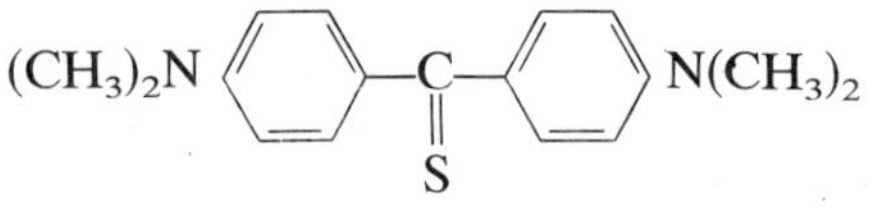

$$(CH_3)_2N \longbenzene{}-\underset{\|}{\underset{S}{C}}-\longbenzene{} N(CH_3)_2$$

The placement of this reagent is uncertain; the structure of its metal complexes is unknown (as far as we are aware). It contains no H replaceable by metals and is described as forming colored addition compounds with Hg(II,I), Ag, Cu(I), Au(III), Pd(II), and Pt(II).[180] This list is the same as that of metals reacting with dimethylaminobenzylidenerhodanine and presumably S is a bonding element. (Thioketones R_2CS are known to form compounds in the ratio 1:1 and 2:1 with Cu_2Cl_2 and $HgCl_2$.) The reagent (yellow in feebly acidic solution) gives an orange-pink complex with Cu(I) and purplish ones with the other metals in acetate-buffered medium. Metals mentioned as not reacting include Bi, Cd, Co, Fe, Ni, Os, Sb, Sn, Tl, U, and Zn. In the presence of mineral acid, the reagent (yellow) is oxidized (as by free chlorine in water) to blue disulfide. In addition to Hg, Ag[181] and Au[182] have been determined with this reagent by extraction methods.

5. 2-(o-Hydroxyphenyl)benzthiazoline

Condensation of salicylaldehyde with o-aminothiophenol yields this compound

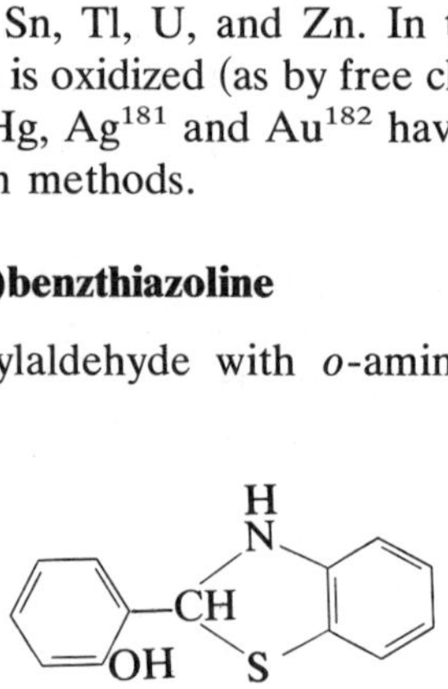

$$\underset{\displaystyle OH}{\underset{\displaystyle \|}{\text{benzene}}}-\underset{\displaystyle S}{CH}\underset{\displaystyle \overset{H}{N}}{\text{benzene}}$$

which forms chelates with sulfophilic metals,[183]

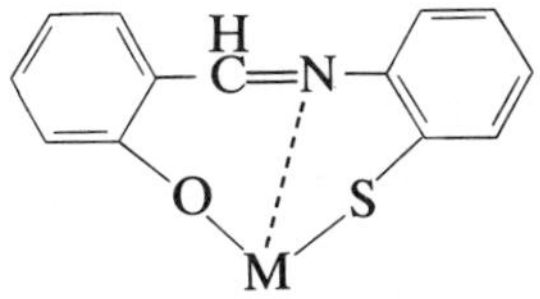

in which metal is bound to O as well as to S and N. Tin (irrespective of its original valence) forms a yellow Sn(IV) chelate extractable into xylene or toluene at pH 2.[183,184] Lead is extractable at pH 11.

6. 2-Amino-1-cyclopentene-1-dithiocarboxylic Acid

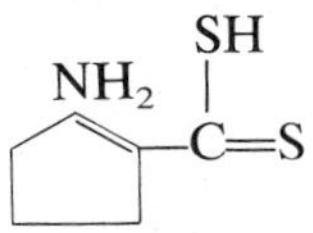

This reagent reacts sensitively with nickel in the pH range 2–8 to form a red chelate extractable into chloroform.[185] Fe(II and III), Mn, Zn, Cu, and Co also react, but less sensitively than Ni. The reacting ratio Ni:reagent is 1:3.

7. Bis(thioantipyrinyl)methane[186]

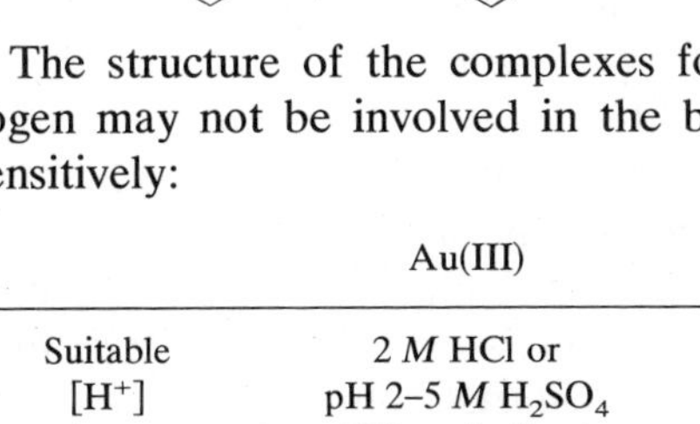

Readily soluble in mineral and organic acids, acetone, $CHCl_3$ and alcohols. M. p. 237.5°.

The structure of the complexes formed is apparently unknown. Nitrogen may not be involved in the bonding. The following metals react sensitively:

	Au(III)	Bi(III)	Mo(VI)
Suitable [H$^+$]	2 M HCl or pH 2–5 M H$_2$SO$_4$	pH 3–0.5 M H$_2$SO$_4$	0.6 M H$_2$SO$_4$ (critical)
λ_{max}	370 nm (yellow)	535 nm (crimson-red)	540 nm
ϵ	3.5×10^4	1.26×10^4	8×10^3
M:L	1:4	1:2	—

8. 2-Thiophene-*trans*-aldoxime

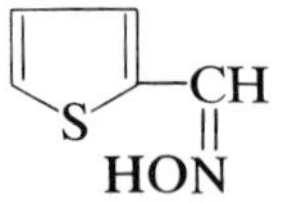

This is not an established photometric reagent,[186a] but is mentioned here because of the composition of its insoluble Pd chelate, formed in HCl solution. H of the –NOH group of the reagent is not replaced by metal, and the precipitate has the composition $Pd(C_5H_5ONS)_2Cl_2$. Apparently heterocyclic S and presumably N are bonded to Pd.

IV. PHOSPHORODITHIOATES

O,O-Dialkyl- and *O,O*-diaryl phosphorodithioic acids resemble diethyldithiocarbamic acid in their reactions with sulfophilic elements.[187] Extraction with di-*n*-butyl phosphorodithioic acid is described in Chapter 9C. Ionization and partition constants of dialkyl phosphorodithioic acids are given in Table 9C–5.

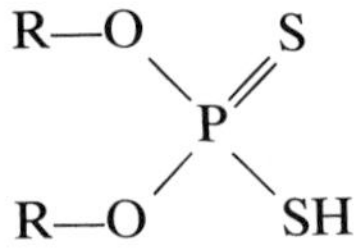

Sodium diethyl phosphorodithioate (m.p. 186°–187°) is hygroscopic and forms a trihydrate. It reacts with the following elements in dilute HCl or H_2SO_4 solutions to form carbon tetrachloride-extractable complexes: Cu(II), Ag, Au(III), Cd, Hg(II), In, Pb, As(III), Sb(III), Bi, Se(IV), Te(IV), and Pd(II). The following metals can be determined spectrophotometrically with Na, K, or Ni diethyl phosphorodithioate after extraction with carbon tetrachloride: Cu ($\lambda_{max} = 420$ nm, $\epsilon = 1.5 \times 10^4$), Pb (295 nm, 7.5×10^3), Bi (329 nm, 1.6×10^4), and Pd (295 nm, 3.0×10^4; a wavelength of 340 nm can also be used for determination, $\epsilon = 2.5 \times 10^3$). The nickel complex is soluble in water and in organic solvents.

Zinc *O,O*-diisopropyl phosphorodithioate (m.p. 145°) has been used for the determination of copper.[188] Copper is extracted from 4 *M* hydrochloric acid solution with a carbon tetrachloride solution of the reagent. The absorption maximum of the complex lies at 420 nm. Bismuth interferes seriously.

O,O-Diphenyl phosphorodithioic acid forms 1:2 complexes with

Cu(II) in dilute sulfuric acid solution[189] and with Pd(II) in dilute hydrochloric acid solution,[190] and the complexes can be extracted into dichloroethane or carbon tetrachloride ($\epsilon_{420} = 1.3 \times 10^4$ for Cu and $\epsilon_{295} = 3.0 \times 10^4$ for Pd).

Diphenyldithiophosphinic acid $(C_6H_5)_2PSSH$, forms CCl_4-extractable complexes with Ru, Os(IV), Rh(III), and Pt(II).[191]

Potassium diphenylselenophosphate has been suggested as a reagent for Bi in methanol or ethanol solution.[192]

ADDENDA

2-Aminobenzenethiol

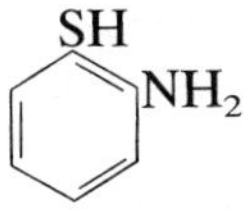

in excess reacts with Mo(VI) at pH $\sim$2 to form a green complex $MoO(OH)L_3$, which is extractable into chloroform; $\mathscr{S} = 0.0075$ μg Mo/cm^2 at 700 nm.[193] Small amounts of W(VI) can be tolerated. Various other sulfophilic metals also react to a greater or less extent. Presumably Mo is bonded to S and N, forming a five-membered ring system with the reacting group

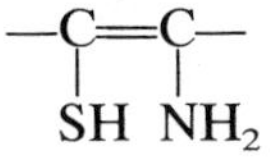

2,2'-Diaminodiphenyldisulfide is a reagent for Pt(IV), which forms a 1 Pt:2 L complex at pH 5–6. $\mathscr{S} = 0.0035$ μg Pt/cm^2 in 50% ethanol.[194] Apparently the other platinum metals may be present in 10- to 50-fold amounts.

NOTES

1. Geochemists refer to elements forming sulfides at high temperatures, in melts for example, as chalcophilic. Chromium(III) has chalcophilic tendencies in the essential absence of water, as in meteorites. But in aqueous systems, the oxyphilic character predominates (Cr sulfide would be hydrolyzed).

2. A. I. Busev and V. V. Evsikov, *Vest. Mosk. Gos. Univ. Ser. Khim.* (5), 96 (1969); A. I. Busev and V. V. Evsikov, *Zh. Anal. Khim.*, **25**, 953 (1970); M. G. Voronkov and F. P. Tsiper, *Zh. Anal. Khim.*, **6**, 331 (1951).

3. A. I. Busev, V. V. Evsikov, and F. A. Khromova, *Vest. Mosk. Gos. Univ. Ser. Khim.* (1), 99 (1969).

4. O. P. Ryabushko, A. T. Pilipenko, and N. L. Emchenko, *Ukr. Khim. Zh.*, **40**, 190 (1974). Mo(VI or V) can also be determined, as a yellow complex: A. I. Busev, F. Chang, and Z. P. Kuzyaeva, *Anal. Abst.*, **9**, 2288.

5. J. Segall, M. Ariel, and L. M. Shorr, *Analyst*, **88**, 314 (1963); A. Corsini and E. Nieboer, *Talanta*, **20**, 291 (1973).

6. T. W. Gilbert, Jr., and E. B. Sandell, *J. Am. Chem. Soc.*, **82**, 1087 (1960), prepared the solid Mo complex. E. I. Stiefel and others, *ibid.*, **88**, 2956 (1966), prepared the Mo, W, and Re dithiolates, and also the Mo, W, and Re *tris* complexes of benzene-1,2-dithiol and *cis*-stilbenedithiol. These authors list wavelength maxima and molar absorptivities of the complexes in chloroform solution. Concerning the electronic structure of square-planar metal dithiolates, see H. B. Gray and E. Billig, *ibid.*, **85**, 2019 (1963). F. J. Miller and P. F. Thomason, *Anal. Chem.*, **33**, 404 (1961), found at least two complexes (rose and golden colored in CCl_4) to be formed in the reaction of Tc(VII) with dithiol in acid solution.

7. R. E. D. Clark, *Analyst*, **82**, 177 (1957).

8. R. E. D. Clark, *Analyst* **82**, 760 (1957); **84**, 16 (1959).

9. R. E. D. Clark, *Analyst*, **82**, 182 (1957).

10. H. G. Hamilton and H. Freiser, *Anal. Chem.*, **41**, 1310 (1969). Various constants for the extraction system are evaluated and qualitative observations are made on the extraction of Cu, Bi, Pd, Fe, Co, and others at pH ~ 4.

11. E. Hoyer, H. Mueller, and H. Wagler, *Anal. Abst.*, **23**, 2442 (1972).

12. A. K. Chakrabarti and S. P. Bag, *Talanta*, **19**, 1187 (1972).

13. J. A. W. Dalziel and A. K. Slawinski, *Talanta*, **15**, 367 (1968), use *S*-2-(3-mercaptoquinoxalinyl)thiuronium chloride as reagent in the simultaneous determination of Ni and Co. It hydrolyzes to quinoxaline-2,3-dithiol in ammoniacal solution. Zinc is added to increase the stability of the reagent and decrease the blank absorbance of the analytical solution.

14. J. A. W. Dalziel and A. K. Slawinski, *Talanta*, **19**, 1240 (1972).

15. D. A. Skoog, M.-G. Lai, and A. Furst, *Anal. Chem.*, **30**, 365 (1958).

16. R. W. Burke and J. H. Yoe, *Anal. Chem.*, **34**, 1378 (1962). Simultaneous determination in 80% dimethylformamide: G. H. Ayres and R. R. Annand, *ibid.*, **35**, 33 (1963); they conclude that polymeric Co and Ni complexes are formed, having chains consisting of alternating metal ion and quinoxaline dithiol links.

17. G. H. Ayres and H. F. Janota, *Anal. Chem.*, **31**, 1985 (1959).

18. Janota and S. B. Choy, *Anal. Chem.*, **46**, 670 (1974).

19. Ayres and R. W. McCrory, *Anal. Chem.*, **36**, 133 (1964). Simultaneous Pd and Pt determination: Janota and Ayres, *ibid.*, **36**, 138 (1964).

20. J. A. W. Dalziel and A. K. Slawinski, *Talanta*, **19**, 1190 (1972). Citrate can

be added to prevent precipitation of Fe, Al and Mn. Most sulfophilic elements interfere.

21. L. I. Chernomorchenko, T. V. Chuiko, and A. G. Akhmetshin, *Zh. Anal. Khim.*, **27**, 2262 (1972).

22. L. I. Chernomorchenko and G. A. Butenko, *Zav. Lab.*, **39**, 1448 (1973).

23. See E. M. Donaldson, *Talanta*, **23**, 417 (1976), for a review of extraction of metal xanthates, wherein composition of complexes, determination, and separations are covered. The red-violet Mo chelate has the composition $Mo_2O_3L_4$. These determination forms are infrequently used. Xanthate with thiocyanate gives a mixed chelate having $\mathscr{S}_{380} = 0.0016\,\mu g$ Mo (in acetophenone), as shown by M. K. Arunachalam and M. K. Kumaran, *Talanta*, **21**, 355 (1974).

24. E. M. Donaldson, *Talanta*, **23**, 411 (1976), has shown that the ethyl xanthates of As(III), Se(IV), Te(IV), Pd, and possibly Au can be extracted quantitatively from strong (10 M) HCl into chloroform and thus separated from all other elements tested except Ge, Re, Pt, and Cu. Elements not extracted from acid solutions include Mn, Zn, Rh(III), Ir(IV), Ru(III), Os(IV), Cr, and Ce. Other elements are extracted more or less at intermediate acidities. A method for determination of As in Cu, Ni, Mo, Pb, and Zn sulfides entails its separation from most of these metals by ammonia precipitation with $Fe(OH)_3$ as collector, followed by separation from Fe, Pb, and other elements by xanthate extraction from 11 M HCl: Donaldson, *Talanta*, **24**, 105 (1977). An old method for the separation of Mo from W involves extraction of Mo(V) ethyl xanthate into chloroform from ~2 M HCl containing tartrate. Selenium, Te, and As are also extracted. This separation illustrates the sulfophilic character of molybdenum. V. Yatirajam and J. Ram, *Talanta*, **21**, 439 (1974), reduce Mo(VI) to Mo(V) before ethyl xanthate–chloroform extraction, and effect separations from various metals with the aid of masking agents. Earlier review of xanthate separations: A. K. De, *Sepn. Sci.*, **3**, 103 (1968).

25. Reviews: A. Hulanicki, *Talanta*, **14**, 1371 (1967); D. J. Halls, *Mikrochim. Acta*, 62 (1969). G. D. Thorn and R. A. Ludwig, *The Dithiocarbamates and Related Compounds*, Elsevier, Amsterdam, 1962, is not primarily concerned with analytical aspects. D. Coucouvanis, *Progr. Inorg. Chem.*, **11**, 233 (1970). Xanthates and dithiocarbamates: R. J. Magee, *Rev. Anal. Chem.*, **1**, 333 (1973). Substoichiometric extraction of dithiocarbamates: A. Wyttenbach and S. Bajo, *Anal. Chem.*, **47**, 2 (1975).

26. E. Gagliardi and W. Haas, *Mikrochim. Acta*, 864 (1955); A. Musil and Haas, *ibid.*, 756 (1958). In contrast to the dialkyldithiocarbamates, the monoalkyldithiocarbamates are unstable in basic as well as acid solutions. Concerning the decomposition of monoalkyldithiocarbamates, see S. J. Joris, K. I. Aspila, and C. L. Chakrabarti, *Anal. Chem.*, **42**, 647 (1970). In acid solution, one decomposition pathway leads to formation of CS_2 and amine, the other to H_2S and isothiocyanate; other products are also formed.

In alkaline medium, the decomposition products include alkyl isothiocyanate and sulfide. The formation of H_2S is very objectionable analytically.

27. In metal determinations, an excess of dithiocarbamate is used and $M^{II}L_2$ is formed, but with a deficiency of reagent, $M^{II}L^+$ can be formed. T. E. Cullen, *Anal. Chem.*, **36**, 221 (1964), gives absorption curves for the 1:1 and 1:2 chelates of Cu(II) with N,N-bis-(2-hydroxyethyl)dithiocarbamic acid.

28. H. Bode, *Z. Anal. Chem.*, **142**, 414 (1954).

29. K. Vadasdi et al., *Anal. Chem.*, **43**, 318 (1971).

30. Kinetic studies: Bode[28]; H. Bode, K. J. Tusche and H. F. Wahrhausen, *Z. Anal. Chem.*, **190**, 48 (1962); A. Hulanicki, *Chem. Anal. (Warsaw)*, **11**, 1081 (1966); R. Zahradnik and P. Zuman, *Coll. Czech. Chem. Commun.*, **24**, 1132 (1959); R. Zahradnik, *ibid.*, **21**, 1111 (1956); P. Zuman and R. Zahradnik, *Z. Phys. Chem.*, **208**, 135 (1957); K. I. Aspila, V. S. Sastri, and C. L. Chakrabarti, *Talanta*, **16**, 1099 (1969); S. J. Joris, K. I. Aspila, and C. L. Chakrabarti, *J. Phys. Chem.*, **74**, 860 (1970); K. I. Aspila, V. S. Sastri, and C. L. Chakrabarti, *Anal. Chem.*, **45**, 363 (1973). (This list is not intended to be complete.)

31. F. Kukula, M. Krivanek, and M. Kyrs, *J. Radioanal. Chem.*, **3**, 43 (1969). The two compounds are almost colorless and yellow, respectively. Also see M. Bobtelsky and J. Eisenstadter, *Anal. Chim. Acta*, **16**, 479 (1957). Mixed .chloro complexes are known for Fe, Hg, As, Sb, Bi, Re, and other metals (see Wyttenbach and Bajo[25]). These are preferentially, or only, formed with an excess of the metal (as in substoichiometric extraction) and most of them are of no great interest to the photometric analyst, who will usually be working with an excess of the dithiocarbamate.

32. pH seems to be the important factor. Stary worked at pH 2–2.6. In approximately neutral or alkaline solutions, Co(III) is extracted.

33. G. St. Nikolov, N. Jordanov, and I. Havezov, *J. Inorg. Nucl. Chem.*, **33**, 1055 (1971).

34. S. Åkerström, *Arkiv Kemi*, **14**, 387 (1959).

35. Åkerström, *Acta Chem. Scand.*, **18**, 824 (1964).

36. W. Kemula, A. Hulanicki, and W. Nawrot, *Roczniki Chem.*, **36**, 1717 (1962); **38**, 1065 (1964).

37. R. R. Scharfe, V. S. Sastri, and C. L. Chakrabarti, *Anal. Chem.*, **45**, 413 (1973). Formation constants of the chelates of these metals with tetramethylene-, pentamethylene-, and hexamethylenedithiocarbamate are also given. The constants for diethyldithiocarbamate and tetramethylenedithiocarbamate chelates are much the same.

38. J. Stary and K. Kratzer, *Anal. Chim. Acta*, **40**, 93 (1968). Values of K_{ext}^* for H_2O–$CHCl_3$ are given by Still, *loc. cit.*, and Usatenko et al., *loc. cit.* I. M. Grekova, *J. Anal. Chem. USSR*, **27**, 1034 (1972), reports some values of K_{ext} for H_2O–$CHCl_3$.

39. A. Ringbom, *Complexation in Analytical Chemistry*, Interscience, New York, 1963, Table A5, p. 352.

40. M. Sadilkova, *Mikrochim. Acta*, 934 (1968). A recent procedure for the separation of traces of Cd from most metals, worked out by S. Bajo and A. Wyttenbach, *Anal. Chem.*, **49**, 158 (1977), calls for the extraction of Cd from 5 M NaOH with a chloroform solution of HDt, followed by back-extraction with 2 M HCl. Tl(I) is said to be the only interfering element. Because its amphoteric properties are not strong, Cd can be extracted from solutions of high hydroxyl ion concentration, as demonstrated long ago with dithizone as the extraction reagent.

41. R. C. Rooney, *Anal. Chim. Acta*, **19**, 428 (1958).

42. V. M. Byrko, *Tr. Kommis. Anal. Khim. Akad. Nauk SSR*, **14**, 191 (1963).

43. K. J. Hahn, D. J. Tuma, and J. L. Sullivan, *Anal. Chem.*, **40**, 974 (1968).

44. H. Bode, *Z. Anal. Chem.*, **143**, 182 (1954); **144**, 165 (1955). Absorption curves are given in the latter paper. For the use of disubstituted dithiocarbamates in extraction separations, particularly by displacement reactions, see G. Eckert, *Z. Anal. Chem.*, **155**, 23 (1957).

45. T. V. Cherkasina, E. I. Petrova, and G. P. Goryanskaya, *Zh. Anal. Khim.*, **23**, 1338 (1968).

46. A. K. Babko, S. V. Freger, M. I. Ovrutskii, and G. S. Lisetskaya, *J. Anal. Chem. USSR*, **22**, 580 (1967).

47. S. V. Freger, M. I. Ovrutskii, and G. S. Lisetskaya, *Zh. Anal. Khim.*, **29**, 19 (1974). The rate and degree of extraction of metal (Cu, Pb, etc.) diethyldithiocarbamates into chloroform at pH 6 increase appreciably in the presence of sodium acetate (0.1–0.3 M), $NaNO_3$ or NaCl.

48. R. J. Lacoste, M. H. Earing, and S. E. Wiberley, *Anal. Chem.*, **23**, 871 (1951).

49. R. Näsänen and V. Tamminen, *Suomen Kemistilehti*, **23B**, 28 (1950). J. M. Chilton, *Anal. Chem.*, **25**, 1274 (1953); **26**, 940 (1954).

50. G. Eckert, *Z. Anal. Chem.*, **155**, 23 (1957); H. Bode and K. J. Tusche, *ibid.*, **157**, 414 (1957); H. Bode and F. Neumann, *ibid.*, **172**, 1 (1960). Also see R. Wickbold, *ibid.*, **152**, 259, 262, 266, etc.; G. Gottschalk, *ibid.*, **194**, 321 (1963); A. Wyttenbach and S. Bajo, *Anal. Chem.*, **47**, 1813 (1975). The last two authors found the diethyldithiocarbamates of Se(IV), In, Co(III), and Fe(III) to be kinetically inert. The rate of extraction decreases roughly in the order Hg(II), Ag, Bi, Sb(III), Cu(II), Te (IV), Cd, Mo(VI), Se(IV), As(III), and In. log $(1-f)$, f being the fraction of metal extracted, varies linearly with extraction time. Shaking for 2 minutes is usually long enough. J. Bajo and A. Wyttenbach, *Anal. Chem.*, **48**, 902 (1976), use a chloroform solution of Bi diethyldithiocarbamate to extract Cu(II) in a radiochemical method for Cu. Extraction of Ni is not appreciable because of slow kinetics.

51. See a list in Hulanicki[25], p. 1384, and B. Ya. Spivakov and Yu. A. Zolotov, *J. Anal. Chem. USSR*, **25**, 529 (1970).

52. H. Bode and F. Neumann, *Z. Anal. Chem.*, **169**, 410 (1959).

53. A difference in rates is possible, inasmuch as the diethyldithiocarbamate concentration will initially be greater in the aqueous phase, other factors being the same, when Na diethyldithiocarbamate is the reagent.

54. G. C. Goode, J. Herrington, and J. K. Bundy, *Analyst*, **91**, 719 (1966).

55. O. Grossmann and H. Grosse-Ruyken, *Z. Anal. Chem.*, **233**, 14 (1968).

56. R. J. Everson and H. E. Parker, *Anal. Chem.*, **46**, 1966 (1974). This half life at pH 3 differs considerably from the value 270 min mentioned earlier.

57. H. Malissa and S. Gomiscek, *Z. Anal. Chem.*, **169**, 402 (1959); precipitation reactions are described by H. Malissa and E. Schöffmann, *Mikrochim. Acta*, **1955**, 187. Ti and Cr dithiocarbamates can be precipitated under certain conditions (Ti in the absence of acetate and other organic anions) as shown by H. Malissa and H. Kotzian, *Talanta*, **9**, 997 (1962). In concentrated HCl, Nb forms an orange-red complex extractable into $CHCl_3$, as found by I. M. Gibalo, I. P. Alimarin, and P. Davaadorzh, *Dokl. Akad. Nauk SSR*, **149**, 1326 (1963). Ta, Ti, Zr, and W do not react under these conditions. J. B. Willis, *Anal. Chem.*, **34**, 614 (1962), isolated Pb, Hg, Bi, and Ni from urine by methyl-*n*-amyl ketone extraction of the chelates at pH 2–3 before determination by atomic absorption spectroscopy. Since that time, ammonium tetramethylenedithiocarbamate has become a popular extraction reagent for traces of heavy metals before atomic absorption spectrometry. It has also found use in the isolation of V, Cr, Mn, Fe, Co, Ni, Cu, and Zn from sea water by continuous extraction with MIBK at pH 2–3 before determination by X-ray fluorescence. See A. W. Morris, *Anal. Chim. Acta*, **42**, 397 (1968). Cr(VI) is extractable into MIBK at pH ~ 2. Probably reduction to Cr(III) occurs, and this is extracted as formed. Direct extraction of Cr(III) is unsuccessful because of its well-known inertness.

58. K. A. Uvarova, Yu. I. Usatenko, and Zh. G. Klopova, *Zav. Lab.*, **38**, 1431 (1972), found that in a 1 *M* $HClO_4$–6 *M* H_2SO_4 solution containing tartaric acid, the reagent forms an insoluble complex containing Nb, reagent and ClO_4^- in the ratio 1:4:1. The complex is extractable into $CHCl_3$ ($\epsilon_{435} =$ 18,200). Ti, Zr, and F give essentially no interference; Fe (reduced with Al), V, and Ta, small interference; Cu, Co, and Bi, much. Nb is usually not an S-bonding metal so that its reaction is unexpected. Butyldithiocarbamate (*ibid.*, **36**, 909) reacts similarly in 4.5–7 *M* H_2SO_4.

59. This conclusion is borne out by the results of E. Kovacs and H. Guyer, *Z. Anal. Chem.*, **186**, 267 (1962), who used sodium tetramethylenedithiocarbamate to determine Bi, Cu, Sn, and Sb in zinc. In CCl_4, λ_{max} for Bi = 360 nm, $\epsilon = 9860$; for Cu, $\lambda_{max} = 436$, $\epsilon = 13,950$, and the Cu absorbance curve is much the same as that of Cu diethyldithiocarbamate. The reagent itself absorbs at the blue end of the spectrum. Thus an unspecified volume of $CHCl_3$ shaken with about 35 ml of aqueous solution of pH 5.2 containing 0.14% of the reagent shows an absorbance of ~0.02 in a 1 cm cell at 400 nm; at pH 8.5 the absorbance is reduced to ~0.01 at 400 nm (slightly

less at 450 and 500 nm) and ~0.02 at 350 nm. The color of a CCl_4 solution of the Bi chelate is reported to fade in light; the light sensitivity varies with the particular reagent product used.

60. M. B. Kalt and D. F. Boltz, *Anal. Chem.*, **40,** 1086 (1968).

61. W. Likussar and D. F. Boltz, *Anal. Chem.*, **43,** 1273 (1971). These values do not accurately predict the extractability of Co or Zn in 0.1 M HCl (Table 6F-8).

62. K. Hayashi, Y. Sasaki, M. Nakanishi, and S. Ito, *Japan Analyst*, **19,** 1673 (1970); K. Hayashi and K. Ito, *ibid.*, **20,** 1550 (1971).

63. I. M. Gibalo, I. P. Alimarin, G. V. Eremina, and L. A. Stanovova, *Zh. Anal. Khim.*, **23,** 1821 (1968).

64. F. M. Tulyupa, V. S. Barkalov, and Yu. I. Usatenko, *J. Anal. Chem. USSR,* **22,** 347 (1967).

65. J. T. Pyle and W. D. Jacobs, *Anal. Chem.*, **36,** 1796 (1964).

66. T. Yamane, T. Mukoyama, and T. Sasamoto, *Anal. Chim. Acta*, **69,** 347 (1974). The Pb chelate is dissociated by shaking the organic phase with 3.5 M HCl, whereas the Bi chelate is not; microgram amounts of Bi can be determined in the presence of 1 g Pb.

67. E. J. Serfass and W. S. Levine, *Chem.-Anal.*, **36,** 55 (1947). E. Geiger and H. G. Müller, *Helv. Chim. Acta*, **26,** 996 (1943).

68. Kh. K. Kirspuu and A. I. Busev, *J. Anal. Chem. USSR*, **23,** 291 (1968).

69. H. Yoshida, M. Taga, and S. Hikime, *Japan Analyst*, **16,** 605 (1967); J. Michal and J. Zyka, *Zh. Anal. Khim.*, **14,** 422 (1959).

70. K. Lesz and T. Lipiec, *Chemia Anal.*, **11,** 523 (1966).

71. Y. I. Usatenko, A. M. Arishkevich, and A. G. Akhmetshin, *J. Anal. Chem. USSR*, **20,** 429 (1965).

72. Y. I. Usatenko, A. M. Arishkevich, and A. A. Moroz, *Zh. Anal. Khim.*, **22,** 1823 (1967).

73. A. I. Busev, M. E. Dzintarnieks, and G. P. Rudzit, *Zh. Anal. Khim.*, **29,** 1353 (1974); R. Pribil and J. Adam, *Talanta*, **18,** 349 (1971).

74. E. L. Abramova, L. L. Talipova, and N. A. Parpiev, *Anal. Abst.*, **17,** 3440 (1969). Pink complex, 1 Re:2, in presence of $SnCl_2$.

75. E. Jacobsen and W. Lund, *Anal. Chim. Acta*, **36,** 135 (1966). $Cr(SCH_2COO)_3^{3-}$ (green) is formed in alkaline solution; the sensitivity is too low for the reaction to be of interest in trace determinations. See this paper for references to the determination of other metals. Also, V. M. Bhuchar, *Nature*, **191,** 489 (1961); **194,** 835 (1962).

76. A. T. Pilipenko and N. N. Maslei, *Ukr. Khim. Zh.*, **33,** 730 (1967).

77. E. Jacobsen and P. Ström, *Anal. Chim. Acta*, **53,** 309 (1971); M. Ziegler and W. Rittner, *Z. Anal. Chem.*, **164,** 310 (1958).

78. A. J. Cameron and N. A. Gibson, *Anal. Chim. Acta*, **24,** 360 (1961).

79. J. Weyers and T. Gancarczyk, *Z. Anal. Chem.*, **235,** 418 (1968).

80. E. Uhlemann and H. Müller, *Anal. Chim. Acta*, **41**, 311 (1968).

81. E. Uhlemann, B. Schuknecht, K. D. Busse, and V. Pohl, *Anal. Chim. Acta*, **56**, 185 (1971). $\epsilon_{410} = 23,700$.

82. E. Uhlemann and B. Schuknecht, *Anal. Chim. Acta*, **63**, 236 (1973).

83. E. Uhlemann and H. Müller, *Anal. Chim. Acta*, **48**, 115 (1969).

84. H. Tanaka, N. Nakanishi, Y. Sugiura, and A. Yokoyama, *Japan Analyst*, **17**, 1428 (1968). E. Uhlemann and B. Schuknecht, *Anal. Chim. Acta*, **69**, 79 (1974) (also determination of Tl). A. Yokoyama et al., *Chem. Pharm. Bull.* (*Tokyo*), **20**, 1856 (1972), have studied the extraction of the mercury(II) chelate with $CHCl_3$. $\log K_{ext} = 19.3$, $\log P = 3.4$.

85. B. Schuknecht, G. Röbisch, and E. Uhlemann, *Anal. Chim. Acta*, **69**, 329 (1974). $\mathscr{S}_{400}$ (benzene) $= 0.004 \ \mu g/cm^2$. Zn may be present up to Zn/Cd ~ 100.

86. V. M. Shinde and S. M. Khopkar, *Anal. Chem.*, **41**, 342 (1969). For V(IV), K. R. Solanke and Khopkar, *Talanta*, **21**, 245 (1974). Use in thin-layer chromatography: H. Müller and H. Rother, *Anal. Chim. Acta*, **66**, 49 (1973).

87. H. Hashitani and K. Katsuyama, *Japan Analyst*, **19**, 355 (1970). Extraction range, pH 5–18 $N \, H_2SO_4$. K. Itsuki and H. Komuro, *ibid.*, **19**, 1214 (1970).

88. T. Honjyo and T. Kiba, *Bull. Chem. Soc. Japan*, **45**, 185 (1972).

89. R. S. Barratt, R. Belcher, W. I. Stephen, and P. C. Uden, *Anal. Chim. Acta*, **58**, 107 (1972).

90. M. Katyal, V. Kushwaha, and R. P. Singh, *Analyst*, **98**, 659 (1973).

91. J. N. Srivastava and R. P. Singh, *Talanta*, **20**, 1210 (1973).

92. Reviews: C. Rozycki, *Chemia Anal.*, **11**, 447 (1966); **15**, 3 (1970). Z. Kh. Sultanova et al., *Zh. Anal. Khim.*, **28**, 413 (1973).

93. For a review, see S. Z. Lewin and R. Seider, *J. Chem. Educ.*, **30**, 445 (1955). See also the citations on p. 526 of *Colorimetric Determination of Traces of Metals*, 3rd ed.

94. W. D. Kingery and D. N. Hume, *J. Am. Chem. Soc.*, **71**, 2393 (1949).

95. Concerning thiocyanato complexes of W(V), which also have OH and O groups in the anionic species, see H. Böhland and E. Niemann, *Z. anorg. allgem. Chem.*, **336**, 225 (1965); H. Funk and Böhland, *ibid.*, **318**, 169 (1962).

96. L. P. Greenland and E. G. Lillie, *Anal. Chim. Acta*, **69**, 335 (1974). Earlier work on the system Mo(VI)–SCN⁻-reducing agent is discussed in Sandell, *Colorimetric Determination of Traces of Metals*, 3rd ed., p. 644. This work is still of significance, but is not examined here because it is chiefly of interest in connection with molybdenum. Also see A. M. Wilson and O. K. McFarland, *Anal. Chem.*, **36**, 2488 (1964). In a recent method for Mo, ascorbic acid and titanous chloride are used as reducing agents and the Mo–thiocyanate anionic complex is extracted as an ion pair complex with tetraphenylarsonium into chloroform (A. G. Fogg, J. L. Kumar, and D. T. Burns, *Analyst*, **100**, 311 (1975)). The sensitivity ($\mathscr{S}_{470} = 0.006 \ \mu g \ Mo/cm^2$)

is better than in most Mo–thiocyanate methods. Again it is found that iron must be present for full development of the Mo color. The role of iron is discussed in Sandell, *op. cit.* Its function has not yet been unequivocally established. Oxalic acid largely prevents reaction of tungsten in this method. In strong HCl ($\geqq 6\,M$), stannous chloride reduces Mo(VI) to Mo(III) and W(VI) to W(V). The anionic thiocyanate complex of the latter can be extracted into $CHCl_3$ as a partner of Zephiramine cation, whereas Mo(III) is not appreciably extracted, as shown by C. Yonezawa and H. Onishi, *J. Radioanal. Chem.*, **36**, 133 (1977). An extraction chromatographic method can also be used for separating the two metals (Zephiramine on Teflon powder).

97. G. Gottschalk, *Z. Anal. Chem.*, **187**, 164 (1962).

98. H. E. Affsprung and J. W. Murphy, *Anal. Chim. Acta*, **30**, 501 (1964); **32**, 381. Organic cations other than tetraphenylarsonium can be used in the determination of W. E. M. Donaldson, *Talanta*, **22**, 837 (1975), used diantipyrinylmethane in chloroform, after reducing W to the V state with stannous chloride. Niobium is complexed with fluoride and Mo is separated with xanthate if need be.

99. R. P. Yaffe and A. F. Voigt, *J. Am. Chem. Soc.*, **74**, 2500 (1952). W. L. Belew, G. R. Wilson, and L. T. Corbin, *Anal. Chem.*, **33**, 886 (1961), determined Ru in this way by extracting RuO_4 into CCl_4 and shaking the extract with $0.3\,M$ KSCN to form the blue complex in that phase; maximum color intensity is attained within 30 minutes. Recently, U. Muralikrishna, K. V. Bapanaiah, N. S. N. Prasad, and P. Kannarao, *Indian J. Chem.*, A, **14**, 291 (1976), studied the reaction of SCN^- with OsO_4 in acid solution. The orange complex has 1:1 stoichiometry and is believed to contain Os(VI). $\mathscr{S}$ at 440–470 nm $= 0.005\ \mu g\ Os/cm^2$. The reaction was used earlier by a number of other workers.

100. Y. Oka and T. Kato, *J. Chem. Soc. Japan Pure Chem. Sec.*, **87**, 580 (1966).

101. N. Jordanov and M. Pavlova, *Zh. Anal. Khim.*, **19**, 221 (1964).

102. C. Djordjevic and B. Tamhina, *Anal. Chem.*, **40**, 1512 (1968), isolated yellow crystals of isoperthiocyanic acid, $C_2H_2N_2S_3$, from mineral acid solutions that had stood for some time. This substance is soluble in organic solvents. Absorption spectra in water, acetone, alcohol, and ether show a maximum in the region 360–380 nm, which is where the Nb thiocyanate complex absorbs maximally. Tautomeric forms of isoperthiocyanic acid can be represented as follows:

This paper also discusses the thiocyanato complexes formed by Nb.

103. C. E. Crouthamel and C. E. Johnson, *Anal. Chem.*, **24,** 1780 (1952); **26,** 1284 (1954); C. E. Crouthamel, B. E. Hjelte, and C. E. Johnson, *ibid.*, **27,** 507 (1955).

104. Sulfolane (tetrahydrothiophene-1,1-dioxide) is better than acetone, as lately shown by H. Flaschka and R. Barnes, *Anal. Chim. Acta*, **63,** 489 (1973).

105. F. I. Lobanov, V. M. Zatonskaya, and I. M. Gibalo, *Zh. Anal. Khim.*, **29,** 826 (1974).

106. Silver and Cu_2^{2+} complexes have high stability ($\log \beta_3 \sim 13$ and $\log \beta_4 = 15.4$, respectively), the Pb complex low ($\log \beta_4 = 2.0$). Thiourea is often used as a masking agent in acid solution for $Cu(II \rightarrow I)$, $Hg(II)$, $Au(III \rightarrow I)$ and Pt. Stability constants of Ni, Co, and Bi complexes in K. Swaminathan and H. Irving, *J. Inorg. Nucl. Chem.*, **28,** 171 (1966).

107. W. Nielsch and L. Giefer, *Z. Anal. Chem.*, **145,** 347 (1955).

108. L. Chugaev, *Compt. rend.*, **167,** 235 (1918); R. D. Sauerbrunn and E. B. Sandell, *J. Am. Chem. Soc.*, **75,** 3554 (1953), who showed osmium to be present as Os(III).

109. R. P. Yaffe and A. F. Voigt, *J. Am. Chem. Soc.*, **74,** 2503 (1952).

110. In the presence of Cl^- or Br^-, the formation of complexes having Bi : thiourea : halide = 1 : 2 : 2 has been reported (*Anal. Abst.*, **23,** 3117). $\epsilon_{410} = 8700$.

111. K. Hayashi, Y. Sasaki, D. Araki, and S. Ito, *Japan Analyst*, **19,** 1370 (1970).

112. In 50% dioxane, however, K_a is reported as 0.01.

113. J. H. Yoe and L. G. Overholser, *Ind. Eng. Chem. Anal. Ed.*, **14,** 435 (1942).

114. S. B. Knight, R. L. Parks, S. C. Leidt, and K. L. Parks, *Anal. Chem.*, **29,** 570 (1957). Absorbance curves of other platinum metals are given. It may be mentioned in passing that diphenylthiourea has been used as a group extraction reagent for the Pt metals, Ag, and Au (G. A. Vorob'eva et al., *Zh. Anal. Khim.*, **29,** 497). Reaction of di-*o*-tolylthiourea with Rh and other Pt metals in presence of $SnCl_2$: E. E. Rakovskii et al., *Zh. Anal. Khim.*, **29,** 2250, 2263 (1974).

115. H. Yoshida and S. Hikime, *Talanta*, **11,** 1349 (1964).

116. E. N. Pollock and L. P. Zopatti, *Anal. Chim. Acta*, **32,** 418 (1965); **47,** 367 (1969).

117. Busev et al., *Zh. Anal. Khim.*, **24,** 1833 (1969).

118. A. I. Busev, N. V. Shvedova, and V. K. Akimov, *Zh. Anal. Khim.*, **24,** 1679 (1969).

119. V. P. Maklakova, *Zh. Anal. Khim.*, **25,** 257 (1970). Determination of Cu in steel is described.

120. V. P. Maklakova and I. P. Ryazanov, *Zav. Lab.*, **34,** 1049 (1968). In steel.

121. W. Geilmann and R. Neeb, *Z. Anal. Chem.*, **152,** 96 (1956).

122. T. Hara and E. B. Sandell, *Anal. Chim. Acta*, **23,** 65 (1960). Selenium

reagent: S. G. Sushkova and V. I. Murashova, *Zh. Anal. Khim.*, **21,** 1475 (1966); Cu, Au, Pt, Pd, Rh, Cd, Ag, and Hg must be removed. Also a Te reagent in Br⁻ solution.

123. A. N. Chechneva and V. N. Podchainova, *Anal. Abst.*, **16,** 3040. A. V. Radushev and L. A. Statina, *Zh. Anal. Khim.*, **28,** 2360 (1973), use the reagent for the simultaneous determination of Pt (PtIV → PtII) and Pd: $\epsilon_{Pt, 747} = 3.2 \times 10^4$, $\epsilon_{Pd, 555} = 1 \times 10^4$. A. I. Tolubara and Yu. I. Usatenko, *Zav. Lab.*, **32,** 807 (1966), extract the green 1:1 complex of Pt(II) into butanol ($\epsilon_{750} = 28,000$).

124. L. V. Borisova, E. I. Plastinina, and A. N. Ermakov, *Zh. Anal. Khim.*, **29,** 743 (1974).

125. K. Ballschmiter, *Z. Anal. Chem.*, **263,** 203 (1973). 1,3-Cyclohexanedione bisthiosemicarbazone: J. J. Berzas Nevado, J. A. Muñoz Leyva, and M. Roman Caba, *Talanta*, **23,** 257 (1976).

126. J. M. Cano Pavon and F. Pino, *Talanta*, **19,** 1659 (1972). For reactions of Re(IV and V) with methyl-2-pyridylketone and 8-quinolinaldehyde thiosemicarbazones, see L. V. Borisova, E. I. Plastinina, and A. N. Ermakov, *Zh. Anal. Khim.*, **29,** 1362 (1974).

127. B. W. Budesinsky and J. Svec, *Anal. Chim. Acta*, **55,** 115 (1971). Bipyridylglyoxal dithiosemicarbazone has been suggested as an extraction reagent for Ni by J. L. Bahamonde, D. Peréz Bendito, and F. Pino, *Analyst*, **99,** 355 (1974).

128. S. C. Shome and P. K. Gangopadhyay, *Anal. Chim. Acta*, **65,** 216 (1973).

129. *Ibid.*, **66,** 460 (1973).

130. W. D. Jacobs and J. H. Yoe, *Anal. Chim. Acta*, **20,** 332 (1959). Cu(I) forms a brown charged species: A. Paul, *Anal. Chem.*, **35,** 2119 (1963).

131. J. Xavier and P. Ray, *J. Indian Chem. Soc.*, **35,** 432 (1958), use an ethanolic pyridine solution in determining Co, Ni, and other metals (the dithiooxamides are kept in solution). A procedure for Pd is described in which its dithiooxamide chelate is extracted into isoamyl alcohol.

132. Z. Kolarik and C. Konecny, *Coll. Czech. Chem. Commun.*, **25,** 1775 (1960). The sensitivity with dithiooxamide is about 2.5 times that with thiourea.

133. S. K. Bhowal, *Anal. Chim. Acta*, **69,** 465 (1974).

134. W. D. Jacobs and J. H. Yoe, *Anal. Chim. Acta*, **20,** 435 (1959).

135. W. D. Jacobs, *Anal. Chem.*, **32,** 512 (1960). For Pd and Pt(II) simultaneously: *ibid.*, **33,** 1279 (1961) The high concentration of HCl required is notable. Maximal color intensity in the Pd reaction is attained above 3.5 *M* HCl and remains the same at 7.5 *M*. The yellow water-soluble complex formed has the composition 1 Pd:2 reagent. Presumably Pd is bonded to S and N in a five-membered ring system (as in the chelate of the parent reagent). The solubility of the complex is explained by the protonation of the amine N. Especially Ru, Os, Cu, and Fe(III) interfere, even in small amounts.

136. W. D. Jacobs and J. H. Yoe, *Talanta*, **2**, 270 (1959).

137. W. D. Jacobs, *Anal. Chem.*, **32**, 514 (1960). Heating is required to form the yellow complex in strong HCl. Unselective.

138. A. Goeminne, M. Herman, and Z. Eeckhaut, *Anal. Chim. Acta*, **28**, 512 (1963). Dissociation constants of reagent: *ibid.*, **30**, 569; **33**, 229.

139. A. A. Janssens, G. L. van de Cappelle, and M. A. Herman, *Anal. Chim. Acta*, **31**, 325 (1964).

140. L. C. van Poucke and M. A. Herman, *Anal. Chim. Acta*, **42**, 467 (1968).

141. DMABR: R. Borissova, M. Koeva, and E. Topalova, *Talanta*, **22**, 791 (1975); DEABR: E. B. Sandell and J. J. Neumayer, *J. Am. Chem. Soc.*, **73**, 654 (1951).

142. W. I. Stephen and A. Townshend, *J. Chem. Soc.*, 3738 (1965), suggested this structure for the rhodanine compound of silver. F. G. Moers and J. P. Langhout, *Rec. Trav. Chim. Pays-Bas*, **89**, 1237 (1970), found rhodanine to form ML_2 complexes with Pd(II) and Pt(II) having a polymeric structure; M is coordinated to N and the thiocarbonyl group of the ligand. Palladium and Pt(II) form ML_2X_2 (not polymeric) with 3-methyl-, 3-butyl- and 3-phenylrhodanines (X = Cl, Br, I); the ligand is bonded to the metal through the thiocarbonyl group. F. G. Moers and J. J. Steggerda, *J. Inorg. Nucl. Chem.*, **30**, 3217 (1968), found rhodanine, 3-methyl- and 3-butylrhodanines to form CuL_3X, CuL_2X, and $CuLX$ with Cu(I). A polymeric structure is suggested for these, in which ligand, halogen, or both act as bridges; Cu is linked to the thiocarbonyl group of the ligand.

143. For reactions of DMAR with Ag, see Borissova, Koeva and Topalova[141]; R. Doicheva and E. Topalova, *Ann. L'Ecole Superieure Chim. Industrielle Sofia*, **18**, 191 (1971); O. Navratil and L. Kotas, *Coll. Czech. Chem. Commun.*, **30**, 2736 (1965); H. Lux, T. Niedermaier, and K. Petz, *Z. Anal. Chem.*, **171**, 173 (1959); M. Castagna and J. Chauveau, *Bull. Soc. Chim. France*, 1165 (1961). Reactions with Pd: Borissova et al.[141]; F. Pantani, *Gazz. Chim. Ital.*, **90**, 999 (1960); G. H. Ayres and B. D. Narang, *Anal. Chim. Acta*, **24**, 241 (1961). Reactions with Au: R. Borissova, *Talanta*, **22**, 797 (1975).

144. Earlier procedures (still valid), with consideration of the separations of these metals required in practical work, may be found in Sandell, *Colorimetric Determination of Traces of Metals*, 3rd ed. Succinct procedures are given by Borissova et al.[141] and Borissova.[143] I. E. Lichtenstein, *Anal. Chem.*, **47**, 465 (1975), has proposed the determination of Au in pyridine (20%)–H_2O solution, in which the Au dimethylaminobenzylidenerhodanine complex is soluble. The final pH should be 7.8–8.6. The sensitivity is $\mathscr{S}_{510} = 0.005$ μg Au/cm^2. The color develops slowly (95% after 30 min) but is stable. EDTA is used as a general masking agent. Platinum(IV) can be tolerated up to a ratio of 6:1 Au, but Pd interferes if Pd/Au > 2. Chloride does not mask Ag, but I$^-$ allows five times as much Ag as Au to be present. Base metals may be present in small amounts. The pyridine procedure has advantages and

disadvantages compared to the acid medium procedure. The reagent itself absorbs more strongly at pH ~8 than in acidic medium.

145. N. S. Poluektov, *Trudy Vsesoyuz. Konferentsii Anal. Khim.*, **2**, 393 (1943), extracted the Au(I) chelate of DMABR into a mixture of $CHCl_3$ and C_6H_6. T. M. Cotton and A. A. Woolf, *Anal. Chim. Acta*, **22**, 192 (1960), considered $CHCl_3$ less suitable than isoamyl acetate as extractant. The $CHCl_3$ solution of the Au complex is fluorescent, as observed by N. K. Podberezskaya, E. A. Shilenko, and D. P. Shcherbov, *Zav. Lab.*, **36**, 661 (1970). Nitrobenzene has been used as extractant for the Au and Pd complexes by R. Borissova, P. Mosheva, Z. Ivanova, and E. Topalova, *Z. Anal. Chem.*, **274**, 31 (1975); R. Doicheva, E. Topalova, and P. Mosheva, *Compt. Rend. Acad. Bulg. Sci.*, **24**, 1675 (1971).

146. W. I. Stephen and A. Townshend, *Anal. Chim. Acta*, **33**, 257 (1965).

147. R. F. Propistsova, S. B. Savvin, and Yu. G. Rozovskii, *Zh. Anal. Khim.*, **26**, 2424 (1971), where other references may be found; **27**, 1554 (1972); N. N. Basargin et al., *Anal. Abst.*, **23**, 317, 3191 (1972); R. F. Propistsova and S. B. Savvin, *Zh. Anal. Khim.*, **29**, 2097 (1974); Rh: *ibid.*, **28**, 1768 (1973); Ag: *ibid.*, **31**, 660 (1976).

148. Bankovskii et al., footnote *d*, Table 6F-11.

149. Theoretical aspects of 8-mercaptoquinoline fluorescence: S. G. Schulman, *Anal. Chem.*, **44**, 400 (1972).

150. General survey: V. I. Kuznetsov, Yu. A. Bankovskii and A. F. Ievin'sh, *J. Anal. Chem. USSR*, **13**, 299 (1959). See further the extensive series of papers by Bankovskii and coworkers, mostly in *Zh. Anal. Khim.* (*J. Anal. Chem. USSR*), 1958– .

151. Yu. A. Bankovskii, A. F. Ievin'sh, and Z. E. Liepina,. *J. Anal. Chem. USSR*, **15**, 1 (1960). Formation constants of 1:1 Pb, Mn, Ni, and Zn complexes in 50% dioxane are given by A. Corsini, Q. Fernando, and H. Freiser, *Anal. Chem.*, **35**, 1424 (1963); the values are comparable to those of the corresponding oxinates.

152. *Anal. Abst.*, **25**, 2946.

153. V. A. Nazarenko and E. N. Poluektova, *Zh. Anal. Khim.*, **26**, 1331 (1971).

154. Yu. I. Usatenko, V. I. Suprunovich, and V. V. Velichko, *Zh. Anal. Khim.*, **29**, 807 (1974).

155. O. E. Veveris et al., *J. Radioanal. Chem.*, **9**, 47 (1971). These values are to be looked upon as minimum values. Solubilities of mercaptoquinolates in $CHCl_3$ are: $HgTx_2$, 0.14 M; $ZnTx_2$, 0.07 M; $FeTx_3$, 0.017 M.

156. C. Yoshimura, H. Noguchi, and H. Hara, *Japan Analyst*, **13**, 1249 (1964). Even $(HTx \cdot H_2Tx)SbCl_6$ has been reported. In this connection it is of interest to note that 8,8'-diquinolyl disulfide forms a cation in acid solution by adding H^+, and this gives slightly soluble ion associates with anionic complex iodides of Sb, Bi, and Hg.

157. N. Nakamura and E. Sekido, *Talanta*, **17**, 515 (1970); E. Sekido, I.

Fujiwara, and Y. Masuda, *ibid.*, **19**, 479 (1972); E. Sekido and I. Fujiwara, *ibid.*, **19**, 647 (1972).

158. Yu. M. Dedkov and M. G. Slotintseva, *Zh. Anal. Khim.*, **28**, 2367 (1973).

159. Yu. M. Dedkov, L. V. Lozovskaya, and M. G. Slotintseva, *Zh. Anal. Khim.*, **27**, 512 (1972).

160. A. T. Pilipenko, I. P. Sereda, and E. P. Semenyuk, *Zh. Anal. Khim.*, **25**, 1958 (1970).

161. B. C. Bera and M. M. Chakrabartty, *Anal. Chem.*, **38**, 1419 (1966). Somewhat similar reagents, substituted pyrimidinethiols, have been used as extraction reagents for Pd and Os in acid solution by A. K. Singh, M. Katyal, R. P. Singh, and N. K. Ralhan, *Talanta*, **23**, 851 (1976).

162. A. K. Majumdar and M. M. Chakrabartty, *Anal. Chim. Acta*, **20**, 379 (1959).

163. L. N. Lomakina and I. P. Alimarin, *Vest. Mosk. Gos. Univ., Ser. Khim.* (3), 109 (1966).

164. K. A. Uvarova, Yu. I. Usatenko, and T. S. Chagir, *Zh. Anal. Khim.*, **28**, 693 (1973).

165. J. G. Sen Gupta, *Talanta*, **8**, 729, 785 (1961).

166. B. C. Bera and M. M. Chakrabartty, *Mikrochim. Acta*, 1094 (1966).

167. R. F. Wilson and P. Merchant, Jr., *J. Inorg. Nucl. Chem.*, **29**, 1993 (1967).

168. B. C. Bera and M. M. Chakrabartty, *Microchem. J.*, **11**, 420 (1966). Determination of Ir (extraction into dichloroethane after heating): E. E. Rakovskii and Yampolskii, *Anal. Abst.*, **28**, 2B 179 (1975). Mercaptobenzothiazole also finds use as a separatory reagent. As shown by A. Diamantatos, *Anal. Chim. Acta*, **66**, 147 (1973), it extracts Pd(II) into $CHCl_3$ from HCl solution at room temperature, Pt(IV) partially, and Rh(III), Ir(IV), Os(VIII), and Ru not at all (kinetic inertness). In the presence of iodide, both Pt and Pd are extracted and separated from Rh and Ir; Rh is separated from Ir by adding $SnCl_2$ and extracting the mercaptobenzothiazole complex of the former into $CHCl_3$ [*ibid.*, **67**, 317 (1973)]. The reactions occurring are not well understood.

169. E. G. Walliczek, *Talanta*, **11**, 573 (1964). Purification of the reagent is described.

170. Review: H. Yoshida and S. Hikime, *J. Spectrosc. Soc. Japan*, **14**, 131 (1966).

171. H. Tomioka and K. Terajima, *Japan Analyst*, **22**, 264 (1973), have determined palladium(II) by extraction of its reddish brown complex into tributyl phosphate. The molar absorptivity at 450 nm is 4.6×10^3.

172. A. K. Majumdar and S. K. Bhowal, *Anal. Chim. Acta*, **62**, 223 (1972).

173. A. I. Busev, L. N. Simonova, and A. A. Arutyunyan, *Zh. Anal. Khim.*, **25**, 1880 (1970).

174. A. I. Busev, L. N. Simonova, and E. J. Gaponyuk, *Zh. Anal. Khim.*, **23**, 59 (1968). Later for Pd: A. I. Busev, L. N. Simonova, A. N. Shishkov, and A. D. Toleva, *ibid.*, **29**, 1134 (1974). 5-Mercapto-3-(2-pyridyl-1,3,4-thiadiazol-2-thione reacts with numerous metals; a reagent for Bi, $\epsilon_{340} = 29{,}000$ in $CHCl_3$. See *Zh. Anal. Khim.*, **29**, 1986 (1974).

175. A. V. Radushev and E. N. Prokhorenko, *Zh. Anal. Khim.*, **27**, 2209 (1972); L. P. Romanenko and A. V. Radushev, *ibid.*, **28**, 1908 (1973).

176. S. C. Srivastava, *Anal. Chem.*, **35**, 1165 (1963); Srivastava and M. L. Good, *Anal. Chim. Acta*, **32**, 309 (1965). *Cf.* R. W. Burke and J. H. Yoe, *Talanta*, **10**, 1267 (1963).

177. R. Berg and W. Roebling, *Berichte*, **68**, 403 (1935); *Angew. Chem.*, **48**, 430, 597 (1935). R. Berg and E. S. Fahrenkamp, *Z. Anal. Chem.*, **109**, 305 (1937); **112**, 162 (1938).

178. Pd has been determined directly (375 nm) by extraction of the thionalide complex into $CHCl_3$–isopentyl alcohol from 2–3 M HCl. Pt, Rh, Ir, Os, Ni, and Co do not interfere, but Fe(III), Bi, Zn, Pb, Sn(IV), and Mo(VI) do. See A. I. Busev and A. Naku, *Zh. Anal. Khim.*, **18**, 1479 (1963).

179. H. K. L. Gupta and N. C. Sogani, *Anal. Chem.*, **31**, 918 (1959).

180. B. Gehauf and J. Goldenson, *Anal. Chem.*, **22**, 498 (1950). G. Ackermann and H. Röder, *Talanta*, **24**, 99 (1977), describe determination of Hg(II) in water-n-propanol mixture (red color). The sensitivity is good ($\mathcal{S} = 0.0013 \, \mu g \, Hg/cm^2$), but many metals interfere; pH ~6.

181. K. L. Cheng, *Mikrochim. Acta*, 820 (1967). $\mathcal{S}_{520} = 0.001$.

182. H. W. Lakin and H. M. Nakagawa, *Prof. Paper U. S. Geol. Surv.*, **525–C**, 168 (1965); K. L. Cheng and B. L. Goydish, *Microchem. J.*, **10**, 158 (1966).

183. E. Uhlemann and V. Pohl, *Anal. Chim. Acta*, **65**, 319 (1973).

184. G. R. E. C. Gregory and P. G. Jeffery, *Analyst*, **92**, 293 (1967); S. Maekawa and K. Kato, *Japan Analyst*, **20**, 474 (1971). This reagent has also been used for extraction-photometric determination of Ni [H. Ishii and H. Einaga, *J. Chem. Soc. Japan Pure Chem. Sec.*, **90**, 175 (1969)] and Cu [H. Ishii and H. Einaga, *ibid.*, **91**, 734 (1970)]. Both Ni and Cu form 1:1 complexes.

185. M. Yokoyama and T. Takeshima, *Anal. Chem.*, **40**, 1344 (1968).

186. A. V. Dolgorev and Ya. G. Lysak, *Zh. Anal. Khim.*, **29**, 1766 (1974).

186a. Proposed by S. G. Tandon and S. C. Bhattacharya, *Anal. Chem.*, **32**, 194 (1960), for the gravimetric determination of Pd. The chelate is soluble in chloroform ($\epsilon_{306} = 3.44 \times 10^4$).

187. A. I. Busev and M. I. Ivanyutin, *Trudy Komis. Anal. Khim. Akad. Nauk SSSR*, **11**, 172 (1960); *Chem. Abstr.* **55**, 24381 (1961); H. Bode and W. Arnswald, *Z. Anal. Chem.*, **185**, 99, 179 (1962).

188. W. A. Forster, P. Brazenall, and J. Bridge, *Analyst*, **86**, 407 (1961).

189. A. I. Busev and A. N. Shishkov, *Zh. Anal. Khim.*, **23,** 181 (1968). Bi, As(III), Sb(III), and Sn(II) interfere, but not Fe, Ni, Co, Zn, or Cd.

190. A. I. Busev and A. N. Shishkov, *Zh. Anal. Khim.*, **23,** 1675 (1968).

191. L. L. Kabanova and S. V. Usova, *Zh. Anal. Khim.*, **29,** 2248 (1974).

192. N. T. Yatsimirskaya, A. Rafee, and A. I. Busev, *Talanta*, **23,** 722 (1976).

193. A. K. Chakrabarti and S. P. Bag, *Talanta*, **23,** 736 (1976).

194. S. P. Bag and S. K. Chakrabarti, *Talanta*, **24,** 128 (1977).

6G

ORGANIC PHOTOMETRIC REAGENTS
DIPHENYLTHIOCARBAZONE

I. PROPERTIES OF THE REAGENT

Diphenylthiocarbazone (3-mercapto-1,5-diphenylformazan),

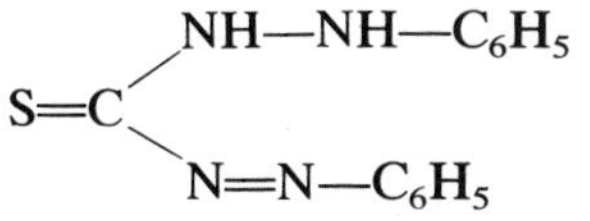

$$S=C \overset{\displaystyle NH-NH-C_6H_5}{\underset{\displaystyle N=N-C_6H_5}{\Bigg<}}$$

Violet-black solid.
Mol. wt. 256.3.

(thione form)

more familiarly known as dithizone, is an outstanding reagent for heavy metals having sulfophilic properties. It was first prepared by Emil

Fischer,[1] who noted its reaction with heavy metals to give brilliantly colored products. Curiously, it did not attract the attention of analytical chemists and lay unused for almost 50 yr until 1925 when Hellmut Fischer[2] showed its great value for the detection and determination of trace metals. Since that time, up to the present, dithizone reactions have been studied extensively. Dithizone is probably best known as a reagent for separation of lead and its photometric determination, but it is also valuable for Zn, Cd, Hg, Cu, and Ag, and occasionally other reacting metals. Even if the final determination is made by methods other than the photometric, it finds use in trace metal separations. The great sensitivity of dithizone reactions arises from the high absorptivity of organic solvent solutions of most dithizonates in the region 500–550 nm where the green reagent does not absorb strongly. ϵ_{max} of dithizonates is usually greater than ϵ_{max} of dithizone, which is understandable from the greater than 1 molar ratio of dithizone:metal in most dithizonates.

A. SOLUBILITY AND SPECTRAL PROPERTIES

Dithizone (molecular form) is essentially insoluble in water ($\sim 2 \times 10^{-7}\,M$). It dissolves to various extents in organic solvents. Solubility values in the literature are not always in good agreement. The following list is not complete, but contains many if not most values that have been reported:

Solvent	Temperature (°C)	Solubility	
		g/liter	M
$CHCl_3$			6.8×10^{-2}
	20	23.5	9.2×10^{-2}
	20		9.5×10^{-2}
	30	20.3	7.9×10^{-2}
CH_2Cl_2	30	12.6	4.9×10^{-2}
CCl_4			2.5×10^{-3}
	20		3.1×10^{-3}
	25		2.27×10^{-3}
	25		2.1×10^{-3}
	30	0.74	2.9×10^{-3}
C_6H_6	20	4.2	1.64×10^{-2}
	30	3.6	1.4×10^{-2}
C_6H_5Cl	30	2.5	9.8×10^{-3}
Toluene		4.5	1.8×10^{-2}
Isoamyl alcohol	20		9.5×10^{-2}
Ethyl alcohol			8.4×10^{-4}
Isopentyl acetate		24.3	9.5×10^{-2}

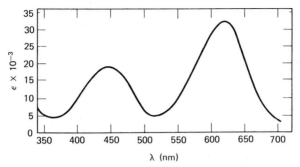

Fig. 6G-1. Molar absorptivity curve of dithizone in CCl_4 (K. Marhenke).

Chloroform and carbon tetrachloride are commonly used as solvents for dithizone in photometric determinations and for separations. Occasionally, other solvents are used. Ethyl propionate has been suggested as a good solvent for extraction preceding determination by atomic absorption. Saturated hydrocarbons are poor solvents for dithizone and dithizonates (e.g., 0.04 g dithizone/liter of n-heptane at room temperature).

Organic solvent solutions of dithizone are green, but the hue varies more or less from one solvent to another. Thus a CCl_4 solution is grass green; a $CHCl_3$ solution is bluish green. The wavelength–absorbance curve of dithizone in an organic solvent shows two peaks (Fig. 6G-1), whose relative heights vary from one solvent to another. The two peaks are believed to correspond to two forms of dithizone in tautomeric equilibrium with each other in organic solvents:

$$S=C \underset{N=N-C_6H_5}{\overset{NH-NH-C_6H_5}{<}} \rightleftharpoons HS-C \underset{N=N-C_6H_5}{\overset{N-NH-C_6H_5}{<}}$$

thione form   thiol form

The peak at longer wavelengths corresponds to the thione form.

More concentrated solutions of dithizone in organic solvents are dichroic (red).

The two absorption maxima and the minimum (in the visible range)[3] of dithizone in $CHCl_3$ and CCl_4 may be taken to lie at the following wavelengths (nm):

	λ_{max1}	λ_{max2}	λ_{min}
$CHCl_3$	442	605	504
CCl_4	450	620	515

TABLE 6G-1
Reported Values of ϵ_{max} of Dithizone in Carbon Tetrachloride

Reference	$\epsilon_{450} \times 10^{-4}$	$\epsilon_{620} \times 10^{-4}$	$\epsilon_{620}/\epsilon_{450}$
Clifford, *J. Assoc. Offic. Agr. Chem.*, **26**, 26 (1943)	1.87	3.11	1.66
Cooper and Sullivan, *Anal. Chem.*. **23**, 613 (1951)	2.03	3.46	1.70
Liebhafsky and Winslow, *J. Am. Chem. Soc.*, **59**, 1966 (1937)	1.90	3.04	1.60
Weber and Vouk, *Analyst*, **85**, 40 (1960)	2.14	3.64	1.70
Geiger, Ph.D. Thesis, Univ. of Minn., 1951	1.99	3.18	1.60
Mathre, Ph.D. Thesis, Univ. of Minn., 1956	1.90	3.17	1.67
Marhenke, Ph.D. Thesis, Univ. of Minn., 1965	1.99	3.27	1.64
Brock, M.S. Thesis, Univ. of Minn., 1969	1.92	3.18	1.66
Bidleman, Ph.D. Thesis, Univ. of Minn., 1970	1.93	3.25	1.68
King and Pruden, *Analyst*, **96**, 146 (1971)		3.31	1.59
Marczenko and Mojski, *Chim. anal.*, **54**, 29 (1972)	1.94	3.20	1.65
Average, excluding first four entries	1.945	3.22	1.656

Different workers agree quite well in their values for λ_{max}, but not so well in their values of ϵ_{max}. As is well known, the accurate determination of molar absorptivity is not easy. With dithizone, there is the additional problem of purity of the reagent and the solvent. In Table 6G-1, values of ϵ_{450} and ϵ_{620} for dithizone in CCl_4, as reported by various workers, are listed. The values of ϵ_{450} and ϵ_{620} and of the ratio $\epsilon_{620}/\epsilon_{450}$ are of importance in judging the purity of dithizone (see p. 632). In $CHCl_3$, the values $\epsilon_{442} = 16,000$ and $\epsilon_{605} = 41,490$ may be accepted.[4] Dithizone dissolved in CCl_4 having low contents of $CHCl_3$ as an impurtity gives markedly higher ϵ_{620} values than in pure CCl_4; ϵ_{450} is less affected[5]:

$CHCl_3(\%)$	ϵ_{620}	ϵ_{450}	$\epsilon_{620}/\epsilon_{450}$
<0.1	32,000	19,400	1.65
0.3	33,000	19,400	1.70
0.6	33,400	19,200	1.74
1.9	34,400	19,200	1.78

ϵ_{max} values are much the same for benzene and toluene as for CCl_4: benzene, $\epsilon_{450} = 20,800$, $\epsilon_{620} = 38,800$; toluene, $\epsilon_{455} = 18,400$, $\epsilon_{620} = 33,300$.

In aqueous solution also, dithizone shows two absorption peaks (Fig. 6G-2), which lie at shorter wavelengths (420, 580 nm) than in organic solvents.[6] The molar absorptivities are $\epsilon_{420} = 1.50 \times 10^4$, $\epsilon_{580} = 4.22 \times 10^4$, $\epsilon_{580}/\epsilon_{420} = 2.8$.

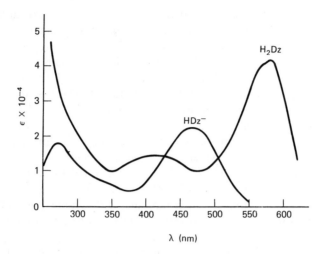

Fig. 6G-2. Molar absorptivity curves of H$_2$Dz and HDz$^-$ in water. After Komar' and Manzhelii, *loc. cit.*

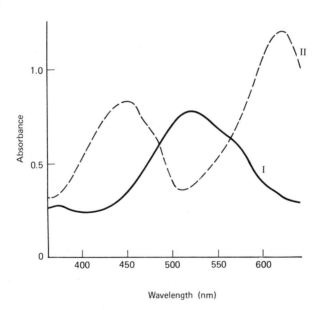

Fig. 6G-3. Absorbance curve of dithizone $(3.7 \times 10^{-5}\ M)$ in 60% H$_2$SO$_4$(I). From Akaiwa and Kawamoto, *loc. cit.* Absorbance curve (II) of dithizone $(3.7 \times 10^{-5}\ M)$ in CCl$_4$ is also shown.

583

We have observed (ca. 1940) that when a CCl_4 solution of dithizone is shaken with concentrated hydrochloric acid, dithizone partially passes into the acid layer and imparts a violet-red color to it (dithizone is not significantly extracted into 6 M hydrochloric acid). This behavior indicates that in strongly acid aqueous solutions, one of the nitrogens in dithizone is protonated and dithizonium ion (H_3Dz^+) is formed. The absorption curve of dithizone in 60% sulfuric acid (Fig. 6G-3) shows a single peak at 520 nm.[7]

Dithizone dissolves in basic aqueous solutions, forming primary dithizonate ion (HDz^-), which has a yellow color (Fig. 6G-2). The absorption peak lies at 475 nm and $\epsilon_{475} = 2.25 \times 10^4$ (average value).[8]

Dithizonate ion combines with organic cations of large size in basic solution to give ion-association compounds extractable into chloroform and other immiscible organic solvents. Thus tetraphenylarsonium dithizonate $(C_6H_5)_4As(HDz)$ is readily extractable into chloroform from 0.01 M NaOH solution. The orange solution has an absorption peak at 490 nm (T. Bidleman).

B.   DITHIZONE EQUILIBRIA

Dithizone is a dibasic acid:

$$H_2Dz \rightleftharpoons H^+ + HDz^-$$
$$HDz^- \rightleftharpoons H^+ + Dz^{2-}$$

At a specified ionic strength, the primary ionization constant in aqueous solution is

$$K_{pd} = \frac{[H^+][HDz^-]}{[H_2Dz]} \tag{6G-1}$$

or if, as is usual in dealing with weakly acidic or basic solutions, aH^+ is used in place of $[H^+]$

$$K_{pd} = \frac{(aH^+)[HDz^-]}{[H_2Dz]} \tag{6G-2}$$

At 0.1 M ionic strength, K_{pd} based on aH^+ has a value of about 2×10^{-5} at room temperature.[9] The secondary ionization constant

$$K_{sd} = \frac{[H^+][Dz^{2-}]}{[HDz^-]} \tag{6G-3}$$

is so small that it has not yet been determined. From the observation that the absorption curve of dithizone in 1.5 M NaOH is not significantly different from that in 0.1 M NaOH, the conclusion[10] may be drawn that K_{sd} is unlikely to be greater than 10^{-15}.

The basic properties of dithizone are extremely weak. An approximate value has been reported[7] for the dissociation of H_3Dz^+:

$$\frac{[H_2Dz][H^+]}{[H_3Dz^+]} = 4 \times 10^4 \qquad (6G\text{-}4)$$

The distribution of dithizone between an organic solvent such as CCl_4 or $CHCl_3$ and an aqueous solution is given by its *extraction coefficient*

$$E_d = \frac{[H_2Dz]_{org}}{\Sigma\,[H_2Dz]} \qquad (6G\text{-}5)$$

(Concentrations without subscripts refer to aqueous phase.) Except in very strongly acid and very strongly basic solutions the only dithizone species present in aqueous solutions are H_2Dz and HDz^-, and we have

$$E_d = \frac{[H_2Dz]_{org}}{([H_2Dz]+[HDz^-])} \qquad (6G\text{-}6)$$

Designate the *partition constant* of molecular dithizone by P_d:

$$\frac{[H_2Dz]_{org}}{[H_2Dz]} = P_d \qquad (6G\text{-}7)$$

No polymerization of H_2Dz occurs in CCl_4, $CHCl_3$, and similar solvents. Therefore, from Eqs. (6G-2) and (6G-7)

$$E_d = \frac{(aH^+)P_d}{K_{pd}+aH^+} \qquad (6G\text{-}8)$$

at specified ionic strength. (In extractions from mineral acid solutions, E_d may be based on $[H^+]$ instead of aH^+.)

For basic solutions in which $[H_2Dz]$ is small compared to $[HDz^-]$, and $[H^+]$ is small compared to K_{pd}, Eq. (6G-8) may be written

$$E_d \sim \frac{[H_2Dz]_{org}}{[HDz^-]} = \frac{(aH^+)P_d}{K_{pd}} \qquad (6G\text{-}9)$$

The value of

$$\frac{P_d}{K_{pd}} = \frac{[H_2Dz]_{org}}{(aH^+)[HDz^-]} = \frac{[H_2Dz]}{(aH^+)[HDz^-]} \times \frac{[H_2Dz]_{org}}{[H_2Dz]}$$

$$\cong \frac{\text{molar solubility of } H_2Dz \text{ in } org}{K_{sp} \text{ of } H_2Dz \text{ in } H_2O \text{ satd. with } org} \qquad (6G\text{-}10)$$

has been determined for CCl_4 and $CHCl_3$ as the organic solvents. For CCl_4–0.1 M HCl, $P_d/K_{pd} = 7 \times 10^8$ (25°),[11] and for $CHCl_3$–H_2O ($\mu \sim 0.15\ M$) it is 4×10^{10} (20°).[12] It may be noted that P_d/K_{pd} is the equilibrium constant for the reaction

$$HDz^- + H^+ \rightleftharpoons (H_2Dz)_{org}$$

For CCl_4–0.1 M HCl, $P_d = 1.1 \times 10^4$ (25°).[9] Since the solubility of dithizone in carbon tetrachloride at room temperature is $2.1 \times 10^{-3}\ M$, the calculated solubility of dithizone in water ($\mu = 0.1$) saturated with carbon tetrachloride is $2 \times 10^{-7}\ M$. A direct determination of the solubility of dithizone in 0.1 M perchloric acid has given $2.5 \times 10^{-7}\ M$ (Dyrssen and Hök[13]). For $CHCl_3$–H_2O ($\mu = 0.25$), P_d has been reported as 8×10^5 at 25°.

With a change in ionic strength, the values of P_d and K_{pd} change. Thus denoting the thermodynamic constant P_d by P_d^0, we have

$$P_d^0 = \frac{[H_2Dz]_{org}(f_d)_{org}}{[H_2Dz]f_d}$$

or

$$P_d = P_d^0 \times \frac{f_d}{(f_d)_{org}} \qquad (6\text{G-}11)$$

Since the organic solvent solution will usually be very dilute, Eq. (6G-11) may be written

$$P_d = P_d^0 \times f_d \qquad (6\text{G-}12)$$

The foregoing expression holds only if the change in the solubility of the organic solvent in the aqueous phase resulting from the change in electrolyte concentration is so small that the solubility of dithizone in water is not changed appreciably. The value of P_d should increase with increasing electrolyte concentration, because f_d should then increase (decrease in the solubility of dithizone in the aqueous phase by the salting out effect). Data are not available for testing this equation for dithizone.

Since

$$K_{pd} = \frac{K_{pd}^0}{f_{d^-}} \qquad (6\text{G-}13)$$

$$\frac{P_d}{K_{pd}} = \frac{P_d^0}{K_{pd}^0} \times f_d f_{d^-} \qquad (6\text{G-}14)$$

(f_{d^-} is the activity coefficient of primary dithizonate ion, and K_{pd} is based on aH^+.)

Koroleff[14] has determined the distribution of dithizone between basic

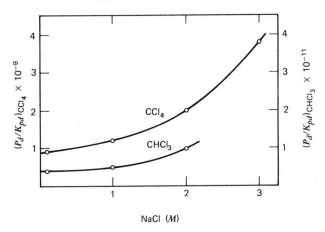

Fig. 6G-4. The value of P_d/K_{pd} ($= [H_2Dz]_{org}/(aH^+[HDz^-])$) for carbon tetrachloride and chloroform as a function of the sodium chloride concentration in the aqueous phase at room temperature (20°–22°C). (From data of F. Koroleff, Ph.D. Thesis, p. 21, Helsingfors, 1950.)

aqueous solutions with variable sodium chloride concentration and carbon tetrachloride or chloroform. His results are plotted in Fig. 6G-4. As the electrolyte concentration is increased, less dithizone partitions into the aqueous phase at constant aH^+, but there is little change up to 1 M. According to Komar' and Manzhelii, the activity coefficient of HDz^- is given by

$$\log f_{d^-} = -(0.28 \pm 0.05)\mu \qquad (25°)$$

The solubility product constant of dithizone is

$$(aH^+)[HDz^-] = K_{sp} = 2 \times 10^{-5} \times 2 \times 10^{-7} = 4 \times 10^{-12}$$

(25°, in water saturated with CCl_4, $\mu \sim 0$).

C. OXIDATION OF DITHIZONE

Dithizone is subject to oxidation by air, by products formed in photodecomposition of organic solvents, and by oxidants in the sample solution. Several dithizone oxidation products can be formed, most of which are yellow in dilute solution. In analytical practice, the major trouble makers are oxidants in the sample [e.g., Fe(III)] in cyanide solution, giving ferricyanide) and those derived from decomposition products of the solvent (CCl_4, $CHCl_3$). Addition of a reducing agent, such as NH_3OHCl, is of help in removing oxidizing agents from the sample solution.

Two analytically important oxidation products[15] are the disulfide

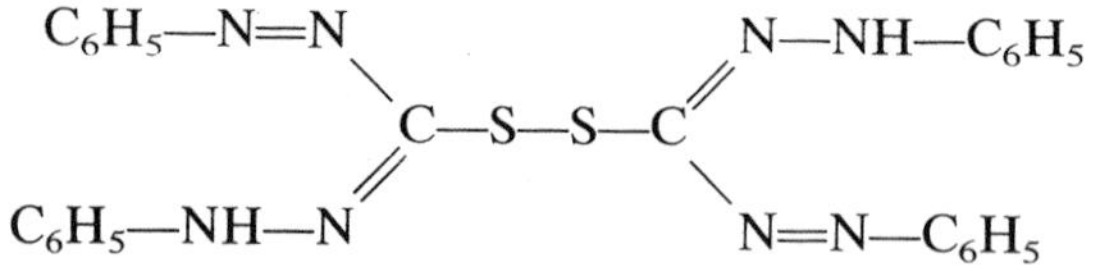

$$
\begin{array}{l}
C_6H_5-N\!=\!N \\
\qquad\qquad\searrow \\
\qquad\qquad C\!-\!S\!-\!S\!-\!C \\
\qquad\qquad\nearrow \qquad\qquad\quad \\
C_6H_5-NH-N \qquad\qquad N\!=\!N-C_6H_5
\end{array}
$$

C_6H_5—N=N \ C—S—S—C / N—NH—C_6H_5
C_6H_5—NH—N /   \ N=N—C_6H_5

Bis-1,5-diphenylformazan-3-yl-disulfide

and the meso-ionic compound[16]

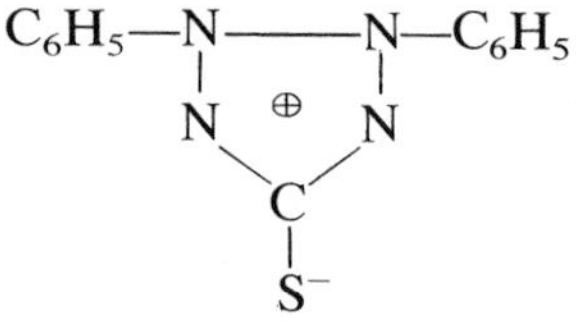

$$
\begin{array}{c}
C_6H_5-N\!-\!\!-\!\!-\!N-C_6H_5 \\
\quad | \qquad\qquad | \\
\quad N \quad \oplus \quad N \\
\qquad\searrow \quad\swarrow \\
\qquad\quad C \\
\qquad\quad | \\
\qquad\quad S^-
\end{array}
$$

2,3-Diphenyl-2,3-dihydrotetrazolium-5-thiolate

When a dilute solution of H_2Dz in chloroform reacts with an equivalent amount of I_2 in the presence of water, the disulfide is formed stoichiometrically according to the equation

$$2\,H_2Dz + I_2 \rightarrow (HDz)_2 + 2\,HI$$

The disulfide is not stable in chloroform (or other organic solvent) solution; it undergoes thermal fission into equimolar amounts of dithizone and the meso-ionic compound:

$$
\left(
\begin{array}{l}
C_6H_5N\!=\!N \\
\qquad\qquad\searrow \\
\qquad\qquad C\!-\!S\!- \\
\qquad\qquad\nearrow \\
C_6H_5NH\!-\!N
\end{array}
\right)_2
\longrightarrow
\begin{array}{l}
C_6H_5N\!=\!N \\
\qquad\qquad\searrow \\
\qquad\qquad C\!-\!SH\; + \\
\qquad\qquad\nearrow \\
C_6H_5NH\!-\!N
\end{array}
\begin{array}{c}
C_6H_5N\!-\!NC_6H_5 \\
\quad | \;\oplus\; | \\
\quad N \qquad N \\
\qquad\searrow\swarrow \\
\qquad\; C \\
\qquad\; | \\
\qquad\; S^-
\end{array}
$$

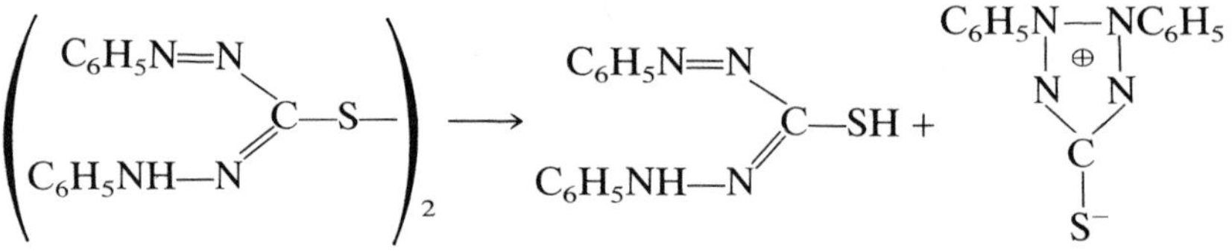

(The thiol form of dithizone is represented here, but the thione form will also be present in tautomeric equilibrium with the former.) The rate of fission depends upon the organic solvent polarity; the half life of the disulfide is 86 min in $CHCl_3$ (still longer in CCl_4), but only 10.5 min in alcohol or acetone at 25°.

The disulfide shows a strong absorption band in the visible:

415 nm ($\epsilon = 48{,}500$), CCl_4 solution

420 nm ($\epsilon = 42{,}000$), $CHCl_3$ solution

and two bands in the UV:

302, ? nm, CCl_4 solution

305, 257 nm, $CHCl_3$ solution

In chloroform solution, the meso compound has a characteristic absorption spectrum with

$\lambda_{max} = 467$ nm, $\epsilon_{467} = 1170$; 266 nm, $\epsilon_{266} = 19,700$

$\lambda_{min} = 360$ nm

The disulfide is easily reduced back to dithizone by sulfurous acid, thiosulfate, hypophosphite and other reducing agents. The meso-ionic compound is not affected by reducing agents. The disulfide in organic solvent solution reacts with metal ions to form primary dithizonate and the meso compound; Hg(II) combines with the latter in aqueous solution, forming $Hg\,(meso)_2^{2+}$.

More than an equivalent amount of iodine in aqueous solution and other oxidizing agents such as H_2O_2 in ammoniacal solution, Fe(III) in basic solution, and Tl(III) in acid, form the meso-ionic compound or a mixture of this and the disulfide. Ferricyanide in basic solution oxidizes dithizone in $CHCl_3$ to the meso compound. Under drastic oxidizing conditions, such as excess H_2O_2 in 0.5 M hydroxide solution (hardly analytical conditions), a sulfonic acid

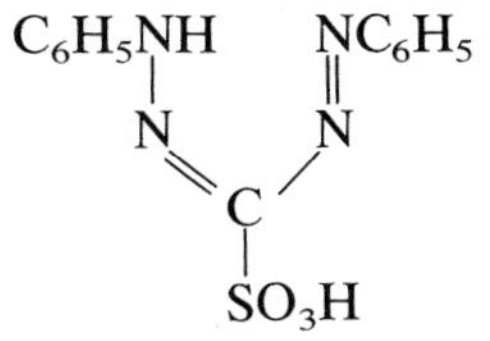

$$C_6H_5NH \qquad NC_6H_5$$

is formed.[17] In hot concentrated acetic acid, oxidation by air forms a purple compound as the main product[18]:

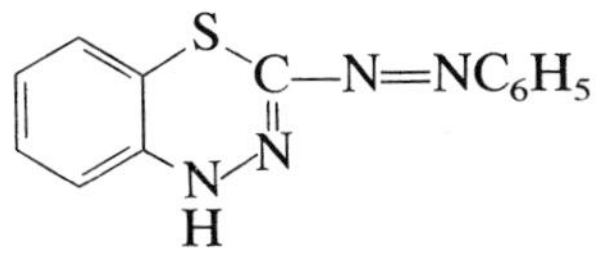

3-Phenylazobenzo-1,3,4(4H)-thiadiazole

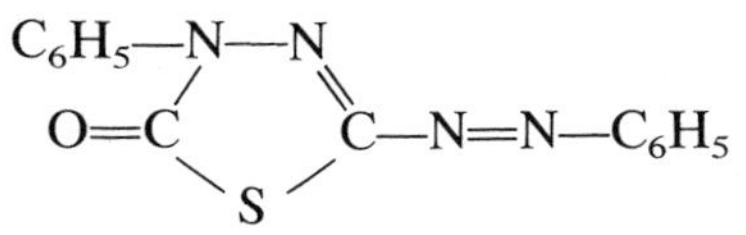

Exposure of a dry chloroform solution of dithizone to UV radiation forms 5-phenylazo-3-phenyl-1,3,4-thiadiazole-2-one by condensation of dithizone with phosgene produced by photochemical oxidation of the solvent.[19] This compound is yellow to brownish red in solution ($\lambda_{max} = 375$, $\epsilon_{375} = 19,300$ in CCl_4). A carbon tetrachloride solution of dithizone exposed to sunlight to convert dithizone completely to the yellow oxidation product(s) shows maximum absorption at 390 nm; a solution initially containing 1 mg H_2Dz in 100 ml gives $A_{390} = 0.254$.[20] Absorbance at 620 nm is negligible. Judging from the similarity in λ_{max} with that of irradiated chloroform–dithizone, this compound may also be present in deteriorated carbon tetrachloride solutions of dithizone. The composition of such solutions has not yet been fully investigated.

Irving and coworkers were unable to obtain diphenylthiocarbadiazone ($(C_6H_5N{=}N)_2C{=}S$) earlier postulated as an oxidation product by H. Fischer. If formed, it is assumed to isomerize rapidly to the meso-ionic compound.

For stability of dithizone in purified CCl_4 and $CHCl_3$, see Section III.C of this chapter. In dilute alkaline solution, for example 0.1 M NaOH, dithizone shows little change in absorbance after 2 or 3 hr if the solution is made 0.05 M in NH_3OHCl.

II. THE METAL DITHIZONATES

A. STRUCTURE

Dithizone contains two active hydrogen atoms, each of which can be replaced by an equivalent of metal. When one hydrogen of the dithizone molecule is replaced, a primary dithizonate is formed; when both are replaced, a secondary dithizonate is formed. All dithizone-reacting metals form primary dithizonates, but not all form secondary dithizonates. Dithizone–metal equilibria are considered in detail later. Here we discuss the composition of dithizone chelates and their structure.

The structure of a primary dithizonate of the metal M(II) can be represented as follows:

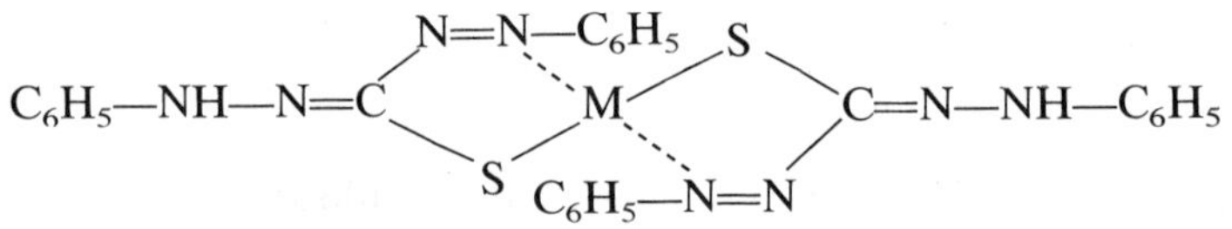

The failure of S-methyl dithizone to form chelates with dithizone-reacting metals points to the presence of metal–sulfur bonds in dithizonates.[21] This structure is supported by X-ray study of metal dithizonate crystals.[22] X-ray structure determination of zinc dithizonate has shown that the

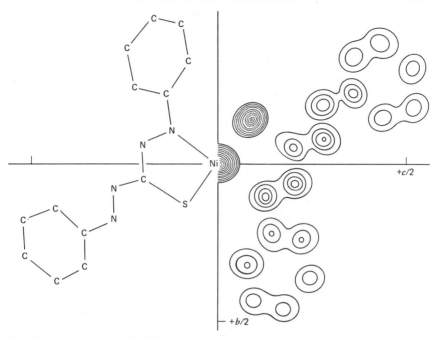

Fig. 6G-5. Structure of solid Ni(HDz)$_2$: projection down the a-axis. The Ni atom is bonded in approximately square planar configuration to S and one N. From Laing and Alsop, *loc. cit.*

molecule consists of two bidentate dithizone residues coordinated tetrahedrally to zinc through two S and two N atoms; one phenyl group of each ligand is associated with the chelate ring and the other phenyl group is extended as far as possible from the central atom with two intervening N atoms holding it in *trans* configuration.[23] The geometries of the dithizonate residues are much the same in primary copper(II) and mercury(II) dithizonates. Likewise in Ni(HDz)$_2$, Ni is bonded to S and one N atom (Fig. 6G-5).[24] Ni(HDz)$_2$, Pd(HDz)$_2$, and Pt(HDz)$_2$ are closely iso-structural, and their solutions in organic solvents show multiple absorption bands.

The following structures are proposed for the secondary dithizonates of Cu(I), Ag, and Hg(I)[25]:

Secondary Cu(II) dithizonate is formulated as a dimer:

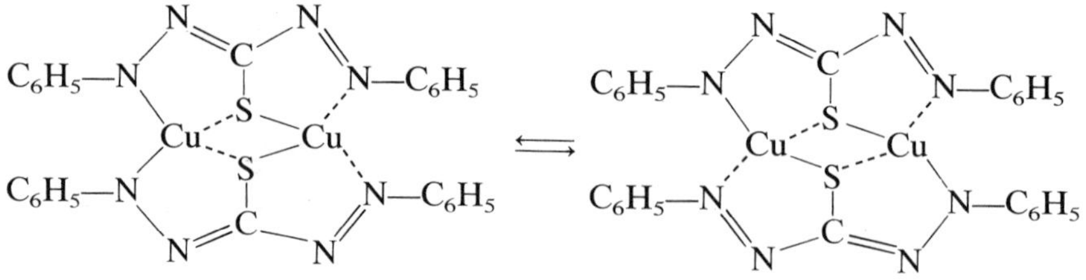

B. DITHIZONE-REACTING METALS AND THEIR CHELATES

The elements known at present (1974) to form dithizonates extractable into organic solvents immiscible with water are indicated in Table 6G-2. The total number is 27; probably a few more will eventually be added to the list. However, a number of elements react with such difficulty or under such special conditions that they are virtually nonreactive from the analytical viewpoint.

Table 6G-3 summarizes general properties of the dithizonates. Dithizone can form mixed as well as simple metal dithizonates; some mixed dithizonates are mentioned in the table. Most of the analytically important dithizonates are simple.

1. Solubility

The chelates are usually less soluble than dithizone in $CHCl_3$ and CCl_4. They are more soluble in $CHCl_3$ than in CCl_4, as illustrated by some of the data in Table 6G-4. Almost always, the primary dithizonate is more soluble than the secondary dithizonate. For example, Ag(HDz) is roughly

TABLE 6G-2

Elements Reacting with Dithizone (extracted into CCl_4, $CHCl_3$, and similar solvents as dithizonates from aqueous solution)[a]

H																	
Li	Be											B	C	N	O	F
Na	Mg											Al	Si	P	S	Cl
K	Ca	Sc	Ti	V	Cr	Mn	Fe	Co	Ni	Cu	Zn	Ga	Ge	As	Se	Br
Rb	Sr	Y	Zr	Nb	Mo	Tc	Ru	Rh	Pd	Ag	Cd	In	Sn	Sb	Te	I
Cs	Ba	La	Hf	Ta	W	Re	Os	Ir	Pt	Au	Hg	Tl	Pb	Bi	Po	
	Ra	Ac	Th	Pa	U											

[a] Elements enclosed by solid lines form definite dithizonates; those enclosed by broken lines probably form dithizonates.

Survey of Dithizonates[a]

Element	P or S[b]	Color[c]	Acidity for Extraction[d]	Remarks
Ag	P	Yellow	Dilute mineral acid	Solid reported as $Ag(HDz) \cdot H_2O$.
	S	Red-violet (solid)	Basic soln.	Slightly soluble in $CHCl_3$ (red solution), practically insoluble in CCl_4. Ag_2Dz has been isolated as solid.
As(III)				According to M. Hranisavljevic-Jakovljevic and I. Peykovic-Tadic, *Mikrochim. Ichnoanal. Acta*, **1965**, 936, As(III) can be extracted from 6 M HCl into benzene; dithizonate gives orange-yellow color on cellulose chromatographic plate. Incomplete extraction from aqueous solution of pH 2–3. H. Fischer does not mention reaction of As(III) with CCl_4 or $CHCl_3$ soln. of H_2Dz. Mixed chelate of As (and Sb) with dithizone-dithiocarbamate has been reported (see note 31, this chapter), namely AsDzDt. T. Kamada, *Talanta*, **23**, 835 (1976), found As(III and V) to be unextractable by dithizone in CCl_4 over the pH range 1–10; As(III) is poorly extracted from 3–10 M HCl (as $AsCl_3$?).
Au(I)	P	Red-brown ($CHCl_3$)	Dilute mineral acid	Apparently a *P* chelate of Au(I) from ratio $H_2Dz:Au = 1$ (H_2Dz in excess). Beardsley, Briscoe, Ruzicka, and Williams, *Talanta*, **13**, 328 (1966).
Au(I)	S	Yellow		With excess gold, chelate has composition 1 $H_2Dz:2$ Au (Beardsley et al.)
Au(III)			Dilute mineral acid	Conflicting statements made about Au(III)–H_2Dz reaction. Difficulties arise from reduction of Au(III), and possibly of Au(I), by H_2Dz. Cox and Servant, *Anal. Chim. Acta*, **66**, 123 (1973), found oxidation products of H_2Dz responsible for several peaks in absorption curve of $CHCl_3$ extract. One peak due to bis-3,3′-(1,5-diphenylformazan) disulfide. A transitory peak at 490 nm doubtfully attributed to Au dithizonate. Zolotov, Demina and Petrukhin, *Zh. Anal. Khim.*, **25**, 2315 (1970), found no evidence for formation of $Au(HDz)_3$ or Au_2Dz_3 ($CHCl_3$). From 5 M H_2SO_4 containing Cl^-, $Au(HDz)Cl_2$ is extracted. In presence of trichloroacetic acid (HT), $Au(HDz)_2T$ is extracted. $Au(HDz)Cl_2$ and $Au(HDz)(OH)_2$ have been isolated as solids by Meriwether et al. H. Fischer reported yellow *P* dithizonate from 0.5 M H_2SO_4 and red *S* (insol. in CCl_4) from basic solns., but reduction to Au(I) may have occurred.

Element	P or S^b	Color^c	Acidity for Extraction^d	Remarks
Bi	P	Orange-yellow	pH > 1(CCl$_4$)	Solid Bi(HDz)$_3$ isolated. No evidence for secondary dithizonate. Solid Bi(HDz)$_2$Cl·2 HCl also reported (not formed under analytical conditions).
Cd	P	Red	Basic soln.	Cd(HDz)$_2$ in CCl$_4$ or CHCl$_3$ is stable toward 1 M NaOH or saturated H$_2$S soln.
Co(II)	P	Violet-red	pH > 5 (CCl$_4$)	Extraction easier from basic (ammoniacal tartrate or citrate) solution, but oxidation to Co(III) and extraction of Co(HDz)$_3$ seems to occur, at least partially. Once extracted, Co dithizonate is quite resistant to dilute mineral acid solutions, especially if present as Co(HDz)$_3$. Co(HDz)$_2$ has been obtained in solid form. Some observations indicate incomplete extraction of Co from ammoniacal solutions, which presumably is connected with oxidation of Co(II) to Co(III). Extraction should be made in the presence of a reducing agent to keep Co in the II state.
Co(III)	P	Brownish		From very nearly neutral soln. in presence of air. Duncan and Thomas, *J. Chem. Soc.*, 2814 (1960).
Cu(I)	P	Orange-brown (CCl$_4$)	Dilute mineral acid (e.g., 0.1 M HCl, CCl$_4$)	Kiwan and Irving, *Anal. Chim. Acta*, **57**, 59 (1971).
Cu(I)	S	Violet (CHCl$_3$)	Basic soln.	Insol. in CCl$_4$.
Cu(II)	P	Violet-red	pH $\lesssim$ 1	With pure CCl$_4$, extn. equilibrium reached in a minute or two of shaking. Geiger and Sandell, *Anal. Chim. Acta*, **8**, 197 (1953).
Cu(II)	S	Yellow-brown (CCl$_4$)	Basic soln.	Dimeric in CHCl$_3$, Irving and Kiwan, *Anal. Chim. Acta*, **56**, 435 (1971). Can also be formed in slightly acidic soln. if Cu > H$_2$Dz. Solid chelate isolated, but hydration uncertain or variable; CuDz·2 H$_2$O claimed. Brand and Freiser, *Anal. Chem.*, **46**, 1147 (1974), maintain that this compound is a Cu(I) complex.
Fe(II)	P	Violet-red	pH > 6.5 (CCl$_4$)	Easily air-oxidized (color fades). Marczenko and Mojski, *Chim. anal.*, **53**, 529 (1971).
Fe(III)				Fe(III) has generally been considered not to form a dithizonate, but Zolotov, Kiseleva, Shakhova, and Lebedev, *Anal. Chim. Acta*, **79**, 237 (1975), report the formation of the cationic complex Fe(HDz)$_2^+$, extracted into dichloroethane at pH ~6 in the presence of capric acid ($\lambda_{max} = 520$ nm). Acetate, tetraphenylborate, and ReO$_4^-$ also serve as anionic partners for extraction into chloroform and other solvents. The relatively small size of Fe^{3+} may explain the nonformation of Fe(HDz)$_3$.

594

Element	*P* or *S*[b]	Color[c]	Acidity for Extraction[d]	Remarks
				Fe(III) oxidizes dithizone, especially in the presence of cyanide.
Ga	*P*	Red-violet	pH 2.5–3 (CCl$_4$, optimum)	Extraction incomplete. Ga in large amounts can interfere with determination of other metals in feebly acidic soln. Ga dithizonate extd. at pH 2.8 (CCl$_4$) has Ga:H$_2$Dz = 1:2; presumably contains Cl or OH. Tartrate and citrate inhibit extn. Iwantscheff and Jörrens, *Anal. Chim. Acta*, **38**, 470 (1967); Pierce and Peck, *ibid.*, **27**, 392 (1962).
Ga alkyl cations	*P*	Red-violet	1–10	Various Ga alkyls form RGa(HDz)$_2$ and R$_2$Ga(HDz). Iwantscheff and Jörrens, *Z. Anal. Chem.*, **264**, 131 (1973).
Ge(II)				GeCl$_2$ in dry CCl$_4$, CHCl$_3$, or C$_6$H$_6$ gives red color with H$_2$Dz, perhaps due to Ge(HDz)$_2$ (Iwantscheff and Jörrens).
Ge(IV)				GeCl$_4$ in dry benzene gives red-brown color with H$_2$Dz (Iwantscheff and Jörrens).
Hg(I)	*P*	Orange	Dil. mineral acid	
Hg(I)	*S*	Purplish red	Basic soln.	Only slightly soluble in CCl$_4$.
Hg(II)	*P*	Orange-yellow	Mineral acid	Formed in strongly acid soln. (in absence of much chloride). Light sensitive.
Hg(II)	*S*	Purplish red	Basic soln.	Also formed in feebly acidic soln. with Hg in excess of H$_2$Dz.
In	*P*	Red	pH 5–6 (optimum) CCl$_4$; pH 8.5–9.5 (optimum), CHCl$_3$	Reacts in presence of cyanide, but citrate hinders extn.
Mn(II)	*P*	Violet	pH 10–11 (CHCl$_3$), incomplete extn.; pH 9–10 (CCl$_4$), incomplete extn.	Mn(HDz)$_2$ unstable; Mn(II) oxidized in basic solution. Mn(HDz)$_2$·H$_2$O isolated as red-brown solid. Marczenko and Mojski, *Anal. Chim. Acta*, **54**, 469 (1971).
Ni	*P*	Brown or brownish violet	Weakly basic	Once extracted into CCl$_4$, Ni(HDz)$_2$ is not easily dissociated by slightly acidic aqueous solutions. When Ni is extracted from strongly basic soln., a gray CCl$_4$ extract is obtained. Ni(HDz)$_2$ isolated as greenish black solid.
Os(III)				Os(HDz)$_3$ said to have been isolated as solid.
Pb	*P*	Cinnabar red	pH 7–11 (CCl$_4$)	Reacts in presence of cyanide.

Element	P or S^b	Color^c	Acidity for Extraction^d	Remarks
Pd(II)	P	Greenish brown	Mineral acid	Extraction is slow but quantitative in strongly acid solutions. Chelate converted to blue $Pd(HDz)_2(OH)_2^{2-}$ by shaking with 0.1 M NaOH or 2 M ammonia [Minczewski, Krasiejko, and Marczenko, *Chemia Anal.*, **15,** 43 (1970)]. $Pd(HDz)_2$ isolated as violet-blue solid. Mixed chelate(s) formed with Cl^-.
	S	Red? Gray-violet?	pH < 1, excess Pd (CHCl$_3$)	The secondary, 1:1, chelate is converted to $PdDz(NH_3)_2(OH)_2^{2-}$ on shaking with 10 M NH_4OH (Minczewski et al.). This mixed species also formed when Pd(II) reacts with H_2Dz in aqueous NH_3. Solid $PdDz \cdot 2 H_2O$ has been obtained. The secondary dithizonate is only slightly soluble in organic solvents. Beardsley, Briscoe, Ruzicka, and Williams, *Talanta*, **13,** 328 (1966), report PdDz as gray-violet and soluble in CHCl$_3$ and CCl$_4$, contrary to earlier findings.
Po			pH 0–5 (CHCl$_3$)	Bouissierès and Ferradini, *Anal. Chim. Acta*, **4,** 610 (1950). According to Bagnall and Robertson, *J. Chem. Soc.*, 509 (1957), complex is probably $PoO(HDz)_2$.
Pt(II)	P	Yellow	Mineral acid	From 1:3 HCl in presence of SnCl$_2$ [Young, *Analyst*, **76,** 49 (1951)]. H. Fischer records different observations. Formation of mixed Cl^- chelate(s) not excluded. Pt(IV) does not react. $Pt(HDz)_2$ has been obtained as solid. Miyamoto, *Japan Analyst*, **9,** 925 (1960), extracts Pt(II) from 1 M HCl containing SnCl$_2$ into CCl$_4$. Also see *Anal. Abst.*, **19,** 209.
Re			9 M HCl	Ryabchikov and Borisova, *Talanta*, **10,** 13 (1963). $\lambda_{max} = 530$ nm (CHCl$_3$).
Sb(III)				Sb has been considered nonreactive with H_2Dz but Hranisavljevic-Jakovljevic and Peykovic-Tadic (*loc. cit.*) claim extraction at pH 4 with benzene soln. of H_2Dz. Carlton, Bradburn, and Kruh, *Anal. Chim. Acta*, **12,** 101 (1955), claim Sb is extracted with H_2Dz in benzene from ~1 M HCl (red-brown color).
Se(IV)	P	Yellow	HCl soln. (CCl$_4$)	Reacts in ratio 1 Se:4 H_2Dz [Stary, Marek, Kratzer, and Sebesta, *Anal. Chim. Acta*, **57,** 393 (1971)]. $Se(HDz)_2$ may be formed [reduction of Se(IV) by H_2Dz]. Ramakrishna and Irving, *ibid.*, **49,** 9 (1970), claim no chelate formed, only oxidation products. Mabuchi and Nakahara, *Bull. Chem. Soc. Japan*, **36,** 151 (1963), extract Se from 6 M HCl (CCl$_4$). Kasterka and Dobrowolski, *Chemia Anal.*, **15,** 303 (1970), extract Se into CCl$_4$ from 6–6.5 M HCl.

596

TABLE 6G-3 (*Continued*)

Element	P or S[b]	Color[c]	Acidity for Extraction[d]	Remarks
Sn(II)	P	Red	pH > 4 (optimum 6–9 in CCl$_4$)	Unstable, air oxidation. Solid Sn$_2$O(HDz)$_2$ · 2 H$_2$O prepared. For supposed dithizonate of Sn(IV) in isopentyl alcohol, see Vancea and Volusniuc, *Anal. Abst.*, **11**, 5418 (1964).
Te(IV)	P		Mineral acid (CCl$_4$)	Mabuchi, *Bull. Chem. Soc. Japan*, **29**, 842 (1956): Mabuchi and Okada, *ibid.*, **38**, 1478 (1965). Marhenke and Sandell, *Anal. Chim. Acta*, **38**, 421 (1967), found the combining ratio 1 Te : 4 H$_2$Dz, and consider Te(HDz)$_2$ as the extracted species. *E* increases with decreasing Te(IV) concentration.
Tl(I)	P	Red	pH 3–4 (CCl$_4$)	Tl(III) seems to react also but is quickly reduced to Tl(I). Solid Tl(HDz) has been prepared.
V(IV)?		Violet		Bloch and Lazare, *Bull. Soc. Chim. France*, 1148 (1960), claim formation of complex having molar ratio H$_2$Dz : V = 2, soluble in water and oxygen-containing organic solvents (butanone).
Zn	P	Purplish red	Feebly acidic to weakly basic solns.	Zn(HDz)$_2$ isolated as solid.

[a] Based on early observations of H. Fischer, *Angew. Chem.*, **47**, 685 (1934); **50**, 919 (1937), supplemented by later work, to much of which references are given. The solid dithizonates mentioned are mostly those prepared by Meriwether, Breitner, and Sloan, *J. Am. Chem. Soc.*, **87**, 4441 (1965).

[b] Primary (*P*) or secondary (*S*).

[c] In CCl$_4$ usually, but similar in CHCl$_3$.

[d] Acidity for quantitative extraction depends on dithizone concentration in organic solvent at equilibrium. Figures, when given, are approximate and pertain to ~0.001% dithizone in CCl$_4$. More exact values can be calculated from K_{ext} in Table 6G-7. For upper pH limits of extraction, see p. 608.

1000 times more soluble than Ag$_2$Dz in CCl$_4$—though this is an extreme example. The secondary dithizonates are of slight importance analytically. Carbon tetrachloride is usually preferred to chloroform as solvent in dithizone separations and photometric determinations, and the solubility of the primary dithizonates in it is almost always sufficient to allow its use in trace analysis.

The solubilities of a few dithizonates in water have been determined[26]: Cu(HDz)$_2$, 8×10^{-9} M; CuDz, $\approx 6 \times 10^{-8}$; Pb(HDz)$_2$, 3×10^{-9}; Bi(HDz)$_3$, 2.5×10^{-8}; Zn, $\sim 5 \times 10^{-9}$ (calculated).

2. Spectral Properties

Molar absorptivity values of dithizonates in CCl$_4$ and CHCl$_3$ at the wavelengths of maximum absorption are collected in Table 6G-5. It has

Solubility of Dithizonates in Organic Solvents

Dithizonate	Solvent	Temp. (°C)[a]	Solubility (M)	Reference
Ag(HDz)	$CHCl_3$	30	1.6×10^{-2}	Dyer and Schweitzer, *Anal. Chim.*
	CCl_4	30	3.7×10^{-3}	*Acta*, **23**, 1 (1960).
	C_6H_6	30	8.2×10^{-3}	
Ag_2Dz	CCl_4		$< 10^{-6}$	
$Au(HDz)_3$	CCl_4		$\sim 10^{-5}$	Iwantscheff, *Das Dithizon*,
				1958 (2nd ed., 1972).
$Au_2Dz_3(?)$	CCl_4		"insol."	
$Bi(HDz)_3$	CCl_4		1.3×10^{-5}	Iwantscheff.
	CCl_4	25	5.5×10^{-4}	Bidleman, Thesis, Univ. of Minn.
$Cd(HDz)_2$	$CHCl_3$		1.3×10^{-4}	Iwantscheff.
	CCl_4		1×10^{-5}	Klein and Wichmann, *J. Assoc.*
				Offic. Agr. Chem., **28**, 257 (1945).
$Co(HDz)_2$	$CHCl_3$		1.4×10^{-3}	Iwantscheff.
	CCl_4		1.6×10^{-4}	Iwantscheff.
$Cu(HDz)_2$	CCl_4	25	5.5×10^{-4}	Geiger, Thesis, Univ. of Minn.,
				1951.
CuDz	CCl_4	25	1.2×10^{-3}	Geiger, Thesis, Univ. of Minn.,
(Cu_2Dz_2)				1951.
$Hg(HDz)_2$	CCl_4		6.6×10^{-5}	Iwantscheff.
HgDz	CCl_4		1.3×10^{-3}	Iwantscheff.
	$CHCl_3$		2.8×10^{-3}	Iwantscheff.
$In(HDz)_3$	$CHCl_3$		1.1×10^{-3}	Iwantscheff.
	CCl_4		8×10^{-4}	
$Pb(HDz)_2$	$CHCl_3$	25	4.3×10^{-4}	Iwantscheff.
	CCl_4	25	5.7×10^{-6}	Mathre, Thesis,[b] Univ. of Minn.,
				1956.
$Pd(HDz)_2$	CCl_4		4.5×10^{-4}	Iwantscheff.
Tl(HDz)	$CHCl_3$		$> 1 \times 10^{-5}$	Iwantscheff.
$Zn(HDz)_2$	$CHCl_3$	30	3.6×10^{-2}	Schweitzer and Honaker, *Anal. Chim.*
	CCl_4	30	1.25×10^{-3}	*Acta*, **19**, 224 (1958).
	CH_2Cl_2	30	2.8×10^{-2}	
	C_6H_6	30	4.0×10^{-3}	
	$CHCl_3$		6.1×10^{-2}	Stary and Ruzicka, *Talanta*, **15**,
	CCl_4		7.2×10^{-3}	513 (1968).

[a] Room temperature if not specified.

[b] Mathre made a careful study of the solubility of $Pb(HDz)_2$ in CCl_4 in the presence and absence of aqueous phase. Recrystallization of finely divided $Pb(HDz)_2$ took place on shaking, more rapidly in the presence of aqueous phase, so that the equilibrium value was $\sim 1/4$ the initial solubility.

λ_{max} and ϵ_{max} of Dithizonates in Carbon Tetrachloride and Chloroform[a]

Dithizonate	λ_{max} (nm)		ϵ_{max} [(liter/mole cm) $\times 10^{-4}$]	
	CCl_4	$CHCl_3$	CCl_4	$CHCl_3$
Ag, P	426		3.05	
	460		2.91	
	462		2.72	
		470		3.1
Ag, S		490		
Au(III), P(?)		450	2.4	4.4
Au(HDz)Cl$_2$		540, 570		
Au(HDz)(OH)$_2$		460		
Bi, P	490	498, 490	8.0	9.0 (490 nm)
	490		7.9	
	490		8.46	
Cd, P	520	520	8.8	8.6, 8.76
	520		8.42	
		518		
Co(II), P	542		5.92	
Co(III), P	~540		~5	
Cu(I), P	480	490	2.13	2.8
Cu(II), P	550	545	4.52	
	545		4.5	
	548		4.50	
Cu(II), S	450	445	2.27	
Fe(II), P	~560			
Ga, P	530		~5	
Ge(II)	~530		>5.2	
Hg(I), P		490		3.06
Hg(II), P	485		7.1	
	490		7.2, 6.85	
S	515		2.36	
	490			
	526		3.2	
In, P	510		8.7, 6.1	
Mn(II), P		515		
Ni, P	665	670	1.92	2.0
	480, 565,	340, 475, 565, 675	3.04 (480 nm)	2.7 (475 nm)
	665	290, 480, 560		2.6 (480 nm)
Os(III)(?)		390, 520		
Pb, P	520	510	6.63 (7.0)	6.36
	520		6.9–7.2	
	520		6.65	
Pd, P	620	450, (570), 640	4.25	
	450		3.4	
Pd, S	340, 550			
	485		2.6	
	425, 490, 730		3.0 (730)	

TABLE 6G-5 (Continued)

Dithizonate	λ_{max} (nm)		ϵ_{max} [(liter/mole cm) $\times 10^{-4}$]	
	CCl$_4$	CHCl$_3$	CCl$_4$	CHCl$_3$
Pt(II), P	490		3.16	
	490 (C$_6$H$_6$)		2.6 (C$_6$H$_6$)	
	(430), 488,			
	710			
	720 (C$_6$H$_6$)		2.7 (C$_6$H$_6$)	
Se(IV → II?); Se(HDz)$_2$(?)	420		7.4, 7.0	
Sn(II), Sn$_2$-O(HDz)$_2$ · 2 H$_2$O	520			
Sn(II), P	520		5.4	
	508			
Te(IV → II); probably Te(HDz)$_2$	~450		~4.0	
	427		4.1	
Tl(I), P	510	505	3.32	3.36
		515		
		508		
Zn, P	535	530	9.6	8.8
	538	532		
	536		9.2	
	530		9.84	

[a] From various authors. Each λ_{max} or ϵ_{max} entry represents an independent source.

been noted[27] that tetrahedral and octahedral dithizonates, which include most dithizonates, show a single strong absorption band in the visible range, the planar dithizonates (Ni dithizonate, for example), several.

Dithizonates can be divided into four groups on the basis of very similar infrared spectra of the solids in each group[27]:

1. Mn(HDz)$_2$ · H$_2$O, Zn(HDz)$_2$, Cd(HDz)$_2$, Hg(HDz)$_2$, Sn$_2$O(HDz)$_2$ · 2 H$_2$O, Pb(HDz)$_2$, and Bi(HDz)$_3$ (tetrahedral and octahedral complexes).

2. Ag(HDz) · H$_2$O, Au(HDz)(OH)$_2$, Tl(HDz), and "brown" Cu(HDz)$_2$ (formed by irradiation of violet Cu(HDz)$_2$).

3. Ni(HDz)$_2$, Pd(HDz)$_2$, "red" Pd(HDz)$_2$ (formed by irradiation of greenish Pd(HDz)$_2$), PdDz · 2 H$_2$O, Pt(HDz)$_2$, and Cu(HDz)$_2$ (probably square planar).

4. Os(HDz)$_3$ and Co(HDz)$_2$.

CuDz · 2 H$_2$O, Ag$_2$Dz, and "violet" Pd(HDz)$_2$ (a form absorbing at 470 and 520 nm) have IR spectra unlike other dithizonates.

3. Photochromism[28]

The effect of light on organic solvent solutions of metal dithizonates can be reversible or irreversible. The latter effect involves oxidative decomposition (p. 630). The reversible effect (photochromism) is considered here.

On exposure to light of sufficient intensity in the absorbing region (λ usually > 450 nm), solutions of some dithizonates show a change in color (hue), with more or less rapid return to the original color when irradiation is cut off. The activated form is usually blue or violet. Secondary dithizonates do not seem to be photochromic. Photochromism is especially marked with $Hg(HDz)_2$ in benzene, chloroform, or carbon tetrachloride. When its solution is strongly irradiated, the color changes from orange ($\lambda_{max} = 490$ nm) to blue ($\lambda_{max} = 605$ nm). The color returns to orange some minutes after irradiation is stopped. The half life of activated $Hg(HDz)_2$ is ~ 0.5 min at $25°$. The return rates depend on the metal dithizonate and increase as follows: $Hg < Pd < Ag < Pt \sim Zn < Bi < Cd \sim Pb$.

Analytically, photochromism is not of importance for the dithizonates of metals other than Hg, Ag, Pd, or Pt. Even at high illumination levels, such as would not be encountered in analytical work, only $\sim 0.1\%$ of lead or cadmium dithizonates in benzene solution at room temperature is in the activated form, compared to 80–90% of $Hg(HDz)_2$ under the same conditions. The return rates decrease with decreasing temperature:

$$\text{normal form} \underset{\Delta}{\overset{h\nu}{\rightleftarrows}} \text{active form}$$

The strongest photochromic effects are shown in dry nonpolar solvents. Hydroxylic solvents and organic acids and bases hinder photochromism by accelerating the return reaction. A small amount of acetic acid in the chloroform or carbon tetrachloride solution of $Hg(HDz)_2$ prevents its photochromism (*Colorimetric Determination of Traces of Metals*, 3rd ed., p. 627).

The activated form of $Hg(HDz)_2$ is believed[29] to have the structure

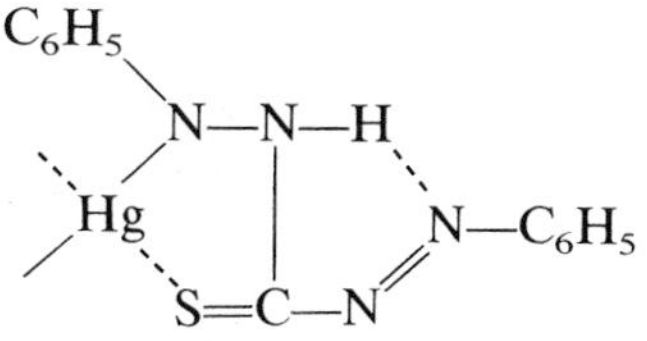

4. Mixed Dithizone Complexes

These are of three types: mixed ligand, mixed metal, and adducts.

Mixed ligand chelates may have either inorganic or organic anions as the second anionic component. A number of mixed ligand chelates are mentioned in Table 6G-3. Chloride (and other halide ions) and OH⁻, especially the former, form such complexes. In a nonaqueous medium, such as ether, Hg(II) yields Hg(HDz)Cl.[30] Au(HDz)Cl$_2$ is extracted into CHCl$_3$ from sulfuric acid solutions containing chloride. Gallium forms Ga(HDz)$_2$Cl. Pd(HDz)$_2$(OH)$_2^{2-}$ is formed when a CHCl$_3$ solution of Pd(HDz)$_2$ is shaken with dilute sodium hydroxide solution.

An example of mixed ligand complexes having an organic component is provided by dithizonate–dithiocarbamate chelates of Cu(II), As(III), Sb(III), Ni, and Fe(II).[31] As(III) forms AsDzDt (Dt = diethyldithiocarbamate anion); the other mixed chelates contain the primary dithizonate anion, HDz⁻. No mixed chelates of dithizone and oxine are known.

The reactions of organomercury(II) compounds with dithizone are of special interest. Thus RHgX (X = Cl, etc.) in water reacts with H$_2$Dz in CCl$_4$ to give yellow RHg(HDz), the Hg–X bond being cleaved, and the organomercury dithizonate is extracted into the organic solvent.[32] (In a sense RHg⁺ is a mixed cation, and the reaction is conveniently considered here.) Organomercurials can be determined photometrically in this way. Organotin and organo Ga compounds react similarly with dithizone. Phenylmercury(II) and other mercurials form both primary and secondary dithizonates.[33] In very alkaline solution, the magenta secondary dithizonate ion is formed:

$$PhHg(HDz) \rightleftharpoons PhHgDz^- + H^+$$

      yellow   magenta

In 53% ethanol–47% H$_2$O(v/v), pK_a for this reaction has the value 11.46. Evidence was obtained for the formation of the secondary dithizonate (PhHg)$_2$Dz in CHCl$_3$. This solution shows two maxima at 510 and 272 nm, whereas PhHg(HDz) in CHCl$_3$ shows maxima at 476 and 270 nm. Other organomercurials behave similarly to phenylmercury(II). The secondary dithizonates are less stable and less soluble in organic solvents than the corresponding primary dithizonates.

The mixed cation secondary dithizonate Hg(AgDz)$_2$ is formed by shaking a chloroform solution of Hg(HDz)$_2$ with aqueous silver nitrate solution.[34] Its solution has a magenta color, with λ_{max} at 512 nm. A possible structure is

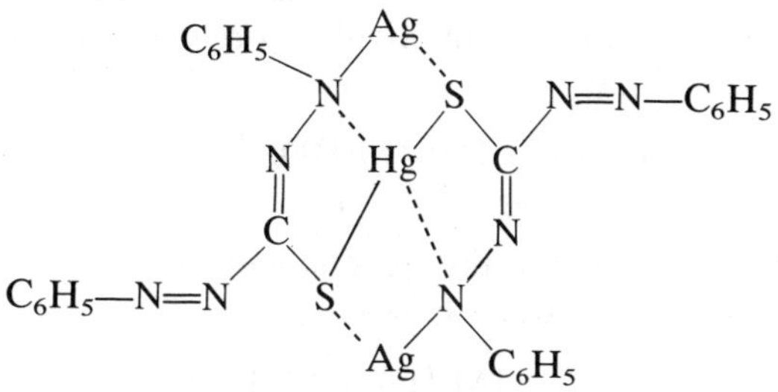

$$C_6H_5-N \underset{N}{\overset{Ag}{\diagup}} S \quad N{=}N-C_6H_5$$

The relatively easy formation of the mixed chelate from the primary chelate in this way is explained by the increased tendency of imino-H to ionize if the S atom is bound to a metal atom or an alkyl group. Mixed complexes are also formed by silver and phenylmercury(II) cations: PhHg(AgDz) and Ag(PhHgDz).[35]

NH_3 is a neutral component of some mixed dithizonate complexes [as of palladium (II)].

Heterocyclic nitrogen bases (pyridine, dipyridyl, phenanthroline, and others) form adducts with dithizonates.[36] Some analytical use has been made of such adduct formation. In the presence of pyridine and NH_3OHCl, Mn(II) can be reproducibly extracted at pH ~ 9.5 with a CCl_4 solution of dithizone and determined photometrically at 530 nm.[37] The sensitivity is high ($\mathscr{S} = 0.0014$ μg Mn/cm^2). The extraction of Fe(II) is also synergized by pyridine. This method is, of couse, unselective for Mn, but a preliminary dithizone extraction at pH 6.5 in the presence of pyridine removes small amounts of Fe(II), Cu, Pb, Ni, Co, and Zn. Tl interferes. The rate of extraction of Co and Ni dithizonates is increased in the presence of pyridine.[38]

Nickel is extracted quite rapidly as the 1,10-phenanthroline adduct with a chloroform solution of dithizone and phenanthroline. Excess dithizone can be removed from the extract with dilute sodium hydroxide, leaving the pink adduct ($\mathscr{S}_{520} = 0.0012$ μg Ni/cm^2).[39]

Addition of N-bases to a chloroform solution of Ni(HDz)$_2$ converts the multiple-peaked absorption curve to one having a single peak in the range 520–535 nm ($\epsilon \sim 5 \times 10^4$). The Ni absorbance is thus approximately doubled. The adducts have a pink color.

C. METAL–DITHIZONE EQUILIBRIA

1. General Aspects

The reaction for the extraction of a metal from an aqueous solution as primary dithizonate with an immiscible organic solvent solution of dithizone may be written

$$M^{n+} + n H_2 Dz \rightleftharpoons M(HDz)_n + n H^+$$
$$\qquad\quad org \qquad\qquad org$$

The equilibrium constant for this reversible[40] reaction (extraction constant) at a specified ionic strength is

$$K_{ext} = \frac{[M(HDz)_n]_{org}[H^+]^n}{[M^{n+}][H_2Dz]_{org}^n} \qquad (6G\text{-}15)$$

This equation holds at any pH and in the presence of substances complexing M^{n+}, but it is useful analytically only when species other than those written are present in such low concentrations that they can be neglected. This is often true for an acidic aqueous solution free from substances complexing or otherwise reacting with M^{n+}.

The extraction coefficient of M is the ratio of the analytical concentration of the metal in the organic solvent phase to its analytical concentration in the aqueous phase (Chapter 9A):

$$E_M = \frac{\Sigma[M]_{org}}{\Sigma[M]} \qquad (6G\text{-}16)$$

When $M(HDz)_n$ is the only metal species in the organic phase (usually true) and M^{n+} is the only significant metal species in the aqueous phase, the following approximation is valid:

$$E_M \sim \frac{[M(HDz)_n]_{org}}{[M^{n+}]} \sim \frac{K_{ext}[H_2Dz]_{org}^n}{[H^+]^n}\text{[from Eq. (6G-15)]} \quad (6G\text{-}17)$$

As shown elsewhere (Chapter 9B), K_{ext} is a composite constant, whose value at specified ionic strength is given by

$$K_{ext} = \frac{P_{pm}}{K_{pm}} \times \frac{K_d^n}{P_d^n} \qquad (6G\text{-}18)$$

where P_{pm} is the partition constant of the primary metal dithizonate (org/aqueous), K_{pm} the ionization constant of the primary metal dithizonate, and K_d and P_d are the (primary) ionization constant and partition constant of dithizone.

When the acidity and concentration of dithizone in the organic phase are such that little M^{n+} remains in the aqueous phase at equilibrium, E_M approaches P_{pm}.

If K^0 is the extraction constant for zero ionic strength,

$$E_M = \frac{K^0 [H_2Dz]_{org}^n f_{M^{n+}}}{[H^+]^n f_{H^+}^n} \qquad (6G\text{-}19)$$

At constant aH^+ and $[H_2Dz]_{org}$, E_M varies as $f_{M^{n+}}$.

When metals are extracted by dithizone from solutions containing substances forming complexes with the metals or from solutions of pH such that hydrolysis of metal ions occurs, the following expression for the extraction coefficient of the metal can be applied[41]:

$$E_M = \frac{K_{ext}[H_2Dz]_{org}^n}{[H^+]^n \left\{ 1 + \dfrac{k_1}{[H^+]} + \dfrac{k_2}{[H^+]^2} + \cdots + \beta_1[X^-] + \beta_2[X^-]^2 + \cdots \right\}} \qquad (6G\text{-}20)$$

where the metal ion hydrolysis constants are defined as

$$k_1 = \frac{[MOH^{n-1}][H^+]}{[M^{n+}]} \qquad (6G\text{-}21a)$$

$$k_2 = \frac{[M(OH)_2^{n-2}][H^+]^2}{[M^{n+}]}, \text{ and so on} \qquad (6G\text{-}21b)$$

and the metal complex formation constants are defined as

$$\beta_1 = \frac{[MX^{n-1}]}{[M^{n+}][X^-]} \qquad (6G\text{-}22a)$$

$$\beta_2 = \frac{[MX_2^{n-2}]}{[M^{n+}][X^-]^2}, \text{ and so on.} \qquad (6G\text{-}22b)$$

It is assumed that polymerized and mixed species of M, L, and X are not formed. If metal–dithizone species such as $M(HDz)^+$ are present in the aqueous phase in significant amounts, their concentrations (in terms of $[HDz^-]$ and the formation constants) will need to be added within the braces of Eq. (6G-20). (See p. 607.) One metal–dithizone species always present in the aqueous phase is $M(HDz)_n$, the uncharged chelate, but its concentration there will always be small compared to its concentration in the organic phase (CCl_4 or $CHCl_3$). However, P_{pm} sets a limit to the maximum attainable value of E_M.

If the masking anion X^- forms a weak acid HX, $[X^-]$ will, of course, be a function of $[H^+]$. This is true for CN^-, PO_4^{3-}, and almost all anionic organic masking agents. The masking ability of uncharged complexing agents forming cationic metal complexes (e.g., NH_3, 1,10-phenanthroline) is also usually a function of acidity. But thiourea, being a very weak base, is hardly affected by $[H^+]$.

When the extraction equation (6G-20) is used to calculate E in extractions from aqueous solutions sufficiently basic to extract dithizone essentially completely from the organic phase, it is convenient to replace the numerator by

$$(7 \times 10^8)^n K_{ext}[H^+]^n[HDz^-]^n \qquad (CCl_4, 25°, \mu \sim 0.1\ M, pH > 11)$$

or

$$(4 \times 10^{10})^n K_{ext}[H^+]^n[HDz^-]^n \qquad (CHCl_3, 25°, \mu \sim 0.1\ M, pH > 13)$$

Metal–Dithizone Complexes in the Aqueous Phase. When exact values for the extraction coefficients of metals with dithizone are required, metal–dithizone species in the aqueous phase may need to be

TABLE 6G-6
Partition Constants (CCl_4/H_2O) and Aqueous Stability Constants of Some Metal Dithizonates

Dithizonate	$P = \dfrac{[M(HDz)_n]_{CCl_4}}{[M(HDz)_n]_{H_2O}}$	β_1^a	β_n (25°, $\mu = 0.1$) B. and S.[a]	Earlier Values
Ag(HDz)			9.5×10^6	
Bi(HDz)$_3$	2.5×10^4		1.3×10^{32}	1.2×10^{33b}
Cu(HDz)$_2$	7.3×10^4	2.2×10^9	1.5×10^{19}	2.5×10^{23c}
				2.1×10^{23d}
Hg(HDz)$_2$	3.6×10^3	4.4×10^{20}	2×10^{40}	2×10^{40e}
Pb(HDz)$_2$	2×10^3	2×10^7	1.4×10^{14}	1.4×10^{15f}
Pd(HDz)$_2$		2.5×10^{11}	6×10^{21}	
Zn(HDz)$_2$	2×10^{5g}	8.5×10^6	9×10^{13}	3×10^{14g}

[a] Budesinsky and Sagat, *loc. cit.*
[b] Bidleman, Ph.D. Thesis, University of Minnesota, 1970. From K_{ext} and P ($\beta_3 = (7 \times 10^8)^3 K_{ext}/P$).
[c] Geiger and Sandell [reference under Cu(II), Table 6G-3]. Ethanol solutions, with extrapolation to H_2O.
[d] Geiger and Sandell. From K_{ext} and $P_{Cu(HDz)_2}$.
[e] Duncan and Thomas, *J. Chem. Soc.*, 2814 (1960). In 0.64–2.7 M HCl. If the P value of D. and T. is combined with K_{ext} of Yokoyama et al., $\beta_2 = 2 \times 10^{39}$ for $\mu = 0.1$.
[f] Mathre and Sandell, *Talanta*, **11**, 295 (1964). From K_{ext} and P.
[g] Komar, Manzhelii, and Gorodilova, *Zh. Anal. Khim.*, **29**, 2391 (1974). Honaker and Freiser, *J. Phys. Chem.*, **66**, 127 (1962), found $[Zn(HDz)^+]/\{[Zn^{2+}][HDz^-]\} = 5.6 \times 10^7$; $[Zn(HDz)_2]/\{[Zn(HDz)^+][HDz^-]\} = 1.4 \times 10^7$.

taken into account. For a divalent metal forming a primary dithizonate these species are $M(HDz)^+$ and $M(HDz)_2$ (assuming no anionic species formed). The concentration of $M(HDz)_2$ in the aqueous phase will always be small compared to its concentration in the organic phase (CCl_4 or $CHCl_3$) because of the large values of $P_{M(HDz)_2}$. P values for CCl_4/H_2O have been determined for a few metal dithizonates (Table 6G-6).

The stability constants β_1 and β_n ($\beta_n = 1/K_{pm}$, $n = 2$ for a divalent metal) of a dozen dithizonates in aqueous solution have recently (1973) been reported.[42] Some of these values are listed in Table 6G-6, together with (mostly) earlier values of β_n (agreement is not striking).

It is of interest to compare the concentrations of $MHDz^+$ and $M(HDz)_2$ in the aqueous phase with the concentration of M^{2+} after a dithizone extraction. Lead dithizonate will be taken as an example. Suppose a 0.001% (0.00004 M) dithizone solution in CCl_4 is shaken to equilibrium with an equal volume of 5×10^{-6} M Pb^{2+} solution at pH 5 and 6. What are the final concentrations of $PbHDz^+$, $Pb(HDz)_2$, and Pb^{2+} in the aqueous phase?

$$\frac{[Pb(HDz)_2]_{CCl_4}[H^+]^2}{[Pb^{2+}][H_2Dz]_{CCl_4}^2} = 5.6$$

$$\frac{[PbHDz^+]}{[Pb^{2+}][HDz^-]} = 2 \times 10^7$$

$$\frac{[Pb(HDz)_2]_{CCl_4}}{[Pb(HDz)_2]} = 2 \times 10^3$$

Using these constants, we find the following values:

	$[Pb^{2+}]$	$[PbHDz^+]$	$[Pb(HDz)_2]$	$\dfrac{[Pb(HDz)_2]_{CCl_4}}{[Pb^{2+}]}$	E
pH 5	1.0×10^{-7}	8.4×10^{-9}	2.5×10^{-9}	50	45
pH 6	1.0×10^{-9}	8.4×10^{-10}	2.5×10^{-9}	~5000	1150^a

^{a} ~1000 if $PbOH^+$ is taken into account.

At pH 6, the concentrations of Pb^{2+} and $PbHDz^+$ are comparable, but $Pb(HDz)_2$ accounts for more than half of the unextracted lead. With a stronger H_2Dz–CCl_4 solution $PbHDz^+$ could outweigh Pb^{2+}, but the concentrations of both would decrease. For example, at pH 5 with a 4×10^{-4} M H_2Dz solution in CCl_4, $[Pb^{2+}] = 6 \times 10^{-10}$, $[PbHDz^+] = 7 \times 10^{-10}$, $[Pb(HDz)_2]_{CCl_4}/[Pb^{2+}] = 1 \times 10^4$, $E = 1.3 \times 10^3$. Often in approximate calculations for analytical purposes, $MHDz^+$ and $M(HDz)_2$ in the aqueous phases can be neglected.

Extraction of Secondary Dithizonates. At least five metals form secondary dithizonates: Ag, Au(I), Cu(I and II), Hg(I and II), and Pd. The secondary chelates are preferentially formed in alkaline solutions or with a deficiency of dithizone compared to metal. A secondary dithizonate can be transformed into the primary by treatment with acid and dithizone or with acid alone:

$$M^{II}Dz + H_2Dz \rightleftharpoons M^{II}(HDz)_2$$
$$\quad\text{org}\qquad \text{org}\qquad\qquad \text{org}$$

$$2\,M^{II}Dz + 2\,H^+ \rightleftharpoons M^{II}(HDz)_2 + M^{2+}$$
$$\quad\text{org}\qquad\qquad\qquad\quad \text{org}$$

Analytically, the secondary dithizonates are not of great interest. They are usually less soluble in organic solvents than the corresponding primary dithizonate (secondary cupric dithizonate is an exception) and may be virtually insoluble in CCl_4 (Ag, Au, Pd). Almost always the analyst wishes to avoid their formation. Extraction constants have been reported for secondary Ag and Cu(II) dithizonates.

Transformations of primary dithizonates into the secondary, and vice versa, are not always rapid. $Cu(HDz)_2$ is converted into $CuDz$ (or Cu_2Dz_2) by shaking a CCl_4 solution with base.[43] On the other hand, $Hg(HDz)_2$, once formed, is said not to be affected by 2 M NaOH. HgDz is extracted from neutral or basic aqueous solutions or even from weakly acidic medium with a deficiency of H_2Dz. It is readily converted to $Hg(HDz)_2$ by H_2Dz.

Extraction of Dithizonates from Basic Solutions. If a metal does not form a slightly dissociated or slightly soluble hydroxide, an anionic hydroxo complex or an anionic dithizonate, its extractability with increasing pH will eventually reach an approximately constant value, namely when substantially all excess dithizone has been transformed into dithizonate ion in the aqueous phase. With a CCl_4 solution of dithizone, $[H_2Dz]_{org}/[HDz^-]$ becomes equal to ~0.01 at pH 11 (i.e., $7 \times 10^8 \times 10^{-11}$), and further increase in pH will not significantly increase the extractability of any metal.

Most metals show decreasing extractability at higher pH values after having attained a plateau in their E–pH curves. Metals quantitatively extracted in strongly basic solutions (0.1–1 M OH$^-$) include Cu(II), Cd, Hg(II), and Co. These metals are not significantly amphoteric. Other metals are extracted less completely above the following pH values (approximate values with a CCl_4 solution of dithizone): Ag, 9; Bi, 11; In, 7; Pb, 11–12; Sn(II), 9–10; Zn, 10–11. With a chloroform solution of dithizone, the pH range of quantitative extraction is shifted to higher pH values in accordance with lower K_{ext} of a metal with $CHCl_3$ than with CCl_4 solutions of dithizone, usually by 1–2 units.

When the hydrolysis constants of a metal cation are known, E can be calculated in the alkaline range by making use of the general extraction expression [Eq. (6G-20)].[44] Dithizone does not show a marked tendency to form anionic metal dithizonates,[45] and it appears that for most metals such species are not of importance in calculating extractability in alkaline solutions.

The extraction of cadmium dithizonate from strongly basic solution is put to use in separation of Cd from Zn, In, and other metals whose dithizonates are only slightly extracted under such conditions.

2. Extraction Rate

The kinetics of dithizone extractions from the general standpoint is considered in Chapter 9B. Some observations of practical importance are noted here. The extraction of Bi, Cd, Cu(II), In, Pb, Hg, Ag, and Zn is very rapid from weakly basic or more strongly basic aqueous solutions with CCl_4 solutions of dithizone (Fig. 6G-6). Bi and Cu are extracted less

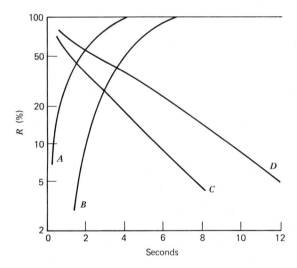

Fig. 6G-6. Percentage of metal extracted (R,%) as a function of shaking time. A, extraction of Cd from pH 13 aqueous solution (tartrate + NaOH) into 25×10^{-6} M H_2Dz in CCl_4; B, extraction of In at pH 5.5 into 25×10^{-6} M H_2Dz in CCl_4; C, back-extraction of Cd from CCl_4–dithizone into aqueous phase of pH 3; D, back-extraction of In from CCl_4 –dithizone into aqueous phase of pH 2.5. After H. Münzel and L. Koch, Radiochim. Acta, **4**, 188 (1965). Relative phase volumes similar, possibly the same.

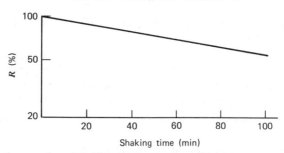

Fig. 6G-7. Back-extraction of Co(II?) dithizonate from CCl_4–H_2Dz into aqueous solution of pH 3. R is the percentage of Co remaining in CCl_4. Münzel and Koch, *loc. cit.*

rapidly from dilute mineral acid solution. However, equilibrium is attained in less than 2 min of shaking if sufficiently pure CCl_4 is used, namely CCl_4 that has been treated with Cl_2 or Br_2 in the purification process. Most metals are extracted more slowly from acidic than from basic solutions (Chapter 9B). The extraction of Ni and Co is slow (CCl_4) from faintly acidic solution and even faintly basic solution. Stripping Co^{46} and Ni from CCl_4 dithizonate solutions likewise is slow. Mercury(II) is extracted rapidly from acid solutions. Back-extraction of Cd and In (and other metals also) is quite rapid as shown by Fig. 6G-6, but that of Co (Fig. 6G-7) is not.

Extractions with $CHCl_3$ solutions of H_2Dz are slower than with CCl_4 solutions (Chapter 9B). Extraction of most metals from slightly basic solution is fast enough with $CHCl_3$, but the difference in rates is evident in acid solutions (Cu, Bi, but not Hg). The extraction of Ni and Co from weakly acidic or neutral medium with $CHCl_3$ solutions of dithizone is very slow.[47] Separations of Ni, Co, and Zn from each other based on differences in extraction rates is not satisfactory, however.

3. Extraction Constants of Dithizonates

Values of this constant for various metals in the CCl_4–H_2O and $CHCl_3$–H_2O systems are collected in Table 6G-7. Not all of these values are for the true constant K_{ext} defined by Eq. (6G-15). Some are effective constants (which can be denoted K'_{ext}) and are based on experimental values of $[M(HDz)_n]_{org}/\sum[M]$ instead of $[M(HDz)_n]_{org}/[M^{n+}]$. They hold only for specified $[H^+]$, specified concentrations of complexing substances, and other specifications, but are useful for calculating E_M under these conditions if E_M does not depend on the metal concentration (this has not always been established). The context (Table 6G-7) will usually make evident whether K_{ext} or K'_{ext} is intended.

If temperatures are not noted, 20°–25° may be assumed. A temperature variation of 5° may make an appreciable difference in the value of K_{ext}.[48]

Dithizonates can react with diethyldithiocarbamic acid (HDt) in an organic solvent, such as CCl_4:

$$M(HDz)_n + nHDt \rightleftharpoons M(Dt)_n + nH_2Dz$$
$$\quad\;\; org \qquad\quad org \qquad\quad org \qquad\quad org$$

Constants for this exchange reaction for various metals have been determined.[49] Some metals form mixed dithizonate–dithiocarbamate complexes (p. 602). Diethyldithiocarbamates are almost all more stable than the corresponding dithizonates, so that the equilibrium tends to lie to the right in the above reaction. Nickel and cobalt dithizonates react very slowly, the latter metal forming Co(III) diethyldithiocarbamate. Exchange constants for dithizone–oxine are also available.[50] Oxinates are less stable than the corresponding dithizonates.

III. PHOTOMETRIC DETERMINATION OF METALS

A. GENERAL ASPECTS

Most metal dithizonates in CCl_4, $CHCl_3$ and other organic solvents show an absorption maximum in the region ~500–550 nm (Table 6G-8 and Figs. 6G-8 to 6G-12) where dithizone itself absorbs minimally ($\epsilon_{min} = 4900$ at 515 nm in CCl_4). Commonly, a metal is determined extractively[51] by shaking the sample solution with a fixed volume of an organic solvent solution of dithizone of known concentration at a pH for quantitative extraction. With CCl_4 as the extracting solvent, most of the excess dithizone remains in the organic phase when the pH < 8 and a mixed color (of dithizone and dithizonate) results. The absorbance of the organic phase is then measured[52] at a suitable wavelength, usually in the range mentioned. The standard (calibration) curve will usually be a straight line (as long as sufficient dithizone is present to convert the metal to the dithizonate), whose intercept on the absorbance axis is the absorbance of the standard dithizone solution. If the pH for metal extraction lies in the basic range, some of the excess dithizone distributes into the aqueous phase, and the intercept absorbance decreases. Unless the pH is so high that practically all the excess dithizone is transferred to the aqueous phase (pH >11, CCl_4), a buffer is required to maintain the pH, and therewith the distribution of dithizone, constant. It is desirable, though not necessary, that the pH be high enough to bring most of the excess dithizone into the aqueous phase. Especially if the metal is amphoteric, the pH must not be raised to a point where the aquo cation is

TABLE 6G-7
Extraction Constants of Metal Dithizonates

Dithizonate	Conditions	K_{CCl_4}	K_{CHCl_3}	Reference
AgHDz	$[H^+] = 4$–$10(CCl_4)$; 1.5–2 $(CHCl_3)$	1.5×10^7	1.0×10^6	F. Koroleff, Thesis, University of Helsingfors, 1950. (This source is identified as Koroleff in entries below.)
	$6\ M\ H_2SO_4([H^+]$ taken as 6) Temp. $18°$	1×10^7 4×10^7		Sandell, unpublished, 1940. B. Tremillon, *Bull. Soc. Chim.*, 1156, 1160 (1954).
		9×10^8		A. T. Pilipenko, *Zh. Anal. Khim.*, **8**, 286 (1953).
	$0.25\ M\ H_2SO_4(+Cl^-)$, $24°$	2.1×10^7		N. Suzuki, *Japan Analyst*, **8**, 283 (1959).
	$1\ M\ HClO_4(30°)$	3×10^6	6×10^5	G. K. Schweitzer and F. F. Dyer, *Anal. Chim. Acta*, **22**, 172 (1960); F. F. Dyer and G. K. Schweitzer, *ibid.*, **23**, 1 (1960).
	$6\ M\ HClO_4(30°)$		1×10^5	
Ag_2Dz	$\dfrac{aH^+}{[AgHDz]_{CCl_4}[Ag^+]}$ (Ag_2Dz insoluble), $18°$	5×10^5		Tremillon, 1954.
Au(III?)			$>10^{27}$	
$Bi(HDz)_3$	Perchlorate soln., $\mu = 0.1$ $(25°)$	8.2×10^{10}		T. Bidleman, Thesis, Univ. of Minn.; *Anal. Chim. Acta*, **56**, 221 (1971).
	Perchlorate soln., $\mu = 1.0$ Perchlorate soln., $0.5\ M$ pH 0.9–2.4 (CCl_4); 2.4–3.7 $(CHCl_3)$	5.8×10^{10} 4.4×10^{10} 5.6×10^9	2.3×10^5	D. Brock, Univ. of Minn., 1969. Koroleff, 1950. Following values from same source. All these values are effective constants.
	$1\ M\ NaCl$; pH 3.6–3.9 (CCl_4), 3.8–4.8 $(CHCl_3)$	8×10^4	3.5×10^2	
	$0.1\ M\ KCN$; pH 1.4–1.9 (CCl_4), 3.2–3.9 $(CHCl_3)$	3.5×10^9	1.6×10^5	

Compound	Conditions			Reference
	0.1 M Na citrate; pH 6.6–7.6 (CCl$_4$), 8.3–8.5 (CHCl$_3$)	3.9×10^{-6}	2.7×10^{-11}	A. I. Busev and L. A. Bazhanova, *Zh. Neorg. Khim.*, **6**, 2210 (1961).
	Perchlorate soln., 0.2 M	9.4×10^9	5×10^8	A. T. Pilipenko, *Zh. Anal. Khim.* **8**, 286 (1953).
	$\mu = 0.5$	7×10^{10}		Koroleff, 1950.
Cd(HDz)$_2$	pH 3.5–5.6 (acetate buffer)	39	3.4	A. K. Babko and A. T. Pilipenko, *Zh. Anal. Khim.*, **2**, 33 (1947).
	1 M NaCl; pH 5.2–6.2 (CCl$_4$), 8.3–8.5 (CHCl$_3$)	0.27	0.25	G. K. Schweitzer and F. F. Dyer, *Anal. Chim. Acta*, **22**, 172 (1960).
	0.2 M Na citrate; pH 6.6 (CCl$_4$), 7.85 (CHCl$_3$)	0.12	0.14	Koroleff, 1950
		1.4×10^2		
Co(HDz)$_2$	HClO$_4$–NaClO$_4$, $\mu=0.1$ (30°)		3	
	HClO$_4$–NaClO$_4$, $\mu=1.0$ (30°)		10	
	pH 3.5–5.1 (CCl$_4$), 5–6.3 (CHCl)$_3$. Shaken 1.5 hr; equilibrium very slowly attained	39	0.32	Koroleff, 1950
	Acidic soln., shaken 10 days		0.08	B. E. McClellan and P. Sabel, *Anal. Chem.*, **41**, 1077 (1969). Koroleff.
Co(HDz)$_3$	1 M NaCl; pH 4.9 (CCl$_4$), 5.6–7.1 (CHCl$_3$)	13	8×10^{-3}	J. F. Duncan and F. G. Thomas, *J. Chem. Soc.*, 2814 (1960).
	0.2 M Na citrate; pH 6.3–6.5 (CCl$_4$), 7.6–8.3 (CHCl$_3$)	5×10^{-4}	3×10^{-7}	

$$\frac{[Co(HDz)_3]_{CCl_4}[H^+]^2[H_2O]^{1/2}P_d^3}{[Co^{2+}][H_2Dz]^3_{CCl_4}[O_2]^{1/4}K_d^2 P_{Co(HDz)_3}} = 2\times10^{24}$$

(For H$_2$O–CCl$_4$, $P_{Co(HDz)_3} = 1\times10^4$). P_d and K_d are the partition constant (CCl$_4$) and ionization constant of H$_2$Dz.

TABLE 6G-7 (Continued)

Dithizonate	Conditions	K_{CCl_4}	K_{CHCl_3}	Reference
Cu(HDz)	0.5 M NH$_3$OH·Cl, 0.5 M NaClO$_4$, pH 2.7–6.5. K'_{ext} decreases with [Cl⁻]		4×10^3	A. M. Kiwan and H. M. N. H. Irving, *Anal. Chim. Acta*, **57**, 59 (1971).
Cu(HDz)$_2$	[H⁺] ~ 1, [Cl⁻] ~ 1	2×10^{10}		S. Anonsen, Univ. of Minn., 1942. R. W. Geiger and E. B. Sandell, *Anal. Chim. Acta*, **8**, 197 (1953).
	[H⁺] = 0.25–1, [Cl⁻] = 1.0 (25°)	1.0×10^{10}		
	[H⁺] = 2.0, [Cl⁻] = 2.0 (25°)	0.7×10^{10}		
	[H⁺] = 3.1, [Cl⁻] = 3.1 (25°)	0.32×10^{10}		
	[H⁺] = 0.6, [ClO$_4^-$] = 0.6 (25°)	1.9×10^{10}		
	[H⁺] = 1.15, [ClO$_4^-$] = 1.15 (25°)	1.5×10^{10}		
	[HClO$_4$] = 0.1 M (calc.) (25°)	3×10^{10}		
	[H⁺] = 0.4–4 (CCl$_4$); pH 1–2 (CHCl$_3$)	3.4×10^{10}	3.2×10^6	Koroleff, 1950; equilibrium not reached in CHCl$_3$?
	0.05 M H$_2$SO$_4$; corrected for sulfate complexing. Equil. after several hours shaking		3×10^8	J. Stary, *Talanta*, **16**, 360 (1969).
	$\dfrac{[CuDz]_{CCl_4}[H^+]^2}{[Cu^{2+}][H_2Dz]_{CCl_4}}$			Geiger and Sandell, *loc. cit.*
CuDz		$\sim3\times10^2$		T. B. Pierce and P. F. Peck, *Anal. Chim. Acta*, **27**, 392 (1962).
Ga(HDz)$_3$		~0.05		A. T. Pilipenko, *Zh. Anal. Khim.*, **8**, 286 (1953); T. Kato, S. Takei, and A. Okagami, *Japan Analyst*, **5**, 689 (1956).
Hg(HDz)$_2$		6×10^{26}		M. Breant, *Bull. Soc. Chim.*, **1956**, 948.
	$\mu=0.5$		1.6×10^{25}	Yokoyama et al., *Chem. Pharm. Bull. (Tokyo)*, **20**, 1856 (1972).
	$\mu=0.1$	$\sim7\times10^{26}$	6×10^{23}	T. B. Pierce and P. F. Peck, *Anal. Chim. Acta*, **26**, 557 (1962).

Species	Conditions			Reference
HgDz	$\dfrac{[HgDz]^2_{CHCl_3}[H^+]^2}{[Hg(HDz)_2]_{CHCl_3}[Hg^{2+}]} = K$ pH 4.5	7×10^4	10^{-3}	Breant, *loc. cit.*
In(HDz)$_3$				A. T. Pilipenko, *Zh. Anal. Khim.*, **5**, 14 (1950).
Mn(HDz)$_2$·2 Py	$\dfrac{[Mn(HDz)_2\cdot2\,Py]_{CCl_4}[H^+]^2}{[Mn^{2+}][H_2Dz]^2_{CCl_4}[Py]^2_{CCl_4}}$ $=10^{-6.5}$ ($\mu = 0.2$). (Py = pyridine)			Z. Marczenko and M. Mojski, *Anal. Chim. Acta*, **54**, 469 (1971).
Ni(HDz)$_2$	Acetate buffer, pH 4.8–5.9 (CCl$_4$), 5.7–6.8 (CHCl$_3$). Equilibrium slowly attained, shaken 1.5 hr	6.4×10^{-2}	1×10^{-3}	Koroleff, 1950. Values may be low, but they serve for extraction under same conditions.
				Pilipenko, 1953.
		0.2	0.6	B. E. McClellan and P. Sabel, *Anal. Chem.*, **41**, 1077 (1969). After 10 days.
Pb(HDz)$_2$	pH 1.5–4, $\mu = 0.1$ (25°)	5.6		O. B. Mathre and E. B. Sandell, *Talanta*, **11**, 295 (1964).
	pH 4.1–5.6 (CCl$_4$), 4.8–5.8 (CHCl$_3$); acetate buffer	2.4	0.13	Koroleff, 1950.
	1 M NaCl, pH 5.1–5.9 (CCl$_4$), 5.6–6.9 (CHCl$_3$)	0.12	0.01	
	0.1 M KCN, pH 4.8 (CCl$_4$), 4.9–5.7 (CHCl$_3$)	4.5	0.9	
	0.2 M Na citrate, pH 6.7–7.6 (CCl$_4$), 7.7–7.8 (CHCl$_3$)	5.4×10^{-5}	7×10^{-6}	
Pd(HDz)$_2$	Indirect method. Pd(HDz)$_2$ not affected by oxine	18	$>10^{27}$	Babko and Pilipenko, 1947. J. Stary, *Talanta*, **16**, 362 (1969).
Pt(HDz)$_2$ Sn(HDz)$_2$			$>10^{27}$	
Te(IV $\to$ II)	$\dfrac{[Te(HDz)_2]^2_{CCl_4}[H^+]}{[TeO(OH)^+][H_2Dz]^4_{CCl_4}} \approx 10^{10}$	1×10^{-2}		Pilipenko, 1953.
				Marhenke and Sandell, *Anal. Chim. Acta*, **38**, 421 (1967).

TABLE 6G-7 (Continued)

Dithizonate	Conditions	K_{CCl_4}	K_{CHCl_3}	Reference
Tl(HDz)	pH 9.7–10	5×10^{-4}	2×10^{-4}	Pilipenko, 1950. G. K. Schweitzer and A. D. Norton, *Anal. Chim. Acta*, **30**, 119 (1964).
Zn(HDz)$_2$	Acetate buffer, pH 3.1–4.2 (CCl$_4$), 3.7–5.1 (CHCl$_3$). Shaken 5 min (CCl$_4$) or 1 hr (CHCl$_3$)	2×10^{2}	4.4	Koroleff, 1950.
	1 M KCl, pH 4.1 (CCl$_4$), 4.4–5.2 (CHCl$_3$)	1.0×10^{2}	2	
	0.1 M Na$_2$S$_2$O$_3$, pH 4.3–4.6 (CCl$_4$), 4.7–5.3 (CHCl$_3$)	20	1.2	
	Acidic soln., $\mu = 0.025$ M		8.7	McClellan and Sabel, *loc. cit.*
	0.005% Na diethyldithiocarbamate, pH 5.4–6.5 (CCl$_4$), 7.2–7.3 (CHCl$_3$)	2.4×10^{-2}	4×10^{-4}	
	Biphthalate buffer, pH 3–4	200	3.2–4.5 ~9 (calc., $\mu = 0$)	I. M. Kolthoff and E. B. Sandell, *J. Am. Chem. Soc.*, **63**, 1906 (1941).
	Biphthalate buffer, 0.005 M	97 130 (calc., $\mu = 0$) 66	8 11 (calc., $\mu = 0$)	K. Buch, *Finska Kemistsamfundets Medd.*, **53**, 25 (1944). H. Irving, C. F. Bell, and R. J. P. Williams, *J. Chem. Soc.*, 357 (1952).
	Acetate buffer, 0.1 M	5×10^{2}		Babko and Pilipenko, 1947.
	$\mu = 0.10$ (perchlorate), pH 2–4 (30°)	50	0.03	G. K. Schweitzer and C. B. Honaker, *Anal. Chim. Acta*, **19**, 224 (1958). With benzene, $K_{ext} = 8$.
	0.1 M citrate, pH ~6		5×10^{-7}	
	0.1 M citrate, pH ~2	30		

TABLE 6G-8

Wavelengths for Determination of Metals as Dithizonates and Values of ϵ[a]

	CCl$_4$		CHCl$_3$	
Dithizonate	λ (nm)	$\epsilon \times 10^{-3}$	λ (nm)	$\epsilon \times 10^{-3}$
Ag(HDz)	~470			
	462 (*max*)	29 (*max*)	470 (*max*)	31 (*max*)
Bi(HDz)$_3$	490–500		495	90
	490 (*max*)	79 (*max*)		
Cd(HDz)$_2$	515–520	88	517 (*max*)	87 (*max*)
Cu(HDz)$_2$	535			
	545–550 (*max*)	45 (*max*)		
Hg(HDz)			490 (*max*)	31 (*max*)
Hg(HDz)$_2$	485–490			
	485 (*max*)	71 (*max*)	485 (*max*)	68 (*max*)
In(HDz)$_3$	510	87	510	
Pb(HDz)$_2$	520 (*max*)	66.5 (*max*)	510 (*max*)	63.5 (*max*)
Tl(HDz)	510 (*max*)	33 (*max*)	505 (*max*)	33.5 (*max*)
Zn(HDz)$_2$	520–530			
	535 (*max*)	96	530 (*max*)	

[a] From various authors. Values of ϵ are to be looked upon as approximate only, the effective values depending upon the procedure, particularly whether a mono or mixed color method is used. In a mixed color method, the effective ϵ is less than ϵ of the dithizonate. For example, taking ϵ_{545} of H$_2$Dz in CCl$_4$ as ~10,000, the change in absorbance in determination of Cu(II) by the mixed color dithizone method corresponds to $\epsilon = 45,000 - (2 \times 10,000) = 25,000$. (Similar considerations apply in reversion procedures.) If extraction of metal is complete, the sensitivity index, $\mathscr{S}$, can be calculated from ϵ values in the table and ϵ of dithizone (all excess remaining in organic solvent) at the particular λ.

converted into anionic hydroxy species. The extractability of lead as a function of pH is illustrated in Fig. 6G-13.

There is another possibility in the photometry of a mixed solution of metal dithizonate and free dithizone. The absorbance of the solution may be measured at two suitable wavelengths, namely one at which dithizone absorbs strongly and the dithizonate as little as possible, and another at which the reverse is true. The advantage of this method lies in the possibility of using dithizone solutions whose exact strengths need not be known, that is, it is not necessary to prepare a new standard curve every time a new batch of reagent solution is made up. The metal concentration is calculated from the absorbance values at the two wavelengths and the previously determined absorptivities of dithizone and the metal dithizonate at these two wavelengths. There must be a linear relation between metal concentration and absorbance (which is not always true when filters

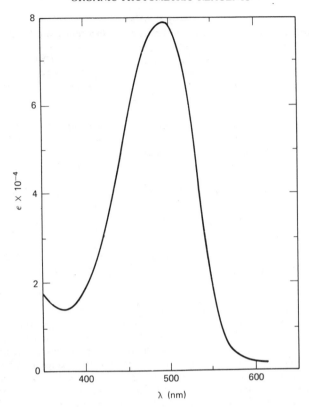

Fig. 6G-8. Molar absorptivity curve of $Bi(HDz)_3$ in CCl_4. From T. Bidleman, Thesis.

are used) if the usual formula is to be applied. Formation of oxidation products of dithizone may lead to error so that decomposition of the dithizone solution on standing must be guarded against. This method has been applied in the determination of copper, lead, and mercury.

Still another procedure for the determination of metals with dithizone (and other organic reagents giving colored complexes) is based on the principle of reversion, that is, on the change in absorbance of a mixture of free dithizone and a metal dithizonate in an organic solvent produced by the addition of a substance forming a stable complex with the metal, which thus liberates an equivalent amount of dithizone.[53] A plot of the increase in absorbance at ~620 nm (where dithizone absorbs strongly and most dithizonates only slightly) against the concentration of metal will generally give a straight line. An advantage of this procedure is that the dithizone solution need not be of known or constant strength.[54] Mineral acid can be used as reversion reagent for lead and zinc dithizonates; iodide and acid can be used for silver and mercury. 2,3-Dimercapto-propanol is a general reversion agent for dithizonates over a wide pH range. The reversion method probably does not have any advantage over the ordinary method when the metal determined is present in a solution entirely free from other substances reacting with dithizone.[55] Two absorbance readings are required in the reversion method compared to one in the usual method. Reversion should have a real value when small amounts of foreign reacting substances are present and a reversion agent is available that will convert the dithizonate being determined into the equivalent amount of dithizone without affecting the others.[56] For example, lead dithizonate can be reverted with $\sim 0.01\,M$ hydrochloric acid

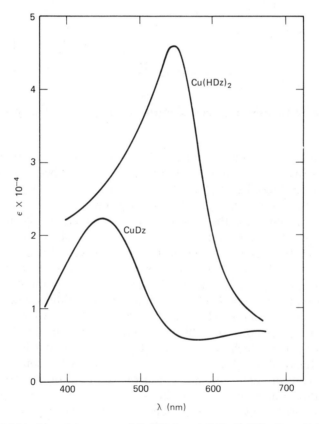

Fig. 6G-9. Molar absorptivity curves of $Cu(HDz)_2$ and CuDz in CCl_4. From R. W. Geiger, Thesis.

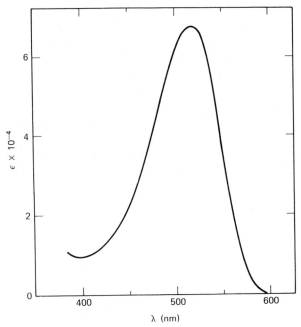

Fig. 6G-10. Molar absorptivity curve of Pb(HDz)$_2$ in CCl$_4$. From O. B. Mathre, Thesis.

without affecting primary cupric dithizonate. Reversion also would be advantageous if a small amount of dithizone has been converted to oxidation products by mild oxidizing agents (such as ferric iron), and these products are unchanged during reversion.

Finally, we consider the determination of metals by the mono-color method, in which the organic extract of the metal is shaken with an alkaline solution such as ammonia to remove excess dithizone and (presumably) leave the metal dithizonate whose absorbance is then measured at its absorption peak.[57] This is a dubious method. As the concentration of dithizone in the organic phase is reduced, metal dithizonate is increasingly dissociated, and metal enters the aqueous phase. A balance must be struck between removal of dithizone and retention of metal dithizonate in the washing step, which may be difficult or impossible to achieve for some metals. Removal of 99% of the dithizone from a CCl$_4$ solution requires shaking with an equal volume of aqueous solution of pH 10.85. If the original dithizone concentration were 4×10^{-5} M (0.0010% w/v), its concentration after shaking would then be 4×10^{-7} M. The latter solution would give an absorbance of $4 \times 10^{-7} \times 4700 = 0.002$ in a 1 cm cell at 510 nm, almost meeting the requirement of no absorption by dithizone. If

K_{ext} of the metal as well as the necessary hydrolysis and complex formation constants are known, one can calculate the extraction coefficient of the metal at pH ~ 10.85 when $[H_2Dz]_{CCl_4} = 4 \times 10^{-7}$ M, and learn whether the metal is present quantitatively (say >99%) in the organic phase. Such a calculation for lead shows that it is quantitatively extracted under the conditions specified (pH 10.85, original concentration of H_2Dz in $CCl_4 = 4 \times 10^{-5}$ M) without an appreciable amount of H_2Dz being left in the CCl_4 phase. Lead can be quantitatively extracted at lower pH values (roughly down to pH 6), but it is advantageous to carry out the determination at pH ~ 10.8, where a slight variation (±0.2) in pH of the buffer would have a negligible effect on the absorbance resulting from a variation in the amount of dithizone extracted.

Washing an organic solvent solution of a primary dithizonate with an alkaline aqueous solution may transform it into the secondary dithizonate (but equilibrium is not always rapidly attained). For example, it is hardly possible to wash all dithizone out of a CCl_4 solution of primary Cu(II) dithizonate without forming some of the secondary dithizonate, as is made evident by a slight change in hue of the originally violet solution (CuDz is brown).

In brief, it is not necessary to remove excess dithizone from an extract of a dithizonate to determine a metal spectrophotometrically, though it

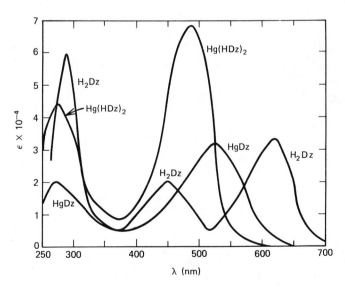

Fig. 6G-11. Molar absorptivity curves of Hg(HDz)$_2$, HgDz, and H$_2$Dz in CCl$_4$. From T. Kato, S. Takei, and A. Okagami, *Japan Analyst*, **5**, 689 (1965).

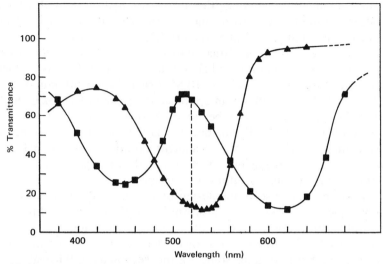

Fig. 6G-12. Transmission curves of Zn(HDz)$_2$ and H$_2$Dz in CCl$_4$. Concentration of Zn(HDz)$_2$ not specified, but calculated to be $\sim 1 \times 10^{-5}$ M (1 cm cell). ▲, Zn(HDz)$_2$; ■, H$_2$Dz. Reprinted with permission from E. T. Verdier, W. J. A. Steyn, and D. J. Eve, *Agr. Food Chem.*, **5**, 356 (1957). Copyright by the American Chemical Society.

may be of some advantage to do so, if this can be done more or less incidentally as by raising the pH without producing harmful effects.

Preliminary Dithizone Extraction. A metal frequently cannot be determined directly with dithizone in the solution of the sample: Oxidizing agents attacking dithizone to a greater or less extent may be present, it may be difficult to adjust the acidity of the solution to the exact value required, and there may be interference from other heavy metals or even neutral salts. When such conditions exist, a preliminary extraction with dithizone (e.g., 0.01% w/v) is made to separate completely the metal in question. The dithizonate in the organic solvent layer may next be dissociated or decomposed and the metal converted into ionic form for final determination in a number of ways:

1. The dithizonates of lead, zinc, cadmium, and thallium are readily dissociated and the metal transferred to the aqueous phase when the organic solvent is shaken with dilute acid.[58] This procedure also provides a separation from such metals as copper whose dithizonates are not appreciably attacked by dilute acids (cf. p. 624). Nickel and cobalt dithizonates are more or less resistant to the action of dilute acids because of kinetic inertness, and they (especially cobalt) remain in CCl$_4$ to a large extent after shaking with ~ 0.01 M HCl.

2. The organic solvent phase is evaporated to dryness, and the dithizonate residue is heated with a few drops of concentrated sulfuric acid and perchloric acid to destroy organic matter completely.[59] This method may be used for copper and other metals forming dithizonates not dissociated by dilute acids.

3. The organic solvent is shaken with an acid solution of a strong oxidizing agent such as potassium permanganate or ceric sulfate to destroy the dithizonate radical and any free dithizone. The metal enters the aqueous phase and the excess of the oxidizing agent can be destroyed by a suitable reducing agent such as hydroxylammonium hydrochloride. This procedure has been used to decompose mercuric dithizonate, which cannot be treated by either of the two preceding methods. (Nitrite has also been used to destroy dithizone, but nitrous acid must then be expelled.)

4. The organic solvent solution of the dithizonate is shaken with an aqueous solution of a reagent forming a stable complex with the metal. This method has been applied in decomposing silver and mercuric dithizonates. Before the metal in the aqueous layer can be determined with dithizone, the complex must be destroyed or at least made much weaker compared to the dithizone complex by a change in conditions, for example, a decrease in the acidity.

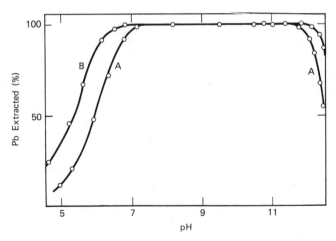

Fig. 6G-13. Extraction of lead dithizonate from sodium citrate–potassium cyanide solutions. The solid curves are those calculated for the extraction of lead (0.0625 mg in 25 ml of aqueous solution, 0.01 M in sodium citrate and 0.02 M in potassium cyanide, $\mu = 0.1$) with 25 ml of 0.001% (w/v) dithizone in carbon tetrachloride (A), and with 25 ml of 0.002% (w/v) dithizone (B). The circles represent experimental values. From O. B. Mathre and E. B. Sandell, *Talanta*, **11**, 295 (1964).

B. SELECTIVITY OF DITHIZONE REACTIONS

Dithizone reacts with more than a score of metals, but selectivity can be obtained by:

1. Adjusting the pH of the solution to be extracted.
2. Altering the oxidation state of interfering metals (of minor importance).
3. Adding a complex-forming agent that will mask other reacting metals.

Separation of metals by the adjustment of the pH of the sample solution can be successfully carried out only with certain combinations of metals, namely those whose dithizonates have extraction constants differing markedly from each other. Extraction from a fairly acid (0.1–0.5 N) solution serves to separate such metals as palladium, silver, mercury, and copper from other metals.[60] Lead and bismuth can be separated by extraction of bismuth from 0.05 M HClO$_4$ solution with dithizone in CCl$_4$. By making a solution strongly basic (1 N in sodium hydroxide) after adding citrate or tartrate to prevent precipitation of metal hydroxides, cadmium can be quantitatively extracted with a carbon tetrachloride solution of dithizone and separated from metals such as lead and zinc, but not from the noble metals, mercury, copper, cobalt, and nickel.

Instead of extracting a metal from a solution of adjusted pH, it is sometimes preferable with a low total content of reacting metals to extract under such conditions (for example, from a faintly ammoniacal solution in the presence of citrate or sulfosalicylate to prevent hydroxide precipitation) that other metals also are extracted, and then to shake the separated organic solvent containing the dithizonates with dilute acid of such concentration that the desired metal either remains in the organic solvent alone or goes into the aqueous phase while the other dithizonates remain undissociated. Thus if a carbon tetrachloride solution containing the dithizonates of copper, zinc, and lead is shaken with one or two portions of dilute acid (0.05 N or even weaker), copper remains in the carbon tetrachloride, whereas zinc and lead go into the aqueous phase.

Since this procedure is frequently applied, it may be worthwhile to consider how sharply separations of metals can be made in this way. As an illustration, we may take the case of two divalent metals whose dithizonates have the extraction constants 10^{10} (metal M) and 100 (metal N).[61] Suppose that the metals are extracted with a 4×10^{-4} M (approximately 0.01% weight by volume) solution of dithizone in an organic solvent, and the resulting solution of dithizonates is shaken with an equal volume of 0.01 N hydrochloric acid. First, let us find the extent of dissociation of a pure $2 \times 10^{-4}\ M$ solution of $M(HDz)_2$, containing no excess dithizone, under these conditions. It will be assumed that the equilibrium may be expressed by

$$\frac{[M(HDz)_2]_{org}}{[M^{2+}]} = K_{ext} \frac{[H_2Dz]^2_{org}}{[H^+]^2}$$

Since the volumes of organic solvent and aqueous phases are equal:

$$[H_2Dz]_{org} = 2[M^{2+}]$$

$$\frac{[M(HDz)_2]_{org}}{[M^{2+}]} = K_{ext} \frac{4[M^{2+}]^2}{[H^+]^2}$$

$$[M^{2+}] = \left\{ \frac{[M(HDz)_2]_{org}[H^+]^2}{4\ K_{ext}} \right\}^{1/3}$$

Since the value of K_{ext} is very large, the concentration of $M(HDz)_2$ in the organic phase will change little on shaking with dilute acid, and it may be set equal to $2 \times 10^{-4}\ M$; the concentration of metal in the aqueous phase when equilibrium is established is given by

$$[M^{2+}] = \left\{ \frac{2 \times 10^{-4} \times 10^{-4}}{4 \times 10^{10}} \right\}^{1/3} = 8 \times 10^{-7}$$

(Any change in $[H^+]$ is neglected.) The percentage of M transferred to the aqueous phase is, therefore, about 0.4. In any actual separation, more or less free dithizone would be present and the amount of M^{2+} going into the acid solution would be enormously reduced. Thus if the dithizone concentration in the organic solvent phase were 4×10^{-5} M, $[M(HDz)_2]_{org}/[M^{2+}] = 1.6 \times 10^5$. It is evident that the organic phase could be shaken successively with many fresh portions of 0.01 N acid without the loss of any significant amount of M^{2+}, provided a little free dithizone is present.

To find the extent of dissociation of $N(HDz)_2$ in the organic phase on shaking with acid, we may first assume that virtually all of N goes into the aqueous phase, and we then find

$$\frac{[N(HDz)_2]_{org}}{[N^{2+}]} = \frac{10^2 \times (4 \times 10^{-4})^2}{10^{-4}} = 0.16$$

This value indicates that an appreciable amount of $N(HDz)_2$ remains in the organic solvent and that the method of successive approximations should be applied in the calculation to obtain a more exact value. The exact value of N^{2+} is found to be $1.78 \times 10^{-4}\ M$, which corresponds to 89% dissociation of the dithizonate by acid. Evidently the organic solvent must be shaken with a fresh portion of 0.01 N acid in order to decompose most of $N(HDz)_2$. If again an equal volume of acid is used, $[N(HDz)_2]_{org}/[N^{2+}] = \sim 0.16$, and $[N^{2+}] = 0.19 \times 10^{-4}$. Accordingly, two shakings with acid transfers $(1.78 + 0.19)$ $100/2$ or 98.5% of N to the aqueous phase. In other words, M and N can be separated in this way without much difficulty.[62] As a matter of fact, N can be separated from a metal whose dithizonate has an extraction constant of 10^5 provided the free dithizone concentration during the shaking with acid is fairly high. Thus if $[H_2Dz]_{org} = 4 \times 10^{-4}$, less than 1% of the metal whose dithizonate constant is 10^5 would be brought into the aqueous phase when shaken with an equal volume of $0.01\ N$ acid, since the ratio of metal in organic solvent and acid will be $[10^5 \times (4 \times 10^{-4})^2]/10^{-4} = 160$. It appears then that divalent metals whose extraction constants differ by the factor 1000 can be separated fairly satisfactorily in the way described by the proper choice of acid and free dithizone concentration when present in comparable quantities, but a larger difference in constants is desirable.

As already noted, iron, tin, and platinum do not form complexes with dithizone when present as Fe(III), Sn(IV), and Pt(IV). (Ferric iron apparently can react slightly.)

Reagents forming stable complexes with interfering metals play an important role in dithizone procedures. It will be seen from Table 6G-9, referring especially to CCl_4 as organic solvent, that the number of metals reacting with dithizone is greatly reduced in the presence of such complex-forming agents as cyanide, thiocyanate, and thiosulfate. An example is provided by the determination of lead. The only metals reacting in slightly basic medium containing cyanide are lead, univalent thallium, divalent tin, and bismuth.[63] Since stannic tin does not react, and bismuth can be separated from lead by extraction with dithizone in slightly acidic medium, the only other metal that will interfere in the lead determination is thallium; a special procedure has been worked out for the determination of lead in its presence.

At a pH of approximately 5 in the presence of a sufficient quantity of thiosulfate, the only metals reacting appreciably with dithizone in carbon tetrachloride solution are zinc, stannous tin, and palladium.[64] Large amounts of nickel and cobalt will react slightly under these conditions, but can be made harmless with cyanide, which also prevents the reaction of palladium. Sodium diethyldithiocarbamate has also been used as a general complex former in the determination of zinc after removal of copper. Both thiosulfate and diethyldithiocarbamate complex zinc weakly and interfere to some extent with the extraction of zinc dithizonate. A better masking agent in the determination of zinc is bis(carboxymethyl)-dithiocarbamate.[65] It forms, predominantly, $M^{II}L_2$ complexes whose $\log \beta_2$ constants are: Zn, <4.5; Cu, 21.5; Pb, 15.5; Cd, 11.25; Ni, 7.95.

TABLE 6G-9
Complex-Forming Agents in Dithizone Reactions

Conditions	Metals Reacting
Basic solution containing cyanide	Pb, Sn(II), Tl(I), Bi, In
Slightly acid solution containing cyanide	Pd, Hg, Ag, Cu (not Au)
Dilute acid solution containing thiocyanate	Hg, Au, Cu
Dilute acid solution containing thiocyanate + cyanide	Hg, Cu
Dilute acid solution containing bromide or iodide[a]	Pd, Au, Cu and others
Slightly acid solution (pH 5) containing thiosulfate (CCl_4 solution of dithizone)	Pd, Sn(II), Zn (Cd, Co, Ni)
Slightly acid solution (pH 4–5) containing thiosulfate + cyanide	Sn(II), Zn
Ethylenediaminetetraacetate, pH ~4.5[b]	Au, Ag, Hg
Citrate and tartrate in basic medium	Usually do not interfere with extraction of reacting metals

[a] V. Concialini, P. Lanza, and M. T. Lippolis, *Anal. Chim. Acta*, **52**, 529 (1970), were able to determine ppm quantities of Cu, Zn, Cd, Pb, Ni, and Co in AgCl (1 g) dissolved in 4.5 M I^- solution by dithizone extraction at pH 8.
[b] According to V. Vasak and V. Sediveč, *Coll. Czech. Chem. Commun.*, **15**, 1076 (1950). Copper does not appear to be complexed very firmly. Also see V. G. Goryushina and Y. Y. Gailis, *Zav. Lab.*, **22**, 905 (1956). S. S. Yamamura, *Anal. Chem.*, **32**, 1897 (1960), separates Hg from milligram amounts of Cu in an Al matrix by extracting from EDTA–citric acid solution at pH 2.75; U, Fe, Cr, Mn, and Ni are without effect at low levels.

The extraction of Zn as dithizonate is not appreciably affected by this complexer, whereas the extraction of Cd, Pb, and Cu is shifted into the alkaline region (Fig. 6G-14), so that the determination of Zn in the presence of these metals (in small to moderate amounts) should be possible.

Other sulfur-containing masking agents are thiosemicarbazide and thiourea, whose complexing ability is little affected by pH. Thiourea forms firm complexes with Cu(I), Ag, and Hg(II), but its utility (and that of thiosemicarbazide) in increasing the selectivity of dithizone reactions remains to be studied; the complex constants of most metal–thiourea chelates are not known [those of Ag, Bi, Cd, Cu(I), Hg(II), and Pb have been determined].

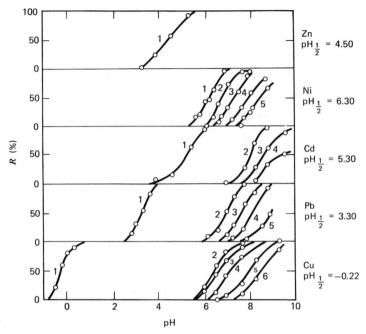

Fig. 6G-14. Extractability (percentage recovery) of metals with dithizone in CCl$_4$ in the presence of bis(carboxymethyl)dithiocarbamate (DTC). From Hulanicki and Minczewska, *loc. cit.* Curve 1 is nonmasked extraction for each metal. Other curves as follows:

Ni; [DTC] $2—2 \times 10^{-4}$ M; $3—10^{-3}$ M; $4—2 \times 10^{-3}$ M· $5—10^{-2}$ M
Cd; [DTC] $2—10^{-3}$ M; $3—2 \times 10^{-3}$ M; $4—10^{-2}$ M
Pb; [DTC] $2—10^{-4}$ M; $3—3.3 \times 10^{-4}$ M; $4—10^{-3}$ M; $5—3.3 \times 10^{-3}$ M
Cu; [DTC] $2—6 \times 10^{-5}$ M; $3—2 \times 10^{-4}$ M; $4—6 \times 10^{-4}$ M; $5—2 \times 10^{-3}$ M; $6—6 \times 10^{-3}$ M

When a metal dithizonate does not absorb much at the peak absorbance of dithizone (620 nm, CCl$_4$), which is usually the situation, the absorbance of the dithizone–dithizonate mixture may be measured at this wavelength; the decrease is proportional to the metal concentration. This method requires that the aqueous solution be acidic, or only weakly basic, so that the excess of dithizone remains in the organic solvent phase. For most metals, the sensitivity is comparable to that obtainable by measuring the absorbance of the metal dithizonate. For example, for Bi, neglecting the small absorbance of Bi(HDz)$_3$ at 620 nm, the reaction of 1 mg Bi with H$_2$Dz in 1 liter of CCl$_4$ decreases the absorbance by $(3 \times 32,500 \times 0.001)/209 = 0.47$ (1 cm) at this wavelength, compared to an increase of 0.42 at 490 nm (λ_{max}).

C. SOURCES OF ERROR IN DITHIZONE DETERMINATIONS

It is clear enough that in applied analysis the greatest errors are likely to arise from the presence of other metals that also react with dithizone and, less often, may impede the extraction of the desired metal. Avoidance of such errors requires a good knowledge of dithizone chemistry and of the composition of the sample. For the routine analyst—the term is not used disparagingly—the choice of a sound procedure, in which possible interferences have been provided for, is important.

We mention here some other sources of error. The formation of a precipitate or a turbidity in an aqueous solution to be extracted with dithizone must be regarded with suspicion. Formation of precipitates is most likely to occur in weakly acidic, neutral, or basic solutions. Thus solutions of samples high in phosphorus (biomaterial) may give a precipitate of calcium phosphate on addition of ammonia, even if citrate is present, and such a precipitate will carry down the greater part of lead in the solution. Metastannic acid will also retain lead. Other examples could be cited. Even if the aqueous solution appears clear, hydrolysis products of metals may hinder extraction of dithizonates. Relatively small amounts (5 mg) of titanium prevent complete extraction of lead from an ammoniacal citrate solution (pH 7–11).[66] A comparatively high concentration of aluminum under the same conditions makes the extraction of lead more difficult and a large number of portions of dithizone is then required for quantitative extraction. Preliminary isolation of lead as sulfide has been recommended in these cases. H. Fischer noted that aluminum prevented the complete extraction of zinc in faintly acidic solution. The effect of titanium and aluminum is probably due to a hydrolysis that results in the formation of a colloidal dispersion having strong adsorptive properties. Adsorption of lead and bismuth on glass from basic solutions also requires consideration in dithizone extractions.

The solubility of some dithizonates in carbon tetrachloride is low (Section II.B), and if the metal concentration is relatively high, the dithizonate may be precipitated, and it then forms flocs in the organic solvent or gathers as a scum at the interface of the two liquids. This can lead to losses of metals in separations and errors in determinations. The solubility of $Pb(HDz)_2$ in CCl_4 at $25°$ is $5.7 \times 10^{-6}\,M$, corresponding to ~ 1.2 mg Pb in a liter of CCl_4 or $12\,\mu g$ in 10 ml. This concentration is sufficient for maximal photometric precision. Dithizonates are more soluble in $CHCl_3$ than in CCl_4, and the former solvent can be used for larger amounts of metals when necessary.

Like dithizone itself, solutions of dithizonates in organic solvents are subject to decomposition on standing. With purified solvents (see next section) the decrease in absorbance is very slow in refrigerated solutions kept in the dark. Usually the analyst is not concerned with the long-period stability of dithizonate solutions. Stability at room temperature is sufficiently good to allow measurement of absorbance without undue haste after extraction if solutions are not exposed to strong light.[67] Bidleman (thesis, 1970) found bismuth dithizonate solutions in CCl_4 to decrease in concentration on standing, especially at room temperature in light, as illustrated by the following figures:

Conditions	Decrease in Concentration after 20 hr (%)
6.3×10^{-6} M solution in Pyrex in dark at ~4°	5.4
5.7×10^{-6} M solution in Pyrex in dark at ~25°	11
5.7×10^{-6} M solution in Pyrex under fluorescent light at ~25°	35

These decreases include adsorption on glass (68% and 59% of total decrease for first two solutions, 13% for the third).

The irreversible decomposition of metal dithizonates in carbon tetrachloride and chloroform seems to be largely due to oxidizing substances (e.g., $COCl_2$) formed by the action of light on the moist solvents. Benzene solutions (0.001%) of Ag(HDz), Hg(HDz)$_2$, Pb(HDz)$_2$, Zn(HDz)$_2$, and H_2Dz are not decomposed irreversibly by light of wavelengths greater than 400 nm, and, with the exception of Ag(HDz) and H_2Dz, are unaffected by light >360 nm.[68] However, decomposition begins at 340 nm and is rapid at 300–320 nm. The relative stabilities of the dithizonates to radiation >340 nm, compared to H_2Dz as 1, are approximately: Cd(HDz)$_2$, 0.5 < Bi(HDz)$_3$, 0.6 < Pb(HDz)$_2$, 1.5 < Pt(HDz)$_2$, 4.3 < Ag-(HDz), 4.5 < Zn(HDz $_2$, 6.0 < Pd(HDz)$_2$, 6.5 < Hg(HDz)$_2$, 7.0 < Ni(HDz)$_2$, 12.0

Precision. Results obtained in the determination of lead (pH 10.8, citrate–cyanide solution) illustrate the precision attainable in dithizone procedures. In our experience, 5 μg of Pb can be determined with a standard deviation of ~1% (10 ml CCl_4–dithizone solution, 1 cm cell, 520 nm). As expected, photometric sensitivity is not the limiting factor in the precision of the determination. The sensitivity index is $\mathscr{S}_{520} =$ 0.003 μg Pb/cm^2 for $A = 0.001$. The absolute standard deviation of the determination of 5 μg Pb is ~$0.01 \times 5/10 = $ ~0.005 μg Pb/cm^2, which is approximately twice the photometric deviation if this is taken as 0.001 A.

If ~ 1 μg Pb is determined by extraction into 25 ml of dithizone–CCl$_4$ solution and the absorbance of the extract is measured in a 10 cm cell, the standard deviation is found to be $\sim 1\%$ (± 0.01 μg Pb).

D. PURIFICATION OF DITHIZONE AND SOLVENTS

Purification of dithizone for analytical purposes is not always necessary. Commercial products vary in quality, and some may be satisfactory for use in *some* trace analyses without purification. If a 0.01% (w/v) solution in CCl$_4$ yields a CCl$_4$ layer that is only faintly yellow after being shaken with dilute (1:100) metal-free ammonia, the product may usually be used without further treatment.[69]

The following procedure is recommended for the preparation of dithizone of the highest purity.[70] Dissolve 2.0 g of commercial dithizone (e.g., Eastman "White Label") in 200 ml of chloroform, and filter through a fritted glass crucible to remove any insoluble material. Extract the filtered solution with four 50 ml portions of pure 1:100 ammonium hydroxide. Shake the combined extracts with 10 ml of chloroform and discard the latter. Pass sulfur dioxide into the aqueous solution to neutralize the ammonia and precipitate dithizone.[71] Extract the dithizone with two 50 ml portions of pure chloroform. To the combined extracts add an equal volume of carbon tetrachloride. Pass a rapid stream of pure dry nitrogen through the solution heated to about 40° in a water bath, until its volume has been reduced to one-third. Cool (preferably in ice), filter off the dithizone on a fritted disk, and wash with a few milliliters of pure carbon tetrachloride. Finally, dry to constant weight over potassium hydroxide in a vacuum desiccator.

The dithizone content of the product can be obtained by preparing a solution of known concentration in CCl$_4$ and titrating a known amount of silver in 0.05 M H$_2$SO$_4$.[72]

A simpler and more rapid (but less effective) purification procedure than that just described involves crystallization from chloroform. An almost saturated, filtered solution of dithizone in chloroform is evaporated at about 40° in a stream of filtered air (nitrogen is better) until one-third to one-half of the dithizone has crystallized out. The precipitate is collected in a sintered-glass filter crucible, washed with a few small portions of carbon tetrachloride, and air-dried. An Eastman product of 94% purity (as determined by silver nitrate titration) yielded a 98% material by this purification procedure. The deficit of 2% is at least in part accounted for by the presence of water or organic solvent. None of the yellow oxidation product was present.

Chromatography has also been applied in the purification of

dithizone.[73] Large-scale chromatography on Whatman SG 81 paper (silica-impregnated) allows 50 mg of dithizone to be treated; benzene is used as the developing solvent and acetone for elution. A low-grade reagent was found to give spots of the following colors: dull green ($R_f = 0$), bright green (H_2Dz, 0.49), yellow-green (0.55), pink (apparently due to metal, 0.67), and yellow (0.72). Larger amounts of dithizone can be purified by column chromatography. The most satisfactory column packing is a 1:1 mixture of acid-washed silica gel (Kieselgel N) and acid-washed Celite 545. Benzene is the preferred eluting agent. About 280 g of the mixture is required for purifying ~0.1 g H_2Dz. ϵ_{620} in CCl_4 of H_2Dz purified chromatographically is 33,100 compared to 32,700 for H_2Dz purified by extraction–crystallization, or ~1% higher. The difference is large enough to require explanation, not available at present.

Purification of CCl_4 and $CHCl_3$. There are at least two reasons for purifying these solvents for use in dithizone work. First, the stability of the solutions is increased by using solvents free of oxidizing substances; second, the rate of extraction of such metals as Cu(II) and Bi from acid solutions is greater with properly purified solvent[74] than with unpurified.

The following procedure for the purification of CCl_4 has been found effective by us and other workers.[75] Add a few drops of liquid bromine to 1 liter of reagent-grade carbon tetrachloride, and allow to stand for a few days. Then reflux with 500 ml of 10% sodium hydroxide solution for 4 hr, wash free of alkali, shake for 5 min with 10% hydroxylamine hydrochloride solution, and finally with water. Dry the carbon tetrachloride with a desiccant such as Drierite and distill. Carbon tetrachloride thus treated gives dithizone solutions that are stable and that react rapidly with copper in mineral acid solution. A 4×10^{-5} M dithizone solution in such carbon tetrachloride has been found to reach equilibrium in 1.5 min when shaken at 300 cycles/min with copper in 0.1 M hydrochloric acid.

Mathre (*loc. cit.*) modified the foregoing procedure slightly, in preparing carbon tetrachloride for general dithizone use, by refluxing for 0.5 hr with 10% hydroxylamine hydrochloride solution, after washing out sodium hydroxide with water, and finally distilling from a mixture of Drierite and calcium oxide. A dithizone solution (1.3×10^{-5} M) made up in carbon tetrachloride thus purified, showed the following absorbances (1 cm) when freshly prepared and after 1 week in the dark at room temperature (23°), respectively: 450 nm, 0.256 and 0.259; 510 nm 0.066 and 0.067; 620 nm, 0.422 and 0.416. A 3×10^{-5} M dithizone solution stored in a refrigerator (5°C) showed practically no change after 3 weeks. Mathre reclaims used carbon tetrachloride by shaking with 0.1 volume portions of concentrated sulfuric acid until the latter remains colorless, washing with water to remove acid, shaking with 1% sodium hydroxide, then with water, and finally distilling from a mixture of Drierite and calcium oxide. Carbon tetrachloride reclaimed as many as five times showed no deterioration in quality.

It is of interest to note that analytical-grade CCl_4 may contain up to 2% $CHCl_3$, which is enough to affect ϵ.[4] The $CHCl_3$ content can be decreased by distillation. Up to 2% CCl_4 in $CHCl_3$ has little effect on ϵ.

For the purification of chloroform, Clifford[76] recommends the procedure of Biddle.[77] This procedure may be used both for the recovery of chloroform from used dithizone solutions and for improving the quality of U.S.P. chloroform. It is carried out as follows. Separate the chloroform from any aqueous phase and wash with one-tenth its volume of concentrated sulfuric acid (commercial quality will do) until colorless. Add lime to the separated chloroform to remove sulfuric acid and distill slowly in the presence of lime. Treat the distillate with 1% its volume of ethanol (preservative).

Dithizone solutions prepared from chloroform thus purified are quite stable if stored at 40°F in the dark. A chloroform solution containing 4 mg dithizone/liter lost 5% in strength after 6.5 months storage under these conditions (Clifford). Overlaying a chloroform–dithizone solution with an aqueous solution of sulfur dioxide or hydroxylamine hydrochloride does not increase the stability of the solution.

The Biddle–Clifford procedure for recovery of chloroform from used dithizone has been criticized as being inadequate, especially when the solvent is badly contaminated with photochemical oxidation products (e.g., phosgene), and the following procedure is recommended.[78] Distill the contaminated chloroform through a layer of 10% aqueous sodium sulfite solution. Shake the distillate either with 3% hydroxylamine hydrochloride or 10% sodium sulfite solution until the aqueous phase remains colorless. Redistill the product from lime or active charcoal into about 1% its volume of ethyl alcohol. A second redistillation will give a more stable product if the original chloroform was badly contaminated, but usually this is not necessary. A product redistilled twice from lime gave a dithizone solution entirely stable for 168 hr when stored in the dark at 5°C.

IV. SUBSTITUTED DITHIZONES AND DITHIZONE HOMOLOGS AND ANALOGS

Numerous derivatives of dithizone have been prepared and studied more or less thoroughly as heavy metal reagents. The following discussion is not intended to be exhaustive. Some differences in the behavior of the derivatives from that of the parent compound are pointed out, which may occasionally make the derivatives preferable as reagents.

A. SUBSTITUTED DITHIZONES

2,2'-Dichlorodithizone ($o,o,'$-dichlorodithizone) has absorption maxima at 462 nm ($\epsilon = 2.98 \times 10^4$) and 644 nm ($\epsilon = 3.34 \times 10^4$) in carbon tetrachloride.[79] $pK_a = 4.75 \pm 0.5$ and $P_{CCl_4} = 6 \times 10^4$. Log K_{ext} values (CCl$_4$) are:

Cu(HL)$_2$	6.88	Cd(HL)$_2$	1.47
Zn(HL)$_2$	0.4	Pb(HL)$_2$	1.85
Hg(HL)$_2$	26.18		

$P_{Cu(HL)_2} = 3.5 \times 10^4$, $P_{Zn(HL)_2} = 1.5 \times 10^4$. K_{ext} for Cu is appreciably smaller than K_{ext} with H$_2$Dz as reagent. This decrease is explained by the assumption of square planar coordination of copper, in which greater steric hindrance might be expected than in tetrahedral coordination (as with zinc).
The λ_{max} and ϵ_{max} values of 2,2'-dichlorodithizonates in CCl$_4$ are:

	λ_{max} (nm)	$\epsilon_{max} \times 10^{-4}$		λ_{max} (nm)	$\epsilon_{max} \times 10^{-4}$
Cu(HL)$_2$	541	6.58	AgHL	460	3.9
Zn(HL)$_2$	533	9.72	Bi(HL)$_3$	490	10.47
Hg(HL)$_2$	485	8.63	Co(HL)$_2$	539	5.57
Cd(HL)$_2$	527	11.1	Ni(HL)$_2$	428	4.13
Pb(HL)$_2$	514	7.08		545	3.74

ϵ_{max} for the Cu chelate is considerably larger than for Cu(HDz)$_2$.

p,p'-Dichlorodithizone has also been prepared.[80] It has $P_{CCl_4}/K_a = [H_2L]_{CCl_4}/([H^+][HL^-]) = 1.1 \times 10^8$ (compared to 7×10^8 for dithizone). Its solubility in CCl$_4$ is 1.3×10^{-3} M. (In general, halo-substituted dithizones are less soluble than H$_2$Dz in organic solvents and are slightly stronger acids.) K_{ext} of Bi (CCl$_4$) has been reported as 2×10^{11}.

o,o'-Dibromodithizone[81] in CCl$_4$ has its two absorption maxima at 470 and 645 nm. ϵ_{max} of its Zn, Cd, Pb, Cu, Hg, and Ag chelates in CCl$_4$ are almost the same as for the dithizonates; ϵ_{max} of Cu(HL)$_2 = 51,300$, of Hg(HL)$_2 = 53,300$, and the respective K_{ext} values are 1.15×10^7 and 2×10^{26} (CCl$_4$). p,p'-Dibromodithizone[82] gives chelates whose absorption maxima in CCl$_4$ are usually shifted ~20 nm toward longer wavelengths compared to dithizonates, and whose ϵ_{max} are appreciably larger: Cu(HL)$_2$, 57,600; Hg(HL)$_2$, 72,900; Zn(HL)$_2$, 112,000. K_{ext} (CCl$_4$) = 1×10^9 for Cu(HL)$_2$ and 8×10^{26} for Hg(HL)$_2$. The chelates (primary) of Cd and Ag are very slightly soluble in CCl$_4$. The reagent has $P_{CCl_4}/[H^+][HL^-] = 2 \times 10^8$ Monobromodithizone (p) in CCl$_4$ has $\lambda_{max} = 456$ and 625 nm and $\epsilon_{625} = 34,500$, thus much the same as the corresponding constants for dithizone.

Di-iodo (o,o'- and p,p') and methyl-iodo derivatives of dithizone have been prepared, and ϵ_{max} of the reagents and their Cu, Hg, and Zn chelates have been determined.[83] λ_{max} of the reagents lie at longer wavelengths than the dithizone maxima (by as much as 30–40 nm in CCl_4) and λ_{max} of the chelates usually at longer wavelengths than the dithizonate maxima, but the shift is usually smaller than with the reagents. ϵ_{max} of the o-I derivative is smaller (23,500) than that of dithizone (30,900) in the >600 nm region, whereas the opposite is true for the p-I derivative (44,300), CCl_4 as solvent. The Cu(II) chelates of the derivatives have greater ϵ_{max} than Cu(HDz)$_2$, whereas ϵ_{max} of the Hg(II) chelates of the derivatives can be larger or smaller than ϵ_{max} of Hg(HDz)$_2$. K_{ext} of the Cu(II) chelates of the derivatives are smaller than of Cu(HDz)$_2$ by as much as 100 times. It is not evident that these reagents have any analytical importance. They are more subject to oxidative decomposition than dithizone.

Absorptivity data, acid dissociation constants, and formation constants (in dioxane–water) of some chelates of di-p Cl, Br, I, and F derivatives of dithizone are available.[84] In 50% dioxane, K_a of the di-p-F-phenyl derivative is 1×10^{-5} (H$_2$Dz, 1.6×10^{-6}), of the di-m-CF$_3$-phenyl derivative, 2.7×10^{-3}.

The reactions of S-methyldithizone (5-methylmercapto-1,5-diphenyl-formazan) with certain metals are not of analytical interest.[85]

o,o'(or 2,2')-Dimethyldithizone[86] in CCl_4 has λ_{max} at 628 nm ($\epsilon =$ 38,100) and 460 nm. It is less acidic than dithizone. K_{ext} of the Hg and Cu chelates (CCl_4) is 4×10^{22} and 5×10^7. It would seem to be of little interest as a reagent. p,p'-Dimethyldithizone[87] in CCl_4 has λ_{max} at $\sim$630 nm ($\epsilon = 46,400$) and 456 nm. Its solubility in CCl_4 corresponds to $\sim$6 $\times 10^{-3}$ M, and $[H_2L]_{CCl_4}/([H^+][HL^-]) = 4 \times 10^{10}$. ϵ_{max} of its chelates is a little larger than ϵ_{max} of dithizonates. K_{ext} of the Cu(II) and Hg(II) chelates is not greatly different from that of the dithizonates.

Various asymmetrical derivatives of dithizone have been prepared.[88]

Doubling the two phenyl groups in dithizone to give di-(o-diphenyl)thiocarbazone or di-(p-diphenyl)thiocarbazone,[89] shifts λ_{max} of the reagents and most of their chelates some 10–30 nm toward longer wavelengths (CCl_4). ϵ_{max} of the Ag, Hg, Cu, and Cd chelates of the o-diphenyl reagent ($\epsilon_{650} = 33,900$) remains much the same as ϵ_{max} of the corresponding dithizonates or decreases. Neither is there much change in the values of K_{ext}, if the Hg and Cu chelates are representative.

B. DINAPHTHYLTHIOCARBAZONES

The compounds mentioned previously are rarely used as reagents in practice. But di-β-naphthylthiocarbazone has demonstrated value in applied analysis and may have advantages over dithizone for some elements under some conditions. First, we examine the α-compound briefly.

Di-α-naphthylthiocarbazone

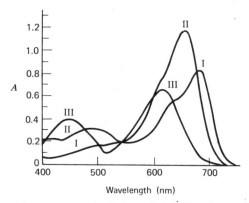

This compound is apparently more acidic than dithizone[90] and $[H_2L]_{CCl_4}/([H^+][HL^-]) = 1 \times 10^{11}$ according to A. S. Grzhegorzhevskii.[91] P_{H_2L} must be much larger than P_{H_2Dz} (~ 1000 times larger). At pH 11, the compound is then equally distributed between aqueous and CCl_4 phases. The CCl_4 solution shows a single absorption peak at 685 nm ($\epsilon = 4.93 \times 10^4$).[92] The Ag and Cd chelates are very slightly soluble in CCl_4. In general, metals will be extracted by this reagent at higher pH values than by dithizone; at the same pH (in the acidic or weakly alkaline range), the concentration of HL^- in the aqueous phase will be less than $\frac{1}{100}$ that of HDz^-, with equal concentrations of reagent in CCl_4. According to Takei, Cu(II) can be quantitatively extracted above pH 1.3 ($\lambda_{max} = 560$ nm,

Fig. 6G-15. Absorption spectra of di-(α-naphthyl)thiocarbazone(I), di-(β-naphthyl)-thiocarbazone(II) and dithizone(III) in CCl_4 according to S. Takei, *Japan Analyst,* **9,** 290 (1960).

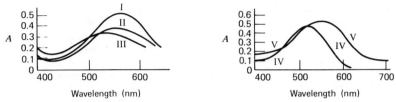

Fig. 6G-16. Absorption spectra of metal complexes of di-(α-naphthyl)thiocarbazone: I, Zn; II, Pb; III, Bi; IV, Hg(II); V, Cu((II). CCl$_4$ solutions (metal concentration not specified). Takei, *loc. cit.*

$\epsilon_{560} = 66,300$ in CCl$_4$, $K_{ext} = 2 \times 10^8$), Hg(II) above pH 0.5 (λ_{max} in CCl$_4$ = 525 nm, $\epsilon_{525} = 51,500$, $K_{ext} = 1.4 \times 10^{22}$). ϵ_{max} for Cu(HL)$_2$ is larger than for Cu(HDz)$_2$, but for Hg(HL)$_2$, it is distinctly less than for Hg(HDz)$_2$. The impression is obtained that for these metals at least no advantage is gained by using this reagent for extraction. Indeed, H$_2$Dz is to be preferred because it enables metals to be extracted from more acid solutions.

For absorption spectra of some metal complexes of di-(α-naphthyl)thiocarbazone see Fig. 6G-16.

Di-β-naphthylthiocarbazone

Mol. wt. 356.4

This reagent has $P_{CCl_4}/K_a = [H_2L]_{CCl_4}/[H^+][HL^-] = 5 \times 10^{12}$, compared to 7×10^8 for dithizone.[93] K_a seems to be much the same as for dithizone.[94] No doubt P_{H_2L} is much larger than P_{H_2Dz}.

The absorption peaks of the reagent and of its chelates in organic solvents lie at longer wavelengths than do those of dithizone and the dithizonates, and the ϵ_{max} values are larger for the reagent and its chelates. The reagent solution shows a single absorption peak, which in CHCl$_3$ falls at 650 nm and $\epsilon_{650} = 67,000$ (λ_{max} of dithizone in CHCl$_3$ is 605 nm and $\epsilon_{max} = 41,490$). The absorption peaks of the di-β-naphthylthiocarbazonates are shifted some 20–30 nm toward longer wavelengths compared to the dithizonates. This shift is sufficient to produce a noticeable change in hue (Hubbard):

	Di-β-naphthylthiocarbazone (CHCl$_3$)	Dithizone (CHCl$_3$)
Hg(II)	Red (blue shade)	Yellow-orange
Pb	Purple	Rose red
Bi	Magenta	Orange

Reported values of ϵ_{max} of the di-β-naphthylthiocarbazonates are uniformly higher than corresponding values of the dithizonates.[95] For some metals the difference is not large enough to be of much analytical significance, whereas for others it is considerable. The differences are largest for Ag, Cu(II), Ni, Au(III), Pd(II). Suprunovich[96] claimed greater sensitivity in the determination of lead, but Hubbard[97] found no significant difference in the determination of lead and bismuth.

K_{ext} has been determined[98] for a small number of di-β-naphthylthiocarbazonates: Bi (CCl$_4$), 6×10^6; Ni (CCl$_4$), ~ 2; Zn (CCl$_4$), 3×10^4; Cd (CHCl$_3$), 40. Because naphthylthiocarbazone continues to be extracted into the aqueous phase as HL$^-$ with a rise in pH until the solution is quite alkaline (at pH 13, $[H_2L]_{CCl_4}/[HL^-] = 10^{-13} \times 5 \times 10^{12} = 0.5$), the optimum pH range for the extraction of a metal extends further into the basic range than with dithizone in CCl$_4$ or even in CHCl$_3$. As shown by Cholak et al.,[99] the extraction of zinc with a dithizone solution in CHCl$_3$ from an ammonium citrate solution decreases above pH 8.3, whereas with a di-β-naphthylthiocarbazone solution of comparable concentration in CHCl$_3$, the extraction of zinc is still $\sim 100\%$ at pH 10 or 10.5 (range of quantitative extraction pH 8–>10). Bismuth can still be extracted quantitatively at pH 12 or higher. Di-β-naphthylthiocarbazone may have advantages over dithizone when a metal is to be extracted in the basic range, whether for separation or determination. There seem to be no reasons for preferring this reagent for extractions from acid medium. It may be noted that the naphthylthiocarbazone does not, as far as known, form secondary complexes.

Procedures for the photometric determination of Hg, Zn, Pb, and Cd in biomaterial with naphthylthiocarbazone have been described. It has also been studied as a reagent for the determination of trace metals in natural waters.[100]

Di-β-naphthylthiocarbazone in pure form is not readily obtained in good yield.[101] The products on the market are of low purity, sometimes containing as little as 30% of the active reagent. Chromatography of a CHCl$_3$ solution on an aluminum oxide column has been found to give a product of 94% average purity.[102]

C. SELENAZONE[103]

This is the trivial name given to the selenium analog of dithizone, 3-seleno-1,5-diphenylformazan

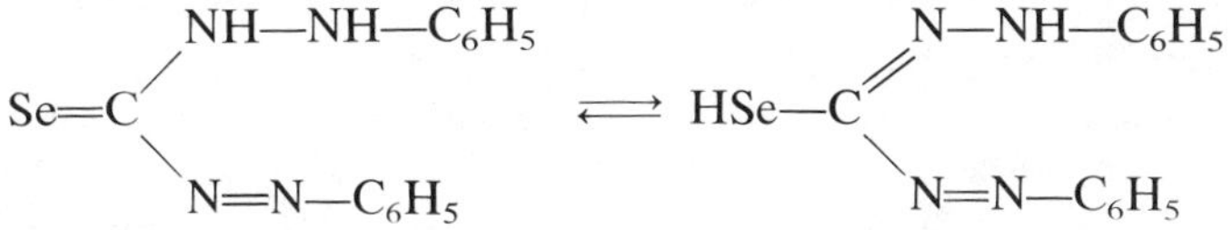

It is a black powder, m.p. 116°, soluble with a green color in organic solvents:

	$\lambda_{max\,1}$	$\lambda_{max\,2}$	λ_{min}	A_2/A_1
CHCl$_3$	482	622	550	0.66
CCl$_4$	487	648	560	0.38

Like dithizone, it is a weak acid and dissolves in alkaline solutions to give red HDsz$^-$. It reacts with heavy metals in much the same manner as dithizone to give strongly colored chelates. Selenazone would seem to be of little interest as a photometric trace reagent because of its easy air-oxidation in solution and because no advantages over dithizone are evident in its analytical reactions.

Oxidation of selenazone yields the diselenide

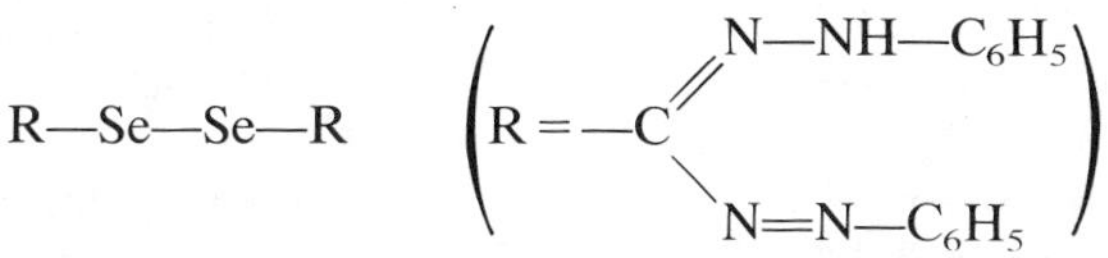

Interestingly, the diselenide forms metal complexes extractable into organic solvents. The metals reacting appear to be limited to Hg, Ag, Pd, and Ni. Possibly the reaction occurring is of the type

$$R—Se—Se—R + HgCl_2 \rightleftharpoons R—Se—HgCl + RSeCl$$

NOTES

1. E. Fischer, *Annalen,* **190,** 118 (1878); E. Fischer and E. Besthorn, *ibid.,* **212,** 316 (1882).
2. H. Fischer, *Wiss. Veröffentl. Siemens-Konzern,* **4,** 158 (1925); *Angew. Chem.,* **47,** 685 (1934); **50,** 919 (1937).
3. There is another maximum in the ultraviolet (at ~280 nm in CCl$_4$). The UV minimum lies at ~370 nm in CCl$_4$, at 365 nm in CHCl$_3$.
4. R. S. Ramakrishna and H. M. N. H. Irving, *Anal. Chim. Acta,* **48,** 252

(1969). Later values by Z. Marczenko and M. Mojski, *Chim. anal.*, **54,** 29 (1972), are $\epsilon_{445} = 16,400$, $\epsilon_{605} = 40,800$, $\epsilon_{605}/\epsilon_{445} = 2.49$.

5. Marczenko and Mojski.[4]

6. N. P. Komar' and L. S. Manzhelii, *J. Anal. Chem. USSR*, **28,** 1293 (1973).

7. H. Akaiwa and H. Kawamoto, *J. Inorg. Nucl. Chem.*, **29,** 541 (1967).

8. Bidleman found $\epsilon_{475} = 2.26 \times 10^4$, Geiger and Sandell, 2.24×10^4; Mathre obtained a higher value, 2.40×10^4. The former values are in good agreement with the peak value $\epsilon_{470} = 2.26 \times 10^4$ obtained later by N. P. Komar' and L. S. Manzhelii, *loc. cit.* These authors find another maximum in the UV at 270 nm. See Fig. 6G-2.

9. E. B. Sandell, *J. Am. Chem. Soc.*, **73,** 4660 (1950), found the value 1.5×10^{-5} at 25° ($\mu = 0.1 \, M$) from measurement of the partitioning of dithizone between basic aqueous solutions and CCl_4, combined with the aqueous molecular solubility of dithizone. At zero ionic strength, the value becomes 2×10^{-5} ($pK_{pd} = 4.7$). New determinations by Komar' and Manzhelii, *loc. cit.*, give the values $pK_{pd} = 4.70 \pm 0.08$ (by spectrophotometry) and 4.80 ± 0.05 (by solubility) at zero ionic strength.

10. R. W. Geiger and E. B. Sandell, *Anal. Chim. Acta*, **8,** 197 (1953). They found that in $\sim 10 \, M$ potassium hydroxide the yellow color of primary dithizonate ion changes to red-violet. This may be due to formation of Dz^{2-}. (Cyanide was added to mask Ag in KOH.)

11. Sandell.[9] Marczenko and Mojski[4] found the value 7×10^8 at $\mu = 1$, and 5×10^8 at $\mu = 0.2$ (temperature not specified).

12. H. Irving, S. J. H. Cooke, S. C. Woodger, and R. J. P. Williams, *J. Chem. Soc.*, 1847 (1949); H. Irving and C. F. Bell, *ibid.*, 1216 (1952). F. Koroleff also found the value 4×10^{10} ($\mu = 0.1$). J. S. Oh and H. Freiser, *Anal. Chem.*, **39,** 295 (1967), give 1.5×10^{10} ($\mu = 0.1$, 25°). Marczenko and Mojski, *loc. cit.*, report 4.5×10^{10} ($\mu = 1$, temperature not specified).

13. D. Dyrssen and B. Hök, *Svensk Kem. Tids.*, **64,** 80 (1952).

14. F. Koroleff, Thesis, University of Helsingfors, 1950, p. 21.

15. H. M. N. H. Irving, A. M. Kiwan, D. C. Rupainwar, and S. S. Sahota, *Anal. Chim. Acta*, **56,** 205 (1971). Thin-layer chromatography of dithizone and its oxidation products on silica gel is discussed also.

16. Also called dehydrodithizone. Synthesis and structure: J. W. Ogilvie and A. H. Corwin, *J. Am. Chem. Soc.*, **83,** 5023 (1961).

17. H. M. N. H. Irving, D. C. Rupainwar, and S. S. Sahota, *Anal. Chim. Acta*, **45,** 249 (1969).

18. H. M. N. H. Irving, U. S. Mahnot, and D. C. Rupainwar, *Anal. Chim. Acta*, **49,** 261 (1970); J. W. Ogilvie and A. H. Corwin, *J. Am. Chem. Soc.*, **83,** 5023 (1961).

19. H. M. N. H. Irving and D. C. Rupainwar, *Anal. Chim. Acta*, **48,** 187 (1969). This substance was obtained by M. Preund and F. Kuh, *Berichte*, **23,** 2821 (1890), by reaction of phosgene with dithizone.

20. King and Pruden, *loc. cit.*

21. H. Irving and C. Bell, *J. Chem. Soc.*, 4253 (1954).

22. Hg(II) and Cu(II) primary dithizonates: M. M. Harding, *J. Chem. Soc.*, 4136 (1958); R. F. Bryan and P. M. Knopf, *Proc. Chem. Soc.*, 203 (1961). Earlier, H. Fischer postulated a keto structure for primary dithizonates in which the metal was bonded to N atoms only, an implausible bonding. The secondary dithizonates were assigned an enol structure

which is stereochemically improbable.

23. A. Mawby and H. M. N. H. Irving, *Anal. Chim. Acta*, **55,** 269 (1971); *J. Inorg. Nucl. Chem.*, **34,** 109 (1972). K. S. Math and H. Freiser, *Talanta*, **18,** 435 (1971).

24. M. Laing and P. A. Alsop, *Talanta*, **17,** 242 (1970). The structure proposed by K. S. Math, Q. Fernando, and H. Freiser, *Anal. Chem.*, **36,** 1762 (1964), in which Ni is bonded to two N atoms, thus receives no support.

25. H. M. N. H. Irving and A. M. Kiwan, *Anal. Chim. Acta*, **56,** 435 (1971).

26. Workers at Univ. of Minnesota; cf. Table 6G-6.

27. Meriwether, Breitner and Sloan, *loc. cit.*

28. L. S. Meriwether, E. C. Breitner, and C. L. Sloan, *J. Am. Chem. Soc.*, **87,** 4441, 4448 (1965).

29. Meriwether et al., *J. Am. Chem. Soc.*, **87,** 4448 (1965).

30. This mixed chelate is also formed in aqueous chloride solution if Hg(II) is in excess of H_2Dz, as shown by B. G. Cooksey, *Proc. Soc. Anal. Chem.*, **3,** 143 (1966); G. B. Briscoe and B. G. Cooksey, *J. Chem. Soc. A*, 205 (1969). Apparently Br^-, I^-, and other inorganic anions can form mixed dithizone complexes with Hg(II): R. Litman et al., *Anal. Chem.*, **49,** 983 (1977).

31. J. Stary and J. Ruzicka, *Talanta*, **15,** 505 (1967).

32. H. M. N. H. Irving and A. M. Kiwan, *Anal. Chim. Acta*, **45,** 255 (1969), in which the reaction of a wide variety of organomercurials with dithizone is described. Extractabilities, absorption spectra, and molar absorptivities are reported. See this paper for earlier work. Extraction constants (CCl_4) are given in *Anal. Chim. Acta*, **45,** 447 (1969).

33. H. M. N. H. Irving and A. M. Kiwan, *Anal. Chim. Acta*, **45,** 271 (1969).

34. H. M. N. H. Irving and T. Nowicka-Jankowska, *Anal. Chim. Acta*, **54,** 55 (1971).

35. A. M. Kiwan and H. M. N. H. Irving, *Anal. Chim. Acta*, **54,** 351 (1971).

36. Nickel adducts: K. S. Math and H. Freiser, *Anal. Chem.*, **41**, 1682 (1969). The 1:1 1,10-phenanthroline adduct is the most stable among those of a number of N bases (log $K \sim 6$). Nickel is bonded octahedrally to 2 S and 2 N of dithizone and 2 N of the heterocyclic base.

37. K. Akaiwa and H. Kawamoto, *Anal. Chim. Acta*, **40**, 407 (1968). Z. Marczenko and M. Mojski, *ibid.*, **54**, 469 (1971), showed that the adduct has the composition $Mn(HDz)_2 \cdot 2 Py$, and find $\mathscr{S}_{510} = 0.001$.

38. K. Akaiwa, H. Kawamoto, and M. Hara, *J. Chem. Soc. Japan Pure Chem. Sec.*, **90**, 186 (1966). The change in the absorption curves of the dithizonates produced by pyridine is noted.

39. K. S. Math, K. S. Bhatki, and H. Freiser, *Talanta*, **16**, 412 (1969); B. S. Freiser and H. Freiser; *ibid.*, **17**, 540 (1970).

40. Not all metal ion–dithizone reactions are easily reversible. For example, Co(II) and Ni do not react to an appreciable extent with dithizone in dilute acid medium (e.g., at pH 2), but when either dithizonate has once been formed (extraction from basic medium), it is not easily dissociated by dilute acids. The seeming irreversibility is the result of very low reaction rate. As well known, nickel and cobalt [especially Co(III)] tend to form inert complexes.

41. See p. 892, Chapter 9B.

42. B. W. Budesinsky and M. Sagat, *Talanta*, **20**, 228 (1973).

43. The $P \rightleftharpoons S$ equilibrium has been studied by Geiger and Sandell, reference in Table 6G-3.

44. O. Mathre and E. B. Sandell, *Talanta*, **11**, 295 (1964), found the experimental course of the lead extraction curve to agree with the calculated.

45. However, a few mixed $OH^- – HDz^-$ complexes are known, as of Pd(II).

46. The literature does not clearly distinguish between Co(II) and Co(III) dithizonates. Apparently—as would be expected—Co(III) dithizonate is not much affected by acids and remains in the CCl_4 phase when it is shaken with acids, for example, 0.1 *M* HCl.

47. B. E. McClellan and P. Sabel, *Anal. Chem.*, **41**, 1077 (1969).

48. According to F. F. Dyer and G. K. Schweitzer, *Anal. Chim. Acta*, **23**, 1 (1960), log K_{ext} of Ag(HDz) varies linearly with the reciprocal of the absolute temperature:

$$\log K_{ext} = (3500/T) - 5.9 \text{ (for } CHCl_3)$$
$$\log K_{ext} = (5300/T) - 11.0 \text{ (for } CCl_4)$$

K_{ext} for $CHCl_3 – H_2O$ decreases by $\sim 8\%$ for a 1° rise in temperature at 27°.

49. J. Ruzicka and J. Stary, *Talanta*, **14**, 909 (1967); J. Stary and J. Ruzicka, *ibid.*, **15**, 505 (1968).

50. J. Stary, *Talanta*, **16**, 359 (1969).

51. Metals can be determined without extraction by adding a water-miscible organic solvent to the aqueous sample solution in sufficient amount to keep

dithizone and the dithizonate in solution. Such a method has been described for silver, zinc, and lead. Usually, a method of this type is not advantageous.

52. Seldom in the laboratory, but often in the field, the determination is made by comparing the hue of the organic phase with the hues of a series of standards, all prepared with the same volume of dithizone solution as the sample. The standards exhibit a series of hues ranging from the unchanged green of dithizone to the color of the metal dithizonate alone, if the pH is such that most of the dithizone remains in the organic solvent. The comparison is conveniently made in glass-stoppered tubes (p. 210), and the organic solvent layers are examined transversely against a white background. The dithizone solution should not be stronger than 0.001% w/v, else the solutions will be too strongly colored for comparison at a depth of 1–2 cm. If the approximate amount of the metal is not known, the dithizone solution is added in small portions from a buret with shaking after each addition until a suitable mixed color is obtained. A carbon tetrachloride or chloroform solution containing somewhere near equal amounts of dithizone and dithizonate gives the most satisfactory hue for precise comparison. A grayish mixed color is then obtained with red or violet-red dithizonates, and the eye can detect differences of about 3% in the metal concentration. If the reaction between metal and dithizone proceeds sufficiently rapidly, the colorimetric titration technique can be applied, that is, a standard solution of the metal is added to a solution of the same pH as the unknown and containing the same volume of dithizone, until after thorough shaking the hues of the organic solvents match.

53. H. Irving, E. J. Risdon, and G. Andrew, *J. Chem. Soc.*, 537 (1949).

54. But the requirement of known or constant strength need not be fulfilled in the usual method either, if the absorbance of the dithizone–dithizonate solution is measured against the standard dithizone solution.

55. Owen Mathre, University of Minnesota, 1956, found that the reversion method had no advantage over the usual method in the determination of lead.

56. Successive determination of Ag, Bi, and Hg(II) by the reversion method: R. A. Chalmers and D. M. Dick, *Anal. Chim. Acta,* **32,** 117 (1965). Ag, 1 M NaCl, pH 1; Bi, 0.2 M HCl; Hg, 1 M KI, pH 1.

57. Separating the phases and washing dithizone out of the organic extract is not necessary, if the pH of the sample solution can be raised sufficiently high in the extraction of the metal to transfer almost all the dithizone to the aqueous phase in this step.

58. Zinc and cadmium dithizonates in carbon tetrachloride or chloroform may be dissociated with 0.02 M hydrochloric acid (double back-extraction if the dithizone concentration is high). Lead dithizonate is dissociated at even lower acidities.

59. Alternatively, the residue is heated with a few drops of concentrated nitric acid and 30% H_2O_2 and evaporated to dryness. The treatment is repeated if

necessary, so that an essentially colorless residue is obtained. This procedure was shown by M. Nishimura to decompose $Ni(HDz)_2(+H_2Dz)$ readily, and presumably should be effective for other dithizonates.

60. A metal, such as zinc, not readily extractable from a fairly acid solution, can give an appreciable absolute amount of extracted metal if present in large amounts, so that difficulties may arise when, for example, a trace of copper is to be determined in zinc metal. A perceptible amount of zinc may be extracted with the copper. Thus when a solution of zinc in 0.1 N (mineral) acid is extracted with 0.001% w/v dithizone in CCl_4 $(4 \times 10^{-5}\ M)$, $[Zn(HDz)_2]_{CCl_4}/[Zn^{2+}]$ at equilibrium is equal to $100 \times (4 \times 10^{-5})^2/0.1^2 = 1.6 \times 10^{-5}$, which with equal volumes of phases and 1 g Zn corresponds to 16 μg Zn in the CCl_4 extract; copper is extracted quantitatively (if the concentration of dithizone remains much the same). If the amount of coextracted zinc is not too great, the carbon tetrachloride or chloroform phase containing the two dithizonates may be drawn off and shaken with dilute acid, which will decompose the zinc complex, thus leaving the copper dithizonate alone in the organic solvent (see p. 624). Another difficulty in a case of this kind may be the practical impossibility of the initial extraction of the trace metal on account of the reaction of most of the dithizone with the other metal present in much higher concentration.

61. These values are practically those for the extraction constants of copper(II) and zinc dithizonates in carbon tetrachloride.

62. If M and N are initially present in equal or comparable amounts. If not, obtaining the lesser constituent substantially free from the greater constituent becomes more difficult. See the discussion in Chapter 9D.

63. At pH 10, with a CCl_4 solution of H_2Dz, $In(HDz)_3$ is extracted in the presence of cyanide. However, citrate or tartrate greatly hinders the extraction, and one of these is always used in the Pb determination. Much In may, nevertheless, cause trouble in the Pb determination.

64. H. Fischer and G. Leopoldi, *Z. Anal. Chem.*, **107**, 241 (1936). In 2% (v/v) H_2SO_4 solution, thiosulfate prevents the reaction of Hg but not of Cu, so that the two metals are separable under these conditions.

65. A. Hulanicki and M. Minczewska, *Talanta*, **14**, 677 (1967). Bis(2-hydroxyethyl)dithiocarbamate is a similar masking agent: D. W. Margerum and F. Santacana, *Anal. Chem.*, **32**, 1157 (1960), in which various dithizone methods for Zn are compared; A. Galik, *Talanta*, **14**, 731 (1967). β-Dithiocarbaminopropionic acid is another possibility: E. Russeva and O. Budevsky, *Talanta*, **20**, 1329 (1973).

66. J. Schultz and M. A. Goldberg, *Ind. Eng. Chem. Anal. Ed.*, **15**, 155 (1943).

67. As already pointed out, exposure of primary mercury(II) dithizonate to strong light results in a reversible change in hue from orange to blue (with intermediate hues). The photochromism is inhibited by addition of acetic acid.

68. Meriwether et al., note 28.

69. H. G. C. King and G. Pruden, *Analyst*, **96**, 146 (1971), determined dithizone in eight commercial samples of dithizone (most of them supposedly laboratory-reagent or analytical-reagent grade) by measuring A_{620} and found surprisingly low contents of active reagent. The worst sample contained ~38% H_2Dz, the best 94%. One milligram pure H_2Dz in 100 ml CCl_4 should give $A_{620} = 1.27$ (1 cm) if $\epsilon_{620} = 32,500$.

70. The purity of dithizone prepared in this way is indicated by the following assay values (titration of silver) obtained by various workers at the University of Minnesota: 99.8–100.2% (Mathre); 99.4–99.8% (Marhenke); $99.4 \pm 0.5\%$ (Bidleman). The small deficit is probably largely due to solvent trapped in the crystals.

71. Hydrochloric acid may be used for acidification, but sulfur dioxide is preferable because of its reducing action and because no heavy metals are introduced. The solid dithizone, which is obtained at this point, probably can be filtered off and used without further treatment, as far as most requirements are concerned. It should be washed with pure water containing a little sulfur dioxide and dried at room temperature (finally over a desiccant such as potassium hydroxide if an anhydrous product is desired).

72. H. Fischer, G. Leopoldi, and H. von Uslar, *Z. Anal. Chem.*, **101**, 1 (1935). The titration can be made as described by these authors, with withdrawal of each increment of H_2Dz–CCl_4 that has reacted, until the last increment shows a yellow-green color, or more conveniently by photometric titration of H_2Dz solution (CCl_4) with a silver nitrate solution at 620 nm. At this wavelength, Ag(HDz) does not absorb and a plot of A versus volume of $AgNO_3$ gives a straight line whose intersection with the volume axis marks the equivalence point.

73. King and Pruden, *loc. cit.*

74. Our observations pertain to CCl_4. Extractions from slightly basic solutions are so rapid that the presence of substances inhibiting extraction from acid solutions is usually of no importance.

75. R. W. Geiger, Ph.D. Thesis, University of Minnesota, 1951.

76. P. A. Clifford, *J. Assoc. Off. Agr. Chem.*, **21**, 695 (1938).

77. D. A. Biddle, *Ind. Eng. Chem. Anal. Ed.*, **8**, 99 (1936).

78. J. B. Mullin and J. P. Riley, *Analyst*, **80**, 316 (1955).

79. These and other data are from A. M. Kiwan and A. Y. Kassim, *Talanta*, **22**, 931 (1975).

80. A. I. Busev and L. A. Bazhanova, *Zh. Neorg. Khim.*, **6**, 2805 (1961). This paper deals with Bi complexes of dithizone derivatives. It is concluded that electron-donating substituents decrease the partition coefficient and increase the pH for 50% extraction of the Bi complexes (opposite effect with electron-accepting substituents).

81. S. Takei, *Japan Analyst*, **9**, 402 (1960). Concerning di-(o-bromo-p-tolyl)-thiocarbazone and di-(p-bromo-o-tolyl)thiocarbazone and their chelates, see S. Takei, *ibid.*, **10**, 708 (1961).

82. Takei, *loc. cit.*; Busev and Bazhanova, *loc. cit.*; Irving and Bell, *J. Chem. Soc.*, 3538 (1953).

83. S. Takei, *Japan Analyst*, **10**, 715 (1961).

84. A. R. Al-Salihy and H. Freiser, *Talanta*, **17**, 182 (1970).

85. H. M. N. H. Irving, A. H. Nabilsi, and S. S. Sahota, *Anal. Chim. Acta*, **67**, 135 (1973).

86. S. Takei, *Japan Analyst*, **6**, 630 (1957); **10**, 708 (1961); S. Takei and T. Kato, *Tech. Rept. Tohoku Univ.*, **21**, 135 (1957); Synthesis: Suprunovich and D. L. Shamshin, *Zh. Anal. Khim.*, **1**, 198 (1946); said to be more resistant than H_2Dz to oxidation. B. E. McClellan and H. Freiser, *Anal. Chem.*, **36**, 2263 (1964), give λ_{max} in $CHCl_3$ as 610 nm and ϵ_{610} as 4.95×10^4. Solutions are stable for one month.

87. S. Takei and K. Shibuya, *Japan Analyst*, **5**, 695 (1956); Busev and Bazhanova, *loc. cit.*; Hubbard and Scott (see note 101).

88. L. S. Pupko and P. S. Pelkis, *Zh. Obshchei Khim.*, **24**, 1640 (1954).

89. S. Takei, *Japan Analyst*, **9**, 409 (1960).

90. K. S. Math, Q. Fernando, and H. Freiser, *Anal. Chem.*, **36**, 1762 (1964), found $K_a = 5.9 \times 10^{-6}$ in 50% dioxane ($\sim 2 \times 10^{-4}$ in H_2O). The purity of the compound is uncertain.

91. *Zh. Anal. Khim.*, **11**, 689 (1956); *Trudy Kom. Anal. Khim.*, **11**, 165 (1960). With $CHCl_3$, B. E. McClellan and H. Freiser, *Anal. Chem.*, **36**, 2262 (1964), find the value 2×10^{12} for this constant.

92. S. Takei, *Japan Analyst*, **9**, 288, 294 (1960). See Fig. 6G-15. McClellan and Freiser report ϵ_{685} in $CHCl_3$ as 2.15×10^4, which seems low.

93. Grzhegorzhevskii, *Zh. Anal. Khim.*, **11**, 689 (1956).

94. Math et al.[90] found K_a for the two compounds to be about the same in 50% dioxane–water, but the purity of the di-β-naphthylthiocarbazone is uncertain.

95. R. J. DuBois and S. B. Knight, *Anal. Chem.*, **36**, 1316 (1964), tabulate λ_{max} and ϵ_{max} values for $CHCl_3$ solutions and compare with corresponding values for dithizonates in CCl_4. These authors report ϵ_{max} for Bi and Zn naphthylthiocarbazonate in $CHCl_3$ as 114,000 and 121,000, whereas Grzhegorzhevskii reports 170,000 for both in CCl_4. It is of interest that DuBois and Knight report reaction of di-β-naphthylthiocarbazone with metals that do not react, or only react difficultly, with dithizone. Thus Rh(III) and Ir(III) give complexes having $\epsilon_{515} = 58,300$ and $\epsilon_{545} = 138,500$, respectively. Os(III) forms unstable complexes absorbing maximally at 475 nm (acid solution) and 510 nm (basic solution). Mo(VI) is said to give an unstable violet complex; Sb(III), a red-violet complex in basic solution; and Sb(V), a questionable red-violet complex in basic solution. As(III) is said to react in acid solution; As(V), in basic.

96. I. B. Suprunovich, *J. Gen. Chem. USSR*, **8**, 839 (1938). This seems to be the earliest analytical paper on this reagent.

97. D. M. Hubbard, *Ind. Eng. Chem. Anal. Ed.*, **12,** 768 (1940).

98. Busev and Bazhanova, *loc. cit.*, Bi; Grzhegorzhevskii, *loc. cit.*, Ni, Cd, and Zn. According to the latter author, K_{ext} of Bi is 8×10^8.

99. J. Cholak, D. M. Hubbard, and R. E. Burkey, *Ind. Eng. Chem. Anal. Ed.*, **15,** 754 (1943).

100. O. W. Lombardi, *Anal. Chem.*, **36,** 415 (1964).

101. The most feasible method of preparation is described by D. M. Hubbard and E. W. Scott, *J. Am. Chem. Soc.*, **65,** 2390 (1943). See also D. M. Hubbard, note 97; M. Preund, *Berichte*, **24,** 4178 (1891). S. S. Cooper and V. K. Kofron, *Anal. Chem.*, **21,** 1135 (1949), describe the purification of naphthylthiocarbazone. An improved method of purification by recrystallization has been given by Hubbard, *ibid.*, **28,** 1802 (1956).

102. R. J. DuBois and S. B. Knight, *Anal. Chem.*, **36,** 1313 (1964). As many as four bands were obtained, including those of red and blue impurities. None of the impurities absorbed appreciably at 650 nm, and they did not react with metals. The purified reagent is reported to be unstable in $CHCl_3$, decomposing to a blue product whose spectrum is identical with that of the blue component in the impure commercial product. Decomposition is more rapid in the UV than in the visible spectral range.

103. R. S. Ramakrishna and H. M. N. H. Irving, *Anal. Chim. Acta*, **48,** 251 (1969).

ORGANIC PHOTOMETRIC REAGENTS
ION-ASSOCIATION REAGENTS

I. GENERAL CONSIDERATIONS

Almost always in trace analysis, ion associates are of interest because of their extractability into organic solvents. An ion-association reagent can supply either the cationic or anionic part of an absorbing or fluorescent ion-association compound for photometric determination. Usually the spectral characteristics of an ion associate are very similar to those of its component ions. Even in an organic solvent, the color of the ion associate may not differ greatly from the color of its absorbing ion in aqueous solution. For example, Rhodamine B chloroantimonate $(RHSbCl_6)$ in benzene is red-violet, similar in color to Rhodamine B (RH^+) in acidic aqueous solution. However, benzene does not appreciably extract Rhodamine B as the chloride from hydrochloric acid solution of the proper concentration, so there is no interference from the excess of reagent. Less often in practice the cationic reagent is colorless, and the constituent to be determined is a colored anion. Tetraphenylarsonium cation extracts MnO_4^- into chloroform as $[(C_6H_5)_4As^+][MnO_4^-]$, but the only sensitivity gain results from the concentration of MnO_4^- into a smaller volume of solvent, since ϵ_{MnO_4} is much the same in water and in chloroform. There may be a gain in selectivity, however, because other colored constituents may not be extracted from the aqueous solution (compare Chapter 9C).

One of the components of an ion associate can be a chelated metal ion. Such a combination is likely to give an extractable product of high, or fairly high, absorptivity. For example, Cu(I) can be chelated with 2,2'-biquinoline to form violet $CuBq_2^+$, extractable into amyl alcohol as $CuBq_2X$, where X^- is Cl^- or other singly charged anion. Color production occurs in the chelation step, but ion-pair formation is required for the extraction of the cationic chelate. A metal can also be chelated into an anion, which can be extracted after the addition of a cationic partner of large ionic volume. A sulfonic acid group in the chelate may supply the negative charge. Thus Co chelated with Nitroso-R salt (sulfonated 1-nitroso-2-hydroxynaphthalene) can be extracted into chloroform with diphenylguanidinium cation.

Extractable ion-association (often ion-pair) compounds are preferentially formed from cations and anions of large ionic volume, which is why at least one component and preferably both are complex inorganic ions or large organic ions.[1] The extractability into an immiscible organic solvent is the result of the slight solubility of the ion associate in water and its solubility (due to the organic part) in organic solvents (Chapter 9C). The organic solvent must be one that does not extract the excess reagent paired with ions of opposite charge other than the ion to be determined.

There are exceptions to the general rule that no dramatic color changes occur when ions composing an ion associate are brought together or the latter is extracted. For example, when silver ion, excess 1,10-phenanthroline and Brompyrogallol Red (a polyphenolic triarylmethane dye) are mixed, a blue color appears,[2] which is due to the formation of

$$[AgPhen^+]_2R^{2-} \qquad (H_2R = Brompyrogallol\ Red)$$

The absorption maximum of the dye is shifted from 560 to 635 nm. No protons are released. The ion associate is only slightly soluble in water and precipitates at once if concentrated solutions are mixed, or on standing if dilute solutions are used. The product is extractable into sundry organic solvents. Apparently, the association of $AgPhen_2^+$ with BPR (Brompyrogallol Red) produces a change in the molecule of the latter that is closely allied to release of the bonded protons (fully ionized BPR absorbs maximally at ~630 nm). It is believed that an electron is transferred from an ionized phenolic group to $AgPhen_2^+$, thus weakening the bonding of the protons. The aromatic chromophore (the resorcinol part of the molecule) thus assumes a form very similar to that of the completely ionized BPR molecule. In other words, a charge transfer takes place. The mechanism is illustrated as follows[3]:

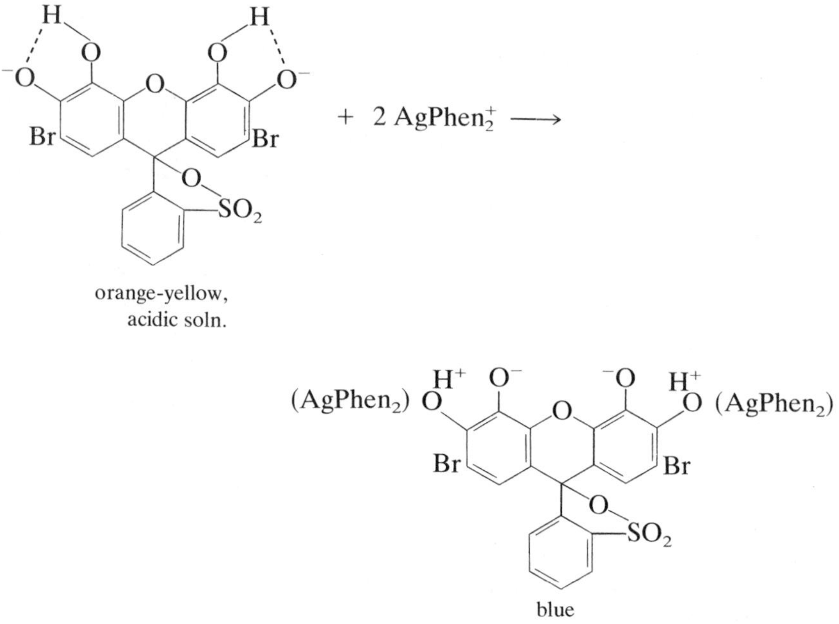

orange-yellow,
acidic soln.

blue

The mechanism of reaction and color change in the system Sn(IV)–Catechol Violet–cetyltrimethylammonium bromide (CAB) are similar.[4] Catechol Violet forms a red-purple 2 CV: 1 Sn complex, the color intensity of which increases, with a shift in λ_{max} to ~660 nm, when CAB is added. CAB is a surfactant and forms micelles above a certain critical concentration. The cationic CAB micelles combine with the anionic

Sn(IV)–Catechol Violet chelate to form the ion associate:

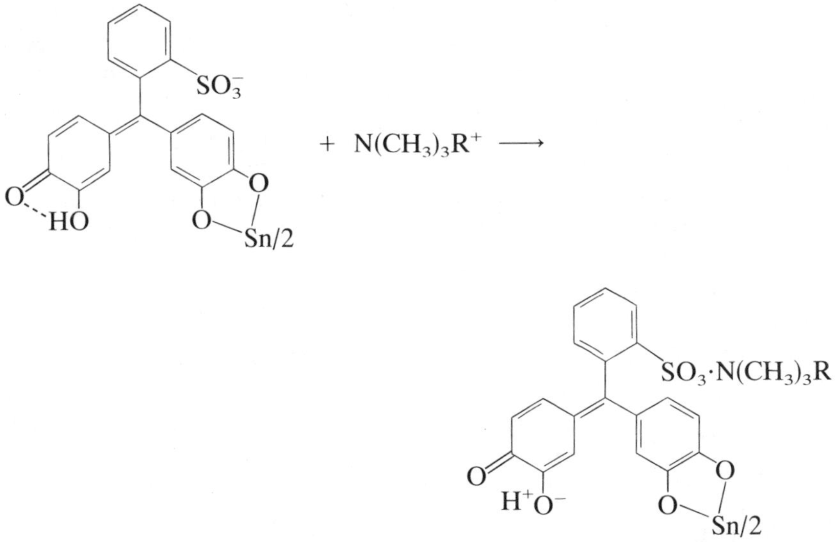

$$+ \; N(CH_3)_3 R^+ \longrightarrow$$

(The anionic associate represented above can combine with another molecule of CAB to form an uncharged association complex extractable into organic solvents.) Electrophilic reaction of the quaternary ammonium group at the site of the sulfonic group promotes electron withdrawal from the remaining phenolic groups of the tin chelate, thus leading to ionization that forms an absorbing species similar to blue, fully ionized Pyrocatechol Violet. It is considered that micelle formation is necessary for formation of the colored species. Quaternary ammonium salts such as tetraethyl- and tetrabutylammonium bromide do not produce a color change. This tin reaction is very sensitive ($\epsilon = 9.6 \times 10^4$) but unselective. Similar reactions are given by Mo and Sb. Again the sensitivity is high. Differential complexing is of some aid in improving the selectivity, but for the reaction to be useful in general practice, careful separations are needed.[5]

Other reagents and other metals give analogous reactions involving cetyltrimethylammonium and similar ions, for example, Al, Eriochrome Cyanine R, and cetyltrimethylammonium chloride.[6] Hexadecylpyridinium bromide intensifies the color given by Nb and Catechol Violet and stabilizes it.[7] A very sensitive reaction of rare earth elements with Xylenol Orange and cetylpyridinium chloride[8] probably has a mechanism similar to that of the preceding tin reaction.

Although ion-association compounds are not limited to the 1:1 type (ion pairs), these form more readily and are most often encountered. This tendency to form ion pairs is of importance in enhancing the selectivity of ion-association reactions, or at least of those with some reagents of this class. For example, with Rhodamine B as reagent in 1:1 hydrochloric acid solution, the only metals extracted easily by benzene and similar solvents are those forming singly charged chloroanions, namely $SbCl_6^-$, $AuCl_4^-$, $GaCl_4^-$, $FeCl_4^-$, and $TlCl_4^-$ (these are also extracted readily by ethers and other oxygenated solvents as $HAuCl_4$, etc.). This reaction becomes selective for gallium if Sb(V) is reduced to Sb(III), Au(III) to Au(0), Fe(III) to Fe(II), and Tl(III) to Tl(I). Tungsten(VI) gives Rhodamine B tungstate, insoluble in benzene.

The Rhodamine B reaction illustrates an advantage possessed by some ion-association reactions: They can be carried out in strongly acid solutions, in fact may require such solutions (to form RH^+). The situation is especially favorable when a high halide ion concentration is needed to form the complex metal anion, so that both H^+ and X^- are supplied by the acid HX. Of course, not all ion-association reactions can be carried out in strongly acid solutions. If the metal is chelated in the cation or anion, a low-acidity buffered solution may be needed, as in formation of ferrous 4,7-diphenyl-1,10-phenanthroline cation.

A number of metals, including Ga, Sb, and Ta, are generally better determined by ion-association complexing than by chelate complexing. For Ag, Au, and Tl, ion-association color methods are also important and, at least, competitive with chelate methods. It will be realized that the composition of the sample being analyzed is an important factor in the choice of a method, and a particular method is not always the best for all samples.

Even if chelate formation is the primary color reaction, ion association may be necessary for the extraction of a colored or fluorescent species and then is an integral part of the method.

II. CATIONIC REAGENTS

By far the larger number of reagents used in photometric determinations based on formation of ion associates are of value because they are electrolytes having a cation conferring a required or desirable property on the ion associate. As already noted, some ion-association reagents (colorless) do not contribute to the color of the product but make possible the extraction of a colored species into a suitable extractant, the purpose of this step being the concentration of the colored substance or its separation from other colored substances in the sample. A few reagents of this type are considered in Section II.A.[9] The more useful and interesting reagents are those that have a color of their own and thus form a colored (and sometimes fluorescent) extractable ion associate. Occasionally, as we have seen, the dye cation interacts with the anion being determined to give a color distinctly different from its own, but most of the useful reagents do not behave in this way, though differences in the respective absorption curves are to be expected, if for no other reason than the difference in solvents.

A. SOME COLORLESS CATIONIC REAGENTS

1. Tetraphenylarsonium Chloride and Other Onium Salts

The general extractive properties of this reagent and its analogs are considered in Chapter 9C. It is an example of a colorless cationic reagent useful for the photometric determination of strongly absorbing oxo- or complex anions. Chloroform is the usual extracting solvent. A chloroform solution of $(C_6H_5)_4AsCl$ absorbs maximally at 264 nm, so that there is a wide wavelength range in the visible and much of the ultraviolet for measurement of absorption by the anionic component. The extractive determination of $AuCl_4^-$ and $IrCl_6^{2-}$ is of some interest. The latter is extracted in spite of its double charge. Its extraction is aided by the formation of an adduct $[(C_6H_5)_4As]_2IrCl_6 \cdot 2(C_6H_5)_4AsCl$. Rhodium, as $RhCl_6^{3-}$, is not extracted at all (because of its high charge), so that Ir is readily determinable in the presence of much Rh. However, some of the other Pt metals interfere by being extracted more or less, and the sensitivity of the Ir determination is low. Bismuth is determinable as BiI_4^-.

Tetraphenylarsonium chloride has also been used for the extraction of the anionic thiocyanates of Co, Mo(V), W(V), Nb, and Pd into chloroform. These complex thiocyanates are extractable as the acids into oxygen-containing solvents, but chloroform has advantages as the extractant. The extracted Mo complex appears to have the composition $[(C_6H_5)_4As]_2[MoO(SCN)_5]$.

Tetraphenylarsonium cation has found occasional use in the extractive determination of anionic chelates. For example, the yellow anionic complex of sulfosalicylic acid with titanium can be extracted into chloroform in the presence of tetraphenylarsonium cation.

Tetraphenylphosphonium and triphenylmethylarsonium cations are also used as extraction reagents.[10]

2. Diantipyrinylmethane and Derivatives[11]

In acid medium, 4,4'-diantipyrinylmethane (diantipyrylmethane)

H₃C—C══════C—CH₂—C══════C—CH₃
H₃C—N   C═O O═C   N—CH₃
     N   N
     C₆H₅   C₆H₅

Mol. wt. ($+H_2O$) 406.5.
M. p. 155°–157° (decomp.)
pK_a in methyl ethyl
ketone = 9.96.

is protonated through the carbonyl groups with one or two H^+. Photometrically, the reagent is chiefly of importance for its selective reaction with titanium, which in 0.5–1 M HCl forms a soluble yellow complex. The bonding of titanium is obscure, but it seems reasonable to suppose that binding to oxygen of –C═O at least occurs. The formation of the species TiR_3^{4+} has been reported.[11] Other oxyphilic metals react similarly. Uranium(VI) gives a yellow color; Fe(III), an orange-red, and Mo(VI), a weak yellow. Iron(III) and V(V) do not react if reduced with ascorbic acid or hydroxylamine hydrochloride. Iron(III) is stated to form FeR_n^{3+} ($n = 1$, 2, or 3). The sensitivity of the Ti reaction in aqueous solution is $\mathscr{S} = 0.003$ μg Ti/cm² for $A = 0.001$. If stannous chloride is added to the HCl solution, an ion associate, believed to be $TiR_3(SnCl_3)_4$, can be extracted into $CHCl_3$. The absorption maxima of the aqueous and $CHCl_3$ solutions lie at much the same wavelength, but in $CHCl_3$, $\mathscr{S}_{395} = 0.0007$ μg Ti/cm², an exceptional sensitivity.[12] Stannous chloride also reduces Fe(III) and prevents its interference.

The diantipyrinylmethane cation (RH^+ or RH_2^{2+}) forms ion associates with anionic complexes of Cl^-, Br^-, I^-, and SCN^- with various metals. These reactions, then, are different from those mentioned in the preceding paragraph in that the absorption is due to the metal in the anion and not to a new complex formed by the metal in the cation. The normal ion associates, as they may be called, are sparingly soluble in water and can be extracted into benzene, chloroform, and dichloroethane. Such extraction methods have been proposed for determination of Au, Bi, Co, Cr, Ir, Os, Pd, Pt, Rh, Mo, Sb, Te, Ti, Tl, U, and other metals. Bi, Sb, and Tl are determined as iodo complexes, most of the other metals as chloro or bromo complexes. Pd forms RH_2PdBr_4. Separations by extraction are briefly described in Chapter 9C, Section I.C.3.

Diantipyrinylmethylmethane (1,1-diantipyrinylethane) has been used to determine Bi and Sb(III) in iodide media. Diantipyrinylpropylmethane (1,1-diantipyrinylbutane)[13] has been proposed as a reagent for Au, Bi, Ir,

Os, Pt, Re, Sb, Te, and Tl.[14] Diantipyrinylphenylmethane (α,α-diantipyrinyltoluene) also has been used to determine Sb(III). The molar absorptivity at 400 nm is approximately 4700, less than that (approximately 6600) for the iodo complex with diantipyrinylmethylmethane or diantipyrinylpropylmethane.[15]

3. Chlorpromazine

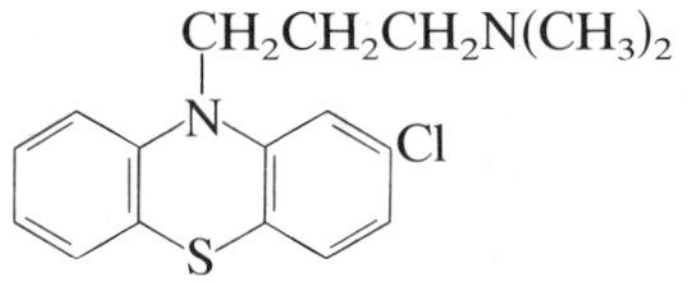

This compound, 2-chloro-10-(3-dimethylaminopropyl)phenothiazine, forms orange $(RH)_3BiI_6$, which is extracted by $CHCl_3$.[16] Various metals interfere. The formation and extraction of an ion associate having a triply charged anion is interesting; it rarely occurs.

4. 1,3-Diphenylguanidine

Anionic chelates and 1,3-diphenylguanidinium cations form ion associates that are extractable by organic solvents. The titanium (IV)-Tiron chelate is extracted at pH 3.0–4.0 into butanol in the presence of 1,3-diphenylguanidinium ion.[17] The combining ratio Ti:Tiron:DPG is 1:2:4. The absorbance maximum (385 nm) and molar absorptivity (1.4×10^4) are much the same as those in aqueous medium. The uranium (VI)-Alizarin Red S-1,3-diphenylguanidinium ion-associate can be extracted at pH 5 with butanol.[18] Its probable structure is

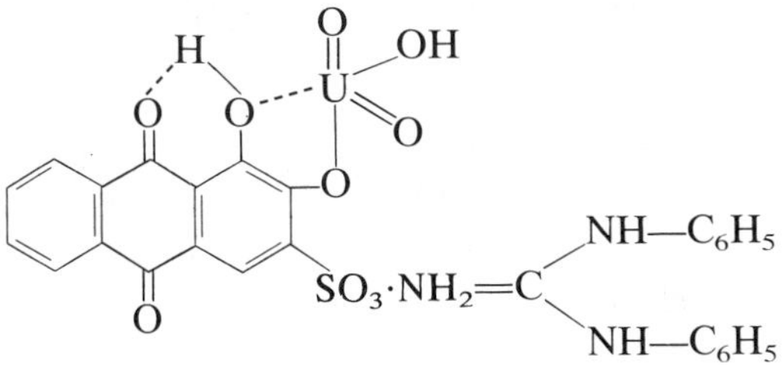

Alizarin Red S and alizarin chelates of germanium form chloroform-extractable associates with diphenylguanidinium ion at pH 6–7.[19]

Uranium has been determined by extracting the uranium(VI)–Arsenazo III–diphenylguanidinium associate into butanol.[20] EDTA is used as masking agent. The Indoferron complexes of Fe(III), Mo, Th, and Ti are extracted by chloroform or butanol in the presence of diphenyl-guanidinium ion.[21] Determination methods based on these reactions are unselective.

5. Amines of High Molecular Weight

A tertiary amine (Alamine 336-S) solution in chloroform has been used to extract a molybdenum–mercaptoacetate complex (ϵ at 370 nm = 1.95×10^4).[22] A stable colored compound is formed by combining a commercial polycyclic ketoamine with the aluminum–Eriochrome Cyanine R complex.[23] The absorbance maximum shifts toward a longer wavelength (bathochromic effect), and the molar absorptivity increases (hyperchromic effect); $\lambda_{max} = 595$ nm and $\epsilon = 1 \times 10^5$.

Various quaternary ammonium salts have been used to extract colored complexes or to increase sensitivity of color reactions in aqueous media (sensitized reactions). Some of the quaternary ammonium salts that have been used are

cetyltrimethylammonium bromide or chloride
hydroxydodecyltrimethylammonium bromide
tetradecyldimethylbenzylammonium chloride (Zephiramine)
dodecyloctylmethylbenzylammonium chloride
cetylpyridinium bromide
Aliquat 336-S

These salts are also cationic surfactants.[24] The system Sn(IV)–Catechol Violet–cetyltrimethylammonium bromide has already been discussed. Addition of cetyltrimethylammonium bromide or chloride to Al and Fe(III)–Eriochrome Cyanine R systems and to Al, Be, Fe, Ga, Th, and U-Chrome Azurol S systems brings both bathochromic and hyperchromic effects. Probably micelles of the surfactant play an important role in the formation of ion associates. However, both bathochromic and hyperchromic effects are observed in the Al and Fe–Eriochrome Cyanine R–cetyltrimethylammonium chloride systems with a surfactant concentration less than the critical micelle concentration.[25] The complexes formed are probably $Fe(ECR)_3(CAC)_3$ and $Al(ECR)_3(CAB)_3$.

Zephiramine has been used to extract many thiocyanate complexes and iodide complexes before photometry. Addition of Zephiramine to aqueous solutions containing Chrome Azurol S and metal ions produces both bathochromic and hyperchromic effects.[26] In the presence of cetylpyridinium bromide, the rare earths and nickel can be determined sensitively with Xylenol Orange.[27] Some metal–Xylenol Orange complexes are extractable by a solution of Aliquat 336-S in chloroform.[28]

B. BASIC DYES

1. General Aspects

Numerous dyes furnishing the ions R^+ or RH^+ form extractable ion associates with anionic metal complexes of the halides, of thiocyanate, and, less often, of nitrate, cyanide, organic acids, and chelating agents. Extraction of the ion associate with a suitable immiscible organic solvent separates it from excess dye. Ion associates of a few oxo anions, particularly ReO_4^-, are also easily extractable.[29] Hundreds of procedures involving basic dyes have been described. For the most part, they have been developed empirically. Comparative studies are few, and it is not always certain that the best dye has been selected for a particular determination. The survey following touches on some of the more commonly used reagents of this class. It is intended to be illustrative, not exhaustive.[30]

All these methods furnish high or very high sensitivity (ϵ of ion associate as large as 10^5). As we see, the basic dyes are quite selective in strong (1:1) HCl solution, reacting principally with Au(III), Sb(V), Ga, and Tl(III). At lower HCl concentrations, a few more metals react, for example, Hg(II) and Sn(IV), but these reactions are of lesser analytical value. If HBr and HI are used in place of HCl as complexing anions, a number of other metals react (In, Te, etc.). In HF medium, Ta can be determined sensitively as TaF_6^-, and B as BF_4^-. The single negative charge on all these ions is the predominant reason for the formation and extraction of their dye complexes.[31] In the presence of SCN^-, Zn can be extracted and determined, but many metals interfere. Uranium is determinable in the presence of excess benzoic acid or other organic acids or chelating agents forming $UO_2L_3^-$. The anionic chelate species formed by 8-hydroxyquinoline, salicylic acid, benzoic acid, and others with the alkaline earths, the rare earths, and others can be used for their determination by reaction with basic dyes to give ion associates; compare Rhodamine B. Some of these determinations can be made fluorimetrically as well as absorptiometrically with a large gain in sensitivity, if the dye itself is fluorescent (e.g., rhodamines). In general, xanthene dyes are fluorescent; triarylmethane dyes are not. The fluorescence of the former, as exemplified by the rhodamine dyes and their ion associates, is generally attributed to the rigidity of the respective molecules resulting from the O atom joining the two benzene rings, but coplanarity of the rings may be as important.

The choice of extracting solvent is an important factor in this class of determinations. Everything depends on finding a solvent and working under such conditions that the metal ion associate is largely extracted without simultaneous extraction of more than a trace of the excess dye and this be present in colorless form. Benzene and its analogs, sometimes mixed with a small fraction of an oxygenated solvent, are often suitable. The choice of organic solvent may depend on the particular dye used. See p. 667 for the extractability of Rhodamine B chloroantimonate by various immiscible solvents. The molar absorptivity of a cationic dye associate may vary considerably with the organic solvent. The reasons are not well understood. The molar absorptivity of Safranine O chloroaurate $(RAuCl_4)$ in various organic solvents as a function of dielectric constant is plotted in Fig. 6H-1. Liquids of low dielectric constant are seen to give solutions of low ϵ.[32]

Dyes of the triarylmethane series and the xanthene group (rhodamines and others) were the first to be used as photometric cationic reagents for metals in anionic forms and are still standard reagents, but other types of dyes have also been tried and found to be useful. Almost any dye convertible to an ion fulfills the requirement of large ionic volume. An essential requirement for the present purpose is a dye structure that leads to the formation of a singly charged, intensely colored (or fluorescent) organic cation. This cation should be capable of existence in fairly strongly acid aqueous solutions. Dyes of the triarylmethane and xanthene groups give such ions. However, the nitrogen atom of the amino and substituted amino groups of these dyes can be protonated, so that with increasing hydrogen-ion concentration, ions of charge greater than one

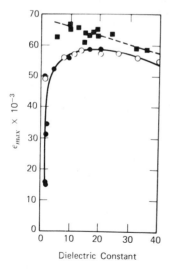

Fig. 6H-1. ϵ_{max} of Safranine O chloroaurate in various solvents according to Burgess, Fogg, and Burns; ■, alcohols (unspecified); ●, solvents other than alcohols (unspecified); ○, dioxane–H_2O mixtures. Reproduced with permission from The Chemical Society, from C. Burgess, A. G. Fogg, and D. T. Burns, *Analyst,* **98,** 607 (1973).

(for example, RH^{2+} and RH_2^{3+}) can be formed. These organic cations of higher charge are generally much less effective than singly charged cations in forming ion associates with metal complex anions.

It is important that the dye used be stable in fairly acid solutions. The ion associate with the metal complex should have a favorable extraction coefficient in some immiscible solvent of the inert type and there should be very little tendency for the excess dye to be extracted. (General expressions for extraction of ion associates are considered in Chapter 9C.)

Broadly, the cationic dyes are quite similar in their reactions with metals, but differences in the properties mentioned may make some dyes preferable to others. Substituents in the dye molecule play an important role. Attempts have been made, and continue to be made, to correlate the structure of the dye cation with properties that are analytically important, for example, stability of an ion associate (size of formation constant), its solubility, and absorbance or fluorescence. Many factors are involved, relations are complex, and definite correlations are difficult, though general trends may be apparent.

The π-electron density distribution in some basic dyes has been calculated by the molecular orbital method and the relationship between the structure of these dyes, and the stability of their associates with metal anions has been traced.[33] The chief type of interaction between dye cation and complex metal anion is believed to be dipole–dipole interaction; local dipole moment in the cation may be more important than the overall dipole moment. The polarizability of the electron shell of the dye cation is an important factor. The stability of an ion associate varies with the dipole moment of the dye cation, which depends on the electron density distribution, assessable by quantum-chemical calculations. However, the extractability of an ion associate depends on the organophilic properties of the dye as well as on the stability of the complex. Introduction of NO_2- and $CN-$ groups (strong electron acceptors) into positions of highest electron density in the cation should increase the stability of ion associates.

Purification of Basic Dyes. Commercial dyes are often very impure, containing inorganic salts (salting-out agents in precipitation) and foreign-colored organic substances that not only dilute the product but can interfere in procedures. Even products marketed as purified may leave much to be desired. Some of the conflicting results reported in the literature are likely to be due to the use of dyes having inadequate purity. Especially in proposed methods, dyes of established purity should be used.

Two generally applicable methods for purifying organic reagents can be applied to basic dyes—crystallization from solution and adsorption chromatography. Crystallization (recrystallization) can be brought about by cooling a hot solution, adding a miscible solvent in which the dye has a lower solubility than in the first solvent and by precipitation as a less soluble salt by adding a suitable anion. Usually, a basic dye will be available as the chloride, or can be converted to the chloride by dissolving in dilute HCl, which may be crystallizable by cooling a warm or hot filtered saturated solution (say 60°–70°). Rhodamine S can be purified in this way (p. 672). If the dye salt is only slightly soluble in aqueous solution (e.g., Rhodamine B in 1 M HCl has a solubility of 2.4×10^{-3} M), it may be dissolved in a suitable organic solvent, from which it may be crystallized by adding another organic solvent. Thus Rhodamine B chloride can be crystallized from ethanolic solution by adding ethyl ether. Crystallization of a dye from a hot organic solvent solution is also a possibility.[34]

Recrystallization may have to be repeated one or more times to obtain a product of satisfactory purity. Some impurities are difficult to eliminate by recrystallization, especially if they are less soluble than the dye itself. Even if freely soluble, an impurity may be concentrated in the solid phase instead of in the solution on crystallization if it forms mixed crystals, or more generally a solid solution, with the main component. However, more often than not, crystallization from hot solution or by change of solvent is an effective purification method and provides large amounts of purified reagent without the expenditure of much time or labor, though with considerable waste of dye at times. Dye perchlorates are usually much less soluble than the chlorides and can often be precipitated from solutions of the latter by addition of perchlorate ion. Brilliant Green has been purified in this way (p. 672). The presence of perchlorate in a solid organic compound raises doubts about its stability. Also, dye perchlorates may have very low solubility in aqueous solutions, and the presence of perchlorate may be objectionable as already noted.

For chromatographic purification, columns of silica gel and aluminum oxide are commonly used, and the dye is dissolved in an organic solvent such as alcohol or acetone. Specific directions are available for some dyes (Brilliant Green, for example). The dye is recovered from the eluate by evaporation of the solvent to crystallize the dye. Chromatography may give indication of the presence of colored impurities. Other types of chromatography—paper[35] and thin layer—may be useful for revealing the presence of colored impurities in a dye.

The purity of a dye can be evaluated more or less successfully from its chemical composition, usually by obtaining the content of some constituent such as chloride, which can be determined accurately enough to indicate significant deviation from stoichiometric composition. Titration of dye with Ti(III) (Knecht–Hibbert) can be useful. Precipitation with tungstosilicic acid is also recommended. Inorganic salts in a dye can be determined by destroying organic matter (ignition in air) and weighing the sulfated residue. If the molar absorptivity of the pure dye has been reported, comparison can be made with this value. A check on the possible presence of colored components other than the main constituent by chromatography can be valuable.

It has been reported that water-free dye samples are hard to obtain. Thus a 100% sample of Brilliant Green recrystallized from nonanhydrous acetone showed 93% of dye, even if dried at 110°. Such behavior is not universal. In our experience, Rhodamine B chloride recrystallized from ethanol–ether and air dried contains ~99% of the dye.

2. Rhodamine B[36] (Tetraethylrhodamine) (C. I. Basic Violet 10; C. I. 45170)

The usual form of the reagent is the chloride:

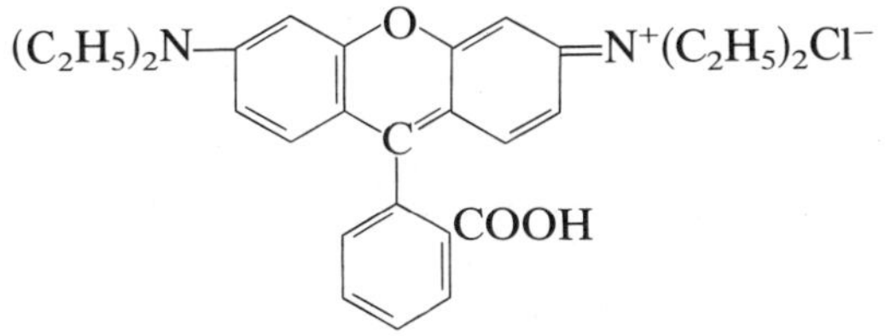

$(C_2H_5)_2N$ ... $=N^+(C_2H_5)_2Cl^-$

COOH

Tetraethylrhodaminium Chloride

Mol. wt. 479.0. Can be purified by adding 10 volumes of anhydrous ethyl ether to 1 volume of absolute ethanol saturated with the dye, and washing crystals with ether.

Equilibria. Rhodamine B equilibria in a benzene (or other immiscible organic solvent)-aqueous solution system may be represented as follows:

$R^°$ (benzene)

$\Updownarrow$

$R^°$ (aqueous)

$\Updownarrow$

$$(R^{+-})_2 \rightleftharpoons R^{+-} \rightleftharpoons RH^+ \rightleftharpoons RHX \rightleftharpoons (RHX)_2 \quad \text{(aqueous)}$$

violet   violet   violet   violet   violet

$\Updownarrow$

$$RH_2^{2+} \rightleftharpoons RH_2X^+ \rightleftharpoons RH_2X_2 \rightleftharpoons (RH_2X_2)_2 \quad \text{(aqueous)}$$

orange   orange   orange   orange

$\Updownarrow$

$$RH_3^{3+} \rightleftharpoons \text{etc.}$$

yellow

(X is Cl, Br, ClO$_4$, etc.). The constants for most of these equilibria have been determined.[37]

In benzene and ether solutions, Rhodamine B exists as the colorless lactone[38] form R°:

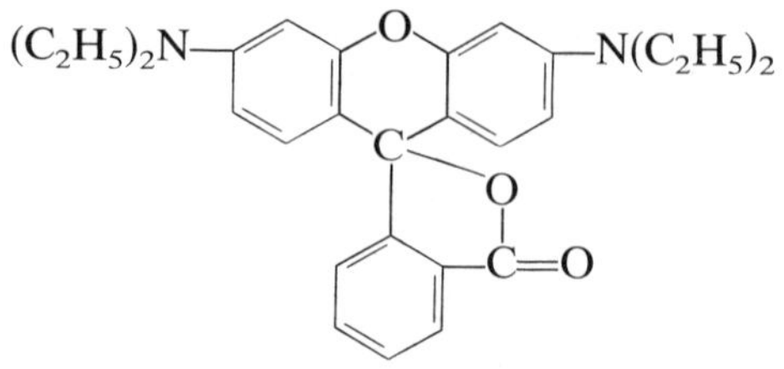

$(C_2H_5)_2N$———O———$N(C_2H_5)_2$
C
O
C=O

Little or no color would be expected, since only benzenoid rings are present. In a polar solvent such as alcohol, acetone, or water, the lactone ring opens and a zwitterion R^{+-} is formed. This is a resonance hybrid, one of whose structures is

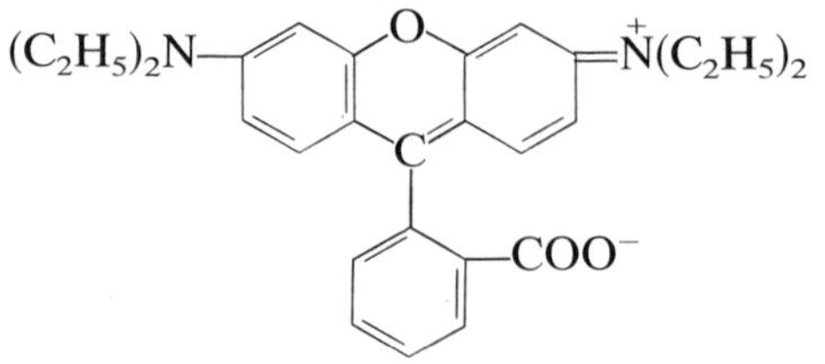

$(C_2H_5)_2N$———O———$=\overset{+}{N}(C_2H_5)_2$
C
COO$^-$

Because of the resonance possibilities, an intense color would be expected, and in fact the maximum molar absorptivity, at 553 nm, is 110,000. As the pH of an aqueous solution is lowered to about 3, a proton is added to the –COO$^-$ group and the form RH$^+$ is produced. The structure of this ion may be written in the same extreme forms as the zwitterion except for the carboxyl group. A shift of only 3 nm occurs in the absorption maximum (556 nm) and the molar absorptivity remains much the same.

As the acidity is increased further, a second proton adds to the cation, forming RH$_2^{2+}$. Since the second proton blocks resonance in one ring, the color of RH$_2^{2+}$ should be, and is, markedly different from that of RH$^+$. In concentrated sulfuric acid, a third proton adds, probably to the second nitrogen, and RH$_3^{3+}$ is formed. The cationic forms associate with anions, such as Cl$^-$, in the solution. Moreover, R^{+-}, RHCl and RH$_2$Cl$_2$ dimerize. The optical properties of these Rhodamine B species are summarized below:

Species	Color	Absorption Maximum (nm)	ϵ_{max}	pH Range	Fluorescence (Hg lamp)
$R°$	None	316	1.8×10^4		Weak blue
R^{+-}	Violet	553	1.1×10^5	>3	Strong yellow
RH^+	Violet	556	1.1×10^5	3 to 1	Strong yellow
RH_2^{2+}	Orange	494	1.5×10^4	0 to -1	Weak yellow[a]
RH_3^{3+}	Yellow	366	3.6×10^4	<-1	None
$(R^{+-})_2$	Violet	520			
$(RHCl)_2$	Violet	520			
$(RH_2Cl_2)_2$	Orange				

[a] Unless due to a very small amount of RH^+.

Absorbance curves of Rhodamine B solutions at various pH values are shown in Figs. 6H-2 and 6H-3. The curves of RH^+ and $RHCl$ are believed to be essentially identical, similarly those of RH_2^{2+} and RH_2Cl_2. Dimeric species have absorption curves distinctly different from those of the corresponding monomeric species (λ_{max} of dimers are shifted to lower wavelengths).

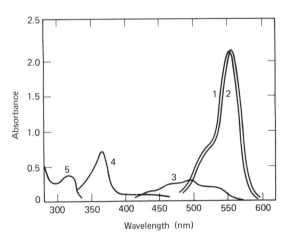

Fig. 6H-2. Absorption spectra of $1.0 \times 10^{-5} M$ Rhodamine B (cell path 2 cm). Curve 1, $0.03 M$ NH$_3$, $1.0 M$ KCl (R^{+-}); curve 2, $0.004 M$ HCl, $1.0 M$ KCl (RH^+); curve 3, 5 M HCl (RH_2^{2+}); curve 4, concentrated H$_2$SO$_4$ (RH_3^{3+}); curve 5, benzene ($R°$). In these dilute solutions, dimeric species can be neglected. Reprinted with permission from R. W. Ramette and E. B. Sandell, *J. Am. Chem. Soc.*, **78**, 4873 (1956). Copyright by the American Chemical Society.

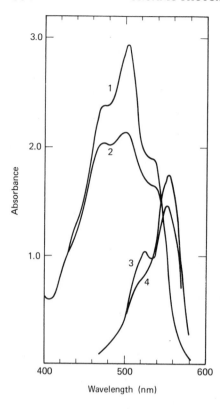

Absorbance

Wavelength (nm)

Fig. 6H-3. Absorption spectra of Rhodamine B showing effect of dye concentration. Curve 1, $2 \times 10^{-5} M$ dye, 4 M HCl (cell path 10.0 cm); curve 2, $2 \times 10^{-3} M$ dye, 4 M HCl (cell path 0.1 cm); curve 3, $2 \times 10^{-4} M$ dye, pH 8 (cell path 0.1 cm); curve 4, $2 \times 10^{-5} M$ dye, pH 8 (cell path 1 cm). Curve 1 represents ~95% of the monomer RHCl, curve 2 roughly equal concentrations of monomer and dimer (53% RH_2Cl_2, 47% $(RH_2Cl_2)_2$). Curve 3 represents approximately 85% R^{+-}–15% $(R^{+-})_2$, curve 4 essentially R^{+-}. Reprinted with permission from R. W. Ramette and E. B. Sandell, *J. Am. Chem. Soc.*, **78**, 4873 (1956). Copyright by the American Chemical Society.

At 25°, the values of various equilibrium constants in the Rhodamine B system are the following. These values refer to solutions that are 1.0 M in Cl^- (HCl + KCl), with the exception of K_2 and K_3 which are thermodynamic values.

$$P' = \frac{P}{Z} = \frac{[R°]_{C_6H_6}}{[R°]} \times \frac{[R°]}{[R^{+-}]} = \frac{[R°]_{C_6H_6}}{[R^{+-}]}$$

$$= 2.8 \times 10^3$$

(Concentrations without subscripts refer to aqueous phase.)

$$K_1 = \frac{[R^{+-}]^2}{[(R^{+-})_2]} = 1.0 \times 10^{-3}$$

$$K_2 = \frac{[R^{+-}][H^+]}{[RH^+]} = 6 \times 10^{-4}$$

$$K_3 = \frac{[RH^+][Cl^-]}{[RHCl]} = 3.4 \times 10^{-3}$$

$$K_4 = \frac{[RHCl]^2}{[(RHCl)_2]} = 1.9 \times 10^{-3}$$

$$K_5' = \frac{[RHCl][H^+][Cl^-]}{[RH_2Cl_2]} = 2.1 \times 10^{-1}$$

$$K_8 = \frac{[RH_2Cl]^2}{[(RH_2Cl_2)_2]} = 8.0 \times 10^{-4}$$

The equilibrium constants for

$$RH_2^{2+} \rightleftharpoons RH^+ + H^+$$

and

$$RH_3^{3+} \rightleftharpoons RH_2^{2+} + H^+$$

have not been determined, but the constant K_5' indicates that $RH^+ +$ RHCl predominates over $RH_2^{2+} + RH_2Cl^+ + RH_2Cl_2$ above pH 1 (Fig. 6H-4). R^{+-} can be neglected below pH 2. In other words, RH^+ and its anion associates are the important forms in the approximate acidity range 0.01–$0.1\ M\ H^+$. RH_3^{3+} and its anionic associates are important only in strongly acid solutions.

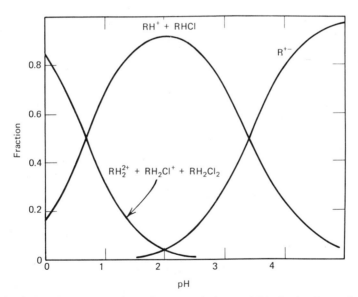

Fig. 6H-4. Relative concentrations of monomeric forms of Rhodamine B as a function of pH in 1.0 M Cl$^-$ solution. R. W. Ramette, Thesis.

Values of K_5' as a function of the chloride (LiCl) concentration are available (Ramette, Thesis):

$[Cl^-]$	K_5'
0.05	1.5
0.1	0.77
0.5	0.27
1.0	0.16
2.0	0.055
3.0	0.018
4.0	0.006
5.0	0.002

Dimeric Rhodamine B species can usually be neglected at analytical concentrations of 10^{-5} M or less.[39]

Addition of a polar solvent, which is characterized by high electronegativity of the OH group, to the colorless solution of Rhodamine B (lactone form) in a nonpolar solvent produces a violet solution having solvated R^{+-}:

$$R^{\circ} + nS \rightleftharpoons R^{+-}S_n$$

(S = polar solvent). Depending on the polar and nonpolar solvent, n varies from 1 to 3. Equilibrium constants for various polar–nonpolar systems have been obtained.[40] For example, for propanol–benzene, $n = 3$ and

$$K = \frac{[R^{+-}S_3]}{[R^{\circ}][S]^3} = 4.7 \times 10^{-3}$$

Chloroform alone is sufficiently polar (dipole moment = 1.06) to give a weakly colored solution when the reagent is extracted from neutral solution.

Metal Reactions in Chloride Solution. Rhodamine B is best known for its sensitive reactions with metals forming singly charged chloro anions MCl_{n+1}^-. The red-violet ion-pair complexes have the composition $(RH^+)-(MCl_{n+1}^-)$. They are extractable into benzene and similar solvents in which they can be determined absorptiometrically or fluorimetrically. The excess reagent is not extracted appreciably into nonpolar solvents from sufficiently acidic solutions. The chief metals determined in this way are Sb(V), Ga(III), Au(III), and Tl(III).[41] Iron(III) also reacts, as $FeCl_4^-$, and is partially extracted. Tungsten(VI) forms Rhodaminium B tungstate, insoluble in organic solvents. Molybdenum(VI) and tantalum react insensitively. Apparently, Sc can react.

The reactions with MCl^-_{n+1} are carried out at high Cl^- concentrations in general. Conversion of Sb(V) and Ga into the chloro complexes requires a high Cl^- concentration, and especially for Sb(V) a high acidity as well to minimize hydrolysis, so that these two metals are determined in ~6 M HCl. As already noted, acidities above ~0.1 M, and chloride, reduce the concentration of the reacting ion RH^+. Nevertheless, Sb(V) is extracted well from 6 M HCl and Ga moderately well, an indication of large values for the formation constant. Also the partition coefficient of $RHSbCl_6$ into benzene must have a large value; no doubt, 6 M HCl exerts a considerable salting-out effect.

For a stable chloro anion such as $AuCl_4^-$, a high Cl^- concentration is not needed, but procedures for Au call for moderately high HCl concentrations, for example, ~1 M. Higher HCl concentrations minimize reaction of some other metals. For example, Hg(II) interferes to some extent at relatively low chloride concentrations at which $HgCl_3^-$ is present. Its reactivity is less at higher Cl^- concentrations, where $HgCl_3^-$ is decreased in concentration, forming $HgCl_4^{2-}$. Another reason for making Rhodamine B extractions from moderately acid solutions is to prevent extraction of the lactone form of Rhodamine B, which extracts in increasing amounts as the acidity is decreased below 3 M H^+. On standing, there is a tendency for the lactone to react with HCl in the separated benzene phase to give the violet forms of the dye and the blank increases.

A few elements other than those mentioned in the preceding discussion have been determined in more dilute HCl solutions. Thus Sn(IV) can be determined in 2 M HCl if ethyl acetate is used as extractant.[42] Interfering elements include Hg(II), Sb, Zn, Bi, and Fe(III); there is little interference from Cu, Co, Pb, and Cd. Tellurium(IV) has been determined fluorimetrically by extraction of its Rhodamine B complex from 5–7% HCl (~1.7 M) into benzene–ether (2:1).[43] Presumably the reacting Te species is $TeCl_5^-$.

The effect of solvent on extractability of Rhodamine B ion associates can be illustrated with $(RH)SbCl_6$, which is extracted by benzene, ethyl benzene, toluene, xylene, chloroform, and isopropyl ether, without extraction of appreciable amounts of excess reagent from HCl solution. Ethyl ether and carbon tetrachloride do not extract $(RH)SbCl_6$. Isoamyl alcohol extracts reagent as well as complex.

Some organic solvent solutions of Rhodamine B ion associates are fluorescent[44] as are aqueous solutions of the dye itself (R^{+-} and RH^+ or RHCl show strong yellow fluorescence). In benzene, Rhodamine B as $R°$ shows only a weak blue fluorescence. In ethanol, Rhodamine B (RHCl) is reported to absorb maximally at 544 nm and emit maximally at 571 nm. Thallium(III), Ga, and Au(III) have been determined fluorimetrically after extraction from HCl solutions with Rhodamine B. These complexes show a yellow-orange fluorescence. We are not too well informed about the fluorescence of Rhodamine B association complexes in organic solvents. It is known that Rhodamine B chloroaurate in isopropyl ether is best determined fluorimetrically by excitation at 550 nm with measurement of fluorescence at 575 nm.[45] Probably other chloro complexes will have similar wavelengths of excitation and emission.

Reviews of Rhodamine B and other basic dye reactions are available.[46]

Other Rhodamine B Reactions. In the presence of HBr, Sn(IV) can be extracted into benzene, toluene, or xylene as the bromostannate of Rhodaminium B.[47] The extract is fluorescent. Other fluorescing metals include In, Tl(III), and W(VI); Cr(VI) and Au(III) produce negative interference. Thallium(III), In, and Hg(II) have also been determined as bromo complexes. Thallium can be determined in Zn or Cd metal without interference from In by extracting Tl(III) from 1 M HBr solution into isopropyl ether and then forming the Rhodamine B complex in this solvent. Bismuth can be determined absorptiometrically as BiI_4^-.

Zinc can be determined fluorimetrically by extraction into ethyl ether as $(RH)_2Zn(SCN)_4$ from weakly acidic solutions.[48]

Ion-association compounds of Rhodamine B and other basic dyes with complex halo-anions are the best known, but are not limited to these. Nitrate ion complexes a few metals sufficiently strongly to yield anionic nitrates, which form ion associates with cationic basic dyes. Uranium(VI) with nitrate and Rhodamine B gives an ion-pair compound having the composition 1 U:3 NO_3^-:1 RhB, which is extractable into benzene–acetone.[49] Presumably this is $[RH^+][UO_2(NO_3)_3^-]$.

Rhodamine B (and analogs) forms a germanomolybdate stable in HCl solutions, not extractable into the usual organic solvents.[50] With molybdophosphate it gives a fluorescent product extractable into chloroform–butanol.[51] Molybdosilicate can also be determined.[52] The complex $(RH)_4(SiMo_{12}O_{40})$ can be separated from excess reagent by flotation at a water–isopropyl ether interface and dissolved in ethanol, in which $\epsilon =$ 500,000, an usually high value, explained by 4 RH in the molecule.

Numerous chelating agents combine with various metals to form anionic complexes, which (especially when they have a single negative charge) form ion associates with Rhodamine B, or other basic dye cations, that can be extracted into inert organic solvents. For example, Rhodamine B and 5,7-dibromo-oxine extract the rare earth metals into benzene as $RH(ML_4)$.[53] Y and Th can also be extracted. Benzoic acid and other organic acids form anionic complex species with U(IV), extractable with Rhodamine B as cationic partner into benzene (strongly fluorescent solution).[54] Beryllium can be determined similarly.[55] A 1:1:1 complex is formed in basic solution by Ca (Sr, Mg), Rhodamine B, and 2-thenoyltrifluoroacetone, which is extractable into benzene.[56] Mg can be determined in the presence of Ca and Sr by masking these with oxalate. $\epsilon_{Mg} = 30,000$, $\epsilon_{Ca} = 56,000$. Phenylanthranilic acid or 5,7-dibromo-oxine with Rhodamine B form complexes with Mg, Ca, and Sr having the composition 3:1:1, extractable into benzene from basic solution.[57] These examples are only a small sampling of many to be found in the literature. The anionic metal–chelate species can be looked upon as being formed by ionization of a reagent adduct, $ML_2 \cdot HL \rightarrow ML_2 \cdot L^-$. The extractability of a metal should then depend on its ability to form adducts.

3. Other Rhodamines

Rhodamine 6G (C. I. Basic Red 1; C. I. 45160)

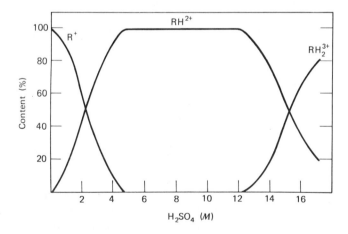

This is the ethyl ester of methylated[58] diethylrhodamine. A proton adds to the R^+ ion (chloride salt above is RCl) in relatively dilute acid solution and a second proton adds in strongly acid solution as shown in Fig. 6H-5.

Fig. 6H-5. Relative contents of R^+, RH^{2+}, and RH_2^{3+} in sulfuric acid solutions of Rhodamine 6G. After Zorov, Golovina, Alimarin, and Khvatkova, *loc. cit.*

The first protonation constants of Rhodamine 6 G and other rhodamines have been determined in sulfuric acid solutions[59]:

	K
Rhodamine 6G	0.08
Rhodamine 3GO	0.40
Rhodamine 4G	0.61
Rhodamine 3C	0.95

The composition of the last three rhodamines is the following:

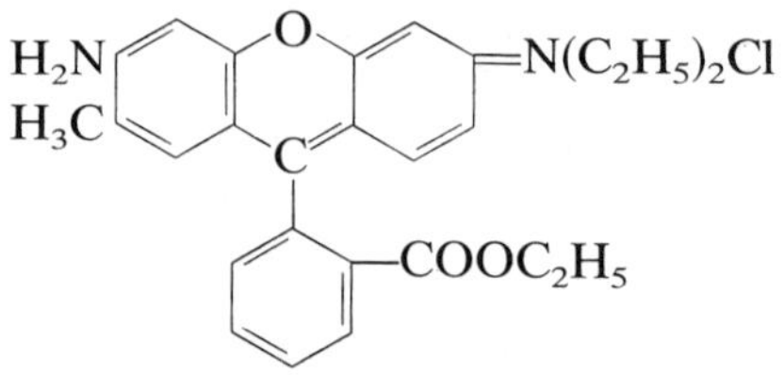

Rhodamine 3GO (C.I. Basic Red 4; C.I. 45215)

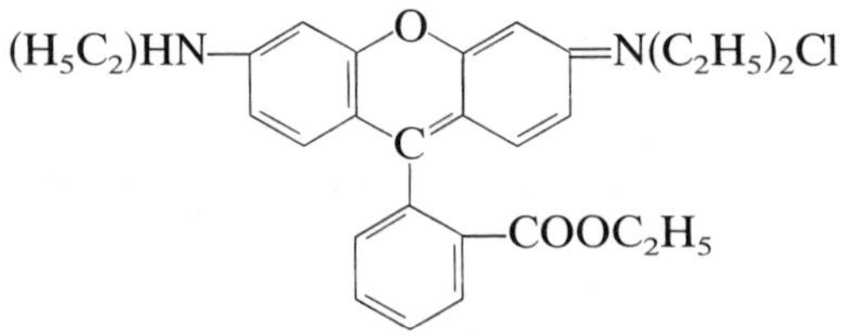

Rhodamine 4G (C. I. 45166)

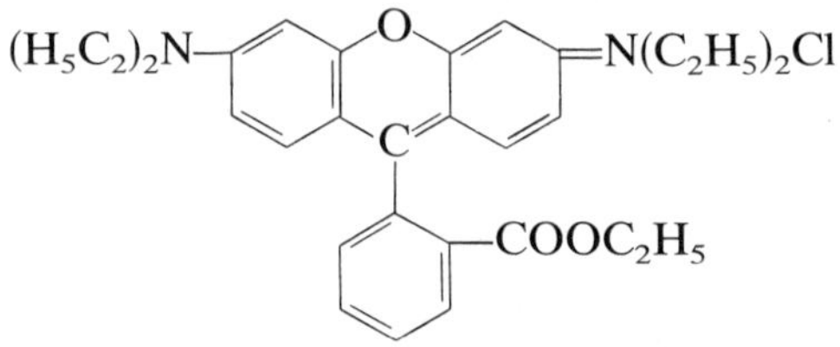

(ethyl ester of Rhodamine B)
Rhodamine 3C (Rhodamine 3B) (C. I. Basic Violet 11; C. I. 45175)

The reactions of Rhodamine 6G and its photometric properties are much the same as those of Rhodamine B. It does not seem to have any marked advantages over the latter in most determinations. The ester should be more hydrophobic than Rhodamine B. When this brings advantages, the butyl ester should be better than the ethyl ester. There is indeed a difference between Rhodamine B and its esters at very low acidities, where the former is present largely as R^{+-}, the ester as R^+ in aqueous solutions.

Rhodamine 6G has been used as a reagent for ReO_4^-, which is extracted into benzene (fluorescent solution) from phosphoric acid solution. It has also been used to determine Ga but is said to be less selective than Rhodamine B; the fluorimetric sensitivity is 5–10 times better than with Rhodamine B. Gallium can be determined fluorimetrically in the presence of much In by benzene extraction from chloride solution.[60] On the other hand, In can be determined fluorimetrically in the presence of Ga by benzene extraction from bromide solution. Te is determinable as $TeBr_6^{2-}$.

The singly charged *tris*-(5,7-dichloro-8-hydroxyquinoline)manganese(II) anion forms an ion pair with Rhodamine 6G cation which is extracted by benzene at pH 7.5 ($\epsilon_{540} = 70,000$).[61]

The butyl ester is also used as reagent in place of Rhodamine B, and is preferred by some workers.[62] It has found use in the determination of $GaCl_4^-$ (said to be more sensitive than Rhodamine B), $TlBr_4^-$, $InBr_4^-$, $TeBr_5^-$, $AuCl_4^-$, and ReO_4^-. Silver can be determined fluorimetrically after extraction into benzene from H_2SO_4–Br^- solution.

The formation of the fluotantalate ion-association complexes of Rhodamine 6G and of the ethyl (Rhodamine 3B) and butyl esters of tetraethylrhodamine has been studied.[63] The butyl ester and Rhodamine 3B give tantalum complexes extracted into benzene with favorable extraction coefficients (more favorable than with Crystal Violet). Under comparable conditions, the extent of extraction with the butyl ester of Rhodamine B, Rhodamine 3B, and Rhodamine 6G is 97, 98, and 79%, respectively. The Ta complex can be determined fluorimetrically as well as absorptiometrically. These determinations rest on the formation of the singly charged complex TaF_6^-. Niobium shows a smaller tendency to form an analogous complex, but conditions for its extraction have been established.[64]

The stability of benzene extracts of various complex ion associates of the ethyl and butyl esters of Rhodamine B is increased by the addition of acetone.[65]

In the presence of bromide, the butyl ester forms an ion-association complex with Te(IV), extractable into benzene–acetone.[66]

The butyl or ethyl ester and oxine give complexes with Ni, Zn, and U(VI) extractable into benzene from aqueous solutions of pH 6–9.[67]

Butylrhodamine B has also been used in the exchange determination of Au in benzene after extraction into this solvent as Crystal Violet chloroaurate.[68]

Rhodamine S (C. I. Basic Red 11; C. I. 45050)

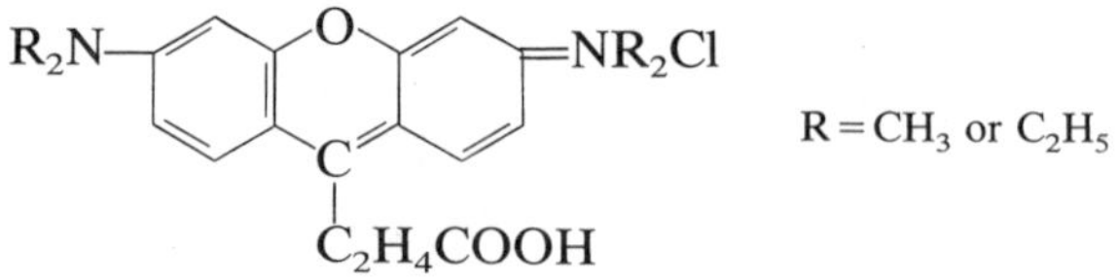

$R = CH_3$ or C_2H_5

This reagent (not a true rhodamine) is claimed to be superior to Rhodamine B for the determination of Sb.[69] Isopropyl ether is used as extractant; $\epsilon_{Sb} = 83,700$. The dye is purified by recrystallization. This reagent is also said to be superior to triphenylmethane dyes, which are stated to be unstable in solution and to yield unstable antimony complexes. Determination of Hg(II) in 0.2 M Cl^- by fluorimetry has been described.

In passing, it may be noted that compounds having structural similarity to the rhodamines react similarly to them with Sb(V), Ga, and others. The presence of strongly basic alkyl or dialkyl amino groups is a requisite, so that a high concentration of cationic dye can be formed. Flaveosine, in which bridging –O– of the rhodamines is replaced by –N=, but which otherwise is like Rhodamine B, extracts Ga (and presumably like-reacting elements) from 6 M HCl into benzene.[70] Likewise, thiopyronine, in which –S– takes the place of –O–, reacts with Ga. The *o*-carboxyphenyl group is not essential for reaction (it is not present in thiopyronine or Acridine Red, both reactive). 3,6-Diamino-9-(*o*-carboxyphenyl)acridine hydrochloride, having a primary amine group, is nonreactive:

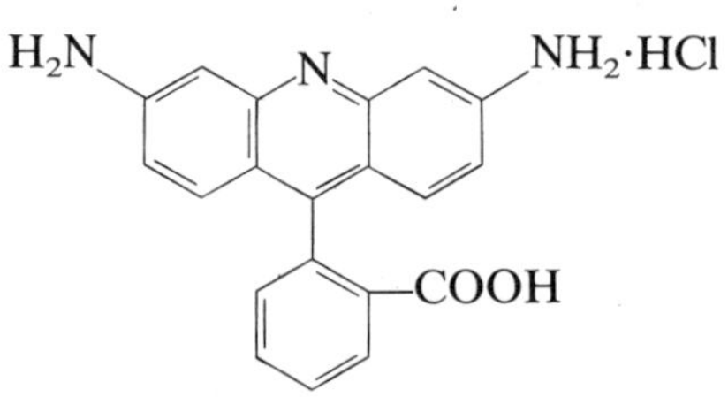

4. Brilliant Green (C. I. Basic Green 1; C. I. 42040)

The solid dye is usually in the form of the chloride, RCl. Commercial products are likely to have low purity and purification is required.[71]

In approximately neutral solution, the dye form R^+ predominates; in acid solution, RH^{2+}; and in alkaline, the colorless carbinol form ROH:

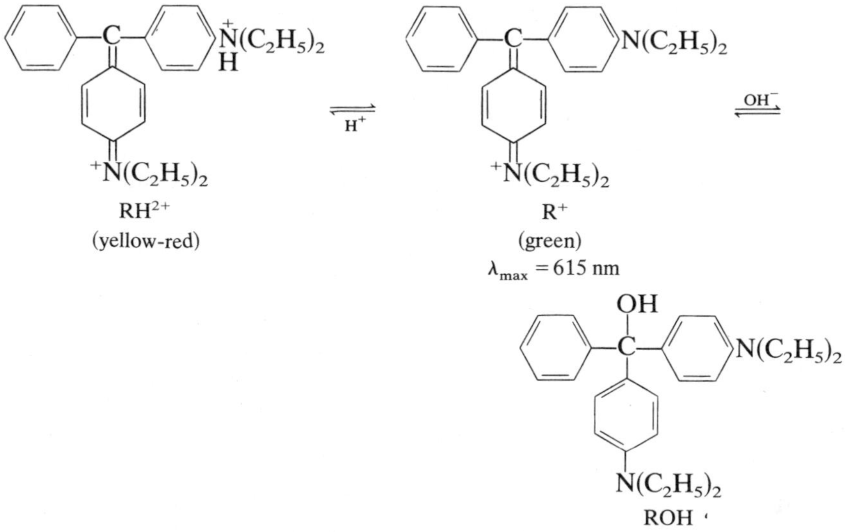

The equilibrium constant[72] for the reaction $R^+ + OH^- \rightleftharpoons ROH$ has been reported as 7.9. The formation constant of RH^+ apparently has not been determined.[73] The equilibrium between the forms is described as being slowly attained.[71]

The dye cation R^+, green in color, forms slightly soluble ion-association complexes with $SbCl_6^-$, $GaCl_4^-$, $TlCl_4^-$, $AuCl_4^-$, $CuCl_2^-$, AgI_2^-, TaF_6^-, ReO_4^-, and other singly charged complex anions. The ion pairs are extractable into various organic solvents, benzene and toluene often being used. The general behavior of Brilliant Green is much like that of Rhodamine B. The optimal range of HCl concentration for the extraction of chloro anions depends upon the metal. Often 1:1 HCl is satisfactory. Mercury(II) is extractable from HCl of lower concentrations as a 1:1 complex,[74] which must contain $HgCl_3^-$. Antimony(V) and Tl(III) can be determined satisfactorily in HCl solution with toluene as extractant ($\epsilon = 1 \times 10^5$). Gallium cannot be determined satisfactorily. Benzene or toluene does not extract $RGaCl_4$ well, and other solvents or mixtures (e.g., 1,1,1-trichloroethane) extract the reagent and give high blank values.[75] For the same reason, the determination of In as $InBr_4^-$ is unsatisfactory.[76]

Thallium(III) has been determined in the isopropyl ether extract obtained in extractive separation of $HTlCl_4$ from HCl solution (optimal HCl concentration in aqueous Brilliant Green solution 0.15 M).[77] $\epsilon_{630} = 1.06 \times 10^5$.

It has been found that the extractability of ReO_4^- depends on the length of time that the reagent remains in the buffered sample solution before benzene extraction is made (Fig. 6H-6). A pH of 6 is optimal. Extractability from a solution in which the dye forms are in equilibrium clearly parallels the curve in Fig. 6H-6a giving the proportion of R^+ at equilibrium. Almost complete extraction is obtained over a wider pH range (2–8 approximately) when the extraction is made within a minute after adding the alcoholic Brilliant Green solution. This behavior is explained by the slow conversion of R^+ to RH^{2+} and ROH.

The extraction of $AuCl_4^-$ (from 0.5 M HCl with toluene) should likewise be made immediately after the addition of Brilliant Green and a purified product (>96%) should be used; a double extraction is best made, with further addition of reagent. The purity of the reagent is less important when the extraction is made from approximately neutral solution (ReO_4^-).

Brilliant Green is better than Malachite Green for the determination of Re because of greater color stability of the complex and greater sensitivity.[78]

Tantalum can be determined as TaF_6^-. Chloroform is a better extractant than benzene.[79]

The anionic species formed by reaction of Ge with 3,5-dinitrocatechol is extractable into CCl_4 with Brilliant Green and other basic dye cations.[80] No doubt many other organic-metal anions show similar behavior.

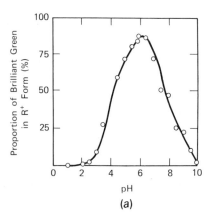 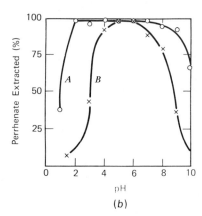

(a)   (b)

Fig. 6H-6. a. Variation of proportion of R^+ with pH (after 5 hr) in Brilliant Green solutions.

b. Extractability of ReO_4^- as a function of pH: A, after 1 min; B, after 5 hr. Reproduced with permission from The Chemical Society, from A. G. Fogg, C. Burgess, and D. T. Burns, *Analyst*, **95**, 1013 (1970).

Silver has been determined in cyanide solution (as $Ag(CN)_2^-$).[81] Brilliant Green as well as other triphenylmethane dyes form ion pairs with iodotellurite anion (said to be TeI_6^{2-}), and these can be used to determine Te. Silver has been determined by extracting $R[AgI_2]$ into benzene.[82] $\mathscr{S}_{640} = 0.002 \; \mu g \, Ag/cm^2$.

Malachite Green (C. I. Basic Green 4; C. I. 42000) is the tetramethyl homolog of Brilliant Green and gives the same reactions but is perhaps less often used as reagent. It has been used to determine Ga in HCl solution. It provides a sensitive determination of tantalum as fluotantalate in the presence of Nb ($\mathscr{S} = 0.002 \; \mu g \, Ta/cm^2$). Malachite Green as well as other triphenylmethane dyes have been proposed for the determination of iodotellurite. Gold as $Au(CN)_2^-$ has been determined by ether extraction of Malachite Green dicyanoaurate(I).[83] Tin has been determined in thiocyanate–sulfuric acid solution and also in HCl solution as $SnCl_3^-$.[84] Determination of In in Br^- and I^- solutions has also been described. Ga does not react in I^- solution (CCl_4). With benzoic acid, Malachite Green gives a sensitive reaction with U(VI); $\epsilon_{630} = 130,000$ in $CHCl_3$ + benzene.

Zinc is extracted into benzene and other solvents from neutral thiocyanate solutions as $R_2Zn(SCN)_4$ ($\epsilon_{630} \sim 100,000$).[85] Ni, Mn, Al, Cr(III), Fe(II), and other metals not readily forming anionic thiocyanates can be present in considerable amounts, and even Cd and Fe(III) are permissible in 100:1 Zn ratio. Cu, Pb, and Au(III) can be masked with thiourea.

Malachite Green, unlike the structurally similar Rhodamine B, is not fluorescent. The difference in behavior of the two compounds has been attributed to the greater molecular rigidity of Rhodamine B, but coplanarity of its rings could also be the explanation.

5. Crystal Violet (C. I. Basic Violet 3; C. I. 42555)

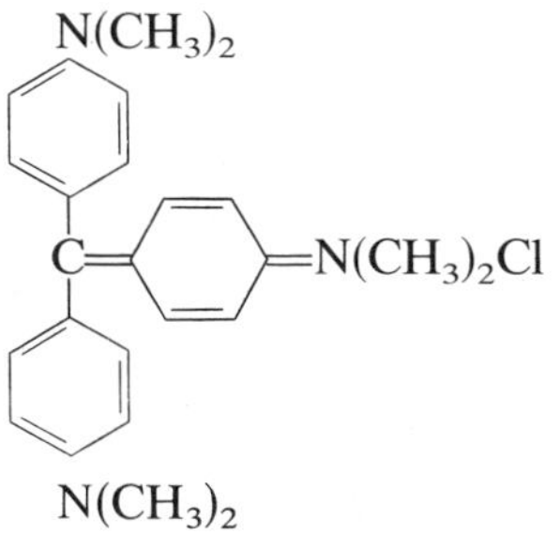

Mol. wt. 408.0. M. p. 194°–196°.

The presence of six CH_3 groups in this molecule results in a stronger color than with less highly substituted derivatives. In water, $\lambda_{max} = 591$, 541 nm. In concentrated sulfuric acid, two of the dimethylamino groups are protonated (orange-brown solution, $\lambda_{max} \sim 425$ nm):

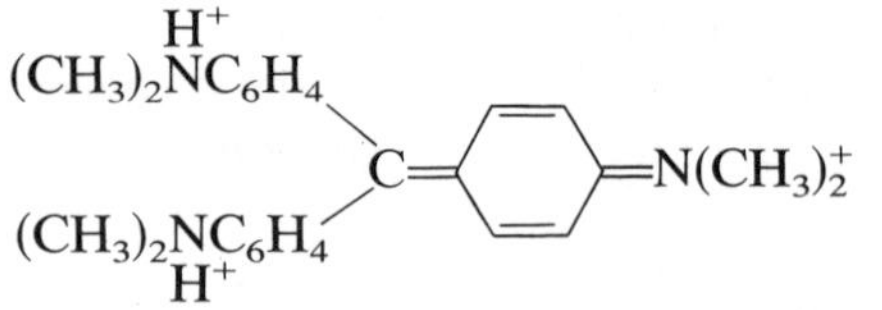

$$(CH_3)_2\overset{H^+}{N}C_6H_4$$
$$(CH_3)_2\underset{H^+}{N}C_6H_4$$
$$C=\hspace{-4pt}\big\langle\hspace{-6pt}=\hspace{-6pt}\big\rangle\hspace{-4pt}=N(CH_3)_2^+$$

Crystal Violet has been used for the determination of Hg (in the presence of Cl⁻, Br⁻, or I⁻), Au (+I⁻ or Br⁻) and Tl(III) (+Cl⁻ or Br⁻).[86] It has also found application in the determination of Sb, Re, and U. In the presence of thiocyanate, Ru and Os form insoluble ion associates with Crystal Violet, determinable photometrically as suspensions. Platinum(IV) reacts similarly, forming $R_2Pt(SCN)_6$. Molybdenum(V) thiocyanate anion forms an insoluble ion associate with Crystal Violet, which can be gathered in toluene, followed by solution in ethanol for photometric determination.[87] Crystal Violet also reacts with Sn(as $SnCl_3^-$) but seems to be a poorer reagent than Malachite Green.

In the presence of benzoic acid, Crystal Violet forms the complexes $R[UO_2(C_6H_5COO)_3]$ and $R_2[(ReO_4)(C_6H_5COO)]$, which are extractable into toluene at pH 4.1; at pH 2.5, only the Re complex is extracted.[88]

Ta is determined sensitively as TaF_6^- with Crystal Violet; chlorobenzene is the preferred extractant.[89] A comparative study of various triphenylmethane dyes for determination of TaF_6^- has been made.[90] With Crystal Violet, Malachite Green, and Brilliant Green as reagents, 400 times as much Nb as Ta may be present in the determination of the latter (benzene as extractant).

Crystal Violet has also found use in the extractive determination of molybdophosphate in the presence of Si and As.[91]

The extraction of alkylbenzenesulfonate perchlorate and periodate ions has been studied.[92] Significant extraction of dye occurs in 0.1 M NaCl solution. Polyvalent ions are not extracted.

Sulfite and cyanide react with Crystal Violet (color weakened or destroyed).

The nature of associative and aggregative processes in Crystal Violet solutions has been examined.[93]

6. Methyl Violet (C. I. Basic Violet 1)

The commercial product is a mixture of the hydrochlorides of the more highly methylated pararosanilines, containing principally N-tetra-, penta-, and hexamethyl derivatives. Determination of Au, Ga, Re,[94] Sb, Ta[95], and Tl with this reagent has been described, but other basic dyes are usually preferred to Methyl Violet. Ethyl Violet has also found use as reagent.

Rosaniline (Fuchsin) (C. I. Basic Violet 14; C. I. 42510)

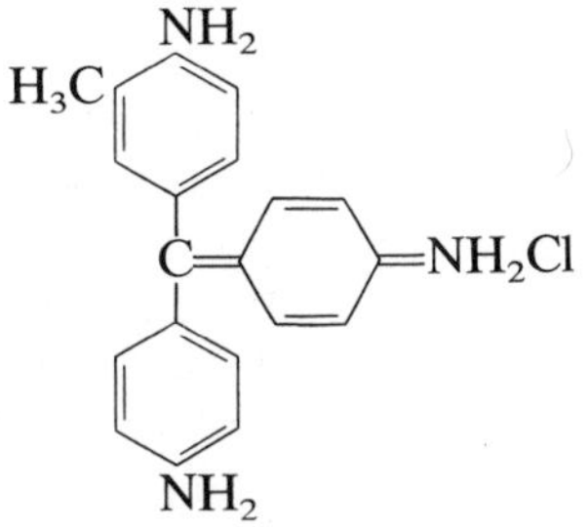

Mol. wt. 337.9.

as the chloride, gives a red solution in water (λ_{max} = 547 and 499 nm). In approximately neutral solution, the reagent cation forms an ion-pair compound with ReO_4^-, which is extractable into chloroform and oxonium-type solvents and serves as a photometric determination form for Re. Rosaniline has also been used to determine Au and Tl.

7. Methyl Green (C. I. Basic Green 5; C. I. 42590)

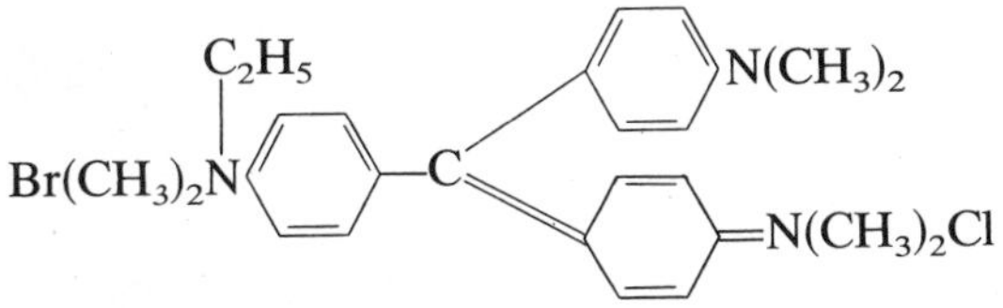

A sensitive reagent for ReO_4^- at pH 4.4 (ϵ = 1 × 10⁵). It has also been proposed for the determination of Bi as the anionic iodide, with benzene-nitrobenzene as extractant; Au, Te, and Hg interfere.[96] Like other basic dyes it can be, and has been, used to determine Au, Ga, Sb, Ta (and B), and Tl. Mercury(II) can be determined in 0.1 M Br⁻ solution (ϵ = 100,000), but the reaction is not selective. It is of interest that the benzene-extracted complex has the composition 1 Hg:2 dye,[97] namely $RHgBr_3 \cdot RBr$.

8. Victoria Blue B (C. I. Basic Blue 26; C. I. 44045)

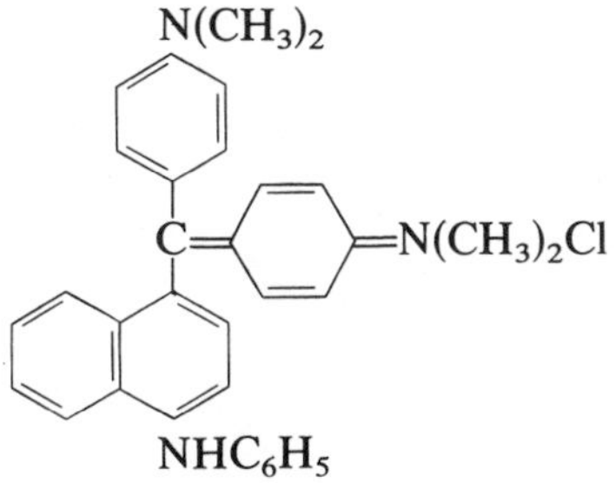

Above pH ~1, the dye exists as blue R^+ in water ($\lambda_{max} = 635$ nm) and at lower pH as RH^{2+} (yellow, $\lambda_{max} = 475$ nm).

Anionic Hg(II) complexes of the type HgX_3^- ($X = Cl^-$, Br^-, I^-, SCN^-) give benzene-extractable ion-pair compounds with Victoria Blue B cation, R^+. Iodide and bromide provide the best sensitivity ($\epsilon_{RHgI_3} = 78,000$).[98] Zinc can be determined as the anionic thiocyanate at pH 7, with benzene containing TBP as the extractant. Au(III) and Pb especially interfere, but they can be masked with thiourea. Indium is determinable sensitively ($\epsilon = 120,000$ in benzene at 630 nm) as $InBr_4^-$. This reagent has also been used to determine Ta as fluotantalate in ~2 M H_2SO_4. The benzene-extractable complex is said to be $(RH^{2+})(TaF_7^{2-})$. Mercury(II) and Th have been reported to interfere, but Nb in small amounts does not.

Victoria Blue 4R (C. I. Basic Blue 8; C. I. 42563)

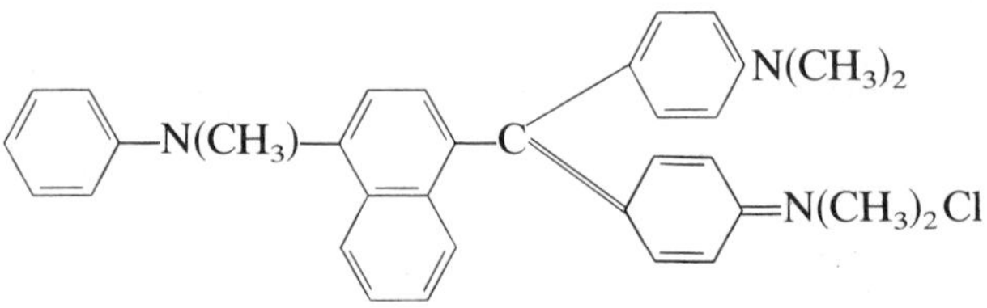

has found use in the determination of bromotellurite ($\epsilon = 80,000$ in benzene). In and Hg interfere.[99] It is also a reagent for Re in dilute H_2SO_4 solution, and for $AuCl_4^-$ and $SbCl_6^-$.

9. Meldola Blue (C. I. Basic Blue 6; C. I. 51175) and Related Dyes

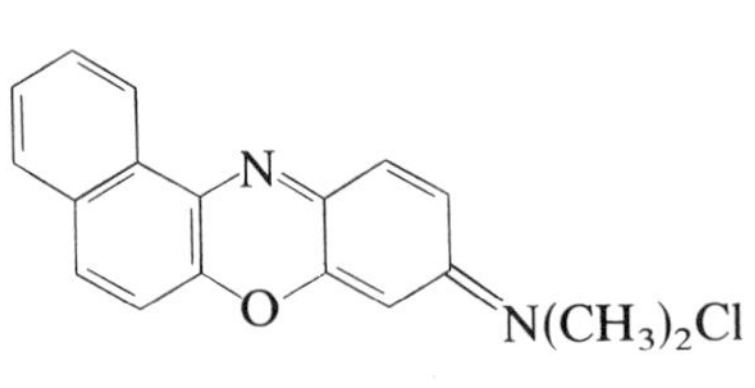

In F^- solution, Meldola Blue is a reagent for Ta.[100] Chlorobenzene is used as extractant for $RTaF_6$. Apparently, Nb does not have much effect, since the method is used to determine Ta in Nb. In Br^- solution, a reagent for $TlBr_5^{2-}$.

Nile Blue is an oxazine dye, similar to Meldola Blue, that has been used to determine B as BF_4^-, Tl as $TlCl_4^-$, and Ta as TaF_6^- and probably could be used for other complex anions having a single charge.

Another oxazine dye, Capri Blue (C. I. 51015)

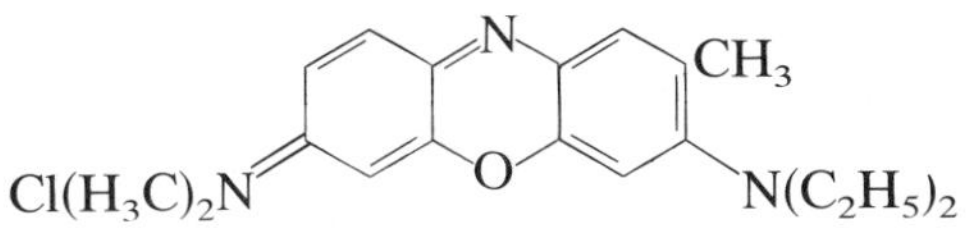

also has been used for the determination of Ta,[101] with chloroform as extractant.

10. Safranine T (or O) (C. I. Basic Red 2; C. I. 50240)

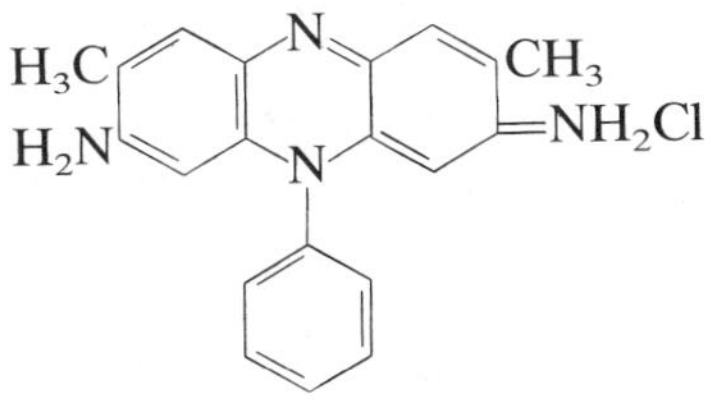

This compound, belonging to the phenylphenazonium group of dyes, has been used for the determination of Sb,[102] and can be applied to the determination of other singly charged chloro anions. $RSbCl_6$ is formed in 3.5 M HCl and $R_2H(SbCl_6)_3$ in 1 M HCl according to Pilipenko and Shinh. Benzene is a suitable extractant. Thallium(III) can be determined by benzene extraction from bromide solution in the presence of up to 5 mg of Ga or In.[103]

At low acidities ($<1\ M\,H^+$), the unprotonated form of the dye R^+ predominates in solution. This form has a red color, $\lambda_{max} = 520$ nm. As the acidity is increased, one of the amino nitrogens is protonated and RH^{2+} forms.

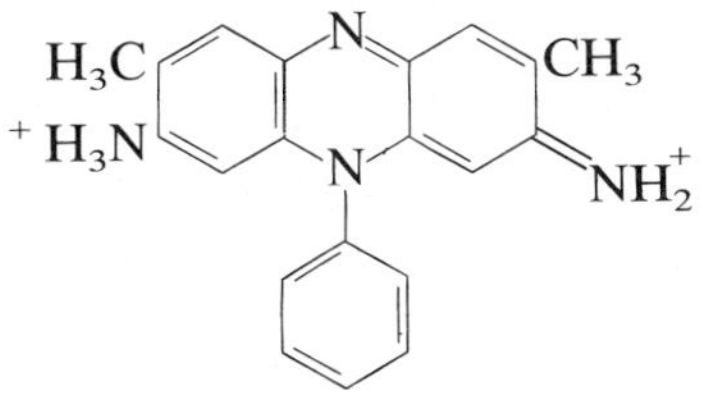

Blue, $\lambda_{max} = 593$ nm.

The relative content of RH^{2+} reaches a maximum in ~7 M HCl and decreases as the acidity is further increased, as a consequence of formation of RH_2^{3+}:

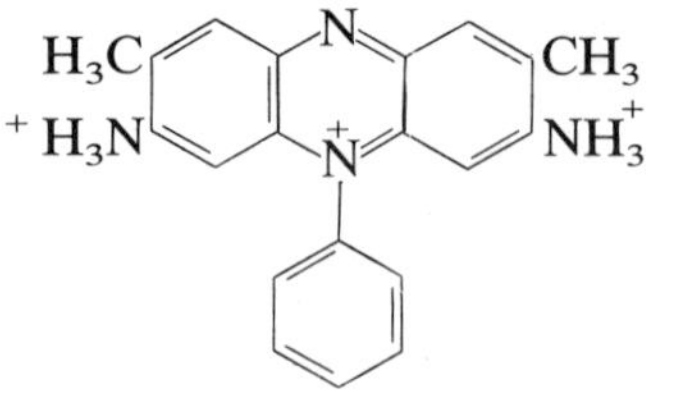

Colorless

(Burgess et al).[102] The protonation constants are reported[104] as

$$R^+ + H^+ \rightleftharpoons RH^{2+} \qquad pK_1 = -0.42$$
$$RH^{2+} + H^+ \rightleftharpoons RH_2^{3+} \qquad pK_2 = -3.96$$

Benzene is the best extractant in the antimony-determination (less interference from other ions). Toluene gives extracts not stable with time. Safranine T chloroantimonate extracted with benzene from 2 M HCl gives only weak fluorescence. However, addition of acetone (30% of the resultant mixture) to the benzene extract greatly increases the fluorescence intensity and fluorimetric determination of antimony is then possible.[105]

$RAuCl_4$ in various solvents shows λ_{max} (522–538 nm) similar to λ_{max} of R^+ in water, but ϵ_{max} varies considerably.

11. Methylene Blue (C. I. Basic Blue 9; C. I. 52015)

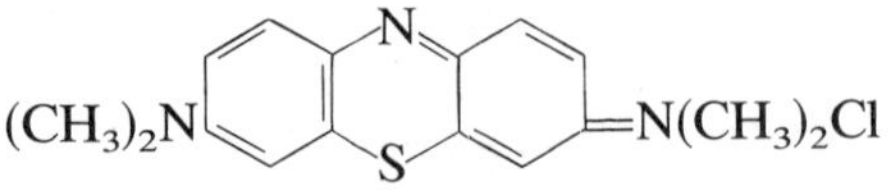

A reagent for Re(VII).[106] Dichloroethane is used as extractant for $RReO_4$ from H_2SO_4 solution. Much iron and moderate amounts of B, Mo, W, and Ta do not interfere. Methylene Blue has also been used to determine Au, B, Ga, Hg, In, Sb, Ta, Tl,[107] and U by methods similar to those with other basic dyes. Palladium can be determined unselectively in weakly acidic solution by chloroform extraction of its anionic azide complex with Methylene Blue as cationic partner.[108]

The determination of zinc by precipitation as $R_2Zn(SCN)_4$, resulting in a decrease in the absorbance of a Methylene Blue solution, has been described.[109]

12. Acridine Orange (C. I. Solvent Orange 15)

The dye cation forms a perrhenate extractable into 1,2-dichloroethane (fluorescent solution).[110] Molybdenum (100 mg) and small amounts of W

(8 μg) and V do not interfere with the fluorimetric determination of 1 μg Re. Chloro complexes of Au(III), Hg(II), Sb(V), and Tl(III) are also extracted.

13. Janus Green B (C. I. 11050)

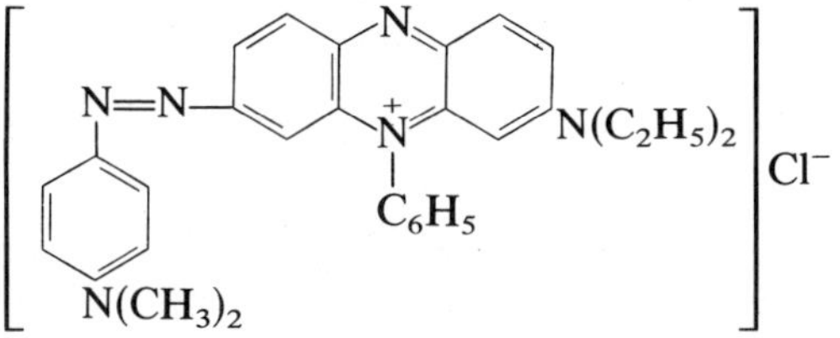

The singly charged dye cation is a reagent for $AuCl_4^-$ or $AuBr_4^-$, benzene–acetone as extractant.[111] It has also been used to determine $TeBr_6^{2-}$ as a blue 1 Te:2 dye complex in 1.4 M HBr, extractable into benzene–acetone.[112]

14. Antipyrine Dyes

These are triphenylmethane analogs in which one of the phenyl groups is replaced by the antipyrine radical. They have the composition

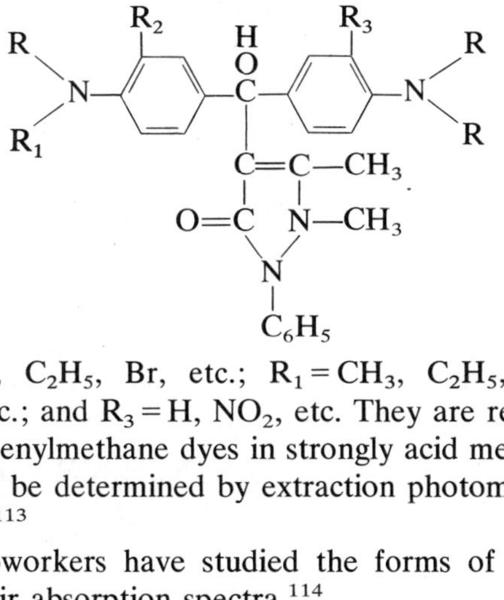

where R = CH_3, C_2H_5, Br, etc.; R_1 = CH_3, C_2H_5, CH_2–C_6H_5, etc.; R_2 = H, NO_2, etc.; and R_3 = H, NO_2, etc. They are reported to be more stable than triphenylmethane dyes in strongly acid medium. No less than 12 elements can be determined by extraction photometric methods with antipyrine dyes.[113]

Busev and coworkers have studied the forms of antipyrine dyes in solution and their absorption spectra.[114]

Chromepyrazole I

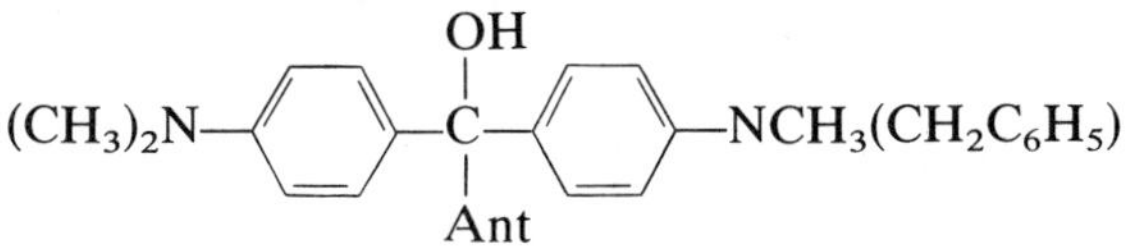

(Ant = antipyrine radical = antipyrinyl)

in aqueous solution at pH 2, gives a spectrum similar to that of Malachite Green, with the main absorption band at ~600 nm and subsidiary bands at 480 and ~320 nm. A toluene extract of the dye (chloride form) shows the major absorption peak at 560 nm. Addition of a proton gives

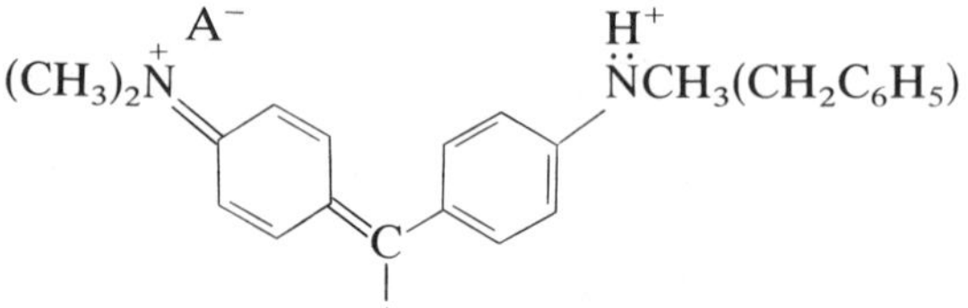

$$5\,M\;H_2SO_4$$
$$\lambda_{max} = 490\;nm,$$
$$\epsilon \sim 2 \times 10^4.$$

(A⁻ = anion)

A second proton is added to the other amino group in concentrated H_2SO_4, and the absorption peak is shifted to 430 nm. The protonated species are not extracted by nonpolar solvents.

Ethyl ether extracts colorless carbinol base ($\lambda_{max} = 260$ nm) from aqueous solutions of pH 11–12

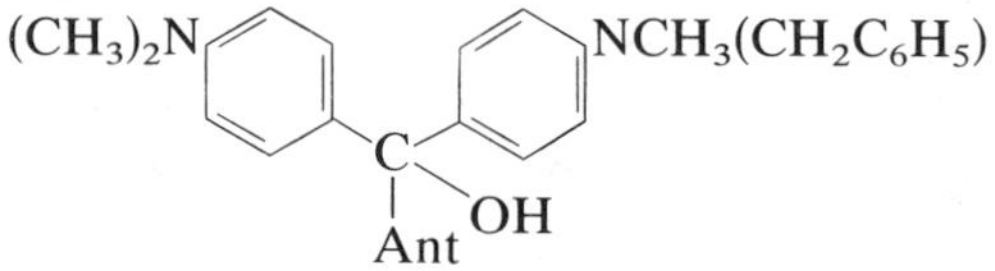

From acidic solutions (pH 1–2), ether extracts another colorless form ($\lambda_{max} = 340$ nm), which is believed to be an oxonium salt

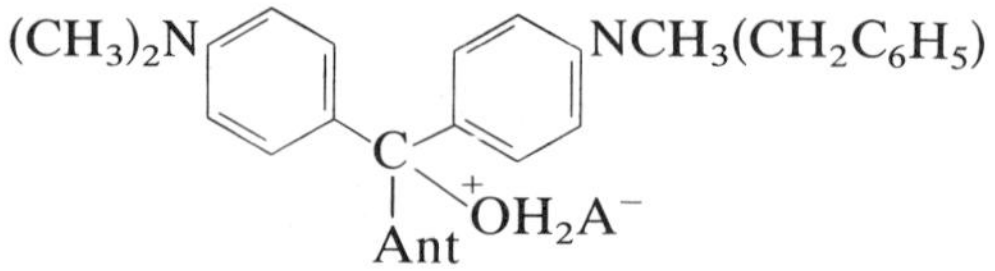

Ether extracts colored base (absorption peaks in the visible range at ~400 and ~480 nm, but main peak at ~260 nm) from strongly alkaline solutions, which is tentatively formulated as

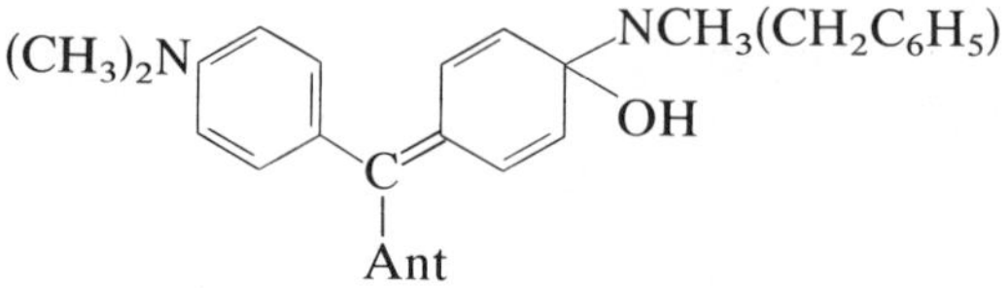

After several hours, the ether solution becomes colorless. This change is attributed to isomerization, which gives the colorless carbinol base represented above.

Antipyrine dyes have found their chief use in the photometric determination of Tl, Ga, Sb, and Cd. Some half-dozen of these dyes have been examined as reagents for Sb.[115] The most selective was 4,4'-bis(dimethylamino)-3'-nitrodiphenylantipyrinylcarbinol. The blue 1:1 chloroantimonate complex is extracted into benzene from 1 M HCl solution. 4,4'-Bis(methylbenzylamino)phenylantipyrinylcarbinol reacts with Fe(III), Cd(II), Sn(IV), Zn, Pb, and Bi in 1 M HCl solution to give blue or violet precipitates, not extracted by toluene; 4,4'-bis(dimethylaminophenyl)antipyrinylcarbinol reacts only with Sn(IV) among these.

A number of antipyrine dyes have been examined as reagents for Cd in bromide solution.[116] The slightly soluble products are stated to have the composition R_2CdBr_4. The effect of substituents in 4,4'-bis(dimethylaminophenyl)antipyrinylcarbinol (R_2 and $R_3 = H$) on the precipitation sensitivity of Cd has been studied particularly.[116] Introduction of the electronegative group $-NO_2$ increases sensitivity up to a point. The presence of a large number of electronegative groups weakens the basic properties of the dye, and it then no longer readily forms ion associates. An increase in sensitivity resulting from the introduction of strongly negative groups is evident only in weakly acidic solutions. Introduction of benzyl or ethyl groups has little effect on the basic character of the dye, but makes it, and more especially its ion-association complex, more hydrophobic, that is, less soluble in water, and thus more sensitive. The increase in sensitivity thus brought about remains much the same in strongly acid solutions.

15. Some Miscellaneous Dyes

3-(4-*Diethylaminophenylazo*)-1,4-*dimethyl*-1,2,4-*triazolium chloride*

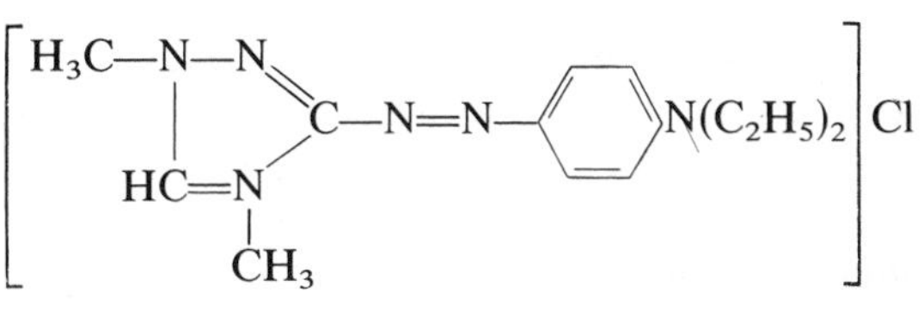

has been used for extraction-photometric determination of Sb(V),[117] Bi (from H_2SO_4–KI solution),[118] In,[119] and Tl(III).[120]

6-*Methoxy*-3-*methyl*-2-[4-(*N*-*methylanilino*)*phenylazo*]*benzo*-*thiazolium chloride*

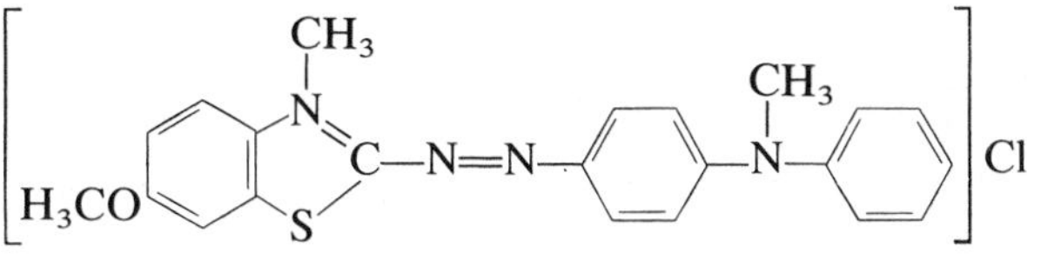

gives blue ion associates with $FeCl_4^-$,[121] $HgBr_3^-$,[122] $SbCl_6^-$,[123] and $Zn(SCN)_4^{2-}$,[124] which can be extracted by organic solvents. Sensitivities are high (ϵ for Fe = 70,600 at 640 nm and for Hg = 75,000 at 642 nm).

III. ANIONIC REAGENTS

Compared to anionic (and uncharged) chelated metal species, the cationic are fewer and generally less important, and so anionic reagents that allow their extraction play a less important role in analysis than the cationic reagents discussed in the preceding section. As already noted, the formation of an ion associate is usually not accompanied by a marked color change (though there are striking exceptions), so that the extractive determination of a cationic species usually requires the extraction of a colored (or fluorescent) associate under conditions that do not permit the extraction of appreciable amounts of the anionic partner if this is colored. Of course, chelated cationic species can be determined without extraction if the complex has a color different from (or more intense than) the reactants. However, extraction may be advantageous because of the increase in sensitivity or selectivity that it brings.

One type of cationic metal chelate is formed by the combination of an uncharged organic molecule with a simple metal ion.[125] The cation complex may be colored. For example, colorless 1,10-phenanthroline combines with weakly colored Fe^{2+} to give red $FePhen_3^{2+}$, which is extracted into amyl alcohol and other oxygenated solvents as Fe Phen$_3$ X$_2$ in the presence of such colorless, singly charged anions as Cl^-, Br^-, I^-, and ClO_4^-. Other cationic chelates of this type may not be extracted as easily as this compound and may require the addition of large hydrophobic and organophilic anions (SCN^-, salicylate, etc.). A solvent of high dielectric constant, such as nitrobenzene, is sometimes used as extractant, alone or in admixture. On the other hand, $CuPy_2(SCN)_2$ is extractable into $CHCl_3$. The presence of pyridine in the cation and thiocyanate in the anion makes this ion associate markedly organophilic. Sulfate and phosphate are usually much less effective than singly charged anions as counter ions for the extraction of cationic metal chelates, because of their greater charge and hydration.

There is a possibility of cationic chelate species being formed as the result of steric effects in the combination of a metal ion with chelating anions. For example, Sn(IV), having a coordination number of 6, cannot form an uncharged chelate by combining with four singly charged bidentate chelating anions. Thus $SnOx_4$ (HOx = 8-hydroxyquinoline) is not known, but the mixed chelate $SnOx_2Cl_2$ can be prepared. In the latter compound, the deficiency of Ox^- is made up by Cl^-. Since Cl^- presumably occupies coordination positions of tin, it is unlikely to be an ion associate, $(SnOx_2)^{2+}Cl_2^{2-}$. However, it is conceivable that some mixed complexes can have an ionic character. Mixed complexes having suitable co-anionic components are extractable into various organic solvents. $SnOx_2Cl_2$ is extractable into $CHCl_3$. If chloride is replaced by acetate and other organic acid anions, the resulting molecule can be extracted more easily into organic solvents. The strongly colored red-violet complex of diphenylcarbazide with chromium, $Cr(HL)_2^+$ ($H_2L =$ diphenylcarbazide), is extractable into chloroform in the presence of a long chain alkyl sulfate.

Steric hindrance may prevent arrangement of a sufficient number of ligands around a cation to neutralize its charge. It is believed that this is the reason for the failure of dimethylglyoxime (H_2Dx) to give a precipitate with Co(III). However, brown $Co(HDx)_2(H_2O)_2^+$ is formed and can be extracted into isoamyl alcohol–chloroform after addition of anions of large ionic volume, especially if H_2O is replaced by pyridine.[126]

Simple metal cations are occasionally extracted into organic solvents by combination with large anions, most successfully when both ions have a single charge and the extractant has a high dielectric constant (e.g., nitrobenzene). These extractions are applied more for separations than for determinations (Chapter 9C). Examples of such extractions with colored anions are Cs^+ with BiI_4^- and with reineckate ion $Cr(NH_3)_2(SCN)_4^-$ into nitrobenzene and nitromethane.

It would be expected that acidic dyes without chelating groups (e.g., Methyl Orange) might be of use in the extraction of colorless cation chelates for determination, and in fact the (colored) chelate cation $FePhen_3^{2+}$ can be determined in this way as mentioned earlier. Another example is provided by Bromcresol Green anion (RSO_3^-), which forms an

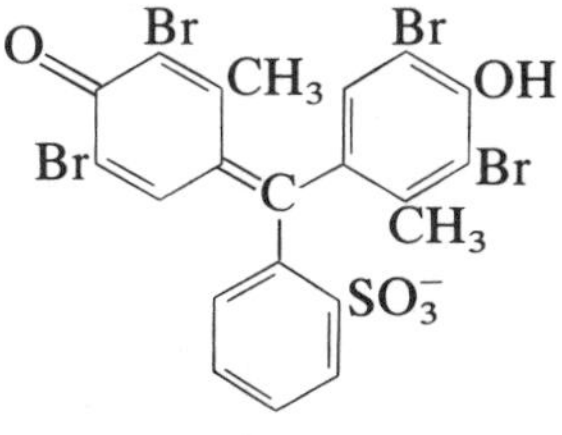

ion associate with colorless $ZnPhen_3^{2+}$, namely $ZnPhen_3(RSO_3)_2$ (Phen = 1,10-phenanthroline), extractable into 1,2-dichloroethane at pH 7.[127] Because of its lack of selectivity, this method is of little practical importance for the determination of Zn.

Fluorescein (phthalein) dyes,

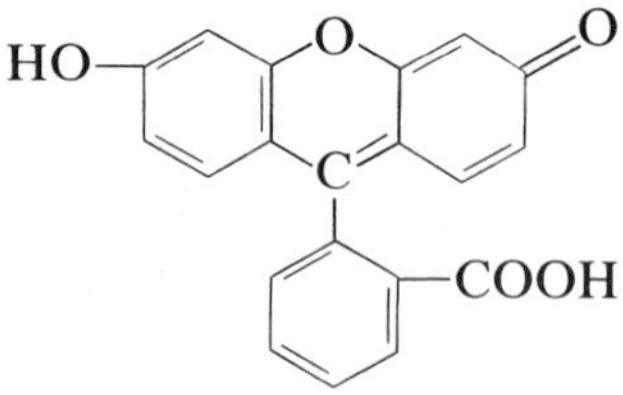

Fluorescein

especially halogenated derivatives, find use in the determination of metals that have been converted to cationic chelates by reaction with 1,10-phenanthroline and other reagents of that type. Some examples were given earlier in this chapter. Eosin, Erythrosine, and Rose Bengal are some of the anionic dyes used in this way. Determinations are often made fluorimetrically. Ag, Cd, Cu, Pb, and Zn can be determined.[128] Aromatic and chlorinated hydrocarbons are used as extractants. Addition of acetone to chloroform (for example) increases the fluorescence intensity. This effect has been explained by dissociation of the ion-association complex in the mixed solvent, which results from the increase in the dielectric constant ($CHCl_3$, 4.8; acetone, 20.7).

NOTES

1. Fe(II)-1,10-phenanthroline cation provides an illustration of this generalization. This cation (red) has been proposed for the photometric determination of $AuCl_4^-$, $Ni(CN)_4^{2-}$, $Sn(C_2O_4)_3^{2-}$, and other complex anions after nitrobenzene extraction. See papers by Y. Yamamoto and coworkers in *Bull. Chem. Soc. Japan, Japan Analyst,* and *J. Chem. Soc. Japan Pure Chem. Sec.,* 1964–1971; L. Ducret, *Anal. Chim. Acta.,* **20,** 565 (1959). Conversely, the determination of Fe(II) can be made more sensitive—though less selective—by extracting its 1,10-phenanthroline complex $FePhen_3^{2+}$ with a strongly colored organic anionic partner such as the alkaline form of Methyl Orange, $(CH_3)_2NC_6H_4N = NC_6H_4SO_3^-$, and measuring the absorbance of the ion associate $(FePhen_3)(RSO_3)_2$. The determination of Cu with 2,2′-biquinoline can be made about four times more sensitive by extracting the ion associate with Bromphenol Blue (RSO_3^-) and measuring the absorbance of $(CuBq_2)(RSO_3)$ in 3-methyl-1-butanol at 605 nm instead of $(CuBq_2)Cl$ in the solvent at 546 nm (K. Sekine and H. Onishi, *Anal. Lett.,* **7,** 187).

2. R. M. Dagnall and T. S. West, *Talanta,* **11,** 1533 (1964). A similar type of color reaction occurs with 1,10-phenathroline, tetrachloro(P)tetraiodo(R)fluorescein, and the metals Cd, Co, Cu(II), Mn, Ni, Pb, and Zn (but not Fe(II)), as shown by B. W. Bailey, R. M. Dagnall, and T. S. West, *ibid.,* **13,** 753 (1966). The reactions are very sensitive, ϵ of the solutions in ethyl acetate or chloroform ranging from 65,000 to 95,000. A similar reaction has

been suggested for the determination of Pd by R. M. Dagnall, M. T. El-Ghamry, and T. S. West, *Talanta*, **15**, 1353 (1968). As shown by El-Ghamry and R. W. Frei, *Talanta*, **16**, 235 (1969), Pt can be determined as $[Pt(NH_3)_6^{4+}][Eos^-]_4$, where Eos is 2,4,5,7-tetrabromofluorescein. A color change from orange-yellow ($\lambda_{max} = 525$ nm) to red ($\lambda_{max} \sim 565$ nm) occurs when the ion associate (sparingly soluble and not extractable by the usual organic solvents) is formed at pH ~ 10; $\mathscr{S} = 0.00024 \mu g\ Pt/cm^2$. Fe(III), Au, and Rh interfere. The color changes occurring when anionic halofluoresceins react with bulky cations containing Pt metals, Au, and some other metals having strong affinity for halogens point to interaction of some kind between them and halogen atoms. Halogenated fluoresceins, for example eosin, in conjunction with 1,10-phenanthroline have been used for the fluorimetric determination of Cd by M. A. Matveets and D. P. Shcherbov, *J. Anal. Chem. USSR*, **26**, 723 (1971). Chloroform is the extractant. A similar (unselective) reaction is applied for the absorptiometric determination of Cd or Zn: M. M. Tananaiko and N. S. Bilenko, *Zh. Anal. Khim.*, **30**, 689 (1975). The complex formed has the composition 1 Cd(Zn):2 phenanthroline:1 Rose Bengal B.

3. B. W. Bailey, J. E. Chester, R. M. Dagnall, and T. S. West, *Talanta*, **15**, 1359 (1968).

4. R. M. Dagnall, T. S. West, and P. Young, *Analyst*, **92**, 27 (1967).

5. A. Ashton, A. G. Fogg, and D. T. Burns, *Analyst*, **98**, 202 (1973).

6. Y. Shijo and T. Takeuchi, *Japan Analyst*, **17**, 323 (1968).

7. R. Nakashima, S. Sasaki, and S. Shibata, *Japan Analyst*, **22**, 723 (1973).

8. V. Svoboda and V. Chromy, *Talanta*, **13**, 237 (1966); **12**, 431 (1965). W. J. de Wet and G. B. Behrens, *Anal. Chem.*, **40**, 200 (1968).

9. Diantipyrinylmethane is a colorless reagent, but it forms a yellow complex with Ti, which is cationic and can be extracted with a suitable anionic partner.

10. Review of analytical applications of onium compounds: A. J. Bowd, D. T. Burns, and A. G. Fogg, *Talanta*, **16**, 719 (1969).

11. A book in Russian is available: S. I. Gusev, Ed., *Diantipyrylmethane and Its Homologs as Analytical Reagents*, Permsk. Gos. Univ., Perm, 1974, 280 pp. V. K. Akimov and A. I. Busev, *Zh. Anal. Khim.*, **26**, 134 (1971). Other reviews: V. P. Zhivopistsev, *Zav. Lab.*, **31**, 1043 (1965); V. K. Akimov and A. I. Busev, *Zh. Anal. Khim.*, **26**, 964 (1971) (antipyrine and derivatives); H. Ishii, *Japan Analyst*, **21**, 665 (1972).

12. V. N. Podchainova and A. V. Dolgorev, *J. Anal. Chem. USSR*, **20**, 1117 (1965). However, other workers, for example, L. Ya. Polyak, *Zh. Anal. Khim.*, **29**, 1338 (1974), find $\mathscr{S}$ much the same in H_2O and $CHCl_3$ so that the difference may be due to different reaction conditions.

13. Infrared spectra of reagent and its ion associates: V. K. Akimov, B. E. Zaitsev, A. I. Busev, and I. A. Emel'yanova, *Zh. Obshch. Khim.*, **40**, 2711 (1970).

14. Review: V. K. Akimov and A. I. Busev, *Zav. Lab.*, **38**, 3 (1972).

15. A. I. Busev and E. S. Bogdanova, *Zh. Anal. Khim.*, **19**, 1346 (1964).

16. H. Basinska and M. Tarasiewicz, *Chemia Anal.*, **13**, 1287 (1968). K.-T. Lee, *Anal. Chim. Acta*, **26**, 285, 478 (1962), has used chlorpromazine to determine gold. The red color formed is attributed to an oxidation product of the reagent.

17. Y. Wakamatsu and M. Otomo, *Bull. Chem. Soc. Japan*, **45**, 2764 (1972).

18. M. Otomo, *J. Chem. Soc. Japan Pure Chem. Sec.*, **92**, 171 (1971).

19. V. A. Nazarenko, G. V. Flyantikova, and T. N. Selyutina, *Zh. Anal. Khim.*, **27**, 2369 (1972).

20. V. I. Kuznetsov and S. B. Savvin, *Radiokhim.*, **2**, 682 (1960).

21. References in Y. Wakamatsu and M. Otomo, *Japan Analyst*, **21**, 1650 (1972).

22. R. Pribil and J. Adam, *Talanta*, **18**, 349 (1971).

23. V. T. Hill, *Anal. Chem.*, **38**, 654 (1966).

24. Review: K. Ueno, *Japan Analyst*, **20**, 736 (1971).

25. Y. Shijo and T. Takeuchi, *Japan Analyst*, **20**, 980 (1971).

26. For example, see H. Nishida, *Japan Analyst*, **20**, 410 (1971) (Fe(III)–CAS–Zephiramine).

27. Earlier,[8] the composition $[RE][(XO)(CPB)_2]_2$ was assigned to the extracted RE complexes. However, M. Otomo and Y. Wakamatsu, *Japan Analyst*, **17**, 764 (1968), give RE:XO:CPB = 1:2:2. For determination of U with Chrome Azurol S and cetylpyridinium bromide, see C. L. Leong, *Anal. Chem.*, **45**, 201 (1973).

28. R. Pribil and V. Vesely, *Talanta*, **17**, 801 (1970). The same extractant has been used to extract the Co–Nitroso-R salt chelate by J. Adam and R. Pribil, *Talanta*, **18**, 733 (1971).

29. The reaction of basic dyes with nonmetal anions of large volume should be borne in mind by the trace-metal analyst. Perchloric acid should not be used to decompose samples. Sulfonated detergents, possible contaminants from washing of glassware, can give extractable colored ion-association compounds with basic dyes.

30. Further treatment (in Russian): I. A. Blyum, *Extraction-Photometric Analysis with the Use of Basic Dyes*, Nauka, Moscow, 1970. Review of use of dyes in fluorimetric determinations: P. R. Haddad, *Talanta*, **24**, 1 (1977).

31. Exceptions to the 1:1 combining ratio occur, especially with bromo- and iodo complex anions. Thus butyl Rhodamine S gives extractable R_2TeBr_6 ($\epsilon = 150,000$) as well as $RTeBr_5$ ($\epsilon = 78,000$) (Murashova and Skripchuk, *Zh. Anal. Khim.*, **27**, 340 (1972)). Moreover, there is evidence that in 6 M HCl, $SbCl_6^-$ is extracted to a small extent as $RH_2(SbCl_6)_2$ in addition to the main complex $RHSbCl_6$ (RH^+ = Rhodaminium B cation). In the presence of thiocyanate, zinc is extracted as $Zn(SCN)_4^{2-}$. Methyl Green is reported to form a 1:2 complex with $AuCl_4^-$.

32. Are true solutions formed?

33. A. T. Pilipenko, L. I. Savranskii, and N. M. Shin, *J. Anal. Chem. USSR,* **24,** 337 (1969); S. A. Lomonosov, *ibid.,* **22,** 949 (1967); S. A. Lomonosov, M. K. Zvezdin, and V. D. Inishev, *ibid.,* **24,** 890 (1969). Triarylmethane dyes and rhodamines are considered. Also see *ibid.,* **28,** 1469, 1475 (1973).

34. Compare A. G. Fogg, C. Burgess, and D. T. Burns, *Talanta,* **18,** 1190 (1971); C. Burgess, A. G. Fogg, and D. T. Burns, *Lab. Pract.,* **22,** 472 (1973). The most effective purification method is considered to be by hot recrystallization in a Soxhlet extractor with a poor dye solvent. See p. 691 of this chapter.

35. A. G. Fogg, A. Willcox, and D. T. Burns, *J. Chromatogr.,* **77,** 237 (1973). Electrophoresis: S. B. Savvin and T. G. Akimova, *Zh. Anal. Khim.,* **27,** 1693 (1972).

36. In Russian translations often designated Rhodamine C or S.

37. R. W. Ramette and E. B. Sandell, *J. Am. Chem. Soc.,* **78,** 4872 (1956).

38. Early in the century, E. Noelting and K. Dziewonski, *Berichte,* **38,** 3516 (1905), concluded that the two principal forms of the dye were the lactone and a salt form RHCl.

39. Dimerization constants of Rhodamine B, butyl Rhodamine B, and other basic dyes are reported by N. S. Poluektov, S. V. Bel'tyukova, and S. B. Meshkova, *J. Anal. Chem. USSR,* **26,** 935 (1971), but the dimerizing species are not specified.

40. N. S. Poluektov, V. N. Drobyazko, and S. B. Meshkova, *J. Anal. Chem. USSR,* **28,** 1253 (1973).

41. The reactions of these metals, with pertinent literature references, are discussed in the respective chapters on the individual metals in *Colorimetric Determination of Traces of Metals.* Reactions of this type were discovered by Eegriwe (1927), who noted the formation of a colored precipitate by Sb(V) in the presence of Rhodamine B. The reaction was misinterpreted at first as an oxidation of the dye by Sb(V), and not until considerably later was the ion-association nature of the reaction demonstrated. In passing, it may be noted that ion-pair compounds such as Rhodamine B chloroantimonate(V) are sometimes called ternary complexes. This is dubious practice, which does not contribute to clarity. Nominally they may be ternary, but their behavior is binary. They should be distinguished from mixed complexes, which may be, and mostly are, ternary as far as the number of components is concerned.

42. R. T. Arnesen and A. R. Selmer-Olsen, *Anal. Chim. Acta,* **33,** 335 (1965). H. Nagai and H. Onishi were unable to obtain satisfactory results because of high reagent blanks.

43. D. P. Shcherbov and A. I. Ivankova, *Zav. Lab.,* **24,** 1346 (1958).

44. $RHSbCl_6$ fluoresces in $CHCl_3$ and isopropyl ether, but not (or very little) in benzene. $RHAuCl_4$ fluoresces in isopropyl ether but not in benzene. On the other hand, $RHGaCl_4$ and $RHTlCl_4$ fluoresce in benzene solution. In

benzene, the Ga detection limit is ~1:10.[9] Addition of acetone to benzene quite generally increases the fluorescence of xanthene dye ion-associates; ethyl and methyl alcohols behave similarly. This effect is ascribed to dissociation of the ion associate in the mixture: I. A. Blyum, F. P. Kalupina, and T. I. Tsenskaya, *Zh. Anal. Khim.*, **29**, 1572 (1974). The solvating effect of polar solvents also enters in here (see note 40). The favorable effect of acetone in fluorimetric determinations must be weighed against the increase in blank fluorescence it produces. Moreover, the use of acetone can make fluorimetric determinations less selective.

45. J. Marinenko and I. May, *Anal. Chem.*, **40**, 1137 (1968). The method described can be used to determine as little as 1–2 parts Au in 10^9 of solid sample.

46. Especially A. G. Fogg, C. Burgess, and D. T. Burns, *Talanta*, **18**, 1175 (1971). I. A. Blyum and N. N. Pavlova, *Zav. Lab.*, **29**, 1407 (1963); I. A. Blyum and L. I. Oparina, *ibid.*, **36**, 897 (1970). T. Matsuo, *Japan Analyst*, **21**, 671 (1972).

47. Y. Nishikawa, K. Hiraki, T. Naganuma, and S. Niina, *Japan Analyst*, **19**, 1224 (1970).

48. A. K. Babko and Z. I. Chalaya, *Zh. Anal. Khim.*, **16**, 268 (1961); **17**, 286 (1962).

49. L. M. Burtnenko and N. S. Poluektov, *Zh. Anal. Khim.*, **23**, 700 (1968).

50. G. Popa and I. Paralescu, *Talanta*, **16**, 315 (1969).

51. G. F. Kirkbright, R. Narayanaswamy, and T. S. West, *Anal. Chem.*, **43**, 1434 (1971).

52. A. Golkowska, *Chemia Anal.*, **15**, 59 (1970); A. Golkowska and L. Pszonicki, *Talanta*, **20**, 749 (1973).

53. N. S. Poluektov and V. T. Mishchenko, *Zh. Anal. Khim.*, **24**, 1434 (1969); diiodo-oxine: **29**, 2396 (1974). Salicylic acid as chelant: N. S. Poluektov and M. A. Sandu, *ibid.*, **24**, 1828 (1969). Sc with Rhodamine B and cinchophen: S. V. Bel'tyukova and N. S. Poluektov, *Ukr. Khim. Zh.*, **37**, 1277 (1971). (These references are illustrative, not complete.)

54. N. R. Anderson and D. M. Hercules, *Anal. Chem.*, **36**, 2138 (1964).

55. N. S. Poluektov, S. B. Meshkova, S. V. Bel'tyukova, and E. I. Tselik, *Zh. Anal. Khim.*, **27**, 1721 (1972).

56. N. S. Poluektov and S. V. Bel'tyukova, *Zh. Anal. Khim.*, **25**, 2106 (1970). U(VI) with Rhodamine B and HTTA: *ibid.*, **23**, 1647 (1968).

57. S. V. Bel'tyukova and N. S. Poluektov, *Zh. Anal. Khim.*, **25**, 1714 (1970).

58. Rhodamine 6G is also represented without the two methyl groups, and sometimes with one methyl group. In the Russian literature, 6Zh = 6G.

59. N. B. Zorov, A. P. Golovina, I. P. Alimarin, and Z. M. Khvatkova, *J. Anal. Chem. USSR*, **26**, 1310 (1971).

60. I. P. Alimarin, A. P. Golovina, N. B. Zorov, and E. P. Tsintsevich, *Izv. Akad. Nauk SSSR Ser. Khim.*, 2678 (1968).

61. J. Minczewski, J. Chwastowska, and E. Lachowicz, *Chemia Anal.*, **18,** 199 (1973).

62. A. T. Pilipenko, P. P. Kish, and Yu. K. Onishchenko, *Anal. Abst.*, **23,** 217 (1972), compared 18 basic dyes for the extraction determination of $SbCl_6^-$. The best reagents are reported to be the butyl ester of Rhodamine B, Crystal Violet, and 1,4-dimethyl-3-(4-dimethylaminophenylazo)-1,2,4-triazoline. The best extractants are benzene, chlorobenzene, amyl acetate, and mixtures of benzene or carbon tetrachloride with nitrobenzene or dichloroethane. Also see *Zh. Anal. Khim.*, **26,** 514 (1971) (xanthene dyes for Sb). Ethyl ester of Rhodamine B for determination of bromo complexes of many metals: I. A. Bochkareva and I. A. Blyum, *ibid.*, **30,** 874 (1975).

63. S. V. Markarova and I. P. Alimarin, *Zh. Anal. Khim.*, **19,** 564, 847 (1964).

64. N. N. Pavlova and V. G. Sayapin, *Zh. Anal. Khim.*, **20,** 1016 (1965).

65. I. A. Blyum and N. N. Pavlova, *Zh. Anal. Khim.*, **20,** 898 (1965); Blyum et al., **29,** 1572 (1974). The effect of acetone on the fluorescence of the complexes in benzene is also examined. N. K. Podberezskaya and V. A. Sushkova, *Zav. Lab.*, **36,** 1197 (1970), also add acetone in the fluorimetric determination of Au.

66. L. S. Alekseeva, V. I. Murashova, L. I. Bakunina, and V. G. Skripchuk, *Zav. Lab.*, **37,** 1299 (1971). For the use of asymmetric rhodamines in the determination of Te in Cl^- and Br^- solution, see *Anal. Abst.*, **25,** 815 (1973). Rhodamine 4G is better than Rhodamine 3GO.

67. Yu. A. Zolotov, I. V. Seryakova, G. A. Vorob'eva, and M. S. Sapragonene, *Zh. Anal. Khim.*, **25,** 1845 (1970).

68. I. A. Blyum, N. N. Pavlova, and F. P. Kalupina, *Zh. Anal. Khim.*, **26,** 55 (1971). Other extractable metals can be determined similarly (earlier papers). The object of the exchange is to extract with a more selective reagent and to determine with a reagent having superior photometric properties.

69. W. Z. Jablonski and C. A. Watson, *Analyst*, **95,** 131 (1970).

70. R. J. Argauer and C. E. White, *Anal. Chim. Acta*, **32,** 596 (1965).

71. As shown by A. G. Fogg, C. Burgess, and D. T. Burns, *Analyst*, **95,** 1012 (1970), purification is easily effected by crystallization from acetone. The impure material is extracted with acetone in a Soxhlet extractor. As the acetone becomes saturated, crystals of Brilliant Green separate in the Soxhlet flask and can be filtered off. Titanium(III) chloride titration of the recrystallized dye indicates a purity near 100%. Another purification method, due to G. O. Kerr and G. R. E. C. Gregory, *Analyst*, **94,** 1036 (1969), involves precipitation of the dye perchlorate from aqueous solution, followed by recrystallization from ethanol–water (1:4). The dye perchlorate is stable for some weeks in the solid state and is reported to be stable to heat and shock. Chromatography of an acetone solution on a silica gel column has also been used for purification.

72. R. J. Goldacre and J. N. Phillips, *J. Chem. Soc.*, 1724 (1949). Recently, A. G. Fogg, A. Willcox, and D. T. Burns, *Analyst*, **101,** 67 (1976), have called

attention to dimerization in Brilliant Green solutions. The dye is dimerized in more concentrated solutions, approximately 80% in an aged $10^{-3} M$ solution. Dilution of such a solution results in de-dimerization, but the reaction is slow, 18 hr or more being required for equilibrium. Since the monomer is the active form (giving extractable ion-pair species with metal anions), the slowness in reaching equilibrium is of some analytical importance, especially if only a slight excess of Brilliant Green is present. Dimerization is also a slow process. Ethanol solutions of the dye are not dimerized. Slow dimerization and de-dimerization has not been observed with Rhodamine B and other rhodamines, Methylene Blue, and Crystal Violet, as well as other cationic dyes. For dimerization of these dyes at higher concentrations, see Poluektov et al., *Zh. Anal. Khim.*, **26**, 1042 (1971).

73. An estimate from Fig. 6H-6a gives 1 to 2×10^4.

74. V. M. Tarayan, E. N. Ovsepyan, and N. S. Karimyan, *Anal. Abst.*, **19**, 4764 (1970). $\epsilon_{645} = 1.4 \times 10^5$. Mercury is also extractable as HgI_4^{2-} (pH 0.7), as shown by T. Sawaya, H. Ishii, and T. Odashima, *Japan Analyst*, **22**, 318 (1973). In benzene, $\mathscr{S}_{640} = 0.0017 \,\mu g \, Hg \, cm^{-2}$.

75. A. G. Fogg, C. Burgess, and D. T. Burns, *Analyst*, **98**, 347 (1973).

76. Brilliant Green tetrabromoindate is not extracted at all by benzene, toluene, and other low-polarity solvents. Polar solvents on the other hand (ketones, alcohols, and nitro compounds and, in addition, chloroform and dichloroethane) readily extract the simple dye salt and for this reason are not suitable as extractants in extraction-photometric methods. However, mixtures of nonpolar and oxygen-containing solvents may find use as extractants of the tetrabromoindate. See Yu. A. Zolotov, P. P. Kish, V. V. Bagreev, and I. I. Pogoida, *Zh. Anal. Khim.*, **29**, 221 (1974).

77. Z. Marczenko, H. Kalowska, and M. Mojski, *Talanta*, **21**, 93 (1974).

78. N. Bausova and E. M. Lebedeva, *Anal. Abst.*, **18**, 2402 (1970). They use 0.5 N H_3PO_4 as reaction medium.

79. A. N. Nevzorov, N. S. Onoprienko, and S. N. Mordvinova, *Zav. Lab.*, **36**, 1176 (1970). Tantalum is extracted over a wide acidity range (pH 3–8 N in H_2SO_4). Nb is not extracted from $>3 \, N$ H_2SO_4. Only traces of Mn, Fe, Sn, Ti, Zr, Ni, Cr, Cu, Co, and Al are reported to be extracted.

80. V. A. Nazarenko, N. V. Lebedeva, and L. I. Vinarova, *Zh. Anal. Khim.*, **27**, 128 (1972). Nazarenko et al., *Zh. Anal. Khim.*, **28**, 1966 (1973), determine W by flotation (light petroleum) of the 1:2:2 complex with 3,5-dinitrocatechol and Brilliant Green, dissolving in chloroform and measuring absorbance at 646 nm ($\mathscr{S} = 0.0014 \,\mu g \, W/cm^2$). The precipitate is formed at pH ~ 2. Also see *ibid.*, **28**, 101 (1973).

81. For example, N. V. Markova et al., *Anal. Abst.*, **21**, 68 (1971).

82. A. I. Busev and N. L. Shestidesyatnaya, *Zh. Anal. Khim.*, **29**, 1138 (1974).

83. V. Armeanu, L. M. Baloiu, and M. Damian, *Revue roum. Chim.*, **13**, 1617 (1968). Astraviolet 3R is used similarly with benzene as extractant (*Anal.*

Chim. Acta, **44,** 230). Many other complex cyanides also react, but $HAu(CN)_2$ can first be separated by extraction into amyl alcohol at pH < 1.

84. G. Ackermann and J. Koethe, *Chemia Anal.*, **17,** 445 (1972).

85. P. P. Kish, I. I. Zimomrya, and Yu. A. Zolotov, *Zh. Anal. Khim.*, **28,** 252 (1973).

86. E. L. Kothny, *Analyst*, **94,** 198 (1969).

87. L. I. Ganago and I. F. Ivanova, *Zh. Anal. Khim.*, **27,** 713 (1972).

88. L. Sokolova, P. N. Kovalenko, and G. G. Shchemeleva, *Anal. Abst.*, **19,** 3787. The anionic complex of PAN with V(V) can be extracted as an ion pair with Crystal Violet into a mixture of benzene and MIBK (J. Minczewski, J. Chwastowska, and Pham thi Hong Mai, *Analyst*, **100,** 708 (1975)). $\mathscr{S}_{585} = 0.0005$ μg V/cm^2. The blank absorbance is high, many anions interfere, and the color is not stable.

89. I. P. Alimarin and S. V. Makarova, *Zh. Anal. Khim.*, **19,** 90 (1964).

90. S. V. Makarova and I. P. Alimarin, *Zh. Anal. Khim.*, **19,** 564 (1964).

91. A. K. Babko, Yu. F. Shkaravskii, and E. M. Ivashkovich, *Zh. Anal. Khim.*, **26,** 854 (1971).

92. C. E. Hedrick and B. A. Berger, *Anal. Chem.*, **38,** 791 (1966).

93. S. A. Lomonosov et al. (8 coauthors), *Zh. Anal. Khim.*, **28,** 1653 (1973). For dimerization constants of Crystal Violet, Methylene Blue, and others in aqueous solution, see also E. M. Zadorozhnaya, B. I. Nabivanets, and N. N. Maslei, *Zh. Anal. Khim.*, **29,** 993 (1974).

94. A. T. Pilipenko and V. A. Obolonchik, *Ukr. Khim. Zh.*, **24,** 506 (1958). In toluene, Methyl Violet perrhenate has $\epsilon_{330} = 2.3 \times 10^5$.

95. A. T. Pilipenko and V. A. Obolonchik, *Anal. Abst.*, **8,** 4597 (1961). Under the optimal extraction conditions for Ta as fluotantalate with toluene (pH 1.9–2.3), Nb does not react and does not interfere. In this solvent, Methyl Violet fluotantalate shows absorption maxima at 330, 550, and 600 nm (ϵ, respectively, 1.5×10^5, 2.27×10^4, and 3.09×10^4). Optimal concentration of F^- is 5 mg/ml.

96. N. L. Shestidesyatnaya and P. P. Kish, *Ukr. Khim. Zh.*, **38,** 489 (1972); *Zh. Anal. Khim.*, **25,** 1547 (1970).

97. S. P. Lebedeva, *Anal. Abst.*, **25,** 74 (1973). See also *Zh. Anal. Khim.*, **26,** 1745 (1971); **29,** 2372 (1974).

98. A. T. Pilipenko, P. P. Kish, and G. M. Vitenko, *Ukr. Khim. Zh.*, **37,** 1149 (1971); determination as bromide, *ibid.*, **38,** 479 (1972).

99. P. P. Kish and S. G. Kremeneva, *Zh. Anal. Khim.*, **25,** 2200 (1970).

100. A. T. Pilipenko and N. D. Tu, *Ukr. Khim. Zh.*, **34,** 1291 (1968). General reactions: *ibid.*, **34,** 703.

101. S. V. Elinson, A. N. Nevzorov, M. V. Belogortseva, N. A. Mirzoyan, and S. N. Mordvinova, *Zh. Anal. Khim.*, **29,** 1234 (1974).

102. A. T. Pilipenko and N. M. Shinh, *Ukr. Khim. Zh.*, **34,** 1286 (1968);

determination of ReO_4^-: *ibid.*, **32,** 1211 (1966). C. Burgess, A. G. Fogg, and D. T. Burns, *Analyst,* **98,** 605 (1973). Fluorescence determination of Re: Pilipenko, A. I. Volkova, and T. L. Shevchenko, *Zh. Anal. Khim.,* **28,** 1524 (1973).

103. A. T. Pilipenko and Nguen Mong Shin, *Zh. Anal. Khim.,* **23,** 934 (1968).

104. A. T. Pilipenko, A. I. Volkova, and T. L. Shevchenko, *Zh. Anal. Khim.,* **29,** 983 (1974).

105. M. A. Matveets, D. P. Shcherbov, and S. D. Akhmetova, *Zh. Anal. Khim.,* **29,** 740 (1974).

106. H. Nagai and H. Onishi, *Japan Analyst,* **21,** 1590 (1972).

107. Methylene Blue (or Methylene Green) is preferred over Methyl Violet and Brilliant Green by V. M. Tarayan, E. N. Ovsepyan, and V. Zh. Artsruni, *Zav. Lab.,* **35,** 1435 (1969), because of greater sensitivity and wider acid range.

108. R. Kuroda, N. Yoshikuni, and Y. Kamimura, *Anal. Chim. Acta,* **60,** 71 (1972).

109. I. Noerdin, *Proc. Inst. Tekn. Bandung,* **3,** 63 (1963); **3,** 187 (1965).

110. L. A. Grigoryan, A. G. Gaibakyan, and V. M. Tarayan, *Zav. Lab.,* **40,** 136 (1974).

111. I. K. Guseinov and A. B. Abdullaeva, *Anal. Abst.,* **22,** 2113 (1972).

112. Guseinov et al., *Anal. Abst.,* **24,** 119 (1973).

113. See references in A. I. Busev, N. N. Yurzhenko, Yu. A. Mittsel', and V. M. Ivanov, *J. Anal. Chem. USSR,* **26,** 1144 (1971).

114. Note 113 and *J. Anal. Chem. USSR,* **26,** 1149 (1971).

115. A. I. Busev, E. S. Bogdanova, and V. G. Tiptsova, *J. Anal. Chem. USSR,* **20,** 542 (1965).

116. V. P. Zhivopistsev and M. N. Chelnokova, *Zh. Anal. Khim.,* **18,** 148, 717 (1963).

117. P. P. Kish and Yu. K. Onishchenko, *Zh. Anal. Khim.,* **25,** 112, 500 (1970).

118. A. I. Busev, N. L. Shestidesyatnaya, and P. P. Kish, *Zh. Anal. Khim.,* **27,** 298 (1972).

119. P. P. Kish and I. I. Pogoida, *Zh. Anal. Khim.,* **28,** 1923 (1973).

120. P. P. Kish, S. G. Kremeneva, and E. E. Monich, *Zh. Anal. Khim.,* **29,** 1741 (1974).

121. L. I. Kotelyanskaya, P. P. Kish, and E. P. Gabrilets, *Anal. Abst.,* **23,** 3857 (1972).

122. P. P. Kish and G. M. Vitenko, *Zav. Lab.,* **38,** 5 (1972). Zn, Cd, Al, Ni, Co, Fe, and Bi do not interfere.

123. P. P. Kish and Yu. K. Onishchenko, *Zav. Lab.,* **36,** 520 (1970).

124. Kish and I. I. Zimomrya, *Zav. Lab.,* **35,** 541 (1969). No interference from Al, Cr, Mn, Ni; only limited amounts of Fe(III), Co, Cd, Hg may be present.

125. Rarely, a cationic metal chelate can be formed by protonation of =N— or other strongly basic group in an uncharged chelate. Such protonation is of rare occurrence because almost always the chelate is dissociated at the hydrogen-ion concentration required for protonation of the basic group.

126. Yu. A. Zolotov and G. E. Vlasova, *Zh. Anal. Khim.*, **24,** 1542 (1969).

127. M. Tsubouchi, *Japan Analyst,* **19,** 1400 (1970).

128. The following papers are illustrative: D. N. Lisitsyna and D. P. Shcherbov, *Zh. Anal. Khim.*, **25,** 2310 (1970) (determination of Ag); M. A. Matveets and Shcherbov[2] (Cd); B. Kasterka and J. Dobrowolski, *Chemia Anal.*, **16,** 619 (1971) (Cu).

6I

ORGANIC PHOTOMETRIC REAGENTS
ORGANIC OXIDIZING
AND REDUCING AGENTS

Suitable organic compounds can be oxidized to strongly colored products in inorganic systems of high oxidation potential, and sometimes indirect photometric methods of practical value for metals can be based on such reactions. These methods are likely to be lacking in specificity and must be applied with caution. Reactions may not be stoichiometric. Another drawback of these methods is the frequent poor stability of the colored organic product. In some of these reactions, the metal being determined functions as an oxidation catalyst. For example, manganese(II) catalyzes the oxidation of N,N-diethylaniline by potassium iodate to a yellow product.

Some illustrations of oxidation reactions are mentioned here. Several aromatic amines have been used for indirect photometric determinations. o-Tolidine in acid solution is oxidized by gold(III) and cerium(IV) to a yellow product. N,N'-Tetramethyl-o-tolidine reacts similarly but requires a higher oxidation potential. o-Dianisidine gives a red color with gold(III) and vanadium(V). 3,3'-Diaminobenzidine is oxidized by vanadium(V) to a red-brown product; the molar reaction ratio is 1:1. This reaction serves well for the determination of vanadium(V). 3,3-Dimethylnaphthidine also has been used for the determination of vanadium. N,N'-Di-(2-naphthyl)-p-phenylenediamine is oxidized by iridium(IV) to a red product that can be extracted into chloroform.[1] Gold(III), iron(III), and H_2O_2 all react identically. 8-Aminoquinoline gives a red-violet oxidation product with iron(III) and manganese(VII), which is extractable into a mixture of benzyl alcohol and chloroform.

The leuco compound of Malachite Green is oxidized to blue-green Malachite Green by gold(III) and manganese(VII). The dye can be extracted with chloroform before photometry. The leuco compound of Crystal Violet is oxidized to Crystal Violet ($\lambda_{max} = 590$ nm) by iridium(IV); gold(III) interferes. Cerium(IV) can be determined with the iron(II)–1,10-phenanthroline complex by measuring the decrease in absorbance at 510 nm.

Thio–Michler's ketone, 4,4'-bis(dimethylamino)thiobenzophenone, has been used for the determination of gold(III). A possible course of reaction is reduction of gold(III) to gold(I) by the reagent and complex formation (red color) between gold(I) and the unconsumed reagent. The oxidized reagent is also colored (p. 560). Hg, Pd, and Ag also react.

Antimony has been determined by oxidation of diphenylaminesulfonic acid with Sb(V).[2]

In all reactions of this class, the excess of the oxidizing agent used to convert the metal to the higher oxidation state must, in general, be removed or be made harmless in some way.

Chemiluminescence methods have been proposed for determination of low concentrations of metals. These methods are based on the catalytic effect of metal ions on the oxidation of Luminol (5-amino-2,3-dihydro-phthalazine-1,4-dione)[3] and Lucigenin (10,10'-dimethyl-9,9'-biacridinium dinitrate)[4] by peroxide or oxygen.

The reduction of an organic compound to a strongly colored product by an inorganic substance is rarely applied in photometry. Cacotheline, the nitro compound of brucine, is reduced by tin(II) and other strong reducing agents[5] to a soluble violet-colored substance. This reaction has been applied to the determination of tin in iron and steel.[6]

NOTES

1. F. G. Nasouri and A. S. Witwit, *Anal. Chim. Acta*, **50,** 163 (1970); M. D. Booth, *ibid.*, **59,** 304 (1972).

2. L. Naruskevicius, R. Kazlauskas, and J. Skadauskas, *Zav. Lab.*, **38,** 1435 (1972).

3. Determination of Cr(III): W. R. Seitz, W. W. Suydam, and D. M. Hercules, *Anal. Chem.*, **44,** 957 (1972) (see Chapter 4); Fe(II): W. R. Seitz and D. M. Hercules, *ibid.*, **44,** 2143 (1972). The former cites a number of Russian papers.

4. Determination of Mn: L. I. Dubovenko and A. P. Tovmasyan, *Zh. Anal. Khim.*, **25,** 940 (1970); Pd and Os: A. K. Babko and A. V. Terletskaya, *Zav. Lab.*, **35,** 1046 (1969); Tl: L. I. Dubovenko and A. P. Tovmasyan, *Ukr. Khim. Zh.*, **37,** 845 (1971); some other metals: A. K. Babko, L. I. Dubovenko, and A. V. Terletskaya, *Ukr. Khim. Zh.*, **32,** 1326 (1966).

5. L. Rosenthaler, *Mikrochim. Acta*, **3,** 190 (1938).

6. H. Goto and Y. Kakita, *Sci. Rept. Res. Inst. Tohoku Univ.*, **A5,** 554 (1953); *ibid.*, **A6,** 12 (1954).

ANALYTICAL SEPARATIONS

INTRODUCTORY NOTE

In a separation, the constituents A, B, ... of a sample are isolated into separate regions of space by suitable processes. The primary need for separations in analysis obviously arises from the lack of selectivity of most methods of determination, so that it becomes necessary to free a given constituent from accompanying interfering constituents. A secondary aspect of separation processes is the concentration or enrichment of a trace constituent, as from a large weight of sample or a large volume of solution, so that it is obtained in a small volume of solution permitting a sensitive determination or one more precise. Such concentration of a trace constituent may be accompanied by separation from constituents interfering in the final determination.

Two factors are of fundamental importance in analytical separations:

1. The extent of recovery of the desired constituent A in the sample which has been carried through the separation process.
2. The extent of separation of constituent A from the other constituents B, C, ... of the sample.

The completeness of recovery of A in a specified procedure is expressed by the recovery factor R_A:

$$R_A = \frac{Q_A}{(Q_A)_0}$$

where $(Q_A)_0$ is the quantity of A in the sample taken and Q_A is the quantity isolated. If desired, this quotient may be multiplied by 100 to give the percentage recovery.

The values of R_A required in quantitative analysis usually depend on the relative amounts of A in the sample. When A is a major constituent, R_A should be ≥ 0.999. When it is a minor constituent (0.01–1%), R_A should usually be > 0.99, and when it is a trace constituent, a similar value is desirable, but $R_A > 0.95$ may be acceptable if $A < 10^{-3}$ or $10^{-4}\%$.

The extent of separation of two constituents A and B from each other

699

is given by the *separation factor.*[1] We define the separation factor for B, the undesired constituent, with respect to A, as the value which multiplied by the initial ratio of B to A gives the final ratio:

$$\frac{Q_B}{Q_A} = S_{B/A} \frac{(Q_B)_0}{(Q_A)_0}$$

Therefore,

$$S_{B/A} = \frac{(Q_A)_0}{(Q_B)_0} \times \frac{Q_B}{Q_A} = \frac{R_B}{R_A}$$

In quantitative work $R_A \sim 1$, so that

$$S_{B/A} \sim R_B$$

The extent of separation required depends on:

1. The initial ratio of B to A, $(Q_B)_0/(Q_A)_0$.
2. The maximum permissible value of Q_B/Q_A, which is determined by the extent of interference of B in the determination of A and the accuracy required.

Suppose that in the photometric determination of A, B gives the same absorbance as an equal weight of A, so that 1 μg of B will be counted as 1 μg of A. Then, if the error due to B is not to exceed 1% (10 ppt), the separation factors will need to have the following values for various initial ratios of B to A:

$(Q_B)_0/(Q_A)_0$	$S_{B/A} \ (= R_B)$
1	$<10^{-2}$
10^2	$<10^{-4}$
10^4	$<10^{-6}$
10^6	$<10^{-8}$

(The separation factor as defined is the reciprocal of the *decontamination factor* used by radiochemists.)

Separation factors must often have very small values in trace analysis. Of course, the final determination of a given constituent is made as selective as possible by choosing a reagent that reacts sensitively with it and not with others that are likely to be present in the sample. The use of differential complexing agents (masking), adjustment of pH, and sometimes a change in oxidation state of one or more of the elements present help to reduce the stringent separation requirements in trace analysis.

Direct and Indirect Separations. As a rule, in photometric analysis, the sample is brought completely into aqueous solution, and required separations of constituents are carried out in either of two ways. In what may be called a direct separation, the constituent to be determined is isolated in, or as, a new phase (e.g., by precipitation, coprecipitation, extraction, etc.), and the other constituents are left in the original solution. A selective separation reaction is required. An advantage of this approach is the possibility of keeping the final volume small so that sensitivity will not be impaired. In an indirect separation, the interfering constituents are brought into, or constitute, a new phase, and the constituent being determined is left in the original solution. Although this approach sometimes has advantages, it is usually inferior to the first. Especially when the interfering constituents are major ones, severe requirements are put on the completeness of their removal. Precipitation is suspect because of the possibility of coprecipitation of the desired trace constituent, even if the major constituents form precipitates that are sufficiently insoluble. When applicable, which is seldom in inorganic analysis, volatilization methods are perhaps superior to other methods for removal of a major constituent, provided the volatility is sufficiently great. Immiscible solvent extraction has limitations imposed by limited solubility of the extracted species in organic solvents or extraction coefficients that are not as large as desirable, so that more than a few extractions will be needed. The extraction reagent should be one that is unselective (e.g., cupferron, diethyldithiocarbamate), so that as many metals as possible are removed at one time. The relatively large amount of reagent required may increase the blank value. Nevertheless, indirect separations have value, especially for those elements, such as the alkaline earth metals, which are deficient in selective direct separation reactions. The disadvantage of a relatively large volume of final aqueous solution is mitigated by extractive photometric methods. Indirect separations may be attractive when many trace elements are to be determined in a matrix element, such as a relatively pure metal.

In the following chapters we consider separations in inorganic trace analysis, with special reference to metals, according to the following classification:

1. Methods involving precipitation and coprecipitation (Chapter 7).
2. Chromatographic methods, with emphasis on ion exchange, which is the most important type of chromatography in inorganic trace analysis (Chapter 8).
3. Liquid–liquid extraction (Chapters 9A–9D).
4. Volatilization methods (Chapter 10).

A great deal of work has been done during the last three or four decades on extraction and chromatographic methods of separation[2] and these are likely to be given first consideration in separation problems involving traces. Methods based on precipitation and coprecipitation are of more restricted application, but are not yet passé. Volatilization methods are applicable to As, Sb, Sn, Ge, Hg, Re, Ru, Os, and many nonmetals, and are valuable because of the sharp separations they often provide.

The relative extent of use of the various types of separation methods in inorganic trace analysis in the early or mid 1970s may be judged from the replies received from 188 laboratories in response to a questionnaire circulated by the Commission on Microchemical Techniques and Trace Analysis of the Union of Pure and Applied Chemistry[3]:

Type of Method	Number of Laboratories Using
Liquid–liquid extraction	79
Ion exchange	21
Precipitation	18
Distillation and sublimation	8
Electrodeposition	5

NOTES

1. This is not the separation factor (so called) of some workers who apply this term to α, the ratio of distribution coefficients (extraction coefficients E in liquid–liquid extraction). The extent of separation of A and B from each other cannot be formulated or calculated from α, and the term is a misnomer leading to obfuscation (see further Chapter 9D, note 1). We need a factor that can be used to obtain the change in the ratio of the amounts of A and B occurring in *any* (single or multistage) separation. The separation factor S defined here does this. The S factor was used as early as 1950 by E. B. Sandell, *Anal. Chim. Acta*, **4,** 504 (1950), in dealing with liquid–liquid extraction separations. G. Günzler, C. Fischer, and P. Mühl (in Dyrssen, Liljenzin, and Rydberg, *Solvent Extraction Chemistry*, p. 620) call α (β in their symbology) the *conventional* separation factor, and use the term *integral* separation factor for the ratio (percentage extraction of A)/(percentage extraction of B), A being more easily extractable than B. We go a step beyond this and reject the term separation factor for α. Though not the separation factor, α is of obvious importance in separations. If, in multistage separations, the distribution coefficients can be given the proper values, a separation will be easier the more α differs from 1. See Chapter 9D.

2. Methods for the actinide and rare earth elements, Li, Rb, Cs, Fr, Be, Ra, Ga, In, Tl, Ge, Se, Te, Po, Ag, Au, Ti, Zr, Hf, V, Nb, Ta, Mo, W, Tc, Re, and the

platinum metals are treated in J. Korkisch, *Modern Methods for the Separation of Rarer Metal Ions*, Pergamon, Oxford, 1969. Coprecipitation and distillation methods are also considered. With 4000 references, this is a book for the practicing analyst. E. W. Berg, *Physical and Chemical Methods of Separation*, McGraw-Hill, New York, 1963, is a course textbook. For other recent books on separations, see J. I. Dinnin, *Anal. Chem.*, **49**, 34R (1977) (review, inorganic and geosamples).

3. This is the same questionnaire mentioned in Chapter 1, p. 40.

PRECIPITATION AND COPRECIPITATION

I. PRECIPITATION

Precipitation (without the use of a collector) is occasionally applied in the separation of trace constituents from major (or interfering minor) constituents of a sample. As already indicated, there are two possibilities:

1. The trace constituent is precipitated, leaving the major constituents in solution.

2. The major constituents are precipitated, leaving the trace constituent in solution.

The first approach is more desirable, but may not be applicable because of appreciable solubility of the precipitation form. The solubility of an ionic precipitate is the concentration sum of three types of species: component ions, undissociated molecules (representing the molecular or intrinsic solubility), and complex ions formed, for example, by an excess of the precipitating ion. It may be possible to reduce the concentration of a component ion of a precipitate in a saturated solution to a very low value by the addition of an excess of a common ion, but the molecular solubility remains the same at constant ionic strength in the same solvent. At the concentration of excess common ion required to bring the ionic solubility to a negligible value, there may be a large solubility increase due to the formation of complex ions. The solubility product constant of a precipitate may give an erroneous impression of its insolubility. For example, though K_{sp} of HgS (metacinnabar) is 10^{-51}, the concentration of mercury in an aqueous solution of pH 0–5 saturated with H_2S at 20° (all temperatures are in °C) and $\mu = 1$ in equilibrium with solid HgS is $\sim 10^{-7}$ M or about 20 μg Hg/liter.[1] In such a solution, mercury is present almost entirely as $Hg(HS)_2$ (which can be regarded as H_2HgS_2). The concentration of Hg^{2+} is negligible (one ion in a ton of solution) and the intrinsic solubility of HgS is not greater than $\sim 1 \times 10^{-9}\ M$. The solubility of HgS in acid solution can be reduced by decreasing the H_2S concentration. If the concentration of hydrogen sulfide in the solution (pH 0–5) is reduced to $10^{-3}\ M$ (about 0.01 of the concentration of H_2S in a saturated solution at room temperature), the solubility of HgS ($=[Hg(HS)_2]$) becomes $\sim 0.2\ \mu g$ Hg/liter.

The intrinsic solubilities of a few precipitates at 20°–25° are listed below (from various sources):

	Solubility	
	M	μg metal/liter
AgCl	2.25×10^{-7}	24
Hg (metal)	3×10^{-7}	60
$Fe(OH)_3$	$< 2 \times 10^{-9}$	< 0.1
$Bi(OH)_3$	5×10^{-6}	1450
$Co(OH)_3$	$< 5 \times 10^{-8}$	< 3
$Pb(OH)_2$	6×10^{-5}	12,500
$Ni(HDx)_2$ (dimethylglyoximate)	1×10^{-6}	60
$Cu(HDz)_2$ (dithizonate)	8×10^{-9}	0.5
CuDz	$< 6 \times 10^{-8}$	< 3.7
$ZnOx_2 \cdot 2\,H_2O$ (oxinate)	3×10^{-8}	2
$CuOx_2$	3×10^{-7}	19
$FeOx_3$	1.6×10^{-7}	9

Assuming a precipitation and wash volume of 100 ml, it is clear that a few micrograms of Ag could not be isolated as the chloride, even if the excess of Cl^- were controlled so that a minimum of $AgCl_2^-$ were formed and Ag^+ were negligible. Nor would metallic mercury be a suitable precipitation form for traces of mercury. The effective solubility would need to be reduced with a suitable collector (p. 709). Metal chelates sometimes have such low intrinsic solubility that precipitation can be made quantitative for microgram amounts,[2] but the use of a collector is advisable. However, most chelates are extractable into organic solvents and precipitation is not often used for them.

When the major constituents are precipitated in a separation, solubility considerations are less important—but by no means unimportant. Possible loss of a trace constituent by coprecipitation with a large mass of insoluble material now becomes the critical factor. This type of precipitation separation must be looked upon with suspicion. It is true that sometimes it is satisfactory. Some precipitates—the silver halides, for example—do not have a pronounced tendency to carry down most other metals, so that the method cannot be universally condemned. Precipitation of a major constituent can at times be used with advantage when it is the chief component of a fairly pure metal or alloy; the possibility of eliminating the matrix element, thus allowing a host of trace elements to be determined, is an attractive one. Thus precipitation of silver as chloride has been applied in the determination of Cu, Cd, Zn, Ni, and Co in silver metal.[3]

In determination of traces of cobalt in nickel metal, the bulk of the nickel can be precipitated as Ni hexammine perchlorate. Cobalt(III) ammine remains in solution. Jackwerth and Graffmann[4] have studied the coprecipitation of many heavy metals in the precipitation of silver as $AgCl$, $AgBr$, AgI, $AgSCN$, $AgCN$, $AgBrO_3$, and $AgIO_3$. From the analytical standpoint, precipitation of Ag as $AgCl$ is of the greatest interest and importance, and the findings with that separation form will be summarized here. A 0.25 M HNO_3 solution of silver (~1 g) was treated with various amounts of NaCl solution, less and greater than the equivalent volume, at room temperature. The final volume was 60 ml. In each precipitation, 250 μg of trace metal was present. The trace metals were determined in the centrifugate. The following elements are not appreciably coprecipitated, >95% remaining in solution:

Ni	Cu	In	Ce(III and IV)
Co	U(VI)	Cr(III)	Sn(IV)
Mn	Fe(III)	Rh	Th
Zn	Ga	La	

Coprecipitated elements, with the percentage remaining in solution after the addition of 50% excess chloride, are:

Pb 13	Hg(II) 9
Cd 36	Tl(I and III) 16 and 2
Pd(II) 21	Bi 4

These metals are considered to be adsorbed, because addition of excess Ag^+ releases them with a recovery of 95% except for

Pd 70	Tl(III) 75
Tl(I) 90	Bi 88

(figures indicate percentage recovery). $AuCl_4^-$ is strongly adsorbed in the vicinity of the equivalence point; a slight excess of Cl^- releases ~90% of it. $AgIO_3$ differs from $AgCl$ in carrying down most of the elements named above, and addition of Ag^+ does not release them to any large extent from the precipitate. Coprecipitation does not seem to be due to mixed crystal formation (Doerner–Hoskins type of incorporation).

A large number of trace metals have been determined spectrographically in lead metal after removal of most of the lead as sulfate.[5] Mercury has been analyzed for trace elements by precipitating metallic mercury from nitric acid solution with formic acid.[6] Metals such as Au, Ag, platinum metals, Cu, and Hg are readily electrodeposited, and it may be expected that metals that are not electroreduced under the conditions will not greatly contaminate the deposited metals. See p. 708 for the use of the mercury cathode.

Precipitation of water-soluble salts by change of solvent offers a means of removing much of one or more of the matrix elements of a sample. For example, by saturating an ammonium chloride solution with HCl at 0°C, most of the titanium is precipitated as $(NH_4)_2TiCl_6$ and most of the aluminum as $AlCl_3 \cdot 6 H_2O$. Large amounts of lead can be precipitated as $Pb(NO_3)_2$ by adding HNO_3 to an aqueous solution. Addition of a miscible organic solvent sometimes provides a way of removing the bulk of an undesired element in the form of a suitable salt. Methods of this kind may be useful only for removing most of a matrix element and another separation may be needed to remove the milligram (or greater) amounts remaining. Coprecipitation of the adsorption–occlusion type may be expected to be small with the relatively coarse crystals, but loss of a micro constituent by mixed crystal formation is a possibility that requires checking.

When appreciable coprecipitation of trace elements occurs with a matrix element, reprecipitation can be applied, but the greater expenditure of time and the increase in the volume of solution are objectionable.[7] Precipitation from homogeneous solution may be of aid in obtaining a purer precipitate.[8] Trace elements must not form solid solutions with the matrix precipitate.

Fusions may be superior to precipitations in eliminating matrix elements without serious loss of trace constituents. The insoluble material is usually obtained in more compact form with weaker adsorptive power. Thus fusion of a silicate with sodium carbonate followed by leaching brings Cr, V, and Mo into solution without important retention by the residue containing Fe, Ca, Mg, and other metals.

Organic reagents find less frequent use than inorganic reagents in precipitating major or matrix constituents, but can be of value at times.[9]

Concentration of very dilute aqueous solutions by freezing out most of the water has been tried.[10]

Electrodeposition.[11] Precipitation without reagents, that is by electrolysis, sometimes has advantages in direct or indirect separations, particularly the latter, in trace analysis. Very small amounts of Cu,[12] Pb,[12] Hg,[13] Zn,[13] Ag,[14] and Au have been isolated by deposition on a platinum or other metal cathode. Hg is not completely deposited because of its slight solubility in water unless amalgamation with the cathode occurs. A special apparatus for the deposition of traces of metals from large volumes of solution (250 ml) allows the separation of 0.1 mg of Zn from 5 g of Al, 0.1 mg Cu from 5 g Ni, and so on.[15] Lead can be anodically deposited as PbO_2 but there seem to be difficulties in obtaining quantitative recovery.[16] Tl is depositable as Tl_2O_3 on a platinum anode.[17]

Large amounts of many heavy metals that are not wanted can frequently be effectively removed by electrolysis with a mercury cathode.[18] In dilute sulfuric acid solution, without special control of the cathode potential, Fe, Cr, Ni, Co, Zn, Cd, Ga, In, Ge, Cu, Sn, Mo, Re, Bi, Tl, Ag, and the platinum metals (except Ru and Os) are deposited in the mercury, while such elements as Ca, Mg, Al, Ti, W, Zr, P, rare earth metals, V, and U are quantitatively left in solution.[19] The method is particularly valuable in the determination of traces of the latter elements in metallurgical materials. Thus electrolysis with a mercury cathode provides an excellent separation of interfering iron in a determination of aluminum in steel. It is not always easy to remove the last traces of the deposited elements; microgram quantities can remain, even when considerable care is taken.[20] Similar amounts of elemental mercury will be introduced into the solution because of its atomic solubility. Provision must be made for the presence of such traces when they interfere in the subsequent photometric method. In general, the deposition of matrix elements takes place without codeposition of elements that should remain in solution under the conditions of electrolysis, that is, coprecipitation, occurring when a (solid) precipitate is formed, is not observed.

By applying controlled potential electrolysis, other useful separations become possible with the mercury cathode. Thus Pb and Cd (and Cu) can be deposited in mercury from a hydrochloric acid solution of a zinc-base alloy; anodic stripping transfers Pb and Cd to the aqueous solution, leaving Cu in the mercury.[21] But separations such as these are made more for the purpose of polarographic determination than as an end in themselves.

Without electrolysis, noble metals such as Ag, Au, and Pt in trace concentrations can be collected in mercury metal.[22] This is a separation from copper. Some other metals find sparing use in plating out reducible metals.

Various forms of mercury cathode electrolysis cells for removal of large amounts of matrix elements have been described. Stirring of mercury and the aqueous solution, required for rapid deposition of metals, can be done

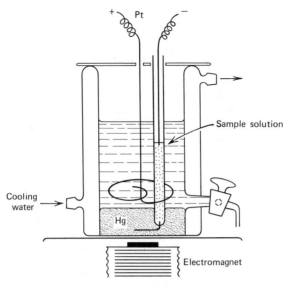

Fig. 7-1. Mercury cathode electrolysis cell with electromagnetic stirring. According to Mizuike (Morrison, *Trace Analysis,* p. 141).

mechanically or electromagnetically. The use of a cell cooled by water is advantageous. The cell illustrated schematically in Fig. 7-1 is available commercially.[23] It enables 5 g of iron to be removed from 100 ml of 0.1 N sulfuric acid in about 1.5 hr.

II. COPRECIPITATION AND GATHERING

Coprecipitation is usually understood to mean the carrying down by a precipitate of substances normally soluble under the conditions. The carrying down of an insoluble substance by a precipitate may be called *gathering.* Thus a precipitate of aluminum hydroxide can be used to gather a small quantity of ferric hydroxide. As we have seen, quantitative precipitation of a trace substance by itself may be impossible. But it may be possible to *collect* (coprecipitate or gather) the trace constituent in a small amount of a suitable slightly soluble substance. Usually the *collector* is formed in the solution by addition of a precipitant, but occasionally it is added preformed. A collector can function in various ways, as by adsorbing the trace constituent, forming a mixed crystal (solid solution) with it, or combining with it to produce an insoluble compound.

The chief desideratum in the choice of a collector is adequate recovery of the trace constituent. The collector should not interfere in the subsequent determination of the trace substance, or if it does, it should be easily eliminated (as by volatilization) or easily masked. The amount of collector required should be small (of the order of milligrams), so that the carrying down of substances other than the trace constituent is minimized. Selectivity of collecting action is highly desirable. Ease of filtration and ease of dissolution of the collector are other desiderata.

The mechanism of collection in a given case is not always clear, and indeed several mechanisms may be involved. It may not even be possible to draw a line between coprecipitation and gathering. These processes are still far from being well understood. So far as coprecipitation is concerned, two different types of processes are of special interest to the trace analyst:

1. Adsorption of a trace constituent at the surface of a precipitate consisting of extremely fine particles ("amorphous" precipitates) and therefore having a very large surface area per unit weight.

2. Incorporation of a trace constituent in the interior of the particles (often micro crystals) of a precipitate by solid solution, more explicitly by mixed crystal formation.

Generally speaking, mixed crystal formation is the only type of occlusion (enclosure in the particles of a precipitate) important in trace analysis. Occlusion in micro crystals resulting from irregular incorporation of adsorbed substances by rapid crystal growth, a nonequilibrium phenomenon that is a nuisance in gravimetric analysis, is of little value for collecting traces. Conceivably this type of occlusion could play a more important role when the precipitate is "amorphous."

A. GATHERING, ADSORPTION, AND OCCLUSION

In some collecting, an important mechanism is doubtless *gathering*. The precipitate of the trace constituent formed by the precipitant is very insoluble, and the main function of the collector may be to drag down the trace constituent that might otherwise remain in supersaturated solution or that might form a colloidal suspension. For example, barium sulfate has been used to gather silver halides. However, something more than mere physical action must be involved in most collecting. Titanium in concentrations as low as 1 μg/liter can be collected in a precipitate of aluminum hydroxide, and many like examples could be mentioned.[24] We would expect that titanium hydroxide would be adsorbed on aluminum hydroxide from a solution of titanium hydroxide and that the concentration of titanium could be reduced below that existing in a saturated solution.[25]

Hydroxide Collectors. In the precipitation of trace heavy metals as hydroxides, $Fe(OH)_3$, $Al(OH)_3$, and sometimes other hydroxides are used as collectors. The collecting hydroxides have huge surface development and adsorption must play a major role. Compound formation and solid solution can occur. Not all hydroxides are equally effective as collectors for a given metal and the choice often rests on empirical findings. Ferric hydroxide is often an effective collector and can be used over a wide pH range. The iron introduced may be objectionable in some photometric determinations (but reduction to Fe^{2+} or complexing is sometimes possible), and $Al(OH)_3$ or $La(OH)_3$[26] may be preferred collectors. The precipitation is, of course, made in the pH range of minimum solubility of the metal hydroxide or other metal compound being collected.

Because most metals form insoluble hydroxides, hydroxide precipitation is likely to be carried out more for concentration than for separation of a particular metal. The quantitative recovery of the metal is the primary requirement. The accompanying elements, if small in amount, can often be provided for in other steps of the determination. The nature of the sample has an important bearing on the applicability of precipitation methods. They are not likely to be applied if large amounts of other precipitable elements are present. Precipitation methods may be suitable for dilute solutions found in nature (sea water, fresh waters), in which the major constituents will not be precipitated to a significant extent in the procedure. If the pH of precipitation of collector and of the collected metal differs sufficiently from that of the major elements, pH adjustment may suffice to keep the latter in solution. Since major elements may be coprecipitated to some extent, the amount of collector should be restricted to the amount actually required. The possible presence in the sample of substances (such as organic material) interfering with hydroxide precipitation should be borne in mind.

The final pH may be very important. Suppose, the collector is $Fe(OH)_3$ (more correctly termed hydrous ferric oxide) or some other hydroxide whose precipitation begins in faintly acidic solution and that is not significantly amphoteric so that it remains undissolved at high pH. For a metal whose hydroxide is precipitated at a higher pH than $Fe(OH)_3$, maximum collection will in general be attained when the ionic solubility has become smaller than the intrinsic (molecular) solubility. If the metal is not amphoteric, so that anionic species are not formed, further increase in pH would be expected to have relatively little effect on the collection, barring any changes in the structure of the $Fe(OH)_3$ precipitate with pH or of compound formation as a function of pH. If the metal is present as an oxo-anion (e.g., CrO_4^{2-}), a decrease in the amount collected will occur at higher pH values as the adsorption of the oxo-anion meets increasing competition from the adsorption of OH^-.[27] Likewise the adsorption of amphoteric metals will decrease above some optimal pH or pH range. In the acidic range, hydrogen ions compete with simple cations or positively charged hydroxo complexes to be adsorbed at OH^- sites at the surface of an insoluble hydroxide.

In general, $Fe(OH)_3$ and $Al(OH)_3$ are good collectors for such metals as Bi, Sn, Pb, Ga, In, Cr(III), Ti, Te, and Ge in neutral or faintly basic medium.[28] Ammonia may not be a good precipitant when ammine-forming metals (Cu, Zn, Ni, etc.) are to be collected. Masking reactions are of value in some hydroxide separations. Mercaptoacetic (thioglycolic) acid complexes iron and other heavy metals in weakly basic solution, while allowing Al, Be, and other metals to be precipitated.

Especially metal ions hydrolyzing in acidic solution can be collected with hydrous manganese dioxide[29] (Table 7-1) precipitated in such solutions by interaction of Mn^{2+} and MnO_4^-. This collection is of interest because of its more selective character, compared to hydroxide collection in the neutral or slightly basic range, which brings down most of the heavy metals and often is of greater value for concentration than for separation. Collection with $MnO_2 \cdot xH_2O$ is particularly effective with Sb, Bi, and Sn, which in trace amounts can be recovered quantitatively, or almost so, at acidities up to $\sim 0.5\ M\ HNO_3$.[30]   See Figs. 7-2 and 7-3. Lead is completely carried down at low acidities (0.01–$0.1\ M\ HNO_3$). Iron(III) is strongly coprecipitated, Cu(II) much less so. Nickel is not adsorbed at low acidities ($< 1\ M\ HNO_3$).

Presumably the adsorption of the heavy metals mentioned involves replacement of H in $-\overset{|}{\underset{|}{Mn}}OH$ by the metal ion or a cationic hydroxo species; at low acidities uncharged species could be adsorbed and carried down. Iron(III) is almost completely coprecipitated from $6\ M\ HClO_4$, in which it must be present as Fe^{3+}. The coprecipitation of metals can be looked upon as the local formation of insoluble manganites.

Metastannic acid is a collector for W in $1.5\ M\ HNO_3$.[31]

Sulfide Collectors.   Quantitative precipitation of traces of metals as sulfides without the use of collectors is hardly possible. The equilibria involved in sulfide precipitations may be quite complex. Solubility in metal–sulfide systems may be determined by species other than simple metal ions. Calculations based only on solubility product constants, as sometimes presented in textbooks, can lead to grossly erroneous conclusions. For at least three metals, Hg, Ag, and Zn, complex sulfide anions and the corresponding acids (e.g., H_2HgS_2, AgHS, $ZnHS_2^-$) need to be taken into account in a realistic calculation of solubilities. As we have seen, the solubility of HgS is actually reduced if the solid is in equilibrium

TABLE 7-1
$MnO_2 \cdot xH_2O$ as Collector[a]

Metal Collected	Matrix	Conditions	Reference
1 As	Pb	$\sim 0.1\ M$ HNO_3 ($\sim 90\%$ recovery of As). Second pptn. in filtrate after removing Pb as sulfate and making ammoniacal. Milligram amounts of As.	C. L. Luke, *Ind. Eng. Chem. Anal. Ed.*, **15,** 626 (1943).
2 As	Pb	0.03–$0.5\ M$ HNO_3.	M. Tsuiki, *J. Electrochem. Soc. Japan*, **29,** 42 (1961).
3 Bi	Pb	pH ~ 5. Double pptn.	N. Kameyama and S. Makishima, *J. Chem. Soc. Japan*, **36,** 346 (1933).
4 Bi	Cu	$0.02\ M$ HNO_3. Double pptn.	Y. Yao, *Anal. Chem.*, **17,** 114 (1945).
5 Bi	Cu		Y. Yamasaki, *Japan Analyst*, **6,** 422 (1957).
6 Bi	Pb		See No. 2.
7 Bi	—	Recovery $\sim 96\%$ from $0.06\ M$ HNO_3 (10–20 mg Bi in 200 ml) and $\sim 10\%$ (10 mg Bi in 200 ml) from $1.2\ M$ HNO_3.	D. Ogden and G. F. Reynolds, *Analyst*, **89,** 538 (1964).
8 Bi	Cu, Pb, Pb ore	$< 0.07\ M$ HNO_3. Greater than trace amounts.	S. Kallmann and F. Pristera, *Ind. Eng. Chem. Anal. Ed.*, **13,** 8 (1941).
9 Bi	Ni	Collection of 100 μg Bi complete from 0.01 to $0.1\ N$ HNO_3, $HClO_4$ or H_2SO_4 (200 ml volume).	K. E. Burke, *Anal. Chem.*, **42,** 1536 (1970).
10 Ce	Cast iron	$\sim 1\ M$ HNO_3	A. A. Amsheev, *Zav. Lab.*, **34,** 789 (1968); *Anal. Abst.*, **17,** 2740.
11 Co	Sea water	Nanogram amounts of Co in 1 liter collected in MnO_2 formed by UV irradiation of ~ 0.1 mg Mn^{2+}.	B. R. Harvey and J. W. R. Dutton, *Anal. Chim. Acta*, **67,** 377 (1973).
12 Fe(III)		pH 7.	H. N. Stokes and J. R. Cain, *J. Am. Chem. Soc.*, **29,** 409, 443 (1905).
		With 30 mg Mn(IV), $\sim 98\%$ Fe copptd. in $0.01\ N$ HNO_3 and 89% in $0.1\ N$ (1 mg Fe). With more Fe, percentage of copptn. decreases. In $HClO_4$, 0.01–$6\ M$, 99% Fe copptd.	K. E. Burke, *loc. cit.* (see No. 9).
13 Fe	NaCl	$\sim 0.003\ M$ HCl.	S. Hirano and E. Igami, *Japan Analyst*, **7,** 106 (1958).
14 Ga	Al, Zn	pH 1.5.	V. S. Biskupsky, *Anal. Chim. Acta*, **46,** 149 (1969).
15 Mo	Cu	$0.016\ M$ HNO_3.	B. Park, *Ind. Eng. Chem. Anal. Ed.*, **6,** 189 (1934).

713

TABLE 7-1 (Continued)

Metal Collected	Matrix	Conditions	Reference
16 Mo	—	Recovery ~100% at pH 2.0–5.3.	M. Tanaka, *Mikrochim. Acta*, 204 (1958).
17 Mo	Sea water	pH 3.8.	K. Sugawara, M. Tanaka, and S. Okabe, *Bull. Chem. Soc. Japan*, **32**, 221 (1959).
18 Nb	Biomaterial (plants)		N. A. Tyutina, V. B. Aleskovsky, and P. I. Vasil'ev, *Geochem. Internat.*, **6**, 668 (1959).
19 Nb	Ti, Cr, Fe, etc.	Dilute H_2SO_4.	V. M. Dorosh, *Zh. Anal. Khim.*, **16**, 250 (1961).
20 Pb		With <10 mg Pb, recovery is 97% or better in 0.008 M HNO_3.	G. F. Reynolds and F. S. Tyler, *Analyst*, **89**, 579 (1964).
21 Pb	Cu, Zn	Milligram amounts of Pb; recovery somewhat low (tin present). 0.008 M HNO_3. Cu coprecipitated ~0.2% with 100 mg present, proportionately more with smaller amounts. Copptn. of Zn very small.	C. M. Pyburn and G. F. Reynolds, *Analyst*, **93**, 375 (1968).
22 Pb	Ni	Copptn. of Pb (0.1 mg) with 30 mg Mn (IV) 98–100% in 0.008–0.1 M HNO_3 or $HClO_4$. Copptn. of Ni very small, but $MnO_2 \cdot xH_2O$ contaminated with Fe and Cu.	Burke, *loc. cit.* (see No. 9).
23 Pb(Bi)	Ni	pH ~4. Pptn. in cold, then heated. Two pptns. for Pb.	S. J. H. Blakeley, A. Manson, and V. J. Zatka, *Anal. Chem.*, **45**, 1941 (1973).
24 Sb	Cu	Dilute HNO_3.	H. Blumenthal, *Z. Anal. Chem.*, **74**, 33 (1928).
25 Sb	Cu	Weakly acidic.	B. Park and E. J. Lewis, *Ind. Eng. Chem. Anal. Ed.*, **5**, 182 (1933); Park, *ibid.*, **6**, 189 (1934).
26 Sb	Cu, Pb, ores	<0.07 M HNO_3. Greater than trace amounts	See No. 8.
27 Sb	Cu	1:20 HNO_3. Microgram quantities of Sb quantitatively recovered in presence of 1 g Cu. Reprecipitation can be made from 0.4 N H_2SO_4 to remove copptd. elements.	H. Goto and Y. Kakita, *Sci. Rept. Res. Inst. Tohoku Univ. Ser. A*, **4**, 589 (1952).
28 Sb	Pb	1.2 M HNO_3. Mg amounts of Sb.	See No. 20.

714

TABLE 7-1 (Continued)

Metal Collected	Matrix	Conditions	Reference
29 Sb	Pb	From $\sim0.1\ M$ HNO$_3$, recovery of Sb is $\sim95\%$; a second pptn. is made.	See No. 1.
30 Sb	Bi	1.2 M HNO$_3$. Slightly low recoveries of 0.5–1 mg Sb in presence of 10–20 mg Bi.	See No. 7.
31 Sb		1.2 M HNO$_3$. Microgram amounts Sb recovered satisfactorily. Au(III) said not to be copptd.	B. J. McNulty and L. D. Woollard, *Anal. Chim. Acta*, **13,** 64 (1955).
32 Sb		pH 1–7.	A. K. Babko and M. I. Shtokalo, *Zav. Lab.*, **21,** 767 (1955).
33 Sb	Ni	Recovery of 0.1 mg Sb with 30 mg Mn (IV) 100% in 0.01–0.1 M HNO$_3$ and 0.01–0.6 M HClO$_4$; recovery decreases above these ranges, e.g., 83% in 1.2 M HNO$_3$. ~10 mg Mn(IV) in 200 ml 0.1 M HNO$_3$ enough for 0.1 mg Sb. 10 M HCl $(+H_2O_2)$ required to dissolve Sb in MnO$_2$. 1 g Fe or Cu does not hinder collection.	See No. 9.
34 Sb	Cast iron	pH 1 (HNO$_3$)	R. C. Rooney, *Analyst*, **82,** 619 (1957).
35 Sb	Pb		See No. 2.
36 Sb, Sn	Cd	For sepn. of Sb from Cd, $<0.06\ M$ HNO$_3$; partial copptn. of Bi.	E. Temmerman and F. Verbeek, *Anal. Chim. Acta*, **43,** 263 (1968).
37 Sn	Cu, Pb, or Ag	0.5 N HNO$_3$ or H$_2$SO$_4$.	See No. 8.
38 Sn	Cu	Weakly acidic.	See No. 25.
39 Sn	Pb	1.2 M HNO$_3$. Mg amounts of Sn. Some copptn. of Pb. Sn incompletely recovered from 0.008 M HNO$_3$.	See Nos. 21 and 20.
40 Sn	Pb	$\sim0.1\ M$ HNO$_3$: Sn in mg amounts 99% recovered.	See No. 1.
41 Sn	Ni	Recovery of 0.1 mg Sn 98% from 0.01–1.2 M HNO$_3$, 95% from 2 M HNO$_3$. Essentially no copptn. of Ni. As little as ~10 mg Mn(IV) sufficient. 10 M HCl $(+H_2O_2)$ required to dissolve Sn in MnO$_2$. 1 g Fe or Cu does not hinder collection.	See No. 9.

TABLE 7-1 (Continued)

Metal Collected	Matrix	Conditions	Reference
42 Sn	Bi	1.2 M HNO₃. Some copptn. of Bi.	See No. 7.
43 Sn	Cast iron	~1 M HNO₃	See No. 10.
44 Tl	Pb	Microgram quantities of Tl quantitatively recovered from HNO₃ solutions as strong as 1 (HNO₃): 9.	C. L. Luke, *Anal. Chem.*, **31**, 1680 (1959).
45 W	U alloys	0.1–0.2 M H₂SO₄.	A. V. Vinogradov et al., *Zav. Lab.*, **39**, 150 (1973).

[a] This table is not intended to be complete. It should be noted that many of these procedures apply to milligram and centigram quantities of metals, not to microgram quantities. Generally, the collection of microgram quantities will be relatively better, or at least not poorer, than the collection of larger quantities. For adsorption on preformed MnO₂ by batch or column method, see C. Bigliocca et al., *Anal. Chem.*, **39**, 1634 (1967).

with an acidic solution that is less than saturated with H_2S. Collectors also help to reduce the solubility. The mechanism of collection is not always known. There is good evidence for mixed crystal formation and compound formation in some trace metal–metal collector–sulfide systems. Adsorption phenomena presumably also play a role, more in some systems than in others. To a large extent the choice of a collector is still rather empirical. The rate of precipitation in some sulfide systems is very slow (possibly because of the low concentration of S^{2-} in acidic solutions). A collector may hasten precipitation.

Sulfide precipitation in mineral acid solution has some degree of selectivity, which may make it useful for separating traces of sulfophilic elements from samples composed chiefly of such common elements as Ca, Mg, Ti, Cr, Mn, Fe, Al, and P. Biomaterials fall in this class, as do silicate and carbonate rocks, ferrous metals and alloys, some nonferrous metals and alloys, and like materials. Extraction methods, as with dithizone, are likely to be superior to precipitation methods, but some trace metals can still be isolated advantageously at times by sulfide precipitation.

Copper sulfide is a common collector for metals whose sulfides are insoluble in dilute mineral acid solutions.[32] As little as 0.02 μg Hg/liter can be satisfactorily recovered with CuS. Cadmium sulfide forms mixed crystals with HgS and should be a good collector for Hg; it is known that 1 μg Cd in 100 ml can be recovered satisfactorily in this way. HgS as a collector has the advantage that it can be volatilized at a low temperature.

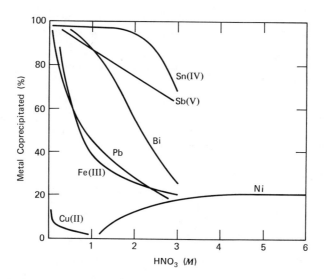

Fig. 7-2. Collection of various metals by $MnO_2 \cdot xH_2O$ ($MnO(OH)_2$) in HNO_3 solution. Potassium permanganate solution (2.5 ml, 1.25%) added to 200 ml of boiling sample solution containing 5 ml 5% $MnSO_4$ (~31 mg Mn precipitated). 100 μg Sb, Bi, Sn, and Pb; 1000 μg Ni, Cu, Fe. Curves have been plotted from values tabulated by Burke (reference in note 30).

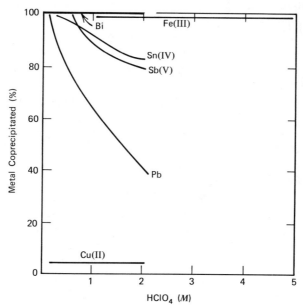

Fig. 7-3. Collection of metals by $MnO_2 \cdot xH_2O$ in $HClO_4$ solution. Conditions as in Fig. 7-2. Coprecipitation of Ni < 1% at all acidities. From Burke, *loc. cit.*

717

Quantitative sulfide precipitation of gold from sea water at the extreme dilution of 1 μg Au in 1 metric ton $(1:10^{12})$ with the aid of PbS has been reported (Haber).

Zinc sulfide can be used to collect gallium by isomorphous replacement in the host lattice.

Collectors are also used in the precipitation of trace metals as sulfides with thioacetamide. Thus in the separation of ~ 1 ppm of Sn and Cd in Nb, and like amounts of Pb and Bi, HgS was used as collector for the first two metals and CuS for the last two.[33]

Masking by complex formation is of value in some sulfide precipitations. When lead is precipitated as sulfide,[34] a dilute mineral acid solution cannot be used, even with a collector. The precipitation can be made at pH 3 (CuS as collector) if citrate is added to keep iron(III) and other phosphates in solution. Citrate is also useful in preventing the coprecipitation of W with MoS_3. Sn can be separated from Cu in acid solution if sufficient thiourea is added to keep Cu in solution.

When a photometric reagent has considerable tolerance to a major constituent in a sample, a small fraction of that constituent may be precipitated to collect a less soluble trace sulfide, for example, enrichment of CdS in ZnS. The principle may, of course, be used for precipitates other than sulfides.

Other Inorganic Collectors. Collectors are used in various precipitations with phosphates,[35] fluorides, sulfates, etc., as for example $AlPO_4$ for U, CaF_2 for the rare earth metals,[36] $BaSO_4$ for Th, and so on. Some selectivity is obtained by pH adjustment and masking. Thus Be is collected by $AlPO_4$ at pH 4.4, and Ca is left in solution. At about the same pH, $FePO_4$ can be used to collect Al, leaving Ca and Mg in solution.[37]

Organic Collectors. An organic reagent only slightly soluble in water may serve as a collector as well as a precipitant. The insoluble reagent, as it crystallizes from a solution, may not only act as a gathering agent, but may also adsorb on its surface and occlude in the interior of its particles otherwise unprecipitable quantities of a reacting element. In this way, the concentration of the element may even be brought below that corresponding to the normal solubility of its insoluble compound. Microgram quantities of Au, Ag, Pd, Cu, and other metals can be collected on solid dithizone[38] and gold on p-dimethylaminobenzylidenerhodanine or 2-mercaptobenzothiazole.[38] In general, however, if the metal–organic reagent compound can be extracted with an immiscible organic liquid, the precipitation procedure has no advantage. PAN (Chapter 6E) has been used as precipitant and collector for Ti, V, Mn, Fe, Co, and other metals.[39]

Slow precipitation (crystallization) of an organic reagent, as by cooling a hot aqueous solution or by boiling to remove the volatile organic solvent, has been studied as a means of concentrating tracer concentrations of metals in the separated crystals.[40] Solid solution formation occurs in some of these crystallizations and the distribution expressions holding for such systems are obeyed more or less closely.[41]

5,7-Dibromo-8-hydroxyquinoline serves well for cocrystallization of Cu, Zn, Fe(III), Co, and Cr(III) from sea water.[42] Microgram quantities of these metals per liter can be coprecipitated essentially completely at pH 8. The recovery of Mn, Pb, Ag, and Ce(III) is unsatisfactory. The reagent is added in acetone solution, and the acetone is boiled off. A similar procedure is used to coprecipitate As(III) (As(V) is reduced with ascorbic acid), Sb, and Cu with thionalide in the determination of these elements in sea water by neutron activation.[43]

Chitosan is a polymer having an amino group on each anhydroglucose unit. It is insoluble in water but soluble in dilute formic or acetic acid. Addition of such a reagent solution to a solution of transition (and other) metals precipitates chitosan (easily ashed), which by its chelating properties carries down these metals. However, the recovery is not quantitative.[44]

Another type of organic collector is the ion-association product formed between cationic and anionic organic species, which carries down a slightly soluble metal complex formed with one or the other of the organic components. Thus Methyl Violet thiocyanate is a collector for Mo(VI),[45] Methyl Violet tannate for thorium[46] and other metals.[47] Traces of Ti, Zr, and Th can be collected in pyridinium molybdophosphate.[48]

A slightly soluble metal chelate can also serve as a collector for another metal chelate.[49] But chelate collectors are infrequently used in photometric determinations, because most chelates are soluble in organic solvents immiscible with water and extraction provides a more convenient and better separation method as already mentioned.

Preformed Collectors. The best coprecipitating and gathering action is usually obtained by forming the collector in the sample solution. However, a collector that has strong adsorptive properties may be added as such to the solution. Frequently, the collector also functions as a reagent. For example, gold, platinum, palladium, selenium, tellurium, and arsenic present in ionic form at very low concentrations can be removed from a solution by shaking with solid mercurous chloride.[50] The reduced metal is held on the surface of the mercurous chloride particles. Lead, copper, zinc, and other heavy metals can be adsorbed on solid calcium carbonate from tap water. Heavy metal sulfides precipitate and collect less soluble sulfides. For example, when a solution containing copper in very low concentration is passed through a disk of filter paper impregnated with

CdS, CuS is precipitated on the paper because of its lower solubility:

$$CdS + Cu^{2+} \rightleftharpoons CuS + Cd^{2+}$$

Because K_{sp} of CuS is 6×10^{-36}, compared to K_{sp} of 8×10^{-27} for CdS, this reaction can be made to run to practical completion from left to right under the proper conditions of acidity and flow rate.[51] An asbestos pad impregnated with CdS is useful for isolating traces of mercury from solution as HgS. Some selectivity in sulfide separations can be obtained in this way. Copper can be separated from much lead ($K_{sp} = 2 \times 10^{-27}$) and silver from lead (Pb/Ag = 10^4).[52]

Silver can be separated from much Cu by percolating the sample solution (0.1 M HNO$_3$) through dithizone-impregnated paper.[53]

As little as 0.02 μg Ag in 2 liters 0.2 M HNO$_3$ can be isolated by adding a suspension of p-dimethylaminobenzylidenerhodanine (separation of Ag from Bi, Pb, Cu, and Fe).[54]

A collector falling in a different category is cellulose and its modifications. Because of the presence of carboxyl groups, cellulose (as a wad of cotton, for example) adsorbs heavy metals such as lead from neutral or faintly acidic solutions by ion exchange. Unbleached sulfite pulp, containing sulfonic acid groups, adsorbs copper strongly from distilled water.[55] Controlled oxidation of cellulose introduces carboxyl groups in large numbers, producing "oxycellulose," which fixes heavy metals and finds use in the isolation of metal traces.[56]

Carbon, especially as activated carbon, has found some use as an adsorbent for metals from aqueous solution.[56a] The adsorption of metals existing in ionic form in solution is usually not effective except for some easily reducible metals (Au); adsorption of chelated metals is stronger.

A number of insoluble inorganic substances function as adsorbers by ion exchange, for example, salts of heteropoly acids, complex cyanides of transition metals, and zirconium phosphate. Ammonium phosphomolybdate (ammonium 12-molybdophosphate, $(NH_4)_3[P(Mo_{12}O_{40})]$) adsorbs especially K, Rb, and Cs, the adsorption increasing in that order, with marked selectivity for Cs. The adsorbed ions enter the lattice of the salt by exchange with NH_4^+. Mixed crystal formation occurs. The salt lattice consists of an array of large spherical $P(Mo_{12}O_{40})^{3-}$ anions with NH_4^+ cations fitted in between them. The interspaces are large so that NH_4^+ is readily replaced by large Cs$^+$ ions. In fact, a more stable structure results when the cations are large and the interspaces are better filled, which probably explains why Cs is more strongly adsorbed than other alkali metal ions. From acidic solution, in addition to the alkali metals, the following metals, forming slightly soluble phosphomolybdates, are adsorbed: Fr, Tl(I), Hg(I), and Ag(I). Multivalent ions are exchanged to some extent in approximately neutral solution. The adsorption of Na is so weak that Cs can be adsorbed from sea water

$$D = \frac{[\text{Cs}]_{solid}}{[\text{Cs}]_{soln}} \sim 1500$$

The distribution coefficient for Cs decreases with acidity but still has large values of 3500 and 2000 in 0.1 and 1 M HCl.[57] The solubility of ammonium phosphomolybdate can be decreased by adding NH_4^+ (common ion) and HNO_3 to the solution. Ammonium ion cuts down the adsorption of K. In 0.1 M ammonium nitrate solution, the distribution coefficients of the alkali metals have the following values[58]:

Na	K	Rb	Cs
<1	3.4	230	6000

Because of its large distribution coefficient, cesium can be adsorbed on a relatively small amount of ammonium phosphomolybdate by a batch process, but a chromatographic column can also be used. Cesium and other adsorbed metals can be eluted with ammonium nitrate solution.

Separation by Froth Flotation. This type of separation is based on differences in surface activity.[59] A substance (as ions, molecules, colloid, or precipitate) is adsorbed or attached at the surfaces of bubbles rising through a liquid, and is thereby separated. A substance which is not surface active itself can often be made so through union with or adherence to a surface active agent (surfactant). Froth flotation involves the removal or separation of particulate material by frothing (foaming). It has many subdivisions. From the standpoint of trace analysis, ion flotation and precipitate flotation are important. If an insoluble product results from interaction between the ion to be separated and a surfactant, the process is called ion flotation. If the ion is first precipitated and the precipitate is then floated with or without the addition of a surfactant, the process is called precipitate flotation.[60] Adsorbing colloid flotation, in which the ion to be separated is adsorbed on colloidal particles and then floated, may be included in precipitate flotation.

An apparatus for batch flotation is relatively simple (Fig. 7-4). Flotation is accomplished in a flotation cell which is a cylindrical glass vessel having a sintered-glass disk to break the gas stream into small bubbles. Nitrogen or air is passed into the flotation cell.

Ion Flotation. Important parameters that affect both ion and precipitate flotation are the concentrations of the ion to be separated and of the surfactant (including the selection of surfactant), pH, ionic strength, and

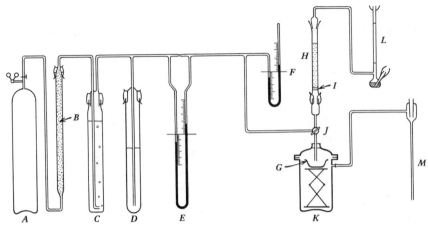

Fig. 7-4. Apparatus for flotation. (*A*) Nitrogen tank; (*B*) glass wool; (*C*) gas washing bottle; (*D*) regulator; (*E*) flowmeter; (*F*) manometer; (*G*) silica dish; (*H*) flotation cell; (*I*) sintered-glass filter; (*J*) T-bore stop-cock; (*K*) bell jar; (*L*) soap film meter; (*M*) aspirator. From K. Sekine and H. Onishi.[60]

flow rate. Temperature and the surface area of bubbles may affect the results.

Ion flotation has been used as a preconcentration technique for Ag, Au, Co, Cu, and Fe.[61] Anionic complexes of these elements are formed by adding oxalate, cyanide, or thiosulfate, and then floated from the solution by using a cationic surfactant and nitrogen. Submicrogram amounts of the trace elements can be separated with recoveries greater than 90% from 0.5–3 g of matrix elements such as Mg, Na, and Zn. The concentration factors are 30, 10^3, and 3×10^4 for single-, double, and triple-stage separations, respectively.

Milligram amounts of uranium have been separated by converting UO_2^{2+} to $UO_2(CO_3)_3^{4-}$ and floating the anionic complex with a cationic surfactant and nitrogen.[62] Microgram amounts of uranium can be separated by flotation of the UO_2-Arsenazo III–Zephiramine (a cationic surfactant) complex.[63] This method has been applied in the separation of U from sea water.[64]

Precipitate Flotation. The following examples illustrate uses of precipitate flotation. Traces of silver have been separated from copper by precipitating the former with *p*-dimethylaminobenzylidenerhodanine and floating the precipitate with sodium dodecylbenzenesulfonate and air.[65] As little as 0.02–1 μg of Ag can be concentrated with recoveries greater than 95% from 100–500 ml of 0.1–1 *M* nitric acid solution within 5 min. When the sample solution contains 1 μg of Ag and 0.5 g of Cu, the concentration factor of Ag with respect to Cu is approximately 10^3.

In some cases, no surfactants are required, but the solid formed when the ion to be separated and precipitant react must have a hydrophobic surface. Some of the precipitants that have been used, and the elements that have been floated by them, are as follows: 1,2-cyclo-hexanedionedioxime: Ni,[66] Pd[67]; 1-nitroso-2-naphthol: Ag,[68] U[68]; phenyl-2-pyridyl ketoxime: Au.[68]

Adsorbing colloid flotation has been used for the separation of Mo,[69] U,[70] Zn,[71] and Cu[71] from sea water. As little as 0.02 μg Hg/1 liter sea water can be recovered satisfactorily by coprecipitation (mixed crystal formation) with CdS at pH 1 and flotation of the latter with the aid of octadecyl-trimetylammonium chloride on N_2 bubbles.[72] The flotation collection is a substitute for filtration in this method. Molybdenum as MoO_4^{2-} and uranium as $UO_2(CO_3)_3^{4-}$ are adsorbed on positively charged iron(III) hydroxide at pH 4.0 and 5.7, respectively, and the colloids are floated with an anionic surfactant (sodium dodecyl sulfate) and air. On the other hand, zinc and copper present as cations are adsorbed on negatively charged iron(III) hydroxide at pH 7.6, and the colloid is floated with a cationic surfactant (dodecylamine) and air. Traces of tin(IV) can be separated from zinc by coprecipitation with Fe(III) hydroxide and flotation of the precipitate at pH 6 with the aid of a hot ethanolic paraffin solution and nitrogen.[73]

B. MIXED CRYSTAL (SOLID SOLUTION) FORMATION

Replacement of an ion M in a crystal lattice by an ion N of the same sign and similar size forms a mixed crystal (M, N)X.[74] It is not necessary that M and N have the same charge. A difference in charge can be compensated by suitable cosubstitutions (e.g., $KMnO_4$ in $BaSO_4$) or lattice vacancies. Y^{3+} replaces Ca^{2+} in the lattice of CaF_2, the resulting excess + charge being compensated by cation vacancies in the lattice. Mixed crystal formation is important in the collection of traces because it offers a way of circumventing the limitation imposed by the solubility of the micro component in a precipitation.

The two most important factors in mixed crystal formation in precipitation or crystallization from aqueous solution are the relative ionic sizes of M and N and the relative solubilities of MX and NX. Other factors are the manner of precipitation (e.g., rate) and the treatment of the precipitate after its formation. For mixed crystal formation to occur at all, the sizes (usually we are concerned with ionic sizes) of M and N must not differ too much. The permissible difference cannot be precisely specified. As a matter of fact, the radius of a cation depends on its coordination number, and the ionicity of its bond to anions of different electronegativities. The empirical rule of V. M. Goldschmidt states that mixed crystal formation between MX and NX may be expected to occur if the ionic radii of M and N agree within ~ 15%.[75] The extent of mixed crystal formation—the miscibility of NX in MX—can be quite limited and still allow incorporation of N in MX when N is a micro component and N/M is < 0.001 or more likely < 0.0001 in solution (see p. 731).

It will be understood that if NX is much more soluble than MX, N will not be incorporated in MX to a significant extent. Although the cations Na^+ and Ca^{2+} and the anions NO_3^- and CO_3^{2-} have similar ionic radii, respectively, little $NaNO_3$ will be found in $CaCO_3$ precipitated from a $Ca^{2+}-Na^+$ solution. We shall return to this point shortly.

Homogeneous and Inhomogeneous Distribution. The distribution of the microcomponent N between a solution and a mixed crystal is a great deal more complicated than would be its distribution between two immiscible liquids. The difference results in large measure from the limited mobility of N in MX. At the end of the precipitation, the interior of the crystals of MX will not, in general, be in equilibrium with the external solution, and a drastic recrystallization may be needed to bring this about. To be sure, the existence of nonequilibrium may be to the advantage of the trace analyst.

If the precipitation or crystallization process is such that the trace component is distributed homogeneously through the crystals of the macrocomponent, that is, if the crystals as a whole are in equilibrium with the solution, the following relation holds ideally (activities should actually be written instead of concentrations)[76]:

$$\frac{[N]_{cryst}}{[N]_{soln}} = D\frac{[M]_{cryst}}{[M]_{soln}} \tag{7-1}$$

Ideally,[77] the homogeneous distribution coefficient D is equal to the ratio of the solubility product constants of the two precipitates MX and NX (1 : 1 type):

$$D = \frac{(K_{sp})_{MX}}{(K_{sp})_{NX}} \tag{7-2}$$

The larger the value of D the greater will be the fraction of N in the crystals. This means that it is desirable to have MX more soluble (within limits) than NX. However, the solubility of MX should be small enough to enable $[M]_{cryst}/[M]_{soln}$ to be made 100 or greater. The state of affairs represented by Eq. (7-1) is approached if the precipitate is rapidly formed so that it is as fine as possible and it is then allowed to recrystallize in the mother liquor to produce a more or less homogeneous distribution of N through MX.

If, during precipitation, equilibrium exists between the solution and an infinitesimally thin surface layer of the crystals, the relation holding is[78]:

$$\log \frac{N_{tot}}{N_{soln}} = \lambda \log \frac{M_{tot}}{M_{soln}}$$

or

$$\log \frac{[N]_{initial}}{[N]_{final}} = \lambda \log \frac{[M]_{initial}}{[M]_{final}} \tag{7-3}$$

(concentrations in solution)

where λ is the logarithmic distribution coefficient. This expression represents heterogeneous distribution of NX through MX. Equation (7-3) has been shown to hold very well for the coprecipitation of radium chromate with barium chromate formed by slow precipitation brought about by gradual neutralization of a nitric acid solution with urea or cyanate.[79] The distribution coefficient for the *surface layer* in the heterogeneous precipitation process should ideally be D of Eq. (7-1).

Precipitation represented by Eq. (7-3) is more favorable for collecting traces than precipitation represented by Eq. (7-1). Compare the calculted values in Table 7-2. In practice, it is usually found that the distribution does not follow either of the two limiting equations but falls somewhere between the two. The rate of precipitation exerts a strong effect on the value of λ. If λ is greater than 1 (enrichment system), it becomes smaller as the rate of precipitation increases, tending toward 1. The data in Table 7-3 illustrate this. Hermann correlates the decrease in λ with an increase in supersaturation during precipitation (Table 7-3).

In the system rare earth element–calcium–dimethyl oxalate (hydrolyzing in hot acetic acid medium to give oxalate), there is good correlation

TABLE 7-2

Illustrating Collection By Homogeneous and Heterogeneous Mixed Crystal Formation

M pptd. (%)	N pptd. (%)					
	Homogeneous D			Heterogeneous λ		
	0.5	1	2	0.5	1	2
90	81.8	90	94.8	68.5	90	99.0
95	90.5	95	97.3	77.7	95	99.75
99	98.1	99	99.5	90.0	99	99.99

TABLE 7-3

λ for Coprecipitation of Am in $La_2(C_2O_4)_3$ as a Function of Rate of Precipitation of $La_2(C_2O_4)_3$[a]

Average precipitation rate (%La/hr)	0.01	0.07	1.2	1.9	4.1	53
λ		6.6	6.1	5.8	5.0	3.8	1.5

[a] J. A. Hermann, Ph.D. Thesis, University of New Mexico, 1955. Precipitation from homogeneous solution by hydrolysis of methyl oxalate.

between the quotient of the apparent rate constants of precipitation of the rare earths (R.E.) and Ca ($k_{R.E.}/k_{Ca} \sim \lambda_{calc}$) and observed λ.[80] This correlation fails when trace amounts of Eu and Y are coprecipitated with other rare earth oxalates as collectors. In the system CaC_2O_4–SrC_2O_4, $\lambda = 0.35$ when CaC_2O_4 is the collector and SrC_2O_4 is the micro component (0.1 and 10^{-6} M, respectively) and $\lambda = 2.1$ for the converse case. It is claimed that here λ is determined by the K_{sp} values rather than the precipitation rates. (Evidently our understanding of the dependence of λ or D on solubilities of macro and micro components, conditions of precipitation, and ionic sizes is far from satisfactory.)

It may be noted that even if λ (or D) is only 1, collection may be adequate provided that the collector is sufficiently slightly soluble. Thus if $\lambda = 1$ and $M_{tot}/M_{soln} = 100$ ($\sim 99\%$ M precipitated), $N_{tot}/N_{soln} = 100$ also; the same result is obtained if $D = 1$. If mixed crystal formation did not occur, and K_{sp} of NX were not exceeded, all of N would remain in solution except for such amounts as might be adsorbed on MX or be occluded by processes other than mixed crystal formation.

A limited number of distribution coefficients λ or D are available in the literature. The trace analyst may wish to find a collector functioning by mixed crystal formation for trace constituents not previously studied. With the aid of a table of ionic radii and a table of solubilities he may be able to select some possibilities that will shorten the experimental testing. To be able to calculate D is too much to expect, but at least it may be possible to select an MX for a given N that possibly will give a D larger than unity. Equation (7-3) may be expected to hold better when M and N are very similar in size (and have the same charge). We seek an anion giving MX of low solubility, and, if possible, an NX of still lower solubility. Ionic radii can be taken, for many purposes, from a table or chart (Fig. 7-5) of Goldschmidt's empirical values. More sophisticated sets of ionic radii are also available (for example, Table 7-4).

As an illustration, we may consider a case in which the conclusions can be tested by experimental results. This is the isolation of $PbSO_4$ by mixed crystal formation with another slightly soluble sulfate. $SrSO_4$ and $BaSO_4$ immediately suggest themselves for this purpose. The pertinent data are:

	K_{sp}	Ionic Radius (Å) (Goldschmidt)	
$PbSO_4$	$10^{-7.8}$	Pb^{2+}	1.32
$SrSO_4$	$10^{-6.6}$	Sr^{2+}	1.27
$BaSO_4$	$10^{-10.0}$	Ba^{2+}	1.43

$SrSO_4$ is the choice for the collector. The ionic radius of Pb^{2+} is apparently closer to that of Sr^{2+} than to that of Ba^{2+}. The solubility

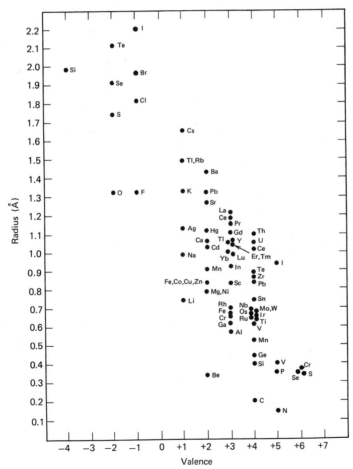

Fig. 7-5. Empirical ionic radii according to V. M. Goldschmidt.

TABLE 7-4
Ionic Radii (Å)[a]

Ag(I)	1.23	Mn(III)	L 0.66
Al(III)	0.61		H 0.73
Am(III)	1.08	Mn(IV)	0.62
Am(IV) (8)	1.03	Mn(VI) (4)	0.35
As(V)	0.58	Mn(VII) (4)	0.34
Au(III) (4, square)	0.78	Mo(III)	0.75
Ba(II)	1.44	Mo(IV)	0.73
Be(II) (4)	0.35	Mo(V)	0.71
Bi(III)	1.10	Mo(VI)	0.68
Ca(II)	1.08	Na(I)	1.10
Cd(II)	1.03	Nb(III)	0.78
Ce(III)	1.09	Nb(IV)	0.77
Ce(IV)	0.88	Nb(V)	0.72
Co(II)	L 0.73	Nd(III)	1.06
	H 0.83	Ni(II)	0.77
Co(III)	L 0.61	Ni(III)	L 0.64
	H 0.69		H 0.68
Cr(III)	0.70	Np(IV) (8)	1.06
Cr(VI) (4)	0.38	Os(IV)	0.71
Cs(I)	1.78	Pa(IV) (8)	1.09
Cu(I) (2)	0.54	Pa(V) (8)	0.99
Cu(II)	0.81	Pb(II)	1.26
Dy(III)	0.99	Pb(IV)	0.86
Er(III)	0.97	Pd(II)	0.94
Eu(II)	1.25	Pd(IV)	0.70
Eu(III)	1.03	Pm(III)	1.04
Fe(II)	L 0.69	Po(IV) (8)	1.16
	H 0.86	Pr(III)	1.08
Fe(III)	L 0.63	Pr(IV)	0.86
	H 0.73	Pt(IV)	0.71
Ga(III)	0.70	Pu(III)	1.09
Gd(III)	1.02	Pu(IV)	0.88
Ge(IV)	0.62	Ra(II) (8)	1.56
Hf(IV)	0.79	Rb(I)	1.57
Hg(II)	1.10	Re(IV)	0.71
Ho(III)	0.98	Re(V)	0.60
In(III)	0.88	Re(VI)	0.60
Ir(III)	0.81	Re(VII)	0.65
Ir(IV)	0.71	Rh(III)	0.75
K(I)	1.46	Ru(III)	0.76
La(III)	1.13	Ru(IV)	0.70
Li(I)	0.82	Sb(III) (5)	0.88
Lu(III)	0.94	Sb(V)	0.69
Mg(II)	0.80	Sc(III)	0.83
Mn(II)	L 0.75	Se(−II)	1.88
	H 0.91	Se(VI) (4)	0.37

728

TABLE 7-4 (Continued)

Si(IV) (4)	0.34	Ti(IV)	0.69
(6)	0.48	Tl(I)	1.58
Sm(III)	1.04	Tl(III)	0.97
Sn(II) (8)	1.30	Tm(III)	0.96
Sn(IV)	0.77	U(IV) (7)	1.06
Sr(II)	1.21	U(VI)	0.81
Ta(III)	0.75	V(III)	0.72
Ta(IV)	0.74	V(IV)	0.67
Ta(V)	0.72	V(V)	0.62
Tb(III)	1.00	W(IV)	0.73
Tb(IV)	0.84	W(VI)	0.68
Tc(IV)	0.72	Y(III)	0.98
Te(IV) (3)	0.60	Yb(III)	0.95
Th(IV)	1.08	Zn(II)	0.83
Ti(III)	0.75	Zr(IV)	0.80

[a] Selected from E. J. W. Whittaker and R. Muntus, *Geochim. Cosmochim. Acta,* **34,** 945 (1970). Values are for sixfold coordination, except where indicated by arabic numerals in parentheses. L and H = low and high spin. For radii for other coordination numbers, see this paper.

relations are also favorable, $PbSO_4$ being a little less soluble than $SrSO_4$. The absolute solubility of $SrSO_4$ is also favorable. Neglecting activity coefficients, we calculate the solubility of $SrSO_4$ in 0.1 M sulfate solution as $10^{-6.6+0.1} = 10^{-6.5}$ M or ~0.003 mg Sr in 100 ml. It will be advantageous to make the precipitation of $SrSO_4$ in faintly acidic solution so that its solubility will be greater than this value, but it should not exceed 1 mg Sr in 100 ml.

The calculated value of D [Eq. (7-2)] at 25° is

$$D = \frac{(K_{sp})_{SrSO_4}}{(K_{sp})_{PbSO_4}} = 10^{-6.6+7.8} = 10^{1.2} \sim 10$$

The magnitude of calculated D is such that we may expect that it is at least greater than 1, and enrichment of Pb in $SrSO_4$ will occur. Assuming that D actually is 10, and only 50% of the strontium in the solution is precipitated as sulfate, the ratio of lead in the $SrSO_4$ precipitate to lead left in solution will be, if the homogeneous distribution law (the less favorable case) is followed:

$$\frac{[Pb^{2+}]_{cryst}}{[Pb^{2+}]_{soln}} = 10 \times 1 = 10$$

(With heterogeneous distribution, $[Pb^{2+}]_{cryst}/[Pb^{2+}]_{soln} = 1000$, and 99.9%

of the lead would then be in the precipitate.) Hahn[81] found that when 50% of the strontium in solution had been precipitated, ~99% of total lead (as ThB) originally present in the solution was in the $SrSO_4$ precipitate. It is not difficult to precipitate 99% of the strontium (say 0.1–0.2 g total Sr in 100 ml), so that the coprecipitation of lead can be made quantitative.

The lead in the strontium sulfate precipitate can be brought into solution by fusing the precipitate with sodium carbonate, leaching, filtering, washing, and dissolving the insoluble carbonates in hydrochloric acid. The isolation of lead by coprecipitation with strontium sulfate is not often applied in practice, because it is generally more advantageous to extract lead with dithizone.[82] Ion-exchange methods are also available.

Evidently the collector must be a substance that does not interfere subsequently in the determination of the trace constituent or that allows easy separation of the two before the determination. If these conditions cannot be fulfilled, isolation of a constituent by mixed crystal formation is of no advantage, even if recovery is ~100%.

Foreign substances may have a marked effect on the isolation of traces of metal by mixed-crystal formation. Any substance converting N to a complex ion not fitting into the lattice of MX will, of course, decrease the distribution coefficient of N. Chloride forms chloroplumbite ions and reduces the amount of lead incorporated in strontium sulfate. Thus in a 1 M KCl solution, 82% of the lead is found in the precipitate when 50% of the strontium is precipitated, and only 30% when the precipitation is made in 2.5 M KCl.[81]

When a collector for a given metal is being selected, one should be chosen that has approximately the same cationic radius, provided that the corresponding compound is sufficiently slightly soluble. Thus we have found that $Co(OH)_2$ is a good collector for $Sc(OH)_3$ (Goldschmidt's ionic radii: Co^{2+} 0.83 Å, Sc^{3+} 0.83 Å). In the sulfide precipitation of Mo(VI), Sb(V) is an effective collector (Mo^{6+} 0.68 Å, Sb^{5+} 0.69 Å for sixfold coordination).[83] Antimony(III) cannot be used in place of Sb(V), nor is Cu^{2+} effective. Presumably the excess positive charge resulting from the replacement of Mo(VI) by Sb(V) is compensated by cation vacancies: 5 Mo(VI) + 1 cation "hole" ≡ 6 Sb(V).

Lathanum fluoride is a good collector for tetrapositive Ce and U.[84] Many other examples could be mentioned in which the replacing and replaced ions do not have the same charge. It might even be expected that, for example, an ion N^{3+} could be held more firmly than the replaced ion M^{2+} in the lattice of MX_2 if other factors are favorable.

Barium Sulfate for Collection of Radioactive Elements. An interesting and useful method for the separation of small amounts of uranium and transuranium elements uses barium sulfate as collector.[85] Trace concentrations of tri- and quadrivalent cations having Goldschmidt radii larger than ~ 1.0 Å can be quantitatively coprecipitated with $BaSO_4$ in the presence of K^+ (following a $K_2S_2O_7$ fusion of the sample). With the exception of At, Fr, and Rn, all elements from Pb to at least Cf are coprecipitated efficiently with $BaSO_4$, as are La and the lighter lanthanides. The elements U through Am are prevented from coprecipitating by oxidation to the highest oxidation states, in which the oxygenated cations of low charge are too large to fit into the lattice of $BaSO_4$. The conditions of the precipitation of $BaSO_4$ are very important. If $BaSO_4$ precipitates locally where Ba^{2+} first encounters SO_4^{2-}, the collection of the metals named may not be quantitative and the presence of Th in more than very small amounts leads to their loss. On the other hand, if Ba^{2+} is distributed throughout the solution before precipitation of $BaSO_4$, these unfavorable effects are minimized. When precipitation is delayed for a few seconds, it then occurs in a homogeneous solution and the whole volume of a $BaSO_4$ crystal is available for the incorporation of coprecipitable trace constituents (heterogeneous mixed crystals). Delay of barium sulfate precipitation can be brought about by the use of sufficiently dilute barium solutions, dropwise addition with good mixing and precipitation in boiling solution. Under such conditions 1 mg of Th (one of the worst interferences) produces a loss of only 0.3% Am.

When the coprecipitation of U is desired, U(VI) can be reduced to U(IV) with Ti(III). Differential oxidation enables various separations to be made in the group of coprecipitable elements. Oxidation with peroxydisulfate in the presence of Ag^+ keeps U $\cdots$ Am in solution when Ba^{2+} is subsequently added. Mono- and divalent cations (except those forming insoluble sulfates) and trivalent cations in moderate amounts produce little error.

Soluble Salts. Although soluble salts find little use as collectors, the possibility is worth consideration at times. For example, lead is enriched in certain alkali halides crystallizing from aqueous solution. When 18.4% of KCl in a solution is crystallized out at 0°, 92.8% of the lead (as ThB) in the solution is incorporated in the crystals; the distribution coefficient D has a value of 57.[86] Lead is also incorporated in NaCl.[87] TlCl can be accumulated in KCl but not in NaCl. Crystallization of 5% of the KCl in a solution removes 99% of the thallium.[88] Cadmium in concentrations below 1 ppm is carried down almost completely in the first 5% of the NaCl crystals separating from aqueous solution.[89] Zinc, bismuth and manganese are reported to remain in solution.

Limited Miscibility. An example of mixed crystal formation occurring to a limited extent seems to be provided by Ga_2S–HgS. Gallium sulfide is

TABLE 7-5
Coprecipitation of Ga with HgS[a]

Ga Present (mg)	R (%)		Molar Ratio[b] (Hg/Ga)
	0.05 M HCl	0.2 M HCl	
5×10^{-7}	70	67	1.4×10^7
5×10^{-6}	84	81	
5×10^{-5}	93	78	1.4×10^5
2×10^{-3}	95	79	
5×10^{-2}	95	64	
0.1	93	35	70
0.21	90	17	33
0.5	—	6.5	
0.7	78	—	
1.0	75	—	7
2.1	65	3.3	3.3
63	12	—	0.11

[a] Selected values from N. A. Rudnev, A. A. Loginov, and G. I. Malofeeva, *J. Anal. Chem. USSR*, **22**, 703 (1967). H_2S passed for 3 min at a rate of 350 ml/min into 10 ml of solution, either 0.05 or 0.2 M in HCl, containing 20.0 mg Hg(II); room temperature. R = recovery.
[b] In solution, before precipitation.

incompletely precipitated from 0.05 M HCl solution (cf. last row of Table 7-5), but is coprecipitated with As_2S_3, HgS, and other sulfides. Apparently, a solid solution is formed between HgS and Ga_2S_3 over a limited range. With a molar ratio of Hg/Ga in the precipitate between $\sim 10^5$ and ~ 70, the percentages of Ga coprecipitated from 0.05 M HCl remain essentially constant at 94 as the Ga concentration in the solution varies from 0.005 to 10 ppm (Table 7-5). When the molar ratio Hg/Ga in the precipitate falls below ~ 70, the percentage of coprecipitated Ga decreases. An odd feature of the HgS–Ga_2S_3 system is the apparent failure of Ga to be incorporated in HgS to the expected extent when its concentration in the solution is less than ~ 0.005 ppm. This behavior has not been satisfactorily explained.

C. COMPOUND FORMATION

There is little doubt that coprecipitation or gathering sometimes involves compound formation between the trace constituent and the collector. Coprecipitation phenomena are complex, and it is not always possible to distinguish the processes operating. Adsorption may be a step along the way to compound formation. The carrying down of As(V) by ferric hydroxide may be a combination of adsorption of arsenate on the precipitate and the formation of ferric arsenate. If, at equilibrium, ferric arsenate is present as a solid phase, the concentration of arsenic in the solution would be that existing in a saturated solution of ferric arsenate under the conditions. There is evidence that ferric hydroxide can form ferrites with certain divalent metals such as Zn, Ni, and Co. On digestion, in ammoniacal solution containing these ions, their content in precipitated ferric hydroxide increases. Manganites may be formed when $MnO_2 \cdot xH_2O$ is used as collector.

Tellurium, formed by reduction of tellurite with $SnCl_2$ or SO_2 in HCl solution, is an excellent collector for traces of Au, Ag, Pd, Pt, Po, and some other easily reducible metals.[90] No doubt tellurides of these metals are formed,[91] although how the tellurides occur in the precipitate is apparently not known. Perhaps there is an atomic dispersion (solid solution) of the reduced metals in the tellurium matrix. Selenium has also been used as a collector for some easily reducible metals. Because of its greater selectivity, collection of the metals mentioned with Te is superior to collection by various sulfides, which, moreover, may not be as effective in their collecting action. Elemental arsenic, precipitated by hypophosphite in acid solution, is an effective collector for tellurium.

Metal sulfide collectors may, at times, form compounds with trace element sulfides. Complex sulfides (called sulfosalts by mineralogists) are of common occurrence in nature. The coprecipitation of Tl with As_2S_3, SnS_2, and MoS_3 and of In with CuS and Ag_2S has been attributed to compound formation, for example, $CuInS_2$.[92]

NOTES

1. G. Schwarzenbach and M. Widmer, *Helv. Chim. Acta*, **46**, 2613 (1963).

2. Thionalide (p. 558), which is similar to H_2S as a metal precipitant, has been used to precipitate Cu, Sn, Sb, Bi, and As in amounts as small as 1 ppm (in alloys) from such metals as Al, Be, Mg, Mn, Zn, Fe, Ni, Co, Ti and Cr in acid solution. See Z. S. Mukhina, I. A. Zhemchuzhnaya and G. S. Kotova, *J. Anal. Chem. USSR*, **17**, 169 (1962).

3. Z. Marczenko and K. Kasiura, *Chemia Anal.*, **9**, 87 (1964). E. Jackwerth and W. Schmidt, *Z. Anal. Chem.*, **225**, 352 (1967), also separate silver as a major constituent by chloride precipitation; by a two-stage precipitation (90 and 10%, at 100°), they were able to recover 98% of the micro constituents Pb, Cd, Tl, Cu, Ni, Zn, Fe, and Mn. E. Jackwerth and G. Graffmann, *ibid.*, **241**, 96 (1968), show that the coprecipitation of Hg and Pd by AgCl is prevented by EDTA. Iridium(IV) is known to be carried down strongly by AgCl (K. Beyermann, *Z. Anal. Chem.*, **200**, 191). Au(III) is coprecipitated less strongly, Pt(IV) less than Au.

4. E. Jackwerth and G. Graffmann, *Z. Anal. Chem.*, **251**, 81 (1970).

5. A. G. Karabash, L. S. Bondarenko, G. G. Morozova, and Sh. I. Peyzulayev, *Zh. Anal. Khim.*, **15**, 623 (1960). O. F. Degtyareva and M. F. Ostrovskaya, *ibid.*, **20**, 814 (1965), removed lead as the chloride, as did D. T. Englis and B. B. Burnett, *Anal. Chim. Acta*, **13**, 574 (1955).

6. J. Meyer, *Z. Anal. Chem.*, **219**, 147 (1966).

7. Digestion of the precipitate may be helpful. Thus digestion of $PbSO_4$ with boiling HNO_3 (1:1) allows Li, Na, Mg, Ca, Al, Ti, V, Cr, Mo, Mn, Fe, Co, Ni, Cu, Zn, Cd, In, Sn, Sb, As, and Bi to be determined spectrographically in $>99.999\%$ Pb. Ba, Sr, Ag, and Au are not quantitatively recovered from the precipitated $PbSO_4$ in this way. See L. S. Bondarenko in I. P. Alimarin, Ed., *Analysis of High-Purity Materials*, translated by J. Schmorak, Israel Program for Scientific Translations, Jerusalem, 1968, p. 348.

8. R. P. Hair and E. J. Newman, *Analyst*, **89**, 42 (1964), found homogeneous precipitation of CuSCN superior to conventional precipitation when determining traces of Fe and Pb in Cu metal.

9. Zirconium in amounts greater than 100 mg can be separated from 5–50 μg of U(VI) by precipitation with bromomandelic acid without serious loss of uranium (<2–3%), as shown by R. F. Rolf, *Anal. Chem.*, **36**, 1398 (1964).

10. Salts such as NaCl, NaH_2PO_4 and $Mg(NO_3)_2$ have been collected in a small volume with good recoveries if the solution was shaken while being frozen. See G. H. Smith and M. P. Tasker, *Anal. Chim. Acta*, **33**, 559 (1965); S. Kobayashi and G. F. Lee, *Anal. Chem.*, **36**, 2197 (1964); J. Shapiro, *Science*, **163**, 2063 (1961). M. Murozumi, T. J. Chow, and C. Patterson, *Geochim. Cosmochim. Acta*, **33**, 1247 (1969), and Murozumi, S. Nakamura, and C. Patterson, *Japan Analyst*, **19**, 1057 (1970), had limited success in concentrating Na and Mg by freezing melted snow and ice. The tendency for relatively much Na and Mg to remain in the ice phase may be, in part, related to the presence of colloidal or suspended solids in the water.

11. A review is presented by A. Mizuike in G. H. Morrison, Ed., *Trace Analysis*, Interscience, New York, 1965, p. 136–142.

12. R. Lucas, F. Grassner, and E. Neukirch, *Mikrochemie (Emich Festschrift)*, 199 (1930) (Cu); 201 (1930) (Pb). J. W. Arden and N. H. Gale, *Anal. Chem.*, **46**, 2 (1974), have made a careful study of the separations of Pb from silicates and stone meteorites by electrolytic deposition. For specificity, Pb is first deposited cathodically, and after solution, anodically (as PbO_2). Lead in amounts greater than 0.01 μg can be 90–95% recovered. Special measures must be taken to minimize introduction of Pb.

13. See especially A. Stock, *Berichte*, **71**, 550 (1938).

14. *Cf.* L. B. Rogers, *Anal. Chem.*, **22**, 1386 (1950).

15. B. L. Clarke and H. W. Hermance, *J. Am. Chem. Soc.*, **54**, 877 (1932).

16. Such was the conclusion of G. von Hevesy and R. Hobbie, *Z. Anal. Chem.*, **88**, 1 (1932) (isolation of Pb from silicate rocks). But work by I. L. Barnes, T. J. Murphy, J. W. Gramlich, and W. R. Shields, *Anal. Chem.*, **45**, 1881 (1973) have established conditions for quantitative recovery of Pb in the absence of

much iron and some other elements. Electrolysis with a Pt wire anode at 1.8–2 volts in a 0.025 M HNO_3 or $HClO_4$ solution for 16 hr deposits >98% of the lead (5 μg) if the volume is not greater than 20 ml. Up to 10 mg of iron may be present.

17. Large amounts of Tl have been removed from solution in this form. See V. A. Nazarenko et al. in I. P. Alimarin, Ed., *Analysis of High-Purity Materials*, p. 236 (determination of zinc in high purity Tl with dithizone).

18. The isolation of small amounts of certain trace metals (that are to be determined) by the use of a mercury cathode is generally not attractive because of the necessity for removing large amounts of mercury by volatilization. Nevertheless, this procedure is occasionally of value, and the use of a small mercury cathode for this purpose has been described by N. H. Furman, C. E. Bricker, and B. McDuffie, *J. Wash. Acad. Sci.*, **38,** 159 (1948). An alternative method for removing certain metals from the mercury is by anodic stripping, in which the mercury is made the anode and maintained at a potential a little lower (by a few tenths of a volt) than its solution potential in a suitable salt solution.

19. W. F. Hillebrand, G. E. F. Lundell, H. A. Bright, and J. I. Hoffman, *Applied Inorganic Analysis*, Wiley, New York, 1953, p. 138; G. E. F. Lundell and J. I. Hoffman, *Outlines of Methods of Chemical Analysis*, Wiley, New York, 1938, p. 94; J. A. Maxwell and R. P. Graham, *Chem. Rev.*, **46,** 471 (1950); J. A. Page, J. A. Maxwell, and R. P. Graham, *Analyst*, **87,** 245 (1962). For the removal of large amounts of heavy metals prior to the determination of such elements as Al, Ca, and Na, see H. O. Johnson, J. R. Weaver, and L. Lykken, *Anal. Chem.*, **19,** 481 (1947); T. D. Parks, H. O. Johnson and L. Lykken, *ibid.*, **20,** 148 (1948). These papers contain data on the quantities of heavy metals left in solution.

20. When large quantities of an element are being removed, it is helpful to use one or two portions of fresh mercury. In this way, 5–10 g of Cd can be removed from solution, leaving only ~0.1 mg/liter.

21. J. K. Taylor and S. W. Smith, *J. Res. Natl. Bur. Stand.*, **56,** 301 (1956).

22. S. Hirano and A. Mizuike, *Japan Analyst*, **8,** 746 (1959); A. Mizuike, *Talanta*, **9,** 948 (1962).

23. E. J. Center, R. C. Overbeck, and D. L. Chase, *Anal. Chem.*, **23,** 1134 (1951).

24. V. T. Chuiko, *Zav. Lab.*, **8,** 950 (1939).

25. Compounds having an ion forming a slightly soluble or slightly dissociated product with an ion of opposite sign in the lattice of a precipitate will be more or less strongly adsorbed by the precipitate (Paneth–Fajans–Hahn rule). To be sure, if the adsorbed hydroxide (or other compound) exists as a solid phase in the solution, the concentration of the adsorbed metal will be that in the saturated solution at equilibrium, so that a decrease below this concentration requires the absence of such a phase.

26. Ammonia precipitation at pH 9–10 with $La(OH)_3$ as collector has been used

to separate Bi, Sb, Sn, Pb, Fe, Te, Se, and As from much copper (W. Reichel and B. G. Bleakley, *Anal. Chem.*, **46,** 59). Copper is also coprecipitated so that this is really a concentration method for the elements named. Coprecipitation of metals with $Mg(OH)_2$ has been studied by A. I. Novikov and V. I. Ruzankin, *Sov. Radiochem.*, **15,** 647 (1973). $Al(OH)_3$ serves as a collector for traces of Fe in its determination in Cu and Ni if excess ammonia is added to keep these metals in solution for the most part. In connection with hydroxide collectors, it may be mentioned that nonamphoteric oxides can be used as collectors in sodium carbonate or hydroxide fusions intended to bring amphoteric metals into solution. For example, F. S. Grimaldi, *Anal. Chem.*, **32,** 119 (1960), when separating microgram amounts of Nb from milligram amounts of Mo, W, V, and Re by sodium hydroxide fusion, used a few milligrams of Fe_2O_3 and MgO (better than Fe_2O_3 alone) to retain Nb in the residue on leaching the melt with water.

27. For example, the optimum pH for coprecipitation of Cr(VI) with metal hydroxides is 5–6.5 and coprecipitability increases in the order Be, Ti, Al, Fe, In, Zr, Y, Th, La, and Pb (Bi, Ce(IV)) hydroxides according to V. I. Plotnikov, V. L. Kochetkov, and E. G. Gibova, *Zh. Anal. Khim.*, **22,** 86 (1967). Cr(III) can be quantitatively recovered by coprecipitation with hydroxides of Fe(III), Zr, Ti, Th, and Ce in the pH range 5.5–10.5. Adsorption of Mo on ferric hydroxide (from sea water) is maximal at pH ~ 4, as found by Y. S. Kim and H. Zeitlin, *Anal. Chim. Acta*, **46,** 1 (1969). M. Ishibashi and coworkers have published a series of papers on the coprecipitation of traces of metals in sea water with fresh, externally prepared ferric hydroxide. They have shown that W is adsorbed at pH 8.2–8.3, but Mo is not (M. Ishibashi, T. Fujinaga, T. Kuwamoto, M. Koyama, and S. Sugibayashi, *J. Chem. Soc. Japan Pure Chem. Sec.*, **81,** 392 (1960)). Mo is adsorbed at pH 3.5–4.0 (M. Ishibashi, T. Fujinaga, and T. Kuwamoto, *ibid.*, **79,** 1496 (1958)). Coprecipitation of As(V) with $Fe(OH)_3$ is virtually complete up to pH 8.5; at pH 10, 15% desorption occurs (V. I. Plotnikov and L. P. Usatova, *J. Anal. Chem. USSR*, **19,** 1101 (1964)). Incidentally, coprecipitation increases in the order Al, Ti, Fe, In, Zr, and Ce(IV). Coprecipitation of As(III) is less complete than of As(V). Te(IV, VI) and Se(IV) can also be collected with $Fe(OH)_3$. P. N. W. Young, *Analyst*, **99,** 588 (1974), was able to separate microgram quantities of Ca, Fe, Mn, Si, Ti, V (partially), and Zn from much Al by coprecipitation with $Zr(OH)_4$ at pH > 10. Aluminum shows little tendency to be carried down. The coprecipitation of V and Zn, which is unexpected at this pH, seems to indicate compound formation. Zirconium hydroxide was found to be superior to Ce and Th hydroxides as coprecipitating agent.

28. Ferric hydroxide adsorbs UO_2^{2+} strongly in the pH range 5.5–8; uranium is desorbed by an alkaline solution (pH > 8) containing carbonate. P. Strohal, K. Molnar, and I. Bacic, *Mikrochim. Acta*, 586 (1972), found Co, Zn, and Cu to be completely coprecipitated with $Al(OH)_3$ at pH 6–10. Hg(II) was very incompletely carried down (attributed to nonionized $HgCl_2$). P. Strohal and D. Nothig-Hus, *Mikrochim. Acta*, 899 (1974), have studied the coprecipitation of various metals with hydroxides of Be, Zn, Fe, La, and Ti at pH 3–11.

29. Although called manganese dioxide, the precipitate is of indefinite composition and contains Mn(III) as well as Mn(IV). It is very insoluble in acid.

30. K. E. Burke, *Anal. Chem.*, **42**, 1536 (1970). Data on collection from H_2SO_4 solution, not shown here, are also reported. Coprecipitation of Sn for atomic absorption determination: P. N. Vijan and C. Y. Chan, *ibid.*, **48**, 1788 (1976).

31. H. Nishida, *Japan Analyst*, **13**, 760 (1964). A. Mizuike, H. Kawaguchi, K. Fukuda, Y. Ochiai, and Y. Nakayama, *Mikrochim. Acta*, 915 (1974), obtained quantitative collection of Fe, Co, Cu, Zn, and Cd on stannic hydroxide at pH 4–8. Sn can mostly be removed by volatilization as the bromide.

32. H. Kamada, Y. Ujihira, and K. Fukuda, *Radioisotopes*, **14**, 206 (1965), for example, obtained good recoveries of Au, Ag, Hg, and Pt by sulfide precipitation from 0.1–$1\,M$ HCl solution with CuS as collector; Zn, Co, and Tl remain in solution. As and Sb are not quantitatively recovered. As and Hg sulfides were also found to be satisfactory collectors for Ag and Au in $0.1\,M$ HCl. Good recoveries of zinc with CuS as collector requires H_2S precipitation from acetic acid–acetate solution at pH ~ 4. CuS is also a satisfactory collector for Sb_2S_3 in $0.5\,N$ HCl or H_2SO_4 solution. It has been used for Pb, as by R. L. Lucas and F. Grassner, *Mikrochemie (Emich Festschrift)*, 203 (1930). CuS has been recommended for collecting MoS_3, but, in our experience, is not satisfactory. Ag_2S has been used for PbS (the two sulfides are not isomorphous).

33. P. Ya. Yakovlev, G. P. Razumova, R. D. Malinina, and M. S. Dymova, *J. Anal. Chem. USSR*, **17**, 89 (1962).

34. As from a solution of bone ash, to which dithizone extraction cannot be applied.

35. For example, Cr(III), Mn, Fe, and Zn in tracer concentrations are quantitatively collectable by coprecipitation with Sr or Al phosphate in the pH range 4–10, as shown by I. Bacic, N. Radakovic, and P. Strohal, *Anal. Chim. Acta*, **54**, 149 (1971).

36. H. Onishi and C. V. Banks, *Talanta*, **10**, 399 (1963). LaF_3 as collector: R. W. Perkins and D. R. Kalkwarf, *Anal. Chem.*, **28**, 1989 (1956); F. L. Moore, *ibid.*, **30**, 1368 (1958); H. J. Gräbner, *Z. Anal. Chem.*, **201**, 401 (1964). ThF_4: B. J. Bornong and J. L. Moriarty, *Anal. Chem.*, **34**, 871 (1962). YF_3 and SrF_2 should also be useful collectors. CeF_3 has been used to collect La and Sm in their separation from U by B. Helger and R. Rynninger, *Svensk Kem.Tids.*, **64**, 224 (1952). With a sufficient excess of HF, fluoride precipitation leaves Fe(III), Ti, Zr, Al, Nb, Ta, and U(VI) in solution.

37. See Sandell, *Colorimetric Determination of Traces of Metals*, 3rd ed., for the use of various collectors in the isolation of individual metals.

38. Y. Ujihira, *Japan Analyst*, **14**, 399 (1965); *J. Chem. Soc. Japan Pure Chem. Sec.*, **84**, 642 (1963); Au and Pd: E. Jackwerth and P. G. Willmer, *Talanta*, **23**, 197 (1976).

39. R. Püschel, E. Lassner, W. Martin, and D. Martin, *Mikrochim. Acta*, 145 (1969).

40. H. V. Weiss and coworkers have published a series of papers in this field, some of which are mentioned here. Cobalt has been isolated from sea water by cocrystallization with 1-nitroso-2-naphthol: H. V. Weiss and J. A. Reed, *J. Marine Res.*, **18**, 185 (1960); molybdenum has been cocrystallized with α-benzoinoxime (sea water): H. V. Weiss and M. G. Lai, *Talanta*, **8**, 72 (1961); Sn, Hg, Ag, Ta, and Au are carried down, with good recovery, at pH 1 with 2-mercaptobenzimidazole: H. V. Weiss and M. G. Lai, *Anal. Chim. Acta*, **28**, 242 (1963). 1-Nitroso-2-naphthol is also a collector for U, thionalide for Ag (both from sea water).

41. R. Dams, *Anal. Chim. Acta*, **33**, 349 (1965), found that cocrystallization of Co with 1-nitroso-2-naphthol (evaporation of an ethanolic solution) followed the logarithmic distribution law (p. 725).

42. J. P. Riley and G. Topping, *Anal. Chim. Acta*, **44**, 234 (1969).

43. S. Gohda, *Bull. Chem. Soc. Japan*, **45**, 1704 (1972).

44. R. A. A. Muzzarelli, *Anal. Chim. Acta*, **54**, 133 (1971).

45. V. I. Kuznetsov and G. V. Myasoedova, *Trudy Komiss. Analit. Khim.*, **9**, 89 (1958).

46. V. I. Kuznetsov, T. G. Akimova, and O. P. Eliseeva. *Radiokhimiya*, **4**, 188 (1962).

47. For a review of organic collectors, see G. V. Myasoedova, *J. Anal. Chem. USSR*, **21**, 533 (1966). They are classified according to the following types: (organic cation)(MX_n); (M)(organic anion); (MY)(organic anion), where Y = pyridine, 1,10-phenanthroline, and others; (organic cation)(ML_n); ML_n; colloid–chemical compounds, such as M tannates.

48. K. Murata and S. Ikeda, *Anal. Chim. Acta*, **51**, 489 (1970).

49. E. E. Pickett and B. E. Hankins, *Anal. Chem.*, **30**, 47 (1958), have shown that cobalt and molybdenum are effectively collected with aluminum 8-hydroxyquinolate at pH 5; Cu is incompletely collected. Indium hydroxyquinolate is inferior to aluminum hydroxyquinolate for collection of Mo and Co.

50. G. G. Pierson, *Ind. Eng. Chem. Anal. Ed.*, **6**, 437 (1934). Later work has shown that Rh(III) can be reduced to metal by Hg_2Cl_2 in dilute HCl containing bromide and thus sepatated from Ir(III), which remains in solution. See K. W. Lloyd and D. F. C. Morris, *Talanta*, **8**, 16 (1961); J. P. Francois, R. Gijbels, and J. Hoste, *Radiochem. Radioanal. Lett.*, **15**, 267 (1973). Pt is also precipitated.

51. B. L. Clarke and H. W. Hermance, *Ind. Eng. Chem. Anal. Ed.*, **10**, 591 (1938). S. Alexandrov, *Talanta*, **23**, 684 (1976), was able to obtain better than 95% recovery of Hg(II) and Au(III) added in trace concentrations to sea water at pH 1–1.5 by percolation of the latter through a 3-cm column of PbS (0.5 g). A sample of Black Sea water was found (radiometrically) to contain 0.8 ± 0.2 μg Hg and 0.006 ± 0.003 μg Au per liter.

52. M. Ziegler and M. Gieseler, *Z. Anal. Chem.*, **191**, 122 (1962); Bi from Pb: Ziegler and M. M. Tha, *ibid.*, **196**, 81 (1963). Addition of Cd^{2+} to the solution prevents precipitation of PbS.

53. K. Fukuda and A. Mizuike, *Japan Analyst*, **18**, 1130 (1969). The use of an ultrasonic field speeds up the adsorption of metals on dithizone and other solids: K. Fukuda and A. Mizuike, *Anal. Chim. Acta*, **51**, 77, 527 (1970); **67**, 207 (1973).

54. A. Mizuike and K. Fukuda, *Anal. Chim. Acta*, **44**, 193 (1969).

55. C. Kullgren, *Svensk Kem. Tids.*, **43**, 99 (1931).

56. E. Schulek and coworkers have studied extensively the adsorption of metals by oxycellulose. See *Talanta*, **9**, 529 (1962); **10**, 821 (1963); **11**, 941 (1964). It is of interest that long ago, G. Witz and F. Osmond, *Bull. Soc. Chim.*, [2], **45**, 309 (1886), were able to detect 0.001 μg of V in a liter of solution by adsorption on treated cotton fibers. The adsorbed vanadium was revealed by the catalytic formation of aniline black; a rough estimate of the amount could be made by comparison against a series of standards. In several respects, this work was ahead of its time.

56a. Charcoal was used by C. K. Kropachev, *Sovet. Zolotoprom.*, **1935,** No. 8, 46, for isolation of Au in its determination in stream waters; the charcoal was ashed after adsorption and Au brought into solution. E. Jackwerth, J. Lohmar, and G. Wittler, *Z. Anal. Chem.*, **266**, 1 (1973), studied the adsorption of chelated metals on activated carbon, and B. M. Vanderborght and R. E. Van Grieken, *Anal. Chem.*, **49**, 311 (1977), were able to adsorb oxinates of heavy metals on the same adsorbent for their determination in natural waters.

57. M. W. Wilding, USAEC Report IDO-14544 (1961).

58. J. R. van Smit, *Nature*, **181**, 1530 (1958).

59. F. Sebba, *Ion Flotation*, Elsevier, Amsterdam, 1962; R. Lemlich, Ed., *Adsorptive Bubble Separation Techniques*, Academic Press, New York, 1972.

60. Actually, there are cases for which demarcation is difficult. For example, it is not certain that the UO_2^{2+}–Arsenazo III–Zephiramine system (K. Sekine and H. Onishi, *Anal. Chim. Acta*, **62**, 468 (1972)) belongs to ion flotation.

61. A. Mizuike, K. Fukuda, and J. Suzuki, *Japan Analyst*, **18**, 519 (1969).

62. C. Jacobelli-Turi, A. Barocas, and F. Salvetti, *Gazz. Chim. Ital.*, **93**, 1493 (1963); A. Barocas, C. Jacobelli-Turi, and F. Salvetti, *J. Chromatogr.*, **14**, 291 (1964).

63. Sekine and Onishi, *loc. cit.* Thorium can be concentrated similarly. See K. Sekine, *Z. Anal. Chem.*, **273**, 103 (1975).

64. K. Sekine, *Mikrochim. Acta*, **I**, 313 (1975).

65. K. Fukuda and A. Mizuike, *Japan Analyst*, **17**, 319 (1968).

66. E. J. Mahne and T. A. Pinfold, *J. Appl. Chem.*, **18**, 52 (1968).

67. E. J. Mahne and T. A. Pinfold, *J. Appl. Chem.*, **18**, 140 (1968).

68. E. J. Mahne and T. A. Pinfold, *J. Appl. Chem.*, **19**, 57 (1969).

69. Y. S. Kim and H. Zeitlin, *Sepn. Sci.*, **6**, 505 (1971).

70. Y. S. Kim and H. Zeitlin, *Anal. Chem.*, **43**, 1390 (1971) (iron(III) hydroxide); G. Leung, Y. S. Kim, and H. Zeitlin, *Anal. Chim. Acta*, **60**, 229 (1972) (thorium hydroxide).

71. Y. S. Kim and H. Zeitlin, *Sepn. Sci.*, **7**, 1 (1972).

72. D. Voyce and H. Zeitlin, *Anal. Chim. Acta*, **69**, 27 (1974).

73. A. Mizuike and M. Hiraide, *Anal. Chim. Acta*, **69**, 231 (1974).

74. A mixed crystal is a solid solution, but not all solid solutions are mixed crystals. Solid solutions can also be formed by incorporation of the component N between lattice ions (interstitial solid solution).

75. An exception immediately coming to mind is the mixed crystals formed by replacement of NH_4^+ in ammonium phosphomolybdate by the alkali metal ions whose radii (Goldschmidt's empirical) range from 0.98 (Na) to 1.65 Å (Cs), with $NH_4 \sim 1.3$ (approximately the same as K). But in the phosphomolybdates, the packing of the anions leaves vacancies large enough to be filled by cations of widely varying size.

76. L. M. Henderson and F. C. Kracek, *J. Am. Chem. Soc.*, **49**, 738 (1927). The relation follows from the Berthelot–Nernst distribution law.

77. For nonideal systems see F. Vaslow and G. E. Boyd, *J. Am. Chem. Soc.*, **74**, 4691 (1952); A. P. Ratner, *J. Chem. Phys.*, **1**, 789 (1933).

78. H. A. Doerner and W. M. Hoskins, *J. Am. Chem. Soc.*, **47**, 662 (1925).

79. M. L. Salutsky, J. G. Stites, and A. W. Martin, *Anal. Chem.*, **25**, 1677 (1953).

80. M. Munakata, S. Toyomasu, and T. Shigematsu, *Anal. Chem.*, **44**, 2057 (1972).

81. O. Hahn, *Applied Radiochemistry*, Cornell Univ. Press, Ithaca, 1936, p. 105.

82. J. T. Rosenqvist, *Am. J. Sci.*, **240**, 356 (1942), isolated lead from silicate rocks by coprecipitation with strontium sulfate, using samples of 50–300 g. The object of this work was not only the determination of the lead content but also the determination of the isotopic composition of the lead. The $SrSO_4$ coprecipitation of lead is of value when the dithizone extraction cannot be directly applied. V. A. Nazarenko et al., *Zav. Lab.*, **26**, 131 (1960), used this method to separate traces of lead occurring in high purity indium. B. C. Flann and J. C. Bartlet, *J. Assoc. Off. Anal. Chem.*, **51**, 719 (1968), applied $SrSO_4$ coprecipitation of Pb as a separation before dithizone extraction.

83. R. B. Henrickson and E. B. Sandell, *Anal. Chim. Acta*, **7**, 57 (1952).

84. K. Schlyter and L. G. Sillén, *Acta Chem. Scand.*, **4**, 1323 (1950); Schlyter, *Arkiv Kemi*, **5**, 61 (1952). For coprecipitation of rare earth iodates with $Th(IO_3)_4$, see K. J. Shaver, *Anal. Chem.*, **28**, 2015 (1956).

85. C. W. Sill and R. L. Williams, *Anal. Chem.*, **41**, 1624 (1969), and earlier papers. The method is applied in the radiochemical determination of these elements in process solutions and environmental samples.

86. O. Hahn, *Applied Radiochemistry*, p. 103. H. Käding, *Z. physik. Chem.*, A**162**, 174 (1932).

87. A. C. Gillet, *Bull. Soc. Chim. Belge*, **77**, 109 (1968), cocrystallizes microgram amounts of Pb with NaCl after adding KI. Values of D or λ for incorporation of Pb in various soluble salts are given by M. Kahn in A. C. Wahl and N. A.

Bonner, Eds., *Radioactivity Applied to Chemistry*, Wiley, New York, 1951, p. 394.

88. P. Pringsheim, *Rev. Mod. Phys.*, **14,** 132 (1942).

89. A. H. Booth, *Nature*, **165,** 968 (1950).

90. Tellurium coprecipitation seems to have been first applied by S. K. Hagen, *Mikrochemie*, **20,** 180 (1936), in isolating Pt for qualitative purposes. For quantitative trace collection: Au, E. B. Sandell, *Anal. Chem.*, **20,** 253 (1948); Ag, E. B. Sandell and J. J. Neumayer, *ibid.*, **23,** 1863 (1951); Pt, L. E. Salyards, unpublished work, 1955 (see *Colorimetric Determination of Traces of Metals*, 3rd ed., p. 721); Pd, E. R. R. Marhenke and E. B. Sandell, *Anal. Chim. Acta*, **28,** 259 (1963). N. M. Potter, *Anal. Chem.*, **48,** 531 (1976), has confirmed the quantitative collection of Pt (99.7%) and Pd (100%) with Te. W. B. Pollard, *Analyst*, **62,** 597 (1937), was able to collect gold quantitatively at a dilution of 1 μg/liter by coprecipitation in Te formed by SO_2 reduction. Stannous chloride is a better reducing agent than SO_2 or hypophosphite when Au, Ag, Pt, and Pd are present with much Fe(III). Ag is not coprecipitated when SO_2 or hydrazine is the reducing agent: K. Kudo, *J. Chem. Soc. Japan Pure Chem. Sec.*, **81,** 570 (1960). Rhodium(III) is not completely collected in Te precipitated by $SnCl_2$, but is satisfactorily collected when Ti(III) is the reductant. Palladium can also be used as collector and has the advantage of easy separation from Rh by extraction of Pd dimethylglyoximate into chloroform. H. Stein (M.S. Thesis, University of Minnesota, 1953), working with 4–16 μg of Rh and 200 μg of Pd as collector, obtained average recoveries of 94% Rh (dimethylglyoxime extraction included).

91. H. Bode and E. Hettwer, *Z. Anal. Chem.*, **173,** 287 (1960), showed that in 0.5 *M* HCl solution, with SO_2 as reductant, mercury forms the chlorotelluride, $2\,HgTe \cdot HgCl_2$.

92. N. A. Rudnev and G. I. Malofeeva, *Talanta*, **11,** 531 (1964). Coprecipitation of Ga with In_2S_3: G. I. Malofeeva, A. A. Loginov, and N. A. Rudnev, *J. Anal. Chem. USSR*, **23,** 1140 (1968).

CHROMATOGRAPHY
ION EXCHANGE

The two most important types of chromatography in the separation of inorganic trace constituents (especially from major constituents) are:

1. Liquid (solution)–solid, in which the solid is an ion exchanger used in column form.

2. Liquid–liquid (immiscible), in which one of the liquids is held by an inert solid support in column form; this is partition chromatography.

Chromatography in which columns are not used (paper, thin layer) is generally less useful in inorganic trace analysis, but is sometimes applied, as in the separation of trace constituents from each other.

I. RATIONALE OF COLUMN CHROMOTOGRAPHY

We assume that the reader already has some knowledge of chromatographic principles. The following brief simplified treatment of column chromotography is intended as a kind of review or summary of principles, without derivations and with a minimum of explanation. The expressions given should be of aid to the analyst in deciding whether chromotographic separations described in the literature, with distribution coefficients stated, are applicable to his samples or what modifications are needed to make them so.

Ideally, the conclusions arrived at in the following treatment are independent of the mechanism of the distribution of a solute between stationary and mobile phases so that they hold for liquid–liquid distribution as well as for ion exchange. Ideal behavior assumes that D, the distribution coefficient in ion-exchange adsorption, or E, the extraction coefficient in liquid–liquid distribution, are independent of the solute concentration in the course of the separation (linear adsorption isotherm).

Consider first the elution of a single species (ion or molecule) from a column of finely divided solid (ion-exchange resin, cellulose powder, etc.). The species is assumed present initially as a thin adsorbed (or sorbed) layer at the very top of the column; the pore space of the column is filled with some solution. Now a suitable eluting solution is added slowly to

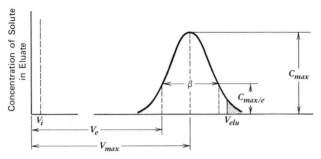

Volume of Effluent (= volume of eluant)

Fig. 8-1. Gaussian elution curve. V_{max} is the volume of eluting solution required to bring the center of the solute band to the base of the column; V_e is the volume of the eluate at which the concentration of the solute is $1/e$ (=0.368) of that at the maximum; V_i is the interstitial volume of the column of adsorbent (initially occupied by solvent). C_{max} is given by $(m/V_{max})(N/2\pi)^{1/2}$, where m is the total amount of eluted ion. V_{elu} is some elution volume; the stippled area under the curve represents uneluted solute.

move the adsorption band down the column. The eluting solution contains a replacing species that is less strongly adsorbed than the species being eluted. In its passage down the column, the adsorption band becomes broader and the axial distribution of the adsorbed substance is more or less Gaussian. A plot of the concentration of the eluted species in a small volume of effluent versus the total volume of eluant added up to that point gives the elution curve (Fig. 8-1), which ideally is the error curve of Gauss. The quantities indicated have the following meaning:

V_{max} = volume of eluant required to bring the center of the adsorbed band to the base of the column. When this volume of eluting solution has been added, one-half of the adsorbed species remains in the column, the other half has been eluted.

C_{max} = concentration of the species in a small volume of eluate corresponding to volume V_{max}.

V_e = volume of eluant corresponding to a concentration of species equal to $C_{max}/e = C_{max}/2.71$ or $0.37 C_{max}$.

β = width of the elution band = $2(V_{max} - V_e)$.

According to the plate theory of chromatography, elution equilibrium takes place in (fictive) cells, usually called plates. The height of a theoretical plate (h) is not constant, but depends on the particle size of the adsorbent, the flow rate of the eluant, and other factors. It is not necessarily the same for all solutes. The number of theoretical plates in a column, N, when the elution band is Gaussian, is given by

$$N = \frac{2 V_{max} V_e}{(V_{max} - V_e)^2} \tag{8-1}$$

$$\sim \frac{2 V_{max}^2}{(V_{max} - V_e)^2}$$

(when difference between V_{max} and V_e is small) or

$$\sim \frac{8 V_{max}^2}{\beta^2} \tag{8-3}$$

If m is the total weight of the adsorbed species

$$C_{max} = \frac{m}{V_{max}} \left(\frac{N}{2\pi} \right)^{1/2} \tag{8-4}$$

V_{max} and V_e, and therewith N, can be found experimentally by collecting successive small increments of effluent when the elution band is near the base of the column and determining the concentration of the species in each. Or the concentration of the species in question can be continuously monitored by suitable instrumentation, for example, by measuring the absorption of the eluant in the visible or UV or by measuring its fluorescence. In forced flow chromatography (in which gravity flow is replaced by pressure flow), the eluate can be mixed continuously with a solution of absorptiometric or fluorimetric reagent and the absorbance or fluorescence then recorded as a function of the volume of eluant. If the distribution coefficient of the species A

$$D_A = \frac{[A]_{solid}}{\Sigma[A]}$$

(Concentrations without subscripts refer to *solution*; Σ indicates that A may be present in various forms in solution.)

is known under the conditions (it may have been reported), V_{max} can be calculated:

$$V_{max} = V_{col}\{i + D_A(1 - i)\} \tag{8-5}$$

where V_{col} is the geometrical column volume (cross-sectional area of adsorbent $\times$ length of cylindrical column) and i is the fractional interstitial volume in column of adsorbent (fractional void space, filled with solution). If the distribution coefficient is based not on the weight of the adsorbent but on a unit volume (ml) of adsorbent column (designate this coefficient D_v, here $D_{A,v}$):

$$D_{A,v} = D_A(1-i)$$
$$V_{max} = V_{col}\{i + D_{A,v}\} \tag{8-6}$$
$$\sim V_{col}D_{A,v}$$

when $D_{A,v}$ is large.

It may be noted from Eq. (8-3) that β varies as $N^{1/2}$, so that doubling the length of an adsorption column ideally increases the elution band width by the factor 1.4.

Extent of Elution. The volume of eluting solution required to remove a specified fraction of an adsorbed substance from a chromatographic column is readily calculated if the elution curve is Gaussian, if V_{max} and β are known, and if the substance initially forms a thin band at the column top. Not all elution curves are Gaussian. The theoretical form is approached more closely in some types of chromatography than in others. In ion-exchange chromatography, Gaussian curves are often approximated if certain conditions are fulfilled (see later). The results of calculations to be described are to be looked upon as approximate, and subject to large errors if the assumptions on which they are based are not fulfilled.

The recovery factor (Analytical Separations, Introductory Note) for an eluted species A is given by the ratio of the area under the elution curve corresponding to V_{elu} (volume of eluant) to the whole area ($= 1$). (In Fig. 8-1, the stippled area represents noneluted A.) From a table of areas under the Gaussian error curve, the recovery factor R_A, corresponding to V_{elu}, is quickly found. Table 8-1 is a condensed table of this kind.

TABLE 8-1
Areas Under Gaussian Curve Giving Elution Fractions

$\phi = \dfrac{V_{elu} - V_{max}}{V_e - V_{max}}$	Fraction Eluted (recovery factor)	Fraction Not Eluted
0	0.500	0.500
1	0.921	0.079
1.5	0.983	0.017
2	0.998	0.002
2.19	0.999	0.001
2.5	0.9998	0.0002
2.64	0.9999	0.0001
3	0.99999	0.00001

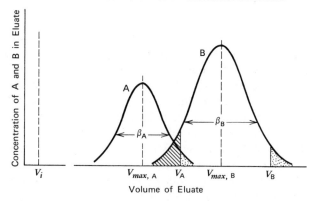

Fig. 8-2. Illustrating separation of A and B by elution. The stippled area under curve A represents the fraction of A not eluted when the volume V_A of effluent is collected (this fraction is collected subsequently with B) and the stippled area under curve B represents the fraction of B not eluted when the additional volume of effluent $V_B - V_A$ is collected. The fraction of B represented by the cross-hatched area under B is collected with the major portion of A in the volume V_A of effluent and thus escapes recovery. The interstitial volume is represented by V_i. The elution curves are shown as being symmetrical, but this is not quite correct, because the band width increases as the band moves down the column. The error is usually not important if β, in volume units, is small compared to V_{max}.

Separations. Two constituents A and B can be separated by differential elution if their elution curves do not overlap too much. The problem is illustrated in Fig. 8-2. The elution curves of A and B are represented for a particular chromatographic column and eluting solution. The curves overlap considerably. If the volume V_A of effluent is collected (see legend of figure), the recovery factor for A, R_A, is the fraction represented by $\phi_A = \{V_A - (V_A)_{max}\}/\{(V_A)_e - (V_A)_{max}\}$ (Table 8-1). The fraction of B contained in this volume of effluent (= volume of eluant added) is

$$1 - \phi_B = 1 - \frac{(V_B)_{max} - V_B}{(V_B)_e - (V_B)_{max}} \tag{8-7}$$

The weights of B and A in the volume V_A of effluent are given by $R_B(Q_B)_0$ and $R_A(Q_A)_0$, $(Q_B)_0$ and $(Q_A)_0$ being the weights of B and A in the column at the beginning of elution. The ratio of the two constituents in the volume V_A of effluent is then

$$\frac{Q_B}{Q_A} = \frac{R_B(Q_B)_0}{R_A(Q_A)_0} \cong \frac{R_B(Q_B)_0}{(Q_A)_0} \cong S_{B/A}\frac{(Q_B)_0}{(Q_A)_0} \tag{8-8}$$

$S_{B/A}$ is the separation factor under the specified conditions (p. 700).[1]
The value of R_A is fixed by the recovery considered satisfactory in the

determination of A. For a trace constituent, R_A might have a value as low as 0.95. The elution volume V_A is, therefore, essentially fixed for a given column. The permissible amount of B that may accompany A will, of course, depend on the extent to which B interferes in the determination of A. With a given adsorbent and eluting solution, a better separation of B and A can be obtained by increasing the length of the column. The distance between the elution peaks (V_{max}) increases directly in proportion to the length of the column, whereas, as already pointed out, the width of the elution band ideally increases as the square root of the length. There is thus a net gain in separability with column length, but at a cost of larger volumes of eluant and longer elution times.

In the preceding discussion, it was (necessarily) assumed that h and the distribution coefficients were constant. In practice, the analyst can vary h by choosing different adsorbents, or even with a particular adsorbent h can be changed to some extent by altering the particle size (as of a resin) and the flow rate. The degree of overlap of two elution curves depends not only on the separation of the peaks but also on the narrowness of the peaks. The separation of the peaks depends on the difference in D values, and these can be varied by changing the chemical composition of the eluting solution(s). These matters are discussed further under ion-exchange chromatography (Section II. B.3.). Additional information can readily be obtained from the many available treatments of chromatography.[2]

Column Length. Expressions can be derived that give the column length and eluant volume required for a specified recovery of A and a specified degree of separation from B, if the distribution coefficients and band widths of each are known. Again, Gaussian elution curves are assumed. Even if the values obtained by the use of these expressions are only very approximate, they serve a useful purpose. They are likely to be most useful in ion-exchange chromatography, where conditions for their applicability can often be met.

The problem is represented graphically in Fig. 8-3, in which the top of the adsorption column lies to the left and the concentrations of A and B in the adsorption band in the column are shown. The band of B lies almost entirely within the column, and the band of A almost entirely outside the column.[3] It is advantageous to introduce a quantity called the *elution constant* $\mathscr{E}$, which is the distance (cm) traversed by the band maximum in a column of adsorbent of 1 cm² cross section when 1 ml of eluant is added:

$$\mathscr{E} = \frac{al}{V} = \frac{al_{col}}{V_{max}} = \frac{V_{col}}{V_{max}} \tag{8-9}$$

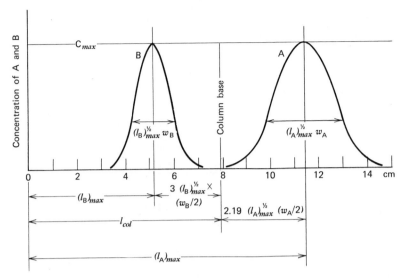

Fig. 8-3. Diagram illustrating the effect of column length on the separation of A and B. Curves are Gaussian. Concentrations of A and B are in terms of C_{max} ($=1$) for each. The adsorption band of A lying to the right of "column base" is hypothetical since it lies outside the column and is represented as if in the exchanger to correspond to the tail in the column; the ratio of the area under curve A to the left of "column base" to the whole area as shown gives the fraction of A not eluted from the column (1×10^{-3}). Lengthening the exchanger column and therewith increasing the elution volume increases the distance between the peaks of A and B. At the same time the widths of the bands increase, but only in proportion to the square root of the distance traversed by the peaks, so that a net gain in separability results. The geometry of the diagram is readily convertible to algebra that gives R_A and R_B as a function of column length and elution volumes and therewith the separation factor R_B/R_A ($R_A \sim 1$).

where a is the cross-sectional area (cm^2), l is the distance traversed by band maximum (cm), and l_{col} is the length of adsorbent column (cm).

$$\mathscr{E} = \frac{1}{D_v + i} \sim \frac{1}{D_v} \qquad (8\text{-}10)$$

when D_v is large, say >10. The ratio of the elution constants (>1) is denoted by ϵ:

$$\epsilon = \frac{\mathscr{E}_A}{\mathscr{E}_B} \qquad (8\text{-}11)$$

Other quantities needed are:

ϕ_A, ϕ_B = factors providing specified values of R_A and R_B (Table 8-1);

they are the factors by which the half-band widths must be multiplied to give the stipulated recoveries.

w_A, w_B = band widths (cm) of A and B in the adsorbent when $l_{max} = 1$.

In terms of $(l_B)_{max}$ (see Fig. 8-3), the required column length is given by

$$l_{col} = (l_B)_{max} + \phi_B (l_B)_{max}^{1/2} \left(\frac{w_B}{2}\right) \tag{8-12}$$

It can be shown that

$$(l_B)_{max} = \frac{(\phi_B w_B + \phi_A w_A \epsilon^{1/2})^2}{4(\epsilon - 1)^2} \tag{8-13}$$

The volume of added eluant (or collected effluent) at which the cut is to be made is given by

$$V_A = \frac{a(\phi_B w_B + \phi_A w_A \epsilon^{1/2})^2}{4\mathscr{E}_B(\epsilon - 1)^2} \tag{8-14}$$

When the band widths (at 1 cm) of A and B are equal ($w_A = w_B$)

$$(l_B)_{max} = \frac{2h(\phi_B + \phi_A \epsilon^{1/2})^2}{(\epsilon - 1)^2} \tag{8-15}$$

Example

Two constituents A and B in the ratio $(Q_B)_0/(Q_A)_0 = 1000$ are to be separated, with the requirements $R_A = 0.999$ and $Q_B/Q_A = 0.01$. With a $1 \text{ cm}^2 \times 10 \text{ cm}$ column of adsorbent, $(V_A)_{max} = 50$ ml, $(V_A)_e = 58$ ml, $(V_B)_{max} = 110$ ml and $(V_B)_e = 125$ ml. What minimum length of column is theoretically required for the separation, and what volume of eluant is needed to elute 0.999 of A?

Answer

$$w_A = \frac{2\{(58/50 \times 10) - 10\}}{(11.6)^{1/2}}$$

$$= 0.94 \text{ cm}$$

(transforming width of elution peak in milliliters to width of elution band in column in centimeters; when $(l_A)_e = 10$, $(l_A)_{max} = 11.6$ cm).

$w_B = 0.804$ cm
$\phi_A = 2.19$
$\phi_B = 3.0$ (corresponding to $R_B = 0.01/1000 = 1 \times 10^{-5}$)
$\epsilon = \dfrac{\mathscr{E}_A}{\mathscr{E}_B} = \dfrac{10/50}{10/110} = 2.2$

$(l_B)_{max} = 5.2$ (5.19) cm [from Eq. (8-13)]
  $l_{col} = 5.2 + 2.75 = 7.95$ cm [from Eq. (8-12)]
  $V_A = 57.1$ ml [from Eq. (8-14)]

For illustration, these calculations have been carried out more precisely than reality warrants. In practice, the elution peaks may not be Gaussian. When elution is begun, the adsorbed substances may not form a very narrow band at the very top of the column. Especially if the volume of the sample solution is comparatively large, or if the amounts of the adsorbed substances are not small, the adsorption band may be rather broad and may be some distance from the top of the column. Also the values of V_{max} and V_e for A and B may not be as accurately known as implied. There may be appreciable dead space at the bottom of the column. Therefore, in practice the analyst would use a column reasonably longer than the calculated minimum, possibly as long as 10 cm in this example. Even then an experimental test of the separation should be made.

II. ION-EXCHANGE SEPARATIONS

Our aim in this section is to touch on the basic principles involved in ion exchange and then, in more detail, to survey applications in the separation of metals, particularly those separations pertaining to traces. More likely than not, the reader is already familiar with ion-exchange materials and their use or can obtain more detailed information elsewhere.[4]

A. ION-EXCHANGE MATERIALS

Strong-acid cation-exchange resins and strong-base anion-exchange resins are the most frequently used ion-exchange materials. The former consist of crosslinked polystyrene having $-SO_3H$ as the ionogenic group and the latter consist of crosslinked polystyrene having quaternary ammonium groups such as $-NMe_3^+OH^-$. Weak-acid cation-exchange resins with $-COOH$ as the ionogenic group and weak-base anion-exchange resins with primary, secondary, or tertiary amine groups are used infrequently.

Chelating resins contain a chelating group such as iminodiacetate

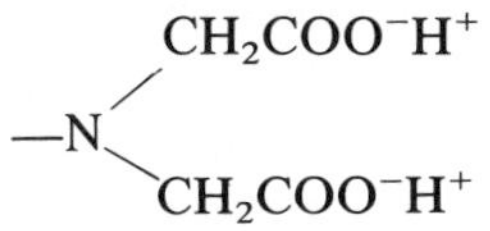

$$-N{\Large\langle}{{CH_2COO^-H^+}\atop{CH_2COO^-H^+}}$$

or mercapto. They preferentially adsorb metals bonding to these groups and are more selective than the common ion-exchange resins mentioned previously.[5]

When cellulose is treated so that ionogenic groups are introduced, cellulosic ion exchangers are formed. They include both cation and anion exchangers. For example, cellulose phosphate and sulfoethylcellulose are strong-acid cation exchangers, carboxycellulose is a weak-acid cation exchanger, and aminoethylcellulose is a weak-base anion exchanger. For some uses of cellulosic ion exchangers see Section II.C.3.

Inorganic ion exchangers have found some use in trace metal separations, but they do not have the versatility of resin exchangers. They are mostly gels, such as hydrous zirconium oxide or zirconium phosphate (cation exchangers). Ammonium molybdophosphate is a cation exchanger, especially for the alkali metal ions, which replace NH_4^+ in the crystal lattice. Acid-washed alumina acts as an anion exchanger. See Section II.C.4.

For "liquid ion exchangers" (the designation is dubious), see p. 756 and Chapter 9C, Section I.C.2.

B. GENERAL PRINCIPLES

To a large extent, the following discussion applies to ion exchangers in general but especially to ordinary ion-exchange resins.

1. Ion-Exchange Equilibria

Denote a cation-exchange resin by R^-A^+. If B^+ is the cation in solution exchanging with A^+ (A^+ and B^+ are known as counter ions, R^- as the fixed ion), the equilibrium constant (ion-exchange constant) for the reaction

$$R^-A^+ + B^+ \rightleftharpoons R^-B^+ + A^+$$

is given by

$$\frac{[B]_r[A^+]}{[A]_r[B^+]} = K$$

(Concentrations with subscript r refer to resin phase, without subscript to aqueous phase.)

K is a constant for a specified ionic strength and loading of the resin (and of course for a particular resin).

Rearrangement of the preceding expression gives the distribution coefficient D_{B^+} of the ion B^+ between the resin and the aqueous solution under the specified conditions:

$$D_{B^+} = \frac{[B]_r}{[B^+]} = K\frac{[A]_r}{[A^+]}$$

If forms of B other than B^+, such as hydrolyzed B ions or complex ions of B, are present in solution, the distribution coefficient of B is given by

$$\frac{[B]_r}{\Sigma[B]} = D_B$$

D_{B^+} will be approximately constant, independent of the concentration of B^+, if $[A]_r$ remains essentially constant ($[A]_r \gg [B]_r$) and likewise $[A^+]$ ($[A^+] \gg [B^+]$). These conditions are fulfilled if the original concentration of B^+ is low—as when B is a trace constituent in a sample—and $[A^+]$ is not too small at the beginning. Similarly, D_B remains essentially constant if $[A]_r \gg \Sigma[B]$.

Concentrations of the exchanging ion can be expressed as meq/ml in the external solution (N) and as meq/g in the dry resin. The resulting D is sometimes symbolized D_g.[6]

In many papers on ion-exchange separations, values of D for the elements involved are stated. These values are based on the total concentration of the element in the aqueous phase, with acidity and concentration of complexing agents specified.

Equilibria involved in the use of liquid ion exchangers do not seem to call for discussion here. They are extraction equilibria and the general principles of immiscible solvent extraction apply (Chapter 9C).

2. Batch Operations

It would be convenient if a gram or two of solid ion exchanger could be added to an aqueous solution to adsorb a particular ion and leave others in solution. With the common ion exchangers, the possibilities are rather limited, however. The usual cation-exchange resins are not sufficiently selective to offer much promise unless complexing agents are used, and even then the approach is not likely to be especially fruitful, although possibly of use at times. Regulation of the hydrogen-ion concentration in the use of cation-exchange resins as a means of increasing selectivity in batch operations is of limited effectiveness. As an illustration, we can compare the adsorption of Na^+ and Ca^{2+} on the cation-exchange resin Dowex 50-X8 (8% nominal crosslinking, that is, 8% divinylbenzene content).

Let 100 ml of a 0.1 M HCl solution containing a few milligrams of a sodium salt be shaken with 2.0 g of resin in the H^+-form. The ion-exchange capacity of this resin may be taken as $\sim$5 meq/g of dry resin, so that even if all the Na^+ is exchanged, $[H]_r$ will remain almost the same. From Table 8-2, the ion-exchange constant for H^+–Na^+ is 1.6 for this resin, and

$$D_{Na^+} = \frac{[Na]_r}{[Na^+]} = 1.6 \times \frac{5}{0.1} = 80$$

The ratio of weights of sodium in solution and in the exchanger is $(100 \times 1)/(2 \times 80) \sim 0.6$ at equilibrium, so that the recovery (adsorption) of Na is poor.

A similar calculation for Ca^{2+}, again assuming a few milligrams are

TABLE 8-2
Ion-Exchange Equilibrium Constants for Dowex 50 Resin $(K_H = 1)^a$

$$K_{nH}^M = \frac{[M]_r [H^+]^n}{[M^{n+}][H]_r^n}$$

	4% DVB	8% DVB
H	1	1
Na	1.2	1.6
K	1.7	2.3
Rb	1.9	2.5
Cs	2.0	2.6
Ag	3.6	6.7
Mg	1.0	1.3
Ca	1.9	3.2
Sr	2.5	5.2
Ba	6.3	16.2
Zn	1.1	1.5
Cu	1.2	1.8
Ce, La	6.9	22

[a] Based on values given by A. Ringbom, *Complexation in Analytical Chemistry*, Wiley-Interscience, New York, 1963, p. 202. Concentrations in the resin are expressed as meq of exchanging ions per gram of dry H-form resin, in the solution as meq per milliliter, that is, as N. These constants are to be looked upon as quite approximate, because they vary with electrolyte concentration and the loading of the resin. Also, commercial resins are not uniform in composition and (as can be seen from this table) a relatively small variation in the extent of crosslinking can have a comparatively large effect on the exchange constant. Nevertheless, constants such as those given here are useful for calculations at low loadings, and especially when complexing of cations is involved and accurate exchange constants are not really required. See Ringbom for exchange constants for anions with Dowex 1 and Dowex 2. The range of constants for anions tends to be greater than for cations.

present so that $[H]_r$ remains essentially constant, gives

$$D_{Ca^{2+}} = \frac{[Ca]_r}{[Ca^{2+}]} = 3.2\left(\frac{5}{0.1}\right)^2 = 8000$$

Ca in solution/Ca in resin $= (100 \times 1)/(2 \times 8000) = 1/160$, and the adsorption of Ca is complete as far as trace analysis is concerned. The much better recovery of Ca is due not so much to its larger ion-exchange constant as to the squaring of the term $[H]_r/[H^+]$ required by the mass-action law.

If the preceding calculations are carried out with $[H^+] = 1$ instead of 0.1, the ratio {metal in solution/metal in resin} is 6 for Na and $1/1.6 \sim 0.6$ for Ca (these values can be only approximate because of changes in activity coefficients). Adjustment of hydrogen-ion concentration is not a possible means of effecting a separation by a single batch operation. A remaining possibility is converting calcium into an uncharged or anionic complex, not adsorbed by the exchanger, at a pH where sodium would be quantitatively taken up. However, a simpler solution, and the one that would be adopted, is to use the resin in a chromatographic column. We have seen (p. 750) that even with $D_B/D_A \sim 2$ a relatively short column will suffice for a good separation if $h \sim 0.1$ cm. We may expect that a separation of Ca and Na can be obtained without much difficulty, because the D ratio is 10 in 1 M HCl.

Batch operations may have greater significance for chelating resins, whose ion-exchange constants vary over a much greater range than those of the common cation-exchange resins.

Determination of D can conveniently be carried out by shaking a solution of appropriate concentration with a suitable weight of resin and analyzing the solution before and after. Determination of D by the use of a column (p. 762) is more realistic analytically (errors from nonattainment of equilibrium tend to cancel) but is more laborious to carry out, although column characteristics (especially h) can be obtained at the same time.

The relative ability of an ion exchanger to select one ion over another (in the same solution) is expressed by the selectivity coefficient. For $Ca^{2+} - Na^+$, for example,

$$(RH)_r + Na^+ \rightleftharpoons (RNa)_r + H^+$$
$$2(RH)_r + Ca^{2+} \rightleftharpoons (R_2Ca)_r + 2\,H^+$$
$$\frac{[Na]_r[H^+]}{[H]_r[Na^+]} = K_{Na}$$
$$\frac{[Ca]_r[H^+]^2}{[H]_r^2[Ca^{2+}]} = K_{Ca}$$

the selectivity coefficient $k_{Ca,Na}$ is given by

$$k_{Ca,Na} = \frac{K_{Ca}}{K_{Na}^2} = \frac{[Ca]_r[Na^+]^2}{[Ca^{2+}][Na]_r^2}$$

When the counter ions do not have the same charge, as above, the concentration units must be specified.

Liquid Ion Exchangers. Analytical use of these is commonly by batch operation, although they can also be used in columns (extraction chromatography). They are dissolved in organic liquids immiscible with water, and this solution is shaken with an aqueous solution of the sample to extract cationic or anionic species. An example of a liquid ion exchanger extracting anions is provided by the salt of a tertiary high-molecular-weight amine such as tri-*n*-octylamine. When a benzene solution of the salt is shaken with an aqueous solution containing a reacting anion, a reaction of the following type occurs:

$$R_3NH^+A^- + B^- \rightleftharpoons R_3NH^+B^- + A^-$$

org   H_2O   org   H_2O

The reaction occurring is similar to that of anion-exchange resins under the same conditions and hence the amine is sometimes called a liquid anion exchanger.[7] B^- can be a complex anion such as $FeCl_4^-$. Differences in the distribution (extraction) coefficients of similar metals can be so large that a separation can be obtained in one or two extractions. Thus with an 8% solution of tribenzylamine in $CHCl_3$, D for Nb in 11 M HCl is 74 and 0.002 for Ta in the same acid.[8]

An example of a liquid cation exchanger is di-(2-ethylhexyl)phosphoric acid (HDEHP, p. 1004), in which H of the OH group is replaced by metal. Other substances, dissolved in a liquid immiscible with water, functioning in this way are:

Dinonylnaphthalene sulfonic acid, $-SO_3H$
Naphthenic acids

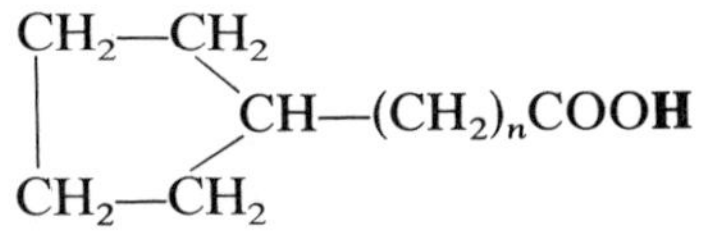

$$\begin{array}{c} CH_2-CH_2 \\ | \qquad\qquad \diagdown \\ \qquad\qquad\qquad CH-(CH_2)_n COOH \\ | \qquad\qquad \diagup \\ CH_2-CH_2 \end{array}$$

3. Ion-Exchange Chromatography

Before considering the variables involved in separations by column fractionation, it may be noted that an ion-exchange column may be used for separations without fractionation. A cation A^+ may be adsorbed on a column and held, while another constituent B, present as an uncharged or anionic species, is not adsorbed and is washed out of the column with a suitable solution, namely one that does not elute A (see p. 763).

a. Distribution Coefficients and Elution Constants

A favorable feature of ion-exchange chromatography is often the relative ease of varying distribution coefficients of metals by changing the hydrogen-ion concentration of solutions or by using differential complexing agents. Other factors being favorable, the separation of A and B will be easier the more $\mathscr{E}_A/\mathscr{E}_B$ or D_B/D_A differs from 1. But the absolute values of $\mathscr{E}_A$ and $\mathscr{E}_B$ are also important. If the $\mathscr{E}$ values are too small (D values too large), A and B are held so tightly on the exchange column that large volumes of eluant are needed to elute them. If eventually they are eluted, the resulting volumes may be so large as to be analytically unsatisfactory; moreover, the separation becomes so time consuming that no one will want to use it.

Suppose A is the ion held more strongly on the column. The aim of the analyst is to adjust the variables determining $\mathscr{E}_A$ so that it is given a value enabling A to be held quantitatively on the column while B is eluted with reasonable speed, and the column length is no greater than really necessary. If A and B are cations of the same charge, adsorbed on a cation-exchange column, $\mathscr{E}_A/\mathscr{E}_B$ (or D_B/D_A) will be approximately independent of the hydrogen-ion concentration. If K_{AH} and K_{BH} are the respective ion-exchange constants for a particular H^+–resin

$$\frac{D_B}{D_A} = \frac{K_{BH}}{K_{AH}}$$

If A has the charge $a+$ and B, the charge $b+$,

$$\frac{D_B}{D_A} = \frac{K_{BH}}{K_{AH}}[H]_r^{(b-a)}[H^+]^{(a-b)}$$

D_B/D_A increases with a decrease in $[H^+]$ if $b > a$. In the previous Na^+-Ca^{2+} illustration, $D_{Ca}/D_{Na} = 8000/80 = 100$ in 0.1 N HCl and $80/8 = 10$ in 1 N HCl.

Suppose 0.1 N HCl is used to elute Na^+ from the narrow band of Na and Ca at the top of the column. The volume of acid required to bring the peak of the Na band to the base of the column will be

$$(V_{Na})_{max} = V_{col}(D_v + i) = V_{col}\{80(1-0.4)+0.4\}$$
$$= 48.4\ V_{col} \qquad (i \text{ taken to be } 0.4)$$

Nearly 50 geometrical column volumes of 0.1 N HCl will be required to elute one-half of the sodium from the column. This is an impossibly large volume. Try 1 N HCl as eluting agent (the D_{Ca}/D_{Na} ratio is still favorable: 10). Now

$$(V_{Na})_{max} = V_{col}\{8(1-0.4)+0.4\} = 5.2 \ V_{col}$$

which is reasonable.

For the elution of Ca with 1 N HCl:

$$(V_{Ca})_{max} \sim 50 \ V_{col}$$

If V_e for both Na and Ca is taken to be 1.2 V_{max} (see the example on p. 750, where $V_e \sim 1.15 \ V_{max}$ and $h \sim 0.1$ cm), the volume theoretically needed to elute 0.9998 of the sodium (this value is chosen to provide a reasonable factor of safety over 0.999 recovery that is aimed for) is

$$V_{max} + 2.5(V_e - V_{max})$$

We may expect (with the illustration on p. 750 as a guide) that a column length of 5 cm will provide a good separation of Ca and Na if $h \sim 0.1$ cm. If the column is 1 cm^2 in cross section, the eluant volume for Na is

$$V_{elu} = (5.2 \times 5) + 2.5 \times 0.2 \times 5.2 \times 5 = 26 + 13 = 39 \ ml$$

When 39 ml of 1 N HCl has passed through the column, the Ca peak has moved $39/250 \times 5 = 0.8$ cm and its width is $(39/250)^{1/2} \times (0.2 \times 5) = 0.4$ cm. The value of ϕ_{Ca} is so large $\{(5-0.8)/0.4 \sim 10\}$ that theoretically essentially no Ca has been eluted. Even with a large disparity between the calculated and actual value, we would expect the latter to be very small and the prediction that a 5 cm column is long enough is verified. If the theoretical plate height is increased four times (to ~ 0.4 cm), the band width is increased twofold $(4^{1/2})$, and essentially all of the Ca will still be retained on the column, since $\phi_{Ca} = (5-0.8)/0.8 \sim 5$.

It is sometimes stated that if D_B/D_A is constant and the individual values of D can be varied, subject to this condition, the product $D_B D_A$ should be made ~ 1 in separations. A more fundamental relation in quantitative separations requires that R_A and R_B have *acceptable* values ($R_A \geq$ some specified value and $R_B \leq$ some other specified value). Suppose A is held more strongly than B on the column (B is to be eluted). If the column length is specified (and h is constant), D_A must be made such that $R_A \sim 1$ under the conditions. Once D_A is fixed, D_B is fixed also if their ratio is constant, and accordingly their product is fixed and so is S

(separation factor, p. 747). If the separation is not satisfactory (the required S depends on the ratio of B to A in the sample), the column length must be increased or h decreased. When this is done, D_A should not be made larger than necessary (allowing a margin for error), that is, not larger than needed for a satisfactory R_A. An unnecessary increase in D_A increases D_B also and makes the separation poorer.

Complexing. A powerful means of varying D of metals is by converting cations into anionic (less often uncharged) species by the use of complexing agents. In this way, the adsorbability of some metals on cation exchangers can be varied over a wide range so that separation by differential adsorption or elution becomes possible. More important in practice are the separations that can be made by adsorbing the anionic metal species on anion exchangers. An outstanding example is conversion of heavy metals into anionic chloro complexes with hydrochloric acid, which can be held by Dowex 1 and other anion-exchange resins. Some of the common metals (alkalies, alkaline earths, Al, Ni) are not adsorbed at any chloride concentration, others hardly at all at low or moderate chloride concentrations (Ti, Mn, Fe(II), Zr, Co, and others), whereas still others have distribution coefficients as large as 10^5 at optimal chloride concentrations (p. 775). An important advantage of hydrochloric acid for adsorption and elution is the general harmlessness of Cl^- in subsequent absorptiometric or fluorimetric determination of metals, although the hydrogen-ion concentration will have to be adjusted. Other inorganic anions also have value in separations. Organic complexing agents are occasionally used, for example EDTA, oxalic acid, and sulfosalicylic acid.

The extent of complexing can, in general, be increased by adding a water-miscible organic solvent to the aqueous solution. Such an organic solvent has a lower dielectric constant than water (e.g., methanol 33, ethanol 24, acetone 21, 2-methoxyethanol 16, dioxane 2.5, compared to water 80) and the dissociation of a complex ion is decreased in the mixture (p. 769).

Water-miscible organic solvents can also alter distribution coefficients of simple ions by decreasing their hydration. Thus the adsorbability of Na^+ by a cation exchanger is increased by addition of ethanol to an aqueous solution, whereas that of K^+, less hydrated at the beginning, is less affected.[9]

b. Loading, Flow Rate, and Plate Height

The preceding discussion on chromatographic separation has assumed that elution curves are Gaussian. This is an ideal that can only be approached. If the approach is not close, predictions of separability can be badly in error. There is often a tendency for tailing in elution curves, that is, the adsorbed solute is eluted less rapidly than expected. The closer the approach to equilibrium, the closer will be the approach of the real elution curve to the ideal.

Distribution coefficients depend upon the loading of an ion-exchange resin (the fraction of exchangeable ion that has been replaced), especially with highly crosslinked resins. If the loading is not greater than ~0.01, distribution coefficients should remain approximately constant and a good approach to symmetrical elution peaks should result. Dowex 50-X8 has a capacity of ~5 meq/g of dry H-form resin or ~2 meq/1 ml of resin bed (fully swollen resin). For Dowex 1 or Amberlite IR-400 (6–8% DVB), the corresponding values are ~3 meq/ml and ~1 meq/ml. A loading of 0.01 for Cu^{2+} referred to 1 ml of Dowex 50-X8 resin bed corresponds to ~1.0 mg. For trace constituents in a sample the permissible loading is unlikely to be exceeded. The matter is otherwise if a major constituent is to be exchanged. Where larger amounts of constituents are to be separated by chromatography the cross-sectional area of a column can be increased, but only at the cost of larger volumes.

The rate of ion exchange is determined either by diffusion of ions through the Nernst film around the resin particles or by diffusion of ions in the resin particles. In column operation, particle diffusion is likely to be the rate-determining step. An increase in crosslinking decreases the diffusion rate in the resin grains. Decreasing the particle size reduces the time required for attaining equilibrium, but there is a limit beyond which the decrease cannot be carried without making the flow rate through the column prohibitively slow unless pressure is applied. Other factors remaining the same, the height of a theoretical plate h depends on the resin particle size and the flow rate. As the flow rate approaches zero, $h \rightarrow \sim 4r$ (r is the particle radius in centimeters). For a 100-mesh resin ($r \sim$ 0.07 mm), a rough average value of h is 0.5 mm for a flow rate of 0.1 ml/cm^2/min (or 0.1 cm/min), 1 mm for 1 ml/cm^2/min and 3 mm for 5 ml/cm^2/min.[10]

In most inorganic chromatography, resins of 100–200 mesh size are suitable; difficult separations may require 200–400 mesh resins. A flow rate of 1 ml/cm^2/min is often satisfactory (if it can be obtained[11]), but sometimes a lower rate is required. In simple percolation procedures, in which one or more constituents are to be retained on the resin and the others pass through, without differential adsorption, coarser resins (e.g., 50–100 mesh) may be suitable and a flow rate of 5 ml/cm^2/min may be reasonable.

With suitable equipment, the flow rate in long columns or beds of fine adsorbent can be increased by applying pressure (forced flow chromatography).

C. APPLICATIONS

In this section we survey general applications of ion-exchange methods in separations of metals, especially from the standpoint of trace analysis when such information is available. Ion-exchange resins, especially the nonchelating, are the most widely applicable ion-exchange materials, and receive most attention here.

1. Nonchelating Resins

a. Some Practical Aspects

Whether or not fractional separation—true chromatography—is involved, ion exchangers are usually used in columns.[12] The analyst must decide what exchange material is best suited for the required separation. Purification and other treatment (sizing) may then be required. Finally the column is prepared. In the following discussion, we assume the ion-exchange material to be a polystyrene resin of the strong-acid or strong-base type.

i. PRELIMINARY TREATMENT

This consists of purification—removal of inorganic and organic contaminants—and sizing, the latter usually restricted to removing fine particles. In critical chromatographic separations, uniform particle size is important; it is then usually advantageous for the analyst to purchase resins that have been specially sized rather than attempt this himself. Also, such resins have been purified. Some resins sold as "reagent" or "analyzed" are actually standard commercial resins and are far from pure as may be seen from the following comparison[13]:

	Maximum Content (ppm)	
	Analytical-grade (purified) Dowex 1	Typical Specification Dowex 1 or Dowex 50
Fe	0.5	100
Cu	0.2	20
Ni	0.05	15
Pb	0.005	20
Al	5	800
Aliphatic amines	5	—
Minimum percent in size range specified	95	50

Cation-exchange resins can be purified by soaking in warm 4 M HCl and washing with pure water (Buchner funnel); this treatment can be repeated once or twice if thought necessary. In this operation, the fines can be removed by pouring off the supernatant liquid after allowing the bulk of the material to settle for a minute or two in a beaker and repeating. Anion-exchange resins can be purified by treating successively with 1 M HCl, water, 0.5 M NaOH, water, HCl, and water and repeating if necessary. Anion-exchange resins may contain organic impurities in amounts that can interfere in separations of metals or in their subsequent photometric determination. Such organic matter can be removed by percolating alcohol or acetone through the resin and finally washing with warm water (the anion exchanger should be in the Cl$^-$ form). When necessary, organic impurities can be removed from cation-exchange resins in the same way.[14]

Anion-exchange resins are not very stable in the OH$^-$ form and are best used in the Cl$^-$ form. Strong oxidizing agents such as permanganate attack polystyrene cation-exchange resins in acid and basic solutions. At room temperature, there is little attack by HNO$_3$ below 3 M. Cr(VI), V(V), Au(III), Ir(IV), and apparently Mo(VI) are partly reduced by resin exchangers. Hydrogen peroxide slowly attacks resins, especially in the presence of Fe, Mn, and Cu, which act as catalysts.

Complexing substances derived from resins in operations with strongly acid solutions may complex some metals so firmly in subsequent photometric determinations that destruction of organic matter with a strong oxidizing agent (KMnO$_4$) is required.[15]

ii. COLUMN PREPARATION

When a column is to be used for the adsorption of a single constituent (or group of constituents) without fractionation, its preparation need not be as painstaking as when it is to be used chromatographically. Nevertheless, some thought and care are needed. The dimensions of the tube will be determined, among other factors, by the size of sample, the percentage of the constituent(s) to be retained on the column, and the value of D. The tube is half-filled with water (or eluant) and a slurry of 1 resin to 2 of water, free from air bubbles, is added to give a uniform resin bed, water being drawn off and slurry added until the desired length of column has been built up. The water level should not be allowed to drop below the top of the resin. Before use, it may be necessary to replace the water in the column with a solution similar to the sample solution. Columns that have stood for any length of time (overnight) should be washed before use to remove traces of organic matter that may have been released from the resin grains.

A chromatographic column will usually require a finer resin than a percolation column (preceding paragraph), so that h is requisitely small. Resins of 100–200 mesh size are often satisfactory, but the more difficult separations may require 200–400 mesh resin. As mentioned earlier, uniformity of grain (bead) size is important. A relatively small fraction of larger size beads will increase h markedly. The column packing must be uniform: no air bubbles or channels. Channelling can ruin a separation. Some workers backwash a resin column to obtain more uniform packing: Water is passed upward through the column at a rate that breaks up the column and forms a suspension, which then is allowed to settle. A peristaltic pump may be used in this operation.

The required column length and flow rate in chromatographic separations have already been discussed. A margin of safety must be provided in the column length, but unnecessarily long columns should be avoided because of larger solution volumes and the longer time required. If, for a given column length and resin size, the flow rate needs to be regulated by adjusting the stopcock, it may be advantageous to use a finer resin to reduce the flow rate.

Different lots of the same kind of exchange resin may show variations in properties (e.g., crosslinking), so it should not be assumed that columns prepared from different batches of resin will all have precisely the same characteristics (D, for example, determining V_{max}).

b. Nonchromatographic Separations

A sharp differentiation between chromatographic and nonchromatographic separations cannot be made nor is this necessary in practice. Ions not adsorbed or having small values of D can easily be washed out of the column and separated from ions that are at least moderately strongly held. The ideal situation is that in which the trace constituent is adsorbed and the major constituents are not. At the same time that the separation is made, the trace constituent may be concentrated, if the sample volume is large and the elution volume for the constituent is small. It may be advantageous from the standpoint of keeping volumes small to remove the trace constituent at the end by running the displacing solution through the column in the reverse direction, since the trace constituent will be at the top of the column. Nonchromatographic separations are most likely to be applicable when the respective constituents are ions of opposite charge. This can often be brought about by using a complexing agent (see separations in HCl solution, p. 774). However, some separations of cations from large amounts of others can be effected by using chelating resins having special affinity for the former. In a sense, this is utilizing complexing in the resin phase. See Table 8-3 for some examples of adsorption of trace constituents on ion exchangers.

The second possibility in separations is the use of an ion exchanger in column form to retain major constituents while letting the trace constituent pass through. Obviously, this is a less favorable case than the previous one, but nevertheless one sometimes useful in trace analysis. The amount of resin should in general be such that there are at least three times as many equivalents of replaceable ions in it as there are ions to be held on the column; moreover, the column length must be adequate (see discussion that follows). Because an ion-exchange resin provides only 1 to

TABLE 8-3
Examples of Nonchromatographic Separations by Use of Ion Exchangers

Element	Separated from	Exchanger	Conditions	Reference
Cu	(A few ppb Cu in 8–10 liters of water concentrated)	Cation	Cu eluted with 5 M HCl	B. Tuck and E. M. Osborn, *Analyst*, **85**, 105 (1960).
Rh, Ir, Pd, Pt	Fe, Cu, Ni	Cation	Base metals adsorbed from chloride solution of pH 1.5	G. G. Tertipis and F. E. Beamish, *Talanta*, **10**, 1139 (1963); *Anal. Chem.*, **34**, 108 (1962); J. G. Sen Gupta, *Anal. Chem.*, **39**, 18 (1967).
Th	Phosphate	Cation	Th adsorbed from 0.1 M HCl and eluted with 10 M HCl	O. Menis, D. L. Manning, and G. Goldstein, *Anal. Chem.*, **29**, 1426 (1957).
Au	Cu	Anion	Au adsorbed from HCl–HNO$_3$ solution; resin containing Au ashed	A. Mizuike, Y. Iida, K. Yamada, and S. Hirano, *Anal. Chim. Acta*, **32**, 428 (1965).
Au, Bi	Sea water	Anion	Both metals adsorbed from sea water (250 liters) made 0.1 M in HCl, and eluted with 0.25 M HNO$_3$	R. R. Brooks, *Analyst*, **85**, 745 (1960).
Co	Th	Anion	Co adsorbed from 10 M HCl and eluted with 4 M HCl	G. Goldstein, D. L. Manning, and O. Menis, *Anal. Chem.*, **31**, 192 (1959).
Rare earths	Th	Anion	Th adsorbed from 8 M HNO$_3$	B. J. Bornong and J. L. Moriarty, *Anal. Chem.*, **34**, 871 (1962).
U	Isolation from ores	Anion	U adsorbed from sulfate solution of pH 1.0–1.5 and eluted with 1 M HClO$_4$	S. Fisher and R. Kunin, *Anal. Chem.*, **29**, 400 (1957).

2 meq of exchangeable ion/ml of bed, a rather large amount of resin may be needed. For removal of 0.18 g of Fe(III) ($\sim$10 meq) by exchange with H^+, about $10/2 = 5$ ml cation-exchange resin is needed (Dowex 50 furnishes $\sim$2 meq of replaceable H^+/ml of bed), so that a column bed volume of $3 \times 5 = 15$ ml is called for. If Dowex 1 resin is used ($FeCl_4^-$ replacing Cl^-) a bed volume of $3 \times 0.18/56 \times 1000 \times 1 \sim 10$ ml should be allowed (the capacity of this resin is $\sim$1 meq/ml bed).

When a macro constituent is separated from a trace constituent in this way, the separation factor (ratio of recovery factors) must be much more favorable than in macro analysis (of major constituents). In the latter, 0.1 mg of unseparated Fe might be tolerated, whereas in trace analysis no more than 0.1 μg Fe might be allowed, if the photometric reagent is sensitive to Fe. Therefore, the analyst must consider whether his exchanger column is long enough to hold the major constituent. Two factors are involved here: the capacity of the column and the number of theoretical plates N. If N is too small (h relatively large and column length too short), the width of the band may be so great that on washing to remove the micro constituent from the column, a significant amount of the major constituent may also be removed. The value of D_v is also important. If this is small, the adsorption band moves rapidly down the column so that little or much of the major constituent escapes before the trace constituent has been completely rinsed out of the column. In the iron example above, the fraction of iron held in the column must be at least

$$\frac{(0.18 \times 10^6) - 0.1}{0.18 \times 10^6} \sim 0.9999994.$$

Consider the possibilities, using Dowex 1 anion-exchange resin and Fe(III) in 2.5 M HCl, D_v being $\sim$100. Suppose the analyst uses a resin column 10 cm long and 1 cm^2 in cross section. If Cl^- in the resin is quantitatively replaced by $FeCl_4^-$, 3.3 ml or 3.3 cm of resin will ideally be transformed from the Cl^- to the $FeCl_4^-$ form (0.18 g of Fe). Suppose that 50 ml of 2.5 M HCl is used to rinse the micro constituent out of the column. This should be more than enough if D_v for the micro constituent is essentially zero; some of the micro constituent will have been removed in the effluent while the sample solution (say 50 ml in volume) was being added. Simple column theory assumes that the adsorbed substance is present in a narrow band at the very top of the column. Here, however, the iron is distributed over a column segment at least 3.3 cm long. To get an answer to the question of feasibility of separation, assume that the band maximum is at 3.3 cm (from the column top). This assumption makes the situation worse than it really is, because most of the iron is behind the 3.3 cm point. When the column is washed with 50 ml of 2.5 M HCl, the assumed band will move 0.5 cm down the tube, because V_{max} (from the top to the base of the column) is equal to V_{col} $(D_v + i) \sim$ $10 \times 100 \sim 1000$ and hence $50/1000 \times 10 = 0.5$ cm. The assumed band peak is, therefore, $3.3 + 0.5 = 3.8$ cm from the top of the column when the 50 ml of 2.5 M HCl has been added. Take h as 0.3 cm, a plausible value in practice (see p. 760). At 3.8 cm, the half-band width in the resin is $(2 \, hl_{max})^{1/2} = (2 \times 0.3 \times 3.8)^{1/2} = 1.5$ cm. Accordingly, $\phi = (10 - 3.8)/1.5 \sim 4$, corresponding to a fractional elution of $< 5 \times 10^{-7}$ of the iron. On the basis of ideal behavior, the stipulated separation can be carried out. Practically, it *may* be possible, especially if h is reduced.

Note that if $D_v = 10$ (h the same as before), the separation will fail, because 50 ml of 2.5 M HCl will carry the assumed band maximum $3.3 + (50/100 \times 10) = 8.3$ cm from the column top and much iron will be eluted. The column length would have to be increased.

Occasionally, electron-exchange resins are used in trace separations. For example, Ag^+ can be reduced to Ag^0 by passage down a column of quinol–quinone type resin, followed by elution with 3 M HNO_3 (separation of Ag from Cu, Pb, Bi, etc.).[16]

c. Chromatographic Separations

In this section, ion-exchange separations of elements are surveyed on the basis of the composition of the solution from which the adsorption is effected or of the eluting solution. Solutions of the common acids are the most important and these are roughly arranged in the order of increasing complexing ability (oxygen acids and halogen acids, followed by miscellaneous reagents).

i. PERCHLORIC ACID

Adsorption of elements from perchloric acid solutions (0.2–12 M) by a cation-exchange resin, Dowex 50-X4, has been studied by Nelson et al. (Fig. 8-4). Distribution coefficients of metal ions on AG 50 W-X8 in perchloric acid solutions (0.2–4.0 M) are shown in Table 8-4.

In dilute perchloric acid, D decreases with acidity as a result of increasing adsorption of H^+. But, with the exception of the alkalies and a few other metals, the decrease in D comes to a stop at 4-6 M $HClO_4$ and D then increases as the concentration of $HClO_4$ increases further. The minima are more pronounced with the cations of higher charge. For cations of the same charge, the difference in D at high $HClO_4$ concentrations probably results from differing hydration strengths. The more strongly adsorbed cations part more easily with coordinated water molecules and form a stronger bond with the sulfonate group of the resin.

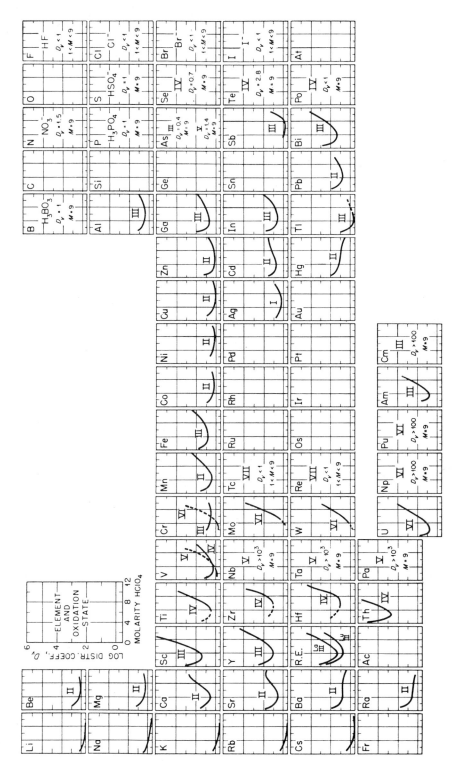

Fig. 8-4. Adsorption of elements from $HClO_4$ solutions by cation-exchange resin Dowex 50-X4. From F. Nelson, T. Murase, and K. A. Kraus, *J. Chromatogr.*, **13**, 533 (1964).

TABLE 8-4
Distribution Coefficients of Metal Ions on AG 50 W-X8 Resin in Perchloric Acid Solutions[a]

Metal Ion	D_g				
	0.2 M	0.5 M	1.0 M	2.0 M	4.0 M
Ag(I)	90	39	20	10	5.8
Al(III)	5250	516	106	31	11
Ba(II)	2280	429	127	44	19
Be(II)	206	49	14	4.4	1.9
Bi(III)	$>10^4$	935	243	76	42
Ca(II)	639	147	50	18	7.7
Cd(II)	423	101	36	13	6.3
Ce(III)	$>10^4$	2380	459	114	53
Co(II)	378	92	31	11	4.8
Cr(III)	8410	585	120	29	11
Cu(II)	378	90	30	10	4.5
Dy(III)	$>10^4$	1370	258	63	39
Fe(II)	389	95	32	12	5.2
Fe(III)	7470	562	119	30	12
Ga(III)	5870	556	112	29	11
Hg(I)	4160	548	147	38	9.0
Hg(II)	937	222	85	38	23
In(III)	6620	619	128	32	14
La(III)	$>10^4$	2470	475	118	58
Mg(II)	312	76	24	7.9	3.1
Mn(II)	387	94	32	11	4.7
Mo(VI)	22	12	5.5	4.4	4.5
Mo(VI)[b]	0.7	0.5	0.4	0.3	1.3
Ni(II)	387	91	32	11	5.3
Pb(II)	1850	368	117	39	17
Sn(IV)	ppt	ppt	ppt	12	7.5
Sr(II)	870	198	67	24	10
Th(IV)	$>10^4$	$>10^4$	5780	844	686
Ti(IV)	549	88	19	7.6	5.7
Tl(I)	131	52	23	8.2	2.7
Tl(III)	1550	548	176	57	41
U(VI)	276	75	29	15	18
V(IV)	201	50	18	6.9	4.4
V(V)	9.8	4.5	2.2	1.3	0.8
V(V)[b]	9.3	6.4	3.0	1.1	1.0
W(VI)[b]	0.4	0.3	0.4	0.2	0.4
Y(III)	$>10^4$	1390	246	59	24
Yb(III)	$>10^4$	1120	205	51	20
Zn(II)	361	88	30	9.9	5.3
Zr(IV)	$>10^4$	$>10^4$	$>10^4$	1960	333

[a] Selected data from F. W. E. Strelow and H. Sondorp, *Talanta*, **19**, 1113 (1972). D_g is based on weight of dry resin.
[b] Hydrogen peroxide present.

Sc, Y, U(VI), Th, Ti and some other metals are adsorbed strongly from concentrated $HClO_4$ solutions, but such solutions attack the resin.

ii. NITRIC ACID

Cation Exchange. Distribution coefficients of metal ions on a strong-acid cation-exchange resin, AG 50 W-X8, in nitric acid solutions are shown in Table 8-5. With very few exceptions, adsorption characteristics are much the same as in $HClO_4$ solutions, as would be expected. Chromatographic separations of the following mixtures are possible: Na(elution with 0.6 M HNO_3)–Be(1.2 M HNO_3)–Ba(2.0 M HNO_3)–Y(3.0 M HNO_3)–Zr(5.0 M HCl); Hg(II)(0.2 M HCl)–Cd(0.5 M HCl)–Be (1.2 M HNO_3)–Fe(III)(1.75 M HCl)–Ba(2.5 M HNO_3)–Zr(5.0 M HCl).

The cation-exchange behavior of 19 elements in nitric acid–organic solvent media has been described.[17] In 90% tetrahydrofuran–10% 6 M nitric acid medium, uranium(VI) has a very low distribution coefficient, whereas many other metals have coefficient values of the order of 10^3–10^4 and consequently their separation from uranium is readily achieved.[18]

Anion Exchange. Distribution coefficients of elements on a strong-base anion-exchange resin in 0.1–14 M nitric acid are summarized in Fig. 8-5. Thorium is strongly adsorbed from 8 M nitric acid, whereas uranium is slightly adsorbed. Separation of thorium from the rare earths is also effected (Table 8-3). (There is an explosion hazard in the use of anion-exchange resins in strong HNO_3. For their safe use, HNO_3 concentrations should be kept as low as possible and should not exceed 7 or 8 M.[19]) Distribution coefficients of Fe(III), Bi, Pb, Hg(II), Mg, U(VI), Be, Na, and Li (decreasing in that order) on AG 50W-X8 in (0.5–2.0) M HNO_3–(0–80)% methanol are available.[20] Traces of lithium in silicate rocks can be separated by elution with 1 M HNO_3 in 80% methanol. Adsorption from methanol–HNO_3 solutions has also been applied in the isolation of lead from silicate rocks (age determinations).

Anion-exchange behavior of metal ions in organic solvent mixtures containing nitric acid has been extensively studied.[21] The presence of high percentages of organic solvents increases distribution coefficients of many metal ions. For example, both thorium and uranium(VI) are adsorbed strongly on Dowex 1-X8 from 96% 1-propanol–4% 5 M nitric acid solution.[22,23] Procedures have been described for the separation of thorium and uranium from each other and from many other elements by anion exchange from nitric acid–organic solvent media.[22] Other applications of anion exchange from nitric acid–organic solvent (mainly alcohols) media include separations of Mg–Ca–Sr,[24] Cd from Al, Cu, and Zn,[25] of Pb and Bi from other elements,[26] and of Re(VII) from Mo(VI).[27]

TABLE 8-5
Distribution Coefficients of Metal Ions on AG 50W-X8 Resin in Nitric Acid Solutions[a]

Metal Ion	D_g				
	0.2 M	0.5 M	1.0 M	2.0 M	4.0 M
Ag(I)	86	36	18	7.9	4.0
Al(III)	3900	392	79	17	5.4
As(III)	<0.1	<0.1	<0.1	<0.1	<0.1
Ba(II)	1560	271	68	13	3.6
Be(II)	183	52	15	6.6	3.1
Bi(III)	305	79	25	7.9	3.0
Ca(II)	480	113	35	9.7	1.8
Cd(II)	392	91	33	11	3.4
Ce(III)	>10^4	1840	246	44	8.2
Co(II)	392	91	29	10	4.7
Cr(III)	1620	418	112	28	11
Cu(II)	356	84	27	8.6	3.1
Fe(III)	4100	362	74	14	3.1
Ga(III)	4200	445	94	20	5.8
Hg(I)	7600	640	94	34	14
Hg(II)	1090	121	17	5.9	2.8
In(III)	>10^4	680	118	23	5.8
La(III)	>10^4	1870	267	47	9.1
Mg(II)	295	71	23	9.1	4.1
Mn(II)	389	89	28	11	3.0
Mo(VI)	5.2	2.9	1.6	1.0	0.6
Ni(II)	384	91	28	10	7.3
Pb(II)	1420	183	36	8.5	4.5
Pd(II)	62	24	9.1	3.4	2.5
Rh(III)	45	20	7.8	4.1	1.0
Sc	3300	500	116	23	7.6
Th(IV)	>10^4	>10^4	1180	123	25
Ti(IV)	461	71	15	6.5	3.4
Tl(I)	91	41	22	9.9	3.3
U(VI)	262	69	24	11	6.6
V(V)	11	4.9	2.0	1.2	0.5
Y(III)	>10^4	1020	174	36	10
Zn(II)	352	83	25	7.5	3.6
Zr(IV)	>10^4	>10^4	6500	652	31

[a] Selected data of F. W. E. Strelow, R. Rethemeyer, and C. J. C. Bothma, *Anal. Chem.*, **37,** 106 (1965). Distribution coefficients (D_g) were determined at a total amount of metal ion/total resin capacity ratio of 0.4. For another set of distribution coefficients in 0.15–1.2 M HNO_3, see J. Korkisch, F. Feik, and S. S. Ahluwalia, *Talanta*, **14,** 1069 (1967).

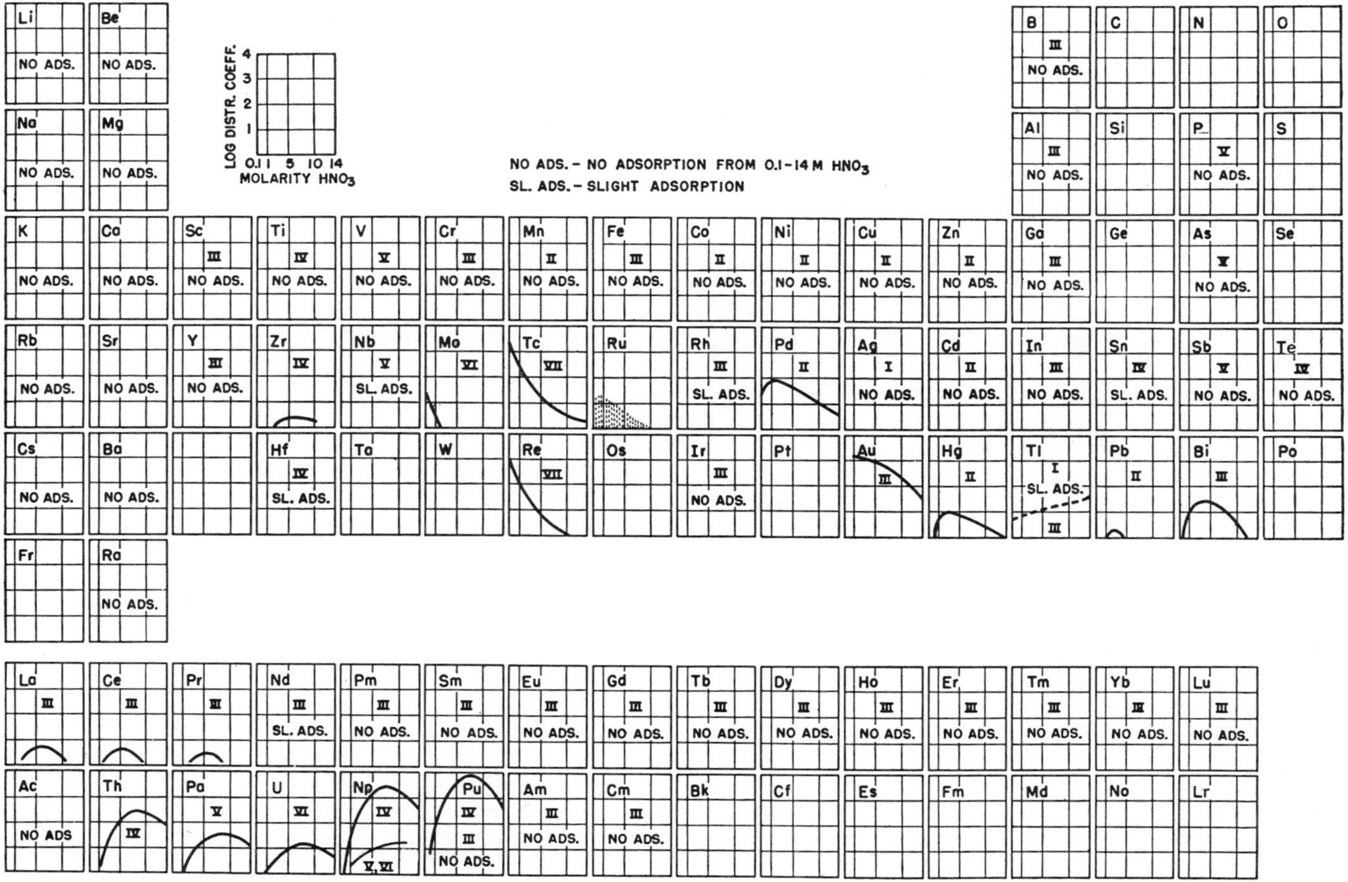

Fig. 8-5. Adsorption of elements from HNO_3 solutions by strongly basic anion-exchange resin. Reprinted with permission from J. P. Faris and R. F. Buchanan, *Anal. Chem.*, **36**, 1158 (1964). Copyright by the American Chemical Society.

771

Incidentally, a selective method for separation of thorium is based on its adsorption on DEAE-cellulose from a 20:1 methanol–8 M nitric acid medium and its elution with 1 M hydrochloric acid.[28]

iii. SULFURIC ACID

Cation Exchange.   Distribution coefficients of metal ions on AG 50W-X8 resin, 100–200 mesh, in 0.1–2.0 M sulfuric acid solutions are listed in

TABLE 8-6
Distribution Coefficients of Metal Ions on AG 50W-X8 in Sulfuric Acid Solutions[a]

Metal Ion	D_g				
	0.1 M	0.25 M	0.5 M	1.0 M	2.0 M
Al(III)	8300	540	126	28	4.7
As(III)	<0.1	<0.1	<0.1	<0.1	<0.1
Be(II)	305	79	27	8.2	2.6
Bi(III)	>10^4	6800	235	32	6.4
Cd(II)	540	144	46	15	4.3
Ce(III)	>10^4	1800	318	66	12
Co(II)	433	126	43	14	5.4
Cr(III)	176	126	55	19	0.2
Cu(II)	505	128	42	13	3.7
Fe(III)	2050	255	58	14	1.8
Ga(III)	3500	618	137	27	4.9
Hg(II)	1790	321	103	35	12
In(III)	3190	376	87	17	3.8
La(III)	>10^4	1860	329	68	12
Mg(II)	484	124	42	13	3.4
Mn(II)	610	165	59	17	5.5
Mo(VI)	5.3	2.8	1.2	0.5	0.2
Ni(II)	590	140	46	17	2.8
Pd(II)	71	33	14	6.0	2.7
Rh(III)	49	29	16	4.5	1.3
Sc(III)	1050	141	35	8.5	3.4
Th(IV)	3900	263	52	9.0	1.8
Ti(IV)	225	46	9.0	2.5	0.4
Tl(I)	236	97	50	21	8.7
Tl(III)	1490	205	47	12	5.2
U(VI)	118	29	9.6	3.2	1.8
V(V)	15	6.7	2.8	1.2	0.4
Y(III)	>10^4	1380	253	50	9.4
Zn(II)	550	135	43	12	4.0
Zr(IV)	474	98	4.6	1.4	1.0

[a] Selected data of F. W. E. Strelow, R. Rethemeyer, and C. J. C. Bothma, *Anal. Chem.*, **37,** 106 (1965). Distribution coefficients D_g were determined at a ratio of total amount of metal ion to total resin capacity, $q = 0.4$; for Bi, $q = 0.06$.

Table 8-6. Scandium has been separated from the rare earths by elution of the former with 1 M sulfuric acid.[29] Ti has been separated from Al, Be, Fe, Mg, Mn, Ni, rare earths, Th, and other metals by elution of Ti with 0.5 M sulfuric acid containing 1% hydrogen peroxide; the other ions stay on the column.[30] Hydrogen peroxide is added as stabilizing agent to suppress hydrolysis of Ti. Mo, Nb, and V can be eluted with 0.25 M sulfuric acid containing 1% hydrogen peroxide before the elution of Ti.

The distribution coefficients of some 20 metal ions on Dowex 50W-X8 in 0.1–1.0 M $(NH_4)_2SO_4$–0.025 M H_2SO_4 solutions have been determined.[31] Beryllium has been separated from a number of foreign metal ions, and uranium, thorium, and rare earths have been separated from one another.

Anion Exchange. Anion-exchange distribution coefficients of metal ions in sulfuric acid media are given in Table 8-7.[32] As examples of

TABLE 8-7
Distribution Coefficients of Metal Ions on AG 1-X8 in Sulfuric Acid Solutions[a]

Metal Ion	D_g					
	0.05 M	0.1 M	0.25 M	0.5 M	1.0 M	2.0 M
As(III)	0.9	0.6	<0.5	<0.5	<0.5	<0.5
Bi(III)	18	4.7	2.1	0.9	0.5	<0.5
Cr(III)	2.1	0.7	0.5	<0.5	<0.5	<0.5
Cr(VI)	12000	7800	4400	2100	800	302
Fe(III)	16	9.1	3.6	1.4	0.9	0.5
Ga(III)	0.6	<0.5	<0.5	<0.5	<0.5	<0.5
Hf(IV)	4700	701	57	12	3.2	1.2
In(III)	2.4	0.8	<0.5	<0.5	<0.5	<0.5
Mo(VI)	533	671	484	232	52	4.6
Mo(VI)[b]	2560	1400	451	197	74	33
Nb(V)[b]	120	96	3.4	<0.5	<0.5	<0.5
Sc(III)	22	11	4.8	2.6	1.5	0.5
Ta(V)[b]	1860	1070	310	138	50	3.9
Th(IV)	35	21	8.3	3.7	2.0	0.6
Ti(IV)[b]	hydrol	0.5	<0.5	<0.5	<0.5	<0.5
U(VI)	521	248	91	27	9.3	2.9
V(V)	6.5	3.3	1.6	0.7	<0.5	<0.5
V(V)[b]	45	11	4.6	2.5	2.1	1.9
W(VI)[b]	528	457	337	222	127	110
Yb(III)	0.6	<0.5	<0.5	<0.5	<0.5	<0.5
Zr(IV)	1350	704	211	47	11	2.9

[a] Selected data of F. W. E. Strelow and C. J. C. Bothma, *Anal. Chem.*, **39**, 595 (1967).
[b] 0.3% H_2O_2 present.

possible separations, elution curves have been presented for the multicomponent systems Y–Th–U–Mo, Th–Hf–Zr–Mo and Cr(III)–V(V)–Mo–W. Adsorption of uranium(VI) from sulfate solution is useful for its isolation from ores (Table 8-3). Traces of Th, Zr, and U in silicate rocks have been separated by anion-exchange chromatography in sulfate media.[33]

The adsorption behavior of some 20 metals on a weak-base anion-exchange resin, Amberlite CG-4B, in sulfuric acid media (0.005-2 M) has been examined.[34] Titanium is adsorbed on CG-4B from 0.1 M H_2SO_4–0.15% H_2O_2, whereas Mg, Al, Cu, Ga, and rare earths are not. Titanium is eluted with 2 M sulfuric acid.

The anion-exchange behavior of metal ions in miscible organic solvent–sulfuric acid media has been studied.[35] In 80–90% methanol–sulfuric acid solutions having a final acidity of 0.05 M, Cu and In have high distribution coefficients, and Cd, Co, Fe, Mg, Mn, Ni, and Zn can be separated from Cu and In.

The distribution coefficients of uranium(VI) and thorium on Dowex 1 in sulfuric acid–alcohol mixtures increase markedly with increasing alcohol concentration, and they decrease with increasing acid concentration in the range 0.1–0.8 M.[22]

iv. HYDROCHLORIC ACID

Anion Exchange. A number of factors combine to make separations of metals with strong-base anion-exchange resins in hydrochloric acid solutions of outstanding value. Metals vary greatly in their extent of complexing with chloride and it is easy to vary the degree of complexing of a particular metal, and therewith D, over a wide range by adjustment of the hydrochloric acid concentration. Often metals can be eluted from the column merely by using hydrochloric acid of the proper concentration. Chloride ion is not a trouble maker in subsequent photometric determinations, although the hydrogen-ion concentration of the eluate will have to be adjusted after an elution. The chloro anions are often strongly adsorbed on a strong-base type of anion-exchange resin so that D_v may have values of 100 or 1000 or even higher. The specific adsorbabilities of the anionic chloro complexes are orders of magnitude greater than that of Cl^-.

The many separation possibilities will be evident from Fig. 8-6, in which D_v is plotted logarithmically versus HCl concentration.

The [HCl]–D_v curves of one-half or more of the metals show a maximum or an approach to a maximum in the range of hydrochloric acid concentrations investigated. Other metals—those forming stable chloro complexes—giving steadily decreasing D_v values with an increase in the

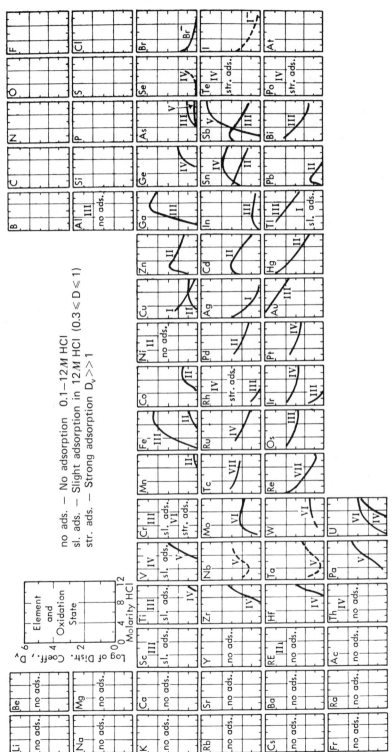

Fig. 8-6. Adsorption of elements from HCl solutions by strongly basic anion-exchange resin. From K. A. Kraus and F. Nelson, *Proc. Int. Conf. Peaceful Uses Atom. Energy,* **7**, 118 (1956).

acid concentration, doubtless have a maximum at very low [HCl]. Kraus and Nelson[36] pointed out that D_v of a metal reaches its maximum value when the average charge of all the metal species in solution is zero. An equivalent statement is[37]: The distribution coefficient of a metal adsorbed on an anion exchanger saturated with a univalent ligand attains its maximum value when the average ligand number is equal to the charge on the free metal ion. Accordingly, one can estimate the ligand concentration at which maximum metal adsorption occurs, if the formation constants of the metal complexes are known for the ligand. Metals such as Au, Ag, Cu(I), and the platinum metals, forming slightly dissociated chloro anions, are adsorbed strongly at low Cl^- concentrations. The decrease of D_v at high hydrochloric acid concentrations, shown by most metals, is partly due to increased competitive adsorption of Cl^-. For some metals, for example Au, the distribution coefficient is not the same for HCl and alkali metal chloride solutions. At least in part, this may be due to incomplete ionization of the metal chloro acid.

The intrinsic adsorbability of a complex chloro anion (or chloro acid) is also an important factor determining the distribution coefficient of a metal. It appears that adsorption of undissociated chloro acid on the resin matrix can also contribute to the overall adsorption of a metal. It is interesting to note (Fig. 8-6) that those chloro complexes, which as acids are readily extracted into ether and other oxonium solvents, namely of Fe(III), Ga, Sb(V), Au(III), and Tl(III), have unusually high distribution coefficients ($\sim 10^5$) on quaternary amine resins. The single charge on the chloro anion probably explains, in large measure, this strong adsorption.

When chromatographic separations are to be made, the sample solution should have as small a volume as conveniently possible and the HCl concentration should be such that the metals to be separated are initially adsorbed in a narrow band at the top of the column. Elution with successive solutions of suitable HCl concentration may enable a number of metals to be separated from each other (Fig. 8-7). The procedure is easier when only one metal, or group of metals, is to be retained on the column and others are run through, provided the first can be given D_v values that are high, and the latter have small values of D_v. Sometimes, if metals are held strongly at low HCl concentrations, another acid can be used for elution. Nitric acid is sometimes useful, because nitrate ion is adsorbed quite strongly by anion-exchange resins. Perchloric acid (dilute) is even more effective.

Strong adsorption in hydrochloric acid media has been applied to separations of many elements, for example, Ag from Cu,[38] Co in the analysis of iron and steel,[39] and Zn in the analysis of meteorites,[40] rocks,[41] and plant materials.[42] Also see Table 8-3. Nonadsorption has

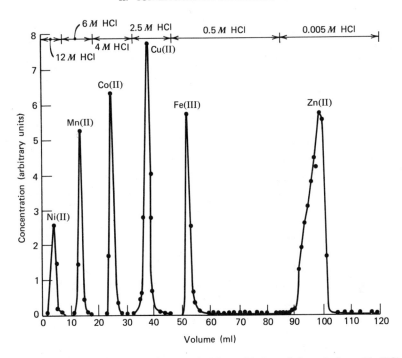

Fig. 8-7. Separation of the transition metals Mn · · · Zn by echelon elution with HCl of different concentrations. Reprinted with permission from K. A. Kraus and G. E. Moore, *J. Am. Chem. Soc.*, **75**, 1462 (1953). Copyright by the American Chemical Society.

been utilized for separations of Al from Fe[43] and U,[44] and Ni in the analysis of blood.[45]

The anion-exchange behavior of metal ions in organic solvent mixtures containing hydrochloric acid has been studied extensively.[46] Alcohols and acetone are favored solvents.[47] Some interesting examples follow. Both uranium(VI) and thorium are adsorbed on Dowex 1-X8 from hydrochloric acid solutions containing high percentages of ethanol, 1-propanol, 2-propanol, 1-butanol, or 1-pentanol. The distribution coefficients for uranium are always higher than those for thorium. After adsorption, uranium and thorium can be separated from each other with suitable eluants. Uranium can be separated from large amounts of iron(III) by adsorption of uranium(VI) on Dowex-1X8 from a medium consisting of 90% methyl glycol (2-methoxyethanol)–10% 6 *M* hydrochloric acid.[48] Uranium is eluted with 1 *M* hydrochloric acid. An equally effective medium is 50% tetrahydrofuran–40% methyl glycol–10% 6 M hydrochloric acid.[49] Distribution coefficients of uranium and other elements are:

% THF	% MG	% 6 M HCl	U(VI)	Fe(III)	Cu(II)	Co(II)	Mn(II)
0	90	10	32,000	17	260	9500	310
50	40	10	3600	1.3	83	5900	160

On Dowex 1-X8, Ga, In, and Al are separated by elution with 80% acetone–20% 3 M HCl, 90% acetone–10% 6 M HCl and 70% acetone–30% 2 M HCl, respectively.[50] Pb, Sn(IV), and Cu are adsorbed on a strong-base anion-exchange resin from 2.5 M HCl–65% ethanol solution, whereas Al is not. This behavior is utilized for the separation of microgram amounts of aluminum from large amounts of Pb and Sn.[51]

The use of tributyl phosphate mixtures has been studied.[52]

The anion-exchange behavior of a number of metals in hydrochloric acid on a weak-base anion-exchange resin (phenol condensation type, Amberlite CG 4B) has been described.[53] Separations of rare earths and Th, Te(IV) and Bi, and Sn(IV) and Sb(III), which are difficult with a strong-base anion-exchange resin in hydrochloric acid media, can be effected with Amberlite CG 4B.

The anion-exchange behavior of the following metal ions has been studied in hydrochloric acid–perchloric acid mixtures: Au, Bi, Cr(VI), Hg(II), Pd, Pt, Ru(IV), Sb(III), Sn(IV), and Tl(III) (on a macroreticular strong-base resin Amberlyst A-26).[54] The following separations are obtainable by using HCl and HCl–HClO$_4$ mixtures as eluants: As(III)–Bi–Sb, Ni–Pd–Pt, Pb–Cu–Fe(III)–Hg–Sn.

Cation Exchange. This is much less important than anion exchange in hydrochloric acid but has some points of interest. Metals that are not strongly complexed by chloride will be adsorbed from dilute HCl solution. With increasing HCl concentration adsorption should decrease because of increasing competition from H$^+$ and possibly also, for some metals, because of increasing formation of anionic chloro complexes. Hydrochloric acid at higher concentrations might, therefore, have value as an eluant of metals adsorbed on a strong-acid cation exchanger. At least, it should be a better eluant than nitric or sulfuric acid for metals complexed by chloride. In general, experiment confirms these expectations.

Distribution coefficients of metal ions on a strong-acid cation-exchange resin, AG 50W-X8, 100–200 mesh, in 0.2–4.0 M hydrochloric acid solutions (found by a static method) are given in Table 8-8. Volume distribution coefficients D_v in more concentrated hydrochloric acid solutions are given by Nelson et al.[55] In 9 M hydrochloric acid, Au(III),

TABLE 8-8

Distribution Coefficients of Metal Ions on AG 50W-X8 in Hydrochloric Acid Solutions[a]

Metal Ion	D_g				
	0.2 M	0.5 M	1.0 M	2.0 M	4.0 M
Al(III)	1900	318	61	13	2.8
As(III)	1.6	2.2	3.8	2.2	
Au(III)	0.1	0.4	0.8	1.0	0.2
Ba(II)	2930	590	127	36	12
Be(II)	117	42	13	5.2	2.4
Bi(III)	ppt	<1.0	1.0	1.0	1.0
Ca(II)	790	151	42	12	5.0
Cd(II)	84	6.5	1.5	1.0	
Ce(III)	10^5	2460	265	48	11
Co(II)	460	72	21	6.7	3.0
Cr(III)	262	73	27	7.9	2.7
Cu(II)	420	65	18	4.3	1.8
Fe(III)	3400	225	35	5.2	2.0
Ga(III)	3040	260	43	7.8	0.4
Hg(II)	0.9	0.5	0.3	0.3	0.2
La(III)	10^5	2480	265	48	10
Mg(II)	530	88	21	6.2	3.5
Mn(II)	610	84	20	6.0	2.5
Ni(II)	450	70	22	7.2	3.1
Pt(IV)			1.4		
Sb(III)	ppt	ppt	ppt	2.8	
Sn(IV)	45	6.2	1.6	1.2	
Sr(II)	1070	217	60	18	7.5
Th(IV)	>10^5	~10^5	2050	239	67
Ti(IV)	297	39	12	3.7	1.7
U(VI)	860	102	19	7.3	3.3
V(V)	7.0	5.0	1.1	0.7	0.3
Y(III)	>10^4	1460	145	30	8.6
Zn(II)	510	64	16	3.7	1.6
Zr(IV)	>10^5	~10^5	7250	489	15

[a] Selected data of F. W. E. Strelow, *Anal. Chem.*, **32,** 1185 (1960). Distribution coefficients D_g were determined at a ratio (total amount of metal ion/total resin capacity) of 0.4. For another set of distribution coefficients see J. Korkisch and S. S. Ahluwalia, *Talanta*, **14,** 155 (1967). The Pt metals as a group can be separated from base metals by adsorbing these from 0.01–0.02 M HCl solution; the Pt metals, as anionic chloro complexes, are not retained.

Fe(III), Ga(III), and Sb(V) have $D_v > 100$; Sc(III) and Th(IV) have $10 < D_v < 100$. The increase in D at high hydrochloric acid concentrations has been attributed to dehydration of metal ions and invasion of resin by co-ions. The effect of resin crosslinking (2 to 16% DVB) on the cation-exchange elution of Cd, Cu, Fe(II and III), Hg(II), Mn, Ni, and Zn by 0.5–12 M hydrochloric acid has been studied.[56] D in 1 M HCl decreases with crosslinking up to 12% DVB, except for Cd and Hg, which are completely complexed by Cl^-. Fe(III) behaves abnormally because of reaction of $FeCl_4^-$ with the resin matrix.

Cation-exchange behavior of metal ions in miscible organic solvent–hydrochloric acid media has been studied by many workers.[57] In these mixtures more possibilities exist for separation of various metal ions from one another than in pure aqueous hydrochloric acid media. For example, the following separations may be effected by using ethanol–hydrochloric acid eluants: Fe(III)–U(VI)–Ca–La; Cd–Zn–Fe(III)–Ca–Ba; In–Ga–Be–Al–Y.[58] Uranium can be separated from large amounts of bismuth by adsorption of the former on Dowex 50 from 90% 2-propanol–10% 6 M hydrochloric acid solution.[59] The constituents of biological ash are separated into four groups by using acetone–hydrochloric acid[60]: MoO_4^{2-}, SO_4^{2-}, PO_4^{3-}; Fe^{3+}, Zn, Pb, Cu; Co, Mn; Na, K, Ca, Mg. Small amounts of Cu, Ni, Co, and Mn can be separated from large amounts of Cd, Zn, and Fe(III) by using acetone–hydrochloric acid eluants.[61] The metal ions in low concentrations are eluted with 3 M hydrochloric or nitric acid. For selecting possible separations with acetone–hydrochloric acid, tables of distribution coefficients presented by Strelow et al. (1971)[57] are useful.

Preconcentration of trace elements by precipitation ion exchange has been proposed.[62] This technique is based on the adsorption of matrix and traces on a cation-exchange column, followed by immobilization of the matrix on the column by precipitation and nonselective elution of as many trace elements as possible before breakthrough of the matrix by its dissolution. Many trace elements are concentrated from NaCl, KCl, $BaCl_2$, $SrCl_2$, and AgCl by using a Dowex 50X-8 column and concentrated hydrochloric acid as precipitant–eluant. This technique has been extended by using miscible organic solvent–hydrochloric acid systems.[63]

V. HYDROBROMIC ACID

Distribution coefficients of Bi, Cd, Hg(II), Mn, Pb, and Zn on Dowex 50W-X8 cation-exchange resin in 0.3–0.6 M hydrobromic acid solutions have been reported.[64] Hg, Bi, Cd, and Pb can be separated from each other as well as from other metals by using different concentrations of hydrobromic acid as eluants.

Thorium can be separated from rare earths, zirconium, and other elements on a strong-acid cation-exchange resin by eluting the latter elements with 5.5 M hydrobromic acid. Thorium is then eluted with 5 M nitric acid.[65]

The anomalous adsorption of $AuBr_4^-$ on the cation exchanger Dowex 50W-X8 from 6 M HBr (0.0035 M in free bromine to prevent reduction of AuIII) serves for the separation of Au from the Pt metals, which are not adsorbed.[66] Gold is eluted with acetylacetone. The adsorption of anionic halide complexes of metals by cation-exchange resins is not well understood. The authors mentioned suggest that undissociated $HAuBr_4$ partitions into the resin by forming a molecular (charge transfer) complex with the aromatic rings of the resin network.

The cation-exchange behavior of 19 elements in hydrobromic acid–organic solvent (alcohols, acetone, tetrahydrofuran, etc.) media has been studied.[67] These media can be used for the separations of Zn–Cd, Mo–V, and U–Fe.

Anion-exchange distribution coefficients of elements in hydrobromic acid–organic solvent mixtures have been determined.[68] Gallium has been separated from the constituents of bauxite by using 90% methanol–10% 4.5 M hydrobromic acid mixture. Several other possible separations are indicated.

Cadmium is adsorbed much more strongly than Zn ($D \sim 3$) on Dowex 1-X8 from 0.15 M HBr, so that this solution can be used as eluant to remove Zn from a column on which both metals have been adsorbed. This separation has been applied preparatory to dithizone determination of Cd in natural waters.[69]

vi. HYDROFLUORIC ACID

Because of metal fluoride solubility problems, hydrofluoric acid is usually combined with a mineral acid (H_2SO_4, HNO_3, HCl), but has been used alone.

Cation Exchange. The elution of metal ions, adsorbed on Dowex 50W resin, with 0.1 and 1 M HF has been studied.[70] Metals forming stable anionic fluoride complexes are readily eluted with 0.1 M HF: Al, Cd(?), Mo, Nb, Sc, Sn(IV), Ta, Ti, U(VI), W, Zr, [As(III), Sb(III)]. With 1 M HF, additional elements are eluted: As, Ba, Fe(III), Ga, Hg, Mn(II)(?), Sb(III), Sr, V(IV), Zn(?), (Bi, Ca, Pb). The insolubility of Ca and other fluorides does not cause trouble if adsorption takes place in the absence of fluoride.

The adsorption of Co (not forming a fluoride complex) follows quite closely the relation $\log D = 2.00\,\mathrm{pH} + 1.25$ in 0.2–7.0 M HF.[71]

Distribution coefficients of elements on Dowex 50-X8 in HF–organic solvent mixtures are available.[72]

Anion Exchange. Figure 8-8 shows the distribution coefficients of metals on Dowex 1-X8 as a function of the hydrofluoric acid concentration. D decreases continuously with the HF concentration (difference from the behavior of many metals in HCl solution).

Anion exchange is a valuable method for the separation of trace amounts of Mo, W, Nb, Ta, Ti, and Zr from much iron (steel analysis). Danielsson[73] found that D for these elements depends slightly on the iron concentration, for example for Mo, $D = 3.4 \times 10^3$, 1.3×10^3, and 1.0×10^3 with 0, 20, and 40 mg Fe/ml in 1 M HF. D for Fe(III) also varies slightly with its concentration (from 11 to ~3 as Fe varies from 0 to 40 mg/ml). The elution of iron is better done with 1 M HF–0.05 M HNO$_3$ than with HF alone.

Anion exchange in fluoride solution has been applied for separations in the neutron-activation determination of impurities in titania.[74] Elution with 1 M HF removes such elements as alkalies and alkaline earths, Fe, Co, Ni, As, Zn, Ga, Cu, and In; elution with 22 M HF removes Sc, U, Th, etc. as well as Ti itself; and finally, elution with 6 M HCl removes Ta and the rare earth elements. Sb, Au, W, and Mo are left in the resin.

The chromatographic behavior of macro amounts of V, Zr, Ti, Mo, W, and Nb in strong HF solution has been studied.[75]

vii. HYDROFLUORIC ACID WITH NITRIC OR SULFURIC ACID

Cation Exchange. In mixtures of HF with HNO$_3$ or H$_2$SO$_4$, D values of adsorbable elements are smaller than in HF alone at the same concentration. Low concentrations of HNO$_3$ and H$_2$SO$_4$ are, therefore, of aid in sharpening the separation of elements such as Fe(III) and Al, which would be adsorbed too strongly from pure HF solutions. For most elements there is not much difference between the use of HF–HNO$_3$ and HF–H$_2$SO$_4$, if the hydrogen-ion concentrations are about the same. Values of D for 0.1 and 1 M HF with 0.1 M HNO$_3$ or 0.05 M H$_2$SO$_4$ are presented in Table 8-9.[76] The elements studied can be classified as follows (1 M HF, 0.1 M HNO$_3$, or 0.05 M H$_2$SO$_4$):

Weakly adsorbed ($D < 10$)
Al, As(III, V), Cr(III), Fe(III),* Mo(VI), Nb(V), Sb(III, V), Sn(IV), Ta, Ti, V(V), W, Zr
Moderately adsorbed ($D = 10$–100)
   Bi,* Fe(II),* V(IV)*
Strongly adsorbed ($D > 100$)
   Ag, Cd, Co, Cu(II), Mg, Mn(II), Ni, Pb, Zn

* Strongly adsorbed if concentration of HF is decreased to 0.1 M.

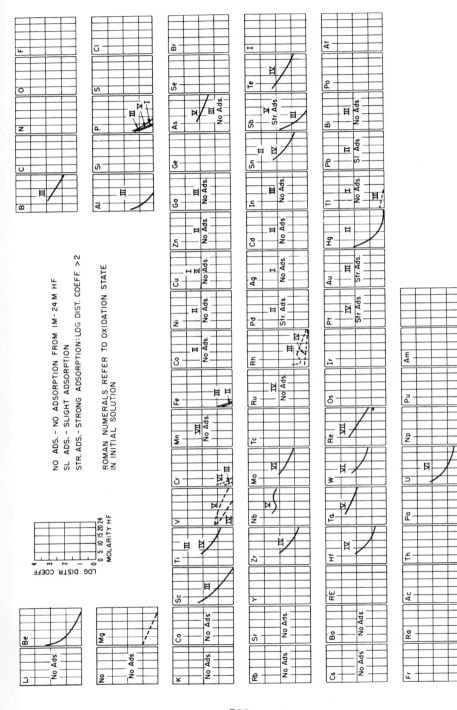

Fig. 8-8. Adsorption of elements from 1 to 24 M HF solution with anion-exchange resin Dowex 1-X10 (200-mesh). Reprinted with permission from J. P. Faris, *Anal. Chem.*, **32**, 521 (1960). Copyright by the American Chemical Society.

TABLE 8-9
Distribution Coefficients of Metal Ions on Dowex 50W-X8 in HF–HNO$_3$ and HF–H$_2$SO$_4$ Solutions^a

Metal	log D			
	0.1 M HF–0.1 M HNO$_3$	0.1 M HF–0.05 M H$_2$SO$_4$	1 M HF–0.1 M HNO$_3$	1 M HF–0.05 M H$_2$SO$_4$
Ag	2.3	2.6	2.25	2.3
Al	0.4	0.4	−1	−1
Bi	2.7		1.2	
Cd	3.3	3.5	3.1	3.1
Co	3.3	3.5	3.1	3.3
Cr(III)	0.6	0.6	0.4	0.3
Cu(II)	3.3	3.5	3.15	3.2
Fe(II)		2.8		1.8
Fe(III)	2.0	2.0	0.3	0.3
Mg	3.3	3.4	3.0	3.0
Mn	3.4	3.5	3.1	3.3
Ni	3.35	3.5	3.2	3.3
Pb	4.0		3.9	
V(IV)		2.5		1.1
V(V)	0.7	0.9	−0.7	<0
Zn	3.3	3.5	3.1	3.2

As(III, V), Mo(VI), Nb(V), Sb(III, V), Sn(IV), Ta(V), Ti(IV), W(VI), Zr: $D < 1(0.1\ M$ HF–0.1 M HNO$_3$)

^{a} Danielsson, *Acta Chem. Scand.*, **19,** 1859 (1965).

Because of slowness of attainment of equilibrium between its species, Cr(III) shows variable adsorption behavior. It is strongly adsorbed from solutions that have not been heated.

Separation of elements in the first group from those in third group should be easy, but separations of elements within groups would be difficult in fluoride solutions. In the separation of adsorbable elements from much iron, this element can be washed out of the column with 1 M HF and the former elements can then be eluted as a group with 4 M HNO$_3$.

Anion Exchange. Distribution coefficients of the metals mentioned under cation exchange have been determined for Dowex 1-X8 in HF–HNO$_3$ and HF–H$_2$SO$_4$ solutions.[76] The adsorptive behavior may be summarized as follows for 1 M HF–0.1 M HNO$_3$ or 0.05 M H$_2$SO$_4$:

Weakly adsorbed ($D < 10$)
   Ag, Al, As(III), Bi, Cd, Co, Cr(III), Cu, Fe(II, III), Mg, Mn(II), Ni, Pb, Sb(III), V(IV), Zn

Moderately adsorbed ($D = 10–100$)

As(V), V(V)

Strongly adsorbed ($D > 100$)

Mo(VI), Nb(V), Sb(V), Sn(IV), Ta(V), Ti, W, Zr

In these mixed acid systems, adsorption is due almost entirely to anionic fluoride complexes. If the H_2SO_4 concentration is increased to 2.5 M, Sn, Ti, and Zr are easily eluted (HF concentration is not critical), and they can be separated from the more strongly held Mo and W.[77]

Elution with 1 M HF–4 M HNO$_3$ easily removes Zr, Sn, Ti, W, Nb, and Mo as a group, but not Ta.[78]

A summary of distribution coefficients of 19 elements between Dowex 1-X4 and HNO_3–HF solutions has been presented.[79] Strong adsorption of tantalum from these media has been applied in the separation of trace impurities from tantalum.

viii.   HYDROFLUORIC–HYDROCHLORIC ACIDS

The distribution coefficients of Co, Cr(III), Fe(III), and V(IV) between a strong-acid cation-exchange resin, Zeo-Karb 225, and 1 M HF–HCl solutions have been determined.[80] Co, Cu, Mn, and Ni are strongly adsorbed at concentrations of hydrochloric acid less than 0.5 M in the presence of 1 M hydrofluoric acid, whereas Al, Cr(III), and Fe(III) are scarcely adsorbed.

Volume distribution coefficients between an anion-exchange resin (Dowex 1-X10) and HCl–HF solutions are shown in Fig. 8-9. The distribution coefficients of Co, Fe(III), Mo, Nb, Ta, Ti, V(V), W, and Zr between De-Acidite FF resin and 1 M HF–HCl solutions have been determined.[80] Ti, Zr, Nb and Ta in steel can be separated by anion exchange using fluoride–chloride mixtures.[81]

The anion-exchange behavior of Zr, Nb, Ta, and Pa in HF–HCl solutions, with and without organic solvents (acetic acid, methanol, i-propanol), has been examined.[82] The bahavior is sufficiently different to allow separation of all these elements. For example, Nb is adsorbed much less strongly than Ta on Dowex 1-X8 from 3–9 M HCl and 0.05–0.2 M HF solutions (Fig. 8-10).

ix.   THIOCYANATE

Cation- and anion-exchange separations of various metal ions with thiocyanate have been studied.[83] Scandium is separated from other metals by cation exchange[84] or anion exchange,[85] using NH$_4$SCN–0.5 M HCl solutions.

Distribution coefficients of several metal ions on Dowex 50-X8 cation-exchange resin[86] and Dowex 1-X8 anion-exchange resin[87] in aqueous

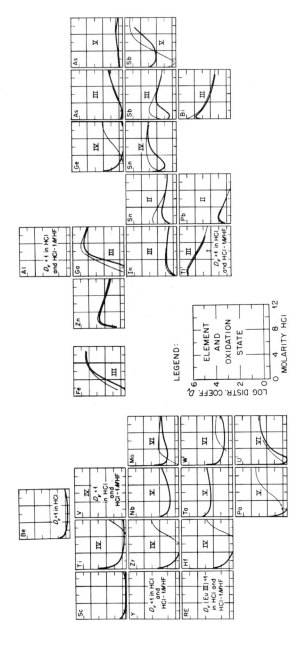

Fig. 8-9. Adsorption of elements from HF–HCl solutions by Dowex 1-X10 resin. ——, distribution coefficients in absence of HF; ——, distribution coefficients in HF–HCl mixtures (usually 1 M HF except Zr, Hf, Nb, Ta, and Pa where [HF] = 0.5 M). Reprinted with permission from F. Nelson, R. M. Rush, and K. A. Kraus, *J. Am. Chem. Soc.*, **82**, 346 (1960). Copyright by the American Chemical Society.

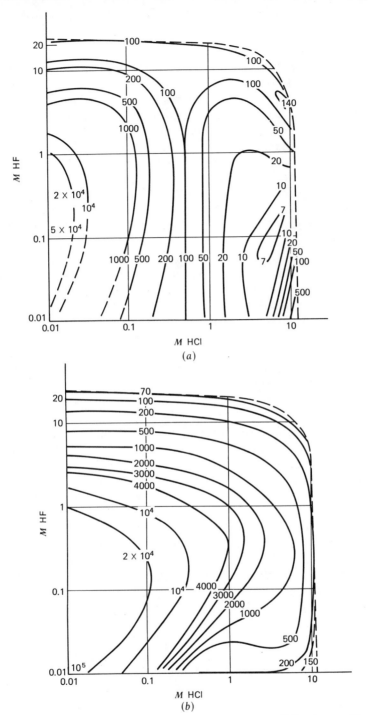

Fig. 8-10. Distribution coefficients of Nb (*a*) and Ta (*b*) in HF–HCl mixtures (Dowex 1-X8). From J. I. Kim, H. Lagally, and H.-J. Born, *Anal. Chim. Acta,* **64,** 29 (1973).

NH$_4$SCN–organic solvent mixtures have been determined. Zinc and cadmium can be separated from each other by cation exchange, with aqueous NH$_4$SCN–methanol mixtures or aqueous NH$_4$SCN–acetone mixtures as eluants. Rare earths as a group can be separated from thorium and scandium by anion exchange.

An anion-exchange method has been proposed for the separation of Re(VII) from Mo(VI) or Tc(VII).[88] After adsorption on Dowex 1-X8, Re is eluted with 0.5 M NH$_4$SCN–0.5 M HCl solution while Mo and Tc remain on the column.

Distribution coefficients of metal ions on a macroreticular weak-base anion-exchange resin (Amberlyst A-21) in 0.03–1.0 M KSCN–0.5 M HCl solutions have been determined.[89] Separations of Ni from V(IV), rare earths from Fe(III) and Th from Ti are effected. Examples of separations with DEAE-cellulose in thiocyanate media are given on p. 793. The adsorption behavior of metal ions on DEAE-cellulose in methanol–thiocyanate–hydrochloric acid media has been investigated.[90] The distribution coefficients of Bi, Cu, Zn, In, and U(VI) increase with increasing concentration of methanol. Several possible separations are indicated.

Adsorption of Co as the SCN$^-$ complex on Dowex 1-X8 is a suitable isolation method for 1 μg/liter or less of Co in natural waters.[91]

X.   ACETIC ACID

Adsorption of many elements on a strong-acid cation-exchange resin, AG 50W-X8, from 1–17 M acetic acid solutions has been studied.[92] Separations of Fe(III) from Co and Mn and of Au from Ag and Cu are described.

Distribution coefficients of many elements on a strong-base anion-exchange resin, Dowex 1-X8, in 2–17 M acetic acid solutions are available.[93] The following separations are suggested: Cu–Zn, Mn–Cu–Zn, and Na–As(III)–Cu. Later it was found that the presence of Cl$^-$ is necessary.[94]

Cation-exchange behavior[95] of metal ions on Amberlite IR-120 in 0.01–1.2 M NH$_4$OAc–0.1 M HNO$_3$ solutions and anion-exchange behavior[96] on DEAE-cellulose in acetic acid–HNO$_3$ media have been investigated. Differences in the anion-exchange distribution coefficients allow separations such as Fe–Mo–W and U–Sm–Mo–Bi–Th to be achieved.

xi.   SUNDRY COMPLEXING AGENTS

The adsorption behavior of some 50 elements on Dowex 1-X8 resin from H$_3$PO$_4$ solution has been recorded.[97] Au, Hg, and Pt are strongly adsorbed; Ag and some rare earths moderately.

Distribution coefficients of elements on a strong-base anion-exchange resin in oxalic acid,[98] oxalic acid–hydrochloric acid[99] and tartrate medium[100] have been determined. By sequential elution with oxalic acid–hydrochloric acid mixtures, the following mixtures can be resolved: Mn–Ni–V(V)–Nb–U(VI), Mg–Al–Ga–In, Co–Cu–Ti–Mo, and Cu–Al.

Ion exchange in formic acid and in citric acid solutions has received some attention.

Separations with EDTA[101] and diethylenetriaminepentaacetic acid[102] have been described. Elimination of these complexing agents before determination is problematic.

A 0.01 M solution of dithizone in tetrahydrofuran–HNO_3 has been used as eluant for the separation of Ag, Cu, and Bi from numerous metals adsorbed on Dowex 50 from tetrahydrofuran–HNO_3 solution.[103] Other organic chelating agents can be used similarly.[104]

xii. SODIUM HYDROXIDE

The distribution of 35 inorganic species between sodium hydroxide solutions (0.1–10 M) and Amberlyst A-29 (a macroreticular strong-base anion-exchange resin) has been reported.[105] The following ions show high distribution coefficients: Re(VII), Tc(VII), Mo(VI), and Cr(VI).

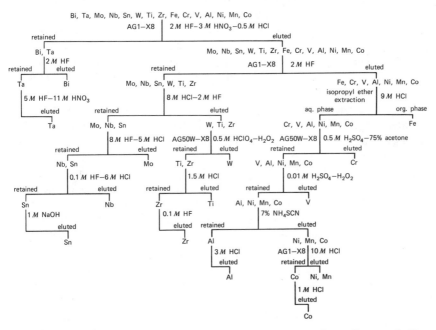

Fig. 8-11. A separation scheme for the analysis of steel. From P. de Gelis, *Chim. anal.*, **53**, 673 (1971).

Multicomponent samples, for example, a mixture of radionuclides, can sometimes be analyzed by selecting a suitable ion exchanger and eluants. However, they can be more easily analyzed by combination of cation and anion exchange. For example, an ion-exchange analysis of silicate rocks is based on the anion- and cation-exchange behavior of the constituent metals in oxalic acid–hydrochloric acid media.[106] (A separation scheme for the analysis of multicomponent samples by ion exchange and reversed-phase partition chromatography is mentioned in Section III.B.) An extensive separation scheme has been presented for the analysis of steel (Fig. 8-11).[107] The 14 elements, present as trace or minor constituents, were determined spectrophotometrically. Results of analyses of standard steel samples were satisfactory.

2. Chelating Resins

As we have seen, ordinary ion-exchange resins show mediocre selectivity for ions of the same sign, especially when the charge is the same. This weakness is overcome to a large extent for most metals by differential complexing in the aqueous phase. Another way of gaining the same end is to introduce chelating groups in the resin, so that depending on their tendency to be complexed, metals will be adsorbed to different extents and have widely different distribution coefficients.

Resins with iminodiacetate groups are commercially available (Dowex A-1). Close correlation between the complexing of metals by EDTA in water and their distribution coefficients on Dowex A-1 is hardly to be expected, but there is a rough correspondence.[108] The distribution coefficient of Hg(II) is ~10^4 as large as that of Ca and that of Cu(II), ~500 times as large as that of Ca. Such a resin should, therefore, have value in collecting traces of strongly complexing metals from solutions containing salts of the alkali metals and alkaline earths.[109] Elements incompletely retained include Ba, Cr(III), and Tl(I). Those not retained include Al, As(V), Cs, P, Se(VI), U(VI). HNO_3 (2 M) is suitable for eluting most cations; $HClO_4$ (2 M) for Bi, H_2SO_4 (1 M) for Th, HCl (2 M) for Co, and ammonia for V, Mo, W and Re. No reagent was found that would completely elute Ag, Hg, Cr, and Sn.

The metal-binding ability of the chelating resins depends strongly on the acidity, since the anionic groups combine with H^+ as well as with metal.

Chelating resins of the iminodiacetate type are effective in isolating Ag, Bi, Cu, Cd, Pb, Mo, Sc, Zn, and other heavy metals from sea water[110] Two chelating resins in the H^+ form were tested: Chelex-100 (purified

Dowex A-1) and Permutit S1005. These gave similar results except the latter resin did not completely retain Mn; Chelex-100 (50–100 mesh) tends to change its particle size by swelling and contracting as the counter ion is changed, with slowing down of flow. The exchange rate is slow so that flow should not exceed 5 ml/min (12 mm diameter bed). The optimal pH for sea water is 7.5–8 for most elements; Mo, W, V, and Re are retained most effectively at pH 5–6, Bi, Ce(III), In, Y, and Mn at pH 9. The elements adsorbed ~100% are:

Bi	In	Mo	Ag
Cd	Pb	Ni	Sn(IV) (85%)
Ce(III)	Mn	Re (90%)	Th
Co	Hg(II) (85%)	Sc	W
Cu			V
			Y
			Zn

Uranium(VI) has been separated from other metals by adsorption on Dowex A-1 from a solution (pH 6.5–8) containing EDTA and elution with 4 M hydrochloric acid.[111] Adsorption of the first-row transition elements by Dowex A-1 in hydrochloric acid media has been studied.[112] Adsorption decreases in the order Zn, Co, Fe(III), Cu, Mn(II); there is negligible adsorption of Cr(III), and none of V(IV) or Ni.

Chelating resins containing

$$-COCH_2N(CH_2COOH)_2 \quad \text{(resin 1)},$$

$$-COCH_2N(CH_2CH_2COOH)_2 \quad \text{(resin 2)},$$

$$HO-\underset{\underset{CH_2N(CH_2COOH)_2}{|}}{\overset{|}{C}}-PO(OH)_2 \quad \text{(resin 3)},$$

and

$$\underset{\underset{CH_2N(CH_2COOH)_2}{|}}{\overset{\underset{|}{C(SCH_2COOH)_2}}{}} \quad \text{(resin 4)}$$

have been synthesized.[113] Compared with Dowex A-1, resins 1 and 2 have a keto group as an extra coordinating site. Distribution coefficients of metal ions at pH 3.0 (0.25 M ammonium acetate) show the following order:

Resin 1: Th > Pb > Ni > U(VI) > La > Zn > Cd > Co > Ca > Sr > Mg
Resin 2 (4% DVB): Th > U(VI) > Pb > La > Zn > Ni > Co > Ca > Sr > Mg
Resin 3: Th ≫ U(VI) > Pb > Fe(III) > La > Zn > Ni > Co > Cd > Ca > Sr > Mg
Resin 4: Th > Zn > Cu

Separations of binary mixtures have been carried out at pH 3.0 or 2.0.

Chelating resins selective for gold and the platinum metals have been prepared by introducing resonating amino groups attached to the same C atom into a styrene–divinylbenzene copolymer matrix.[114] These groups,

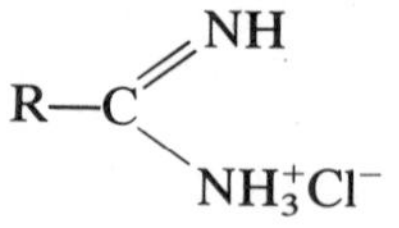

$$R-C \overset{\displaystyle \diagup NH}{\diagdown NH_3^+Cl^-}$$

bond the metals mentioned in square-planar complexes. The metals are supposedly bound in valences stabilized in a d^8 electronic configuration: Os^0, Rh^+, Ir^+, Pd^{2+}, Pt^{2+}, Au^{3+}. The bonding of Pd is illustrated by

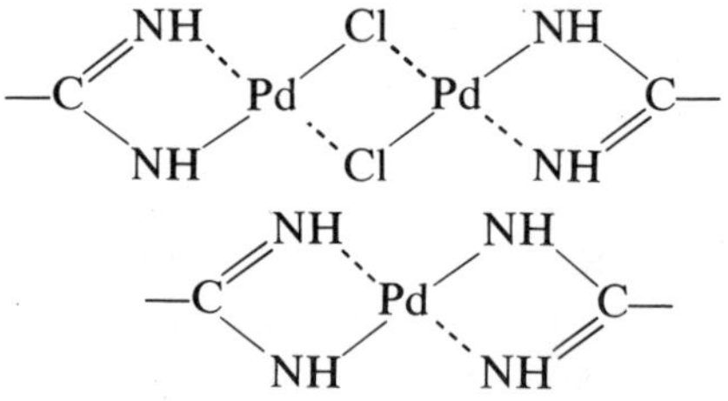

$$-C\overset{NH}{\underset{NH}{}}Pd\overset{Cl}{\underset{Cl}{}}Pd\overset{NH}{\underset{NH}{}}C-$$

$$-C\overset{NH}{\underset{NH}{}}Pd\overset{NH}{\underset{NH}{}}C-$$

A solution of pH 0.5–2.5 is suitable for the adsorption. Selective adsorption of Pt in the presence of Ni, Cu, and Fe(III) has been demonstrated. Base metals can be washed out of the column with 0.05 M HCl. The Pt metals can be eluted with acidic thiourea solution (or the resin can be ignited). Adsorption of the Pt metals is slow.

A chelating resin with mercapto groups is said to adsorb Au, Ag, and Hg specifically and has been used to isolate a microgram or so of Au from 100 liters of sea water.[115]

A resin obtained by condensing naphthalene-1,8-diol with chloroacetic acid is useful for the separation of Zr from many elements.[116]

A chelating resin based on 8-hydroxyquinoline has been used to separate U and fission products.[117] Various oxine-containing resins have been prepared and their properties studied.[118] The only one of these likely to find practical application is a lightly crosslinked, macroporous, polystyrene-azo-oxine resin, which has reasonable capacity and shows a much faster equilibration rate than the other resins examined. The minimal pH values for quantitative retention of cations are

2.8 Fe(III), Cu, U(VI)
4.0 Hg(II), Ni, Co
4.8 Al, Zn
6.1 Cd
10.0 Mg

Uranium(VI) can be separated from other metals by adsorption at pH 2.8 and elution from the column with Na_2CO_3 solutions.

What has been called an immobilized chelate, Controlled Pore Glass-8-Hydroxyquinoline, is prepared by azo-linkage to an arylamine coupled to Controlled Pore Glass through γ-aminopropyltriethoxysilane.[119] It is stable in acidic and neutral solutions. Metals are adsorbed by exchange with H^+. The adsorption pattern is much the same as extraction with HOx in $CHCl_3$. The adsorptive capacity is small. Isolation of traces of Cu(II) and Fe(III) from water (distilled) has been demonstrated.

A crosslinked polystyrene-azo-PAR resin has been synthesized and its ion-exchange properties have been examined.[120]

Dithiocarbamate resins have been prepared.[121]

Dithizone-coupled cellulose completely collects Cu, Zn, and Ag from 1 liter of a solution (pH 5) containing 1 ppm each of these elements.[122] Chelating resins based on catechol or pyrogallol[123] and chitosan (a natural chelating polymer)[124] also serve for collection of metals from dilute solutions.

A monograph on the chelating resins is available.[125]

3. Cellulosic Ion Exchangers

A few papers illustrating the use of these ion exchangers are mentioned here, without any attempt at extensive treatment.

The distribution coefficients of Li, Cs, Ca, Sr, Ba, La, Pr, Yb, Al, Cr, Co, and Ni ions on cellulose phosphate in $0.01-10\ M$ hydrochloric acid solutions have been measured.[126] Separations of rare earths and alkaline earths, alkali metals and alkaline earths, and aluminum and alkaline earths have been successfully carried out. Traces of Cu, Pb, Cd, and Zn can be separated from 0.5 g uranium by adsorption of uranium(VI) on cellulose phosphate from $1\ M$ hydrochloric acid solution.[127]

Carboxycellulose, a weak-acid cation exchanger, collects heavy metals from solutions (pH 3–4) of 0.001–0.1 ppm concentrations.[128]

Diethylaminoethylcellulose (DEAE-cellulose) is a weak-base anion exchanger. Separations of Hg(II) from other metals,[129] Pt and Pd from base metals,[130] and Re(VII), Mo(VI), and W(VI) from one another[131] have been effected by using DEAE-cellulose.

Gold(III) can be separated from Cd, Co, Cu, Fe(III), Ni, Zn, and Se(IV) by adsorption of the former on DEAE-cellulose from $0.01\ M$ HCl and elution with $1\ M$ HCl.[132] This is in contrast to strong-base anion-exchange resins, from which the elution of gold is difficult. Gold can be separated from Pd(II), Pt(IV), and Hg(II) on a DEAE-cellulose column by elution with $8:2$ methanol–$1\ M$ HCl mixture. Gold is eluted easily, but the other metals are retained on the column.

4. Inorganic Ion Exchangers

A large number of inorganic substances have ion-exchange properties, including hydrous oxides, acidic salts of multivalent metals (e.g., zirconium phosphate), salts of heteropolyacids (e.g., ammonium molybdophosphate), ferrocyanides, and synthetic aluminosilicates (zeolites).[133] The inorganic ion exchangers do not have the wide analytical applicability of resin exchangers, but they have their special uses and can be valuable at times in trace separations. For example, ammonium molybdophosphate is useful for separating the alkali metals,[134] zirconium phosphate for isolating Cs from silicate rocks, and zirconium molybdate for separating Mg, Ca, and Sr. Alumina, which has been washed with perchloric acid, adsorbs the anionic Nitroso-R salt complex of Co and finds use in the determination of Co in iron and steel.

III. PARTITION CHROMATOGRAPHY

Partition chromatography (liquid–liquid) is based on the difference in distribution of different solutes between two immiscible liquid phases. The method was developed for organic compounds by Martin and Synge in 1941. In normal partition chromatography, the stationary phase is water (aqueous solution) held in or on a solid support (silica gel, cellulose, etc.), and the mobile phase is an organic solvent. In reversed-phase chromatography,[135] the stationary phase is an organic solvent, and the mobile phase is an aqueous solution.

Partition chromatography is a powerful separation method: it can make use of all the varied separations of immiscible solvent extraction as carried out by the batch process, with the advantage that constituents having D values not greatly different can be separated more conveniently and more rapidly. The general principles of column chromatography (Section I of this chapter) apply to column partition chromatography and close approach to ideal behavior has been demonstrated in a number of inorganic separation systems. If liquid–liquid partitioning is the only, or predominant, process (no adsorption or ion exchange by the solid material holding one of the liquid phases), D can be equated with E, the extraction coefficient, whose value may be known under the conditions or which can be calculated. Partitioning of chelates, ion-pair compounds, and so on can be made the basis of this type of chromatography, which thus supplements ion-exchange chromatography. Some separations are better carried out by partition chromatography than by ion exchange. The possibilities in partition chromatography have not as yet been as fully explored as in ion exchange.[136] To be sure, there are limitations. Extractions that proceed slowly (as some with HTTA, p. 924) cannot conveniently be applied. Capacity may be limited. The two liquid phases should not be too soluble in each other, even if mutually saturated beforehand.

Reversed-phase partition chromatography is generally more useful than ordinary partition chromatography in trace analysis. Usually the sample is dissolved in an aqueous solution. It is easier to find suitable supports for organic liquids than for aqueous solutions. For the latter, hydrophilic materials such as cellulose are necessary, and these can have adsorptive or ion-exchange properties that obscure or spoil the partition process being exploited. Many inert hydrophobic materials are available for holding the organic phase, for example, polyethylene, polytrifluorochloroethylene (Kel-F), polyurethane foam, and Teflon. The particle size is important; usually the mesh size (U.S.) lies in the range 100–200. Theoretical plate heights of 1 mm are obtainable in practice. Column preparation is critical; the bed must be homogeneous.

A. EXAMPLES OF PARTITION CHROMATOGRAPHY WITH STATIONARY AQUEOUS PHASE

Useful separations have been obtained with cellulose columns. Several examples follow. In the analysis of steel for Cu, Co, and Ni,[137] a dilute hydrochloric acid solution of the sample is placed on the top of the column. Iron is first eluted with a mixture of methyl propyl ketone, acetone, water, and hydrochloric acid (100:30:4:1). Copper follows. Cobalt is eluted with methyl propyl ketone containing 2% hydrochloric acid. Nickel is finally eluted with 2 M hydrochloric acid. Each element is determined colorimetrically. Aluminum is adsorbed on a cellulose column from a methyl ethyl ketone–hydrochloric acid solution and separated from large amounts of iron(III).[138] Aluminum is eluted with 2 M hydrochloric acid.

Niobium and tantalum have been separated from other metals by elution with a mixture of methyl ethyl ketone and hydrofluoric acid.[139] Tantalum can be separated from niobium by elution with methyl ethyl ketone saturated with water. In the determination of traces of uranium in silicate rocks, uranium is separated by selective elution from activated cellulose with ethyl ether containing 5% of nitric acid.[140]

Uranium has been separated from other metals, including thorium and zirconium, by elution of the former from a silica-gel column with methyl isobutyl ketone previously equilibrated with 6 M nitric acid.[141] The other metals remaining on the column can be eluted with 6 M nitric acid and other eluants.

B. EXAMPLES OF REVERSED-PHASE PARTITION CHRO-MATOGRAPHY[142]

An early use of reversed-phase partition chromatography was made in isolation of Cd, Co, Cu, Mn, Pb, and Zn from natural waters by dithizone extraction.[143] The sample, adjusted to pH 7.0, is passed through a column of cellulose acetate holding a carbon tetrachloride solution of dithizone as the stationary phase. Cd, Mn, Pb, and Zn can be eluted with $1 M$ hydrochloric acid, and Co and Cu with concentrated ammonia. A better support matrix is the open-cell type of polyurethane foam in which plasticizers are used as solvents for the chelating agent.[144]

The use of high-molecular-weight amines, di-(2-ethylhexyl)phosphoric acid, tributyl phosphate and tri-n-octylphosphine oxide as stationary phases has been extensively studied.[145] Separations of rare earths are noteworthy.

Ni, Mn, Co, Cu, Fe, Zn, and Cd can be separated by reversed-phase chromatography with tri-n-octylamine as the stationary phase and applying a gradient elution technique.[146] A separation of uranium in $5.5 M$ HNO_3 makes use of its retention by a column of Kel-F supporting tributyl phosphate.[147] Other elements (except Th, Zr, and Ce^{4+}) are eluted with $5.5 M$ nitric acid, and uranium is eluted with water. Small amounts of zirconium have been separated from niobium by using methylene-bis(di-n-hexylphosphine oxide) as the stationary phase and $10 M$ hydrofluoric acid as the mobile phase.[148]

A Teflon-6 support impregnated with methyl isobutyl ketone has been used for the separation of W, Mo, Nb, and Ta.[149] W is eluted with $7 M$ HCl–$2 M$ HF, Mo with $6 M$ HCl–$6 M$ HF, Nb with $3 M$ HCl–$1 M$ HF, and Ta with methyl isobutyl ketone. A method for the quantitative separation of zirconium and hafnium has been described, in which Teflon-6 serves as support and methyl isobutyl ketone equilibrated with $3 M$ thiocyanic acid as stationary phase.[150] Zirconium is eluted from the column with $4 M$ NH_4SCN–$0.2 M$ $(NH_4)_2SO_4$ solution and hafnium with $1.2 M$ $(NH_4)_2SO_4$ or $1 M$ H_2SO_4.

A reversed-phase chromatographic procedure for the separation of iron has been described.[151] Iron(III) is extracted from a mobile $7 M$ hydrochloric acid phase into a stationary 2-octanone (methyl hexyl ketone) phase supported on a column of Fluoropak. Iron is eluted with 2-propanol and determined spectrophotometrically with 1,10-phenanthroline.

A scheme has been proposed for quantitative group separation of 19 metal ions by reversed-phase partition chromatography (Fig. 8-12).[152] In another separation scheme,[153] five different chromatographic and

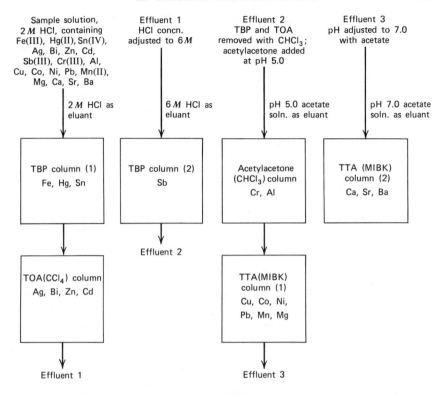

Fig. 8-12. Group separations by reversed-phase partition chromatography. Organic solvent is absorbed on polytrifluorochloroethylene. TBP: tributyl phosphate, TOA: tri-*n*-octylamine, TTA: thenoyltrifluoroacetone. From Akaza, Tajima, and Kiba, *loc. cit.*

ion-exchange columns are used:

1. Isopropyl ether on Amberlyst XAD-2 column	Sb(V), Ga, Fe(III) as a group
2. MIBK on XAD-2 column	Mo–Sn(IV) (– means separable)
3. Amberlyst A-26 column	Anion-exchange separation of Co–Cu–U(VI)–Zn–Cd–Bi
4. Trioctylphosphine oxide (cyclohexane) on XAD-2 column	Ti, Sc–Th–Zr, Hf
5. Dowex 50W-X8 column	Cation-exchange separation of V(IV)–Pb–Mn(II)–Al–Ni, Mg–Ca, Ba, Sr, rare earths

IV. MISCELLANEOUS CHROMATOGRAPHIC METHODS

In this final section of the chapter we collect a number of methods that are less often applied in inorganic trace analysis than those considered in the preceding discussion, but which can be useful at times. Formally they can be characterized as solid–liquid chromatographic methods but cannot be simply classified by mechanism.

A. ADSORPTION CHROMATOGRAPHY

The mechanism of solute distribution here is primarily surface adsorption in which ion exchange is not an essential feature of the process. The distribution process can, however, be complex, and depending on the adsorbent and the solutions, ion exchange and liquid–liquid distribution can be involved. For trace separations, adsorption chromatography is carried out with columns.

Alumina is the most popular adsorbent. Some work has been done on the separation of metals by running an organic solvent solution of their chelate complexes through an adsorption column. Separation of metal dithizonates on an alumina column has been applied to the determination of Co, Hg, Ni, Pd, and Pt. Many other adsorbents have been studied.[154] Chromium has been separated from other metals by passing a chloroform–benzene solution of their 8-hydroxyquinaldine complexes through a column of activated alumina. Chromium is eluted with chloroform–benzene; the other metals remain on the alumina.

Traces of metals in natural waters have been concentrated on a mixture of activated charcoal and chlorinated lignin.[155]

Alumina and other adsorption columns also find use in the purification of organic reagents, as of dyes.

B. PAPER CHROMATOGRAPHY

In paper chromatography, a small volume (0.002–0.02 ml) of solution is placed near the end of a strip of suitable paper and dried, and the chromatogram is developed by allowing a suitable solvent to ascend the paper by capillary action or to descend. This process is carried out in a closed chamber. The solvent is usually a mixture of organic solvent (or solvents) and aqueous solution. The movement of a solute is characterized by its R_f value:

$$R_f = \frac{\text{distance traveled by a particular solute band}}{\text{distance traveled by the solvent front}}$$

(R_f values obtained under standardized conditions are of aid in identification of the solutes in a chromatogram.) The separated constituents in the bands or spots are usually determined by treating the paper with a suitable reagent forming colored or fluorescent products (p. 230). This technique is usually not applied for separations alone.

Paper chromatography usually involves both adsorption and partitioning (between the mobile external phase and fixed aqueous layer at or near the cellulose interface), the latter process probably being the more important.

Paper-chromatographic separations may find use in trace analysis if they have been preceded by other separations which remove the bulk of the major constituents. Larger amounts of sample can be handled in a cellulose column.

A few examples of applications of paper chromatography follow.[156] Microgram amounts of Rh, Ir, Pd, and Pt have been separated by descending development with a mixture of methyl isobutyl ketone, pentanol, and hydrochloric acid.[157] Uranium in soils, rocks, natural waters, and vegetation has been separated by means of paper chromatography with a mixture of ethyl acetate, nitric acid, and water and determined by comparison against standard spots of brown uranyl ferrocyanide on the paper.[158] Traces of Fe, Mn, Zn, and Cu in plant material have been separated by paper chromatography with a mixture of butanol, hydrochloric acid, and water and determined by reflectance measurements on the chromatogram.[159]

Occasionally the section of the strip holding the metal in question is cut out and leached with water or acid to allow the metal to be determined more exactly in solution.

C. THIN-LAYER CHROMATOGRAPHY

The technique of thin-layer chromatography is very similar to that of paper chromatography. Instead of paper, a thin layer of adsorbent adhering to a glass plate is used as the stationary phase. Silica gel, alumina, cellulose, or ion-exchange material (e.g., DEAE-cellulose) serve as adsorbents. R_f values are expressed in the same way as in paper chromatography, and they provide a measure of possible separations. In thin-layer chromatography, development is very rapid as compared with paper chromatography. Moreover, larger samples can be used.

Microgram amounts of Cu, Ni, and Zn have been separated on a thin layer of MN cellulose powder with a 3:1 mixture of 1-butanol and concentrated hydrochloric acid as solvent (ascending technique). After treatment with color-forming reagents, the three elements on the chromatogram were determined by reflectance spectroscopy.[160]

Separations by thin-layer chromatography have been described for diethyldithiocarbamates,[161] dithizonates,[162] and other metal chelates. Elution with ethanol–benzene and acetone–chloroform has been used for the separation of Co, Cu, Ni, and Fe 1-(2-pyridylazo)-2-naphtholates on "Silufol" plates, followed by reflectance scanning photometry of the chromatograms.[163] The chelates are first extracted by chloroform.

A theory of thin-layer chromatographic separation of chelates has been offered, which has been tested with dithizonates and diethyldithiocarbamates.[164]

Individual rare earth elements have been separated by developing the chromatogram with $0.1\ M$ HSCN in methyl ethyl ketone on a layer composed of silica gel, NH_4NO_3, and starch.[165]

D. GEL CHROMATOGRAPHY

Gel chromatography is a form of liquid–solid chromatography in which gels are used as the stationary phase.[166] Dextran gels (Sephadex) and polyacrylamide gels (Bio-Gel P) are frequently used. The separation of substances is based primarily on molecular or ionic size, that is, on differential exclusion from the gel, although adsorption and other effects are also frequently involved.[167] Large molecules are eluted faster than smaller ones. There are many names for this branch of chromatography, for example, molecular-sieve chromatography, gel permeation chromatography, and gel filtration.

Gel chromatography has been applied predominantly to complex biochemical and highly polymerized molecules. However, it is not confined to organic molecules. The distribution behavior of inorganic species on columns of polyacrylamide and dextran gels has been studied.[168] Separations of alkali metals[169] and of alkaline earth metals[170] on dextran gel columns have been achieved to some extent. The use of complexing agents may improve the separation of metal ions. At present, gel chromatography is of little importance for separation of traces of metals.

E. ELECTROPHORESIS

Differences in the mobility of charged particles (ions and colloids) in liquid media under the influence of an electric field may be used for their separation.[171] Components are separated into zones with the aid of paper, gels, etc. as supporting or stabilizing media. The position of the zones is detected by means of radioactivity, color reaction, and other properties.

Factors that affect the mobility of charged species are:

State of migrating species—charge and mass (molecular weight).

Properties of background electrolytic solution—viscosity, ionic strength, kind and concentration of electrolyte (and complexing agent), pH, and temperature.

Properties of stabilizing (supporting) medium—electroosmosis and sorption[172] of migrating species.

Electric field–voltage gradient.

The large majority of electrophoretic applications lie in the general realm of biochemistry. Inorganic analytical applications are of minor importance. Examples include separations of alkali metals, alkaline earth metals, platinum metals, and rare earths.[173] The size of sample that can ordinarily be taken for paper electrophoresis is very small. Much larger samples can be handled if electrophoresis is carried out through many chambers with cellulose as stabilizing medium.[174] The following separations and determinations have been effected. Pt and Pd (50 mg each); Os and Ru (50 mg each); Mn (0.8 mg) from Fe (83 mg) and Ni (88 mg); Pt (0.3 mg) from Sb (240 mg) and Co (59 mg); Pd (0.07 mg) from Ni (89 mg) and Pt (20 mg); Os (0.2 mg) from Hg (200 mg) and Cu (64 mg); Ni (0.2 mg) from As (320 mg) and Os (40 mg). After electrophoresis the metal in question is eluted with 3 M hydrochloric acid and determined gravimetrically or photometrically.

Electrophoresis also finds use in the testing of photometric reagents for impurities. For example, ~1% of colored impurities can be detected in sulfonephthaleins by paper electrophoresis.[175] Moreover, it is of use in elucidation of the composition of determination and separation forms in photometric analysis.

NOTES

1. Note that $S_{B/A}$ is not what is often (and erroneously) called the separation factor in the literature, namely the ratio of distribution coefficients (α). From this ratio alone, the extent of separation cannot be calculated. That requires the use of recovery factors.

2. For example, Chapter 5 in B. L. Karger, L. R. Snyder, and C. Horvath, *An Introduction to Separation Science*, Wiley, New York, 1973; J. J. Kirkland, Ed., *Modern Practice of Liquid Chromatography*, Wiley-Interscience, New York, 1971.

3. Specifically, the figure represents the solution to the illustrative problem on p. 750.

4. O. Samuelson, *Ion Exchange Separations in Analytical Chemistry*, Wiley, New York, 1963. W. Rieman and H. F. Walton, *Ion Exchange in Analytical*

Chemistry, Pergamon, Oxford, 1970. F. Helfferich, *Ion Exchange*, McGraw-Hill, New York, 1962. Y. Marcus and A. S. Kertes, *Ion Exchange and Solvent Extraction of Metal Complexes*, Wiley, New York, 1969.

5. Reviews of chelating resins: G. Nickless and G. R. Marshall, *Chromatogr. Rev.*, **6**, 154 (1964); G. Schmuckler, *Talanta*, **12**, 281 (1965); G. V. Myasoedova, O. P. Eliseeva, and S. B. Savvin, *Zh. Anal. Khim.*, **26**, 2172 (1971).

6. As already indicated, in column operation, it is customary to express the concentration of the exchanging ion in the exchanger in terms of the column (bed) volume. D_v is then obtained. If ρ is the number of grams of dry resin in 1 ml of geometrical column volume, $D_v = \rho D_g$.

7. The designation is not entirely a happy one, because many metal extractions, cationic and anionic, are of the ion-exchange type, for example, $H_2Dz(org) + Cu^{2+}$ or $Zn(HDz)_2(org) + Cu^{2+}$, and the resemblance to the behavior of an ion-exchange resin is only formal.

8. J. Y. Ellenburg, G. W. Leddicotte, and F. L. Moore, *Anal. Chem.*, **26**, 1045 (1954).

9. The effect of various factors involved in the use of organic solvents in ion exchange are examined in the review article "Inorganic Ion Exchange in Organic and Aqueous-Organic Solvents" by G. J. Moody and J. D. R. Thomas, *Analyst*, **93**, 557 (1968) (31 pp., 282 references).

10. F. W. Cornish, *Analyst*, **83**, 634 (1958). E. Glueckauf, *Ion Exchange and Its Applications*, Society for Chemical Industry, London, 1955, p. 34, has proposed an expression for the calculation of h, involving particle radius, flow rate, diffusion coefficients in resin and solution, D_v, and i. Also see Chapter 9 in Helfferich.[4]

11. Approximately 30 min is required for 10 ml of water to percolate through a 6.5 cm column of 200–400 mesh Dowex 1-X8 resin 1 cm^2 in cross section.

12. Ion-exchange material in the form of discs finds use in collecting minute amounts of metals from aqueous solutions. For example, H. James, *Analyst*, **98**, 274 (1973), has shown that heavy metals such as Co, Ni, Zn, Cu, Fe, and Cr at very low concentrations in reactor cooling waters can be isolated by passing the sample through an ion-exchange membrane consisting of finely divided ion-exchange resin dispersed in a polymer film. Three membrane discs are used in series. The collected metals are brought into solution by digesting the discs in HCl–HNO$_3$. With a sample of 250 liters, the recovery of 0.01 μg/liter of Co is ~85%, of 0.1 μg/liter 99%.

13. Bio-Rad Laboratories Catalog, May 1965.

14. Ion exchangers based on copolymers of styrene and divinylbenzene have a tendency to deteriorate slowly on long contact with solutions. The nature of these changes is not well understood. Carboxylic groups seem to be formed: G. M. Armitage and S. J. Lyle, *Talanta*, **20**, 315 (1973). Concerning the physical and chemical stability of ion-exchange resins, see G. J. Moody and J. D. R. Thomas, *Lab. Pract.*, **21**, 632 (1972).

15. J. Korkisch and D. Dimitriadis, *Talanta*, **20**, 1199 (1973) (determination of Th).

16. N. F. Podorvanova, *J. Anal. Chem. USSR*, **26**, 718 (1971).

17. J. Korkisch, F. Feik, and S. S. Ahluwalia, *Talanta*, **14**, 1069 (1967).

18. J. Korkisch and S. S. Ahluwalia, *Anal. Chem.*, **38**, 497 (1966).

19. I. L. Jenkins, *Actinides Rev.*, **1**, 187 (1969).

20. F. W. E. Strelow, C. H. S. W. Weinert, and T. N. van der Walt, *Anal. Chim. Acta*, **71**, 123 (1974).

21. G. J. Moody and J. D. R. Thomas, *Analyst*, **93**, 557 (1968); C. W. Walter and J. Korkisch, *Mikrochim. Acta*, 81 (methanol), 137 (ethanol, 2-propanol, methyl glycol), 158 (tetrahydrofuran and acetone), 181 (acetic acid and dimethyl sulfoxide), 194 (discussion and theoretical treatment) (1971). Separation of rare earth metals, Y, and Sc in HNO_3–CH_3OH: J. P. Faris and J. W. Warton, *Anal. Chem.*, **34**, 1077 (1962).

22. J. Korkisch and G. E. Janauer, *Talanta*, **9**, 957 (1962); J. Korkisch, *Mikrochim. Acta*, 816 (1964).

23. According to L. W. Marple, *J. Inorg. Nucl. Chem.*, **26**, 635 (1964), the adsorption of thorium and uranium from alcoholic hydrochloric and nitric acid solutions entails distribution of the neutral complexes between the two phases. The main factor that affects the distribution is the change in the free energy of solvation of the neutral complex with a change in alcohol content. Cf. C. W. Walter and J. Korkisch, *Mikrochim. Acta*, 194 (1971).

24. J. S. Fritz, H. Waki, and B. B. Garralda, *Anal. Chem.*, **36**, 900 (1964).

25. J. Korkisch and F. Feik, *Anal. Chim. Acta*, **32**, 110 (1965).

26. S. S. Ahluwalia and J. Korkisch, *Z. Anal. Chem.*, **208**, 414 (1965).

27. J. Korkisch and F. Feik, *Anal. Chim. Acta*, **37**, 364 (1967).

28. R. Kuroda, T. Ono, and K. Ishida, *Japan Analyst*, **20**, 1142 (1971).

29. F. W. E. Strelow and C. J. C. Bothma, *Anal. Chem.*, **36**, 1217 (1964).

30. F. W. E. Strelow, *Anal. Chem.*, **35**, 1279 (1963).

31. K. Kawabuchi, T. Ito, and R. Kuroda, *J. Chromatogr.*, **39**, 61 (1969).

32. L. Danielsson, *Acta Chem. Scand.*, **19**, 670 (1965), determined D for 26 elements in H_2SO_4 up to 5 M on Dowex 1-X8 resin. The two sets of results are in substantial agreement. For separation of Zr and Hf from each other by elution with H_2SO_4, see J. L. Hague and L. A. Machlan, *J. Res. Nat. Bur. Stand.*, **A65**, 75 (1961).

33. T. Kiriyama and R. Kuroda, *Anal. Chim. Acta*, **71**, 375 (1974).

34. R. Kuroda, K. Oguma, N. Kono, and Y. Takahashi, *Anal. Chim. Acta*, **62**, 343 (1972).

35. J. Korkisch and S. S. Ahluwalia, *Z. Anal. Chem.*, **215**, 86 (1966).

36. K. A. Kraus and F. Nelson in W. J. Hamer, Ed., *The Structure of Electrolytic Solutions*, Wiley, New York, 1959, chapter 23.

37. S. Fronaeus, *Svensk Kem. Tids.*, **65**, 1 (1953); *Acta Chem. Scand.*, **8**, 1174 (1954).

38. M. Miyamoto, *Japan Analyst*, **10**, 321 (1961).

39. D. Monnier, W. Haerdi, and J. Vogel, *Anal. Chim. Acta*, **23**, 577 (1960); S. Hirano, A. Mizuike, Y. Iida, and N. Kokubun, *Japan Analyst*, **10**, 326 (1961).

40. M. Nishimura and E. B. Sandell, *Anal. Chim. Acta*, **26**, 242 (1962).

41. C. Huffman, H. H. Lipp, and L. F. Rader, *Geochim. Cosmochim. Acta*, **27**, 209 (1963).

42. R. H. Maier and J. S. Bullock, *Anal. Chim. Acta*, **19**, 354 (1958).

43. A. D. Horton and P. F. Thomason, *Anal. Chem.*, **28**, 1326 (1956).

44. A. D. Horton, P. F. Thomason, and M. T. Kelley, *Anal. Chem.*, **29**, 388 (1957).

45. M. L. Cluett and J. H. Yoe, *Anal. Chem.*, **29**, 1265 (1957).

46. Moody and Thomas.[21] J. S. Fritz and D. J. Pietrzyk, *Talanta*, **8**, 143 (1961); Korkisch and Janauer[22]; Korkisch[22]; J. Korkisch and I. Hazan, *Talanta*, **11**, 1157 (1964); C. Cleyrergue, N. Deschamps, and P. Albert in J. R. DeVoe and P. D. LaFleur, Eds., *Modern Trends in Activation Analysis*, NBS Spec. Publ. **312**, 1969, p. 646; Z. K. Karalova, B. F. Myasoedov, L. M. Rodionova, and Z. I. Pyzhova, *Zh. Anal. Khim.*, **29**, 1332 (1974) (distribution coefficients of Ac, Th, R.E. elements, Fe, Bi, and Pb on anion-exchange resin AV-17X8 from various organic solvent–HCl solutions). For adsorption benavior ot metals on a weak-base anion exchanger, DEAE-cellulose, in methanol–hydrochloric acid mixtures, see R. Kuroda and N. Yoshikuni, *Talanta*, **18**, 1123 (1971).

47. For anion-exchange separation of metal ions in dimethyl sulfoxide–methanol-hydrochloric acid, see J. S. Fritz and M. L. Gillette, *Talanta*, **15**, 287 (1968).

48. J. Korkisch and I. Hazan, *Anal. Chem.*, **36**, 2464 (1964).

49. J. Korkisch and I. Steffan, *Mikrochim. Acta*, 837 (1972) (U in geo materials).

50. J. Korkisch and I. Hazan, *Anal. Chem.*, **36**, 2308 (1964). Separation by using 2-methoxyethanol–hydrochloric acid mixtures is also described. Distribution coefficients in acetone–HCl: J. M. Peters and G. del Fiore, *Radiochem. Radioanal. Lett.*, **21**, 11 (1975).

51. S. Onuki, K. Watanuki, and Y. Yoshino, *Japan Analyst*, **15**, 924 (1966).

52. For example, separation of U: W. Koch and J. Korkisch, *Mikrochim. Acta*, 225, 245, 263, 865 (1973).

53. R. Kuroda, K. Ishida, and T. Kiriyama, *Anal. Chem.*, **40**, 1502 (1968).

54. M. D. Seymour and J. S. Fritz, *Anal. Chem.*, **45**, 1394 (1973).

55. F. Nelson, T. Murase, and K. A. Kraus, *J. Chromatogr.*, **13**, 503 (1964).

56. C. K. Mann and C. L. Swanson, *Anal. Chem.*, **33**, 459 (1961). HETP and

normalized band widths are given. For an elution study of Mg with HCl and HClO$_4$, see C. K. Mann, *ibid.*, **32,** 67 (1960).

57. G. J. Moody and J. D. R. Thomas, *Analyst,* **93,** 557 (1968) (review); J. Korkisch and S. S. Ahluwalia, *Talanta,* **14,** 155 (1967); T. Cummings and Korkisch, *ibid.,* **14,** 1185 (1967); J. S. Fritz and T. A. Rettig, *Anal. Chem.,* **34,** 1562 (1962) (acetone); F. W. E. Strelow, A. H. Victor, C. R. van Zyl, and C. Eloff, *ibid.,* **43,** 870 (1971) (acetone); G. E. Janauer, H. E. Van Wart, and J. T, Carrano, *ibid.,* **42,** 215 (1970) (dimethyl sulfoxide); M. N. Renault and N. Deschamps, *Radiochem. Radioanal. Lett.,* **9,** 199 (1972) (dioxane). Separation of Au (acetone): Strelow et al., *Talanta,* **23,** 173 (1976).

58. F. W. E. Strelow, C. R. van Zyl, and C. J. C. Bothma, *Anal. Chim. Acta,* **45,** 81 (1969).

59. J. Korkisch and S. S. Ahluwalia, *Anal. Chem.,* **37,** 1009 (1965).

60. P. C. van Erkelens, *Anal. Chim. Acta,* **25,** 42 (1961).

61. J. S. Fritz and J. E. Abbink, *Anal. Chem.,* **37,** 1274 (1965).

62. F. Tera, R. R. Ruch, and G. H. Morrison, *Anal. Chem.,* **37,** 358 (1965).

63. R. R. Ruch, F. Tera, and G. H. Morrison, *Anal. Chem.,* **37,** 1565 (1965).

64. J. S. Fritz and B. B. Garralda, *Anal. Chem.,* **34,** 102 (1962).

65. F. W. E. Strelow and M. D. Boshoff, *Anal. Chim. Acta,* **62,** 351 (1972). Distribution coefficients for 54 cations in HBr–acetone solution: F. W. E. Strelow, M. D. Hanekom, A. H. Victor, and C. Eloff, *ibid.,* **76,** 377 (1975).

66. R. Dybczynski and H. Maleszewska, *Analyst,* **94,** 527 (1969).

67. J. Korkisch and E. Klakl, *Talanta,* **16,** 377 (1969). The use of acetone in U separations is described by F. W. E. Strelow and C. H. S. W. Weinert, *Talanta,* **20,** 1127 (1973).

68. J. Korkisch and I. Hazan, *Anal. Chem.,* **37,** 707 (1965); E. Klakl and J. Korkisch, *Talanta,* **16,** 1177 (1969).

69. J. Korkisch and D. Dimitriadis, *Talanta,* **20,** 1295 (1973). J. Korkisch and A. Sorio, *ibid.,* **22,** 273 (1975), isolate Pb (0.001–0.01 ppm) from natural waters by adsorption on Dowex 1-X8 from 0.15 *M* HBr, followed by elution with 6 *M* HCl. An easier ion-exchange separation of Cd from Zn (and some other metals also) can be obtained by using iodide in preference to bromide medium. For example, S. Kallmann, H. Oberthin, and R. Liu, *Anal. Chem.,* **32,** 59 (1960), pass the iodide-containing sample solution through a Dowex 50 or Amberlite IR-120 resin column and wash with iodide solution (Zn is left in the column). Cadmium can be determined directly with dithizone in the eluate.

70. J. S. Fritz, B. B. Garralda, and S. K. Karraker, *Anal. Chem.,* **33,** 882 (1961). M. K. Nikitin, *Dokl. Akad. Nauk SSR,* **148,** 595 (1963), determined distribution coefficients of a large number of elements using the resin KU-2X6. Rare earth elements are strongly adsorbed. Also see C. De

Wispelaere, J. P. Op de Beeck, and J. Hoste, *Anal. Chim. Acta*, **70,** 1 (1974).

71. L. Danielsson and T. Ekström, *Acta Chem. Scand.*, **21,** 1173 (1967).

72. J. Korkisch and A. Huber, *Talanta*, **15,** 119 (1968).

73. L. Danielsson, *Arkiv Kemi*, **27,** 453 (1967).

74. R. Neirinckx, F. Adams, and J. Hoste, *Anal. Chim. Acta*, **46,** 165 (1969).

75. T. A. Ferraro, *Talanta*, **16,** 669 (1969).

76. L. Danielsson, *Acta Chem. Scand.*, **19,** 1859 (1965). Values are also given for 1.0 M HNO$_3$ and 0.5 M H$_2$SO$_4$ (0.1 M and 1 M HF), which are omitted here. As would be expected, adsorbability decreases with the concentration of HNO$_3$ and H$_2$SO$_4$.

77. L. Danielsson, *Arkiv Kemi*, **27,** 459 (1967), where the adsorptive and elutional properties of Nb, Ta, V(V), and Sb(V) are also discussed.

78. L. Danielsson, *Arkiv Kemi*, **27,** 453 (1967).

79. E. A. Huff, *Anal. Chem.*, **36,** 1921 (1964).

80. J. B. Headridge and E. J. Dixon, *Analyst*, **87,** 32 (1962).

81. J. L. Hague and L. A. Machlan, *J. Res. Nat. Bur. Stand.*, **62,** 11 (1959); E. J. Dixon and J. B. Headridge, *Analyst*, **89,** 185 (1964) (separation of Mo and W is included.); L. Danielsson, *Arkiv Kemi*, **27,** 467 (1967). Nb and Ta in ores and minerals: S. Kallmann, H. Oberthin, and R. Liu, *Anal. Chem.*, **34,** 609 (1962). (These references are not intended to be exhaustive.)

82. J. I. Kim, H. Lagally, and H.-J. Born, *Anal. Chim. Acta*, **64,** 29 (1973). The relation of adsorbability to the kinds of complex species in solution is discussed.

83. A. K. Majumdar and B. K. Mitra, *Z. Anal. Chem.*, **208,** 1 (1965).

84. H. Hamaguchi, R. Kuroda, K. Aoki, R. Sugisita, and N. Onuma, *Talanta*, **10,** 153 (1963).

85. H. Hamaguchi, N. Onuma, M. Kishi, and R. Kuroda, *Talanta*, **11,** 495 (1964).

86. D. J. Pietrzyk and D. L. Kiser, *Anal. Chem.*, **37,** 233 (1965).

87. D. J. Pietrzyk and D. L. Kiser, *Anal. Chem.*, **37,** 1578 (1965).

88. H. Hamaguchi, K. Kawabuchi, and R. Kuroda, *Anal. Chem.*, **36,** 1654 (1964).

89. J. S. Fritz and E. E. Kaminski, *Talanta*, **18,** 541 (1971).

90. R. Kuroda, T. Kondo, and K. Oguma, *Talanta*, **19,** 1043 (1972).

91. J. Korkisch and D. Dimitriadis, *Talanta*, **20,** 1287 (1973); J. Korkisch and L. Gödl, *Anal. Chim. Acta*, **71,** 113 (1974). Also for U and Cd: J. Korkisch and L. Gödl, *Talanta*, **21,** 1035 (1974); Zn: J. Korkisch, L. Gödl, and H. Gross, *ibid.*, **22,** 281 (1975).

92. S. K. Jha, F. De Corte, and J. Hoste, *Anal. Chim. Acta*, **62,** 163 (1972).

93. P. Van den Winkel, F. De Corte, A. Speecke, and J. Hoste, *Anal. Chim.*

Acta, **42**, 340 (1968); P. Van den Winkel, F. De Corte, and J. Hoste, *J. Radioanal. Chem.*, **10**, 139 (1972); P. Van den Winkel, F. De Corte, and J. Hoste, *Anal. Chim. Acta*, **56**, 241 (1971). Protactinium and homologs in acetic–nitric acid: J. I. Kim, H. Lagally, and H.-J. Born, *J. Inorg. Nucl. Chem.*, **33**, 3547 (1971).

94. F. De Corte, P. Van Acker, and J. Hoste, *Anal. Chim. Acta*, **64**, 177 (1973).

95. A. Mahan, A. K. Ghose, and A. K. Dey, *Sep. Sci.*, **6**, 781 (1971).

96. R. Kuroda, T. Kondo, and K. Oguma, *Talanta*, **20**, 533 (1973). The metals are probably adsorbed as nitrato complexes.

97. H. Polkowska-Motrenko and R. Dybczynski, *J. Chromatogr.*, **88**, 387 (1974). Adsorption on Dowex 50: A. Dadone et al., *Chromatographia*, **7**, 258 (1974).

98. F. De Corte, P. Van den Winkle, A. Speecke, and J. Hoste, *Anal. Chim. Acta*, **42**, 67 (1968). Cation exchange: T. Nozaki, O. Hiraiwa, C. Henmi, and K. Koshiba, *Bull. Chem. Soc. Japan*, **42**, 245 (1969).

99. F. W. E. Strelow, C. H. S. W. Weinert, and C. Eloff, *Anal. Chem.*, **44**, 2352 (1972).

100. G. F. Pitstick, T. R. Sweet, and G. P. Morie, *Anal. Chem.*, **35**, 995 (1963); Morie and Sweet, *ibid.*, **36**, 140 (1964) and *J. Chromatogr.*, **16**, 210 (1964). Cation exchange has been applied by A. Dadone, F. Baffi, and R. Frache, *Talanta*, **23**, 593 (1976); J. C. Rouchaud and G. Revel, *J. Radioanal. Chem.*, **16**, 221 (1973). Lead can be separated from Sn, Sb, Nb, Ta, Mo, and W on a cation-exchange resin in tartaric–nitric acid solution as shown by F. W. E. Strelow and T. N. van der Walt, *Anal. Chem.*, **47**, 2272 (1975).

101. For example, J. S. Fritz and G. R. Umbreit, *Anal. Chim. Acta*, **19**, 509 (1958) (cation-exchange resin); J. Vanderdeelen, *ibid.*, **49**, 360 (1970) (anion-exchange resin); V. Kratochvil, P. Povondra, and Z. Sulcek, *Chem. Listy*, **63**, 1185 (1969) (review). Isolation of Be: J. R. Merrill, M. Honda, and J. R. Arnold, *Anal. Chem.*, **32**, 1420 (1960).

102. M. Zukriegelova, V. Sixta, and Z. Sulcek, *Coll. Czech. Chem. Commun.*, **37**, 68 (1972); M. Zukriegelova, Z. Sulcek, and P. Povondra, *ibid.*, **37**, 560 (1972).

103. K. A. Orlandini and J. Korkisch, *Anal. Chim. Acta*, **43**, 459 (1968).

104. Cupferron: J. Korkisch and M. M. Khater, *Talanta*, **19**, 1654 (1972).

105. R. F. Hirsch and J. D. Portock, *Anal. Chim. Acta*, **49**, 473 (1970); T. Fukasawa, S. Kano, H. Ono, and A. Mizuike, *Japan Analyst*, **19**, 1417 (1970).

106. F. W. E. Strelow, C. J. Liebenberg, and F. von S. Toerien, *Anal. Chim. Acta*, **47**, 251 (1969).

107. P. de Gelis, *Chim. anal.*, **53**, 673 (1971).

108. R. Rosset, *Bull. Soc. Chim. France*, 59 (1966).

109. Examples: separation of Cu from much NH_4Cl, R. Turse and W. Rieman,

Anal. Chim. Acta, **24,** 202 (1961); Mn from much NaCl, H. Imoto, *Japan Analyst*, **10,** 124 (1961). Metals are eluted with 1 *M* mineral acid.

110. J. P. Riley and D. Taylor, *Anal. Chim. Acta*, **40,** 479 (1968). But see T. M. Florence and G. E. Batley, *Talanta*, **23,** 179 (1976). Another type of chelating resin for isolating Cd, Co, Cu, Mn, Ni, and other heavy metals from sea water and other natural waters at pH ~5 (Th at pH 1) has arsonic acid groups in a macroporous matrix: R. F. Hirsch, E. Gancher, and F. R. Russo, *Talanta*, **17,** 483 (1970); J. S. Fritz and E. M. Moyers, *ibid.*, **23,** 590 (1976). (The metals are eluted with perchloric acid.) A resin of this type is superior to Dowex A-1 in being faster in action and having lesser affinity for Ca and Mg. The macroporous matrix is conducive to more rapid reaction kinetics. The macroporous (or macroreticular) resins have a network of wider pores (50 nm or more in diameter) than the gel-type (microreticular) resins such as Dowex A-1. Macroporous resins shrink and swell less than gel-type resins.

111. Z. Sulcek and P. Povondra, *Coll. Czech. Chem. Commun.*, **32,** 3140 (1967).

112. D. G. Birney, W. E. Blake, P. R. Meldrum, and M. E. Peach, *Talanta*, **15,** 557 (1968).

113. M. Marhol and K. L. Cheng, *Talanta*, **21,** 751 (1974).

114. G. Koster and G. Schmuckler, *Anal. Chim. Acta*, **38,** 179 (1967).

115. E. Bayer, *Angew. Chem.*, **76,** 76 (1964). A poly-Schiff base prepared by condensing 1,4-diamino-2,5-dimercaptobenzene is effective in isolating pp10[9] of Au from chloride solutions [A. Zlatkis, W. Bruening and E. Bayer, *Anal. Chem.*, **41,** 1692 (1969)]. For other chelating resins having sulfur-containing groups see G. V. Myasoedova et al., *Zh. Anal. Khim.*, **28,** 1550 (1973).

116. E. Blasius and G. Kynast, *J. Radioanal. Chem.*, **2,** 55 (1969).

117. H. Bernhard and F. Grass, *Mikrochim. Acta*, 426 (1966).

118. F. Vernon and H. Eccles, *Anal. Chim. Acta*, **63,** 403 (1973). Earlier literature is summarized. J. R. Parrish and R. Stevenson, *Anal. Chim. Acta*, **70,** 189 (1974), have also prepared oxine resins, and arrive at conclusions different from Vernon and Eccles.

119. K. F. Sugawara, H. H. Weetall, and G. D. Schucker, *Anal. Chem.*, **46,** 489 (1974).

120. H. Eccles and F. Vernon, *Anal. Chim. Acta*, **66,** 231 (1973).

121. J. F. Dingman, Jr., K. M. Gloss, E. A. Milano, and S. Siggia, *Anal. Chem.*, **46,** 774. (1974).

122. A. J. Bauman, H. H. Weetall, and N. Weliky, *Anal. Chem.*, **39,** 932 (1967).

123. G. V. Myasoedova and L. I. Bol'shakova, *Zh. Anal. Khim.*, **23,** 504 (1968). Phenol resin based on salicylic acid (for FeIII): F. Vernon and H. Eccles, *Anal. Chim. Acta*, **72,** 331 (1974).

124. R. A. A. Muzzarelli, and O. Tubertini, *Talanta*, **16,** 1571 (1969); R. A. A. Muzzarelli, G. Raith, and O. Tubertini, *J. Chromatogr.*, **47,** 414 (1970).

125. R. Hering, *Chelatbildende Ionenaustauscher*, Akad.-Verlag, Berlin, 1967. Review: G. Schmuckler, *Talanta*, **12**, 281 (1965).

126. D. H. Schmitt and J. S. Fritz, *Talanta*, **15**, 515 (1968).

127. G. C. Goode and M. C. Campbell, *Anal. Chim. Acta*, **27**, 422 (1962).

128. E. Schulek, Zs. Remport-Horváth, A. Lásztity, and E. Körös, *Talanta*, **16**, 323 (1969). A. J. Bauman, H. H. Weetal, and N. Weliky, *Anal. Chem.*, **39**, 932 (1967), have diazo-coupled dithizone and other ligands to a modified carboxymethylcellulose matrix (used for isolating reacting trace metals from sea water).

129. R. Kuroda, T. Kiriyama, and K. Ishida, *Anal. Chim. Acta*, **40**, 305 (1968).

130. K. Ishida, T. Kiriyama, and R. Kuroda, *Anal. Chim. Acta*, **41**, 537 (1968).

131. K. Ishida and R. Kuroda, *Anal. Chem.*, **39**, 212 (1967).

132. R. Kuroda and N. Yoshikuni, *Mikrochim. Acta*, 653 (1974).

133. C. B. Amphlett, *Inorganic Ion Exchangers*, Elsevier, Amsterdam, 1964. The literature in the period 1965–70 has been comprehensively reviewed by V. Vesely and V. Pekarek, *Talanta*, **19**, 219, 1245 (1972). Also see M. J. Fuller, *Chromatogr. Rev.*, **14**, 45 (1971). A more recent, extensive review: M. Abe, *Japan Analyst*, **23**, 1254, 1561 (1974).

134. Recently, ammonium molybdophosphate has been used for the separation of Th from Y, and cesium molybdophosphate for the separation of Nb from Ti. See V. F. Barkovskii, T. V. Velikanova, L. K. Neudachina, and L. N. Yudina, *Zh. Anal. Khim.*, **29**, 1231 (1974) (Th–Y); V. F. Barkovskii, T. V. Velikanova, L. P. Kharchenko, and N. V. Semchuk, *Zh. Anal. Khim.*, **29**, 1014 (1974).

135. Reversed-phase partition chromatography is also called extraction chromatography or reversed-phase extraction chromatography.

136. For surveys of column partition chromatography see G. S. Katykhin, *J. Anal. Chem. USSR*, **20**, 570 (1965); **27**, 758 (1972).

137. F. H. Burstall, G. R. Davies, and R. A. Wells, *Disc. Faraday Soc.*, **7**, 179 (1949).

138. J. R. Bishop, *Analyst*, **81**, 291 (1956).

139. R. A. Mercer and R. A. Wells, *Analyst*, **79**, 339 (1954).

140. J. A. S. Adams and W. J. Maeck, *Anal. Chem.*, **26**, 1635 (1954). For separation of traces of metals from uranium, see R. A. A. Muzzarelli and L. C. Bate, *Talanta*, **12**, 823 (1965).

141. J. S. Fritz and D. H. Schmitt, *Talanta*, **13**, 123 (1966).

142. Reviews of reversed-phase partition chromatography: E. Cerrai and G. Ghersini in J. C. Giddings and R. A. Keller, Eds., *Advances in Chromatography*, vol. 9, Dekker, New York, 1970, p. 3; V. Spevackova, *Chem. Listy*, **62**, 1194 (1968); H. Eschrich and W. Drent, *Eurochemic Technical Report ETR*-271 (1971). Book: T. Braun and G. Ghersini, Eds., *Extraction Chromatography*, Elsevier, New York, 1975, 566 pp.

143. D. E. Carritt, *Anal. Chem.*, **25**, 1927 (1953). For a tabulation of various methods of partition chromatography with dithizone, see note 144.

144. T. Braun and A. B. Farag, *Anal. Chim. Acta*, **69**, 85 (1974), in which references to earlier use of polyurethane foam may be found. This paper deals with the collection of silver on such a matrix impregnated with zinc dithizonate (more stable than dithizone). The recovery of 1 μg Ag in 1 liter of solution was greater than 95% for a flow rate of 50 ml/min with 5 g foam. For collection of Hg, see *ibid.*, **71**, 133. Also see Y. Sekizuka, T. Kojima, T. Yano, and K. Ueno, *Talanta*, **20**, 979 (1973).

145. E. Cerrai, *Chromatogr. Rev.*, **6**, 129 (1964); Cerrai and Ghersini.[142]

146. B. Neef and H. Grosse-Ruyken, *J. Chromatogr.*, **79**, 275 (1973).

147. A. G. Hamlin, B. J. Roberts, W. Loughlin, and S. G. Walker, *Anal. Chem.*, **33**, 1547 (1961); T. J. Hayes and Hamlin, *Analyst*, **87**, 770 (1962); E. A. Huff, *Anal. Chem.*, **37**, 533 (1965); J. W. Arden and N. H. Gale, *ibid.*, **46**, 687 (1974).

148. G. J. Kamin, J. W. O'Laughlin, and C. V. Banks, *J. Chromatogr.*, **31**, 292 (1967).

149. J. S. Fritz and L. H. Dahmer, *Anal. Chem.*, **40**, 20 (1968).

150. J. S. Fritz and R. T. Frazee, *Anal. Chem.*, **37**, 1358 (1965).

151. M. A. Wade and S. S. Yamamura, *Anal. Chem.*, **36**, 1861 (1964). Compare J. S. Fritz and C. E. Hedrick, *ibid.*, **34**, 1411 (1962).

152. I. Akaza, T. Tajima, and K. Kibà, *Bull. Chem. Soc. Japan*, **46**, 1199 (1973).

153. J. S. Fritz and G. L. Latwesen, *Talanta*, **17**, 81 (1970). For separation of Ga, In, and Tl by reversed-phase chromatography, see J. S. Fritz, R. T. Frazee, and G. L. Latwesen, *Talanta*, **17**, 857 (1970).

154. A. I. Busev, I. P. Kharlamov, and O. V. Smirnova, *Anal. Lett.*, **3**, 177, 191 (1970).

155. N. I. Brodskaya, I. P. Vychuzhanina, and A. D. Miller, *Zh. Prikl. Khim.*, **40**, 802 (1967).

156. For a summary of paper-chromatographic separations see L. Meites, Ed., *Handbook of Analytical Chemistry*, McGraw-Hill, New York, 1963, p. 10-48. Review: G. M. Varshal, *Zh. Anal. Khim.*, **27**, 904 (1972).

157. N. F. Kember and R. A. Wells, *Analyst*, **80**, 735 (1955). Cf. G. G. Tertipis and F. E. Beamish, *Talanta*, **10**, 1139 (1963).

158. F. N. Ward et al., *U.S. Geol. Surv. Bull.*, **1152**, 79 (1963). Cf. D. Purushottam, *Z. Anal. Chem.*, **185**, 214 (1962).

159. R. A. Webb, D. G. Hallas, and H. M. Stevens, *Analyst*, **94**, 794 (1969).

160. M. M. Frodyma, D. F. Zaye, and V. T. Lieu, *Anal. Chim. Acta*, **40**, 451 (1968). For reviews of thin-layer chromatography in inorganic analysis, see D. I. Ryabchikov and M. P. Volynets, *Zh. Anal. Khim.*, **21**, 1348 (1966); M. P. Volynets, *ibid.*, **27**, 871 (1972).

161. H.-J. Senf, *J. Chromatogr.*, **21**, 363 (1966); silica gel D as adsorbent. K. Ballschmiter, *Z. Anal. Chem.*, **254**, 348 (1971).

162. M. Hranisavljevic-Jakovlevic, J. Pejkovic-Tadic, and J. Miljkovic-Stojanovic, *Mikrochim. Acta*, 141, 936, 940 (1965); T. Takeuchi and Y. Tsunoda, *J. Chem. Soc. Japan Pure Chem. Sec.*, **88**, 176 (1967); Z. Gregorowicz, J. Kulicka, and T. Suwinska, *Chemia Anal.*, **16**, 169 (1971).

163. A. Galik and A. Vincourova, *Anal. Chim. Acta*, **46**, 113 (1969).

164. A. Galik, *Anal. Chim. Acta*, **67**, 357 (1973); **57**, 399 (1971).

165. Kh. S. Babayan and G. M. Varshal, *Zh. Anal. Khim.*, **28**, 921 (1973).

166. H. Determann, *Gel Chromatography*, 2nd ed., Springer-Verlag, Berlin, 1969; Reviews: D. M. W. Anderson, I. C. M. Dea, and A. Hendrie, *Talanta*, **18**, 365 (1971); N. Yoza, *Japan Analyst*, **20**, 505 (1971).

167. Separation of traces of alkali and alkaline earth metals from molybdenum on a dextran gel column is based on the formation of an Mo–NH$_3$–glucose complex. See S. Karajannis, H. M. Ortner, and H. Spitzy, *Talanta*, **19**, 903 (1972). For the separation of these metals from tungsten, see H. Krainer, H. M. Ortner, K. Müller, and H. Spitzy, *ibid.*, **21**, 933 (1974).

168. D. Saunders and R. L. Pecsok, *Anal. Chem.*, **40**, 44 (1968); P. A. Neddermeyer and L. B. Rogers, *ibid.*, **40**, 755 (1968); B. Z. Egan, *J. Chromatogr.*, **34**, 382 (1968); Y. Ueno, N. Yoza, and S. Ohashi, *ibid.*, **52**, 321 (1970).

169. H. Ortner and H. Spitzy, *Z. Anal. Chem.*, **221**, 119 (1966); **238**, 167, 251 (1968). Concentration of lithium from Dead Sea brines has been described by M. Rona and G. Schmuckler, *Talanta*, **20**, 237 (1973).

170. N. Yoza and S. Ohashi, *J. Chromatogr.*, **41**, 429 (1969); N. Yoza, T. Ogata, and S. Ohashi, *ibid.*, **52**, 329 (1970).

171. M. Lederer, *An Introduction to Paper Electrophoresis and Related Methods*, Elsevier, Amsterdam, 1955; T. Wieland and K. Dose, *Electrochromatography*, in W. G. Berl, Ed., *Physical Methods in Chemical Analysis*, vol. III. Academic Press, New York, 1956, pp. 29–70; Ch. Wunderly, *Principles and Applications of Paper Electrophoresis*, Elsevier, New York, 1961; R. D. Strickland, "Electrophoresis" in F. J. Welcher, Ed., *Standard Methods of Chemical Analysis*, 6th ed., vol. 3, part A, Van Nostrand, Princeton, N.J., 1966; E. Heftmann, Ed., *Chromatography*, 2nd ed., Van Nostrand Reinhold, New York, 1967; M. Bier in B. L. Karger, L. R. Snyder, and C. Horvath, *An Introduction to Separation Science*, J. Wiley, New York, 1973. Reviews: H. H. Strain and T. R. Sato, *Anal. Chem.*, **28**, 687 (1956); H. H. Strain, *ibid.*, **30**, 620 (1958), **32**, 3R (1960); R. D. Strickland, *ibid.*, **34**, 31R (1962); **36**, 80R (1964); **38**, 99R (1966); **40**, 74R (1968); **42**, 32R (1970).

172. E. K. Korchemnaya, A. N. Ermakov, and V. I. Naumova, *Zh. Anal. Khim.*, **26**, 26 (1971).

173. Reviews: D. I. Ryabchikov, E. K. Korchemnaya, and V. I. Naumova, *Zh. Anal. Khim.*, **23**, 741 (1968); E. K. Korchemnaya, A. N. Ermakov, and V. I. Naumova, *ibid.*, **25**, 705 (1970).

174. H. Meier, R. Dieberg, E. Zimmerhackl, W. Albrecht, D. Bösche, W. Hecker, P. Menge, A. Ruckdeschel, E. Unger, and G. Zeitler, *Mikrochim. Acta*, 102 (1970).

175. E. P. Krysin et al., *Anal. Abst.*, **29**, 3A10 (1975).

LIQUID–LIQUID EXTRACTION
GENERAL CONSIDERATIONS—INORGANIC
SUBSTANCES

The term extraction as used in chemical analysis usually refers to liquid–liquid (immiscible solvent) extraction. Extraction is the process of distributing a dissolved constituent, which may be present as more than one species, between two contacting liquids that are immiscible or, more accurately expressed, have limited miscibility.[1] Almost always in conventional inorganic trace separations, the sample is initially dissolved in an aqueous solution, and the second solvent is an organic liquid, which in addition to being of limited miscibility, must extract the constituent in question to a larger extent than the others—or, less desirably, vice versa.

The greatly different extractabilities of metal compounds, especially of metal–organic complexes, into the same organic solvent or into various organic solvents makes extraction an effective and widely applicable means of separation of metals. All metals and semimetals except the alkali metals—and even these to some extent—can be extracted in one form or another. All things considered, if a choice is offered, extraction–separation methods will generally be chosen in preference to other methods for most trace metals (the alkalies and alkaline earths are exceptions). Usually, by adjustment of pH and with the aid of differential complexing agents, most metals can be separated extractively without resort to extensive fractionation procedures. When fractionation is required, ion exchange or one of the other chromatographic methods allowing more rapid and convenient fractionation is usually better than extraction. Otherwise, extraction methods will generally be chosen in preference to ion-exchange methods for separating metals on the basis of rapidity, the small volume of solvent in which the constituent in question can be obtained, and, often, the exclusion of masking agents from the final solution. An extraction may be an integral part of an absorptiometric or fluorimetric determination, and the determination step may then automatically supply a separation from some elements that would otherwise interfere.

LITERATURE ON EXTRACTION

For the convenience of the reader, we collect here a list of books and review articles on general extraction, with special reference to inorganic analytical applications or analytical possibilities. This list is not intended to be exhaustive. Some books on specialized extractions are noted elsewhere in this chapter.

1. Monographs and the Like

G. H. Morrison and H. Freiser, *Solvent Extraction in Analytical Chemistry*, Wiley, New York, 1957.

J. Stary, *The Solvent Extraction of Metal Chelates*, Pergamon, London, 1964.

Y. Marcus and A. S. Kertes, *Ion Exchange and Solvent Extraction of Metal Complexes*, Wiley-Interscience, New York, 1969. (Physicochemical.)

A. K. De, S. M. Khopkar, and R. A. Chalmers, *Solvent Extraction of Metals*, Van Nostrand-Reinhold, New York, 1970.

Yu. A. Zolotov (transl. by J. Schmorak), *Extraction of Chelate Compounds*, Ann Arbor–Humphrey Science Publishers, Ann Arbor, Mich., 1970.

Yu. A. Zolotov and N. M. Kuz'min, *Extractive Concentration* [in Russian], Khimiya, Moscow, 1970.

V. V. Bagreev, Yu. A. Zolotov, N. A. Kurilina, and G. F. Kalinina, *Ekstraktsiya neorganicheskikh soedinenii*, Moscow, 1970. Two volumes of bibliography (1945–1962 and 1963–1967).

S. Hartland, *Counter-Current Extraction*, Pergamon, Oxford, 1971.

In addition, information on various aspects of extraction will be found in O. G. Koch and G. A. Koch-Dedic, *Handbuch der Spurenanalyse*, Springer, Berlin, 2nd ed., 1974; D. D. Perrin, *Organic Complexing Reagents*, Wiley-Interscience, New York, 1964 (Chapter 10); A. Ringbom, *Complexation in Analytical Chemistry*, Interscience, New York, 1963 (Chapter VII); J. Korkisch, *Modern Methods for the Separation of the Rarer Metal Ions*, Pergamon, Oxford, 1969.

2. Summarizing and Review Articles

R. M. Diamond and D. G. Tuck, "Extraction of Inorganic Compounds into Organic Solvents," in F. A. Cotton, Ed., *Progress in Inorganic Chemistry*, Vol. 2, p. 109, Interscience, New York, 1960.

H. Irving and R. J. P. Williams, "Liquid–Liquid Extraction," in I. M. Kolthoff, P. J. Elving, and E. B. Sandell, Eds., *Treatise on Analytical Chemistry*, Part I, Vol. 3, Chapter 31, Wiley, New York, 1961.

F. A. von Metzsch, "Solvent Extraction," in W. G. Berl, Ed., *Physical Methods in Chemical Analysis*, Vol. 4, p. 317, Academic, New York, 1961.

L. I. Katzin, "Solvent Extraction of Inorganic Species," in J. J. Lagowski, Ed., *Chemistry of Non-Aqueous Solvents*, Vol. 1, Chapter 5, Academic, New York, 1966.

D. F. Peppard, "Liquid–Liquid Extraction of Metal Ions," in H. J. Eméleus and A. G. Sharpe, Eds., *Advances in Inorganic Chemistry and Radiochemistry*, Vol. 9, p. 1, Academic, New York, 1966.

C. Hanson, Ed., *Recent Advances in Liquid–Liquid Extraction*, Pergamon, Oxford, 1971.

S. J. Lyle, "Solvent Extraction in Inorganic Analytical Chemistry," in *Selected Annual Reviews of the Analytical Sciences*, Vol. 3, Chemical Society, Letchworth, England, 1973. (661 references.)

Yu. A. Zolotov, "Extraction Methods of Separation in Analytical Chemistry," in A. I. Tugarinov, Ed., *Recent Contributions to Geochemistry and Analytical Chemistry*, Wiley-Halsted, New York, 1975.

Among journal articles may be mentioned: H. Irving, *Quart. Rev. (London)*, **5**, 200 (1951); F. S. Martin and R. J. W. Holt, *ibid.*, **13**, 327 (1959); Y. Marcus, *Chem. Rev.*, **63**, 139 (1963).

Comprehensive reviews of extraction: G. H. Morrison and H. Freiser, *Anal. Chem.*, **30**, 632 (1958); **32**, 37R (1960); **34**, 64R (1962); **36**, 93R (1964); H. Freiser, *ibid.*, **38**, 131R (1966); **40**, 522R (1968).

J. A. Marinsky and Y. Marcus, Eds., *Ion Exchange and Solvent Extraction*, Dekker, New York. (Vol. 6, 1974, and earlier volumes.)

Y. Marcus, Ed., *Solvent Extraction Reviews*, Vol. 1, Dekker, New York, 1971.

3. Proceedings of Conferences

H. A. C. McKay, T. V. Healy, I. L. Jenkins, and A. Naylor, Eds., *Solvent Extraction Chemistry of Metals*, Macmillan, London, 1966. (Mainly concerned with nuclear fuel reprocessing.)

D. Dyrssen, J.-O. Liljenzin, and J. Rydberg, Eds., *Solvent Extraction Chemistry*, North-Holland, Amsterdam, 1967.

A. S. Kertes and Y. Marcus, Eds., *Solvent Extraction Research*, Wiley-Interscience, New York, 1969.

Proceedings of the International Solvent Extraction Conference, The Hague, 1971, 2 vols., Society for Chemical Industry, London, 1971.

G. V. Jeffreys, Ed., *Proceedings of the International Conference on Solvent Extraction, Lyon, 1974*, Vols. 1–4, Society for Chemical Industry, London, 1974.

ORGANIC SOLVENT HAZARDS

Among the common organic solvents, carbon tetrachloride and benzene are especially toxic.[2] As little as 50–100 ppm CCl_4 (100 ppm ≡ 0.32 mg/liter) produces toxic effects in a relatively short period of time. The maximum allowable concentration in air has been set (1963) at 10 ppm of CCl_4. Benzene is equally harmful, chloroform rather less so. Carbon tetrachloride, $CHCl_3$, benzene, and halo ethers are described as, or suspected to be, carcinogenic. The maximum allowable concentration for benzene in air has recently been set at 1 ppm (USA, 1977). Other chlorinated hydrocarbons are suspect. It is evident that anyone working with these (and other) volatile solvents needs to exercise care. They should not be spilled nor allowed to evaporate into the laboratory air from open vessels. It is best to work in a hood.

Volatile solvents that are combustible, such as ethyl ether, can be hazardous if used incautiously.

Ethers, especially isopropyl ether, can form peroxides on long standing, which can result in explosions if redistillation is attempted without their removal. Explosions without heating have been reported.

I. GENERAL CONSIDERATIONS

A. TERMINOLOGY[3]

In extraction separations the analyst is primarily interested in the ratio of the analytical concentrations of an element M in the two phases at (presumed) equilibrium

$$\frac{\sum[M]_{org}}{\sum[M]_w} \tag{9A-1}$$

where *org* and *w* refer to the organic and aqueous (water) phases.[4] This ratio allows the recovery factor R for an element to be calculated in an extraction or series of extractions. Especially in the aqueous phase of an extraction system a number of species of a metal may exist, so that the total metal concentration is their sum. If polymerized or polynuclear metal species are present, for example of the types $(ML)_2$ and M_2L_3', the molar concentrations of each must be multiplied by the proper coefficient to furnish the analytical concentration $\sum[M]$.

We call this ratio (9A-1) the *extraction coefficient* and symbolize it as E. It is often designated D and is then called the distribution coefficient (or ratio), a term not free from ambiguity. E is usually a function of such system variables as reagent concentration, acidity, and sometimes others, and may be calculable. In most systems, E is largely independent of the amount or concentration of the constituent (M) present. The usual ease of varying E accounts in large measure for the power of liquid–liquid extraction as a separation method.

In any liquid–liquid extraction, the species extracted into the organic solvent phase is also to be found in the aqueous phase at equilibrium, albeit often in very low concentration and possibly not contributing significantly to the total metal concentration in that phase. For example, in an extraction of a metal chelate ML, its partitioning is given by a constant

$$\frac{[\text{ML}]_{org}}{[\text{ML}]_w} = P \qquad (9\text{A-2})$$

(For a given solvent system, at a specified ionic strength in the aqueous phase; temp. = const.) Similarly, in other extraction systems there is a P for some other distributing molecule. We designate this constant as the *partition ratio* or *partition constant*, avoiding the use of the term distribution ratio or coefficient because of its ambiguity. The corresponding thermodynamic partition constant (sometimes designated K_P^0) is, for example,

$$P^0 = K_P^0 = \frac{f_{org}[\text{ML}]_{org}}{f[\text{ML}]}$$

where f_{org} and f are the activity coefficients of the partitioning substance in the organic solvent and water phases (see p. 826). (Here and later, concentrations without subscripts refer to the aqueous phase.) In the very early days of immiscible solvent extraction, the substances studied were mostly those not involved in complex equilibria, the systems consisted essentially of a single substance distributing between the two phases, and P was sufficient to describe the extraction. This constant was called *coefficient de partage* by Berthelot and Jungfleisch, and there is thus historical precedent for the use of the term partition constant and the symbol P.

As we see later, for example in the discussion of metal chelate extraction, P and the constants of the various equilibria (especially in the aqueous phase) determine E in a particular extraction system. P is the pivot around which the extraction revolves, so to speak.

The extent of extraction (extractability) does not need a new symbol. It is designated by R, the recovery factor, which can be expressed as a percentage if desired. If it is necessary to indicate that R refers to extraction, that can be done by using R_{ext}.

Extractant refers to the organic solvent, usually containing a dissolved reagent (*extraction reagent*) that makes extraction possible. At times the extracting organic solvent is also the reagent. The physical properties (viscosity and density, for example) of an organic solvent can sometimes be improved by adding another organic solvent, called *diluent.* The diluent may also improve extractability or the selectivity of an extraction. The *extract* is (usually) the organic phase after an extraction, but an aqueous extract of an organic phase is also possible.

Successive extraction refers to the use of consecutive portions of an extractant, which are combined, to improve recovery. *Continuous extraction* involves passage of drops of organic solvent through an aqueous solution, by an automatic recirculating arrangement, to enable a (difficultly) extractable substance to be extracted into a small total volume.

Washing of an (organic) extract involves shaking the extract with water, or more commonly with a suitable aqueous solution, to remove mechanically included foreign material in the form of aqueous droplets without removing a significant amount of the extracted substance. (Similarly, a separated aqueous phase can be washed with organic solvent.) This operation may be carried out with the intent of removing coextracted "foreign" substances and the term *scrubbing* is then used.

Back-extraction (*retrograde extraction*) refers to the transfer of a substance, in whole or in part, from an organic extract to an aqueous phase by shaking the two together. The substance may be the one being determined, or it may be a foreign, unwanted substance. When the back-extraction involves the desired substance and the intent is to remove it entirely from the organic phase, the operation is known as *stripping.* Often back-extraction entails fractionation and is carried out to improve the separation effected in direct extraction (Chapter 9D). *Backwashing* is often used in this sense.

Coextraction is used in two different senses. Usually it implies the simultaneous extraction, to a greater or less extent, of substances other than the one desired. However, it is sometimes used in a narrower sense to indicate the extraction of a foreign substance B brought about by the presence of the desired extractable substance A, that is, B would not be extracted in the absence of A. In the latter sense, coextraction is analogous to coprecipitation in precipitation processes. In this sense, coextraction results from chemical interaction of A and B to give an extractable substance. Prediction of separation possibilities from the behavior of individual constituents is not always safe.

Synergism in its general sense is defined as the joint action of two agents to produce an effect greater than the sum of the independent effects. As used at present in liquid–liquid extraction, the term refers chiefly to the enhanced extraction of an uncharged metal chelate produced by an uncharged ligand (such as an organophosphorus ester or a heterocyclic base), which forms an adduct having an extraction coefficient greater than that of the chelate.[5] In a broader sense, synergism in the extraction field embraces any increase in the extractability of an uncharged species resulting from combination with one or more other species. Thus a chelate ML_n may be convertible into a more easily extractable ion associate $(B^+)(ML_n A^-)$ by combination with A^- and B^+.

The increase in extractability or rate of extraction produced by the introduction of a neutral ligand into the coordination sphere of a metal ion must in large measure be due to the displacement of water molecules.

Separation factor is discussed later (p. 1031).

B. CLASSIFICATION OF METHODS

It is difficult, if not impossible, to assign all extraction systems to sharply defined categories. Nevertheless, some kind of classification scheme is necessary, even for a cursory survey of extraction methods. The following classification is based primarily on the type of the extracted compound. Four types are distinguished, but the distinction between some types is sometimes rather arbitrary. Systems in which solvated extracted species are formed offer difficulties in classification. A given reagent may function in different ways depending on the conditions of extraction or on the metal. Some reagents can, at the same time, be chelating and solvating in their action.

1. *Systems having simple, nonpolar species* (*molecules, rarely atoms*) *in both the organic solvent and aqueous phases.*

Examples are I_2 and other halogens, OsO_4, covalent halides such as $AsCl_3$ and SnI_4, and Hg metal. That these substances are all more or less volatile is not a coincidence, but an indication of their nonpolar character. In such systems, the organic solvent is often largely inert: It dissolves the substance but usually does not react with it chemically to a significant extent. A covalent molecule may be accommodated more easily within the disordered structure of an inert organic solvent, such as CCl_4, than within the ordered structure of water (hydrogen bonding of molecules), for which it has no particular affinity. Accordingly, it partitions preferentially into the organic solvent. Polar, oxygenated solvents are less desirable than inert solvents for extracting these substances, because the former are likely to extract other substances as well. An ether would be a bad choice for extracting $GeCl_4$ or $AsCl_3$ from a hydrochloric acid solution, because it would also extract certain metal chloro acids that might be present.

2. *Systems in which uncharged chelate complexes of metals with organic reagents are extracted into the organic solvent.*

These comprise a large and important group (Chapter 9B). Dithizonates, 8-hydroxyquinolates and most other uncharged chelates are extractable into a variety of organic solvents. In a chelate complex, the metal cation is more or less surrounded by hydrophobic organic radicals, and the uncharged molecule is generally very slightly soluble in water, but much more soluble in an organic solvent. The solubility may be greater in an inert solvent than in an oxygen-containing organic solvent, and the former are usually preferred (p. 906).

The metal cation in some chelates is not coordinatively saturated by the anionic reagent, with the result that water molecules are attached to the metal. The presence of water strikingly diminishes the solubility in most organic solvents, especially of the inert type. A divalent metal of coordination number 6 forming the chelate ML_2 with the bidentate reagent HL will have two coordination positions unoccupied by the reagent, which, in the absence of other substances to fill these sites, are occupied by two water molecules:

$$H_2O$$
$$L \mathbin{\cdot\!\!-} \overset{\text{II}}{M} \mathbin{-\!\!\cdot} L \qquad \text{(schematic representation)}$$
$$H_2O$$

M(II) could be Ca^{2+}, Sr^{2+}, Ba^{2+}, Mg^{2+}, Mn^{2+}, Zn^{2+}, Co^{2+}, Fe^{2+}, UO_2^{2+}, and some other cations. In general, coordination unsaturation results when the coordination number of the metal (central atom) is greater than twice its valency. (Multidentate ligands may then give better extraction.) Among triply charged ions, the rare earths can give coordinatively unsaturated chelates with bidentate reagents. On the other hand, Fe^{3+}, Al^{3+}, Ga^{3+}, In^{3+}, and Cr^{3+} among the trivalent ions, and Zr^{4+}, Th^{4+}, and U^{4+} among the quadrivalent, all having a coordination number of 6 or 8, respectively, form coordinatively saturated chelates (with bidentate reagents):

$$L \mathbin{\cdot\!\!-} \overset{\text{III}}{M} \mathbin{-\!\!\cdot} L$$
$$L$$

Copper(II), usually having a coordination number of 4, often forms chelates that are coordination-saturated.

The hydrated chelates are virtually nonextractable by such solvents as CCl_4 and $CHCl_3$. However, they can be extracted by suitable oxygen-containing solvents (higher alcohols), as we shall see (p. 909). Moreover, it may be possible to replace H_2O in the molecule by uncharged molecules of reagent or by other substances conferring solubility in organic solvents in general, so that inert solvents may then be suitable for extraction.

Except for the nature of the extracted species, metal salts of fatty acids, or more generally of organic acids, fall in this group of extractions. The chelate extraction expression holds for these salts, at least at low concentrations.

Extraction with Acidic Organic Phosphorus Compounds. The chelate extractions briefly discussed in the foregoing paragraphs are, in the main, simple. Extractions with organophosphorus acids are often more complex. These acids, in many cases, give definite compounds in which an acidic H is replaced by an equivalent of metal. These compounds are chelates, but not of the simple type formed by the more conventional chelating agents. Acidic organophosphorus compounds will be considered at greater length later (Chapter 9C), but here we may note, as an illustration of the products formed, the reaction occurring under some conditions in the extraction of a metal ion with a solution of a dialkylphosphoric acid in a solvent such as chloroform, in which the reagent is dimeric:

$$M^{n+} + n(HX)_2 \rightleftharpoons M(X \cdot HX)_n + nH^+$$
$$\quad\quad\quad org \quad\quad\quad\quad org$$

HX is $(RO)_2PO(OH)$. One H of $(HX)_2$ is replaced by an equivalent of metal. The resulting chelate can be represented as

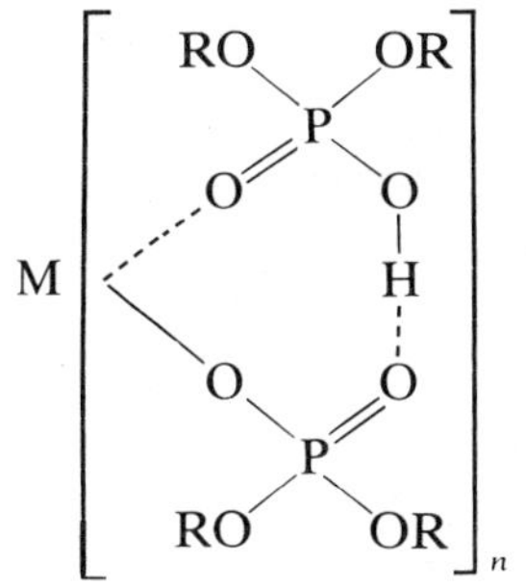

3. *Systems in which ion-association compounds are formed with organic reagents.*

The compound formed can be an organic cation combined with an inorganic anion, a metal cation combined with an organic anion, or both cation and anion can be organic. Some examples are

$[(C_6H_5)_4As^+](ReO_4^-)$

$[Co^{III}(HDx)_2 \cdot 2\,Py^+]Cl^-$ H_2Dx = dimethylglyoxime, Py = pyridine

$Cs^+[B(C_6H_5)_4^-]$

$[(C_6H_5)_4N^+](UO_2Ox_3^-)$ Ox^- = 8-hydroxyquinolate anion

An ion-pair compound having a large cation and a large anion tends to be only slightly soluble in water. Large ions are but little solvated as a rule and do not fit well into the water structure. The mutual charge neutralization brought about by bringing together a bulky cation and a bulky anion results in the formation of an ion pair of limited water solubility. If, in addition, one of the ions is organophilic, the ion pair will be soluble in organic solvents of different types. The combination of solubility in immiscible organic solvents and low water solubility means that the partition constant will have a large or fairly large value.

In general, ion associates in which the cation–anion ratio is 1:1 are more easily extracted than those in which the ratio is 1:2 or 2:1. This behavior is commonly explained by the greater hydration of ions of higher charge and resulting tendency to remain in the aqueous phase. Steric factors also play a role. The most used ion-association reagents are cationic, $(C_6H_5)_4As^+$, $(R_4N)^+$ and dye cations of the type RH^+, for example, which combine with singly charged oxo- or complex metal anions such as $AuCl_4^-$ or $SbCl_6^-$ to give ion pairs effectively extracted by $CHCl_3$ and other solvents.

Positively charged metal chelates, functioning as the cation of ion-pair compounds, are encountered occasionally. They can be formed by organic chelating reagents that coordinate to a metal without replacement of hydrogen. An example is the cation

$$Fe^{II}(1,10\text{-phenanthroline})_3^{2+}$$

in which molecular phenanthroline occupies the six coordination positions of Fe^{2+}. In the presence of perchlorate ion, an ion associate is formed, which is sparingly extracted by chloroform. Better extraction is obtained with long-chain alkyl sulfates and sulfonates, thiocyanate, salicylate, and the like as the anion.

Amines. The cations of high-molecular-weight amine salts (R_3NH^+ for example) and of quaternary ammonium salts (R_4N^+) extract complex metal anions (halides, nitrates and others) as well as metal oxoanions into various organic solvents. These extraction systems are not always simple. The electrolyte concentration in the aqueous phase may be high (salting out), and polymerization of reagent can occur in both phases so that extractability may not be related to system parameters in a simple way. See further Chapter 9C where a general account of amine extractants is presented.

4. *Systems in which coordination (solvation) complexes of inorganic species with the extraction reagent molecules are formed.*

This class is not as well defined as the previous ones. It comprises systems of rather diverse nature, and some are not easily distinguishable from ion-association systems. In fact, many of these systems are, or can be looked upon, as ion-associate systems. The molecules of the reagent (which sometimes are the extracting solvent itself) can coordinate to the hydrogen ion of a complex metallo acid (e.g., $HFeCl_4$) or a mineral acid, or to the cation of a salt or an ion-association compound. The strength of the bonding can vary over a wide range. In some systems, definite compounds are formed, which can be isolated. In others, there is an indefinite solvation of the extracted species. An important factor is always the displacement, not necessarily complete, of water molecules around the species to be extracted.

Extraction reagents falling in this class include oxygen-containing organic solvents (ethers, ketones, alcohols, and esters) and neutral organophosphorus compounds, the latter usually used in an "inert"[6] diluent. Generally the coordinating ability of C—O solvents increases in the order esters $<$ ethers $<$ ketones $<$ alcohols (it depends on the steric availability of O as well as on the basicity). Solvents having P–O bond more strongly than C–O solvents to metal salts and complex metallo acids.

One of the earliest uses of immiscible organic solvent extraction in inorganic analysis was in the separation of iron(III) as chloride by extraction into ethyl ether from $1:1$ HCl. The extraction of iron and some other trivalent metal chlorides by ether involves the formation of a coordination complex between the complex metal acid and the ether. The hydrogen of $HMCl_4$ (or more generally HMX_4) bonds to the basic oxygen atom of the ether

$$\begin{array}{c} | \\ -C \\ \diagdown \\ O\text{-}\text{-}\text{-}H^+(M^{III}X_4)^- \\ \diagup \\ -C \\ | \end{array}$$

The hydrogen ion has associated water molecules (not shown) accompanying iron into the extract. This is an ion-pair compound but not of the essentially unsolvated type 3 (p. 821). Actually, more than one ether molecule (apparently 2 or 3) solvates H^+.

Other oxonium solvents form similar coordination complexes. Esters and ketones extract hydrated $HFeCl_4$ and analogous complexes very well.

Tributyl phosphate (TBP) and other neutral organophosphorus compounds extract metal chlorides also, but they are of greater interest for the extraction of metal nitrates. Extraction with these reagents is usually carried out with their solutions in a solvent such as benzene. The solvation of inorganic species need not take place through the agency of an acidic hydrogen; sometimes coordination of the reagent with the metal ion takes place. The strong solvating power of TBP and related compounds is explained by the basicity of the oxygen atom in the phosphoryl group

$$\equiv P{=}O \qquad (\text{more accurately} \equiv P \to O)$$

and its ready steric availability. From strong hydrochloric acid solution the species $H[FeCl_4(TBP)_2]$ is extracted, from dilute hydrochloric acid, $FeCl_3(TBP)_3$. Nitrates are usually extracted as the solvated neutral salt with TBP.

See Chapter 9C for a general account of organophosphorus reagents.

II. EXTRACTION OF INORGANIC SUBSTANCES: ELEMENTS, OXIDES, OXYGEN ACIDS, HALIDES, NITRATES, AND OTHERS

It will be convenient to discuss here in one place the extraction of these metal compounds, though different types of extraction are involved. In general, the anionic component of the inorganic compound accompanies the metal into the organic solvent. Solvation or compound formation with the organic extracting solvent almost always occurs—if not always to some degree. Excepting simple covalent compounds, the extracted species can mostly be assigned to our class 4. With an acidic organo-P compound dissolved in an "inert" organic solvent, even chelate formation may occur. But extraction of chelates is reserved mainly for discussion in Chapter 9B, and of ion associates and complexes of organo-P reagents for Chapter 9C.

A. ELEMENTS

The only metal perceptibly extractable as such into organic solvents is mercury. That mercury should dissolve in organic liquids is not unexpected. It has an appreciable vapor pressure at room temperature, and all gases dissolve in liquids. A rough value for the solubility of mercury in an organic solvent can even be estimated on the basis of comparison with the solubility of the monatomic inert gases. The extraction of metallic mercury has not hitherto found any important application in trace separations, in contrast to the separation of mercury by volatilization (Chapter 10).

The solubility of mercury in water is 0.06 mg/liter at 25°. In cyclohexane the solubility is 30 times as great so that the partition constant P of the metal is ~30. This will also be the extraction coefficient E in the presence of a strong reducing agent to keep mercury in the elemental form. The solubility in benzene at room temperature has been reported as 1.5–2 mg Hg/liter.

Dry or Fire Assaying.   This is a high-temperature liquid–liquid extraction process which is of great importance in the isolation of gold, silver, and the platinum metals. In a conventional assay, a mixture of sample, lead oxide, flux, and reducing agent is heated to a high temperature. The molten lead that is formed dissolves the noble metals and sinks to the bottom of the crucible. Lead is removed from the button by cupellation (oxidation to the oxide, which is absorbed by the bone-ash *cupel*), leaving a globule (bead) of noble metal. The bead is treated further to separate silver and gold. Classically, the determination is finished by weighing the metals, but determination by other methods is also possible and may be advantageous.

Other high temperature liquid–liquid systems find limited analytical use.

Molten 8-hydroxyquinoline (m.p. 76°) has been used to extract hydroxyquinolates from water, and naphthalene (m.p. 80°) and other substances melting below 100° to extract dithizonates.

B. OXIDES

Extraction of OsO_4 into inert solvents, for example, CCl_4, provides an excellent, perhaps generally the best, method for the separation of traces of osmium from almost all elements. Similarly, extraction of RuO_4 is valuable for separating ruthenium. Details are given in the respective chapters on these metals in *Colorimetric Determination of Traces of Metals*, but some points of general interest may be touched upon here. The extractions can be carried out from HNO_3 or H_2SO_4 (and no doubt $HClO_4$) solutions containing the elements in the 8+ state. The extractions are highly selective when an inert solvent is used. The oxides have hardly any basic character and the extraction coefficients are much the same in acidic as in neutral solutions.[7] OsO_4, as perosmic acid, is a very weak acid ($K_a \sim 1 \times 10^{-12}$), so that it is extracted less easily from basic than from acidic or neutral solutions, but this is a point of no importance, since basic solutions would never be used in practice.

The extraction coefficient of OsO_4 is appreciably larger for $CHCl_3$ than for CCl_4 (19.1 versus 13.0, 25°, $\mu = 0$). An increase in the coefficient is obtainable by salting out. As un-ionized substances, OsO_4 and RuO_4 are salted out by indifferent electrolytes. The quantitative effect of electrolytes on the solubility of a nonelectrolyte is expressed by the Setchenow[8] equation:

$$\log f = \log \frac{\text{solubility of nonelectrolyte in water}}{\text{solubility of nonelectrolyte in electrolyte solution}}$$

$$= k\mu$$

where f is the activity coefficient of nonelectrolyte in electrolyte solution, k is a constant whose value depends on the nonelectrolyte and the electrolyte (Setchenow salting coefficient), and μ is the ionic strength of electrolyte in aqueous solution. If the electrolyte does not affect the solubility of the nonelectrolyte in the organic solvent, that is, if it does not affect the mutual solubilities of the aqueous or organic phases or dissolve in the organic solvent, P of the nonelectrolyte will increase in proportion to the decrease of its solubility in the aqueous phase, so that

$$P_{elect} = fP_w$$

where P_{elect} is the partition constant for an electrolyte solution of specified concentration and P_w is the partition constant for water. In Fig. 9A-1a, $\log f$ of RuO_4 for various salts (H_2O–CCl_4) is plotted against the salt concentration. The Setchenow relation holds up to rather high electrolyte concentrations. Salting out can be ascribed to the lowering of water

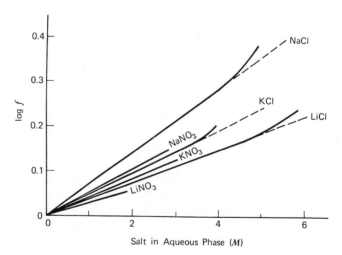

Fig. 9A-1. **a.** Variation of $\log f$ ($= \log P_{elect}/P_w$) of RuO_4 in salt solutions according to Martin and Holt. Reproduced with permission from The Chemical Society, from F. S. Martin and R. J. W. Holt, *Quart. Rev.*, **13**, 331 (1959).

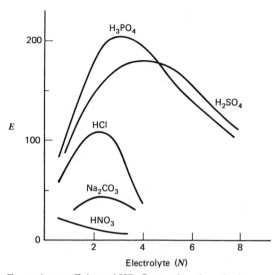

Fig. 9A-1. b. Extraction coefficient of $HReO_4$ as a function of mineral acid concentration ($NaReO_4$ in presence of Na_2CO_3). With NaOH as the electrolyte, E is 12–13 over the range 2–9 N. Jordanov, Pavlova, and Bojkova, *loc. cit.*

activity by the binding of water molecules into the hydration spheres of the added electrolyte. A modest increase in P of RuO_4 can be obtained by adding an electrolyte, but P is already quite large in pure water (58.4 at 25°), so that usually this is not worthwhile.[9]

Osmium tetroxide is easily returned to aqueous solution by shaking the organic solvent extract with an aqueous reducing solution converting Os(VIII) to Os(IV). For example, thiourea serves both as reducing agent and photometric reagent. Ruthenium tetroxide reacts with SCN^- to give blue $Ru(SCN)^{2+}$ ($\mathscr{S}_{590} = 0.0025$ μg Ru/cm²) when CCl_4–RuO_4 is shaken with aqueous thiocyanate solution.

Other (soluble) oxides, except possibly Ir in a high oxidation state, are not extractable by inert organic solvents because they are acidic or basic and combine with water to give products usually insoluble in these organic solvents.

C. OXYGEN ACIDS

The extraction of simple oxygen acids as such is of little importance in metal separations except for Cr(VI) and Re. (Extraction of certain oxoanions—MnO_4^-, ReO_4^-, and some others—can be brought about by forming ion-pair compounds with bulky, little-solvated, organic cations such as

Ph_4As^+ and extracting into chloroform and other essentially nonsolvating solvents; these extractions are not considered in this section.)

Re(VII) is extractable with favorable coefficients into cyclohexanone and other ketones from mineral acid solutions as solvated and hydrated $HReO_4$.[10] The common acids differ in their salting-out effect, the more hydrated H_2SO_4 and H_3PO_4 being more effective than the less hydrated HNO_3 and HCl (Fig. 9A-1b). Presumably the decrease in extraction of Re at higher acid concentrations is due to competition with the corresponding anions. In the presence of NaOH and Na_2CO_3, Re is extracted as solvated, but not hydrated, $NaReO_4$. Molybdenum is extracted very little from sulfuric and phosphoric acid solutions, so that cyclohexanone extraction has been suggested for the separation of Re from Mo.[10] This separation is not as sharp as that with tetraphenylarsonium.

Chromic acid is extractable into MIBK and other ketones from dilute HCl and H_2SO_4.[11] The optimal concentration for HCl is 2 M. From HCl solutions, the extracted species is said to be $HCrO_3Cl$.

The extraction of heteropoly acids (as of Mo) is of greater interest. These acids are extractable into oxygenated solvents (especially alcohols and esters)[12] and such extractions are of importance in the photometric determination of P, As, and Si, for example, but less often for the metals (e.g., V) involved. Hardly any use of these extractions primarily for metal separations has been made.[13] The extraction of heteropoly acids into oxygenated solvents is a consequence of their solvation by these (hydrogen bonding). The solvation varies from one acid to another and with the extracting solvent so that separations become possible. A nonsolvating organic solvent would be useless.

Extraction of Salts from Strongly Alkaline Solution. Re(VII), as well as Tc(VII), can be extracted from strong NaOH or KOH solution into pyridine,[14] quinoline, and ketones such as methyl ethyl ketone. The extracted species must be the alkali metal perrhenate. The extraction coefficient of Re increases with the sodium hydroxide concentration, thus $E = 24.2$ with pyridine as extractant when the original NaOH concentration is 1.2 M, 391 at 5.5 M NaOH and 2.26×10^3 at 9.0 M NaOH.[15] Molybdenum(VI) is extracted very slightly ($E = 0.002$ at 1.2 M NaOH, and ~0.001 at 8.4 M NaOH), and W(VI) to much the same extent as Mo. In the presence of carbonate, U(VI) remains in the aqueous phase. Manganese(VII) is extracted. Technetium is extracted better than Re. It is of interest that Tc(VII) can be extracted with a 0.1 M solution of pyridine in cyclohexane. Nitrate decreases the extraction coefficient.

Rhenium(VII) is extractable into methyl ethyl ketone from 5 M NaOH containing $(NH_3OH)Cl$.[16] This is a separation primarily from Tc, but also from Fe, Ga, Mo, and other metals.

D. HALIDES[17]

This is a large and varied class of extractable substances. Elements are extractable from halide solutions as covalent halides, as solvated halides and as complex halide anions forming halo acids or ion associates with suitable cationic species. These classes cannot always be sharply distinguished from each other. For a given element the extracted species can vary with the solvent and the halide ion concentration. Evidently the two important factors in metal halide extraction are the degree of formation of the halide and its partitionability.[18]

1. Chlorides

a. *Inert Solvents*

The covalent chlorides of As(III) and Ge(IV), especially the latter, are extractable from strong HCl solutions into such solvents as benzene, chloroform, and carbon tetrachloride (particulars are given in the respective element chapters of *Colorimetric Determination of Traces of Metals*). For the latter solvent, E_{As} ~2.7 with 9.5 M HCl and E_{Ge} ~500 with 9 M HCl.[19] These extractions are highly selective. Other metals, if complexed at all by chloride, form anionic species that are not significantly extracted into inert solvents. Some thousandths or hundredths of a percent of such elements as Te, Se, Sb(III),[20] Sn, and Hg are extracted by CCl_4 from 9 M HCl, but are removed when the extract is washed with HCl of the same concentration. Such washing is needed in any event to remove entrained droplets of aqueous phase. Tc is reported to be extractable as TcO_3Cl into $CHCl_3$, CCl_4, and hexane.[21] Arsenic(V), not forming a chloride, is not extractable, so that Ge and As can be separated after oxidizing As(III).

A mixture of benzene and isoamyl alcohol extracts $AsCl_3$ to a much larger extent than either alone (Fig. 9A-2).[22] The optimum volume ratio is about 1 of the alcohol to 9 of benzene. The synergic effect has been explained by the breaking down by benzene of the hydrogen-bonded structure of isoamyl alcohol, thus supposedly increasing the concentration of "free solvent." The effect is interesting, but the use of such a mixture has little practical importance. The addition of isoamyl alcohol to benzene destroys the highly selective arsenic extraction by the latter, because the alcohol extracts complex chloro acids.

Chloro anions of metals can be extracted into *inert* solvents if they are combined with bulky organic cations such as $(C_6H_5)_4As^+$ or $(C_2H_5)_4N^+$ to form unsolvated, unhydrated ion pairs. See Chapters 6H and 9C.

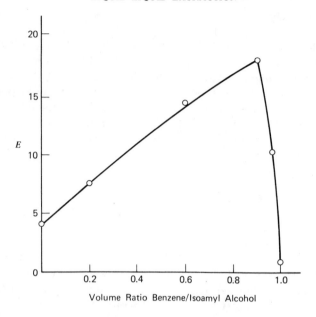

Fig. 9A-2. Extraction coefficient of As(III) ($[As]_{org}/[As]_{HCl}$) for 6 M HCl as a function of ratio benzene/isoamyl alcohol in the organic solvent phase. After Kuz'min, Popova, Kuzovlev, and Solomatin (reference in note 22).

b. Oxygen-type Solvents

These solvents extract metal chloro acids ($HMCl_m$). Ethyl ether was the first organic solvent used to extract iron(III) and some other metals from ~6 M HCl. It has a number of disadvantages: low boiling point (but this helps when extracts are to be evaporated for recovery of metals), flammability, marked solubility in HCl, and somewhat mediocre extracting power for metals. Although it is still used, other oxygen-type solvents have advantages and are often substituted. Isopropyl ether is superior to ethyl ether in lower volatility and lesser solubility and it gives somewhat larger extraction coefficients, but it is subject to marked peroxide formation on standing. Amyl acetate is a better extractant for iron at high HCl concentrations. Ketones, as exemplified by methyl isobutyl ketone (MIBK), are better metal chloro-acid extractants than ethers. Tributyl phosphate is an especially powerful extractant and is usually dissolved in an inert diluent. The more powerful extractants are usually less selective than the less powerful, and are not always the best choice.

i. ETHERS

Ethyl Ether. Table 9A-1 summarizes the extractability of metals from HCl solutions by diethyl ether.

The ether extraction of iron (and other elements) has been much studied,[23] but a quantitative interpretation of the extractability as a function of HCl and Fe concentrations is not simple, and this system

TABLE 9A-1
Extraction of Various Metals as Chlorides with Ethyl Ether[a]

Metal	1% HCl (0.3 M)	5% HCl (1.4 M)	10% HCl (2.9 M)	15% HCl (4.4 M)	20% HCl (6.0 M)
	Percentage of Metal in Ether Layer				
Au(III)	84	98	98		95
Fe(III)	Trace	0.1	8	92	99
Tl(I)(0.3 mg/ 100 ml)					~10
Tl(III)		~98		~99	~98
Sb(III)	0.3	8	22	13	6
Sb(V)	Trace	2.5	6	22	81
As(III)	0.2	0.7	7	37	68
Te(IV)	Trace	0.2	3	12	34
Sn(IV)	0.8	10	23	28	17
In (3 mg/ 100 ml)					~5
Hg(II)	13		0.4		0.2
Cu(II)	Trace		0.05		0.05
Zn	Trace		0.03		0.2
Ni	Trace		Trace		Trace
Pt(IV)	Trace		Trace		Trace
Ir(IV)	Trace		0.02		5
Pd(II)	Trace		0.02		

[a] A hydrochloric acid solution containing 1 g of the metal as chloride in 100 ml was shaken with an equal volume of ethyl ether saturated with water. Data mostly from Mylius, *Z. Anorg. Chem.*, **70**, 203 (1911); Mylius and Hüttner, *Berichte*, **44**, 1315 (1911). Data obtained by various other workers show that the following elements are also extracted from HCl solutions (~1:1): As(V), slightly (a few percent from 6 M HCl, more from 8 M); Ga, 97%; Ge, ~50%; Mo(VI), 80–90%; Sn(II), 15–30%; Te(IV) slightly ($E = 0.08$ from 8 M HCl). The following metals are not extracted appreciably: Al, Be, Bi, Ca, Cd, Cr(III), Co, Fe(II), Pb, Mg, Mn, Ni, Os, rare earths, Re(<1%), Rh, Sc, Ag, Ti, Th, W, U, Zr. V(V) is extracted slightly, V(IV) hardly at all. Se(IV) seems to be extracted very slightly. Nb(V) (tracer concentrations) is extracted ~30% from 6 M HCl. Au(I) and Cu(I) are reported to be extracted from HCl solutions. Data on the extraction of metals into 2-chloroethyl ether are available: P. I. Artyukhin et al., *Dokl. Akad. Nauk. SSSR*, **169**, 98 (1966). Interestingly, platinum(IV) is almost completely extracted at HCl concentrations greater than 7 M by this solvent.

illustrates the difficulties that arise when aqueous and organic solvent phases are soluble in each other to a large extent and the electrolyte concentration is high and variable. The extracted species is solvated and hydrated $HFeCl_4$, formulated as $H[xH_2O, y(C_2H_5)_2O]FeCl_4$. The course of the extraction curve of Fe(III) as a function of the HCl concentration (Fig. 9A-3) is explained as follows. The extraction of iron increases with the HCl concentration up to 6.5 M in consequence of increasing formation of $FeCl_4^-$ and decreasing dissociation of $HFeCl_4$ in the aqueous phase. Complex iron species higher than $FeCl_4^-$ are not formed. Above ~6.5 M HCl, the extraction of iron decreases because of increasing solubility of ether in the aqueous phase. The two phases approach each other in composition and the extraction coefficient of iron must decrease.

From the analytical standpoint, an important feature of the iron(III) extraction (and also that of gallium and trivalent gold) is the increase in E at higher metal concentrations.[24] (Cf. Fig. 9A-4 for iron–isopropyl ether, which shows the same behavior as ethyl ether.) The reason for this increase has been much argued. At least in part it is due to increased polymerization of $HFeCl_4$ (formation of $\{HFeCl_4\}_2$, etc.) in the organic

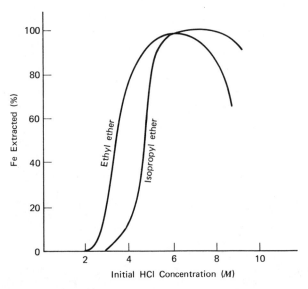

Fig. 9A-3. Extraction of Fe(III) from HCl solution by ethyl and isopropyl ether; 25 ml $FeCl_3$–HCl solution (250 mg Fe with ethyl ether, 125 or 250 mg Fe with isopropyl ether) shaken with 25 ml of dry ether. Two isopropyl ether phases form above 7.5 M HCl. After R. W. Dodson, G. J. Forney, and E. H. Swift, *J. Am. Chem Soc.*, **58**, 2573 (1936).

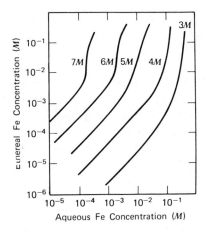

Fig. 9A-4. Distribution of Fe(III) between isopropyl ether and aqueous HCl solution as a function of the concentration of the latter and of the iron concentration. Note that below a certain iron concentration (which varies with HCl concentration), E becomes constant. From N. H. Nachtrieb and R. E. Fryxell, *J. Am. Chem. Soc.*, **70**, 3557 (1948).

phase, but not in the aqueous phase, with increasing concentration of iron. Ethyl and isopropyl ether favor such polymerization because of their low dielectric constants ($\epsilon \sim 4$).[25] Another, entirely different, reason for incomplete removal of small amounts of iron is reduction of Fe(III) to Fe(II), especially if the extraction takes some time, as it would if continuous extraction is applied; the reduction is promoted by light.

Examination of Table 9A-1 reveals that the metals forming singly charged MCl_4^- anions are the ones extracted to large extents. (Ge and As(III) are exceptions, but their extractability is explained by the formation of covalent chlorides that are extracted by active as well as inert solvents; Mo, another exception, is extracted, it is believed, as MoO_2Cl_2.) No divalent metals are extracted appreciably from 6 M HCl, since they do not form singly charged chloro complexes under the conditions. The extractable metals are those forming $HMCl_4$ ion pairs, similar in some respects to the ion pairs of Class 3 (p. 821). The bulky cation here is hydronium ion with solvating $(C_2H_5)_2O$ as well as H_2O.

Gold(III) is extracted well from dilute HCl because $AuCl_4^-$ is stable at very low Cl^- concentrations.

Isopropyl and Other Ethers. With diisopropyl ether, the maximum extraction coefficient of $\sim 10^3$ for Fe(III) is reached in 8 M HCl, but the extractability is good over the approximate range 7–8.5 M.[26] It may be expected that for other elements the optimum extraction acidity will be slightly higher than with ethyl ether, and that the extraction coefficients will be larger. It is known that the optimal acidity for the extraction of gallium is 7 M ($E \sim 200$). Ether (ethyl or isopropyl) extraction of Ga as $HGaCl_4$ is perhaps the best way of separating Ga because Fe, Sb, and Au are not extracted if reduced. Especially because of the simultaneous extraction of Fe(III), isolation of Sb[27] and Au by ether extraction is not of much interest; there are better methods. With the exception of Ga [and possibly Tl(III)], then, the chief use of ether extraction is for removal of large amounts of iron.[28] Cu, Co, Mn, Ni, Zn, Al, Cr, Ti, and S (as SO_4^{2-}) are some of the elements left in solution. See further Table 9A-2. V(V) is extracted to a considerable extent, V(IV) much less so. Mo(VI) and H_3PO_4 accompany iron to a large extent. Vanadium(V) can be reduced to V(IV) by SO_2 in 8–9 M HCl medium (ferric iron is not reduced), thus enabling V to be separated from much Fe; the small amounts of V and Cr extracted by isopropyl ether can be washed out with 8 M HCl.[29]

Isopropyl ether extraction of Tl(III) from 6 M HCl has been used to separate Tl from much In.[30] The small amount of In extracted can be removed from the ether phase by a double wash with 3 M HCl, which also removes small amounts of Fe(III) and Ga. For 6 M HCl, log E_{Tl} = 1.2, log E_{In} = −2.5. Gold and Sb can be removed before ether extraction by deposition on Cu metal.

TABLE 9A-2
Extraction of Elements with Diisopropyl Ether from 8 M HCl[a]

Element	Extraction (%)	Element	Extraction (%)
As(III)	67[b]	Sb(III)	~2
Au(III)	99[b]	Sb(V)	≧99
Fe(III)	99.9	Tl(III)	99
Ga	99.9	V(V)	22
Mo(VI)	21[c]		

Roughly equal phase volumes. Not appreciably extracted from 6 M HCl (<1%, often much less): Ag, Al, Be, Bi, Ca, Cd, Co, Cr(III), Cu, Fe(II), In, Mg, Mn, Ni, Pb, Pd(II), Pt(IV), Ru(III,IV), Sn(IV), Te(IV), Ti, U, Zn, Zr, rare earths. P is extracted slightly when alone, but ~50% in presence of Fe(III). $S(SO_4^{2-})$ is very slightly extracted, usually <1%. V(IV) is extracted 0.4% in presence of Fe(III); V(V) ~40% in presence of Fe(III).

[a] R. W. Dodson, G. J. Forney, and E. H. Swift, *J. Am. Chem. Soc.*, **46**, 668 (1926); F. C. Edwards and A. F. Voigt, *Anal. Chem.*, **21**, 1204 (1949); N. H. Nachtrieb and R. E. Fryxell, *J. Am. Chem. Soc.*, **71**, 4035 (1949); F. A. Pohl and W. Bonsels, *Z. Anal. Chem.*, **161**, 108 (1958).
[b] 6 M HCl.
[c] More in presence of Fe(III) (45%).

Ethers of higher molecular weight, such as β,β'-dichlorodiethyl, di-n-butyl, and di-n-amyl, extract iron maximally above 8 M HCl.[31] It is not evident that they have any advantages over diisopropyl ether. Fe(III), Sb(V), Ga, and Tl(III) are readily extracted by β,β'-dichlorodiethyl ether from concentrated HCl (equal volumes of phases), but these metals are accompanied by 20–55% of tracer amounts of Sn(IV), In, Se(IV), Te(IV), Mo, W, Pt, Pd, Os(VI), and Ir(IV), as well as smaller amounts of other heavy metals.[32] The following extraction coefficients have been reported for 11.4 M HCl[33]:

	Fe(III)	Zn	In	Te(IV)	Sn(IV)	Co	Cu(II)	As
E	169	0.0064	0.28	0.6	0.63	0.06	0.006	0.0085
E with 50 mg Fe(III)/ml in aqueous phase		0.0022	0.027	0.086	0.25	0.03		

ii. KETONES

Chloroferric acid is extracted much better by ketones than by ethers, often by a factor of 10 or 100 (Fig. 9A-5).[34] An E in the vicinity of 10^4 is

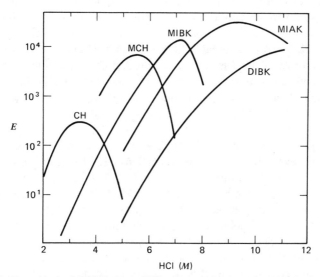

Fig. 9A-5. Extraction of Fe(III) from HCl solutions by ketones. Equal phase volumes, 200 mg Fe in 25 ml aqueous solution. From E. Bankmann and H. Specker, *Z. Anal. Chem.*, **162**, 18 (1958). CH, cyclohexanone; MCH, methylcyclohexanone; MIBK methyl isobutyl ketone; MIAK, methyl isoamyl ketone; DIBK, diisobutyl ketone.

obtainable in the range 6–10 M HCl for many of the ketones. As with the ethers, those ketones dissolving in HCl solutions to a considerable extent show a maximum in the E–[HCl] curves.

Methyl isobutyl ketone (MIBK, hexone) is the ketone most often used in extractions (vapors are toxic). It extracts Fe(III) maximally from 7 M HCl, $E \sim 10^4$. It is not very volatile (b.p. 116°), but it has a tendency to form emulsions. A 1:1 mixture with amyl acetate is free from the emulsifying tendency[35] and extracts iron as well as MIBK alone. Other elements extracted to a large or considerable extent by MIBK include As, Au(III), Ga, Ge, In, Mo, Sb, Sn, Te, Tl(III) (Tables 9A-3 and 9A-4). The same elements are extracted by ethers, but less strongly. It is of interest that Nb is extracted from 8 M HCl by MIBK though E is not large ($\sim$3), whereas Ta is hardly extracted ($E < 10^{-4}$). Molybdenum(V) is extracted from HCl solutions by O-containing solvents, especially ketones. The extracted species is hydrated and solvated $MoOCl_3$.[36] Evidence has been presented that both H_2TeCl_6 and $H_2Te(OH)_2Cl_4$ are the solvated species in Te extractions. If the aqueous solution is made 7 M in LiCl instead of HCl, the extraction of iron(III) remains much the same, but that of some other elements is reduced (e.g., As(V), Cd, Mo, and V). With equal volumes of phases, Ga can be extracted $\sim$99.9% from a mixture 4 M in HCl and 2 M in HNO_3 or $HClO_4$ (H_2SO_4 causes emulsification).[37]

A mixture of methyl ethyl ketone and CCl_4 in the volume ratio 3:2 has been suggested for the extraction of Fe(III). It has the advantage of being denser than water; moreover the phases separate easily.[38] The extraction of Fe(III) is 97% from 4 M HCl, 98.1% from 5 M and 99.9% from 6 M. The extractability of iron is the same whether 3 or 100 mg is present.

In using ketones as extracting solvents, it should be borne in mind that they may give decomposition and condensation products and even chloro derivatives in strong HCl. Ketones are reported not to be subject to peroxide formation.

Observations on the rate of extraction by oxygen-type solvents are scarce. In the extraction of Se(IV) from 7 M HCl by addition of methyl ethyl ketone in small amount followed by $CHCl_3$ extraction, it was noted that the reaction between Se (10^{-4} M) and MEK (0.35 M) in the aqueous phase was slow.[39] A period of 15 min was allowed for this reaction, before extraction with chloroform, which appears to proceed rapidly. Once formed, the Se complex is not easily decomposed. Selenium cannot be returned to the aqueous phase by shaking the $CHCl_3$ extract with water or dilute HCl solution. It is necessary to evaporate the extract with nitric acid to decompose MEK.

The extraction of Au, Hg, and other metals from HCl (and HNO_3 and H_2SO_4) solutions with mesityl oxide has been investigated.[40]

**Extraction of Elements with MIBK and MIBK+ Amyl Acetate from HCl and LiCl
Solutions**

Element	Extraction (%)		
	7 M HCl–MIBK[a]	7 M HCl–1:1 MIBK+amyl acetate[b]	7 M LiCl–MIBK[c]
As(III)	~88	80	33.4
As(V)	~3.5	25	0
Au(III)		99	
Cd	~12	2.9	0
Cr(VI)	~98[d]		
Cu(II)	4	0.3	1.0
Fe(III)	99.996	>99.9	99.944
Ga	>99.9	90	>99.9
Ge		97	
Hg(II)		5	
In	~94	16	~60
Mo(VI)	~96	74	0
P(H_3PO_4)		0.5–4	
Sb(III)	69	46	47
Sb(V)		94	
Se(IV)		5.8	
Se(VI)		4.0	
Sn(IV)	~93	30	90.8
Te(IV)	much	42	
Te(VI)		0.6	
Ti(IV)		1.0	0
Tl(III)		>99	
U(VI)	~22	<0.05	4.5
V(V)	~81	28	0
W(VI)		49[e]	
Zn	~5	1.8	0.8

Elements not extracted appreciably (usually <0.01%) by MIBK or MIBK–amyl acetate: Ag, Al, Ba, Be, Bi, Ca, Cr(III), K, and other alkali metals, Mg, Mn, Ni, Pb, Pd, Th, Tl(I), rare earths, Zr. MIBK + amyl acetate extracts 0.3% B as H_3BO_3, 0.1–0.2% Co, ~0.5% Pt.

[a] H. Specker, *Arch. Eisenhüttenw.*, **29**, 467 (1958); W. Doll and Specker, *Z. Anal. Chem.*, **161**, 354 (1958). Also see C. R. Boswell and R. R. Brooks, *Mikrochim. Ichnoanal. Acta*, **1965**, 814. For Te(IV) in 4 M HCl, $E \approx 81$ has been reported.
[b] Claassen and Bastings, *loc. cit.* They also give extraction values for a 2:1 mixture of MIBK + amyl acetate, which are not greatly different except for: Ga, 99; In 48; Mo, 92; Sb(III), 59; Sb(V), >99; Sn(IV), 78; Te(IV), 97; Te(VI), 4; W, 20; Zn, 4%.
[c] Specker, and Doll and Specker (footnote a). According to N. Ichinose, *Talanta*, **18**, 21 (1971), U(VI) is extracted quantitatively from a solution 5–8 M in HCl and 10 M in LiCl or from a solution 7–8 M in HCl and 1 M in $MgCl_2$. Th is essentially not extracted.
[d] Some reduction of Cr(VI) occurs.
[e] H_3PO_4 present (1 ml conc. acid/25 ml).

TABLE 9A-4
Extraction of Metals from HCl Solutions by MIBK[a]

Element	log E				
	HCl (M)				
	1	2	4	6	8
Au(III)[b]	2.6	1.9	1.8	1.8	1.7
Cr(VI)	0.1	0.1	0.05	0.0	−0.3
Fe(III)	−0.8	0.2	0.6	0.6	0.6
Ga	−0.9	1.0	2.5	2.8	2.75
In	−0.9	−0.4	0.7	1.7	1.85
Mo(VI)	−2.2	−1.4	0.3	1.1	1.3
Nb				−1.4	0.5
Pt	0.1	0.1	0.1	0.1	0.1
Re	0.5	0.6	0.65	0.7	0.7
Se(IV)	−1.5	−1.5	−0.8	0.3	1.75
Ta				−4.5	−4.5
U(VI)				−0.8	0.2

[a] Ishimori et al., *JAERI Rept.*, **1106** (1966). Tracer concentrations of metals. For coefficients of less easily extractable metals see the original.
[b] Higher values were reported by F. W. E. Strelow et al., *Anal. Chem.*, **38,** 115 (1966), for HCl solutions containing ~100 mg Au in 100 ml.

Au(III). The extraction of gold with methyl ethyl ketone has been studied with the view of elucidating the mechanism.[41] It has been concluded that MEK solvates both the cation and the anion of $MAuCl_4$, M being H^+ or alkali metal. The extracted species is assigned the general formula

$$M^+(H_2O)_{3-14}(MEK)_7AuCl_4^-$$

The extraction coefficient varies as follows with the cation (1 M solutions of MCl):

Cation	H^+	Li^+	Na^+	K^+	NH_4^+
E	900	220	60	35	30

The lesser extractability of the alkali metal chloroaurates, compared to $HAuCl_4$, is understandable from the smaller tendency of the cations to be solvated owing to the larger ionic radii. With $HAuCl_4$, E is independent of the gold concentration in the range 5×10^{-5}–2×10^{-3} M. At higher gold concentrations, E increases because of association in the organic solvent phase.

When MEK is diluted with an inert solvent, the course of E with the composition of the extractant is given by the curves in Fig. 9A-6.

Au(III) is slowly reduced, eventually to the metal, by MEK. The reduction is prevented by free chlorine, obtained by adding hypochlorite to the acid solution.

iii. ALCOHOLS

The extractability of many elements from HCl solutions by 2-ethyl-hexanol has been determined.[42] See Fig. 9A-7. There is, as would be expected, a general similarity to ether and ketone (MIBK) extractability. E of Fe and V(V) decreases above 1 mg/ml of HCl solution and E of Mo(VI) and Zr remains substantially constant.

Cyclohexanol and cyclohexanone have been tested as metal extractants

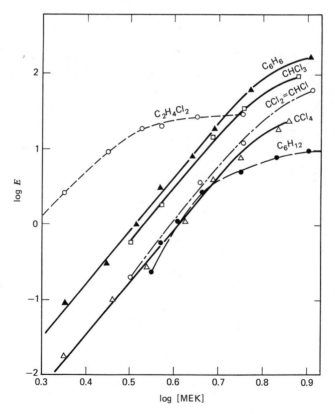

Fig. 9A-6. E as a function of diluent–methyl ethyl ketone composition in extraction of HAuCl$_4$ from 1 M HCl. After Jordanov and Havesov (reference in **note** 41).

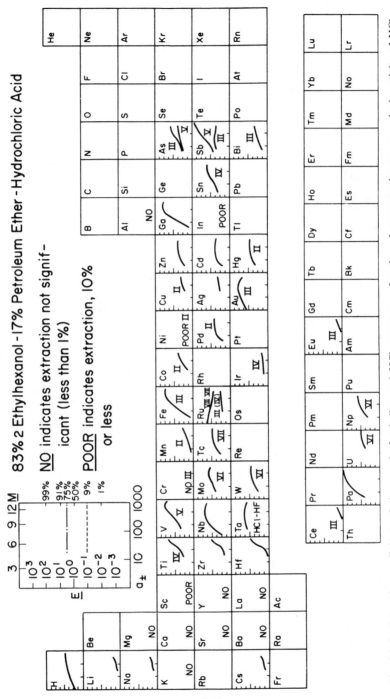

Fig. 9A-7. Extraction coefficients of elements in 2-ethylhexanol–HCl system as a function of concentration or mean ionic activity of HCl. Reprinted with permission from K. A. Orlandini, M. A. Wahlgren, and J. Barclay, *Anal. Chem.*, **37**, 1149 (1965). Copyright by the American Chemical Society.

840

in HCl solution.[43] The latter is generally the better extractant for Zn, Ga, Cd, In, Sb, Hg, Pb, and Bi (for Sn and Tl the two are equally good), but only Ga, In, Sn, and Tl are extracted really well (from 2–3 M HCl); the extraction of Zn and Hg amounts to 90–95% with equal volumes of phases. It does not seem that cyclohexanol has much value as a metal chloride extractant, though it can be used to separate Tl from In in 0.5 M HCl, the latter element hardly being extracted. The behavior of Fe(III) is not mentioned, but doubtless it is extracted to a large extent.

Such alcohols as butyl, amyl, and hexyl readily extract the Nb chloro complex from HCl solutions stronger than 9 M.[44] Ta is not readily extracted, even from >10 M HCl. Synergic effects are shown in the extraction of Nb by mixtures of alcohols with inert diluents such as $CHCl_3$ and CCl_4.

2,6-Dimethyl-4-heptanol (diisobutylcarbinol) in n-heptane solution has found use in extraction of Sb(V) from 7 M HCl–6 M H_3PO_4 in radiochemical separations.[45] As, Hg, Fe, Mo, Po, Se, and Tc in appropriate oxidation states are also extracted to large or moderate extents.

Molybdenum(VI) is extractable into amyl alcohol from HCl solutions.[46] E is about 13 when the aqueous solution is 2.0 M in HCl. Each extracted MoO_2Cl_2 molecule is solvated with two molecules of alcohol. Molybdenum(V) is extracted less well. For 2 M HCl and 0.004 M $SnCl_2$, E is roughly 6, and it decreases with higher $SnCl_2$ concentrations.

The formation of solvates of H_2TeCl_6 and $HTeCl_5$ has been demonstrated in the alcohol extraction of Te(IV) from HCl solution. Alcohols, being stronger H^+ donors than ethers, are better extractants for Te.

Attention needs to be drawn to the common presence of peroxides in immiscible alcohols (e.g., amyl, hexanol), which may adversely affect the extraction of certain metals.

iv. ESTERS (NON-P)

Esters, excluding phosphate esters, are not as good extractants of $HFeCl_4$ as are the ketones, but are better than the ethers (Fig. 9A-8). Most of them are at their best in 8 M or stronger HCl. Amyl acetate has been recommended for iron.[47] From 8 M HCl, the extraction of Fe(III) is essentially complete and that of Sb(V) and Mo(VI) can be made so by salting agents (inert chlorides). Sn(IV) and V(V) also accompany iron.

Ethyl acetate is used for the extraction of $HAuCl_4$. From 2 M HCl, Au(III) is effectively extracted, hardly accompanied by Ir(IV).

Isopentyl acetate extracts 98% Nb from 11 M HCl (equal volumes of phases) and is as effective an extractant as di-isopropyl ketone, tribenzylamine in $CHCl_3$, and di-n-octylmethylamine in $CHCl_3$.[48] It has the

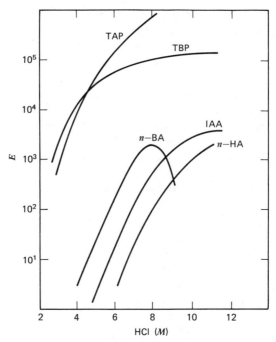

Fig. 9A-8. Extraction coefficients of Fe(III) in HCl–ester systems. From E. Bankmann and H. Specker, *Z. Anal. Chem.*, **162**, 18 (1958). TAP, triamyl phosphate; TBP, tributyl phosphate; *n*-BA, *n*-butyl acetate; IAA, isoamyl acetate; *n*-HA, *n*-heptyl acetate.

advantage of not forming emulsions as do the amine extractants. At this high acidity, there is no interference from HF, H_3PO_4, and organic acids such as tartaric and citric. Iron(III) accompanies Nb; it cannot easily be reduced to Fe(II) under the conditions.

Amyl and isoamyl acetates are also good extractants for Nb in $\sim 9\ M$ HCl.[44]

An unconventional method for the separation of Fe(III) involves its extraction from 5 M HF–8 M HCl into pentyl acetate; Nb and Ta remain in the aqueous phase if complexed with H_2O_2.[49]

c. *Organophosphorus Reagents*

i. TRI-*n*-BUTYL PHOSPHATE (TDP)

This is one of a family of powerful extraction reagents having phosphoryl, $\equiv P = O$, as the solvating group. The general extraction properties of TBP are discussed in Chapter 9C.

TBP is ordinarily used in an inert solvent such as benzene or carbon tetrachloride. If used undiluted (it is viscous), it gives extraordinarily high extraction coefficients for Fe(III). From 5–11 M HCl, more than 99.997% Fe is extracted. (Triamyl phosphate extracts >99.9999% Fe from 9–11 M HCl.[34]) Undiluted TBP extracts many metal chlorides from HCl solutions, including those of Mo, Sc, Zr, Nb, and Ta, most of which are not appreciably extracted by ethers or ketones (Table 9A-5).[50] It even extracts Pt, but poorly. TBP, therefore, is less selective than ethers and ketones for such metals as Fe(III) and Ga, and does not, in general, have advantages for the extraction of these metals. In fact, TBP is much more important for the extraction of nitrates than chlorides.

From a mixture of HCl (4 M) and LiCl (4 M), an equal volume of 50% TBP in xylene extracts (%): As 71, Ag 85, Co 61, Hg(II) 83, Sc 94, Zn 95.[51] Such extractions are too unselective to be of much value, even if they could be made quantitative.

A scheme for separating Ga, In, and Tl(III) by TBP, based on varying the HCl concentration, has been proposed.[52]

TBP has been suggested for the separation of U from Th.[53] The latter is extracted very little, whereas U is extracted quite well from ~6 M HCl, as the following extraction coefficients indicate (30% TBP by volume in CCl_4)[54]:

HCl (M)	5	6	7	8
E { U(IV)	0.5	15	50	102
U(VI)	3.6	9.0	15	21

Pu(IV and VI) is also extracted well at these HCl concentrations. With undiluted TBP, $E_U = 40$ and $E_{Th} = 0.006$ when the aqueous phase is 5.2 M HCl.[55] Washing the extract with a small volume of HCl of the same concentration will remove most of any extracted Th without seriously affecting the recovery of U. HCl below 6 M is not significantly extracted into TBP.

Titanium is extracted from ~6 M HCl salted with $MgCl_2$ by a xylene solution of TBP. The extracted species is apparently $TiOCl_2 \cdot 2$ TBP. Zirconium is extracted well from >8 M HCl solution by TBP in kerosene.[56] From 11 M HCl Ti is extracted as $TiCl_4 \cdot 2$ TBP by 0.1 M TBP in CCl_4.[57]

According to Gibalo et al.,[44] Nb is extracted satisfactorily by TBP from >9 M HCl, Ta above 10 M HCl.

Extraction of Pt and Pd from 6 M HCl solutions by TBP in various solvents has been studied.[58]

Extraction of Metals into Undiluted Tributyl Phosphate from Hydrochloric Acid Solutions[a]

Element	log E				
	HCl (M)				
	1	4	6	9	12
Ag	1.1	0.6	−0.25	−1.0	−1.4
As(III)	−0.75	0.4	1.3	1.6	1.5
Au(III)	3.9	4.1	4.3	4.6	4.3
Bi	1.3	0.3	−0.3	−1.0	−1.5
Cd	0.2	1.3	1.25	0.75	0.0
Ce	−3.0	−2.8	−2.4		
Co		−1.8	−1.3	0.1	−0.2
Cu(II)	−2.0	−0.4	−0.1	−0.1	−0.4
Fe(III)	1.5	3.75	3.8	3.8	3.9
Ga	0.1	2.8	3.1	3.15	2.5
Ge	−1.5	0.4	1.8	1.7	1.9
Hg	1.9	1.9	1.5	0.9	0.2
In	0.6	2.3	2.2	1.6	0.8
Mo	0.5	2.5	2.5	2.3	2.3
Nb	0.4	1.1	2.7	3.3	3.3
Os	0.6	0.9	0.8	0.5	0.25
Pa(V)		0.75	2.7	3.0	2.5
Pd	−0.3	0.4	0.25	−0.1	−0.3
Pt	0.75	1.3	1.25	0.5	0.0
Re	2.1	2.5	1.8	1.0	0.4
Sb		1.6	2.7	3.0	2.5
Sc		−1.4	0.4	2.5	3.1
Sn(II)	1.0		0.4	0.4	0.5
Ta	2.8	2.2	2.5	3.5	3.7
Tc	1.75	1.8	2.0	1.6	1.4
Te(IV)[b]	−0.25	2.7	2.7	2.75	2.8
Tl(III)	3.7	3.9	4.0	3.9	3.2
U(VI)	−0.5	1.5	1.9	1.8	
V(V)[c]	−1.4	0.8	1.4	1.75	1.5
W	2.5	2.6	2.2	2.0	1.8
Zn	0.8	1.4	1.0	0.4	−0.1
Zr	−2.0	−1.0	0.8	3.6	3.4

[a] From T. Ishimori and E. Nakamura, *JAERI Rept.*, **1047** (1963). Tracer concentrations. I. M. Gibalo, D. S. Al'badri, and G. V. Eremina, *Zh. Anal. Khim.*, **22,** 816 (1967), found 83% extraction of Nb from 9 M HCl and 82% of Ta from 10 M HCl (tartrate present) with 5 mg oxides in 30 ml aqueous phase and 10 ml TBP. For extraction of Sc, see A. R. Eberle and M. W. Lerner, *Anal. Chem.*, **27,** 1551 (1955).

[b] According to M. Haissinsky and A. Raitz, *Anal. Chim. Acta*, **45,** 143 (1969), Te(IV) but not Te(VI), is extractable from 4 M HCl. K. Inarida, *Japan Analyst*, **7,** 449 (1958), found Te(IV), but not Te(VI), to be extracted from 2–10 M HCl.

[c] I. V. Vinarov, N. P. Kirichenko, and G. M. Popelyukh, *Ukr. Khim. Zh.*, **41,** 742 (1975), found the molar ratios H:V:Cl:TBP in the extracted complex to be 2:1:1:1 or 2:1:1:2. S. K. Majumdar and A. K. De, *Anal. Chem.*, **33,** 297 (1961), report the extracted species as $VOCl_3 \cdot 2\,TBP$.

ii. TRI-n-OCTYLPHOSPHINE OXIDE (TOPO)

See Chapter 9C for a brief general account of this reagent.

The extraction coefficients of the metals more readily extractable from HCl solution by a 5% solution of TOPO in toluene are listed in Table 9A-6. These values hold for tracer concentrations of the elements. Thorium is also known to be extracted (maximally at 5–7 M HCl,

TABLE 9A-6

Extraction of Metals from HCl Solution by a 5% Solution of TOPO in Toluene^a

Element	log E HCl (M)			
	1	4	8	12
As(III)		−0.1	1.3	1.3
Au(III)	4.8	4.8	4.9	4.4
Bi	2.1	−0.1	−1.5	
Fe(III)	0.3	1.25	1.2	1.2
Ga	−1.2	3.3	3.8	3.3
Hg(II)	1.1	1.6	1.1	−0.5
In	1.5	1.8	1.9	0.4
Mo(VI)	1.6	2.8	1.4	0.75
Nb	−0.2	1.8	2.6	2.5
Pa	0.3	0.2	1.3	1.3
Pt	−0.1	1.25	0.8	−1.2
Re	1.0	1.3	0.9	−0.4
Sc	−1.3	1.3	1.9	−0.2
Se	−1.5	−1.2	0.0	1.1
Sn(IV)^b	2.0	2.0	2.3	1.5
Tc(VII)	1.5	2.2	1.6	1.0
Ti		−0.7	1.4	2.0
U(VI)	3.1	4.3	3.75	1.3
V(V)	−1.0	−0.1	0.8	0.8
W	0.2	0.8	1.0	0.8
Zn	1.8	2.1	0.6	−1.1
Zr	−0.75	2.3	2.0	2.4

^{a} T. Ishimori and E. Nakamura, *JAERI Rept.*, **1047** (1963). For additional data (benzene), see T. Sato and M. Yamatake, *Z. Anorg. Allg. Chem.*, **391**, 174 (1972); C. K. Mann and J. C. White, *Anal. Chem.*, **30**, 989 (1958). Extracted very little from 1–4 M HCl: Al, Ba, and other alkaline earths, Co, Cr(III), Cu(II), Mg, Mn(II), Ni, Pb, Ta ($E \sim 0.5$), Y, and rare earth metals.

^{b} W. J. Ross and J. C. White, *Anal. Chem.*, **33**, 424 (1961), state that tris(2-ethylhexyl)phosphine oxide is a more selective extraction reagent than TOPO for tin. Determination of Sn with Catechol Violet is described; cf. *ibid.*, **33**, 421 (1961).

TABLE 9A-7
Extraction of Metals with Di-2-ethylhexylphosphoric Acid (50% in Toluene) from Hydrochloric Acid Solution[a]

Element	log E HCl (M)		
	0.01	0.1	1.0
Ac	2.5	−0.6	
Ag	0.1	−1.0	
Am	4	1.1	
Bi		2.3	
Cr	−1.0		
Cu	0.0	−1.8	< −3
Fe(III)	4	3.3	0.4
Ga		2.2	−0.8
Hf		4	4
Hg		−0.3	
In		3.4	−0.4
Ir	−0.9	−1.5	
La		1.0	−2.7
Mg	−1.2		< −4
Mn	−0.25	−1.4	
Mo		1.5	1.6
Nb			2
Np		4	3.6
Os	1.0	0.9	0.8
Pa	4	4	4
Pb	1.5	−0.5	−2.9
Pm	4	1.7	−1.3
Sb		1.3	0.5
Sc	4	4	
Sn		0.7	0.0
Ta		−1.2	−0.9
Th	4	4	4
Ti			−0.2
Tm	4	4	1.4
U		4	3.2
V		−0.7	−1.6
W		−0.5	−0.7
Y		4	1.5
Zn	2.0	0.5	−1.5
Zr	2	2	2

[a] From K. Kimura, *Bull. Chem. Soc. Japan,* **33,** 1038 (1960). Tracer concentrations. See original for E of less easily extractable elements. The oxidation state of some elements is uncertain.

$\log E = 1\text{--}2$), as are Sb [Sb(III) at $3\,M$ HCl, Sb(V) $>5\,M$ HCl] and Cr(VI).[59]

A separation of Nb from Ta is reported to be obtained by extracting the former from $10\,M$ HCl.[60]

iii. DI-(2-ETHYLHEXYL)-o-PHOSPHORIC ACID (HDEHP)

This reagent (Chapter 9C) finds more use in HNO_3 than HCl medium. Separation possibilities in HCl solution are evident from Table 9A-7.

Yttrium is quantitatively extracted from an HCl solution of bone ash, having a pH of 1, with $0.45\,M$ HDEHP in n-heptane.[61] The extract is washed with $0.5\,M$ HCl, and Y is stripped from the organic phase with $9\,M$ HCl. Decontamination factors (the inverse of separation factors, S) greater than 1000 are obtained for such metals as Co, Zn, Mn, Fe, Pb, and Nb. Zr, U, and Th accompany Y, but can be separated by extracting the HCl solution with the quaternary ammonium salt Aliquat 336. The decontamination factor for Th is ~200.

From $0.5\text{--}1.5\,M$ HCl salted with $CaCl_2$, Ga is extracted into heptane as $HGaCl_4(HR)_2$.[62]

d. Long-Chain Amines

See Chapter 9C for a brief general account of these extraction reagents.

Much information is available on the extraction of chlorides with tertiary amines and quaternary ammonium salts. Triisooctylamine may be taken as representative of tertiary amine extractants. Extractabilities of metals with this reagent in xylene from HCl solutions of various concentrations are shown in Table 9A-8. Such metals as Au(III), Fe(III), and Ga, which are extracted well by the solvents considered in the foregoing sections, are also extracted well by this reagent in an inert solvent, but many other metals are also extracted to a large extent. This extraction lacks selectivity, but, nevertheless, it can be useful for certain samples having as major constituents those metals that are extracted poorly, for example, Ca, Al, Ni.

Secondary and tertiary amine salts extract Nb into CCl_4 from $7\text{--}11\,M$ HCl, in which the species $NbOCl_5^{2-}$ is present.[63]

Only aliphatic amines are effective extractants of Fe(III) and Co from HCl solution.

Primary amines are reported to be better extractants than the secondary and tertiary for Cu(II) and Mn(II) from HCl solution.[64]

n-Octylaniline, $C_8H_{17}\cdot C_6H_4NH_2$, in toluene extracts Pt(II, IV), Ir(III, IV), Rh(III), Pd(II), and Ru(III, IV) from $1\text{--}3\,M$ HCl.[65] Extraction coefficients are mostly >40 (for Rh, 6 in $1\,M$ HCl, 20 in $3\,M$; Ru(III), 1 in $1\,M$ HCl, 160 in $3\,M$). Extraction coefficients of Ni, Co, Cu(II), and

Extraction of Metals from Hydrochloric Acid Solutions with Tri-isooctylamine in Xylene[a]

Element	$\log E$			
	HCl (M)			
	1	4	8	12
Ag	1.8	0.4	−1.0	−2.4
Au(III)	4.3	4.75	4.4	3.7
Bi	2.9	1.8	0.1	−1.9
Cd	2.1	2.3	1.6	−0.4
Co	−3.0	−0.6	1.2	−0.2
Cu	−1.7	0.5	0.8	−0.6
Fe(III)	1.0	1.5	1.9	1.4
Ga	−0.25	2.9	2.0	0.8
In	0.5	1.8	1.9	0.2
Mo	−0.9	1.2	1.5	0.4
Nb	−2.5	−1.7	2.0	0.8
Os	2.9	2.8	2.8	2.8
Pa		1.0	3.2	3.6
Pb	0.6	0.3	−1.3	−2.6
Pt	2.3	2.4	1.8	0.1
Re	2.7	2.2	1.3	−0.4
Ru	−0.7	−0.25	−0.2	−1.2
Se(IV)	−1.1	−0.8	1.0	1.0
Ta	−2.5	−2.8	−2.3	−1.3
Tc	2.4	2.3	1.8	1.3
Te(IV)	−0.7	1.7	1.5	1.2
Tl(I)	−0.5	−0.5	−0.5	2.8
U(VI)	1.1	1.5	1.9	1.4
V	−0.9	0.6	1.9	0.3
W	0.9	0.75	0.5	0.3
Zn	2.2	2.6	1.75	0.2
Zr	−1.0		0.0	0.9
Hg(II)	extracted			

The following elements have $\log E < -1$ at all acidities: Al, Ba, Ca, Ir, Mg, Ni, rare earths, Sc, Sr, Th, Y.

[a] From T. Ishimori et al., *JAERI Rept.*, **1106** (1966). Tracer amounts of elements were used. Concentration of triisooctylamine in xylene = 0.11 M. For trans-U elements see F. L. Moore, *Anal. Chem.*, **30,** 908 (1958).

Fe(III) are much less than 1. Scrubbing the extract with 3 M HCl largely removes the base metals. Silver and Au are also extracted. M. M. L. Khosla, S. R. Singh, and S. P. Rao, *Talanta*, **21**, 411 (1974), separate Ga, In, and Tl with *N*-benzylaniline in chloroform.

For metal extraction with tetraphenylarsonium chloride—a reagent analogous to quaternary ammonium chlorides—see Chapter 9C.

e. Miscible Organic Solvents

At this point, we introduce a separatory process that is sometimes called solid–liquid extraction, but that may as well be designated by the homelier, more accurate term *leaching*. It is illustrated by a solid metal chloride mixture but, of course, is not restricted to chlorides. From such a mixture, certain chlorides that are soluble in organic solvents (predominantly oxygenated solvents) can be dissolved out, leaving other chlorides largely undissolved. Particularly those chlorides (or other salts) that are deliquescent—indicating a strong tendency to be solvated by oxygen-containing organic solvents—will be dissolved.

An old example is leaching of lithium chloride from a dry mixture of the other alkali metal chlorides by anhydrous organic solvents, such as ethyl ether + ethyl alcohol (Rammelsberg, 1845). In such processes, there is danger that the large amount of insoluble material will occlude some of the soluble and prevent its complete solution. Nevertheless, the method is sometimes of value, especially for concentration of a trace element. The trace component must not form a solid solution (mixed crystal) with the matrix.

The separation of microgram quantities of Fe(III), Co, Cu, and Zn from gram quantities of NaCl, KCl, $NiCl_2$, $CdCl_2$, $BaCl_2$, and $PbCl_2$ by leaching has been studied.[66] The solvents used included ethanol, *n*-butanol, and acetone, with and without a small proportion of 12 M HCl. Agglomerates of the solid were broken up by trituration, and the leaching was carried out in an ultrasonic field, usually for 30 min. This reduces the particle size to $\sim 2\,\mu$m. As a rule, recoveries of 95–98% could be obtained with one of the solvents mentioned (1 g of matrix chloride) under the most favorable conditions. The amount of matrix element accompanying the trace component varied from milligrams to less than a milligram. It was found that if less than a milligram or two of the matrix element dissolved in the leaching the recovery of the trace element was poor. This finding agrees with what might be expected: The minor components are present at the surface of the crystals of the matrix or in the interstitial space between agglomerated crystals, and a certain amount of the matrix must be dissolved to bring most of the trace component into solution.

Though not pertinent to the preceding examples, it may be noted that chemical reaction may be involved in a leaching process. For example, Ag and Au, which have been coprecipitated with elemental Te can be separated by treating the precipitate with nitric acid: Ag (and Te) dissolves, leaving Au behind. This process is perhaps better described as differential solution, since the bulk of the material does not remain undissolved. This separation is similar to the parting of assay buttons containing Ag in considerable excess over Au by treatment with HNO_3 or H_2SO_4: Ag dissolves.

2. Bromides

a. Inert Solvents

These extract covalent $SbBr_3$, $AsBr_3$, $GeBr_4$, $SnBr_4$, and, less well, $HgBr_2$ and $SeBr_4$ from bromide solutions of high acidity (H_2SO_4, $HClO_4$). A high acidity is needed to form appreciable amounts of the bromides of these easily hydrolyzed cations in aqueous solution; in addition, a high acid concentration increases extractability by salting out of the bromide. The percentage extractions by benzene from a 10 M H_2SO_4–0.03 M HBr solution (0.1 mg element in 1 ml), equal volumes of phases, have been reported[67] as

Sb(III)	99.4	Sn(II?)	95.7	Nb	0.9
As(III)	99.4	Se(IV)	95.6		
Ge(IV)	98.2	Hg(II)	74.1		

A sulfuric acid concentration of 8 M and a Br^- concentration of 0.01 M are sufficient for good extraction of Sb(III). The extractability of Sb does not vary a great deal with its concentration (apparently it decreases slightly below 1 ppm Sb). Antimony in the benzene phase is easily stripped with water. Extraction equilibrium is rapidly attained.

Tellurium(IV) extracts readily into nitrobenzene from Br^- solutions of high acidity.[68]

Inert solvents can also be used to extract certain metals forming anionic bromo complexes (especially singly charged) if a bulky organic cation is added. Metals not forming such complex anions remain in aqueous solution. These separations have not been much studied, but behavior similar to that of complex chlorides can be expected (see p. 829). Tri-n-octylamine as cationic partner extracts bromobismuthite ($BiBr_4^-$) into benzene. Bismuth can thus be determined in the presence of Ni, Co, Cu, and Pb by measuring the absorbance of the yellow extract at 380 nm.[69] Nitrate and perchlorate decrease the color intensity. N-Benzylaniline has been used to extract Tl(III) into $CHCl_3$ from HBr solution.[70]

b. Oxygen-type Solvents

Ethyl Ether. The data of several authors are summarized in Table 9A-9. Some apparent disagreements may be noted, which, partly at least, arise from different conditions, particularly differences in element concentrations. For many metals there is a rather close similarity between the extraction of bromides and chlorides, but there are some differences also. Probably the differences of greatest analytical interest occur with In and Tl(III). Whereas In is extracted only slightly with ethyl ether from HCl solution, it can be extracted >99% from 4–5 M HBr. Tl(III) is extracted 99% or better over a wide range of HBr concentrations, from 0.1 to 6 M. Au(III) is also extracted more easily from HBr than from HCl solutions. Tl(III) and Au(III) can be extracted easily from 0.5 M HBr, a concentration at which there is little extraction of Fe(III). On the other hand, there is greater extraction of divalent metals (Cd, Cu, Zn) in HBr solution, presumably because the bromo complex anions are more stable than the chloro anions.

Extraction from HBr solution into 2-chloroethyl ether has been studied.[71]

Isopropyl Ether. Data for the extractability of some 25 elements with this solvent from 5 M HBr are available and are recorded in Table 9A-10. Isopropyl ether is useful for removing much gold, so that trace elements associated with it can be determined.

MIBK. The extraction of Fe, Sn, Zn, and Cu from mixtures of HBr and NH_4Br has been studied.[72] From 2.0 M NH_4Br + 3.8 M HBr, ~98% Sn(II or IV) and ~51% Zn are extracted.

The extraction of In from bromide solutions by MIBK has been thoroughly examined.[73] Indium can be extracted as $HInBr_4$, $InBr_3$, and $NaInBr_4$ (if much Na is present), depending on the acidity and bromide concentration. The extraction coefficient is independent of the indium concentration below 10^{-5} M (< ~1 ppm). With macro concentrations of indium, E decreases with increasing indium concentration, the reverse of the effect shown when ethyl ether is the extracting solvent.[74] These effects are in part explained by polymer formation. Dimeric species such as In_2Br_6 and $In_2Br_7^-$ are thought to be formed in the aqueous phase. The main extracted species is hydrated and solvated $HInBr_4$.

Os(III), but not Ru(III), is extractable from 6 M HBr by MIBK.

Butyl Acetate. This extracts more than 99% Tl(III) from an equal volume of 1 M HBr containing Br_2; E is ~400.[75] Most other elements are extracted only slightly (Table 9A-11, in which some of the values have been rounded). Indium and Ga are extracted less well by butyl acetate than by ethyl ether. It will be noticed that some foreign elements are

TABLE 9A-9
Extraction of Elements with Ethyl Ether from Hydrobromic Acid Solutions[a]

Element	Extraction (%) HBr Concentration (M)							
	0.1	0.5	1	2	3	4	5	6
As(III)			3.0		6.7	22.8	63.1	72.9
Au(III)		99[b]	99.5; 99[b]		>99.9	99[b]		73[b]
Cd			0.4					0.9
Co			0.01					0.08
Cu(II)			0.5		1.5		4.2	6.2
Fe(III)	0[b]	0[b]	<0.1	0.2[b]; 1.4	55; 9.5[b]	97; 70[b]	97; 72.5[b]	95; 60[b]
Ga	0[b]	0[b]	0[b]	0.9; 0[b]	1.5; 0.5[c]	54.8; 16.9[c]; 3.6[b]	96.7; 28.9[c]; 53[b]	95; 57[b]; 8.9[c]
Hg	58.3	3[b]; 4.9	1[b]; 3.4	0.2[b]	2.3	0.3[b]		0.3[b]; 1.5
In		1.7; 1[b]	15; 3.2[b]	85.2; 19.2[b]	98.6; 89.3[b]	99.9; 99[b]	99.4	99[b]; 93.5
Mo(VI)				0.16		2.8	25.0	54.1
Sb(III)				37.9	22.3	14.9	9.0	6.1
Sb(V)							95.4	79.6
Se(IV)			0.3			3.5	18.3	31.0
Sn(II)			32	64	79	84	78	36
Sn(IV)			11.5	45.2	73.6	85.4	77.4	45.1
Te(IV)		0.3[b]	0.7; 2[b]	3[b]		3[b]		3[b]; 2
Tl(I)	11.7[c]	61.7[c]	99.9[c]		99.8[c] (3.2 M)		77.3[c]	6.2[c] (6.4 M)
Tl(III)	99[b]	99[b]	>99.9	99.7[b]	99.7[b]	>99.9	99.6[b]	99.0; 99.2[b]
Zn			1.4[d]; 1.3	5.0; 3.3[d]	10.0[d]	4.9; 8.9[d]	9.2[d]	3.6

Ni is not appreciably extracted (<0.03% in 1 and 6 M HBr); V(IV) is not extracted. Re(VII) is extracted to the extent of several percent (Bode).

[a] From R. Bock, H. Kusche, and E. Bock, *Z. Anal. Chem.*, **138**, 167 (1953), except as indicated. Equal volumes aqueous solution and ether (latter not previously shaken with HBr).

[b] I. Wada and R. Ishii, *Bull. Inst. Phys. Chem. Res. Tokyo*, **13**, 264 (1934). 20 ml of aqueous solution, 30 ml of ether previously shaken with HBr solution of corresponding concentration. 100 mg Tl(III), 250 mg Fe(III), 83 mg Ga, 100 mg In, 60 mg Au(III), 100 mg Te(IV), 100 mg Hg(II).

[c] H. Irving and F. J. C. Rossotti, *Analyst*, **77**, 801 (1952). Ether not previously shaken with HBr. 2.7 mg Tl/liter and 14.9 mg Ga/liter. Equal volumes of ether and aqueous phase for both Ga and Tl extractions.

[d] L. Kosta and J. Hoste, *Mikrochim. Acta*, 790 (1956). Conditions as in footnote *a*.

Extraction of Elements with Diisopropyl Ether from 5 M HBr^a

Element	Extraction (%)	Element	Extraction (%)
As(III)	31	Pd(II)	0.16^b
Au(III)	~99	Pt(II)	0.10^b
Fe(III)	35; 0.14–1^b	Sb(V)	6.5
Ga(III)	29	Sn(IV)	6.5
In(III)	~99	Te(IV)	1.4
Ir	0.03^b	Tl(III)	~99
Os	25 or more^b	Zn^c	

Extracted less than 1% from 5 M HBr: Ag, Al, Bi, Ca, Cd, Co, Cr(III), Cu, Fe(II), Mg, Mn, Mo(VI), Ni, Pb, Rh (0.07%), Ru (0.00%), Ti(IV), V(V), Zn, Zr.

^{a} F. A. Pohl and W. Bonsels, *Z. Anal. Chem.*, **161**, 108 (1958). Equal volumes acid and ether phases (ether not previously shaken with HBr).
^{b} 2.5–3 M HBr. W. A. E. McBryde and J. H. Yoe, *Anal. Chem.*, **20**, 1094 (1948).
^{c} T. A. Collins, Jr. and J. H. Kanzelmeyer, *Anal. Chem.*, **33**, 245 (1961), found E for Zn (extracted from a solution 6.0 M in HBr and 0.3 M in HClO$_4$) to be 0.0072 (0.3 M Zn). $E_{Pb} = 0.016$ and $E_{Cd} = 0.014$.

extracted to a somewhat greater extent in the presence of Tl than in its absence.

TBP. Extractions with this solvent from HBr (and HClO$_4$) solutions have been examined.[76]

3. Iodides

a. *Inert Solvents*

A group of adjacent elements in the periodic table

		Ge	As	(Se)
	(In)	Sn	Sb	Te
Hg	—	(Pb)	(Bi)	

are extractable as covalent iodides into nonpolar solvents such as benzene or toluene from aqueous solutions. Under optimal conditions, Ge, As, Sn, Sb, Te, and Hg are easily extracted quantitatively. In, Pb, and Bi are extracted to a small extent only, and this extraction is of no value for their isolation. Selenium(IV) is reduced to elemental Se by iodide, and amorphous Se dissolves in toluene or benzene, but the solution is unstable and Se separates on standing.

TABLE 9A-11

Extraction Coefficients of Metals into Butyl Acetate from 1 M HBr[a]

Metal	Concentration (M)	E	
		Tl Absent	2×10^{-4} M Tl(III) Present
Al	0.5	6×10^{-3}	5×10^{-3}
V	0.1	1×10^{-6}	5×10^{-4}
Bi	0.001	7×10^{-3}	8×10^{-3}
Ga	0.007	3×10^{-7}	7×10^{-7}
Dy	0.003	7×10^{-3}	2×10^{-3}
Fe(III)	0.45	8×10^{-4}	4×10^{-3}
Au(III)	5×10^{-4}	3.6	4.5
Ca	0.13	1.3×10^{-3}	
Cd	0.04	7×10^{-5}	
Mg	0.2	8×10^{-3}	2.4×10^{-2}
Mn(II)	0.09	3×10^{-4}	2.4×10^{-3}
Cu(II)	0.08	5×10^{-4}	4×10^{-4}
Mo(VI)	0.05	7×10^{-5}	1×10^{-4}
Na	0.2	5×10^{-5}	3×10^{-5}
Ni	0.09	5×10^{-4}	3×10^{-4}
Sn(II)	0.04	3×10^{-4}	4×10^{-4}
Hg(II)	0.002	small	
Pb	0.02	8×10^{-5}	4×10^{-4}
Ag	0.005	$<2 \times 10^{-5}$	5×10^{-5}
Sb(V)	0.004	3.5×10^{-3}	3×10^{-3}
Ti	0.1	8×10^{-5}	2×10^{-4}
Zn[b]	0.01	5×10^{-3}	2×10^{-3}

[a] Each phase 20 ml in volume; free bromine in aqueous phase. Organic phase not washed. Foreign elements do not affect extractability of Tl(III).
[b] P. P. Kish, I. I. Zimomrya, I. S. Balog, V. V. Roman, and V. P. Ugrin, *Zh. Anal. Khim.*, **29**, 1539 (1974), found 80% of Zn to be extracted from 7 M H_2SO_4–1 M Br^- by an equal volume of isobutyl acetate (90% extraction of Cd). The metals are extracted as $HMBr_3$.

Mercuric iodide extracts into benzene, with $P \sim 50$. Even a slight excess of iodide spoils the extraction by forming nonextractable anionic mercuric iodide complexes. However, Hg(II) can be extracted to the extent of $\sim 98\%$ into toluene from 5×10^{-4} M I^- in 6 N H_2SO_4 (equal phase volumes).[77] The acidity is not important.

The acidity as well as the iodide concentration play an important role in the extraction of As, Sb, Ge, (Te), and Sn. Sulfuric acid is often used in practice (HCl and HNO_3 are obviously unsuitable). The extractability of these metals as iodides increases with the sulfuric acid concentration for two reasons. First, a high acidity is needed to form the iodide in the aqueous phase by increasing the cation concentration and, second, the iodide is salted out by sulfuric acid and P increases. In general, the higher the acidity, the lower the iodide concentration required for quantitative extraction.

If we take 6 N as a reasonably high H_2SO_4 concentration:

As(III) is extracted almost 100% into toluene at $[I^-] = 1.0\ M$ and ~65% at $[I^-] = 0.5$ (equal phase volumes).[78]

Ge is not extracted even when $[I^-] = 1.0$ (~100% extraction requires 12 $N\ H_2SO_4$ with 0.5 $M\ I^-$). Sb(III) is extracted almost 100% at $[I^-] = 0.05$. Sn(IV) is extracted almost 100% when $[I^-] = 1.0$ and ~98% at $[I^-] = 0.5$.[79]

Iodide extraction with nonpolar solvents is of special importance for separations of Sb and Sn and some further particulars are given in the following paragraphs (benzene as extractant).

The extraction of SbI_3 from sulfuric and perchloric acid solutions has been studied (Fig. 9A-9).[80] From 5 $M\ H_2SO_4$, maximal E is about 3000, which is attained in 0.01 $M\ I^-$ solutions. From 2.0 $M\ HClO_4$, maximal E is a little greater than 70 at an iodide concentration of ~0.06 M. In mineral acid solutions, Sb(III) is present almost entirely as SbO^+ and the concentration of Sb^{3+}, which determines the formation of SbI_3 and the extractability of Sb, rises with the acidity. A moderately high acidity is, therefore, favorable for the extraction of Sb. An increase in acidity also increases the salting out of SbI_3. All six Sb(III)–I^- species are known to

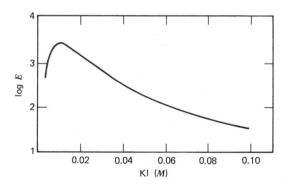

Fig. 9A-9. Extraction of SbI_3 with benzene from 5 $M\ H_2SO_4$: E as a function of iodide concentration. Based on data of Ramette (reference in note 80).

be formed (Julin): SbI^{2+}, SbI_2^+, SbI_3, SbI_4^-, SbI_5^{2-}, and SbI_6^{3-}. The positively charged species predominate to the left of the maximum in E–$[I^-]$ plots, the negatively charged to the right. In 5 M H_2SO_4, E does not fall below ~100 in the range ~0.002–~0.06 M iodide, so the iodide concentration need not be closely controlled for quantitative extraction of Sb. E is independent of the antimony concentration. Formation of I_2 by air oxidation of iodide or by oxidation with Fe(III) and the like is prevented by adding ascorbic acid. Bismuth iodide is essentially insoluble in benzene (as well as in $CHCl_3$), and if formed, is suspended in the extract. There may be some extraction of AsI_3 and possibly of GeI_4 at higher acidities. In other words, the acidity should be no higher than needed to give an E of 50–100, which is high enough for quantitative recovery of Sb in two successive extractions. The presence of small amounts of As and Ge does no harm in the usual photometric methods for Sb.

The successful extraction of Sn(IV) as iodide requires a comparatively high concentration of I^- and H^+ and the presence of a salting-out agent. Benzene is a good extracting solvent. n-Hexane can also be used. From a 2.5 M $NaClO_4$ solution (salting agent), 1 M $HClO_4$ and 1.5 M iodide give an E of 380, and higher values can be obtained by increasing their concentrations (Table 9A-12). E is somewhat dependent on the tin concentration in the aqueous phase, decreasing with increasing concentration of tin, pointing to polymerization in that phase. There is little indication of the formation of anionic tin(IV) iodide species in 5 M I^- solutions. It seems likely that Sb(III) will be present mostly as anionic iodide species when $[I^-]$ is 1 or higher, so that little Sb will be extracted, and this can be largely removed from the benzene phase by washing with

TABLE 9A-12
Extraction Coefficients (25°) for Sn(IV) in
NaI, $NaClO_4$, $HClO_4$–Benzene System[a]

$HClO_4$ (M)	NaI (M)	$NaClO_4$ (M)	E
1.0	1.5	2.5	380
1.0	4.0	—	1020
4.0	1.0	—	1950
2.5	2.5	—	2100

[a] D. D. Gilbert and E. B. Sandell, *Microchem. J.*, **4**, 491 (1960). Ionic strength 5.0 M in all solutions. Equilibrium tin concentrations in the aqueous phase approximately 1×10^{-6}–4×10^{-6} M.

a solution that is (for example) 1 M in $HClO_4$ and 3 or 4 M in I^-. Some As is likely to accompany Sn.

The extraction of both SnI_4 and SbI_3 is rapid, equilibrium being reached within a minute of shaking. The foregoing results hold for 25°. The extraction of Ge and Sb has been reported to increase above 25° (toluene as extractant).

When toluene is the extractant, In is extracted ~2% from 6 N H_2SO_4 which is 1 M in I^-.[77] A like amount of Bi is extracted from 0.05 M I^- and high (12 N) H_2SO_4 concentration; its extractability decreases with iodide concentration. Under like conditions, 5% Pb extracts, but its extractability also decreases with iodide concentration and greater formation of anionic iodide complex species. These elements are removed from the organic phase by backwashing. Elements not extracted include Cd, Cu, Ga, Au, Mo, W, and Zn.

Sulfuric acid solutions are also suitable for the extraction of SnI_4 and probably find more use than perchloric acid in practice.

The extraction of Te(IV) from I^-–H_2SO_4 solutions has been studied.[81] Tellurium is extracted as TeI_4 with widely varying extraction coefficients depending on the nonpolar solvent:

	E (0.05 M KI, 3 M H_2SO_4)
n-C_5H_{12}	0.007
CCl_4	0.033
$CHCl_3$	0.040
CH_2Cl_2	0.050
$C_2H_4Cl_2$	0.012
C_6H_6	3.0
$C_6H_5CH_3$	4.0
1,2-$(CH_3)_2C_6H_4$	3.8

As a result of π-bonding, the aromatic solvents are much better extractants than the other nonpolar solvents. For a given acidity, E rises with increasing iodide concentration to a maximum and then falls as anionic Te-iodide species increase in concentration. E decreases above a Te concentration of ~10^{-4} M (with acetophenone and 0.025 M KI–4 M H_2SO_4, E ~ 1000 in the range 10^{-6}–10^{-4} M Te, then drops to ~40 at 10^{-3} M Te). This behavior indicates formation of less easily extractable TeI_4 polymers in the aqueous solution at higher Te concentrations.

Acetophenone is a better extractant than xylene for TeI_4, which is extracted as the solvate $TeI_4 \cdot 3$ acetophenone. The large iodine atoms in TeI_4 are easily polarizable, as by strongly polar keto groups, but acetophenone is a better extractant than aliphatic ketones (effect of benzene ring).

Inert Solvent + Pyridine. As a matter of convenience, we include this combination here, although it really belongs elsewhere.

Pyridine coordinates with a number of heavy metal cations, which then can form ion-association complexes with halide ions. For example, with Cu(I) the following complexes are reported to be formed[82]:

$$CuPy_2X, CuPy_3X, CuX \cdot HXPy_{1-2} \qquad (X = halide, best\ Br^- \ or\ I^-)$$

They are extractable into inert solvents such as $CHCl_3$. Pyridine plays much the same role as TBP, for example, and $CHCl_3$ is the diluent. Hydroxylammonium chloride or ascorbic acid can be used as reductants for Cu(II). EDTA prevents extraction of metals such as Fe and Zn.

Extraction of metals from iodide solution into benzene–pyridine has been studied.[83] The aqueous solution (0.02 M in KI, 1% in ascorbic acid and 0.02–0.3 M in HCl or HNO_3) is extracted with an equal volume of a 5% (v/v) solution of pyridine in benzene. Cadmium is extracted as $CdPy_2I_2$, Cu probably as $Cu_2Py_2I_2$, and Hg possibly as $HgPy_2I_2$. In the pH range 5.3–6.1, the extraction of Cd is quantitative. In addition to serving as a reducing agent to prevent formation and extraction of I_2, ascorbic acid prevents extraction of Bi(III). Other metals not extracted include Co, Ni, Al, Fe(III), U(VI), Th, Mn(II), Mg, Cr(III), Tl(I), La, As, Sb(III), In, Sn(II). Partial extraction of Pb, Zn, and Ag occurs. The extraction of Cd and Cu is still quantitative from 0.002 M I^-.

b. Oxygen-type Solvents

i. ETHYL ETHER. Available data on the extraction of iodides (mostly as solvated hydrated iodo acids) are summarized in Table 9A-13. In contrast to their behavior in HCl and HBr solutions, Cd, Hg, Sb(III), and Sn(II) can be extracted quantitatively from HI solution.

The extraction of In and the nonextraction of Ga from iodide solutions is one of the more interesting analytical characteristics of this system. The extraction of In from 1.5 M HI solution is not disturbed by 0.5 M Br^-, F^-, cyanide, citrate, or phosphate; 0.5 M Cl^- has little effect (Irving and Rossotti).

HI solutions are air oxidized on standing and solid KI and H_2SO_4 can be substituted.

ii. MIBK AND METHYL ISOPROPYL KETONE. The extractability of elements from a solution 1 M in KI and 2 M in H_2SO_4 with an equal volume of

TABLE 9A-13
Extraction of Metals as Iodides with Ethyl Ether

Metal	From 6.9 M HI[a] (%)	From 1.5 M KI–0.75 M H$_2$SO$_4$[b] (%)
As(III)	63	
Au(III)	~100	
Bi	34	<10
Cd	~100	100
Hg	~100	33
In	~8	99.8[c]
Mo(VI)	6.5	<1
Sb(III)	~100	<50
Sn(II)	~100	~100
Te(IV)	5.5	
Tl(I)	>99.5[d]	
Tl(III)	>99.5[e]	
Zn	10.6	33

Not extracted appreciably: Al, Ba, Be, Ca, Co, Cr(III), Fe(II), Ga, Ir, Mn(II), Ni, Os, Pb, Pd, Pt, Ru, Th, Ti, U, V, W, Zr.

[a] S. Kitahara, *Rept. Sci. Res. Inst. Tokyo*, **24**, 454 (1948). Volume ratio ether:aqueous phase = 4.
[b] H. M. Irving and F. J. C. Rossotti, *Analyst*, **77**, 801 (1952); equal phase volumes.
[c] 0.5–2.5 M HI.
[d] 0.5–2.0 M HI.
[e] 0.05–2.0 M HI.

MIBK may be summarized as follows[84]:

Extracted	Partly Extracted	Not Extracted
Ag, As, Bi, Cd, Cu, Hg, Pb, Sb, Sn, Te	Zn, Mo	Al, Ca, Co, Cr, Fe, Mg, Mn, Ni, Th, Ti, V, W

With equal volumes of phases under the conditions indicated, percentage extraction is:

Bi, 99.7; 3 M H$_2$SO$_4$, 0.25 M KI
Cd, 99.9; >2 M H$_2$SO$_4$, ~0.25 M KI
Cu(II), 99.7; 3 M H$_2$SO$_4$, 0.25 M KI
Pb, 94; 2 M HCl, 0.5–0.75 M KI
Sb, 99.4; 0.5–3 M H$_2$SO$_4$, ~0.25 M KI
Te, 99.4; 0.5–4 M H$_2$SO$_4$; 0.25–2.5 M KI

Most of these results were obtained with 10 mg of metal in 10 ml of solution.

Microgram amounts of Pb, Cd, In, Bi, Cu, and Sb can be extracted quantitatively, or almost so, as iodides from 5% HCl solutions of such matrix elements as Fe, Al, Mg, and Zn.[85] Good recoveries of the first-named metals can also be obtained in the presence of 0.5 g each of Co, Ti, Cr, Mn, Ga, U, Th, V and As. Some of the latter elements, for example, Ga, accompany Pb, Cd, and others to a small extent.[86]

MIBK extracts Pd(II) readily from 20% v/v H_2SO_4 solution containing 10% KI (separation from Nb and Zr).[87]

Methyl isopropyl ketone would be expected to have much the same extractive properties for metal iodides as MIBK. Qualitative observations are available.[88]

iii. CYCLOHEXANONE. Fractional extraction of metal iodides by cyclohexanone has been studied, with special reference to removal of much Hg and Bi from aqueous solution.[89] The order of decreasing extraction is $Hg > Bi > Cd > Zn$. This solvent has also been used in the separative extraction of In from such metals as Al and Fe.[90] A $4+1$ mixture of cyclohexanone and tetrahydrofuran extracts Zn but hardly any Cd from weakly ammoniacal iodide solution.[91] Cyclohexanone alone extracts In but not Ga from dilute H_2SO_4–KI solution.[92] Ag in the presence of excess iodide is extracted from acid or ammoniacal solution.[93]

iv. ALIPHATIC ALCOHOLS. 2-Ethyl-1-butanol, 4-methyl-2-propanol and 1-butoxy-2-propanol extract In, Tl(III), Sn(II), Sb, Bi, Te, Cu, Ag, Au, Hg(II), Cd, Pd, and Pt to a large extent (98–100%) from H_2SO_4 solutions (1 to 8 N, depending on metal) containing KI (0.07–0.9 M, depending on metal).[94] Al, Ga, Cr, Mn, Fe, Co, Ni, Rh, and U are not extracted. Other metals are extracted to varying extents.

Mercury(II) is extracted into n-octanol from sulfuric acid solutions of low iodide concentration,[95] probably mostly as HgI_3^-. It can be stripped from the alcohol solution with neutral 3 M iodide, probably as HgI_4^{2-}.

v. ISOAMYL ACETATE. Only meager information is available on the extraction of iodides by alcohols and esters (except TBP). The extraction of BiI_3 by isoamyl acetate and isoamyl alcohol has been examined in a preliminary way with the view of application to the separation of Bi from samples containing much Ca, Mg, and phosphate, which are not amenable to dithizone extraction.[96] Extraction of Bi as BiI_3 was considered to be preferable to extraction as $HBiI_4$, inasmuch as conditions favoring the extraction of the latter would also lead to extraction of other iodo acids.

With isoamyl acetate as extractant, extractability of Bi shows a maximum at $0.0065\ M\ I^-$ $(R = 97\%$ in $3\ N\ H_2SO_4$, 98.9% in $5\ N\ H_2SO_4$, and 99.2% in $3\ N\ HClO_4$, org/aqueous volume ratio = 0.25). The appearance of the maximum is interpreted as signifying predominant extraction of Bi as BiI_3. On the other hand, isoamyl alcohol extracts Bi equally well (99%, $3\ N\ H_2SO_4$) over the iodide range 0.015–$0.10\ M\ I^-$, pointing to extraction of $HBiI_4$. Partial extraction of Hg(II), Sb(III), Pb(II), and possibly of Tl may be expected when isoamyl acetate is used as extractant. Experimentally, Pb is found to be slightly extracted $(R \sim 2\%)$ when the conditions are optimal for the extraction of Bi.

Isobutyl acetate extracts 98% of Zn from 5–$7\ M\ H_2SO_4$–$1\ M\ I^-$ and 97% of Cd from an equal volume of 4–$7\ M\ H_2SO_4$–$1\ M\ I^-$ and $4\ M\ H_2SO_4$–$(0.01$–$1)\ M\ I^-$. Cd $(\sim 10^{-4}\ M)$ can be separated from Zn $(\sim 10^{-4}\ M)$ by isobutyl acetate extraction from $5\ M\ H_2SO_4$–$0.01\ M\ I^-$; E_{Cd}/E_{Zn} is 8000.[97]

c. High-Molecular-Weight Amines and Organophosphorus Compounds

Tri-n-octylamine (5% solution in benzene) has been used for separative extraction of Au, Ag, Bi, Cd, Hg, and Pb from U(VI) in $0.01\ M\ KI$–$0.4\ M\ HCl$.[98] Some U is also extracted and the effectiveness of the separation seems doubtful.

Indium is quantitatively extracted from $2\ M\ H_2SO_4$–$1.5\ M\ KI$ solution by a solution of N-benzylaniline in $CHCl_3$, and can be stripped with $2\ M\ HCl$.[99] Many other metals are also extracted but some selectivity results from the stripping step.

Amines of high molecular weight in xylene quite generally extract CdI_4^{2-} well from acidic (sometimes slightly basic) solution. One of the best reagents is Aliquat 336-S (quaternary amine).[100] Other metals forming negatively charged iodide complexes would be expected to be extracted also. Cadmium can be stripped from the xylene extract with $10\ M\ NH_4OH$ (difference from Hg).

Tribenzylamine in $CHCl_3$ extracts Cd readily from $0.5\ M\ H_2SO_4$–$0.3\ M\ I^-$ and provides an excellent separation from much Zn.[101]

TOPO (p. 1003) in MIBK extracts Sb, Bi, Pb, Sn, and Ag.[102]

TBP (+diluent) is not often used in the extraction of iodides, but it has received some attention.[103] In TBP extractions, the neutral iodide may be solvated. Thus BiI_3 forms a neutral complex.

Platinum(IV) and Pd(II) have been separated from Rh and Ir(III) by extraction of the former metals from $6\ M\ HCl$ containing iodide with a TBP solution in hexane.[104]

d. High-Dielectric Constant Solvents: Extraction of Polyiodides

Iodine forms complex anions of the type $X(I_2)_n^-$ (X = halide), I_3^- being the most important one. This is a large anion, and it can form ion pairs with the alkali metal cations, which can be extracted into oxygen-type solvents of high dielectric constant, for example, nitro compounds (nitromethane, nitrobenzene). In such solvents, the ion pairs are dissociated. Extractability of the alkali metals increases with the ionic radius: Li < Na < K < Rb < Cs.[105] The extraction coefficient rises with the concentration of free iodine in the organic solvent. The extraction of MI_3 and other alkali metal polyiodides (e.g., $CsCl \cdot I_2$) by a variety of organic solvents has been systematically studied.[106] By way of illustration, with nitromethane as the extractant, $E_K = 8.8$ when 2 g KI and 10 g I_2 are present in 100 ml aqueous solution, and $E_{Cs} = 68$ when 2 g CsI and 10 g I_2 are present in 100 ml. If I_2 is reduced from 10 to 2 g, $E_K = 0.86$ and $E_{Cs} = 3.44$. It is evident that with such values of E a satisfactory separation of Cs and K would require extensive fractionating extractions. A set of two or three successive extractions at an E_{Cs} sufficiently large to give satisfactory recovery of Cs would also extract much K so that only an enrichment of Cs in K, and not a good one at that, would result. Even a more serious deficiency in this extraction system is the decrease in E with a decrease in MI (or MX) concentration. Separation of the alkali metals by ion-exchange chromatography is much to be preferred.

4. Fluorides

Anionic metal complexes formed by fluoride in excess tend to have charges greater than -1 (FeF_6^{3-}, for example), and this is probably the main reason for the poor extractability of most metals from hydrofluoric acid solutions by oxygen-type solvents. In practice, the use of fluoride solutions is unpopular because of the precipitation of many metals (alkaline earths, Bi, Mg, Pb, Th, rare earths, and others) and the need to use vessels other than glass. However, fluoride extraction with ketones is of real value in separations of Nb and Ta.

a. Ethyl Ether

See Table 9A-14 for extractabilities with this solvent with equal volumes of phases.[107] Earlier work gave the following figures for the percentage extraction of metals with a 4:1 volume ratio of ethyl ether/aqueous phase[108]:

Sn(II, IV)	100	Se(IV)	3
As(III)	62	Sb(III)	0.4
Mo(VI)	10		

TABLE 9A-14
Extraction of Elements with Ethyl Ether from HF Solutions

Element	Extraction (%)				
	HF (M)				
	1	5	10	15	20
Be	<0.05	<0.05	0.5	1.9	4.0
Hg(II)	<0.05	<0.05	<0.05	0.9	2.7
Ge	<0.2	<0.2	0.5	2.7	6.7
Sn(II)		2.0	3.8		4.9
Sn(IV)		0.6	5.0		5.2
Zr	0.4	0.5	0.5	1.2	2.9
P(V)	<0.1	1.1	3.1	9.9	14.8
As(III)	10.0	18.5	30.2	34.6	37.7
As(V)	<0.1	1.7	4.6	10.8	13.6
Sb(III)	<0.05	0.3	1.9	3.8	6.3
V(V)	<0.1	0.4	1.7	5.3	8.5
Nb(V)	0.6	4.2	32.4	51.9	65.8
Ta	1.2	43.6	79.2	79.3	79.3
Se(IV)	0.06	2.2	7.4	11.6	12.9
Te(IV)	0.01	2.0	6.6	19.2	23.0
Mo(VI)	0.7	1.8	3.0	5.8	9.3
Re(VII)	0.05	10.8	61.2	64.0	61.8

Not appreciably extracted: Ag, Al, Ba, Bi, Ca, Cr(III), Cu(II), Fe(II, III), Ga, In, Mg, Mn(II), Ni, Pb, Pd(II), Rh, Sb(V), Sr, Th, Ti, Tl, U(VI, ~1% from 20 M HF), Zn (~1% from 20 M HF), rare earth metals.

These values are for 4.6 M HF, except 3.5 M for Sb and Mo. The apparently complete extraction of Sn(II or IV) is suspect.

There is little in Table 9A-14 that would encourage the use of extraction of fluorides by ethyl ether for separations. The situation is more favorable with other solvents, especially ketones, of which MIBK finds the most use.

b. MIBK

From a solution 10 M in HF, 6 M in H_2SO_4, and 2.2 M in NH_4F, MIBK extracts the following percentages of metals when the phase ratio MIBK/aqueous is 1[109]:

Mo(VI)	9.1	Nb	96
W(VI)	26	Ta	99.6

Not extracted: Al, Fe(III), Ga, Mn, Sn, Ti, U, and Zr. For extraction from $10\ M$ HF–$6\ M$ H_2SO_4 the extraction coefficients are[110]:

Nb   67   Zr   0.025
Ta   >250

From $1.6\ M$ HF-$6.3\ M$ H_2SO_4 solution, MIBK extracts 96% Nb,[111] and from $0.4\ M$ HF–$6\ M$ H_2SO_4, 98% Ta.[112] The presence of much Cl^- ($6\ M$ HCl) leads to extraction of Fe, Ga, Sb, As, Mo, etc.

Tantalum is extracted as $HTaF_6$.

c. Other Solvents

Alcohols (diisobutylcarbinol, pentanol, isobutanol) and esters (ethyl acetate, TBP) have also found use as extractants for Nb and Ta in fluoride solutions. Pentanol has been used to extract Ta (separation from Nb).

A $0.1\ M$ solution of n-dodecylamine in $CHCl_3$ extracts FeF_6^{3-} at an optimal pH of 5.5.[113]

Cyclohexanone extracts Nb from $3\ M$ H_2SO_4–$4\ M$ HF and Ta from $1\ M$ H_2SO_4–$0.5\ M$ HF, leaving Ti in the aqueous phase.[114]

If a suitable organic cation is added to a TaF_6^- solution, the ion-pair compound can be extracted by inert solvents. Thus chloroform extracts Brilliant Green fluotantalate from pH 3 to $8\ N$ H_2SO_4; Nb is not extracted if the concentration of H_2SO_4 is greater than $3\ N$.[115] Ordinarily, a strongly colored reagent such as Brilliant Green is avoided for separations because of the necessity for destroying it subsequently. It can be replaced by a colorless cationic reagent, for example, amines of high molecular weight (p. 989).

E.   THIOCYANATES[116]

Metal thiocyanates are extractable into "active" organic solvents as the solvated salts or as solvated thiocyanato acids or salts (alkali metal) of these acids. It appears that with the more active solvents the number of thiocyanate ions bound to a metal tend to be smaller than with a less active solvent and definite compounds can be formed. For example, $NH_4[Co(SCN)_3]\cdot 3$ TBP and $(NH_4)_2[Co(SCN)_4]\cdot x$ cyclohexanone are believed to be the species extracted into TBP and cyclohexanone.[117] Of course, the concentration of thiocyanate and the acidity are important factors with a particular solvent. With ethyl ether as extractant at low acidities, the simple thiocyanates $Zn(SCN)_2$, $Sc(SCN)_3$, $Ga(SCN)_3$, $In(SCN)_3$, $Fe(SCN)_3$, and $UO_2(SCN)_2$ are extracted at low thiocyanate concentrations, whereas complex thiocyanates of some metals (Fe, Sc, Ga, for example) are extracted at higher thiocyanate concentrations.[118] The compounds $(NH_4)_2Zn(SCN)_4\cdot 4\ H_2O$ and $(NH_4)_2Co(SCN)_4\cdot 4\ H_2O$ can be isolated from the ether phase. Beryllium and Al are extracted as simple thiocyanates (possibly as ion pairs) at thiocyanate concentrations as high as 7 M. The decrease in the extractability of U(VI) with increasing thiocyanate concentration is attributed to the formation of less easily extractable complex uranyl thiocyanates.

The extraction of thiocyanates of Zn, Cd, Pb, Bi, Fe, Al, Ga, In, and Co into TBP, MIBK, isoamyl alcohol, ethyl ether, isobutyl acetate, and TBP in $CHCl_3$ has been studied systematically.[119] Extractability at higher acidities (HCl) decreases, partly because of increasing extraction of HCNS. A pH of 2–3 is suitable for most of these metals. For most metals, extractability increases with thiocyanate concentration up to a plateau value; for a few, maxima are reached; optimal values depend on the extractant. TBP is the best extractant, and for that reason is not always the best choice for separations.

1. Ethyl Ether

The extractabilities of metals as thiocyanates by diethyl ether are listed in Table 9A-15.[120] It is evident that thiocyanate–ethyl ether extractions lack selectivity, and they are incomplete in a single equilibration except for Mo(V) and Sn(IV) at low (1–2 M) thiocyanate concentrations and Ga and Sn(IV) at high concentrations. The extraction of Sc is of analytical interest, but it must be combined with other separations. It should be borne in mind that thiocyanic acid undergoes slow decomposition and polymerization in mineral acid solution. HCN can be formed in strongly acid solutions. HCNS is subject to attack by oxidizing agents.

Thiocyanic acid is extracted from acid aqueous solutions into ether (and other oxygenated solvents, but little into inert solvents), and the residue left on evaporation of the latter requires oxidative treatment before determination of extracted elements.

The presence of anionic complexing agents such as phosphate, fluoride, and sulfate are detrimental in thiocyanate extractions, especially at low acidities.

A 2:3 mixture of ethyl ether and tetrahydrofuran extracts Ga well from 4 N H_2SO_4 thiocyanate solutions and enables it to be separated from Al.[121]

2. MIBK

The extraction of some metals is improved by using MIBK in place of ethyl ether.[122] With equal volumes of phases, $R = 99.9\%$ for Co with 4 M SCN^- in 0.5 M HCl, and 97% for Fe(III) with 1 M or stronger SCN^- and 0.1 M HCl. Titanium thiocyanate is extracted by MIBK.[123] Zinc is

TABLE 9A-15
Extraction of Elements as Thiocyanates with Ethyl Ether from 0.5 M HCl Solutions

	Extraction (%)				
Element	NH$_4$SCN Concentration (M)				
	1	2	3	5	7
Ag					0.1[a]
Al		1.1	9.0		48.9
As(III)	0.4[b]				0.4[b]
As(V)	0.1				
Be	3.8	24.3	49.7	84.1	92.2
Bi	0.3				0.1
Cd	0.1				0.2
Co(II)	3.6	37.7	58.2	74.9	75.2
Cr(III)	0.06[c]				3.4[c]
Cu(I)		2.9		0.4	
Fe(III)	88.9		83.7	75.5	52
Ga	65.4		90.5		99.3
Ge(IV)	<0.3				<0.5
Hg(II)	0.15[a,c]				0.65[a,c]
In	51.5	75.1	75.3	68.3	48, 59
Mn(II)					0.17
Mo(V)	99.3		97.2		97.3
Pd(II)	1.7				<0.1
Sb(III)					2.2
Sc	12.7	55.4	79.8		89, 94
Sn(IV)	99.3		99.9	>99.9	>99.9
Ti(III)	58.8	80.5	84.0	79.8	76.3
Ti(IV)					1, 13, 2.4
Th					0.13
U(VI)	45.1	41.4	29.4	13.8	7, 5, 2
V(IV)	15.0	13.1	8.7		2.2
Y					0.07
Zn	96.0		97.4	94.8	92.8

Not extracted (<0.1%): Ca, La, Li, Mg, Ni, Y, Zr. A few values above have been rounded. Equal volumes of phases except as indicated.

[a] Neutral solution.
[b] 0.8 M HCl.
[c] 10 ether:1 aqueous by volume.

extracted more easily than Cd at low SCN^- concentration (0.1 M) and separation should be feasible.[119] The nonextraction of Pb should make various separations possible.

V(III) is extractable from HCl solution (greenish-yellow).

Extraction of a solution 0.5 M in HCl, 0.5 M in KSCN, and 8% in ascorbic acid with an equal volume of MIBK transfers the following percentages of metal to the MIBK phase[124]:

Sb(V)	5	Al, As(V), Bi, Ca,	0
Co(II)	67	Cr(III), Pb, Mg,	
Fe(III)	6	P, Se, S, Te	
Mn(II)	1		
Mo(VI)	78		
Ni	1		
Nb	2		
Sn(IV)	99.9		
Ti	5		
V(V)	27		
Zn	99		

Thus extraction under these conditions provides good recoveries of Sn and Zn with separation from sundry metals.

3. Other Solvents

Alcohols and esters are used occasionally in extractions of thiocyanates, but more often in determination of colored thiocyanate complexes (Mo, W, Nb, Re, U) than in separations.[125] Rare earth elements can be separated (more or less) by extensive fractional extraction of their thiocyanates by butanol, but this is hardly a separation of interest to the trace analyst.

Isobutyl acetate extracts 84% of Zn from 1 M H_2SO_4–1 M SCN^-; E_{Zn}/E_{Cd} is 1100.[126] Zinc is extracted as $HZn(SCN)_3$.

Hafnium is extracted to a larger extent than Zr by cyclohexanone from a solution 3 M in ammonium thiocyanate, 2 M in ammonium sulfate and 1.5 M in HCl.[127] $E_{Hf}/E_{Zr} = 110$ at room temperature. Evidently a fractionating extraction procedure must be applied, even if the initial ratio Hf/Zr = 1.

Tributyl phosphate extracts $Fe(SCN)_3$ better than ether does, and some workers have proposed to remove much Fe in this way in determining nonextractable elements (Al) in steel.[128] TBP extracts Ca from 2 M thiocyanate solution (0.1 M H^+), and separates it from Fe, Ni, and rare earths if EDTA is present.[129] Magnesium is also extracted to a large extent.[130] Data have been reported for the extraction of Mn(II), Cr(III), Cu(I, II), Y, La, Mg, Ca, and Sr by TBP from weakly acidic thiocyanate solutions.[131] Be is extractable from dilute HCl with TBP–toluene. Good recoveries of most of these elements can be obtained at sufficiently high thiocyanate concentrations, but separation possibilities seem few.

Tri-n-octylphosphine oxide–cyclohexane extracts titanium thiocyanate (yellow color) from an acidic chloride or sulfate solution.[132] This extraction has been examined only from the standpoint of determination of Ti, but since U, Fe, and Co are not extracted to an appreciable extent, it may have separation possibilities. Niobium, W, and Mo are extracted. Extraction of Ti thiocyanate with diantipyrinylmethane–CHCl₃ is primarily for determination.

Scandium is reported to be extracted with CHCl₃ from a thiocyanate or iodide solution containing diantipyrinylmethane (separation from rare earths, Al, Cr, Ni, Co, Be, Mg).[133]

Trioctylamine has been used in the extraction of Ga (pH 2, SCN⁻ concentration 0.2–8 M).[134] The extraction of Co as $Co(SCN)_4^{2-}$ by trioctylamine in various diluents has been studied.[135] Manganese can be separated from Ca and Mg by extraction of the $Mn(CNS)_6^{4-}$ associate with trioctylmethylammonium cation into benzene.[136] The extraction of Cr(III) and divalent transition metals has been examined.[137] Amberlite LA-1 in CHCl₃ has found use in extraction of Mo, W, and Re, reduced with $SnCl_2$, from HCl solution.

Beryllium is extracted from 5 M SCN⁻ solution, 0.5 M in HCl, by mesityl oxide.[138] Unselective.

Acetophenone has found use in the extraction of Zr and Hf thiocyanates from HCl solution.

Various organic base cations are useful for extractions of complex thiocyanate anions into CHCl₃.[139]

Cyanate finds no use in quantitative extractive metal separations, but complex cyanides offer possibilities (see p. 876 and note 46 on p. 1019).

F. NITRATES

The extraction of nitrates is of great importance in separations of uranium and other actinide elements. The literature on nitrate extractions is voluminous and will not be reviewed thoroughly here,[140] but some main features of these extractions will be touched upon.

1. Ethyl Ether

Peligot in 1842 found solid uranyl nitrate to be soluble in ethyl ether and made use of this property in preparing this salt from pitchblende. The extraction of uranyl nitrate into ether from an aqueous phase was a later development.

Table 9A-16 summarizes the extractability of metal nitrates with an equal volume of ethyl ether from 8 M HNO$_3$.[141] Gold(III) and cerium(IV) are the only metals extracted really well.

From neutral or slightly acidic solutions, metals are extracted as hydrated and ether-solvated nitrates, for example,

$$UO_2(NO_3)_2 \cdot 2\, H_2O \cdot 2(C_2H_5)_2O$$

At higher concentrations of nitric acid, anionic nitrate complexes can be formed, which are extracted as solvated nitrato acids, for example, $(C_2H_5)_2O \cdot HUO_2(NO_3)_3$. The extractability of U and other metals can be increased by adding salting agents, namely nitrates of high or fairly high aqueous solubility, which by the common-ion effect increase the concentration of the species to be extracted. At the same time, by decreasing the solubility of the extracted species in the aqueous phase by the electrolyte effect (see p. 826), they further increase the extraction coefficient. Fig. 9A-10 illustrates the salting out of $UO_2(NO_3)_2$ by ammonium nitrate in HNO$_3$ solution. Generally, nitrates having a strongly hydrated cation are the most effective. Thus Al(NO$_3$)$_3$ and Ca(NO$_3$)$_2$ are better salting agents than NH$_4$NO$_3$. The inhibiting effect of such substances as F$^-$ and PO$_4^{3-}$ on the extraction is minimized or eliminated by using a salting agent whose cation (e.g., Al) complexes them. It may be better to use a salting agent

TABLE 9A-16
Diethyl Ether Extraction of Elements from 8 M Nitric Acid

Element	Extraction (%)	Element	Extraction (%)
Ag	2.4	Mn(II)	0.2
As(V)	14.4	Mn(VII)^a	<0.5
Au(III)	97	Mo(VI)	0.6
Be	1.4	P(PO$_4^{3-}$)	20.4
Bi	6.8	Pb	0.5
Cd	0.3	Sb(V)	<1
Ce(IV)	96.8	Th	34.6
Co	0.2	Ti	<0.5
Cr(VI)^a	>15	Tl(I)	<0.5
Fe(III)	0.13	Tl(III)	7.7
Ga	<0.2	U(VI)	65
Ge	2.2	V(V)	2.0
Hg(II)	4.7	Zr	~8

Not extracted (0.1% or less): Al, In, La, Sc, Y, Zn, Ni.

^{a} Decomposed.

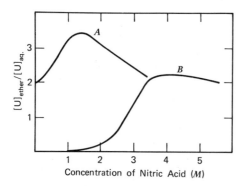

Fig. 9A-10. Distribution of U(VI) between diethyl ether and aqueous HNO_3 solutions: A, in the presence of NH_4NO_3 (saturated solution); B, in the absence of NH_4NO_3. From A. Norström and L. G. Sillén, *Svensk Kem. Tids.*, **60**, 232 (1948).

with a low concentration of HNO_3, than HNO_3 alone at a higher concentration, to avoid the greater extraction of undesired elements that may result from the latter condition. Because, even with salting agents, the extraction coefficient of U(VI) and other extractable metals is not very high, continuous extraction may be preferred. The continuous extraction may require several hours for completion. Ether reduces U(VI) in strong light, so that the extraction should not be carried out under such conditions. Nitric acid by itself is extracted into ether, and the acidity is decreased by continued extraction. The extraction coefficient is substantially independent of the uranium concentration. It is evident that the presence of chloride is objectionable.

Zinc nitrate has been recommended as the salting agent in the extraction of Th.[141] With a $3\ M\ HNO_3$ solution saturated with $Zn(NO_3)_2$, ~90% recovery of Th is obtained on extraction with an equal volume of ether. Lithium nitrate has also been used for salting out. It allows Sc to be extracted to the extent of 84% from $1\ M\ HNO_3$ (saturated with $LiNO_3$).

Ethers other than diethyl have been tried for the extraction of U and Th. Except for lesser solubility in water, they do not seem to have marked advantages.

TcO_4^- is extracted into ethyl ether from strong HNO_3, possibly as $(C_2H_5)_2O \cdot H^+TcO_4^-$ or similar species; E ~10 in 6–10 $M\ HNO_3$.[142]

2. Ketones

MIBK. From HNO_3 alone, this solvent apparently does not show any unusual extractive properties (Table 9A-17). However, it extracts U(VI) well from 2.3–2.7 $M\ Al(NO_3)_3$ containing 5 vol. % of concentrated ammonia (pH 0.7).[143] The recovery of U(VI) is a trifle better than 99% when 16 ml of such solution is extracted with 20 ml of MIBK. This extraction was investigated from the standpoint of the determination of

TABLE 9A-17
MIBK Extraction of Elements from HNO₃ Solution[a]

Element	log E				
	HNO₃(M)				
	2	4	6	8	10
Hg(II)	0.7	0.5	0.2	0.0	−0.2
Tc	−0.05	0.0	0.0	−0.2	−0.4
U(VI)	−0.5	0.0	0.2	0.4	0.3

log E of other elements <0.0. Tracer concentrations.

[a] T. Ishimori et al., *JAERI Rept.*, **1106** (1966).

U, so that the amounts of other elements extracted were not determined, but the simultaneous extraction of Th, Pu, Ce(IV), and Zr may be expected.

MIBK is used as solvent in the important extraction of tetrapropyl-ammonium uranyl trinitrate (p. 993).

Evaporation of MIBK solutions containing HNO_3, that is, those that have been shaken with HNO_3-containing solutions, should be avoided. It has been observed by one of us (H.O.) that small explosions can result on evaporation.

Mesityl Oxide. This solvent, isopropylidene acetone,

$$(CH_3)_2C{=}CH{-}CO{-}CH_3$$

has been used by a number of workers for the extraction of U(VI) and Th. With $Al(NO_3)_3$-salted (2.5 M) solutions of HNO_3, good recoveries of both elements are possible, for example, ~99.9% of Th (equal volumes of phases). Other elements extracted include Zr (to a large extent), V (~10%), Al, Fe(III) (~1%), Sn(II), Pt(IV), and In. Metals extracted very slightly include Be, Mg, Ti, Mn, Co, Ni, Zn, Mo, Pb, and the rare earths. Y and Ce are slightly extracted. It is seen that a separation of Th and U can be obtained from the common elements and from some of the less common, especially if the extract is washed with a solution having the same composition as that from which the extraction was made.

Mesityl oxide is subject to oxidation by air (brown color) and more rapidly by HNO_3. It is toxic.

3. Ethyl Acetate

This solvent has been a popular extractant for U(VI) nitrate. The extraction coefficient depends more on the concentration of the salting nitrate than of HNO_3, pointing to the extraction of solvated neutral salt. Equilibrium is reached in a minute or less of shaking, and U is easily returned to the aqueous phase by shaking the extract with water. Aluminum nitrate is a good salting-out agent, but Ca and Mg nitrates have also been used. The required concentration of $Al(NO_3)_3 \cdot 9\,H_2O$ is about 1 g/ml of final solution. The recovery of U depends on the relative phase volumes and conditions of washing, but in practice a recovery of ~99% may be expected, especially if a double extraction is made. Thorium is extracted to a large extent (although not as well as U), as is Ce(IV), and Zr appears to be extracted partially. Most other metals, including Al, As, Ca, Co, Cu, Fe(III), Mg, Mn, Mo(VI), and V(V), are hardly extracted (<1%, some much less).[144]

Because of its hydrolysis, ethyl acetate is not suitable for continuous extraction.

A 2:1 mixture of ethyl acetate and acetone extracts Th from solutions salted with $Al(NO_3)_3$ (separation from rare earth elements).

4. Tri-*n*-butyl Phosphate (TBP)

Undiluted TBP is a good, or moderately good, extractant for Au(III), Ce(IV), Pu(VI), Sc, Th, U(VI), and Y from HNO_3 solutions.[145] See Table 9A-18. It is chiefly of interest for separations of U, Th, Sc, and sometimes of the rare earth elements. It can be diluted with an inert solvent to give solutions that are less viscous than undiluted TBP. The resulting decrease in extractive power can be more than compensated by salting the aqueous solution with various nitrates. Sodium nitrate and $Al(NO_3)_3$ are effective salting-out agents in the extraction of U. With 60 g $Al(NO_3)_3 \cdot 9\,H_2O$ in 100 ml of 2 M HNO_3, an extraction coefficient of ~1000 is attained with 20% TBP in hexane. Diluents commonly used include kerosene, CCl_4, C_6H_6, hexane, isooctane, and MIBK. At low HNO_3 concentrations, U is extracted mostly as $UO_2(NO_3)_2 \cdot 2\,TBP$. Nitric acid is extracted by TBP, especially at high concentrations, a fact that may need to be taken into account in extraction of metals from HNO_3 solutions not containing salting nitrates.

Masking agents are of aid in improving selectivity in TBP extractions. EDTA reduces the amount of Th and Zr extracted, as well as such elements as Fe(III), Cu, Ni, Mn(II), and Cr(III), otherwise reportedly extracted in small amounts, if the acidity is not too great (pH ~1).[146] EDTA prevents the hydrolysis and possible precipitation of Bi, which would otherwise lead to emulsion formation. Washing the TBP extract with suitable solutions is also helpful in removing undesired elements. Saturated oxalic acid solution has been used to remove Zr from a 100% TBP extract.[147]

TABLE 9A-18
TBP Extraction of Elements from Nitric Acid^a

Element	log E				
	HNO$_3$(M)				
	1	3.5	6	11	14
Au(III)	3.9	3.2	2.25	0.6	0.3
Bi	0.3	−0.1	−0.7	−1.5	−1.3
Cu(II)	−2.8		−3.8	−3.7	−2.8
Fe(III)	−2.2	−2.1	−2.0	−0.3	1.2
Hf		−0.6	0.9		
Hg	1.0	0.0	−0.4	−0.6	−0.5
Lu		−0.5		1.5	2.3
Mo	−1.7	−1.8	−1.9	−0.8	0.5
Nb	−0.7	0.5	0.7	1.3	1.75
Np(IV)	0.8	1.5	2.0	1.9	
Os	1.5	0.8	0.7	0.5	0.5
Pa	0.25	1.25	1.7		
Pd	0.5	−0.3	−1.3	−2.2	−1.8
Po	−0.3	−1.0	−1.7	−1.3	−0.8
Pt	0.25	−0.7	−1.7	−2.4	−2.3
Pu(VI)	1.7	1.8	2.0	1.8	
Re	1.2	0.3	−0.5	−1.0	−1.1
Ru	0.8	0.3	−0.5	−1.9	
Sc	−0.5	−0.05	0.3	1.6	2.5
Th	0.7	1.5	1.9	2.6	2.7
Tl(III)	−0.5	−0.9	−1.2	−2.0	−1.9
Ti	−2.2	−2.0	−1.5		0.6
U(VI)	1.5	2.0	2.4		
V	−1.6	−1.9		−1.2	−0.4
W	1.5	0.6	−0.4	−1.1	−0.8
Y	−1.0	−0.3	0.0	1.1	1.9
Zr			0.75		

For other elements, log $E < -1$ for most HNO$_3$ concentrations. Tracer concentrations of elements used; some elements in micro and macro concentrations may not show the same behavior.

^{a} T. Ishimori and E. Nakamura, *JAERI Rept.*, **1047** (1963).

As already mentioned for some of the other solvents, phosphate and fluoride do not interfere in the presence of $Al(NO_3)_3$.

Recovery of U from a TBP phase is not as easy as from an ether or ethyl acetate phase. Evaporation of TBP is troublesome because of its high boiling point and creeping; the residue is likely to contain carbonaceous matter. Stripping of U with a carbonate solution (after a water wash to remove HNO_3) or NH_4OH–H_2O_2 solution has been recommended.

TBP in MIBK has been suggested for the extraction of Th from HNO_3–$Al(NO_3)_3$ solution, mesotartaric acid being used to prevent extraction of Zr.[148] TBP alone has also been used for HNO_3 solutions.

Undiluted TBP is useful for removal of the bulk of uranium from $6\,M\,HNO_3$ preliminary to the determination of Al, Cd, Ca, Cr, Co, Cu, Fe, Pb, Mg, Mn, Ni, K, Na and Zn (atomic absorption).[149]

5. Tri-*n*-octylphosphine Oxide (TOPO)

A solution of TOPO (Chapter 9C) in an inert solvent such as cyclohexane, benzene, toluene, or carbon tetrachloride extracts the actinide elements (IV and VI) from 1–$6\,M\,HNO_3$ (Table 9A-19). Other elements extracted under these conditions include Au, Bi, Hf, Hg, Sc, Sn, and Zr, as well as others in lesser amounts, depending on the HNO_3 concentration. Fe(II) is not extracted, Fe(III), weakly. Other metals not much extracted include the rare earths, Al, Mg, Co, Zn, Mn, Cu, Cr(III). Masking with fluoride (+excess Al) has been used to hinder extraction of such metals as Zr, Ti and Th. Back-extraction of U from the organic phase has been effected with carbonate.

6. Di-(2-ethylhexyl)-*o*-phosphoric Acid (HDEHP)

The analytical usefulness of HDEHP lies largely in its ability to extract elements of the IVA, VA, and VIA groups of the periodic table from strong HNO_3 and other acid solutions. This is illustrated by the following figures giving the percentage extraction when a $10\,M\,HNO_3$ solution is extracted with an equal volume of $0.75\,M$ HDEHP in *n*-heptane[150]:

Zr	>99	Ce(III)	<0.1	Tc, Te	~0
Th	>98	Ce(IV)	>99	Mo(VI)	much
U(VI)	>98	Ru	<2	R.E.	~0

When the extract is washed with $10\,M\,HNO_3$ containing H_2O_2, Ce, and Mo are transferred to the aqueous phase, and Zr remains in the organic

TABLE 9A-19
Extraction of Elements from Nitric Acid Solution with
TOPO (5% Solution in Toluene)[a]

Element	log E				
	HNO$_3$(M)				
	1	3.5	6	11	15
Au(III)	2.8	2.0	0.3	−0.5	−0.8
Bi	0.7	−2.5	−3.1		
Fe(III)	−2.4	−2.5	−2.2	−1.2	−0.2
Hf	1.3	1.1	1.0	1.0	0.9
Hg	0.1	−0.5	−0.7	−0.4	−0.2
Mo	−0.6	−1.7	−1.6	−0.5	0.8
Nb	−0.6	−0.2	0.0	0.4	0.7
Os	−0.2	−0.7	−0.6	−0.6	0.4
Pa	2.3	1.7	2.2	2.1	2.0
Sc	0.3	−0.1	−0.4	−0.3	−0.3
Sn	−0.7	−0.5	−0.3	−0.1	0.5
Th	1.9	1.8	1.7	1.2	0.1
Ti	−1.6	−1.8	−1.2	−0.5	0.7
Tl	−1.1	−1.0	−1.0		
U(VI)	3.3	2.1	2.5	1.5	−0.2
Y	−1.3	−2.8	−3.0		
Zr	1.0	1.4		1.75	1.7

Tracer concentrations of elements. See the original for extraction coefficients of elements extracted to lesser extents. V(IV) is hardly extracted.

[a] T. Ishimori and E. Nakamura, *JAERI Rept.*, **1047** (1963). Also see C. A. Horton and J. C. White, *Anal. Chem.*, **30,** 1779 (1958).

phase with most of the U and Th. Nb is extracted readily from solutions >3 M in HNO$_3$ (or 5 M in HClO$_4$), Mo almost completely from >5 M HNO$_3$, and Sc almost completely from 1–11 M HNO$_3$ (and HCl).[151] Ti is almost completely extracted from 1–11 M HNO$_3$ (and HClO$_4$ and HCl).

HDEHP can be used to extract large amounts (0.5 g) of In and Bi (R ~98%), while leaving most of a number of micro elements in the aqueous phase. Extraction of nitric acid solutions, with an equal volume of a 1:2 mixture of commercial HDEHP (70%) and CCl$_4$, left the following amounts of macro elements in the aqueous solution at the nitric acid concentration indicated[152]:

In	0.5 M, <10 mg
Ga	pH 3, 20 mg
Bi	0.5 M, 10 mg
Sb	4.5 M H_2SO_4, 10 mg (2 extractions)

The amount of metal taken was 0.5 g (in 30 ml of aqueous phase). The percentages of micro elements extracted from 0.5 M HNO_3 were:

Cr(III)	5	Zn	2
Cr(VI)	50	As(III)	0.1
Fe(III)	99	As(V)	0.2
Mn(II)	<1	Se	3
Co	<1	Cd	7
Ni	8	Sb(III)	30
Cu	2	Sb(V)	2
		Te	3

G. SULFATES

Included among the relatively few reagents used for extraction of metals from sulfuric acid solution are primary, secondary, and tertiary long-chain amines and quaternary ammonium salts. The extraction of uranium in this way is of considerable interest. See Table 9A-20 for extraction coefficients of metals with triisooctylamine in xylene. Coefficients have also been determined for the extraction of a score of elements from ammonium sulfate (and ammonium carbonate) solution into toluene with trioctylmethylammonium chloride.[153]

Extraction of Re(VII) by tertiary amines in $CHCl_3$ from H_2SO_4 and other acid solutions has been investigated.[154] Tribenzylamine ($CHCl_3$) is preferred for extraction of Re from H_2SO_4 or H_3PO_4 solution (incomplete separation from Mo).

Tributyl phosphate (undiluted) extracts Re, Tc, Os(VIII), and Hg(II) well from 0.1–10 N H_2SO_4 ($E > 10$).[155] Niobium is extractable from 10 M H_2SO_4 by TBP in benzene.[156] Palladium(II) has been reported to be extracted from 20% v/v H_2SO_4 solution by MIBK ($E \sim 4$).[87] Niobium and Zr are not extracted.

H. CYANIDES

Extractive separation of metals as complex cyanides is unlikely to become a favored separation method, but it can have value at times. Tetrahexylammonium ion has been studied as a cationic reagent for extraction of cyanides into MIBK, more from the theoretical than the

TABLE 9A-20
Extraction of Metals from Sulfuric Acid
Solution with Triisooctylamine in Xylene
(0.11 M)[a]

Element	log E			
	$H_2SO_4(N)$			
	0.01	0.1	1.0	10.0
Au(III)	1.5	3.1	3.7	
Ba	−0.7	−0.3	−0.3	
Bi	0.0	−0.3	−1.75	−2.4
Hf	2.1	1.9	−0.3	
Mo	0.6	1.9	1.3	−0.4
Nb	0.7	0.8	1.2	−0.2
Np(IV)	1.75	1.2	0.25	−0.9
Os	1.4	1.1	0.7	
Pa	1.4	1.7	1.2	
Pt	0.8	0.8	0.6	0.4
Re	3.2	3.6	3.25	
Ru	0.1	0.3	−0.3	−1.4
Sb	−0.3	−0.2	−0.2	
Se(IV)	1.5	1.5	1.6	
Ta	0.2	−0.3	−0.9	
Tc	2.0	2.0	2.2	2.7
Th	−0.5	0.0	−1.6	
U(VI)	2.4	2.8	1.8	
V		2.9		−0.3
W	0.7	1.5	0.8	
Zr	0.6	0.9	1.3	−0.6

[a] From T. Ishimori, E. Akatsu, W. Cheng, K. Tsukuechi, and T. Osakabe, *JAERI Rept.*, **1062** (1964). Tracer concentrations of metals. See the original for extraction coefficients of metals having log values <0.1. Also see the original for coefficients for extraction with Primene JM-T in xylene and Amberlite LA-1 in xylene.

practical viewpoint.[157] Singly charged $Au(CN)_4^-$, $Au(CN)_2^-$, $Ag(CN)_2^-$, and $Hg(CN)_3^-$ are extracted to a much greater extent than doubly charged $M(CN)_4^{2-}$ (M = Zn, Cd, Hg, Ni, Pd, and Pt) (Table 9A-21). The influence of the geometry of the complex ions is seen in the greater extractability of $Au(CN)_4^-$ (4-coordinate square planar) than of $Au(CN)_2^-$ (linear). In the series of $M(CN)_4^{2-}$ complex ions having the same geometry, extractability

TABLE 9A-21
Constants for Extraction of Complex Cyanides
with $[N(C_6H_{13})_4^+][Co(NH_3)_2(NO_2)_4^-](=NR_4A)$ in
MIBK at $25^{\circ a}$

	K_{ext}		$K_{ext} \times 10^4$
$Ag(CN)_2^-$	0.38	$Zn(CN)_4^{2-}$	4.5
$Au(CN)_2^-$	15.8	$Cd(CN)_4^{2-}$	7.8
$Au(CN)_4^-$	165	$Hg(CN)_4^{2-}$	8.0
$Hg(CN)_3^-$	0.41	$Ni(CN)_4^{2-}$	0.11
		$Pd(CN)_4^{2-}$	1.22
		$Pt(CN)_4^{2-}$	3.56

^{a} $M(CN)_4^{2-} + 2NR_4A \rightleftharpoons (NR_4)_2M(CN)_4 + 2A^-$ and similar reaction for singly charged cyanide complexes.

increases with ionic volume (and atomic number):

$Zn < Cd \ll Hg$ (tetrahedral ions)
$Ni \ll Pd < Pt$ (square-planar ions)

Tetrahedral ions are more readily extracted than comparable square-planar ions (explained by greater energy gain in transporting the larger ions into the organic phase from water). As expected, triply charged $Cu(CN)_4^{3-}$ and $Fe(CN)_6^{3-}$, and $Fe(CN)_6^{4-}$, are not appreciably extracted.

A number of workers have shown that singly charged cyanide complexes are extractable with basic dye cations (for determination).

NOTES

1. Extractions are necessarily distributions, but not all liquid–liquid distributions are extractions. A gaseous species such as NH_3 or a volatile one such as acetic acid can, in a closed system, be distributed between two liquid phases in communication but not in contact.
2. N. I. Sax, *Dangerous Properties of Industrial Materials*, 4th ed., Van Nostrand Reinhold, New York, 1975.
3. Compare H. M. N. H. Irving, Recommended Nomenclature for Liquid–Liquid Distribution, *Pure Appl. Chem.*, **21**, 111 (1970); A. W. Ashbrook and G. M. Ritcey, *Can. Mining J.*, **1972**, 70. Liquid–liquid extraction is frequently called solvent extraction, a term to which purists will object as a repetition of words. Apparently, those who use this term associate solvent with organic solvent. Leaching of a component from a solid mixture is sometimes called extraction of a solid (p. 849).

4. Typists and typesetters are prone to ignore the difference between o and 0, and we, therefore, use *org* to indicate the organic solvent phase.

5. The effect produced by neutral molecules of the chelating agent itself (e.g., forming $ML_2 \cdot 2\,HL$) is usually not considered synergism, though it is a very similar phenomenon, nor is the increase in extractability sometimes produced by using a mixture of two solvents instead of either alone. It may be noted that compounds other than chelates can be extracted synergically. Thus the extraction of HgI_2 into hexane is greatly increased in the presence of TBP, as shown by T. Sekine and Y. Hasegawa in Kertes and Marcus, *Solvent Extraction Research*, p. 295.

6. Inert is used here in a relative sense. Probably few organic solvents are entirely inert with respect to their solutes. Those having weak solvating power are conveniently described as inert.

7. Ru: F. S. Martin, *J. Chem. Soc.*, 2564 (1954). Os: R. D. Sauerbrunn and E. B. Sandell, *J. Am. Chem. Soc.*, **75**, 4170 (1953); *Anal. Chim. Acta*, **9**, 86 (1953).

8. J. Setchenow, *Z. phys. Chem.*, **4**, 117 (1889). Salting out also comes into play in extraction of electrolytes into polar solvents, but other effects are also involved in such extractions so that the state of affairs is less simple than for nonpolar substances extracted into more or less nonpolar organic solvents (see Fig. 9A-1*b*).

9. However, W. L. Belew, G. R. Wilson, and L. T. Corbin, *Anal. Chem.*, **33**, 886 (1961), use $Al(NO_3)_3$ ($0.2\,M\ HNO_3 - 1.2\,M\ Al(NO_3)_3$) to assure quantitative extraction of RuO_4 after AgO oxidation. Shaking for 2 min is sufficient.

10. N. Jordanov, M. Pavlova, and D. Bojkova, *Talanta*, **23**, 463 (1976).

11. S. A. Katz, W. M. McNabb, and J. F. Hazel, *Anal. Chim. Acta*, **25**, 193 (1961); **27**, 405 (1962). Both chromic and perchromic acids are extracted by TBP (D. G. Tuck, *ibid.*, **27**, 296). Cr(VI) can be determined absorptiometrically in the organic solutions.

12. For example, see R. B. Heslop and E. F. Pearson, *Anal. Chim. Acta*, **39**, 209 (1967), who find that *n*-butyl acetate extracts molybdophosphoric acid, but little molybdoarsenic acid; the latter is extracted well by cyclohexanone at pH 2. T. M. Malyutina, S. B. Savvin, V. A. Orlova, V. A. Mineeva, and T. I. Kirillova, *Zh. Anal. Khim.*, **29**, 925 (1974), found that with a 1:3 mixture of isobutyl alcohol + $CHCl_3$ the extraction coefficients E of molybdophosphoric and molybdoarsenic acid were 190 and 0.05, and that with 1:1:2 isobutyl alcohol + ethyl acetate + $CHCl_3$ the respective E values were 0.04 and 23. For other differential separation schemes see Chapter 5, Section V. F. Alt and F. Umland, *Z. Anal. Chem.*, **274**, 103 (1975), studied the extraction of Mo heteropoly acids of P, As, Si, and Ge with di- and trioctylamine in chloroform. These substances, as cations in acid solution, extract the heteropoly acids in anionic form as ion associates.

13. Separation (incomplete) of Mo from Re by isoamyl acetate extraction of the

former as molybdophosphoric acid has been suggested by V. Yatirajam and L. R. Kakkar, *Anal. Chim. Acta*, **54**, 152 (1971). Phosphomolybdenum blue, obtained by reduction of molybdophosphoric acid with hydrazine sulfate, is extractable into MIBK, but the extraction does not seem to be applicable to trace quantities of Mo (V. Yatirajam and J. Ram, *Talanta*, **20**, 885 (1973)).

14. W. Goishi and W. F. Libby, *J. Am. Chem. Soc.*, **74**, 6109 (1952). Extraction of Tc (and Re) by pyridine and methyl ethyl ketone: S. J. Rimshaw and G. F. Malling, *Anal. Chem.*, **33**, 751 (1961). See note 10.

15. Iv. Boyadjov, R. de Neve, and J. Hoste, *Anal. Chim. Acta*, **40**, 373 (1968).

16. A. D. Matthews and J. P. Riley, *Anal. Chim. Acta*, **51**, 455 (1970). A. A. Samadi, P. Ailloud, and M. Fedoroff, *Anal. Chem.*, **47**, 1847 (1975), use acetone and 7 *M* NaOH to separate Re from Mo and W.

17. Yu. A. Zolotov, B. Z. Iofa, and L. K. Chuchalin, *Extraction of Metal Halide Complexes* [in Russian], Izdat. Nauka, Moscow, 1973. 378 pp., ~2000 references. Chapter contents: 1. Theoretical principles; 2. Extraction of halide, thiocyanate, and cyanide complexes; 3. Extraction properties of 40 elements; 4. Practical use.

18. Review of metal halide extraction, particularly from the standpoint of the stability of the halide complexes: B. Ya. Spivakov, O. M. Petrukhin, and Yu. A. Zolotov, *J. Anal. Chem. USSR*, **27**, 1435 (1972). Extraction of covalent halides of As, Sb, and Bi with CCl_4: J. Chwastowska and W. Podgorny, *Chemia Anal.*, **30**, 53 (1975).

19. As: W. Fischer and W. Harre, *Angew. Chem.*, **66**, 165 (1954). Ge: W. A. Schneider, Jr. and E. B. Sandell, *Mikrochim. Acta*, 263 (1954).

20. Schweitzer and Storms (reference in note 27) report the extraction of 1 to 2% Sb(V) into an equal volume of $CHCl_3$ from 6–11 *M* HCl. Also see note 94, Chapter 9B. According to J. Bartura and W. Bodenheimer, *Israel J. Chem.*, **6**, 61 (1968), extraction of an acid chloride solution of Al^{3+} with $CHCl_3$ leads to a loss of a few percent of Al. This observation requires confirmation.

21. C. L. Rulfs, R. A. Pacer, and R. F. Hirsch, *J. Inorg. Nucl. Chem.*, **29**, 687 (1967).

22. N. M. Kuz'min, G. D. Popova, I. A. Kuzovlev, and V. S. Solomatin, *Zh. Anal. Khim.*, **24**, 899 (1969).

23. For our purpose, a review of all this work is not necessary. Those wishing to look into this subject, especially from the physicochemical standpoint, should consult Marcus and Kertes, *op. cit.* (especially p. 887, 634, 467) and Diamond and Tuck, *op. cit.*, p. 158, for analyses of experimental results and further references.

24. The effect disappears at very low concentrations such as those involved in trace analysis. D. C. Grahame and G. T. Seaborg, *J. Am. Chem. Soc.*, **60**, 2524 (1938), found $E = 17.5–18.1$ for 10^{-12} *M* Ga and 16.9 for 0.0016 *M* Ga, both in 6 *M* HCl. G. K. Schweitzer and W. N. Bishop, *J. Am. Chem.*

Soc., **75,** 6330 (1953), found the percentage extraction of Au(III) from 1.5 *M* HCl into an equal volume of ethyl ether to vary as follows with gold concentration: tracer concentrations, 88; 10^{-7} *M*, 91; 10^{-4} *M*, 94; 0.1 *M*, 98, all at room temperature.

25. See references in note 23 for an examination of this effect.

26. The extraction coefficient given holds for 5–10 mg $FeCl_3$/ml. Lower concentrations of iron are extracted less well, but still well enough, so that as little as 1 mg Fe can be recovered by successive extractions. Cf. Fig. 9A-4.

27. G. K. Schweitzer and L. E. Storms, *Anal. Chim. Acta*, **19,** 154 (1958), testing several ethers, found Sb(V) to be extracted 98% or better from 3 to 8 *M* HCl with an equal volume of isopropyl ether.

28. If microgram quantities of iron are to be isolated, other extraction methods (or nonextraction) will usually be preferred.

29. G. A. Dean and J. F. Herringshaw, *Analyst*, **86,** 106 (1961).

30. Z. Marczenko, H. Kalowska, and M. Mojski, *Talanta*, **21,** 93 (1974).

31. As ethers become less soluble in HCl solution, the maximum in *E* as a function of HCl concentration becomes less pronounced or disappears, as already evident with isopropyl ether.

32. I. G. Yudelevich et al., *J. Anal. Chem. USSR*, **21,** 1296 (1966).

33. P. I. Artyukhin, E. N. Gil'bert, and V. A. Pronin, *J. Anal. Chem. USSR*, **22,** 92 (1967).

34. E. Bankmann and H. Specker, *Z. Anal. Chem.*, **162,** 18 (1958). The effectiveness of ketones as extractants was demonstrated earlier by V. I. Kuznetsov, *J. Gen. Chem. (USSR)*, **17,** 175 (1947).

35. A. Claassen and L. Bastings, *Z. Anal. Chem.*, **160,** 403 (1958).

36. A. I. Busev and V. A. Frolkina, *Zh. Neorg. Khim.*, **14,** 1289 (1969).

37. H. Goto, Y. Kakita, and N. Ichinose, *Sci. Rept. Res. Inst. Tohoku Univ.*, **A19,** 219 (1967).

38. E. Gagliardi and H. P. Wöss, *Z. Anal. Chem.*, **248,** 302 (1969). The same authors, *Anal. Chim. Acta*, **48,** 107 (1969), use CCl_4 as diluent for methyl propyl ketone and ethyl butyl ketone, and report ~100% extraction of Fe(III), Ga, As(III), Mo (with 2-butanone), In, Te(IV), Au(III), and Tl(III) from 6 *M* HCl. Metals partly extracted include V(V→IV), Cu, Zn, Cd, Sn(IV), Sb, Pt, and Hg; not extracted: Ti, Zr, Hf, Nb, Ta, Cr(III), Mn, Ni, Co, Pb, Bi, Th, Al, Pd. Mo(VI) is extracted as $HMoO_2Cl_3$ and $HMoOCl_5$.

39. N. Jordanov and L. Futekov, *Talanta*, **12,** 371 (1965).

40. A. Alian, *Mikrochim. Acta*, 988 (1968); gold: V. M. Shinde and S. M. Khopkar, *Anal. Chim. Acta*, **43,** 146 (1968).

41. N. Jordanov and I. Havesov, *Z. Anal. Chem.*, **244,** 176 (1969).

42. K. A. Orlandini, M. A. Wahlgren, and J. Barclay, *Anal. Chem.*, **37,** 1149 (1965).

43. C. R. Boswell and R. R. Brooks, *Anal. Chim. Acta.*, **33,** 117 (1965).

44. I. M. Gibalo, D. S. Al'badri, and G. V. Eremina, *Zh. Anal. Khim.*, **22,** 816 (1967). Lately, G. E. Baiulescu and I. Craciun, *Revta Chim.*, **25,** 578 (1974), have studied the extraction of Au(III) from chloride solutions into butyl alcohols at pH 1–5. They found extractability of gold to decrease in the order butanol, *s*-butyl alcohol, and *t*-butyl alcohol. The diluents benzene, toluene, and *p*-xylene (20–40%) improved extraction.

45. R. W. Lowe, S. H. Prestwood, R. R. Rickard, and E. I. Wyatt, *Anal. Chem.*, **33,** 874 (1961).

46. L. F. Greenland and E. G. Lillie, *Anal. Chim. Acta*, **69,** 335 (1974).

47. J. E. Wells and D. P. Hunter, *Analyst*, **73,** 671 (1948).

48. T. V. Ramakrishna, S. A. Rahim, and T. S. West, *Talanta*, **16,** 847 (1969).

49. T. M. Malyutina, V. A. Orlova, and B. Ya. Spivakov, *Zh. Anal. Khim.*, **29,** 790 (1974). Various oxygen-type solvents were tried. Solvents of relatively high dielectric constant or donor capacity (dichlorodiethyl ether, MIBK, TBP, isoamyl alcohol) extract $FeCl_3$ poorly in the presence of HF, whereas those having low dielectric constants (isopropyl ether, amyl acetate, dibutyl ether) extract iron better in the presence of HF. It is suggested that the extraction of iron under these conditions is due to coextraction of $HFeCl_4$ with H_nF_n molecules.

50. According to H. Specker and R. Shirodker, *Z. Anal. Chem.*, **214,** 401 (1965), a 3.4 M solution of TBP in benzene or isooctane extracts (%) from an equal volume of 7 M HCl: Fe(III) 99.9, Zn 3, Cd 1, and negligible amounts of Cu, Pb, Co, and Ni (from 10^{-2} to 10^{-5} M solutions). Up to a concentration of 0.1 M, Fe(III) has practically no effect on the distribution of Zn, Cd, Cu, Pb, Co, and Ni, but above that concentration, it has a suppressing effect on the extraction of Zn and Cd. See Yu. A. Zolotov and V. I. Golovanov, *Zh. Anal. Khim.*, **26,** 1880 (1971). M. E. M. S. de Silva, *Analyst*, **100,** 517 (1975), found undiluted TBP to extract Mo(VI) well from 5–8 M HCl.

51. A. Alian, R. Shabana, W. Sanad, B. Allam, and K. Khalifa, *Talanta*, **15,** 262 (1968).

52. A. K. De and A. K. Sen, *Talanta*, **14,** 629 (1967).

53. D. Ishii and T. Takeuchi, *Japan Analyst*, **10,** 1125 (1961). A. R. Eberle and M. W. Lerner, *Anal. Chem.*, **29,** 1134 (1957), use a 1 : 1 mixture of TBP and MIBK and 7 M HCl for the U–Th separation.

54. R. P. Larsen and C. A. Seils, *Anal. Chem.*, **32,** 1863 (1960).

55. D. F. Peppard and M. V. Gergel, *U.S. AEC*, ANL-**4490,** 64 (1950); Peppard, G. W. Mason, and M. V. Gergel, *J. Inorg. Nucl. Chem.*, **3,** 370 (1957).

56. T. Sato, *Anal. Chim. Acta*, **52,** 183 (1970). (HDEHP extraction of Zr is also discussed in this paper.) Hf is also easily extracted: N. Ichinose, *Talanta*, **19,** 1644 (1972).

57. G. Roland and B. Gilbert, *Anal. Chim. Acta*, **60,** 57 (1972).

58. R. Stella and N. Genova, *Radiochem. Radioanal. Lett.*, **16**, 273 (1974).

59. J. C. White and W. J. Ross, NAS-NS **3102** (1961).

60. H. Saisho, *Bull. Chem. Soc. Japan*, **35**, 514 (1962).

61. H. G. Petrow, *Anal. Chem.*, **37**, 584 (1965).

62. I. S. Levin, N. A. Balakireva, and L. A. Novosel'tseva, *Zh. Anal. Khim.*, **29**, 1095 (1974).

63. N. A. Ivanov, I. P. Alimarin, I. M. Gibalo, and G. F. Bebikh, *Izv. Akad. Nauk SSSR Ser Khim.*, 2664 (1970); *Anal. Abst.*, **21**, 2537 (1971). This is a separation from Ti, which is not extracted from 8.5 M HCl (*Zh. Anal. Khim.*, **24**, 1521).

64. B. E. McClellan and V. M. Benson, *Anal. Chem.*, **36**, 1985 (1964).

65. A. A. Vasilyeva et al., *Talanta*, **22**, 745 (1975).

66. A. Mizuike, K. Fukuda, and Y. Ochiai, *Talanta*, **19**, 527 (1972).

67. A. P. Grimanis and I. Hadzistelios, *Anal. Chim. Acta*, **41**, 15 (1968). Values for extractability of elements from 5 M H_2SO_4–0.01 M I^- are also given. K. Studlar, *Coll. Czech. Chem. Commun.*, **31**, 1999 (1966), has shown that As(V) is readily extracted from 8 M H_2SO_4–~0.1 M Br^- by CCl_4.

68. K. Tanaka, *Japan Analyst*, **18**, 315 (1969).

69. I. Tsukahara and T. Yamamoto, *Anal. Chim. Acta*, **64**, 337 (1973).

70. M. M. L. Khosla and S. P. Rao, *Anal. Chim. Acta*, **68**, 470 (1974); **58**, 389 (1972).

71. P. I. Artyukhin et al., *Dokl. Akad. Nauk SSSR*, **197**, 337 (1971).

72. A. R. Denaro and V. J. Occleshaw, *Anal. Chim. Acta*, **13**, 239 (1955), who also used methyl ethyl ketone.

73. H. Irving and F. J. C. Rossotti, *J. Chem. Soc.*, 1927 (1955). For extraction of indium halides by various organic solvents, see the same authors, *ibid.*, 1946 (1955).

74. H. Irving and F. J. C. Rossotti, *J. Chem. Soc.*, 1938 (1955).

75. T. V. Gurkina and E. Y. Litvinova, *J. Anal. Chem. USSR*, **24**, 267 (1969).

76. A. Alian, *Mikrochim. Acta*, 981 (1968). The use of trilaurylamine was also studied.

77. A. R. Byrne and D. Gorenc, *Anal. Chim. Acta*, **59**, 81 (1972). CCl_4 extraction: M. D. Morris and L. R. Whitlock, *Anal. Chem.*, **39**, 1180 (1967).

78. Byrne and Gorenc[77] give curves of extractability as a function of H_2SO_4 and I^- concentrations. K. Tanaka, *Japan Analyst*, **9**, 574 (1960), finds $E_{As} \cong 700$ for benzene, 1.0 M KI and 6 M H_2SO_4. Also see *ibid.*, **10**, 1087 (1961). CCl_4 is almost as good an extractant and more convenient in practice. It is used by V. Stara and J. Stary, *Talanta*, **17**, 341 (1970). For extractive separation of As and Sb (bio samples) see A. R. Byrne, *Anal. Chim. Acta*, **59**, 91 (1972).

79. Similar results were obtained by K. Tanaka, *Japan Analyst*, **11**, 332 (1962).

80. R. W. Ramette, *Anal. Chem.*, **30**, 1158 (1958); B. G. Julin, Ph.D. Thesis, University of Minnesota, 1965.

81. I. Havezov and M. Stoeppler, *Z. Anal. Chem.*, **258**, 189 (1972).

82. F. Jancik and J. Körbl, *Talanta*, **1**, 55 (1958).

83. L. E. Mattison and J. C. Wolford, *Anal. Chem.*, **38**, 1675 (1966), with reference to isolation of cadmium in its determination by a radiotracer method (iodine-131 in inactive KI solution).

84. Y. Kakita and H. Goto, *Sci. Rept. Res. Inst. Tohoku Univ.*, **A15**, 133 (1963).

85. C. L. Luke, *Anal. Chim. Acta*, **39**, 447 (1967).

86. They may be removed by cupferron–CHCl₃ extraction or by other means before the trace elements are determined with H_2Dz or other photometric reagent.

87. J. F. Duke and W. Stawpert, *Analyst*, **85**, 671 (1960).

88. P. W. West and J. K. Carlton, *Anal. Chim. Acta*, **6**, 406 (1952), who also tested other solvents. Rh and Ru were found to be extracted quite well in the presence of Pb, hardly at all in its absence. C. K. Hanson, *Anal. Chem.*, **29**, 1204 (1957), extracted iodotellurous acid from 1 *M* HCl, 0.6 *M* in iodide, into *n*-amyl alcohol.

89. E. Jackwerth, *Z. Anal. Chem.*, **202**, 81 (1964); **206**, 269 (1964); **211**, 254 (1965).

90. H. Hartkamp and H. Specker, *Talanta*, **2**, 67 (1959).

91. H. Hartkamp and H. Specker, *Naturwiss.*, **43**, 421 (1956).

92. H. Hartkamp and H. Specker, *Angew. Chem.*, **68**, 678 (1956).

93. H. Specker and H. Hartkamp, *Naturwiss.*, **43**, 516 (1956); *Angew. Chem.*, **69**, 397 (1957).

94. E. Gagliardi and P. Tümmler, *Talanta*, **17**, 93 (1970).

95. I. Kressin, *Talanta*, **19**, 197 (1972).

96. H. A. Mottola and E. B. Sandell, *Anal. Chim. Acta*, **24**, 301 (1961).

97. Kish et al., *loc. cit.*

98. S. de Moraes and A. Abrão, *Anal. Chem.*, **46**, 1812 (1974).

99. M. M. L. Khosla and S. P. Rao, *Anal. Chim. Acta*, **58**, 389 (1972).

100. C. W. McDonald and F. L. Moore, *Anal. Chem.*, **45**, 983 (1973); extraction of Zn and separation from Cd: C. W. McDonald and T. Rhodes, *ibid.*, **46**, 300 (1974). J. R. Knapp, R. E. Van Aman, and J. H. Kanzelmeyer, *ibid.*, **34**, 1374 (1962), effected a good separation of Cd from Zn by extracting the former from I⁻–H_2SO_4 solution with Amberlite LA-2 in xylene; Cd was stripped with Na_2CO_3.

101. P. V. Marchenko and A. I. Voronina, *Anal. Abst.*, **19**, 1065 (1970).

102. K. E. Burke, *Talanta*, **21**, 417 (1974); *Analyst*, **97**, 19 (1972).

103. For example, E. Jackwerth and H. Specker, *Z. Anal. Chem.*, **177**, 327 (1960), who deal with the fractional extraction of Hg, Bi, Cd, Zn, and In.

104. G. H. Faye and W. R. Inman, *Anal. Chem.*, **35,** 985 (1963); A. Hofer, *Z. Anal. Chem.*, **238,** 183 (1968).

105. The extraction of alkali metal polyiodides into nitrobenzene was noted (and investigated) in the early 1900s by Dawson and coworkers.

106. R. Bock and T. Hoppe, *Anal. Chim. Acta*, **16,** 406 (1957).

107. R. Bock and M. Herrmann, *Z. Anorg. Allgem. Chem.*, **284,** 288 (1956).

108. S. Kitahara, *Rept. Sci. Res. Inst. Tokyo*, **25,** 165 (1949); *Chem. Abst.*, **45,** 3743 (1951).

109. G. W. C. Milner, G. A. Barnett, and A. A. Smales, *Analyst*, **80,** 380 (1955).

110. G. W. C. Milner and J. W. Edwards, *Anal. Chim. Acta*, **13,** 230 (1955). Compare E. Gagliardi and E. Fuesselberger, *Mikrochim. Acta*, 700 (1972), who mention extraction of Re, Nb, and Ta from 5 M HF–5 M H_2SO_4 into ethyl methyl ketone.

111. G. R. Waterbury and C. E. Bricker, *Anal. Chem.*, **30,** 1007 (1958).

112. G. R. Waterbury and C. E. Bricker, *ibid.*, **29,** 1474 (1957). Diisopropyl ketone likewise extracts 98% Ta.

113. I. A. Shevchuk, N. A. Skripnik, and V. I. Martsokha, *J. Anal. Chem. USSR*, **22,** 752 (1967). Tri-n-decylamine and tri-n-octylamine do not extract iron (steric hindrance); neither do secondary amines.

114. A. Ya. Romanova, *Zav. Lab.*, **34,** 1444 (1968).

115. A. N. Nevzorov, N. S. Onoprienko, and S. N. Mordvinova, *Zav. Lab.*, **36,** 1176 (1970).

116. Review of thiocyanate extractions: Z. Kh. Sultanova et al., *Zh. Anal. Khim.*, **28,** 413 (1973).

117. H. Specker and G. Werding, *Z. Anal. Chem.*, **200,** 337 (1964). Compare *ibid.*, **177,** 10 (1960); **197,** 109 (1963).

118. R. Bock, *Z. Anal. Chem.*, **133,** 110 (1951).

119. C. Rozycki, *Chemia Anal.*, **14,** 755 (1969).

120. Based on Bock, *loc. cit.*, and W. Fischer and R. Bock, *Z. Anorg. Allgem. Chem.*, **249,** 146 (1942).

121. H. Specker and E. Bankmann, *Z. Anal. Chem.*, **149,** 97 (1956). Iron(III) and Al are separated similarly: H. Specker and H. Hartkamp, *ibid.*, **140,** 353 (1953).

122. M. Namiki, Y. Kakita and H. Goto, *Sci. Rept. Res. Inst. Tohoku Univ.*, **A14,** 239 (1962).

123. S. Tribalat and J. M. Caldero, *Bull. Soc. Chim. France*, 3187 (1964).

124. J. B. Headridge and A. Sowerbutts, *Analyst*, **97,** 442 (1972).

125. E. A. Mari, *Anal. Chim. Acta*, **29,** 303, 312 (1963), has described a separation (and determination) method for Ti, Nb, and Ta based on ethyl acetate as extractant for the thiocyanates. Titanium is essentially not extracted from a solution having low concentrations of H_2SO_4 and NH_4SCN, whereas Nb and Ta are extracted. Back-extraction with HF + HCl removes

Nb from the ethyl acetate extract, leaving Ta therein. Extraction coefficients are independent of the metal concentrations. J. Minczewski and C. Rozycki, *Z. Anal. Chem.*, **239,** 158 (1968), use isoamyl alcohol to extract Zn thiocyanate (separation from much Fe in presence of $S_2O_3^{2-}$ and F^-). Mo(VI) thiocyanate is extracted very well $(E \sim 10^4)$ by amyl alcohol from $\sim 1\,M$ SCN^- solution (0.3 *M* HCl). The extracted species is thought to be $MoO_2(SCN)_2$.

126. Kish et al., *loc. cit.*

127. Y. Hoshino, *J. Chem. Soc. Japan Pure Chem. Sec.*, **81,** 1574 (1960); *Japan Analyst*, **11,** 1032 (1962). Various other solvents were used by earlier workers, for example, W. Fischer et al., *Z. Anorg. Chem.*, **255,** 79 (1947); **255,** 277 (1948).

128. M. Aven and H. Freiser, *Anal. Chim. Acta*, **6,** 412 (1952). Also see *Anal. Chem.*, **25,** 856 (1953), in which, however, separations are not discussed.

129. F. P. Gorbenko et al., *J. Anal. Chem. USSR*, **23,** 1148 (1968).

130. E. D. Kuchkina et al., *Zh. Anal. Khim.*, **28,** 595 (1973).

131. C. Rozycki and E. Lachowicz, *Chemia Anal.*, **15,** 255 (1970). Results with MIBK, isoamyl alcohol, ethyl ether, and isobutyl acetate are also given. They are much poorer extractants than TBP. For extraction of Ni, Sn, Th, Sc, Be, Ba, Nd, Sm, and Er by TBP and other solvents see C. Rozycki and W. Suszczewski, *ibid.*, **17,** 1209 (1972).

132. J. P. Young and J. C. White, *Anal. Chem.*, **31,** 393 (1959).

133. V. P. Zhivopistsev and I. S. Kalmykova, *Uch. Zap. Permsk. Univ.*, **25,** 120 (1963); *Zh. Anal. Khim.*, **19,** 69 (1964).

134. N. B. Mikheev et al., *Anal. Abst.*, **21,** 2457 (1971).

135. H. Watanabe and K. Akatsuka, *Anal. Chim. Acta*, **38,** 547 (1967).

136. R. Pribil and J. Adam, *Talanta*, **20,** 49 (1973).

137. B, E. McClellan, M. K. Meredith, R. Parmelee, and J. P. Beck, *Anal. Chem.*, **46,** 306 (1974).

138. P. V. Dhond and S. M. Khopkar, *Anal. Chem.*, **45,** 1937 (1973).

139. An example: Diphenylguanidine, M. M. Tananaiko and F. V. Mirzoyan, *Ukr. Khim. Zh.*, **38,** 699 (1972).

140. See J. Korkisch, *Modern Methods for the Separation of Rarer Metal Ions*, Pergamon, Oxford, 1969, p. 131 *et seq.* In addition to the common solvents or solvating reagents mentioned in the text, some others have been studied, for example, sulfoxides (in CCl_4) by P. Markl, *Z. Anal. Chem.*, **271,** 7 (1974). N-Oxides in inert solvents have also been tried. M. Ejaz, *Talanta*, **23,** 193 (1976), found an *E* of almost 10 for extraction of $Th(NO_3)_4$ from 0.1–0.2 *M* HNO_3 with 0.1 *M* 5-(4-pyridyl)nonane N-oxide in xylene. A disolvate of thorium nitrate is formed. The rare earth elements are hardly extracted. Uranium (VI) is extracted.

141. R. Bock and E. Bock, *Z. Anorg. Allgem. Chem.*, **263,** 146 (1950).

142. M. Attrep, *Anal. Chem.*, **34,** 1349 (1962).

143. O. A. Nietzel and M. A. De Sesa, *Anal. Chem.*, **29,** 756 (1957).

144. R. J. Guest and J. B. Zimmerman, *Anal. Chem.*, **29,** 931 (1955).

145. B. Bernström and J. Rydberg, *Acta Chem. Scand.*, **11,** 1173 (1957), studied the extraction of various metals with undiluted TBP from HNO_3–$Ca(NO_3)_2$ solutions.

146. H. H. Gill, R. F. Rolf, and G. W. Armstrong, *Anal. Chem.*, **30,** 1788 (1958), use EDTA and F^- as masking agents. C. A. Francois, *ibid.*, **30,** 50 (1958), also uses EDTA.

147. C. J. Rodden, *Analysis of Essential Nuclear Reactor Materials*, USAEC, Washington, 1964, p. 50.

148. D. A. Everest and J. V. Martin, *Analyst*, **84,** 312 (1959).

149. C. R. Walker and O. A. Vita, *Anal. Chim. Acta*, **43,** 27 (1968).

150. J. J. McCown and R. P. Larson, *Anal. Chem.*, **32,** 597 (1960). $KBrO_3$ is used to oxidize Ce(III) to Ce(IV). D. A. Becker and P. D. LaFleur, *Anal. Chem.*, **44,** 1508 (1972), found extraction of U(VI) to be essentially complete (99% or better) from solutions up to 16 M in HNO_3, up to 12 M in $HClO_4$, up to 4 M in HCl, and up to 0.5 M in HF when phase volumes were equal and a 0.75 M HDEHP solution in petroleum ether was used.

151. I. H. Qureshi, L. T. McClendon, and P. D. LaFleur, *Radiochim. Acta*, **12,** 107 (1969). Tracer concentrations of metals.

152. V. G. Goryushina and E. Y. Biryukova, *J. Anal. Chem. USSR*, **24,** 443 (1969).

153. K. Ueno and A. Saito, *Anal. Chim. Acta*, **56,** 427 (1971).

154. V. Yatirajam and L. R. Kakkar, *Anal. Chim. Acta*, **52,** 555 (1970).

155. T. Ishimori et al., *JAERI Rept.*, **1106,** p. 21 (1966).

156. N. N. Kuznetsova, *Zh. Anal. Khim.*, **23,** 1485 (1968).

157. H. M. N. H. Irving and A. D. Damodaran, *Anal. Chim. Acta*, **53,** 267 (1971). In a subsequent paper, *ibid.*, **53,** 277, the extraction of $Hg(CN)_4^{2-}$ and $Hg(CN)_2OH \cdot solv^-$ is treated. In these extractions, an MIBK solution of tetrahexylammonium erdmannate was shaken with the aqueous solution of the complex cyanide and $Co(NH_3)_2(NO_2)_4^-$ displaced into the aqueous solution was determined.

LIQUID–LIQUID EXTRACTION CHELATES

I. EXTRACTION OF METAL CHELATES

A. GENERAL CONSIDERATIONS

1. Simple Chelates

Most chelating reagents for metals are of the type HL, which can react with metal cations, M^{n+}, by replacement of the acidic hydrogen to form

888

an uncharged chelate ML_n. Products of this type may be called simple chelates in contradistinction to chelate adducts or mixed chelate complexes. The neutral chelate is likely to be very slightly soluble in water, but much more soluble in some immiscible organic solvent, so that on extraction with the latter, much of the metal will be transferred to the organic phase. Suppose the extraction reaction is written as

$$M^{n+} \quad + \quad n\,HL \rightleftharpoons ML_n \quad + \quad n\,H^+$$

$$\phantom{M^{n+}}\quad w \qquad\qquad org \qquad org \qquad\qquad w$$

$$\text{(aqueous phase)} \qquad \text{(organic phase)}$$

The equilibrium constant for this two-phase reaction can be written (just as any other equilibrium constant is written) on the basis of the mass action law:

$$K_{ext} = \frac{[ML_n]_{org}[H^+]_w^n}{[M^{n+}]_w[HL]_{org}^n} \tag{9B-1}$$

K_{ext} is the *extraction equilibrium constant* of the reaction as written, at a specified temperature and ionic strength of the two phases.

It will be realized that Eq. (9B-1) may not tell the full story of the extraction. Metal and reagent species other than those indicated may be present in significant concentrations. There may be hydrolysis of the metal ion in the aqueous solution, forming various charged and uncharged hydroxo complexes; complex species of metal with anions such as the halides and tartrate may be present; charged complexes of metal with the chelating agent as well as ML_n may be present in the aqueous phase; more than one chelate may be extracted into the organic solvent phase (e.g., an adduct with HL may be formed). The analyst is concerned with the ratio of total metal concentrations in the organic and aqueous phases (E), and this may not be obtainable directly from Eq. (9B-1). Nevertheless, in spite of other equilibria, Eq. (9B-1) holds and serves as the basis for finding E.

A better insight into a metal chelate extraction is obtained by considering the individual equilibria, which combined, give K_{ext}[1]:

$$ML_n \rightleftharpoons M^{n+} + nL^- \qquad K_m = \frac{[M^{n+}][L^-]^n}{[ML_n]} \tag{9B-2}$$

Concentrations without subscripts refer to the aqueous phase. K_m is the ionization constant of the metal chelate in the aqueous phase.

$$ML_n \rightleftharpoons (ML_n)_{org} \qquad P_m = \frac{[ML_n]_{org}}{[ML_n]} \tag{9B-3}$$

P_m is the partition constant of the metal chelate (org/H$_2$O).

$$HL \rightleftharpoons H^+ + L^- \qquad K_l = \frac{[H^+][L^-]}{[HL]} \tag{9B-4}$$

$$HL \rightleftharpoons (HL)_{org} \qquad P_l = \frac{[HL]_{org}}{[HL]} \tag{9B-5}$$

From Eqs. (9B-2) and (9B-3),

$$\frac{[ML_n]_{org}}{[M^{n+}]} = \left(\frac{P_m}{K_m}\right)[L^-]^n \tag{9B-6}$$

From Eqs. (9B-4) and (9B-5),

$$[L^-] = \left(\frac{K_l}{P_l}\right)\left(\frac{[HL]_{org}}{[H^+]}\right) \tag{9B-7}$$

Accordingly, substituting Eq. (9B-7) into (9B-6), we have

$$K_{ext} = \frac{P_m K_l^n}{K_m P_l^n} = \frac{[ML_n]_{org}[H^+]^n}{[M^{n+}][HL]_{org}^n} \tag{9B-8}$$

for a specified ionic strength and temperature. Comparing different chelate extraction systems, we see that the extraction equilibrium constant is larger the larger the partition constant of the metal chelate and the ionization constant of the reagent, and the smaller the ionization constant of the metal chelate and the partition constant of the reagent. For a given reagent, P_m and P_l depend on the organic solvent, of course, but K_m and K_l will usually remain much the same from one extraction system to another, since the solubility of the organic solvent in the aqueous phase is small. In a series of organic solvents having but slight solubility in water (nonoxygen solvents), P_l and P_m will vary with the solubilities of the chelating reagent and the chelate in the solvents.

Under some conditions, the *extraction coefficient* of a metal can be obtained from Eq. (9B-8):

$$E = \frac{\sum[M]_{org}}{\sum[M]} \sim \frac{[ML_n]_{org}}{[M^{n+}]} = K_{ext}\frac{[HL]_{org}^n}{[H^+]^n} \tag{9B-9}$$

The approximation arises, as already pointed out, because species other than those written may be present in the system in appreciable amounts. There may be a vast difference between $\sum[M]$ and $[M^{n+}]$. It may be worthwhile to set down the various possibilities, or at least the more important ones, rendering the approximation invalid:

1. ML_n may not be the only metal species in the organic phase.

Ionization of ML_n may occur in some solvents of high dielectric constant, but this possibility may be dismissed for such solvents as CCl_4, $CHCl_3$, and the like. Polymerization of ML_n in the same solvents is unlikely but sometimes does take place. With some metals and reagents there is a possibility of a chelate adduct, of the type $ML_n \cdot aHL$, (p. 904), or a mixed chelate (p. 906) being formed.

2. There will always be some ML_n in the aqueous phase. Usually this will be small compared to M^{n+} or hydrolyzed or complexed species of M, and can be neglected. As the concentrations of M^{n+} and its hydrolyzed or complexed species approach zero, E_m approaches P_m as a limit.

3. Charged complexes of M and L may be present in the aqueous phase, for example ML_{n-1}^+ and others of this type. The extent of their formation varies with M and HL, and, for a given combination, on $[L^-]$, which is a function of total ligand concentration and the pH. Some chelating reagents are amphiprotic (amphoteric), forming H_2L^+, thus reducing the concentration of L^- in the aqueous phase (see discussion of 8-hydroxyquinoline). In some solutions, such a cation can form ion pairs with suitable ions (e.g., acetate) and be extracted into the organic solvent phase.

4. Frequently, species of M other than M^{n+} are present in the aqueous phase. If the acidity is low, hydrolysis of the heavy metal cation M^{n+} will occur, with the formation of one or more of a series of metal–hydroxo complexes. Because heavy metals tend to be complexed by inorganic anions, such as chloride, one or more complex species may be formed with them. These species will usually not be extractable into chloroform, carbon tetrachloride and like solvents.

The approximation

$$E \cong K_{ext} \frac{[HL]_{org}^n}{[H^+]^n}$$

may be close to the truth if the aqueous solution is sufficiently acidic to prevent appreciable hydrolysis of the metal, and ions complexing the cation are absent.

When an appreciable fraction of the total metal is present as hydrolyzed species, as species formed by foreign complexing anions, or as charged metal–reagent species, the equilibrium constants of the reactions forming these three types of species must be known if E is to be calculated. Suppose that charged metal–ligand complexes can be neglected, so that only hydrolysis and complex formation need be considered. The hydrolyzed metal species will be

$$MOH^{n-1}, M(OH)_2^{n-2}, \ldots, M(OH)_n, M(OH)_{n+1}^-, \text{ etc.}$$

Polymerization of hydrolyzed metal species is assumed not to occur. The complexing anion X is assumed to be univalent. It forms the complex species

$$MX^{n-1}, MX_2^{n-2}, \ldots, MX_n, MX_{n+1}^{-}, \text{etc.}$$

(not extractable).

The hydrolysis constants are defined as follows

$$k_1 = \frac{[MOH^{n-1}][H^+]}{[M^{n+}]} \tag{9B-10a}$$

whence $[MOH^{n-1}] = k_1[M^{n+}]/[H^+]$

$$k_2 = \frac{[M(OH)_2^{n-2}][H^+]^2}{[M^{n+1}]} \tag{9B-10b}$$

and similarly for k_3 and the others. The complex formation constants are defined as

$$\beta_1 = \frac{[MX^{n-1}]}{[M^{n+}][X^-]} \tag{9B-11a}$$

whence $[MX^{n-1}] = \beta_1[M^{n+}][X^-]$

$$\beta_2 = \frac{[MX_2^{n-2}]}{[M^{n+}][X^-]^2} \tag{9B-11b}$$

and similarly for β_3 and the others. If HX is a weak acid, $[X^-]$ will be a function of pH.

The concentration of M in the aqueous phase is

$$\sum [M] = [M^{n+}] + [MOH^{n-1}] + \cdots + [MX^{n-1}] + \cdots$$

$$= [M^{n+}] + \frac{k_1[M^{n+}]}{[H^+]} + \cdots + \beta_1[M^{n+}][X^-] + \cdots$$

$$= [M^{n+}]\left\{1 + \frac{k_1}{[H^+]} + \cdots + \beta_1[X^-] + \cdots\right\} \tag{9B-12}$$

$$E = \frac{[ML_n]_{org}}{\sum [M]} = \frac{[ML_n]_{org}}{[M^{n+}]\{1 + k_1/[H^+] + \cdots + \beta_1[X^-] + \cdots\}} \tag{9B-13}$$

Since

$$\frac{[ML_n]_{org}}{[M^{n+}]} = \frac{K_{ext}[HL]_{org}^n}{[H^+]^n},$$

$$E = \frac{K_{ext}[HL]_{org}^n}{[H^+]^n\{1 + k_1/[H^+] + \cdots + \beta_1[X^-] + \cdots\}} \tag{9B-14}$$

If metal–ligand species occur in the aqueous phase, terms giving the concentration of these must be added to the denominator (within the braces). Likewise, additional terms will be needed if polymerized or mixed species (e.g., of M, L, and X) are formed, or if a species other than ML_n is extracted into the organic phase. ML_n will always be present in the aqueous phase and will need to be taken into account when E is large (approaching its maximum) and an accurate value of E is required. When the only metal species in the aqueous phase are M^{n+} and ML_n,

$$E = \frac{K_{ext}[HL]_{org}^n}{[H^+]^n + (K_{ext}[HL]_{org}^n/P_m)} \qquad (9B\text{-}15)$$

As $[M^{n+}] \to 0$, $E \to P_m$.

Sometimes the ligand forms a series of complex species with the metal cation in the aqueous phase. Other factors being the same, this is more likely to occur when the partition constant of the reagent is relatively small and the solution is weakly acidic or basic so that the reagent anion can attain an appreciable concentration in the aqueous phase. Let N represent the maximal number of ligands that can combine with the cation M^{n+}. In the general case, the following metal–reagent complex species will be formed in the aqueous phase:

$$ML^{n-1}, ML_2^{n-2}, \ldots, ML_n^0, \ldots, ML_N^{n-N}$$

The formation constants of these species are

$$\beta_{l1} = \frac{[ML^{n-1}]}{[M^{n+}][L^-]} \qquad (9B\text{-}16a)$$

$$\beta_{l2} = \frac{[ML_2^{n-2}]}{[M^{n+}][L^-]^2} \qquad (9B\text{-}16b)$$

(and similar expressions for other β constants)

$$\beta_{lN} = \frac{[ML_N^{n-N}]}{[M^{n+}][L^-]^2} \qquad (9B\text{-}16c)$$

Then, if only the uncharged species is extracted (and neglecting species other than M–L),

$$E = \frac{[ML_n]_{org}}{[M^{n+}] + [ML^{n-1}] + \cdots + [ML_N^{n-N}]}$$

$$= \frac{[ML_n]_{org}}{[M^{n+}] + \beta_{l1}[M^{n+}][L^-] + \beta_{l2}[M^{n+}][L^-]^2 + \cdots + \beta_{lN}[M^{n+}][L^-]^N}$$

$$\qquad (9B\text{-}17a)$$

Dividing numerator and denominator by $[M^{n+}]$:

$$E = \frac{[ML_n]_{org}/[M^{n+}]}{1 + \beta_{l1}[L^-] + \beta_{l2}[L^-]^2 + \cdots + \beta_{lN}[L^-]^N}$$

$$= \frac{K_{ext}([HL]_{org}/[H^+])^n}{1 + \beta_{l1}[L^-] + \beta_{l2}[L^-]^2 + \cdots \beta_{lN}[L^-]^N} \qquad (9B-17b)$$

$[L^-]$ as a function of $[HL]_{org}$ and $[H^+]$ is found from

$$[L^-] = \frac{[HL]_{org}}{[H^+]} \times \frac{K_l}{P_l}$$

If HL is associated (polymerized) in the organic phase, constants must be known that will allow calculation of monomeric $[HL]_{org}$ from its total concentration.

Application of Eq. (9B-9) or (9B-14) is simplest when the amount of extracted metal is small, so that the concentration of the added reagent is not appreciably changed, as with tracer concentrations of metal.

Usually, at a specified pH, a few metal species in the aqueous phase will predominate to the extent that others can be ignored, so that in application, Eq. (9B-14) for E will be simpler than it appears. Values for hydrolysis constants and complex formation constants are conveniently available, as far as they are known, in compilations.[2]

An important question in the extraction of metal traces is whether the extractability is independent of the total metal concentration in the system, other variables remaining constant. E will be independent of the metal concentration if no species of M in either phase contains more than one atom of M per charged or uncharged molecule, in other words if there is neither real nor virtual association of M in any species. E would not be constant if the extracted species were M_2L_3 (derived from H_2L), for example, because this species represents what may be called *virtual* association [as compared to real association, for example $(ML_2)_2$]. Neglecting hydrolyzed and complexed metal species, we would then have

$$\frac{[M_2L_3]_{org}}{[M^{3+}]^2} = const.$$

at specified acidity and reagent concentration and

$$E = 2 \times const. \times [M^{3+}]$$

Other factors remaining the same, E would increase with the amount of M in the system. Such cases are not often encountered, in chelate extractions, because usually the ligand anion has a single negative charge L^-. However, charged metal hydrolysis products can undergo association and form polynuclear species, and this could result in a variation of E with metal concentration (likewise if any other metal species in the aqueous phase associates). Assume all the metal in the aqueous phase to be present as a polynuclear complex containing m atoms of the metal, represented as M_m, and HL is monobasic:

$$E = \frac{[ML_n]_{org}}{m[M_m]} = \frac{const.}{m[ML_n]_{org}^{m-1}}$$

The extraction coefficient decreases with the amount of metal in a system in which $[H^+]$ and $[HL]_{org}$ are fixed. The value of m has been considered to remain the same, a questionable assumption, because m may vary with the metal concentration. Moreover, the polynuclear metal complex may dissociate completely into mononuclear species at sufficiently low total metal concentrations. But the qualitative trend of E is evident.

If the concentration of the complexing substance X is large compared to the concentration of M in the system (and this is usually true in the extraction of a trace constituent), so that its excess remains essentially constant, we may write (disregarding hydrolyzed forms of M and also possible charged metal-chelate species):

$$E = \frac{[ML_n]_{org}}{\sum [M]} \cong K'_{ext} \frac{[HL]_{org}^n}{[H^+]^n}$$

where K'_{ext} is a constant whose value depends on the concentration of X in the aqueous phase. Again it is assumed that real or virtual association of any species of M does not occur in either phase. The use of such effective constants is convenient in practical work.[3]

The effect of indifferent electrolytes on extractability of metals as chelates is complex, especially at high concentrations. First of all, the value of K_{ext} is a function of the ionic strength of the aqueous solution. Most electrolytes salt out nonelectrolytes. P_l and P_m will, in general, have larger values at higher electrolyte concentrations. For example, the intrinsic solubility of Ni dimethylglyoximate corresponds to 57 μg Ni/liter in 0.05 M NaCl and to 30 μg in 2.0 M NaCl, and the molar solubility of dimethylglyoxime decreases as follows with NaCl concentration (M)[4]: 0, 5.48×10^{-3}; 0.05, 5.37×10^{-3}; 0.5, 4.44×10^{-3}; 5.0, 6.1×10^{-4}. In 2.0 M NaCl, P_m/P_l^2 is about one-third of that in water. At the same time, the change in ionic strength brings about a change in the values of K_l and K_m. At high ionic strengths the change is difficult to estimate. For low ionic strengths, activity coefficients can be estimated and approximate values of K_{ext} can be obtained (see discussion of dithizone in Chapter 6G).

In addition to producing a change in the value of K_{ext}, an electrolyte can alter the extractability of a metal by anion complexing of a metal, as by halides. It would be useful to have apparent values of K_{ext} for various electrolyte concentrations in a given system. An apparent or effective extraction constant K'_{ext} is of the form

$$K'_{ext} = \frac{[ML_n]_{org}}{\sum [M]} \times \frac{[H^+]^n}{[HL]^n_{org}}$$

for a specified electrolyte concentration (or complexing anion concentration).

K'_{ext} would be constant with varying $[H^+]$ only if the extent of complexing of the metal were independent of the acidity. If $[H^+]$ were also specified, another constant would be obtained:

$$K''_{ext} = \frac{[ML_n]_{org}}{\sum [M]} \times \frac{1}{[HL]^n_{org}}$$

Values of K'_{ext} could be plotted against the complexing electrolyte concentration to enable interpolation of K'_{ext} for intermediate values of the latter. Figure 9B-1 shows a plot of K'_{ext} for the extraction of ferric cupferrate by chloroform at various concentrations of HCl. In this system, there is a linear (but inverse) relation between $\log K'_{ext}$ and [HCl]. If [Cl⁻]

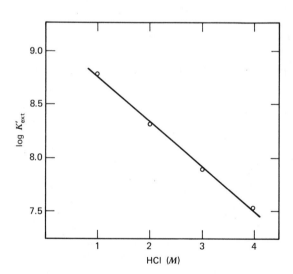

Fig. 9B-1. Effective extraction equilibrium constant of Fe(III) cupferrate as a function of HCl concentration with CHCl₃ as extracting solvent. Reprinted with permission from E. B. Sandell and P. F. Cummings, *Anal. Chem.*, **21**, 1358 (1949). Copyright by the American Chemical Society.

is maintained constant at 4 M, K'_{ext} remains fairly constant as $[H^+]$ is varied from 1 to 4 M. Reproducibility in this system is not easy to attain because of rapid decomposition of the reagent in mineral acid solution (Table 9B-1). Not many effective extraction constants for high electrolyte concentrations are available in the literature. However, values of $pH_{1/2}$ (pH for 50% extraction) or E for a number of reagents in systems of fairly high salt concentration have been determined.[5] Extraction at high electrolyte concentrations may be easier or more difficult than at low electrolyte concentrations. Unless the electrolyte is one complexing the metal, it is difficult to predict its effect at high concentrations. In general, extraction of metals can be carried out at high electrolyte concentrations, except when the electrolytes are strong acids or when they complex the metal. It may be noted that if an oxonium-type solvent is used, the electrolyte may lead to an extraction of ion-pair compounds.

pH–E Curves. In considering chelate extraction equilibria, it is advantageous to plot log E against pH at a constant excess of HL in the organic solvent phase. The simplest possible system has only the species M^{n+}, ML_n (in both aqueous and organic solvent phases), HL (in both aqueous and organic solvent phases), and H^+ as concentration variables. When, at equilibrium, M^{n+} outweighs ML_n in the aqueous phase, Eq. (9B-9) applies, which can be written in logarithmic form

$$\log E = \log K_{ext} + n \log [HL]_{org} + n pH$$

If the amount of metal is minute, so that the initial concentration of HL in the organic solvent hardly changes, we can write (for a specified reagent concentration)

$$\log E = n pH + const.$$

n is the slope of the straight line. A series of straight lines is obtained for

TABLE 9B-1
Extraction of Fe(III) Cupferrate by Chloroform from an Aqueous Solution of Constant Chloride Ion Concentration (4 M)[a]

Concentration of HCl (M)	Concentration of NaCl (M)	$K'_{ext} \times 10^{-7}$
4.0	0	3.2
3.0	1.0	2.9
2.0	2.0	3.5
1.0	3.0	3.4

[a] E. B. Sandell and P. F. Cummings, *Anal. Chem.*, **21**, 1356 (1949).

various initial concentrations of reagent. The straight-line relation can hold only when ML_n in the aqueous phase can be neglected. As $E \to P_m$, its concentration needs to be taken into account [Eq. (9B-15)]. Finally, if no other factors supervene, the $\log E$–pH line bends and eventually runs parallel to the pH-axis when $[ML_n] \gg [M^n]$.

The extraction of $Cu(HDz)_2$, cupric dithizonate, from acid solutions, in which there is no hydrolysis of Cu^{2+}, illustrates the preceding discussion. In Fig. 9B-2, the value of $\log E_{Cu}$ is plotted against pH for three different concentrations of H_2Dz in the organic solvent (CCl_4).

Most experimental pH–E curves will deviate more or less from ideal curves such as in Fig. 9B-2. When hydrolyzed and complexed metal species are taken into account, the equation is

$$\log E = \log K_{ext} + n \log [HL]_{org} + n\mathrm{pH}$$
$$- \log \{ 1 + k_1/[H^+] + \cdots + \beta_1[X^-] + \cdots \}$$

Many chelating reagents are used in a pH range where appreciable hydrolysis of M^{n+} occurs or with aqueous solutions containing species X complexing M^{n+}. The value of the last log term in the preceding expression increases with pH because of hydrolysis and also because of increased complexing if X^- is the anion of a weak acid. Therefore, the slope of the pH–E curve is less than n, and the deviation becomes greater with increasing pH. Formation of charged metal-chelate species also reduces the theoretical value of the slope.

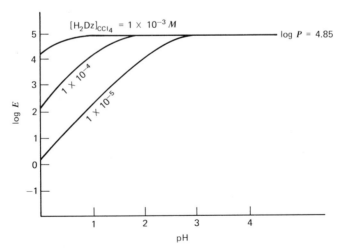

Fig. 9B-2. Extractability of Cu(II) as dithizonate in CCl_4 as a function of $[H_2Dz]_{CCl_4}$ (equilibrium concentration) and pH. $K_{ext} = 1.5 \times 10^{10}$ ($\mu = 1$). $P_{Cu(HDz)_2} = 7 \times 10^4$.

Especially if a metal is amphoteric, the pH–E curve (or the pH–R curve, p. 900), after running parallel to the pH axis for some distance, begins to turn downward in the higher pH ranges. This behavior is well shown by Pb and Zn dithizonates, as a result of formation of biplumbite and zincate ions. Charged complexes of a metal and reagent may be increasingly formed at higher pH values, where the concentration of reagent anion in the aqueous phase increases. Another possibility is the formation of hydroxo chelate complex anions in basic medium. Thus ferric iron forms $FeOx_2(OH)_2^-$ in 0.01–0.07 M OH^- with 8-hydroxyquinoline (see Chapter 6D). The flat portion of the pH–E curve in some systems may be quite short so that a maximum appears instead.

When E is determined over a wide range of pH values, it is not unusual to find a minimum (or even two) in the pH–E curve. Such minima may have several causes. The minimum may represent nonequilibrium conditions, as when the metal ion hydrolyzes to give polymeric species or even colloidal hydroxide. These species may react slowly with the organic reagent in an organic solvent and equilibrium may not be attained even after hours of shaking. Thus an aluminum solution that is approximately neutral or slightly basic reacts very incompletely with a chloroform solution of 8-hydroxyquinoline. But if 8-hydroxyquinoline is added to the aluminum solution before the pH is adjusted to approximate neutrality, aluminum is readily extracted when the solution is shaken with chloroform because hydrolyzed aluminum species are not formed. Such solutions as Zr, Nb, Ta, Ti, and Fe(III) hydrolyze readily and the hydrolysis products impede extraction. Addition of a mild complexing agent, for example, citrate or tartrate, may be helpful.

Extraction minima can also occur under equilibrium conditions, and are then likely to be encountered when a complexing anion A^- is present. The concentration of the latter will often be a function of pH, since HA (or H_nA) frequently is a weak acid. It may happen that in some intermediate pH range the complexing of the metal by A^- outweighs the complexing by the reagent L^-, with a reversal of the situation in a higher pH range. The nature of the extracted metal species may change at higher pHs (mixed chelates containing OH may be formed), and the major anion species may vary with pH if its parent acid is polybasic. Moreover, the extraction reagent may undergo pH-dependent tautomeric transformations.

Extraction minima have been noted in the following systems (among others): W–8-hydroxyquinoline, Ce(III)–acetylacetone, U(VI)–thenoyltrifluoroacetone, Nb and Ta–cupferron or benzoylphenylhydroxylamine, Ti–diethyldithiocarbamate. The minima may become more or less pronounced (or even disappear) if the metal concentration is drastically varied (difference between "micro" and tracer concentrations). They may depend on the extracting solvent.

pH–R Curves. It is sometimes useful to see at a glance how the extractability of a metal varies with the pH at a specified concentration of a chelating agent in the organic solvent phase. For this purpose, the extractability expressed as the recovery factor R in percent is plotted against pH. The recovery factor for a single extraction with a volume ratio r of organic solvent to the aqueous phase is given by (p. 1026):

$$R_{\%} = \frac{100rE}{rE + 1}$$

If $r = 1$,

$$R_{\%} = \frac{100E}{E + 1}$$

A pH–R curve is not fundamentally different from a pH–E curve, but it presents information in a more direct, analytically useful way.

The part of a pH–R curve lying in an acidic range where there is little metal ion hydrolysis and where charged metal chelates are less likely to be formed may be calculable easily enough from K_{ext}. Calculation of the curve in the weakly acidic and basic range is not always possible because all the hydrolysis and complex formation constants may not be known. Any calculated curve should be checked experimentally, because unforeseen factors (e.g., nonequilibrium) may invalidate a theoretical curve. Figure 9B-3 shows calculated pH–R curves for Cu, Bi, and Pb dithizonates in the acidic range (equal phase volumes).

If the extraction of a metal follows Eq. (9B-9), so that the metal ion M^{n+} is the only significant species of M in the aqueous phase, the pH range over which the recovery (one extraction, equal phase volumes) varies from 1 to 99% is

4 for M^+
2 for M^{2+}
1.33 for M^{3+}
1.0 for M^{4+}

The pH for 50% extraction of M^{2+} is

$$pH_{1/2} = -\tfrac{1}{2} \log K_{ext} - \log [HL]_{org}$$

For M^{n+},

$$pH_{1/2} = -\frac{1}{n} \log K_{ext} - \log [HL]_{org}$$

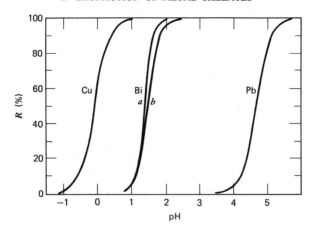

Fig. 9B-3. pH–R curves for extraction of Cu(II), Bi, and Pb as dithizonates in CCl$_4$. Curve a for Bi is calculated curve neglecting hydrolysis of Bi^{3+}. In curve b, hydrolysis to BiOH^{2+} has been taken into account. Equilibrium concentration of H$_2$Dz in CCl$_4$ = 1×10^{-5} M.

A change in $[HL]_{org}$ shifts pH$_{1/2}$ along the pH-axis of a pH–R plot. For example, a tenfold increase in $[HL]_{org}$ decreases pH$_{1/2}$ by 1 unit for all cations irrespective of charge.

A quick estimate of metal separabilities is obtained by examining a pH–R plot, or simply a plot of metal pH$_{1/2}$ values. It is evident from Fig. 9B-3 that only a limited separation of Cu and Bi from each other by dithizone extraction on the basis of pH adjustment alone is possible. With comparable amounts of the two metals present, a fair separation is possible but difficulties arise when the ratio Cu/Bi is large. (See Chapter 9D.) On the other hand, Cu and Pb are easily separable, as are Bi and Pb.

Magnitude of K_{ext}. The extraction equilibrium constants for various chelating agents and metals vary over an enormous range from zero to 10^{27} or greater for a divalent metal (10^{27} is the extraction constant for primary mercuric dithizonate in CCl$_4$). It is evident from Fig. 9B-3 that a constant of 10^{10} for a divalent metal enables it to be extracted quantitatively from a rather acid solution with a low concentration of reagent. Even a K_{ext} of 1 (for a divalent metal) may be large enough for satisfactory extraction. Suppose we require that 99% of a divalent metal be extracted in a single extraction with equal volumes of phases. In a system free from hydrolysis and complexing complications,

$$\frac{[ML_2]_{org}}{[M^{2+}]} = 99 = \frac{[HL]_{org}^2}{[H^+]^2} K_{ext}$$

and

$$\frac{[HL]_{org}}{[H^+]} \sim \frac{10}{K_{ext}^{1/2}}$$

The recovery is made larger by increasing the reagent concentration or decreasing the acidity. There is a limit to the solubility of both chelating reagent and chelate in an organic solvent, which often is reached quite soon, whereas $[H^+]$ can be varied over a very wide range, although at a risk of complications when $[H^+]$ is small. Suppose that $[HL]_{org} \sim 0.001$ (equilibrium concentration), corresponding at most to an original concentration perhaps a few times larger than this, a solubility attained by many (though not all) chelating reagents. Then, if $K_{ext} = 1$, for 99% extraction of M

$$[H^+] = \frac{1 \times 0.001}{10} = 1 \times 10^{-4}$$

Thus even with K_{ext} as small as 1, quantitative extraction can be obtained at pH 4. If hydrolysis of the metal ion occurs, the situation is less favorable. The formation of insoluble hydrolysis products must always be prevented. The addition of a mild complexing agent such as tartrate or citrate may be needed for this purpose. The pH of extraction will then need to be increased, and it is usually found that there is a net increase in extractability on doing so.

   In general, it is desirable to make extractions in acid solutions if possible. One of the advantages of separation by extraction is that reagents exist that provide large values of K_{ext}, allowing the use of acid solutions. For some metals, K_{ext} differ sufficiently for separations to be made largely on the basis of acidity. However, the possibilities of using differential complex formation in acid solution to separate metals are less than in weakly acidic or basic solutions. Most complexing anions are anions of weak acids and do not function well, if at all, when the aqueous solution is too acid. For example, cyanide ($K_{HCN} \sim 10^{-9}$) cannot be used effectively as complexer for most metals in the acid range. Halides (Cl^-, Br^-, I^-) and thiocyanate are anions of strong acids and can be used for complexing in acid solutions with organic solvents not extracting the complexes.

   K_{ext} for a given chelating agent and metal can be varied to some extent by a change in the organic solvent. As may be seen from the relation

$$K_{ext} = \left(\frac{P_m}{P_l^n}\right)\left(\frac{K_l^n}{K_m}\right)$$

changes in P_m and P_l in the same direction tend to oppose each other.

However, if $n \neq 1$, the same relative increase in P_m and P_l would result in a decrease of K_{ext}. In general, an increase in P_l is usually accompanied by an increase in P_m, but this is by no means invariably true. Much depends on the chelating reagent and the solvent. Thus a reagent containing free –OH groups might not be easily soluble in an inert solvent such as chloroform (low P_l), whereas the chelate might be easily soluble in the same solvent, as a result of the blocking of –OH groups in the chelate by hydrogen bonding.[6] Even if both constants change in the same sense, the relative changes can be quite different and there can be a marked change in K_{ext}. K_{ext} for divalent metal dithizonates are higher by a factor of 10–1000, sometimes more, for CCl_4 than for $CHCl_3$. A carbon tetrachloride solution of dithizone allows extraction of metals from more acid solutions than does a chloroform solution. For a divalent metal, $pH_{1/2}$ may be shifted downward by one to two units on going from a chloroform to a carbon tetrachloride dithizone solution.

Values of 100–1000 are common for P_l of chelating reagents in various organic solvents, inert and oxygen-type. Larger values are possible, for example, $\sim 10^4$ for dithizone in carbon tetrachloride. But for dimethylglyoxime in chloroform, $P_l \sim 0.1$. P_m is usually larger than P_l and usually ranges from 10^2 to 10^5 or even 10^6. K_l usually lies in the range 10^{-3} to 10^{-11}. K_m has an extremely wide range, and its variation is mainly responsible for the selectivity of chelate extraction reagents. K_m for aluminum dithizonate may not be ∞, but its value certainly is very large (formation or stability constant is essentially zero). For $Cu(HDz)_2$, K_m is 1×10^{-23}, and for $Hg(HDz)_2$, it is on the order of 10^{-34}.

Alternative Extraction Expression. An extraction reaction can be written in terms of the chelating ion concentration in the *aqueous* phase:

$$M^{n+} + nL^- \rightleftharpoons ML_n$$
$$\phantom{M^{n+}} _{H_2O} \quad _{H_2O} \quad\quad _{org}$$

The equilibrium constant for this two-phase reaction is given by

$$\frac{[ML_n]_{org}}{[M^{n+}][L^-]^n}$$

which may be symbolized K_{ext}^* to distinguish it from K_{ext} discussed previously. The relation between these two constants is

$$K_{ext}^* = K_{ext} \frac{P_l^n}{K_l^n}$$

The use of K_{ext}^* may have an advantage when the extraction reagent is present mostly in the aqueous phase. (See, further, discussion of sodium diethyldithiocarbamate in Chapter 6F.)

Metal Exchange Extraction.[7]   If an organic solvent solution of the chelate ML_m is shaken with an aqueous solution containing the metal ion N^{n+}, more or less of the latter is transformed into NL_n:

$$mN^{n+} + nML_m \rightleftharpoons nM^{m+} + mNL_n$$
$$\text{org} \qquad\qquad\qquad \text{org}$$

and at equilibrium

$$\frac{[M^{m+}]^n[NL_n]_{org}^m}{[N^{n+}]^m[ML_m]_{org}^n} = K_{exch}$$

(If $m = n$, K_{exch} is taken as $([M^{m+}][NL_n]_{org})/([N^{n+}][ML_m]_{org})$.) The equilibrium constant for the reaction written is K_{exch}, the exchange constant for the two metals with the common ligand L. The value of K_{exch} is given by

$$K_{exch} = \frac{(K_{ext})_N^m}{(K_{ext})_M^n} \qquad \left(\text{or } \frac{(K_{ext})_N}{(K_{ext})_M} \text{ if } m = n\right)$$

If $(K_{ext})_N$ is large compared to $(K_{ext})_M$, most N^{n+} is converted to $(NL_n)_{org}$. (In practice, effective extraction constants may be needed in place of K_{ext}.)

The use of an organic solvent solution of ML_m instead of HL may have advantages in the extractive separation or determination of N, the primary advantage being a possible increase in selectivity. By proper choice of M, the extraction of other metals (that would react with HL) may be largely prevented. To be sure the same result may be achieved by adjustment of the hydrogen-ion concentration when HL is used as reagent, but this is not always possible, as when the required acidity lies outside the attainable range. In separations, introduction of the metal M (since ML_m is used in excess) may be objectionable. Exchange equilibrium may not always be established rapidly. Inert complexes [e.g. of Co(III)] in the organic solvent may not exchange appreciably, when, on equilibrium considerations, exchange would be expected.

For further discussion of chelate exchange extraction, see "Diethyldithiocarbamate," p. 523).

2.   Chelate Adducts

Reference has already been made to the formation of chelates containing uncharged components S, of the type ML_nS_s. S may be H_2O, neutral reagent molecules or molecules of some uncharged organic donor. Especially when S is a substance other than H_2O, these compounds are termed *chelate adducts.* S usually occupies coordination positions of M, but it appears that sometimes it can be bonded to L. Some examples of chelate adducts are

SrOx$_2$·2 HOx   (HOx = 8-hydroxyquinoline)
MgOx$_2$·RNH$_2$ or MgOx$_2$·RNH$_3$Ox   (RNH$_2$ = aliphatic amine)
Co(TTA)$_2$·2 Py   (HTTA = thenoyltrifluoroacetone; Py = pyridine)
UO$_2$(TTA)$_2$·TOPO   (TOPO = tri-n-octylphosphine oxide)
Ca(TTA)$_2$·CsX   (X = inorganic anion)

Most adduct-forming agents contain a basic oxygen or nitrogen.

Looked at in another way, most of these compounds are mixed ligand chelates, at least if the adducting component occupies a coordination position in the coordination sphere of the cation. Ca(TTA)$_2$·CsX is a mixed metal, mixed ligand chelate. Some adducts can be looked upon as ion-association compounds, and indeed are (MgOx$_2$·RNH$_3$Ox = MgOx$_2$·RNH$_3^+$Ox$^-$).

As we have already seen, the presence of H$_2$O molecules in a chelate molecule has an unfavorable effect on its extractability into organic solvents, especially of the inert type. Replacement of H$_2$O by organophilic molecules results in extractability of the chelate adduct into CHCl$_3$, CCl$_4$, C$_6$H$_6$, and similar solvents. Occasionally, uncharged molecules of the chelating reagent itself at comparatively high concentrations can displace H$_2$O, but more commonly a suitable organic addend (uncharged ligand) is chosen for the purpose. Often it is found that the extraction coefficient of the adduct is greater than that of the simple chelate, even if replacement of H$_2$O is not involved. That this should be so is not surprising. In general, the addition of further (organic) addends makes the chelate more organophilic and less hydrophilic and, therewith, more easily extractable into organic solvents. This enhancement of extraction by adduct formation is commonly called synergism (p. 819), a term which may evoke the notion of action by mysterious forces. A neutral ligand S can form an extractable complex with the ion M^{n+} only when the charge of the latter has been neutralized by anionic L or inorganic anions. The extent of enhancement of the extraction of M by formation of ML$_n$·sS (or ML$_n$S$_s$ as it also is written), compared to extraction of ML$_n$, depends upon the increase in P brought about by the formation of the adduct and the extent of formation of the latter. It is not uncommon to find that ML$_n$·sS is a much more stable complex than ML$_n$. With the same metal and a series of similar chelating agents, the adduct formed by a given S tends to be more stable the less stable ML$_n$ is.

Neutral organophosphorus compounds form adducts with chelates (those of β-diketones have been studied extensively), showing marked enhancement of metal extraction (see p. 1012). Nonphosphorus adducts display much the same general behavior.

Synergic effects, usually not strong, are shown by some mixed adducts, i.e., $ML_n \cdot s_I S_I \cdot s_{II} S_{II}$, as well as by mixed chelates (L_I and L_{II}).

Knowing the composition of the chelate adduct formed, one can write the expression for the equilibrium extraction constant of the reaction, and therefrom the expression for the extraction coefficient. This expression is analytically useful, if it includes the concentrations of all the major species in both phases. The effective concentration of S may be a function of pH and other variables. There is a possibility of interaction between S and the chelating agent (destruction of synergism). (See p. 1014.)

3. Mixed Chelates

The formation of mixed chelates (Chapter 6A) is of considerable importance in extraction separations. A mixed chelate may be more easily (larger P) or less easily extractable than the corresponding simple chelate. The formation of these chelates is not always readily predictable and may bring the analyst unpleasant surprises. It is unsafe to deduce extraction separabilities from the behavior of individual constituents. The formation of chelates containing OH is not uncommon with some chelating agents (8-hydroxyquinoline, for example), rare with others (dithizone). The hydroxo chelates may be formed only in a certain pH range. The presence of groups such as OH and Cl in a chelate is, in general, not favorable for its extraction into nonoxygen solvents. On the other hand, the presence of another organophilic anionic component can increase the solubility of a chelate in organic solvents.

Mixed cation chelates are encountered less often than mixed anion chelates in extractions. An example is the complex citrate or malate of In and Cr(III), which contains the two metals in a $1:1$ ratio (p. 937), and the similar Cu(II)–Cr(III) citrate (p. 287). The formation of the latter hinders the extraction of copper in the biquinoline method for its determination. The coextraction of metals expected to remain in the aqueous phase in certain extractions has been attributed to mixed metal chelate formation (Fig. 9B-4). Coextraction can, however, also result from the formation of ion association-type complexes, for example, $Ca(ScOx_4)_2$, in which Ca is present as cation and Sc in the complex anion ($Ox^- = $ oxinate anion).

B. CHOICE OF EXTRACTING SOLVENT

It was noted earlier that a metal chelate in which all the coordination positions are occupied by anionic or neutral organic reagent, that is, in which water molecules are not present, is soluble in organic solvents of the inert and noninert (oxygen-containing) types. Solubilities of chelates in organic solvents vary over a wide range, as illustrated by the following

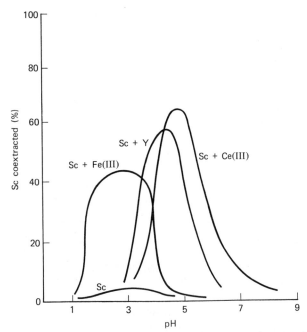

Fig. 9B-4. Coextraction of Sc (as radioactive isotope) with Y, Ce(III), and Fe(III) chelates of Lumogallion (Chapter 6E, Section I.A). Diethyl ether as extractant. According to Alimarin, Gibalo, and Pigaga, *Dokl. Akad. Nauk SSSR*, **186**, 1323 (1969). Mixed metal complexes have composition 1 Sc:1 M:2 Lumogallion.

figures for $CHCl_3$ and CCl_4 (room temperature):

	Soly. in $CHCl_3$ (M)	Soly. in CCl_4 (M)
Pb dithizonate	4.3×10^{-4}	2.5×10^{-5}
Zn dithizonate	3.7×10^{-2}	1.3×10^{-3}
Al 8-hydroxyquinolate	4.5×10^{-2}	
Fe(III) 8-hydroxyquinolate	1.9×10^{-2}	
Cu(II) 8-hydroxyquinolate	2.7×10^{-3}	3.4×10^{-4}
Al cupferrate	~ 1.4	
Bi diethyldithiocarbamate	0.22	

Cupferrates tend to be very soluble in organic liquids, some dithizonates sparingly soluble, and most other chelates fall between these two extremes.

When there is no special advantage to be gained by using an oxygen-containing solvent, CCl_4 or $CHCl_3$ is often chosen as the extractant. Carbon tetrachloride is frequently used. Its solubility in water is very low

908   LIQUID–LIQUID EXTRACTION

(0.05 ml/100 ml H_2O) and it is not unduly volatile (b.p. 77°). Moreover, it is considerably denser (1.58) than water and separates well from an aqueous solution after being shaken with it. Chloroform is chosen when the chelate is not sufficiently soluble in CCl_4. It is more polar than CCl_4, tends to form hydrogen bonds with the ligand, and usually dissolves chelates to a larger extent, sometimes by a factor of 10 or more, as with the dithizonates. Although $CHCl_3$ is more soluble in water and more volatile than CCl_4, it is still a satisfactory extractant. Other inert solvents than these are sometimes used. Saturated hydrocarbons are often found to be poor solvents for chelates. Organic solvents that are lighter than water are inconvenient when more than one extraction of an aqueous phase is necessary. Solvents tending toward emulsion formation (more common among oxygenated solvents) obviously are to be avoided.

Oxygenated solvents are usually shunned in the extraction of normal (saturated) chelates. Their physical properties may be appreciably inferior to solvents of the inert type, and the solubilities of HL and ML_n may be no greater in them than in selected inert solvents (see Table 9B-2). Moreover, the oxygen-type solvents may extract undesired species from the aqueous solution, even metals that would not be extracted by inert solvents. They may extract ionic forms of the reagent as certain ion-pairs. Thus amyl alcohol extracts 8-hydroxyquinoline, HOx, to some extent from mineral acid medium as $HOx \cdot H^+$ combined with a suitable anion or as Ox^- with a cation (Na^+) from sodium hydroxide solution.[8] But there are times when an oxygen-type solvent will be needed for chelate extractions. Chelates with free –OH groups (chelates of morin, PAN, etc.) are more easily extracted by the active solvents (alcohols, esters, ketones) than by the inert. When the chelate is hydrated (for reasons mentioned earlier), CCl_4, $CHCl_3$, and like solvents tend to extract it poorly, sometimes hardly at all. For example, the solubility of $ZnOx_2 \cdot 2H_2O$ in $CHCl_3$ is only $\sim 2 \times 10^{-6}\ M$ at room temperature; the

TABLE 9B-2
Solubility of Pd(II) Dimethylglyoximate in Various Organic Solvents (mg Pd(HDx)$_2$ in 100 ml, 22°)[a]

Toluene	2.8	Isoamyl alcohol	12.5
Isobutanol	4.4	CCl_4	13.4
Ethyl ether	5.4	$C_2H_4Cl_2$	14.6
n-Butanol	6.7	$CHCl_3$	18.8
Benzene	10.8		

[a] P. B. Mikhel'son and L. V. Kalabina, J. Anal. Chem. USSR, **24,** 160 (1969).

anhydrous oxinate dissolves readily in $CHCl_3$. Hydrated chelates can often be extracted into oxygen-type solvents. Such a solvent is likely either to replace coordinated water molecules or to hydrogen-bond to them. The chelate molecule then becomes more organophilic and prefers the organic phase to the aqueous. An immiscible alcohol, isoamyl alcohol for example, is often a good extractant; esters are also often effective.[9]

As we have seen, extraction of hydrated chelates can also be brought about by replacing H_2O with hydrophobic or organophilic substances such as the reagent itself at relatively high concentrations (e.g., transformation of $SrOx_2 \cdot 2\,H_2O$ to $SrOx_2 \cdot 2\,HOx$) or with a foreign organic coordinating agent such as n-butylamine (see discussion of 8-hydroxyquinoline). The resulting complex can then be extracted into chloroform and possibly other inert solvents.

When charged (anionic or cationic) chelate species are to be extracted, a suitable ion of opposite sign is needed to form an ion pair, or more generally an ion associate. This type of extraction is discussed in Chapter 9C.

A mixture of two organic solvents may extract a chelate better than either solvent alone, but usually the synergic effect is not strong. The effect is stronger and more often observed in the extraction of nitrates, chlorides, and others (see p. 830 for an example). A solvent may be mixed with another organic solvent to improve its physical properties, for example, density and viscosity, without unduly decreasing its extractive capability. It may even be advantageous at times to reduce the extractive power of an organic solvent by dilution with a solvent of much lower extractive power. As we shall see later, high extractive power is not necessarily advantageous in separations.

The rate of extraction of a particular chelate often depends on the organic solvent in which the reagent is dissolved. The differences in rate may be so great that this factor may play a decisive role in the choice of solvent. See the section following.

Finally, in choosing a solvent, thought should be given to its stability under the extraction conditions (oxidizability, photochemical decomposition, etc.).

Physical properties of common organic solvents are listed in Table 9B-3.

C. EXTRACTION RATE

1. General

The rate of extraction of metals as chelates from aqueous solutions by organic solvents varies over a wide range.[10] In most such extractions, the

TABLE 9B-3
Properties of Some Extraction Solvents[a]

Solvent	Solubility (wt. %)[b] Solv. in H_2O	H_2O in solv.	Density (g/ml)	Boiling Point (°C)	Dielectric Constant	Dipole Moment (Debye units)
n-Amyl acetate	0.20	0.9	0.88	149	4.8	1.9
iso-Amyl alcohol	2.3	9.1	0.81	131	14.7	1.8
n-Amyl alcohol	2.2 (25°)		0.81	138	13.9	
Benzene	0.18 (25°)	0.06 (25°)	0.88	80	2.3	0.0
n-Butyl acetate	1.0	1.37	0.88	126	5.0	
iso-Butyl alcohol	8.5	16.4	0.81	108	18	1.7
n-Butyl alcohol	7.8	20.0	0.81	118	17	1.7
Carbon tetrachloride	0.08 (25°)	0.09 (25°)	1.58 (25°)	77	2.2	0.0
Chloroform	0.80	0.97	1.49	61	4.9	1.2
Cyclohexane	0.01	0.01	0.78	81	2.0	
Cyclohexanone	2.3	8.0	0.95	157	18.2	2.8
Diamyl ether	6.9	1.26	0.78	187	3.1	
Dibutyl ether	0.77	0.3	0.77	142	3.1	1.2
o-Dichlorobenzene			1.30	178	9.9	
Diethyl ether	7.4	1.26	0.71	35	4.3	1.2
Diisopropyl ether	0.90	0.60	0.73	68	3.9	1.2
Ethyl acetate	8.5	3.1	0.90	77	6.0	1.8
Heptane	0.005		0.68	98		
Hexane	0.014		0.66	69	1.9	0.0
1-Hexanol	0.70		0.82	~155	13.3	
Mesityl oxide	3.2	3.1	0.85	129	15.6	
Methyl isobutyl ketone (MIBK)	1.7 (25°)	1.9 (25°)	0.80	116	13.1	
4-Methyl-2-pentanol	1.6	6.4	0.80	132	13.3	
Nitrobenzene	0.21	0.22	1.21	211	34.8	4.0
Nitromethane	~10		1.13	101	35.9	
Toluene	0.05	0.06	0.87	111	2.4	
Tributyl phosphate (TBP)	0.6		0.97	(289)	8.0	
m-Xylene	0.02	0.04	0.87	139	2.4	

[a] Compiled from various sources. For salting-out coefficients of some organic solvents by salts, see T. Groenewald, *Anal. Chem.*, **43,** 1678 (1971).
[b] Usually 20° unless otherwise specified.

equilibrium state is reached in 1 or 2 min, often in less time, when the two phases are shaken together vigorously, but some extractions are very slow, requiring hours, less often days, for attainment of equilibrium. The analyst does not willingly use extractions that need more than a few minutes of shaking time. Also, slow extractions can lead to the decomposition of some chelating agents and may introduce an element of uncertainty. In some extractions, the extraction coefficient E apparently has a smaller value at very low metal concentrations than at higher concentrations (Chapter 9D, Section II). The difference has been explained by incomplete reaction at the low concentrations because of low reaction rate.

When a metal is initially present as a simple (hydrated) cation (or a soluble complex species)[11] in the aqueous phase and the chelating reagent is dissolved in the organic solvent, the extraction may be looked upon as consisting of a number of steps as schematically represented in Fig. 9B-5. Briefly described, they are the following:

1.   Metal ions are brought to the border of the aqueous diffusion layer (film) adjacent to the aqueous–organic interface by convectional mixing and to a lesser extent by diffusion. They diffuse into the film toward the interface.

2.   Chelating molecules in the organic phase are transported to the interface by processes mentioned in 1.

3.   Chelating molecules cross the interface into the aqueous phase and are more or less ionized depending on the pH and K_a.

4.   Chelating anions in the aqueous phase react with hydrated M^{n+} to form ML_n. If $n > 1$, there may be more than one reaction step. Reaction of the reagent in the form HL is not excluded.

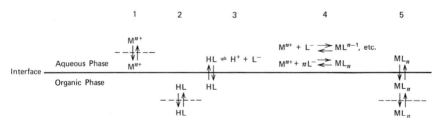

Fig. 9B-5. Schematic representation of extraction of a metal cation M^{n+} in aqueous solution by a chelating reagent HL in an organic solvent. The broken lines indicate hypothetical diffusion zones (films). HL and L^- are shown as being present only in the diffusion zone in the aqueous phase, but if reaction with metal ions is slow they can also be present outside this zone, as can ML_n.

5. ML_n molecules cross the interface into the organic solvent and are distributed in it by diffusion and convection.

Any one of these steps conceivably can be markedly slower than the others and thus can be rate determining. To differentiate one step from another may not be easy. Each step is more or less complex. No attempt will be made here to examine in detail the processes involved, but some observations may be useful. Roughly, the processes can be divided into the physical and the chemical, over which (especially the physical) the analyst has some control. The former class include mass transport of reactants and products by molecular (or ionic) diffusion and eddy diffusion, the first being more important over very short distances, as at the interface. Transfer of material into or from the bulk of the two phases is effected largely by turbulent mixing (eddy diffusion). Molecular (ionic) diffusion evidently determines the rate of steps 1, 2, and 5 and may be involved in the others. Diffusion-controlled extraction rates are examined in more detail later (p. 916).

The important chemical factor is the rate of reaction between reagent species and metal species in the aqueous phase. The analyst may be able to change this rate by varying pH, temperature, or chemical composition. Reaction rate is considered further later.

Passage of HL from organic to aqueous phase and of ML_n from aqueous to organic phase is not necessarily a physical process (diffusion across the interface) only. It may be necessary for HL to shed solvating molecules to enter the aqueous phase (and conversely), and for ML_n to shed hydrating water molecules to dissolve in the organic phase. These reactions are not instantaneous and can conceivably be rate determining in some extractions. Transport across the interface by pure diffusion should (initially) be rapid because of the very small distance involved (*cf.* equation on p. 917). The foregoing concepts assume that reaction between chelating agent and metal takes place in the aqueous phase close to the interface. There is evidence, considered later, that this is true for common chelating extraction reagents such as dithizone. However, reaction at the interface is a possibility and the rate of such reactions may determine the rate of extraction.[12] Conceivably, more than one reaction mechanism may be operative.

Important factors in convectional and diffusional mixing are the rate of movement of one phase past the other, extent of turbulent mixing within the phases and the interfacial area per unit volume of organic phase. If, as often is true, the volume of the organic phase is smaller than that of the aqueous phase, droplets of the former will be distributed in an aqueous matrix on vigorous shaking. Increasing the vigor of shaking decreases the average droplet size of the organic solvent and increases the interfacial area per unit volume in proportion to $1/r$ (r = droplet radius). The rate of mass transfer of chelating reagent from organic solvent to aqueous solution by molecular diffusion will increase with decreasing droplet size. Eddy diffusion will be important for mixing in the aqueous phase, thus bringing M^{n+} to the diffusion layer; presumably it will be less important for mixing in the drops of organic solvent. Mass transfer increases with the velocity of movement of one phase past the other. When organic solvent droplets become very small, they tend to be carried along with the surrounding aqueous solution. But, in general, a net gain in mass transfer is to be expected by increasing the rate of shaking the two phases together.

Clearly, when hours of mixing, even if not particularly rapid, are needed for attainment of equilibrium in extractions at room temperature, the reaction between the chelating reagent and the metal species is very slow. An extreme example is provided by $Cr(H_2O)_6^{3+}$, an ion of the inert type, which does not react appreciably with 8-hydroxyquinoline or acetylacetone at room temperature. Other inert-type metal species include Co(III) complexes and certain inorganic complexes of the platinum metals, for example, complex halides. Some of these reactions can be speeded up by raising the temperature to $\sim 100°$, or by converting an inert species into a less inert or a labile one by an inorganic ligand. Thus Cr(III) is extracted more rapidly by HTTA if F^- or acetate is added. If once formed by a suitable reaction, a chelate of the inert type cannot be dissociated easily, if at all, as by shaking the organic solvent solution with a strongly acid aqueous solution. Cobalt(III) nitrosonaphtholate is not appreciably decomposable by strong sulfuric acid.

Some extractions with dithizone, acetylacetone, and thenoyl-trifluoroacetone (especially the latter two reagents) are slow. Some particulars are given under the respective reagents. In general, chelate extractions are more rapid in less acidic, or basic, solutions than in acid solutions. This behavior is easy to understand: The less acidic the solution, the greater the concentration of L^-, and the greater the rate of formation of the chelate. Also, other factors remaining the same, the smaller P_l (of HL) the greater is the extraction rate since more HL and, therefore, L^- is present in the aqueous phase than when P_l is large. Zinc dithizonate is extracted more rapidly with a CCl_4 than with a $CHCl_3$ solution of dithizone (for H_2Dz, $P_{CCl_4} = 1.1 \times 10^4$, $P_{CHCl_3} = 3 \times 10^5$).[13] This behavior is taken as evidence for the reaction of the metal in the aqueous solution and not with the reagent on the organic solvent side of the interface.

Substances forming water soluble complexes with a metal cation can decrease or increase its reaction rate with a soluble chelating agent. If the cation reacts as such, it might be expected, in general, that reducing its concentration by complexing would reduce its reaction rate. EDTA slows down the extraction of Cu(II) with diethyldithiocarbamate in organic solvents or of Fe(III) with 8-hydroxyquinoline. But it is possible for the metal complex to react directly with the chelating agent, and such reactions need not be particularly slow. In a sense, a simple metal cation in water alone is already complexed, having water molecules occupying coordination positions around the metal. The rate of formation of the chelate complex in the aqueous phase will be determined by the net rate of exchange of chelating anions with water molecules. Replacement of coordinated water molecules by other uncharged molecules may result in more rapid formation of chelate. The extraction of zinc into a chloroform solution of dithizone is markedly accelerated by acetate (25-fold) and thiocyanate.[14] Aquo Ni^{2+} does not react rapidly with dithizonate ion (sluggish replacement of H_2O by HDz^-). If coordinated H_2O is replaced by a nitrogen-base molecule such as pyridine, dithizonate ion substitutes much more rapidly, and therewith the rate of extraction of Ni is increased. Actually, in this case, the extracted species is a mixed chelate, nickel pyridine dithizonate, not all of the pyridine molecules being replaced. In such exchange reactions, the lability of the metal ion complex, not its thermodynamic stability, may be the decisive rate factor.

Masking agents such as citrate or tartrate increase the extraction rate of such metals as Nb, Ta, and Zr, which otherwise would form colloidal or polymeric hydrolyzed species impeding extraction. Citrate and tartrate are also used in the extraction of heavy metals from basic solution with dithizone. These reactions proceed rapidly.

If we compare the extraction rate of a series of metals with the same reagent in the same solvent we may find a rough correlation between the rate and the formation constant of the chelate. This relation is observed with the dithizonates. Ag, Hg(II), and Au(III, chloro complex) are extracted very rapidly with a CCl_4 solution of dithizone from acid aqueous solution, Cu(II) more slowly but still quite rapidly, and such metals as Zn, Ni, and Co (in less acid solutions) markedly less rapidly; the extraction of Tl(I) at pH 7 has been reported to require an hour ($CHCl_3$ solution of H_2Dz). This order is the order of decreasing stability of the dithizonates.

The occurrence of very rapid metal extractions enables us to conclude that mass transport (eddy and molecular or ionic diffusion) is rapid under the conditions of some extractions. For example, if a separatory funnel half full of very dilute acidic aqueous solution of Hg^{2+} and a CCl_4 solution of dithizone in slight excess is shaken vigorously (3–4 shakes/sec), the color change signifying essentially complete reaction of mercury takes place within a few seconds at most. The diffusion and convectional processes are rapid enough to transport dithizone from the CCl_4 phase and Hg^{2+} from the bulk of the aqueous solution to the reaction zone and $Hg(HDz)_2$ from the latter to the bulk of the CCl_4 phase in a matter of seconds. When extractions are slow under conditions of vigorous mixing, the slowness is likely to be due to sluggish reaction between metal and chelating reagent, or slow transfer of the latter across the interface.

Extraction rates obtained for the same metal–reagent system in different laboratories are not easily compared because shaking rates and conditions vary and are difficult to specify. Nevertheless, it may be concluded that quite different rates have been found for the extraction of the same metal with the same extraction reagent by different workers. Cupric dithizonate is an example. We have found that the extraction of $Cu(HDz)_2$ by dithizone–CCl_4 in mineral acid medium is much more rapid than reported by others. Likewise, we have found that $Bi(HDz)_3$ is extracted more rapidly from acid solution than indicated by others. In the Minnesota laboratories, it was found that equilibrium in the extraction of bismuth from 1 M perchloric acid was reached in less than 2 min of hand shaking,[15] whereas others have reported that a 0.5 hr shaking period is needed to reach equilibrium in extraction with similar concentrations of dithizone from 0.1 M acid. We believe that such differences may be due to a trace impurity in the solvent (CCl_4). In our experience, CCl_4 purified by treatment with chlorine or bromine (Chapter 6G) can give faster extraction of copper and bismuth dithizonates than reagent-grade CCl_4, the quality of which, from the extraction rate, varies. The nature of the inhibiting impurity is unknown. In this connection, it may be noted that the extraction rate of Fe(III) by diisopentylorthophosphoric acid is influenced by the presence of very small amounts of impurities, especially monoalkyl dihydrogen phosphates.[16] The inhibiting impurities may exert their effect at the interface of the two phases, rather than in the bulk of the solution. An effect on droplet size does not seem excluded.

Apparently, step 5 can sometimes be the rate-determining step in an extraction. Schweitzer and Rimstidt[17] concluded that with fast stirring, the rate of extraction of zinc acetylacetonate into chloroform was determined by the rate of transfer of the chelate from the aqueous to the organic phase. Sometimes in extractions the chelating reagent is added as a salt to the sample solution, for example, sodium diethyldithiocarbamate and cupferron, and the chelate is formed in an aqueous phase having a high concentration of the reagent. The rate of extraction under these conditions is necessarily the rate of transfer of the chelate from the aqueous to the immiscible organic solvent. It is often found that a metal can be extracted more rapidly in this way than when the reagent is initially present in the organic solvent. The metal concentration may be sufficiently high to give rise to the formation of a precipitate. It has been found that in one such system—diethyldithiocarbamate added to an aqueous solution of copper and other metals—the extraction rate varies with the extracting organic solvent, being much slower with $CHCl_3$, CCl_4, and C_6H_6 (>40 min for equilibrium) than with ethyl acetate or butyl acetate.[18] The slow extraction with nonpolar or weakly polar solvents is believed to be due to the low wetability of Cu(II) diethyldithiocarbamate by these.

Rates of back-extraction (stripping) can also vary greatly for metals complexed by the same reagent. Not all extractions are reversible. For example, Ni and Co dithizonates (especially the latter) in CCl_4 (and doubtless other solvents) are not easily back-extracted into an acid aqueous phase. This behavior seems to be connected with the formation of inert complexes. On the other hand, Pb and Zn as dithizonates can be returned to an aqueous phase quite rapidly (in a few minutes or less) by shaking a CCl_4 solution with 0.02 M HCl; a second portion of acid is used to assure dissociation of these dithizonates (equilibrium effect). Fe(III), Sr, Ca and Zn thenoyltrifluoroacetonates in benzene are dissociated into the metal ions in a relatively short time (within 15 min) by shaking the organic phase with concentrated hydrochloric acid. The Al complex requires a number of hours, and the Be complex is incompletely dissociated after ~ 2 days.[19] The reversion of Fe(III) thenoyltrifluoroacetonate by *dilute* HCl is slow.

When slowness of extraction is due to slowness of the reaction between a metal ion and the chelating agent in the aqueous phase, it may be possible to add a substance that brings about a more rapid extraction. For example, the reaction between Fe(III) and thenoyltrifluoroacetone (HTTA) is very slow, so that extraction of the Fe chelate into benzene is very slow. If thiocyanate is added to the system, the extraction rate is increased.[20] Thiocyanate reacts rapidly with Fe(III) to form a series of complexes, of which $Fe(SCN)_3$ extracts into benzene. In benzene, the SCN^- of the complex is replaced by TTA^- from HTTA in the organic phase. The extraction rate is further increased by using benzene–MIBK as the extracting solvent, in which $Fe(CNS)_3$ is more soluble.[21] Thiocyanate is an extraction catalyst for Fe–HTTA. Zirconium behaves similarly to iron.

Diffusion-controlled Extraction Rates. The rate of extraction of a metal as a chelate from an aqueous phase into an organic solvent phase may be limited by the diffusion rate of the reactants, especially of the metal cation in the aqueous phase. We examine the effect of this factor in an elementary way.

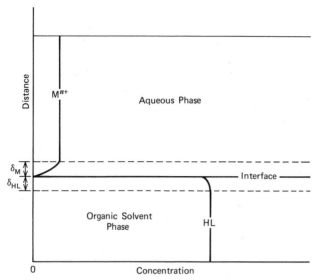

Fig. 9B-6. Schematic representation of concentration profiles of reactants M^{n+} and HL according to the two-film theory. Rate of diffusion of M^{n+} across aqueous film is taken to be the slow step in extraction reaction.

Nernst in 1904 postulated the existence of a stagnant zone (usually called *film*) around a dissolving solid particle, in which transport of material takes place only by molecular (or ionic) diffusion. The film theory can be extended to liquid–liquid extraction (Fig. 9B-6).[22] The thickness of the films on both sides of the interface decreases with increasing flow velocity of one phase past the other. Actually, the stationary film is a fiction, and laminar or turbulent flow takes place up to the interface, with diminished velocity as the latter is approached. However, the concept of a diffusion layer in which convection does not occur is very useful and on its basis diffusion processes can be described quite accurately.

It is assumed that equilibrium is rapidly established at the interface, so that the rate-determining step in a diffusion-controlled extraction is diffusion across the aqueous or organic solvent film indicated in Fig. 9B-6. In what follows, we assume that the rate of diffusion of metal ion across the aqueous film is the slow step in the extraction. If diffusion of the chelating agent across the organic film is the rate-controlling step, conclusions similar to those drawn here can be arrived at.

According to Fick's first law the flux across a unit area of the aqueous film is given by

$$Q = \frac{D([M^{n+}]-[M^{n+}]_i)}{1000\delta}$$

where Q is the moles of M^{n+} diffusing per second across 1 cm^2 of film, D is the diffusion coefficient (cm^2/sec) of M^{n+}, $[M^{n+}]$ and $[M^{n+}]_i$ are the concentrations (moles/liter) of metal cation at aqueous border of film and at organic–aqueous interface, and δ is the aqueous film thickness (cm). It is assumed that stirring or agitation keeps $[M^{n+}]$ at the aqueous side of the film the same as in the bulk of the solution, and because of rapid reaction at the surface of organic solvent $[M^{n+}]_i$ is essentially 0.

A question of interest is the length of time required for the extraction of a specified fraction of M if the extraction rate is limited by diffusion. First, obtain an approximate value of Q, knowing D and very approximately δ. D may be taken[23] as $\sim 10^{-5}$. Rough values of δ for what the analyst would call mild mixing are available. Assume that $\delta \sim 5 \times 10^{-3} \text{ cm}$ for such conditions.[24] With a concentration gradient

$$[M^{n+}] - [M^{n+}]_i = [M^{n+}] - 0 = [M^{n+}] = 1,$$

$$Q \sim \frac{10^{-5} \times 1}{1000 \times 5 \times 10^{-3}} \sim 2 \times 10^{-6} \text{ mole/cm}^2 \text{ sec}$$

Consider the extraction of M^{n+} from V ml of aqueous solution having an initial metal concentration of $[M^{n+}]_0$ into an organic solvent solution of HL (in excess), the area of the interface being 1 cm^2. The rate of extraction is assumed to be determined by the rate of diffusion of the cation

$$-\frac{d[M^{n+}]}{dt} \times \frac{V}{1000} = 2 \times 10^{-6}[M^{n+}]$$

or

$$-\frac{d[M^{n+}]}{[M^{n+}]} = \frac{2 \times 10^{-3}}{V} dt$$

The decrease in metal cation concentration in the time interval $t - t_0 = t$ is

$$\int_{[M^{n+}]_0}^{[M^{n+}]_t} -\frac{d[M^{n+}]}{[M^{n+}]} = \int_0^t \left(\frac{2 \times 10^{-3}}{V}\right) dt$$

or

$$-\ln \frac{[M^{n+}]_t}{[M^{n+}]_0} = \frac{(2 \times 10^{-3})t}{V}$$

$$-2.30 \log \frac{[M^{n+}]_t}{[M^{n+}]_0} = \frac{(2 \times 10^{-3})t}{V}$$

With the value of δ already given and $V = 1$ ml (i.e., 1 ml of aqueous solution for 1 cm^2 interface), the time required to reduce the concentration of M^{n+} to 0.1 of its initial value is

$$t = -2.30 \log 0.1 \times \frac{1}{2 \times 10^{-3}} = 1150 \text{ sec}$$

For 99% extraction, $2 \times 1150 = 2300$ sec would be required. With $V = 10$ ml and the initial metal concentration the same, 23,000 sec would be required for 99% extraction. The initial concentration of the metal in the aqueous phase is immaterial theoretically, but the ratio aqueous volume/interface area is important. If the extraction rate is actually determined by the rate of metal ion diffusion, the concentration of reagent (if always in excess of metal) and the organic solvent volume (if the interface area remains constant) as well as the aqueous volume for a constant weight of metal are theoretically immaterial. Actually in analytical extractions, the interfacial area will depend on organic solvent volume (droplet formation on shaking).

The extraction rates in the foregoing example are inadmissibly slow for analytical extractions. Actually, under analytical conditions, extraction rates are much more rapid than this for two reasons: The ratio aqueous volume/interface area is much smaller and δ is also smaller. When an extraction is made in a separatory funnel, the two phases are shaken together vigorously, and the liquids are broken into droplets. In vigorous shaking, it is reasonable to believe that the organic phase is broken into droplets having a diameter of less than 1 mm (or the same for the aqueous phase). One milliliter of either phase reduced to droplets of 1 mm diameter suspended in a matrix of the other will give an interface of ~ 60 cm^2 compared to 1 cm^2 in the previous example. The rate of mass transfer by diffusion increases by the same factor. But the vigorous mixing also decreases δ. A tenfold decrease to 5×10^{-4} cm at least does not seem unreasonable. The combined effect of these two factors would reduce the extraction time by $1/60 \times 1/10$ or say 1/500. The extraction of 99% metal from 1 ml of aqueous phase by 1 ml of organic solvent (its volume now becomes significant) would then require 2300/500 ~ 5 sec. Accepting this figure only as an order of magnitude, we can still conclude that if other steps in an extraction are faster than the diffusion step, the latter will not in general require shaking the two phases together for more than 0.5–1 min (even if V_w/V_{org} is as large as 5/1), provided agitation is vigorous. In fact, some extractions are known to be 90% or more complete in 1–2 sec (4–8 shakes). If, on vigorous shaking, an extraction is not within 90–99% of the equilibrium value in 0.5 min, we may suspect that diffusion is not the rate-limiting process, whether in the organic or aqueous phase.

In the following pages the extraction rates of chelates of some common reagents are discussed, illustrating general principles and supplying information of practical interest.

2. Dithizone and Dithizonates

Dithizone. Data on the rate of mass transport of a molecular chelating reagent from an organic solvent phase to an aqueous phase are sparse. The results obtained by Geiger[25] on the transport rate of dithizone from a CCl_4 to an acidic aqueous solution are of interest.

The following expression can be derived:

$$\log \frac{1}{1-f} = \log \frac{100}{100-f_{\%}} = 0.434 \frac{k_{org}}{r}(1+rP_{H_2Dz})t$$

$$= \text{const.} \times t$$

in which

f = fraction (or $f_{\%}$ = percentage) of equilibrium (i.e., final) concentration of dithizone in aqueous phase at time t after beginning of extraction (shaking of water with H_2Dz solution in organic solvent).

k_{org} = rate constant for transfer of dithizone from organic solvent phase to aqueous phase under specified conditions of shaking. Depends on interfacial area for unit volume of organic solvent (taken to be smaller than aqueous volume), therefore on drop size, which varies with vigor of shaking.

r = volume of organic solvent phase/volume of aqueous phase.

$P_{H_2Dz} = [H_2Dz]_{org}/[H_2Dz]_{H_2O}$ (equilibrium concentrations in bulk of solution).

$[H_2Dz]_{org}$ is taken to remain essentially constant, since $[H_2Dz]_{org} \gg [H_2Dz]_{H_2O}$ at equilibrium and phase volumes are not too disproportionate.

Provided that k_{org} actually remains constant, $\log[1/(1-f)]/t$ should not vary much with a change in r, since rP_{H_2Dz} in practice is $\gg 1$ (P large for CCl_4 and $CHCl_3$). Mixing is assumed to keep the concentration of dithizone on the aqueous side of the hypothetical film the same as the concentration in the bulk of the aqueous solution.

In Fig. 9B-7, $\log(100-f_{\%})$ is plotted against t for 0.001 and 0.1 M $HClO_4$. It is seen that the rate of transfer of dithizone to the aqueous phase is slow. In spite of vigorous shaking (small drop size), ~9 min is required to bring the dithizone concentration in the aqueous phase to 90% of the equilibrium concentration (25°). Conversely, the rate of

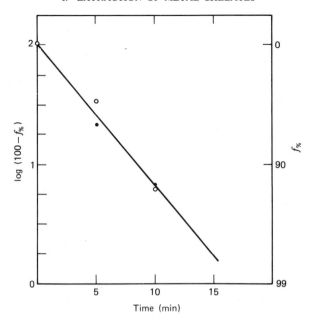

Fig. 9B-7. Transport of dithizone from CCl_4 to an acidic aqueous phase as a function of time of shaking (25 °C). 25 ml of 1.3×10^{-3} M H_2Dz in CCl_4 was shaken with 100 ml $HClO_4$ solution in a 250 ml bottle at 300 shakes/min. Bottle placed 29 cm from pivot of shaker arm, which moved through an arc of 15°. $f_{\%}$ is the percentage of equilibrium concentration of dithizone in aqueous phase. ○ 0.10 M $HClO_4$, ● 0.001 M $HClO_4$.

transfer from an aqueous phase to a carbon tetrachloride phase is rapid. We would expect the initial rate $H_2O \rightarrow CCl_4$ to be 1.1×10^4 $(= P_{CCl_4/H_2O})$ times the initial rate $CCl_4 \rightarrow H_2O$. Within the experimental error, the rate of transfer is the same into 0.10 M $HClO_4$ as into 0.001 M $HClO_4$, which would be expected, since dithizone is present essentially completely as undissociated H_2Dz in both solutions $(K_a = 2 \times 10^{-5})$.

If the acidic aqueous solution of Fig. 9B-7 is replaced by an alkaline solution, for example, 0.1 M NaOH, dithizone is removed practically completely from the CCl_4 phase in a few shakes. Conversion of H_2Dz immediately adjacent to the interface on the aqueous side to HDz^- practically stops the passage of H_2Dz molecules back into the organic phase. Likewise, transfer of H_2Dz from the organic phase is very rapid when Hg(II) is present in the aqueous solution; equilibrium is reached in a few seconds of shaking.

Copper Dithizonate. A preliminary investigation of the rate of copper extraction by a carbon tetrachloride solution of dithizone[26] showed that the extraction rate is affected relatively little by the hydrogen-ion concentration in acid solutions. Thus in Fig. 9B-8*a*, the initial rate of copper extraction is roughly 1.5 times as great in 0.01 *M* H$^+$ as in 0.1 *M* H$^+$ (ionic strength 1 *M* in both). Also, the rate in 1 *M* HCl is much the same as in 0.1 *M* HCl (Geiger). The rate of copper extraction is greater than would be expected from the rate of transfer of H$_2$Dz from CCl$_4$ to the aqueous solution, so that the latter rate does not seem to limit the copper

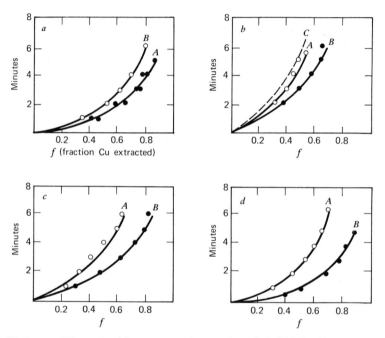

Fig. 9B-8.   *a.* Effect of acidity on rate of extraction of Cu(HDz)$_2$. Equal volumes of 3×10^{-5} *M* dithizone in carbon tetrachloride and 1.0×10^{-5} *M* Cu^{2+} at 1.0 *M* ionic strength (chloride). *A*, 0.01 *M* HCl, *B*, 0.1 *M* HCl. *Note:* Only curves on the same figure are comparable.

     b. Effect of dithizone concentration on rate of extraction of Cu(HDz)$_2$ from 0.01 *M* HCl. Equal volumes of phases; 1.0×10^{-5} *M* Cu^{2+}. *A*, 3.0×10^{-5} *M* H$_2$Dz (CCl$_4$), *B*, 5.0×10^{-5} *M* H$_2$Dz. *C*, The theoretical curve for second order reaction [Cu^{2+}] [H$_2$Dz]$_o$ for *A* based on *B*.

     c. Effect of dithizone concentration on rate of extraction of Cu(HDz)$_2$ from 0.10 *M* HCl. Equal volumes of phases; 1.0×10^{-5} *M* Cu^{2+}. *A*, 3.0×10^{-5} *M* H$_2$Dz (CCl$_4$), *B*, 5.0×10^{-5} *M* H$_2$Dz.

     d. Effect of dithizone concentration on rate of extraction of Cu(HDz)$_2$ from 0.10 *M* HCl + 0.90 *M* NaCl. Equal volumes of phases; 1.0×10^{-5} *M* Cu^{2+}. *A*, 3.0×10^{-5} *M* H$_2$Dz (CCl$_4$), *B*, 5.0×10^{-5} *M* H$_2$Dz.

extraction rate. Since the rate is so little dependent on the acidity, it seems unlikely that HDz^- is kinetically important in extraction of copper from acid solutions. Apparently, H_2Dz plays a more important role. With the same initial concentration of copper, the rate rises with an increase in the dithizone concentration, but the relationship is not a simple one. In 0.01 M HCl, the initial rate does not seem to increase quite in proportion to the dithizone concentration in CCl_4, in 0.1 M HCl it increases approximately so, whereas in 0.1 M HCl + 0.9 M NaCl it increases almost as the square of the dithizone concentration (Fig. 9B-8b,c,d). The rate is found to increase with the copper concentration but not in direct proportion. Reproducibility is hard to obtain in experiments of this kind; more exact and more extensive data will be needed to elucidate the kinetics of the copper extraction. It may turn out that $Cu(HDz)_2$ is formed by more than one path. At least, the preliminary data show that a variation in HCl concentration has a relatively small effect on the extraction rate of copper.

The results depicted in Fig. 9B-8 were obtained under conditions less favorable than those usually employed in separations; the shaking was less vigorous than usually used (so that the reaction could be followed) and the dithizone concentration was smaller. Possibly the moderate agitation (~1 cycle/sec) has obscured some trends by allowing diffusion control. Though not permitting answers to fundamental questions, the results are of interest to the applied analyst.

Zinc Dithizonate. In weakly acidic solutions, zinc tends to be extracted quite slowly by dithizone. It has been concluded that with sufficiently vigorous shaking the rate of extraction of zinc with a CCl_4 or $CHCl_3$ solution of dithizone is determined by the rate of formation of $ZnHDz^+$ by reaction of HDz^- with hydrated zinc ion.[27] The extraction rate is first order with respect to zinc and (approximately) to dithizone and approximately inverse first order with respect to the hydrogen-ion concentration (the latter varied from 6×10^{-6} to 5×10^{-4} M). A mechanism has been proposed in which the rate-controlling step is the reaction involving the change in configuration of the monodithizonate complex from a 6- to 4-coordinate zinc ion, accompanied by replacement of water.[28] The mechanism is represented by the following equations:

$$Zn(H_2O)_6^{2+} + HDz^- \rightarrow Zn(H_2O)_4HDz^+ + 2\,H_2O$$
$$Zn(H_2O)_4HDz^+ \rightarrow Zn(H_2O)_2HDz^+ + 2\,H_2O$$
$$Zn(H_2O)_2HDz^+ + HDz^- \rightarrow Zn(HDz)_2 + 2\,H_2O$$

The same mechanism holds for the extraction of Cd, Co, and Ni with H_2Dz (and di-o-tolylthiocarbazone). Cd is extracted more rapidly than

Zn; Co and Ni are extracted less rapidly. The reaction of zinc with di-α-naphthylthiocarbazone is second order in the reagent anion, and approximately inverse second order with respect to hydrogen-ion concentration. The higher extraction rate with di-o-tolylthiocarbazone than with dithizone is attributed to a steric effect, the former displacing water molecules more readily.

3. β-Diketones

Reagents of this class are notorious for their slow extraction of some metals.

Acetylacetone (HAA) in CCl_4, $CHCl_3$, benzene, and other organic solvents extracts Fe(III), Mg, Ni, Co, and Mo from acidic solution so slowly that hours, even days, are needed for attainment of equilibrium. Most other metals seem to be extracted in a matter of minutes, anyhow in less than an hour. Extraction time depends on the reagent concentration and the acidity. As we have already seen, extraction rate varies with the solvent in which the reagent is dissolved. In the extraction of Fe(III) at pH 1.3 with a 0.1 M solution of HAA in CCl_4, extraction equilibrium is reached within an hour, whereas with $CHCl_3$ as solvent, equilibrium is not quite reached in 5 hr (Zolotov). $\log P_{HAA} = 0.5$ for CCl_4/H_2O and 1.4 for $CHCl_3/H_2O$.

Extractions with benzoylacetone (HBA), as of Be, Mg, and Ni, may be even slower than with acetylacetone, which is not surprising considering that P_{HBA} is some thousand times larger than P_{HAA} (CCl_4, $CHCl_3$, benzene). The difference in rates is not as great, however, as might be expected from the difference in P values.

Dibenzoylmethane extractions of Be, Ni, Pd, Hg, and Al are reported to require days, a behavior that eliminates this reagent as a possible one for the practical extraction of these elements. Iron(III), Cu, and U(VI), on the other hand, can be extracted in some minutes of shaking.

Thenoyltrifluoroacetone (in benzene) likewise gives some very slow extractions, as with Be, Fe(III), Al, Ga, and In. In fact, extraction equilibrium for HTTA itself requires hours. The kinetics of formation of the 1:1 complex ion, $M(TTA)^{n-1}$, of a number of metals in acidic aqueous solution has been studied.[29] The rate-determining step for the formation of the 1:1 complexes of Sc, Cu, Zn, and Mg is the rate of enolization of HTTA, whereas for Fe(III) it is the rate of the reaction of the metal ion with the enolate ion. For Cu(II) and Sc the rate constants (25°) are greater than $10^6 \, l \, mole^{-1} \, min^{-1}$ and probably also for Mg and Zn. The rate constants for 1:1 complex formation have been correlated inversely with the Born charging energy of the central ion, $(Ze)^2/r$, which varies from 5.4 to 11.0 for the fast reacting ions Zn, Cu, Mg, and Sc and from 12.5 to 18.0 for the slow reactors Be, Cr(III), Fe(III), and Al. The rate of formation of 1:1 complexes is faster than that of higher complexes.

When a chelating extraction reagent reacts slowly with a metal ion, an obvious way to obtain better extraction is to add the reagent to the aqueous solution in such a way that the reaction proceeds in a homogeneous phase, and then to extract the chelate. This may be less easily done than said, but there are possibilities. It might be possible to add the reagent (usually water-insoluble) in a comparatively small volume of water-miscible organic solvent, for example, alcohol. Such an approach will not always be successful because the reagent may not be sufficiently soluble in the miscible organic solvent–water mixture, or the solvent may interfere in the subsequent extraction. Another approach is to use a solution of reagent in an organic solvent of limited miscibility with water at room temperature but wholly miscible at higher temperatures. Such a solvent is propylene carbonate, which dissolves to the extent of 2.1 ml in 10 ml of water at 25°, but to the extent of 10 ml in the same volume at 73°. It has been proposed to extract $Fe(TTA)_3$ by adding an equal volume of propylene carbonate containing HTTA to an Fe(III) solution in 0.01 M $HClO_4$, heating to homogenization (~80°), and then cooling to room temperature to separate the propylene carbonate.[30] Equilibrium is reached in not more than 5 min heating at 80°. It appears that the high temperature of the reaction is as important as the homogenization, if not more so, for rapid extraction of iron. The extraction of iron by the customary extraction method at room temperature is also much more rapid with a propylene carbonate solution of HTTA (0.02 M) than with a benzene solution, equilibrium being reached with the former solution in 5–10 min, though only with ~90% recovery. It may be noted that species such as $Fe(TTA)_2ClO_4$ can be extracted into propylene carbonate, so that formation of $Fe(TTA)_3$ is not essential.

II. SOME CHELATING REAGENTS FOR EXTRACTION SEPARATIONS

In this section, we discuss reagents that are primarily useful in separations by extraction. Some of them are of subsidiary interest as absorptiometric reagents. 8-Hydroxyquinoline and dithizone, which are important as determination as well as separation reagents, are treated in Chapters 6D and 6G. Dithiocarbamate reagents are discussed in Chapter 6F. A number of other reagents mentioned in earlier chapters on photometric reagents are also important in separations (for example, nitrosonaphthols for Co).

A. β-DIKETONES

The parent compound of this group of extraction reagents is acetylacetone, which in its enolic form has an acidic hydrogen replaceable by metals

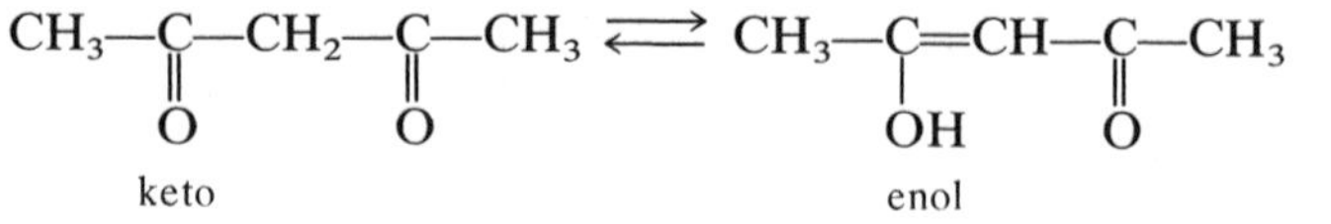

$$\underset{\text{keto}}{CH_3-\underset{\underset{O}{\|}}{C}-CH_2-\underset{\underset{O}{\|}}{C}-CH_3} \rightleftharpoons \underset{\text{enol}}{CH_3-\underset{\underset{OH}{|}}{C}=CH-\underset{\underset{O}{\|}}{C}-CH_3}$$

to form chelates of the type

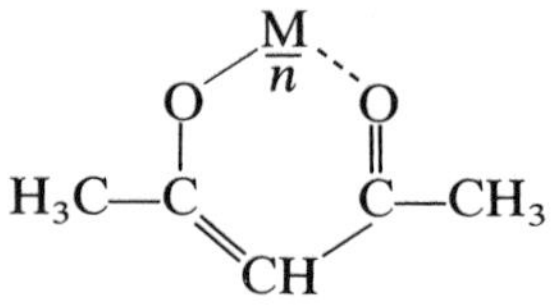

$$\underset{\substack{\\ CH}}{H_3C-\underset{\underset{O}{}}{C} \overset{\overset{\displaystyle M}{\overset{\displaystyle n}{\diagup}\,\diagdown}}{} \underset{\underset{O}{}}{C}-CH_3}$$

Acetylacetone is very weakly acidic in aqueous solution, with a pK_a of 8.9. Substitution of fluorine for hydrogen atoms increases the acidity. Trifluoroacetylacetone has a pK_a of 6.7, thenoyltrifluoroacetone of 6.2 (see the following discussion). Other factors remaining the same, the more acidic reagents permit metal extractions in more acidic solutions. An increase in the acid character of the reagent may, however, lead to lesser stability (greater dissociability) of the metal chelates.

Physical constants for some analytically useful β-diketones are listed in Table 9B-4.

1. Acetylacetone (2,4-Pentanedione, Diacetylmethane)

Colorless liquid, density 0.976 (25°), b.p. 135°–137° at 745 mm Hg; mol. wt. 100.1. Solubility in water, 173 g/liter at 25°.[31] Commercial acetylacetone contains acetic acid (sometimes 10% or more). Can be purified by shaking with 1:10 ammonia, washing with water, drying with Na_2SO_4 and distilling. A benzene solution (20% v/v) can be freed from acetic acid by shaking several times with equal volumes of water. Stable, undiluted, and in organic solvents.

Most metals are chelated by acetylacetone (HAA) and can be extracted by acetylacetone itself or by its solution in various organic solvents. The solubility of the acetylacetonates in common organic solvents is generally high (reported to be of the order of grams per liter). The use of undiluted HAA offers the advantage of a high concentration of the reagent in the aqueous phase to (presumably) increase the extraction coefficient of metals. However, the relatively high solubility of HAA in water may lead to an appreciable solubility of the metal chelate in the aqueous phase and

TABLE 9B-4
Some Properties of Acetylacetone and Common Derivatives[a]

	m.p. (°C)	b.p. (°C)	pK_a	log P (organic/aqueous)		
				C_6H_6	CCl_4	$CHCl_3$
Acetylacetone		135–137 (745 mm)	8.9	0.76	0.52 0.15 (1.9 M $HClO_4$)	1.4 1.0 (1.9 M $HClO_4$)
Benzoylacetone	59–60		8.7	3.1	2.8	3.4
Dibenzoylmethane	77–78		9.35	5.35	4.5	5.4
Thenoyltrifluoroacetone	43		6.2, 6.4	1.6 2.12 $(\mu = 1.0)^b$	1.66	1.8
Trifluoroacetylacetone			6.7, 6.6	0.28	0.32	0.53, 0.3
Hexafluoroacetylacetone			4.35			

[a] From various sources. Temperature 25° or 20° for pK_a and log P; $\mu = 0.1 M$ or 0. I. M. Korenman and M. I. Gryaznova, *Zh. Anal. Khim.*, **29**, 964 (1974), give the following log P values (20°±2°) for C_6H_6, CCl_4, and $CHCl_3$: acetylacetone, 0.72, 0.61, 1.40; benzoylacetone, 3.03, 2.79, 3.53. For P of various alkyl-substituted β-diketones see H. Koshimura and T. Okubo, *Anal. Chim. Acta*, **67**, 331 (1973).

[b] HTTA is slightly associated in benzene so that E varies from 40 at very low concentrations to 58.7 for a 1.79 M solution in benzene, when the aqueous phase is 0.12 M HCl (King and Reas, *J. Am. Chem. Soc.*, **73**, 1804 (1951)).

therewith to a decrease in the extraction coefficient. For this reason, and for others, HAA is often diluted with benzene, carbon tetrachloride, or chloroform. The partition constants (20°, $\mu = 0.1$), $[HAA]_{org}/[HAA]_w$, for these solvents are[32]: C_6H_6, 5.8; CCl_4, 3.3; $CHCl_3$, 25. Benzene is often used as diluent, but carbon tetrachloride and chloroform, being denser than the aqueous phase, are more convenient solvents when successive extractions are required—and this is the rule.

The benzene–H_2O extraction equilibrium constants of some metal acetylacetonates are listed in Table 9B-5.[33] The chelates are of the normal type, except that of U(VI), which is $UO_2(AA)_2$ (HAA).[34] Molybdenum seems to give $MoO_2(AA)_2$. These values are offered only for the purpose of rough calculations. They are, for the most part, effective constants, that is, they are based on the total concentration of metal in the aqueous phase. This is advantageous from the analytical standpoint, but it must be remembered that the constants may hold only for relatively constant conditions. Especially when extractions are carried out in weakly

TABLE 9B-5
Extraction Equilibrium Constants for Some Metal β-Diketonates (Benzene–Water)[a]

Metal	$-\log K_{ext}$		
	Acetylacetonate	Benzoyl-acetonate	Dibenzoyl-methanate
Be	2.8	3.9	3.45
Ca		18.3	18.0
Cd		14.1	14.0
Co		11.1	10.8
Cu	(~3.9)	4.15	3.8
Fe(III)	1.4[b]	0.5	1.95
Ga	5.5	6.35	5.75
In	7.2	9.3	7.6
La		20.45	19.45
Mg		16.65	14.7
Mn(II)		14.6	13.7
Ni			11.0
Pb		9.6	9.45
Pd	<−2	−1.2	
Sc	5.85	6.0	6.05
Th	12.15	7.7	6.4
U(VI)[c]		4.7	4.1
Zn		10.8	10.65

[a] Stary and Hladky, reference in note 37. Values rounded to nearest 0.05 log unit. Constants are given only for metals having more or less normal sigmoid extraction curves, with 100% extraction attained at a higher pH.
[b] With undiluted HAA, $-\log K_{ext} = 3.2$.
[c] Chelate is $UO_2L_2 \cdot HL$.

acidic solutions, and even more so in basic solutions, hydrolyzed metal species may not be taken into account and may upset the analytical calculations. Likewise, formation of intermediate, charged, acetyl-acetone–metal species may be important.[35] Even the neglect of uncharged chelate in the aqueous phase may lead to serious error. The solubility of some uncharged metal acetylacetonates in the aqueous phase may be appreciable, especially when undiluted acetylacetone is used as extractant. Under the latter conditions, $P_{M(AA)_n}$ ($[M(AA)_n]_{org}/[M(AA)_n]_w$) is 41 for Be and ~7 for Cu(II).[36] Thus the extraction coefficient of copper can never exceed 7 and with one extraction, equal volumes of HAA and aqueous solution, the recovery of copper under the most favorable conditions is only ~87%. With a benzene solution of HAA slightly better recoveries of copper can be obtained.

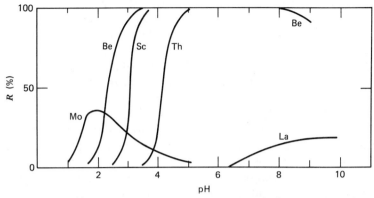

Fig. 9B-9. Extractability (= R,%) of Be, La, Sc, Th, and Mo(VI) by 0.1 M acetylacetone in benzene as a function of pH. Stary and Hladky, *Anal. Chim. Acta*, **28,** 227 (1963).

Because of such complications, more generally useful information can be imparted to the analyst by plotting the extractability (as the percentage recovery of the element under specified conditions of reagent concentration and phase volume ratios) versus pH. Such curves for HAA and its derivatives are to be found in the literature, and two samples are shown in Figs. 9B-9 and 9B-10. Even such curves, however, should be used with caution in the weakly acidic range, if information on the manner of adjustment of the pH is lacking.

Only a few metals can be extracted easily ($R > 90\%$ with an equal volume of 0.1 M HAA in benzene) from solutions more acid than pH 3.

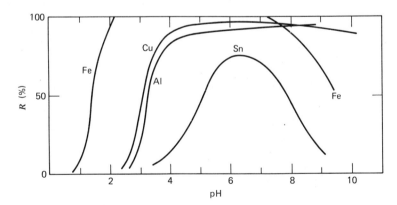

Fig. 9B-10. Extractability (= R,%) of Fe(III), Al, Cu(II), and Sn(II) by 0.1 M acetylacetone in benzene as a function of pH. Stary and Hladky, *loc. cit.*

These include Pd(II), Fe(III), and Tl(III). The approximate pH values required for >90% extraction of other metals are Sc, 3.5; Th, 4.5; Ga, 3.5; In, 4.5.[37] Metals for which 90% extraction is not reached at any pH include Mg, Mo, Ti, Zr, U, La, Ni, Co, Mn, Hg, Zn, Sn, and Pb. Some of these show a narrow peak in the R–pH curve, others a plateau extending over several pH units. Aluminum gives a slowly rising curve in the pH 4–9 range, with a maximum near or at 90% at pH 9. The unfavorable behavior of the metals mentioned is due largely to the presence of hydrolyzed species or to appreciable intrinsic solubility of the metal acetylacetonate in the aqueous phase. Magnesium is not extracted below pH 8.5 and incompletely above; Ca, Sr, Ba, Ag, As(V), Cd, Bi, and Cr(III, at room temperature) are not extracted at any pH; Sb(III) and Ge are extracted hardly at all from HCl solution by undiluted HAA.

Some metals are extracted very slowly by HAA (p. 924). Chromium(III) does not readily form an acetylacetonate at room temperature and must be boiled with the reagent.[38] $Co(AA)_2$ is known to form a dihydrate, so that its poor extraction by benzene is to be expected. It can be extracted by immiscible ketones and alcohols (e.g., butanol).

Although acetylacetone extractions have received considerable study, the useful analytical applications are limited. Probably the most interesting extraction is that of beryllium. Somewhat unexpectedly, it is extracted from rather acidic solutions (pH >3 with 0.1 M benzene solution of HAA), but pH adjustment alone does not provide sufficient selectivity. However, most metals can be masked with EDTA, which does not mask Be, so that with its aid, useful separations of Be can be made. When EDTA is used to mask Al, time should be allowed for formation of the EDTA complex (which is slow at room temperature); boiling for a few minutes in neutral or slightly acidic solution after addition of EDTA is advisable. The extraction of Fe(III) from acidic solutions (~0.01 M H^+) is of mild interest. It allows removal of fairly large amounts of iron from solutions of metals extracted only at definitely higher pH. However, many reagents yield iron(III) chelates or association complexes extractable from acid solution. Acetylacetone extraction in approximately neutral solution has been applied with the intent to rid an aqueous solution of most metals before the determination of calcium and magnesium. Too much reliance should not be placed on this separation; it may have value for samples of favorable composition.

Acetylacetone has been suggested for the extraction of Re in basic solution.[39] In 0.5 M NaOH, $E_{Re} \sim 2$, $E_{Mo} \sim 2 \times 10^{-4}$. Tungsten(VI) and V(V) are extracted slightly less than Mo.

A chloroform solution of acetylacetone extracts Mo but not W from a citric acid solution.[40]

If acetylacetone extraction is carried out in the presence of pyridine (pH ~5), the extraction of Ni and Cr(III) can be made ~98% complete.[41] The extracted Ni species is considered to be $Ni(AA)_2Py_2$. Elements not extracted include Sb(V), As(V), Ca, Mg, P, SO_4^{2-}, Te(VI), and probably Tl(I). In addition to Ni, Cr(III), and Fe(III), elements extracted to a large extent include Al, Bi, Ce(III), Co, Cu, Sn(IV), and V(V).

Data are available on the extraction of sundry metals by benzene solutions of a number of alkyl-substituted β-diketones.[42] Their behavior does not differ significantly from that of acetylacetone.

2. Benzoylacetone

Mol. wt. 162.2. Water solubility $2.4 \times 10^{-3} M$ (25°); CCl_4, 1.9 M; $CHCl_3$, 3.9 M; C_6H_6, 2.5 M.

This reagent has a pK_a of 8.7, much the same as that of acetylacetone, but its P values are larger (thus 1150 for benzene, 660 for carbon tetrachloride, and 2500 for chloroform) and the same is true of the metal chelates. The pH ranges for the extraction of the benzoylacetonates are much the same, for most metals, as those of the acetylacetonates. But a number of the benzoylacetonates are extracted more completely in the optimal pH range than are the same acetylacetonates, and, in fact, some metals not extractable as the latter can be extracted as the former.[43] For example, with a 0.10 M benzoylacetone solution in benzene quantitative extraction of Mg can be obtained at pH >10; U(VI) >5; Ti >3.5; Zn >7.5; Cd >9; Pb >7. Bismuth is extracted to a considerable extent in basic solution, Ca almost completely at pH 11–11.5, Sr ~50% and Ba not at all at this pH.

Extraction equilibrium is reached more slowly with benzoylacetone than with acetylacetone. The extraction of Be, Mg, Mo (always incomplete) and Ni requires shaking for several hours, that of other metals 10–30 min.

3. Dibenzoylmethane

Mol. wt. 224.25. Solubility $6 \times 10^{-6} M$ in 0.1 M $NaClO_4$, 1.3 M in CCl_4, 1.8 M in C_6H_6.

See Table 9B-4 for K_a and other constants.

The extractability of metals with this reagent in benzene and the like is very similar to that with benzoylacetone (Table 9B-5). One characteristic of this reagent will not endear it to analysts—the rate of attainment of extraction equilibrium is usually slow or even very slow. Several days of shaking are needed for reaching equilibrium with Pd, Hg, Ni, Mo, and Al. Other metals may require hours, although Fe(III), Cu, and U(VI) can be extracted in a few minutes. The U(VI) chelate is of some interest as a determination form of uranium ($\epsilon_{400} = 20,000$ or $\mathscr{S} = 0.012 \ \mu g \ U/cm^2$ in benzene).

4. 2-Thenoyltrifluoroacetone

The more precise name is 4,4,4-trifluoro-1-(2-thienyl)-1,3-butane-dione:

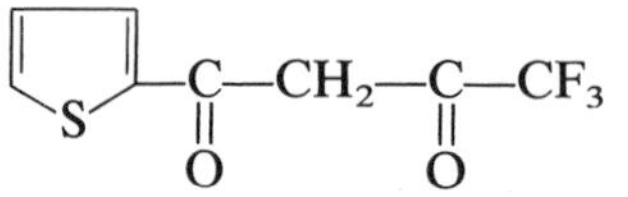

Mol. wt. 222.2. M. p. 43°. Very slightly soluble in water, ~0.001 M. Light sensitive; solid unstable on storage. Can be purified by vacuum distillation. $K_a = 7 \times 10^{-7}$. Sufficiently soluble in most common organic solvents to enable preparation of 0.5 M or stronger reagent solutions.

This fluorinated diketone (HTTA) is more acidic than acetylacetone and partly for this reason allows metal extractions to be made at lower pH values than does that reagent. Benzene, xylene, toluene, chloroform, carbon tetrachloride, and occasionally oxonium solvents (e.g., MIBK) are used to prepare solutions (see Table 9B-4 for some P values). HTTA exists almost entirely as the keto hydrate in aqueous solution[44]:

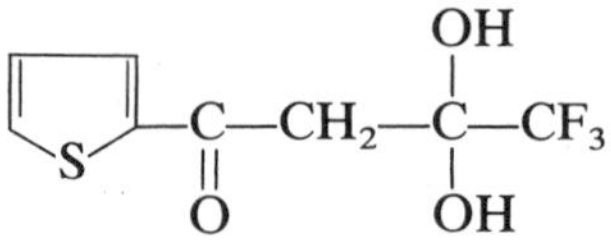

Approximately 2% is present in the enol form. In dry benzene, HTTA is present mostly in the enol form. In benzene solution in equilibrium with an aqueous solution, 11% of HTTA is present as the hydrate. The final distribution equilibrium is slowly established (over a period of days). Figure 9B-11 gives the extraction coefficient as a function of the ionic strength in the aqueous phase. An $E_{benzene}$ of 40 is indicated for zero ionic strength and of ~400 for $\mu = 6.0 \ M$ (perchlorate).

In weakly alkaline solutions, HTTA is converted to enolate ion and E decreases according to the relation

$$E = \frac{[H^+]P}{K_a + [H^+]}$$

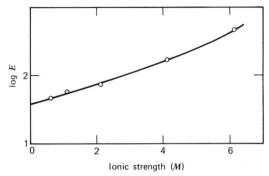

Fig. 9B-11. Distribution of HTTA between benzene and aqueous solutions of NaClO$_4$ (25 °C). Benzene 0.50 M in HTTA; aqueous solution 0.09 M in HClO$_4$, with NaClO$_4$ as remainder of electrolyte. From data of E. L. Zebroski, H. W. Alter, and F. K. Heumann, *J. Am. Chem. Soc.*, **73**, 5646 (1951).

so that $E \sim 1$ for benzene–H$_2$O at pH ~ 8. Above pH ~ 9, HTTA decomposes into trifluoroacetic acid and acetylthiophene.

HTTA has found its greatest use in separations of the actinide elements. In general, chelates of the type ML$_n$ are formed, having six-membered rings, that are extractable into the solvents mentioned.[45] The extraction equilibrium constants, as far as known, are given in Table 9B-6, from which separation possibilities can be deduced (if there are no complications, $E = 1$ at pH$_{1/2} = -\log K_{ext}/n - \log [\text{HTTA}]_{org}$). HTTA has an undesirable characteristic: equilibrium may be reached very slowly in some extractions.[46] This is especially likely when the reagent anion concentration is low (in more acid solutions, low total reagent concentration), and the effect is more pronounced with some metals than with others. Beryllium, for example, is extracted slowly, and it has been reported[47] that 12 hr of shaking are required to reach equilibrium in the extraction of Fe(III) from hydrochloric acid solutions with 0.2 M benzene HTTA solution. (See further, p. 924.)

Niobium is poorly extracted by HTTA in xylene from HCl solutions. If 10% (v/v) butanol is added to the xylene solution, the extraction of Nb (<1 μg/ml aqueous solution) rises to $>99\%$ from strong HCl (7 M or higher).[48] The extraction also succeeds in HCl–H$_2$SO$_4$ solutions and is quite rapid. Ta, Zr, and Fe(III) are also extracted to a large extent. This is another example of the enhancement of extraction of a coordinatively unsaturated chelate by replacement of water molecules by organophilic molecules. Lower alcohols are much less effective than butanol. Co(TTA)$_2$, a coordinatively unsaturated chelate, is also readily extracted into butanol.

**Extraction Equilibrium Constants of Metal Thenoyltrifluoro-
acetonates**[a]

Metal	$\log K_{ext}$	
	Benzene	Other Solvents
Al	−5.25	
Am(III)	−7.45	−8.6, toluene
		−6.6, cyclohexane
Ba	−14.4	
Be	−3.2	
Bi	−3.3	
Bk		−7.5, toluene
Ca	−12	
Cd		−11.4, $CHCl_3$
Ce(III)	−9.45	
Ce(IV)		3.3, xylene
Cf(III)		−7.8, toluene
Cm(III)		−8.6, toluene
Co(II)	−6.7	
Cu(II)	−1.3	
Dy	−7.05	
Es(III)		−7.9, toluene
Eu	−7.65	
Fe(III)	3.3	
Fm(III)		−7.7, toluene
Gd	−7.6	
Hf	7.8 to	7.3, dichlorobenzene
	8.2	
Ho	−7.25	
In	−4.35	
La	−10.5	
Lu	−6.75	
Mn(II)		−6.5, acetone–benzene
		−5.5, ethyl acetate
Mn(III)		3.0, xylene
Nd	−8.55	
Np(IV)	5.6	
Pa(IV)	6.7	
Pb	−5.2	
Pm(III)	−8.05	
Pr	−8.5,	
	−8.85	
Pu(III)	−4.45	−4.7, cyclohexane
Pu(IV)	6.85,	5.0, CCl_4
	6.35	6.35, cyclohexane
Pu(VI)	−1.8	−1.55, cyclohexane
Sc[b]	−0.75	

| Metal | $\log K_{ext}$ | |
	Benzene	Other Solvents
Sm	-7.7	
Sr	-14	
Tb	-7.5	
Th	-0.7,	-1.0, CCl_4
	-1.4	-1.0, MIBK
Tl(I)	-5.2	
Tl(III)	-5.7	
Tm	-6.95	
U(IV)	5.3 .	
U(VI)	-2.25,	-2.8, cyclohexane
	-2.0	
Y	-7.4	
Yb	-6.7	-7.3, toluene
Zr	9.2 to	8.5, dichlorobenzene
	9.6	

[a] Values (rounded to nearest 0.05 log unit) from various sources, including F. Hagemann, *J. Am. Chem. Soc.*, **72**, 768 (1950), for Bi, Pb, Tl(I,III) and Th; R. A. Bolomey and L. Wish, *ibid.*, **72**, 4483 (1950), for Al, Be, Ca, Sr, and Cu; A. M. Poskanzer and B. M. Foreman, *J. Inorg. Nucl. Chem.*, **16**, 323 (1961), for Ba, Fe, In, Sc, U, Y, and rare earths; G. K. Schweitzer and D. R. Randolph, *Anal. Chim. Acta*, **26**, 567 (1962), for Cd; G. Goldstein, O. Menis, and D. L. Manning, *Anal. Chem.*, **32**, 400 (1960) for Th (CCl_4); H. Yoshida, H. Nagai, and H. Onishi, *Japan Analyst*, **15**, 513 (1966), for Ce(IV), and *Talanta*, **13**, 37 (1966), for Mn(II,III). For more complete references see J. Stary, *Solvent Extraction of Metal Chelates*. These constants are to be looked upon as effective constants, and are useful for rough calculations in the vicinity of $pH_{1/2}$ (equal concentrations of metals in both phases). The ionic strength and, more importantly, the presence of complexing ions (including OH^-) will, of course, affect the values, and the original sources should be consulted for this information. Also see p. 905 (formation of mixed complexes). Sn(IV) is extracted from H_2SO_4–HCl into MIBK: J. R. Stokely and F. L. Moore, *Anal. Chem.*, **36**, 1203 (1964). Nb(V) is extractable from *ca.* neutral solutions by higher alcohols.

[b] Extraction into sundry esters: N. Suzuki, K. Akiba, and T. Kanno, *Anal. Chim. Acta*, **43**, 311 (1968). Extraction into C_6H_6, $CHCl_3$, and CCl_4 has been studied by Yu. A. Zolotov, N. V. Shakhova, and I. P. Alimarin, *Zh. Anal. Khim.*, **23**, 1321 (1968). Partition constants of HTTA and its Sc chelate have been determined for various ethers and water by N. Suzuki, K. Akiba, and H. Asano, *Anal. Chim. Acta*, **52**, 115 (1970).

The organic solvent solutions of some metal–HTTA chelates are colored (e.g., ferric iron) and have been proposed for colorimetric determination, but are of minor value in trace analysis.

Regulation of the acidity is the most important means of obtaining selectivity in HTTA extraction separations. For example, thorium can be extracted from ~0.1 M HNO$_3$ solution by a 0.5 M HTTA benzene solution, whereas little uranium(VI) is extracted. By washing the benzene extract with 0.1–0.2 M HNO$_3$, one can remove the small amount of uranium in one or two steps. Of course, when the ratio U/Th is very unfavorable this simple procedure fails. Uranium(VI) can be extracted quantitatively at pH ~3. Many metals will accompany uranium as can be seen from Table 9B-6. Some use has been made of masking agents to increase selectivity, for example, as of EDTA.[49] Hydrogen peroxide complexes Nb and Pa in 2 M HNO$_3$ (separation from Zr).

The extraction of Group II and III elements by thenoyltrifluoroacetone in MIBK (0.1 M solution) has been studied as a function of the pH.[50] In the pH range indicated, the extraction coefficient of the first named metal is at least 1000 times greater than that of the second metal:

pH	Metal Pair
2.0–2.2	Sc–Y
2.0–2.3	Sc–In
2.0–2.8	Sc–Zn
2.0–2.8	Sc–La
2.0–4.0	Sc–Cd
2.0–4.3	Sc–Tl
2.0–4.6	Sc–Ca
2.0–4.8	Sc–Sr
2.0–5.3	Sc–Ba
3.8–4.0	Y–Cd
3.8–4.3	Y–Tl
3.8–4.6	Y–Ca
3.8–4.8	Y–Sr
3.8–5.3	Y–Ba
3.8–4.2	In–Cd
3.8–4.3	In–Tl
4.0–4.2	La–Cd
4.2–4.3	La–Tl
4.2–4.6	La–Ca
4.2–4.8	La–Sr
4.2–5.3	La–Ba

The extraction coefficient of the first metal of each pair is typically ~10^2–10^3, of the second metal 10^{-3}–10^{-2}, in the pH range indicated. Equilibrium is usually attained in 10-min shaking (Ga is an exception);

the reverse extraction is slow for Be and Ga. Similar data have been reported for other elements[51]:

pH	Metal Pair
1.6–2.0	Th–Pb Th–Y
1.6–2.2	Th–La
1.6–4.0	Th–Ca
2.8–3.0	Cu–Pb
3.8–4.0	Fe(III)–Mg
4.4–4.6	Y–Ra

Fe(III) is said to be irreversibly extracted.

A solution of HTTA in MIBK is reported to be more stable than in other solvents, especially when used in basic and strongly acid solutions.

Iron(III), which otherwise would be troublesome in HTTA separations, can be reduced to Fe(II), not extracted in acid medium, as with NH_3OHCl.

Many metals can be stripped from the organic solvent by shaking with mineral acid (2 M HNO_3 for Th) to dissociate the thenoyltrifluoro-acetonates. Not all back-extractions proceed rapidly; some are very slow (p. 916).

It has been observed that the extraction of In from acidic citrate medium by HTTA is inhibited by Fe(III), Al, and Cr(III). The effect of Cr(III) has been studied especially.[52] Some hydroxycarboxylic acids, in the presence of Cr(III), decrease the extraction of In, others do not. Isocitric acid, tartaric acid, and malic acid inhibit the extraction, whereas such acids as tricarballylic, tartronic, and glutaric do not. The conclusion has been drawn that the inhibiting acids must contain two or more carboxyl groups and one hydroxyl group, with the carboxyl groups bonded to adjacent carbons, one of which must also be bonded to a hydroxyl group. No doubt mixed complexes of In and Cr are formed with an inhibiting acid and the following structure has been suggested for the 1:1:2 In, Cr, malic acid complex:

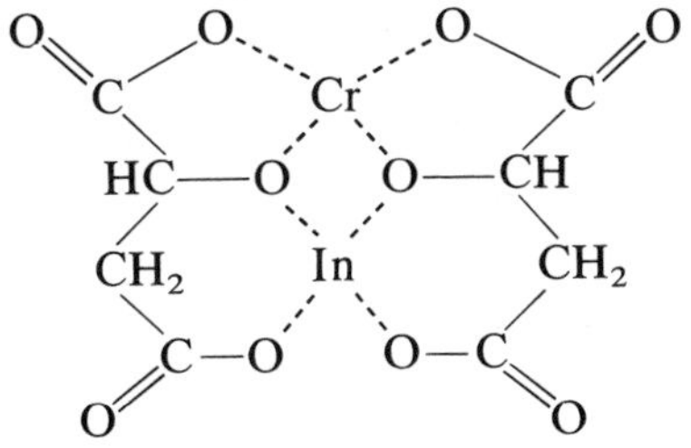

Similar structures can be written for the In, Cr, citrate complex. The extraction coefficient (benzene/aqueous) of In can be decreased as much as tenfold in the presence of Cr and citrate (tartrate shows a lesser, but still definite, effect).

HTTA and other β-diketones have been much used in studies of synergism (see p. 1013). An unusual type of synergism occurs in acetate solutions of HTTA and $(C_6H_5)_4AsCl$. From such solutions at pH 3.8, Co is extracted much better into benzene than in the absence of acetate, though the effect does not seem to have analytical importance. The synergic complex is $\{(C_6H_5)_4As^+\}\{Co(TTA)_2Ac^-\}$.[53]

Heterocyclic bases, 1,10-phenanthroline for example, increase P of lanthanide–HTTA chelates by a factor of 10^6 and decrease extraction time from hours to 1 minute.[54] The effect is due to adduct formation: displacement of water molecules in the coordination sphere by organophilic organic base.

Thiothenoyltrifluoroacetone. This compound has been synthesized, and spectral data for sundry chelates in organic solvents have been obtained.[55] It exists almost entirely in the thio-enolic form. The chelates, soluble in polar and nonpolar organic solvents, are more strongly colored than those of thenoyltrifluoroacetone. Divalent ions chelate more readily than the trivalent. Extraction takes place from weakly acidic solution. This reagent may have more value for determination than for extractive separation. At the present time it finds little use for either.

5. Other β-Diketones

Among these may be mentioned trifluoroacetylacetone,[56] furoyltrifluoroacetone, pyrroltrifluoroacetone, selenoylacetone, and selenoyltrifluoroacetone. It is not evident that they have any special analytical utility in connection with photometric analysis. Hexafluoroacetylacetone in isoamyl acetate extracts Ba well at pH 10–11. Its chelates with various metals form stable adducts with TOPO.[57]

Dipivaloylmethane (2,2,6,6-tetramethyl-3,5-heptanedione)

$$CH_3-\underset{\underset{CH_3}{|}}{\overset{\overset{CH_3}{|}}{C}}-\underset{\overset{||}{O}}{C}-CH_2-\underset{\overset{||}{O}}{C}-\underset{\underset{CH_3}{|}}{\overset{\overset{CH_3}{|}}{C}}-CH_3$$

Soly. in $H_2O \sim 9 \times 10^{-4}$ M.
$pK_a = 11.75$ ($\mu = 0.1$).

has attracted some attention as forming a chelate with lithium that is (poorly) extractable by ethyl ether from strongly basic solution. Its use for the extraction of the lanthanides (itself as the solvent) has been studied.[58]

Two other β-diketones whose extractive properties have been studied are 7-methyl-2,4-octanedione ($pK_a = 10.0$) and 1,1,1-trifluoro-7-methyl-2,4-octanedione ($pK_a = 7.14$).[59] Both are liquids and the former has a b.p. of 88° at 4 mm Hg. The maximal E values for various metals, the reagents being the extracting solvents, are larger than with acetylacetone. Some $pH_{1/2}$ values (pH at which $E = 1$) for metals are:

	7-Methyl-2,4-octanedione	1,1,1-Trifluoro-7-methyl-2,4-octanedione
Th	1.4	<0
In	1.5	0.6
Sc	1.1	0.1
Zn	4.8	~3.1

1-Phenyl-3-methyl-4-benzoylpyrazolone-5.[60]   This reagent (PMBP)

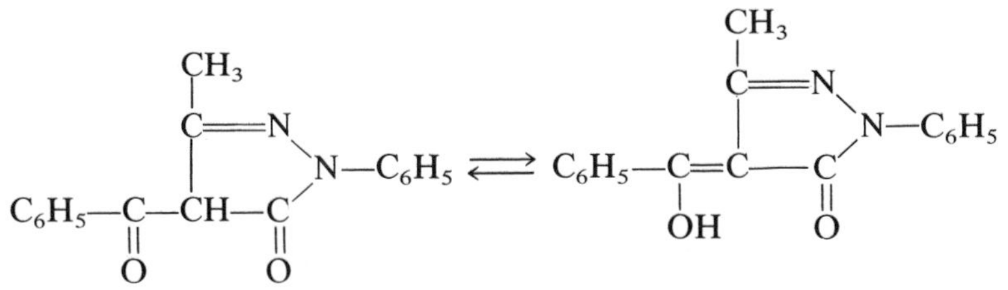

is a general extractant for metals. It appears to be a promising reagent (Table 9B-7). It has advantages over HTTA, being more easily prepared and purified (by recrystallization from alcohol) and being more stable. A considerable number of elements are extracted at comparatively low pH (Sc and PuIV in a more acid medium than with HTTA as reagent). The chelates formed may contain water of hydration. A suitable solvent is a 1:1 mixture of isoamyl alcohol and chloroform.

The enol form of PMBP (yellow-green) is obtained by recrystallization from nonpolar solvents, the keto form (colorless) from polar solvents. The ratio of the two forms depends on the particular solvent. The chelate is formed as a six-membered ring from the enol form.

The extractive separation of Zr, Hf, and Nb with this and similar reagents has been studied.[61] The extraction of Nb from HCl solution by a benzene solution of a number of 4-acylpyrazolones is attributed to formation of solvated ion-association complexes.[62]

Metal Extractions with 1-Phenyl-3-Methyl-4-Benzoylpyrazolone-5

Subject	Reference[a]
Soly. in H_2O and 7 organic solvents. $pK_{HA} = 4.04$ (20°).	N. T. Sizonenko and Yu. A. Zolotov, ZAK, **24,** 1305 (1969).
Extn. of Cu and Ni.	Zolotov and Sizonenko, ZAK, **25,** 54 (1970).
Extn. of Zn and Cu.	V. G. Lambrev and V. S. Vlasov, ZAK, **25,** 1638 (1970).
Extn. of Fe from HNO_3, H_2SO_4 and HCl solns.	M. K. Chmutova and N. E. Kochetkova, ZAK, **24,** 216 (1969).
Extn. of Fe from HCl solns.	M. K. Chmutova, N. E. Kochetkova, and Yu. A. Zolotov, ZAK, **24,** 711 (1969).
Extn. of Pa from HCl, HNO_3 and H_2SO_4 solns.	B. F. Myasoedov and N. P. Molochnikova, ZAK, **24,** 702 (1969).
Sepn. of Nd and Nb. Nd extd. with 0.05 M PMBP in $CHCl_3$ at pH 4–9, Nb then extd. from 8 M HCl.	N. T. Sizonenko and Yu. A. Zolotov, ZAK, **24,** 1341 (1969).
Extn. of Cm and Eu at pH 1–2 with PMBP in isoamyl, butyl and benzyl alcohols.	M. K. Chmutova and N. E. Kochetkova, ZAK, **24,** 1757 (1969).
Extn. of actinides and Eu with PMBP + TBP or TOPO.	M. K. Chmutova and N. E. Kochetkova, ZAK, **25,** 710 (1970).
Extn. of Th.	Z. K. Karalova and Z. I. Pyzhova, ZAK, **23,** 1564 (1968).
Extn.–sepn. of Pu.	M. K. Chmutova, P. N. Palei, and Yu. A. Zolotov, ZAK, **23,** 1476 (1968).
Extn. of rare earth elements into $CHCl_3$.	I. P. Efimov, L. G. Tomilova, L. S. Voronets, and V. M. Peshkova, ZAK, **28,** 267 (1963).
Extn. determination of Nb. PMBP + catechol or SCN^- extd. from 6 M HCl into benzene. $\epsilon_{410} \sim 10^4$. (Mixed ligand complexes.)	O. D. Savrova, I. M. Gibalo, S. S. Spiridonova, and F. I. Lobanov, ZAK, **28,** 817 (1973).
Separations of Am, Cm, and certain rare earths.	V. M. Vdovenko, M. P. Koval'skaya, and E. A. Smirnova, Radiokhim., **15,** 316 (1973).
Synergic extraction of trivalent transplutonium elements from HNO_3 solution into benzene in presence of TOPO.	M. K. Chmutova, G. A. Pribylova, and B. F. Myasoedov, ZAK, **28,** 2340 (1973).

[a] ZAK = Zh. Anal. Khim.

B. β-ISOPROPYLTROPOLONE

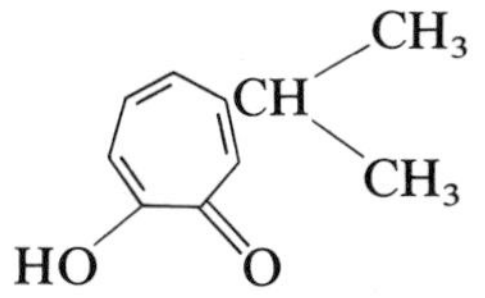

Slightly soluble in water, readily soluble in $CHCl_3$. $pK_a = 7.04$ ($\mu = 0.1$, 25°). $\log P_{CHCl_3/H_2O} = 3.37$ ($\mu = 0.1$, 25°). $CHCl_3$ solutions in dark bottles stable at room temperature for at least 1 week.

The parent compound tropolone is not very stable and is more difficult to prepare than this compound, which, moreover, has more favorable extraction properties.

β-Isopropyltropolone (HIPT) forms the following types of extractable chelates[63]:

ML_n with Ni, Cu, Zn, Sc, Fe(III), In, Pr, Th.
$ML_n \cdot HL$ with Ca, Sr, Ba, Cd, U(VI), Eu, Ho, Yb, Lu.
$NaML_3$ with Ni and Zn.
ML_2ClO_4 with Al and Ga (possibly).

H of the OH group is replaced by M/n to form a five-membered ring system in the chelate.

With a 0.1 M solution of HIPT in chloroform, the following metals are extracted to a large extent below a pH of 1:

Fe(III)[64]	Cu(II)
U(VI)	In
Th	Al and Ga (probably)
Sc	

Some of these metals may be separable in practice from those extractable in the pH range 2–6 [Co, Ni, Zn, Cd, Y, lanthanides, Am(III)] or at higher pH values (alkaline earths, Ag). See Figs. 9B-12 and 9B-13, from which separation possibilities may be deduced.

The solubilities of the chelates in chloroform and other common solvents are low (10^{-4}–10^{-3} M), but this is not important in trace analysis.

Extractions apparently are quite rapid, except of Fe(III), which requires 1 to 4 days to reach equilibrium in 0.5–1 M $HClO_4$. The formation of FeL_3 is the slow step.

β-Isopropyltropolone has advantages over HTTA as metal extractant. It does not decompose in alkaline solution, and it extracts metals as well or better than HTTA, that is, its K_{ext} values are as large or larger than those of HTTA for most metals. It is comparable to 8-hydroxyquinoline in the extraction of some metals from acidic solution. The HIPT complexes of Cu(II), Fe(III), and U(VI) absorb strongly at 400–450 nm, where the reagent itself does not absorb.

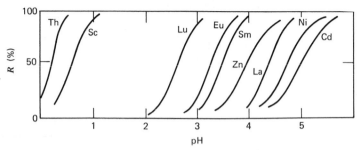

Fig. 9B-12. Extraction of some metals by 0.1 M HIPT in CHCl$_3$ as a function of pH. Dyrssen, *loc. cit.*

C. CUPFERRON

As a precipitant, nitrosophenylhydroxylamine, which is almost always used as its ammonium salt, cupferron,

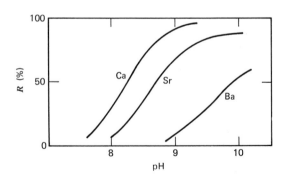

Mol. wt. 155.2. M. p. 163°–164°. Soluble in water (~6%). Can be recrystallized from ethanol.

is known for its ability to form very slightly soluble chelates in mineral acid solution (~1 M H$^+$) with Fe(III), Ti(IV), Zr(IV), V(V), U(IV), Sn(IV), and a few other metals. These metals can be quantitatively precipitated and separated from many other metals that are precipitated only in weakly acidic solution. In trace analysis, the separation of the metals named is usually carried out by extraction with chloroform or other organic solvent.[65] Most of the metal cupferrates are quite soluble in

Fig. 9B-13. Extraction of Ca, Sr, and Ba by 0.1 M HIPT in CHCl$_3$ as a function of pH. Dyrssen, *loc. cit.*

organic solvents so that sometimes a major constituent, such as Fe(III), can be removed by extraction, leaving various trace constituents in the aqueous phase. In general, extractabilities of metals parallel the insolubilities of their cupferrates, so that metals precipitated in mineral acid medium can be extracted therefrom.

Cupferrates usually have the composition MCp_n. Uranium(VI) forms $UO_2Cp_3^-$, which can give slightly soluble heavy metal salts in weakly acidic or neutral solutions. $UO_2Cp_3^-$ forms ion associates with large organic cations, for example, $(C_6H_5)As^+$, that are not chloroform extractable.

1. Properties of Nitrosophenylhydroxylamine (HCp)

It is a weak acid, having $K_a = 7 \times 10^{-5}$ at 25° ($\mu = 0.1$)[66] and 4×10^{-5} at 0°. The solubility of HCp in water is limited, being 0.02–0.025 M at 25°.[67] In organic solvents, it is very soluble, for example, ~5.5 M in chloroform at 25°.[67]

Acid aqueous solutions of cupferron, as well as its organic solvent solutions, are very unstable and appreciable decomposition into nitrosobenzene and other products occurs in a period of a few minutes.[68] Neutral solutions are also unstable, but not to the extent of acidic solutions. Even the solid is unstable, but it can be preserved by placing a lump of ammonium carbonate in the reagent bottle. Solutions to be extracted should be cool, certainly not above 25°.

The following partition constants have been reported for HCp:

	P
$CHCl_3/H_2O$[69]	151 ± 4 (room temp.)
$CHCl_3/H_2O$[70]	210 (room temp.)
$CHCl_3$ [1% ethanol/ H₂O (equal volume)][71]	150 (25°)
$CHCl_3/H_2O$[72]	145, 120–135 (20°)
CCl_4/H_2O[73]	2300 (15°)
$MIBK/H_2O$[71]	85 (25°)

2. Extraction of Metal Cupferrates

In general, those metals forming cupferrates coordinatively saturated by the reagent tend to be more soluble in inert solvents than in oxonium solvents and are extracted better by the former. Thus gallium is extracted better by chloroform and benzene than by isoamyl alcohol from mineral acid solution. Chloroform is the preferred solvent for most extractions.

Cupferrates only slightly soluble in chloroform include those of Ag, Cd, Mo, Nb, and Ta. Better extraction of W, Mo, Nb, and Ta (the latter two elements in tartrate medium) can be obtained with isoamyl alcohol than with chloroform. A number of cupferrates, for example of La, U(VI), Y, and Mn, are quite soluble in water, so that their P values (with chloroform) will be rather low. As a consequence, their extractability will be poor even when [Cp$^-$] is high and a favorable extracting solvent is used. Thus LaCp$_3$ has a solubility of $\sim 4 \times 10^{-4}$ M in water, and its P with MIBK as solvent is 4, so that the maximum extraction of lanthanum is 80% with equal volumes of phases (Dyrssen).

A survey of cupferron extractability of metals, mostly with chloroform as solvent, is contained in Table 9B-8; elements easily extracted quantitatively are marked with an asterisk. Of course, the concentration of cupferron is important, so that by increasing this the extractability of the less easily extractable elements can be increased, and possibly made essentially complete, especially if successive extractions are made. Thus copper can be extracted quantitatively from 1 N mineral acid if need be, but 0.1 N acid medium allows easier extraction. Some of the metals not extracted from 0.1 M H$^+$ solution can be extracted in neutral or alkaline solutions,[74] but the chief interest in the use of cupferron extraction lies in the separation of metals listed in the 1 N and 0.1 N acid columns.

The theoretical treatment of cupferron extractions is complicated by the formation of intermediate (positively charged) metal–cupferrate species[75] that play a significant role in the extraction of some metals, especially at low acidities. Practical consideration of extraction separations under such conditions is simplified by plotting the extraction coefficient E or the percentage extraction R of a metal for a fixed ratio of the two phases against the aqueous cupferrate ion concentration on a log scale.[76] The latter concentration is, of course, a function of the hydrogen-ion concentration and total cupferron concentration, so that E or R can also be plotted against pH for a constant quantity of cupferron. For equal volumes of aqueous and MIBK phases, Th is extracted to the extent of 50% at p[Cp$^-$] ~ 7.6, U(VI) at p[Cp$^-$] $= 6.1$ and La at p[Cp$^-$] $= 3.75$.[76]

Because of the instability of cupferron and cupferrate solutions, the high ionic strength in many extractions and the formation of intermediate complexes, accurate values of K_{ext} for cupferrates would be difficult to determine and would not be of great practical value. Effective extraction constants are of some interest:

$$K'_{ext} = \frac{[\mathrm{MCp}_n]_{org}[\mathrm{H}^+]^n}{\sum [\mathrm{M}][\mathrm{HCp}]^n_{org}}$$

General Extractability of Certain Elements as Cupferrates by Chloroform as a Function of Acidity

(~0.05 M cupferron in aqueous phase before extraction; starred elements can be extracted quantitatively.)

Extracted from 1 N Mineral Acid	Extracted from 0.1 N Mineral Acid	Not Appreciably Extracted from 0.1 N Mineral Acid[b]	
Cu(II)	Bi*(H_2SO_4, HNO_3)	Ag	Pb
Fe(III)*	Cu*(II)	Al	Pt(IV)?
Mo(VI)*[a]	Ga*	As(III,V)	R.E.
Nb, Ta	Hg(II)	Be	Sb(V)
Pd*	In	Ca, Sr, Ba	Se(IV)
Sb(III)*(H_2SO_4, HNO_3)	Sc	Cd	Te(IV)
Sn(IV)*,[c] Sn(II)*	Th*(ethyl acetate)	Ce(III)	Tl(I)
Tc		Co	U(VI)
Ti*		Cr(III)	Y
U(IV)*		Ge	Zn
V(V)*[d]		Mg	
W		Mn	
Zr*, Hf*		Ni	

[a] A trace remains in the aqueous phase; Mo(VI) possibly partly reduced. More than 98% of Mo (~0.1 μg) can be extracted from 50 ml of 2 M HCl + 1 ml 6% cupferron with 10 ml $CHCl_3$ if phases are separated within 5 min (W. B. Healy and W. J. McCabe, *Anal. Chem.*, **35**, 2117).

[b] Under conditions not precisely defined, a 1:1 mixture of benzene and amyl alcohol extracts 50% of the metal at the following approximate pH values: Cu(II), 0.1; Pb, 1.9; Zn, Co, Ni, 2.5–2.8; Mn(II), 3.9; Mg, 5 (J. S. Fritz, M. J. Richard, and A. S. Bystroff, *Anal. Chem.*, **29**, 578 (1957)). T. Kiba and M. Kanetani, *Bull. Chem. Soc. Japan*, **31**, 1013 (1958), when extracting an aqueous solution twice with about one-fourth its volume of $CHCl_3$ (5% HCp) found 50% extraction of various metals at the following approximate pH values: Cu, 0; Pb, 1.7; Zn, 2.9; Co, Ni, 3.0–3.1; Mn, 4.5; Mg, 5.2.

[c] According to S. J. Lyle and A. D. Shendrikar, *Anal. Chim. Acta*, **36**, 286 (1966), Sn(IV) in 1 M HCl is only 50% extracted by an equal volume of 5% cupferron–$CHCl_3$. Above the minima at HCl = 4.5 and 7 M, respectively, increasing extraction of Sn(IV) and Sn(II) is shown, attributed to $CHCl_3$ alone. The decomposition products of cupferron are reported to inhibit the extraction of Sb(III and V) at high HCl concentrations by $CHCl_3$ alone.

[d] V(V) cupferrate is unstable in $CHCl_3$. The original red color changes to yellow, and the absorption curve of this solution is much like that of V(IV) cupferrate. The reduction of V(V) is attributed by P. Crowther and D. M. Kemp, *Anal. Chim. Acta*, **29**, 97 (1963), to $CHCl_3$ rather than to cupferron. The reduction is largely complete after ~1 hr. V(IV) cupferrate is not entirely stable in $CHCl_3$. The $CHCl_3$ used contained alcohol.

K'_{ext} depends on the ionic strength and, more importantly, on the concentration of complexing anions such as Cl^- and SO_4^{2-}, as well as on $[HCp]_{org}$ if charged cupferrate complexes are formed; at sufficiently high acidities, hydrolysis of metal cations can be neglected. Rough calculations of separability can be made with the aid of effective extraction constants, but caution is needed (see the subsequent discussion of the extraction of aluminum cupferrate). Effective extraction constants are available for a score of metals (Table 9B-9).

At pH 5.5–5.7 in the presence of EDTA and citrate, the elements precipitated (not necessarily quantitatively) and extracted by MIBK include Be, Al, Ti, V, Zr, Hf, Nb, Ta, Sn(II,IV), U(IV,VI), Ce(III,IV), and the rare earth elements.[77] Titanium cupferrate can be extracted by chloroform from slightly alkaline solutions containing tartrate (not citrate)

TABLE 9B-9
Effective Extraction Equilibrium Constants ($CHCl_3/H_2O$) for Metal Cupferrates at Room Temperature[a]

(Initial aqueous concentration of cupferron 0.005 M for Bi, Ga, Mo, Sc, Th, Tl, Y and 0.05 M for Be, Co, Cu, Hg, La, Pb; perchlorate solutions; μ usually 0.1 M.)

Metal	$\log K'_{ext}$	Metal	$\log K'_{ext}$
Al[b]		Hf	>8
Be	−1.5	Hg(II)	0.9
Bi	5.1	In	2.4
Ce(IV)	4.6	La	−6.2
(butyl acetate)		Pb	−1.5
Co	−3.6	Pu(IV)	7
Cu(II)	2.7	Sb(III)	~7
Fe(III), 1 M $HClO_4$	9.8	Sc	3.3
1 M HCl	8.8	Sn(II)	~6
2 M HCl	8.3	Th	4.4; 6 (MIBK)
3 M HCl	7.9	Tl(III)	~3
4 M HCl	7.5	U(IV) (ethyl ether)	~8
Ga	4.9[c]	Y	−4.7

[a] Values for Fe from reference a, Table 9B-1; for most of other metals from J. Stary and J. Smizanska, *Anal. Chim. Acta*, **29**, 545 (1963); Ce(IV) from Z. Hagiwara, *J. Chem. Soc. Japan Ind. Chem. Sec.*, **57**, 266 (1954) (0.1 M H_2SO_4); Hf from Dyrssen and Dahlberg, *op. cit.*; Th in MIBK from same authors.

[b] See p. 947.

[c] A larger value is indicated for H_2SO_4 solution, from which at 5 N concentration (from Russian data), ~90% Ga can be extracted with $CHCl_3$, more with C_6H_6, much less with $(C_2H_5)_2O$ and hardly any with isoamyl alcohol.

and EDTA and thus separated from W, Mo (partially), V, Cu, Ni, Co, Cd, Bi, and Mn (sulfite added).[78] Other complexing agents than EDTA find use in cupferron separations.

Iron(III) and copper(II) cupferrates have been used as extraction reagents, as for zirconium and gallium (exchange extraction).[79]

Recovery of extracted metal usually requires evaporation of the organic solvent solution and oxidative destruction of excess cupferron and cupferrate in the residue, as with $HNO_3 + H_2O_2$.

Chloroform–Cupferron Extraction of Aluminum. A preliminary investigation of the chloroform extraction of aluminum cupferrate has shown that K'_{ext} depends on the pH and the time of standing before extraction.[67] K'_{ext} was found to decrease with increasing pH and cupferrate ion concentration when the extraction was begun soon after the addition of cupferron to the aqueous aluminum solution. For example, when $3.9 \times 10^{-4}\,M$ Al solution was treated with about a 20-fold excess of cupferron ($2.36 \times 10^{-2}\,M$) and the extraction was begun after 30 sec with an equal volume of chloroform, the apparent (effective) value of K (all Al in the aqueous phase counted as Al^{3+}) was 2.3 at pH 1.23 and 9×10^{-3} at pH 2.40. The first figure corresponds to 13% extraction of Al, the second to 62%. The longer the time elapsing between the addition of cupferron and the extraction, the greater was the amount of Al extracted under otherwise the same conditions, and the larger the apparent value of K.[80] Thus at pH 1.65 with $[HCp]_{CHCl_3} \sim 1 \times 10^{-2}$, the percentage of Al extracted increases from 70% after 5 min standing to 96% after 1 hr (apparent $K_{ext} = 26$ and 296). These values of K'_{ext} may be compared with the value 3×10^{-4} reported by Stary to hold when $10^{-4}\,M$ Al perchlorate solution (pH ~ 2.5) is made $0.05\,M$ in cupferron and shaken with an equal volume of chloroform for 5 min. Evidently, apparent K_{ext} for aluminum is sensitive to small changes in conditions, and only rough calculations can be based upon it.

In spite of the variability of K'_{ext}, its magnitude is such that the feasibility of the separation of iron from aluminum by cupferron extraction is indicated. The aqueous solution should be $1\,M$ or stronger in hydrochloric acid for this separation, and the extraction should be made immediately or very soon after cupferron has been added. When a solution of $2.5 \times 10^{-3}\,M$ Al in $1\,M$ hydrochloric acid is shaken with an equal volume of $3 \times 10^{-2}\,M$ cupferron–chloroform solution, $\sim 0.1\%$ of the aluminum is extracted. Herzog found that aluminum cupferrate in chloroform was not completely transformed to Al^{3+} and transferred to the aqueous phase by shaking with $1\,M$ hydrochloric acid. Whether this behavior is due to slowness in attainment of equilibrium or whether some other effect is involved is not known (presumably the former).

Neocupferron. This is the ammonium salt of nitrosonaphthylhydroxyl-amine. It is a slightly stronger acid than its phenyl analog ($K_a = 8 \times 10^{-5}$) and forms more insoluble chelates, but has no features of special interest. At room temperature, $P_{CHCl_3/H_2O} = 1.31 \times 10^3$.

N-Nitroso-N-cyclohexylhydroxylamine. This compound differs from nitrosophenylhydroxylamine only in having a saturated ring in place of the benzene ring. The sodium salt ("hexahydrocupferron") is more stable in solution (a 10^{-4} M solution in 6 M HCl is ~50% decomposed after 4 days) than cupferron, but its metal reactions are very similar.[81] $pK_a = 5.58$.

The pH ranges for maximal extraction of metal hexahydrocupferrates into CHCl$_3$, and some K_{ext} values have been reported.[82] K_{ext} is smaller than for cupferrates, thus log $K_{ext} = -0.3$ for Cu(II) and -2.84 for Pb. By way of illustration, pH ranges for quantitative extraction are: Cu(II), 1.7–11; Pb, 3.7–11; Bi, 1.2–9.3; Sb(III), 0–11; Ti, 0–11; Fe(III), 0–10.4; V(V), 0–3.85. These values pertain to 0.05 M solutions of hexahydro-cupferron in CHCl$_3$.

D. N-BENZOYL-N-PHENYLHYDROXYLAMINE

This reagent (HBPHA),[83] also known as N-phenylbenzohydroxamic acid,

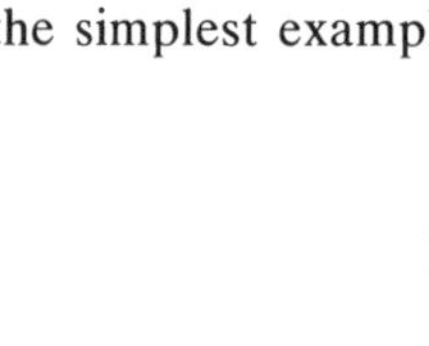

Mol. wt. 213.1. M. p. 120°–122°. Purified by re-crystallization from hot water.

is one of the simplest examples of a hydroxamic acid

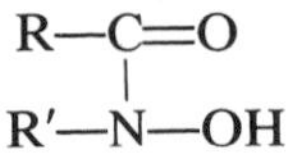

$$\begin{matrix} R{-}C{=}O \\ | \\ R'{-}N{-}OH \end{matrix}$$

Its value in the separation of traces of metals lies in its ability to form very slightly soluble chelates that can be extracted into nonoxygen solvents, especially chloroform, from mineral acid solutions. HBPHA is an analog of cupferron, the nitroso group of the latter being replaced by benzoyl. Its reactions with metals are similar, but it is superior to cupferron in a number of respects, particularly in its much greater stability in mineral acid solution. It is reported to resist oxidation by nitric acid up to about 4 M.

Some constants of analytical interest (from various sources) for HBPHA follow:

Solubility, in

H_2O (25°, pH 2.3, $\mu = 0.1$)	1.95×10^{-3} M
$CHCl_3$ (25°)	0.74 M
CCl_4	0.026 M
C_6H_6	0.15 M
pK_a (25°, $\mu = 0.1$, stoichiometric)	8.15
P_{CHCl_3/H_2O} (25°, $\mu = 0.1$)	214
$P_{CHCl_3/1.0\ M\ HClO_4}$	137 ± 5
$P_{CHCl_3/6\ M\ HClO_4}$	24 ± 1
$P_{C_6H_6/1.0\ M\ HClO_4}$ (25°)	
$\quad [\Sigma\ HBPHA]_{C_6H_6} = 1 \times 10^{-3}$	23
$\quad [\Sigma\ HBPHA]_{C_6H_6} = 6 \times 10^{-2}$	37

Extractions from Weakly Acidic or Dilute ($< \sim 1\ M\ H^+$) *Acid Solutions.* Leaving to a later section the extractability of metals from strongly acid solutions, we consider here extraction from solutions of relatively low acidity. As might be expected, the extraction of metal–HBPHA chelates into chloroform shows considerable similarities to the extraction of cupferrates. Metals of the IVA, VA, and VIA groups, as well as some of the IVB and VB groups, are extractable from dilute mineral acid medium. Thus a 0.5% solution of HBPHA in chloroform will extract 90% or more of the following metals from an equal volume of 0.1–1 M HCl solution: Zr, Hf, V(V), Nb, Cr(VI), Mo(VI), W(VI),[84] Sn(IV),[85] Sb(III), Bi, Ce(IV), and Fe(III). Ta and Ti(IV) are also extracted to a large extent from dilute mineral acid solutions. Most of the other elements are not extracted to a large or even significant extent from such solutions. Molybdenum(VI) is extracted very little above pH 10. Information gleaned from the literature indicates that extensive extraction of other metals occurs at roughly the following acidities[86]:

pH 3–4		pH 5	pH > 6	
Al	In	Be	Cd	Nd
Cu(II)	Pd(II)	Fe(II)	Ce(III)	Pb
Ga (also pH 1)	Th	Ni	Co	Pr
	U(VI)	Sc	Gd	Y
			La	Zn
			Mn	

If hydrolysis of metal cations is allowed for, it seems likely that the general extraction expression can be applied to predict extractability.

Dyrssen[87] reported the following values of $\log K_{ext}$ (CHCl$_3$): La, -14.4; U(VI), -3.14; Th, -0.68. These values indicate that with a $0.1\,M$ solution of the reagent in chloroform, the extraction of Th should begin at pH ~1 and be completed at pH ~2.5, whereas the corresponding pH values for La should be ~5 and ~7. These two elements should, therefore, be easily separable by extraction with HBPHA.

Extraction constants for a considerable number of other metals have been determined by Riedel (Table 9B-10). Note that La and Co(II) form adducts with the reagent. It is of interest that the partition constants of the molecular chelates (CHCl$_3$/H$_2$O) are not very large (Table 9B-10), that is, some of the chelates are appreciably soluble in water, or more accurately in water saturated with CHCl$_3$.

Extractions from basic solutions can be useful at times if most metals can be masked by suitable complexing agents. This is so for aluminum, which can be extracted from ammonium carbonate solution as its

TABLE 9B-10
K_{ext} and P of HBPHA Chelates (CHCl$_3$–H$_2$O)[a]
($\mu = 0.1, 20 \pm 2°$)

Chelate	$\log K_{ext}$	$\log P$
CdL$_2$	-12.1	1.73
CeL$_3$	-13.2	
CoL$_2$·HL	1.0	
CuL$_2$	-0.7	2.35
FeL$_3$	5.3	
GdL$_3$	-12.6	2.75
InL$_3$	-1.8	3.03
LaL$_3$·HL	-13.6	
ThL$_4$	-0.7	3.45
TlL	-7.3	0.30
YL$_3$	-12.4	
ZnL$_2$	-10.0	1.93

[a] A. Riedel, *J. Radioanal. Chem.*, **13**, 125 (1973). Values of $\log K_{ext}$ have been rounded to nearest 0.1 above. Values of K_{ext}^* and stability constants of some chelates in aqueous solution are also given in the original. F. G. Zharovskii and M. S. Ostrovskaya, *Zh. Anal. Khim.*, **29**, 1646 (1974), determined the solubility of chelates of Cu, Al, Ga, In, and Fe(III) in various solvents. Solubility in nonoxygenated solvents follows the order chloroform $>$ 1,2-dichloroethane $>$ benzene $>$ toluene $>$ carbon tetrachloride. Solubility in oxygen-type solvents generally decreases from ketones to alcohols, esters, and ethers.

HBPHA chelate into benzene, most other metals being masked with hexametaphosphate, mercaptoacetate, cyanide, and hydrogen peroxide.[88] Aluminum is readily back-extracted from the benzene phase into 0.2 M HCl.

An exact treatment of the extraction of HBPHA chelates requires a knowledge of the values of the formation constants of the charged complexes as well as of the uncharged complex formed by the reagent with a metal ion. These and other constants have been determined for Fe(III).[89] Denoting benzoylphenylhydroxylamine anion, BPHA[-], by L[-], we have

β_1 (formation constant of FeL^{2+}) $\sim 2 \times 10^{11}$

β_2 $\qquad\qquad\qquad\qquad\qquad \sim 4 \times 10^{20}$ } Variable μ, 0.01–0.5 M

β_3 $\qquad\qquad\qquad\qquad\qquad \sim 1 \times 10^{29}$

K_{sp} of $FeL_3 = 3 \times 10^{-36}$ (1 min shaking)

$\qquad\qquad\quad 7.5 \times 10^{-38}$ (6 hr shaking)

P_{CHCl_3/H_2O} of $FeL_3 = 3.5 \times 10^8$

(Note the very large value of P, one of the highest recorded for a metal chelate.) From the extraction coefficient E of 1.5 given for Fe(III) in a system in which $[L^-]$ is 5.5×10^{-13} ($\sim$0.001 M HL in $CHCl_3$) and pH = 1.3, we calculate a rough value of 2×10^5 for (effective) K_{ext}. At a pH of $\sim$1, in such an extraction, the concentration of Fe^{3+} in the aqueous phase greatly exceeds the concentration of Fe–L species at this HL concentration. At lower acidities, hydrolysis of Fe^{3+} would need to be considered.

Although extraction of a number of metals by HBPHA–$CHCl_3$ from (say) 0.1–1 M H[+] solutions is the basis of useful separations, extraction at still higher acidities is of special interest because of increased selectivity. For example, V(V) can be determined spectrophotometrically by extraction of its violet chelate with $CHCl_3$–HBPHA from $\sim$4 M HCl, an acidity at which little Fe(III) is extracted.

The rate of extraction of metals by HBPHA from weakly acidic solutions apparently has not been much studied, but some observations have been recorded. Extraction of 10^{-3} M Fe(III) with a 0.01 M solution of HBPHA in chloroform from a solution of pH 1.7 requires 10–15 min for attainment of equilibrium; with a CCl_4 solution of HBPHA, equilibrium is reached in 1–2 min.[90] The difference in rates agrees with the difference in the reagent P values. Extraction rate increases with pH and reagent concentration.

Extractions from Strongly Acid Solutions. As long as the extracted metal complex is ML_n and the predominant species in the aqueous phase are HL, L^-, M^{n+}, ML^{n-1}, and the like, it would be expected in general that E of the metal in solutions having $[H^+] > 1$ would continue to decrease with rising acidity, although increasing ionic strength leading to greater salting out of reagent and metal complex, not necessarily to the same extent, could cause some irregularities. Some metals, for example, Fe(III), Bi(III), Sb(III),[91] and Sn(IV), do show decreasing extraction of their HBPHA complexes into chloroform as the HCl concentration is increased to 6 M and higher (complexing of the metal by Cl^- can account for some decrease). However, at least eight metals show continued high, and even higher, extractability as the acidity is increased to 10 M or more in HCl (Figs. 9B-14 through 9B-16). The extraction of Ti increases smoothly to high HCl concentrations, that of Zr and Hf shows a shallow minimum at ~5 M HCl, and that of others (Nb, Ta, V, Mo, W), while showing a more or less complex relationship to acid concentration, remains high at 8–11 M HCl. Extensive extraction of some metals also takes place from 12–14 M H_2SO_4, namely of Nb, Ta, V, Zr, Sn(IV), Sb(III), and Ga; as much as 75% of Ga can be extracted by chloroform from 11 M H_2SO_4.[92] With the exception of Ga, Sb(III), and Sn(IV), these

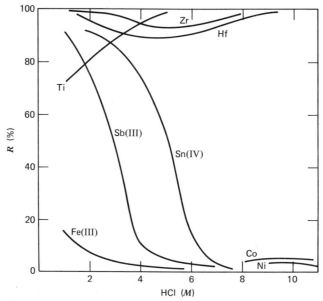

Fig. 9B-14. Percentage extraction (R) of Ti, Zr, Hf, Sn(IV), Sb(III), Fe(III), Ni, and Co by HBPHA in $CHCl_3$. From 0.5 to 1 mg of metal sulfate dissolved in 25 ml of HCl solution extracted with 25 ml of 0.5% (~0.025 M) HBPHA for 15 min. O. A. Vita, W. A. Levier, and E. Litteral, *Anal. Chim. Acta*, **42**, 87 (1968).

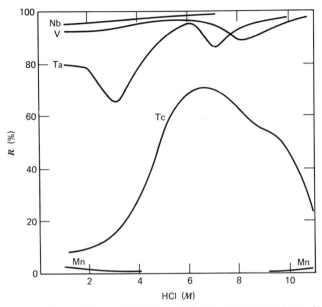

Fig. 9B-15. Extraction of Nb, Ta, V(V), Tc, and Mn(II) with HBPHA in CHCl$_3$. Conditions as in Fig. 9B-14. Vita, Levier, and Litteral, *loc. cit.*

are also the elements that are extracted to a large extent from 6–10 M HCl.

Almost all elements extractable from HCl solution by chloroform reach equilibrium in 5 min (Vita et al.); Nb and Ta were found to require 10 min of mixing.[93] Solutions of the two latter elements allowed to stand for 2 days before extraction give poorer recoveries than freshly prepared solutions. Both V(V) and Cr(VI) are reduced on standing in strong HCl and extractability decreases, because V(IV) and Cr(III) are not extractable.

The reasons for continued, or even higher, extractability of some metals with HBPHA at high acidities are uncertain.[94] The mechanism may not be the same for all the elements showing this behavior. If ML$_n$ is postulated to be the only extracted species, continued extraction at high acidities might be imagined to result if the metal cation is extensively hydrolyzed at, say, 1 M H$^+$ and increasing acidity then increases the metal cation concentration at a rate keeping pace with the decrease in [L$^-$] brought about by decreasing ionization of HL. The good extraction at high acidities of metals forming oxo-cations may be explainable in this way. HBPHA is known to react with V(V) in HCl solutions to form a

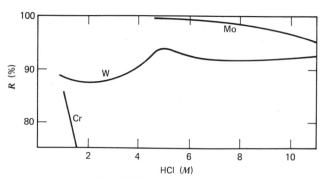

Fig. 9B-16. Extraction of Mo(VI), W(VI), and Cr(VI) with HBPHA in CHCl₃. Conditions as in Fig. 9B-14. Cr(VI) is reduced to Cr(III), not extractable, in strong HCl. Vita, Levier, and Litteral, *loc. cit.*

purple chloroform-extractable complex having the ratio HBPHA/V = 2. The absorbance of the extract has been reported to remain constant from 5 to 9 M HCl (by others constant to 4.3 M HCl, with a slow decrease above this acidity). The absorbance peak remains at the same wavelength. Although the structure of the extracted complex is not known, a reaction of the type

$$V_2O_5 + 4\,HL \rightleftharpoons 2\,VO(OH)L_2 + H_2O$$

in which H^+ is not released, would be in accord with the reaction of V(V) with the reagent at high acidities. Possibly the extraction of Mo and W from strongly acid solutions can be explained similarly.

An important factor in increasing extraction at high acidities might well be the extraction of species other than, or in addition to, the simple chelate ML_n. Such species could include mixed chelates, addition complexes of the type $MX_n \cdot mHL$ (X = inorganic anion), $ML_n \cdot mHL$, or neutral ion-pair or association compounds of protonic HBPHA and metal anion.[95] It may be noted that the solubility of HBPHA increases in mineral acid solutions, so that protonated HBPHA seems to be formed to some extent. The extraction coefficient of HBPHA for $CHCl_3$–H_2O decreases with a rise in $HClO_4$ or HCl concentration, but the decrease is more marked with $HClO_4$ (Fig. 9B-17). At constant $HClO_4$ concentration (2 M), an increase in $NaClO_4$ concentration increases E of HBPHA (salting-out effect).

The extraction of Zr and Hf from strongly acid solutions of HCl and $HClO_4$ has been attributed to the formation of the neutral complexes $MX_4 \cdot 2\,HBPHA$, in which $X = ClO_4$ or Cl.[96] Contrarily, the mechanism of extraction of Hf has been held to be the same in strongly acid as in weakly acid solution, the rise in extraction past the minimum being attributed to an increase in $[BPHA^-]$ in the aqueous phase resulting from the decrease in E of the reagent, which more than compensates for the repression of the ionization of HBPHA at higher acidities.[97] However, this explanation can only be valid if the decrease in E of the reagent is not due to formation of new reagent species such as $HBPHA \cdot H^+$ at high acidities. Moreover, metals other than Hf and Zr do not show a minimum at 4-5 M H^+, although other effects could conceivably supervene.

Whatever the mechanisms turn out to be for the extraction of certain metals as HBPHA[98] complexes from strongly acid solutions, this extraction is of practical value for separations of a number of elements. For example, traces of Hf, Zr, Nb, Ta, Ti, and V can be separated from U, Th, Np, and Pu (see Figs. 9B-14 through 9B-16). In the HCl range 1-11 M, E for U is $\sim 10^{-6}$ and 10^{-4} for Np and Pu.[99] Other elements having very low E values in strongly acid medium include Be, Mg, Ca, Sr, Ba, Sc, Y, La and other R. E., Ru, Zn, Cd, Hg, B, Al, Ga, In, Tl, Pb, Bi. Fe(III) is slightly extracted at high acidities (HCl) as are Co and Ni, and seemingly Mn(II).

Approximate separation factors for systems of analytical interest have been recorded (CHCl₃ solutions of HBPHA).[100] For example:

	S
Sn(IV) from Bi, 4 M HClO₄, 1% HBPHA	10^{-4}
Sn(IV) from Pb, 0.8 M HCl, 1% HBPHA	10^{-3}
Nb from Ta, 1 M HCl, 0.05 M HF, 0.2% HBPHA	10^{-2}
Nb from Zr, 1 M HCl, 0.05 M HF, 0.2% HBPHA	10^{-4}

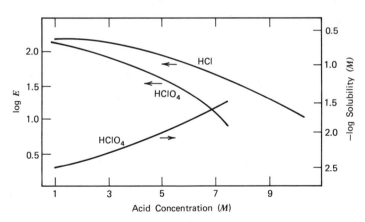

Fig. 9B-17. Solubility and E_{CHCl_3/H_2O} of HBPHA as a function of HCl and HClO₄ concentrations. K. F. Fouché, *J. Inorg. Nucl. Chem.*, **30**, 3057 (1968).

S is defined as on p. 700, being the recovery factor for the undesired element (the second element in each of the above pairs).

HBPHA has also been used as an absorptiometric reagent, as for V and Nb.[101] The latter element forms a mixed complex with HBPHA and thiocyanate ($\mathscr{S}_{365} = 0.0029$ μg Nb/cm^2) in toluene.[93] Ta interferes.

Thiobenzoylphenylhydroxylamine. The thio analog of HBPHA has been prepared and tested as an analytical reagent in a preliminary way.[102]

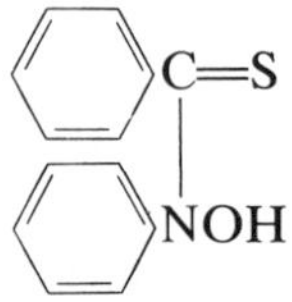

M. p. 102°–103°. Recrystallizable from alcohol–water.

Its reactions are similar to those of HBPHA and most of its chelates are extractable into chloroform. It is a stronger acid than HBPHA ($pK_a = 8.0$ in 1:1 dioxane–water compared to 10.45 for HBPHA in the same solvent), and its chelates are less dissociated.

The solid reagent is indefinitely stable and a 1% alcoholic solution shows no decomposition after several weeks.

E. OXIMES

1. α-Dioximes

Dimethylglyoxime and other α-dioximes are excellent extraction reagents (chloroform and other solvents) for Pd(II) and Pt(II) from dilute mineral acid solution and for Ni from weakly acidic or weakly basic solutions. The separability of minute amounts of Ni from much Fe(III) and Co in this way is especially important. Copper(II) is partially extracted. See Chapter 6C.

2. α-Benzoinoxime

C$_6$H$_5$—CH—C—C$_6$H$_5$
          | ‖
         OH   NOH

Mol. wt. 227.3. M. p. 149°–151° (152°–154°). Recrystallizable from ethanol. Alcoholic solutions stable for a week or more (protect from light). Solubility in water 0.30 g/liter (26°), in CHCl$_3$ 6.0 g/liter (25°). E_{CHCl_3/H_2O} varies from 14.1 at $[H_2Bx]_{CHCl_3} = 0.0012$ to 19.9 at $[H_2Bx]_{CHCl_3} = 0.026$ ($\mu = 0.3\ M$), indicating slight association in chloroform. Unstable in mineral acid solution. pK_a ~6.

This compound was proposed as a specific reagent for Cu(II) by F. Feigl in 1923. Actually, it is of no great interest as a copper reagent and more likely than not it would have fallen into a state of innocuous desuetude had not Knowles (1932) found that it gives selective precipitation of Mo from mineral acid solution. α-Benzoinoxime forms slightly soluble and extractable chelates with a block of elements in the periodic table:

23	24
V	Cr
41	42
Nb	Mo
73	74
Ta	W

and also with Pd(II). Precipitation and extraction can be carried out in dilute mineral acid medium. With the exception of Pd, the reacting metals form oxocations in their highest oxidation state. Chromium and vanadium do not react if reduced to Cr(III) and V(IV); in their highest oxidation state, they oxidize the reagent (though the reagent has been used for the extraction of V from faintly acidic solution).

Extraction is superior to precipitation for the separation of traces of Mo, W, and Nb(Ta). Good separations from iron and many other elements can be obtained. Chloroform is a good extracting solvent.

The extraction of W has received special attention, partly because of the lack of good extraction reagents for this element. The α-benzoinoxime extraction of W has some features of general interest that seem to merit discussion here.

The composition of the tungsten chelate is $WO_2(HBx)_2$ ($H_2Bx = \alpha$-benzoinoxime),[103] and the structure probably is

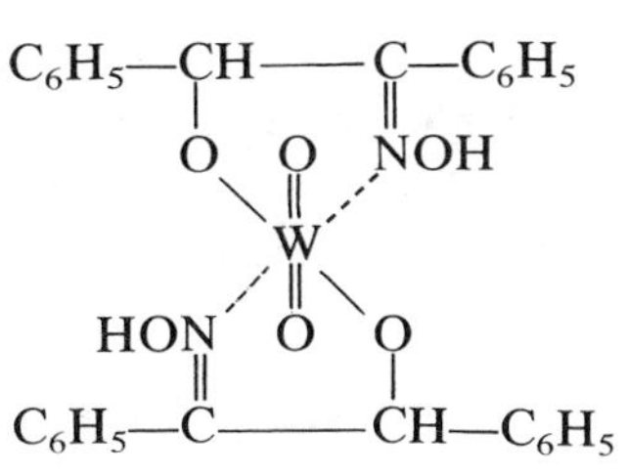

The molybdenum complex has an analogous composition.

The best extracting solvents (highest P) for the tungsten chelate are benzene, chlorobenzene, cyclohexane, and toluene.[104] Chloroform and carbon tetrachloride give slightly lower partition constants but are preferable because of their greater density (heavier than water). The following

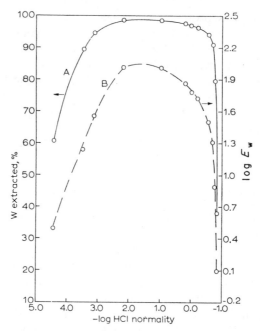

Fig. 9B-18. Percentage extraction of (*A*) W and (*B*) log E_W as functions of [HCl]. Volume of aqueous phase 20 ml, of CHCl$_3$ phase 10 ml (containing 15 mg α-benzoinoxime); 20 μg W. Peng and Sandell, *loc. cit.*

discussion refers to chloroform as extractant. The effect of hydrogen-ion concentration on the extraction is of chief interest.[105] The extractability increases with acidity, reaching a maximum at 0.01–0.1 *M* HCl, is still favorable in 1 *M* HCl, but decreases markedly at higher acidities (Fig. 9B-18). The rising portion of the log *E*–pH curve reflects the presence of anionic tungsten species. If at a particular pH only one tungsten species were present (an obvious simplification, even if polymeric species are neglected), the slope of the curve at that point would have an integral value. Thus

$$WO_4^{2-} + 2\,H^+ + 2H_2Bx \rightleftharpoons WO_2(HBx)_2 + 2\,H_2O$$
$$\quad\quad\quad\quad\quad\quad org \quad\quad\quad\quad org$$

$$HWO_4^- + H^+ + 2\,H_2Bx \rightleftharpoons WO_2(HBx)_2 + 2\,H_2O$$
$$\quad\quad\quad\quad\quad org \quad\quad\quad\quad org$$

If the species HWO$_4^-$ predominates over other W species in a certain pH range, the extraction coefficient is given by

$$E \sim \frac{[\text{WO}_2(\text{HBx})_2]_{org}}{[\text{HWO}_4^-]} \sim const. \times [\text{H}^+][\text{H}_2\text{Bx}]_{org}^2$$

When $[\text{H}_2\text{Bx}]_{org}$ is constant,

$$E \sim const'. \times [\text{H}^+]$$

$$\log E \sim \log const'. + \log [\text{H}^+] \sim const''. - pH$$

and the slope of the $\log E - pH$ curve in the figure will be ~ 1. If the predominant species were WO_4^{2-}, the slope would be ~ 2, and so on. In Fig. 9B-18, the curve between pH 4.5 and 3 is approximately linear, with a slope of ~ 1.2. In this pH region, anionic W species are present, together with some WO_3 or similar uncharged species, the average charge being a little greater than 1.

Between pH 1 and 2, E is almost constant, and the major species is WO_3 (or other uncharged species such as $\text{WO}_2(\text{OH})_2$). This conclusion receives support from the distribution of W species as a function of pH reported by Nazarenko and Poluektova.[106] See Fig. 9B-19. Below pH 1, WO_2OH^+ and WO_2^{2+} become the dominant species, and E decreases with increasing acidity in accordance with the extraction reaction

$$\text{WO}_2^{2+} + 2 \text{H}_2\text{Bx} \rightleftharpoons \text{WO}_2(\text{HBx})_2 + 2 \text{H}^+$$
$$\quad\quad\quad\quad org \quad\quad\quad\quad\quad\quad org$$

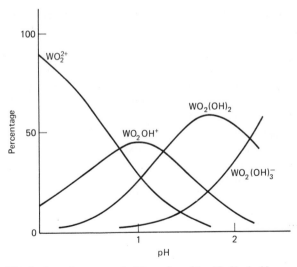

Fig. 9B-19. Distribution of monomeric W species with pH. V. A. Nazarenko and E. N. Poluektova, *Zh. Neorg. Khim.*, **14,** 209 (1969).

The sharp drop in strong HCl solutions may be aided by the formation of chloro complexes in the aqueous phase and decomposition of α-benzoinoxime (including, or limited to, the formation of the β-oxime, reported to occur at high acidities).

The chloroform-α-benzoinoxime extraction is suitable only for relatively small amounts of W because of the limited solubility of the chelate in chloroform and of WO_3 in water. As much as 25 μg W in 10 ml of water can be extracted satisfactorily. There is little decrease in E with an increase in tungsten concentration. Thus under otherwise constant conditions, 5 μg W in 10 ml aqueous solution was found to be 97% extracted, 25 μg, 95.3%. Apparently, formation of polymeric W species is not important at these low W concentrations.

It may be expected that Mo will give a pH extraction curve similar to that of W.[107] The extraction of Mo is known to be favorable at $\sim 0.5 M H^+$, at which acidity both W and Mo can be quantitatively extracted for subsequent photometric determination.

Niobium (tracer amounts) is extracted maximally ($\sim 100\%$) by α-benzoinoxime in chloroform at about 5 $M H^+$ (HCl).[108] Low concentrations of oxalic or tartaric acid are permissible. This is a good way of separating Nb from Zr and Hf. Nb can be back-extracted into H_2SO_4–H_2O_2 or into oxalic acid (0.1 M).

Ta is also extractable by α-benzoinoxime into organic solvents, but apparently less easily.[109]

Vanadium(V) is best extracted into $CHCl_3$ in the approximate pH range 2–4, but many other metals are also extracted (even Cu, Ni, Co, etc.), so that the extraction is of little interest.

F. THIOGLYCOLIC-β-AMINONAPHTHALIDE (THIONALIDE)

This compound is best known as a precipitation reagent. As far as we know, no systematic study has been made of the extraction of its chelates, which are of the type ML_n. However, qualitative observations indicate that some of these are soluble in chloroform (As, Sb, Bi, Pd, Cd, Co), so that thionalide may find use as an extraction reagent, and deserves mention here. The chelates of Ag, Au, Cu, Hg(II), Pd, Sn(II, IV), and Ni have been reported to be chloroform insoluble.

The following dissociation constants have been reported for thionalide (HL)[110]:

$$\frac{[H^+][L^-]}{[HL]} = 7 \times 10^{-11} \text{ (ionization of } -SH)$$

$$\frac{[H^+][HL]}{[H_2L^+]} = 2 \times 10^{-3}$$

Values for the water solubility are contradictory, for example, 9×10^{-5} M and $\sim 5 \times 10^{-4}$ M, and $\sim 2 \times 10^{-4}$ M in 0.1 M HCl at 20°. Thionalide is readily soluble in the usual organic extracting solvents and in acetic acid. Its solutions are very unstable (air oxidation to disulfide), decomposing in a few hours. It is destroyed by even weak oxidizing agents such as ferric iron; hydroxylammonium sulfate has been used to maintain a reducing environment.

Thionalide precipitates most metals of the hydrogen sulfide analytical group from dilute mineral acid solution (e.g., 0.2 N),[111] including Ag, Au, Cu, Hg, Sn, As, Sb, Bi, Pt, Pd, Rh, and Ru.[112] Unlike H_2S, it does not precipitate Pb and Cd from mineral acid solution. Thionalide has been used to separate (precipitation) traces of Cu, Sb, Sn, Bi, and As in Ni–Co–Fe–Cr alloys.[113] Various separations are also possible in basic solutions containing tartrate, cyanide, and the like. Precipitation sensitivities in 0.2 M H^+ solution are $1:10^7$ and even $1:10^8$, so that extraction into immiscible organic solvents would be expected to provide large E values.

Indirect colorimetric determination of metals by reduction of phosphotungstomolybdic acid by the precipitated metal thionalate has been described, but is of minor interest (Chapter 6F).

G. AZO-AZOXY BN (AZO-AZOHYDROXY BN, 2-HYDROXY-2′-(2-HYDROXY-1-NAPHTHYLAZO)-5-METHYLAZOXY-BENZENE)[114]

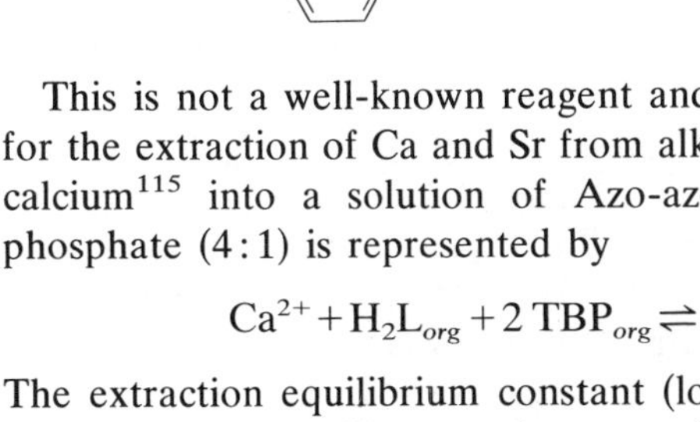

$K_{a1} = 3.3 \times 10^{-11}$ and $K_{a2} = 5 \times 10^{-14}$. Practically insoluble in water. Solubility 2×10^{-7} M in 0.05 M NaOH, 6.5×10^{-6} M in 0.5 M NaOH, 6×10^{-3} M in $CHCl_3$, 4.6×10^{-4} M in TBP. $P_{CCl_4/H_2O} = 2.6 \times 10^5$ and $P_{CCl_4-TBP(4:1)/H_2O} = 4.0 \times 10^5$ ($\mu = 1$).

This is not a well-known reagent and is a specialized one. It is useful for the extraction of Ca and Sr from alkaline solutions. The extraction of calcium[115] into a solution of Azo-azoxy BN (H_2L) in CCl_4–tributyl phosphate (4:1) is represented by

$$Ca^{2+} + H_2L_{org} + 2\,TBP_{org} \rightleftharpoons CaL \cdot 2\,TBP_{org} + 2\,H^+$$

The extraction equilibrium constant (log K_{ext}) for this reaction is -20.0 ($\mu = 1$). When the diluent carbon tetrachloride is replaced by cyclohexane and chloroform, log K_{ext} becomes -18.4 and -22.14, respectively. The probable formula[116] of the extracted complex is:

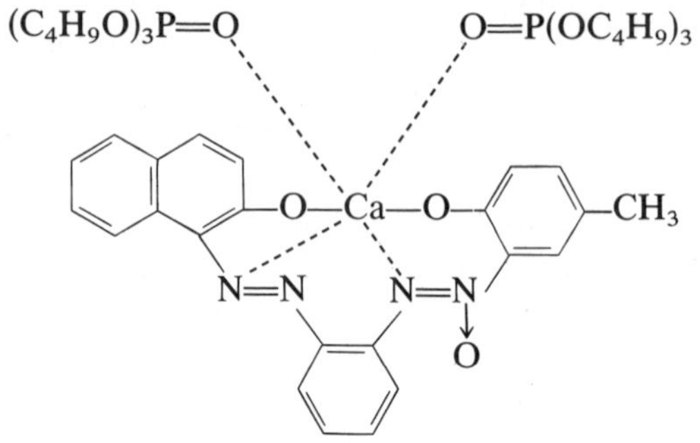

Microgram amounts of calcium are extracted quantitatively with a 0.04% solution of Azo-azoxy BN in CCl_4–tributyl phosphate (4:1) from an equal volume of 0.05–5 M NaOH solution.[117] Shaking for 1–2 min is sufficient. About 90% strontium is extracted into an equal volume of the Azo-azoxy BN solution from 0.8–2 M NaOH solution. Barium is not appreciably extracted. Therefore, calcium can be separated from strontium by extracting the former into the Azo-azoxy BN solution from 0.05 M NaOH solution. Strontium is then extracted from 0.8 M NaOH solution. Calcium and strontium can be back-extracted into dilute (e.g., 0.1 M) HCl. Large amounts of Na, K, Al, Cr(III), and Pb do not interfere with either extraction. Magnesium is not extracted from 0.8 M NaOH solution, and separation of calcium from magnesium is effected.[118] The Azo-azoxy BN extraction has been applied to the separation of calcium from a variety of matrix elements.

Traces of Li are separable from NaCl and KCl by extraction with a CCl_4–TBP solution of the reagent from NaOH medium.[119]

Zinc also is extracted by a 0.02% solution of Azo-azoxy BN in CCl_4-tributyl phosphate (4:1) from 0.4 M NaOH solution.[120] Azo-azoxy BN has been used for the extraction–photometric determination of copper[121] and cobalt.[122]

Azo-azoxy AN (8-acetamido-1-[2-(2-hydroxy-5-methylphenylazoxy)-phenylazo]-2-naphthol) in CCl_4-tributyl phosphate quantitatively extracts calcium (from 0.2–0.5 M NaOH solution) and copper, and partially extracts strontium (from 1.5–2.5 M NaOH solution).[123] Calcium or copper can be determined by measuring the absorbance of the extract, or absorptiometrically after back-extraction into 0.1 M HCl.

Azo-azoxy FMP (Azo-azoxy PMP; 4-[2-(2-hydroxy-5-methylphenyl-azoxy)phenylazo]-3-methyl-1-phenyl-2-pyrazolin-5-one) in CCl_4-tributyl

phosphate

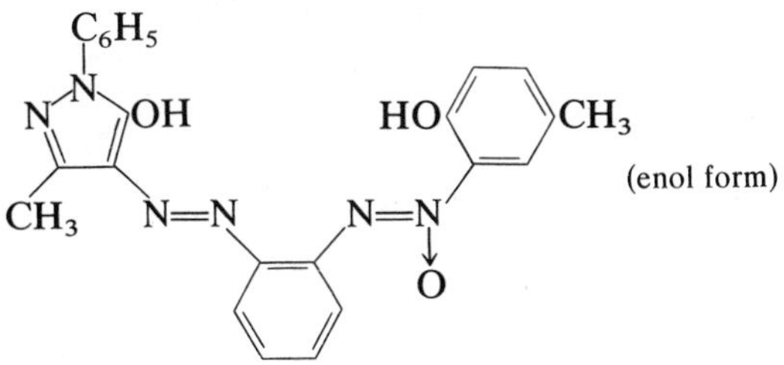

(enol form)

quantitatively extracts strontium at pH 13–14.[124] The composition of the extracted complex is analogous to that of the calcium-Azo-azoxy BN–TBP complex. Large amounts of Na, K, Al, Pb, Sb, and Cr do not interfere with the extraction, but Ca does. Barium is extracted slightly. Strontium in the organic phase can be back-extracted into 0.1 M HCl.

H. CARBOXYLIC ACIDS

Metal carboxylates are not chelates, but it is convenient to consider their extraction here. Formally, the extraction of metals as carboxylates is much the same as the extraction of metal chelates, account being taken of the polymerization (often dimerization) of the reagent and of the carboxylate in the organic solvent and of the formation of adducts and mixed complexes. The extraction of metals by various carboxylic acids (aliphatic, aromatic, naphthenic, etc.) has been studied fairly extensively.[125] Though some facts of undoubted analytical interest have emerged, the general impression one obtains is that carboxylic acid extraction is of subsidiary importance in analytical separations. Some examples follow.

Phenylacetic acid $C_6H_5CH_2COOH$ in chloroform solution extracts Cu(II), Fe(III), and U(VI) from hexamine solution, leaving Mn, Ni, Co, Mo(VI), and V(V) in the aqueous phase.[126]

Butyric acid has been used for the chloroform extraction of Be in the presence of Fe(III) and Al, with EDTA as masking agent.[127] Salicylic, cinnamic, and other carboxylic acids, best in MIBK, have been used to separate Th and U(VI) from the rare earths (La).[128]

In early work of preliminary nature, scandium was shown to be extracted as benzoate at pH ~7 by butyl or amyl alcohol and ethyl acetate.[129]

Extraction of metals with α-hydroxy-α-dibutylphosphineoxide-propionic acid (or 2-dibutylphosphinyl-2-hydroxypropionic acid)[130] and α-hydroxy-α-dibutylphosphineoxidephenylacetic acid[131]

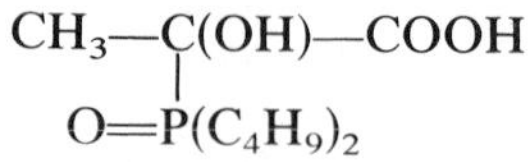

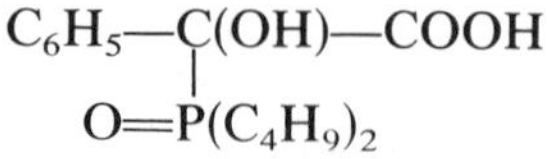

$$CH_3—\underset{\underset{O=P(C_4H_9)_2}{|}}{C}(OH)—COOH$$

$$C_6H_5—\underset{\underset{O=P(C_4H_9)_2}{|}}{C}(OH)—COOH$$

in chloroform and isoamyl alcohol has been studied as a function of aqueous-phase pH. With the former, Al can be separated from Fe(III), Ti, Ga or In. With the latter, extraction of Fe(III), Cu, Ga, and In (isoamyl alcohol) and Ga, Al, In, and Cu (chloroform) proceeds quantitatively. Separation of these metals from Co and Ni is possible.

NOTES

1. This approach was used by E. B. Sandell in 1939 (University of Minnesota, unpublished) to derive expression (9B-1), using dithizone as the model reagent. The first test of the extraction expression, somewhat preliminary in character, was carried out with zinc and dithizone by I. M. Kolthoff and E. B. Sandell, *J. Am. Chem. Soc.*, **63,** 1906 (1941), and the results showed conformity with theory. A more thorough test was made by H. Irving, C. F. Bell, and R. J. P. Williams, *J. Chem. Soc.*, 356 (1952).

2. L. G. Sillén and A. E. Martell, *Stability Constants of Metal-Ion Complexes,* The Chemical Society, London, 1964. A supplement giving data appearing in the years 1963–1968 has been published: *Stability Constants of Metal-Ion Complexes. Supplement No. 1.* Part 1, Inorganic Ligands, L. G. Sillén. Part 2, Organic Including Macromolecule Ligands, A. E. Martell. Chemical Society, London, 1971. xx, 866 pp. Special Publication No. 25, Supplement to Special Publication No. 17.

    A useful source of analytical complexing data is Table A5 in A. Ringbom, *Complexation in Analytical Chemistry*, Interscience, New York, 1963.

3. See Ringbom, *Complexation in Analytical Chemistry*, Chapter 7, in which the use of effective constants in extractions is systematically developed. See Chapter 6F, p. 520, of the present book for an illustration.

4. H. Christopherson and E. B. Sandell, *Anal. Chim. Acta*, **10,** 1 (1954).

5. A summary is given in Zolotov, *Extraction of Chelate Compounds*, pp. 48–59.

6. Yu. A. Zolotov and V. V. Bagreev, *Trudy Komiss. Anal. Khim.*, **17,** 251 (1969), have examined the relations between P_l and P_m for homologous reagents and various solvents. See also, Zolotov (transl. by J. Schmorak), *Extraction of Chelate Compounds*, Humphrey Science Publ., Ann Arbor, Mich., 1970.

7. Review: B. Ya. Spivakov and Yu. A. Zolotov, *J. Anal. Chem. USSR*, **25,** 529 (1970).

8. V. M. Krasikova, cited in Zolotov, *Extraction of Chelate Compounds*, p. 21. D. Dyrssen, *Svensk Kem. Tidskr.*, **64,** 213 (1952), studied the extraction of HOx·H$^+$ by MIBK.

9. G. K. Schweitzer, R. B. Neel, and F. R. Clifford, *Anal. Chim. Acta*, **33,** 514 (1965), concluded that zinc oxinate is extracted as ZnOx$_2$·2S (S = solvent molecule) into the oxygenated solvents 2-methyl-2-pentanone, 1-butanol, and 4-methyl-2-pentanol; as ZnOx$_2$·HOx·S into CHCl$_3$ and C$_6$H$_6$; as ZnOx$_2$·S·H$_2$O into butyl ether and pentyl ether; as ZnOx$_2$·HOx·H$_2$O into carbon tetrachloride. The extraction of zinc oxinate into hexane and cyclohexane is so poor ($E = 0.01$–0.0025 at pH 2–7, 30°) that the extracted species is thought to be ZnOx$_2$·2 H$_2$O. The replacement of only one water molecule by butyl ether, pentyl ether, and carbon tetrachloride is believed to be related to the small dielectric constants and dipole moments of these solvents. The organic solvent molecules are thought to occupy coordination sites around the zinc ion when the ketone and the two alcohols are the extractants. The sites of the chloroform and benzene molecules are uncertain; attachment to O or N of Ox or HOx by hydrogen bonding is a possibility. In these experiments, the initial Zn concentration was 5×10^{-7} *M* and the aqueous solutions were usually 0.1 *M* in perchlorate.

10. For general reviews of analytical extraction kinetics, see Yu. A. Zolotov, I. P. Alimarin, and V. A. Bodnya, *J. Anal. Chem. USSR*, **19,** 23 (1964); I. P. Alimarin, *Pure Appl. Chem.*, **34,** 1 (1973).

11. If it is present as colloidal hydroxide or basic salt, or polymerized hydrolysis products, extraction may be very slow. Zirconium, niobium, tantalum, and like elements especially may give such products, even when the solution is quite acidic.

12. The extraction rate of some solvation coordination complexes of metals (TBP) has been shown to be determined by chemical reactions at the interface. (See p. 1002).

13. H. M. Irving and R. J. P. Williams, *J. Chem. Soc.*, 1844 (1949); H. M. Irving, C. F. Bell, and R. J. P. Williams, *ibid.*, 356 (1952). Yu. A. Zolotov, *Doklady Akad. Nauk SSSR*, **162,** 577 (1965), studying the extraction of ferric acetylacetonate, ferric benzoylphenylhydroxylaminate, scandium dibenzoylmethanate, and other compounds with CHCl$_3$, C$_6$H$_6$, and CCl$_4$, found that the extraction rate decreased in the order of increase of P_i ($P_{CHCl_3} > P_{C_6H_6} > P_{CCl_4}$). It appears that in rather strongly acid solution (see rate of extraction of Cu(II) dithizonate, p. 921) the reagent may react with a metal ion without ionizing, and the extraction rate is less dependent on acidity than in weakly acidic solutions.

14. P. R. Subbaraman, Sr. M. Cordes, and H. Freiser, *Anal. Chem.*, **41,** 1878 (1969).

15. T. F. Bidleman, *Anal. Chim. Acta*, **56,** 221 (1971).

16. Yu. B. Kletenik and V. A. Navrotskaya, *Zh. Neorg. Khim.*, **12,** 3114 (1967).

17. G. K. Schweitzer and J. R. Rimstidt, *Anal. Chim. Acta*, **27,** 389 (1962).

18. V. Sedivec and J. Fleck, *Coll. Czech. Chem. Commun.*, **29,** 1310 (1964). See further "Sodium Diethyldithiocarbamate," Chapter 6F.

19. R. A. Bolomey and L. Wish, *J. Am. Chem. Soc.*, **72,** 4483 (1950).

20. H. L. Finston and Y. Inoue, *J. Inorg. Nucl. Chem.*, **29,** 199 (1967).

21. H. L. Finston and Y. Inoue, *ibid.*, **29,** 2431 (1967).

22. W. G. Whitman, *Chem. Met. Eng.*, **29,** 147 (1923); K. F. Gordon and T. K. Sherwood, *AIChEJ*, **1,** 129 (1955).

23. Values of D for electrolytes, nonelectrolytes, and a few ions may be found in *American Institute of Physics Handbook*, 3rd ed., p. 2–221. At 25°, in aqueous solution, few electrolytes have values of D greater than 2×10^{-5}, or much less than 1×10^{-5}. Nonelectrolytes tend to have values of 1×10^{-5} or smaller. D increases with a rise in temperature and a decrease in viscosity of the solution (of greater significance for organic solvent solutions than for aqueous).

24. This is the value reported for a system in which dextromethorphanium cation was extracted as an ion associate into chloroform from water, in which the two phases were stirred in opposite directions at 40 rpm. The interface remained effectively quiescent under these stirring conditions. See T. Higuchi and A. F. Michaelis, *Anal. Chem.*, **40,** 1925 (1968). In passing, it may be noted that extractions have occasionally been carried out by allowing the two immiscible phases to remain in contact for a long time without mixing or stirring (static extraction). For example, R. Ko, *Anal. Chem.*, **39,** 1903 (1967), found that when 10 ml of aqueous phase (0.1 mg Ga/liter) was allowed to stand with 2 ml of organic phase (CCl₄–chlorobenzene) containing Rhodamine B, Ga was completely extracted (~99%) as Rhodamine B chlorogallate after 20 hours. Extraction of U(VI) (4 g/liter) as the chloro complex into a xylene solution of tri-*n*-octylamine was found to be almost complete after 48 hours. Resort to static extraction may be advantageous if emulsions are formed in conventional extraction (as when finely divided material is suspended in the aqueous phase, or for other reasons) and the time factor is unimportant. Gentle movement of the phases past each other, as by propeller stirring, without droplet formation should reduce the time required for attainment of equilibrium.

25. R. W. Geiger, Ph. D. Thesis, University of Minnesota, 1951.

26. R. W. Geiger and E. B. Sandell, unpublished work, University of Minnesota, 1951–1952. The curves reproduced in Fig. 9B-8 were obtained by one of the authors.

27. C. B. Honaker and H. Freiser, *J. Phys. Chem.*, **66,** 127 (1962).

28. B. E. McClellan and H. Freiser, *Anal. Chem.*, **36,** 2262 (1964).

29. R. W. Taft and E. H. Cook, *J. Am. Chem. Soc.*, **81,** 46 (1959).

30. K. Murata, Y. Yokoyama, and S. Ikeda, *Anal. Chem.*, **44,** 805 (1972).

31. The solubility changes only slightly in 0.1 M acid solutions, as shown by J. F. Steinbach and H. Freiser, *Anal. Chem.*, **26**, 375 (1954). The salting out effect of the acid may be counteracted by slight formation of $(HAA)H^+$.

32. N. P. Rudenko and J. Stary, *Trudy Kom. Anal. Khim.*, **9**, 28 (1958).

33. There is evidence that HAA can function as a cation in mineral acid solution. H. Jaskolska and J. Minczewski, *Chemia Anal.*, **12**, 1147 (1967), found that In is extracted from ~1 M HBr and HI solutions into HAA or benzene. Presumably an ion-association complex of $(HAA)H^+$ and $InBr_4^-$ or InI_4^- is formed. Fe(III) behaves similarly.

34. J. Rydberg, *Arkiv Kemi*, **8**, 113 (1955).

35. A. Krishen and H. Freiser, *Anal. Chem.*, **31**, 923 (1959).

36. J. F. Steinbach and H. Freiser, *Anal. Chem.*, **25**, 881 (1953). Also see *Anal. Chem.*, **31**, 923 (1959).

37. These and other values are taken from curves given by J. Stary and E. Hladky, *Anal. Chim. Acta*, **28**, 227 (1963), who used 10^{-3}–10^{-4} M metal solutions.

38. H. E. Hellwege and G. K. Schweitzer, *Anal. Chim. Acta*, **29**, 46 (1963), using a $CHCl_3$ solution of HAA, found induction times varying from 90 min at pH 4 to 5 min at pH 7. The extraction kinetics as a function of Cr and reagent concentration and pH were studied.

39. V. Yatirajam and L. Kakkar, *Anal. Chim. Acta*, **44**, 468 (1969).

40. H. Grubitsch and T. Heggeboe, *Monatsh. Chem.*, **93**, 274 (1962).

41. J. B. Headridge and J. Richardson, *Analyst*, **94**, 968 (1969). The extraction is used to separate Fe, Ni, and Cr in the atomic absorption determination of Ca in stainless steel. Also see K. Tanino and S. Kitahara, *Sci. Pap. Inst. Phys. Chem. Res. Tokyo*, **61**, 35 (1967).

42. H. Koshimura and T. Okubo, *Anal. Chim. Acta*, **49**, 67 (1970).

43. Stary and Hladky, reference in note 37. See Table 9B-5.

44. E. L. King and W. H. Reas, *J. Am. Chem. Soc.*, **73**, 1806 (1951).

45. Intermediate species such as $Th(TTA)_3^+$ can be formed in the aqueous phase. At higher reagent concentrations, U(VI) seems to form $UO_2(TTA)_2 \cdot HTTA$ and Sr gives $Sr(TTA)_2 \cdot HTTA$. Mixed complexes can also be formed, for example $Ga(OH)(TTA)_2$, $GaBr(TTA)_2$ [*J. Inorg. Nucl. Chem.*, **32**, 483 (1970)].

46. R. A. Bolomey and L. Wish, *J. Am. Chem. Soc.*, **72**, 4483 (1950), made observations on the kinetics of the forward and back extraction.

47. A. M. Poskanzer and B. M. Foreman, *J. Inorg. Nucl. Chem.*, **16**, 323 (1961).

48. A. Jurriaanse and F. L. Moore, *Anal. Chem.*, **39**, 494 (1967).

49. S. M. Khopkar and A. K. De, *Analyst*, **85**, 376 (1960), state that Ag, Th, Zr, Cu, and Fe can be masked by EDTA in HTTA extraction of U; however, H. Onishi and K. Sekine, *Japan Analyst*, **19**, 547 (1970), could not mask Th by EDTA. P. Crowther and F. L. Moore, *Anal. Chem.*, **35**, 2081

(1963), found that EDTA made their extraction method for Cs highly selective. In this method, Cs is extracted from an aqueous solution containing LiOH (synergist) and Na_2CO_3 by HTTA in nitromethane or nitrobenzene. This is not a separation from other alkali metals, however.

50. W. M. Jackson, G. I. Gleason, and P. J. Hammons, *Anal. Chem.*, **42**, 1242 (1970). For extraction of alkaline earths, see I. Akaza, *Bull. Chem. Soc. Japan*, **39**, 971 (1966). Lithium is extractable to the extent of 90% or so at pH 9.5–10 with HTTA in MIBK, as shown by H. T. Delves, G. Shepherd, and P. Vinter, *Analyst*, **96**, 260 (1971). For separation of Sc from U, Th, V, and Ti, see A. W. Ashbrook, *ibid.*, **88**, 113 (1963).

51. W. M. Jackson and G. I. Gleason, *Anal. Chem.*, **45**, 2125 (1973). Only a few figures are given above.

52. T. W. Gilbert, L. Newman, and P. Klotz, *Anal. Chem.*, **40**, 2123 (1968).

53. M. S. Rahaman and H. L. Finston, *Anal. Chem.*, **41**, 2023 (1969). Under similar conditions, Ga is extracted as $\{(C_6H_5)_4As^+\}\{Ga(TTA)_2AcCl^-\}$ (*ibid.*, **40**, 1709).

54. E. F. Kassierer and A. S. Kertes, *J. Inorg. Nucl. Chem.*, **34**, 3221 (1972).

55. E. W. Berg and K. P. Reed, *Anal. Chim. Acta*, **36**, 372 (1966); V. M. Shinde and S. M. Khopkar, *Chem. Ind.*, 1785 (1967).

56. W. G. Scribner, W. J. Treat, J. D. Weis, and R. W. Moshier, *Anal. Chem.*, **37**, 1136 (1965); G. P. Morie and T. R. Sweet, *ibid.*, **37**, 1552 (1965), where earlier literature is mentioned. Al, Fe(III), and Cu are extracted from acetate-buffered solution by $CHCl_3$. Benzene extraction of Al, Ga, and In chelates is combined with gas chromatography. Ni, Co, Cu, Zn, and other metals are readily extracted into $CHCl_3$ by trifluoroacetylacetone if isobutylamine is present, which forms adducts, as shown by W. G. Scribner and A. M. Kotecki, *Anal. Chem.*, **37**, 1304 (1965). M. L. Lee and D. C. Burrell, *Anal. Chim. Acta*, **62**, 153 (1972), extract Fe, In, Co, and Zn from sea water with a toluene solution of the reagent; isobutylamine must be added to extract Co and Zn.

57. See J. W. Mitchell and R. Ganges, *Anal. Chem.*, **46**, 503 (1974).

58. T. R. Sweet and H. W. Parlett, *Anal. Chem.*, **40**, 1885 (1968). Formation constants of ML^{2+}, $MLOH^+$, ML_2^+, ML_2OH, ML_3, and the like are given for Pr, Sm, Eu, Gd, and Tb.

59. G. K. Schweitzer and W. Van Willis, *Anal. Chim. Acta*, **36**, 77 (1966).

60. Yu. A. Zolotov and V. G. Lambrev, *Zh. Anal. Khim.*, **20**, 659 (1965); Yu. A. Zolotov, M. K. Chmutova, and P. N. Palei, *ibid.*, **21**, 1217 (1966); Yu. A. Zolotov et al., *Dokl. Akad. Nauk SSSR*, **165**, 117 (1965); references in Table 9B-7.

61. O. Navratil and B. S. Jensen, *J. Radioanal. Chem.*, **5**, 313 (1970).

62. O. Navratil, *Coll. Czech. Chem. Commun.*, **38**, 1333 (1973).

63. D. Dyrssen, *Trans. Roy. Inst. Tech. Stockholm*, No. 188 (1962); *Acta Chem. Scand.*, **15**, 1614 (1961).

64. Formation constants of Fe(III) and V(V) complexes: O. Menis, B. E. McClellan, and D. S. Bright, *Anal. Chem.*, **43**, 431 (1971); kinetics and mechanism of Fe(III) extraction: B. E. McClellan and O. Menis, *ibid.*, **43**, 436 (1971).

65. N. H. Furman, W. B. Mason, and J. S. Pekola, *Anal. Chem.*, **21**, 1325 (1949), have reviewed precipitation and extraction of cupferrates.

66. D. Dyrssen, *Svensk Kem. Tidskr.*, **64**, 213 (1952). I. V. Pyatnitskii, *Zh. Anal. Khim.*, **1**, 135 (1946), found $K_a = 5 \times 10^{-5}$ (25°).

67. H. Herzog, M. S. Thesis, University of Minnesota, 1951. Instability of the solutions makes exact solubility determination difficult.

68. S. J. Lyle and A. D. Shendrikar, *Anal. Chim. Acta*, **36**, 286 (1966), state that in solutions up to 6–8 M in HCl, the main decomposition product is nitrosobenzene, whose amount increases with acidity. At higher acidities the main product, not identified, is not nitrosobenzene. Others have mentioned benzenediazonium nitrate and 4,4'-dinitrodiphenylamine as other decomposition products. For further study of cupferron decomposition, see A. Yu. Grigorovich, F. I. Lobanov, and V. M. Savostina, *Zh. Anal. Khim.*, **26**, 265 (1971). They found the decomposition to be autocatalytic, with the induction period longer in H_2SO_4 than in HCl or $HClO_4$. Even traces of degradation products (initially nitrosobenzene) strongly catalyze the decomposition. The decomposition of cupferron in H_2SO_4 is a second-order reaction.

69. N. P. Komar and I. G. Perkov, *J. Anal. Chem. USSR*, **19**, 1329 (1964).

70. Bunce, *Anal. Chem.*, **21**, 1325 (1949).

71. Dyrssen, reference in note 66. The value for $CHCl_3$ is much the same at 0°.

72. D. M. Kemp, *Anal. Chim. Acta*, **27**, 480 (1962). The value 145 was determined at pH 3.5–5, the value 120–135 at pH 1–2 (approximately).

73. Z. Hagiwara, *Tech. Rept. Tohoku Univ.*, **18**, 16 (1953).

74. For example, if cupferron is added to the aqueous solution and this is made alkaline, Al and Th can be extracted to a large extent up to a pH of ~9, Ga to a pH of 9, and Bi to pH 11 at least. On the other hand, Mo(.VI) and V(V) are extracted little above pH 7, W(VI) little above pH 4; their nonextraction from basic solution is understandable from the absence of cationic species of these metals.

75. Negatively charged species can also be formed by some metals. A few have been isolated as alkali metal complexes in the solid state. Uranium(VI) forms $NH_4UO_2Cp_3$ as shown by W. S. Horton, *J. Am. Chem. Soc.*, **78**, 897 (1956).

76. D. Dyrssen and V. Dahlberg, *Acta Chem. Scand.*, **7**, 1186 (1953).

77. K. L. Cheng, *Anal. Chem.*, **30**, 1941 (1958). Ti is determined photometrically.

78. E. M. Donaldson, *Talanta*, **16**, 1505 (1969).

79. Yu. A. Zolotov, B. Ya. Spivakov, and G. N. Gavrilina, *Zh. Anal. Khim.*, **24,** 1168 (1969).

80. At least in part this behavior is due to more complete precipitation of Al on standing, in spite of the decomposition of nitrosophenylhydroxylamine. Other organic chelates show the same behavior (Al oxinate). On the other hand, examples could be given showing that a metal chelate is extracted more readily by an organic solvent if it is not first precipitated (Cu diethyldithiocarbamate, p. 522). This last behavior is not common.

81. F. Buscarons and J. Canela, *Anal. Chim. Acta*, **67,** 349 (1973). For cyclooctyl and cyclodecyl analogs, see *ibid.*, **71,** 468 (1974).

82. F. Buscarons and J. Canela, *Anal. Chim. Acta*, **70,** 113 (1974).

83. A. K. Majumdar, *N-Benzoylphenylhydroxylamine and Its Analogues*, Pergamon, New York, 1972. Also see H. Förster, *J. Radioanal. Chem.*, **6,** 11 (1970).

84. Not extracted in presence of tartrate or citrate (separation from Mo).

85. Tin forms a mixed chelate $Sn(BPHA)_2Cl_2$ when precipitated or extracted from dilute HCl solution, as shown by D. E. Ryan and G. D. Lutwick, *Can. J. Chem.*, **31,** 9 (1953); Lutwick and Ryan, *ibid.*, **32,** 949 (1954); S. J. Lyle and A. D. Shendrikar, *Anal. Chim. Acta*, **36,** 286 (1965).

86. According to Shendrikar,[100] quantitative extraction of Cd and Mn occurs at pH 10, of Ni at pH 9. Lyle and Shendrikar, *Anal. Chim. Acta*, **32,** 575 (1965), using a 1% solution of BPHA in $CHCl_3$, found quantitative extraction of Ga at pH 3.1, In at pH 5.3, Pb at pH 9.0, and Tl(I) at pH 10.5 (equal phase volumes); Sn(IV) is extracted ~95% from 0.8 M HCl or 4 M $HClO_4$. H. Förster, *J. Radioanal. Chem.*, **6,** 11 (1970), has studied the extraction and back-extraction of many metals with HBPHA. Extraction of Th: N. P. Rudenko and L. A. Samarai, *Anal. Abst.*, **18,** 3035 (1970).

87. D. Dyrssen, *Svensk Kem. Tidskr.*, **68,** 212 (1956).

88. R. Villarreal, J. R. Krsul, and S. A. Barker, *Anal. Chem.*, **41,** 1420 (1969).

89. I. G. Per'kov, N. Komar', and V. V. Mel'nik, *J. Anal. Chem. USSR*, **22,** 429, 567 (1967).

90. Zolotov, reference in note 13.

91. E. E. Rakovskii and O. M. Petrukhin, *J. Anal. Chem. USSR*, **18,** 465 (1963), found that quantitative extraction of Sb(III) was obtained from 10 N H_2SO_4 with 0.4 M HBPHA in $CHCl_3$. Lately, J. Chwastowska, K. Lissowska and E. Sterlinska, *Chemia Anal.*, **19,** 671 (1974), have found that W, Zr, Nb, Ta, Ti, and Sb can be extracted quantitatively into chloroform from 11.5 M HCl. As shown by A. T. Pilipenko, E. A. Shpak, and O. S. Zul'figarov, *Zh. Anal. Khim.*, **30,** 1009 (1975), addition of phenylfluorone to the extract gives a mixed chelate, $\mathscr{S}_{540} = 0.0005$ μg Ti/cm^2.

92. I. P. Alimarin and O. Petrukhin, *Zh. Neorg. Khim.*, **7,** 1191 (1962). Alimarin, E. V. Smolina, I. V. Sokolova, and T. V. Firsova, *Zh. Anal. Khim.*, **25,** 2287 (1970), found Ge, Sn(IV), Ti, and Zr to be extracted 90%

or more from >2 M HClO$_4$ by an equal volume of 0.1 M HBPHA in benzene.

93. But R. Villarreal and S. A. Barker, *Anal. Chem.*, **41**, 611 (1969), using toluene as extracting solvent, found the extraction of tracer amounts of Nb from 9–12 M HCl to be complete in 1 min.

94. It should be noted that a few elements can be extracted into CHCl$_3$ from HCl solution in the absence of HBPHA. The extraction of As(III) as AsCl$_3$ into CHCl$_3$ and other nonoxygen solvents is well known. S. J. Lyle and A. D. Shendrikar, *Anal. Chim. Acta*, **36**, 286 (1966), found Sb(III) to be extracted (presumably as SbCl$_3$) by chloroform from HCl solutions >9 M, and Sb(V) to be extracted above 6 M HCl (as much as 50% at 9–10 M). Sb(V) apparently does not extract into CHCl$_3$ alone from HClO$_4$ solution.

95. According to L. P. Varga et al., *Anal. Chem.*, **37**, 1003 (1965), Ta is apparently extracted as HTaF$_4$(BPHA)$_2$ by HBPHA in CHCl$_3$ from 3 M HClO$_4$ solution having low concentrations of F. In a later paper (J. S. Erskine, M. L. Sink, and L. P. Varga, *ibid.*, **41**, 70), the extracted species is stated to be TaF$_3$·2 BPHA (mixed anion chelate). Niobium is reported to be extracted as H$_2$NbF$_5$·2 BPHA. Separation possibilities are discussed.

96. J. Hála, *J. Inorg. Nucl. Chem.*, **29**, 187 (1967). The trend of the Zr and Hf E versus acid concentration curves is the same as found by Vita et al., with a minimum at 4-5 M HCl or HClO$_4$ (CHCl$_3$ as extractant). Extraction of these metals by ion-pair formation involving MCl$_5^-$ (or MCl$_6^{2-}$) is considered unlikely.

97. K. F. Fouché, *J. Inorg. Nucl. Chem.*, **30**, 3057 (1968); *Talanta*, **15**, 1295 (1968).

98. The phenomenon of metal extraction from very strongly acid solution is not limited to HBPHA. Some other reagents containing N show a similar behavior. Thus a chloroform solution of ammonium tetramethylene-dithiocarbamate (pyrrolidinedithiocarbamate) extracts niobium from strong acid ($\sim 9\,M\,H^+$). New reagent species resulting from protonation of N may explain the metal extraction.

99. According to M. K. Chmutova, O. M. Petrukhin, and Yu. A. Zolotov, *J. Anal. Chem. USSR*, **18**, 507 (1963), a 0.4 M solution of HBPHA in CHCl$_3$ extracts Pu quantitatively from 1–3 M HNO$_3$, but not from HCl or H$_2$SO$_4$ solutions. Pa is extractable even from 12 M HCl [Pal'shin et al., *ibid.*, **18**, 566 (1963)].

100. A. D. Shendrikar, *Talanta*, **16**, 51 (1969); also see *Anal. Chim. Acta*, **36**, 292 (1966).

101. T. Shigematsu, Y. Nishikawa, S. Goda, and H. Hirayama, *Bull. Inst. Chem. Res. Kyoto Univ.*, **43**, 347 (1965).

102. G. A. Brydon and D. E. Ryan, *Anal. Chim. Acta*, **35**, 190 (1966). Formation constants of Mn, Zn, and Ni chelates are reported. Also see R. M. Cassidy and D. E. Ryan, *ibid.*, **41**, 319 (1968).

103. H. J. Hoenes and K. G. Stone, *Talanta*, **4**, 250 (1960).

104. V. Pfeifer and F. Hecht, *Z. Anal. Chem.*, **177**, 175 (1960).

105. P. Y. Peng and E. B. Sandell, *Anal. Chim. Acta*, **29**, 325 (1963).

106. V. A. Nazarenko and E. N. Poluektova, *Zh. Neorg. Khim.*, **14**, 204 (1969).

107. J. S. Fritz and D. R. Beuerman, *Anal. Chem.*, **44**, 692 (1972), present such a curve for 10-hydroxy-eicosan-9-one oxime ($C_9H_{19}CH(OH)C-(NOH)C_9H_{19}$). With a toluene solution of the reagent, the $\log E$–$\log[H^+]$ curve shows a maximum at $[H^+] = 0.1$–1 M. An E of $\sim$600 can be attained.

108. V. Pfeifer and H. Bildstein in Dyrssen, Liljenzin, and Rydberg, *Solvent Extraction Chemistry*, p. 142 (1967). Also see H. Yoshida and C. Yonezawa, *J. Radioanal. Chem.*, **5**, 201 (1970).

109. C. Yonezawa (Japan Atomic Energy Research Institute) found 45% of ^{182}Ta extracted from 5 M HCl by an equal volume of $CHCl_3$.

110. P. Ya. Yakovlev and R. D. Malinina, *Zh. Anal. Khim.*, **23**, 1296 (1968).

111. See reference in note 177, Chapter 6F.

112. W. J. Rogers, F. E. Beamish, and D. S. Russell, *Ind. Eng. Chem. Anal. Ed.*, **12**, 561 (1940).

113. Z. S. Mukhina, I. A. Zhemchuzhnaya, and G. S. Kotova, *J. Anal. Chem. USSR*, **17**, 169 (1962).

114. F. P. Gorbenko and V. V. Sachko, *Zh. Anal. Khim.*, **24**, 15 (1969).

115. F. P. Gorbenko and V. V. Sachko, *Zh. Anal. Khim.*, **25**, 1884 (1970).

116. F. P. Gorbenko and V. V. Sachko, *Zh. Anal. Khim.*, **18**, 1198 (1963). For the extraction of calcium with Azo-azoxy BN in carbon tetrachloride in the presence of alkylamines, see F. P. Gorbenko, I. A. Shevchuk, Yu. K. Tselinskii, and V. V. Sachko, *ibid.*, **18**, 1397 (1963).

117. F. P. Gorbenko and E. V. Lapitskaya, *Zh. Anal. Khim.*, **23**, 1139 (1968); F. P. Gorbenko and A. A. Nadezhda, *Ukr. Khim. Zh.*, **34**, 625 (1968). Calcium has been determined by measuring the absorbance of the extract at 520 nm, but this method does not appear very satisfactory. See F. P. Gorbenko and V. V. Sachko, *Ukr. Khim. Zh.*, **30**, 402 (1964).

118. Z. Marczenko and M. Mojski, *Chemia Anal.*, **12**, 1155 (1967).

119. Z. M. Vashun', F. P. Gorbenko, and I. A. Shevchuk, *Ukr. Khim. Zh.*, **38**, 1058 (1972).

120. F. P. Gorbenko and L. I. Degtyarenko, *Anal. Abst.*, **15**, 3196 (1968).

121. E. Wieteska, *Chemia Anal.*, **13**, 413 (1968).

122. E. Wieteska and M. Kamela, *Chemia Anal.*, **17**, 85 (1972).

123. F. P. Gorbenko and L. Ya. Enal'eva, *Anal. Abst.*, **15**, 3123 (1968).

124. F. P. Gorbenko and A. A. Nadezhda, *Zh. Anal. Khim.*, **24**, 671 (1969); A. A. Nadezhda et al., *Anal. Abst.*, **23**, 104, 105 (1972).

125. Review: F. Miller, *Talanta*, **21**, 685 (1974). About 140 references. Also see R. Pietsch and H. Sinic, *Anal. Chim. Acta*, **49**, 51 (1970). The latter authors studied the extraction of 20 metals with aliphatic carboxylic acids, C_3–C_{10}, in $CHCl_3$. A general survey of carboxylate extraction may be found in Marcus

and Kertes, *Ion Exchange and Solvent Extraction of Metal Complexes*, p. 551 *et seq.*

126. J. Adam and R. Pribil, *Talanta*, **19**, 1105 (1972); **20**, 1344; **21**, 113 (1974), the latter including derivatives. J. Adam, R. Pribil, and V. Vesely, *ibid.*, **19**, 825 (1972).

127. S. Banerjee, A. K. Sundaram, and H. D. Sharma, *Anal. Chim. Acta*, **10**, 256 (1954); A. K. Sundaram and S. Banerjee, *ibid.*, **8**, 526 (1953). L. L. Galkina and L. A. Glazunova, *J. Anal. Chem. USSR*, **21**, 941 (1966), separated the rare earth elements from Fe, Al, Ti, Nb, Zr, and others by extraction with butyric acid in the presence of sulfosalicylic acid.

128. B. Hök-Bernström, *Acta Chem. Scand.*, **10**, 163 (1956); *Svensk Kem. Tidskr.*, **68**, 34 (1956).

129. S. E. J. Johnsen, unpublished observations, University of Minnesota, 1942. Ethyl acetate extraction of aluminum benzoate has been used by J. E. Chester, R. M. Dagnall, and T. S. West, *Talanta*, **17**, 13 (1970), for the separation of Al from Fe (in presence of 1,10-phenanthroline and hydroxylammonium chloride), Cr(VI), Sb(III), Ti(IV), and Hg(II) in a photometric method for Al; Zr, V, and the rare earth elements, as well as Be, are not separated.

130. V. F. Toropova, A. Kh. Miftakhova, M. G. Zimin, and Z. A. Sabitova, *Zh. Anal. Khim.*, **27**, 1836 (1972).

131. A. Kh. Miftakhova, M. G. Zimin, N. I. Bairamova, and V. F. Toropova, *Zh. Anal. Khim.*, **29**, 1771 (1974).

LIQUID–LIQUID EXTRACTION ION-ASSOCIATION REAGENTS— ORGANOPHOSPHORUS REAGENTS

Ion-association reagents finding use in photometric determinations are treated in Chapter 6H. Those useful in separations are considered here. Organophosphorus reagents form both chelates and coordinatively solvated metal complexes, but it is convenient to discuss these reagents here, separately (for the most part) from the nonphosphorus reagents.

I. ION-ASSOCIATION REAGENTS

In this section, we amplify the generalities briefly stated in the classification of extraction systems (Chapter 9A) that pertain to this category and discuss the use of some common ion-association reagents in analytical separations.

A. EXTRACTION EXPRESSIONS

It is always advantageous to be able to connect the parameters of an extraction system in an equation allowing calculation of the extractability of a constituent, even if the results are only approximate. Ion-association extraction systems vary a great deal in their amenability to formulation in relatively simple expressions. Some approach conformity to ideal expressions that can be derived in a simple way. For example, extractions with tetraphenylarsonium cation as the reagent often lend themselves quite well to simple theoretical treatment. On the other hand, extractions with high-molecular-weight amine salts and quaternary ammonium salts may not be formulatable in simple expressions.

An important general case in practice is presented by the formation of an ion-pair compound between a singly charged organic cation R^+ and a bulky singly charged metal anion, which can be extracted into an inert organic solvent. Although extractability of ion-association compounds is not limited to the 1:1 type, these are usually extracted more easily than 1:2 or 2:1 types. The singly charged metal anion can be an oxo-anion such as MnO_4^- or ReO_4^- or, more often, a complex metal halide ion of the type $(M^{n+}X_{n+1})^-$. X represents Cl^-, Br^-, or I^- (less often F^-), or nonhalides (SCN^-, CN^-, OH^-, NO_3^-). The organic cation R^+ can be Ph_4As^+, $(alkyl)_4N^+$, a dye cation $R'H^+$ or the like. The charge of a complex anion depends on the charge and coordination number of the central metal cation and on the ligand also (charge, size). The geometry of the complex anion, as well as its charge and size, affects its extractability (Chapter 9A, Section II.H). The organophilic properties of the coordinating ligand also influence the extractability of the ion associate. Thiocyanate ion sometimes forms doubly charged metal anions that give ion associates with singly charged R^+ extractable into chloroform and other nonsolvating liquids.

The anionic metal may occur as a chelate complex, for example, $UO_2Ox_3^-$, $Fe^{III}Y^-$ ($H_4Y = EDTA$), and complex anions with hydroxy acids.[1] The presence of an $-SO_3H$ group in a chelating agent converts a normal uncharged chelate, ML_n into an anion, $M(LSO_3^-)_n$; the $-SO_3H$ group in the reagent and its chelate is dissociated in quite strongly acid solutions (pH ~ 1). Such species may be extractable into organic solvents (especially of the active type) as ion associates with bulky cationic species. At present, extracted species of this type are not of great importance in separations.

Derivations of extraction expressions for simple ion-association systems follow. The existence of equilibria other than those postulated obviously invalidate the expression for the extraction coefficient. For example, polymerization of the reagent or the ion associate in either phase is not considered here. The composition of extracted products may not always be as simple as represented here; adducts may be formed.

The hydrogen-ion concentration does not enter explicitly into the extraction expressions that follow, but it may affect the extractability. Thus hydrolysis of metal species at low acidities would lead to reduced extraction, as would the formation of slightly ionized $HMCl_{n+1}$, for example. The composition of the extracted product may depend on the acidity. Also, we have assumed that $[R^+]$ is independent of the hydrogen-ion concentration. Actually, many cations of ion pairs are of the type RH^+, formed by the addition of a proton (less often, of more than one proton) to a base R. The concentration of RH^+ may be a function of H^+—if R is a weak base. If R is a charged chelate, for example $Fe(Phen)_3^{2+}$, its concentration will almost always depend on the acidity, since the ligand can also combine with H^+.

REAGENT IS CATIONIC, R^+. Take the reagent to be added as the chloride, RCl, and other metals to be present as chlorides. Let A^- be the anion to be extracted (as RA).

1. *RCl and RA are essentially completely dissociated into their component ions in water, but not at all in the organic extracting solvent.*

This case is often encountered in practice. Solvents such as $CHCl_3$, CCl_4, and C_6H_6 are nondissociating. The dissociation of RCl and RA in water may be so extensive that the molecular forms (ion pairs) can be neglected. Consider first the extraction of the reagent itself:

$$R^+ + Cl^- \rightleftharpoons \underset{org}{RCl}$$

The extraction equilibrium constant is given by

$$\frac{[RCl]_{org}}{[R^+][Cl^-]} = K_{extR}$$

Concentrations without subscripts refer to aqueous phase. The extraction coefficient of the reagent is

$$E_R = \frac{[RCl]_{org}}{[R^+]} = K_{extR}[Cl^-]$$

If there is no extra chloride in solution, that is, if

$$[R^+] = [Cl^-]$$

the extraction coefficient increases with reagent concentration

$$E_R = \frac{[RCl]_{org}}{[R^+]} = K_{extR}[R^+]$$

The extraction equilibrium constant of the constituent A^- corresponding to the extraction reaction

$$A^- + R^+ \rightleftharpoons \underset{org}{RA}$$

is given by

$$\frac{[RA]_{org}}{[A^-][R^+]} = K_{ext}^*$$

(K_{ext}^* is the extraction constant based on reagent concentration in the *aqueous* phase) and

$$K_{ext}^* = \frac{K_{extR}[RA]_{org}[Cl^-]}{[A^-][RCl]_{org}}$$

in which $[Cl^-]$ is independent of reagent concentration.

$$E_A = \frac{[RA]_{org}}{[A^-]} = \frac{K_{ext}^*[RCl]_{org}}{K_{extR}[Cl^-]} = \frac{K_{ext}[RCl]_{org}}{[Cl^-]} \qquad (9C\text{-}1)$$

(K_{ext} is the conventional extraction constant based on $[RCl]_{org}$.) Cl^- competes with A^- to be extracted and E_A varies inversely as $[Cl^-]$. Ionic strength effects have been disregarded.

When the extracted species is R_2A,

$$E_A = \frac{[R_2A]_{org}}{[A^{2-}]} = K_{ext}\frac{[RCl]_{org}^2}{[Cl^-]^2} \qquad (9C\text{-}1a)$$

When A^- is MCl_{n+1}^-. Equation (9C-1) applies when a trace metal M^{n+} is converted quantitatively to MCl_{n+1}^- at the existing Cl^- concentration. Then

$$E_A = E_M = \frac{[RMCl_{n+1}]_{org}}{[MCl_{n+1}^-]} = \frac{K_{ext}^*[RCl]_{org}}{K_{extR}[Cl^-]} = K_{ext}\frac{[RCl]_{org}}{[Cl^-]} \qquad (9C\text{-}1b)$$

If the conversion of M^{n+} to MCl_{n+1}^- is not complete, M^{n+}, MCl^{n-1},···, MCl_{n+1}^- are present (it is assumed that MCl_{n+1}^- is the highest chloro complex anion formed). Then

$$E_M = \frac{[RMCl_{n+1}]_{org}}{\sum [M]} = \frac{[RMC]_{n+1}]_{org}}{[M^{n+}]+[MCl^{n-1}]+ \cdots +[MCl_{n+1}^-]+[RMCl_{n+1}]}$$

If $\beta_1 \cdots \beta_{n+1}$ are the formation constants (cumulative) of the chloro complexes, and taking $[RMCl_{n+1}]$ to be small compared to the concentrations of the other metal species in the aqueous solution, the extraction coefficient of the metal is given by

$$E_M \sim \frac{K_{ext}\beta_{n+1}[Cl^-]^n[RCl]_{org}}{1+\beta_1[Cl^-]+ \cdots +\beta_{n+1}[Cl^-]^{n+1}} \qquad (9C\text{-}1c)$$

When $[MCl_{n+1}^-]$ becomes small compared to $[RMCl_{n+1}]$, $E_M \rightarrow P_M$.

2. *RCl and RA are essentially completely dissociated in both phases.*

This case is less often encountered, but can occur when the extracting solvent has a high dielectric constant (nitrobenzene, nitromethane, and the like).

The extraction constant of the reagent is

$$\frac{[R^+]_{org}[Cl^-]_{org}}{[R^+][Cl^-]} = K_{extR}$$

from which

$$[R^+]_{org} = \frac{K_{extR}[R^+][Cl^-]}{[Cl^-]_{org}}$$

If

$$[Cl^-]_{org} = [R^+]_{org}$$

(each Cl^- entering *org* has R^+ as a partner)

$$[R^+]_{org} = K_{extR}^{1/2}[R^+]^{1/2}[Cl^-]^{1/2}$$

($[Cl^-]$ can be varied, that is, it is not equivalent to $[R^+]$.)

The extraction equilibrium constant of the reaction

$$A^- + R^+ \rightleftharpoons A^- + R^+$$
$$\qquad\qquad\quad org \quad org$$

is

$$\frac{[A^-]_{org}[R^+]_{org}}{[A^-][R^+]} = K_{ext}$$

and

$$E_A = \frac{[A^-]_{org}}{[A^-]} = \frac{K_{ext}[R^+]}{[R^+]_{org}}$$

$$= \frac{K_{ext}[R^+]^{1/2}}{K_{extR}^{1/2}[Cl^-]^{1/2}}$$

$$= const. \times \frac{[R^+]^{1/2}}{[Cl^-]^{1/2}} \tag{9C-2}$$

The assumption made above that $[R^+]_{org} = [Cl^-]_{org}$ will be approximately true if $[A^-]_{org}$ is small compared to $[Cl^-]_{org}$ and will often be valid if A is a trace constituent.

3. *RCl and RA are incompletely dissociated in water and not at all in the organic solvent.*

Denote the dissociation constants of RCl and RA by K_{RCl} and K_{RA}:

$$\frac{[R^+][Cl^-]}{[RCl]} = K_{RCl}$$

$$\frac{[R^+][A^-]}{[RA]} = K_{RA}$$

Now

$$E_A = \frac{[RA]_{org}}{[RA] + [A^-]}$$

$$= \frac{[RA]_{org}}{[RA]\left(1 + \dfrac{K_{RA}}{[R^+]}\right)}$$

$$= \frac{P_{RA}[R^+]}{[R^+] + K_{RA}} \tag{9C-3}$$

If RCl is completely dissociated in water, and RCl is not extracted into the organic solvent phase, this expression gives E_A directly, $[R^+]$ being the reagent concentration in the aqueous phase ($=$ initial concentration less that extracted as RA). When RCl must be taken into account

$$E_A = \frac{P_{RA}K_{RCl}[RCl]}{K_{RCl}[RCl] + K_{RA}[Cl^-]} \quad \text{or} \quad \frac{P_{RA}[RCl]}{[RCl] + (K_{RA}/K_{RCl})[Cl^-]} \tag{9C-3a}$$

The analyst requires the values of three constants to calculate E_A, two of which appear in the preceding equation, the third, P_{RCl}, being needed to find the distribution of RCl between the two phases. Alternatively, K_{RA}/K_{RCl} can be expressed in terms of K_{ext} and $K_{ext\ R}$.

4. *RCl and RA are incompletely dissociated in water and completely dissociated in the organic solvent.*

Define the extraction equilibrium constant on the basis of the reaction

$$A^- + R^+ \rightleftharpoons A^- + R^+$$
$$\qquad\qquad org \quad org$$

$$K_{ext} = \frac{[A^-]_{org}[R^+]_{org}}{[A^-][R^+]}$$

$$K_{ext\ R} = \frac{[R^+]_{org}[Cl^-]_{org}}{[R^+][Cl^-]}$$

and K_{RCl} and K_{RA} are defined as earlier.
Then

$$E_A = \frac{[A^-]_{org}}{[A^-] + [RA]}$$

$$= \frac{K_{ext}[A^-][R^+]/[R^+]_{org}}{K_{RA}[RA]/[R^+] + [RA]}$$

$$= \frac{K_{ext}K_{RA}}{(K_{RA}/[R^+] + 1)[R^+]_{org}}$$

$$= \frac{K_{ext}K_{RA}}{(K_{RA}/[R^+] + 1)(K_{ext\ R}^{1/2}[R^+]^{1/2}[Cl^-]^{1/2})} \qquad (9C\text{-}4)$$

As earlier, it is assumed that $[Cl^-]_{org} = [R^+]_{org}$, that is, that $[A^-]_{org} \ll [Cl^-]_{org}$. In the aqueous phase, $[Cl^-]$ is independent of $[R^+]$.

5. *RCl and RA are completely dissociated in water, but incompletely dissociated in the organic solvent.*

$$E_A = \frac{[A^-]_{org} + [RA]_{org}}{[A^-]}$$

$$= \frac{K_{RA,org}[RA]_{org}/[R^+]_{org} + [RA]_{org}}{[A^-]}$$

$$= \frac{[RA]_{org}}{[A^-]}\left\{\frac{K_{RA,org}}{[R^+]_{org}} + 1\right\}$$

$$= K_{ext}[R^+]\left\{\frac{K_{RA,org}}{[R^+]_{org}} + 1\right\}$$

Since $[R^+]_{org} = K_{RCl,org}^{1/2}[RCl]_{org}^{1/2}$

$$E_A = K_{ext}[R^+]\left\{\frac{K_{RA,org}}{K_{RCl,org}^{1/2}[RCl]_{org}^{1/2}} + 1\right\} \tag{9C-5}$$

6. *RCl and RA are incompletely dissociated in both phases.*

$$E_A = \frac{[A^-]_{org} + [RA]_{org}}{[A^-] + [RA]}$$

$$= \frac{[RA]_{org}\{K_{RA,org}/[R^+]_{org} + 1\}}{[RA]\{K_{RA}/[R^+] + 1\}}$$

$$= \frac{P_{RA}\{K_{RA,org}/[R^+]_{org} + 1\}}{\{K_{RA}/[R^+] + 1\}}$$

$$= \frac{P_{RA}\{K_{RA,org}/[K_{RCl,org}^{1/2}[RCl]_{org}^{1/2}] + 1\}}{\{K_{RA}/K_{RCl} \times [Cl^-]/[RCl] + 1\}}$$

$$= \frac{P_{RA}\{K_{RA,org}/K_{RCl,org}^{1/2}P_{RCl}^{1/2}[RCl]^{1/2} + 1\}}{\{K_{RA}/K_{RCl} \times [Cl^-]/[RCl] + 1\}} \tag{9C-6}$$

REAGENT IS ANIONIC, R^-. We suppose that the reagent is NaR and that the aqueous solution may contain other sodium salts so that R is not equivalent to Na. Designate the metal to be extracted as M and assume it is singly charged. The only species extracted into the organic solvent are NaR and MR, and no interaction between R^- and other metals that may be present takes place.

The extraction expressions derived for the cation reagent R^+ will hold for the anionic reagent if R^+ is replaced by R^- and Cl^- by Na^+.

B. EXTRACTION RATE

The possible rate-determining steps in the extraction of ion-association compounds are the same as, or analogous to, those in chelate extractions (p. 911). As a rule, the extraction rate of ion-association complexes is not limited by the formation rate of the complex, as sometimes is true in chelate reactions. As a matter of fact, the formation rate is not an important factor in many, if not most, of the ion-association reactions discussed in this section, because the reactants are brought together in the aqueous phase and, unless unusually slow, the reaction is complete by the time the extracting solvent is added and extraction is made. To be sure, no reaction takes place in the aqueous phase if the metal ion associate is completely dissociated in the aqueous phase (and no precipitation occurs), and the formation of the complex would then take place at the organic solvent–aqueous solution interface. This step could conceivably be the rate-determining one for the extraction. However, it appears that usually the extraction rate of the complexes considered here is diffusion controlled. Both molecular and eddy diffusion are involved under analytical conditions, for example, when the two phases are shaken together in a separatory funnel.

Although not dealing with the extraction of metals, a study of the extraction rate of dextromethorphanium ion is of interest here.[2] This bulky cation (RH^+) is a protonated amine. The rate of transfer of R from an aqueous solution to $CHCl_3$, both phases smoothly stirred, was found to be proportional to the stirring rate up to 50 r.p.m., above which marked turbulent mixing presumably occurred. The extraction rate did not depend on the nature of the anion (bromide, iodide, perchlorate, trichloroacetate) and increased only slightly with the anion (bromide) concentration. The conclusion was drawn that R diffused as RH^+, not as an ion associate in the aqueous phase, and that the ion associate is formed at the phase boundary. (This conclusion is not generally valid, since many of the cationic reagents considered in this section and in Chapter 6H form ion pairs, RCl for example, in the aqueous phase; tetraphenylarsonium chloride and Rhodamine B chloride are largely undissociated in aqueous chloride solution.) Finally, the extraction rate was found to be smaller for a mixture of chloroform–carbon tetrachloride than for chloroform (R^+ ion pairs are less soluble in carbon tetrachloride than in chloroform).

The authors mentioned consider their results to be in accord with the classical film theory (p. 917), and calculate an aqueous film thickness of 0.005 cm at a stirring speed of 40 r.p.m. Since extraction in a separatory funnel involves much more vigorous agitation and a more favorable geometric factor (surface area/volume of aqueous phase) than in these experiments, analytical extraction should be very rapid.

C. SOME COMMON ION-ASSOCIATION EXTRACTION REAGENTS

Most of these are cationic in nature and form extractable compounds with metals in anionic form, that is, oxo-anions or complex anions having large size. Anionic organic ion-association reagents are less important analytically. Their use generally requires conversion of the metal into a bulky cation with an organic reagent (chelating agent). Thus Fe^{2+} is converted to $Fe(Phen)_3^{2+}$ with an excess of 1,10-phenanthroline, and this cation is extracted into chloroform after adding a long-chain alkyl sulfate or sulfonate.

1. Quaternary Arsonium, Phosphonium, and Similar Reagents

These include reagents containing the groups Ph_4As^+, Ph_4Sb^+, Ph_4P^+, Ph_3S^+, Ph_3Sn^+, and analogous groups (Ph is phenyl). Tetraphenylarsonium (chloride) is the most used of these reagents and requires discussion. Cationic N reagents are considered under (2). In general, the onium cation concentration is not much affected by hydrogen ion concentration, at least at moderate acidities, but the anion introduced with H^+ may reduce its concentration by forming ion pairs.

a. Tetraphenylarsonium Chloride

i. PROPERTIES OF THE REAGENT. The solubility of $(C_6H_5)_4AsCl\cdot2 H_2O$ in water at 25° is 0.99 M and in $CHCl_3$ (as the anhydrous salt), 0.70 M. The salt can be purified[3] by adding concentrated hydrochloric acid to a saturated aqueous solution to crystallize $(C_6H_5)_4AsCl\cdot HCl\cdot2 H_2O$,[4] dissolving this in water, neutralizing with sodium carbonate, evaporating to dryness, extracting with dry chloroform (to separate from sodium chloride), and finally crystallizing from ethanol by adding ethyl ether.

Tetraphenylarsonium chloride is polymerized in chloroform solution. The variation of the extraction coefficient of the reagent with its concentration in chloroform is best represented by a system consisting of monomer, dimer, and tetramer. The following values have been obtained for the polymerization constants (25°)[5]:

$$\beta_2 = \frac{[(RCl)_2]}{[RCl]^2} = 163 \pm 16 \qquad (R = (C_6H_5)_4As)$$

$$\beta_4 = \frac{[(RCl)_4]}{[RCl]^4} = 4.5 \times 10^5$$

According to these values, a chloroform solution that is 0.010 M in the monomer will be $163 \times (0.01)^2 = 0.0163$ M in the dimer and $4.5 \times 10^5 \times (0.01)^4 = 0.0045$ M in the tetramer. The analytical concentration of tetraphenylarsonium chloride in this solution is $0.010 + 2(0.0163) + 4(0.0045) = 0.061$ M. In other words, a 0.06 M solution of tetraphenylarsonium chloride in chloroform has ~16% of the reagent in monomeric form.

There is no evidence for dissociation of Ph_4AsCl in chloroform solution.[6]

In water, tetraphenylarsonium chloride exists as the monomer, whose ionization constant is

$$K_{mon} = \frac{[R^+][Cl^-]}{[RCl]} = 0.082 \qquad (\mu = 0.1, 25°; \text{ for } \mu = 0, K_{mon} = 0.079)$$

(Thus when $[Cl^-] = 0.1$, $[R^+]/[RCl] = 0.82$ and 45% of the ion pair is dissociated.) The partition constant of the monomer is

$$P = \frac{[RCl]_{CHCl_3}}{[RCl]} = 3.7 \qquad (\mu = 0.1,\ 25°;\ \text{for } \mu = 0,\ P = 3.3)$$

The extraction coefficient of the reagent

$$E_r = \frac{[RCl]_{CHCl_3} + 2[(RCl)_2]_{CHCl_3} + \cdots}{[R^+] + [RCl]}$$

is a function of the RCl concentration as well as the chloride ion concentration. For relatively low concentrations of Cl^- and RCl (up to $0.2\ M$ total in $CHCl_3$ solution), E_r can be calculated from

$$E_r = \frac{P^0_{mon}(1 + 2\beta_2[RCl]_{CHCl_3} + 4\beta_4[RCl]^4_{CHCl_3})}{K^0_{mon}/f_+ f_-[Cl^-] + 1/f}$$

where f_+, f_-, and f are the activity coefficients of R^+, Cl^-, and RCl. This expression may be used for HCl concentrations up to $\sim 0.1\ M$. Above this chloride concentration there is a difference in E_r in NaCl and HCl solutions (Fig. 9C-1). The difference may reasonably be attributed to the formation of such species as $RCl \cdot H^+$ and $RCl \cdot HCl$ in the aqueous phase when HCl is present.[7]

ii. EXTRACTION OF METALS. Most of the tetraphenylarsonium ion-association compounds that are very slightly soluble in water[8] are more soluble in chloroform and similar solvents of low polarity (Table 9C-1). The combination of low water solubility and appreciable solubility in

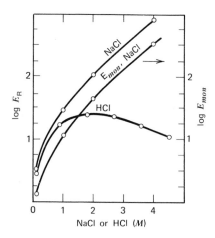

Fig. 9C-1. Extraction coefficient ($CHCl_3/H_2O$) of tetraphenylarsonium chloride in all forms as a function of NaCl and HCl concentrations (initial $\sum[R]_{org} = 0.010$, equal phase volumes), and extraction coefficient of the monomer as a function of NaCl concentration. From J. S. Fok, Z. Z. Hugus, and E. B. Sandell, *Anal, Chim. Acta*, **48**, 248 (1969).

TABLE 9C-1
Solubility of Some Tetraphenylarsonium Compounds in Chloroform and 1,2-Dichloroethane[a]

	Solubility (M)	
	Chloroform	1,2-Dichloroethane
$(Ph_4As)I$	0.2	
$(Ph_4As)AuCl_4$	4.4×10^{-3}	$>2.8 \times 10^{-2}$
$(Ph_4As)MnO_4$	7×10^{-4}	3.1×10^{-3}
$(Ph_4As)ClO_4$	8×10^{-4}	1.1×10^{-2}
$(Ph_4As)ReO_4$	1.1×10^{-3}	6.1×10^{-3}
$(Ph_4As)_2Cr_2O_7$		$>3.5 \times 10^{-2}$
$(Ph_4As)_2IrCl_6$	4.85×10^{-5}	

[a] I. P. Alimarin and G. A. Perezhogin, *Talanta*, **14**, 109 (1967). The value for $(Ph_4As)_2IrCl_6$ is from J. Fok, Ph.D. Thesis, University of Minnesota.

organic solvents results in analytically favorable extraction coefficients. Tetraphenylarsonium illustrates the general rule—to which there are some exceptions—that singly charged bulky cations form the most easily extractable ion-association compounds with large anions of single charge. Chromium(VI) is extracted from acidic solution but hardly at all from basic solution, no doubt because of the presence of $HCrO_4^-$ or $HCr_2O_7^-$ in acidic medium.

The extraction of ReO_4^- as Ph_4AsReO_4 into chloroform is reported to be quantitative over a wide acidity range, from pH 13 to an HCl concentration of 6 M.[9] The extraction expression derived on p. 977 is applicable if account is taken of the polymerization of Ph_4AsCl in chloroform. Tribalat gives the value 10^4 for K in the extraction expression

$$E_m = \frac{[Ph_4AsReO_4]_{CHCl_3}}{[ReO_4^-]} = K \frac{[Ph_4AsCl]_{CHCl_3}}{[Cl^-]}$$

but the true value is smaller because of the polymerization of the reagent. This extraction is of value in separating rhenium from much molybdenum (not singly charged), which is not extracted in the pH range 7–13.

Tantalum as TaF_6^- is extractable with Ph_4As^+ into $CHCl_3$ and can thus be separated from Nb, even when Nb:Ta = 100; this is also a separation from Ti and Zr.[10]

Extraction coefficients of inorganic anions[11] with Ph_4As^+, chloroform as extractant, are listed in Table 9C-2.[12] Most of these values are not of

TABLE 9C-2
Extraction Coefficients (CHCl₃/H₂O) of Anions with Tetraphenylarsonium as Reagent[a]

Anion	E		Anion	E	
	pH 1.5	pH ~12		pH 1.5	pH ~12
F^-	<0.01	<0.01	SO_4^{2-}	<0.001	<0.001
Cl^-	0.22	0.23	SO_3^{2-}		0.015
Br^-	4.6	4.75	$S_2O_3^{2-}$		0.017
I^-	>300	>300	PO_4^{3-}	~0.01	~0.01
SCN^-	34.7	32.3	$P_2O_7^{4-}$		0.003
ClO_3^-	>150	>150	CrO_4^{2-}	71	0.025
BrO_3^-	0.92	0.84	MoO_4^{2-}		<0.005
IO_3^-	<0.004	<0.004	WO_4^{2-}		<0.005
ClO_4^-	>200	>200	VO_3^-		<0.005
MnO_4^-	>300	>300	AsO_3^{3-}	0.006	0.05
ReO_4^-	>200	>200	AsO_4^{3-}	<0.01	<0.01
IO_4^-		0.02	SeO_3^{2-}	<0.01	0.09
NO_3^-	20.3	42	TeO_3^{2-}	<0.01	0.09
NO_2^-		0.22			

[a] R. Bock and G. M. Beilstein, *Z. Anal. Chem.*, **192**, 44 (1963). 50 ml CHCl₃ shaken with 50 ml aqueous solution containing 1/3 meq anion and 2/3 meq $(C_6H_5)_4AsOH$.

direct interest in metal extractions, but are given here because of general interest and because of their bearing on the effect of some common anions on trace metal extractions.

Table 9C-3 lists extraction coefficients of metals in hydrochloric acid solution, again with chloroform as the extractant. These values are based on tracer concentrations of the elements. Most of these metals are extracted as chloro anions, some (Re, Tc) as oxo-anions. It may be noted that E is large (≥ 100) for Ag, Au(III), Hg(II), Re, Tc, and Tl(III) in 2 M HCl. At lower chloride concentrations some improvement in selectivity may be expected. Extraction of Fe(III) is prevented by reduction to Fe(II) [as with Ti(III)].

The extraction of metal halide complexes has not been studied in detail except for $IrCl_6^{2-}$. The extraction of Ir(IV) has some features of interest. First, in spite of its double charge, $IrCl_6^{2-}$ is extractable by Ph_4As^+ into chloroform. Second, $(Ph_4As)_2IrCl_6$ forms an adduct with the reagent,[13] $(Ph_4As)_2IrCl_6·2(Ph_4As)Cl$, in chloroform. The formation of the adduct greatly increases the extractability of $IrCl_6^{2-}$. From the equilibrium constant

$$\frac{[R_2IrCl_6 \cdot 2\,RCl]_{CHCl_3}}{[R_2IrCl_6]_{CHCl_3}[RCl]^2_{CHCl_3}} = 1.6 \times 10^5$$

we find that in a chloroform solution having a monomeric tetraphenylarsonium chloride concentration of $0.01\,M$, the ratio of the two iridium species is

$$\frac{[R_2IrCl_6 \cdot 2\,RCl]_{CHCl_3}}{[R_2IrCl_6]_{CHCl_3}} = 16$$

Additional constants needed for the formulation of the extraction coefficient of iridium (the ionization constant of R_2IrCl_6 in water and its partition constant for $CHCl_3/H_2O$) have been determined.[13] The following expression can be written for the extraction coefficient of iridium from $0.1\,M$ HCl as a function of the monomeric tetraphenylarsonium chloride concentration in the chloroform phase and the total tetraphenylarsonium

TABLE 9C-3
Extraction of Metals from Chloride Solution with Tetraphenylarsonium Chloride in Chloroform $(0.05\,M)^a$

	log E			log E	
	HCl (M)			HCl (M)	
	2	4		2	4
Ag	2.0	1.25	Re	2.2	2.0
Au	2.75	2.75	Ru(III)	−0.6	−0.75
Bi	−0.5	−1.4	Sb(III)	1.6	1.75
Cd	0.3	0.25	Sn(IV)	−0.2	0.2
Fe(III)	0.7	1.4	Tc	2.8	2.4
Ga	1.0	2.75	Tl(I)	−0.9	−1.1
Hg(II)	2.3	1.75	Tl(III)	2.4	2.4
In	−0.3	0.4	V^b	−1.1	−0.2
Mo	−2.4	−0.5	W	0.1	0.1
Os	1.4	1.25	Zn	−0.3	−0.3
Pa	−2.8	−0.4			

^{a} K. Ueno and C. Chang, *J. Atom. Energy Soc. Japan*, **4**, 457 (1962). See the original for values for 8 and 12 M HCl, and for E of elements poorly extracted.

^{b} The chloro complex of V(V) is extracted well from concentrated HCl (chlorate prevents reduction of V) into dichloromethane or -ethane. R. Bock and B. Jost, *Z. Anal. Chem.*, **250**, 358 (1970).

concentration R (RCl + R$^+$) in the aqueous phase[13]:

$$E_{Ir} = \frac{[R_2IrCl_6]_{CHCl_3} + [R_2IrCl_6 \cdot 2\,RCl]_{CHCl_3}}{[IrCl_6^{2-}] + [R_2IrCl_6]}$$

$$= \frac{19\{1 + 1.6 \times 10^5 [RCl]_{CHCl_3}^2\}}{1.4 \times 10^{-8}/\{(\sum [R])^2 + 1\}}$$

E_{Ir} does not depend on the iridium concentration.

The extraction of Ir(IV) is quantitative with a comparatively low concentration of tetraphenylarsonium chloride. With equal phase volumes, the extraction of iridium is >99% complete when a 1% (w/v) chloroform reagent solution is used (0.1 M HCl aqueous phase).

The chloroform extraction of $IrCl_6^{2-}$ with tetraphenylarsonium chloride provides one of the best[14] methods for the separation of iridium and rhodium, not otherwise easily carried out. Because of its high charge, $RhCl_6^{3-}$ is not detectably extracted, so that a trace of iridium is sharply separable from much rhodium.

Iridium(III) as $IrCl_6^{3-}$ is not extractable into chloroform by tetraphenylarsonium chloride. Ascorbic acid reduces $IrCl_6^{2-}$ very rapidly to $IrCl_6^{3-}$, whereas $PtCl_6^{2-}$ is reduced only slowly, so that platinum can be extracted as $(Ph_4As)_2PtCl_6$, whereas iridium is left in the aqueous solution. Microgram quantities of platinum can thus be separated from a hundred times as much iridium.[15]

b. Tetraphenylphosphonium Chloride

$(C_6H_5)_4P^+$ behaves much the same as $(C_6H_5)_4As^+$ in the extraction of anions. For most anions, the extraction coefficients (CHCl$_3$) are very similar to those of the $(C_6H_5)_4As^+$ compounds.[16] The tetraphenylphosphonium ion associates are more soluble in chloroform than the corresponding tetraphenylarsonium associates. The extraction of $Zn(SCN)_3^-$ has been studied.[17] $(C_6H_5)_4PCl$ has found use in the extraction separation of iridium from rhodium. $(C_6H_5)_4PBr$ has been used in the extraction determination of $IrBr_6^{2-}$.

Tetraphenylphosphonium chloride is polymerized in chloroform solution.

Methyltriphenylphosphonium chloride has been recommended for the extractive separation of Ir(IV) and Os(IV) from Rh(III).[18] Ru(IV) is largely extracted. Chloroform is the solvent.

c. Other Onium Reagents

These do not seem to offer much of interest in metal separations. In general, they resemble Ph$_4$As$^+$ in their extraction behavior, but there are occasional marked differences. Thus F^- is extracted into CCl_4 by Ph_4Sb^+ but not by Ph_4As^+; Ph_3Sn^+ does not extract ReO_4^- into benzene from acidic solution. Some references to these reagents are appended.[19]

2. Amines of High Molecular Weight

Since about 1950, salts of high-molecular-weight (long-chain) amines, particularly of secondary and tertiary amines, and quaternary ammonium salts have become important extraction reagents for metals.[20] The bulky substituted ammonium cation of these reagents forms ion-association compounds with complex metal anions (e.g., halo, nitrato, sulfato) and oxo-anions, which are extractable into immiscible organic solvents. Both inert (and semi-inert) and oxygenated solvents such as MIBK find use in these extractions. The extraction of metals by amine reagents cannot always be represented by simple mass-action law expressions. The amine salt is usually aggregated in the organic solvent and often in the aqueous solution (micelles may be formed), the metal salt may be polymerized, and other complexities may be encountered, so that the extractability often cannot be formulated as simply as when $(C_6H_5)_4As^+$, for example, is the extraction reagent. In fact, the composition of the extracted species is not always clear. The extraction coefficient may not be independent of the metal concentration. High ionic strengths may complicate interpretations. Moreover, commercially available long-chain amines are often mixtures of isomers and similar amines and are frequently contaminated with foreign, undesired amines and even nonamine organic compounds. This may sometimes confuse the issue.

Extractions of complex metal anions from aqueous solution by amine salts in an immiscible organic solvent, for example,

$$\underset{org}{R_3NHA} + \underset{H_2O}{MA_{n+1}^-} \rightleftharpoons \underset{org}{R_3NH(MA_{n+1})} + \underset{H_2O}{A^-}$$

(in which A^- could be Cl^-) are analogous to the reactions occurring with anion-exchange resins, and the amine salts are sometimes called liquid ion exchangers, but the analogy should not be pushed too far.[21]

A considerable number of high-molecular-weight amines are commercially available. Some of these are:

Alamine 336	"Tricaprylamine." Tertiary amine with three straight-chain alkyl groups (mainly octyl and decyl). 90–95% tertiary amine, average mol. wt. 392.
Aliquat 336	"Tricaprylmethylammonium chloride." Alkyl groups mainly octyl and decyl, the former predominating; average mol. wt. of $R_3CH_3NCl \sim 450$.

Amberlite LA-1	*N*-Dodecenyl(trialkylmethyl)amine. 24–27 C atoms, average mol. wt. 372.
Amberlite LA-2	*N*-Lauryl(trialkylmethyl)amine. A saturated secondary amine having an average mol. wt. of 374.
Primene	A primary trialkylmethylamine having 18–24 C atoms.
TBA	Tribenzylamine (solid). ~99% pure.
TIOA	Triisooctylamine. Mixture of dimethylhexyl, methylheptyl, etc., chains. Liquid.
TLA	Trilaurylamine.
TNOA	Tri-*n*-octylamine.

Not only are these commercial amines usually mixtures but they contain more or less inert material. Aliquat 336 is reported to contain at least 88% of the active ingredient.

A multitude of factors affect the extractability of metals with high-molecular-weight amines, including the structure of the amine (degree of substitution of H on N, length of carbon chain, extent of branching of chain, polymerization), the nature of the organic solvent (diluent), and, of course, the anion complex of a metal.

Although salts of primary, secondary, and tertiary amines find use in metal extractions, the last and quaternary ammonium salts are more commonly used, and these will be considered in the following few pages. This treatment is in the nature of a general sketch and references are illustrative. The use of some reagents of this class in the extraction of metal chlorides, nitrates, and the like is discussed in Chapter 9A.

Equilibrium is usually attained rapidly in these extractions. Shaking for 1 or 2 min is often sufficient.

The vapors of amines should not be inhaled, nor should they be allowed to come in contact with the skin.

a. Tertiary Amines

These are used as extractants for metals from HCl, H_2SO_4, HNO_3, acetic acid, and other acidic solutions, in which the metals form anionic complex species combining with bulky R_3NH^+ to give ion associates extractable into a variety of organic solvents.

HCl Solution. Metal extractability varies with the amine and with the organic solvent and extensive investigations of various systems have been made. For example, a study of the effect of amine structure on the extractability of zirconium into benzene from HCl solution revealed that tertiary amines were better extractants than secondary amines, and that among tertiary amines extractability increased with the chain length of the alkyl group and with branching of the alkyl chain.[22] The presence of a benzyl group in the molecule was found to increase the extractability of zirconium.

Such metals as Au, Bi, Pt, Ir(IV), Zn, Cd, Co, Ag, Fe(III), Ga, In, and U(VI) are extractable from HCl solutions of suitable concentration. Triisooctylamine is a tertiary amine often used in such extractions. See Table 9A-8 for a listing of extraction coefficients of metals most easily extracted from HCl solution by this reagent into xylene. Many metals show a maximum in the plot of the extraction coefficient versus the HCl concentration. This behavior is illustrated in Fig. 9C-2, in which log E for Zn^{2+} is plotted vs the HCl concentration for triisooctylamine–xylene and also, for comparison, for a primary and secondary amine and for a quaternary ammonium salt. The decrease of E at higher HCl concentrations has been attributed to the combination of HCl with R_3NHCl to form $R_3NHCl \cdot HCl$ or $R_3NHCl_2^-$. Another possibility is the formation of slightly ionized complex chloro acids at higher HCl concentrations. When an alkali chloride is used in place of HCl to supply chloride ion, the maximum disappears, at least with some metals. A maximum could be obtained with increasing alkali metal chloride concentration if a singly charged metal anion were being converted into a less easily extractable doubly charged anion.

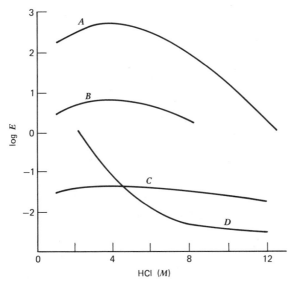

Fig. 9C-2. Extraction coefficients of zinc from hydrochloric acid solutions with Primene JM-T in xylene, 0.28 M (Curve C); Amberlite LA-1 in xylene, 0.20 M (Curve B); triisooctylamine in xylene, 0.11 M (Curve A); and benzyldimethylphenylammonium chloride in CHCl$_3$, 0.1 M (Curve D). Based on T. Ishimori and E. Nakamura, JAERI Rept., **1047** (1963).

Differences in the stability of metal chloride complexes and in the charges on the anions are the chief reasons for differences in metal extractabilities. Thus $\log E$ with triisooctylamine–xylene (0.11 M) and 1 M HCl is 2.2 for Zn (2.1 for Cd) and ~ -3 for Ni and Co, indicating that Zn (and Cd) can easily be separated from Ni and Co.[23] The extractability of anionic metal chlorides is not limited to those having a single negative charge, but these are preferentially extracted. For example, $AuCl_4^-$ has an E lying between 10^4 and 10^5 when extracted from 4 M HCl by triisooctylamine–xylene. $GaCl_4^-$ likewise has a large extraction coefficient. U(VI) can be extracted from 4 M or stronger HCl (separation from Th),[24] but many other metals are also extracted from chloride medium. Niobium is extractable from strong HCl (separation from Ta).[25] Some doubly charged chloro anions are extracted quite well ($IrCl_6^{2-}$, for example), but triply charged anions little if at all (chloro complexes of Rh, for example).

The effect of a variation in the amine salt concentration on the extraction coefficient in extractions from chloride solutions is often not what would be expected at first sight. Thus E for Fe(III), extracted as $FeCl_4^-$, often shows a second-power dependence on the tertiary amine salt concentration. Apparently, species such as $(R_3NH)^+(FeCl_4)^- \cdot (R_3NH)^+Cl^-$ can be extracted.[26] Indium and gallium behave similarly.

As a matter of interest, but hardly of importance from the standpoint of separations, it may be mentioned that nickel can be extracted into a cyclohexane solution of Alamine 336 (and other amines) from aqueous LiCl (LiBr, LiI) solutions of high concentration.[27] The free acidity should be low. The extracted species is $(R_3NH)_2NiCl_4$.

HNO$_3$ Solution. Uranium(VI), Th, Pu(IV, VI) Np(IV), and the rare earth metals are extractable by amines from nitrate solutions.

A study of the extraction of thorium from HNO_3 solution with benzene solutions of long-chain aliphatic amines showed that extractability decreased as follows with the amine: *tris*-(2-ethylhexyl)amine > tri-*n*-dodecylamine $\sim$ tri-*n*-octylamine > di-*n*-octylamine.[28] Branching of the alkyl chain increases the extraction of thorium. Maximum extractability occurs at an initial HNO_3 concentration of ~ 9 M. Extractability depends strongly on the amine solvent. Thus the extraction coefficient has the following values with 0.1 M tri-*n*-octylamine in various solvents when the aqueous phase is 6 M in HNO_3:

	E (rounded)		E (rounded)
Kerosene	4.6	Toluene	0.47
Benzene	0.4	CHCl$_3$	0.028
n-Hexane	3.7		

Investigation of the amine extraction of U(VI) and Pu(IV) from nitrate solutions into MIBK or CCl_4 led to the conclusion that extractability increases with the basicity of the amines and with the polarity of their salts.[29] The effect of the carbon chain length is comparatively small. With tertiary amines, maximal extraction occurs with C_9 amines (with secondary amines, C_{13} or C_{14} is best).

Other Acids. Amine extractions become more selective when carried out in sulfuric acid solution, since fewer metals form anionic sulfate than anionic chloride complexes (see Table 9A-20 for extraction coefficients with triisooctylamine–xylene in sulfuric acid medium). Most of the metals of the periodic groups IVA to VII are extracted well at suitable sulfuric acid concentrations. A number of these metals are no doubt extracted as oxo-anions (Mo, W, V, Re, Tc). Triisooctylamine in xylene extracts U(VI), Zr, and Nb rather well. The extraction of U is of chief interest. Thorium is extracted to a much lesser extent so that U and Th can be separated. Tricaprylamine is also used for this purpose.

Extractions from thiocyanate solutions are also useful. Niobium can be separated from Ta by extraction of the former with tributylamine salt.[30] Isoamyl acetate is the extracting solvent.

Amberlite LA-2 in benzene extracts Ta but not Nb from HCl–HF solution.[31]

b. Quaternary Ammonium Salts

Various reagents of this type are used for the extraction of metals as complex anions from chloride (less often, bromide and iodide), nitrate, thiocyanate, and occasionally other solutions. When necessary, they can be used for extractions from basic medium (the quaternary ammonium salts correspond to a strong-base solid ion exchanger). When the complexing anion is added as an acid, many metals show a maximum in the E versus [acid] curve, which may not appear if the anion is added as a neutral salt.[32] With a particular quaternary ammonium salt, extractability broadly follows the trend of the stabilities of the complex anions. The preferred extraction of singly charged anions is evident, but extraction is not confined to these.[33] The extractability of a given metal varies with the quaternary ammonium cation, that is, with the number of carbons in the chains and the structure of the chains; substitution of aromatic groups for alkyl groups may improve metal extraction. Extracting power generally increases with the molecular weight of the reagent, at least up to a certain point. The nature of the extracting solvent also affects extractability.

Some of the quaternary ammonium salts that have been, or are, used analytically are mentioned. Tetrapropylammonium cation extracts U(VI) well from an $Al(NO_3)_3$ solution into MIBK; the extracted species is (prop)$_4$N[UO$_2$(NO$_3$)$_3$].[34] The extraction is fairly selective. Extensive data have been presented for the extraction of metals by the tetraalkylamine cations (propyl)$_4$N$^+$, (butyl)$_4$N$^+$, and (hexyl)$_4$N$^+$ into MIBK from OH$^-$, NO$_3^-$, SO$_4^{2-}$, F$^-$, and Cl$^-$ solutions.[35] Extraction rises with increasing molecular weight of amine and with increasing dielectric constant of the solvent.

Extraction of U(VI) from acid-deficient Al(NO$_3$)$_3$–tetrapropyl-ammonium nitrate solution into n-amyl acetate containing 2% TBP is useful for its separation from complex mixtures.[36]

Tricaprylmethylammonium ion (the chloride salt is Aliquat 336) finds use in the extraction of U and Th from nitric acid solutions. The extraction coefficient of U(VI) attains a maximum value of about 15 in 7 M HNO$_3$ when a 0.67 M reagent in Solvesso-100 (an aromatic solvent) is used.[37] Thorium is extracted more easily, E being ~50 in 2–5 M HNO$_3$ with 5% tricaprylmethylammonium nitrate in xylene.[38] U(VI) is also extracted ($E \sim 2$, 4 M HNO$_3$), but on back-extraction with 3 M HCl it enters the aqueous phase only slightly ($E \sim 30$), whereas Th is easily stripped ($E < 2 \times 10^{-3}$). Cerium(IV), obtained by bromate oxidation of Ce(III), is extracted well into xylene from 2 M HNO$_3$ as (R$_3$CH$_3$N)$_2$Ce(NO$_3$)$_6$ (separation from Bk, trivalent lanthanides–actinides, Fe, Zr, and Nb).[39] UO$_2^{2+}$, NpO$_2^{2+}$, and PuO$_2^{2+}$ accompany Ce(IV), but they can be separated by stripping Ce from the xylene phase with strong HCl, which reduces Ce(IV). The difference in extraction behavior of Ce(IV) and Bk(IV) is believed to be due to the smaller ionic radius of the latter, which leads to easier hydration. Primary, secondary, and tertiary amines are ineffective extraction reagents for cerium.

Tricaprylmethylammonium cation in acidic thiocyanate solution serves for the separation of the trivalent actinide elements from the lanthanides.[40] The former group of elements is extracted into xylene to a larger extent, as the thiocyanates, than the latter. The thiocyanate system is superior to the HCl–LiCl system (with trioctylamine)[41] used for the same purpose.

Quaternary ammonium salts also form ion-association compounds with anionic complexes of metals (particularly of the trivalent actinide–lanthanide group) with various organic acids, including EDTA (H$_4$Y). Thus tricaprylmethylammonium cation reacts with M^{III}Y$^-$ to give the ion pair (R$_3$CH$_3$N)MY, which can be extracted into xylene and MIBK. These extractions are of potential value in separations.[42] The extraction of the 1:2 complexes formed by Am and Cm with hydroxyethylethylene-diaminetriacetic acid into a benzene solution of tricaprylmethyl-ammonium chloride has been studied with a view to the separation of the two elements.[43]

The anionic complexes of Ti with ascorbic acid can be extracted with trioctylmethylammonium chloride into $CHCl_3$ at pH 3.7 from sulfate solutions.[44] Chloride and nitrate must be absent. Aluminum begins to be extracted at pH 4, iron (III→II) at pH 4.5, so that separation of Ti from these is possible if ratios are not unfavorable. Uranium and more or less V accompany Ti.

Tetra-n-butylammonium cation extracts magnesium in the presence of 8-hydroxyquinoline at pH ~11 into $CHCl_3$. The ion pair $(R_4N)(MgOx_3)$ is the extracted species. UO_2^{2+} forms the analogous species with a quaternary ammonium cation. Butylamine and other organic bases giving cationic species function similarly (Chapter 6D).

Trioctylmethylammonium chloride (Aliquat 336-S) in $CHCl_3$ extracts Cr(VI) from H_2SO_4 solution.[45] Other elements extracted include Mo(VI), U, V(V), Au, and Pt metals. The same reagent in diisobutyl ketone extracts Au (I and III) from basic cyanide solution.[46]

Tetra-n-hexylammonium chloride in ethylene dichloride readily extracts Cu(I) from 1 M Cl^- solution as R_4NCuCl_2.[47] Lead is also extracted as a singly charged chloro complex R_4NPbCl_3. Contrary to the behavior of copper, the extractability of lead is strongly dependent on its concentration. Thus with a phase ratio of 1:1, HCl = 0.58 M and $[R_4NCl]_{org} = 0.1$, the extraction coefficient varies as follows with the lead concentration:

$[Pb]_{tot} \times 10^3$	1.74	0.87	0.058	0.029
E	55	11	0.17	0.003

Tetrahexylammonium erdmannate (the anion is $Co(NH_3)_2(NO_2)_4^-$) in hexone (MIBK) readily extracts the singly charged complex cyanides $Ag(CN)_2^-$, $Au(CN)_4^-$, $Hg(CN)_3^-$; poorly extracts doubly charged $Zn(CN)_4^{2-}$, $Cd(CN)_4^{2-}$, $Hg(CN)_4^{2-}$, $Ni(CN)_4^{2-}$, $Pd(CN)_4^{2-}$, $Pt(CN)_4^{2-}$; and $Cu(CN)_4^{3-}$, $Fe(CN)_6^{3-}$, $Fe(CN)_6^{4-}$, and indium cyanide not at all.[48] See further p. 877.

3. Other Cationic Reagents Derived from Organic Bases

We mention here some reagents that, while not amines, are similar to the reagents in the preceding section and that owe their cationic character to basic N (mostly) and usually are of the type RH^+. The number of such compounds is large, but systematic investigations of their use are small. They often allow extractive separations to be made in mineral acid solutions, which frequently is an advantage. Metals are usually extracted in the form of anionic halo complexes.

In recent years, diantipyrinylmethane (Chapter 6H) and related compounds have been used considerably in separations by ion-associate extraction. Diantipyrinylmethane is a weak base that can add one or two

H^+ to give cationic forms. These cations form ion associates with metal halo and thiocyanato anions that are extractable into $CHCl_3$. Thus Sb(III), Bi, Zn, Cd, Te,[49] Cu(II), Ga, In, and Tl(III) can be extracted from 2.5–3 M HCl, and Sb(III), Bi, Cd, Pb, Sn(IV), and In can be extracted from 3% KI solution acidified with H_2SO_4.[50] Separations of these metals from those (such as Al, Mo, W, Nb, and Ta) not forming halo complexes under the conditions can thus be effected. Fe(III) is reduced with ascorbic acid to prevent its extraction. Anionic metal nitrate complexes, as of Th,[51] are also extractable into $CHCl_3$ by this reagent. Sc and Bi are extracted from 0.2–0.3 M HNO_3, Hf from 6 M HNO_3. Procedures for the separation of Zr and Hf from various elements have been developed.[52]

1,1-Diantipyrinylbutane has been used to separate Pt and Ir. Platinum(IV) is extracted from an acid iodide solution by $CHCl_3$, whereas Ir (reduced to III) remains in the aqueous phase.[53] After removal of iodide, Ir(IV) can be extracted from HCl solution by the reagent into dichloroethane. Diantipyrinylheptane is a better extraction reagent than its homologs, as for Zn and Cd in HCl medium.[54]

Diphenylguanidine, triphenylguanidine, and phenazone (antipyrine) are some other organic bases that have been applied in extractive separations.

The proper choice of base may contribute to the selectivity of ion associate extractions. For example, weak bases such as diantipyrinyl-methane, phenazone, and tribenzylamine (all having $pK_b > 9$) form chloroform-extractable ion associates with Cd but not with Zn in acid solutions containing iodide.[55] A stronger base such as pyridine gives an extractable ion-pair complex with both Zn and Cd in weakly acidic solution. Diphenylguanidine (strong base) forms complexes with Cd in acid iodide and even basic solutions, but with Zn only at pH >6. Thiocyanate is much less effective than iodide for Zn–Cd separations. In acid Cl^- solutions, neither Cd nor Zn is extracted with weak bases; both Cd and Zn are extracted from weakly acidic or basic Cl^- solutions.

4. Anionic Reagents

These are mainly of interest for the separation (or attempted separation) of the alkali metals.

Tetraphenylborate. The extraction of K, Rb, and Cs tetraphenylborates, with Na tetraphenylborate as reagent, into nitromethane, nitroethane, nitrobenzene, MIBK, and TBP has been studied.[56] Nitrobenzene is the best extractant for K, Rb, Cs. The extractability increases in that order. With an aqueous solution that is 0.1 M in $NaClO_4$ and 0.01 M in Na tetraphenylborate, log E is, respectively, 0.89, 1.60, and 2.48 for K, Rb, and Cs, so that there is a possibility of separation by fractional

TABLE 9C-4
Extraction of Alkali Metals as Salts of Tetrathiocyanato-diamminechromate(III) and Tetrathiocyanatodianiline-chromate(III)[a]

(Equal volumes of 10^{-4} M alkali metal ion and 4×10^{-3} M reagent in water and of organic solvent; reagent added as Ca salt.)

Reagent	Solvent	Percentage Extraction			
		Na	K	Rb	Cs
TAnCr⁻	Nitromethane			95.2	98.9
TAmCr⁻	Nitromethane	26.6	82.3	94.6	98.2
TAnCr⁻	Nitrobenzene	5.6	75.1	94.3	99.1
TAmCr⁻	Nitrobenzene	1.5	17.2	70.5	91.6

[a] T. Fujinaga, M. Koyama, and O. Tochiyama, *Anal. Chim. Acta*, **42**, 219 (1968). Also see *J. Radioanal. Chem.*, **3**, 413 (1969).

extraction. The reagent is itself extracted to a large extent. The extraction behavior of these elements indicates that all the tetraphenylborates are completely dissociated in water and in nitrobenzene. E of K, Rb, and Cs is not much affected by the Na tetraphenylborate concentration for low concentrations of these metals. Addition of Na as NaCl, that is, in a form other than Na tetraphenylborate, decreases the extraction of K, Rb, and Cs. Compare the expressions for extraction of ion-pair compounds of the type MR, p. 981.

Other reagents include dipicrylamine[57] (K, Rb, Cs) and 4-*sec*-butyl-2-(α-methylbenzyl)-phenol[58] (cyclohexane solution extracts Cs and Rb from NaOH solutions). For still other alkali metal extraction reagents see Table 9C-4. The extraction of Cs and Rb as polyiodides has already been noted in Chapter 9A. Extraction of tetrachloroaurates and tetra-chloroferrates by nitrobenzene and 2-nitropropane has been tried.[59]

5. Miscellaneous

Cationic chelates rarely find use in ion-pair systems for separatory extractions. Possibilities with 1,10-phenanthroline have been indicated. Nitrobenzene is a good extractant for the ion-association complexes of perchlorate with 1,10-phenanthroline cations of Mn, Fe(II), Co, Ni, Cu(II), Ag, Zn, Cd, and Pb; Tl(I, III) is extracted moderately well, Mg, Ti(III), V(IV), Cr(III), Mo(IV), Sn(II), and Bi poorly or not at all.[60] Extractions are necessarily made from weakly acidic or neutral solutions.

II. ORGANOPHOSPHORUS REAGENTS

Since about 1950, organic extractants for metals based on phosphorus have been investigated extensively, especially in connection with nuclear technology. The analytical applications are important, and they are reviewed here from the general standpoint. The literature is voluminous, and it cannot be examined in detail. Only a brief survey is attempted.[61]

The organophosphorus extraction reagents can be divided into those not containing hydrogen atoms replaceable by a metal atom, and those containing –OH groups, the acidic hydrogen of which is metal replaceable. The former react with metal salts or metallo acids to give coordination complexes extractable into the diluent solvent in which the reagent is usually dissolved. The latter can give chelates or chelates with the uncharged reagent as an adduct.

An example of a reagent of the first type is tri-n-butyl phosphate (TBP), which was used as early as 1949 by Warf for the extraction of U(VI), Th, and Ce(IV) from nitric acid solution. This reagent and others of the same type bond to a metal salt through the phosphoryl group to give a coordinatively solvated salt such as $UO_2(NO_3)_2(TBP)_2$ of definite composition.

Successive replacement of –OG in a trialkyl phosphate by –G (or –G′, –G″) increases the polarity of the O–P bond, and the extractability of a metal generally rises with increasing solvating power in the order

$$\underset{\substack{\text{Dialkyl alkyl} \\ \text{phosphonate}}}{OP{\overset{\displaystyle OG}{\underset{\displaystyle G}{{\Large<}}}}OG} \quad < \quad \underset{\substack{\text{Alkyl dialkyl} \\ \text{phosphinate}}}{OP{\overset{\displaystyle OG}{\underset{\displaystyle G}{{}}}}G} \quad < \quad \underset{\substack{\text{Trialkyl phos-} \\ \text{phine oxide}}}{OP{\overset{\displaystyle G}{\underset{\displaystyle G}{{}}}}G}$$

The trialkyl phosphine oxides are especially powerful extraction reagents for U(VI), Th, and the like from 1 M or stronger nitric and hydrochloric acid solutions (see TOPO below). Bu_3PO gives an extraction coefficient larger by a factor of ~1000 than does $(BuO)_3PO$ for $UO_2(NO_3)_2$ in 1 M HNO_3.

Replacement of an alkyl group by a phenyl or other electronegative group in these compounds usually leads to a decrease in extractability (and possibly an increase in selectivity) because of the resulting decrease in the basicity of the $OP{<}$ group. Steric effects also play a role in extractability and selectivity.

Figure 9C-3 illustrates the wide range of extractability of $UO_2(NO_3)_2$ from 1 M HNO_3 with various neutral organophosphorus compounds.

Acidic organophosphorus reagents can be monobasic or dibasic (if mononuclear):

$$OP \overset{\displaystyle OH}{\underset{\displaystyle Z}{-Z}} \qquad OP \overset{\displaystyle OH}{\underset{\displaystyle Z}{-OH}}$$

where Z represents –OG or –G. The monobasic reagents (di-esters) are more important analytically. Dibutylphosphoric acid

$$OP \overset{\displaystyle OH}{\underset{\displaystyle OBu}{-OBu}}$$

was one of the earliest monobasic acids to be used, but di-(2-ethylhexyl)phosphoric acid (HDEHP)

$$OP \overset{\displaystyle OH}{\underset{\displaystyle OC_6H_{12} \cdot C_2H_5}{-OC_6H_{12} \cdot C_2H_5}}$$

is now more often used as reagent. Both of these dialkyl phosphoric acids are usually dimeric in nonpolar media and monomeric in alcohols. See, further, the following discussion of HDEHP.

The extent of dimerization of such acids as diethyl- and dibutylphosphoric[62] is large in inert solvents such as carbon tetrachloride and kerosene but decreases in polar solvents. In the latter type of solvent (ethers, esters, ketones and alcohols), interaction occurs between the

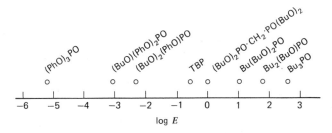

Fig. 9C-3. Extraction coefficients of $UO_2(NO_3)_2$ (tracer concentrations of U) for organophosphorus esters and 1 M HNO_3 according to E. Glueckauf, *Ind. Chim. Belge,* **23,** 1215 (1958).

organophosphoric acid and the solvent. This association increases with the basicity of the solvent and the order is

$$\text{kerosene} < CCl_4 < C_6H_6 < CHCl_3 < \text{ethers} < \text{ketones} < \text{alcohols}$$

The order of increasing dimerization is the reverse. It appears that $1:1$ associates are formed by dialkyl phosphoric acids with $CHCl_3$, ethers, and ketones.

Binuclear acidic organophosphorus reagents of the type

$$(HO)X(O)P—Y—P(O)X(OH) \qquad Y = \text{alkyl chain or O atom}$$

have also received attention as metal extractants. See Section II.D.

Extraction rates with organophosphorus reagents range from slow (hours) to moderately rapid (minutes). Iron(III), aluminum, and beryllium are extracted slowly by chelating organophosphorus esters. A kinetic study of the extraction of Fe(III) with di-isopentylphosphoric acid led to the conclusion that the rate-limiting step in the extraction occurs at the interface of aqueous and organic solvent phases (compare TBP, p. 1002) and that the transfer of reagent to the aqueous phase or the reaction in the aqueous phase are not rate determining.[63] The extraction rate is decreased by very small amounts of impurities. See further, HDEHP (C below).

A. TRIBUTYL PHOSPHATE (TBP)

$$O=P{\overset{\displaystyle O—C_4H_9}{\underset{\displaystyle O—C_4H_9}{—O—C_4H_9}}}$$

Mol. wt. 266.3. Colorless, viscous liquid; b.p. 154°–157° (10 mm Hg), 290° at 1 atm; sp. gr. 0.976 (25°/25°). Solubility in water ~0.5 ml/liter (25°); solubility of H_2O in TBP~7%. Miscible with most organic solvents, and such mixtures (e.g., with CCl_4, kerosene, iso-octane, benzene, and other aliphatic or aromatic hydrocarbons) are often used analytically to vary the extracting power and to improve physical properties of TBP as extractant. Polar diluents are usually less suitable. Chloroform solutions of TBP (and other organophosphorus reagents) have reduced extraction ability because of the formation of hydrogen bridges between the molecules of the two solvents. TBP is not, in general, an ionizing solvent.

Purification. Several procedures have been described for the purification of TBP (removal of organic pyrophosphates, mono- and dibutyl phosphates, and butanol). In one procedure,[64] 0.4% sodium hydroxide solution is added, the mixture is boiled, the TBP is then repeatedly washed with water, and finally dried under vacuum. In another procedure,[65] TBP is stirred with 6 M HCl at 60° for 12 hr and is then washed repeatedly with water, 5% aqueous Na_2CO_3 solution and water. Finally, it is dried at 30° under vacuum.

The use of TBP for the extraction of actinide and other elements from nitric and hydrochloric acid solutions, and occasionally from other media, has been surveyed in Chapter 9A. The extraction of U(VI) from a nitrate solution with an "inert" organic solvent solution of TBP (in which it is taken to be monomeric) is represented by the equation

$$UO_2^{2+} + 2 NO_3^- + 2 TBP \underset{org}{\rightleftharpoons} (UO_2)(NO_3)_2(TBP)_2$$

(*org* indicates the organic solvent phase). The extraction equilibrium constant for this reaction is

$$K_{ext} = \frac{[UO_2(NO_3)_2(TBP)_2]_{org}}{[UO_2^{2+}][NO_3^-]^2[TBP]_{org}^2}$$

If U(VI) is present almost entirely as UO_2^{2+} in the aqueous phase, the ratio of concentrations of uranium in the two phases at equilibrium should be given by

$$E \cong \frac{[UO_2(NO_3)_2(TBP)_2]_{org}}{[UO_2^{2+}]} = K_{ext}[NO_3^-]^2[TBP]_{org}^2$$

under otherwise constant conditions, for example, constant ionic strength. Actually, the distribution of uranium between an aqueous and a CCl_4 or kerosene phase conforms quite well to this expression, E increasing as the square of the nitrate concentration in the aqueous solution and as the square of the TBP concentration in the organic solvent at equilibrium.[66] When, as with thorium, the cation exists to a considerable extent as nitrate complex species in the aqueous solution, E is a more complex function of the nitrate concentration. The rare earth nitrates give trisolvates with TBP.

In nitrate extractions, salting-out nitrates (not extracted into the organic solvent) will have a large effect on E, since, depending on the metal, the nitrate ion concentration enters into the extraction expression as the second, third, or fourth power.

TBP can bond to a metallo acid as well as to a metal salt. For example, Fe(III) forms $H[FeCl_4(TBP)_2]$ (from 6 M HCl), as well as $FeCl_3(TBP)_3$ (from 2 M HCl). A change in the hydrogen-ion concentration will have an appreciable effect on the extractability of a metal in such a case. Acidity can also affect extractability by repression of hydrolysis. Moreover, mineral acids combine with TBP. Nitric acid forms $TBP \cdot HNO_3$, and thus lowers the concentration of TBP available for solvation of the metal complex. Solvated acido complexes such as $H(UO_2)_2(NO_3)_3$ are extracted better than solvated neutral salts such as

$UO_2(NO_3)_2$. In practice, TBP extractions are made from fairly strongly acid solutions containing a salting-out agent (the anion of the salt being the same as the anion of the acid).

Although TBP is quite resistant to hydrolysis in acid solutions, on longer contact hydrolysis products can be formed that alter its extractive properties (synergic effects) or which may make stripping of metals from the organic phase more difficult. Thus dibutyl phosphoric acid, formed by hydrolysis, increases the extraction of zirconium.[67] Nitric acid of 70% concentration is reported to attack TBP slowly at room temperature. A few tenths of a milligram of orthophosphate are said to be formed from 100 ml TBP on extraction of 6 M HNO_3 solutions.

The rates of extraction of U(VI) and Pu(IV) by TBP solutions have been interpreted on the basis of formation of interfacial complexes.[68] These gradually add TBP and NO_3^-. In other words, extraction rate is determined by chemical reaction at the interface. The initial transfer rates of extraction of Pu(IV) from 3 M nitric acid into dodecane solutions of TBP conform well to the relation

$$v = \frac{280 \times 10^{-3}[Pu]_{aq}}{1 + 35.0[TBP]_{org}^2 + 7.1[Pu]_{aq}}$$

v is expressed as moles/cm^{2}sec.

Triphenyl Phosphite.[69] As a neutral organophosphorus reagent, this ester

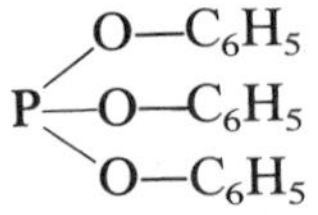

$$P \begin{matrix} O-C_6H_5 \\ O-C_6H_5 \\ O-C_6H_5 \end{matrix}$$

may be included here. Dissolved in CCl_4 it extracts Cu(I) halides by forming addition compounds $(RO)_3P:CuX$. Copper(II) can be reduced with ascorbic acid (60°, dilute HCl). The extraction coefficient for copper is ~100 in 0.1–1 M HCl with 5% triphenyl phosphite in CCl_4 at 25°. Extraction tends to be slow and 30 min shaking should be allowed for HCl solutions, 10 min for alkali chloride solutions. Copper can be stripped from the CCl_4 phase with ammonia (a few minutes shaking is sufficient). The copper extraction is highly selective. Gold(I) is also extracted, but gold can be reduced to metal by ascorbic acid. Thorium is poorly extracted ($E \sim 0.4$), but it and traces of other metals can be washed out of the CCl_4 extract with 1 M HCl (0.1% loss of copper, equal volumes of phases) or water. Although available photometric reactions are selective for copper, separation of traces of Cu from large amounts of such metals as Hg, Zn and the platinum metals may be required at times, and this reagent would appear to be suitable.

B. TRI-*n*-OCTYLPHOSPHINE OXIDE (TOPO)

$$OP \overset{\diagup C_8H_{17}}{\underset{\diagdown C_8H_{17}}{\overline{}} C_8H_{17}}$$

Mol. wt. 386.65, m.p. 51°–52°, b.p. 200° (0.1 mm Hg). Soluble in hydrocarbons and most other organic solvents. Cyclohexane, *n*-octane, kerosene, C_6H_6, and CCl_4 are often used as solvents. Commercial product said to be useable without purification.

TOPO is the most commonly used trialkyl phosphine oxide. Its solution in "inert" solvents extracts many metals in the III, IV, V, and VI oxidation states from hydrochloric and nitric acid solutions, mostly as coordinatively solvated chlorides and nitrates (see Chapter 9A). Typical extracted salts are $UO_2Cl_2(TOPO)_2$ and $UO_2(NO_3)_2(TOPO)_2$. Trivalent lanthanides and actinides are extracted as $M(NO_3)_3(TOPO)_3$.[70] TOPO is of special value in the extraction of U, Th, and other actinides.[71] Extraction of these metals is more selective from HNO_3 than from HCl solutions. From HCl solution some metals are extracted as ion pairs of complex chloride anions and $TOPO \cdot H^+$. Fe(III) is extracted as $(TOPO \cdot H^+)(FeCl_4^-)$ from 7 M HCl by a benzene solution of TOPO when [TOPO]/[Fe] is high.[72] The extraction of $AuBr_4^-$ into chloroform[73] must have an analogous explanation.

Uranium is also extractable from sulfuric acid solution (as $UO_2SO_4(TOPO)_2$), but this extraction is not important analytically. Some metal extractions from thiocyanate and perchlorate solutions have been reported. From 3 M $HClO_4$ containing F^-, tantalum appears to be extracted as $TaF_4ClO_4(TOPO)_2$ into $CHCl_3$.[74]

TOPO in an organic solvent extracts mineral acids from aqueous solution, the order of extraction being $HNO_3 > HClO_4 > HCl$ up to 2 M acid. The formation of $TOPO \cdot HNO_3$ and $TOPO \cdot 2 HNO_3$ leads to a decrease in the extraction coefficient of metals above a certain HNO_3 concentration.[75] Indifferent metal nitrates are, therefore, used to increase the extractability of metals by salting-out and mass-action effects.

The extractive properties of TOPO can be strongly affected by small amounts of organophosphorus impurities (synergic effects).

The synergic and antagonistic effects of carboxylic acids on the extraction of U(VI) with TOPO have been studied.[76]

The extraction of lanthanide nitrates, chlorides and perchlorates by methylenebis(di-*n*-hexylphosphine oxide) in CCl_4 or 1,2-dichlorobenzene has been examined.[77] The nitrates form trisolvates at sufficiently high reagent concentrations.

C. DI-(2-ETHYLHEXYL)PHOSPHORIC ACID (HDEHP)

Viscous liquid, sp. gr. 0.975 (20°), mol. wt. 322. Very slightly soluble in water, easily soluble in organic solvents. $pK_a = 1.4$. The commercial grade material requires purification.[78] It is stirred for 16 hr with 6 M HCl at 60° (destruction of pyro esters by hydrolysis).

HDEHP may be considered typical of the monobasic dialkylphosphates. Dissolved in heptane, toluene, carbon tetrachloride, and the like, it is used principally to extract Zr, Hf, Ti, U(VI), Ce(IV), Th, Nb, Mo(VI), Sc, and trans-U elements[79] from HNO_3 solutions (see Chapter 9A). Extractions are also made from HCl solutions (Chapter 9A). Lanthanide extraction from $HClO_4$ into toluene has been studied.[80]

As already mentioned, HDEHP is usually dimeric in nonpolar organic solvents and monomeric in alcohols. The general extraction reaction in which a simple chelate is formed can then be represented as follows when a solvent such as benzene is used:

$$M^{n+} + n(HR)_2 \rightleftharpoons M(HR_2)_n + nH^+$$
$$\quad\quad\quad org \quad\quad\quad\quad org$$

According to this equation, the extractability of the metal (expressed as E) should be strongly dependent on the acidity of the aqueous phase, varying as $1/[H^+]^n$, but not on the nitrate or chloride ion concentration, except insofar as the ionic strength is affected.[81] Actually, some metals can give more complex reaction products than shown by the preceding reaction. Calcium, strontium, and barium are extracted as adducts by a benzene solution of HDEHP according to the equation[82]

$$M^{2+} + 3(HR)_2 \rightleftharpoons M(HR_2)_2(HR)_2 + 2H^+$$
$$\quad\quad\quad org \quad\quad\quad\quad org$$

E thus varies as $[(HR)_2]^3_{benzene}$ and as $1/[H^+]^2$. Thorium in perchlorate or chloride solutions of low acidity reacts with HDEHP in toluene according to the equation

$$Th^{4+} + 3(HR)_2 \rightleftharpoons ThR_2(HR_2)_2 + 4H^+$$
$$\quad\quad\quad org \quad\quad\quad\quad org$$

and the extractability is inversely proportional to the fourth power of the hydrogen-ion concentration. When the aqueous phase contains nitrate, a mixed ligand complex is predominantly extracted[83]:

$$Th^{4+} + NO_3^- + 3(HR)_2 \rightleftharpoons ThNO_3(HR_2)_3 + 3H^+$$

Mixed complexes of this kind are formed mostly by quadri- and quinquevalent cations.

When thorium is extracted with HDEHP in n-decyl alcohol, in which the reagent is predominantly monomeric, the products ThR_4 and $ThR_2(HR_2)_2$ are obtained, and when U(VI) is extracted, the products are UO_2R_2 and $UO_2(HR_2)_2$, the latter product becoming more important as the reagent concentration is increased. From perchlorate solution, U(VI) is extracted largely as $UO_2R_2(HR)_2$ into hexane. With MIBK as extracting solvent, $UO_2R_2 \cdot MIBK$ is formed.

Gallium is extracted into heptane from ~1 M HCl, salted with Cl^-, as $HGaCl_4(HR)_2$ according to the equation

$$Ga^{3+} + 4\,Cl^- + H^+ + (HR)_2 \rightleftharpoons HGaCl_4(HR)_2$$
$$\qquad\qquad\qquad\qquad\quad org \qquad\qquad\quad org$$

Here HDEHP functions as a solvating agent, not as a chelating agent. The extracted species can also be looked upon as an ion-pair complex, $[(HR)_2H]GaCl_4$.

A considerable number of metals, for example, Zr, Ti, and Sc, are extractable by HDEHP from strongly acid solutions (HCl, HNO_3, $HClO_4$). Probably compounds of the type $MX_n(HR)_2$ are extracted. ($X = Cl^-$, NO_3^-, etc.). U(VI) can be extracted from concentrated HNO_3 and $HClO_4$.

The analytical literature tends to be deficient in comparative results, that is, results by different methods applied to the same separation or determination. More comparative data on organic phosphorus reagents would be helpful to the analyst. Some are available. For example, it has been found that HDEHP is much superior to TBP, thenoyltrifluoroacetone and mesityl oxide in the extraction of thorium from acid phosphate solutions (bone ash). HDEHP extraction is also superior to anion exchange in 7 M HNO_3 solution, which indeed fails with solutions of bone ash.[84]

HDEHP and other organophosphorus compounds find use as the stationary phase in reversed-phase partition chromatography for separations of the actinides.

The rate of some HDEHP extractions is very slow. For example, attainment of equilibrium in the extraction of beryllium by 0.5 M HDEHP in toluene requires hours at room temperature. A number of studies of the kinetics of HDEHP extraction have been made.[85] Some cations, e.g., U, Zn, and Eu, show much the same (fairly high) extraction rates independent of acidity, whereas others, for example, Fe(III), Be and Sr (all slow), show rates decreasing with acidity. Ions of the first group presumably react with un-ionized HL or H_2L_2, those of the second with

L^- or HL_2^-. With quiescent interface, the extraction rate of Fe(III) is controlled by chemical reaction at the interface. Reaction kinetics are complex (rate increases with HDEHP concentration in organic solvent and with a decrease in Fe concentration if $[HDEHP] > \sim 10^{-3}\ M$). A rate equation has been formulated by Coleman and Roddy that represents the experimental results well. The power dependence of rate on HDEHP concentration under certain conditions decreases at high HDEHP concentrations in dispersion mixing (droplet formation). The extraction kinetics of U, Be, and other metals have also been studied.

D. BINUCLEAR ORGANOPHOSPHORUS COMPOUNDS

A considerable number of compounds of this type have been studied analytically. Some of these are mentioned here. Extraction coefficients of metal complexes of binuclear reagents may be greater than those of corresponding mononuclear reagents.

Di-n-butylethane-(1,2)-diphosphonic acid has been examined as an extraction reagent for Eu,[86] and the effect of the solvent is of interest. The reagent is monomeric in benzene up to a concentration of $1.3 \times 10^{-2}\ M$, above which association occurs, with $n \sim 9$ at saturation; addition of 1-pentanol (4%) monomerizes the reagent. In MIBK, the reagent is monomeric to saturation. Eu is extracted into benzene and chloroform as $Eu(AH)_3AH_2$ (reagent $= AH_2$), into MIBK and benzene containing 12% pentanol-1 as $EuA(AH)\cdot xRR'C{=}O$ and $Eu(AH)_3\cdot xROH$. In benzene, the Eu complex exists as

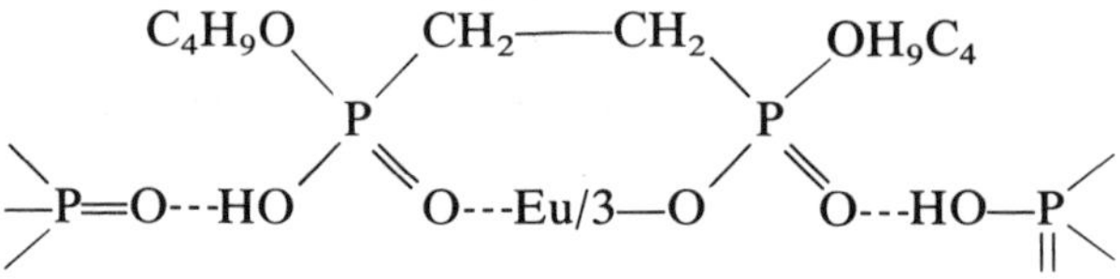

Benzene has low solvating power, and in this solvent the free polar groups in the complex are coordinated to molecules of the reagent.

The extraction of U(VI) from nitrate solution by dioxides of diphosphine in 1,2-dichlorobenzene has been found[87] to increase in the order

$$\text{TOPO} < 1,1\text{-DiPO} < 1,4\text{-DiPO} < 1,3\text{-DiPO} < 1,2\text{-DiPO},$$

where 1,n-DiPO represents

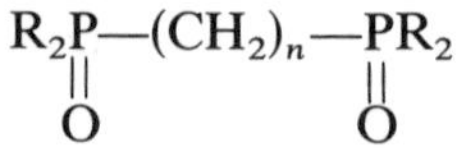

The extractability of trivalent actinides follows the order[88]

$$TBP < TOPO < 1,1\text{-DiPO} < 1,5\text{-DiPO} < 1,4\text{-DiPO}$$

E. ORGANIC PHOSPHORUS–SULFUR REAGENTS

Replacement of one or more O atoms by S in organophosphorus compounds leads to extraction reagents of much greater selectivity. They contain $\equiv P = S$ and $= PS(SH)$ groups and form chelates or association complexes, almost entirely with sulfophilic elements (those forming sulfides insoluble in acidic solution), which are extractable into organic solvents of the inert type.[89] Those compounds not having replaceable H react with a limited number of metals.

1. Di-*n*-butyl Phosphorothioic Acid

Dialkyl phosphorothioic acids dissolved in organic solvents show tautomeric equilibrium:

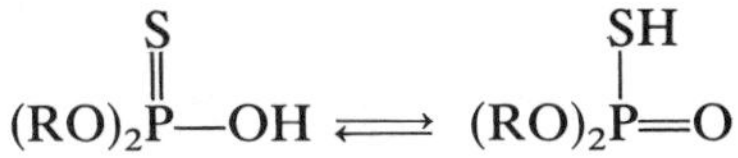

$$\underset{\text{thiono}}{\overset{\displaystyle S \atop \displaystyle \|}{(RO)_2 P - OH}} \rightleftharpoons \underset{\text{thiol}}{\overset{\displaystyle SH \atop \displaystyle |}{(RO)_2 P = O}}$$

The active group for most metals is –SH. The metal complexes formed by replacement of hydrogen can be extracted into inert and oxygenated solvents, the former usually being preferred. Four-membered ring systems are formed.

The extractive properties of di-*n*-butyl phosphorothioic acid have been studied.[90] Figure 9C-4 summarizes the results for extraction from sulfuric acid solutions. (In HCl solutions, most metals show similar behavior, except some show decreasing extraction at high acid concentrations because of Cl^- complexing.) It may be noted that some nonsulfophilic elements [Y, Nb, and Cr(VI)] are extracted to a greater or less extent. But, as expected, sulfide-reacting metals are the ones extracted well. Some of these show smaller E values at higher acidities, others (Pd, Ag, Hg, and Cu), higher values. In general, if simple chelates are formed, extraction should decrease at higher acidities. Coordination complexes not involving replacement of H may be formed by some metals. The complex formed by Zn has been shown to be ZnL_2 (reagent = HL) and those formed by Cd, Ga, In, and Tl(I) must be of analogous composition (chelates). The nature of the Pd, Ag, Hg, and Cu complexes is uncertain.

The reagent is subject to oxidation, disulfide being formed, more so in HCl than H_2SO_4 solution; strong light accelerates decomposition.

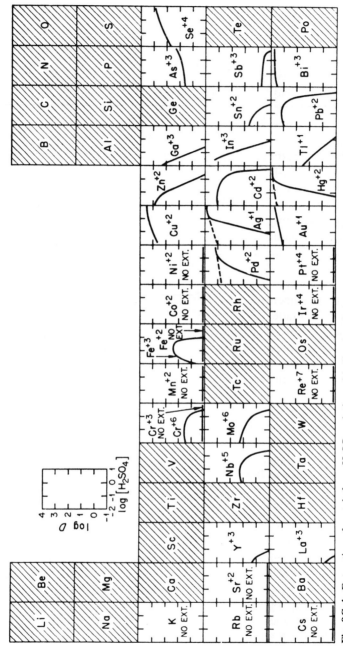

Fig. 9C-4. Extraction of metals from H_2SO_4 solution with di-n-butyl phosphorothioic acid (0.22 M in CCl_4). Equal volumes of organic and aqueous phases. Metal concentration, 0.5 mg/ml. Reprinted with permission from T. H. Handley, *Anal. Chem.,* **35**, 992 (1963). Copyright by the American Chemical Society.

2. Di-n-butyl Phosphorodithioic Acid

Dialkyl phosphorodithioic acids

$$\underset{\text{(RO)}_2\text{P}-\text{SH}}{\overset{\overset{\displaystyle S}{\|}}{}}$$

dissolved in inert or oxygenated organic solvents extract sulfophilic elements from mineral acid solution. The diethyl-[91] and dibutyl-[92] acids have been studied. They are fairly strong acids (Table 9C-5). An increase in the length of the hydrocarbon chain increases the number of metals that can be extracted. Cesium is reported to be extracted ($E = 0.5$) by di-(2-ethyl)hexyl phosphorodithioic acid at pH 8–9. In the following, we briefly survey the use of di-n-butyl phosphorodithioic acid.

The free acid is subject to slow hydrolysis. For this reason it is best used in the form of the sodium or ammonium salt. Figure 9C-5 summarizes the extractability of many elements from HCl solution (extractability from H_2SO_4 solution is much the same) with CCl_4 as the extracting solvent. Equilibrium is usually attained in 10 min. Separation of phases is aided by centrifugation.

The reagent is not rapidly decomposed in acid solution.[93] It is oxidized by strong oxidizing agents, but can be used in <6 M HNO_3 solutions if extractions are made in <10 min. The oxidation product is $(RO)_2P(S)SS(S)P(RO)_2$. Cu(II) is partly reduced to Cu(I), Fe(III) to Fe(II), and Au(III) to Au(I).

We note in Fig. 9C-5 that for some metals, for example, Hg(II), E increases with the acidity to 1 M H^+ or higher. Evidently, simple chelate formation does not tell the whole story of metal extractions with this reagent. For Hg(II) in 0.1 M HCl, oxygenated solvents such as esters and

TABLE 9C-5
Ionization and Partition (CCl_4/aq) Constants of Dialkyl Phosphorodithioic Acids[a]

Alkyl Group	pK_a	$\log P_r$
Ethyl	−0.10	0.45
Isopropyl	0.0	1.90
n-Butyl	0.22	2.52
Isobutyl	0.10	2.63

[a] R. H. Zucal, J. A. Dean, and T. H. Handley, *Anal. Chem.*, **35**, 988 (1963). $\mu = 1.0$ (mostly). The acids are not dimerized in CCl_4 or other organic solvents.

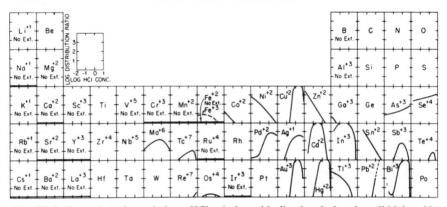

Fig. 9C-5. Extraction of metals from HCl solution with di-n-butyl phosphorodithioic acid (0.21 M in CCl_4). Equal volumes of aqueous and CCl_4 phases. Reprinted with permission from T. H. Handley and J. A. Dean, *Anal. Chem.*, **34**, 1313 (1962). Copyright by the American Chemical Society.

ethers give much higher E than inert solvents, pointing to the formation of association complexes.[94]

Metals can be stripped from the organic phase with H_2O_2 in HCl (destruction of reagent and complexes).

Some of the metal complexes formed must have great stability: HgS and Bi_2S_3 dissolve when shaken with the reagent in CCl_4.

It is reported that extractions at acid concentrations less than 0.2 M are facilitated by the presence of salts in 0.2–0.3 M concentration. In the absence of these electrolytes, some complexes (Bi, Cu, Cd, Hg, Pd, Ag) tend to remain in the aqueous phase as a very fine suspension. Addition of an oxygenated solvent dissolves the suspended material.

The phosphorodithioic acids resemble diethyldithiocarbamic acid in their reactions, but they are more stable and can be used in more strongly acid solutions. There are some differences in the extraction behavior of di-n-butyl phosphorodithioic and di-n-butyl phosphorothioic acids as can be seen by comparing Figs. 9C-4 and 9C-5.

Sodium diethyl phosphorodithioate has found use in the separation of little Cd from much Zn by extraction of the former into CCl_4 from acetate solution.[95]

3. Trialkyl Thiophosphates

$$S{=}P{-}OR$$

with OR groups:

$$S{=}\underset{\diagdown OR}{\overset{\diagup OR}{P{-}OR}}$$

Triisooctyl thiophosphate (O,O,O-triisooctyl thionophosphate) and tri-n-butyl thiophosphate in an inert solvent such as CCl_4 selectively extract Ag and Hg from HNO_3 or $HClO_4$ solution.[96] With a 0.67 M solution of triisooctyl thiophosphate (TOTP) in CCl_4 and an aqueous phase 6 M in HNO_3, $E_{Ag} > 100$ and $E_{Hg(II)} \sim 90$. Salting agents—$NaNO_3$ and $Ca(NO_3)_2$—increase E. Of other elements tested, only Ta and V showed a slight tendency to be extracted. The extracted Ag and Hg species seem to be $AgNO_3 \cdot TOTP$ and $Hg(NO_3)_2 \cdot 2\,TOTP$.

Triisooctyl thiophosphate in MIBK (or benzene) has been used to extract silver from 4–8 M HNO_3 solutions (digests of soils, rocks, etc.) preparatory to its photometric or atomic absorption determination.[97]

Trialkyl- and Hexaalkylphosphorothioic Triamides. Compounds of this type

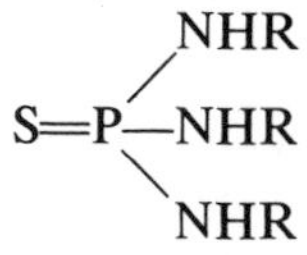

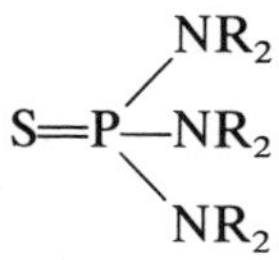

$$S{=}P{\overset{\displaystyle NHR}{\underset{\displaystyle NHR}{-}}}NHR \qquad S{=}P{\overset{\displaystyle NR_2}{\underset{\displaystyle NR_2}{-}}}NR_2$$

as exemplified by N,N',N''-trihexylphosphorothioic triamide and N,N',N''-hexabutylphosphorothioic triamide extract Cu(I), Hg(II), Ag, Pd(II), and Au(III) into $CHCl_3$ and other "inert" solvents from mineral acid solutions.[98] Extraction is rapid from 0.1–6 M HNO_3 or H_2SO_4 solutions and E does not depend on hydrogen-ion concentration. Coordination compounds such as $HgSO_4 \cdot 2$ reagent and $AgNO_3 \cdot$ reagent are the extracted species.

4. Tri-n-octylphosphine Sulfide

This reagent (TOPS)

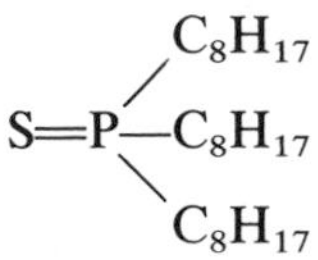

$$S{=}P{\overset{\displaystyle C_8H_{17}}{\underset{\displaystyle C_8H_{17}}{-}}}C_8H_{17}$$

in an inert solvent such as cyclohexane is a good extractant for Hg, Ag and Pd from HNO_3 solution (Table 9C-6).[99] Au and Hg(II) are extracted well from 1–7 M HCl, Pd(II) poorly, Sn(II) and Ga very slightly, and other metals even less.[100] It is seen that TOPS is a much more selective extraction reagent than TOPO (Chapter 9A). The selectivity extends into the sulfophilic group of elements.

TABLE 9C-6
Extraction of Metals from Nitric Acid Solution by TOPS[a]

	$\log E$		
	HNO_3 (M)		
	1	3	5
Hg(II)	4.0	1.96	1.6
Ag	3.6	3.3	2.9
Pd(II)	3.0	2.3	2.0
Mo(VI)	0.5	0.5	0.5

[a] Condensed from Elliott and Banks. Organic phase 0.05 M TOPS in cyclohexane (nitrobenzene for Hg), aqueous phase ~0.01 M in metal. Equal volumes of phases. Se(IV) is extracted slightly ($\log E \sim -1$ in 1–5 M HNO_3); Pb, Zn, Cd, and As(III) less than Se; and Th, U(VI), Bi, and Lu essentially not at all.

F.   SYNERGISM WITH ORGANOPHOSPHORUS REAGENTS

These reagents provide striking instances of synergism. Pronounced synergism seems to have been first observed in the mid 1950s, when it was noted that the extraction of Pr and Nd by thenoyltrifluoroacetone in kerosene was increased 60–100 times when TBP was added.[101] Soon thereafter the synergic effect of a neutral organophosphorus compound of the type G_3PO (G = alkyl or alkoxy group) on the extraction of U(VI) with a dialkylphosphoric acid was observed and investigated systematically.[102] Since then many synergic systems have been studied, but usually not with analytical applications in mind.[103]

The organophosphorus complexes formed in synergic extractions fall into two main classes:

1.   Those in which the neutral (i.e., uncharged) organophosphorus component forms an adduct with a metal non-P chelate. The extracted product can be represented as $ML_n \cdot S_s$, where S is the organophosphorus addend.

2.   Those in which an acidic organophosphorus compound [HX, or $(HX)_2$ if dimerized] forms a chelate with a metal, which undergoes adduct formation with a neutral organophosphorus component. The extracted product can be represented as $MX_n \cdot S_s$, $M(X \cdot HX)_n \cdot S_s$, and the like.

In adducts of the general type $ML_n \cdot S_s$, S can be a non-P addend, but usually the synergism is much stronger when S is a neutral phosphorus compound. In most of the studies that have been made, HL has been a β-diketone, especially thenoyltrifluoroacetone (HTTA). Figure 9C-6 shows the synergic effect of TBP and TBPO on the extraction of U(VI) thenoyltrifluoroacetonate by cyclohexane. The extraction of U is enhanced by a factor of $\sim 10^4$ by TBPO. Enhancement factors of 10^4 or 10^5 are common in this type of synergic extractions. Of course, such factors signify hardly any extraction of the simple chelate. Synergic effects of TBP, TOPO, and other neutral phosphorus addends on the extraction of HTTA (and other β-diketone) complexes of the actinide and lanthanide

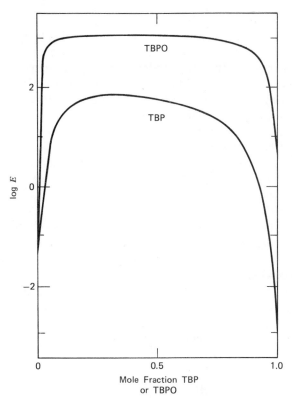

Fig. 9C-6. Extraction of U(VI) into cyclohexane from $0.01\,M$ HNO_3 by thenoyltrifluoroacetone (HTTA) + tributyl phosphate (TBP) or tributylphosphine oxide (TBPO). Concentration of HTTA + TBP and HTTA + TBPO in cyclohexane = $0.02\,M$. From H. Irving and D. N. Edgington, *J. Inorg. Nucl. Chem.*, **15**, 158 (1960).

TABLE 9C-7
Constants for the Reaction CuL$_2$ + S $\rightleftharpoons$ CuL$_2$·S in Benzene^a

HL	TPP	TBP	TPPO	TOPO
			S	
Acetylacetone				45
Trifluoroacetylacetone	3	36	180	940
Thenoyltrifluoroacetone	3	46	220	1700
Hexafluoroacetylacetone	130	4800	12,000	

TPP = triphenyl phosphate, TPPO = triphenylphosphine oxide.
^{a} C. N. Ke and N. C. Li, *J. Inorg. Nucl. Chem.*, **28**, 2255 (1966).

elements, as well of some other elements, have been extensively investigated and equilibrium constants of the adducts have been determined.[104] The synergic effect increases with the base strength of the neutral phosphorus component, the order being phosphate < phosphonate < phosphinate < phosphine oxide. Steric factors also play a role. Synergism increases with decreasing complexing power of HL or roughly with increasing acidity of the latter (see Table 9C-7). Finally, the extracting solvent affects synergism, the enhancement of extraction increasing with a decrease in the polarity of the solvent.

The course of the curves in Fig. 9C-6 imply that an increase in the ratio[S]/[HL] above a certain value, or range of values, results in a decrease in the synergic effect of S. This is more clearly shown in Fig. 9C-7, which demonstrates that the synergic effect of TBP on the extraction of UO$_2$(TTA)$_2$ decreases above a concentration of ~0.3 M under the conditions indicated. The descending limb of the curve corresponds to destruction of synergism, which is due to interaction between HTTA and TBP. An association product is formed between the keto hydrate of HTTA and TBP.[105] Such destruction of synergism is a general phenomenon, occurring also in organophosphate chelate–neutral organophosphorus addend systems, and indeed in non-P systems.[106] The destruction of synergism in the system Zn–hexafluoroacetylacetone–TOPO has been attributed[107] to the formation of

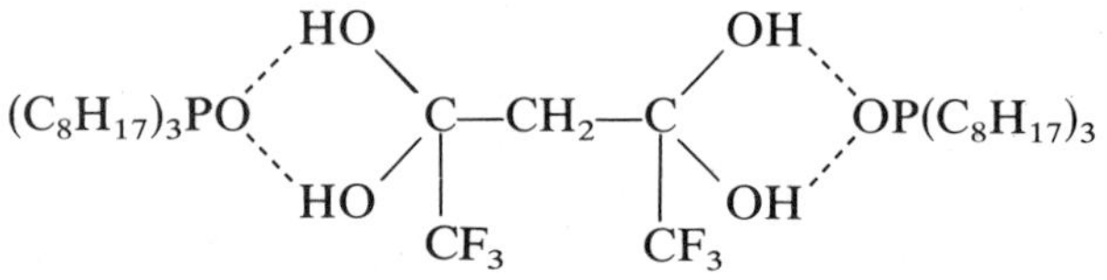

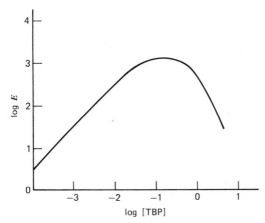

Fig. 9C-7. Extraction of U(VI) into 0.015 M HTTA solution in cyclohexane containing TBP in varying concentrations; aqueous phase 0.01 M HCl. From T. V. Healey, D. F. Peppard, and G. W. Mason, *J. Inorg. Nucl. Chem.*, **24,** 1429 (1962). The slope of +1 for the ascending limb of the curve indicates the extraction of the adduct $UO_2(TTA)_2 \cdot TBP$. The slope of descending limb of the curve is -2.

An analogous compound is believed formed between hexafluoroacetyl-acetone and TBP (in cyclohexane).[108]

The synergism produced by a neutral organophosphorus compound forming an adduct with a metal chelate derived from an acidic or-ganophosphorus compound (class 2 mentioned previously) is less pro-nounced than that of class 1. To a large extent this is the result of moderately large values for E of the metal chelate alone. Dialkyl-phosphoric acids have been largely used as the chelating agent in these synergic extractions, with TBP, TOPO, and the like as the neutral ligand. Dialkylphosphoric acids are largely dimeric in solvents of low polarity, as we have seen. The adducts tend to have the composition $M(X \cdot HX)_n \cdot S_s$ at low metal concentrations or $MX_n \cdot S_s$ at high metal concentrations or in solvents of high dielectric constant. As a rule, the stability of the synergic adduct increases with the basicity of S and the acidity of HX.

NOTES

1. In passing, it may be noted that addition of masking agents such as EDTA and hydroxy acids may lead to the unintentional extraction of anionic metals if suitable cationic partners should be present.
2. T. Higuchi and A. F. Michaelis, *Anal. Chem.*, **40,** 1925 (1968).
3. A product sold by a U.S. supplier as $(C_6H_5)_4AsCl$ was found by us to be

mostly the bromide. The widely varying solubility values obtained by different workers for the chloride point to the use of impure salts.

4. This salt is more accurately formulated $(C_6H_5)_4AsHCl_2\cdot2H_2O$. K. W. Loach, *Anal. Chim. Acta*, **44**, 323 (1969), found a m.p. of 254°–256° for tetraphenylarsonium hydrogen dichloride dihydrate and a m.p. of 256°–257° for $(C_6H_5)_4AsCl\cdot2H_2O$. $(C_6H_5)_4AsHCl_2\cdot2H_2O$ gives the anhydrous salt at 70°–113°, and this loses HCl above 125°; above 250° the anhydrous chloride decomposes rapidly (probably to $(C_6H_5)_3As$ and C_6H_5Cl). $(C_6H_5)_4AsCl\cdot2H_2O$ is converted to the monohydrate at 54°–82° and to the anhydrous salt at 92°–126° according to Loach. $(C_6H_5)_4AsCl$ is very hygroscopic. The dihydrate is stable at relative humidities above 43% at 25°. Careful measurements by K. W. Loach, *Anal. Chim. Acta*, **45**, 93 (1969), give λ_{max} of $(C_6H_5)_4AsCl$ in water as 264.0 nm ($\epsilon = 3348$) and in $CHCl_3$, $\lambda_{max} = 265.0$ (values for other bands also given). The absorption peaks are very sharp. Triphenylarsine oxide is a possible impurity in tetraphenylarsonium salts.

5. J. S. Fok, Z Z. Hugus, and E. B. Sandell, *Anal. Chim. Acta*, **48**, 243 (1969).

6. Dissociation of some quaternary P and As salts can occur in favorable solvents. N. A. Gibson and D. C. Weatherburn, *Anal. Chim. Acta*, **58**, 149 (1972), found that triphenylmethylarsonium chloride shows an increase in E when its concentration in the organic solvent decreases below $2.5\times10^{-3}M$ in dichloroethane and below $6\times10^{-3}M$ in 2,2′-dichloroethyl ether (aqueous phase $= 0.5\,M$ KCl). This behavior is interpreted as resulting from dissociation of the ion pair $(Ph_3CH_3As)^+Cl^-$ in the organic solvents mentioned. Dissociation does not occur in $CHCl_3$.

7. Gibson and Weatherburn[6] ascribe the difference in the effect of HCl and alkali chlorides to greater association of the component ions in HCl than in (say) NaCl solution because of the greater lowering of the dielectric constant of the solution by HCl.

8. The intrinsic (molecular) solubilities of tetraphenylarsonium ion-association compounds in aqueous solution have not been determined except that of $(Ph_4As)_2IrCl_6$, for which J. Fok found the value $2.5\times10^{-6}M$ in 0.1 M HCl (only the electrolyte concentration, not the acidity, is relevant) at 25°.

9. S. Tribalat, *Anal. Chim. Acta*, **3**, 113 (1949).

10. I. P. Alimarin and S. V. Makarova, *Zh. Anal. Khim.*, **17**, 1072 (1962).

11. Anionic complexes of metals with organic chelating reagents can also be extracted as tetraphenylarsonium salts into organic solvents. For example, Yu. A. Zolotov, O. M. Petrukhin, and I. P. Alimarin, *Zh. Anal. Khim.*, **20**, 347 (1965), found iron and thorium to be extracted by tetraphenylarsonium (and diphenylguanidinium) into a 1:1 mixture of isobutyl alcohol and nitromethane from EDTA solution. The mixed anion $Ga(TTA)_2AcCl^-$ (TTA is thenoyltrifluoroacetone anion, Ac acetate) is extracted with $(C_6H_5)_4As^+$ as cationic partner into dichlorobenzene from acetate-buffered

solutions at pH 2.5, as shown by M. S. Rahaman and H. L. Finston, *Anal. Chem.*, **40**, 1709 (1968). (An example of synergism, but of little analytical interest.) The strongly colored complex formed by Nb with PAR(H_2L) in oxalate solution is extractable as $[(C_6H_5)_4As][NbO(C_2O_4)L]$ into $CHCl_3$, as shown by M. Siroki, L. Maric, and M. J. Herak, *Anal. Chem.*, **48**, 55 (1976). Tetraphenylphosphonium can also be used as reagent. Tungsten, Ti, and Zr do not extract; Ta does not interfere seriously, but U(VI) does.

12. N. A. Gibson and D. C. Weatherburn, *Anal. Chim. Acta*, **58**, 159 (1972), determined the extraction coefficients of the salts of 14 phosphonium and arsonium cations with 19 inorganic anions and four solvents. In general, with a given anion, E increases with the size of the cation. The effect of the solvent (chloroform, dichloromethane, 1,2-dichloroethane, and 2,2'-dichlorodiethyl ether) is relatively minor with most salts; there is little dissociation of ion pairs in these solvents. A rough correlation was found between the extractability of an anion and the enthalpy of its hydration: The larger the latter, the better the extraction.

13. J. Fok, Ph. D. Thesis, University of Minnesota, Minneapolis, 1963.

14. Tetraphenylphosphonium chloride can also be used.

15. P. Peng, University of Minnesota, 1965.

16. R. Bock and J. Jainz, *Z. Anal. Chem.*, **198**, 315 (1963); S. Tribalat, *Anal. Chim. Acta*, **4**, 228 (1950); cf. *ibid.*, **5**, 115 (1951).

17. A. I. Busev, G. Rudzitis, and A. K. Kauke, *Zh. Anal. Khim.*, **29**, 1847 (1974). Diphenylguanidine, triphenylguanidine, and S-benzylthiourea were also investigated as cationic reagents.

18. M. A. Ashy and J. B. Headridge, *Analyst*, **99**, 285 (1974).

19. Tetraphenylstibonium: R. Bock and E. Grallath, *Z. Anal. Chem.*, **222**, 283 (1966). Triphenylantimony dihalides (for F^-): H. Chermette et al., *Anal. Chem.*, **44**, 857 (1972). Triphenyl-lead: R. Bock and H. Deister, *Z. Anal. Chem.*, **230**, 321 (1967). Triphenyltin: R. Bock, H. Niederauer, and K. Behrends, *Z. Anal. Chem.*, **190**, 33 (1962); G. K. Schweitzer and S. W. McCarty, *J. Inorg. Nucl. Chem.*, **27**, 191 (1965) (also Ph_3Pb^+, $Ph_3Sb(OH)^+$, $Ph_3As(OH)^+$). Triphenylsulfonium: R. Bock and C. Hummel, *Z. Anal. Chem.*, **198**, 176 (1963).

20. V. S. Shmidt, *Amine Extraction*, trans. by J. Schmorak, Israel Program for Scientific Translations, 1971.

21. For tabulations of liquid ion exchangers see C. F. Coleman, C. A. Blake, and K. B. Brown, *Talanta*, **9**, 297 (1962). Reviews: H. Green, *ibid.*, **11**, 1561 (1964) and **20**, 139 (1973); D. Dyrssen, *Svensk Kem. Tidskr.*, **77**, 387 (1965).

22. T. Sato and H. Watanabe, *Anal. Chim. Acta*, **54**, 439 (1971).

23. In an early paper in this field, H. A. Mahlman, G. W. Leddicotte, and F. L. Moore, *Anal. Chem.*, **26**, 1939 (1954), recorded the extraction of Zn by methyldioctylamine in trichloroethylene from 3 M HCl and the nonextraction of Co, Ni, and Mn. W. L. Ott, H. R. MacMillan, and W. R. Hatch,

Anal. Chem., **36**, 363 (1964), used triisooctylamine–xylene extraction in the determination of Zn in metallic Ni.

24. F. L. Moore, *Anal. Chem.*, **30**, 908 (1958). Apparently, U(VI) is extracted as $(R_3NH)_2UO_2Cl_4$, U(IV) as $(R_3NH)_2UCl_6$ (second-power dependence on amine concentration).

25. G. W. Leddicotte and F. L. Moore, *J. Am. Chem. Soc.*, **74**, 1618 (1952), who used methyldioctylamine in xylene. J. Y. Ellenburg, G. W. Leddicotte and F. L. Moore, *Anal. Chem.*, **26**, 1045 (1954), using 8% tribenzylamine in methylene chloride obtained $E_{Nb}/E_{Ta} = 3.5 \times 10^4$ in 11 M HCl $(80.5/2.3 \times 10^{-3})$.

26. M. L. Good and S. C. Srivastava, *J. Inorg. Nucl. Chem.*, **27**, 2429 (1965).

27. T. M. Florence and Y. J. Farrar, *Anal. Chem.*, **40**, 1200 (1968). Lower concentrations of Li halides suffice if the cyclohexane extraction is made from anhydrous methanol solution.

28. T. Sato, *Anal. Chim. Acta*, **43**, 303 (1968).

29. A. M. Rozen and Z. I. Nagnibeda, *Radiokhim.*, **13**, 284 (1971); *Anal. Abst.*, **23**, 1352, on which this summary is based.

30. M. Ziegler, O. Glemser, and A. von Baeckmann, *Z. Anal. Chem.*, **172**, 105 (1960).

31. Y. Kitano, W. Ishibashi, and S. Sato, *Japan Analyst*, **16**, 922 (1967).

32. A. M. Wilson, L. Churchill, K. Kiluk, and P. Hovsepian, *Anal. Chem.*, **34**, 203 (1962).

33. M. L. Good, S. C. Srivastava, and F. F. Holland, Jr., *Anal. Chim. Acta*, **31**, 534 (1964), found Co to be extracted as $CoCl_4^{2-}$, Fe(III), Ga, and In as MCl_4^-.

34. W. J. Maeck, G. L. Booman, M. C. Elliott, and J. E. Rein, *Anal. Chem.*, **30**, 1902 (1958); **31**, 1130 (1959). Lanthanides can be extracted into nitroethane, as shown by W. J. Maeck, M. E. Kussy, and J. E. Rein, *ibid.*, **37**, 103 (1965).

35. W. J. Maeck, G. L. Booman, M. E. Kussy, and J. E. Rein, *Anal. Chem.*, **33**, 1775 (1961); correction, *ibid.*, **34**, 212. Extraction of TcO_4^- into $CHCl_3$ from NaOH solution: R. E. Foster, Jr., W. J. Maeck, and J. E. Rein, *ibid.*, **39**, 563 (1967).

36. C. R. Walker and O. A. Vita, *Anal. Chim. Acta*, **67**, 119 (1973).

37. G. Koch, *Radiochim. Acta*, **4**, 128 (1965), who found $R_3CH_3N[UO_2(NO_3)_3]$ to be associated in benzene. In a 0.015 M solution, $(MW)_{observed}/(MW)_{theoretical} \sim 2$.

38. M. Cospito and L. Rigali, *Anal. Chim. Acta*, **57**, 107 (1972). They use tricaprylamine in xylene to extract U.

39. F. L. Moore, *Anal. Chem.*, **41**, 1658 (1969).

40. F. L. Moore, *Anal. Chem.*, **36**, 2158 (1964).

41. F. L. Moore, *Anal. Chem.*, **33**, 748 (1961).

42. F. L. Moore, *Anal. Chem.*, **37**, 1235 (1965); **38** 905 (1966). H. M. N. H. Irving and R. H. Al-Jarrah, *Anal. Chim. Acta*, **55,** 135 (1971), found $CrY(H_2O)^-$ extractable at pH 3.6–6.0 into a dichloroethane solution of tetra-n-hexylammonium chloride. $CrY(OH)^{2-}$ and $CrY(OH)_2^{3-}$ are not extractable.

43. N. Zaman, E. Merciny, and G. Duyckaerts, *Anal. Chim. Acta*, **56,** 271 (1971). In an earlier paper, *ibid.*, **56,** 261 (1971), the same authors found tricaprylmethylammonium chloride to be polymerized in benzene, with an average polymerization factor of 2.75.

44. J. Adam and R. Pribil, *Talanta*, **22,** 905 (1975).

45. J. Adam and R. Pribil, *Talanta*, **18,** 91 (1971). Cr can be determined in the extract with diphenylcarbazide (*ibid.*, **21,** 616).

46. T. Groenewald, *Anal. Chem.*, **40,** 863 (1968). (for atomic absorption spectrometry). Many other complex cyanides are also extracted. V. Y. Armeanu and L. M. Baloiu, *Anal. Chim. Acta*, **44,** 230 (1969), report that $HAu(CN)_2$ can be extracted into isoamyl alcohol at $pH < 1$ and thus separated from complex cyanides of other metals. See O. E. Zvyagintsev and O. I. Zakharov-Nartsissov, *Zh. Prikl. Khim.*, **33,** 55 (1960); R. Ripan, M. Marcu, and V. Cordis, *Rev. Chim. (Bucharest)*, **15,** 684 (1964).

47. H. M. N. H. Irving and A. H. Nabilsi, *Anal. Chim. Acta*, **41,** 505 (1968).

48. H. M. N. H. Irving and A. D. Damodaran, *Anal. Chim. Acta*, **53,** 267 (1971).

49. E. N. Pollock, *Anal. Chim. Acta*, **40,** 285 (1968), extracts Te from $\sim 6\,M$ HCl with diantipyrinylmethane in 1,1,2,2-tetrachloroethane. Separation from Cu.

50. V. P. Zhivopistsev et al., *Trudy Kom. Analit. Khim.*, **16,** 80 (1968); **17,** 304 (1969); see A. I. Busev et al., *Talanta*, **19,** 173 (1972).

51. V. P. Zhivopistsev and L. P. Pyatosin, *Zh. Anal. Khim.*, **22,** 70 (1967). With diantipyrinylbutane as reagent, Th can be extracted into chloroform from $6\,M$ HNO_3 (separation from Sc): B. I. Petrov et al., *Zav. Lab.*, **41,** 655 (1975).

52. V. P. Zhivopistsev and B. I. Petrov, *Zh. Anal. Khim.*, **23,** 1634 (1968).

53. N. P. Rudenko and V. O. Kordyukevich, *Zh. Anal. Khim.*, **23,** 1061 (1968). For the rationale of this separation, see Rh–Ir, p. 988.

54. B. I. Petrov et al., *Anal. Abst.*, **29,** 3B48 (1975).

55. P. V. Marchenko and A. I. Voronina, *Ukr. Khim. Zh.*, **35,** 652 (1969).

56. T. Sekine and D. Dyrssen, *Anal. Chim. Acta*, **45,** 433 (1969). Earlier work: K. Haruyama and T. Ashizawa, *Japan Analyst*, **14,** 120 (1965); A. G. Collins, *Anal. Chem.*, **35,** 1259 (1965); C. Feldman and T. C. Rains, *ibid.*, **36,** 405 (1964). Substoichiometric extraction of sodium into $CHCl_3$ by the use of dicyclohexyl-18-crown-6 (a macrocyclic polyether) and tetraphenylborate has been described by J. W. Mitchell and D. L. Shanks, *Anal. Chem.*, **47,** 642 (1975). The crown reagent chelates Na^+ as $Na \cdot Crown^+$, which forms the chloroform-extractable ion pair with $(C_6H_5)_4B^-$. Cf. note 109.

57. J. Benes and M. Kyrs, *Anal. Chim. Acta*, **29**, 564 (1963); T. Iwachido, *Bull. Chem. Soc. Japan*, **44**, 1835 (1971).

58. W. R. Ross and J. C. White, *Anal. Chem.*, **36**, 1998 (1964); R. J. Everett and H. A. Mottola, *Anal. Chim. Acta*, **54**, 309 (1971).

59. Z. Maksimovic and M. Kyrs, *Coll. Czech. Chem. Commun.*, **34**, 3436 (1969).

60. A. A. Schilt, R. L. Abraham, and J. E. Martin, *Anal. Chem.*, **45**, 1808 (1973).

61. A comprehensive review treating the inorganic and physicochemical aspects of these extraction reagents is presented by D. F. Peppard in Eméleus and Sharpe, Eds., *Advances in Inorganic Chemistry and Radiochemistry*, **9**, 25, Academic, New York, 1966. A bibliography of more than 300 references on organophosphorus reagents may be found in Chapter 9 of A. K. De, S. M. Khopkar, and R. A. Chalmers, *Solvent Extraction of Metals*, Van Nostrand Reinhold, New York, 1970. For interactions of acidic organophosphorus reagents in the organic phase, see Z. Kolarik in Y. Marcus, Ed., *Solvent Extraction Reviews*, vol. 1, Dekker, New York, 1971, in which will be found an extensive collection of extraction coefficients of these compounds and their dissociation and self-association constants.

62. D. Dyrssen and D. H. Liem, *Acta Chem. Scand.*, **14**, 1091 (1960); D. Dyrssen, S. Ekberg, and D. H. Liem, *ibid.*, **18**, 135 (1964); D. Dyrssen and D. H. Liem, *ibid.*, **18**, 224 (1964). The dimerization and solvent association of bis(p-chlorophenyl)phosphoric acid have been studied by F. Krasovec, M. Ostanek, and C. Klofutar, *Anal. Chim. Acta*, **36**, 431 (1966).

63. V. A. Navrotskaya and Y. B. Kletenik, *Russ. J. Inorg. Chem.*, **14**, 997 (1969).

64. K. Alcock, S. S. Grimley, T. V. Healey, J. Kennedy, and H. A. C. McKay, *Trans. Faraday Soc.*, **52**, 39 (1956).

65. D. F. Peppard, W. J. Driscoll, R. J. Sironen, and S. McCarty, *J. Inorg. Nucl. Chem.*, **4**, 326 (1957); D. F. Peppard, G. W. Mason, and J. L. Maier, *ibid.*, **3**, 215 (1956).

66. K. Naito, *Bull. Chem. Soc. Japan*, **33**, 363 (1960); D. F. Peppard, G. W. Mason, I. Hucher, and F. A. J. A. Brandao, *J. Inorg. Nucl. Chem.*, **24**, 1387 (1962).

67. R. F. Rolf, *Anal. Chem.*, **33**, 149 (1961); J. Huré, M. Rastoix, and R. Saint-James, *Anal. Chim. Acta*, **25**, 118 (1961).

68. F. Baumgärtner and L. Finsterwalder in Kertes and Marcus, *Solvent Extraction Research*, p. 313.

69. T. H. Handley and J. A. Dean, *Anal. Chem.*, **33**, 1087 (1961).

70. J. Goffart and G. Duyckaerts, *Anal. Chim. Acta*, **46**, 91 (1969).

71. J. C. White and W. J. Ross, NAS-NS **3102** (1961); T. Ishimori and E. Nakamura, *Japan Atom. Energy Res. Inst. JAERI*, **1047** (1963).

72. G. Michel, E. de Ville de Goyet, and G. Duyckaerts, *Anal. Chim. Acta*, **72**, 97 (1974).

73. W. B. Holbrook and J. E. Rein, *Anal. Chem.*, **36**, 2451 (1964). Used for determination of gold.

74. L. P. Varga, W. D. Wakley, L. S. Nicolson, M. L. Madden, and J. Patterson, *Anal. Chem.*, **37**, 1003 (1965).

75. Equilibrium constants for the extraction of HNO_3 into benzene by TOPO and other phosphine oxides have been determined by J. Goffart and G. Duyckaerts, *Anal. Chim. Acta*, **38**, 529 (1967). J. C. White and W. J. Ross, NAS–NS **3102** (1961), found indications of $TOPO \cdot 2 HNO_3$ at high HNO_3 concentrations. J. Goffart and G. Duyckaerts, *Anal. Chim. Acta*, **36**, 499 (1966), reported the extracted species $TBPO \cdot HNO_3$ and $TBPO \cdot 2 HNO_3 \cdot H_2O$ (TBPO = tri-n-butylphosphine oxide).

76. Aliphatic acids: M. Konstantinova, St. Mareva, and N. Jordanov, *Anal. Chim. Acta*, **68**, 237 (1974); aromatic acids: **59**, 319 (1972).

77. J. W. O'Laughlin and D. F. Jensen, *Anal. Chem.*, **41**, 2010 (1969).

78. U.S. Atomic Energy Comm. Report ORNL **3548** (1964).

79. D. F. Peppard, G. W. Mason, J. L. Maier, and W. J. Driscoll, *J. Inorg. Nucl. Chem.*, **4**, 334 (1957). Mono-(2-ethylhexyl)phosphoric acid has also been examined as a metal extraction reagent, but it is of lesser value.

80. T. B. Pierce and P. F. Peck, *Analyst*, **88**, 217 (1963).

81. Be, Fe(III), lanthanides, and actinides at low concentrations in 1 M mineral acid solutions are largely extracted according to this equation by HDEHP and other dialkylphosphoric acids.

82. D. F. Peppard, G. W. Mason, S. McCarty, and F. D. Johnson, *J. Inorg. Nucl. Chem.*, **24**, 321 (1962).

83. D. F. Peppard, G. W. Mason, and S. McCarty, *J. Inorg. Nucl. Chem.*, **13**, 138 (1960).

84. H. G. Petrow and C. D. Strehlow, *Anal. Chem.*, **39**, 265 (1967).

85. C. F. Coleman and J. W. Roddy in Y. Marcus, Ed., *Solvent Extraction Reviews*, vol. 1, Dekker, New York, 1971, p. 63, wherein their own and other work is discussed.

86. M. Jamil, P. zur Nedden, and G. Duyckaerts, *Anal. Chim. Acta*, **55**, 145 (1971).

87. J. E. Mrochek and C. V. Banks, *J. Inorg. Nucl. Chem.*, **27**, 589 (1965).

88. J. Goffart and G. Duyckaerts, *Anal. Chim. Acta*, **48**, 99 (1969). The same authors investigated the extraction of HNO_3 into benzene solutions of diphosphine dioxides (*ibid.*, **39**, 57); the extracted species are $DiPO \cdot HNO_3$ and $DiPO \cdot 2 HNO_3$.

89. Review: T. H. Handley, *Talanta*, **12**, 893 (1965).

90. T. H. Handley, *Anal. Chem.*, **35**, 991 (1963).

91. H. Bode and W. Arnswald, *Z. Anal. Chem.*, **185**, 99, 179 (1962). Early work on the use of these reagents was done by A. I. Busev and coworkers. See for example, *Zh. Anal. Khim.*, **4**, 49, 234 (1949); *Anal. Abst.*, **6**, 1268; **8**, 4449.

92. T. H. Handley and J. A. Dean, *Anal. Chem.*, **34**, 1312 (1962).

93. In 10 *M* HCl the half life of the diethyl ester is 4.8 hr according to H. Bode and W. Arnswald, *Z. Anal. Chem.*, **185**, 99 (1962).

94. In later work, T. H. Handley, *Anal. Chem.*, **37**, 311 (1965), found that $[(BuO)_2P(S)S]_2Hg$ and $[(BuO)_2P(S)SAg$ in CCl_4 extracted Ag from HNO_3 solution. Addition-type complexes are formed in which Ag (as $AgNO_3$) is probably bonded to S=. Zinc is extracted as the normal chelate by dialkyl-phosphorodithioic acids in CCl_4: T. H. Handley, R. H. Zucal, and J. A. Dean, *Anal. Chem.*, **35**, 1163 (1963).

95. H. Bode and K. Wulff, *Z. Anal. Chem.*, **219**, 32 (1966).

96. T. H. Handley and J. A. Dean, *Anal. Chem.*, **32**, 1878 (1960).

97. H. M. Nakagawa and H. W. Lakin, *U.S. Geol. Surv. Profess. Paper*, **525-C**, C172 (1965); T. T. Chao, J. W. Ball, and H. M. Nakagawa, *Anal. Chim. Acta*, **54**, 77 (1971).

98. T. H. Handley, *Anal. Chem.*, **36**, 2467 (1964).

99. D. E. Elliott and C. V. Banks, *Anal. Chim. Acta*, **33**, 237 (1965). The use of the reagent in reversed-phase partition paper chromatography is also described.

100. Tri-*n*-butylphosphine sulfide in CCl_4 behaves much the same as TOPS, as shown by R. B. Hitchcock, J. A. Dean, and T. H. Handley, *Anal. Chem.*, **35**, 254 (1963).

101. J. G. Cuninghame, P. Scargill, and H. H. Willis, AERE C/M 215 (1954).

102. C. A. Blake, C. F. Baes, K. B. Brown, C. F. Coleman, and J. C. White, *Proc. Second Intern. Conf. Peaceful Uses Atomic Energy*, vol. 28, p. 289, Geneva, 1958.

103. For reviews of synergic systems, many of which involve organophosphorus reagents, see H. M. N. H. Irving in D. Dyrssen, J.-O. Liljenzin, and J. Rydberg, Eds., *Solvent Extraction Chemistry*, Wiley, New York, 1967, p. 91, and T. V. Healey in A. S. Kertes and Y. Marcus, Eds., *Solvent Extraction Research*, Wiley-Interscience, New York, 1969, p. 257. A chapter is devoted to synergic extraction in Marcus and Kertes, *Ion Exchange and Solvent Extraction of Metal Complexes*.

104. For listings see Marcus and Kertes, *Ion Exchange and Solvent Extraction of Metal Complexes*.

105. Healy, *op. cit.* (note 103). A solvent such as chloroform and benzene can decrease synergism in this system, but less effectively than an excess of TBP.

106. The destruction of synergism can be put to analytical use. For example, after In has been extracted with mono-2-ethylhexylphosphoric acid or with HDEHP, it can be stripped from the organic phase (hydrocarbon) more

easily if TBP is added. See I. S. Levin and T. G. Azarenko, *J. Anal. Chem. USSR*, **20,** 419 (1965).

107. S. M. Wang, W. R. Walker, and N. C. Li, *J. Inorg. Nucl. Chem.*, **28,** 875 (1966).

108. B. B. Tomazic and J. W. O'Laughlin, *Anal. Chem.*, **45,** 106 (1973). They found sodium to be extracted at higher pH values, possibly as the result of an ion-exchange with a proton in $HHFA \cdot 2 H_2O \cdot 2 TBP$. [For synergic extraction of iron as $Fe(HFA)_2 \cdot (TBP)_2$ and $Fe(HFA)_3 \cdot TBP$, see *ibid.*, **45,** 1519.]

109. H. Sumiyoshi, K. Nakahara, and K. Ueno, *Talanta*, **24,** 763 (1977), determine K in the presence of much Na(blood serum) by adding Bromcresol Green and acetic acid-lithium acetate buffer to the aqueous sample solution, extracting with benzene solution of 18-crown ether-6 (see note 56) and measuring the absorbance of the extract at 410 nm. The extracted species is $(K Crown^+) (BCG^-)$. The pH must be below 4 in the extraction to maintain Bromcresol Green in the singly charged anionic state. The extraction of K is far from complete, so that the standard curve must be established under the same conditions. Apparently, K can be determined in the presence of 100 times as much Na. K^+ fits better than Na^+ into the structure of the crown ether.

9D

LIQUID–LIQUID EXTRACTION
RATIONALE OF SEPARATION

I. GENERAL CONSIDERATIONS

The power of liquid–liquid extraction in the separation of metals lies in the great range of extraction coefficients that can be obtained with their complexes and in the ease with which the analyst can vary them to optimize separability. The extractability of a metal is a function of the fraction of a metal in aqueous solution that has been converted to a species extractable into an organic solvent. For a given metal and reagent, this fraction depends on the concentration of excess reagent forming the extractable substance, more often than not on the hydrogen-ion concentration and on the presence of masking (mostly complexing) agents. The analyst can usually control these variables so that the extraction coefficient of the desired constituent is given a value that will assure quantitative recovery in a small number of successive extractions (usually not more than two in trace analysis), while the extraction coefficients of undesired constituents are kept at such low levels that these remain in the aqueous phase for the most part. As we shall see, further separation can be obtained by backwashing the organic extract.

The separability of two constituents A and B depends on the individual values of E_A and E_B, not on their ratio α (witness the tabular example on p. 1034).[1] The magnitude of α is obviously of importance in separation. If E_A is fixed, E_B under those conditions is determined by α. But being a ratio, α is indeterminate as far as separability is concerned.

In theory, two constituents A and B can be separated by extraction if their extraction coefficients are not equal. In practice, a satisfactory separation will not be feasible if the values E_A and E_B are so close together that the required expenditure of time and labor becomes prohibitive. We shall limit the discussion of extraction separations to very simple multistage schemes in which the number of steps is small and simple apparatus (e.g., separatory funnels) is used. These are more important in usual inorganic analytical practice than extensive fractionation schemes.[2]

Many metals can often be separated adequately from each other in a comparatively small number of extraction steps, even when the ratio of one metal to the other is large, say 10^3, 10^4, 10^5, perhaps higher. This requires a considerable difference in the values of E_A and E_B, that is, α needs to be 10^3–10^4 if $(Q_B)_0/(Q_A)_0$ is large, say $> 10^4$. The use of differential complexing agents, in addition to adjustment of acidity and excess reagent concentration, will often—by no means always—bring this about. A change in the oxidation state of a metal (reduction of Fe^{3+} to Fe^{2+}, for example) can be a valuable aid, but is a less general possibility.

When simple extraction schemes fail, the trace analyst is likely to consider abandonment of liquid–liquid extraction as such and its replacement by chromatographic separation in which the required fractionation may be provided more expeditiously.

II. RECOVERY FACTOR IN EXTRACTIONS

The aim in any analytical separation of the constituents A and B is to obtain a recovery factor Q/Q_0 for one of them (say A) that approaches 1, and a recovery factor for the other that approaches 0, within certain limits of time and effort. The required closeness of approach to 0 depends on the extent that B interferes in the determination of A, and their relative amounts in the sample. As already pointed out, R_A should preferably have a value $\lesssim 0.99$ in trace analysis, but a value $\lesssim 0.95$ may be acceptable in the lower trace ranges, especially if the separation is difficult. For a fixed value of E_A, R_A can be made to approach 1 by making successive extractions with fresh portions of organic solvent solution of the reagent under conditions such that the concentration of the extraction reagent (and of H^+) remains the same, or much the same, through the series.

R as a Function of E and n. If E is independent of the concentration of the extracted constituent, the fraction of the latter in the combined extracts after n extractions $(= R_n)$, in each of which the volumes of the aqueous and organic solvent phases are equal (complete immiscibility assumed) is given by

$$R_n = \frac{Q_n}{Q_0} = 1 - \left(\frac{1}{E+1}\right)^n = \frac{(E+1)^n - 1}{(E+1)^n}$$

Multiplication of R_n by 100 gives the percentage recovery, $R_{n,\%}$.

If the aqueous phase volume is V_w, and each portion of organic phase is V_{org}, and letting $r = V_{org}/V_w$,

$$R_n = 1 - \left(\frac{1}{rE+1}\right)^n = \frac{(rE+1)^n - 1}{(rE+1)^n} = \frac{(E'+1)^n - 1}{(E'+1)^n}$$

where rE ($= E'$) is the effective extraction coefficient. Approximately (exactly if $n = 1$), when $E > {\sim} 10$

$$R_n = \frac{(E')^n}{(E')^n + 1}$$

With a fixed total volume of organic solvent and an infinite number of extractions,

$$R_{max} = 1 - \exp(-r'E)$$

where $r' = (\Sigma V_{org})/V_w$ as $V_{org} \to 0$.

Continuous Extraction. When E has a small value, especially when a large amount of constituent is to be extracted as completely as possible to reduce its concentration, resort may be made to continuous extraction, in which the organic solvent falls through, or rises through, a column of aqueous solution dropwise. The operation is carried out automatically in some form of continuous extraction apparatus. A limited volume of organic solvent is used, its circulation usually being maintained by vaporization and condensation. The effective ratio V_{org}/V_w can thus be made large. By reducing the size of each organic solvent portion, as becomes practicable in continuous extraction, good recoveries can be secured even when E is small, as brought out in the following tabulation:

	Number of Equal Portions for Extraction	R			
		$E = 1$	$E = 2$	$E = 3$	$E = 5$
$V_{org} = V_w$	2	0.56	0.75	0.84	
	4	0.59	0.80	0.89	
	10	0.615	0.84	0.93	
	∞	0.632	0.86	0.95	
$V_{org} = 2V_w$	2	0.75	0.89	0.94	0.97
	4	0.80	0.94	0.974	0.993
	10	0.84	0.966	0.991	0.9990
	100	0.86	0.980	0.9971	$1 - 7.2 \times 10^{-5}$
	∞	0.865	0.982	0.9975	$1 - 5 \times 10^{-5}$
$V_{org} = 3V_w$	100	0.948	0.997	$1 - 2 \times 10^{-4}$	$1 - 9 \times 10^{-7}$
	∞	0.95	0.9975	$1 - 1.2 \times 10^{-4}$	$1 - 3 \times 10^{-7}$
$V_{org} = 5 V_w$	∞	0.993	$1 - 5 \times 10^{-5}$	$1 - 3 \times 10^{-7}$	$1 - 1 \times 10^{-11}$
$V_{org} = 10 V_w$	∞	$1 - 4 \times 10^{-5}$	$1 - 2 \times 10^{-9}$		

Note that there is little difference between extracting with 100 portions and an infinite number, if equilibrium is attained.

Even if E is not particularly small, effective E ($= E'$) may be, as when a large volume of aqueous solution is to be extracted with a small volume of organic solvent. Continuous extraction may then be advantageous. It is applied, for example, in the extractive isolation of trace metals from sea water. Continuous extraction may be indicated if extraction is slow.

Finding E. Tabulations of E for some extraction reagents are available (see Chapter 9A). Values of E for common extraction reagents at specified concentrations of H^+ and complexing anions can often be calculated from known values of the extraction equilibrium constant K_{ext} and complex formation constants as we have seen.[3] Such values are subject to experimental verification. Even when only approximate, they are useful in the selection of experimental conditions and in limiting the amount of necessary experimental work. Some extractions are slow and shaking must be continued long enough for attainment of equilibrium or an approach to it.

It should be noted that E cannot have a value greater than P, the partition constant of the extracted species. The P values of most metal–organic complexes for common extracting solvents tend to be so large that this is usually not an important restriction analytically. Lead dithizonate has $P_{CCl_4/H_2O} = 2 \times 10^3$, one of the smaller chelate values. This is larger than a desirable value of 10–100 for E_A' in extraction separations. In extractions of solvated species with oxonium-type solvents, E_{max} may not be as large as desirable and a salting-out agent may be needed to increase it.

Often it is convenient to use a volume of extracting organic solvent smaller than the aqueous sample solution. In general, there should be no particular difficulty in working with V_{org} as small as $0.1\ V_w$. Thus extraction of 50 ml of aqueous solution with 5 ml of organic solvent is easy enough; extraction of 10 ml of aqueous solution with 1 ml of organic phase would require more care. Care must obviously be taken to prevent mechanical contamination of one phase by the other. Sufficient time must be allowed for the phases to separate after shaking. Centrifugation is sometimes advantageous. Washing of an organic extract by an aqueous solution of the proper composition (pH, reagent concentration) and conversely is required. Suppose $Q_B \gg Q_A$ and A is being extracted into an organic solvent. After the first extraction (organic phase denser than aqueous), washing of the aqueous solution and the separatory funnel with organic solvent to remove surface organic film or droplets adhering to funnel walls will usually not be necessary because this will be done by the second portion of organic extracting solution. But small droplets of aqueous phase in the organic extract should be removed by washing (shaking) with a small volume of aqueous solution of the proper composition. In the back-extraction (p. 818), the organic phase is automatically washed by the second portion of aqueous scrubbing solution.

Stems of separatory funnels should not exceed 1–2 cm in length and the ends should be cut at a slant.

Constancy of E. It is important for the analyst to know whether E remains constant as the concentration of A varies, but the concentration of the extraction reagent and other variables such as $[H^+]$ remain constant.[4]

The most obvious reason for a change in E_A with a change in the concentration of A is the occurrence of A in entities not containing the same number of atoms of A in the two phases. Suppose A occurs as monomer in the aqueous phase and as A_n in equilibrium with the monomer in the organic phase:

$$A_{H_2O} \rightleftarrows A_{org}$$
$$(nA)_{org} \rightleftarrows (A_n)_{org}$$

The partition constant of the monomer is

$$P = \frac{[A]_{org}}{[A]_{H_2O}}$$

and the self-association constant of A in the organic phase is

$$K_{assoc} = \frac{[A_n]_{org}}{[A]_{org}^n}$$

$$\therefore E_A = \frac{\Sigma [A]_{org}}{\Sigma [A]_{H_2O}} = \frac{[A]_{org} + n[A_n]_{org}}{[A]_{H_2O}} = \frac{[A]_{org} + nK_{assoc}[A]_{org}^n}{[A]_{H_2O}}$$

$$= P(1 + nK_{assoc}P^{n-1}[A]_{H_2O}^{n-1})$$

An increase in the equilibrium concentration of A in the aqueous phase results in an increase in E. Conversely, if A is associated in the aqueous phase but not in the organic phase, E will decrease as the concentration of A increases. The association leading to a change in E need not be real, only virtual, as pointed out earlier (e.g., formation of polynuclear complexes).

In most—by no means all—analytical extractions, E is found to be essentially independent of the concentration of the constituent extracted. In not a few systems, the constancy of E has been demonstrated to hold over a wide range of concentrations. From the examples given in Table 9D-1, a 10^8- or even 10^9-fold concentration range of constancy is seen to hold for some extraction forms. However, in a few systems, E definitely

TABLE 9D-1
Constancy of E^a

Metal	Aqueous Solution	Organic Phase	Range (μg/ml) Over Which E Is Constant
Ag	Conc. HCl	$CHCl_3 + n$-octyl-amine	10^{-2}–4×10^2
As(III)	Conc. HCl	Various	$\sim 10^{-3}$–$\sim 10^4$
Au(III)	Conc. HCl	Benzene + TBP	10^{-2}–4×10^2
Au(III)	1.5 M HCl	Ethyl ether	$< 10^{-2}$–20
Au(III)	1 M HBr	Ethyl ether	10^{-3}–2×10^4
Be	4 N H_2SO_4	Di-n-decylphos-phoric acid in toluene	~ 0–10^3
Co	NH_4SCN	sec-Amyl alcohol	$\sim 10^{-8}$–0.4
Co	NH_4SCN	Ph_4AsCl	$\sim 10^{-5}$–1.5
Co		H_2Dz in CCl_4	$\sim 10^{-5}$–~ 1
Fe(III)	7 M HCl	Isopropyl ether	12–100
Ga	6 M HCl	Ethyl ether	$\sim 10^{-7}$–70
Ge	7.5 M HCl	CCl_4, etc.	5×10^{-2}–5×10^3
In	4.5 M HBr	Isopropyl ether	$\sim 10^{-2}$–10^2
In	pH 3.85	8-Hydroxyquinoline in $CHCl_3$	10^{-7}–6
Mo(VI)	6 M HCl	MIBK	10^{-4}–10^3
Nb(IV)	NH_4SCN, 2 M HCl	Ethyl ether	10^{-5}–66
Os(VIII)	H_2O	CCl_4	$\sim 10^{-4}$–$\sim 10^2$
Pa	5 M HCl	Di-iso-butyl-carbinol	10^{-9}–20
Pd(II)	Conc. HCl	$CHCl_3 + n$-octyl-amine	10^{-2}–4×10^2
Pt(IV)	NH_4OH + Na diethyldithio-carbamate	$CHCl_3$	10^{-2}–4×10^2
Re(VII)	Ph_4AsCl, pH 8.5	$CHCl_3$	2×10^{-4}–20
Sc	SCN^-	Ethyl ether	2×10^{-4}–5×10^3
Tl(III)	3 M HCl	Isopropyl ether	2×10^{-2}–5×10^2
U(VI)	6 M HCl	Xylene + tri-i-octyl-amine	~ 0–4×10^3
Zn	SCN^-	Ethyl ether	1×10^{-2}–2×10^2
Zr	2 M HNO_3	Xylene + thenoyl-trifluoroacetone	~ 0–2×10^3

^{a} Selected values from R. Bock and A. Monerjan, *Z. Anal. Chem.*, **226,** 29 (1967), either from their list of literature values or from the experimental values of these authors. The true ranges may be wider than the ranges given. According to R. Litman, E. T. Williams, and H. L. Finston, *Anal. Chem.*, **49,** 983 (1977), Hg(II) in nitric acid solution at a concentration of 0.015 μg/ml is not completely extracted by dithizone in carbon tetrachloride. When 5×10^{-5} M $K_2Cr_2O_7$ is present, extraction is 99% complete.

varies with a comparatively small change in concentration of the extracted metal. This is true for the extraction of zinc with dithizone (0.01% solution in CCl_4) from an aqueous solution of pH 5 (acetate) containing thiosulfate. The extraction coefficient was found[5] to vary as follows with the initial concentration of zinc in the aqueous phase:

μg Zn/ml	1.1×10^{-2}	5.6×10^{-2}	0.56	3.6
E	24	125	2800	5200

Incomplete formation of the chelate at low concentrations must be the reason for the low E. Another system in which E increases with the concentration of the metal is Mo(VI)–dithiol–$CHCl_3$ ($E = 19$ and 160 at 4.3×10^{-2} and 0.94 μg Mo/ml, respectively). It has been known for a long time that E for Fe(III)–HCl–ethyl ether increases with the iron concentration. At least in part this behavior is due to association (polymerization) of $HFeCl_4$ in the ether phase.

The incomplete extraction of Co at low concentrations ($< 1\ \mu g/ml$) by a $CHCl_3$ solution of oxine, even after long shaking, is attributed to slow formation of the chelate.[6] The extraction becomes essentially complete if oxine is added to the aqueous solution, which is then heated to 85° for 20 min (and cooled) before extraction with $CHCl_3$ (Fig. 9D-1). The extraction of Mo(VI) as dithiolate by amyl acetate is incomplete below 1 ppm Mo, even if shaking is continued for several hours. Complete extraction is obtained (even at 0.001 ppm Mo) if the solution is warmed to 75°. In these, and other cases, extraction is rapid if the complex is first

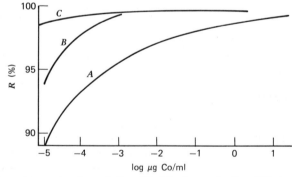

Fig. 9D-1. Percentage extraction of Co with 8-hydroxyquinoline (1% in $CHCl_3$) from aqueous solution of pH 7.8 (equal phase volumes, shaking for 3 min). (A) without heating; (B) after heating to 70°C; (C) after heating to 85°. From R. Bock and K.-D. Freitag, Z. Anal. Chem., **254,** 105 (1971).

completely formed in the aqueous phase. The rate of formation of some chelates is unexpectedly slow at very low metal concentrations.

III. SEPARATION FACTOR IN EXTRACTIONS

As we have already seen, the extent of separation of two constituents A and B from each other, A being the desired constituent, is expressed by the separation factor

$$S_{B/A} = R_B/R_A$$

In an extraction separation with a given reagent and solvent, the R values for A and B will depend on the E values, the relative volumes of the two immiscible phases and the number of successive extractions. For n successive direct extractions in which the phase volumes (org/aq) have the value r (so that $E' = rE$), the separation factor is

$$(S_{B/A})_n = \frac{(E'_B + 1)^n - 1}{(E'_B + 1)^n} \times \frac{(E'_A + 1)^n}{(E'_A + 1)^n - 1}$$

In quantitative work, $R_A \sim 1$, so that

$$(S_{B/A})_n = \frac{(E'_B + 1)^n - 1}{(E'_B + 1)^n}$$

It is assumed that there is no interaction between A and B, that is, E_A and E_B are the same as when A and B are each alone.

E'_A should not be unnecessarily large. If A and B react with the same reagent to give extractable complexes, an increase in E'_A by a change in pH or reagent concentration generally also increases E'_B. With a relatively low value of E'_A, R_A can, of course, be made larger by increasing n, but R_B is increased at the same time. Conveniently n is restricted to 2 or 3. If one direct extraction is sufficient to give an adequately large value of R_A, the separatory funnel would in any event need to be washed with a portion of the extracting solvent, so that this step might as well be made a second extraction with a fresh portion of reagent solution. If desirable, E'_A can be reduced in the second extraction by decreasing the volume of the organic solvent.

In chelate extractions (and other types of reactions as well), extraction coefficients of constituents to be separated may be correlated: a change in E_A may produce the same relative increase or decrease in E_B. This would be true if the two metals A and B are present as simple cations with the same charge in the aqueous phase, and they form analogous extractable complexes. Thus

$$A^{2+} + 2\,HL \underset{org}{\rightleftharpoons} AL_2 + 2\,H^+$$
$$\phantom{A^{2+} + 2\,}_{org}$$

$$B^{2+} + 2\,HL \underset{org}{\rightleftharpoons} BL_2 + 2\,H^+$$

$$E_A = \frac{[AL_2]_{org}}{[A^{2+}]} = \frac{[HL]^2_{org}}{[H^+]^2} \times K_{ext\,A}$$

and similarly for E_B. Therefore,

$$\frac{E_A}{E_B} = \frac{K_{ext\,A}}{K_{ext\,B}} = \alpha$$

Addition of a complexing agent that complexes A and B to different extents destroys the constancy of α with a change in $[H^+]$ and $[HL]_{org}$. Hydroxyl ion, leading to hydrolysis of metal cations, would be included among such complexing agents.

If the cations A and B have different charges, the relation between E_A and E_B is

$$\frac{E_A}{E_B} = \alpha = \frac{K_{ext\,A}}{K_{ext\,B}} \times [H^+]^{b-a}\,[HL]^{a-b}_{org}$$

and α varies with acidity and reagent concentration in the organic solvent phase.

General Separatory Extraction Scheme. As before take $E_A > E_B$. The first requirement in any separatory extraction preliminary to the determination of constituent A is its adequate recovery. As already noted, R_A should preferably exceed 0.99, but the working analyst may be willing to accept 0.95, especially when the content of A in the sample is less than 10 ppm, or even less than 100 ppm, and interfering constituents are present in unfavorably large ratios. Multiple direct extractions assure attainment of an acceptable R_A but at the cost of increasing R_B (if B is extracted at all) as the number of direct extractions increase. By reversing the process—back-extracting the combined organic extracts with an aqueous solution providing suitable values for E'_A and E'_B—B is removed more effectively than A (B is removed in geometric decrements approximately, A in small equal decrements if $E_B < 1$ and $E_A > 1$). By judicious choice of E'_A and E'_B in the backwashing, a large fraction of B can be removed from the organic phase without bringing R_A below the required value.[7] The separation obviously will be easier the greater the spread between E_A and E_B. A further requirement we arbitrarily impose is a small number of steps in the process: two (possibly three) direct extractions and two backwashings. Only two separatory funnels are needed.

This scheme requires that the extraction of B is reversible. As noted in earlier chapters, not all extractions are reversible, but there are more reversible than irreversible extractions. Finally, a relatively rapid attainment of extraction equilibrium is desirable.

What can be accomplished by applying such a scheme will be illustrated by two metals having correlated extraction coefficients, $E_A/E_B = \alpha = 10^4$. It will be assumed that all extractions are reversible and that E'_A can be given[8] values in the range 10 to 10^4 (actually 10 to 10^3 as it turns out) and E'_B values in the range 1 to 10^{-3}. Recoveries of A and B obtained by various combinations of E'_A and E'_B in direct extraction and in backwashing are shown in the table on p. 1034 (from values in Table 9D-2) in part (a).

Without backwashing, the best combination is $E'_A = 10$ and $E'_B = 10^{-3}$, which gives $R_A = 0.992$ and $R_B = 0.002$ ($= S_{B/A}$) after two extractions. If we specify that after separation $(Q_B)_{n=2} < 0.1 (Q_A)_{n=2}$, the original ratio $(Q_B)_0/(Q_A)_0$ could not exceed 50. This is a poor separation for trace analysis. But by backwashing twice with a solution giving $E'_A = 10^2$ and $E'_B = 10^{-2}$, $R_B = 2 \times 10^{-7}$, and $(Q_B)_0/(Q_A)_0$ could be 2×10^6. This separation has been gained with a loss of 3% in A, which is not serious. To be sure, in practice it may not be possible to adjust E'_A and E'_B as precisely as this example requires, so that in a sense this is an ideal result, but then R_B is so small that in many analyses an R_B of 2×10^{-6} or 2×10^{-5} might be allowed, thus permitting a margin for error.[9]

If $\alpha = 1000$ (b in the table), $R_A = 0.97$ and $R_B = 2 \times 10^{-4}$ with $E'_A = 10$ and $E'_B = 10^{-2}$ in the direct extraction ($n = 2$), and $E'_A = 10^2$ and $E'_B = 0.1$ in the back-extraction ($N = 2$). When $\alpha = 100$, separation is poor ($E'_A = 10$, $E'_B = 0.1$ in direct extraction, $n = 2$, and $E'_A = 10^2$, $E'_B = 1$ in back-extraction, $N = 2$), giving $R_A = 0.97$ and $R_B = 0.04$. The situation is not much improved if an attempt is made to recover A from the aqueous backwashings when $E'_A = 10$ and $E'_B = 0.1$ (in both direct and back-extractions, $n = 2$, $N = 2$, and also in one extraction of backwashings). Although R_A now becomes 0.98, $R_B = 0.017$. The slight decrease in R_B is hardly worth the introduction of another step (which moreover violates our limitation of four extractions total).

Denoting the number of direct (d) extractions as n and the number of backwashings (b) as N, the approximate expression for the recovery of B is

$$R_B = nE'_{B,d}E'^N_{B,b} \qquad (E' < 0.1)$$

This is also $S_{B/A}$ when $R_A \sim 1$, as already noted. If

$$E'_{B,d} = E'_{B,b} = E'_B, \qquad R_B = nE'^{N+1} \qquad (E' < 0.1)$$

Direct Extraction

Back-Extraction

	α	E'_A	E'_B	$n=1$ R_A	$n=1$ R_B	$n=2$ R_A	$n=2$ R_B	E'_A	E'_B	$N=1$ R_A	$N=1$ R_B	E'_A	E'_B	$N=2$ R_A	$N=2$ R_B
a	10^4	10^4	1	$1-10^{-4}$	0.50	$1-10^{-8}$	0.75								
	10^4	10^3	10^{-1}	0.999	0.091	$1-10^{-6}$	0.17								
	10^4	10^2	10^{-2}	0.99	0.01	$1-10^{-4}$	0.02	10^2	10^{-2}	0.99	2×10^{-4}				
	10^4	10	10^{-3}	0.91	0.001	0.992	0.002	10^2	10^{-2}	0.98	2×10^{-5}	10^2	10^{-2}	0.98	2×10^{-6}
b	10^3	10	10^{-2}			0.992	0.02	10	10^{-2}	0.90	2×10^{-4}	10^2	10^{-2}	0.97	2×10^{-7}
	10^3	10	10^{-2}			0.992	0.02	10^2	10^{-1}	0.98	2×10^{-3}	10^2	10^{-1}	0.97	2×10^{-4}
c	10^2	10	10^{-1}			0.992	0.17	10^2	1	0.98	0.09	10^2	1	0.97	0.04

TABLE 9D-2

R as a Function of E and Number of Extractions

	R						
	Direct Extraction n			Back-Extraction[a] N			
E'	1	2	3	1	2	3	
10^{-5}	1×10^{-5}	2×10^{-5}	3×10^{-5}	1×10^{-5}	1×10^{-10}	1×10^{-15}	
10^{-4}	1×10^{-4}	2×10^{-4}	3×10^{-4}	1×10^{-4}	1×10^{-8}	1×10^{-12}	
10^{-3}	0.001	0.002	0.003	0.001	1×10^{-6}	1×10^{-9}	
10^{-2}	0.01	0.02	0.029	0.01	1×10^{-4}	1×10^{-6}	
10^{-1}	0.091	0.17	0.25	0.091	0.008	7.5×10^{-4}	
1	0.50	0.75	0.875	0.50	0.25	0.125	
3	0.75	0.938	0.984				
10	0.91	0.992	0.9992	0.909	0.826	0.75	
10^2	0.990	0.99990	$1 - 10^{-6}$	0.990	0.980	0.97	
10^3	0.999	$1 - 10^{-6}$	$1 - 10^{-9}$	0.999	0.998	0.997	
10^4	0.9999	$1 - 10^{-8}$		0.9999	0.9998	0.9997	
10^5	$1 - 10^{-5}$						

[a] Before back-extraction ($N = 0$), $R = 1$.

Separation of Bi from Pb by Dithizone Extraction from Acid Solution. The application of the foregoing conclusions can be illustrated by this extraction system. The extraction constants with a CCl_4 solution of dithizone are

$$K_{ext\ Bi} = \frac{[Bi(HDz)_3]_{CCl_4}[H^+]^3}{[Bi^{3+}][H_2Dz]^3_{CCl_4}} = 8.2 \times 10^{10} \qquad (\mu = 0.1, 25°)$$

$$K_{ext\ Pb} = \frac{[Pb(HDz)_2]_{CCl_4}[H^+]^2}{[Pb^{2+}][H_2Dz]^2_{CCl_4}} = 5.6 \qquad (\mu = 0.1, 25°)$$

It is evident from the magnitude of the constants that Bi should be easily separable from much Pb. The following calculation verifies the expectation.

First, we seek the $[H^+]$ (from noncomplexing $HClO_4$) at which E_{Bi} will have a value of 10–100, taking hydrolysis of Bi^{3+} to $BiOH^{2+}$ into possible account:

$$E_{Bi} = \frac{8.2 \times 10^{10}[H_2Dz]^3_{CCl_4}}{[H^+]^3(1 + \beta_1 K_w/[H^+])} = \frac{8.2 \times 10^{10}[H_2Dz]^3_{CCl_4}}{[H^+]^3(1 + 0.037/[H^+])}$$

A reasonable concentration of the H_2Dz solution is $1 \times 10^{-4}\,M$ (10 ml is equivalent to 70 μg of Bi). We then find (assuming the dithizone concentration does not change much) that $E_{Bi} \sim 10$ at $[H^+] = 0.2$ and ~ 100 at $[H^+] = 0.09$. If hydrolysis of Bi^{3+} is taken into account, E_{Bi} at $[H^+] = 0.09$ is 80. After two dithizone extractions (equal phase volumes), $R_{Bi} = 0.9998+$ at this $[H^+]$. E_{Pb} under the conditions is 7×10^{-6}, so that R_{Pb} after two extractions is 1.4×10^{-5}. $S_{Pb/Bi}$ is the same. The separation factor is so favorable that ordinarily scrubbing for removal of $Pb(HDz)_2$ from the combined CCl_4 extracts would not be done. But there is a very good reason for carrying it out, namely to remove aqueous Pb^{2+} solution carried along as droplets in the CCl_4 phase (0.001 ml of a 1% Pb solution would introduce 10 μg Pb). A double wash of the Bi extract with an equal volume of $\sim 0.1\,M\,HClO_4$ would not remove significant amounts of Bi from a CCl_4 phase $0.0001\,M$ in H_2Dz.

There is little reason to question the above results, but work by Bidleman[10] confirms them. He separated Bi from Pb under essentially the conditions described above and determined Bi with dithizone in basic cyanide solution:

Pb Present (mg)	Bi Taken (μg)	Bi Found (μg)
74	0	0.0
74	3.27	3.2, 3.2, 3.3, 3.2
147	9.8	9.7, 9.9
167	5.23	5.3, 5.2
167	7.84	8.1, 8.0

Any unseparated Pb would have reacted and given high results for Bi.

Separation by Extraction of Undesired Constituents. The trace constituent A is left in aqueous solution, while B, which may be another trace constituent or a minor or major constituent, is extracted. This is usually a less desirable process than the reverse and may be applied more by necessity than by choice. If B is a major (or minor) constituent, a considerable number of extractions may be needed to reduce its concentration to the desired level in the aqueous solution. Not all chelates are sufficiently soluble in organic solvents to permit easy extraction. Constituent A will be left with other nonextracted constituents in the aqueous phase. Nevertheless, extraction of a major or predominating constituent may be useful at times. Thus the bulk of iron(III) can be extracted as $HFeCl_4$ by oxonium solvents; continuous extraction may be used. It is not always necessary that the last traces of constituent B be removed. Another separation may follow, or the concentration of B may be reduced to a level where it can be masked or it does not interfere with the determination of A.

The Bi–Pb dithizonate separation already discussed can be used to extract relatively much Bi from Pb. The solubility of Bi(HDz)$_3$ in CCl$_4$ is 5.4×10^{-4} M at 25°. With equal volumes of phases, two extractions at $[H^+] = 0.09$ would reduce the Bi concentration to about 2×10^{-4} of its original value, if the excess dithizone were 10^{-4} M in each extraction. Other metals removed include Au, Ag, Hg, and Cu.

Separations Based on Extraction Rate. At times, separations can be based on the difference in the rates of extraction or back-extraction of different metals, if pronounced. For example, cobalt forms chelate complexes with a number of organic reagents in weakly acidic or weakly alkaline solution, which, once formed, are not easily decomposed by mineral acid solutions. Cobalt(II) reacts with 1-nitroso-2-naphthol or 2-nitroso-1-naphthol in weakly acidic solution to give a Co(III) chelate extractable into chloroform. When the chloroform extract is shaken with 2 M HCl, the Co chelate is unaffected, but Cu(II) and Ni complexes of the reagent are dissociated into Cu^{2+} and Ni^{2+}. The stability of the Co(III) chelate is due to its inert character in the sense of nonequilibrium stability. Some chelates of Cr, the Pt metals, and other metals also show similar inertness. Thus Ru(VI) reacts with 8-hydroxyquinoline in hot aqueous solution at pH 4–7 to give a green chelate (Ru in the III or IV state), which is extractable into chloroform.[11] The chelate is not dissociated if the extract is shaken with H$_2$SO$_4$, < 2.5 M; HCl, < 4 M; H$_2$C$_2$O$_4$, < 0.5 M; and NaOH, < 4 M. Most other metal complexes of 8-hydroxyquinoline are removed from the extract by shaking with 2 M H$_2$SO$_4$. Exceptions are Co, Cr, Mo, Pd, and Rh, most of which must form inert complexes.

Differences in rate of extraction from aqueous solutions are sometimes of value in separations. Cr(III) is hardly extracted by acetylacetone at room temperature, whereas most other metals are extracted more or less rapidly. The slow extraction of copper and the rapid extraction of mercury(II) by a chloroform solution of dithizone helps to separate the two metals. However, such methods should be viewed with some scepticism, because rates of extraction can be affected by other constituents in a sample and the purity of the organic solvent. We know that such metals as Cu and Bi are extracted more rapidly from acid solution by dithizone in *pure* CCl$_4$ than in reagent-quality CCl$_4$.

Separatory extractions based on inspection of a table of K_{ext} values may be unsuccessful because of rate factors. The metal to be separated must be extracted reasonably rapidly from the aqueous solution. Usually such information is available in the literature.

Also, apart from rate factors, it should be remembered that two metals together may not show the same extraction behavior as when present singly. To repeat an example given earlier, Ca is hardly extracted by 8-hydroxyquinoline in benzene at pH 8–9 when present alone, but is extracted in the presence of Sc as $Ca(ScOx_4)_2$.

NOTES

1. As noted earlier, α should not be called the separation factor for A/B. R_B/R_A is the separation factor, not α [E. B. Sandell, *Anal. Chem.*, **40**, 834 (1968)]. Curiously, there is resistance to denying α the name separation factor, apparently from a disinclination to abandon old cherished definitions even when they are demonstrably faulty. In the early days of liquid–liquid extraction, the term separation factor was incautiously bestowed on α (an example of "terminological inexactitude"). To say that R_B/R_A should be called the decontamination factor is useless, for this is the radiochemists' term for separation factor (actually R_B/R_A is the inverse of the decontamination factor). Likewise, the proposal to call R_B/R_A the enrichment factor is inappropriate.

2. When more extensive fractionation is required, the "push through" method of Peppard may be considered (in EMeléus and Sharpe, *Advances in Inorganic Chemistry and Radiochemistry*, Vol. 9, 1966, p. 8), which is based on multiple extraction and multiple scrubbing to give good recovery of the desired constituent and its separation from others. This method has been examined by T. W. Gilbert, *J. Chem. Educ.*, **51**, 822 (1974), who gives a clear account of its design and general application. Gilbert accepts $R_B/R_A = S_{B/A}$ as the true measure of separability.

3. If K_{ext} values are not listed, E can sometimes be obtained or estimated from experimental pH–E or pH–R curves which may be given in the literature. Such extraction curves hold for a specified reagent concentration, but E values can often be calculated or estimated for other concentrations. Even approximate values may suffice for predicting separability. It should be remembered that K_{ext} is usually expressed in terms of $[M^{n+}]$ in the aqueous phase and that forms of M other than M^{n+} may occur in larger concentrations than the latter. E taking only M^{n+} into account is then erroneous and additional equilibria must be taken into consideration. Experimental pH–E or pH–R curves are of importance when the additional constants are not available. Some chelates are quite soluble in water (or in water saturated with the extracting solvent) or can form negatively charged species with an excess of the chelating reagent. For example, acetylacetonates of many metals are very incompletely extracted (Chapter 9B).

4. In order that the concentration of the reagent may not change appreciably, the concentration of A must be small compared to that of the reagent. This condition is most closely fulfilled when A is present as a radioactive isotope in tracer concentrations. The concentration of H^+ and other species involved

directly or indirectly in the extraction reaction may be inadvertently altered in successive extractions if they partition into the organic solvent. Thus ethyl ether extracts HCl from an aqueous solution unless it has been shaken with a sufficiently large volume of aqueous HCl solution having the same composition as the solution to be extracted. The concentration of an extraction reagent that partitions appreciably into an aqueous phase (or into the organic phase if added to the aqueous phase) is likewise subject to change.

5. Bock and Monerjan, reference in Table 9D-1. The time of shaking was 2–3 min.
6. R. Bock and K.-D. Freitag, *Z. Anal. Chem.*, **254**, 104 (1971).
7. *Anal. Chim. Acta*, **4**, 504 (1950).
8. It will be understood that the desired E values may not lie in the accessible range of acidity and reagent concentration. Not only the ratio of K_{ext} values but their absolute values as well are important.
9. The effect of an error in E'_A on R_A can be illustrated by taking the actual value as 3.33, compared to the assumed value of $E'_A = 10$. Actual R_A after two direct extractions is 0.947 compared to 0.992 that the analyst thinks he is getting. Reduction of E'_A by the factor one-third has brought R_A to the edge of acceptable recovery; reduction by one-half would reduce recovery to 0.97. Thus a reasonable error is acceptable.
10. T. Bidleman, Ph. D. Thesis, University of Minnesota, 1970.
11. H. Hashitani, K. Katsuyama, and K. Motojima, *Talanta*, **16**, 1553 (1969).

VOLATILIZATION

Gaseous and volatile compounds are more often found among the nonmetals than the metals. However, some ten metals and semimetals (metalloids) form compounds of sufficient volatility (or are themselves sufficiently volatile) to make them useful in separations. In general, the volatile compounds of metals and semimetals are those formed with elements that are gaseous at room temperature. These compounds are covalent and the metal or metalloid has an oxidation state of +3 or higher; hydrides are also of importance.

Because of the relatively small number of metals giving volatile compounds, the selectivity of volatilization methods is likely to be better than when other separation methods are applied. However, organic solvents of the inert type (CCl_4, $CHCl_3$) extract covalent volatile compounds, so that extraction competes with volatilization methods in their separation, as with $GeCl_4$, $AsCl_3$ and OsO_4.

I. GENERAL ASPECTS

We use the term volatilization to refer to the passing of a substance in the liquid or solid state, or in solution, into a vapor or gas. Vaporization is the transformation of a substance into vapor. Distillation is the volatilization of a liquid, or of a substance in solution, and its subsequent condensation or absorption. Sublimation involves volatilization of, or from, a solid phase, followed by condensation to a solid. Evolution refers to liberation of a gas, which, since not readily condensable, is absorbed in a solution. Evaporation is vaporization without intentional condensation.

In all volatilization separations, condensation or absorption is as important as volatilization and care should be taken to see there is no loss of constituent in this step.

A. CLASSIFICATION OF METHODS

The following classification of volatilization methods is convenient for practical purposes:

1. The sample is a solution, usually obtained by dissolving a solid in water or an aqueous solution.

   a. The desired constituent is evolved as a gas and absorbed in a suitable solution.

   a'. One or more unwanted constituents are evolved as gases.

   b. The desired constituent is distilled, together with water or other solvent, and condensed or absorbed.

   b'. One or more unwanted constituents (or the solvent itself) are vaporized.

2. The sample is a solid.

   a. The desired constituent is volatilized and condensed or absorbed.

   a'. One or more unwanted constituents are volatilized.

B. GENERAL OBSERVATIONS ON METHODS

In the following discussion, the numbers refer to the previous classification.

1a. Among metals and metalloids, arsenic and antimony are virtually the only elements in category 1a. Arsenic(III) is readily converted to AsH_3, as by the action of zinc and mineral acid, which is swept out of the solution by the evolved hydrogen and absorbed, for example, in a bicarbonate solution of iodine. Quantitative evolution of SbH_3 is less easy and requires adherence to specified conditions. The quantitative separation of other metals as hydrides may be possible but is not established (1975).

1a'. This category is unfilled as far as metals and semimetals are concerned, but some nonmetals can be eliminated in this way, for example, sulfur as S^{2-} in some sulfides by treatment with mineral acid. Complete removal of H_2S requires sweeping with an inert gas or boiling the solution, so that this separation then grades into category 1b'.

An important application of volatilization of unwanted constituents is made in the decomposition of silicates by hydrofluoric acid. If this process is carried out in the presence of a high-boiling acid such as sulfuric or perchloric acid with heating, silica is eliminated as SiF_4 (partly as H_2SiF_6) and excess fluoride is volatilized. Most silicates can be decomposed and brought completely into solution this way. Silicates can, of course, be decomposed in other ways, as by sodium carbonate fusion, but silica is then brought into solution and will cause trouble in most photometric methods unless removed. A few metals are lost in hydrofluoric acid decomposition (Chapter 3, Section II.A).

1b. This is the most important category in volatilization separations. Most inorganic compounds of metals or metalloids that are volatile are liquids or solids at room temperature and their vapor pressures may be less than that of water. Nevertheless, it may be possible to concentrate them in a relatively small volume of distillate by distilling their aqueous solutions. The volatile substance partitions into the vapor phase (steam) to an extent determined by its vapor pressure and (inversely) to its solubility in water, or other solvent, at the boiling temperature. One milliliter of water at room temperature gives about 1700 ml of steam at 100° (all temperatures are in °C) and 1 atm. The large value of the ratio vapor/solution and the large increase of the vapor pressure of a substance on going from room temperature to 100° are the reasons for the effectiveness of distillation with water. The simplicity of the operation is an advantage.

If the volatility of a substance is not sufficiently great at ~100°, addition of sulfuric or perchloric acid may be needed to raise the boiling point of the mixture. Passage of an inert gas helps volatilization, but may also make quantitative condensation or absorption more difficult. The boiling range may need to be controlled to prevent the volatilization of undesired elements that may show some volatility. Mechanical carry-over, as by spray and creeping film, may decrease the effectiveness of volatilization separations, and the diminution of this effect may require attention, especially when a major nonvolatile constituent can interfere in the subsequent determination.

The process of simple distillation of a volatile substance from an aqueous solution is akin to extraction. The volatile substance A is distributed between the boiling liquid and the vapor. At a given boiling temperature, with a constant composition of the distilland, the ratio of the concentration of the substance in the vapor phase to its concentration in the liquid phase is a constant

$$D' = \frac{[A]_{vap}}{[A]_{liq}}$$

when equilibrium is attained. Instead of water vapor, an indifferent gas (often air) can be bubbled through the solution below the boiling point. D' may be expected to be independent of the concentration of A in the system, provided there is no change in species (e.g., association) in either phase. If the vapor is condensed so that both A and H_2O enter the liquid in the ratio that they are found in the vapor phase

$$D = \frac{[A]_{distillate}}{[A]_{liq}}$$

An important analytical question is the volume of distillate that must be collected to recover a specified fraction of A. An answer is obtainable if equilibrium is imagined to take place with successive portions of water vapor (or indifferent gas) that are small compared to the liquid phase. If the volume of solution in the distilling flask (V) remains constant,[1] the weight w of A distilled, corresponding to a volume v of distillate, is ideally given by

$$w = W(1 - e^{-vD/V}), \qquad (10\text{-}1a)$$

W being the original weight of A in the distilling flask. (This expression is analogous to that for liquid–liquid extraction of a substance with a fixed total volume of immiscible extractant used in infinitesimal portions.) Since $w/W = R$ (the recovery factor), Eq. (10-1a) can be written

$$\log(1 - R) = -0.43\, D\left(\frac{v}{V}\right) \qquad (10\text{-}1b)$$

If the volume does not remain constant during distillation, the following expression holds if D is assumed to remain constant:

$$\log\frac{1}{1 - R} = D \log\frac{V}{V - v} \qquad (10\text{-}2)$$

Actually, it is most unlikely that D will remain constant during distillation, because other (nonvolatile) solutes will be present and the boiling point will rise as distillation proceeds and D will change.

Estimates of D' and D can be obtained if the vapor pressure and solubility of the volatile substance are known at a specified temperature and there are no complicating factors. See the discussion on the volatilization of metallic mercury (p. 1046).

An example is of interest in connection with these equations. Manganese(VII) can be distilled as $HMnO_4$ or Mn_2O_7 from sulfuric acid solution containing potassium periodate as oxidizing agent.[2] The optimal concentration of sulfuric acid is $\sim 10\ M$:

$H_2SO_4\ (M)$	Temp. (°C)	$D(Mn_{distillate}/Mn_{soln})$
8.0	142	0.35
8.9	151	0.80
10.2	166	1.08
11.9	190	0.12

According to Eq. (10-1b), if distillation is made from 50 ml of 10.2 M H_2SO_4, maintained at constant volume, 217 ml of distillate must be collected for a manganese recovery of 99% ($R = 0.99$).[3] This volume is too large for practical purposes. $HMnO_4$ is not sufficiently volatile under the conditions for easy separation.

Osmium tetroxide presents a more favorable case. The solid has a vapor pressure of 1 atm at 130°. When a 5 M nitric acid solution of OsO_4 is distilled, the osmium is quantitatively recovered in a volume of distillate equal to 20–25% of the original solution (no water or nitric acid added during distillation).

The composition of the aqueous phase will have a great effect on the volatility of metalloids and metals. Thus easy distillation of arsenic(III) as chloride requires a solution strongly acidified with HCl to convert As_2O_3 into $AsCl_3$.

Masking agents find use now and then in distillation methods, as for example H_3PO_4 to prevent or greatly lessen distillation of tin chloride.

1b′. Evaporation of an aqueous solution serves the purpose of concentrating nonvolatile trace constituents into a smaller volume and removing volatile acids such as HCl, HNO_3, $HClO_4$, and H_2SO_4 that are likely to be objectionable in the subsequent determination. This operation entails dangers: Trace elements may be introduced from the atmosphere or the vessels, or the elements in question may be lost by adsorption on the walls of the vessels, occasionally by being volatilized themselves, and less often lost mechanically by improper evaporation techniques. These hazards have been touched upon earlier (Chapter 2). It is inadvisable to evaporate a solution of very low electrolyte content without first adding a few milligrams of a soluble salt to serve as a collector for the trace constituent.

Elements such as Ge and As are readily volatilized from hydrochloric acid solutions. When compounds of these elements are to be examined for nonvolatile trace constituents, this is usually the best way of eliminating these elements.

2a. Heating at elevated or moderately elevated temperatures may be of value in separating constituents of greater or less volatility from those that are essentially nonvolatile under the conditions. The volatilized substance may be a metal or a metal compound such as a halide (see Section II). The volatilization may involve a chemical reaction, as when a solid sample is heated in a current of hydrogen or chlorine.

The expulsion of a volatile substance from a solid is not always easy. Diffusion through a lattice may be very slow, and the substance may be held strongly on the surface of the sample material. Much depends on the matrix material and the substance. Volatilization of mercury from rocks and minerals requires heating for several hours at 800°. Volatilization is more certain if the solid melts or if fusion is brought about by the use of a suitable flux.

2a'. The destruction of organic matter by dry ashing falls under this heading, the conversion of C, H, and the like of organic substances into volatile products being partly the result of oxidation. The possibility of trace metal loss in this operation has already been touched upon (Chapter 3). The removal of ammonium salts by fuming them off is also included here. Losses of metals forming volatile chlorides can occur when ammonium chloride is volatilized, for example, of vanadium (possibly as $VOCl_3$).

Relatively few metals or compounds of metals or metalloids are sufficiently volatile at even moderately elevated temperatures to enable them, as major constituents, to be removed conveniently from a sample in this way, but occasionally this approach is useful.[4]

II. SEPARATION FORMS IN VOLATILIZATION METHODS

A. METALS

1. Mercury

Mercury (b.p. 357°) is unique among the metals in having an appreciable vapor pressure at room temperature. This property, combined with the easy reducibility of its compounds to the metal, enables mercury to be separated readily from other metals, even by volatilization from aqueous solution. Absorption by mercury vapor at 253.7 nm provides a simple photometric determination. This method has its pitfalls but is the most popular of those in use today for microgram or smaller amounts of mercury.

It is of interest to see how easily mercury can be removed from an aqueous solution by bubbling air or other inert gas through a solution at room temperature. Mercurous or mercuric ion, or more generally mercury compounds, can be reduced to mercury metal by a strong reducing agent such as Sn(II).[5] We consider the case in which all of the mercury is dissolved in an aqueous solution, and it partitions as Hg vapor into the gas (air) phase. An approximate value for D' (p. 1043) is estimated from the vapor pressure $(2 \times 10^{-3}\,\mathrm{mm\,Hg})$ and the solubility of elemental mercury in water $(3 \times 10^{-7}\,M$ or $0.06\,\mu\mathrm{g\,Hg/ml})$, both at 25°:

$$D' = \frac{\text{concentration of Hg in saturated gas phase}}{\text{concentration of Hg in saturated solution}}$$

$$D' = \frac{[(2 \times 10^{-3})/760] \times [(201 \times 10^{6})/22{,}400]}{0.06} = 0.4$$

The mercury in the gas (air) can be absorbed in a strongly oxidizing aqueous solution such as acidified permanganate solution, which converts Hg to Hg(II), or amalgamated with Au or Ag metal. If the volume of the aerated solution is 50 ml, what volume of air must be passed to remove 99% of the mercury? From Eq. (10-1b), $V = 50$ ml:

$$v = {\sim}580\,\mathrm{ml} \qquad (\text{or } {\sim}12 \text{ times the volume of aqueous solution})$$

Attainment of equilibrium may not be complete (and the assumption of successive infinitesimal volumes of air is not quite true) so that this is a minimum figure for the volume.[6] If undissolved mercury metal is also present (if more than $50 \times 0.06 = 3\,\mu\mathrm{g\,Hg}$), additional air is required to volatilize this ($1\,\mu\mathrm{g\,Hg}$ requires 42 ml air). Note that if mercury were absolutely insoluble in water, the volume of air required to volatilize $3\,\mu\mathrm{g}$ mercury would be $3 \times 42 = 126$ ml, assuming the air was mercury-saturated.

A point of interest in this separation is the possibility of absorbing mercury in a small volume of oxidizing solution, so that the sensitivity of the subsequent determination is not impaired. Even if a large volume of air is passed, there is essentially no gain or loss of water in the absorbed solution.[7]

The foregoing considerations on the volatilization of mercury assume no tendency for mercury to be adsorbed on the walls of the vessels used or on solids in the original solution. Mercury is coprecipitated with noble metals precipitated by stannous tin and can then no longer be volatilized. It also tends to be adsorbed on finely divided "inert" solids.

It should be possible to expel mercury, as metal, from an aqueous solution easily by boiling. We estimate D' (now for steam/liquid water) as

5.4, from a vapor pressure of 0.27 mm Hg and a solubility of 0.6 μg Hg/ml, both at 100°. D, corresponding to this, is estimated as 9×10^3, a very large value. If there are no complicating factors, such as a tendency for mercury to be held on glass surfaces, traces of mercury should be almost completely recovered in a small volume of distillate.[8]

2. Other Metals

Metals other than mercury are not often separated as such preliminary to absorption photometric determination. Some possibilities are apparent when metals are arranged in order of the temperatures at which their vapor pressures are equal to 10 mm of Hg:

	Temp. (C°)		Temp. (C°)
Hg	184	Sb	1033
Cs	375	Ba	1049
As	437	Bi	1136
Se	442	Pb	1162
Cd	484	Al	1487
Na	549	Mn	1505
Zn	593	Ga	1541
Te	650	Ag	1575
Po	962	Cu	1879
Tl	983	Ni	2057
		W	4507

On being heated in a stream of hydrogen, compounds of the metals or metalloids Hg, As, Se, Cd, Zn, Te, and Tl are reduced and volatilized at temperatures ($\sim$1000°) easily attained without the use of special furnaces. Sparing use has been made of such volatilization for isolation of these metals.[9] Thus zinc in silicate and oxide minerals has been isolated in this way (1000°–1100°). For As and Se, volatilization of the elements is usually not of much interest, because they are more easily separated as halogen compounds from solution. Lead has been volatilized from stone and iron meteorites and from silicate rocks (samples as large as 25 g) by heating the sample in a graphite crucible in an induction furnace at 1300°–1400°C.[10] A stream of nitrogen is passed, and lead is condensed on cooled silica walls and caught in moist quartz wool (Fig. 10-1). This method—compared to hydrofluoric acid decomposition—has the advantages of speed and lower blanks. Reducible trace metals other than lead will also be volatilized from the sample (see the figures in the tabulation on p. 1047) but not major constituents (Si, Ca, Mg, Fe, Al, etc.). The volatilization of lead from silicate rocks and meteorites appears to be

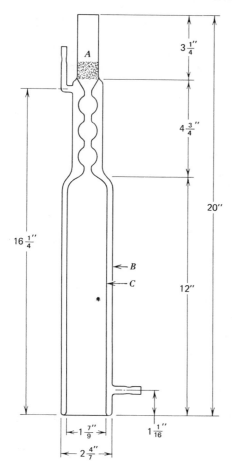

Fig. 10-1. Arrangement of Marshall and Hess for volatilization of lead from silicate rocks and stone and iron meteorites. The inner wall (C) of the tube is fused silica, the outer (B) is Vycor; A is a plug of quartz wool. Water is circulated through the jacket. The graphite crucible (2.625 in. high, 1 in. outer diameter) is placed on a silica pedestal inside the tube, and a 3 kW induction furnace is used for heating. Reprinted with permission from R. R. Marshall and D. C. Hess, *Anal. Chem.*, **32**, 961 (1960); **33**, 22 (1961). Copyright by the American Chemical Society.

complete. In the fusion of silicate rocks, gases are generated, which sweep lead out of the molten silicate. Formation of droplets of iron metal indicates the strongly reducing conditions in the fusion.

Volatilization of certain trace metals by heating in hydrogen is occasionally a useful separation from less volatile matrix elements. Traces of zinc have been separated in this way from Al, Ga, and In.[11] Sometimes vacuum volatilization is used. Conversely, less volatile elements can be concentrated by vaporizing most of a more volatile matrix metal such as zinc or cadmium. In the determination of traces of uranium in sodium, the bulk of sodium is removed by vacuum distillation.[12]

Cadmium metal can be volatilized in a current of steam at 630°. As much as a gram of Cd can be volatilized in 45 min.[13] Metals not volatilized include Pb, Bi, Cu, and Ag. Antimony and Tl are partially volatilized. A small proportion of Cd, 0.1–0.5%, remains as oxide in the unvolatilized residue. Volatilization of Cd in a current of nitrogen was unsatisfactory.

B. OXIDES

A half-dozen adjacent metals in the periodic table form oxides, in their higher oxidation states, that are markedly volatile:

VIIA	VIII	
Mn		
Tc	Ru	
Re	Os	Ir

As we have seen, the volatility of Mn_2O_7 ($HMnO_4$) is not sufficiently great to be of much value in analytical separations. But for the other five metals, the volatile oxides provide good or moderately good separation forms. The tetroxides of Os (b.p. 130°) and Ru (b.p. ~100°, decomposed) are so volatile that they can be distilled with water. In this way, these two metals can readily be separated from other elements and from each other. Oxidation of lower valence states of Ru to the VIII state requires a higher oxidation potential than oxidation of Os to the VIII state, so that the two elements are separable by differential oxidation before or during distillation.

Relatively recently it has been found that Ir in microgram quantities can be distilled from H_2SO_4–$HClO_4$ at 200° if a current of air + Cl_2 is passed through the solution.[14] The volatilization of Ir is slow, and distillation should be continued for 2 hr. For 30 μg Ir, $R \sim 95\%$.

Distillation of Re as Re_2O_7 (b.p. 360°) requires a solution of high boiling point. Steam distillation from sulfuric acid solution at 270°–290° is the standard method. Perchloric acid has also been used.

Technetium heptoxide (b.p. 310°) is a little more volatile than Re_2O_7 and can be distilled from $HClO_4$ or from H_2SO_4[15] + peroxydisulfate solution. The separation of Tc from Re in this way would be difficult, but separation from most other elements should be possible.

Tellurium can be volatilized from solid samples (sulfides, silicate rocks) as TeO_2 by heating to 900° in air.[16] Selenium in copper is easily volatilized as oxide by heating in a stream of oxygen. As_2O_3 sublimes at ~200°.

C. HALIDES

A group of adjacent elements in the periodic table form compounds with the halogens that can be volatilized without much difficulty from acid solutions, if need be from high-boiling solutions such as sulfuric and perchloric acid. These elements are enclosed by solid lines below:

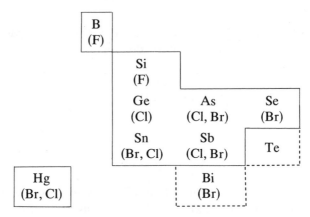

The broken lines, enclosing Bi and Te, indicate elements not easily volatilized from solution but which can be vaporized as halide compounds from a solid. Not shown here is Cr which is volatile in the form of chromyl chloride, CrO_2Cl_2 (b.p. 117°). The symbols below the volatile elements indicate the halides ordinarily used as separation forms.

A more general discussion follows.

1. Fluorides

Volatilization of metallic or submetallic fluorides is of interest as a means of eliminating these elements in a hydrofluoric acid decomposition (or of the possibility of their loss). Cr(VI) can be volatilized as CrO_2F_2 from fuming $HClO_4$–HF solutions.[17] Several additions of HF are needed to remove most of the chromium. The remainder can be volatilized as CrO_2Cl_2. Arsenic(III and V) is completely lost under these conditions, as are B and Si, of course. Other elements volatilized to variable extents include Ge, Sb, Se, and Re.

The anhydrous fluorides of the metals of groups IVA, VA, and VIA of the periodic table are easily vaporized, for example, boiling points[18] are TiF_4, 284°; VF_5, 111°; NbF_5, ~220°; TaF_5, 220°; MoF_6, 35°; WF_6, 20°. These values may be compared with the boiling points of some other fluorides: AlF_3, 1260°; GaF_3, ~1000°; MgF_2, 2239°; PbF_2, 1290°; CdF_2, 1758°; SnF_4, 705°. These differences have been exploited by volatilizing such metals as V, Mo, W, Ti, and Ta as fluorides or oxyfluorides by heating the sample (oxide) at 350° in a current of HF.[19] Obviously the use of gaseous HF is unattractive and the method is likely to be confined to special problems.

2. Chlorides

Two elements coming immediately to mind as being volatilizable as chlorides are arsenic(III) and germanium; the boiling points are 130° for $AsCl_3$ and 84° for $GeCl_4$. They can be distilled from HCl solutions of sufficient strength to convert, to a large extent, As_2O_3 and GeO_2 into the chlorides. Distillation of $AsCl_3$ is often carried out from solutions initially close to concentrated HCl at ~110°, with a stream of N_2 or CO_2 being passed. Arsenic must be present in the tripositive state ($AsCl_5$ is supposedly nonexistent) and hydrazine sulfate is often used as reducing agent. Germanium accompanies As. Antimony(III) may codistill slightly at 110° and in its presence it is best to use a slightly lower distillation temperature ($=107°$).

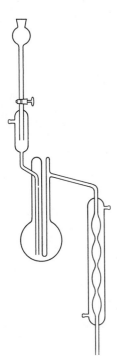

Fig. 10-2. Scherrer distillation apparatus.

TABLE 10-1
**Volatilization of Metallic Compounds from Perchloric and Sulfuric Acid Solutions
at 200–220°[a]**

Procedure 1 ($HClO_4$–HCl): 15 ml of 60% $HClO_4$ added to metal chloride or perchlorate
solution in a Scherrer distilling flask. Distillation made in a stream of CO_2. Temperature
raised to 200°, and hydrochloric acid then added at such a rate that temperature remained at
200–220°. Distillation stopped after 15 ml HCl had been added over a period of 20–30 min.
 Procedure 2 ($HClO_4$–HBr): As in (1) except 40% HBr used in place of HCl.
 Procedure 3 ($HClO_4$–H_3PO_4–HCl): As in (1) except 5 ml 85% H_3PO_4 added to flask
before distillation.
 Procedure 4 ($HClO_4$–H_3PO_4–HBr): As in (3) except HBr substituted for HCl.
 Procedure 5 (H_2SO_4–HCl): As in (1) except H_2SO_4 (sp. gr. 1.84) substituted for $HClO_4$.
 Procedure 6 (H_2SO_4–HBr): As in (2) except H_2SO_4 substituted for $HClO_4$.

	Approximate Percentage of Element Volatilized from 20–100 mg Portions by Distillation with					
Element	HCl– $HClO_4$ Proc. 1	HBr– $HClO_4$ Proc. 2	HCl– H_3PO_4– $HClO_4$ Proc. 3	HBr– H_3PO_4– $HClO_4$ Proc. 4	HCl– H_2SO_4 Proc. 5	HBr– H_2SO_4 Proc. 6.
As(III)	30	100	30	100	100	100
As(V)	5	100	5	100	5	100
Au	1	0.5	0.5	0.5	0.5	0.5
B	20	20	10	10	50	10
Bi	0.1	1	0	1	0	1
Cr(III)	99.7	40	99.8	40	0	0
Ge[b]	50	70	10	90	90	95
Hg(I)	75	75	75	75	75	90
Hg(II)	75	75	75	75	75	90
Mn	0.1	0.02	0.02	0.02	0.02	0.02
Mo	3	12	0	0	5	4
Os[c]	100	100	100	100	0	0
P	1	1	1	1	1	1
Re	100	100	80	100	90	100
Ru	99.5	100	100	100	0	0
Sb(III)	2	99.8	2	99.8	33	99.8
Sb(V)	2	99.8	0	99.8	2	98
Se(IV)	4	2–5	2–5	2–5	30	100
Se(VI)	4	5	5	5	20	100
Sn(II)	99.8	100	0	99.8	1	100
Sn(IV)	100	100	0	100	30	100
Te(IV)	0.5	0.5	0.1	0.5	0.1	10
Te(VI)	0.1	0.5	0.1	1	0.1	10
Tl[d]	1	1	1	1	1	1
V	0.5	2	0	0	0	0

The following elements are not volatilized in any of the procedures: Ag, alkalies (Li, Na,
K, Rb, Cs), Al, Ba, Be, Ca, Cd, Co, Cu, Fe, Ga, Hf, In, Ir, Mg, Nb, Ni, Pb, Pd, Pt, rare earths,
Rh, Si, Ta, Th, Ti, U, W, Zn, Zr.

Germanium can be distilled from ~6 M HCl. Arsenic remains in the distilling flask if oxidized and maintained in the V state, as by passing in Cl_2. Although $SnCl_4$ boils at 114°, it does not pass over, most likely because it forms $SnCl_6^{2-}$.

A standard method for separating macro quantities of As, Sb, and Sn involves precipitation of the sulfides, their solution in $H_2SO_4 + HNO_3$, expulsion of HNO_3, and reduction of As and Sb to the III state with SO_2. Distillation from strong HCl, in an apparatus such as that in Fig. 10-2, follows, with $AsCl_3$ volatilized at 110° and $SbCl_3$ at 155°–165°, with H_3PO_4 being added before the distillation of antimony to prevent distillation of tin. Finally, tin can be distilled at 140° by adding HBr and HCl in the ratio 1:3.[20]

Arsenic(III) can be volatilized quantitatively from concentrated H_2SO_4 at 200°–220° in a stream of CO_2 by dropping in concentrated HCl (Table 10-1, procedure 5). Ge accompanies As, and the high temperature leads to greater or less volatilization of B, Hg (I and II), Re, Sb(III and V), Se(IV and VI), Sn(II, especially IV), and even some Mo(VI) and very little Te and Tl.

Stannic chloride (b.p. 114°) is quantitatively volatilized by passing *dry* HCl gas through concentrated sulfuric acid solution at 200°. Selenium, but not tellurium, accompanies tin. Likewise $HgCl_2$ (b.p. 304°) can be vaporized from concentrated sulfuric acid by gaseous HCl.[21] These methods tend to be inconvenient and are not often used.

Removal of Cr(VI) by vaporization as CrO_2Cl_2 requires repeated addition of concentrated HCl or NaCl to a boiling $HClO_4$ solution. A method for the distillation of Re oxychloride from H_2SO_4–H_3PO_4–HCl has been described.[22]

Under anhydrous conditions, chlorides of metals other than those mentioned above are volatile at moderate temperatures (<500°) and may

---

Footnotes to facing table.

---

[a] J. I. Hoffman and G. E. F. Lundell, *J. Res. Nat. Bur. Stand.*, **22**, 465 (1939). Using tracer amounts of 23 elements, C. Ballaux, R. Dams, and J. Hoste, *Anal. Chim. Acta*, **47**, 397 (1969), obtained good agreement with the results of Hoffman and Lundell in distillation from HBr–H_2SO_4 at 200–220°. More than 99.999% As(III or V), 99.999% Se(IV or VI) and 99.98% Sn(II or IV) could be distilled. The volatilization of a few percent of Au was confirmed.
[b] H_2SO_4 and $HClO_4$ solutions heated to 200° before admitting HCl or HBr.
[c] At 200–220°, no Os is volatilized from H_2SO_4, but at 270–300° it is completely volatilized.
[d] No Tl distils if it is univalent (reduction with SO_2).

serve as separation forms. The solid sample, which should be in a reactive form such as sulfide or oxide is heated in a current of HCl, Cl_2, or a chlorinating agent such as sulfur monochloride or dichloride. Thus niobates, tantalates, and tungstates heated in a stream of S monochloride or the dichloride mixed with chlorine give a sublimate of Ti, Nb, Ta, W, Sn, Mo, Sb, As, and Fe chlorides.[23] Metals such as Al, Ca, Mg, and Pb are not volatilized. This is an old method, intended for decomposition, with the volatilization of the elements somewhat secondary, which is infrequently used. Sulfides heated in a current of Cl_2 give a sublimate containing Sn, Sb, As, Te, Se, Hg, Bi, V, and Ge. WO_3 gives the oxydichloride when heated in dry HCl at redness (separation from nonvolatile Th). MoO_3 gives volatile $MoO_3 \cdot 2\,HCl$ when heated at 250°–300° in HCl. Other possibilities could be mentioned.

Radioactive isotopes of Hg, Pt, Ir, Os, and Re formed by proton bombardment of gold have been separated by sublimation in a stream of argon or chlorine and condensation in a tube having a temperature gradient.[24]

3. Bromides

Although the volatile bromides usually have higher boiling points than the corresponding chlorides, it is found that often it is easier to distill an element from a solution containing HBr than from one containing only HCl. This is particularly true for tin and antimony (compare Procedures 5 and 6 in Table 10-1). Presumably this behavior is due to the presence of a larger fraction of total tin, for example, as $SnBr_4$ than as $SnCl_4$ under corresponding conditions. The greater tendency for $SnBr_4$ than $SnCl_4$ to be formed more than makes up for the lower volatility of the former (b.p. $SnCl_4$ 114°, $SnBr_4$ 202°). The beneficial effect of bromide persists in the presence of chloride (p. 1053). Formation of mixed species is not excluded.

Arsenic is readily distilled from hydrobromic acid solutions, even if present in the V state; $AsBr_5$ is believed to be the species distilled under the latter condition.

Other possibilities of bromide distillation may be seen from Table 10-1, with indications of elements that can be volatilized to a small extent. Mercury (^{203}Hg + carrier) has been recovered to the extent of 96% from a sulfuric acid digest of biomaterial by three additions of HBr.[25]

When the dry bromides are heated for 45 min in a stream of inert gas, As can be volatilized at 150° and Ga, Ge, Mo, and Sn at 200°–230°. At the latter temperature, Ag, Ba, Ca, La, Al, and Ti are not volatilized and other metals are volatilized to the following percentage extents[26]:

Au	15	Pb	10	Cu	15	Bi	70
Be	10	Pt	20	Ni	20	In	70
Mg	20	Tl	10	Co	45	Cr	50
Mn	25	Fe	5	Zn	30		

Much tin can be removed by treating the metal with $HBr + Br_2$ and heating to volatilize $SnBr_4$. Somewhat unexpectedly, bismuth can also be removed in this way (b. p. $BiBr_3$ 461°).[27]

Vanadium as vanadate is completely volatilized when heated with solid ammonium bromide or ammonium chloride.[28]

4. Iodides

A number of metal or metalloid iodides are sufficiently volatile at relatively low temperatures, say in the neighborhood of 500°, to be vaporized (sublimed) from a solid sample (boiling points: SnI_4, 341°; SbI_3, 401°; AsI_3, 403°). Ammonium iodide can be used to convert compounds of many elements to the iodides. It dissociates on heating into NH_3 and HI and the latter attacks metals, oxides, and the like. Moreover, HI is dissociated into H_2 and I_2:

$$2HI \rightleftharpoons H_2 + I_2$$

$$\frac{[H_2][I_2]}{[HI]^2} = 18.9 \text{ at } 327°$$

Consequently, ammonium iodide is a powerful reagent for decomposition. At 400°–500° it readily converts the otherwise refractory SnO_2 into SnI_4, which sublimes.[29] Similarly, compounds of As, Sb, and Te are converted to the volatile iodides.

D. HYDRIDES

The following group of metals and metalloids form gaseous hydrides

Ge	As	Se
Sn	Sb	Te
Pb	Bi	Po

Separation of As by evolution as AsH_3 by action of Zn and acid before its determination is an old, standard method in photometric analysis. The analogous separation of Sb as SbH_3 has been less successful, and evolution of Ge as germane (by electrolysis) is seldom used. Formation of the hydrides of the other elements above, requires (usually) reduction with the powerful reducing agent sodium borohydride, $NaBH_4$. It has been used in connection with atomic-absorption or atomic-fluorescence determination of these elements.[30] This work has not been concerned directly with separation as such and little is known about recoveries attainable. Most of these hydrides are unstable and some are not easily formed.

E. MISCELLANEOUS

Carbonyls. Nickel has been volatilized by conversion to nickel carbonyl by heating at 150 atm CO pressure.[31] Congeneric metals also form carbonyls, but the method is too specialized to be of great interest.

Chelate Complexes. Essentially no use has been made of the volatilization of metal chelate complexes for separations related to absorption photometric methods. However, the volatility of some [e.g., acetylacetonates (and derivatives)] makes them useful in gas chromatography.[32]

NOTES

1. Often in practice, matters are so arranged that this condition is at least approximately fulfilled. A steam distillation is made, or water or acid is added to the distilling flask at a rate equal to the distillation rate.

2. J. D. H. Strickland and G. Spicer, *Anal. Chim. Acta*, **3**, 543 (1949). J. Pijck and J. Hoste, *ibid.*, **26**, 501 (1962), were able to recover 85–90% ^{54}Mn (separation from Fe).

3. This result is in accordance with the procedure of Strickland and Spicer, which calls for collection of 200 ml of distillate, with the volume in the distilling flask being kept constant at 50 ml by addition of 200 ml of 40% v/v HNO_3 (2 g KIO_4 present).

4. K. H. Neeb, *Z. Anal. Chem.*, **200**, 278 (1964), determined Li, Na, K, and Ca in high-purity P, As, and Sb after volatilizing the latter elements in a stream of chlorine.

5. The aeration separation method was applied early by T. Kiba, I. Akaza, and O. Kinoshita, *Bull. Chem. Soc. Japan*, **33**, 329 (1960) and Y. Kimura and V. L. Miller, *Anal. Chim. Acta*, **27**, 325 (1962). Steam distillation of mercury was used by W. L. Miller and L. E. Wachter, *Anal. Chem.*, **22**, 1312 (1950).

6. The calculated result agrees well with experimental values reported by J. Olafsson, *Anal. Chim. Acta*, **68**, 207 (1974). He found that "total recovery" (here equated with 99% recovery) of 0.02 μg Hg in 450 ml of sea water (or just water) required passage of argon at a rate of 140 ml/min for 35 min. The volume of argon required is $(35 \times 140)/450 = \sim 11$ times the volume of the aqueous solution. The exact figures are of no significance, of course.

7. Since 1 ml air saturated with water vapor at 25° contains 23 μg H_2O, quantitative condensation of Hg and H_2O in the issuing air by freezing would give a condensate having a ratio $Hg/H_2O = (3 \times 0.99)/(580 \times 23) = 2.3 \times 10^{-4}$

(50 ml water saturated with mercury metal being used), compared to a ratio of $3/(50 \times 10^6) = 6 \times 10^{-8}$ in the original solution. The increase in the ratio Hg/H_2O comes about because D' for water is lower than D' for mercury. At 25°, $D'_{H_2O} = 0.023/{\sim}1000 = 2.3 \times 10^{-5}$.

8. We do not know of any comprehensive study of the separation of mercury by distillation with water at 100°. W. L. Miller and L. E. Wachter, *Anal. Chem.*, **22,** 1312 (1950), in distilling mercury after stannous sulfate reduction, boil the solution down to half its original volume.

9. Volatilization methods for traces of metals have been studied by W. Geilmann and coworkers: W. Geilmann, *Z. Anal. Chem.*, **160,** 410 (1958) (general); W. Geilmann and R. Neeb, *Angew. Chem.*, **67,** 26 (1955) (Zn); W. Geilmann and K. H. Neeb, *Z. Anal. Chem.*, **165,** 251 (1959) (Tl); W. Geilmann and H. Hepp, *ibid.*, **200,** 241 (1964) (Cd). An advantage of these methods is the low blank, since solid and liquid reagents are not used. Also, relatively large samples can be used. Volatilization methods should, therefore, be especially suitable for nanogram amounts of elements. At least this is true if the elements can be quantitatively expelled (and condensed), which, it seems, has not always been demonstrated. The partial volatilization of other elements may not be objectionable if sufficiently selective determination methods are available. Carbon monoxide has been used as a reducing and sweeping gas in volatilization of As, Se, Zn, Cd, and Hg at ~1180° from ashed organic samples in neutron activation analysis [E. Orvini, T. E. Gills, and P. D. LaFleur, *Anal. Chem.*, **46,** 1294 (1974)].

10. R. R. Marshall and D. C. Hess, *Anal. Chem.*, **32,** 960 (1960). A similar procedure was used earlier by G. Edwards and H. C. Urey, *Geochim. Cosmochim. Acta*, **7,** 154 (1955).

11. K. H. Neeb, *Z. Anal. Chem.*, **194,** 255 (1963).

12. M. Takahashi, Y. Matsuda, Y. Toita, M. Ouchi, and T. Komori, *Japan Analyst*, **20,** 1085 (1971).

13. S. R. Rajic and S. V. Markovic, *Anal. Chim. Acta*, **50,** 169 (1970).

14. R. Gijbels and J. Hoste, *Anal. Chim. Acta*, **36,** 230 (1966).

15. R. E. Meyer, R. D. Oldham, and R. P. Larsen, *Anal. Chem.*, **36,** 1975 (1964).

16. I. Schoenfeld and A. Berman, *Anal. Chem.*, **46,** 1826 (1974).

17. J. I. Dinnin, *U.S. Geol. Surv. Bull.*, **1084B,** 31 (1959).

18. Here and elsewhere, boiling points are given only to indicate possibilities of volatilization separation. Compounds mentioned may be obtained only under special conditions, not always analytically practicable or convenient.

19. Z. G. Fratkin and V. S. Shebunin, *Trudy Kom. Anal. Khim.*, **15,** 127 (1965). Z. G. Fratkin, *Zav. Lab.*, **30,** 170 (1964), analyzed TiO_2 for 10^{-5}–$10^{-6}\%$ of Mn, Mg, Cu, Ni, Sn, Al, Fe, and other metals spectrographically after volatilizing titanium as fluoride.

20. J. A. Scherrer, *J. Res. Nat. Bur. Stand.* **16,** 253 (1936); **21,** 95 (1938). W. D.

Mogerman, *ibid.*, **33**, 307 (1944). H. Onishi and E. B. Sandell, *Anal. Chim. Acta*, **14**, 153 (1956), were able to obtain good recoveries of microgram amounts of Sn by HBr–HCl distillation.

21. E. P. Fenimore and E. C. Wagner, *J. Am. Chem. Soc.*, **53**, 2468 (1931).

22. N. Jordanov and M. Pavlova, *Mikrochim. Ichnoanal. Acta*, 477 (1963).

23. W. F. Hillebrand, G. E. F. Lundell, H. A. Bright, and J. I. Hoffman, *Applied Inorganic Analysis*, Wiley, New York, 1953, p. 597.

24. J. Merinis and G. Bouissières, *Anal. Chim. Acta*, **25**, 498 (1961).

25. K. Samsahl, *Anal. Chem.*, **39**, 1480 (1967). The recovery figure includes an ion-exchange separation.

26. V. V. Malakhov et al., *J. Anal. Chem. USSR*, **24**, 439 (1969).

27. G. W. C. Milner and G. A. Barnett, *Anal. Chim. Acta*, **17**, 223 (1957). The sample (U alloy) is evaporated to dryness with a mixture of Br_2 and HBr. "It is generally necessary to heat the walls of the beaker with a bunsen flame to remove any condensed bismuth from them." By repeating the $HBr–Br_2$ evaporation several times, amounts of Bi up to 1 g can be removed.

28. R. Bock and S. Gorbach, *Mikrochim. Acta*, 593 (1958).

29. E. R. Caley and M. G. Burford, *Ind. Eng. Chem. Anal. Ed.*, **8**, 114 (1936). Ammonium iodide finds important use at the present time more for decomposition of geosamples for tin determination than for the separation of tin.

30. For example, K. C. Thompson and D. R. Thomerson, *Analyst*, **99**, 595 (1974) (all of the elements listed above except Po); K. C. Thompson, *ibid.*, **100**, 307 (1975); J. A. Fiorino, J. W. Jones, and S. G. Capar, *Anal. Chem.*, **48**, 120 (1976) (As, Sb, Se, and Te); M. Bédard and J. D. Kerbyson, *Anal. Chem.*, **47**, 1441 (1975) (Bi). Interferences: A. E. Smith, *Analyst*, **100**, 300 (1975).

31. D. M. Shvarts, *Zav. Lab.*, **26**, 966 (1960).

32. R. W. Moshier and R. E. Sievers, *Gas Chromatography of Metal Chelates*, Pergamon, Oxford, 1965.

INDEX